T0178492

Volume 98
Classical and Quantum Orthogonal Polynomials in One Variable

This is first modern treatment of orthogonal polynomials from the viewpoint of special functions. The coverage is encyclopedic, including classical topics such as Jacobi, Hermite, Laguerre, Hahn, Charlier and Meixner polynomials as well as those, e.g. Askey–Wilson and Al-Salam–Chihara polynomial systems, discovered over the last 50 years: multiple orthogonal polynomials are dicussed for the first time in book form. Many modern applications of the subject are dealt with, including birth and death processes, integrable systems, combinatorics, and physical models. A chapter on open research problems and conjectures is designed to stimulate further research on the subject.

Exercises of varying degrees of difficulty are included to help the graduate student and the newcomer. A comprehensive bibliography rounds off the work, which will be valued as an authoritative reference and for graduate teaching, in which role it has already been successfully class-tested.

ENCYCLOPEDIA OF MATHEMATICS AND ITS APPLICATIONS

All the titles listed below can be obtained from good booksellers or from Cambridge
University Press. For a complete series listing visit
http://publishing.cambridge.org/stm/mathematics/eom/

Classical and Quantum Orthogonal
Polynomials in One Variable

Mourad E.H. Ismail
University of Central Florida

With two chapters by

Walter Van Assche
Catholic University of Leuven

CAMBRIDGE
UNIVERSITY PRESS

CAMBRIDGE
UNIVERSITY PRESS

University Printing House, Cambridge CB2 8BS, United Kingdom

One Liberty Plaza, 20th Floor, New York, NY 10006, USA

477 Williamstown Road, Port Melbourne, VIC 3207, Australia

314-321, 3rd Floor, Plot 3, Splendor Forum, Jasola District Centre, New Delhi - 110025, India

79 Anson Road, #06-04/06, Singapore 079906

Cambridge University Press is part of the University of Cambridge.

It furthers the University's mission by disseminating knowledge in the pursuit of education, learning and research at the highest international levels of excellence.

www.cambridge.org
Information on this title: www.cambridge.org/9780521143479

First published 2005

A catalogue record for this publication is available from the British Library

ISBN 978-0-521-78201-2 Hardback
ISBN 978-0-521-14347-9 Paperback

Transferred to digital printing 2008

Contents

Contents

Contents ix

Foreword

There are a number of ways of studying orthogonal polynomials. Gabor Szegő's book "Orthogonal Polynomials" (Szegő, 1975) had two main topics. Most of this book dealt with polynomials which are orthogonal on the real line, with a chapter on polynomials orthogonal on the unit circle and a short chapter on polynomials orthogonal on more general curves. About two-thirds of Szegő's book deals with the classical orthogonal polynomials of Jacobi, Laguerre and Hermite, which are orthogonal with respect to the beta, gamma and normal distributions, respectively. The rest deals with more general sets of orthogonal polynomials, some general theory, and some asymptotics.

Barry Simon has recently written a very long book on polynomials orthogonal on the unit circle, (Simon, 2004). His book has very little on explicit examples, so its connection with Szegő's book is mainly in the general theory, which has been developed much more deeply than it had been in 1938 when Szegő's book appeared.

The present book, by Mourad Ismail, complements Szegő's book in a different way. It primarily deals with specific sets of orthogonal polynomials. These include the classical polynomials mentioned above and many others. The classical polynomials of Jacobi, Laguerre and Hermite satisfy second-order linear homogeneous differential equations of the form

$$a(x)y''(x) + b(x)y'(x) + \lambda_n y(x) = 0$$

where $a(x)$ and $b(x)$ are polynomials of degrees 2 and 1, respectively, which are independent of n, and λ_n is independent of x. They have many other properties in common. One is that the derivative of $p_n(x)$ is a constant times $q_{n-1}(x)$ where $\{p_n(x)\}$ is in one of these classes of polynomials and $\{q_n(x)\}$ is also. These are the only sets of orthogonal polynomials with the property that their derivatives are also orthogonal.

Many of the classes of polynomials studied in this book have a similar nature, but with the derivative replaced by another operator. The first operator which was used is

$$\Delta f(x) = f(x+1) - f(x),$$

a standard form of a difference operator. Later, a q-difference operator was used

$$D_q f(x) = [f(qx) - f(x)]/[qx - x].$$

Still later, two divided difference operators were introduced. The orthogonal polynomials which arise when the q-divided difference operator is used contain a set of
polynomials introduced by L. J. Rogers in a remarkable series of papers which appeared in the 1890s. One of these sets of polynomials was used to derive what we
now call the Rogers–Ramanujan identities. However, the orthogonality of Rogers's
polynomials had to wait decades before it was found. Other early work which leads
to polynomials in the class of these generalized classical orthogonal polynomials was
done by Chebyshev, Markov and Stieltjes.

To give an idea about the similarities and differences of the classical polynomials and some of the extensions, consider a set of polynomials called ultraspherical
or Gegenbauer polynomials, and the extension Rogers found. Any set of polynomials which is orthogonal with respect to a positive measure on the real line satisfies a three term recurrence relation which can be written in a number of equivalent
ways. The ultraspherical polynomials $C_n^\nu(x)$ are orthogonal on $(-1, 1)$ with respect
to $\left(1 - x^2\right)^{\nu - 1/2}$. Their three-term recurrence relation is

$$2 \left(n + \nu\right) x C_n^\nu(x) = (n + 1) C_{n+1}^\nu(x) + (n + 2\nu - 1) C_{n-1}^\nu(x)$$

The three-term recurrence relation for the continuous q-ultraspherical polynomials of
Rogers satisfy a similar recurrence relation with every $(n + a)$ replaced by $1 - q^{n+a}$.
That is a natural substitution to make, and when the recurrence relation is divided by
$1 - q$, letting q approach 1 gives the ultraspherical polynomials in the limit.

Both of these sets of polynomials have nice generating functions. For the ultraspherical polynials one nice generating function is

$$\left(1 - 2xr + r^2\right)^{-\nu} = \sum_{n=0}^{\infty} C_n^\nu(x)\, r^n$$

The extension of this does not seem quite as nice, but when the substitution $x =
\cos\theta$ is used on both, they become similar enough for one to guess what the left-hand
side should be. Before the substitution it is

$$\prod_{n=0}^{\infty} \frac{\left(1 - 2xq^{\nu+n}r + q^{2\nu+2n}r^2\right)}{\left(1 - 2xq^n r + q^{2n}r^2\right)} = \sum_{n=0}^{\infty} C_n(x; q^\nu \mid q)\, r^n.$$

The weight function is a completely different story. To see this, it is sufficient to
state it:

$$w\left(x, q^\nu\right) = \left(1 - x^2\right)^{-1/2} \prod_{n=0}^{\infty} \frac{\left(1 - \left(2x^2 - 1\right) q^n + q^{2n}\right)}{\left(1 - \left(2x^2 - 1\right) q^{n+\nu} + q^{2n+2\nu}\right)}.$$

These polynomials of Rogers were rediscovered about 1940 by two mathematicians,
(Feldheim, 1941b) and (Lanzewizky, 1941). Enough had been learned about orthogonal polynomials by then for them to know they had sets of orthogonal polynomials,
but neither could find the orthogonality relation. One of these two mathematicians,
E. Feldheim, lamented that he was unable to find the orthogonality relation. Stieltjes
and Markov had found theorems which would have allowed Felheim to work out the
orthogonality relation, but there was a war going on when Feldheim did his work and
he was unaware of this old work of Stieltjes and Markov. The limiting case when

$\nu \to \infty$ gives what are called the continuous q-Hermite polynomials. It was these polynomials which Rogers used to derive the Rogers-Ramanujan identities.

Surprisingly, these polynomials have recently come up in a very attractive problem in probability theory which has no q in the statement of the problem. See Bryc (Bryc, 2001) for this work.

Stieltjes solved a minimum problem which can be considered as coming from one dimensional electrostatics, and in the process found the discriminant for Jacobi polynomials. The second-order differential equation they satisfy played an essential role. When I started to study special functions and orthogonal polynomials, it seemed that the only orthogonal polynomials which satisfied differential equations nice enough to be useful were Jacobi, Laguerre and Hermite. For a few classes of orthogonal polynomials nice enough differential equations existed, but they were not well known. Now, thanks mainly to a conjecture of G. Freud which he proved in two very special cases, and work by quite a few people including Nevai and some of his students, we know that nice enough differential equations exist for polynomials orthogonal with respect to $\exp(-v(x))$ when $v(x)$ is smooth enough. The work of Stieltjes can be partly extended to this much wider class of orthogonal polynomials. Some of this is done in Chapter 3.

Chapter 4 deals with the classical polynomials. For Hermite polynomials there is an explicit expression for the analogue of the Poisson kernel for Fourier series which was found by Mehler in the 19th century. An important multivariable extension of this formula found independently by Kibble and Slepian is in Chapter 4. Chapter 5 contains some information about the Pollaczek polynomials on the unit interval. Their recurrence relation is a slight variant of the one for ultraspherical polynomials listed above. The weight function is drastically different, having infinitely many point masses outside the interval where the absolutely continuous part is supported or vanishing very rapidly at one or both of the end points of the interval supporting the absolutely continuous part of the orthogonality measure.

Chapter 6 deals with extensions of the classical orthogonal polynomials whose weight function is discrete. Here the classical discriminant seemingly cannot be found in a useful form, but a variant of it has been computed for the Hahn polynomials. This extends the result of Stieltjes on the discriminant for Jacobi polynomails. Hahn polynomials extend Jacobi polynomials and are orthogonal with respect to the hypergeometric distribution. Transformations of them occur in the quantum theory of angular momentum and they and their duals occur in some settings of coding theory.

The polynomials considered in the first 10 chapters which have explicit formulas are given as generalized hypergeometric series. These are series whose term ratio is a rational function of n. In Chapters 11 to 19 a different setting occurs, that of basic hypergeometric series. These are series whose term ratio is a rational function of q^n.

In the 19th century Markov and Stieltjes found examples of orthogonal polynomials which can be written as basic hypergeometric series and found an explicit orthogonality relation. As mentioned earlier, Rogers also found some polynomials which are orthogonal and can be given as basic hypergeometric series, but he was unaware they were orthogonal. A few other examples were found before Wolfgang Hahn

wrote a major paper, (Hahn, 1949b) in which he found basic hypergeometric extensions of the classical polynomials and the discrete ones up to the Hahn polynomial
level. There is one level higher than this where orthogonal polynomials exist which
have properties very similar to many of those known for the classical orthogonal
polynomials. In particular, they satisfy a second-order divided q-difference equation
and this divided q-difference operator applied to them gives another set of orthogonal
polynomials. When this was first published, the polynomials were treated directly
without much motivation. Here simpler cases are done first and then a boot-strap
argument allows one to obtain more general polynomials, eventually working up to
the most general classical type sets of orthogonal polynomials.

The most general of these polynomials has four free parameters in addition to the
q of basic hypergeometric series. When three of the parameters are held fixed and
the fourth is allowed to vary, the coefficients which occur when one is expanded
in terms of the other are given as products. The resulting identity contains a very
important transformation formula between a balanced $_4\phi_3$ and a very-well-poised
$_8\phi_7$ which Watson found in the 1920s as the master identity which contains the
Rogers-Ramanujan identities as special cases and many other important formulas.
There are many ways to look at this identity of Watson, and some of these ways
lead to interesting extensions. When three of the four parameters are shifted and this
connection problem is solved, the coefficients are single sums rather than the double
sums which one expects. At present we do not know what this implies, but surprising
results are usually important, even if it takes a few decades to learn what they imply.

The fact that there are no more classical type polynomials beyond those mentioned
in the last paragraph follows from a theorem of Leonard (Leonard, 1982). This
theorem has been put into a very attractive setting by Terwilliger, some of whose
work has been summarized in Chapter 20. However, that is not the end since there
are biorthogonal rational functions which have recently been discovered. Some of
this work is contained in Chapter 18. There is even one higher level than basic
hypergeometric functions, elliptic hypergeometric functions. Gasper and Rahman
have included a chapter on them in (Gasper & Rahman, 2004).

Chapters 22 and 23 were written by Walter Van Assche. The first is on the
Riemann-Hilbert method of studying orthogonal polynomials. This is a very powerful method for deriving asymptotics of wide classes of orthogonal polynomials. The
second chapter is on multiple orthogonal polynomials. These are polynomials in one
variable which are orthogonal with respect to r different measures. The basic ideas
go back to the 19th century, but except for isolated work which seems to start with
Angelesco in 1919, it has only been in the last 20 or so years that significant work
has been done on them.

There are other important results in this book. One which surprised me very much
is the q-version of Airy functions, at least as the two appear in asymptotics. See, for
example, Theorem 21.7.3.

When I started to work on orthogonal polynomials and special functions, I was told by a number of people that the subject was out-of-date, and some even said dead. They were wrong. It is alive and well. The one variable theory is far from finished, and the multivariable theory has grown past its infancy but not enough for us to be able to predict what it will look like in 2100.

Madison, WI   Richard A. Askey
April 2005

Preface

I first came across the subject of orthogonal polynomials when I was a student at Cairo University in 1964. It was part of a senior-level course on special functions taught by the late Professor Foad M. Ragab. The instructor used his own notes, which were very similar in spirit to the way Rainville treated the subject. I enjoyed Ragab's lectures and, when I started graduate school in 1968 at the Univerity of Alberta, I was fortunate to work with Waleed Al-Salam on special functions and q-series. Jerry Fields taught me asymptotics and was very generous with his time and ideas. In the late 1960s, courses in special functions were a rarity at North American universities and have been replaced by Bourbaki-type mathematics courses. In the early 1970s, Richard Askey emerged as the leader in the area of special functions and orthogonal polynomials, and the reader of this book will see the enormous impact he made on the subject of orthogonal polynomials. At the same time, George Andrews was promoting q-series and their applications to number theory and combinatorics. So when Andrews and Askey joined forces in the mid-1970s, their combined expertise advanced the subject in leaps and bounds. I was very fortunate to have been part of this group and to participate in these developments. My generation of special functions / orthogonal polynomials people owes Andrews and Askey a great deal for their ideas which fueled the subject for a while, for the leadership role they played, and for taking great care of young people.

This book project started in the early 1990s as lecture notes on q-orthogonal polynomials with the goal of presenting the theory of the Askey–Wilson polynomials in a way suitable for use in the classroom. I taught several courses on orthogonal polynomials at the University of South Florida from these notes, which evolved with time. I later realized that it would be better to write a comprehensive book covering all known systems of orthogonal polynomials in one variable. I have attempted to include as many applications as possible. For example, I included treatments of the Toda lattice and birth and death processes. Applications of connection relations for q-polynomials to the evaluation of integrals and the Rogers–Ramanujan identities are also included. To the best of my knowledge, my treatment of associated orthogonal polynomials is a first in book form. I tried to include all systems of orthogonal polynomials but, in order to get the book out in a timely fashion, I had to make some compromises. I realized that the chapters on Riemann–Hilbert problems and multiple orthogonal polynomials should be written by an expert on the subject, and

Walter Van Assche kindly agreed to write this material. He wrote Chapters 22 and 23, except for §22.8. Due to the previously mentioned time constraints, I was unable to treat some important topics. For example, I covered neither the theories of matrix orthogonal polynomials developed by Antonio Durán, Yuan Xu and their collaborators, nor the recent interesting explicit systems of Grünbaum and Tirao and of Durán and Grünbaum. I hope to do so if the book has a second edition. Regrettably, neither the Sobolov orthogonal polynomials nor the elliptic biorthogonal rational functions are treated.

Szegő's book on orthogonal polynomials inspired generations of mathematicians. The character of this volume is very different from Szegő's book. We are mostly concerned with the special functions aspects of orthogonal polynomials, together with some general properties of orthogonal polynomial systems. We tried to minimize the possible overlap with Szegő's book. For example, we did not treat the refined bounds on zeros of Jacobi, Hermite and Laguerre polynomials derived in (Szegő, 1975) using Sturmian arguments. Although I tried to cover a broad area of the subject matter, the choice of the material is influenced by the author's taste and personal bias.

Dennis Stanton has used parts of this book in a graduate-level course at the University of Minnesota and kindly supplied some of the exercises. His careful reading of the book manuscript and numerous corrections and suggestions are greatly appreciated. Thanks also to Richard Askey and Erik Koelink for reading the manuscript and providing a lengthy list of corrections and additional information. I am grateful to Paul Terwilliger for his extensive comments on §20.3.

I hope this book will be useful to students and researchers alike. It has a collection of open research problems in Chapter 24 whose goal is to challenge the reader's curiosity. These problems have varying degrees of difficulty, and I hope they will stimulate further research in this area.

Many people contributed to this book directly or indirectly. I thank the graduate students and former graduate students at the University of South Florida who took orthogonal polynomials and special functions classes from me and corrected misprints. In particular, I thank Plamen Simeonov, Jacob Christiansen, and Jemal Gishe. Mahmoud Annaby and Zeinab Mansour from Cairo University also sent me helpful comments. I learned an enormous amount of mathematics from talking to and working with Richard Askey, to whom I am eternally grateful. I am also indebted to George Andrews for personally helping me on many occasions and for his work which inspired parts of my research and many parts of this book. The book by Gasper and Rahman (Gasper & Rahman, 1990) has been an inspiration for me over many years and I am happy to see the second edition now in print (Gasper & Rahman, 2004). It is the book I always carry with me when I travel, and I "never leave home without it." I learned a great deal of mathematics and picked up many ideas from collaboration with other mathematicians. In particular I thank my friends Christian Berg, Yang Chen, Ted Chihara, Jean Letessier, David Masson, Martin Muldoon, Jim Pitman, Mizan Rahman, Dennis Stanton, Galliano Valent, and Ruiming Zhang for the joy of having them share their knowledge with me and for the pleasure of working with them. P. G. (Tim) Rooney helped me early in my career, and was very generous with his time. Thanks, Tim, for all the scientific help and post-doctorate support.

This book was mostly written at the University of South Florida (USF). All the typesettng was done at USF, although during the last two years I was employed by the University of Central Florida. I thank Marcus McWaters, the chair of the Mathematics Department at USF, for his encouragement and continued support which enabled me to complete this book. It is my pleasure to acknowledge the enormous contribution of Denise L. Marks of the University of South Florida. She was always there when I needed help with this book or with any of my edited volumes. On many occasions, she stayed after office hours in order for me to meet deadlines. Working with Denise has always been a pleasure, and I will greatly miss her in my new job at the University of Central Florida.

In closing I thank the staff at Cambridge University Press, especially David Tranah, for their support and cooperation during the preparation of this volume and I look forward to working with them on future projects.

Orlando, FL   Mourad E.H. Ismail
April 2005

1

Preliminaries

In this chapter we collect results from linear algebra and real and complex analysis which we shall use in this book. We will also introduce the definitions and terminology used. Some special functions are also introduced in the present chapter, but the q-series and related material are not defined until Chapter 11. See Chapters 11 and 12 for q-series.

1.1 Hermitian Matrices and Quadratic Forms

Recall that a matrix $A = (a_{j,k})$, $1 \leq j, k \leq n$ is called Hermitian if

$$\overline{a_{j,k}} = a_{k,j}, \quad 1 \leq j, k \leq n. \tag{1.1.1}$$

We shall use the following inner product on the n-dimensional complex space $\mathbb{C}^n$,

$$(\mathbf{x}, \mathbf{y}) = \sum_{j=1}^{n} x_j \overline{y_j}, \quad \mathbf{x} = (x_1, \dots, x_n)^T, \quad \mathbf{y} = (y_1, \dots, y_n)^T, \tag{1.1.2}$$

where A^T is the transpose of A. Clearly

$$(\mathbf{x}, \mathbf{y}) = \overline{(\mathbf{y}, \mathbf{x})}, \quad (a\mathbf{x}, \mathbf{y}) = a(\mathbf{x}, \mathbf{y}), \quad a \in \mathbb{C}.$$

Two vectors $\mathbf{x}$ and $\mathbf{y}$ are called orthogonal if $(\mathbf{x}, \mathbf{y}) = 0$. The adjoint A^* of A is the matrix satisfying

$$(A\mathbf{x}, \mathbf{y}) = (\mathbf{x}, A^*\mathbf{y}). \tag{1.1.3}$$

It is easy to see that if $A = (a_{j,k})$ then $A^* = (\overline{a_{k,j}})$. Thus, A is Hermitian if and only if $A^* = A$. The eigenvalues of Hermitian matrices are real. This is so since $A\mathbf{x} = \lambda\mathbf{x}$, $\mathbf{x} \neq 0$ then

$$\lambda(\mathbf{x}, \mathbf{x}) = (A\mathbf{x}, \mathbf{x}) = (\mathbf{x}, A^*\mathbf{x}) = (\mathbf{x}, \lambda\mathbf{x}) = \overline{\lambda}(\mathbf{x}, \mathbf{x}).$$

Furthermore, the eigenvectors corresponding to distinct eigenvalues are orthogonal. This is the case because if $A\mathbf{x} = \lambda_1\mathbf{x}$ and $A\mathbf{y} = \lambda_2\mathbf{y}$ then

$$\lambda_1(\mathbf{x}, \mathbf{y}) = (A\mathbf{x}, \mathbf{y}) = (\mathbf{x}, A\mathbf{y}) = \lambda_2(\mathbf{x}, \mathbf{y}),$$

hence $(\mathbf{x}, \mathbf{y}) = 0$.

Any Hermitian matrix generates a quadratic form

$$\sum_{j,k=1}^{n} a_{j,k}\, \overline{x_j}\, x_k, \tag{1.1.4}$$

and conversely any quadratic form with $\overline{a_{j,k}} = a_{k,j}$ determines a Hermitian matrix A through

$$\sum_{j,k=1}^{n} a_{j,k}\, \overline{x_j}\, x_k = \mathbf{x}^* A \mathbf{x} = (A\mathbf{x}, \mathbf{x}). \tag{1.1.5}$$

In an infinite dimensional Hilbert space $\mathcal{H}$, the adjoint is defined by (1.1.3) provided it holds for all $x, y \in \mathcal{H}$. A linear operator A defined in $\mathcal{H}$ is called self-adjoint if $A = A^*$. On the other hand, A is called symmetric if

$$(Ax, y) = (x, Ay)$$

whenever both sides are defined.

Theorem 1.1.1 *Assume that the entries of a matrix A satisfy $|a_{j,k}| \le M$ for all j, k and that each row of A has at most ℓ nonzero entries. Then all the eigenvalues of A satisfy*

$$|\lambda| \le \ell M.$$

Proof Take $\mathbf{x}$ to be an eigenvector of A with an eigenvalue λ, and assume that $\|\mathbf{x}\| = 1$. Observe that the Cauchy–Schwartz inequality implies

$$|\lambda|^2 = |(A\mathbf{x}, \mathbf{x})|^2 = \left| \sum_{j,k=1}^{n} a_{j,k}\, \overline{x_j}\, x_k \right|^2 \le \|x\|^2 \sum_{j=1}^{n} \left| \sum_{k=1}^{n} a_{j,k} x_k \right|^2$$

$$\le \ell^2 M^2.$$

Hence the theorem is proved. $\qquad\square$

A quadratic form (1.1.4) is *positive definite* if $(A\mathbf{x}, \mathbf{x}) > 0$ for any nonzero $\mathbf{x}$. Recall that a matrix U is unitary if $U^* = U^{-1}$. The spectral theorem for Hermitian matrices is:

Theorem 1.1.2 *For every Hermitian matrix A there is a unitary matrix U whose columns are the eigenvectors of A such that*

$$A = U^* \Lambda U, \tag{1.1.6}$$

and Λ is the diagonal matrix formed by the corresponding eigenvalues of A.

For a proof see (Horn & Johnson, 1992). An immediate consequence of Theorem 1.1.2 is the following corollary.

Corollary 1.1.3 *The quadratic form* (1.1.4) *is reducible to a sum of squares,*

$$\sum_{j,k=1}^{n} a_{j,k}\,\overline{x_j}\,x_k = \sum_{k=1}^{n} \lambda_k\,|y_k|^2, \qquad (1.1.7)$$

where $\mathbf{y} = U\mathbf{x}$, *and* $\lambda_1, \ldots, \lambda_n$ *are the eigenvalues of* A.

The following important characterization of positive definite forms follows from Corollary 1.1.3.

Theorem 1.1.4 *The quadratic form* (1.1.4)–(1.1.5) *is positive definite if and only if the eigenvalues of* A *are positive.*

We next state the Sylvester criterion for positive definiteness (Shilov, 1977), (Horn & Johnson, 1992).

Theorem 1.1.5 *The quadratic form* (1.1.5) *is positive definite if and only if the principal minors of* A, *namely*

$$a_{1,1},\ \begin{vmatrix} a_{1,1} & a_{1,2} \\ a_{2,1} & a_{2,2} \end{vmatrix}, \ldots, \begin{vmatrix} a_{1,1} & a_{1,2} & \cdots & a_{1,n} \\ a_{2,1} & a_{2,2} & \cdots & a_{2,n} \\ \vdots & \vdots & \vdots & \vdots \\ a_{n,1} & a_{n,2} & \cdots & a_{n,n} \end{vmatrix}, \qquad (1.1.8)$$

are positive.

Recall that a matrix $A = (a_{j,k})$ is called **strictly diagonally dominant** if

$$2\,|a_{j,j}| > \sum_{k=1}^{n} |a_{j,k}|. \qquad (1.1.9)$$

The following criterion for positive definiteness is in (Horn & Johnson, 1992, Theorem 6.1.10).

Theorem 1.1.6 *Let* A *be* $n \times n$ *matrix which is Hermitian, strictly diagonally dominant, and its diagonal entries are positive. Then* A *is positive definite.*

1.2 Some Real and Complex Analysis

We need some standard results from real and complex analysis which we shall state without proofs and provide references to where proofs can be found. We shall normalize functions of bounded variations to be continuous on the right.

Theorem 1.2.1 (Helly's selection principle) *Let* $\{\psi_n(x)\}$ *be a sequence of uniformly bounded nondecreasing functions. Then there is a subsequence* $\{\psi_{k_n}(x)\}$ *which converges to a nondecreasing bounded function,* ψ. *Moreover if for every* n *the moments* $\int_{\mathbb{R}} x^m\, d\psi_n(x)$ *exist for all* m, $m = 0, 1, \ldots$, *then the moments of* ψ *exist and* $\int_{\mathbb{R}} x^m\, d\psi_{n_k}(x)$ *converges to* $\int_{\mathbb{R}} x^m\, d\psi(x)$. *Furthermore if* $\{\psi_n(x)\}$ *does not converge, then there are at least two such convergent subsequences.*

For a proof we refer the reader to Section 3 of the introduction to Shohat and Tamarkin (Shohat & Tamarkin, 1950).

Theorem 1.2.2 (Vitali) *Let $\{f_n(z)\}$ be a sequence of functions analytic in a domain $\mathcal{D}$ and assume that $f_n(z) \to f(z)$ pointwise in $\mathcal{D}$. Then $f_n(z) \to f(z)$ uniformly in any subdomain bounded by a contour C, provided that C is contained in $\mathcal{D}$.*

A proof is in Titchmarsh (Titchmarsh, 1964, page 168).

We now briefly discuss the Lagrange inversion and state two useful identities that will be used in later chapters.

Theorem 1.2.3 (Lagrange) *Let $f(z)$ and $\phi(z)$ be functions of z analytic on and inside a contour C containing the point a in its interior. Let t be such that $|t\phi(z)| < |z - a|$ on the contour C. Then the equation*

$$\zeta = a + t\phi(\zeta), \tag{1.2.1}$$

regarded as an equation in ζ, has one root interior to C; and further any function of ζ analytic on the closure of the interior of C can be expanded as a power series in t by the formula

$$f(\zeta) = f(a) + \sum_{n=1}^{\infty} \frac{t^n}{n!} \left[\frac{d^{n-1} f(x)[\phi(x)]^n}{dx^{n-1}} \right]_{x=a}. \tag{1.2.2}$$

See Whittaker and Watson (Whittaker & Watson, 1927, §7.32), or Polya and Szegő (Pólya & Szegő, 1972, p. 145). An equivalent form is

$$\frac{f(\zeta)}{1 - t\phi'(\zeta)} = \sum_{n=0}^{\infty} \frac{t^n}{n!} \left[\frac{d^n f(x)[\phi(x)]^n}{dx^n} \right]_{x=a}. \tag{1.2.3}$$

Two important special cases are $\phi(z) = e^z$, or $\phi(z) = (1 + z)^\beta$. These cases lead to:

$$e^{\alpha z} = 1 + \sum_{n=1}^{\infty} \frac{\alpha(\alpha + n)^{n-1}}{n!} w^n, \quad w = ze^{-z}, \tag{1.2.4}$$

$$(1 + z)^\alpha = 1 + \alpha \sum_{n=1}^{\infty} \binom{\alpha + \beta n - 1}{n - 1} \frac{w^n}{n}, \quad w = z(1 + z)^{-\beta}. \tag{1.2.5}$$

We say that (Olver, 1974)

$$f(z) = O(g(z)), \quad \text{as } z \to a,$$

if $f(z)/g(z)$ is bounded in a neighborhood of $z = a$. On the other hand we write

$$f(z) = o(g(z)), \quad \text{as } z \to a$$

if $f(z)/g(z) \to 0$ as $z \to a$.

A very useful method to determine the large n behavior of orthogonal polynomials $\{p_n(x)\}$ is Darboux's asymptotic method.

Theorem 1.2.4 *Let* $f(z)$ *and* $g(z)$ *be analytic in* $\{z : |z| < r\}$ *and assume that*

$$f(z) = \sum_{n=0}^{\infty} f_n z^n, \quad g(z) = \sum_{n=0}^{\infty} g_n z^n, \quad |z| < r. \qquad (1.2.6)$$

If $f - g$ *is continuous on the closed disc* $\{z : |z| \le r\}$ *then*

$$f_n = g_n + o\left(r^{-n}\right). \qquad (1.2.7)$$

This form of Darboux's method is in (Olver, 1974, Ch. 8) and, in view of Cauchy's formulas, is just a restatement of the Riemann–Lebesgue lemma. For a given function f, g is called a comparison function. Another proof of Darboux's lemma is in (Knuth & Wilf, 1989).

In order to apply Darboux's method to a sequence $\{f_n\}$ we need first to find a generating function for the f_n's, that is, find a function whose Taylor series expansion around $z = 0$ has coefficients $c_n f_n$, for some simple sequence $\{c_n\}$. In this work we pay particular attention to generating functions of orthogonal polynomials and Darboux's method will be used to derive asymptotic expansions for some of the orthogonal polynomials treated in this work. The recent work (Wong & Zhao, 2005) shows how Darboux's method can be used to derive uniform asymptotic expansions. This is a major simplification of the version in (Fields, 1967). Wang and Wong developed a discrete version of the Liouville–Green approximation (WKB) in (Wang & Wong, 2005a). This gives uniform asymptotic expansions of a basis of solutions of three-term recurrence relations. This technique is relevant, because all orthogonal polynomials satisfy three-term recurrence relations.

The Perron–Stieltjes inversion formula, see (Stone, 1932, Lemma 5.2), is

$$F(z) = \int_{\mathbb{R}} \frac{d\mu(t)}{z - t}, \quad z \notin \mathbb{R} \qquad (1.2.8)$$

if and only if

$$\mu(t) - \mu(s) = \lim_{\epsilon \to 0^+} \int_s^t \frac{F(x - i\epsilon) - F(x + i\epsilon)}{2\pi i} \, dx. \qquad (1.2.9)$$

The above inversion formula enables us to recover μ from knowing its Stieltjes transform $F(z)$.

Remark 1.2.1 *It is clear that if* μ *has an isolated atom* u *at* $x = a$ *then* $z = a$ *will be a pole of* F *with residue equal to* u. *Conversely, the poles of* F *determine the location of the isolated atoms of* μ *and the residues determine the corresponding masses. Formula (1.2.9) captures this behavior and reproduces the residue at an isolated singularity.*

Remark 1.2.2 *Formula (1.2.9) shows that the absolutely continuous component of* μ *is given by*

$$\mu'(x) = \left[F\left(x - i0^+\right) - F\left(x + i0^+\right) \right] / (2\pi i). \qquad (1.2.10)$$

An analytic function defined on a closed disc is bounded and its absolute value attains its maximum on the boundary.

Definition 1.2.1 *Let f be an entire function. The maximum modulus is*

$$M(r;f) := \sup\{|f(z)| : |z| \le r\}, \quad r > 0. \tag{1.2.11}$$

The order of f, $\rho(f)$ is defined by

$$\rho(f) := \limsup_{r\to\infty} \frac{\ln\ln M(r,f)}{\ln r}. \tag{1.2.12}$$

Theorem 1.2.5 ((Boas, Jr., 1954)) *If $\rho(f)$ is finite and is not equal to a positive integer, then f has infinitely many zeros.*

If f has finite order, its type σ is

$$\sigma = \inf\{K : M(r) < \exp(Kr^\rho)\}. \tag{1.2.13}$$

For an entire function of finite order and type we define the Phragmén–Lindelöf indicator $h(\theta)$ as

$$h(\theta) = \overline{\lim_{r\to\infty}} \frac{\ln|f(re^{i\theta})|}{r^\rho}. \tag{1.2.14}$$

Consider the infinite product

$$P = \prod_{n=1}^{\infty}(1+a_n). \tag{1.2.15}$$

We say the P converges to ℓ, $\ell \ne 0$, if

$$\lim_{m\to\infty}\prod_{n=1}^{m}(1+a_n) = \ell.$$

If $\ell = 0$ we say P diverges to zero. One can prove, see (Rainville, 1960, Chapter 1), that $a_n \to 0$ is necessary for P to converge. Similarly, one can define absolute convergence of infinite products. When $a_n = a_n(z)$ are functions of z, say, we say that P converges uniformly in a domain $\mathcal{D}$ if the partial products

$$\prod_{n=1}^{m}(1+a_n(z))$$

converge uniformly in $\mathcal{D}$ to a function with no zeros in $\mathcal{D}$.

Definition 1.2.2 *Given a set of distinct points $\{x_j : 1 \le j \le n\}$, the Lagrange fundamental polynomial $\ell_k(x)$ is*

$$\ell_k(x) = \prod_{\substack{j=1\\j\ne k}}^{n} \frac{(x-x_j)}{(x_k-x_j)} = \frac{S_n(x)}{S_n'(x_k)(x-x_k)}, \quad 1 \le k \le n, \tag{1.2.16}$$

where $S_n(x) = \prod_1^n (x - x_j)$. The Lagrange interpolation polynomial of a function $f(x)$ at the nodes $x_1, \ldots, x_n$ is the unique polynomial $L(x)$ of degree $n - 1$ such that $f(x_j) = L(x_j)$.

It is easy to see that $L(x)$ in Definition 1.2.2 is

$$L(x) = \sum_{k=1}^n \ell_k(x) f(x_k) = \sum_{k=1}^n f(x_k) \frac{S_n(x)}{S'_n(x_k)(x - x_k)}. \tag{1.2.17}$$

Theorem 1.2.6 (Poisson Summation Formula) *Let $f \in L_1(\mathbb{R})$ and F be its Fourier transform,*

$$F(t) = \frac{1}{2\pi} \int_{\mathbb{R}} f(x) e^{-ixt}\, dx, \quad t \in \mathbb{R}.$$

Then

$$\sum_{k=-\infty}^{\infty} f(2k\pi) = \sum_{n=-\infty}^{\infty} \frac{1}{2\pi} \int_{\mathbb{R}} f(x) e^{-inx}\, dx.$$

For a proof, see (Zygmund, 1968, §II.13).

Theorem 1.2.7 *Given two differential equations in the form*

$$\frac{d^2u}{dz^2} + f(z)u(z) = 0, \quad \frac{d^2v}{dz^2} + g(z)v(z) = 0,$$

then $y = uv$ satisfies

$$\frac{d}{dz}\left\{\frac{y''' + 2(f + g)y' + (f' + g')y}{f - g}\right\} + (f - g)y = 0, \; \text{if } f \neq g \tag{1.2.18}$$

$$y''' + 4fy' + 2f'y = 0, \quad \text{if } f = g. \tag{1.2.19}$$

A proof of Theorem 1.2.6 is in Watson (Watson, 1944, §5.4), where he attributes the theorem to P. Appell.

Lemma 1.2.8 *Let $y = y(x)$ satisfy the differential equation*

$$\phi(x)y''(x) + y(x) = 0, \quad a < x < b \tag{1.2.20}$$

where $\phi(x) > 0$, and $\phi'(x)$ is positive (negative) and continuous on (a, b). Then the successive relative maxima of $|y|$, increase (decrease) with x in (a, b) if ϕ increases (decreases) on (a, b).

Proof Let

$$f(x) := \{y(x)\}^2 + \phi(x)\{y'(x)\}^2, \tag{1.2.21}$$

so that $f(x) = \{y(x)\}^2$ if $y'(x) = 0$. Clearly

$$f'(x) = y'(x)\{2y(x) + \phi'(x)y'(x) + 2\phi(x)y''(x)\}$$
$$= \phi'(x)\{y'(x)\}^2.$$

Thus sign $f'(x) = \text{sign } \phi'$ in between the consecutive successive maxima of $|y|$ and the result follows.   □

1.3 Some Special Functions

Standard references in this area are (Andrews et al., 1999), (Bailey, 1935), (Rainville, 1960), (Erdélyi et al., 1953b), (Slater, 1964), (Whittaker & Watson, 1927).

The gamma and beta functions are probably the most important functions in mathematics beyond the exponential and logarithmic functions. Recall that

$$\Gamma(z) = \int_0^\infty t^{z-1} e^{-t}\, dt, \quad \text{Re } z > 0, \tag{1.3.1}$$

$$B(x,y) = \int_0^1 t^{x-1}(1-t)^{y-1}\, dt, \quad \text{Re } x > 0, \quad \text{Re } y > 0. \tag{1.3.2}$$

They are related through

$$B(x,y) = \Gamma(x)\Gamma(y)/\Gamma(x+y). \tag{1.3.3}$$

The functional equation

$$\Gamma(z+1) = z\Gamma(z) \tag{1.3.4}$$

extends the gamma function to a meromorphic function with poles at $z = 0, -1, \ldots,$ and also extends $B(x, y)$ to a meromorphic function of x and y. The Mittag–Leffler expansion for Γ'/Γ is (Whittaker & Watson, 1927, §12.3)

$$\frac{\Gamma'(z)}{\Gamma(z)} = -\gamma - \frac{1}{z} - \sum_{n=1}^{\infty}\left[\frac{1}{z+n} - \frac{1}{n}\right], \tag{1.3.5}$$

where γ is the Euler constant, (Rainville, 1960, §7).

The shifted factorial is

$$(a)_0 := 1, \quad (a)_n = a(a+1)\cdots(a+n-1), \quad n > 0, \tag{1.3.6}$$

hence (1.3.4) gives

$$(a)_n = \Gamma(a+n)/\Gamma(a). \tag{1.3.7}$$

The shifted factorial is also called Pochhammer symbol. Note that (1.3.7) is meaningful for any complex n, when $a + n$ is not a pole of the gamma function. The gamma function and the shifted factorial satisfy the duplication formulas

$$\Gamma(2z) = 2^{2z-1}\Gamma(z)\Gamma(z+1/2)/\sqrt{\pi}, \quad (2a)_{2n} = 2^{2n}(a)_n(a+1/2)_n. \tag{1.3.8}$$

We also have the reflection formula

$$\Gamma(z)\Gamma(1-z) = \frac{\pi}{\sin \pi z}. \tag{1.3.9}$$

We define the multishifted factorial as

$$(a_1, \cdots, a_m)_n = \prod_{j=1}^{m}(a_j)_n.$$

Some useful identities are

$$(a)_m(a+m)_n = (a)_{m+n}, \quad (a)_{N-k} = \frac{(a)_N(-1)^k}{(-a-N+1)_k}. \tag{1.3.10}$$

A hypergeometric series is

$$_rF_s\left(\begin{array}{c} a_1, \ldots, a_r \\ b_1, \ldots, b_s \end{array}\middle| z\right) = {}_rF_s(a_1, \ldots, a_r; b_1, \ldots, b_s; z)$$

$$= \sum_{n=0}^{\infty} \frac{(a_1, \ldots, a_r)_n}{(b_1, \ldots, b_s)_n} \frac{z^n}{n!}. \tag{1.3.11}$$

If one of the numerator parameters is a negative integer, say $-k$, then the series (1.3.11) becomes a finite sum, $0 \le n \le k$ and the $_rF_s$ series is called *terminating*. As a function of z nonterminating series is entire if $r \le s$, is analytic in the unit disc if $r = s+1$. The hypergeometric function $_2F_1(a, b; c; z)$ satisfies the hypergeometric differential equation

$$z(1-z)\frac{d^2y}{dz^2} + [c - (a+b+1)z]\frac{dy}{dz} - aby = 0. \tag{1.3.12}$$

The confluent hypergeometric function (Erdélyi et al., 1953b, §6.1)

$$\Phi(a, c; z) := {}_1F_1(a; c; z) \tag{1.3.13}$$

satisfies the differential equation

$$z\frac{d^2y}{dz^2} + (c - z)\frac{dy}{dz} - ay = 0, \tag{1.3.14}$$

and $\lim_{b \to \infty} {}_2F_1(a, b; cz/b) = {}_1F_1(a; c; z)$. The Tricomi Ψ function is a second linear independent solution of (1.3.14) and is defined by (Erdélyi et al., 1953b, §6.5)

$$\Psi(a, c; x) := \frac{\Gamma(1-c)}{\Gamma(a-c+1)}\Phi(a, c; x) + \frac{\Gamma(c-1)}{\Gamma(a)}x^{1-c}\Phi(a-c+1, 2-c; x). \tag{1.3.15}$$

The function of Ψ has the integral presentation (Erdélyi et al., 1953a, §6.5)

$$\Psi(a, c; x) = \frac{1}{\Gamma(a)}\int_0^{\infty} e^{-xt}t^{a-1}(1+t)^{c-a-1}\, dt, \tag{1.3.16}$$

for $\mathrm{Re}\, a > 0$, $\mathrm{Re}\, x > 0$.

The Bessel function J_ν and the modified Bessel function I_ν, (Watson, 1944) are

$$J_\nu(z) = \sum_{n=0}^{\infty} \frac{(-1)^n (z/2)^{\nu+2n}}{\Gamma(n+\nu+1)\, n!},$$

$$I_\nu(z) = \sum_{n=0}^{\infty} \frac{(z/2)^{\nu+2n}}{\Gamma(n+\nu+1)\, n!}. \tag{1.3.17}$$

Clearly $I_\nu(z) = e^{-i\pi\nu/2}J_\nu\left(ze^{i\pi/2}\right)$. Furthermore

$$J_{1/2}(z) = \sqrt{\frac{2}{\pi z}}\sin z, \quad J_{-1/2}(z) = \sqrt{\frac{2}{\pi z}}\cos z. \tag{1.3.18}$$

The Bessel functions satisfy the recurrence relation

$$\frac{2\nu}{z} J_\nu(z) = J_{\nu+1}(z) + J_{\nu-1}(z). \tag{1.3.19}$$

The Bessel functions J_ν and $J_{-\nu}$ satisfy

$$x^2 \frac{d^2y}{dx^2} + x \frac{dy}{dx} + \left(x^2 - \nu^2\right) y = 0. \tag{1.3.20}$$

When ν is not an integer J_ν and $J_{-\nu}$ are linear independent solutions of (1.3.20) whose Wronskian is (Watson, 1944, §3.12)

$$W\{J_\nu(x), J_{-\nu}(x)\} = -\frac{2 \sin(\nu\pi)}{\pi x}, \quad W\{f, g\} := fg' - gf'. \tag{1.3.21}$$

The function I_ν satisfies the differential equation

$$x^2 \frac{d^2y}{dx^2} + x \frac{dy}{dx} - \left(x^2 + \nu^2\right) y = 0, \tag{1.3.22}$$

whose second solution is

$$K_\nu(x) = \frac{\pi}{2} \frac{I_{-\nu}(x) - I_\nu(x)}{\sin(\pi\nu)}, \tag{1.3.23}$$

$$K_n(x) = \lim_{\nu \to n} K_\nu(x), \quad n = 0, \pm 1, \dots.$$

We also have

$$I_{\nu-1}(x) - I_{\nu+1}(x) = \frac{2\nu}{x} I_\nu(x),$$

$$K_{\nu+1}(x) - K_{\nu-1}(x) = \frac{2\nu}{x} K_\nu(x). \tag{1.3.24}$$

Theorem 1.3.1 *When $\nu > -1$, the function $z^{-\nu} J_\nu(z)$ has only real and simple zeros. Furthermore, the positive (negative) zeros of $J_\nu(z)$ and $J_{\nu+1}(z)$ interlace for $\nu > -1$.*

We shall denote the positive zeros of $J_\nu(z)$ by $\{j_{\nu,k}\}$, that is

$$0 < j_{\nu,1} < j_{\nu,2} < \cdots < j_{\nu,n} < \cdots. \tag{1.3.25}$$

The Bessel functions satisfy the differential recurrence relations, (Watson, 1944)

$$zJ_\nu'(z) = \nu J_\nu(z) - zJ_{\nu+1}(z), \tag{1.3.26}$$

$$zY_\nu'(z) = \nu Y_\nu(z) - zY_{\nu+1}(z), \tag{1.3.27}$$

$$zI_\nu'(z) = zI_{\nu+1}(z) + \nu I_\nu(z), \tag{1.3.28}$$

$$zK_\nu'(z) = \nu K_\nu(z) - zK_{\nu+1}(z), \tag{1.3.29}$$

where $Y_\nu(z)$ is

$$Y_\nu(z) = \frac{J_\nu(z) \cos\nu\pi - J_{-\nu}(z)}{\sin\nu\pi}, \quad \nu \neq 0, \pm 1, \dots,$$

$$Y_n(z) = \lim_{\nu \to n} Y_\nu(z), \quad n = 0, \pm 1, \dots. \tag{1.3.30}$$

The functions $J_\nu(z)$ and $Y_\nu(z)$ are linearly independent solutions of (1.3.20).

The Bessel functions are special cases of $_1F_1$ in the sense

$$e^{-iz} {}_1F_1(\nu + 1/2; 2\nu + 1; 2iz) = \Gamma(\nu + 1)(z/2)^{-\nu} J_\nu(z), \qquad (1.3.31)$$

(Erdélyi et al., 1953b, §6.9.1).

Two interesting functions related to special cases of Bessel functions are the functions

$$k(x) = \frac{\pi}{3} \left(\frac{x}{3}\right)^{\frac{1}{2}} J_{-1/3}\left(2(x/3)^{3/2}\right) = \frac{\pi}{3} \sum_{n=0}^{\infty} \frac{(-x/3)^{3n}}{n!\,\Gamma(n + 2/3)},$$

$$\ell(x) = \frac{\pi}{3} \left(\frac{x}{3}\right)^{\frac{1}{2}} J_{1/3}\left(2(x/3)^{3/2}\right) = \frac{\pi}{9} x \sum_{n=0}^{\infty} \frac{(-x/3)^{3n}}{n!\,\Gamma(n + 4/3)}.$$

$$(1.3.32)$$

Indeed $\{k(x), \ell(x)\}$ is a basis of solutions of the Airy equation

$$\frac{d^2 y}{dx^2} + \frac{1}{3} xy = 0. \qquad (1.3.33)$$

Moreover

$$k(x) = -\ell(x)(1 + o(1)) = \frac{3^{-\frac{3}{4}}}{2} \sqrt{\pi}\, |x|^{-\frac{1}{4}} \exp\left\{2(|x|/3)^{3/2}\right\} (1 + o(1)),$$

as $x \to -\infty$. Thus the only solution of (1.3.33) which is bounded as $x \to -\infty$ is $k(x) + \ell(x)$. Set

$$A(x) := k(x) + \ell(x), \qquad (1.3.34)$$

which has the asymptotic behavior

$$A(x) = \frac{\sqrt{\pi}}{2\,3^{1/4}} |x|^{-\frac{1}{4}} \exp\left\{-2(|x|/3)^{3/2}\right\} (1 + o(1))$$

as $x \to -\infty$. The function $A(x)$ is called the Airy function and plays an important role in the theory of orthogonal polynomials with exponential weights, random matrix theory, as well as other parts of mathematical physics. The function $A(x)$ is positive on $(-\infty, 0)$ and has only positive simple zeros.

We shall use the notation

$$0 < i_1 < i_2 < \cdots, \qquad (1.3.35)$$

for the positive zeros of the Airy function.

The Appell functions generalize the hypergeometric function to two variables.

They are defined by (Appell & Kampé de Fériet, 1926), (Erdélyi et al., 1953b)

$$F_1\left(a;b,b';c;x,y\right) = \sum_{m,n=0}^{\infty} \frac{(a)_{n+m}(b)_m\,(b')_n}{(c)_{m+n}\,m!\,n!}\,x^m y^n,$$ (1.3.36)

$$F_2\left(a;b,b';c,c';x,y\right) = \sum_{m,n=0}^{\infty} \frac{(a)_{n+m}(b)_m\,(b')_n}{(c)_m(c')_n\,m!\,n!}\,x^m y^n,$$ (1.3.37)

$$F_3\left(a,a';b,b';c;x,y\right) = \sum_{m,n=0}^{\infty} \frac{(a)_m\,(a')_n\,(b)_m\,(b')_n}{(c)_{m+n}\,m!\,n!}\,x^m y^n,$$ (1.3.38)

$$F_4\left(a,b;c,c';x,y\right) = \sum_{m,n=0}^{\infty} \frac{(a)_{n+m}(b)_{m+n}}{(c)_m(c')_n\,m!\,n!}\,x^m y^n.$$ (1.3.39)

The complete elliptic integrals of the first and second kinds are (Erdélyi et al., 1953a)

$$\mathbf{K} = \mathbf{K}(k) = \int_0^1 \frac{du}{\sqrt{(1-u^2)(1-k^2 u^2)}},$$ (1.3.40)

$$\mathbf{E} = \mathbf{E}(k) = \int_0^1 \sqrt{\frac{1-k^2 u^2}{1-u^2}}\,du$$ (1.3.41)

respectively. Indeed

$$\mathbf{K}(k) = \frac{\pi}{2}\,{}_2F_1\left(1/2,1/2;1;k^2\right),$$ (1.3.42)

$$\mathbf{E}(k) = \frac{\pi}{2}\,{}_2F_1\left(-1/2,1/2;1;k^2\right).$$ (1.3.43)

We refer to k as the modulus, while the complementary modulus k' is

$$k' = \left(1-k^2\right)^{1/2}.$$ (1.3.44)

1.4 Summation Theorems and Transformations

In the shifted factorial notation the binomial theorem is

$$\sum_{n=0}^{\infty} \frac{(a)_n}{n!}\,z^n = (1-z)^{-a},$$ (1.4.1)

when $|z| < 1$ if a is not a negative integer.

The Gauss sum is

$$_2F_1\left(\begin{matrix}a,b\\c\end{matrix}\bigg|1\right) = \frac{\Gamma(c)\Gamma(c-a-b)}{\Gamma(c-a)\Gamma(c-b)},\quad \mathrm{Re}\{c-a-b\}>0.$$ (1.4.2)

The terminating version of (1.4.2) is the Chu–Vandermonde sum

$$_2F_1\left(\begin{matrix}-n,b\\c\end{matrix}\bigg|1\right) = \frac{(c-b)_n}{(c)_n}.$$ (1.4.3)

A hypergeometric series (1.3.11) is called *balanced* if $r = s + 1$ and

$$1 + \sum_{k=1}^{s+1} a_k = \sum_{k=1}^{s} b_k \qquad (1.4.4)$$

The Pfaff–Saalschütz theorem is

$$_3F_2 \left(\begin{matrix} -n, a, b \\ c, d \end{matrix} \middle| 1 \right) = \frac{(c-a)_n (c-b)_n}{(c)_n (c-a-b)_n}, \qquad (1.4.5)$$

if the balanced condition, $c + d = 1 - n + a + b$, is satisfied. In Chapter 13 we will give proofs of generalizations of (1.4.2) and (1.4.5) to q-series.

The Stirling formula for the gamma function is

$$\operatorname{Log} \Gamma(z) = \left(z - \frac{1}{2} \right) \operatorname{Log} z - z + \frac{1}{2} \ln(2\pi) + O\left(z^{-1} \right), \qquad (1.4.6)$$

$|\arg z| \le \pi - \epsilon, \epsilon > 0$. An important consequence of Stirling's formula is

$$\lim_{z \to \infty} z^{b-a} \frac{\Gamma(z+a)}{\Gamma(z+b)} = 1. \qquad (1.4.7)$$

The hypergeometric function has the Euler integral representation

$$_2F_1 \left(\begin{matrix} a, b \\ c \end{matrix} \middle| z \right)$$

$$= \frac{\Gamma(c)}{\Gamma(b)\Gamma(c-b)} \int_0^1 t^{b-1} (1-t)^{c-b-1} (1-zt)^{-a} \, dt, \qquad (1.4.8)$$

for $\operatorname{Re} b > 0$, $\operatorname{Re} (c - b) > 0$. The Pfaff–Kummer transformation is

$$_2F_1 \left(\begin{matrix} a, b \\ c \end{matrix} \middle| z \right) = (1-z)^{-a} \, _2F_1 \left(\begin{matrix} a, c-b \\ c \end{matrix} \middle| \frac{z}{z-1} \right), \qquad (1.4.9)$$

and is valid for $|z| < 1$, $|z| < |z - 1|$. An iterate of (1.4.9) is

$$_2F_1 \left(\begin{matrix} a, b \\ c \end{matrix} \middle| z \right) = (1-z)^{c-a-b} \, _2F_1 \left(\begin{matrix} c-a, c-b \\ c \end{matrix} \middle| z \right), \qquad (1.4.10)$$

for $|z| < 1$. Since $_1F_1(a; \zeta; z) = \lim_{b \to \infty} {_2F_1}(a, b; c; z/b)$, (1.4.10) yields

$$_1F_1(a; c; z) = e^z \, _1F_1(c - a; c; -z). \qquad (1.4.11)$$

Many of the summation theorems and transformation formulas stated in this section have q-analogues which will be stated and proved in §12.2.

In particular we shall give a new proof of the terminating case of Watson's theorem (Slater, 1964, (III.23))

$$_3F_2 \left(\begin{matrix} a, b, c \\ (a+b+1)/2, 2c \end{matrix} \middle| 1 \right)$$

$$= \frac{\Gamma(1/2)\Gamma(c+1/2)\Gamma((a+b+1)/2)\Gamma(c+(1-a-b)/2)}{\Gamma((a+1)/2)\Gamma((b+1)/2)\Gamma(c+(1-a)/2)\Gamma(c+(1-b)/2)} \qquad (1.4.12)$$

in §9.2. We also give a new proof of the q-analogue of (1.4.12) in the terminating and nonterminating cases. As $q \to 1$ we get (1.4.12) in its full generality. This is just one of many instances where orthogonal polynomials shed new light on the theory of evaluation of sums and integrals.

We shall make use of the quadratic transformation (Erdélyi et al., 1953a, (2.11.37))

$$
{}_2F_1 \left(\begin{matrix} a, b \\ 2b \end{matrix} \middle| z \right) = \left(\frac{1 + (1 - z)^{1/2}}{2} \right)^{-2a}
$$

$$
\times \; {}_2F_1 \left(\begin{matrix} a, a - b + \frac{1}{2} \\ b + \frac{1}{2} \end{matrix} \middle| \left[\frac{1 - (1 - z)^{1/2}}{1 + (1 - z)^{1/2}} \right]^2 \right). \tag{1.4.13}
$$

In particular,

$$
{}_2F_1 \left(\begin{matrix} a, a + 1/2 \\ 2a + 1 \end{matrix} \middle| z \right) = \left(\frac{1 + (1 - z)^{1/2}}{2} \right)^{-2a}. \tag{1.4.14}
$$

Exercises

1.1   Prove that if $c, b_1, b_2, \ldots, b_n$ are distinct complex numbers, then (Gosper et al., 1993)

$$
\prod_{k=1}^{n} \frac{x + a_k - b_k}{x - b_k} = \prod_{k=1}^{n} \frac{c + a_k - b_k}{c - b_k}
$$

$$
+ \prod_{k=1}^{n} \frac{a_k (x - c)}{(b_k - c)(x - b_k)} \prod_{\substack{j=1 \\ j \neq k}} \frac{b_k + a_j - b_j}{b_k - b_j}.
$$

This formula is called the nonlocal derangement identity.

1.2   Use Exercise 1.1 to prove that (Gosper et al., 1993)

$$
\sum_{j=1}^{\infty} \left(1 - (-1)^j \cos \left(\sqrt{j(j + a)} \, \pi \right) \right)
$$

$$
\times \frac{\Gamma \left(\left(j + \sqrt{j(j + a)} \right)/2 \right) \Gamma \left(\left(j - \sqrt{j(j + a)} \right)/2 \right)}{j! \, j} = -\frac{\pi^4 a}{12},
$$

where $\operatorname{Re} a < 4$.

1.3   Derive Sonine's first integral

$$
J_{\alpha + \beta + 1}(z) = \frac{2^{-\beta} z^{\beta + 1}}{\Gamma(\beta + 1)} \int_0^1 x^{2\beta + 1} \left(1 - x^2 \right)^{\alpha/2} J_\alpha \left(z \sqrt{1 - x^2} \right) dx,
$$

where $\operatorname{Re} \alpha > -1$ and $\operatorname{Re} \beta > -1$.

1.4   Prove the identities (1.2.4) and (1.2.5).

1.5   Prove that (Gould, 1962)

$$
f(n) = \sum_{k=0}^{n} (-1)^k \binom{n}{k} \binom{a + bk}{n} g(n)
$$

if and only if

$$\binom{a+bn}{n} g(n) = \sum_{k=0}^{n} \frac{a+bk-k}{a+bn-k} \binom{a+bn-k}{n-k} f(k).$$

Hint: Express the exponential generating function $\sum_{n=0}^{\infty} f(n)(-t)^n/n!$ in terms of $\sum_{n=0}^{\infty} g(n)(-t)^n/n!$ by using (1.2.5).

1.6   Prove the generating function

$$\sum_{n=0}^{\infty} \frac{(\lambda)_n}{n!} \phi_n(x) t^n$$

$$= (1-t)^{-\lambda} {}_{p+2}F_s \left(\begin{matrix} \lambda/2, (\lambda+1)/2, a_1, \ldots, a_p \\ b_1, \ldots, b_s \end{matrix} \middle| \frac{-4tx}{(1-t)^2} \right),$$

when $s \geq p+1$, $\left| tx/(1-t)^2 \right| < 1/4$, $|t| < 1$, where

$$\phi_n(x) = {}_{p+2}F_s \left(\begin{matrix} -n, n+\lambda, a_1, \ldots, a_p \\ b_1, \ldots, b_s \end{matrix} \middle| x \right).$$

Note: This formula and Darboux's method can be used to determine the large n asymptotics of $\phi_n(x)$, see §7.4 in (Luke, 1969a), where $\{\phi_n(x)\}$ are called extended Jacobi polynomials.

2

Orthogonal Polynomials

This chapter develops properties of general orthogonal polynomials. These are polynomials orthogonal with respect to positive measures. An application to solving the Toda lattice equations is given in §2.8.

We start with a given positive Borel measure μ on $\mathbb{R}$ with infinite support and whose moments $\int_{\mathbb{R}} x^n \, d\mu(x)$ exist for $n = 0, 1, \ldots$. Recall that the distribution function $F_\mu(x)$ of a finite Borel measure μ is $F_\mu(x) = \mu((-\infty, x])$. A distribution function is nondecreasing, right continuous, nonnegative, bounded, and $\lim_{x \to -\infty} F_\mu(x) = 0$. Conversely any function satisfying these properties is a distribution function for a measure μ and $\int_{\mathbb{R}} f(x) \, dF_\mu(x) = \int_{\mathbb{R}} f(x) \, d\mu(x)$, see (McDonald & Weiss, 1999, §4.7). Because of this fact we will use μ to denote measures or distribution functions and we hope this will not cause any confusion to our readers.

By a **polynomial sequence** we mean a sequence of polynomials, say $\{\varphi_n(x)\}$, such that $\varphi_n(x)$ has precise degree n. A polynomial sequence $\{\varphi_n(x)\}$ is called **monic** if $\varphi_n(x) - x^n$ has degree at most $n - 1$.

2.1 Construction of Orthogonal Polynomials

Given μ as above we observe that the moments

$$\mu_j := \int_{\mathbb{R}} x^j \, d\mu(x), \quad j = 0, 1, \ldots, \tag{2.1.1}$$

generate a quadratic form

$$\sum_{j,k=0}^{n} \mu_{k+j} \, \overline{x_j} \, x_k. \tag{2.1.2}$$

We shall always normalize the measures to have $\mu_0 = 1$, that is

$$\int_{\mathbb{R}} d\mu = 1. \tag{2.1.3}$$

16

The form (2.1.2) is positive definite since μ has infinite support and the expression in (2.1.2) is

$$\int_{\mathbb{R}} \left| \sum_{j=0}^{n} x_j t^j \right|^2 d\mu(t).$$

Let D_n denote the determinant

$$D_n = \begin{vmatrix} \mu_0 & \mu_1 & \cdots & \mu_n \\ \mu_1 & \mu_2 & \cdots & \mu_{n+1} \\ \vdots & \vdots & \cdots & \vdots \\ \mu_n & \mu_{n+1} & \cdots & \mu_{2n} \end{vmatrix}. \tag{2.1.4}$$

The Sylvester criterion, Theorem 1.1.5, implies the positivity of the determinants $D_0, D_1, \ldots$. The determinant D_n is a Hankel determinant.

Theorem 2.1.1 *Given a positive Borel measure μ on $\mathbb{R}$ with infinite support and finite moments, there exists a unique sequence of monic polynomials $\{P_n(x)\}_0^\infty$,*

$$P_n(x) = x^n + \text{lower order terms}, \quad n = 0, 1, \ldots,$$

and a sequence of positive numbers $\{\zeta_n\}_0^\infty$, with $\zeta_0 = 1$ such that

$$\int_{\mathbb{R}} P_m(x) P_n(x) \, d\mu(x) = \zeta_n \, \delta_{m,n}. \tag{2.1.5}$$

Proof We prove (2.1.5) for $m, n = 0, \ldots, N$, and $N = 0, \ldots$, by induction on N. Define $P_0(x)$ to be 1. Assume $P_0(x), \ldots, P_N(x)$ have been defined and (2.1.5) holds. Set $P_{N+1}(x) = x^{N+1} + \sum_{j=0}^{N} c_j x^j$. For $m < N + 1$, we have

$$\int_{\mathbb{R}} x^m P_{N+1}(x) \, d\mu(x) = \mu_{N+m+1} + \sum_{j=0}^{N} c_j \mu_{j+m}.$$

We construct the coefficients c_j by solving the system of equations

$$\sum_{j=0}^{N} c_j \mu_{j+m} = -\mu_{N+m+1}, \quad m = 0, 1, \ldots, N,$$

whose determinant is D_N and $D_N > 0$. Thus the polynomial $P_{N+1}(x)$ has been found and ζ_{N+1} is $\int_{\mathbb{R}} P_{N+1}^2(x) \, d\mu(x)$. $\qquad \square$

As a consequence of Theorem 2.1.1 we see that

$$P_n(x) = \frac{1}{D_{n-1}} \begin{vmatrix} \mu_0 & \mu_1 & \cdots & \mu_n \\ \mu_1 & \mu_2 & \cdots & \mu_{n+1} \\ \vdots & \vdots & \cdots & \vdots \\ \mu_{n-1} & \mu_n & \cdots & \mu_{2n-1} \\ 1 & x & \cdots & x^n \end{vmatrix}, \quad \zeta_n = \frac{D_n}{D_{n-1}}, \tag{2.1.6}$$

since the right-hand side of (2.1.6) satisfies the requirements of Theorem 2.1.1. The reason is that for $m < n$, $\int_{\mathbb{R}} x^m P_n(x)\, d\mu(x)$ is a determinant whose $n+1$ and $m+1$ rows are equal. To evaluate ζ_n, note that

$$\zeta_n = \int_{\mathbb{R}} P_n^2(x)\, d\mu(x) = \int_{\mathbb{R}} x^n P_n(x)\, dx$$

$$= \frac{1}{D_{n-1}} \begin{vmatrix} \mu_0 & \mu_1 & \cdots & \mu_n \\ \mu_1 & \mu_2 & \cdots & \mu_{n+1} \\ \vdots & \vdots & & \\ \mu_n & \mu_{n+1} & \cdots & \mu_{2n} \end{vmatrix} = \frac{D_n}{D_{n-1}}.$$

Remark 2.1.1 *The odd moments will vanish if μ is symmetric about the origin, hence (2.1.6) shows that $P_n(x)$ contains only the terms x^{n-2k}, $0 \le k \le n/2$.*

We now prove the following important result of Heine.

Theorem 2.1.2 *The monic orthogonal polynomials $\{P_n(x)\}$ have the Heine integral representation*

$$P_n(x) = \frac{1}{n!\, D_{n-1}} \int_{\mathbb{R}^n} \prod_{i=1}^{n} (x - x_i) \prod_{1 \le j < k \le n} (x_k - x_j)^2\, d\mu(x_1) \cdots d\mu(x_n).$$

$$(2.1.7)$$

Proof In the determinant in (2.1.6) write μ_j as $\int_{\mathbb{R}} x_k^j\, d\mu(x_k)$ in row k, $1 \le k \le n$. Thus

$$P_n(x) = \frac{1}{D_{n-1}} \int_{\mathbb{R}^n} \begin{vmatrix} 1 & x_1 & \cdots & x_1^n \\ x_2 & x_2^2 & \cdots & x_2^{n+1} \\ \vdots & \vdots & \cdots & \vdots \\ x_n^{n-1} & x_n^n & \cdots & x_n^{2n-1} \\ 1 & x & \cdots & x^n \end{vmatrix} d\mu(x_1) \cdots d\mu(x_n). \quad (2.1.8)$$

If we choose the integration variables to be $x_{k_1}, x_{k_2}, \ldots, x_{k_n}$ where $(k_1, k_2, \ldots, k_n) = \sigma(1, 2, \ldots, n)$, and σ is a permutation of $(1, 2, \ldots, n)$, then (2.1.8) becomes

$$P_n(x) = \frac{\mathrm{sign}\, \sigma}{D_{n-1}}$$

$$\times \int_{\mathbb{R}^n} \begin{vmatrix} 1 & x_1 & \cdots & x_1^n \\ 1 & x_2 & \cdots & x_2^n \\ \vdots & \vdots & \cdots & \vdots \\ 1 & x_n & \cdots & x_n^n \\ 1 & x & \cdots & x^n \end{vmatrix} x_1^{k_1} \cdots x_n^{k_n}\, d\mu(x_1) \cdots d\mu(x_n),$$

where $\mathrm{sign}\, \sigma$ is (-1) raised to the number of inversions in $\sigma(1, \ldots, n) = (k_1, \ldots, k_n)$, see (Shilov, 1977, Chapter 1). Now sum over all permutations σ of $(1, 2, \ldots, n)$, which are $n!$ in number, then divide by $n!$ we get the factor

$$\sum_\sigma (\text{sign } \sigma) x_1^{k_1} x_2^{k_2} \cdots x_n^{k_n}$$

inside the integrand. The above sum is a Vandermonde determinant hence is equal to $\prod_{1 \le j < k \le n} (x_k - x_j)$. Finally the determinant in the integrand is a Vandermonde determinant with variables $x_1, x_2, \ldots, x_n, x$, and it evaluates to

$$\prod_{i=1}^n (x - x_i) \prod_{1 \le j < k \le n} (x_k - x_j),$$

hence the result. $\qquad\square$

By equating coefficients of x^n on both sides of (2.1.7) we prove the following corollary to Heine's formula.

Corollary 2.1.3 *We have*

$$\int_{\mathbb{R}^{n+1}} \prod_{1 \le j < k \le n+1} (x_k - x_j)^2 \, d\mu(x_1) \cdots d\mu(x_{n+1}) = (n+1)! \, D_n. \qquad (2.1.9)$$

Both Heine's formula and its corollary have been used extensively in the theory of random matrices, (Mehta, 1991), (Deift, 1999). Heine's formula when $\mu'(x) = x^\alpha (1 - x)^\beta$, $x \in [0, 1]$ has been generalized by Selberg to what is known as the Selberg integral

$$\int_{[0,1]^n} \prod_{k=1}^n \left\{ x_k^{\alpha-1} (1 - x_k)^{\beta-1} \right\} \prod_{1 \le i < j \le n} |x_i - x_j|^{2\gamma} \, dx_1 \cdots dx_n$$

$$= \prod_{j=1}^n \frac{\Gamma(\alpha + (j-1)\gamma)\, \Gamma(\beta + (j-1)\gamma)\, \Gamma(1 + j\gamma)}{\Gamma(\alpha + \beta + (n + j - 2)\gamma)\, \Gamma(1 + \gamma)},$$

(Andrews et al., 1999, §8.1).

Another determinant representation for $\{P_n(x)\}$ is

$$P_n(x) = \frac{1}{D_{n-1}} \begin{vmatrix} \mu_0 x - \mu_1 & \mu_1 x - \mu_2 & \cdots & \mu_{n-1} x - \mu_n \\ \mu_1 x - \mu_2 & \mu_2 x - \mu_3 & \cdots & \mu_n x - \mu_{n+1} \\ \vdots & \vdots & \vdots & \vdots \\ \mu_{n-1} x - \mu_n & \mu_n x - \mu_{n+1} & \cdots & \mu_{2n-2} x - \mu_{2n-1} \end{vmatrix}. \qquad (2.1.10)$$

To prove (2.1.10), multiply column j in (2.1.6) by $-x$ and add it to the $j+1$ column for $j = n, n - 1, \ldots, 1$.

It is important to note that the construction of the monic orthogonal polynomials in (2.1.6) and Theorem 2.1.1 depends only on the moments and not on the original measure.

The orthonormal polynomials will be denoted by $\{p_n(x)\}$

$$p_n(x) = P_n(x)/\sqrt{\zeta_n},$$

so that

$$p_n(x) = \frac{1}{\sqrt{D_n D_{n-1}}} \begin{vmatrix} \mu_0 & \mu_1 & \cdots & \mu_n \\ \mu_1 & \mu_2 & \cdots & \mu_{n+1} \\ \vdots & \vdots & \cdots & \vdots \\ \mu_{n-1} & \mu_n & \cdots & \mu_{2n-1} \\ 1 & x & \cdots & x^n \end{vmatrix}. \tag{2.1.11}$$

We shall use the notation

$$p_n(x) = \gamma_n x^n + \text{lower order terms} \tag{2.1.12}$$

with

$$\gamma_n = \sqrt{D_{n-1}/D_n}. \tag{2.1.13}$$

Consequently,

$$D_n = 1/\left[\gamma_1^2 \cdots \gamma_n^2\right]. \tag{2.1.14}$$

Sometimes it is more convenient to use polynomial bases other than the monomials. Let $\{\phi_n(x)\}$ be a polynomial sequence with real coefficients and with $\phi_0(x) = 1$. For a given probability measure with finite moments set

$$\phi_{jk} = \int_{\mathbb{R}} \phi_j(x)\phi_k(x)\, d\mu(x). \tag{2.1.15}$$

Theorem 2.1.4 *The matrices* $\{\phi_{jk} : 0 \le j, k \le n\}$, $n = 0, 1, \ldots$, *are positive definite. With*

$$\tilde{D}_n = \begin{vmatrix} \phi_{0,0} & \phi_{0,1} & \cdots & \phi_{0,n} \\ \phi_{1,0} & \phi_{1,1} & \cdots & \phi_{1,n} \\ \vdots & & & \\ \phi_{n,0} & \phi_{n,1} & \cdots & \phi_{n,n} \end{vmatrix}, \quad n \ge 0, \tag{2.1.16}$$

the polynomials orthonormal with respect to μ are

$$p_n(x) = \frac{1}{\sqrt{\tilde{D}_n \tilde{D}_{n-1}}} \begin{vmatrix} \phi_{0,0} & \phi_{0,1} & \cdots & \phi_{0,n} \\ \phi_{1,0} & \phi_{1,1} & \cdots & \phi_{1,n} \\ & \ddots & & \\ \phi_{n-1,0} & \phi_{n-1,1} & \cdots & \phi_{n-1,n} \\ \phi_0(x) & \phi_1(x) & \cdots & \phi_n(x) \end{vmatrix}. \tag{2.1.17}$$

Proof Clearly $\phi_{jk} = \phi_{kj}$ and the quadratic form $\sum \phi_{jk} x_j x_k$ is $\int_{\mathbb{R}} |\sum x_j \phi_j(x)|^2\, d\mu(x)$ which implies the positive definiteness of the matrices $\{\phi_{jk} : 0 \le j, k \le n\}$, $n = 0, 1, \ldots$. Hence $\tilde{D}_n > 0$, $n \ge 0$. The proof of the orthonormality of $\{P_n\}$ is similar to the proof for $\phi_n(x) = x^n$. $\qquad\square$

It is our opinion that the determinant representations of orthogonal polynomials are underused. For a clever use of these representations see (Wilson, 1991).

A matrix of the form

$$\begin{pmatrix} \phi_{0,0} & \phi_{0,1} & \cdots & \phi_{0,n} \\ \phi_{1,0} & \phi_{1,1} & \cdots & \phi_{1,n} \\ \vdots & \vdots & \cdots & \vdots \\ \phi_{n,0} & \phi_{n,1} & \cdots & \phi_{n,n} \end{pmatrix}$$

is called a **Gram matrix**. A Gram matrix is always positive definite.

Remark 2.1.2 *Observe that the construction of the monic orthogonal polynomials only used the fact that $D_n \neq 0$, so the positivity of D_n was used to construct the orthonormal polynomials. Therefore, one can allow the measure μ to be a signed measure but assume that $D_n \neq 0$. When $\{P_n(x)\}$ are monic polynomials orthogonal with respect to a signed measure μ with $D_n \neq 0$, we shall call $\{P_n(x)\}$ signed orthogonal polynomials.*

Theorem 2.1.5 *Let $f \in L_2(\mu, \mathbb{R})$ and N be a positive integer. If $\{p_j(x) : 0 \leq j \leq N\}$ are orthonormal with respect to μ, then*

$$\inf\left\{\left|\int_{\mathbb{R}} f(x) - \sum_{j=0}^{N} f_j\, p_j(x)\right|^2 d\mu(x) : f_j \in \mathbb{R}, \, 0 \leq j \leq N\right\}$$

is attained if and only if $f_j = \int_{\mathbb{R}} f(x)\, p_j(x)\, d\mu(x)$. The infimum is

$$\int_{\mathbb{R}} |f(x)|^2 d\mu(x) - \sum_{j=0}^{N} |f_j|^2.$$

Proof Clearly

$$\int_{\mathbb{R}} \left| f(x) - \sum_{j=0}^{N} f_j\, p_j(x) \right|^2 d\mu(x)$$

$$= \int_{\mathbb{R}} |f(x)|^2 d\mu - 2\,\mathrm{Re}\left\{\sum_{j=0}^{N} f_j \int_{\mathbb{R}} p_j(x)\, \overline{f(x)}\, d\mu(x)\right\} + \sum_{j=0}^{N} |f_j|^2.$$

Let $a_j = \int_{\mathbb{R}} p_j(x)\, f(x)\, d\mu(x)$. The above expression is

$$\int_{\mathbb{R}} |f(x)|^2 d\mu(x) - \sum_{j=0}^{N} |a_j|^2 + \sum_{j=0}^{N} |f_j - a_j|^2,$$

which is minimized if and only if $f_j = a_j$ for $j = 0, 1, \ldots, N$. $\qquad\square$

Theorem 2.1.6 *Let N be a positive integer and let*

$$q_N(x) = x^N + \sum_{j=0}^{N-1} c_j\, x^j.$$

Then

$$\inf\left\{\int_{\mathbb{R}}|q_N(x)|^2\,d\mu(x):c_0,\dots,c_{N-1}\in\mathbb{R}\right\}$$

is attained if and only if $q_N(x)=p_N(x)/\gamma_N$.

Proof Rewrite $q_N(x)$ as $\sum_{j=0}^{N}d_j\,p_j(x)$ with $d_N=1/\gamma_N$. The rest follows from the orthogonality. □

Since an $N\times N$ Hankel matrix formed by moments of a positive measure is positive definite, it is natural to study the limiting behavior of its smallest eigenvalue, $\lambda_1(N)$. Surprisingly

$$\lim_{N\to\infty}\lambda_1(N)>0$$

if and only if the moments μ_n do not determine a unique measure μ, see (Berg et al., 2002). Berg, Chen and Ismail gave a positive lower bound for the above limit.

2.2 Recurrence Relations

Since $\{P_n\}$ forms a basis for the vector space of polynomials over $\mathbb{R}$ then

$$xP_n(x)=P_{n+1}(x)+\sum_{j=0}^{n}d_jP_j(x),$$

with $d_j\zeta_j=\int_{\mathbb{R}}xP_n(x)P_j(x)\,d\mu(x),0\le j\le n$. On the other hand $xP_n(x)P_j(x)$ is $P_n(x)$ multiplied by a polynomial of degree $j+1$, hence the integrals $\int_{\mathbb{R}}xP_n(x)P_j(x)\,d\mu(x)$ vanish for $j+1<n$, and we find that $d_0=\cdots=d_{n-2}=0$.

Theorem 2.2.1 *A monic sequence of orthogonal polynomials satisfies a three-term recurrence relation*

$$xP_n(x)=P_{n+1}(x)+\alpha_nP_n(x)+\beta_nP_{n-1}(x),\quad n>0,\qquad(2.2.1)$$

with

$$P_0(x)=1,\quad P_1(x)=x-\alpha_0,\qquad(2.2.2)$$

where α_n is real, for $n\ge0$ and $\beta_n>0$ for $n>0$.

Proof We need only to prove that $\beta_n>0$. Clearly (2.2.1) yields

$$\beta_n\zeta_{n-1}=\int_{\mathbb{R}}xP_n(x)P_{n-1}(x)\,d\mu(x)$$

$$=\int_{\mathbb{R}}P_n(x)\left[P_n(x)+\text{lower order terms}\right]d\mu(x)=\zeta_n,$$

hence $\beta_n>0$. □

Note that we have actually proved that

$$\zeta_n = \beta_1 \cdots \beta_n. \tag{2.2.3}$$

Theorem 2.2.2 *The Christoffel–Darboux identities hold for* $N > 0$

$$\sum_{k=0}^{N-1} \frac{P_k(x)P_k(y)}{\zeta_k} = \frac{P_N(x)P_{N-1}(y) - P_N(y)P_{N-1}(x)}{\zeta_{N-1}\,(x-y)}, \tag{2.2.4}$$

$$\sum_{k=0}^{N-1} \frac{P_k^2(x)}{\zeta_k} = \frac{P_N'(x)P_{N-1}(x) - P_N(x)P_{N-1}'(x)}{\zeta_{N-1}}. \tag{2.2.5}$$

Proof Multiply (2.2.1) by $P_n(y)$ and subtract the result from the same expression with x and y interchanged. With $\Delta_k(x,y) = P_k(x)P_{k-1}(y) - P_k(y)P_{k-1}(x)$, we find

$$(x - y)P_n(x)P_n(y) = \Delta_{n+1}(x,y) - \beta_n \Delta_n(x,y),$$

which can be written in the form

$$(x-y)\frac{P_n(x)P_n(y)}{\zeta_n} = \frac{\Delta_{n+1}(x,y)}{\zeta_n} - \frac{\Delta_n(x,y)}{\zeta_{n-1}},$$

in view of (2.2.3). Formula (2.2.4) now follows from the above identity by telescopy. Formula (2.2.5) is the limiting case $y \to x$ of (2.2.4). □

Remark 2.2.1 *It is important to note that Theorem 2.2.2 followed from (2.2.1) and (2.2.2), hence the Christoffel–Darboux identity (2.2.4) will hold for any solution of (2.2.1), with possibly an additional term $c/(x-y)$ depending on the initial conditions. Similarly an identity like (2.2.5) will also hold.*

Theorem 2.2.3 *Assume that α_{n-1} is real and $\beta_n > 0$ for all $n = 1, 2, \ldots$. Then the zeros of the polynomials generated by (2.2.1)–(2.2.2) are real and simple. Furthermore the zeros of P_n and P_{n-1} interlace.*

Proof Let $x = u$ be a complex zero of P_N. Since the polynomials $\{P_n(x)\}$ have real coefficients, then $x = \bar{u}$ is also a complex zero of $P_N(x)$. With $x = u$ and $y = \bar{u}$ we see that the right-hand side of (2.2.4) vanishes while its left-hand side is larger than 1. Therefore all the zeros of P_N are real for all N. On the other hand, a multiple zero of P_N will make the right-hand side of (2.2.5) vanish while its left-hand side is positive.

Let

$$x_{N,1} > x_{N,2} > \cdots > x_{N,N} \tag{2.2.6}$$

be the zeros of P_N. Since $P_N(x) > 0$ for $x > x_{N,1}$ we see that $(-1)^{j-1}P_N'(x_{N,j}) > 0$. From this and (2.2.5) it immediately follows that $(-1)^{j-1}P_{N-1}(x_{N,j}) > 0$, hence P_{N-1} has a change of sign in each of the $N-1$ intervals $(x_{N,j+1}, x_{N,j})$, $j = 1, \ldots, N-1$, so it must have at least one zero in each such interval. This accounts for all the zeros of P_{N-1} and the theorem follows by Theorem 2.2.3. □

Another way to see the reality of zeros of polynomials generated by (2.2.1)–(2.2.2) is to relate P_n to characteristic polynomials of real symmetric matrices. Let $A_n = (a_{j,k} : 0 \leq j, k < n)$ be the tridiagonal matrix

$$a_{j,j} = \alpha_j, \quad a_{j,j+1} = 1,$$
$$a_{j+1,j} = \beta_{j+1}, \quad a_{j,k} = 0, \quad \text{for } |j - k| > 1. \tag{2.2.7}$$

Let $S_n(\lambda)$ be the characteristic polynomial of A_n, that is the determinant of $\lambda I - A_n$. By expanding the determinant expression for S_n about the last row it follows that $S_n(x)$ satisfies the recurrence relation (2.2.1). On the other hand $S_1(x) = x - \alpha_0$, $S_2(x) = (x - \alpha_0)(x - \alpha_1) - \beta_1$, so $S_1 = P_1$ and $S_2 = P_2$. Therefore S_n and P_n agree for all n. This establishes the following theorem.

Theorem 2.2.4 *The monic polynomials have the determinant representation*

$$P_n(x) = \begin{vmatrix} x - \alpha_0 & -1 & 0 & \cdots & 0 & 0 & 0 \\ -\beta_1 & x - \alpha_1 & -1 & \cdots & 0 & 0 & 0 \\ \vdots & \ddots & \ddots & \ddots & \vdots & \vdots & \vdots \\ 0 & 0 & \cdots & -\beta_{n-2} & x - \alpha_{n-2} & -1 \\ 0 & 0 & 0 & \cdots & 0 & -\beta_{n-1} & x - \alpha_{n-1} \end{vmatrix}. \tag{2.2.8}$$

It is straightforward to find an invertible diagonal matrix D so that $A_n = D^{-1}B_n D$ where $B_n = (b_{j,k})$ is a real symmetric tridiagonal matrix with $b_{j,j} = \alpha_j$, $b_{j,j+1} = \sqrt{\beta_{j+1}}, j = 0, \ldots, n - 1$. Thus the zeros of P_n are real.

Theorem 2.2.5 *Let $\{P_n\}$ be a sequence of orthogonal polynomials satisfying (2.1.5) and let $[a, b]$ be the smallest closed interval containing the support of μ. Then all zeros of P_n lie in $[a, b]$.*

Proof Let $c_1, \ldots c_j$ be the zeros of P_n lying inside $[a, b]$. If $j < n$ then the orthogonality implies $\int_{\mathbb{R}} P_n(x) \prod_{k=1}^{j} (x - c_k) \, d\mu = 0$, which contradicts the fact that the integrand does not change sign on $[a, b]$ by Theorem 2.2.3. $\qquad\square$

Several authors studied power sums of zeros of orthogonal polynomials and special functions. Let

$$s_k = \sum_{j=1}^{n} (x_{n,j})^k, \tag{2.2.9}$$

where $x_{n,1} > x_{n,2} > \cdots > x_{n,n}$ are the zeros of $P_n(x)$. Clearly

$$\frac{P_n'(x)}{P_n(x)} = \sum_{j=1}^{n} \frac{1}{x - x_{n,j}} = \sum_{k=0}^{\infty} \frac{1}{x^{k+1}} \sum_{j=1}^{n} (x_{n,j})^k.$$

Thus

$$P_n'(z)/P_n(z) = \sum_{k=0}^{\infty} s_k z^{-k-1}, \tag{2.2.10}$$

for $|z| > \max\{|x_{n,j}| : 1 \le j \le n\}$. Power sums for various special orthogonal polynomials can be evaluated using (2.2.10). If no $x_{n,j} = 0$, we can define s_k for $k < 0$ and apply

$$\frac{P_n'(z)}{P_n(z)} = -\sum_{j=1}^{n} \frac{1}{x_{n,j}} \sum_{k=0}^{\infty} \left(\frac{x}{x_{n,j}}\right)^k,$$

to conclude that

$$P_n'(z)/P_n(z) = -\sum_{k=0}^{\infty} z^k s_{-k-1}, \qquad (2.2.11)$$

for $|z| < \min\{|x_{n,j}| : 1 \le j \le n\}$. Formula (2.2.11) also holds when P_n is replaced by a function with the factor product representation

$$f(z) = \prod_{k=1}^{\infty} (1 - z/x_k).$$

An example is $f(z) = \Gamma(\nu+1)(2/z)^\nu J_\nu(z)$. Examples of (2.2.10) and (2.2.11) and their applications are in (Ahmed et al., 1979), (Ahmed et al., 1982) and (Ahmed & Muldoon, 1983). The power sums of zeros of Bessel polynomials have a remarkable property as we shall see in Theorems 4.10.4 and 4.10.5.

The **Poisson kernel** $P_r(x, y)$ of a system of orthogonal polynomials is

$$P_r(x, y) = \sum_{n=0}^{\infty} P_n(x) P_n(y) \frac{r^n}{\zeta_n}. \qquad (2.2.12)$$

One would expect $\lim_{r \to 1^-} P_r(x, y)$ to be a Dirac measure $\delta(x - y)$. Indeed under certain conditions

$$\lim_{r \to 1^-} \int_{\mathbb{R}} P_r(x, y) f(y) \, d\mu(y) = f(x), \qquad (2.2.13)$$

for $f \in L^2(\mu)$. A crucial step in establishing (2.2.13) for a specific system of orthogonal polynomials is the nonnegativity of the Poisson kernel on the support of μ.

Definition 2.2.1 *The kernel polynomials $\{K_n(x, y)\}$ of a distribution function F_μ are*

$$K_n(x, y) = \sum_{k=0}^{n} \overline{p_k(x)}\, p_k(y) = \sum_{k=0}^{n} \overline{P_k(x)}\, P_k(y)/\zeta_k, \qquad (2.2.14)$$

$n = 0, 1, \ldots$.

Theorem 2.2.6 *Let $\pi(x)$ be a polynomial of degree at most n and normalized by*

$$\int_{\mathbb{R}} |\pi(x)|^2 d\mu(x) = 1. \qquad (2.2.15)$$

Then the maximum of $|\pi(x_0)|^2$ taken over all such $\pi(x)$ is attained when

$$\pi(x) = \zeta K_n(x_0, x)/\sqrt{K_n(x_0, x_0)}, \qquad |\zeta| = 1.$$

The maximum is $K_n(x_0, x_0)$.

Proof Assume that $\pi(x)$ satisfies (2.2.15) and let $\pi(x) = \sum_{k=0}^{n} c_k p_k(x)$. Then

$$|\pi(x_0)|^2 \le \left(\sum_{k=0}^{r} |c_k|^2 \right) \left(\sum_{k=0}^{n} |p_k(x_0)|^2 \right) = K_n(x_0, x_0)$$

and the theorem follows. $\qquad\qquad\qquad\qquad\qquad\qquad\qquad\qquad\qquad\square$

Remark 2.2.2 *Sometimes it is convenient to use neither the monic nor the orthonormal polynomials. If $\{\phi_n(x)\}$ satisfy*

$$\int_{\mathbb{R}} \phi_m(x)\phi_n(x)\, d\mu(x) = \zeta_n \delta_{m,n} \tag{2.2.16}$$

then

$$x\phi_n(x) = A_n \phi_{n+1}(x) + B_n \phi_n(x) + C_n \phi_{n-1}(x) \tag{2.2.17}$$

and

$$\zeta_n = \frac{C_1 \cdots C_n}{A_0 \cdots A_{n-1}} \zeta_0. \tag{2.2.18}$$

The interlacing property of the zeros of P_n and P_{n+1} extend to eigenvalues of general Hermitian matrices.

Theorem 2.2.7 (Cauchy Interlace Theorem) *Let A be a Hermitian matrix of order n, and let B be a principal submatrix of A of order $n-1$. If $\lambda_n \le \lambda_{n-1} \le \cdots \le \lambda_2 \le \lambda_1$ lists the eigenvalues of A and $\mu_n \le \mu_{n-1} \le \cdots \le \mu_3 \le \mu_2$ the eigenvalues of B, then $\lambda_n \le \mu_n \le \lambda_{n-1} \le \mu_{n-1} \le \cdots \le \lambda_2 \le \mu_2 \le \lambda_1$.*

Recently, Hwang gave an elementary proof of Theorem 2.2.7 in (Hwang, 2004). The standard proof of Theorem 2.2.7 uses Sylvester's law of inertia (Parlett, 1998). For a proof using the Courant–Fischer minmax principle, see (Horn & Johnson, 1992).

2.3 Numerator Polynomials

Consider (2.2.1) as a difference equation in the variable n,

$$x\, y_n = y_{n+1} + \alpha_n y_n + \beta_n y_{n-1}. \tag{2.3.1}$$

Thus (2.3.1) has two linearly independent solutions, one of them being $P_n(x)$. We introduce a second solution, P_n^*, defined initially by

$$P_0^*(x) = 0, \quad P_1^*(x) = 1. \tag{2.3.2}$$

It is clear that $P_n^*(x)$ is monic and has degree $n-1$ for all $n > 0$. Next consider the Casorati determinant (Milne-Thomson, 1933)

$$\Delta_n(x) := P_n(x)P_{n-1}^*(x) - P_n^*(x)P_{n-1}(x).$$

From (2.2.1) we see that $\Delta_{n+1}(x) = \beta_n \Delta_n(x)$, and in view of the initial conditions (2.2.2) and (2.3.2) we establish

$$P_n(x)P_{n-1}^*(x) - P_n^*(x)P_{n-1}(x) = -\beta_1 \cdots \beta_{n-1}. \tag{2.3.3}$$

From the theory of difference equations (Jordan, 1965), (Milne-Thomson, 1933) we know that two solutions of a second order linear difference equation are linearly independent if and only if their Casorati determinant is not zero. Thus P_n and P_n^* are linearly independent solutions of (2.3.1).

Theorem 2.3.1 *For $n > 0$, the zeros of $P_n^*(x)$ are all real and simple and interlace with the zeros of $P_n(x)$.*

Proof Clearly (2.3.3) shows that $P_n^*(x_{n,j}) P_{n-1}(x_{n,j}) > 0$. Then P_n^* and P_{n-1} have the same sign at the zeros of P_n. Now Theorem 2.2.4 shows that P_n^* has a zero in all intervals $(x_{n,j}, x_{n,j-1})$ for all j, $2 \leq j < n$. ☐

Definition 2.3.1 *Let $\{P_n(x)\}$ be a family of monic orthogonal polynomials generated by (2.2.1) and (2.2.2). The associated polynomials $\{P_n(x; c)\}$ of order c of $P_n(x)$ are polynomials satisfying (2.2.1) with n replaced by $n + c$ and given initially by*

$$P_0(x; c) := 1, \quad P_1(x; c) = x - \alpha_c. \tag{2.3.4}$$

The above procedure is well-defined when $c = 1, 2, \ldots$. In general, the above definition makes sense if the recursion coefficients are given by a pattern amenable to replacing n by $n + c$. It is clear that $P_{n-1}(x; 1) = P_n^*(x)$. Clearly $\{P_n(x; c)\}$ are orthogonal polynomials if $c = 1, 2, \ldots$.

Theorem 2.3.2 *The polynomials $\{P_n^*(z)\}$ have the integral representation*

$$P_n^*(z) = \int_{\mathbb{R}} \frac{P_n(z) - P_n(y)}{z - y} \, d\mu(y), \quad n \geq 0. \tag{2.3.5}$$

Proof Let $r_n(z)$ denote the right-hand side of (2.3.5). Then $r_0(z) = 0$ and $r_1(x) = 1$. For nonreal z and $n > 0$,

$$r_{n+1}(z) - (z - \alpha_n) r_n(z) + \beta_n r_{n-1}(z)$$
$$= \int_{\mathbb{R}} \frac{P_{n+1}(y) - (z - \alpha_n) P_n(y) + \beta_n P_{n-1}(y)}{y - z} \, d\mu(y)$$
$$= \int_{\mathbb{R}} \frac{y - z}{y - z} P_n(y) \, d\mu(y),$$

which vanishes for $n > 0$. Thus $r_n(z) = P_n^*(z)$ and the restriction on z can now be removed. ☐

2.4 Quadrature Formulas

Let $\{P_n(x)\}$ be a sequence of monic orthogonal polynomials satisfying (2.1.5) with zeros as in (2.2.6).

Theorem 2.4.1 *Given N there exists a sequence of positive numbers $\{\lambda_k : 1 \le k \le N\}$ such that*

$$\int_{\mathbb{R}} p(x)\, d\mu(x) = \sum_{k=1}^{N} \lambda_k p\left(x_{N,k}\right), \qquad (2.4.1)$$

for all polynomials p of degree at most $2N-1$. The λ's depend on N, and $\mu_0, \ldots, \mu_N$ but not on p. Moreover, the λ's have the representations

$$\lambda_k = \int_{\mathbb{R}} \frac{P_N(x)\, d\mu(x)}{P_N'\left(x_{N,k}\right)\left(x - x_{N,k}\right)} \qquad (2.4.2)$$

$$= \int_{\mathbb{R}} \left[\frac{P_N(x)}{P_N'\left(x_{N,k}\right)\left(x - x_{N,k}\right)}\right]^2 d\mu(x). \qquad (2.4.3)$$

Furthermore if (2.4.1) holds for all p of degree at most $2N - 1$ then the λ's are unique and are given by (2.4.2).

Proof Let L be the Lagrange interpolation polynomial of p at the nodes $x_{N,k}$, see (1.2.16)–(1.2.17). Since $L(x) = p(x)$ at $x = x_{N,j}$, for all j, $1 \le j \le N$ then $p(x) - L(x) = P_N(x)r(x)$, with r a polynomial of degree $\le N - 1$. Therefore

$$\int_{\mathbb{R}} p(x)\, d\mu(x) = \int_{\mathbb{R}} L(x)\, d\mu(x) + \int_{\mathbb{R}} P_N(x)r(x)\, d\mu(x)$$

$$= \sum_{k=1}^{N} p(x_{N,k}) \int_{\mathbb{R}} \frac{P_N(x)\, d\mu(x)}{P_N'\left(x_{N,k}\right)\left(x - x_{N,k}\right)}.$$

This establishes (2.4.2). Applying (2.4.1) to $p(x) = P_N(x)^2 / \left(x - x_{N,k}\right)^2$ we establish (2.4.3) and the uniqueness of the λ's. The positivity of the λ's follows from (2.4.3). $\qquad \square$

The numbers $\lambda_1, \ldots, \lambda_N$ are called Christoffel numbers.

Theorem 2.4.2 *The Christoffel numbers have the properties*

$$\sum_{k=1}^{N} \lambda_k = \mu(\mathbb{R}), \qquad (2.4.4)$$

$$\lambda_k = -\zeta_N / \left[P_{N+1}\left(x_{N,k}\right) P_N'\left(x_{N,k}\right)\right], \qquad (2.4.5)$$

$$\frac{1}{\lambda_k} = \sum_{j=0}^{N} P_j^2\left(x_{N,k}\right) / \zeta_j =: K_N\left(x_{N,k}, x_{N,k}\right). \qquad (2.4.6)$$

Proof Apply (2.4.1) with $p(x) \equiv 1$ to get (2.4.4). Next replace N by $N+1$ in (2.2.4) then set $y = x_{N,k}$ and integrate with respect to μ. The result is

$$1 = -\frac{P_{N+1}(x_{N,k})}{\zeta_N} \int_{\mathbb{R}} \frac{P_N(x)}{x - x_{N,k}} \, d\mu(x),$$

and the rule (2.4.2) implies (2.4.5). Formula (2.4.6) follows from (2.4.5) and (2.2.5). □

We now come to the Chebyshev–Markov–Stieltjes separation theorem and inequalities. Let $[a, b]$ be the convex hull of the support of μ. For $N \geq 2$, we let

$$u_k = x_{N,N-k}, \tag{2.4.7}$$

so that $u_1 < u_2 < \cdots < u_N$. Let F_μ be the distribution function of μ. In view of the positivity of the λ's and (2.4.4), there exist numbers $y_1 < y_2 < \cdots < y_{N-1}, a < y_1$, $y_{N-1} < b$, such that

$$\lambda_k = F_\mu(y_k) - F_\mu(y_{k-1}), \quad 1 \leq k \leq N, \quad y_0 := a, \quad y_N := b. \tag{2.4.8}$$

Theorem 2.4.3 (Separation Theorem) *The points $\{y_k\}$ interlace with the zeros $\{u_k\}$; that is*

$$u_k < y_k < u_{k+1}, \quad 1 \leq k \leq N - 1.$$

Equivalently

$$F_\mu(u_k) < F_\mu(y_k) = \sum_{j=1}^{k} \lambda_j < F_\mu(u_{k+1}^-), \quad 1 \leq k < N.$$

An immediate consequence of the separation theorem is the following corollary.

Corollary 2.4.4 *Let I be an open interval formed by two consecutive zeros of $P_N(x)$. Then $\mu(I) > 0$.*

Proof of Theorem 2.4.3 Define a right continuous step function V by $V(a) = 0$, $V(x) = \sum_{j=1}^{N} \lambda_j$ for $x \geq x_{N,N}$ and V has a jump λ_j at $x = u_j$. Formula (2.4.1) is

$$\int_a^b p(x) \, d(\mu - V) = 0.$$

Hence,

$$0 = \int_a^b p'(x) \left[F_\mu(x) - V(x) \right] dx. \tag{2.4.9}$$

Set $\beta(x) = F_\mu(x) - V(x)$. Let $I_1 = (a, u_1)$, $I_{j+1} = (u_j, u_{j+1})$, $1 \leq j < N$, $I_{N+1} = (u_N, b)$. Clearly, $\beta(x)$ is nondecreasing on I_j, V_j. Moreover, $\beta(x) \geq 0$, $\beta(x) \not\equiv 0$, on I_1, but $\beta(x) \leq 0$, $\beta(x) \not\equiv 0$ on I_{N+1}. On I_j, $1 < j \leq N$, β either has a constant sign or changes sign from negative to positive at some point within the interval. Such points where the change of sign may occur must be among the y points defined in (2.4.8). Thus, $[a, b]$ can be subdivided into at most $2N$ subintervals where $\beta(x)$ has a constant sign on each subinterval. If the number of intervals of β constant signs is $< 2N$, then we can choose p' in (2.4.9) to have degree at most $2n - 2$ such that $p'(x)\beta(x) \geq 0$ on $[a, b]$, which gives a contradiction. Thus we must have at least $2N$ intervals where $\beta(x)$ keeps a constant sign. By the pigeonhole principle, we must have $y_j \in I_{j+1}$, $1 \leq j < N$ and the theorem follows. $\square$

Szegő gives two additional proofs of the separation theorem; see (Szegő, 1975, §3.41).

2.5 The Spectral Theorem

The main result in this section is Theorem 2.5.2 which we call the spectral theorem for orthogonal polynomials. In some of the literature it is called Favard's theorem, (Chihara, 1978), (Szegő, 1975), because Favard (Favard, 1935) proved it in 1935. Shohat claimed to have had an unpublished proof several years before Favard published his paper (Shohat, 1936). More interestingly, the theorem is stated and proved in Wintner's book on spectral theory of Jacobi matrices (Wintner, 1929). The theorem also appeared in Marshal Stone's book (Stone, 1932, Thm. 10.27) without attributing it to any particular author. The question of uniqueness of μ is even discussed in Theorem 10.30 in (Stone, 1932).

We start with a sequence of polynomials $\{P_n(x)\}$ satisfying (2.2.1)–(2.2.2), with $\alpha_{n-1} \in \mathbb{R}$ and $\beta_n > 0$, for $n > 0$. Fix N and arrange the zeros of P_N as in (2.2.6). Define a sequence

$$\rho\left(x_{N,j}\right) = \zeta_{N-1}/\left[P_N'\left(x_{N,j}\right) P_{N-1}\left(x_{N,j}\right)\right], \quad 1 \leq j \leq N. \tag{2.5.1}$$

We have established the positivity of $\rho\left(x_{N,j}\right)$ in Theorem 2.2.4. With x a zero of $P_N(x)$, rewrite (2.2.4) and (2.2.5) as

$$\rho(x_{N,r}) \sum_{k=0}^{N-1} P_k\left(x_{N,r}\right) P_k\left(x_{N,s}\right)/\zeta_k = \delta_{r,s}. \tag{2.5.2}$$

Indeed (2.5.2) says that the real matrix U,

$$U = (u_{r,k}), \ 1 \leq r, k \leq N, \quad u_{r,k} := \sqrt{\rho\left(x_{N,r}\right)} \frac{P_{k-1}\left(x_{N,r}\right)}{\sqrt{\zeta_k - 1}},$$

satisfies $UU^T = I$, whence $U^TU = I$, that is

$$\sum_{r=1}^{N} \rho\left(x_{N,r}\right) P_k\left(x_{N,r}\right) P_j\left(x_{N,r}\right) = \zeta_k \delta_{j,k}, \quad j, k = 0, \ldots, N-1. \tag{2.5.3}$$

We now introduce a sequence of right continuous step functions $\{\psi_N\}$ by

$$\psi_N(-\infty) = 0, \quad \psi_N\left(x_{N,j} + 0\right) - \psi_N\left(x_{N,j} - 0\right) = \rho\left(x_{N,j}\right). \tag{2.5.4}$$

Theorem 2.5.1 *The moments* $\int_{\mathbb{R}} x^j \, d\psi_N(x)$, $1 \leq j \leq 2N - 2$ *do not depend on* ζ_k *for* $k > \lfloor (j+1)/2 \rfloor$ *where* $\lfloor a \rfloor$ *denotes the integer part of* a.

Proof For fixed j choose $N > 1 + j/2$ and write x^j as $x^s x^\ell$, with $0 \le \ell, s \le N - 1$. Then express x^s and x^ℓ as linear combinations of $P_0(x), \ldots, P_{N-1}(x)$. Thus the evaluation of $\int_\mathbb{R} x^j \, d\psi_N(x)$ involves only $\zeta_0, \ldots, \zeta_{\lfloor (j+1)/2 \rfloor}$. $\qquad\square$

Theorem 2.5.2 *Given a sequence of polynomials $\{P_n(x)\}$ generated by (2.2.1)–(2.2.2) with $\alpha_{n-1} \in \mathbb{R}$ and $\beta_n > 0$ for all $n > 0$, then there exists a distribution function μ such that*

$$\int_\mathbb{R} P_m(x) P_n(x) \, d\mu(x) = \zeta_n \delta_{m,n}, \qquad (2.5.5)$$

and ζ_n is given by (2.2.3).

Proof Since

$$1 = \zeta_0 = \int_\mathbb{R} d\psi_N(x) = \psi_N(\infty) - \psi_N(-\infty)$$

then the ψ_N's are uniformly bounded. From Helly's selection principle it follows that there is a subsequence ϕ_{N_k} which converges to a distribution function μ. The rest follows from Theorems 2.5.1 and 1.2.1. It is clear that the limiting function μ of any subsequence will have infinitely many points of increase. $\qquad\square$

Shohat (Shohat, 1936) proved that if $\alpha_n \in \mathbb{R}$ and $\beta_{n+1} \ne 0$ for all $n \ge 0$, then there is a real signed measure μ with total mass 1 such that (2.5.5) holds, with $\zeta_0 = 1$, $\zeta_n = \beta_1 \cdots \beta_n$, see also (Shohat, 1938).

The distribution function μ in Theorem 2.5.2 may not be unique, as can be seen from the following example due to Stieltjes.

Example 2.5.3 *Consider the weight function*

$$w(x; \alpha) = [1 + \alpha \sin(2\pi c \ln x)] \exp\left(-c \ln^2 x\right), \quad x \in (0, \infty),$$

for $\alpha \in (-1, 1)$ and $c > 0$. The moments are

$$\mu_n := \int_0^\infty x^n w(x; \alpha) \, dx = \int_{-\infty}^\infty e^{u(n+1)} [1 + \alpha \sin(2\pi cu)] \exp\left(-cu^2\right) du$$

$$= \exp\left(\frac{(n+1)^2}{4c}\right) \int_{-\infty}^\infty \exp\left(-c\left(u - \frac{n+1}{2c}\right)^2\right) [1 + \alpha \sin(2\pi cu)] \, du$$

$$= \exp\left(\frac{(n+1)^2}{4c}\right) \int_{-\infty}^\infty \exp\left(-cv^2\right) [1 - (-1)^n \alpha \sin(2\pi cv)] \, dv.$$

In the last step we have set $u = v + (n+1)/(2c)$. Clearly

$$\mu_n = \exp\left(\frac{(n+1)^2}{4c}\right) \int_{-\infty}^\infty \exp\left(-cv^2\right) dv = \sqrt{\frac{\pi}{c}} \exp\left(\frac{(n+1)^2}{4c}\right)$$

which is independent of α. *Therefore the weight functions* $w(x; \alpha)$, *for all* $\alpha \in (-1, 1)$, *have the same moments.*

We will see that μ is unique when the α_n's and β_n's are bounded. Let

$$\xi := \lim_{n \to \infty} x_{n,n}, \quad \eta := \lim_{n \to \infty} x_{n,1}. \tag{2.5.6}$$

Both limits exist since $\{x_{n,1}\}$ increases with n while $\{x_{n,n}\}$ decreases with n, by Theorem 2.2.4. Theorem 2.2.6 and the construction of the measures μ in Theorem 2.5.2 motivate the following definition.

Definition 2.5.1 *The **true interval of orthogonality** of a sequence of polynomials* $\{P_n\}$ *generated by (2.2.1)–(2.2.2) is the interval* $[\xi, \eta]$.

It is clear from Theorem 2.4.3 that $[\xi, \eta]$ is a subset of the convex hull of $\mathrm{supp}(\mu)$.

Theorem 2.5.4 *The support of every* μ *of Theorem 2.5.2 is bounded if* $\{\alpha_n\}$ *and* $\{\beta_n\}$ *are bounded sequences.*

Proof If $|\alpha_n| < M$, $\beta_n \leq M$, then apply Theorem 2.2.5 to identify the points $x_{N,j}$ as eigenvalues of a tridiagonal matrix then apply Theorem 1.1.1 to see that $|x_{n,j}| < \sqrt{3}\, M$ for all N and j. Thus the support of each ψ_N is contained in $(-\sqrt{3}\, M, \sqrt{3}\, M)$ and the result follows. □

Theorem 2.5.5 *If* $\{\alpha_n\}$ *and* $\{\beta_n\}$ *are bounded sequences then the measure of orthogonality* μ *is unique.*

Proof By Theorem 2.5.4 we may assume that the support of one measure μ is compact. Let ν be any other measure. For any $a > 0$, we have

$$\int_{|x| \geq a} d\nu(x) \leq a^{-2n} \int_{|x| \geq a} x^{2n} d\nu(x) \leq a^{-2n} \int_{\mathbb{R}} x^{2n} d\nu(x)$$

$$= a^{-2n} \int_{\mathbb{R}} x^{2n} d\mu(x).$$

Assume $|x_{N,j}| \leq A$, $j = 1, 2, \ldots, N$ and for all $N \geq 1$. Apply (2.4.1) with $N = n + 1$. Then

$$\int_{|x| \geq a} d\nu(x) \leq a^{-2n} \sum_{k=1}^{n+1} \lambda_k \left(x_{n+1,k}\right)^{2n}$$

$$\leq (A/a)^{2n} \sum_{k=1}^{n+1} \lambda_k = (A/a)^n.$$

If $a > A$, then $\int_{|x| \geq a} d\nu(x) = 0$, hence $\mathrm{supp}\, \nu \subset [-A, A]$. We now prove that $\mu = \nu$.

Clearly for $|x| \geq 2A$, $\sum_{k=0}^{n} t^k x^{-k-1}$ converges to $1/(x-t)$ for all $t \in [-A, A]$. Therefore

$$\int_{\mathbb{R}} \frac{d\mu(t)}{x-t} = \int_{\mathbb{R}} \lim_{n\to\infty} \sum_{k=0}^{n} \frac{t^k}{x^{k+1}} \, d\mu(t)$$

$$= \lim_{n\to\infty} \int_{\mathbb{R}} \sum_{k=0}^{n} \frac{t^k}{x^{k+1}} \, d\mu(t) = \lim_{n\to\infty} \sum_{k=0}^{n} \frac{\mu_k}{x^{k+1}},$$

where in the last step we used the Lebesgue dominated convergence theorem, since $|t/x| \leq 1/2$. The last limit depends only on the moments, hence is the same for all μ's that have the same μ_k's. Thus $F(x) := \int_{\mathbb{R}} d\mu(t)/(x-t)$ is uniquely determined for x outside the circle $|x| = 2A$. Since F is analytic in $x \in \mathbb{C} \setminus [-A, A]$, by the identity theorem for analytic functions F is unique and the theorem follows from the Perron–Stieltjes inversion formula. $\qquad\qquad\square$

Observe that the proof of Theorem 2.5.5 shows that supp μ is the true interval of orthogonality when $\{\alpha_n\}$ and $\{\beta_n\}$ are bounded sequences.

A very useful result in the theory of moment problems is the following theorem, whose proofs can be found in (Shohat & Tamarkin, 1950; Akhiezer, 1965).

Theorem 2.5.6 *Assume that μ is unique and $\int_{\mathbb{R}} d\mu = 1$. Then μ has an atom at $x = u \in \mathbb{R}$ if and only if the series*

$$S := \sum_{n=0}^{\infty} P_n^2(u)/\xi_n \qquad\qquad (2.5.7)$$

converges. Furthermore, if μ has an atom at $x = u$, then

$$\mu(\{u\}) = 1/S. \qquad\qquad (2.5.8)$$

Some authors prefer to work with a positive linear functional $\mathcal{L}$ defined by $\mathcal{L}(x^n) = \mu_n$. The question of constructing the orthogonality measure then becomes a question of finding a representation of $\mathcal{L}$ as an integral with respect to a positive measure. This approach is used in Chihara (Chihara, 1978). Some authors even prefer to work with a linear functional which is not necessarily positive, but the determinants D_n of (2.1.4) are assumed to be nonzero. Such functionals are called regular. An extensive theory of polynomials orthogonal with respect to regular linear functionals has been developed by Pascal Maroni and his students and collaborators.

Boas (Boas, Jr., 1939) proved that any sequence of real numbers $\{c_n\}$ has a moment representation $\int_E x^n \, d\mu(x)$ for a nontrivial finite signed measure μ. In particular, the sequence $\{\mu_n\}$, $\mu_n = 0$ for all $n = 0, 1, \ldots$, is a moment sequence. Here, E

can be $\mathbb{R}$ or $[a, \infty)$, $a \in \mathbb{R}$. For example

$$0 = \int_0^\infty x^n \sin(2\pi c \ln x) \exp\left(-c \ln x^2\right) dx, \quad c > 0 \qquad (2.5.9)$$

$$0 = \int_0^\infty x^n \sin x^{1/4} \exp\left(-x^{1/4}\right) dx. \qquad (2.5.10)$$

Formula (2.5.9) follows from Example 2.5.3. To prove (2.5.10), observe that its right-hand side, with $x = u^4$, is

$$-2i \int_0^\infty u^{4n+3} \left[\exp(-u(1+i)) - \exp(-u(1-i))\right] du$$

$$= \frac{2i(4n+3)!}{(1+i)^{4n+3}} - \frac{2i(4n+3)!}{(1-i)^{4n+3}} = 0.$$

Thus the signed measure in Boas' theorem is never unique. A nontrivial measure whose moments are all zero is called a polynomial killer.

Although we cannot get into the details of the connection between constructing orthogonality measures and the spectral problem on a Hilbert space, we briefly describe the connection. Consider the operator T which is multiplication by x. The three-term recurrence relation gives a realization of T as a tridiagonal matrix operator

$$T = \begin{pmatrix} \alpha_0 & a_1 & 0 & \cdots \\ a_1 & \alpha_1 & a_2 & \cdots \\ & \ddots & \ddots & \ddots \end{pmatrix} \qquad (2.5.11)$$

defined on a dense subset of ℓ^2. It is clear that T is symmetric. When T is self-adjoint there exists a unique measure supported on $\sigma(T)$, the spectrum of T, such that

$$T = \int_{\sigma(T)} \lambda dE_\lambda.$$

In other words

$$(y, p(T)x) = \int_{\sigma(T)} p(\lambda) (y, dE_\lambda x), \qquad (2.5.12)$$

for polynomial p, and for all $x, y \in \ell^2$. By choosing the basis $e_0, e_1, \ldots$ for ℓ^2, $e_n = (u_1, u_2, \ldots)$, $u_k = \delta_{kn}$, we see that $(e_0, dE_\lambda e_0)$ is a positive measure. This is the measure of orthogonality of $\{P_n(x)\}$. One can evaluate $(e_n, dE_\lambda e_m)$, for all $m, n \geq 0$, from the knowledge of $(e_0 dE_\lambda e_0)$. Hence dE_λ can be computed from the knowledge of the measure with respect to which $\{P_n(x)\}$ are orthogonal. The details of this theory are in (Akhiezer, 1965) and (Stone, 1932).

One can think of the operator T in (2.5.11) as a discrete Schrödinger operator (Cycon et al., 1987). This analogy is immediate when $a_n = 1$. One can think of the diagonal entries $\{\alpha_n\}$ as a potential. There is extensive theory known for doubly infinite Jacobi matrices with 1's on the super and lower diagonals and a random potential α_n; see (Cycon et al., 1987, Chapter 9). The theory of general doubly infinite tridiagonal matrices is treated in (Berezans'kiĭ, 1968). Many problems in this area remain open.

2.6 Continued Fractions

A continued J-fraction is

$$\frac{A_0}{A_0z + B_0-} \; \frac{C_1}{A_1z + B_1-} \cdots. \tag{2.6.1}$$

The nth convergent of the above continued fraction is the rational function

$$\frac{A_0}{A_0z + B_0-} \; \frac{C_1}{A_1z + B_1-} \cdots \frac{C_{n-1}}{A_{n-1}z + B_{n-1}}. \tag{2.6.2}$$

Write the nth convergent as $N_n(z)/D_n(z)$, $n > 1$, and the first convergent is
$$\frac{A_0}{A_0z + B_0}.$$

Definition 2.6.1 *The J-fraction* (2.6.1) *is of positive type if* $A_n A_{n-1} C_n > 0$, $n = 1, 2, \ldots$.

Theorem 2.6.1 *Assume that* $A_n C_{n+1} \neq 0$, $n = 0, 1, \ldots$. *Then the polynomials* $N_n(z)$ *and* $D_n(z)$ *are solutions of the recurrence relation*

$$y_{n+1}(z) = [A_n z + B_n]\, y_n(z) - C_n\, y_{n-1}(z), \quad n > 0, \tag{2.6.3}$$

with the initial values

$$D_0(z) := 1, \; D_1(z) := A_0z + B_0, \quad N_0(z) := 0, \; N_1(z) := A_0. \tag{2.6.4}$$

Proof It is easy to check that $N_1(z)/D_1(z)$ and $N_2(z)/D_2(z)$ agree with what (2.6.3) and (2.6.4) give for $D_2(z)$ and $N_2(z)$. Now assume that $N_n(z)$ and $D_n(z)$ satisfy (2.6.3) for $n = 1, \ldots, N - 1$. Since the $N + 1$ convergent is

$$\frac{N_{N+1}(z)}{D_{N+1}(z)}$$

$$= \frac{A_0}{A_0z + B_0-} \; \frac{C_1}{A_1z + B_1-} \cdots \frac{C_{N-1}}{A_{N-1}z + B_{N-1} - C_N/(A_Nz + B_N)} \tag{2.6.5}$$

then $N_{N+1}(z)$ and $D_{N+1}(z)$ follow from $N_N(z)$ and $D_N(z)$ by replacing C_{N-1} and $A_{N-1}z + B_{N-1}$ by $(A_Nz + B_N)\, C_{N-1}$ and $(A_{N-1}z + B_{N-1})\,(A_Nz + B_N) - C_N$, respectively. In other words

$$D_{N+1} = [(A_{N-1}z + B_{N-1})\,(A_Nz + B_N) - C_N]\, D_{N-1}$$
$$- C_{N-1}\,(A_Nz + B_n)\, D_{N-2},$$

which yields (2.6.3) for $n = N$ and $y_N = D_N$. Similarly we establish the recursions for the N_n's. $\qquad\square$

When $A_n = 1$ then D_n and N_n become P_n and P_n^* of §2.3, respectively.

Theorem 2.6.2 (Markov) *Assume that the true interval of orthogonality* $[\xi, \eta]$ *is bounded. Then*

$$\lim_{n \to \infty} \frac{P_n^*(z)}{P_n(z)} = \int_\xi^\eta \frac{d\mu(t)}{z - t}, \quad z \notin [\xi, \eta], \tag{2.6.6}$$

and the limit is uniform on compact subsets of $\mathbb{C} \setminus [\xi, \eta]$.

Proof The Chebyshev–Markov–Stieltjes inequalities and Theorem 2.5.5 imply that μ is unique and supported on E, say, $E \subset [\xi, \eta]$. Since $\int_E x^m d\psi_n(x) \to \int_E x^m d\mu$ for all m then $\int_E f d\psi_n \to \int f d\mu$ for every continuous function f. The function $1/(z-t)$ is continuous for $t \in E$, and $\mathrm{Im}\, z \neq 0$, whence

$$\frac{P_n^*(z)}{P_n(z)} = \int_\xi^\eta \frac{d\psi_n(t)}{z - t} \to \int_\xi^\eta \frac{d\mu(t)}{z - t}, \quad z \notin [\xi, \eta].$$

The uniform convergence follows from Vitali's theorem. □

Markov's theorem is very useful in determining orthogonality measures for orthogonal polynomials from the knowledge of the recurrence relation they satisfy.

Definition 2.6.2 *A solution* $\{u_n(z)\}$ *of* (2.6.3) *is called a minimal solution at* ∞ *if* $\lim_{n \to \infty} u_n(z)/v_n(z) = 0$ *for any other linear independent solution* $v_n(z)$. *A minimal solution at* $-\infty$ *is similarly defined.*

The minimal solution is the discrete analogue of the principal solution of differential equation.

It is clear that if the minimal solution exists then it is unique, up to a multiplicative function of z. The following theorem of Pincherle characterizes convergence of continued fractions in terms of the existence of the minimal solution.

Theorem 2.6.3 (Pincherle) *The continued fraction* (2.6.2) *converges at* $z = z_0$ *if and only if the recurrence relation* (2.6.3) *has a minimal solution* $\left\{ y_n^{(\min)}(z) \right\}$. *Furthermore if a minimal solution* $\left\{ y_n^{(\min)}(z) \right\}$ *exists then the continued fraction converges to* $y_0^{(\min)}(z)/y_{-1}^{(\min)}(z)$.

For a proof, see (Jones & Thron, 1980, pp. 164–166) or (Lorentzen & Waadeland, 1992, pp. 202–203).

Pincherle's theorem is very useful in finding the functions to which the convergents of a continued fraction converge. On the other hand, finding minimal solutions is not always easy but has been done in many interesting specific cases by David Masson and his collaborators. The following theorem whose proof appears in (Lorentzen & Waadeland, 1992, §4.2.2) is useful in verifying whether a solution is minimal.

Theorem 2.6.4 *Let* $\{u_n(z)\}$ *be a solution to* (2.2.1) *and assume that* $u_n(\zeta) \neq 0$ *for all* n *and a fixed* ζ. *Then* $\{u_n(\zeta)\}$ *is a minimal solution to* (2.6.3) *at* ∞ *if and only if*

$$\sum_{n=1}^{\infty} \frac{\prod_{m=1}^{n} \beta_m}{(u_n(\zeta)u_{n+1}(\zeta))} = \infty.$$

2.7 Modifications of Measures: Christoffel and Uvarov

Given a distribution function F_μ and the corresponding orthogonal polynomials $P_n(x)$, an interesting question is what can we say about the polynomials orthogonal with respect to $\Phi(x)\,d\mu(x)$ where Φ is a function positive on the support of μ, and their recursion coefficients. In this section we give a formula for the polynomials whose measure of orthogonality is $\Phi(x)\,d\mu(x)$, when Φ is a polynomial or a rational function. The modification of a measure by multliplication by a polynomial or a rational function can also be explained through the Darboux transformation of integrable systems. For details and some of the references to the literature the reader may consult (Bueno & Marcellán, 2004). In the next section we study the recursion coefficients when Φ is an exponential function.

Theorem 2.7.1 (Christoffel) *Let* $\{P_n(x)\}$ *be monic orthogonal polynomials with respect to* μ *and let*

$$\Phi(x) = \prod_{k=1}^{m} (x - x_k) \tag{2.7.1}$$

be nonnegative on the support of $d\mu$. *If the* x_k's *are simple zeros then the polynomials* $S_n(x)$ *defined by*

$$C_{n,m}\Phi(x)S_n(x) = \begin{vmatrix} P_n(x_1) & P_{n+1}(x_1) & \cdots & P_{n+m}(x_1) \\ P_n(x_2) & P_{n+1}(x_2) & \cdots & P_{n+m}(x_2) \\ \vdots & \vdots & \vdots & \vdots \\ P_n(x_m) & P_{n+1}(x_m) & \cdots & P_{n+m}(x_m) \\ P_n(x) & P_{n+1}(x) & \cdots & P_{n+m}(x) \end{vmatrix} \tag{2.7.2}$$

with

$$C_{n,m} = \begin{vmatrix} P_n(x_1) & P_{n+1}(x_1) & \cdots & P_{n+m-1}(x_1) \\ P_n(x_2) & P_{n+1}(x_2) & \cdots & P_{n+m-1}(x_2) \\ \vdots & \vdots & \vdots & \vdots \\ P_n(x_m) & P_{n+1}(x_m) & \cdots & P_{n+m-1}(x_m) \end{vmatrix}, \tag{2.7.3}$$

are orthogonal with respect to $\Phi(x)\,d\mu(x)$, *and* S_n *has degree* n. *If the zero* x_k *has multiplicity* $r > 1$, *then we replace the corresponding rows of* (2.7.2) *by derivatives of order* $0, 1, \ldots, r-1$ *at* x_k.

Proof If $C_{n,m} = 0$, there are constants $c_0, \ldots, c_{m-1}$, not all zero, such that the polynomial $\pi(x) := \sum_{k=0}^{m-1} c_k P_{n+k}(x)$, vanishes at $x = x_1, \ldots, x_m$. Therefore $\pi(x) =$

$\Phi(x)G(x)$, and G has degree at most $n-1$. But this makes $\int_{\mathbb{R}} \Phi(x)G^2(x)\,d\mu(x) = 0$, in view of the orthogonality of the P_n's. Thus $G(x) \equiv 0$, a contradiction. Now assume that all the x_k's are simple. It is clear that the right-hand side of (2.7.2) vanishes at $x = x_1, \ldots, x_m$. Define $S_n(x)$ by (2.7.2), hence S_n has degree $\leq n$. Obviously the right-hand side of (2.7.2) is orthogonal to any polynomial of degree $< n$ with respect to $d\mu$. If the degree of $S_n(x)$ is $< n$, then $\int_{\mathbb{R}} \Phi(x)S_n^2(x)\,d\mu(x)$ is zero, a contradiction, since all the zero of Φ lie outside the support of μ. Whence S_n has exact degree n. The orthogonality of S_n to all polynomials of degree $< n$ with respect to $\Phi\,d\mu$ follows from (2.7.2) and the orthogonality of P_n to all polynomials of degree $< n$ with respect to $d\mu$. The case when some of the x_k are multiple zeros is similarly treated and the theorem follows. $\qquad\square$

Observe that the special case $m = 1$ of (2.7.2) shows that the kernel polynomials $\{K_n(x, c)\}$, see Definition 2.2.1, are orthogonal with respect to $(x - c)\,d\mu(x)$.

Formula (2.7.2) is useful when m is small but, in general, it establishes the fact that $\Phi(x)S_n(x)$ is a linear combination of $P_n(x), \ldots, P_{n+m}(x)$ and it may by possible to evaluate the coefficients in a different way, for example by equating coefficients of $x^{n+m}, \ldots, x^n$.

Given a measure μ define

$$d\nu(x) = \frac{\prod_{i=1}^{m}(x - x_i)}{\prod_{j=1}^{k}(x - y_j)}\,d\mu(x), \tag{2.7.4}$$

where the products $\prod_{i=1}^{m}(x - x_i)$, and $\prod_{j=1}^{k}(x - y_j)$ are positive for x in the support of μ. We now construct the polynomials orthogonal with respect to ν.

Lemma 2.7.2 *Let $y_1, \ldots, y_k$ be distinct complex numbers and for $s = 0, 1, \ldots, k-1$, let*

$$\frac{1}{\prod_{j=s+1}^{k}(x - y_j)} = \sum_{j=s+1}^{k} \frac{u_j(s)}{x - y_j} \tag{2.7.5}$$

Then we have, for $0 \leq \ell \leq k - s - 1$,

$$\sum_{j=s+1}^{k} u_j(s)\,y_j^{\ell} = \delta_{\ell, k-s-1}. \tag{2.7.6}$$

Proof The case $\ell = 0$ follows by multiplying (2.7.5) by x and let $x \to \infty$. By induction and repeated calculations of the residue at $x = \infty$, we establish (2.7.6). $\qquad\square$

Theorem 2.7.3 (Uvarov) *Let ν be as in (2.7.4) and assume that $\{P_n(x; m, k)\}$ are orthogonal with respect to ν. Set*

$$\tilde{Q}_n(x) := \int\limits_{\mathbb{R}} \frac{P_n(y)}{x - y}\, d\mu(y). \tag{2.7.7}$$

Then for $n \geq k$ we have

$$\left[\prod_{i=1}^{m}(x - x_i)\right] P_n(x; m, k)$$

$$= \begin{vmatrix} P_{n-k}(x_1) & P_{n-k+1}(x_1) & \cdots & P_{n+m}(x_1) \\ \vdots & \vdots & \vdots & \vdots \\ P_{n-k}(x_m) & P_{n-k+1}(x_m) & \cdots & P_{n+m}(x_m) \\ \tilde{Q}_{n-k}(y_1) & \tilde{Q}_{n-k+1}(y_1) & \cdots & \tilde{Q}_{n+m}(y_1) \\ \vdots & \vdots & \vdots & \vdots \\ \tilde{Q}_{n-k}(y_k) & \tilde{Q}_{n-k+1}(y_k) & \cdots & \tilde{Q}_{n+m}(y_k) \\ P_{n-k}(x) & P_{n+1}(x) & \cdots & P_{n+m}(x) \end{vmatrix}. \tag{2.7.8}$$

If $n < k$ then

$$\left[\prod_{i=1}^{m}(x - x_i)\right] P_n(x; m, k)$$

$$= \begin{vmatrix} a_{1,1} & \cdots & a_{1,k-n} & P_0(x_1) & \cdots & P_{n+m}(x_1) \\ \vdots & \vdots & \vdots & \vdots & & \vdots \\ a_{m,1} & \cdots & a_{m,k-n} & P_0(x_m) & \cdots & P_{n+m}(x_m) \\ b_{1,1} & \cdots & b_{1,k-n} & \hat{Q}_0(y_1) & \cdots & \tilde{Q}_{n+m}(y_1) \\ \vdots & \vdots & \vdots & \vdots & \vdots & \vdots \\ b_{k,1} & \cdots & b_{k,k-n} & \tilde{Q}_0(y_k) & \cdots & \tilde{Q}_{n+m}(y_k) \\ c_1 & \cdots & c_{k-n} & P_0(x) & \cdots & P_{n+m}(x) \end{vmatrix}, \tag{2.7.9}$$

where

$$b_{ij} = y_i^{j-1}, \quad 1 \leq i \leq k, \quad 1 < j \leq k - n,$$

$a_{ij} = 0; 1 \leq i \leq m, 1 \leq j \leq k - n, c_j = 0.$

If an x_j (or y_l is repeated r times, then the corresponding r rows will contain $P_s(x_j), \ldots, P_s^{(r-1)}(x_j)$ ($\tilde{Q}_s(x_j), \ldots, \tilde{Q}_s^{(r-1)}(x_j)$), respectively.

Uvarov proved this result in a brief announcement (Uvarov, 1959) and later gave the details in (Uvarov, 1969). The proof given below is a slight modification of Uvarov's original proof.

Proof of Theorem 2.7.3 Let $\pi_j(x)$ denote a generic polynomials in x of degree at most j and denote the determinant on the right-hand side of (2.7.8) by $\Delta_{k,m,n}(x)$. Clearly $\Delta_{k,m,n}(x)$ vanishes at the points $x = x_j$, with $1 \leq j \leq m$ so let $\Delta_{k,m,n}(x)$

be $S_n(x) \prod\limits_{i=1}^{m} (x - x_i)$ with S_n of degree at most n. Moreover, $S_n(x) \not\equiv 0$, so we let

$$S_n(x) = \pi_{n-k}(x) \prod_{i=1}^{k} (x - y_i) + \pi_{k-1}(x),$$

and note the partial fraction decomposition

$$\frac{\pi_{k-1}(x)}{\prod\limits_{i=1}^{k} (x - y_i)} = \sum_{j=1}^{k} \frac{\alpha_j}{x - y_j}.$$

With ν as in (2.7.4) we have

$$\int_{\mathbb{R}} S_n^2(x) \, d\nu(x) = \int_{\mathbb{R}} S_n(x) \prod_{j=1}^{m} (x - x_j) \left\{ \pi_{n-k}(x) + \frac{\pi_{k-1}(x)}{\prod\limits_{i=1}^{k} (x - y_i)} \right\} d\mu(x)$$

$$= \int_{\mathbb{R}} S_n(x) \prod_{j=1}^{m} (x - x_j) \, \pi_{n-k}(x) \, d\mu(x)$$

$$+ \sum_{j=1}^{k} \alpha_j \int_{\mathbb{R}} \frac{\Delta_{k,m,n}(x)}{x - y_j} \, d\mu(x).$$

The term involving the sum in the last equality is zero because the last row in the integrated determinant coincides with one of the rows containing the $\tilde{Q}$ functions. If the degree of S_n is $< n$ then the first term in the last equality also vanishes because now $\pi_{n-k}(x)$ is $\pi_{n-k-1}(x)$ and the term we are concerned with is $\int_{\mathbb{R}} \Delta_{k,m,n}(x) \pi_{n-k-1}(x) \, d\mu(x)$, which obviously is zero. Thus S_n has exact degree n, and $\Delta \neq 0$, where

$$\Delta := \begin{vmatrix} P_{n-k}(x_1) & P_{n-k+1}(x_1) & \cdots & P_{n+m}(x_1) \\ \vdots & \vdots & \vdots & \vdots \\ P_{n-k}(x_m) & P_{n-k+1}(x_m) & \cdots & P_{n+m}(x_m) \\ \tilde{Q}_{n-k}(y_1) & \tilde{Q}_{n-k+1}(y_1) & \cdots & \tilde{Q}_{n+m}(y_1) \\ \vdots & \vdots & \vdots & \vdots \\ \tilde{Q}_{n-k}(y_k) & \tilde{Q}_{n-k+1}(y_k) & \cdots & \tilde{Q}_{n+m}(y_k) \end{vmatrix}.$$

It is evident that from the determinant representation that $S_n(x)$ is orthogonal to any polynomial of degree $< n$ with respect to $d\nu$.

Similarly, we denote the determinant on the right-hand side of (2.7.9) by $\Delta_{k,m,n}(x)$, and it is divisible by $\prod\limits_{j=1}^{m} (x - x_j)$, so we set $\Delta_{k,m,n}(x) = P_n(x; m, k) \prod\limits_{i=1}^{m} (x - x_i)$. To prove that $\int_{\mathbb{R}} x^s P_n(x; m, k) \, d\nu(x) = 0$ for $0 \leq s < n$, it is sufficient to prove that

$$\int_{\mathbb{R}} \prod_{j=1}^{s} (x - y_j) \, P_n(x; m, k) \, d\nu(x) = 0, \text{ for } 0 \leq s < n, \text{ that is}$$

$$\int_{\mathbb{R}} \frac{\Delta_{k,m,n}(x) \, d\mu(x)}{\prod\limits_{j=s+1}^{k} (x - y_j)} = 0, \quad s = 0, 1, \ldots, n.$$

This reduces the problem to showing that the determinant in (2.7.9) vanishes if $P_\ell(x)$ is replaced by

$$\int_{\mathbb{R}} \frac{P_\ell(x) \, d\mu(x)}{\prod\limits_{j=s+1}^{k} (x - y_j)}, \quad \ell = 0, 1, \ldots, n + m.$$

Let D denote the determinant in (2.7.9) with P_ℓ replaced by the above integral. Hence, by expanding the reciprocal of the product as in (2.7.5) we find

$$\int_{\mathbb{R}} \frac{P_\ell(x) \, d\mu(x)}{\prod\limits_{j=s+1}^{k} (x - y_j)} = \sum_{j=s+1}^{k} u_j(s) \tilde{Q}_\ell(y_j).$$

By adding linear combinations of rows to the last row we can replace the last $n + m + 1$ entries in the last row of D to zero. This changes the entry in the last row and column ℓ to $- \sum\limits_{j=s+1}^{k} u_j(s) b_{j,\ell}$, that is $- \sum\limits_{j=s+1}^{k} u_j(s) y_j^{\ell-1}$. This last quantity is $-\delta_{\ell-1,k-s-1}$ by Lemma 2.7.2. The latter quantity is zero since $1 \leq \ell \leq k - n$ and $k - n < k - s$. $\square$

2.8 Modifications of Measures: Toda

In this section we study modifying a measure of orthogonality by multiplying it by the exponential of a polynomial.

The Toda lattice equations describe the oscillations of an infinite system of points joined by spring masses, where the interaction is exponential in the distance between two spring masses (Toda, 1989). The semi-infinite Toda lattice equations in one time variable are

$$\dot{\alpha}_n(t) = \beta_n(t) - \beta_{n+1}(t), \quad n \geq 0, \tag{2.8.1}$$

$$\dot{\beta}_n(t) = \beta_n(t) \left[\alpha_{n-1}(t) - \alpha_n(t) \right], \tag{2.8.2}$$

where we followed the usual notation $\dot{f} = \dfrac{df}{dt}$. We shall show how orthogonal polynomials can be used to provide an explicit solution to (2.8.1)–(2.8.2).

Start with a system of monic polynomials $\{P_n(x)\}$ orthogonal with respect to μ and construct the recursion coefficients α_n and β_n in (2.2.1)–(2.2.2).

Theorem 2.8.1 *Let μ be a probability measure with finite moments, and let α_n and β_n be the recursion coefficients of the corresponding monic orthogonal polynomials. Let $P_n(x, t)$ be the monic polynomials orthogonal with respect to $\exp(-xt) \, d\mu(x)$*

under the additional assumption that the moments $\int_{\mathbb{R}} x^n \exp(-xt)\, d\mu(x)$ exist for all n, $n \geq 0$. Let $\alpha_n(t)$ and $\beta_n(t)$ be the recursion coefficients for $P_n(x,t)$. Then $\alpha_n(t)$ and $\beta_n(t)$ solve the system (2.8.1)–(2.8.2) with the initial conditions $\alpha_n(0) = \alpha_n$ and $\beta_n(0) = \beta_n$.

Proof First observe that the degree of $\dot{P}_n(x,t)$ is at most $n-1$. Let $\beta_0(t) = \int_{\mathbb{R}} e^{-xt}\, d\mu(x)$. Replace ζ_n in (2.1.5) by $\beta_0(t)\beta_1(t)\cdots\beta_n(t)$ then differentiate with respect to t to obtain

$$\beta_0(t)\beta_1(t)\cdots\beta_n(t) \sum_{k=0}^{n} \frac{\dot{\beta}_k(t)}{\beta_k(t)} = 0 - \int_{\mathbb{R}} x P_n^2(x,t) e^{-xt}\, d\mu(x).$$

Formula (2.2.1) implies

$$\alpha_n(t)\zeta_n(t) = \alpha_n(t)\beta_0(t)\beta_1(t)\cdots\beta_n(t) = \int_{\mathbb{R}} x P_n^2(x,t) e^{-xt}\, d\mu(x), \qquad (2.8.3)$$

hence

$$\beta_0(t)\beta_1(t)\cdots\beta_n(t) \sum_{k=0}^{n} \frac{\dot{\beta}_k(t)}{\beta_k(t)} = -\alpha_n(t)\beta_0(t)\beta_1(t)\cdots\beta_n(t), \qquad (2.8.4)$$

and (2.8.2) follows. Next differentiate (2.8.3) to find

$$\dot{\alpha}_n(t)\beta_0(t)\beta_1(t)\cdots\beta_n(t) + \alpha_n(t)\beta_0(t)\beta_1(t)\cdots\beta_n(t) \sum_{k=0}^{n} \frac{\dot{\beta}_k(t)}{\beta_k(t)}$$

$$= -\zeta_{n+1}(t) - \alpha_n^2(t)\zeta_n(t) - \beta_n^2(t)\zeta_{n-1}(t)$$

$$+ 2\beta_n(t) \int_{\mathbb{R}} \dot{P}_n(x,t) P_{n-1}(x,t) e^{-xt}\, d\mu(x).$$

The remaining integral can be evaluated by differentiating

$$0 = \int_{\mathbb{R}} P_n(x,t) P_{n-1}(x,t) e^{-xt}\, d\mu(x),$$

which implies

$$\int_{\mathbb{R}} \dot{P}_n(x,t) P_{n-1}(x,t) e^{-xt}\, d\mu(x) = \int_{\mathbb{R}} x P_n(x,t) P_{n-1}(x,t) e^{-xt}\, d\mu(x).$$

This yields

$$\int_{\mathbb{R}} \dot{P}_n(x,t) P_{n-1}(x,t) e^{-xt}\, d\mu(x) = \beta_n(t)\zeta_{n-1}(t).$$

Combining the above calculations with (2.8.4) we establish (2.8.1). $\qquad\square$

The multitime Toda lattice equations can be written in the form

$$\partial_{t_j} Q = \left[Q, (Q^j)_+ \right], \quad j = 1, \ldots, M, \qquad (2.8.5)$$

where Q is the tridiagonal matrix with entries (q_{ij}), $q_{ii} = \alpha_i$, $q_{i,i+1} = 1$, $q_{i+1,i} = \beta_{i+1}$, $i = 0, 1, 2, \ldots$, and $q_{ij} = 0$ if $|i - j| > 1$. For a matrix A, $(A)_+$ means replace all the entries below the main diagonal by zeros.

We start with a tridiagonal matrix Q formed by the initial values of α_n and β_n and find a measure of the orthogonal polynomials. We form a new probability measure according to

$$d\mu(x; \mathbf{t}) = \frac{1}{\zeta_0(\mathbf{t})} \exp\left(-\sum_{s=1}^{M} t_s x^s\right) d\mu(x), \qquad (2.8.6)$$

where $\mathbf{t}$ stands for $(t_1, \ldots, t_M)$. Let the corresponding monic orthogonal polynomials be $\{P_n(x; \mathbf{t})\}$, and $\{\alpha_n(\mathbf{t})\}$ and $\{\beta_n(\mathbf{t})\}$ be their recursion coefficients. Then the matrix $Q(\mathbf{t})$ be formed by the new recursion coefficients solves (2.8.5).

The partition function is

$$\begin{aligned}
Z_n(\mathbf{t}) := \frac{1}{\zeta_0^n(\mathbf{t})} \int_{\mathbb{R}^n} \exp\left(-\sum_{j=1}^{n}\sum_{s=1}^{M} t_s x_j^s\right) \\
\times \prod_{1 \leq i < j \leq n} (x_i - x_j)^2 \, d\mu(x_1) \cdots d\mu(x_n)
\end{aligned} \qquad (2.8.7)$$

and the tau function is

$$\tau_n(\mathbf{t}) = Z_n(\mathbf{t})/n! \qquad (2.8.8)$$

Formulas (2.1.9) and (2.1.13) establish

$$\tau_{n+1}(\mathbf{t}) = D_n = \prod_{j=1}^{n} \beta_j^{n-j+1}. \qquad (2.8.9)$$

2.9 Modification by Adding Finite Discrete Parts

We consider modification of a measure by adding a finite discrete part and express the polynomials orthogonal with respect to the new measure in terms of the polynomials orthogonal with respect to the old measure. This material is from Uvarov's very interesting paper (Uvarov, 1969).

Let

$$\nu(x) = \mu(x) + \sum_{j=1}^{r} m_j \epsilon_{x_j}, \qquad (2.9.1)$$

where ϵ_u is an atomic measure concentrated at $x = u$. Let $\{P_n(x)\}$ and $\{R_n(x)\}$ be monic polynomials orthogonal with respect to μ and ν, respectively, with $\zeta_n = \int_{\mathbb{R}} P_n^2 \, d\mu$. Set

$$R_n(x) = \sum_{s=0}^{n} C_s P_s(x), \quad C_n = 1. \qquad (2.9.2)$$

Since $\int\limits_{\mathbb{R}} R_n P_s \, d\nu = 0$ for $s < n$, we get

$$\zeta_s C_s + \sum_{j=1}^{r} m_j R_n(x_j) P_s(x_j) = 0, \quad 0 \le s < n.$$

Substituting this evaluation of C_s into (2.9.2) and using $C_n = 1$ we obtain

$$R_n(x) = P_n(x) - \sum_{j=1}^{r} m_j R_n(x_j) a_j(x), \tag{2.9.3}$$

with

$$a_j(x) = \sum_{s=0}^{n-1} \frac{P_s(x) P_s(x_j)}{\zeta_s},$$

so that

$$a_j(x) = \frac{P_n(x) P_{n-1}(x_j) - P_n(x_j) P_{n-1}(x)}{\zeta_{n-1}(x - x_j)}, \quad x \ne x_j$$

$$a_j(x_j) = \frac{P_n'(x_j) P_{n-1}(x_j) - P_n(x_j) P_{n-1}'(x_j)}{\zeta_{n-1}} \tag{2.9.4}$$

upon the use of the Christoffel–Darboux formula (2.2.4). Set

$$a_{ij} = a_j(x_i). \tag{2.9.5}$$

The system of equations (2.9.3) can be written as

$$R_n(x_i) + \sum_{j=1}^{r} a_{ij} m_j R_n(x_j) = P_n(x_i), \quad 1 \le i \le r, \tag{2.9.6}$$

and the values $R_n(x_j)$, $1 \le j \le r$ can be found in terms of the evaluations $\{P_n(x_j)\}_{j=1}^{r}$. Indeed (2.9.6) is

$$\begin{pmatrix} R_n(x_1) \\ \vdots \\ R_n(x_r) \end{pmatrix} = (I + A)^{-1} \begin{pmatrix} P_n(x_1) \\ \vdots \\ P_n(x_r) \end{pmatrix}$$

with

$$A = \begin{pmatrix} m_1 a_1(x_1) & m_2 a_2(x_1) & \cdots & m_r a_r(x_1) \\ m_1 a_1(x_2) & m_2 a_2(x_2) & \cdots & m_r a_r(x_2) \\ \vdots & \vdots & \vdots & \vdots \\ m_1 a_1(x_r) & m_2 a_2(x_r) & \cdots & m_r a_r(x_r) \end{pmatrix}.$$

Thus $R_n(x)$ is given by

$$R_n(x) = P_n(x) - \sum_{j=1}^{r} m_j R_n(x_j) a_j(x). \tag{2.9.7}$$

As an example consider the Jacobi polynomials where $d\mu$ is $(1-x)^\alpha (1+x)^\beta \, dx$, see Chapter 4. Let

$$A d\nu(x) = (1-x)^\alpha (1+x)^\beta \, dx + M_1 \epsilon_1 + M_2 \epsilon_{-1},$$

and A is a normalization constant, as

$$A = 2^{\alpha+\beta+1} \frac{\Gamma(\alpha+1)\Gamma(\beta+1)}{\Gamma(\alpha+\beta+2)} + M_1 + M_2.$$

The monic Jacobi polynomials are (4.1.1) and (4.1.4)

$$
\begin{aligned}
P_n(x) &= \frac{(\alpha+1)_n 2^n}{(\alpha+\beta+n+1)_n} \, {}_2F_1 \left(\begin{array}{c} -n, \alpha+\beta+n+1 \\ \alpha+1 \end{array} \middle| \frac{1-x}{2} \right) \\
&= \frac{(-2)^n(\beta+1)_n}{(\alpha+\beta+n+1)_n} \, {}_2F_1 \left(\begin{array}{c} -n, \alpha+\beta+n+1 \\ \beta+1 \end{array} \middle| \frac{1+x}{2} \right).
\end{aligned}
$$

Hence

$$P_n(1) = \frac{2^n(\alpha+1)_n}{(\alpha+\beta+n+1)_n}, \quad P(-1) = \frac{(-2)^n(\beta+1)_n}{(\alpha+\beta+n+1)_n}.$$

One can then find an explicit formula for $R_n(x)$.

The above example was studied in (Koornwinder, 1984) without the use of Uvarov's theorem.

2.10 Modifications of Recursion Coefficients

We saw in §§2.7–2.9 different examples of modifying a measure of orthogonality through multiplication by a rational function or an exponential function or by adding a finite discrete part to the measure of orthogonality. The question we address here is the effect of changing the recursion coefficients on the orthogonal polynomials.

The following theorem was proved in (Wendroff, 1961), but it was probably known to Geronimus in the 1950s, see (Geronimus, 1977). It is stated without proof as a footnote in (Geronimus, 1946) where the corresponding result for orthogonal polynomials on the circle is proved.

Theorem 2.10.1 (Wendroff) *Given sequences* $\{\alpha_n : n \geq N\}$, $\{\beta_n : n \geq N\}$, $\alpha_n \in \mathbb{R}$, $\beta_n > 0$; *and two finite sequences* $x_1 > x_2 > \cdots > x_N$, $y_1 > y_2 > \cdots > y_{N-1}$ *such that* $x_{k-1} > y_{k-1} > x_k$, $1 < k \leq N$, *then there is a sequence of monic orthogonal polynomials* $\{P_n(x) : n \geq 0\}$ *such that*

$$P_N(x) = \prod_{j=1}^{N} (x - x_j), \quad P_{N-1}(x) = \prod_{j=1}^{N-1} (x - y_j), \tag{2.10.1}$$

and

$$x P_n(x) = P_{n+1}(x) + \tilde{\alpha}_n P_n(x) + \tilde{\beta}_n P_{n-1}(x), \quad n > 0, \tag{2.10.2}$$

and $\tilde{\alpha}_n = \alpha_n$, $\tilde{\beta}_n = \beta_n$, *for* $n \geq N$.

Proof Use (2.10.2) for $n \geq N$ to define $P_{N+j}(x)$, $j > 0$, hence P_n has precise degree n for $n > N$. Define $\tilde{\alpha}_{N-1}$ by demanding that φ_{N-2},

$$\varphi_{N-2}(x) := (x - \tilde{\alpha}_{N-1}) P_{N-1}(x) - P_N(x),$$

has degree at most $N - 2$. Clearly $\operatorname{sgn} \varphi_{N-2}(y_j) = -\operatorname{sgn} P_N(y_j) = (-1)^{j-1}$, hence $\varphi_{N-2}(x)$ has at least $N-1$ sign changes, so it must have degree $N-2$ and its zeros interlace with the zeros of $P_{N-1}(x)$. Choose $\tilde{\beta}_{N-1}$ so that $\varphi_{N-2}(x)/\tilde{\beta}_{N-1}$ is monic. Hence $\tilde{\beta}_{N-1} > 0$. By continuing this process we generate all the remaining polynomials and the orthogonality follows from the spectral theorem. $\qquad\square$

Remark 2.10.1 *It is clear that Theorem 2.10.1 can be stated in terms of eigenvalues of tridiagonal matrices instead of zeros of P_N and P_{N-1}, see (2.2.7). This was the subject of (Drew et al., 2000), (Gray & Wilson, 1976), (Elsner & Hershkowitz, 2003) and (Elsner et al., 2003), whose authors were not aware of Wendroff's theorem or the connection between tridiagonal matrices and orthogonal polynomials.*

Now start with a recursion relation of the form (2.3.1) and define R_N and R_{N-1} to be monic of exact degrees N, $N-1$, respectively, and have real simple and interlacing zeros. Define $\{R_n : n > N\}$ through (2.3.1) and use Theorem 2.10.1 to generate $R_{N-2}, \ldots, R_0 (= 1)$. If $\{P_n(x)\}$ and $\{P_n^*(x)\}$ be as in §2.3. If the continued J-fraction corresponding to (2.3.1) converges then the continued J-fraction of $\{R_n(x)\}$ converges, and we can relate the two continued fractions because their entries differ in at most finitely different places.

It must be noted that the process of changing finitely many entries in a Jacobi matrix corresponds to finite rank perturbations in operator theory.

In general, we can define the kth associated polynomials $\{P_n(x; k)\}$ by

$$P_0(x; k) = 1, \quad P_1(x; k) = x - \alpha_k, \tag{2.10.3}$$

$$xP_n(x; k) = P_{n+1}(x; k) + \alpha_{n+k}P_n(x; k) + \beta_{n+k}P_{n-1}(x; k). \tag{2.10.4}$$

Hence, $P_{n-1}(x; 1) = P_n^*(x)$. It is clear that

$$(P_n(x; k))^* = P_{n-1}(x; k+1). \tag{2.10.5}$$

When $\{P_n(x)\}$ is orthogonal with respect to a unique measure μ then the corresponding continued J-fraction, $F(x)$, say, satisfies

$$\lim_{n\to\infty} \frac{P_n^*(x)}{P_n(x)} = F(x) = \int_{\mathbb{R}} \frac{d\mu(t)}{x-t}. \tag{2.10.6}$$

The continued J-fraction $F(x; k)$ of $\{P_n(x; k)\}$ is

$$\cfrac{1}{x - \alpha_k -} \cfrac{\beta_{k+1}}{x - \alpha_{k+1} -} \cdots.$$

Since

$$F(x) = \cfrac{1}{x - \alpha_0 -} \cfrac{\beta_1}{x - \alpha_1 -} \cdots \cfrac{\beta_k}{x - \alpha_k -} \cdots,$$

then

$$F(x) = \cfrac{1}{x - \alpha_0 -} \cfrac{\beta_1}{x - \alpha_1 -} \cdots \cfrac{\beta_k}{x - \alpha_{k-1} - \beta_k F(x; k)}. \tag{2.10.7}$$

Thus (2.10.7) evaluates $F(x; k)$, from which we can recover the spectral measure of $\{P_n(x; k)\}$.

An interesting problem is to consider $\{P_n(x; k)\}$ when k is not an integer. This is meaningful when the coefficients in (2.10.3) and (2.10.4) are well-defined. The problem of finding $\mu(x; k)$, the measure of orthogonality of $\{P_n(x; k)\}$, becomes highly nontrivial when $0 < k < 1$. If this measure is found, then the Stieltjes transforms of $\mu(x; k)$ for $k > 1$ can be found from the corresponding continued fraction. The parameter k is called the association parameter.

The interested reader may consult the survey article (Rahman, 2001) for a detailed account of the recent developments and a complete bibliography.

The measures and explicit forms of two families of associated Jacobi polynomials are in (Wimp, 1987) and (Ismail & Masson, 1991). The associated Laguerre polynomials are in (Askey & Wimp, 1984). Two families of associated Laguerre and Meixner polynomials are in (Ismail et al., 1988). The most general associated classical orthogonal polynomials are the Ismail–Rahman polynomials which arise as associated polynomials of the Askey–Wilson polynomial system. See (Ismail & Rahman, 1991) for details. The weight functions for a general class of polynomials orthogonal on $[-1, 1]$ containing the associated Jacobi polynomials have been studied by Pollaczek in his memoir (Pollaczek, 1956). Pollaczek's techniques have been instrumental in finding orthogonality measures for polynomials defined by three-term recurrence relations, as we shall see in Chapter 5.

In §5.6 we shall treat the associated Laguerre and Hermite polynomials, and §5.7 contains a brief account of the two families of associated Jacobi polynomials. The Ismail–Rahman work will be mentioned in §15.10.

2.11 Dual Systems

A discrete orthogonality relation of a system of polynomials induces an orthogonality relation for the dual system where the role of the variable and the degree are interchanged. The next theorem states this in a precise fashion.

Theorem 2.11.1 (Dual orthogonality) *Assume that the coefficients $\{\beta_n\}$ in (2.2.1) are bounded and that the moment problem has a unique solution. If the orthogonality measure μ has isolated point masses at α and β (β may $= \alpha$) then the dual orthogonality relation*

$$\sum_{n=0}^{\infty} P_n(\alpha)P_n(\beta)/\zeta_n = \frac{1}{\mu\{\alpha\}} \delta_{\alpha,\beta}, \qquad (2.11.1)$$

holds.

Proof The case $\alpha = \beta$ is Theorem 2.5.6. If $\alpha \neq \beta$, then the Christoffel–Darboux formula yields

$$\sum_{n=0}^{\infty} P_n(\alpha)P_n(\beta)/\zeta_n = \lim_{N\to\infty} \frac{P_{N+1}(\alpha)P_N(\beta) - P_{N+1}(\beta)P_N(\alpha)}{\zeta_N (\alpha - \beta)}$$

which implies the result since $\lim_{N\to\infty} P_N(x)/\sqrt{\zeta_N} = 0$, $x = \alpha, \beta$. $\square$

My friend Christian Berg sent me the following related theorem and its proof.

Theorem 2.11.2 *Let* $\mu = \sum_{\lambda \in \Lambda} a_\lambda \epsilon_\lambda$ *be a discrete probability measure and assume that* $\{p_n(x)\}$ *is a polynomial system orthonormal with respect to* μ *and complete in* $L^2(\mu; \mathbb{R})$. *Then the dual system* $\{p_n(\lambda)\sqrt{a_\lambda} : \lambda \in \Lambda\}$ *is a dual orthonormal basis for* ℓ^2, *that is*

$$\sum_{n=0}^{\infty} p_n(\lambda_1) p_n(\lambda_2) = \delta_{\lambda_1, \lambda_2}/a_{\lambda_1}, \forall \lambda_1, \lambda_2 \in \Lambda.$$

The proof will appear elsewhere.

Dual orthogonality arises in a natural way when our system of polynomials has finitely many members. This happens if β_n in (2.2.1) is positive for $0 \leq n < N$ but $\beta_{N+1} \leq 0$. Examples of such systems will be given in §6.2 and §15.6.

Let $\{\phi_n(x) : 0 \leq n \leq N\}$ be a finite system of orthogonal polynomials satisfying

$$\sum_{x=0}^{N} \phi_m(x)\phi_n(x)w(x) = \delta_{m,n}/h_n. \tag{2.11.2}$$

Then

$$\sum_{n=0}^{N} \phi_n(x)\phi_n(y)h_n = \delta_{x,y}/w(x) \tag{2.11.3}$$

holds for $x, y = 0, 1, \ldots, N$.

de Boor and Saff introduced another concept of duality which we shall refer to as the deB–S duality (de Boor & Saff, 1986). Given a sequence of polynomials satisfying (2.2.1)–(2.2.2), define a deB–S dual system $\{Q_n(x) : 0 \leq n \leq N\}$ by

$$Q_0(x) = 1, \quad Q_1(x) = x - \alpha_{N-1}, \tag{2.11.4}$$

$$Q_{n+1}(x) = (x - \alpha_{N-n-1}) Q_n(x) - \beta_{N-n}Q_{n-1}(x), \tag{2.11.5}$$

for $0 \leq n < N$. From (2.2.1)–(2.2.2) and (2.11.4)–(2.11.5) it follows that

$$Q_n(x) = P_n(x; N - n).$$

The material below is from (Vinet & Zhedanov, 2004).

Clearly the mapping $\{P_n\} \to \{Q_n\}$ is an involution in the sense that the deB–S dual of $\{Q_n\}$ is $\{P_n\}$. By induction we can show that

$$Q_{N-n-1}(x)P_{n+1}(x) - \beta_n Q_{N-n-2}(x)P_n(x) = P_N(x), \tag{2.11.6}$$

$0 \leq n < N$. As in the proof of Theorem 2.6.2, we apply (2.5.1) to see that $\{P_n(x) : 0 \leq n < N\}$ are orthogonal on $\{x_{N,j} : 1 \leq j \leq N\}$ with respect to a discrete measure with masses $\rho(x_{N,j})$ at $x_{N,j}$, $1 \leq j \leq N$. Moreover

$$\frac{P_N^*(x)}{P_N(x)} = \sum_{k=1}^{N} \frac{\rho(x_{N,j})}{x - x_{N,j}}.$$

Using the fact that $P_N(x) = Q_N(x)$ and $Q_N^*(x) = P_{N-1}(x)$ it follows that the masses $\rho_Q(x_{N,j})$ defined by

$$\frac{P_{N-1}(x)}{P_N(x)} = \frac{Q_N^*(x)}{Q_N(x)} = \sum \frac{\rho_Q(x_{N,j})}{x - x_{N,j}},$$

so that the numbers $\rho_Q(x_{N,j}) = P_{N-1}(x_{N,j})/P_N'(x_{N,j})$, have the property

$$\sum_{j=1}^{N} \rho_Q(x_{N,j}) = 1, \quad \sum_{j=1}^{N} \rho_Q(x_{N,j}) Q_m(x_{N,j}) Q_n(x_{N,j}) = \zeta_n(Q)\delta_{m,n},$$

(2.11.7)

and

$$\zeta_n(Q) = \beta_{N-1}\beta_{N-2}\cdots\beta_{N-n}, \quad \text{or} \quad \zeta_n(Q) = \zeta_{N-1}/\zeta_{N-n-1}. \quad (2.11.8)$$

Vinet and Zhedanov refer to $\rho_Q(x_{N,j})$ as the dual weights.

Theorem 2.11.3 *The Jacobi, Hermite and Laguerre polynomials are the only orthogonal polynomials where $\rho_Q(x_{N,j}) = \pi(x_{N,j})/c_N$ for all N, where π is a polynomial of degree at most 2 and c_N is a constant.*

For a proof, see (Vinet & Zhedanov, 2004).

A sequence of orthogonal polynomials is called **semiclassical** if it is orthogonal with respect to an absolutely continuous measure μ, where $\mu'(x)$ satisfies a differential equation $y'(x) = r(x)\,y(x)$, and r is a rational function.

Theorem 2.11.4 ((Vinet & Zhedanov, 2004)) *Assume that $\{Q_n(x)\}$ is the deB–S dual to $\{P_n(x)\}$ and let $\{x_{n,i}\}$ be the zeros of $P_n(x)$. Then $\{P_n(x)\}$ are semiclassical if and only if, for every $n > 0$, the polynomials $\{Q_i(x) : 0 \le i < n\}$ are orthogonal on $\{x_{n,i} : 1 \le i \le n\}$ with respect to the weights*

$$\frac{q(x_{n,i})}{\tau(x_{n,i}, n)},$$

where $q(x)$ is a polynomial of fixed degree and its coefficients do not depend on n, but $\tau(x,n)$ is a polynomial of fixed degree whose coefficients may depend on n.

Exercises

2.1   Let $\{P_n(x)\}$ satisfy (2.2.1)–(2.2.2). Prove that the coefficient of x^{n-2} in $P_n(x)$ is

$$\sum_{0\le i<j<n-1} \alpha_i\alpha_j - \sum_{k=1}^{n-1} \beta_k.$$

2.2   Assume that $\alpha_n = 0$, $n \ge 0$, $\beta_n < 0$ for $n > 0$. If $\{P_n(x)\}$ satisfies (2.2.1)–(2.2.2), prove that $\{i^{-n}P_n(ix)\}$ is a real sequence of orthogonal polynomials.

2.3   Consider an absolutely continuous measure μ on $\mathbb{R}$ with $\mu'(x) = e^{-x^2}dx/\sqrt{\pi}$. Prove that $\mu_{2n+1} = 0$ and $\mu_{2n} = (1/2)_n$. Evaluate the Hankel determinants D_n, $n > 0$ and find an explicit formula for the orthonormal polynomials.

2.4   Repeat Exercise 2.3 for μ supported on $[0, \infty)$ with

$$\mu'(x) = x^\alpha e^{-x}dx/\Gamma(\alpha+1).$$

2.5   Assume that $\{P_n(x)\}$ satisfies (2.2.1)–(2.2.2) and $Q_n(x) = a^{-n}P_n(ax + b)$. Show that if $\{P_n(x)\}$ are orthogonal with respect to a probability measure with moments $\{\mu_n\}$ then $\{Q_n(x)\}$ are orthogonal with respect to a probability measure whose moments m_n are given by

$$m_n = a^{-n} \sum_{k=0}^{n} \binom{n}{k} (-b)^{n-k} \mu_k.$$

2.6   Suppose that $P_n(x)$ satisfies the three-term recurrence relation

$$P_{n+1}(x) = (x - \alpha_n) P_n(x) - \beta_n P_{n-1}(x), \quad n > 0,$$
$$P_0(x) = 1, \quad P_{-1}(x) = 0.$$

Assume that $\beta_n > 0$, and that L is the positive definite linear functional for which P_n is orthogonal and $L(1) = 1$, that is $L(p_m(x)p_n(x)) = 0$ if $m \neq n$. Suppose that S is another linear functional such that $S(P_k P_\ell) = 0$ for $n \geq k > \ell \geq 0$, $S(1) = 1$. Show that S has the same jth moments as L, for j at most $2n - 1$.

2.7   Explicitly renormalize the polynomials $P_n(x)$ to $P_n^*(x)$ in Exercise 2.6 so that the eigenvalues and eigenvectors of the related real symmetric tridiagonal n by n matrix are explicitly given by the zeros $x_{n,k}$ of $P_n(x)$ and the vectors $\left(P_0^*(x_{n,k}), \ldots, P_{n-1}^*(x_{n,k})\right)$. Find these results. Next, using orthogonality of the eigenvectors, rederive the Gaussian quadrature formula

$$L_n(p) = \sum_{k=1}^{n} \Lambda_{nk} p(x_{nk}).$$

Then use Exercise 2.6 to conclude that L_n and L have the identical moments up to $2n - 1$.

2.8   Let $\Delta(x) = \prod_{k>j} (x_k - x_j)$ be the Vandermonde determinant in n variables $x_1, \ldots, x_n$. Let $w(x) = x^a(1-x)^b$ on $[-1, 1]$, where $a, b > -1$. Evaluate

   (a) $\int_{[0,1]^n} [\Delta(x)]^2 w(x_1) w(x_2) \ldots w(x_n) \, dx_1 \ldots dx_n$

   (b) $\int_{[0,1]^n} [\Delta(x)]^2 w(x_1) w(x_2) \ldots w(x_n) e_k(x_1, \ldots, x_n) \, dx_1 \ldots dx_n,$

   where e_k is the elementary symmetric function of degree k.

2.9   Let $\{r_n(x)\}$ and $\{s_n(x)\}$ be orthonormal with respect to $\rho(x)\,dx$ and $\sigma(x)\,dx$, respectively, and assume that $\int_{\mathbb{R}} \rho(x)\,dx = \int_{\mathbb{R}} w(x)\,dx = 1$. Prove that if

$$r_n(x) = \sum_{k=0}^{n} c_{n,k}\, s_k(x),$$

then

$$\sigma(x)s_k(x) = \sum_{n=k}^{\infty} c_{n,k}\, \rho(x)r_n(x).$$

The second equality holds in $L^2(\mathbb{R})$.

2.10   Assume that $y \in \mathbb{R}$ and does not belong to the true interval of orthogonality of $\{P_n(x)\}$. Prove that the kernel polynomials $\{K_n(x, y)\}$ are orthogonal with respect to $|x - y| \, d\mu(x)$, μ being the orthogonality measure of $\{P_n(x)\}$.

3

Differential Equations, Discriminants and Electrostatics

In this chapter we derive linear second order differential equations satisfied by general orthogonal polynomials with absolutely continuous measures of orthogonality. We also give a closed form expression of the discriminant of such orthogonal polynomial in terms of the recursion coefficients. These results are then applied to electrostatic models of a system of interacting charged particles.

3.1 Preliminaries

Assume that μ is absolutely continuous and let

$$d\mu(x) = w(x)\, dx, \quad x \in (a, b), \tag{3.1.1}$$

where $[a, b]$ is not necessarily bounded. We shall write

$$w(x) = \exp(-v(x)), \tag{3.1.2}$$

require v to be twice differentiable and assume that the integrals

$$\int_a^b y^n \frac{v'(x) - v'(y)}{x - y}\, w(y)\, dy, \quad n = 0, 1, \ldots, \tag{3.1.3}$$

exist. We shall also use the orthonormal form $\{p_n(x)\}$ of the polynomials $\{P_n(x)\}$, that is

$$p_n(x) = P_n(x)/\sqrt{\zeta_n}. \tag{3.1.4}$$

Rewrite (2.2.1) and (2.2.2) in terms of the p_ns. The result is

$$p_0(x) = 1, \quad p_1(x) = (x - \alpha_0)/a_1, \tag{3.1.5}$$

$$xp_n(x) = a_{n+1}p_{n+1}(x) + \alpha_n p_n(x) + a_n p_{n-1}(x), \quad n > 0. \tag{3.1.6}$$

The discriminant D of a polynomial g,

$$g(x) := \gamma \prod_{j=1}^n (x - x_j) \tag{3.1.7}$$

is defined by, (Dickson, 1939),

$$D(g) := \gamma^{2n-2} \prod_{1 \le j < k \le n} (x_j - x_k)^2.$$ (3.1.8)

An alternate representation for the discriminant is

$$D(g) := (-1)^{n(n-1)/2} \gamma^{n-2} \prod_{j=1}^{n} g'(x_j),$$ (3.1.9)

as can be seen by direct substitution of g in (3.1.9) and using (3.1.7), see (Dickson, 1939). The resultant of g and a polynomial f of degree m is

$$\text{Res}\{g, f\} = \gamma^m \prod_{j=1}^{n} f(x_i).$$ (3.1.10)

Clearly

$$D(g) = (-1)^{\binom{n}{2}} \gamma^{-1} \text{Res}\{g, g'\}.$$ (3.1.11)

We shall consider the case of logarithmic potential on the line. In this model the potential energy at x of a point charge e located at c is $-e \ln|x - c|$. Consider the system of n movable unit charged particles in the presence of the external potential $V(x)$. The particles are restricted to lie in $[a, b]$. Let

$$\mathbf{x} := (x_1, x_2, \ldots, x_n),$$ (3.1.12)

where $x_1, \ldots, x_n$ are the positions of the particles arranged in decreasing order. The total energy of the system is

$$E(\mathbf{x}) = \sum_{k=1}^{n} V(x_k) - 2 \sum_{1 \le j < k \le n} \ln|x_j - x_k|.$$ (3.1.13)

Let

$$T(\mathbf{x}) := \exp(-E(\mathbf{x})).$$ (3.1.14)

In §3.5 we shall describe how the zeros of p_n describe the equilibrium position of the particles in such a system.

3.2 Differential Equations

Define $A_n(x)$ and $B_n(x)$ via

$$\frac{A_n(x)}{a_n} = \frac{w(y) p_n^2(y)}{y - x} \Big|_a^b + \int_a^b \frac{v'(x) - v'(y)}{x - y} p_n^2(y) w(y) \, dy,$$ (3.2.1)

$$\frac{B_n(x)}{a_n} = \frac{w(y) p_n(y) p_{n-1}(y)}{y - x} \Big|_a^b$$

$$+ \int_a^b \frac{v'(x) - v'(y)}{x - y} p_n(y) p_{n-1}(y) w(y) \, dy.$$ (3.2.2)

We shall tacitly assume that the boundary terms in (3.2.1) and (3.2.2) exist.

Theorem 3.2.1 *Let $v(x)$ be a twice continuously differential function on $[a, b]$. Then the polynomials $\{p_n(x)\}$ orthonormal with respect to $w(x)dx$, $w(x) = \exp(-v(x))$ satisfy*

$$p_n'(x) = -B_n(x)p_n(x) + A_n(x)p_{n-1}(x), \qquad (3.2.3)$$

where A_n and B_n are given by (3.2.1) and (3.2.2).

Proof Since $p_n'(x)$ is a polynomial of degree $n - 1$, it can be expanded as

$$p_n'(x) = \sum_{k=0}^{n-1} c_{n,k}\, p_k(x). \qquad (3.2.4)$$

Applying the orthogonality relation and integration by parts we get

$$c_{n,k} = \int_a^b p_n'(y)\, p_k(y)\, w(y)\, dy$$

$$= p_n(y)p_k(y)w(y)\big|_a^b - \int_a^b p_n(y)\left[p_k'(y) - p_k(y)v'(y)\right]w(y)\, dy,$$

hence the term involving p_k' vanishes. It follows that the right-hand side of (3.2.4) is

$$w(y)p_n(y) \sum_{k=0}^{n-1} p_k(x)p_k(y)\Bigg|_{y=a}^{y=b} + \int_a^b p_n(y)\left[\sum_{k=0}^{n-1} p_k(x)p_k(y)\right]v'(y)w(y)\, dy.$$

The above integral would vanish if $v'(y)$ is replaced by $v'(x)$, hence the right-hand side of (3.2.4) is

$$w(y)p_n(y) \sum_{k=0}^{n-1} p_k(x)p_k(y)\Bigg|_{y=a}^{y=b}$$

$$+ \int_a^b \left[\sum_{k=0}^{n-1} p_k(x)\, p_k(y)\right]\left[v'(y) - v'(x)\right]p_n(y)\, w(y)\, dy.$$

Formula (3.2.3) now follows from the Christoffel–Darboux formula (2.2.4) with $P_n(x) = \sqrt{\xi_n}\, p_n(x)$. $\square$

Theorem 3.2.1 was proved for symmetric polynomials in (Mhaskar, 1990), see also (Bonan & Clark, 1990), (Bauldry, 1990). It was rediscovered in a slightly more general form in (Chen & Ismail, 1997). The treatment and terminology presented here is from (Chen & Ismail, 1997).

Lemma 3.2.2 *We have*

$$B_n(x) + B_{n+1}(x) = \frac{x - \alpha_n}{a_n}\, A_n(x) - v'(x). \qquad (3.2.5)$$

Proof It is clear that the left-hand side of (3.2.5) is

$$\frac{w(y)\,(y-\alpha_n)\,p_n^2(y)}{y-x}\bigg|_a^b + \int_{\mathbb{R}} p_n(y)\,\frac{v'(x)-v'(y)}{x-y}\,[y-\alpha_n]\,p_n(y)\,w(y)\,dy.$$

By writing $y-\alpha_n$ as $y-x+x-\alpha_n$ we see that the left-hand side of (3.2.5) is

$$= w(y)p_n^2(y)\big|_a^b + \frac{x-\alpha_n}{a_n}\,A_n(x) + \int_{\mathbb{R}} [v'(y)-v'(x)]\,p_n^2(y)\,w(y)\,dy$$

$$= \frac{x-\alpha_n}{a_n}\,A_n(x) - v'(x),$$

after an integration-by-parts and the use of the orthonormality of $\{p_n(x)\}$. $\qquad\square$

Rewrite (3.2.3) as

$$L_{1,n}p_n(x) = A_n(x)p_{n-1}(x), \qquad (3.2.6)$$

where $L_{1,n}$ is the differential operator

$$L_{1,n} = \frac{d}{dx} + B_n(x). \qquad (3.2.7)$$

In this form $L_{1,n}$ is a lowering or annihilation operator for p_n. In a few cases it is independent of n, except possibly for a multiplicative constant.

Introduce the weighted inner product

$$(f,g)_w := \int_a^b f(x)\,\overline{g(x)}\,w(x)\,dx, \qquad (3.2.8)$$

and consider the function space where $(f,f)_w$ is finite and $f(x)\sqrt{w(x)}$ vanishes at $x=a,b$. It is easy to see that with respect to the inner product (3.2.8)

$$L_{2,n} := L_{1,n}^* = -\frac{d}{dx} + B_n(x) + v'(x). \qquad (3.2.9)$$

Now the adjoint relation

$$\left(-\frac{d}{dx} + B_n(x) + v'(x)\right)p_{n-1}(x) = A_{n-1}(x)\,\frac{a_n}{a_{n-1}}\,p_n(x), \qquad (3.2.10)$$

follows from (3.2.3) and (3.1.5)–(3.1.6). Therefore $L_{2,n}$ is a creation or raising operator for $\{p_n\}$. For a discussion of raising and lowering operators for specific one-variable polynomials as they relate to two-variable theory, we refer the reader to a recent article (Koornwinder, 2005a).

By combining (3.2.6) and (3.2.10) we establish the next theorem after using (3.2.5)

Theorem 3.2.3 *Under the assumptions in Theorem 3.2.1 the p_n's satisfy the factored equation*

$$L_{2,n}\left(\frac{1}{A_n(x)}\,(L_{1,n}p_n(x))\right) = \frac{a_n}{a_{n-1}}\,A_{n-1}(x)p_n(x). \qquad (3.2.11)$$

Equivalently (3.2.11) is

$$p_n''(x) + R_n(x)p_n'(x) + S_n(x)p_n(x) = 0, \qquad (3.2.12)$$

where

$$R_n(x) := -\left[v'(x) + \frac{A_n'(x)}{A_n(x)}\right], \qquad (3.2.13)$$

$$
\begin{aligned}
S_n(x) := {}& A_n(x)\left(\frac{B_n(x)}{A_n(x)}\right)' - B_n(x)\left[v'(x) + B_n(x)\right] \\
& + A_n(x)A_{n-1}(x)\frac{a_n}{a_{n-1}} \\
= {}& B_n'(x) - B_n(x)\frac{A_n'(x)}{A_n(x)} - B_n(x)\left[v'(x) + B_n(x)\right] \\
& + \frac{a_n}{a_{n-1}}A_n(x)A_{n-1}(x).
\end{aligned}
\qquad (3.2.14)
$$

The so-called Schrödinger form of (3.2.11)–(3.2.12) is

$$\Psi_n''(x) + V(x;n)\Psi_n(x) = 0, \qquad (3.2.15)$$

where

$$\Psi_n(x) := \frac{\exp[-v(x)/2]}{\sqrt{A_n(x)}}\, p_n(x), \qquad (3.2.16)$$

and

$$
\begin{aligned}
V(x,n) = {}& A_n(x)\left(\frac{B_n(x)}{A_n(x)}\right)' - B_n(x)\left[v'(x) + B_n(x)\right] \\
& + A_n(x)A_{n-1}(x)\frac{a_n}{a_{n-1}} + \frac{v''(x)}{2} + \frac{1}{2}\left(\frac{A_n'(x)}{A_n(x)}\right)' \\
& - \frac{1}{4}\left[v'(x) + \frac{A_n'(x)}{A_n(x)}\right]^2.
\end{aligned}
\qquad (3.2.17)
$$

Observe that $A_n(x) > 0$ if we assume that $v(x)$ is convex.

The differential equation (3.2.12) was derived in (Shohat, 1939) for $v(x) = x^4 + c$, (Bauldry, 1985) for general polynomials v of degree 4 and in (Sheen, 1987) for $v(x) = x^6/6 + c$.

Shohat gave a procedure to derive (3.2.3) when $w'(x)/w(x)$ is a rational function, (Shohat, 1939). Another derivation of the same result is in (Atkinson & Everitt, 1981), who also established (3.2.3) when $\int_{\mathbb{R}} (x - t)^{-1} d\mu(t)$ satisfies a linear first order differential equation, without assuming that μ is absolutely continuous.

It is important to note that (3.2.9)–(3.2.10) lead to

$$a_n p_n(x) = [\mathbb{L}_n \mathbb{L}_{n-1} \cdots \mathbb{L}_1]\, 1 \qquad (3.2.18)$$

where

$$(\mathbb{L}_n f)(x) := \frac{1}{A_{n-1}(x)}\left[-\frac{d}{dx} + B_n(x) + v'(x)\right] f(x). \qquad (3.2.19)$$

For the special cases of Legendre, Hermite, Laguerre or Jacobi polynomials formula (3.2.19) is referred to as the Rodrigues formula. It is interesting to note that the general differential equations approach described here extends it for all polynomials orthogonal with respect to absolutely continuous measures when the weight function e^{-v} satisfies the assumptions in Theorem 3.2.1.

As an example, consider the weight function $w(x) = ce^{-x^4+2tx^2}$, $x \in \mathbb{R}$, and $\int_{\mathbb{R}} w(x)\, dx = 1$. Thus,

$$\frac{A_n(x)}{a_n} = \int_{\mathbb{R}} 4\left(x^2 + y^2 + xy - t\right) p_n^2(y)w(y)\, dy.$$

By Remark 2.1.2, we see that $p_n^2(y)w(y)$ is an even function, hence $\alpha_n = 0$. Thus

$$\frac{A_n(x)}{4a_n} = x^2 - t + \int_{\mathbb{R}} \left[a_{n+1}p_{n+1}(y) + a_n p_n(y)\right]^2 w(y)\, dy,$$

so that

$$A_n(x) = 4a_n \left[x^2 - t + a_n^2 + a_{n+1}^2\right].$$

Similarly,

$$B_n(x) = 4a_n^2 x.$$

Thus we find

$$R_n(x) = -4x^3 - \frac{2x}{x^2 + a_n^2 + a_{n+1}^2}$$

$$S_n(x) = 4a_n^2 - \frac{8a_n^2 x^2}{x^2 + a_n^2 + a_{n+1}^2} + 16x^2 a_n^2 \left(a_{n-1}^2 + a_n^2 + a_{n+1}^2\right)$$
$$+ 16a_n^2 \left(a_n^2 + a_{n+1}^2\right)\left(a_n^2 + a_{n-1}^2\right).$$

One important application of (3.2.3) is to derive the so-called string equation. To do so, set

$$p_n(x) = \frac{x^n}{a_1 \cdots a_n} + u_n x^{n-2} + \cdots,$$

and substitute in (3.2.3). From (3.1.6) it follows that

$$u_n = a_{n+1}u_{n+1} + \frac{a_n}{a_1 \cdots a_n - 1}.$$

The result of equating the coefficients of x^{n-1} is the string equation

$$n = 4a_n^2 \left(a_n^2 + a_{n+1}^2 - t\right) + 4a_n^2 a_{n-1}^2. \tag{3.2.20}$$

Additional nonlinear relations can be obtained by equating coefficients of other powers of x in (3.2.3). Freud proved (3.2.20) when $t = 0$ and derived a similar nonlinear relation for the weight $C \exp\left(-x^6\right)$ in (Freud, 1976). This was part of Freud's study of asymptotics and largest zeros of polynomials orthogonal with respect to $C \exp\left(-|x|^m\right)$. Freud conjectured that the recursion coefficients of such polynomials have the limiting behavior

$$\lim_{n\to\infty} n^{-1/m} a_n = [\Gamma(m/2)\Gamma(1+m/2)/\Gamma(m+1)]^{1/m}, \tag{3.2.21}$$

in (Freud, 1976) and (Freud, 1977). He used (3.2.20) and tricky manipulations involving lim sup and lim inf to prove his conjecture for $m = 4$, and applied the same technique to settle his conjecture for $m = 6$. The general case, when m is positive but not necessarily an integer, was proved by Rakhmanov. A stronger conjecture of Freud's states that the largest zero of $p_n(x)$, $x_{n,1}$, has the limiting property

$$\lim_{n \to \infty} n^{-1/m} x_{n,1} = 2[\Gamma(m/2)\Gamma(1 + m/2)/\Gamma(m + 1)]^{1/m}, \qquad (3.2.22)$$

which was proved by Lubinsky, Mhaskar and Saff. For details, references and applications, see (Lubinsky, 1987) and (Lubinsky, 1993).

Motivated by the case $t = 0$ of (3.2.20), Lew and Quarles (Lew & Quarles, Jr., 1983) studied asymptotics of solutions to the nonlinear recursion

$$c_n^2 = x_n \left(x_{n+1} + x_n + x_{n-1}\right), \ n > 0, \text{ with } x_0 \geq 0, \qquad (3.2.23)$$

where $\{c_n\}$ is a given sequence of real numbers.

We now derive recursion relations for $A_n(x)$ and $B_n(x)$. This material is based on the work (Ismail & Wimp, 1998).

Theorem 3.2.4 *The A_n's and B_n's satisfy the string equation*

$$B_{n+1}(x) - B_n(x)$$
$$= \frac{a_{n+1}A_{n+1}(x)}{x - \alpha_n} - \frac{a_n^2 A_{n-1}(x)}{a_{n-1}(x - \alpha_n)} - \frac{1}{x - \alpha_n}. \qquad (3.2.24)$$

Proof We set

$$a_{n+1}A_{n+1}(x) - \frac{a_n^2 A_{n-1}(x)}{a_{n-1}} = I + BT, \qquad (3.2.25)$$

where I and BT stand for integrals and boundary terms on the left-hand side of (3.2.24). The recursion relation (3.1.6) gives

$$I = \int_a^b \frac{v'(x) - v'(y)}{x - y} \left[a_{n+1}p_{n+1}(y) - a_n p_{n-1}(y)\right] (y - \alpha_n) \, p_n(y) \, w(y) \, dy$$

$$= (x - \alpha_n) \left[B_{n+1}(x) - B_n(x)\right]$$

$$+ (x - \alpha_n) \left[\frac{w(a^+) \, p_n(a^+)}{x - a} \left\{a_n p_{n-1}(a^+) - a_{n+1}p_{n+1}(a^+)\right\}\right.$$

$$+ \frac{w(b^-) \, p_n(b^-)}{b - x} \left\{a_n p_{n-1}(b^-) - a_{n+1}p_{n+1}(b^-)\right\}\Bigg]$$

$$- \int_a^b [v'(x) - v'(y)] \left[a_{n+1}p_{n+1}(y) - a_n p_{n-1}(y)\right] p_n(y) \, w(y) \, dy.$$

The remaining integral in I can be dealt with in the following way:

$$\int_a^b [v'(x) - v'(y)] [a_{n+1}p_{n+1}(y) - a_n p_{n-1}(y)] p_n(y) w(y) \, dy$$

$$= \int_a^b [a_{n+1}p_{n+1}(y) - a_n p_{n-1}(y)] p_n(y) w'(y) \, dy$$

$$= \{a_{n+1}p_{n+1}(y) - a_n p_{n-1}(y)\} w(y) p_n(y)|_{y=a}^b$$

$$- \int_a^b [a_{n+1}p_n(y)p'_{n+1}(y) - a_n p_{n-1}(y)p'_n(y)] w(y) \, dy.$$

Since the coefficient of x^n in $p_n(x)$ is $(a_1 \cdots a_n)^{-1}$, we get

$$\int_a^b a_{n+1}p_n(y)p'_{n+1}(y)w(y) \, dy \qquad (3.2.26)$$

$$= (n + 1) \int_a^b p_n^2(y) \, w(y) \, dy = n + 1,$$

and after putting all this information in one pot, we establish

$$I = (x - \alpha_n) [B_{n+1}(x) - B_n(x)]$$

$$- \{a_{n+1}p_{n+1}(y) - a_n p_{n-1}(y)\} \, w(y) \, p_n(y)]_{y=a}^{y=b}$$

$$+ (x - \alpha_n) \left[\frac{w(a^+) p_n(a^+)}{x - a} \{a_n p_{n-1}(a^+) - a_{n+1}p_{n+1}(a^+)\} \right.$$

$$\left. + \frac{w(b^-) p_n(b^-)}{b - x} \{a_n p_{n-1}(b^-) - a_{n+1}p_{n+1}(b^-)\} \right] + 1.$$

Now the boundary terms above combine and when compared with the terms BT in (3.2.25) we find

$$I + BT = (x - b_n) [B_{n+1}(x) - B_n(x)] + 1,$$

and the theorem follows. □

In the special case $w(x) = ce^{-x^4 + 2tx^2}$, formula (3.2.24) gives

$$4a_{n+1}^2 (a_{n+1}^2 + a_{n+2}^2) - 4a_n^2 (a_n^2 + a_{n+1}^2) = 4t (a_{n+1}^2 - a_n^2) - 1,$$

which implies (3.2.20). The following theorem deals with the special cases when v is a general polynomial of degree 4 and describes how the string equation characterizes the orthogonal polynomials.

Theorem 3.2.5 ((Bonan & Nevai, 1984)) *Let $\{p_n(x)\}$ be orthonormal with respect to a probability measure μ. Then the following are equivalent*

(i) *There exist nonnagative integers j and k two sequences $\{e_n\}$ and $\{c_n\}$, $n =$ $1, 2, \ldots$, such that $j < k$ and*

$$p_n'(x) = e_n p_{n-j}(x) + c_n p_{n-k}(x), \quad n = 1, 2, \ldots .$$

(ii) *There exists a nonnegative constant c such that*

$$p_n'(x) = \frac{n}{a_n} p_{n-1}(x) + c a_n a_{n-1} a_{n-2} p_{n-3}(x), \quad n = 1, 2, \ldots,$$

where $\{a_n\}$ are as in (3.1.6).

(iii) *There exist real numbers c, b and K such that $c \geq 0$, if $c = 0$ then $K > 0$ and the recursion coefficients in (3.1.6) satisfy*

$$n = c a_n^2 \left[a_{n+1}^2 + a_n^2 + a_{n-1}^2 \right] + K a_n^2, \quad n = 1, 2, \ldots,$$

and $\alpha_n = b$, for $n = 0, 1, \ldots$.

(iv) *The measure μ is absolutely continuous $\mu' = e^{-v}$ with*

$$v(x) = \frac{c}{4}(x - b)^4 - \frac{K}{2}(x - b)^2 + d, \quad b, c, d, K \in \mathbb{R}.$$

Moreover, $c \geq 0$ and if $c = 0$ then $K > 0$.

Theorem 3.2.6 ((Bonan et al., 1987)) *Let $\{p_n(x)\}$ be orthonormal with respect to μ. They satisfy*

$$p_n'(x) = \sum_{k=n-m+1}^{n-1} c_{k,n} p_k(x). \tag{3.2.27}$$

for constant $\{c_{k,n}\}$ if and only if μ is absolutely continuous, $\mu'(x) = e^{-v(x)}$, and v is a polynomial of exact degree m.

The interested reader may consult (Bonan & Nevai, 1984) and (Bonan et al., 1987) for proofs of Theorems 3.2.5 and 3.2.6.

We now return to investigate recursion relations satisfied by $A_n(x)$ and $B_n(x)$. It immediately follows from (3.2.5) and (3.2.24) that

$$\begin{aligned} 2B_{n+1}(x) = {}& \frac{x - b_n}{a_n} A_n(x) + \frac{a_{n+1} A_{n+1}(x)}{x - \alpha_n} \\ & - \frac{a_n^2 A_{n-1}(x)}{a_{n-1}(x - \alpha_n)} - v'(x) - \frac{1}{x - \alpha_n}, \end{aligned} \tag{3.2.28}$$

and

$$\begin{aligned} 2B_n(x) = {}& \frac{x - b_n}{a_n} A_n(x) - \frac{a_{n+1} A_{n+1}(x)}{x - \alpha_n} \\ & + \frac{a_n^2 A_{n-1}(x)}{a_{n-1}(x - \alpha_n)} - v'(x) + \frac{1}{x - \alpha_n}. \end{aligned} \tag{3.2.29}$$

The compatibility of (3.2.28)–(3.2.29) indicates that A_n must satisfy the inhomogeneous recurrence relation

$$
\begin{aligned}
\frac{a_{n+2}A_{n+2}(x)}{x - \alpha_{n+1}} &= \left[\frac{x - \alpha_{n+1}}{a_{n+1}} - \frac{a_{n+1}}{x - \alpha_n} \right] A_{n+1}(x) \\
&\quad + \left[\frac{a_{n+1}^2}{a_n \left(x - \alpha_{n+1} \right)} - \frac{(x - \alpha_n)}{a_n} \right] A_n(x) \\
&\quad + \frac{a_n^2}{a_{n-1} \left(x - \alpha_n \right)} A_{n-1}(x) \\
&\quad + \frac{1}{x - \alpha_n} + \frac{1}{x - \alpha_{n+1}}, \quad n > 1.
\end{aligned}
\tag{3.2.30}
$$

We extend the validity of (3.2.30) to the cases $n = 0$, and $n = 1$ through

$$
a_0 := 1, \quad p_{-1} := 0,
$$

$$
\frac{A_0(x)}{a_0} := \frac{w\left(a^+\right)}{x - a} + \frac{w\left(b^-\right)}{b - x} + \int_a^b \frac{v'(x) - v'(y)}{x - y} w(y)\, dy.
\tag{3.2.31}
$$

Thus

$$
B_0 = A_{-1}(x) = 0.
$$

Eliminating $A_n(x)$ between (3.2.5) and (3.2.24), and simplifying the result we find that the B_n's also satisfy the inhomogeneous four term recurrence relation

$$
\begin{aligned}
B_{n+2}(x) &= \left[\frac{(x - \alpha_n)(x - \alpha_{n+1})}{a_{n+1}^2} - 1 \right] B_{n+1}(x) \\
&\quad + \left[\frac{a_n^2 (x - \alpha_{n+1})}{a_{n+1}^2 (x - \alpha_{n-1})} - \frac{(x - \alpha_n)(x - \alpha_{n+1})}{a_{n+1}^2} \right] B_n(x) \\
&\quad + \frac{a_n^2 (x - \alpha_{n+1})}{a_{n+1}^2 (x - \alpha_{n-1})} B_{n-1}(x) + \frac{(x - \alpha_{n+1})}{a_{n+1}^2} \\
&\quad + \left[\frac{a_n^2 (x - \alpha_{n+1})}{a_{n+1}^2 (x - \alpha_{n-1})} - 1 \right] v'(x), \quad n > 1.
\end{aligned}
\tag{3.2.32}
$$

Theorem 3.2.7 *For all* $n \geq 0$, *the functions* $A_n(x)$ *and* $B_n(x)$ *are linear combinations of* $A_0(x)$ *and* $v'(x)$ *with rational function coefficients.*

Proof The statement can be readily verified for $n = 0, 1$. The theorem then follows by induction from the recurrence relations (3.2.30) and (3.2.32). $\qquad \square$

Define $F_n(x)$ by

$$
F_n(x) := \frac{a_n}{a_{n-1}} A_n(x) A_{n-1}(x) - B_n(x) \left[v'(x) + B_n(x) \right].
\tag{3.2.33}
$$

Theorem 3.2.8 *The* F_n's *have the alternate representation*

$$
F_n(x) = \sum_{k=0}^{n-1} A_k(x)/a_k.
\tag{3.2.34}
$$

Proof First express $F_{n+1}(x) - F_n(x)$ as

$$\frac{a_{n+1}}{a_n} A_n(x) A_{n+1}(x) - \frac{a_n}{a_{n-1}} A_n(x) A_{n-1}(x)$$

$$+ [B_n(x) - B_{n+1}(x)] [B_n(x) + B_{n+1}(x) + v'(x)].$$

Then eliminate $B_n(x)$ using (3.2.5) and (3.2.24) to obtain

$$F_{n+1}(x) - F_n(x) = \left[\frac{a_{n+1}}{a_n} A_{n+1}(x) - \frac{a_n}{a_{n-1}} A_{n-1}(x) \right] A_n(x)$$

$$+ \frac{x - b_n}{a_n} A_n(x) \left[\frac{1}{x - \alpha_n} + \frac{a_n^2 A_{n-1}(x)}{a_{n-1}(x - \alpha_n)} - \frac{a_{n+1} A_{n+1}(x)}{x - \alpha_n} \right], \tag{3.2.35}$$

which simplifies to $A_n(x)/a_n$ when $n > 0$. When $n = 0$ the relationship (3.2.35) can be verified directly, with $F_0(x) := 0$. This gives $F_n(x)$ as a telescoping sum and the theorem follows.   □

Theorem 3.2.9 ((Ismail, 2000b)) *Let $\mu = \mu_{ac} + \mu_s$, where μ_{ac} is absolutely continuous on $[a, b]$, $\mu'_{ac}(x) = e^{-v(x)}$, and μ_s is a discrete measure with finite support contained in $\mathbb{R} \setminus [a, b]$. Assume that $\{p_n(x)\}$ are orthonormal with respect to μ and let $[A, B]$ be the true interval of orthogonality of $\{p_n(x)\}$. Define functions*

$$\frac{\mathcal{A}_n(x)}{a_n} = \frac{w(y) p_n^2(y)}{y - x} \Bigg]_{y=a}^{b} - \int_A^B \frac{d}{dy} \left[\frac{p_n^2(y)}{x - y} \right] d\mu_s(y)$$

$$+ v'(x) \int_A^B \frac{p_n^2(y)}{x - y} d\mu_s(y) + \int_a^b \frac{v'(x) - v'(y)}{x - y} p_n^2(y) w(y) \, dy, \tag{3.2.36}$$

$$\frac{\mathcal{B}_n(x)}{a_n} = \frac{w(y) p_n(y) p_{n-1}(y)}{y - x} \Bigg]_{y=a}^{b} - \int_A^B \frac{d}{dy} \left[\frac{p_n(y) p_{n-1}(y)}{x - y} \right] d\mu_s(y)$$

$$- v'(x) \int_A^B \frac{p_n(y) p_{n-1}(y)}{x - y} d\mu_s(y) \tag{3.2.37}$$

$$+ \int_a^b \frac{v'(x) - v'(y)}{x - y} p_n(y) p_{n-1}(y) w(y) \, dy,$$

and assume that all the above quantities are defined for all n, $n = 1, 2, \ldots$. Then $\{p_n(x)\}$ satisfy (3.2.3), (3.2.10) and (3.2.11) with A_n, B_n replaced by $\mathcal{A}_n$ and $\mathcal{B}_n$.

The proof is similar to the proof of Theorem 3.2.1 and will be omitted.

It is useful to rewrite (3.2.1) and (3.2.2) in the monic polynomial notation

$$P'_n(x) = \tilde{A}_n(x) P_{n-1}(x) - \tilde{B}_n(x) P_n(x), \tag{3.2.38}$$

where

$$\frac{\tilde{A}_n(x)}{\beta_n} = \frac{w(y)P_n^2(y)}{(y-x)\zeta_n}\Big|_a^b + \int\limits_a^b \frac{v'(x) - v'(y)}{(x-y)\zeta_n} P_n^2(y)w(y)\,dy \qquad (3.2.39)$$

$$\tilde{B}_n(x) = \frac{w(y)P_n(y)P_{n-1}(y)}{(y-x)\zeta_{n-1}} + \int\limits_a^b \frac{v'(x) - v'(y)}{(x-y)\zeta_{n-1}} P_n(y)P_{n-1}(y)w(y)\,dy$$

$$(3.2.40)$$

and ζ_n is given by (2.2.3).

3.3 Applications

We illustrate Theorems 3.2.1 and 3.2.3 by considering the cases of Laguerre and Jacobi polynomials. The properties used here will be derived in Chapter 4.

Laguerre Polynomials. In the case of the (generalized) Laguerre polynomials $\left\{ L_n^{(\alpha)}(x) \right\}$ we have

$$w(x) = \frac{x^\alpha e^{-x}}{\Gamma(\alpha + 1)}, \quad p_n(x) = (-1)^n \sqrt{\frac{n!}{(\alpha + 1)_n}} L_n^{(\alpha)}(x). \qquad (3.3.1)$$

The orthogonality relation of these polynomials as well as the Jacobi polynomials will be established in Chapter 4. We first consider the case $\alpha > 0$.

We shall evaluate a_n as a byproduct of our approach. We first assume $\alpha > 0$. Clearly integration by parts gives

$$x\frac{A_n(x)}{a_n} = \int\limits_0^\infty p_n^2(y)\frac{\alpha y^{\alpha-1}}{\Gamma(\alpha+1)} e^{-y}\,dy = \int\limits_0^\infty p_n^2(y)\frac{y^\alpha e^{-y}}{\Gamma(\alpha+1)}\,dy = 1.$$

Similarly

$$x\frac{B_n(x)}{a_n} = \int\limits_0^\infty p_n(y)p_{n-1}(y)\frac{\alpha y^{\alpha-1}}{\Gamma(\alpha+1)} e^{-y}\,dy = -\int\limits_0^\infty p_{n-1}p_n'(y)w(y)\,dy.$$

Thus

$$A_n(x) = \frac{a_n}{x}, \quad B_n(x) = \frac{-n}{x}. \qquad (3.3.2)$$

Substitute from (3.3.2) into (3.2.24) to get

$$a_{n+1}^2 - a_n^2 = \alpha_n. \qquad (3.3.3)$$

Similarly (3.2.5) and (3.3.2) imply

$$\alpha_n = \alpha + 2n + 1. \qquad (3.3.4)$$

Finally (3.3.3) and (3.3.4) establish

$$a_n^2 = n(n + \alpha). \qquad (3.3.5)$$

Therefore

$$A_n(x) = \frac{\sqrt{n(n+\alpha)}}{x}, \quad B_n(x) = \frac{-n}{x}. \tag{3.3.6}$$

Now remove the assumption $\alpha > 0$ since (3.2.3) is a rational function identity whose validity for $\alpha > 0$ implies its validity for $\alpha > -1$.

This general technique is from (Chen & Ismail, 2005) and can also be used for Jacobi polynomials and for polynomials orthogonal with respect to the weight function $C(1-x)^\alpha (1+x)^\beta (c-x)^\gamma, c > 1$.

In view of (3.3.1), the differential recurrence relation (3.2.3) becomes

$$-x \frac{d}{dx} L_n^{(\alpha)}(x) = (n+\alpha) L_{n-1}^{(\alpha)}(x) - n L_n^{(\alpha)}(x).$$

Finally (3.2.12) becomes

$$x \frac{d^2}{dx^2} L_n^{(\alpha)}(x) + (1 + \alpha - x) \frac{d}{dx} L_n^{(\alpha)}(x) + n L_n^{(\alpha)}(x) = 0.$$

The restriction $\alpha > 0$ can now be removed since (3.3.5), (3.3.6) and the above differential relations are polynomial identities in α.

Jacobi Polynomials. As a second example we consider the Jacobi polynomials $P_n^{(\alpha,\beta)}(x)$ of Chapter 4. Although we can apply the technique used for Laguerre polynomials we decided to use a different approach, which also works for Laguerre polynomials. For Jacobi polynomials

$$w(x) = \frac{(1-x)^\alpha (1+x)^\beta \Gamma(\alpha+\beta+2)}{2^{\alpha+\beta+1} \Gamma(\alpha+1) \Gamma(\beta+1)}, \quad x \in [-1,1], \tag{3.3.7}$$

$$p_n(x) = \frac{P_n^{(\alpha,\beta)}(x)}{\sqrt{h_n^{(\alpha,\beta)}}}, \tag{3.3.8}$$

with

$$h_n^{(\alpha,\beta)} = \frac{(\alpha+\beta+1)(\alpha+1)_n (\beta+1)_n}{(2n+\alpha+\beta+1) n! (\alpha+\beta+1)_n}. \tag{3.3.9}$$

Moreover

$$P_n^{(\alpha,\beta)}(x) = \frac{(\alpha+1)_n}{n!} {}_2F_1 \left(\begin{array}{c} -n, \alpha+\beta+n+1 \\ \alpha+1 \end{array} \middle| \frac{1-x}{2} \right), \tag{3.3.10}$$

$$P_n^{(\alpha,\beta)}(1) = (\alpha+1)_n/n!, \quad P_n^{(\alpha,\beta)}(-1) = (-1)^n (\beta+1)_n/n!, \tag{3.3.11}$$

see (4.1.1) and (4.1.6).

Here again we first restrict α and β to be positive then remove this restriction at the end since the Jacobi polynomials and their discriminants are polynomial functions of α and β.

It is clear that

$$\frac{A_n(x)}{a_n} = \frac{\alpha}{1-x} \int_{-1}^{1} p_n^2(y) \frac{w(y)\,dy}{1-y} + \frac{\beta}{1+x} \int_{-1}^{1} p_n^2(y) \frac{w(y)\,dy}{1+y},$$

and the orthogonality gives

$$
\frac{A_n(x)}{a_n} = \frac{\alpha \, p_n(1)}{1-x} \int_{-1}^{1} p_n(y) \frac{w(y) \, dy}{1-y}
$$

$$
+ \frac{\beta \, p_n(-1)}{1+x} \int_{-1}^{1} p_n(y) \frac{w(y) \, dy}{1+y}.
$$

(3.3.12)

We apply (3.3.7)–(3.3.11) to the above right-hand side and establish the following expression for $A_n(x)/a_n$

$$
\frac{(\alpha + \beta + 1)_n \, (2n + \alpha + \beta + 1) \, \Gamma(\alpha + \beta + 1)}{2^{\alpha + \beta + 1} \, (\beta + 1)_n \, \Gamma(\alpha + 1) \Gamma(\beta + 1)} \sum_{k=0}^{n} \frac{(-n)_k \, (\alpha + \beta + n + 1)_k}{(\alpha + 1)_k \, k!}
$$

$$
\times \int_{-1}^{1} \left[\frac{\alpha(\alpha + 1)_n}{n! \, (1-x) \, 2^k} \, (1-y)^{k+\alpha-1} (1+y)^{\beta} \right.
$$

$$
\left. + \frac{\beta(\beta + 1)_n \, (-1)^n}{n! \, (1+x) \, 2^k} \, (1-y)^{k+\alpha} (1+y)^{\beta-1} \right] dy.
$$

By the evaluation of the beta integral in (1.3.3) and (1.3.7) we find that the integral of quantity in square brackets is

$$
\frac{2^{\alpha+\beta} \, \Gamma(\alpha + 1) \Gamma(\beta + 1)}{n! \, \Gamma(\alpha + \beta + k + 1)} \left\{ \frac{(\alpha)_k \, (\alpha + 1)_n}{1-x} + \frac{(-1)^n \, (\alpha + 1)_k \, (\beta + 1)_n}{1+x} \right\}.
$$

Thus we arrive at the evaluation

$$
\frac{A_n(x)}{a_n} = \frac{(\alpha + \beta + 1)_n \, (2n + \alpha + \beta + 1)}{2 \, n! \, (\beta + 1)_n}
$$

$$
\times \left[\frac{(\alpha + 1)_n}{(1-x)} \, {}_3F_2 \left(\begin{matrix} -n, \alpha + \beta + n + 1, \alpha \\ \alpha + 1, \alpha + \beta + 1 \end{matrix} \middle| 1 \right) \right.
$$

$$
\left. + \frac{(-1)^n \, (\beta + 1)_n}{1+x} \, {}_2F_1 \left(\begin{matrix} -n, \alpha + \beta + n + 1 \\ \alpha + \beta + 1 \end{matrix} \middle| 1 \right) \right].
$$

Now use (1.4.3) and (1.4.5) to sum the ${}_2F_1$ and ${}_3F_2$ above. The result is

$$
\frac{A_n(x)}{a_n} = \frac{(\alpha + \beta + 1 + 2n)}{1 - x^2}.
$$

(3.3.13)

The a_n's are given by

$$
a_n = \frac{2}{\alpha + \beta + 2n} \sqrt{ \frac{n(\alpha + n)(\beta + n)(\alpha + \beta + n)}{(\alpha + \beta - 1 + 2n)(\alpha + \beta + 1 + 2n)} }.
$$

(3.3.14)

Now $B_n(x)/a_n$ is given by the right-hand side of (3.3.12) after replacing $p_n(\pm 1)$ by

$p_{n-1}(\pm 1)$, respectively. Thus

$$\frac{B_n(x)}{a_n} = \frac{(\alpha+\beta+1)_n (2n+\alpha+\beta+1)}{2(n-1)!(\beta+1)_n} \sqrt{\frac{h_n^{(\alpha,\beta)}}{h_{n-1}^{(\alpha,\beta)}}}$$

$$\times \left[\frac{(\alpha+1)_{n-1}}{(1-x)} \,_3F_2\left(\begin{matrix} -n, \alpha+\beta+n+1, \alpha \\ \alpha+1, \alpha+\beta+1 \end{matrix} \middle| 1 \right) \right.$$

$$\left. - \frac{(-1)^n (\beta)_n}{\beta(1+x)} \,_2F_1\left(\begin{matrix} -n, \alpha+\beta+n+1 \\ \alpha+\beta+1 \end{matrix} \middle| 1 \right) \right].$$

Therefore

$$\frac{B_n(x)}{a_n} = -\frac{n(\alpha+\beta+1+2n)}{2(1-x^2)}$$

$$\times \frac{\beta-\alpha+x(2n+\alpha+\beta)}{(n+\alpha)(n+\beta)} \sqrt{\frac{h_n^{(\alpha,\beta)}}{h_{n-1}^{(\alpha,\beta)}}}. \qquad (3.3.15)$$

This shows that (3.2.3) becomes

$$\frac{d}{dx} P_n^{(\alpha,\beta)}(x) = A_n(x) \sqrt{\frac{h_n^{(\alpha,\beta)}}{h_{n-1}^{(\alpha,\beta)}}} \, P_{n-1}^{(\alpha,\beta)}(x) - B_n(x) P_n^{(\alpha,\beta)}(x).$$

Finally (3.3.13)–(3.3.15) and the above identity establish the differential recurrence relation

$$(x^2-1)(2n+\alpha+\beta)\frac{d}{dx} P_n^{(\alpha,\beta)}(x)$$

$$= -2(n+\alpha)(n+\beta)P_{n-1}^{(\alpha,\beta)}(x) + n[\beta-\alpha+x(2n+\alpha+\beta)]P_n^{(\alpha,\beta)}(x) \quad (3.3.16)$$

It is worth noting that the evaluation of the integrals in (3.3.12) amounts to finding the constant term in the expansion of $p_n(x)$ in terms of the polynomials orthogonal with respect to $w(y)/(1\pm y)$. This is a typical situation when v' is a rational function.

Koornwinder Polynomials. Koornwinder considered the measure μ where

$$w(x) = \frac{2^{-\alpha-\beta-1}\Gamma(\alpha+\beta+2)}{\Gamma(\alpha+1)\Gamma(\beta+1)T} (1-x)^\alpha (1+x)^\beta, \quad x \in (-1,1), \quad (3.3.17)$$

$$\mu_s(x) = \frac{M}{T} \delta(x+1) + \frac{N}{T} \delta(x-1), \qquad (3.3.18)$$

where δ is a Dirac measure and

$$T = 1 + M + N, \qquad (3.3.19)$$

see (Koornwinder, 1984), (Kiesel & Wimp, 1996) and (Wimp & Kiesel, 1995). One can show that

$$A_n(x)/a_n = \phi(x)/(1-x^2)^2, \qquad (3.3.20)$$

with

$$\phi_n(x) = c_n + d_n x + e_n x^2, \qquad (3.3.21)$$

$$c_n = \frac{M}{T}(\alpha - \beta - 1)p_n^2(-1) + \frac{N}{T}(\beta - \alpha - 1)p_n^2(1)$$

$$+ \frac{4N}{T}p_n(1)p_n'(1) + 2\alpha \int_{-1}^{1} \frac{p_n^2(y)}{1 - y}w(y)\,dy, \tag{3.3.22}$$

$$d_n = 2\frac{M}{T}(1 + \beta)p_n^2(-1) - 2\frac{N}{T}(1 + \alpha)p_n^2(1), \tag{3.3.23}$$

$$e_n = -\frac{\alpha + \beta + 1}{T}\{Np_n^2(1) + Mp_n^2(-1)\} - \frac{4N}{T}p_n(1)p_n'(1)$$

$$- 2\alpha \int_{-1}^{1} \frac{p_n^2(y)}{1 - y}w(y)\,dy. \tag{3.3.24}$$

Some special and limiting cases of the Koornwinder polynomials satisfy higher-order differential equations of Sturm–Liouville type. Everitt and Littlejohn's survey article (Everitt & Littlejohn, 1991) is a valuable source for information on this topic. We believe that the higher-order differential equations arise when we eliminate n from (3.2.12) (with $A_n(x)$ and $B_n(x)$ replaced by $\mathcal{A}_n(x)$ and $\mathcal{B}_n(x)$, respectively) and its derivatives.

3.4 Discriminants

In this section we give a general expression for the discriminants of orthogonal polynomials and apply the result to the Hermite, Laguerre and Jacobi polynomials.

Lemma 3.4.1 ((Schur, 1931)) *Assume that $\{p_n(x)\}$ is a sequence of orthogonal polynomials satisfying a three-term recurrence relation of the form*

$$p_{n+1}(x) = (\xi_{n+1}x + \eta_{n+1})\,p_n(x) - \nu_{n+1}p_{n-1}(x), \tag{3.4.1}$$

and the initial conditions

$$p_0(x) = 1, \quad p_1(x) = \xi_1 x + \eta_1. \tag{3.4.2}$$

If

$$x_{n,1} > x_{n,2} > \cdots > x_{n,n} \tag{3.4.3}$$

are the zeros of $p_n(x)$ then

$$\prod_{k=1}^{n} p_{n-1}(x_{n,k}) = (-1)^{n(n-1)/2} \prod_{k=1}^{n} \xi_k^{n-2k+1}\nu_k^{k-1}, \tag{3.4.4}$$

with $\nu_1 := 1$.

Proof Let Δ_n denote the left-hand side of (3.4.4). The coefficient of x^n in $\rho_n(x)$ is $\xi_1\xi_2\cdots\xi_n$. Thus by expressing ρ_n and ρ_{n+1} in terms of their zeros, we find

$$\Delta_{n+1} = (\xi_1\xi_2\cdots\xi_n)^{n+1} \prod_{k=1}^{n+1}\prod_{j=1}^{n} [x_{n+1,k} - x_{n,j}]$$

$$= (-1)^{n(n+1)}(\xi_1\xi_2\cdots\xi_n)^{n+1} \prod_{j=1}^{n}\prod_{k=1}^{n+1} [x_{n,j} - x_{n+1,k}]$$

$$= \frac{(\xi_1\xi_2\cdots\xi_n)^{n+1}}{(\xi_1\xi_2\cdots\xi_{n+1})^n} \prod_{j=1}^{n} \rho_{n+1}(x_{n,j}).$$

On the other hand the three-term recurrence relation (3.4.1) simplifies the extreme right-hand side in the above equation and we get

$$\Delta_{n+1} = \xi_1\xi_2\cdots\xi_n \left(-\nu_{n+1}\right)^n \Delta_n. \tag{3.4.5}$$

By iterating (3.4.5) we establish (3.4.4).   □

It is convenient to use

$$x\rho_n(x) = A_n\rho_{n+1}(x) + B_n\rho_n(x) + C_n\rho_{n-1}(x) \tag{3.4.6}$$

in which case (3.4.4) becomes

$$\prod_{k=1}^{n} \rho_{n-1}(x_{n,k}) = (-1)^{n(n-1)/2} \left\{ \prod_{k=0}^{n-1} A_k^{k+1-n} \right\} \left\{ \prod_{j=1}^{n-1} C_j^j \right\}. \tag{3.4.7}$$

The next result uses Lemma 3.4.1 to give an explicit evaluation of the discriminant of $p_n(x)$ and is in (Ismail, 1998).

Theorem 3.4.2 *Let $\{p_n(x)\}$ be orthonormal with respect to $w(x) = \exp(-v(x))$ on $[a,b]$ and let it be generated by (3.1.5) and (3.1.6). Then the discriminant of $p_n(x)$ is given by*

$$D(p_n) = \left\{ \prod_{j=1}^{n} \frac{A_n(x_{n,j})}{a_n} \right\} \left[\prod_{k=1}^{n} a_k^{2k-2n+2} \right]. \tag{3.4.8}$$

Proof From (3.1.5) and (3.1.6) it follows that

$$p_n(x) = \gamma_n x^n + \text{lower order terms}, \quad \gamma_n \prod_{j=1}^{n} a_j = 1. \tag{3.4.9}$$

Now apply (3.2.3), (3.1.9), (3.4.4), and (3.4.9) to get

$$D(p_n) = \gamma_n^{n-2} \prod_{k=1}^{n} A_n(x_{n,k}) \xi_k^{n-2k+1} \zeta_k^{k-1}. \tag{3.4.10}$$

Here $\xi_n = 1/a_n$, $\nu_n = a_{n-1}/a_n$, and γ_n is given in (3.4.9). We substitute these values in (3.4.10) and complete the proof of this theorem.   □

Note that the term in square brackets in (3.4.8) is the Hankel determinant since $\beta_k = a_k^2$. Therefore

$$D\left(p_n\right) = D_n \prod_{j=1}^{n} \frac{A_n\left(x_{n,j}\right)}{a_n}. \tag{3.4.11}$$

Theorem 3.4.3 *Under the assumptions of Theorem* 3.2.9, *the discriminant of the monic polynomial* $P_n(x)$ *is given by*

$$D_n = \left\{ \prod_{j=1}^{n} \frac{A_n\left(x_{n,j}\right)}{a_n} \right\} \left[\prod_{k=1}^{n} a_k^{2k} \right]. \tag{3.4.12}$$

Stieltjes (Stieltjes, 1885b), (Stieltjes, 1885a) and Hilbert (Hilbert, 1885) gave different evaluations of the discriminants of Jacobi polynomials. This contains evaluations of the discriminants of the Hermite and Laguerre polynomials. We now derive these results from Theorem 3.4.2.

Hermite polynomials. For the Hermite polynomials $\{H_n(x)\}$,

$$p_n(x) = 2^{-n/2} \frac{H_n(x)}{\sqrt{n!}}, \quad w(x) = \frac{\exp\left(-x^2\right)}{\sqrt{\pi}}, \quad a_n = \sqrt{\frac{n}{2}}. \tag{3.4.13}$$

Hence $A_n(x)/a_n = 2$, $D\left(H_n\right) = \left[2^n\, n!\right]^{n-1} D\left(p_n\right)$ and (3.4.8) gives

$$D(H_n) = 2^{3n(n-1)/2} \prod_{k=1}^{n} k^k. \tag{3.4.14}$$

Laguerre polynomials. We apply (3.3.1) and (3.3.3) and find that $D\left(L_n^{(\alpha)}\right)$ is $\left[(\alpha+1)_n/(n!)\right]^{n-1} D\left(p_n\right)$. Thus (3.4.8) yields

$$D\left(L_n^{(\alpha)}\right) = \prod_{j=1}^{n} \frac{1}{x_{n,j}} \prod_{k=1}^{n} k^{k+2-2n}(k+\alpha)^k.$$

From (3.3.2) we see that $\prod_{j=1}^{n} x_{n,j} = (\alpha+1)_n$ and we have established the relationship

$$D\left(L_n^{(\alpha)}\right) = \prod_{k=1}^{n} k^{k+2-2n}(k+\alpha)^{k-1}. \tag{3.4.15}$$

Jacobi polynomials. The relationships (3.3.8)–(3.3.9) indicate that

$$D\left(P_n^{(\alpha,\beta)}\right) = \left[h_n^{(\alpha,\beta)}\right]^{n-1} D\left(p_n\right).$$

The substitution of $h_n^{(\alpha,\beta)}$ from (3.3.9), a_n from (3.3.14), and $A_n(x)/a_n$ from (3.3.13), into (3.4.8) establishes the following discriminant formula for the Jacobi polynomials

$$D\left(P_n^{(\alpha,\beta)}\right)$$

$$= 2^{-n(n-1)} \prod_{j=1}^{n} j^{j-2n+2}\, (j+\alpha)^{j-1}\, (j+\beta)^{j-1}\, (n+j+\alpha+\beta)^{n-j}. \tag{3.4.16}$$

In deriving (3.4.16) we have used the fact that $\prod_{j=1}^{n} \left[1 - x_{n,j}^2\right]$ is $(-1)^n P_n^{(\alpha,\beta)}(1) P_n^{(\alpha,\beta)}(-1)/\gamma_n$, where γ_n is the coefficient of x^n in $P_n^{(\alpha,\beta)}(x)$, see (4.1.5). We also used the evaluations of $P_n^{(\alpha,\beta)}(\pm 1)$ in (3.3.11).

3.5 An Electrostatic Equilibrium Problem

Recall that the total energy of a system of n unit charged particles in an external field V is given by (3.1.13). Any weight function w generates an external v defined by (3.1.2). We propose that in the presence of the n charged particles the external field is modified to become V

$$V(x) = v(x) + \ln\left(A_n(x)/a_n\right). \tag{3.5.1}$$

Theorem 3.5.1 ((Ismail, 2000a)) *Assume $w(x) > 0$, $x \in (a, b)$ and let $v(x)$ of (3.1.2) and $v(x) + \ln A_n(x)$ be twice continuously differentiable functions whose second derivative is nonnegative on (a, b). Then the equilibrium position of n movable unit charges in $[a, b]$ in the presence of the external potential $V(x)$ of (3.5.1) is unique and attained at the zeros of $p_n(x)$, provided that the particle interaction obeys a logarithmic potential and that $T(\mathbf{x}) \to 0$ as $\mathbf{x}$ tends to any boundary point of $[a, b]^n$, where*

$$T(\mathbf{x}) = \left[\prod_{j=1}^{n} \frac{\exp\left(-v\left(x_j\right)\right)}{A_n(x_j)/a_n}\right] \prod_{1 \le \ell < k \le n} \left(x_\ell - x_k\right)^2. \tag{3.5.2}$$

Before proving Theorem 3.5.1, observe that finding the equilibrium distribution of the charges in Theorem 3.5.1 is equivalent to finding the maximum of $T(\mathbf{x})$ in (3.5.2). The reason is that at interior points of $[a, b]^n$, the gradient of T vanishes if and only if the gradient of E vanishes. Furthermore at such points of vanishing gradients the Hessians of T and E have opposite signs.

There is no loss of generality in assuming

$$x_1 > x_2 > \cdots > x_n, \tag{3.5.3}$$

a convention we shall follow throughout this section.

Proof of Theorem 3.5.1 The assumption $v''(x) > 0$ ensures the positivity of $A_n(x)$. To find an equilibrium position we solve

$$\frac{\partial}{\partial x_j} \ln T(\mathbf{x}) = 0, \quad j = 1, 2, \ldots, n.$$

This system is

$$-v'(x_j) - \frac{A_n'(x_j)}{A_n(x_j)} + 2 \sum_{1 \le k \le n,\, k \ne j} \frac{1}{x_j - x_k} = 0, \quad j = 1, 2, \ldots, n. \quad (3.5.4)$$

Let

$$f(x) := \prod_{j=1}^n (x - x_j). \quad (3.5.5)$$

It is clear that

$$\sum_{1 \le k \le n,\, k \ne j} \frac{1}{x_j - x_k} = \lim_{x \to x_j} \left[\frac{f'(x)}{f(x)} - \frac{1}{x - x_j} \right]$$

$$= \lim_{x \to x_j} \left[\frac{(x - x_j) f'(x) - f(x)}{(x - x_j) f(x)} \right]$$

and L'Hôpital's rule implies

$$2 \sum_{1 \le k \le n,\, k \ne j} \frac{1}{x_j - x_k} = \frac{f''(x_j)}{f'(x_j)}. \quad (3.5.6)$$

Now (3.5.4), (3.5.5) and (3.5.6) imply

$$-v'(x_j) - \frac{A_n'(x_j)}{A_n(x_j)} + \frac{f''(x_j)}{f'(x_j)} = 0, \quad (3.5.7)$$

or equivalently

$$f''(x) + R_n(x) f'(x) = 0, \quad x = x_1, \ldots, x_n,$$

with R_n as in (3.2.13). In other words

$$f''(x) + R_n(x) f'(x) + S_n(x) f(x) = 0, \quad x = x_1, \ldots, x_n. \quad (3.5.8)$$

To check for local maxima and minima consider the Hessian matrix

$$H = (h_{ij}), \quad h_{ij} = \frac{\partial^2 \ln T(\mathbf{x})}{\partial x_i \partial x_j}. \quad (3.5.9)$$

It readily follows that

$$h_{ij} = 2 (x_i - x_j)^{-2}, \quad i \ne j,$$

$$h_{ii} = -v''(x_i) - \frac{\partial}{\partial x_i} \left(\frac{A_n'(x_i)}{A_n(x_i)} \right) - 2 \sum_{1 \le j \le n,\, j \ne i} \frac{1}{(x_i - x_j)^2}.$$

This shows that the matrix $-H$ is real, symmetric, strictly diagonally dominant and its diagonal terms are positive. By Theorem 1.1.6 $-H$ is positive definite. Therefore $\ln T$ has no relative minima nor saddle points. Thus any solution of (3.5.8) will provide a local maximum of $\ln T$ or T. There cannot be more than one local maximum since $T(\mathbf{x}) \to 0$ as $\mathbf{x} \to$ any boundary point along a path in the region defined in (3.5.3). Thus the system (3.5.4) has at most one solution. On the other hand (3.5.3) and (3.5.8) show that the zeros of

$$f(x) = a_1 a_2 \cdots a_n p_n(x), \quad (3.5.10)$$

satisfy (3.5.4), hence the zeros of $p_n(x)$ solve (3.5.4). $\qquad \square$

Theorem 3.5.2 *Let* $T_{\max}$ *and* E_n *be the maximum value of* $T(\mathbf{x})$ *and the equilibrium energy of the* n *particle system. Then*

$$T_{\max} = \exp\left(-\sum_{j=1}^{n} v\left(x_{n,j}\right)\right) \prod_{k=1}^{n} a_k^{2k}, \tag{3.5.11}$$

$$E_n = \sum_{j=1}^{n} v\left(x_{n,j}\right) - 2\sum_{j=1}^{n} j \ln a_j. \tag{3.5.12}$$

Proof Since $T_{\max}$ is

$$\left[\prod_{j=1}^{n} \frac{\exp\left(-v\left(x_{jn}\right)\right)}{A_n\left(x_{jn}\right)/a_n}\right] \gamma^{2-2n} D_n\left(p_n\right), \tag{3.5.13}$$

(3.5.11) follows from (3.4.8) and (3.5.10). We also used $\gamma a_1 \cdots a_n = 1$. Now (3.5.12) holds because E_n is $-\ln\left(T_{\max}\right)$. $\qquad\square$

Remark 3.5.1 *Stieltjes proved Theorem 3.5.1 when* $e^{-v} = (1-x)^\alpha(1+x)^\beta$, $x = [-1,1]$. *In this case, the modification term* $\ln\left(A_n(x)/a_n\right)$ *is a constant* $-\ln\left(1-x^2\right)$. *In this model, the total external field is due to fixed charges* $(\alpha+1)/2$ *and* $(\beta+1)/2$ *located at* $x = 1, -1$, *respectively. The equilibrium is attained at the zeros of* $P_n^{(\alpha,\beta)}(x)$.

Remark 3.5.2 *The modification of the external field from* v *to* V *certainly changes the position of the charges at equilibrium and the total energy at equilibrium. We maintain that the change in energy is not significant. To quantify this, consider the case* $v = x^4 + c$ *and* n *large. Let* $\tilde{E}_n$ *and* E_n *be the energies at equilibrium due to external fields* v *and* V, *respectively. It can be proved that there are nonzero constants* c_1, c_2 *and constants* c_3, c_4 *such that*

$$E_n = c_1 n^2 \ln n + c_2 n^2 + c_3 n \ln n + O(n),$$
$$\tilde{E}_n = c_1 n^2 \ln n + c_2 n^2 + c_4 n \ln n + O(n),$$

as $n \to \infty$. *Thus, the modification of the external field changes the third term in the large* n *asymptotics of the energy at equilibrium.*

Remark 3.5.3 *For a treatment of electrostatic equilibrium problems (without the modification* $v \to V$) *we refer the reader to (Saff & Totik, 1997), where potential theoretic techniques are used.*

Remark 3.5.4 *An electrostatic equilibrium model for the Bessel polynomials was proposed in (Hendriksen & van Rossum, 1988), but it turned out that the zeros of the Bessel polynomials are saddle points for the energy functional considered, as was pointed out in Valent and Van Assche (Valent & Van Assche, 1995).*

3.6 Functions of the Second Kind

Motivated by the definition of the Jacobi functions of the second kind in Szegő's book (Szegő, 1975, (4.61.4)), we defined in (Ismail, 1985) the function of the second kind associated with polynomials $\{p_n(x)\}$ orthonormal with respect to μ satisfying (3.1.1) as

$$q_n(z) = \frac{1}{w(z)} \int\limits_{-\infty}^{\infty} \frac{p_n(y)}{z - y} w(y)\, dy, \quad n \geq 0, z \notin \mathrm{supp}\{w\}. \tag{3.6.1}$$

It is important to note that $w(z)$ in (3.6.1) is an analytic continuation of w to the complex plane cut along the support of w. Therefore

$$zq_n(z) = \frac{1}{w(z)} \int\limits_{-\infty}^{\infty} \frac{z - y + y}{z - y} p_n(y) w(y)\, dy$$

$$= \frac{1}{w(z)} \delta_{n,0} + \frac{1}{w(z)} \int\limits_{-\infty}^{\infty} \frac{w(y)}{z - y} y p_n(y)\, dy,$$

where the orthonormality of the p_n's was used in the last step. The recursive relations (3.1.5)–(3.1.6) then lead to

$$zq_n(z) = a_{n+1}q_{n+1}(z) + \alpha_n q_n(z) + a_n q_{n-1}(z), \quad n \geq 0, \tag{3.6.2}$$

provided that

$$a_0 q_{-1}(z) := \frac{1}{w(z)}, \quad z \notin \mathrm{supp}\{w\}. \tag{3.6.3}$$

Theorem 3.6.1 *Let $\{p_n(x)\}$ are orthonormal with respect to $w(x) = e^{-v(x)}$ on $[a, b]$, and assume $w(a^+) = w(b^-) = 0$. Then for $n \geq 0$ both p_n and q_n have the same raising and lowering operators, that is*

$$q_n'(z) = A_n(z)q_{n-1}(z) - B_n(z)q_n(z), \tag{3.6.4}$$

$$\left(-\frac{d}{dz} + B_n(z) + v'(z) \right) q_{n-1}(z) = A_{n-1}(z) \frac{a_n}{a_{n-1}} q_n(z). \tag{3.6.5}$$

Furthermore $p_n(x)$ and $q_n(x)$ are linear independent solutions of the differential equation (3.2.12) if $A_n(x) \neq 0$.

Proof We first show that the q's satisfy (3.6.4). Multiply (3.6.1) by $w(x)$, differentiate, then integrate by parts, using the fact that $1/(z - y)$ is infinitely differentiable for z off the support of w. The result is

$$w(x)q_n'(x) - v'(x)w(x)q_n(x) = \int\limits_{a}^{b} \frac{p_n'(y) - v'(y)\, p_n(y)}{x - y} w(y)\, dy$$

$$= \int\limits_{a}^{b} \frac{A_n(y)\, p_{n-1}(y) - [B_n(y) + v'(y)]\, p_n(y)}{x - y} w(y)\, dy.$$

Thus we have

$$w(x)q_n'(x) = v'(x)w(x)q_n(x)$$

$$+ \int_a^b \frac{A_n(y)p_{n-1}(y) - [B_n(y) + v'(y)]\,p_n(y)}{x - y}\, w(y)\, dy$$

or equivalently

$$w(x)q_n'(x) = \int_a^b \frac{A_n(y)\,p_{n-1}(y) - B_n(y)\,p_n(y)}{x - y}\, w(y)\, dy$$

$$+ \int_a^b \frac{v'(x) - v'(y)}{x - y}\, p_n(y)w(y)\, dy. \tag{3.6.6}$$

The second integral on the the right-hand side of (3.6.6) can expressed as

$$\int_a^b \frac{v'(x) - v'(y)}{x - y}\, p_n(y)w(y) \int_a^b \left[\sum_{k=0}^{n-1} p_k(y)p_k(t)\right] w(t)\, dt\, dy$$

$$= a_n \int_a^b \int_a^b \frac{v'(x) - v'(y)}{x - y}$$

$$\times \frac{p_n(y)p_{n-1}(t) - p_n(t)p_{n-1}(y)}{y - t}\, p_n(y)w(y)w(t)\, dt\, dy,$$

where we used the Christoffel–Darboux formula (2.2.4). After invoking the partial fraction decomposition

$$\frac{1}{(y - t)(x - y)} = \frac{1}{(x - t)} \left[\frac{1}{y - t} + \frac{1}{x - y}\right],$$

we see that second integral on the the right-hand side of (3.6.6) can be written as

$$\int_a^b \frac{v'(x) - v'(y)}{x - y}\, p_n(y)w(y)\, dy = I_1 + I_2, \tag{3.6.7}$$

where

$$I_1 = a_n \int_a^b \int_a^b \frac{v'(x) - v'(y)}{x - y}\, \frac{p_n(y)p_{n-1}(t) - p_n(t)p_{n-1}(y)}{x - t}$$

$$\times p_n(y)w(y)\, w(t)\, dt\, dy, \tag{3.6.8}$$

and

$$I_2 = a_n \int_a^b \int_a^b \frac{v'(x) - v'(y)}{x - t}\, \frac{p_n(y)p_{n-1}(t) - p_n(t)p_{n-1}(y)}{y - t}$$

$$\times p_n(y)w(y)\, w(t)\, dt\, dy. \tag{3.6.9}$$

Performing the y integration in I_1 and applying (3.2.1) and (3.2.2) we simplify the form of I_1 to

$$I_1 = \int_a^b \frac{A_n(x)p_{n-1}(t) - p_n(t)B_n(x)}{x - t}\, w(t)\, dt \tag{3.6.10}$$

$$= w(x)\left[A_n(x)q_{n-1}(x) - B_n(x)q_n(x)\right].$$

In I_2 write $v'(x) - v'(y)$ as $v'(x) - v'(t) + v'(t) - v'(y)$, so that

$$\frac{I_2}{a_n} = \int_a^b \int_a^b \frac{v'(x) - v'(t)}{x - t}\, \frac{p_n(y)p_{n-1}(t) - p_n(t)p_{n-1}(y)}{y - t}$$

$$\times p_n(y)w(y)\, w(t)\, dt\, dy$$

$$+ \int_a^b \int_a^b \frac{v'(t) - v'(y)}{x - t}\, \frac{p_n(y)p_{n-1}(t) - p_n(t)p_{n-1}(y)}{y - t} \tag{3.6.11}$$

$$\times p_n(y)w(y)\, w(t)\, dt\, dy.$$

In the first integral in (3.6.11) use the Christoffel–Darboux identity again to expand the second fraction then integrate over y to see that the integral vanishes. On the other hand performing the y integration in the second integral in (3.6.11) gives

$$I_2 = \int_a^b \frac{A_n(t)\, p_{n-1}(t) - B_n(t)\, p_n(t)}{x - t}\, w(t)\, dt. \tag{3.6.12}$$

This establishes (3.6.1). Eliminating $q_{n-1}(x)$ between (3.6.2) and (3.6.4) we establish (3.4.5). Thus p_n and q_n have the same raising and lowering operators. The differential equation (3.2.12) now follows because it is an expanded form of (3.2.11). The case $n = 0$ needs to be verified separately via the interpretation $a_0 = 1$, $A_{-1}(x) = B_0(x) = 0$. The linear independence of $p_n(z)$ and $q_n(z)$ as solutions of the three term recurrence relation follows from their large z behavior, since $zw(z)q_n(z) \to 0$ or 1 as $z \to \infty$ in the z-plane cut along the support of w. On the other hand

$$p'_n(x)q_n(x) - q'_n(x)p_n(x)$$
$$= A_n(x)\left[q_n(x)p_{n-1}(x) - p_n(x)q_{n-1}(x)\right] \tag{3.6.13}$$

follows from (3.6.4) and (3.2.3). This completes the proof. □

Observe that (3.6.13) relates the Wronskian of p_n and q_n to the Casorati determinant. There are cases when $A_0(x) = 0$, and q_0 need to be redefined. This happens for Jacobi polynomials when $n = 0$, $\alpha + \beta + 1 = 0$, see (3.3.13). We shall discuss this case in detail in §4.4.

Theorem 2.3.2 implies

$$P_n^*(z) = \int_{\mathbb{R}} \frac{P_n(z) - P_n(y)}{z - y}\, w(y)\, dy,$$

hence

$$P_n^*(z) = P_n(z)w(z)Q_0(z) - w(z)Q_n(z), \quad (3.6.14)$$

with

$$Q_n(z) = \frac{1}{w(z)} \int_{\mathbb{R}} \frac{P_n(y)}{z-y}\, w(y)\, dy.$$

When w is supported on a compact set $\subset [a, b]$, then Theorem 2.6.2 (Markov) and (3.6.14) prove that $Q_n(z)/P_n(z) \to 0$ uniformly on compact subsets of $\mathbb{C} \setminus [a, b]$. Any solution of (3.1.6) has the form $A(z)p_n(z) + B(z)Q_n(z)$. Thus,

$$Q_n(z)/\left[A(z)P_n(z) + B(z)Q_n(z)\right] \to 0$$

if $A(z) \neq 0$. Therefore, $Q_n(z)$ is a minimal solution of (3.1.6), see §2.6.

3.7 Differential Relations and Lie Algebras

We now study the Lie algebra generated by the differential operators $L_{1,n}$ and $L_{2,n}$. In this Lie algebra the multiplication of operators A, B is the commutator $[A, B] = AB - BA$. This algebra generalizes the harmonic oscillator algebra, which corresponds to the case of Hermite polynomials when $v(x) = x^2+$ a constant, (Miller, 1974).

In view of the identities

$$\exp(-v(x)/2)L_{n,1}\left(y \exp(v(x)/2)\right) = \left(\frac{d}{dx} + B_n(x) + \frac{1}{2}v'(x)\right) y$$

$$\exp(-v(x)/2)L_{n,2}\left(y \exp(v(x)/2)\right) = \left(-\frac{d}{dx} + B_n(x) + \frac{1}{2}v'(x)\right) y,$$

the Lie algebra generated by $\{L_{1,n}, L_{2,n}\}$ coincides with the Lie algebra generated by $\{M_{1,n}, M_{2,n}\}$,

$$M_{1,n} := \frac{d}{dx}, \quad M_{2,n}\, y := [2B_n(x) + v'(x)]\, y. \quad (3.7.1)$$

Define a sequence of functions $\{f_j\}$ by

$$f_1(x) := B_n(x) + \frac{1}{2}\, v'(x), \quad f_{j+1}(x) = \frac{df_j(x)}{dx}, \quad j > 0, \quad (3.7.2)$$

let $M_{j,n}$ be the operator of multiplication by f_j, that is

$$M_{j,n}\, y = f_j(x)\, y, \quad j = 2, 3, \dots . \quad (3.7.3)$$

It is easy to see that the Lie algebra generated by $\{M_{n,1}, M_{n,2}\}$ coincides with the one generated by $\{d/dx, f_j(x) : j = 1, 2, \dots\}$. The M's satisfy the commutation relations

$$[M_{1,n}, M_{j,n}] = M_{j+1,n},\ j > 1, \quad [M_{j,n}, M_{k,n}] = 0,\ j, k > 1. \quad (3.7.4)$$

Theorem 3.7.1 *Assume that $v(x)$ is a polynomial of degree $2m$ and w is supported on $\mathbb{R}$. Then the Lie algebra generated by $L_{1,n}$ and $L_{2,n}$ has dimension $2m + 1$ when for all n, $n > 0$.*

Proof The boundary terms in the definition of $A_n(x)$ and $B_n(x)$ vanish. Clearly the coefficient of x^{2m} in $v(x)$ must be positive and may be taken as 1. Hence $B_n(x)$ is a polynomial of degree $2m-3$ with leading term $2ma_n^2 x^{2m-3}$, so $f_1(x)$ has precise degree $2m-1$. Therefore $f_j(x)$ is a polynomial of degree $2m-j, j = 1, 2, \cdots, 2m$ and the theorem follows. $\qquad\qquad\square$

The application of a theorem of Miller (Miller, 1968, Chapter 8), also stated as Theorem 1 in (Kamran & Olver, 1990), leads to the following result.

Theorem 3.7.2 *Let f_1 be analytic in a domain containing $(-\infty, \infty)$. Then the Lie algebra generated by $M_{1,n}$ and $M_{2,n}$ is finite dimensional, say $k+2$, if and only if f_1 and its first k derivatives form a basis of solutions to*

$$\sum_{j=0}^{k+1} a_j\, y^{(j)} = 0, \tag{3.7.5}$$

where $a_0, \ldots, a_{k+1}$ are constants which may depend on n, and $a_{k+1} \neq 0$.

Next consider the orthogonal polynomials with respect to the weight function

$$w(x) = x^\alpha e^{-\phi(x)}, \quad \alpha > 0, \quad x > 0, \tag{3.7.6}$$

where ϕ is a twice continuously differentiable function on $(0, \infty)$. It is clear that if f is a polynomial of degree at most n and w is as in (3.7.6) then

$$\int_0^\infty \frac{f(y)}{x-y} p_n(y)\, w(y)\, dy = f(x) \int_0^\infty \frac{p_n(y)}{x-y}\, w(y)\, dy \tag{3.7.7}$$

since we can write $f(y)$ as $f(y) - f(x) + f(x)$ and apply the orthogonality.

In order to study the Lie algebra generated by $xL_{1,n}$ and $xL_{2,n}$ associated with the weight function (3.7.6) we need to compute the corresponding A_n and B_n.

$$\frac{A_n(x)}{a_n} = \frac{\alpha}{x} \int_0^\infty \frac{p_n^2(y)}{y}\, w(y)\, dy + \phi_n(x), \tag{3.7.8}$$

$$\frac{B_n(x)}{a_n} = \frac{\alpha}{x} \int_0^\infty \frac{p_n(y)p_{n-1}(y)}{y}\, w(y)\, dy + \psi_n(x), \tag{3.7.9}$$

where

$$\phi_n(x) = \int_0^\infty \frac{\phi'(x) - \phi'(x)}{x-y} p_n^2(y)\, w(y)\, dy, \tag{3.7.10}$$

$$\psi_n(x) = \int_0^\infty \frac{\phi'(x) - \phi'(x)}{x-y} p_n(y)p_{n-1}(y)\, w(y)\, dy. \tag{3.7.11}$$

From the observation (3.7.7) it follows that

$$\frac{A_n(x)}{a_n} = \frac{\alpha}{x}\, p_n(0)\lambda_n + \phi_n(x), \tag{3.7.12}$$

$$\frac{B_n(x)}{a_n} = \frac{\alpha}{x}\, p_{n-1}(0)\lambda_n + \psi_n(x), \tag{3.7.13}$$

with

$$\lambda_n := \int_0^\infty \frac{p_n(y)}{y}\, w(y)\, dy. \tag{3.7.14}$$

We now assume

$$\phi(x) \text{ is a polynomial of degree } m. \tag{3.7.15}$$

It is clear from (3.7.10), (3.7.11) and the assumption (4.7.16) that ϕ_n and ψ_n are polynomials of degree $m-2$ and $m-3$, respectively.

From (3.2.7) and (3.2.9) it follows that $xL_{n,1}$ and $xL_{n,2}$ are equivalent to the operators

$$\pm x\,\frac{d}{dx} + \frac{1}{2}\, xv'(x) + xB_n(x),$$

hence are equivalent to the pair of operators $\{T_{1,n}, T_{2,n}\}$,

$$T_{1,n}y := x\frac{d}{dx}\, y, \quad T_{2,n}y := f_{1,n}y, \tag{3.7.16}$$

where

$$f_{1,n} = xv'(x) + 2xB_n(x). \tag{3.7.17}$$

Since $f_{1,n}$ has degree m, the dimension of the Lie algebra generated by $T_{1,n}$ and $T_{2,n}$ is at most $m+1$. We believe the converse is also true, see §24.5.

The Lie algebras generated by $L_{n,1}$ and $L_{n,2}$ for polynomial v's are of one type while Lie algebras generated $M_{n,1}$ and $M_{n,2}$ for polynomial ϕ's are of a different type.

It is of interest to characterize all orthogonal polynomials for which the Lie algebra generated by $\{L_{1,n}, L_{2,n}\}$ is finite dimensional. It is expected that such polynomials will correspond to polynomial external fields $(v(x))$. This problem will be formulated in §24.5.

Exercises

3.1   Prove that $A_n(x)$ and $B_n(x)$ are rational functions if $w(x) = e^{-v(x)}$, $x \in \mathbb{R}$ and $v'(x)$ is a rational function.

3.2   When $v'(x)$ is a rational function, show that there exists a fixed polynomial $\pi(x)$ and constants $\{a_{nj}\}$ such that

$$\pi(x)P_n'(x) = \sum_{j=0}^{M} a_{nj}P_{n+m-j-1}(x),$$

where m is the degree of π and M is a fixed positive integer independent of n. Moreover, π does not depend on n.

3.3   The Chebyshev polynomials of the second kind $\{U_n(x)\}$ will be defined in (4.5.25) and satisfy the recurrence relation (4.5.28).

(a) Prove that $y_n = U_n(x) + cU_{n-1}(x)$ also satisfies (4.5.28) for $n > 0$.

(b) Using Schur's lemma (Lemma 3.4.1), prove that

$$\text{Res}\,\{U_n(x) + kU_{n-1}(x), U_{n-1}(x) + hU_{n-2}(x)\}$$

$$= (-1)^{\binom{n}{2}} 2^{n(n-1)} h^n \left[U_n\left(\frac{1+kh}{2h}\right) - kU_{n-1}\left(\frac{1+kh}{2h}\right) \right],$$

(Dilcher & Stolarsky, 2005). More general results are in preparation in a paper by Gishe and Ismail.

3.4   Derive the recursion coefficients and find the functions $A_n(x)$, $B_n(x)$ for Jacobi polynomials using the technique used in §3.3 to treat Laguerre polynomials (Chen & Ismail, 2005).

4

Jacobi Polynomials

This chapter treats the theory of Jacobi polynomials and their special and limiting cases of ultraspherical, Hermite and Laguerre polynomials. The ultraspherical polynomials include the Legendre and Chebyshev polynomials as special cases.

The weight function for Jacobi polynomials is

$$w(x; \alpha, \beta) := (1-x)^\alpha (1+x)^\beta, \quad x \in (-1, 1). \tag{4.0.1}$$

To evaluate $\int_{-1}^{1} w(x; \alpha, \beta)\, dx$ we set $1 - x = 2t$ and apply the beta integral (1.3.2)–(1.3.3) to see that

$$\int_{-1}^{1} w(x; \alpha, \beta)\, dx = 2^{\alpha+\beta+1} \frac{\Gamma(\alpha+1)\Gamma(\beta+1)}{\Gamma(\alpha+\beta+2)}. \tag{4.0.2}$$

4.1 Orthogonality

We now construct the polynomials $\{P_n^{(\alpha,\beta)}(x)\}$ orthogonal with respect to $w(x; \alpha, \beta)$ and are known as Jacobi polynomials. It is natural to express $P_n^{(\alpha,\beta)}(x)$ in terms of the basis $\{(1-x)^k\}$ since $(1-x)^k w(x; \alpha, \beta) = w(x; \alpha+k, \beta)$. Similarly we can use the basis $\{(1+x)^k\}$. Thus we seek constants $c_{n,j}$ so that

$$P_n^{(\alpha,\beta)}(x) = \sum_{j=0}^{n} c_{n,j} (1-x)^j,$$

such that $\int_{-1}^{1} (1+x)^k P_n^{(\alpha,\beta)}(x) w(x; \alpha, \beta)\, dx = 0$ for $0 \le k < n$, that is

$$\sum_{j=0}^{n} c_{n,j} \int_{-1}^{1} w(x; \alpha+j, \beta+k) = 0.$$

Therefore

$$\sum_{j=0}^{n} \frac{2^j \, \Gamma(\alpha+j+1)}{\Gamma(\alpha+\beta+k+j+2)} c_{n,j} = 0, \quad 0 \le k < n.$$

The terminating summation formulas (1.4.3) and (1.4.5) require the presence of $(-n)_j/j!$, so we try $c_{n,j} = 2^{-j}(-n)_j(a)_j/(b)_j j!$. Applying (1.3.7) we get

$$\sum_{j=0}^{n} \frac{(-n)_j\,(a)_j\,(\alpha+1)_j}{j!\,(b)_j(\alpha+\beta+k+2)_j} = 0, \quad 0 \le k < n.$$

It is clear that taking $b = \alpha+1$ and applying (1.4.3) amounts to choosing a to satisfy

$$(\alpha+\beta+k+2-a)_n/(\alpha+\beta+k+2)_n = 0.$$

This suggests choosing $a = n + \alpha + \beta + 1$.

Observe that the key to the above evaluations is that the factors $(1-x)^j$ and $(1+x)^k$ attach to the weight function resulting in $(1-x)^j(1+x)^k w(x;\alpha,\beta) = w(x;\alpha+j,\beta+k)$, so all the integrals involved reduce to the evaluation of the single integral

$$\int_{-1}^{1} w(x;\alpha,\beta)\,dx.$$

Theorem 4.1.1 *The Jacobi polynomials*

$$P_n^{(\alpha,\beta)}(x) = \frac{(\alpha+1)_n}{n!}\,{}_2F_1\left(\begin{matrix} -n, \alpha+\beta+n+1 \\ \alpha+1 \end{matrix}\bigg| \frac{1-x}{2}\right), \qquad (4.1.1)$$

satisfy

$$\int_{-1}^{1} P_m^{(\alpha,\beta)}(x)P_n^{(\alpha,\beta)}(x)(1-x)^\alpha(1+x)^\beta dx = h_n^{(\alpha,\beta)}\delta_{m,n}, \qquad (4.1.2)$$

where

$$h_n^{(\alpha,\beta)} = \frac{2^{\alpha+\beta+1}\Gamma(\alpha+n+1)\Gamma(\beta+n+1)}{n!\,\Gamma(\alpha+\beta+n+1)(\alpha+\beta+2n+1)}. \qquad (4.1.3)$$

Proof We may assume $m \le n$ and in view of the calculation leading to this theorem we only need to consider the case $m = n$. Using (4.1.1) we see that the left-hand side of (4.1.2) is

$$\frac{(\alpha+1)_n}{n!}\frac{(-n)_n(n+\alpha+\beta+1)_n}{n!\,(\alpha+1)_n 2^n}\int_{-1}^{1}(1-x)^n P_n^{(\alpha,\beta)}(x)w(x;\alpha,\beta)\,dx$$

$$= (-1)^n\frac{(-n)_n(n+\alpha+\beta+1)_n}{(n!)^2 2^n}\int_{-1}^{1}(1+x)^n P_n^{(\alpha,\beta)}(x)w(x;\alpha,\beta)\,dx.$$

Use $(-n)_n = (-1)^n n!$ and (4.1.1) to see that the above expression becomes

$$\frac{(\alpha+1)_n(n+\alpha+\beta+1)_n}{(n!)^2 2^n} \sum_{j=0}^{n} \frac{(-n)_j(n+\alpha+\beta+1)_j}{j!\,(\alpha+1)_j\, 2^j} \int_{-1}^{1} w(x; \alpha+j, \beta+n)\, dx$$

$$= \frac{(\alpha+1)_n(n+\alpha+\beta+1)_n}{(n!)^2}\, 2^{\alpha+\beta+1}\frac{\Gamma(\alpha+1)\Gamma(\beta+n+1)}{\Gamma(\alpha+\beta+n+2)}$$

$$\times {}_2F_1\left(\begin{matrix} -n, \alpha+\beta+n+1 \\ \alpha+\beta+n+2 \end{matrix}\,\middle|\, 1\right),$$

The Chu–Vandermonde sum (1.4.3) evaluates the above ${}_2F_1$ in closed form and we have established (4.1.2). $\qquad\square$

Observe that replacing x by $-x$ in $w(x; \alpha, \beta)$ amounts to interchanging α and β in $w(x; \alpha, \beta)$ while $h_n^{(\alpha,\beta)}$ is symmetric in α and β. After checking the coefficient of x^n, we find

$$P_n^{(\alpha,\beta)}(x) = (-1)^n\, P_n^{(\beta,\alpha)}(-x)$$

$$= (-1)^n\, \frac{(\beta+1)_n}{n!}\, {}_2F_1\left(\begin{matrix} -n, \alpha+\beta+n+1 \\ \beta+1 \end{matrix}\,\middle|\, \frac{1+x}{2}\right). \qquad (4.1.4)$$

Furthermore it is clear from (4.1.1) and (4.1.4) that

$$P_n^{(\alpha,\beta)}(x) = \frac{(\alpha+\beta+n+1)_n}{n!\, 2^n}\, x^n + \text{lower order terms}, \qquad (4.1.5)$$

and

$$P_n^{(\alpha,\beta)}(1) = \frac{(\alpha+1)_n}{n!}, \quad P_n^{(\alpha,\beta)}(-1) = (-1)^n\, \frac{(\beta+1)_n}{n!}. \qquad (4.1.6)$$

4.2 Differential and Recursion Formulas

From the ${}_2F_1$ representation in (4.1.1) and the observation

$$(a)_{k+1} = a(a+1)_k, \qquad (4.2.1)$$

we see that

$$\frac{d}{dx}\, P_n^{(\alpha,\beta)}(x) = \frac{1}{2}\, (n+\alpha+\beta+1)\, P_{n-1}^{(\alpha+1,\beta+1)}(x). \qquad (4.2.2)$$

Now the orthogonality relation, (4.2.2) and integration by parts give

$$\frac{n+\alpha+\beta+2}{2}\, h_n^{(\alpha+1,\beta+1)} \delta_{m,n}$$

$$= \int_{-1}^{1} P_m^{(\alpha+1,\beta+1)}(x)(1-x)^{\alpha+1}(1+x)^{\beta+1}\frac{dP_{n+1}^{(\alpha,\beta)}(x)}{dx}\, dx$$

$$= -\int_{-1}^{1} P_{n+1}^{(\alpha,\beta)}(x)\frac{d}{dx}\left[(1-x)^{\alpha+1}(1+x)^{\beta+1}P_m^{(\alpha+1,\beta+1)}(x)\right] dx$$

and the uniqueness of the orthogonal polynomials imply

$$\frac{d}{dx}\left[(1-x)^{\alpha+1}(1+x)^{\beta+1}P_n^{(\alpha+1,\beta+1)}(x)\right]$$
$$= -\frac{(n+\alpha+\beta+2)h_n^{(\alpha+1,\beta+1)}}{2h_{n+1}^{(\alpha,\beta)}}(1-x)^{\alpha}(1+x)^{\beta}P_{n+1}^{(\alpha,\beta)}(x).$$

Equation (4.1.3) simplifies the above relationship to

$$\frac{1}{(1-x)^{\alpha}(1+x)^{\beta}}\frac{d}{dx}\left[(1-x)^{\alpha+1}(1+x)^{\beta+1}P_{n-1}^{(\alpha+1,\beta+1)}(x)\right]$$
$$= -2nP_n^{(\alpha,\beta)}(x).$$

(4.2.3)

Combining (4.2.2) and (4.2.3) we obtain the differential equation

$$\frac{1}{(1-x)^{\alpha}(1+x)^{\beta}}\frac{d}{dx}\left[(1-x)^{\alpha+1}(1+x)^{\beta+1}\frac{d}{dx}P_n^{(\alpha,\beta)}(x)\right]$$
$$= -n(n+\alpha+\beta+1)P_n^{(\alpha,\beta)}(x).$$

(4.2.4)

Simple exercises recast (4.2.3) and (4.2.4) in the form

$$\left(x^2-1\right)\frac{d}{dx}P_{n-1}^{(\alpha+1,\beta+1)}(x)$$
$$= [\alpha-\beta+x(\alpha+\beta+2)]P_{n-1}^{(\alpha+1,\beta+1)}(x)+2nP_n^{(\alpha,\beta)}(x).$$

(4.2.5)

Observe that (4.2.4) indicates that $y = P_n^{(\alpha,\beta)}(x)$ is a solution to

$$\left(1-x^2\right)y''(x)+[\beta-\alpha-x(\alpha+\beta+2)]y'(x)$$
$$+n(n+\alpha+\beta+1)y(x) = 0.$$

(4.2.6)

Note that (4.1.1) and (4.1.4) follow from comparing the differential equation (4.2.6) and the hypergeometric differential equation (1.3.12).

It is worth noting that (4.2.6) is the most general second order differential equation of the form

$$\pi_2(x)\,y''(x)+\pi_1(x)\,y'(x)+\lambda y = 0,$$

with a polynomial solution of degree n, where $\pi_j(x)$ denotes a generic polynomials in x of degree at most j. To see this observe that λ is uniquely determined by the requirement with polynomial that y is a polynomial of degree n. Thus we need to determine five coefficients in π_1 and π_2. But by dividing the differential equation by a constant we can make one of the nonzero coefficients equal to 1, so we have only four parameters left. On the other hand the change of variable $x \to ax + b$ will absorb two of the parameters, so we only have two free parameters at our disposal. The differential equation (4.2.6) does indeed have two free parameters.

Consider the inner product

$$(f,g) = \int_{-1}^{1} w(x;\alpha,\beta)\,f(x)\,\overline{g(x)}\,dx,$$

defined on space of functions f for which $w(x; \alpha, \beta) f(x) \to 0$ as $x \to \pm 1$. It is clear that

$$\text{if } T = (1 - x^2) \frac{d}{dx}, \text{ then } (T^* f)(x) = \frac{-1}{w(x; \alpha, \beta)} \frac{d}{dx} (w(x; \alpha + 1, \beta + 1) f(x)).$$

Therefore (4.2.4) is of the form $T^* T y = \lambda_n y$.

By iterating (4.2.3) we find

$$2^k (-1)^k n! P_n^{(\alpha, \beta)}(x)$$

$$= \frac{(n - k)!}{(1 - x)^\alpha (1 + x)^\beta} \frac{d^k}{dx^k} \left[(1 - x)^{\alpha+k} (1 + x)^{\beta+k} P_{n-k}^{(\alpha+k, \beta+k)}(x) \right], \quad (4.2.7)$$

In particular the case $k = n$ is the Rodrigues formula,

$$2^n (-1)^n n! P_n^{(\alpha, \beta)}(x)$$

$$= \frac{1}{(1 - x)^\alpha (1 + x)^\beta} \frac{d^n}{dx^n} \left[(1 - x)^{\alpha+n} (1 + x)^{\beta+n} \right]. \quad (4.2.8)$$

We next derive the three term recurrence relation for Jacobi polynomials. We believe that at this day and age powerful symbolic algebra programs make it easy and convenient to use the existence of a three term recurrence relation to compute the recursion coefficients by equating coefficients of powers of x^{n+1}, x^n, x^{n-1}, and x^{n-2}. The result is

$$2(n + 1)(n + \alpha + \beta + 1)(\alpha + \beta + 2n) P_{n+1}^{(\alpha, \beta)}(x)$$

$$= (\alpha + \beta + 2n + 1) \left[(\alpha^2 - \beta^2) + x(\alpha + \beta + 2n + 2)(\alpha + \beta + 2n) \right] \quad (4.2.9)$$

$$\times P_n^{(\alpha, \beta)}(x) - 2(\alpha + n)(\beta + n)(\alpha + \beta + 2n + 2) P_{n-1}^{(\alpha, \beta)}(x),$$

for $n \geq 0$, with $P_{-1}^{(\alpha, \beta)}(x) = 0$, $P_0^{(\alpha, \beta)}(x) = 1$.

To derive (3.2.3) for the Jacobi polynomials, that is to prove (3.3.16), one can apply (4.2.2) then express $P_{n-1}^{(\alpha+1, \beta+1)}(x)$ in terms of $\left\{ P_k^{(\alpha, \beta)}(x) \right\}$ through Christoffel's formula (2.7.2). Thus for some constant C_n we have

$$C_n (1 - x^2) P_n^{(\alpha+1, \beta+1)}(x)$$

$$= \begin{vmatrix} P_n^{(\alpha, \alpha)}(1) & P_{n+1}^{(\alpha, \alpha)}(1) & P_{n+1}^{(\alpha, \alpha)}(1) \\ P_n^{(\alpha, \alpha)}(-1) & P_{n+1}^{(\alpha, \alpha)}(-1) & -P_{n+1}^{(\alpha, \alpha)}(-1) \\ P_n^{(\alpha, \alpha)}(x) & P_{n+1}^{(\alpha, \alpha)}(x) & x P_{n+1}^{(\alpha, \alpha)}(x) \end{vmatrix}, \quad (4.2.10)$$

where we used the existence of a three term recurrence relation. Expand the determinant and evaluate C_n by equating the coefficients of x^{n+2} via (4.1.5). The result now follows from (4.2.2).

We also note that the Christoffel formula (2.7.2) and (4.1.6) imply

$$P_n^{(\alpha+1, \beta)}(x) = \frac{2 \left[(n + \alpha + 1) P_n^{(\alpha, \beta)}(x) - (n + 1) P_{n+1}^{(\alpha, \beta)}(x) \right]}{(2n + \alpha + \beta + 2)(1 - x)}, \quad (4.2.11)$$

$$P_n^{(\alpha, \beta+1)}(x) = \frac{2 \left[(n + \beta + 1) P_n^{(\alpha, \beta)}(x) + (n + 1) P_{n+1}^{(\alpha, \beta)}(x) \right]}{(2n + \alpha + \beta + 2)(1 + x)}. \quad (4.2.12)$$

Of course (4.2.12) also follows from (4.2.11) and (4.1.4).

The following lemma provides a useful bound for Jacobi polynomials.

Lemma 4.2.1 *Let* $\alpha > -1$, $\beta > -1$, *and set*

$$x_0 = \frac{\beta - \alpha}{\alpha + \beta + 1}, \quad M_n := \max\left\{\left|P_n^{(\alpha,\beta)}(x)\right| : -1 \le x \le 1\right\}.$$

Then

$$M_n = \left\{\begin{array}{ll} (s)_n/n! & \text{if } s \ge -1/2 \\ P_n^{(\alpha,\beta)}(x') & \text{if } s < -1/2 \end{array}\right., \tag{4.2.13}$$

where $s = \min\{\alpha, \beta\}$ *and* x' *is one of the two maximum points closest to* x_0.

Proof We let

$$n(n + \alpha + \beta + 1)f(x) = n(n + \alpha + \beta + 1)\left\{P_n^{(\alpha,\beta)}(x)\right\}^2$$
$$+ (1 - x^2)\left\{\frac{d}{dx}P_n^{(\alpha,\beta)}(x)\right\}^2.$$

Hence

$$n(n + \alpha + \beta + 1)f'(x) = 2\{\alpha - \beta + (\alpha + \beta + 1)x\}\left\{\frac{d}{dx}P_n^{(\alpha,\beta)}(x)\right\}^2$$
$$= 2(\alpha + \beta + 1)(x - x_0)\left\{\frac{d}{dx}P_n^{(\alpha,\beta)}(x)\right\}^2.$$

It follows that $x_0 \in (-1, 1)$ if and only if $(\alpha + 1/2)(\beta + 1/2)$. If $\alpha > -\frac{1}{2}$, $\beta > -\frac{1}{2}$, then $f' \le 0$ on $(-1, x_0]$, hence the sequence formed by $\left|P_n^{(\alpha,\beta)}(-1)\right|$, and the successive maxima of $\left|P_n^{(\alpha,\beta)}(x)\right|$ decreases. Similarly $f' \ge 0$ on $[x_0, 1)$ and the successive maxima of $\left|P_n^{(\alpha,\beta)}(x)\right|$ and $\left|P_n^{(\alpha,\beta)}(1)\right|$ increases. On the other hand if $\alpha > -\frac{1}{2}$, $\beta \le -\frac{1}{2}$ then $f' \ge 0$ and the sequence of relative maxima is monotone on $[-1, 1]$. This proves the case $s \ge -\frac{1}{2}$. If $\alpha, \beta \in (-1/2, 1)$ then $x_0 \in (-1, 1)$ and the sequence of relative maxima increase on $(-1, x_0]$ and decrease on $[x_0, 1]$, so the maximum is attained at one of the stationary points closest to x_0. □

Since (4.1.1) and (4.1.4) express $P_n^{(\alpha,\beta)}(x)$ as a polynomial in $(1 \pm x)$ it is natural to invert such representations and expand $(1 + x)^m(1 - x)^n$ in terms of Jacobi polynomials.

Jacobi Polynomials

Theorem 4.2.2 *We have*

$$\left(\frac{1-x}{2}\right)^m \left(\frac{1+x}{2}\right)^n = \frac{(\alpha+1)_m(\beta+1)_n}{\Gamma(\alpha+\beta+m+n+2)}$$

$$\times \sum_{k=0}^{m+n} (\alpha+\beta+2k+1) \frac{\Gamma(\alpha+\beta+k+1)}{(\beta+1)_k} \qquad (4.2.14)$$

$$\times {}_3F_2 \left(\begin{matrix} -k, k+\alpha+\beta+1, \alpha+m+1 \\ \alpha+1, \alpha+\beta+m+n+2 \end{matrix} \bigg| 1 \right) P_k^{(\alpha,\beta)}(x).$$

Proof Clearly we can expand $(1+x)^m(1-x)^n$ in terms of Jacobi polynomials and the coefficient of $P_k^{(\alpha,\beta)}(x)$ is

$$\frac{(\alpha+1)_k}{2^{m+n}\, k!\, h_k^{(\alpha,\beta)}} \int_{-1}^{1} (1-x)^{m+\alpha}(1+x)^{n+\beta}$$

$$\times {}_2F_1 \left(\begin{matrix} -k, k+\alpha+\beta+1 \\ \alpha+1 \end{matrix} \bigg| \frac{1-x}{2} \right) dx,$$

which simplifies by the evaluation of the beta integral to the coefficient in (4.2.14). $\qquad\square$

The special case $m=0$ of Theorem 4.2.2 is

$$\left(\frac{1+x}{2}\right)^n = (\beta+1)_n\, n! \sum_{k=0}^{n} \frac{(\alpha+\beta+2k+1)\,\Gamma(\alpha+\beta+k+1)}{(\beta+1)_k\,\Gamma(\alpha+\beta+n+2+k)\,(n-k)!} \qquad (4.2.15)$$

$$\times P_k^{(\alpha,\beta)}(x),$$

where we used (1.4.3). Similarly, (1.4.5) and $n=0$ give

$$\left(\frac{1-x}{2}\right)^m = (\alpha+1)_m\, m! \sum_{k=0}^{m} \frac{(\alpha+\beta+2k+1)\,\Gamma(\alpha+\beta+k+1)\,(-1)^k}{(\alpha+1)_k\,\Gamma(\alpha+\beta+m+2+k)\,(m-k)!}$$

$$\times P_k^{(\alpha,\beta)}(x).$$

$$(4.2.16)$$

Theorem 4.2.3 *Let*

$$d_{n,k} = \frac{n!\,(\beta+1)_n\,(\alpha+\beta+1)_k\,(2k+\alpha+\beta+1)}{(n-k)!\,(\beta+1)_k\,(\alpha+\beta+1)_{n+k+1}}. \qquad (4.2.17)$$

Then we have the inverse relations

$$u_n = \sum_{k=0}^{n} d_{n,k}\, v_k \qquad (4.2.18)$$

if and only if

$$v_n = (-1)^n \frac{(\beta+1)_n}{n!} \sum_{k=0}^{n} \frac{(-n)_k(\alpha+\beta+n+1)_k}{k!\,(\beta+1)_k}\, u_k. \qquad (4.2.19)$$

Proof The relationships (4.1.4) and (4.2.15) prove the theorem when

$$u_n = ((1+x)/2)^n \quad \text{and} \quad v_n = P_n^{(\alpha,\beta)}(x).$$

This is sufficient because the sequences $\{((1+x)/2)^n\}$ and $\left\{ P_n^{(\alpha,\beta)}(x) \right\}$ form bases for all polynomials. □

Remark 4.2.1 *When $\alpha > -1$, $\beta > -1$ and $\alpha + \beta \neq 0$, the Jacobi polynomials are well-defined through (4.2.9) with*

$$P_0^{(\alpha,\beta)}(x) = 1, \quad P_1^{(\alpha,\beta)}(x) = [x(\alpha + \beta + 2) + \alpha - \beta]/2. \tag{4.2.20}$$

When $\alpha + \beta = 0$, one must be careful in defining $P_1^{(\alpha,\beta)}(x)$. If we use (4.2.20), then $P_1^{(\alpha,-\alpha)}(x) = x + \alpha$. On the other hand, if we set $\alpha + \beta = 0$ then apply (4.2.9) and the initial conditions $P_{-1}(x) = 0$, $P_0(x) = 1$, we will see in addition to the option $P_1 = x + \alpha$ we may also choose $\mathcal{P}_1 = x$. The first choice leads to standard Jacobi polynomials with $\beta = -\alpha$, i.e., $\left\{ P_n^{(\alpha,-\alpha)}(x) \right\}$, while the second option leads to what is called exceptional Jacobi polynomials $\left\{ \mathcal{P}_n^{(\alpha)}(x) \right\}$. This concept was introduced in (Ismail & Masson, 1991). Ismail and Masson proved

$$\mathcal{P}_n^{(\alpha)}(x) = \frac{1}{2} \left[P_n^{(\alpha,-\alpha)}(x) + P_n^{(-\alpha,\alpha)}(x) \right], \tag{4.2.21}$$

and established the orthogonality relation

$$\int_{-1}^{1} \mathcal{P}_m^{(\alpha)}(x) \mathcal{P}_n^{(\alpha)}(x) w(x;\alpha)\, dx = \frac{(1+\alpha)_n (1-\alpha)_n}{(2n+1)(n!)^2} \delta_{m,n}, \tag{4.2.22}$$

where the weight function is

$$w(x;\alpha) = \frac{2\sin(\pi\alpha)}{\pi\alpha} \frac{\left(1 - x^2\right)^\alpha}{(1-x)^{2\alpha} + 2\cos(\pi\alpha)\left(1-x^2\right)^\alpha + (1+x)^{2\alpha}}, \tag{4.2.23}$$

for $-1 < \alpha < 1$. Define the differential operator

$$D(\alpha,\beta;n)$$
$$:= \left(1 - x^2\right) \frac{d^2}{dx^2} + [\beta - \alpha - (\alpha + \beta + 2)x] \frac{d}{dx} \tag{4.2.24}$$
$$+ n(n + \alpha + \beta + 1).$$

Then $\mathcal{P}_n^{(\alpha)}(x)$ satisfies the fourth-order differential equation

$$D(1 - \alpha, 1 + \alpha; n - 1)\, D(\alpha, -\alpha; n)\, \mathcal{P}_n^{(\alpha)}(x) = 0. \tag{4.2.25}$$

One can show that

$$\mathcal{P}_n^{(\alpha)}(x) = \lim_{c \to 0^+} P_n^{(\alpha,-\alpha)}(x;c), \tag{4.2.26}$$

where $\left\{ P_n^{(\alpha,\beta)}(x;c) \right\}$ are the polynomials in §5.7. Using the representation (5.7.5) one can also confirm (4.2.21).

4.3 Generating Functions

A generating function for a sequence of functions $\{f_n(x)\}$ is a series of the form $\sum_{n=0}^{\infty} \lambda_n f_n(x) z^n = F(z)$, for some suitable multipliers $\{\lambda_n\}$. A bilinear generating function is $\sum_{n=0}^{\infty} \lambda_n f_n(x) f_n(y) z^n$. The Poisson kernel $P_r(x, y)$ of (2.2.12) is an example of a bilinear generating function.

In his review of the Srivastava–Manocha treatise on generating functions (Srivastava & Manocha, 1984) Askey wrote (Askey, 1978)

...The present book (Srivastava & Manocha, 1984) is devoted to the question of finding sequences f_n for which $F(z)$ can be found, where being found means there is a representation as a function which occurs often enough so it has a name. The sequences f_n are usually products of hypergeometric functions and binomial coefficients or shifted factorials, and the representation of $F(z)$ is usually as a hypergeometric function in one or several variables, often written as a special case with its own notation (which is sometimes a useful notation and other times obscures the matter). As is usually the case with a book on this subject, there are many identities which are too complicated to be of any use, as well as some very important identities. Unfortunately the reader who is trying to learn something about which identities are important will have to look elsewhere, for no distinction is made between the important results and the rest.

Our coverage of generating functions is very limited and we believe all the generating functions and multilinear generating functions covered in our monograph are of some importance.

We first establish

$$
\sum_{n=0}^{\infty} \frac{(\alpha + \beta + 1)_n}{(\alpha + 1)_n} t^n P_n^{(\alpha,\beta)}(x) = (1 - t)^{-\alpha - \beta - 1}
$$
$$
\times {}_2F_1 \left(\begin{array}{c} (\alpha + \beta + 1)/2, 1 + (\alpha + \beta)/2 \\ \alpha + 1 \end{array} \middle| \frac{2t(x - 1)}{(1 - t)^2} \right).
$$

$$(4.3.1)$$

The proof consists of using (4.3.1) to substitute a finite sum, over k, say, for $P_n^{(\alpha,\beta)}(x)$ then replace n by $n + k$ and observe that the left-hand side of (4.3.1) becomes

$$
\sum_{n,k=0}^{\infty} \frac{(\alpha + \beta + 1)_{n+2k} \, t^{n+k}}{(-1)^k \, 2^k \, n! \, k! \, (\alpha + 1)_k} (1 - x)^k = \sum_{k=0}^{\infty} \frac{(\alpha + \beta + 1)_{2k}}{2^k \, k! \, (\alpha + 1)_k} \frac{t^k (x - 1)^k}{(1 - t)^{\alpha + \beta + 2k + 1}},
$$

then apply the second relation in (1.3.7) and (4.3.1) follows.

The generating function (4.3.1) has two applications. The first is that when $\alpha = \beta$ it reduces to a standard generating function for ultraspherical polynomials, see §4.5. The second is that it is the special case $y = 1$ of a bilinear generating function and this fact is related to a Laplace type integral for Jacobi polynomials. Multiply (4.3.1) by $t^{(\alpha + \beta + 1)/2}$, then differentiate (4.3.1) with respect to t. After simple manipulations we establish

$$\sum_{n=0}^{\infty} \frac{(\alpha+\beta+1)_n}{(\alpha+1)_n} \frac{(\alpha+\beta+1+2n)}{(\alpha+\beta+1)} t^n P_n^{(\alpha,\beta)}(x)$$

$$= \frac{(1+t)}{(1-t)^{\alpha+\beta+2}} \,_2F_1 \left(\begin{array}{c} (\alpha+\beta+2)/2, (\alpha+\beta+3)/2 \\ \alpha+1 \end{array} \middle| \frac{2t(x-1)}{(1-t)^2} \right). \quad (4.3.2)$$

Formula (4.3.2) is closely connected to the Poisson kernel of $\left\{ P_n^{(\alpha,\beta)}(x) \right\}$ and the generalized translation operator associated with Jacobi polynomials, see §4.7.

Another generating function is

$$\sum_{n=0}^{\infty} \frac{P_n^{(\alpha,\beta)}(x) t^n}{(\beta+1)_n (\alpha+1)_n} \quad (4.3.3)$$

$$= \,_0F_1(-;\alpha+1, t(x-1)/2) \,_0F_1(-;\beta+1; t(x+1)/2).$$

Rainville (Rainville, 1960) refers to (4.3.3) as Bateman's generating function. To prove (4.3.3) apply the transformation (1.4.9) to the $_2F_1$ in (4.1.1) to get

$$P_n^{(\alpha,\beta)}(x) = \frac{(\alpha+1)_n}{n!} \left(\frac{1+x}{2} \right)^n \,_2F_1 \left(\begin{array}{c} -n, -n-\beta \\ \alpha+1 \end{array} \middle| \frac{x-1}{x+1} \right). \quad (4.3.4)$$

Next employ the useful formula

$$(c)_{n-k} = (c)_n (-1)^k / (-c-n+1)_n, \quad (4.3.5)$$

to write $(-n-\beta)_k$ and $(-n)_k/k!$ as $(-1)^k (\beta+1)_n/(\beta+1)_{n-k}$, and $(-1)^k/(n-k)!$, respectively. This allows us to express (4.3.4) as a convolution of the form

$$\frac{P_n^{(\alpha,\beta)}(x)}{(\alpha+1)_n(\beta+1)_n} = \sum_{k=0}^{n} \frac{((x-1)/2)^k}{(\alpha+1)_k \, k!} \frac{((x+1)/2)^{n-k}}{(\beta+1)_{n-k} \, (n-k)!}, \quad (4.3.6)$$

which implies (4.3.3).

Formula (4.3.6) is of independent interest. In fact it can be rewritten as

$$P_n^{(\alpha,\beta)}(x)(-1)^n = \sum_{k=0}^{n} \frac{(-n-\beta)_k}{k!} \left(\frac{x-1}{2} \right)^k \frac{(-n-\alpha)_{n-k}}{(n-k)!} \left(\frac{x+1}{2} \right)^{n-k}, \quad (4.3.7)$$

It is clear that when $x \neq \pm 1$ then (4.3.7) leads to the integral representation

$$P_n^{(\alpha,\beta)}(x) = \frac{1}{2\pi i} \int_C [1+(x+1)z/2]^{n+\alpha} [1+(x-1)z/2]^{n+\beta} \frac{dz}{z^{n+1}}, \quad (4.3.8)$$

where C is a closed contour such that the points $-2(x \pm 1)^{-1}$ are exterior to C. Therefore in a neighborhood of $t = 0$ we have

$$\sum_{n=0}^{\infty} P_n^{(\alpha,\beta)}(x) t^n = \frac{1}{2\pi i} \int_C \frac{[1+(x+1)z/2]^\alpha [1+(x-1)z/2]^\beta}{z - t[1+(x+1)z/2][1+(x-1)z/2]} dz. \quad (4.3.9)$$

With

$$R = R(t) = \sqrt{1 - 2tx + t^2}, \quad (4.3.10)$$

and $R(0) = +1$ we see that for sufficiently small $|t|$ the poles of the integrand are

$$z_1 = 2\frac{xt - 1 + R}{(1 - x^2)\,t}, \quad z_1 = 2\frac{xt - 1 - R}{(1 - x^2)\,t}$$

and z_1 is interior to C but z_2 is in the exterior of C. Now Cauchy's theorem gives

$$\sum_{n=0}^{\infty} P_n^{(\alpha,\beta)}(x)t^n = \frac{[1 + (x + 1)z_1/2]^\alpha [1 + (x - 1)z_1/2]^\beta}{(z_1 - z_2)\,(1 - x^2)\,t/4}$$

It is easy to see that $z_1 - z_2 = 4R/\left[(1 - x^2)\,t\right]$ and

$$z_1 - z_2 = \frac{4R}{(1 - x^2)\,t} \quad \text{and} \quad 1 + \frac{1}{2}(x \pm 1)z_1 = \frac{2}{1 \mp t + R}.$$

This establishes the Jacobi generating function

$$\sum_{n=0}^{\infty} P_n^{(\alpha,\beta)}(x)t^n = \frac{2^{\alpha+\beta}R^{-1}}{(1 - t + R)^\alpha(1 + t + R)^\beta}. \tag{4.3.11}$$

Note that the right-hand side of (4.3.11) is an algebraic function when α and β are rational numbers. In fact the generating function (4.3.11) is the only algebraic generating function known for Jacobi polynomials. For another proof of (4.3.11) see Pólya and Szegő (Pólya & Szegő, 1972, Part III, Problem 219) where the Lagrange inversion formula (1.2.3) was used. Rainville (Rainville, 1960) gives a proof identifying the left-hand side of (4.3.11) as an F_4 function then observes that it is reducible to a product of $_2F_1$ functions. A proof using an idea of Hermite was given in (Askey, 1978).

One important application of (4.3.11) is to apply Darboux's method and find the asymptotics of $P_n^{(\alpha,\beta)}(x)$ for large n.

Theorem 4.3.1 *Let $\alpha, \beta \in \mathbb{R}$ and set*

$$N = n + (\alpha + \beta + 1)/2, \quad \gamma = -(\alpha + 1/2)\pi/2.$$

Then for $0 < \theta < \pi$,

$$P_n^{(\alpha,\beta)}(\cos\theta) = \frac{k(\theta)}{\sqrt{n}}\cos(N\theta + \gamma) + O\left(n^{-3/2}\right), \tag{4.3.12}$$

where

$$k(\theta) = \frac{1}{\sqrt{\pi}}\left[\sin(\theta/2)\right]^{-\alpha-1/2}\left[\cos(\theta/2)\right]^{-\beta-1/2}. \tag{4.3.13}$$

Furthermore, the error bound holds uniformly for $\theta \in [\epsilon, \pi - \epsilon]$, and fixed $\epsilon > 0$.

Proof The t-singularities of the generating function (4.3.11) are when $R = 0, -1\pm t$. The only t-singularities of smallest absolute value for $-1 < x < 1$ are $t = e^{\pm i\theta}$. Thus a comparison function is

$$2^{\alpha+\beta}\left[\frac{\left(1 - e^{i\theta}\right)^{-\alpha}\left(1 + e^{i\theta}\right)^{-\beta}}{\left[(1 - e^{2i\theta})(1 - te^{-i\theta})\right]^{1/2}} + \frac{\left(1 - e^{-i\theta}\right)^{-\alpha}\left(1 + e^{-i\theta}\right)^{-\beta}}{\left[(1 - e^{-2i\theta})(1 - te^{i\theta})\right]^{1/2}}\right].$$

The result now follows from Darboux's method, and the binomial theorem $(1 - z)^{-1/2} = \sum_{0}^{\infty}(1/2)_n z^n/n!$, and (1.4.7). $\qquad\square$

Formula (4.3.6) implies another generating function for $P_n^{(\alpha,\beta)}(x)$. To see this, use (4.3.6) to get

$$\sum_{n=0}^{\infty} \frac{(\gamma)_n(\delta)_n t^n}{(\alpha+1)_n(\beta+1)_n} P_n^{(\alpha,\beta)}(x)$$

$$= \sum_{n=0}^{\infty}\sum_{k=0}^{n} \frac{(\gamma)_n(\delta)_n((x-1)/2)^k((x+1)/2)^{n-k}}{(\alpha+1)_k k!(\beta+1)_{n-k}(n-k)!} t^n.$$

Thus, we proved

$$\sum_{n=0}^{\infty} \frac{(\gamma)_n(\delta)_n t^n}{(\alpha+1)_n(\beta+1)_n} P_n^{(\alpha,\beta)}(x)$$

$$= F_4\left(\gamma,\delta;\alpha+1,\beta+1;\frac{t}{2}(x-1),\frac{t}{2}(x+1)\right), \tag{4.3.14}$$

where F_4 is defined in (1.3.39).

Theorem 4.3.2 ((Srivastava & Singhal, 1973)) *We have the generating function*

$$\sum_{n=0}^{\infty} P_n^{(\alpha+\lambda n,\beta+\mu n)}(x)t^n = \frac{(1-\zeta)^{\alpha+1}(1+\zeta)^{\beta+1}}{(1-x)^\alpha(1+x)^\beta}$$

$$\times \frac{(1-x)^\lambda(1+x)^\mu}{(1-x)^\lambda(1+x)^\mu + \frac{1}{2}t(1-\zeta)^\lambda(1+\zeta)^\nu[\mu-\lambda-z(\lambda+\mu+2)]}. \tag{4.3.15}$$

for $\operatorname{Re} x \in (-1,1)$, *where*

$$\zeta = x - t\frac{(1-\zeta)^{\lambda+1}(1+\zeta)^{\mu+1}}{2(1-x)^\alpha(1+x)^\beta}. \tag{4.3.16}$$

Proof The Rodrigues formula (4.2.8) implies

$$\sum_{n=0}^{\infty} P_n^{(\alpha+\lambda n,\beta+\mu n)}(x)\left(-2\tau(1-x)^\lambda(1+x)^\mu\right)^n$$

$$= (1-x)^{-\alpha}(1+x)^{-\beta}\sum_{n=0}^{\infty} \frac{\tau^n}{n!}\frac{d^n}{dx^n}\left[(1-x)^{n+\alpha+\lambda n}(1+x)^{n+\beta+\mu n}\right]$$

The rest follows from Lagrange's theorem with

$$\phi(z) = (1-z)^{\lambda+1}(1+z)^{\beta+1}, \quad f(z) = (1-z)^\alpha(1+z)^\beta. \quad\square$$

Theorem 4.3.3 (Bateman) *We have the functional relation*

$$\left(\frac{x+y}{2}\right)^n P_n^{(\alpha,\beta)}\left(\frac{1+xy}{x+y}\right) = \sum_{k=0}^{n} c_{n,k} P_k^{(\alpha,\beta)}(x)P_k^{(\alpha,\beta)}(y), \tag{4.3.17}$$

with

$$c_{n,k} = \frac{(\alpha+1)_n \, (\beta+1)_n \, (\alpha+\beta+1)_k \, (\alpha+\beta+1+2k) \, k!}{(\alpha+1)_k \, (\beta+1)_k \, (\alpha+\beta+1)_{n+k+1} \, (n-k)!}. \tag{4.3.18}$$

Moreover (4.3.17) has the inverse

$$\frac{(-1)^n \, n! \, n!}{(\alpha+1)_n(\beta+1)_n} \, P_n^{(\alpha,\beta)}(x) \, P_n^{(\alpha,\beta)}(y)$$
$$= \sum_{k=0}^{n} \frac{(-n)_k \, (\alpha+\beta+n+1)_k}{(\alpha+1)_k \, (\beta+1)_k} \left(\frac{x+y}{2}\right)^k P_k^{(\alpha,\beta)}\left(\frac{1+xy}{x+y}\right). \tag{4.3.19}$$

Proof Expand the left-hand side of (4.3.17) as $\sum\limits_{m=0}^{n} c_{n,m}(y) P_m^{(\alpha,\beta)}(x)$. Then

$$c_{n,m}(y) h_m^{(\alpha,\beta)} = \int_{-1}^{1} \left(\frac{x+y}{2}\right)^n P_n^{(\alpha,\beta)}\left(\frac{1+xy}{x+y}\right) P_m^{(\alpha,\beta)}(x)(1-x)^\alpha (1+x)^\beta \, dx.$$

Using the representations (4.3.6) and (4.1.1) to expand $P_n^{(\alpha,\beta)}$ and $P_m^{(\alpha,\beta)}$, respectively, we find that

$$c_{n,m}(y) = \frac{(\alpha+1)_m \, (\alpha+1)_n \, (\beta+1)_n}{m! \, 4^n \, h_m^{(\alpha,\beta)}} \sum_{k=0}^{n} \frac{(1-y)^k \, (1+y)^{n-k}}{k! \, (\alpha+1)_k \, (n-k)! \, (\beta+1)_{n-k}}$$
$$\times \sum_{j=0}^{m} \frac{(-m)_j \, (\alpha+\beta+m+1)_j}{j! \, (\alpha+1)_j \, 2^j} \int_{-1}^{1} (1-x)^{\alpha+k+j}(1+x)^{\beta+n-k} \, dx$$
$$= \frac{(\alpha+1)_n \, (\beta+1)_n \, \Gamma(\alpha+\beta+m+1) \, (\alpha+\beta+2m+1)}{2^{m+n} \, (\beta+1)_m \Gamma(\alpha+\beta+n+2) \, n! \, (1+y)^{-n}}$$
$$\times \sum_{j=0}^{m} \frac{(-m)_j \, (\alpha+\beta+m+1)_j}{j! \, (\alpha+\beta+n+2)_j} \, {}_2F_1\left(\begin{matrix} -n, \alpha+j+1 \\ \alpha+1 \end{matrix} \middle| \frac{y-1}{y+1}\right).$$

Apply the Pfaff–Kummer transformation (1.4.9) to the ${}_2F_1$ to see that the j sum is

$$\left(\frac{2}{y+1}\right)^n \sum_{s=0}^{n} \sum_{j=s}^{m} \frac{(-m)_j \, (\alpha+\beta+m+1)_j}{j! \, (\alpha+\beta+n+2)_j} \frac{(-j)_s \, (-n)_s}{s! \, (\alpha+1)_s} \left(\frac{1-y}{2}\right)^s$$
$$= \left(\frac{2}{y+1}\right)^n \sum_{s=0}^{n} \frac{(-m)_s \, (-n)_s}{s! \, (\alpha+1)_s} \left(\frac{y-1}{2}\right)^s \frac{(\alpha+b+m+1)_s}{(\alpha+\beta+n+2)_s}$$
$$\times {}_2F_1\left(\begin{matrix} s-m, \alpha+\beta+m+s+1 \\ \alpha+\beta+n+s+2 \end{matrix} \middle| 1\right)$$
$$= \frac{(2/(y+1))^n}{(\alpha+\beta+n+2)_m} \sum_{s=0}^{n} \frac{(-m)_s \, (\alpha+\beta+m+1)_s}{s! \, (\alpha+1)_s}$$
$$\times \left(\frac{y-1}{2}\right)^s (-n)_s \, (n-m+1)_s,$$

where we used the Chu–Vandermonde sum in the last step. The above expression simplifies to

$$\left(\frac{2}{y+1}\right)^n \frac{n!\,m!}{(n-m!)\,(\alpha+1)_m} P_m^{(\alpha,\beta)}(y)$$

and (4.3.17) follows. Next write (4.3.17) as

$$\left(\frac{x+y}{2}\right)^n P_n^{(\alpha,\beta)}\left(\frac{1+xy}{x+y}\right) / P_n^{(\alpha,\beta)}(1) = \sum_{k=0}^{n} d_{n,k} \frac{P_k^{(\alpha,\beta)}(x)\,P_k^{(\alpha,\beta)}(y)}{P_k^{(\alpha,\beta)}(1)},$$

$$(4.3.20)$$

and apply the inversion formulas (4.2.18)–(4.2.19). □

Bateman's original proof of (4.3.17) is in (Bateman, 1905). His proof consists of deriving a partial differential equation satisfied by the left-hand side of (4.3.17) then apply separation of variables to show that the equation has solutions of the form $P_k^{(\alpha,\beta)}(x)P_k^{(\alpha,\beta)}(y)$. The principal of superposition then gives (4.3.17) and the coefficients are computed by setting $y = 1$ and using (4.2.15). A proof of (4.3.19) is in (Bateman, 1932).

4.4 Functions of the Second Kind

In §3.6 we defined functions of the second kind for general polynomials orthogonal with respect to absolutely continuous measures. In the case of Jacobi polynomials the normalization is slightly different. Let

$$Q_n^{(\alpha,\beta)}(x) = \frac{1}{2}(x-1)^{-\alpha}(x+1)^{-\beta} \int_{-1}^{1} (1-t)^\alpha (1+t)^\beta \frac{P_n^{(\alpha,\beta)}(t)}{x-t}\,dt. \quad (4.4.1)$$

When $\alpha > 0$, $\beta > 0$, Theorem 3.6.1 shows that $Q_n^{(\alpha,\beta)}(x)$ satisfies (4.2.4) and (4.2.9). This can be extended to hold for $\operatorname{Re}\alpha > -1$, $\operatorname{Re}\beta > -1$ by analytic continuation except when $n = 0$, $\alpha + \beta + 1 = 0$ hold simultaneously. Furthermore, the Rodrigues formula (4.2.7) and integration by parts transform (4.4.1) into the equivalent form

$$Q_n^{(\alpha,\beta)}(x) = \frac{(n-k)!\,k!}{2^{k+1}\,n!} (x-1)^{-\alpha}(x+1)^{-\beta}$$

$$\times \int_{-1}^{1} \frac{(1-t)^{\alpha+k}(1+t)^{\beta+k}}{(x-t)^{k+1}} P_{n-k}^{(\alpha+k,\beta+k)}(t)\,dt. \quad (4.4.2)$$

In particular

$$Q_n^{(\alpha,\beta)}(x) = \frac{(x-1)^{-\alpha}(x+1)^{-\beta}}{2^{n+1}} \int_{-1}^{1} \frac{(1-t)^{\alpha+n}(1+t)^{\beta+n}}{(x-t)^{n+1}}\,dt. \quad (4.4.3)$$

Formulas (4.4.1)–(4.4.3) hold when $\operatorname{Re}\alpha > -1$, $\operatorname{Re}\beta > -1$ and x in the complex plane cut along $[-1, 1]$ and $n + |\alpha + \beta + 1| \neq 0$. In the exceptional case $n = 0$ and $\alpha + \beta + 1 = 0$, $Q_0^{(\alpha,\beta)}(x)$ is a constant. This makes $P_0^{(\alpha,\beta)}(x)$ and $Q_0^{(\alpha,\beta)}(x)$ linear dependent solutions of (4.2.6) and the reason, as we have pointed out in §3.6, is that $A_0(x) = 0$. A non-constant solution of (4.2.6) is

$$Q^{(\alpha)}(x) = \ln(1+x) + \frac{\sin \pi \alpha}{\pi}(x-1)^{-\alpha}(x+1)^{-\beta}$$

$$\times \int_{-1}^{1} \frac{(1-t)^{\alpha}(1+t)^{\beta}}{x-t} \ln(1+t)\, dt. \tag{4.4.4}$$

The function $\left\{ Q_n^{(\alpha,\beta)}(x) \right\}$ is called the Jacobi function of the second kind. In the exceptional case $n = 0$, $\alpha + \beta + 1 = 0$, the Jacobi function of the second kind is $Q^{(\alpha)}(x)$. Note that

$$Q^{(\alpha)}(x) = 2\frac{\sin \pi \alpha}{\pi} \frac{\partial}{\partial \beta} \left. Q_0^{(\alpha,\beta)}(x) \right|_{\beta=-\alpha-1}. \tag{4.4.5}$$

Formula (4.4.3) and the integral representation (1.4.8) lead to the hypergeometric function representation

$$Q_n^{(\alpha,\beta)}(x) = \frac{\Gamma(n+\alpha+1)\Gamma(n+\beta+1)}{\Gamma(2n+\alpha+\beta+2)2^{-n-\alpha-\beta}}(x-1)^{-n-\alpha-1}(x+1)^{-\beta}$$

$$\times \,_2F_1 \left(\begin{array}{c} n+\alpha+1, n+1 \\ 2n+\alpha+\beta+2 \end{array} \middle| \frac{2}{1-x} \right). \tag{4.4.6}$$

Similarly, (4.4.5)–(4.4.6) yield

$$Q^{(\alpha)}(x) = \ln(x+1) + c + \left(1 - \frac{2}{1-x} \right)^{\alpha+1}$$

$$\times \sum_{k=1}^{\infty} \frac{(\alpha+1)_k}{k!} \left(\sum_{k=1}^{k} \frac{1}{j} \right) \left(\frac{2}{1-x} \right)^k, \tag{4.4.7}$$

where

$$c = -\gamma - \frac{\Gamma'(-\alpha)}{\Gamma(-\alpha)} - \ln 2. \tag{4.4.8}$$

Additional properties of the Jacobi functions are in §4.6 of (Szegő, 1975).

4.5 Ultraspherical Polynomials

The ultraspherical polynomials are special Jacobi polynomials, namely

$$C_n^{\nu}(x) = \frac{(2\nu)_n}{(\nu+1/2)_n} P_n^{(\nu-1/2,\nu-1/2)}(x)$$

$$= \frac{(2\nu)_n}{n!} \,_2F_1 \left(\begin{array}{c} -n, n+2\nu \\ \nu+1/2 \end{array} \middle| \frac{1-x}{2} \right). \tag{4.5.1}$$

The ultraspherical polynomials are also known as Gegenbauer polynomials. Rainville (Rainville, 1960) uses a different normalization for ultraspherical polynomials but his Gegenbauer polynomials are $\{C_n^{\nu}(x)\}$. The Legendre polynomials $\{P_n(x)\}$ correspond to the choice $\nu = 1/2$. The ultraspherical polynomials $\{C_n^{\nu}(x)\}$ are the spherical harmonics on $\mathbb{R}^m$, $\nu = -1 + m/2$. In the case of ultraspherical polynomials, the generating function (4.3.1) simplifies to

$$\sum_{n=0}^{\infty} C_n^\nu(x) t^n = (1-t)^{-2\nu} \left(1 - \frac{2t(x-1)}{(1-t)^2}\right)^{-\nu},$$

via the binomial theorem. Thus

$$\sum_{n=0}^{\infty} C_n^\nu(x) t^n = \left(1 - 2xt + t^2\right)^{-\nu}. \tag{4.5.2}$$

In §5.1 we shall give a direct derivation of (4.5.2) from the three-term recurrence relation

$$2x(n+\nu)C_n^\nu(x) = (n+1)C_{n+1}^\nu(x) + (n + 2\nu - 1)C_{n-1}^\nu(x). \tag{4.5.3}$$

Note that (4.5.3) follows from (4.2.9) and (4.5.1). The orthogonality relation (4.1.2)–(4.1.3) becomes, when $\nu \neq 0$ and $\operatorname{Re}\nu > -\frac{1}{2}$,

$$\int_{-1}^{1} \left(1 - x^2\right)^{\nu - 1/2} C_m^\nu(x) C_n^\nu(x)\, dx = \frac{(2\nu)_n \sqrt{\pi}\, \Gamma(\nu + 1/2)}{n!\, (n+\nu)\Gamma(\nu)} \delta_{m,n}. \tag{4.5.4}$$

Formulas (4.2.5), (4.2.10) and (4.5.1) give

$$\frac{d}{dx} C_n^\nu(x) = 2\nu C_{n-1}^{\nu+1}(x), \tag{4.5.5}$$

$$2\nu \left(1 - x^2\right) \frac{d}{dx} C_n^\nu(x) = (n + 2\nu) \, x C_{n-1}^\nu(x) - (n+1) \, C_{n+1}^\nu(x). \tag{4.5.6}$$

Moreover, (4.5.3) and (4.5.6) give

$$4\left(1 - x^2\right) \frac{d}{dx} C_n^\nu(x) = \frac{(n + 2\nu)(n + 2\nu - 1)}{\nu(n+\nu)} C_{n-1}^\nu(x) - \frac{n(n+1)}{\nu(n+\nu)} C_{n+1}^\nu(x). \tag{4.5.7}$$

The ultraspherical polynomials satisfy the differential equation

$$\left(1 - x^2\right) y'' - x(2\nu + 1) \, y' + n(n + 2\nu) \, y = 0, \tag{4.5.8}$$

as can be seen from (4.2.6).

Differentiating (4.5.2) with respect to t we find

$$\sum_{n=1}^{\infty} C_n^\nu(x) \, n t^{n-1} = 2\nu(x - t) \sum_{n=0}^{\infty} C_n^{\nu+1}(x) \, t^n,$$

hence

$$(n+1) C_{n+1}^\nu(x) = 2\nu x C_n^{\nu+1}(x) - 2\nu C_{n-1}^{\nu+1}(x). \tag{4.5.9}$$

Eliminating $x C_n^{\nu+1}(x)$ from (4.5.9) by using (4.5.3) we obtain

$$(n+\nu) C_n^\nu(x) = \nu \left[C_n^{\nu+1} - C_{n-2}^{\nu+1}(x)\right]. \tag{4.5.10}$$

Formula (4.2.7) in the case $\alpha = \beta = \nu - 1/2$ becomes

$$\left(1 - x^2\right)^{\nu - 1/2} C_n^\nu = \frac{(n-k)!\,(-2)^n(\nu)_k}{n!\,(2\nu + n)_k} \frac{d^k}{dx^k} \left[\left(1 - x^2\right)^{\nu + k - 1/2} C_{n-k}^{\nu+k}(x)\right] \tag{4.5.11}$$

and when $k = n$ we get the Rodrigues formula

$$\left(1 - x^2\right)^{\nu - 1/2} C_n^\nu(x) = \frac{(-1)^n(2\nu)_n}{2^n n!\,(\nu + 1/2)_n} \frac{d^n}{dx^n} \left(1 - x^2\right)^{\nu + n - 1/2}. \tag{4.5.12}$$

We used (1.3.8) in deriving (4.5.11) and (4.5.12).

With $x = \cos\theta$, $1 - 2xt + t^2 = \left(1 - te^{i\theta}\right)\left(1 - te^{-i\theta}\right)$, so we can apply the binomial theorem to (4.5.2) and see that

$$\sum_{n=0}^\infty C_n^\nu(\cos\theta) t^n = \sum_{k,j=0}^\infty \frac{(\nu)_k}{k!} t^k e^{ik\theta} \frac{(\nu)_j}{j!} t^j e^{-ij\theta}.$$

Therefore

$$C_n^\nu(\cos\theta) = \sum_{j=0}^n \frac{(\nu)_j(\nu)_{n-j}}{j!\,(n-j)!} e^{i(n-2j)\theta}. \tag{4.5.13}$$

One application of (4.5.13) is to derive the large n asymptotics of $C_n^\nu(x)$ for $x \in \mathbb{C} \setminus [-1,1]$. With $e^{\pm i\theta} = x \pm \sqrt{x^2 - 1}$ and the sign of the square root chosen such that $\sqrt{x^2 - 1} \approx x$ as $x \to \infty$, we see that $\left|e^{-i\theta}\right| < \left|e^{i\theta}\right|$ if $\operatorname{Im} x > 0$. Using

$$\frac{(\nu)_{n-j}}{(n-j)!} = \frac{\Gamma(\nu + n - j)}{\Gamma(\nu)\Gamma(n - j + 1)} = \frac{n^{\nu-1}}{\Gamma(\nu)}\left\{1 + \mathcal{O}\left(\frac{1}{n}\right)\right\},$$

Tannery's theorem and the binomial theorem we derive

$$C_n^\nu(\cos\theta) = \frac{e^{in\theta}}{(1 - e^{-2i\theta})^\nu} \frac{n^{\nu-1}}{\Gamma(\nu)} \left\{1 + \mathcal{O}\left(\frac{1}{n}\right)\right\}, \quad \operatorname{Im}\cos\theta > 0, \tag{4.5.14}$$

with $\theta \to -\theta$ if $\operatorname{Im}\cos\theta < 0$.

The relationships (4.1.1), (4.1.4) and (4.5.1) imply the explicit representations

$$\begin{aligned} C_n^\nu(x) &= \frac{(2\nu)_n}{n!} \, {}_2F_1(-n, n + 2\nu; \nu + 1/2; (1 - x)/2) \\ &= \frac{(2\nu)_n}{n!} (-1)^n \, {}_2F_1(-n, n + 2\nu; \nu + 1/2; (1 + x)/2). \end{aligned} \tag{4.5.15}$$

Another explicit representation for $C_n^\nu(x)$ is

$$C_n^\nu(x) = \sum_{k=0}^{\lfloor n/2 \rfloor} \frac{(2\nu)_n x^{n-2k}\left(x^2 - 1\right)^k}{2^{2k} k!\,(\nu + 1/2)_k (n - 2k)!} \tag{4.5.16}$$

which leads to the integral representation

$$C_n^\nu(x) = \frac{(2\nu)_n \Gamma(\nu + 1/2)}{n!\,\Gamma(1/2)\Gamma(\nu)} \int_0^\pi \left[x + \sqrt{x^2 - 1}\cos\varphi\right]^n \sin^{2\nu-1}\varphi \, d\varphi, \tag{4.5.17}$$

known as the Laplace first integral. To prove (4.5.17), rewrite the right-hand side of (4.5.2) as

$$\left[(1 - xt)^2 - t^2 \left(x^2 - 1\right)\right]^{-\nu} = (1 - xt)^{-2\nu} \left[1 - \frac{t^2 \left(x^2 - 1\right)}{(1 - xt)^2}\right]^{-\nu}$$

$$= \sum_{n=0}^{\infty} \frac{(\nu)_n}{n!} \frac{\left(x^2 - 1\right)^n t^{2n}}{(1 - xt)^{2n+2\nu}}$$

$$= \sum_{n,k=0}^{\infty} \frac{(\nu)_n}{n!} \frac{(2n + 2\nu)_k}{k!} \left(x^2 - 1\right)^n x^k t^{2n+k}$$

$$= \sum_{n,k=0}^{\infty} \frac{(\nu)_n (2\nu)_{2n+k}}{n! \, k! \, (2\nu)_{2n}} x^k \left(x^2 - 1\right)^n t^{2n+k},$$

which implies (4.5.16) upon equation coefficients of like powers of t. To prove (4.5.17) expand $[\ ]^n$ by the binomial theorem then apply the change of variable $y = \cos^2 \varphi$. Thus the right-hand side of (4.5.17) is

$$\frac{(2\nu)_n \Gamma(\nu + 1/2)}{\Gamma(1/2)\Gamma(\nu)} \sum_{k=0}^{n} \frac{x^{n-k} \left(x^2 - 1\right)^{k/2}}{k! \, (n - k)!} \int_0^{\pi} \cos^k \varphi \sin^{2\nu-1} \varphi \, d\varphi$$

$$= \frac{(2n)_n \Gamma(\nu + 1/2)}{\Gamma(1/2)\Gamma(\nu)} \sum_{k=0}^{\lfloor n/2 \rfloor} \frac{x^{n-2k} \left(x^2 - 1\right)^k}{(2k)! \, (n - 2k)!} 2 \int_0^{\pi/2} \cos^{2k} \varphi \sin^{2\nu-1} \varphi \, d\varphi$$

$$= \frac{(2\nu)_n \Gamma(\nu + 1/2)}{\Gamma(1/2)\Gamma(\nu)} \sum_{k=0}^{\lfloor n/2 \rfloor} \frac{x^{n-2k} \left(x^2 - 1\right)^k}{(2k)! \, (n - 2k)!} \frac{\Gamma(k + 1/2)\Gamma(\nu)}{\Gamma(\nu + k + 1/2)},$$

which completes the proof.

The Chebyshev polynomials of the first and second kinds are

$$T_n(x) = \cos(n\theta); \quad U_n(x) = \frac{\sin(n + 1)\theta}{\sin \theta}, \quad x := \cos \theta, \tag{4.5.18}$$

respectively. Their orthogonality relations are

$$\int_{-1}^{1} T_m(x) T_n(x) \frac{dx}{\sqrt{1 - x^2}} = \begin{cases} \frac{\pi}{2} \delta_{m,n}, & n \neq 0, \\ \pi \delta_{0,n} \end{cases} \tag{4.5.19}$$

and

$$\int_{-1}^{1} U_m(x) U_n(x) \sqrt{1 - x^2} \, dx = \frac{\pi}{2} \delta_{m,n}. \tag{4.5.20}$$

Moreover,

$$T_n(x) = \frac{n!}{(1/2)_n} P_n^{(-1/2,-1/2)}(x), \quad U_n(x) = \frac{(n + 1)!}{(3/2)_n} P_n^{(1/2,1/2)}(x). \tag{4.5.21}$$

In terms of ultraspherical polynomials, the Chebyshev polynomials are

$$U_n(x) = C_n^1(x), \quad T_n(x) = \lim_{\nu \to 0} \frac{n + 2\nu}{2\nu} C_n^\nu(x), \quad n \geq 0. \tag{4.5.22}$$

Therefore $\{U_n(x)\}$ and $\{T_n(x)\}$ have the generating functions

$$\sum_{n=0}^{\infty} U_n(x)t^n = \frac{1}{1 - 2xt + t^2}, \tag{4.5.23}$$

$$\sum_{n=0}^{\infty} T_n(x)t^n = \frac{1 - xt}{1 - 2xt + t^2}. \tag{4.5.24}$$

It is clear that

$$T_n(z) = \frac{1}{2}\left[\left(z + \sqrt{z^2 - 1}\right)^n + \left(z - \sqrt{z^2 - 1}\right)^n\right],$$

$$U_n(z) = \frac{\left(z + \sqrt{z^2 - 1}\right)^{n+1} - \left(z - \sqrt{z^2 - 1}\right)^{n+1}}{2\sqrt{z^2 - 1}}. \tag{4.5.25}$$

Formulas (4.5.16) and (4.5.22) yield

$$T_n(x) = \sum_{k=0}^{\lfloor n/2 \rfloor} \frac{(-n)_{2k}}{(2k)!} x^{n-2k} \left(x^2 - 1\right)^k, \tag{4.5.26}$$

$$U_n(x) = (n+1) \sum_{k=0}^{\lfloor n/2 \rfloor} \frac{(-n)_{2k}}{(2k+1)!} x^{n-2k} \left(x^2 - 1\right)^k. \tag{4.5.27}$$

The representations (4.5.26)–(4.5.27) also follow from (4.5.25). Both $U_n(x)$ and $T_n(x)$ satisfy the three-term recurrence relation

$$2xy_n(x) = y_{n+1}(x) + y_{n-1}(x), \quad n > 0, \tag{4.5.28}$$

with $T_0(x) = 1, T_1(x) = x, U_0(x) = 1, U_1(x) = 2x$.

Theorem 4.5.1 *Let E denote the closure of the area enclosed by an ellipse whose foci are at ± 1. Then $\max\{|T_n(x)| : x \in E\}$ is attained at the right endpoint of the major axis. Moreover, the same property holds for the ultraspherical polynomials $C_n^\nu(x)$ for $\nu \geq 0$.*

Proof The parametric equations of the ellipse are $x = a\cos\phi$, $y = \sqrt{a^2 - 1}\sin\varphi$. Let $z = x + iy$. A calculation gives $z \pm \sqrt{z^2 - 1} = \left(a \pm \sqrt{a^2 - 1}\right)e^{\pm i\phi}$. Thus, the first equation in (4.5.25) and the fact that the maximum is attained on the boundary of the ellipse proves the assertion about $T_n(x)$. For C_n^ν, rewrite (4.5.13) as

$$C_n^\nu(z) = \sum_{j=0}^{n} \frac{(\nu)_j (\nu)_{n-j}}{j!\,(n-j)!} T_{|n-2j|}(z),$$

and use the result for T_n to prove it for $C_n^\nu(z)$.   □

4.6 Laguerre and Hermite Polynomials

The weight function for Laguerre polynomials is $x^\alpha e^{-x}$, on $[0, \infty)$. For Hermite polynomials the weight function is e^{-x^2} on $\mathbb{R}$. It is easy to see that the Laguerre weight is a limiting case of the Jacobi weight by first putting the Jacobi weight on

$[0, a]$ then let $a \to \infty$. The Hermite weight is the limiting case $(1-x/\alpha)^\alpha(1+x/\alpha)^\alpha$ as $\alpha \to \infty$. Instead of deriving the properties of Laguerre and Hermite polynomials as limiting cases of Jacobi polynomials, we will establish their properties directly. Furthermore certain results hold for Hermite or Laguerre polynomials and do not have a counterpart for Jacobi polynomials. In the older literature, e.g., (Bateman, 1932), Laguerre polynomials were called Sonine polynomials. Askey pointed out in (Askey, 1975a) that the Hermite, Laguerre (Sonine), Jacobi and Hahn polynomials are not named after the first person to define or use them.

Theorem 4.6.1 *The Laguerre polynomials have the explicit representation*

$$L_n^{(\alpha)}(x) = \frac{(\alpha+1)_n}{n!} {}_1F_1(-n; \alpha+1; x), \tag{4.6.1}$$

and satisfy the orthogonality relation

$$\int_0^\infty x^\alpha e^{-x} L_m^{(\alpha)}(x) L_n^{(\alpha)}(x)\, dx \tag{4.6.2}$$
$$= \frac{\Gamma(a+n+1)}{n!} \delta_{m,n} \quad \mathrm{Re}\,(\alpha) > -1.$$

Furthermore

$$L_n^{(\alpha)}(x) = \frac{(-1)^n}{n!} x^n + lower\ order\ terms. \tag{4.6.3}$$

Proof Clearly (4.6.3) follows from (4.6.1), so we only prove that the polynomials defined by (4.6.1) satisfy (4.6.2). One can follow the attachment procedure of §4.1 and discover the form (4.6.1) but instead we shall verify that the polynomials defined by (4.6.1) satisfy (4.6.2). It is easy to see that

$$\int_0^\infty x^\alpha e^{-x} x^m {}_1F_1(-n; \alpha+1; x)\, dx = \sum_{k=0}^n \frac{(-n)_k}{k!\,(\alpha+1)_k} \Gamma(m+k+\alpha+1)$$
$$= \Gamma(\alpha+m+1)\, {}_2F_1(-n, \alpha+m+1; \alpha+1; 1) = \frac{\Gamma(\alpha+m+1)(-m)_n}{(\alpha+1)_n},$$

by the Chu–Vandermonde sum (1.4.3). Hence the integral in the above equation is zero for $0 \le m < n$. Furthermore when $m = n$ the left-hand side of (4.6.2) is

$$\frac{(\alpha+1)_n}{n!} \frac{(-1)^n}{n!} \frac{\Gamma(\alpha+n+1)(-n)_n}{(\alpha+1)_n},$$

and (4.6.2) follows.

We next establish the generating function

$$\sum_{n=0}^\infty L_n^{(\alpha)}(x) t^n = (1-t)^{-\alpha-1} \exp\left(-xt/(1-t)\right). \tag{4.6.4}$$

To prove (4.6.4), substitute for $L_n^{(\alpha)}(x)$ from (4.6.1) to see that

$$\sum_{n=0}^{\infty} L_n^{(\alpha)}(x)\, t^n = \sum_{n=0}^{\infty} \sum_{k=0}^{n} \frac{(\alpha+1)_n}{n!} \frac{(-1)^k\, n!\, x^k\, t^n}{k!\,(n-k)!\,(\alpha+1)_k}$$

$$= \sum_{k=0}^{\infty} \frac{(-1)^k}{k!}\, x^k\, t^k \sum_{n=0}^{\infty} \frac{(\alpha+k+1)_n}{n!}\, t^n = \sum_{k=0}^{\infty} \frac{(-xt)^k}{k!\,(1-t)^{\alpha+k+1}},$$

which is equal to the right-hand side of (4.6.4) and the proof is complete. □

We now come to the Hermite polynomials $\{H_n(x)\}$. In (4.6.2) let $x = y^2$ to see that

$$\int_{\mathbb{R}} |y|^{2\alpha+1} e^{-y^2} L_m^{(\alpha)}\left(y^2\right) L_n^{(\alpha)}\left(y^2\right) = 0, \quad \mathrm{Re}\,(\alpha) > -1,$$

when $m \neq n$. The uniqueness of the orthogonal polynomials, up to normalization constants, shows that $H_{2n}(x)$ and $H_{2n+1}(x)$ must be constant multiples of $L_n^{(-1/2)}(x^2)$ and $xL_n^{(1/2)}(x^2)$, respectively. In the literature the constant multiples have been chosen as

$$H_{2n}(x) = (-1)^n\, 2^{2n}\, n!\, L_n^{(-1/2)}\left(x^2\right), \tag{4.6.5}$$

$$H_{2n+1}(x) = (-1)^n\, 2^{2n+1}\, n!\, x L_n^{(1/2)}\left(x^2\right). \tag{4.6.6}$$

We now take (4.6.5)–(4.6.6) as the definition of the Hermite polynomials.

It is important to note that the above calculations also give explicit representations for the polynomials orthogonal with respect to $|x|^\gamma e^{-x^2}$ on $\mathbb{R}$ in terms of Laguerre polynomials.

Theorem 4.6.2 *The Hermite polynomials have the representation*

$$H_n(x) := \sum_{k=0}^{\lfloor n/2 \rfloor} \frac{n!\,(-1)^k (2x)^{n-2k}}{k!\,(n-2k)!}, \tag{4.6.7}$$

and satisfy the orthogonality relation

$$\int_{\mathbb{R}} H_m(x) H_n(x) e^{-x^2}\, dx = 2^n n! \sqrt{\pi}\, \delta_{m,n}. \tag{4.6.8}$$

Proof Formula (4.6.5) and the fact $m! = (1)_m$ combined with the duplication formula (1.3.7) yield

$$H_{2n}(x) = \frac{(1/2)_n\, 2^{2n}\, n!}{(-1)^n\, n!} \sum_{k=0}^{n} \frac{(-n)_k}{(1)_k (1/2)_k}\, x^{2k}$$

$$= \frac{(-1)^n\,(2n)!}{n!} \sum_{k=0}^{n} \frac{(-1)^k\, n!}{(n-k)!\,(2k)!}\, (2x)^{2k}.$$

By reversing the above sum, that is $k \to n - k$ we establish (4.6.7) for even n. The odd n similarly follows. Finally (4.6.8) follows from (4.6.5)–(4.6.6) and (4.6.2). □

Formula (4.6.7) leads to a combinatorial interpretation of $H_n(x)$. Let S be a set of n points on a straight line. A perfect matching of S is a one-to-one mapping ϕ of S onto itself. The fixed points of ϕ are those points x for which $\phi(x) = x$. If $\phi(x) \neq x$, we join x and $\phi(x)$ by an edge (arch). Let $PM(S)$ be the set of all perfect matchings of S. It then follows that

$$H_n(x/2) = \sum_{c \in PM(S)} (-1)^{\# \text{ of edges in } c}\, x^{\# \text{ of fixed points in } c}. \tag{4.6.9}$$

Note that

$$L_n^{(\alpha)}(0) = \frac{(\alpha+1)_n}{n!}, \quad H_{2n+1}(0) = 0, \quad H_{2n}(0) = (-1)^n 4^n (1/2)_n. \tag{4.6.10}$$

The generating functions

$$\sum_{n=0}^{\infty} \frac{H_{2n}(x)\, t^n}{2^{2n}\, n!} = (1+t)^{-1/2} \exp\left(x^2 t/(1+t)\right), \tag{4.6.11}$$

$$\sum_{n=0}^{\infty} \frac{H_{2n+1}(x)\, t^n}{x\, 2^{2n+1}\, n!} = (1+t)^{-3/2} \exp\left(x^2 t/(1+t)\right), \tag{4.6.12}$$

are immediate consequences of Theorem 4.6.2 and (4.6.4).

Formula (4.6.1) implies

$$\frac{d}{dx} L_n^{(\alpha)}(x) = -L_{n-1}^{(\alpha+1)}(x). \tag{4.6.13}$$

The idea of proving (4.2.3) leads to the adjoint relation

$$\frac{d}{dx}\left[x^{\alpha+1}e^{-x}L_n^{(\alpha+1)}(x)\right] = (n+1)x^{\alpha}e^{-x}L_{n+1}^{(\alpha)}(x). \tag{4.6.14}$$

Combining (4.6.13) and (4.6.14) we establish the differential equation

$$\frac{d}{dx}\left[x^{\alpha+1}e^{-x}\frac{d}{dx}L_n^{(\alpha)}(x)\right] + nx^{\alpha}e^{-x}L_n^{(\alpha)}(x) = 0. \tag{4.6.15}$$

In other words $y = L_n^{(\alpha)}(x)$ is a solution to

$$xy'' + (1+\alpha-x)\, y' + ny = 0. \tag{4.6.16}$$

It is important to note that (4.6.15) is the Infeld–Hull factorization of (4.6.16), that is (4.6.15) has the form T^*T where T is a linear first order differential operator and the adjoint T^* is with respect to the weighted inner product

$$(f,g) = \int_0^{\infty} x^{\alpha}e^{-x}f(x)\,\overline{g(x)}\, dx. \tag{4.6.17}$$

Another application of (4.6.14) is

$$L_n^{(\alpha)}(x) = \frac{(n-k)!}{n!}\, x^{-\alpha}e^x \frac{d^k}{dx^k}\left[x^{\alpha+k}e^{-x}L_{n-k}^{(\alpha+k)}(x)\right]. \tag{4.6.18}$$

In particular we have the Rodrigues formula

$$L_n^{(\alpha)}(x) = \frac{1}{n!}\, x^{-\alpha}e^x \frac{d^n}{dx^n}\left[x^{\alpha+n}e^{-x}\right]. \tag{4.6.19}$$

Similarly from (4.6.7) one derives

$$\frac{d}{dx} H_n(x) = 2nH_{n-1}(x), \tag{4.6.20}$$

and (4.6.8) gives the adjoint relation

$$H_{n+1} = -e^{x^2} \frac{d}{dx} \left[e^{-x^2} H_n(x) \right], \tag{4.6.21}$$

The Hermite differential equation is

$$e^{x^2} \frac{d}{dx} \left[e^{-x^2} \frac{d}{dx} H_n(x) \right] + 2nH_n(x) = 0, \tag{4.6.22}$$

or equivalently

$$y'' - 2xy' + 2ny = 0, \quad y = H_n(x). \tag{4.6.23}$$

Furthermore (4.6.21) leads to

$$H_n(x) = (-1)^k e^{x^2} \frac{d^k}{dx^k} \left[e^{-x^2} H_{n-k}(x) \right] \tag{4.6.24}$$

and the case $k = n$ is the Rodrigues formula

$$H_n(x) = (-1)^n e^{x^2} \frac{d^n}{dx^n} e^{-x^2}. \tag{4.6.25}$$

The three-term recurrence relations of the Laguerre and Hermite polynomials are

$$xL_n^{(\alpha)}(x) = -(n+1)L_{n+1}^{(\alpha)}(x) + (2n+\alpha+1)L_n^{(\alpha)}(x) - (n+\alpha)L_{n-1}^{(\alpha)}(x), \tag{4.6.26}$$

$$2xH_n(x) = H_{n+1}(x) + 2nH_{n-1}(x). \tag{4.6.27}$$

In the remaining part of this section we derive several generating functions of Hermite and Laguerre polynomials. For combinatorial applications of generating functions we refer the interested reader to (Stanley, 1978) and (Wilson, 1982).

Theorem 4.6.3 *The Hermite polynomials have the generating functions*

$$\sum_{n=0}^{\infty} \frac{H_n(x)}{n!} t^n = \exp\left(2xt - t^2\right), \tag{4.6.28}$$

$$\sum_{n=0}^{\infty} \frac{H_{n+k}(x)}{n!} t^n = \exp\left(2xt - t^2\right) H_k(x - t). \tag{4.6.29}$$

Proof Formula (4.6.28) follows from the representation (4.6.7). Differentiating (4.6.28) k times with respect to t we see that the left-hand side of (4.6.29) is

$$\frac{\partial^k}{\partial t^k} \exp\left(2xt - t^2\right) = e^{x^2} \frac{\partial^k}{\partial (t-x)^k} \exp\left(-(t-x)^2\right),$$

and (4.6.29) follows from (4.6.25). $\square$

Theorem 4.6.4 *The Laguerre polynomials have the generating functions*

$$\sum_{n=0}^{\infty} \frac{L_n^{(\alpha)}(x)}{(\alpha+1)_n} t^n = e^t \, {}_0F_1(-;\alpha+1;-xt) \tag{4.6.30}$$

$$\sum_{n=0}^{\infty} \frac{(c)_n}{(\alpha+1)_n} L_n^{(\alpha)}(x) \, t^n = (1-t)^{-c} \, {}_1F_1\left(\begin{array}{c} c \\ 1+\alpha \end{array} \middle| \frac{-xt}{1-t}\right). \tag{4.6.31}$$

Proof Use (4.6.1) to see that the left-hand side of (4.6.30) is

$$\sum_{0 \le k \le n < \infty} \frac{(-n)_k \, x^k t^n}{k! \, (\alpha+1)_k \, n!} = \sum_{k=0}^{\infty} \frac{(-tx)^k}{k! \, (\alpha+1)_k} \sum_{n=0}^{\infty} \frac{t^n}{n!}$$

and (4.6.30) follows. Similarly the left-hand side of (4.6.31) is

$$\sum_{0 \le k \le n < \infty} \frac{(c)_n(-n)_k \, x^k t^n}{k! \, (\alpha+1)_k \, n!} = \sum_{k=0}^{\infty} \frac{(-tx)^k(c)_k}{k! \, (\alpha+1)_k} \sum_{n=0}^{\infty} \frac{t^n (c+k)_n}{n!}$$

$$= \sum_{k=0}^{\infty} \frac{(-tx)^k(c)_k}{k! \, (\alpha+1)_k} (1-t)^{-c-k},$$

which establishes (4.6.31). $\qquad\qquad\Box$

Theorem 4.6.5 *The following expansion of scaled Laguerre polynomials holds*

$$L_n^{(\alpha)}(cx) = (\alpha+1)_n \sum_{k=0}^{n} \frac{c^k (1-c)^{n-k}}{(n-k)! \, (\alpha+1)_k} L_k^{(\alpha)}(x). \tag{4.6.32}$$

Proof Let $G(x,t)$ denote the generating function in (4.6.30). Clearly

$$G(cx, t) = G(x, ct) \exp(t - ct).$$

The result follows from equating coefficients of t^n in the above formula. One can also expand $L_n^{(\alpha)}(cx)$ as $\sum_{k=0}^{n} c_{n,k} L_k^{(\alpha)}(x)$, then use the orthogonality relation to evaluate $c_{n,k}$. $\qquad\qquad\Box$

The analogue of Theorem 4.6.5 for Hermite polynomials is

$$H_n(cx) = \sum_{k=0}^{\lfloor n/2 \rfloor} \frac{n! \, (-1)^k}{k! \, (n-2k)!} \left(1-c^2\right)^k c^{n-2k} H_{n-2k}(x). \tag{4.6.33}$$

To prove (4.6.33) use (4.6.28). Hence

$$\sum_{n=0}^{\infty} \frac{H_n(cx)}{n!} t^n = \exp\left(2xct - t^2\right) = \exp\left(2xct - c^2t^2 - \left(1-c^2\right)t^2\right)$$

$$= \sum_{k=0}^{\infty} \frac{(-1)^k}{k!} \left((1-c^2)t^2\right)^k \sum_{m=0}^{\infty} \frac{H_m(x)}{m!} c^m t^m,$$

and (4.6.33) follows.

Observe that (4.6.32) implies that for $c \geq 0$, the coefficients of $L_k^\alpha(x)$ in the expansion of $L_n^{(\alpha)}(x)$ are positive if and only if $c < 1$. Formula (4.6.33) has a similar interpretation.

We now extend (4.6.29) to Laguerre polynomials, so we consider the sum S_k

$$S_k := \sum_{n=0}^\infty \frac{(n+k)!}{n!\,k!} L_{n+k}^{(\alpha)}(x) t^n.$$

Clearly with $m = n + k$ we get

$$\sum_{k=0}^\infty S_k s^k = \sum_{m=0}^\infty L_m^{(\alpha)}(x) \sum_{k=0}^m \frac{m!\,s^k t^{m-k}}{k!\,(m-k)!} = \sum_{m=0}^\infty L_m^{(\alpha)}(x)(s+t)^m$$

$$= (1 - s - t)^{-\alpha - 1} \exp\left(-x(t+s)/(1 - s - t)\right)$$

$$= (1 - t)^{-\alpha - 1} \exp\left(-xt/(1 - t)\right) [1 - s/(1 - t)]^{-\alpha - 1}$$

$$\times \exp\left(\frac{-xs}{(1 - t)^2} \frac{1}{1 - s/(1 - t)}\right).$$

This proves the generating function

$$\sum_{n=0}^\infty \frac{(n+k)!}{n!\,k!} L_{n+k}^{(\alpha)}(x) t^n$$

$$= (1 - t)^{-\alpha - 1 - k} \exp\left(-xt/(1 - t)\right) L_k^{(\alpha)}(x/(1 - t)). \tag{4.6.34}$$

We record the effect of translation on Hermite and Laguerre polynomials. Start with (4.6.4) to obtain

$$\sum_{n=0}^\infty L_n^{(\alpha)}(x + y) t^n = (1 - t)^{\alpha + 1} \sum_{k=0}^\infty L_k^{(\alpha)}(x) t^k \sum_{m=0}^\infty L_m^{(\alpha)}(x) t^m,$$

and we have

$$L_n^{(\alpha)}(x + y) = \sum_{k,m=0}^n L_k^{(\alpha)}(x) L_m^{(\alpha)}(y) \frac{(-\alpha - 1)_{n-k-m}}{(n - k - m)!} \tag{4.6.35}$$

Similarly we establish

$$H_n(x + y) = \sum_{k,s=0}^n \frac{H_s(x)}{s!} \frac{H_{n-2k-s}(y)}{(n - 2k - s)!} \frac{1}{k!}, \tag{4.6.36}$$

where the sum is over $k, s \geq 0$, with $2k + s \leq n$. It is straightforward to derive

$$L_n^{(\alpha + \beta + 1)}(x + y) = \sum_{k=0}^n L_k^{(\alpha)}(x) L_{n-k}^{(\beta)}(y). \tag{4.6.37}$$

A dual to (4.6.37) is

$$\frac{L_{m+n}^{(\alpha)}(x)}{L_{m+n}^{(\alpha)}(0)} = \frac{\Gamma(\alpha+1)}{\Gamma(\beta+1)\Gamma(\alpha-\beta)}$$

$$\times \int_0^1 t^\alpha (1-t)^{\alpha-\beta-1} \frac{L_m^{(\alpha)}(xt)}{L_m^{(\alpha)}(0)} \frac{L_n^{(\alpha-\beta-1)}(x(1-t))}{L_n^{(\alpha-\beta-1)}(0)} dt \qquad (4.6.38)$$

for $\operatorname{Re}\alpha > -1$, $\operatorname{Re}\alpha > \operatorname{Re}\beta$, and is due to Feldheim, (Andrews et al., 1999, §6.2). It can be proved by substituting the series representations for Laguerre polynomials in the integrand, evaluate the resulting beta integrals to reduce the right-hand side to a double sum, then apply the Chu–Vandermonde sum. Formulas (4.6.37) and (4.6.38) are analogues of Sonine's second integral

$$x^\mu y^\nu \frac{J_{\mu+\nu+1}\left(\sqrt{x^2+y^2}\right)}{(x^2+y^2)^{(\mu+\nu+1)/2}}$$

$$= \int_0^{\pi/2} J_\mu(x\sin\theta) J_\nu(y\cos\theta) \sin^{\mu+1}\theta \sin^{\nu+1}\theta \, d\theta, \qquad (4.6.39)$$

(Andrews et al., 1999, Theorem 4.11.1), since

$$\lim_{n\to\infty} n^{-\alpha} L_n^{(\alpha)}\left(\frac{x^2}{4n}\right) = (2/x)^\alpha J_\alpha(x). \qquad (4.6.40)$$

A generalization of (4.6.37) was proved in (Van der Jeugt, 1997) and was further generalized in (Koelink & Van der Jeugt, 1998).

Theorem 4.6.6 *The Hermite and Laguerre polynomials have the integral representations*

$$\frac{H_n(ix)}{(2i)^n} = \frac{1}{\sqrt{\pi}} \int_{-\infty}^\infty e^{-(y-x)^2} y^n \, dy, \qquad (4.6.41)$$

$$n! \, L_n^\alpha(x) = x^{-\alpha/2} \int_0^\infty e^{x-y} y^{n+\alpha/2} J_\alpha\left(2\sqrt{xy}\right) dy, \qquad (4.6.42)$$

valid for $n = 0, 1, \ldots$, and $\alpha > -1$.

Proof The right-hand side of (4.6.41) is

$$\frac{1}{\sqrt{\pi}} \int_{-\infty}^\infty e^{-y^2} (y+x)^n \, dy = \frac{1}{\sqrt{\pi}} \sum_{k=0}^{\lfloor n/2 \rfloor} \binom{n}{2k} x^{n-2k} \int_{-\infty}^\infty y^{2k} e^{-y^2} \, dy$$

$$= \sum_{k=0}^{\lfloor n/2 \rfloor} \frac{n! \, x^{n-2k}}{(n-2k)! \, (2k)!} \frac{\Gamma(k+1/2)}{\Gamma(1/2)},$$

which is $H_n(ix)/(2i)^n$, by (1.3.8) and (4.6.7). Formula (4.6.42) can be similarly proved by expanding J_α and using (4.6.1). $\qquad \square$

The Hermite functions $\left\{e^{-x^2/2}H_n(x)\right\}$ are the eigenfunctions of the Fourier transform. Indeed

$$e^{-x^2/2}H_n(x) = \frac{i^{-n}}{\sqrt{2\pi}}\int_{\mathbb{R}} e^{ixy}e^{-y^2/2}H_n(y)\,dy, \qquad (4.6.43)$$

$n = 0, 1, \ldots$.

The arithmetic properties of the zeros of Laguerre polynomials have been studied since the early part of the twentieth century. Schur proved that $\left\{L_m^{(0)}(x)\right\}$ are irreducible over the rationals for $m > 1$, and later proved the same result for $\left\{L_m^{(1)}(x)\right\}$, (Schur, 1929) and (Schur, 1931). Recently, (Filaseta & Lam, 2002) proved that $L_m^{(\alpha)}(x)$ is irreducible over the rationals for all, but finitely many m, when α is rational but is not a negative integer.

4.7 Multilinear Generating Functions

The Poisson kernel for Hermite polynomials is (4.7.6). It is a special case of the Kibble–Slepian formula (Kibble, 1945; Slepian, 1972), which will be stated as Theorem 4.7.2. The proof of the Kibble–Slepian formula, given below, is a modification of James Louck's proof in (Louck, 1981). An interesting combinatorial proof was given by Foata in (Foata, 1981).

Lemma 4.7.1 *We have*

$$\exp\left(-\frac{1}{4}\partial_x^2\right)e^{-\alpha x^2} = [1-\alpha]^{-1/2}\exp\left(-\alpha x^2/(1-\alpha)\right). \qquad (4.7.1)$$

Proof With $y = \sqrt{\alpha}\,x$ the left-hand side of (4.7.1) is

$$\sum_{n=0}^{\infty}\frac{(-4)^{-n}}{n!}\frac{d^{2n}}{dx^{2n}}e^{-\alpha x^2} = \sum_{n=0}^{\infty}\frac{(-4)^{-n}}{n!}\alpha^n\frac{d^{2n}}{dy^{2n}}e^{-y^2}$$

$$= \sum_{n=0}^{\infty}\frac{(-\alpha)^n}{4^n\,n!}e^{-y^2}H_{2n}(y)$$

and we applied the Rodrigues formula in the last step. The result follows from (4.6.11). $\qquad\square$

For an $n \times n$ matrix $S = s_{ij}$ the Euclidean norm is

$$\|S\| = \left(\sum_{i,j=1}^{n}|s_{ij}|^2\right)^{1/2}.$$

Theorem 4.7.2 (Kibble–Slepian) *Let $S = s_{ij}$ be an $n \times n$ real symmetric matrix, and assume that $\|S\| < 1$, I being an identity matrix. Then*

$$[\det(I + S)]^{-1/2} \exp\left(\mathbf{x}^T S(I + S)^{-1}\mathbf{x}\right)$$

$$= \sum_K \left[\prod_{1 \le i \le j \le n} (s_{ij}/2)^{k_{ij}} / k_{ij}! \right] 2^{-\operatorname{tr} K} H_{k_1}(x_1) \cdots H_{k_n}(x_n), \tag{4.7.2}$$

where $K = (k_{ij})$, $1 \le i, j \le n$, $k_{ij} = k_{ji}$, and

$$\operatorname{tr} K := \sum_{i=1}^n k_{ii}, \qquad k_i := k_{ii} + \sum_{j=1}^n k_{ij}, \quad i = 1, \ldots, n. \tag{4.7.3}$$

In (4.7.2) $\sum_K$ denotes the $n(n+1)/2$ fold sum over $k_{ij} = 0, 1, \ldots$, for all positive integers i, j such that $1 \le i \le j \le n$.

Proof The operational formula

$$\exp\left((-1/4)\partial_x^2\right)(2x)^n = H_n(x), \tag{4.7.4}$$

follows from expanding $\exp\left((-1/4)\partial_x^2\right)$ and applying (4.6.7). Let D be an $n \times n$ diagonal matrix, say $\alpha_j \delta_{ij}$ and assume that $I + D$ is positive definite. Therefore with $\tilde{\Delta}_n$ denoting the Laplacian $\sum_{j=1}^n \partial_{y_j}^2$ we obtain from (4.7.1) the relationship

$$\exp\left((-1/4)\tilde{\Delta}_n\right) \exp\left(\mathbf{y}^T D\mathbf{y}\right) = \prod_{J=1}^n (1 + \alpha_j)^{-1/2} \exp\left(\sum_{k=1}^n \alpha_k y_k^2 / (1 + \alpha_k)\right),$$

with $\mathbf{y} = (y_1, \ldots, y_n)^T$. Therefore

$$\exp\left((-1/4)\tilde{\Delta}_n\right) \exp\left(\mathbf{y}^T D\mathbf{y}\right)$$
$$= [\det(I + D)]^{-1/2} \exp\left(\mathbf{y}^T D(I + D)^{-1}\mathbf{y}\right). \tag{4.7.5}$$

We now remove the requirement that $I + D$ is positive definite and only require the positivity of $\det(I + D)$. Given a symmetric matrix S then there is an orthogonal matrix O such that $S = ODO^T$ with D diagonal. Furthermore $\det(I + D) > 0$ if and only if $\det(I + S) > 0$. Indeed, $\det(I + S) > 0$ since $\|S\| < 1$. With $\mathbf{x} = O\mathbf{y}$ we see that

$$\mathbf{y}^T D(I + D)^{-1}\mathbf{y} = \mathbf{y}^T O^T O D O^T O(I + D)^{-1} O^{-1} O\mathbf{y} = \mathbf{x}^T S[I + S]^{-1}\mathbf{x}$$

Since the Laplacian is invariant under orthogonal transformations we then transform (4.7.5) to

$$[\det(I + S)]^{-1/2} \exp\left(\mathbf{x}^T S(I + S)^{-1}\mathbf{x}\right)$$
$$= \exp\left((-1/4)\tilde{\Delta}_n\right) \exp\left(\mathbf{x}^T S\mathbf{x}\right)$$

Now use

$$\mathbf{x}^T S\mathbf{x} = \sum_{i=1}^n s_{ii} x_i^2 + 2 \sum_{1 \le i < j \le n} s_{ij} x_i x_j$$

and expand the exponential of the above quadratic form as

$$\exp\left(\mathbf{x}^T S \mathbf{x}\right)$$

$$= \sum_K \left[\prod_{1 \le i \le j \le n} (s_{ij}/2)^k / k_{ij}! \right] 2^{-\operatorname{tr} K} (2x_1)^{k_1} (2x_2)^{k_2} \cdots (2x_n)^{k_n}.$$

Using (4.7.4) we arrive at (4.7.2).   □

Example 4.7.3 *Consider the case when S is a 2×2 matrix with $s_{i,j} = t(1 - \delta_{i,j})$. Thus $k_{1,1} = k_{2,2} = 0$, and $k_1 = k_2$. Formula (4.7.4) becomes the Mehler formula, see (Rainville, 1960, §111) and (Foata & Strehl, 1981)*

$$\sum_{n=0}^{\infty} \frac{H_n(x) H_n(y)}{2^n \, n!} \, t^n \tag{4.7.6}$$

$$= \left(1 - t^2\right)^{-1/2} \exp\left[\left(2xyt - x^2 t^2 - y^2 t^2\right) / \left(1 - t^2\right)\right].$$

The Mehler formula is the Poisson kernel for Hermite polynomials, except for a factor of $\sqrt{\pi}$, see (2.2.12) and (4.6.8).

Example 4.7.4 *Let $s_{11} = a$, $s_{12} = s_{21} = t$, $s_{22} = b$. Then (4.7.2) reduces to*

$$\sum_{j,k,l=0}^{\infty} \frac{a^j \, t^k \, b^l}{j! \, k! \, l!} \, 2^{-2j-k-2l} H_{2j+k}(x) H_{k+2l}(y) \tag{4.7.7}$$

$$= \frac{\exp\left(x^2 \left(a + ab - t^2\right) + y^2 \left(b + ab - t^2\right) + 2txy\right)}{\sqrt{1 + a + b + ab - t^2}},$$

and of course when $a = b = 0$ then (4.7.7) reduces to (4.7.6).

Remark 4.7.1 *Although the Kibble–Slepian formula has been known since 1945 several special cases of it, like (4.7.7) for example, have been established in the 1970's and 1980's, and in some instances with very complicated proofs. Most of these special cases have been collected, with some proofs, in the treatise (Srivastava & Manocha, 1984) without mentioning the Kibble–Slepian formula.*

It is important to note that the right-hand side of the Mehler formula is nonnegative for all $x, y \in \mathbb{R}$ and $t \in (-1, 1)$. In fact, the left-hand side of the Kibble–Slepian formula is also nonnegative for $x_1, \ldots, x_n \in \mathbb{R}$ and all S in a neighborhood of $S = 0$ defined by $\det(I + S) > 0$.

Remark 4.7.2 *The Kibble–Slepian formula can be proved using (4.6.41). Just replace $H_{k_1}(x_1) \cdots H_{k_n}(x_n)$ by their integral representations in (4.6.41) and evaluate all the sums to see that the right-hand side of (4.7.2) is the integral of the exponential of a quadratic form. Then diagonalize the quadratic form and evaluate the integral.*

We next evaluate the Poisson kernel for the Laguerre polynomials using a previously unpublished method. Assume that $\{\rho_n(x)\}$ is a sequence of orthogonal polynomials satisfying a three term recurrence relation of the form

$$x\rho_n(x) = f_n\rho_{n+1}(x) + g_n\rho_n(x) + h_n\rho_{n-1}(x), \tag{4.7.8}$$

and the initial conditions

$$\rho_0(x) = 1, \quad \rho_1(x) = (x - g_0)/f_0. \tag{4.7.9}$$

We look for an operator A such that

$$AP_r(x, y) = xP_r(x, y), \quad P_r(x, y) := \sum_{n=0}^{\infty} r^n \rho_n(x)\rho_n(y)/\zeta_n, \tag{4.7.10}$$

with

$$\zeta_{n+1}f_n = \zeta_n h_{n+1}. \tag{4.7.11}$$

Here A acts on y and r and x is a parameter. Thus (4.7.8) gives

$$A\sum_{n=0}^{\infty} \rho_n(x)\rho_n(y)\frac{r^n}{\zeta_n} = \sum_{n=0}^{\infty} \rho_n(y)\frac{r^n}{\zeta_n}[f_n\rho_{n+1}(x) + g_n\rho_n(x) + h_n\rho_{n-1}(x)].$$

If we can interchange the A action and the summation in the above equality we will get

$$\sum_{n=0}^{\infty} \rho_n(x)A\left[\rho_n(y)r^n/\zeta_n\right]$$

$$= \sum_{n=0}^{\infty} \rho_n(x)\left[f_{n-1}\rho_{n-1}(y)\frac{r^{n-1}}{\zeta_{n-1}} + g_n\rho_n(y)\frac{r^n}{\zeta_n} + h_{n+1}\rho_{n+1}(y)\frac{r^{n+1}}{\zeta_{n+1}}\right],$$

where ρ_{-1}/ζ_{-1} is interpreted to be zero. This suggests

$$A[\rho_n(y)r^n] = \frac{\zeta_n}{\zeta_{n-1}} f_{n-1}\rho_{n-1}(y)\, r^{n-1}$$

$$+ g_n\rho_n(y)r^n + \frac{\zeta_n}{\zeta_{n+1}} h_{n+1}\rho_{n+1}(y)\, r^{n+1}.$$

In view of (4.7.11) the above relationship is

$$A\left[\rho_n(y)r^n\right] = h_n\rho_{n-1}(y)\, r^{n-1} + g_n\rho_n(y)r^n + \frac{\zeta_n}{\zeta_{n+1}} f_n\rho_{n+1}(y)\, r^{n+1}.$$

The use of the recurrence relation (4.7.8) enables us to transform the defining relation above to the form

$$A\left[\rho_n(y)r^n\right] = h_n\rho_{n-1}(y)\, r^{n-1}\left(1 - r^2\right)$$
$$+ g_n\rho_n(y)r^n(1 - r) + ry\rho_n(y)\, r^n. \tag{4.7.12}$$

For Laguerre polynomials the Poisson kernel is a constant multiple of the function $F(x, y, r)$ defined by

$$F(x, y, r) = \sum_{n=0}^{\infty} \frac{n!\, r^n}{(\alpha + 1)_n} L_n^{(\alpha)}(x)L_n^{(\alpha)}(y). \tag{4.7.13}$$

Now (4.6.26) shows that

$$f_n = -n - 1, \quad g_n = 2n + \alpha + 1, \quad h_n = -n - \alpha,$$

and (4.7.12) becomes

$$A\left[L_n^{(\alpha)}(y)\, r^n\right] = -(n+\alpha)\left(1 - r^2\right) L_{n-1}^{(\alpha)}(y) r^{n-1}$$
$$+ (2n + \alpha + 1)(1 - r) L_n^{(\alpha)}(y) r^n + ry L_n^{(\alpha)}(y) r^n.$$

Now (4.6.26) and the observation $nr^n L_n^{(\alpha)}(y) = r\partial_r[r^n L_n^{(\alpha)}(y)]$ identify A as the partial differential operator

$$A = (r^{-1} - r)\left[y\partial_y - r\partial_r\right] + (1 - r)\left[\alpha + 1 + 2r\partial_r\right] + ry.$$

Therefore the equation $AF = xF$ is

$$F(x, y, r)$$
$$= -(1 - r)^2 \frac{\partial F(x, y, r)}{\partial r} + \frac{y}{r}\left(1 - r^2\right) \frac{\partial F(x, y, r)}{\partial y}. \tag{4.7.14}$$

The equations of the characteristics of (4.7.14) are (Garabedian, 1964)

$$\frac{dF}{F[(\alpha + 1)(r - 1) + x - yr]} = \frac{-dr}{(1 - r)^2} = \frac{r\,dy}{y\,(1 - r^2)}.$$

The second equality gives $yr(1 - r)^{-2} = C_1$, C_1 is a constant. With this solution the first equality becomes

$$\frac{dF}{F} = \left[C_1(1 - r)^2 - x + (\alpha + 1)(1 - r)\right] \frac{dr}{(1 - r)^2},$$

whose solution is

$$F(x, y, r)\,(1 - r)^{\alpha+1} \exp\left(C_2(1 - r) + x/(1 - r)\right) = C_2 \tag{4.7.15}$$

with C_2 a constant. Therefore the general solution of the partial differential equation (4.7.14) is (Garabedian, 1964)

$$F(x, y, r)\,(1 - r)^{\alpha+1} \exp\left((x + yr)/(1 - r)\right) = \phi\left(yr/(1 - r)^2\right), \tag{4.7.16}$$

for some function ϕ, which may depend on x. Let $\phi(z) = e^x g(x, z)$. Thus (4.7.16) becomes

$$F(x, y, r) = (1 - r)^{-\alpha-1} \exp\left(-r(x + y)/(1 - r)\right) g\left(x, yr/(1 - r)^2\right). \tag{4.7.17}$$

The symmetry of $F(x, y, r)$ in x and y implies

$$g(x, yr/(1 - r)^2) = g(y, xr/(1 - r)^2). \tag{4.7.18}$$

The function g is required to have a convergent power series in a neighborhood of $(0, 0)$, so we let

$$g(x, z) = \sum_{m,n=0}^{\infty} g_{m,n} x^m y^n.$$

The symmetry property (4.7.18) shows that

$$\sum_{m,n=0}^{\infty} g_{m,n} \left[x^m y^n w^n - y^m x^n w^n \right] = 0, \quad w := r(1-r)^{-2},$$

which clearly implies $g_{m,n} = 0$ if $m \neq n$, and we conclude that $g(x,z)$ must be a function of xz, that is $g(x,z) = h(xz)$ and we get

$$F(x,y,r) = (1-r)^{-\alpha-1} \exp\left(-r(x+y)/(1-r)\right) h\left(xyr/(1-r)^2\right). \quad (4.7.19)$$

To determine h replace y by y/r and let $r \to 0^+$. From (4.6.3) it follows that $r^n L_n^{(\alpha)}(y/r) \to (-y)^n/n!$ as $r \to 0^+$, hence (4.7.13) and (4.6.30) show that $h(z) = {}_0F_1(-;\alpha+1;z)$ and we have established the following theorem.

Theorem 4.7.5 *The Poisson kernel*

$$\sum_{n=0}^{\infty} \frac{n!\, r^n}{(\alpha+1)_n} L_n^{(\alpha)}(x) L_n^{(\alpha)}(y)$$
$$= (1-r)^{-\alpha-1} \exp\left(-r(x+y)/(1-r)\right) {}_0F_1\left(-;\alpha+1;xyr/(1-r)^2\right),$$
$$(4.7.20)$$

holds.

The bilinear generating function (4.7.20) is called the Hille–Hardy formula. One can also prove (4.7.20) using (4.6.42) and (4.6.30). Indeed, the left-hand side of (4.7.20) is

$$e^y r^{-\alpha/2} y^{-\alpha} \int_0^{\infty} e^u J_\alpha\left(2\sqrt{yu}\right) e^{ru} J_\alpha\left(2\sqrt{yru}\right) du$$

$$= e^y r^{-\alpha/2} y^{-\alpha} \int_0^{\infty} J_\alpha\left(2\sqrt{yu}\right) J_\alpha\left(2\sqrt{yru}\right) \exp\left((1-r)u^2\right) du,$$

and the result follows from Weber's second exponential integral

$$\int_0^{\infty} \exp\left(-p^2 u^2\right) J_\nu(au) J_\nu(bu)\, u\, du = \frac{1}{2p^2} \exp\left(-\frac{a^2+b^2}{4p^2}\right) I_\nu\left(\frac{ab}{2p^2}\right),$$
$$(4.7.21)$$

(Watson, 1944, (13.31.1)).

A general multilinear generating function for Laguerre polynomials and confluent hypergeometric functions was given in (Foata & Strehl, 1981). It generalizes an old result of A. Erdélyi. Other related and more general generating functions are in (Koelink & Van der Jeugt, 1998).

Motivated by the possibility of the Poisson kernels for Hermite and Laguerre polynomials, Sarmanov, Sarmanov and Bratoeva, considered series of the form

$$f(x,y) := \sum_{n=0}^{\infty} \frac{c_n \, n!}{\Gamma(\alpha + n + 1)} \, L_n^{(\alpha)}(x) L_n^{(\alpha)}(y), \quad \{c_n\} \in \ell^2, \; c_0 = 1 \qquad (4.7.22)$$

$$g(x,y) := \sum_{n=0}^{\infty} \frac{c_n}{2^n \, n!} \, H_n(x) H_n(y), \quad \{c_n\} \in \ell^2, \; c_0 = 1 \qquad (4.7.23)$$

and characterized the sequences $\{c_n\}$ which make $f \geq 0$ or $g \geq 0$.

Theorem 4.7.6 ((Sarmanov & Bratoeva, 1967)) *The orthogonal series $g(x,y)$ is nonnegative for all $x, y \in \mathbb{R}$ if and only if there is a probability measure μ such that*

$$c_n = \int_{-1}^{1} t^n \, d\mu(t). \qquad (4.7.24)$$

Theorem 4.7.7 ((Sarmanov, 1968)) *The series $f(x,y)$ is nonnegative for all $x \geq 0$, $y \geq 0$ if and only if there exists a probability measure μ such that $c_n = \int_0^1 t^n \, d\mu(t)$.*

Askey gave a very intuitive argument to explain the origins of Theorems 4.7.6 and 4.7.7 in (Askey, 1970b).

It is clear that the sequences $\{c_n\}$ which make $g(x,y) \geq 0$, for $x, y \in \mathbb{R}$ form a convex subset of ℓ^2 which we shall denote by $\mathcal{C}_1$. Theorem 4.7.6 shows that the extreme points of this set are sequences satisfying (4.7.24) when μ is a singleton, i.e., $c_n = t^n$ for some $t \in (-1, 1)$. In other words, Mehler's formula corresponds to the cases when $\{c_n\}$ is an extreme point of $\mathcal{C}_1$. Similarly, in the Hille–Hardy formula $\{c_n\}$ is an extreme point of the set of $\{c_n\}$, $\{c_n\} \in \ell^2$, and $f(x,y) \geq 0$ for all $x \geq 0$, $y \geq 0$. The bilinear formulas for Jacobi or ultraspherical polynomials have a more complicated structure.

Theorem 4.7.8 *The Jacobi polynomials have the bilinear generating functions*

$$\sum_{n=0}^{\infty} \frac{n! \, (\alpha + \beta + 1)_n}{(\alpha + 1)_n \, (\beta + 1)_n} \, t^n \, P_n^{(\alpha,\beta)}(x) \, P_n^{(\alpha,\beta)}(y)$$

$$= (1 + t)^{-\alpha - \beta - 1} \, F_4 \left(\begin{matrix} (\alpha + \beta + 1)/2, (\alpha + \beta + 2)/2 \\ \alpha + 1, \beta + 1 \end{matrix} \middle| A, B \right), \qquad (4.7.25)$$

and

$$\sum_{n=0}^{\infty} \frac{n! \, (\alpha + \beta + 1)_n}{(\alpha + 1)_n \, (\beta + 1)_n} \, (2n + \alpha + \beta + 1) \, t^n \, P_n^{(\alpha,\beta)}(x) \, P_n^{(\alpha,\beta)}(y)$$

$$= \frac{(\alpha + \beta + 1)(1 - t)}{(1 + t)^{\alpha + \beta + 2}} \, F_4 \left(\begin{matrix} (\alpha + \beta + 2)/2, (\alpha + \beta + 3)/2 \\ \alpha + 1, \beta + 1 \end{matrix} \middle| A, B \right), \qquad (4.7.26)$$

where

$$A = \frac{t(1 - x)(1 - y)}{(1 + t)^2}, \quad B = \frac{t(1 + x)(1 + y)}{(1 + t)^2}. \qquad (4.7.27)$$

Proof From (4.3.19) we see that the left-hand side of (4.7.25) is

$$\sum_{n=0}^{\infty}\sum_{k=0}^{n}\frac{(\alpha+\beta+1)_n\,(-n)_k\,(\alpha+\beta+n+1)_k}{n!\,(\alpha+1)_k\,(\beta+1)_k\,(-t)^{-n}}\left(\frac{x+y}{2}\right)^k P_k^{(\alpha,\beta)}\left(\frac{1+xy}{x+y}\right)$$

$$=\sum_{k,n=0}^{\infty}\frac{(\alpha+\beta+1)_{n+2k}\,(-1)^n\,t^{n+k}}{(n-k)!\,(\alpha+1)_k\,(\beta+1)_k}\left(\frac{x+y}{2}\right)^k P_k^{(\alpha,\beta)}\left(\frac{1+xy}{x+y}\right)$$

$$=\sum_{k=0}^{\infty}\frac{(\alpha+\beta+1)_{2k}\,t^k}{(\alpha+1)_k\,(\beta+1)_k}\left(\frac{x+y}{2}\right)^k P_k^{(\alpha,\beta)}\left(\frac{1+xy}{x+y}\right)(1+t)^{-\alpha-\beta-2k-1}$$

which, in view of (4.3.14), equals the right-hand side of (4.7.25) after applying $(2a)_{2k}=4^k(a)_k(a+1/2)_k$. Formula (4.7.26) follows by applying $2\frac{d}{dt}+\alpha+\beta$ to (4.7.25). $\qquad\square$

The special case $y=-1$ of (4.7.25) and (4.7.26) are (4.3.1) and (4.3.2), respectively.

Remark 4.7.3 *It is important to note that (4.7.26) is essentially the Poisson kernel for Jacobi polynomials and is positive when $t \in [0, 1]$, and $x, y \in [-1, 1]$ when $\alpha > -1$, $\beta > -1$. The kernel in (4.7.25) is also positive for $t \in [0, 1]$, and $x, y \in [-1, 1]$ but in addition to $\alpha > -1$, $\beta > -1$ we also require $\alpha + \beta + 1 \geq 0$. One can generate other positive kernels by integrating (4.7.25) or (4.7.26) with respect to positive measures supported on subsets of $[0, 1]$, provided that both sides are integrable and interchanging summation and integration is justified. Taking nonnegative combinations of these kernels also produces positive kernels.*

A substitute in the case of Jacobi polynomials is the following.

Theorem 4.7.9 *Let $\alpha \geq \beta$ and either $\beta \geq -1/2$ or $\alpha \geq -\beta$, $\beta > -1$, and assume $\sum\limits_{n=0}^{\infty} |a_n| < \infty$. Then*

$$f(x,y)=\sum_{n=0}^{\infty}a_n\frac{P_n^{(\alpha,\beta)}(x)}{P_n^{(\alpha,\beta)}(1)}\frac{P_n^{(\alpha,\beta)}(y)}{P_n^{(\alpha,\beta)}(1)}\geq 0,\quad 1\leq x,\,y\leq 1,\qquad (4.7.28)$$

if and only if

$$f(x,1)\geq 0,\quad x\in[-1,1].\qquad (4.7.29)$$

When $\alpha \geq \beta \geq -1/2$, this follows from Theorem 9.6.1. Gasper (Gasper, 1972) proved the remaining cases when $-1 < \beta < -1/2$. The remaining cases, namely $\alpha = -\beta = 1/2$ and $\alpha = \beta = -1/2$, are easy. When $\alpha = \beta$, Weinberger proved Theorem 4.7.9 from a maximum principle for hyperbolic equations. The conditions on α, β in Theorem 4.7.9 are best possible, (Gasper, 1972). For applications to discrete Banach algebras (convolution structures), see (Gasper, 1971). Theorem 4.7.9 gives the positivity of the generalized translation operator associated with Jacobi series.

In the case of ultraspherical polynomials, the following slight refinement is in (Bochner, 1954).

Theorem 4.7.10 *The inequality*

$$f_r(x,y) := \sum r^n a_n \frac{n+\nu}{\nu} C_n^\nu(x) C_n^\nu(y) \geq 0, \qquad (4.7.30)$$

holds for all $x, y \in [-1,1]$, $0 \leq r < 1$, *and* $\nu > 0$, *if and only if*

$$a_n = \int_{-1}^{1} \frac{C_n^\nu(x)}{C_n^\nu(1)} \, d\alpha(x),$$

for some positive measure α.

In an e-mail dated January 11, 2004, Christian Berg kindly informed me of work in progress where he proved the following generalizations of Theorems 4.7.6 and 4.7.7.

Theorem 4.7.11 *Let* $\{p_n\}$ *be orthonormal with respect to* μ *and assume that* $f(x,y) \geq 0$, $\mu \times \mu$ *almost everywhere, where* $f(x,y) := \sum_{n=0}^{\infty} c_n p_n(x) p_n(y)$:

1. *If the support of* μ *is unbounded to the right and left, then* c_n *is a moment sequence of a positive measure supported in* $[-1,1]$.
2. *If the support of* μ *is unbounded and contained in* $[0,\infty)$, *then* c_n *is a moment sequence of a positive measure suported in* $[0,1]$.

Nonnegative Poisson kernels give rise to positive linear approximation operators. Let $E \subset \mathbb{R}$ be compact and denote the set of continuous functions on E by $C[E]$. Let L_n be a sequence of positive linear operators mapping $C[E]$ into $C[E]$. Assume that $(L_n e_j)(x) \to e_j(x)$, uniformly on E, for $j = 0, 1, 2$, where $e_0(x) = 1$, $e_1(x) = x$, $e_2(x) = x^2$. Korovkin's theorem asserts that the above assumptions imply that $(L_n f)(x) \to f(x)$ uniformly for all $f \in C[E]$. For a proof see (DeVore, 1972).

Theorem 4.7.12 *Let* $p_n(x)$ *be orthonormal on a compact set* E *with respect to a probability measure* μ. *Then*

$$\lim_{r \to 1^-} \int_E P_r(x,y) f(y) \, d\mu(y) = f(x), \qquad (4.7.31)$$

for all $f \in C[E]$. *Moreover for a given* f, *the convergence is uniform on* E.

Proof Define the operators

$$(L_r f)(x) = \int_E P_r(x,y) f(y) \, d\mu(y).$$

A calculation and Parseval's theorem imply $\lim_{r \to 1^-} (L_r e_j)(x) = e_j(x)$, for $j = 0, 1, 2$ uniformly for $x \in E$. Let $\{r_k\}$ be sequence from $(0,1)$ so that $\lim_{k \to \infty} r_k = 1$. Then $(L_{r_k} e_j)(x) \to e_j(x)$, uniformly on E for $j = 0, 1, 2$. Since this holds for all such sequences, then (4.7.31) follows. $\qquad \square$

4.8 Asymptotics and Expansions

In this section we record asymptotic formulas for Jacobi, Hermite, and Laguerre polynomials. We also give the expansion of a plane wave, e^{ixy}, in a series of Jacobi polynomials.

We start with the expansion of a plane wave e^{ixy} in a series of ultraspherical and Jacobi polynomials. Let $\alpha > -1$ and $\beta > -1$ and set

$$e^{xy} \sim \sum_{n=0}^{\infty} c_n P_n^{(\alpha,\beta)}(x).$$

We now evaluate the coefficients c_n.

Lemma 4.8.1 *We have for* $\operatorname{Re}\alpha > -1$, $\operatorname{Re}\beta > -1$,

$$\int_{-1}^{1} e^{xy}(1-x)^{\alpha}(1+x)^{\beta} P_n^{(\alpha,\beta)}(x)\,dx = \frac{y^n}{n!}\,e^{-y}$$

$$\times \frac{2^{\alpha+\beta+n+1}\Gamma(\alpha+n+1)\Gamma(\beta+n+1)}{\Gamma(\alpha+\beta+2n+1)}\,{}_1F_1\left(\begin{array}{c}\beta+n+1\\\alpha+\beta+2n+2\end{array}\bigg|\,2y\right).$$

(4.8.1)

Proof Substitute for $P_n^{(\alpha,\beta)}$ from (4.2.8) in the above integral. The right-hand side of (4.8.1) becomes, after n integrations by parts,

$$\frac{(y/2)^n}{n!}\int_{-1}^{1} e^{xy}(1-x)^{\alpha+n}(1+x)^{\beta+n}\,dx$$

$$= \frac{(y/2)^n e^{-y}}{n!}\sum_{k=0}^{\infty}\frac{y^k}{k!}\int_{-1}^{1}(1-x)^{\alpha+n}(1+x)^{\beta+n+k}\,dx$$

$$= \frac{(y/2)^n}{n!}\,e^{-y}2^{\alpha+\beta+2n+1}\sum_{k=0}^{\infty}\frac{y^k}{k!}\,2^k\frac{\Gamma(\beta+n+k+1)\Gamma(\alpha+n+1)}{\Gamma(\alpha+\beta+2n+k+2)},$$

and the lemma follows.   □

Theorem 4.8.2 *For* $\alpha > -1$, $\beta > -1$, *we have*

$$e^{xy} = \sum_{n=0}^{\infty}\frac{\Gamma(\alpha+\beta+n+1)}{\Gamma(\alpha+\beta+2n+1)}\,(2y)^n e^{-y}$$

(4.8.2)

$$\times\,{}_1F_1\left(\begin{array}{c}\beta+n+1\\\alpha+\beta+2n+2\end{array}\bigg|\,2y\right) P_n^{(\alpha,\beta)}(x).$$

Proof Let $g(x)$ denote the right-hand side of (4.8.2). From (4.2.13) and Theorem 4.3.1 we see that $\Gamma(s)M_n = \Gamma(s+n)/\Gamma(n+1) \approx n^{s-1}$ if $s \geq -\frac{1}{2}$ and $M_n \sim Cn^{-1/2}$ if $s \leq -\frac{1}{2}$. On the other hand, as $n \to \infty$, after using the duplication formula (1.3.8) we get, for fixed y,

$$\frac{\Gamma(\alpha+\beta+n+1)}{\Gamma(\alpha+\beta+2n+1)} \, {}_1F_1\left(\begin{array}{c}\beta+n+1\\\alpha+\beta+2n+2\end{array}\bigg| 2y\right) M_n$$

$$= O\left(\frac{\Gamma(\alpha+\beta+n+1)M_n}{2^{\alpha+\beta+2n}\,\Gamma(n+(\alpha+\beta+1)/2)\,\Gamma(n+1+(\alpha+\beta)/2)}\right)$$

$$= O\left(\frac{n^{(\alpha+\beta)/2}M_n}{2^{\alpha+\beta+2n}\,\Gamma(n+(\alpha+\beta+1)/2)}\right).$$

Thus the series in (4.8.2) converges uniformly in x, for $x \in [-1,1]$. By Lemma 4.2.1, Lemma 4.8.1, and (4.1.2), the function $e^{xy} - g(x)$ has zero Fourier–Jacobi coefficients. The result now follows from the completeness of the Jacobi polynomials in $L_2\left(-1,1,(1-x)^{\alpha}(1+x)^{\beta}\right)$. $\qquad\square$

The expansion (4.8.2) is called the plane wave expansion because with $y \to iy$ it gives the Fourier–Jacobi expansion of a plane wave.

The special case $\alpha = \beta$ of (4.8.2) is

$$e^{ixy} = \Gamma(\nu)(z/2)^{-\nu}\sum_{n=0}^{\infty} i^n(\nu+n)J_{\nu+n}(y)C_n^{\nu}(x). \tag{4.8.3}$$

The orthogonality of the ultraspherical polynomials implies

$$J_{\nu+n}(z) = \frac{(-i)^n n!\,(z/2)^{\nu}}{\Gamma(\nu+1/2)\Gamma(1/2)(2\nu)_n}\int_{-1}^{1} e^{izy}\left(1-y^2\right)^{\nu-1/2}C_n^{\nu}(y)\,dy, \tag{4.8.4}$$

for $\mathrm{Re}\,\nu > -1/2$. Formula (4.8.4) is called "Gegenbauer's generalization of Poisson's integral" in (Watson, 1944). Note that (4.8.4) can be proved directly from (4.5.1) and (1.3.31). It can also be proved directly using the Rodriguez formula and integration by parts. The cases n even and n odd of (4.8.4) are

$$J_{\nu+2n}(z) = \frac{(-1)^n (2n)!\,(z/2)^{\nu}}{\Gamma(\nu+1/2)\Gamma(1/2)(2\nu)_{2n}}$$

$$\times \int_{0}^{\pi}\cos(z\cos\phi)(\sin\varphi)^{2\nu}C_{2n}^{\nu}(\cos\varphi)\,d\varphi, \tag{4.8.5}$$

and

$$J_{\nu+2n+1}(z) = \frac{(2n+1)!\,(z/2)^{\nu}}{\Gamma(\nu+1/2)\Gamma(1/2)(2\nu)_{2n+1}}$$

$$\times \int_{0}^{\pi}\sin(z\cos\varphi)(\sin\varphi)^{2\nu}C_{2n+1}^{\nu}(\cos\varphi)\,d\varphi. \tag{4.8.6}$$

The next theorem is a Mehler–Heine-type formula for Jacobi polynomials.

Theorem 4.8.3 *Let $\alpha, \beta \in \mathbb{R}$. Then*

$$\lim_{n\to\infty} n^{-\alpha}P_n^{(\alpha,\beta)}(\cos(z/n))$$

$$= \lim_{n\to\infty} n^{-\alpha}P_n^{(\alpha,\beta)}\left(1-z^2/2n^2\right) = (z/2)^{-\alpha}J_{\alpha}(z). \tag{4.8.7}$$

The limit in (4.8.7) is uniform in z on compact subsets of $\mathbb{C}$.

Proof From (4.1.1) it follows that the left-hand side of (4.8.7) is

$$\lim_{n\to\infty} \sum_{k=0}^{n} \frac{n^{-\alpha}(\alpha+\beta+n+1)_k}{k!\,\Gamma(\alpha+k+1)} \frac{\Gamma(\alpha+n+1)}{\Gamma(n-k+1)} \left[-\sin^2\left(\frac{z}{2n}\right)\right]^k,$$

and (4.8.7) follows from the dominated convergence theorem. $\qquad\square$

An important consequence of Theorem 4.8.3 is the following

Theorem 4.8.4 *For real* α, β *we let*

$$x_{n,1}(\alpha,\beta) > x_{n,2}(\alpha,\beta) > \cdots > x_{n,n}(\alpha,\beta)$$

be the zeros of $P_n^{(\alpha,\beta)}(x)$ *in* $[-1,1]$. *With* $x_{n,k}(\alpha,\beta) = \cos(\theta_{n,k}(\alpha,\beta))$, $0 < \theta_{n,k}(\alpha,\beta) < \pi$, *we have*

$$\lim_{n\to\infty} n\,\theta_{n,k}(\alpha,\beta) = j_{\alpha,k}, \qquad (4.8.8)$$

where $j_{\alpha,k}$ *is the kth positive zero of* $J_\alpha(z)$.

Theorem 4.8.5 *For* $\alpha \in \mathbb{R}$, *the limiting relation*

$$\lim_{n\to\infty} n^{-\alpha} L_n^{(\alpha)}(z/n) = z^{-\alpha/2} J_\nu\left(2\sqrt{2}\right) \qquad (4.8.9)$$

holds uniformly for z in compact subsets of $\mathbb{C}$.

Theorem 4.8.6 *For* $\alpha, \beta \in \mathbb{R}$, *we have*

$$P_n^{(\alpha,\beta)}(x) = (x-1)^{\alpha/2}(x+1)^{-\beta/2} \left\{(x+1)^{1/2} + (x-1)^{1/2}\right\}^{\alpha+\beta}$$

$$\times \frac{(x^2-1)^{-1/4}}{\sqrt{2\pi n}} \left\{x + (x^2-1)^{1/2}\right\}^{n+1/2} \{1+o(1)\}, \qquad (4.8.10)$$

for $x \in \mathbb{C} \setminus [-1,1]$. *The above limit relation holds uniformly in x on compact subsets of* $\mathbb{C}$.

The proof is similar to the proof of Theorem 4.3.1. We next state several theorems without proofs. Proofs and references are Szegő's book, (Szegő, 1975, §§8.1, 8.2).

Theorem 4.8.7 (Hilb-type asymptotics) *Let* $\alpha > -1$, $\beta \in \mathbb{R}$. *Then*

$$\left(\sin\frac{\theta}{2}\right)^\alpha \left(\cos\frac{\theta}{2}\right)^\beta P_n^{(\alpha,\beta)}(\cos\theta)$$

$$= \frac{\Gamma(n+\alpha+1)}{n!\,N^\alpha} \left(\frac{\theta}{\sin\theta}\right)^{1/2} J_\alpha(N\theta) + \theta^u O\left(n^v\right), \qquad (4.8.11)$$

as $n \to \infty$, *where N is as in Theorem 4.3.1, and*

$$\begin{aligned} u &= 1/2, & v &= -1/2, & \text{if } c/n \le \theta \le \pi - \epsilon, \\ u &= \alpha+2, & v &= \alpha, & \text{if } 0 < \theta \le cn^{-1}; \end{aligned} \qquad (4.8.12)$$

c and ϵ are fixed numbers.

Theorem 4.8.8 (Fejer) *For $\alpha \in \mathbb{R}$, $x > 0$,*

$$L_n^{(\alpha)}(x) = \frac{e^{x/2}}{\sqrt{\pi}} x^{-\alpha/2-1/4} n^{\alpha/2-1/4} \cos\left\{2(nx)^{1/2} - \alpha\pi/2 - \alpha/4\right\}$$
$$+ O\left(n^{\alpha/2-3/4}\right), \tag{4.8.13}$$

as $n \to \infty$. The O bound is uniform for x in any compact subset of $(0, \infty)$.

Theorem 4.8.9 (Perron) *For $\alpha \in \mathbb{R}$,*

$$L_n^{(\alpha)}(x) = \frac{e^{x/2}}{2\pi} (-x)^{-\alpha/2-1/4} n^{\alpha/2-1/4} \exp\left\{2(-nx)^{1/2}\right\}, \tag{4.8.14}$$

for $x \in \mathbb{C} \setminus (0, \infty)$. In (4.8.13), the branches of $(-x)^{-\alpha/2-1/4}$ and $(-x)^{1/2}$ are real and positive for $x < 0$.

Theorem 4.8.10 (Hilb-type asymptotics) *When $\alpha > -1$ and $x > 0$,*

$$e^{-x/2} x^{\alpha/2} L_n^{(\alpha)}(x) = N^{-\alpha/2} \frac{\Gamma(\alpha+n+1)}{n!} J_\alpha\left(2(Nx)^{1/2}\right) + O\left(n^{\alpha/2-3/4}\right), \tag{4.8.15}$$

where $N = n + (\alpha + 1)/2$.

Theorem 4.8.11 *Let c and C be positive constants. Then for $\alpha > -1$ and $c/n \le x \le C$, we have*

$$L_n^{(\alpha)}(x) = \frac{e^{x/2}}{\sqrt{\pi}} x^{-\alpha/2-1/4} n^{\alpha/2-1/4}$$
$$\times \left\{\cos\left[2(nx)^{1/2} - \alpha\pi/2 - \pi/4\right] + (nx)^{-1/2} \mathcal{O}(1)\right\}. \tag{4.8.16}$$

Theorem 4.8.12 *For x real, we have*

$$\frac{\Gamma(n/2+1)}{\Gamma(n+1)} e^{-x^2/2} H_n(x) = \cos\left(N^{\frac{1}{2}}x - n\pi/2\right)$$
$$+ \frac{x^3}{6} N^{-\frac{1}{2}} \sin\left(N^{\frac{1}{2}}x - n\pi/2\right) + \mathcal{O}\left(n^{-1}\right), \tag{4.8.17}$$

where $N = 2n + 1$. The bound for the error term holds uniformly in x on every compact interval.

Theorem 4.8.13 *The asymptotic formula in (4.8.17) holds in the complex x-plane if we replace the remainder term by $\exp\left\{N^{\frac{1}{2}} |\operatorname{Im}(x)|\right\} \mathcal{O}(n^{-p})$. This is true uniformly for $|x| \le R$ where R is an arbitrary fixed positive number.*

Finally we record another type of asymptotic formulas requiring a more elaborate consideration.

Theorem 4.8.14 (Plancherel–Rotach-type) *Let $\alpha \in \mathbb{R}$ and ϵ and ω be fixed positive numbers. We have*

(a) *for $x = (4n + 2\alpha + 2)\cos^2\phi$, $\epsilon \leq \phi \leq \pi/2 - \epsilon n^{-\frac{1}{2}}$,*

$$e^{-x/2}L_n^{(\alpha)}(x) = (-1)^n(\pi\sin\phi)^{-\frac{1}{2}}x^{-\alpha/2-\frac{1}{4}}n^{\alpha/2-\frac{1}{4}}$$
$$\times \left\{\sin[n + (\alpha+1)/2)(\sin 2\phi - 2\phi) + 3\pi/4] + (nx)^{-\frac{1}{2}}\mathcal{O}(1)\right\};$$
(4.8.18)

(b) *for $x = (4n + 2\alpha + 2)\cosh^2\phi$, $\epsilon \leq \phi \leq \omega$,*

$$e^{-x/2}L_n^{(\alpha)}(x) = \frac{1}{2}(-1)^n(\pi\sinh\phi)^{-\frac{1}{2}}x^{-\alpha/2-\frac{1}{4}}n^{\alpha/2-\frac{1}{4}}$$
$$\times \exp\{(n + (\alpha+1)/2)(2\phi - \sinh 2\phi)\}\left\{1 + \mathcal{O}\left(n^{-1}\right)\right\};$$
(4.8.19)

(c) *for $x = 4n + 2\alpha + 2 - 2(2n/3)^{\frac{1}{3}}t$, t complex and bounded,*

$$e^{-x/2}L_n^{(\alpha)}(x) = (-1)^n\pi^{-1}2^{-\alpha-\frac{1}{3}}3^{\frac{1}{2}}n^{-\frac{1}{2}}\left\{A(t) + \mathcal{O}\left(n^{-\frac{2}{3}}\right)\right\} \quad (4.8.20)$$

where $A(t)$ is Airy's function defined in (1.3.32), (1.3.34).

Moreover, in the above formulas the $\mathcal{O}$-terms hold uniformly.

Theorem 4.8.15 *Let ϵ and ω be fixed positive numbers. We have*

(a) *for $x = (2n+1)^{\frac{1}{2}}\cos\phi$, $\epsilon \leq \phi \leq \pi - \epsilon$,*

$$e^{-x^2/2}H_n(x) = 2^{n/2+\frac{1}{4}}(n!)^{\frac{1}{2}}(\pi n)^{-\frac{1}{4}}(\sin\phi)^{-\frac{1}{2}}$$
$$\times \left\{\sin\left[\left(\frac{n}{2} + \frac{1}{4}\right)(\sin(2\phi) - 2\phi) + \frac{3\pi}{4}\right] + \mathcal{O}\left(n^{-1}\right)\right\};$$
(4.8.21)

(b) *for $x = (2n+1)^{\frac{1}{2}}\cosh\phi$, $\epsilon \leq \phi \leq \omega$,*

$$e^{-x^2/2}H_n(x) = 2^{n/2-\frac{3}{4}}(n!)^{\frac{1}{2}}(\pi n)^{-\frac{1}{4}}(\sinh\phi)^{-\frac{1}{2}}$$
$$\times \exp\left[\left(\frac{n}{2} + \frac{1}{4}\right)(2\phi - \sinh 2\phi)\right]\left\{1 + \mathcal{O}\left(n^{-1}\right)\right\};$$
(4.8.22)

(c) *for $x = (2n+1)^{\frac{1}{2}} - 2^{-\frac{1}{2}}3^{-\frac{1}{3}}n^{-\frac{1}{6}}t$, t complex and bounded,*

$$e^{-x^2/2}H_n(x) = 3^{\frac{1}{3}}\pi^{\frac{3}{4}}2^{n/2+\frac{1}{4}}(n!)^{\frac{1}{2}}n^{1/12}\left\{A(t) + \mathcal{O}\left(n^{-\frac{2}{3}}\right)\right\}. \quad (4.8.23)$$

In all these formulas, the $\mathcal{O}$-terms hold uniformly.

For complete asymptotic expansions, proofs and references to the literature, the reader may consult §8.22 in (Szegő, 1975).

Baratella and Gatteschi proved the following uniform asymptotics of $P_n^{(\alpha,\beta)}(\cos\theta)$ using the Liouville-Stekloff method (Szegő, 1975, §8.6).

Theorem 4.8.16 ((Baratella & Gatteschi, 1988)) *Let*

$$N = n + (\alpha + \beta + 1)/2, \quad A = 1 - 4\alpha^2, \quad B = 1 - 4\beta^2,$$

$$a(\theta) = \frac{2}{\theta} - \cot\left(\frac{\theta}{2}\right), \quad b(\theta) = \tan\left(\frac{\theta}{2}\right), \quad f(\theta) = N\theta + \frac{1}{16N}\left[Aa(\theta) + Bb(\theta)\right],$$

$$u_n^{(\alpha,\beta)}(\theta) = \left(\sin\left(\frac{\theta}{2}\right)\right)^{\alpha+1/2} \left(\cos\left(\frac{\theta}{2}\right)\right)^{\beta+1/2} P_n^{(\alpha,\beta)}(\cos\theta),$$

$$F(\theta) = F_1(\theta) + F_2(\theta),$$

$$F_1(\theta) = \frac{1}{2}\frac{Aa'''(\theta) + Bb'''(\theta)}{16N^2 + Aa'(\theta) + Bb'(\theta)}$$
$$- \frac{3}{4}\left[\frac{Aa''(\theta) + Bb''(\theta)}{16N^2 + Aa'(\theta) + Bb'(\theta)}\right]^2,$$

$$F_2(\theta) = \frac{A}{2\theta^2}\frac{\theta Aa'(\theta) + \theta Bb'(\theta) - Aa(\theta) - Bb(\theta)}{16\,N^2\theta + Aa(\theta) + Bb(\theta)}$$
$$\times \left[1 + \frac{1}{2}\frac{\theta Aa'(\theta) + \theta Bb'(\theta) - Aa(\theta) - Bb(\theta)}{16\,N^2\theta + Aa(\theta) + Bb(\theta)}\right]$$
$$+ \frac{[Aa'(\theta) + Bb'(\theta)]^2}{256\,N^2}.$$

With

$$\Delta(t,\theta) := J_\alpha(f(\theta))\,Y_\alpha(f(t)) - J_\alpha(f(t))\,Y_\alpha(f(\theta))$$

$$I^{(\alpha,\beta)} := \frac{\pi}{2}\int_0^\theta \left[\frac{f(t)}{f'(t)}\right]^{1/2} \Delta(t,\theta)\,F(t)\,u_n^{(\alpha,\beta)}(t)\,dt,$$

we have

$$\left[\frac{f'(\theta)}{f(\theta)}\right]^{1/2} u_n^{(\alpha,\beta)}(\theta) = C_1\,J_\alpha(f(\theta)) - I^{(\alpha,\beta)},$$

where

$$C_1 = \frac{\Gamma(n + \alpha + 1)}{\sqrt{2}\,N^\alpha\,n!}\left[1 + \frac{1}{16\,N^2}\left(\frac{A}{6} + \frac{B}{2}\right)\right]^{-\alpha}.$$

Furthermore, for $\alpha, \beta \in (-1/2, 1/2)$, $I^{(\alpha,\beta)}$ has the estimate

$$\left|I^{(\alpha,\beta)}\right| \leq \begin{cases} \frac{\theta^\alpha}{N^4}\binom{n+\alpha}{n}(0.00812\,A + 0.0828\,B), & 0 < \theta < \theta^* \\[2mm] \frac{\theta^{1/2}}{N^{\alpha+7/2}}\binom{n+\alpha}{n}(0.00526\,A + 0.535\,B), & \theta^* \leq \theta \leq \pi/2, \end{cases}$$

where θ^* is the root of the equation $f(\theta) = \pi/2$.

4.9 Relative Extrema of Classical Polynomials

In this section, we study properties of relative extrema of ultraspherical, Hermite and Laguerre polynomials and certain related functions.

Theorem 4.9.1 *Let $\mu_{n,1}, \ldots, \mu_{n,\lfloor n/2 \rfloor}$ be the relative extrema of $\{C_n^\nu(x)\}$ in $(0,1)$ arranged in decreasing order of x. Then, for $n > 1$, we have*

$$1 > \mu_{n,1}^{(\nu)} > \mu_{n,2}^{(\nu)} > \cdots > \mu_{\lfloor n/2 \rfloor}^{(\nu)}, \quad n \geq 2, \tag{4.9.1}$$

when $\nu > 0$. When $\nu < 0$, then

$$\mu_{n,1}^{(\nu)} < \mu_{n,2}^{(\nu)} < \cdots < \mu_{\lfloor n/2 \rfloor}^{(\nu)}. \tag{4.9.2}$$

Proof Let

$$f(x) = n(n+2\nu)\,y^2(x) + \left(1 - x^2\right)(y'(x))^2, \quad y := C_n^\nu(x).$$

Then

$$
\begin{aligned}
f'(x) &= 2y'(x)\left\{\left(1 - x^2\right)y''(x) - 2xy'(x) + n(n+2\nu)\,y(x)\right\} \\
&= 4\nu x\,(y'(x))^2,
\end{aligned}
$$

where we used (4.5.8). Therefore f is increasing for $x > 0$ and decreasing for $x < 0$. The result follows since $f(x) = n(n+2\nu)\left(C_n^\nu(x)\right)^2$ when $(C_n^\nu(x))' = 0$. $\quad\square$

The corresponding result for Hermite polynomials is a limiting case of Theorem 4.9.1. In the case $\nu = 0$, all the inequality signs in (4.9.1) and (4.9.2) become equal signs as can be seen from (4.5.22).

Theorem 4.9.2 *Assume that $n > 1$. The successive maxima of $(\sin\theta)^\nu\,|C_n^\nu(\cos\theta)|$ for $\theta \in (0, \pi/2)$ form an increasing sequence if $\nu \in (0,1)$, and a decreasing sequence if $\nu > 1$.*

Proof Let $u(\theta) = (\sin\theta)^{-\nu}C_n^\nu(\cos\theta)$, $0 < \theta < \pi$. The differential equation (4.5.6) is transformed to

$$\frac{d^2u}{d\theta^2} + \phi(\theta)\,u = 0, \quad \phi(\theta) = \frac{\nu(1-\nu)}{\sin^2\theta} + (\nu+n)^2.$$

Set

$$f(\theta) = u^2(\theta) + \frac{1}{\phi(\theta)}\left(\frac{du}{d\theta}\right)^2.$$

It follows that $f'(\theta) = -\,(u'(\theta))^2\,\phi'(\theta)/\phi^2(\theta)$. Since

$$\phi'(\theta) = 2\nu(\nu-1)\cos\theta(\sin\theta)^{-2},$$

then f increases when $\nu \in (0,1)$ and decreases if $\nu > 1$. But $f(\theta) = (u(\theta))^2$ when $u'(\theta) = 0$. This completes the proof. $\quad\square$

The next theorem compares the maxima $\mu_{n,k}^{(\nu)}$ for different values of n and was first proved in (Szegő, 1950c) for $\nu = 1/2$ and in (Szász, 1950) for general ν.

Theorem 4.9.3 *The relative maxima $\mu_{n,k}^{(\nu)}$ decreases with n for $\nu > -1/2$, that is*

$$\mu_{n,k}^{(\nu)} > \mu_{n+1,k}^{(\nu)}, \quad n = k+1, k+2, \ldots . \tag{4.9.3}$$

Proof Apply (4.5.5) and (4.5.10) to get

$$(1-x)\frac{d}{dx}\,C_n^\nu(x) = 2\nu\,C_{n-1}^{\nu+1}(x) - 2\nu\,C_{n-1}^{\nu+1}(x) - n\,C_n^\nu(x).$$

Therefore

$$(1+x)\frac{d}{dx}\left\{C_{n+1}^\nu(x) - C_n^\nu(x)\right\} = (n+2\nu)\,C_n^\nu(x) + (n+1)\,C_{n+1}^\nu(x), \tag{4.9.4}$$

follows from (4.5.10). In (4.9.4), replace x by $-x$ and use $C_n^\nu(-x) = (-1)^n C_n^\nu(x)$ to obtain

$$\left(1-x^2\right)\left[\left(y_{n+1}'(x)\right)^2 - \left(y_n'(x)\right)^2\right] = (n+2\nu)^2 y_n^2(x) - (n+1)^2 y_{n+1}^2(x),$$

where $y_n = C_n^\nu(x)$. Let

$$1 > z_{n,1} > z_{n,2} > \cdots > z_{n,n-1} > -1,$$

be the points when $y_n'(x) = 0$. By symmetry, it suffices to consider only the non-negative $z_{n,k}$'s. We have

$$(n+1)^2\left(\mu_{n+1,k}^{(\nu)}\right)^2 = (n+2\nu)^2 y_n^2\left(z_{n+1,k}\right) + \left(1 - z_{n+1,k}^2\right)\left(y_n'\left(z_{n+1,k}\right)\right)^2. \tag{4.9.5}$$

Consider the function

$$f(x) = (n+2\nu)^2 y_n^2(x) + \left(1-x^2\right)\left(y_n'(x)\right)^2.$$

The differential equation (4.5.8) implies

$$f'(x) = 2x(2\nu+1)\left(y_n'(x)\right)^2,$$

hence f increases with x on $(0,1)$ and the result follows from (4.9.5). □

It is of interest to note that Theorem 4.9.3 has been generalized to orthogonal Laguerre functions $\left\{e^{-x/2}L_n(x)\right\}$ in (Todd, 1950) and to orthogonal Hermite functions in (Szász, 1951).

Let $\mu_{n,k}^{(\alpha,\beta)}$ be the relative extrema of $\left|P_n^{(\alpha,\beta)}(x)\right|$. Askey conjectured that

$$\mu_{n+1,k}^{(\alpha,\beta)} < \mu_{n,k}^{(\alpha,\beta)}, \quad k = 1, \ldots, n-1, \quad \text{for } \alpha > \beta > -1/2, \tag{4.9.6}$$

in his comments on (Szegő, 1950c), see p. 221 of volume 3 of Szegő's Collected Papers. Askey also conjectured that when $\alpha = 0$, $\beta = -1$, the inequalities in (4.9.6) are reversed. Askey also noted that

$$P_n^{(0,-1)}(x) = \frac{1}{2}\left[P_n(x) + P_{n-1}(x)\right],$$

$\{P_n(x)\}$ being the Legendre polynomials. Both conjectures are also stated in (Askey, 1990). Wong and Zhang confirmed Askey's second conjecture by proving the desired result asymptotically for $n \geq 25$, then established the cases $n \leq 24$ by direct comparison of numerical values. This was done in (Wong & Zhang, 1994b). Askey's first conjecture has been verified for n sufficiently large by the same authors in (Wong & Zhang, 1994a).

4.10 The Bessel Polynomials

In view of (1.3.16)–(1.3.18) we find

$$I_{1/2}(z) = \sqrt{\frac{2}{\pi z}} \sinh z, \quad I_{-1/2}(z) = \sqrt{\frac{2}{\pi z}} \cosh z,$$

and (1.3.23) gives

$$K_{1/2}(z) = K_{-1/2}(z) = \sqrt{\frac{\pi}{2z}} e^{-z}. \tag{4.10.1}$$

Now (4.10.1) and (1.3.24) imply, by induction, that $K_{n+1/2}(z)e^z \sqrt{z}$ is a polynomial in $1/z$. We now find this polynomial explicitly. Define $y_n(x)$ by

$$y_n(1/z) = e^z z^{1/2} K_{n+1/2}(z)/\sqrt{\pi}. \tag{4.10.2}$$

Substitute for $K_{n+1}(z)$ from (4.10.2) in (1.3.22) to see that

$$y_n''(1/z) + 2z(z+1)y_n'(1/z) - n(n+1)z^2 y_n(1/z) = 0,$$

that is

$$z^2 y_n''(z) + (2z+1)y_n'(z) - n(n+1)y_n(z) = 0. \tag{4.10.3}$$

By writing $y(z) = \sum a_k z^k$ we see that the only polynomial solution to (4.10.3) is a constant multiple of the solution

$$y_n(z) = {}_2F_0(-n, n+1; -; -z/2). \tag{4.10.4}$$

The reverse polynomial

$$\theta_n(z) = z^n y_n(1/z), \tag{4.10.5}$$

also plays an inportant role. More general polynomials are

$$y_n(z; a, b) = {}_2F_0(-n, n+a-1; -; -z/b), \quad \theta_n(z; a, b) = z^n y_n(1/z; a, b). \tag{4.10.6}$$

The corresponding differential equations are

$$z^2 y'' + (az+b)y' - n(n+a-1)y = 0, \quad y = y_n(z; a, b),$$
$$z\theta'' - (2n-2+a+bz)\theta' + bn\theta = 0, \quad \theta = \theta_n(z; a, b). \tag{4.10.7}$$

The polynomials $\{y_n(z)\}$ or $\{y_n(z; a, b)\}$ will be called the Bessel polynomials while $\{\theta_n(z)\}$ and $\{\theta_n(z; a, b)\}$ will be referred to as the reverse Bessel polynomials. Clearly, $y_n(z) = y_n(z; 2, 2)$, $\theta_n(z) = \theta_n(z; 2, 2)$.

The notation and terminology was introduced in (Krall & Frink, 1949). However, the same polynomials appeared over 15 years earlier in a different notation in (Burchnal & Chaundy, 1931).

Define w_B by

$$w_B(z; a, b) = \sum_{n=0}^{\infty} \frac{(-b/z)^n}{(a-1)_n} = {}_1F_1(1; a-1; -b/z). \tag{4.10.8}$$

In the case $a = b = 2$, w_B becomes $\exp(-2/z)$.

Theorem 4.10.1 *The Bessel polynomials satisfy the orthogonality relation*

$$\frac{1}{2\pi i} \oint_C y_m(z; a, b) y_n(z; a, b) w_B(z; a, b) \, dz$$

$$= \frac{(-1)^{n+1} b}{a + 2n - 1} \frac{n!}{(a-1)_n} \delta_{m,n}, \tag{4.10.9}$$

where C is a closed contour containing $z = 0$ in its interior and w_B is as in (4.10.8).

Proof Clearly for $j \leq m$, we have

$$\frac{1}{2\pi i} \oint z^j y_m(z; a, b) w_B(z; a, b) \, dz$$

$$= \sum_{n=0}^{\infty} \frac{(-b)^n}{(a-1)_n} \sum_{k=0}^{m} \frac{(-m)_k (m + a - 1)_k}{k! \, (-b)^k} z^{k+j-n}$$

$$= \frac{(-b)^{j+1}}{(a-1)_{j+1}} \sum_{k=0}^{m} \frac{(-m)_k (m + a - 1)_k}{k! \, (a+j)_k} = \frac{(-b)^{j+1}}{(a-1)_{j+1}} \frac{(j + 1 - m)_m}{(a+j)_m},$$

by the Chu–Vandermonde sum. The factor $(j + 1 - m)_m$ is $m! \delta_{m,j}$, for $j \leq m$. Thus for $n \leq m$, the left-hand side of (4.10.9) is $\delta_{m,n}$ times

$$\frac{(-b)^{n+1} n!}{(a-1)_{n-1}(a+n)_n} \frac{(-n)_n(n + a - 1)_n}{n! \, (-b)^n},$$

which reduces to the right-hand side of (4.10.9). □

It is clear from (4.10.6) and (4.1.1) that

$$y_n(z; a, b) = \lim_{\gamma \to \infty} \frac{n!}{(\gamma + 1)_n} P_n^{(\gamma, a - \gamma)}(1 + 2\gamma z/b). \tag{4.10.10}$$

Therefore (3.3.16) gives the differential recurrence relations

$$(a + 2n - 2)z^2 y_n'(z; a, b) = n[(-a + 2n - 2)z - b]y_n(z; a, b) + bny_{n-1}(z; a, b), \tag{4.10.11}$$

and for θ_n (4.10.11) becomes

$$(a + 2n - 2)\theta_n'(z; a, b) = bn\theta_n(z; a, b) - bnz\theta_{n-1}(z; a, b). \tag{4.10.12}$$

Furthermore, (4.2.9) establishes the three term recurrence relation

$$(a + n - 1)(a + 2n - 2)y_{n+1}(z; a, b) - n(a + 2n)y_{n-1}(z; a, b)$$
$$= (a + 2n - 1)[a - 2 - (a + 2n)(a + 2n - 2)z]y_n(z; a, b). \tag{4.10.13}$$

It is clear from (4.10.13) that $\{y_n(z; a, b)\}$ are not orthogonal with respect to a positive measure. Theorems of Boas and Shohat, (Boas, Jr., 1939) and (Shohat, 1939), show that they are orthogonal with respect to a signed measure supported in $[0, \infty)$. The question of finding such a signed measure was a long-standing problem. The first construction of a signed measure with respect to which $\{y_n(x; a, b)\}$ are orthogonal was in (Durán, 1989) and (Durán, 1993). Several other measures were constructed later by various authors; see, for example, (Kwon et al., 1992). A detailed exposition of the constructions of signed orthogonality measures for $\{y_n(z; a, b)\}$ is in (Kwon, 2002).

Theorem 4.10.2 *The discriminant of the Bessel polynomial is given by*

$$D\left(y_n(x; a, b)\right) = (n!)^{2n-2} \left(-b^2\right)^{-n(n-1)/2} \prod_{j=1}^{n} j^{j-2n+2}(n + j + a - 2).$$

Proof Formula (4.10.10) gives the discriminant as a limiting case of (3.4.16). $\square$

The Rodriguez formula is

$$y_n(x; a, b) = b^{-n} x^{2-a} e^{b/x} \frac{d^n}{dx^n} \left(x^{2n+a-2} e^{-b/x}\right). \tag{4.10.14}$$

Proof With $x = 1/z$, it is easy to see that

$$\left(\frac{d}{dx}\right)^n f = (-1)^n \left(z^2 \frac{d}{dz}\right)^n f = (-1)^n z^{n+1} \frac{d^n}{dz^n} \left(z^{n-1} f\right)$$

Hence the right-hand side of (4.10.14) is

$$b^{-1} z^{a-2+n} e^{bz} (-1)^n \frac{d^n}{dz^n} \left(z^{-n-a+1} e^{-bz}\right)$$

$$= b^{-n} z^{n+a-1} e^{bz} (-1)^n \sum_{k=0}^{n} \binom{n}{k} (-b)^{n-k} e^{-bz} (n + a - 1)_k (-1)^k z^{-n-k-a+1}$$

$$= \sum_{k=0}^{n} \frac{(-n)_k (n + a - 1)_k}{k!} (-b/z)^k = y_n(x; a, b). \quad \square$$

The above proof does not seem to be in the literature.
We now discuss the zeros of Bessel polynomials.

Theorem 4.10.3

(a) *All zeros of $y_n(z; a, b)$ are simple.*
(b) *No two consecutive polynomials $y_n(z; a, b)$, $y_{n+1}(z; a, b)$ have a common zero.*
(c) *All zeros of y_{2n} are complex, while y_{2n+1} has only one real zero, $n = 0, 1, 2, \ldots$.*

Proof Part (a) follows from (4.10.7). If y_n and y_{n+1} have a common zero, say

$z = \xi$, then (4.10.12) forces $y'_{n+1}(\xi) = 0$, which contradicts (a). To prove (c), let $\phi_n(z) = e^{-z}z^n y_n(1/z)$. Thus, $\phi_n(z)$ satisfies

$$zy'' - 2ny' - zy = 0. \tag{4.10.15}$$

Clearly, $\phi_n(-z)$ also satisfies (4.10.15). What is also clear is that $\phi_n(z)$ and $\phi_n(-z)$ are linearly independent and their Wronskian is

$$\phi'_n(z)\phi_n(-z) - \phi_n(z)\frac{d}{dz}\phi_n(-z) = Cz^{2n},$$

and by equating coefficients of z^{2n} we find $C = 2(-1)^n$. Since $\phi_n(z) = e^{-z}\theta_n(z)$, we can rewrite the Wronskian in the form

$$\theta_n(z)\theta_n(-z) + \theta'_n(-z)\theta_n(z) - 2\theta_n(z)\theta_n(-z) = 2(-1)^{n+1}z^{2n}. \tag{4.10.16}$$

If θ_n has a real zero it must be negative because θ_n has positive coefficients. Let α and β be two consecutive real zeros of θ_n, then $\theta'_n(\alpha)\theta_n(-\alpha)$ and $\theta'_n(\beta)\theta_n(-\beta)$ have the same sign. But $\theta'_n(\alpha)\theta'_n(\beta) < 0$, hence $\theta_n(-\alpha)\theta_n(-\beta) < 0$, which is a contradiction because α, β must be negative.   □

Observe that (c) also follows from a similar result for K_ν, $\nu > 0$, (Watson, 1944).

Theorem 4.10.4 *Let $\{z_{n,j} : j = 1, \dots, n\}$ be the zeros of $y_n(x)$. Then*

$$\sum_{j=1}^n z_{n,j} = -1, \quad \sum_{j=1}^n z_{n,j}^{2m-1} = 0, \quad m = 2, 3, \dots, n.$$

Proof By (4.10.6) we obtain

$$\theta_n(z) = z^n {}_2F_0(-n, n+1; -, -1/2z) = \frac{(n+1)_n}{2^n} \sum_{k=0}^n \frac{(-1)_k}{k!\,(-2n)_k}(2z)^k$$

$$= \frac{(n+1)_n}{2^n} \lim_{\varepsilon \to 0} {}_1F_1(-n; -2n+\varepsilon; 2z)$$

$$= \frac{(n+1)_n}{2^n} \lim_{\varepsilon \to 0} e^{2z} {}_1F_1(-n+\varepsilon, -2n+\varepsilon; -2z),$$

where (1.4.11) was used in the last step. Thus, $\phi_n(z) := e^{-z}z^n y_n(1/z)$, contains no odd power of z with exponents less than $2n+1$ and

$$\frac{\phi'_n(z)}{\phi_n(z)} + 1 = \frac{\theta'_n(z)}{\theta_n(z)} = -\sum_{j=1}^n \frac{1}{z - 1/z_{n,j}} = -\sum_{j=1}^n z_{n,j} \sum_{k=0}^\infty z^k z_{n,j}^k.$$

The result now follows.   □

The vanishing power sums in Theorem 4.10.4 appeared as the first terms in an asymptotic expansion, see (Ismail & Kelker, 1976). Theorem 4.10.4 was first proved in (Burchnall, 1951) and independently discovered in (Ismail & Kelker, 1976), where an induction proof was given. Moreover,

$$\sum_{j=1}^n z_{n,j}^{2n+1} = \frac{(-1/4)^n}{(3/2)_n^2}, \quad \sum_{j=1}^n z_{n,j}^{2n+3} = \frac{(-1/4)^n}{(2n-1)(3/2)_n^2},$$

were also proved in (Ismail & Kelker, 1976).

Theorem 4.10.5 ((Burchnall, 1951)) *The system of equations*

$$\sum_{j=1}^{n} x_j = -1, \quad \sum_{j=1}^{n} x_j^{2m-1} = 0, \quad m = 2, 3, \ldots, n, \tag{4.10.17}$$

has a unique solution given by the zeros of $y_n(x)$.

Proof We know that (4.10.17) has at least one solution $y_1, y_2, \ldots, y_n$. Assume that $z_1, z_2, \ldots, z_n$ is another solution. Define variables $\{x_j : 1 \leq j \leq 2n\}$ by $x_j = y_j, x_{n+j} = -y_j, 1 \leq j \leq n$. The elementary symmetric functions σ_{2j+1} of $x_1, \ldots, x_{2n}$ vanish for $j = 0, \ldots, n-1$. Therefore $x_1, \ldots, x_{2n}$ are roots of an equation of the form

$$x^{2n} + \sigma_2 x^{2n-2} + \cdots + \sigma_{2n} = 0.$$

Whence n of the x's must form n pairs of the form $(a, -a)$. If $x_j = -x_k$ for some j, k between 1 and n, we will contradict (4.10.16) since none of the x's are zero. Thus $\{z_1, z_2, \ldots, z_n\} = \{y_1, y_2, \ldots, y_n\}$. $\qquad\square$

Theorem 4.10.6 *The Bessel polynomials have the generating function*

$$\sum_{n=0}^{\infty} y_n(z; a, b) \frac{t^n}{n!} = (1 - 4zt/b)^{-\frac{1}{2}} \left(\frac{2}{1 + \sqrt{1 - 4zt/b}} \right)^{a-2}$$

$$\times \exp\left(\frac{2t}{1 + \sqrt{1 - 4zt/b}} \right). \tag{4.10.18}$$

Proof The left-hand side of (4.10.18) is

$$\sum_{n=0}^{\infty} \sum_{k=0}^{n} \frac{(n+a-1)_k}{(n-k)! \, k!} \left(\frac{z}{b} \right)^k t^n = \sum_{n,k=0}^{\infty} \frac{(k+n+a-1)_k}{n! \, k!} \left(\frac{z}{b} \right)^k t^{n+k}$$

$$= \sum_{n,k=0}^{\infty} \frac{(a-1)_{n+2k}}{(a-1)_{n+k}} \frac{(zt)^k}{b^k \, k!} \frac{t^n}{n!}$$

$$= \sum_{n=0}^{\infty} \frac{t^n}{n!} \, {}_2F_1\left(\begin{array}{c} (a+n-1)/2, (a+n)/2 \\ a+n-1 \end{array} \middle| \frac{4zt}{b} \right)$$

$$= \sum_{n=0}^{\infty} \frac{t^n}{n!} \left(\frac{1 + (1 - 4zt/b)^{1/2}}{2} \right)^{-a-n}$$

$$\times \left[1 - \left(\frac{1 - (1 - 4zt/b)^{1/2}}{1 + (1 - 4zt/b)^{1/2}} \right)^2 \right]^{-1},$$

where we used (1.4.13) in the last step. The above simplifies to the right-hand side in (4.10.18). $\qquad\square$

The special case $a = b = 2$ of (4.10.18) gives an exponential generating function for $\{y_n(x)\}$. Another generating function is the following

$$\sum_{n=0}^{\infty} y_n(z) \frac{t^{n+1}}{(n+1)!} = \exp\left(\frac{2t}{1 + (1 - 2zt)^{\frac{1}{2}}}\right) - 1. \tag{4.10.19}$$

To prove (4.10.19), observe that its left-hand side is

$$\sum_{n=0}^{\infty} \sum_{k=0}^{n} \frac{(n+1)_k}{(n-k)!\,k!} \left(\frac{z}{2}\right)^k \frac{t^n}{n+1} = \sum_{n,k} \frac{(n+2k)!}{(n+k+1)!} \left(\frac{zt}{2}\right)^k \frac{t^{n+1}}{n!\,k!}$$

$$= \sum_{n=0}^{\infty} \frac{t^{n+1}}{(n+1)!}\, {}_2F_1\left(\begin{array}{c} \frac{n+1}{2}, \frac{n+2}{2} \\ n+2 \end{array} \middle| 2zt\right)$$

$$= \sum_{n=0}^{\infty} \frac{t^{n+1}}{(n+1)!} \left(\frac{2}{1 + (1 - 2zt)^{1/2}}\right)^{n+1},$$

which is the right-hand side of (4.10.19) after the application of (1.4.13).

The parameter b in $y_n(z; a, b)$ scales the variable z, so there is no loss of generality in assuming $b = 2$.

Definition 4.10.1 *For $a \in \mathbb{R}$, $a + n > 1$, let*

$$C(n, a) := \left\{ z = re^{i\theta} \in \mathbb{C} : 0 < r < \frac{1 - \cos\theta}{n + a - 1} \right\} \bigcup \left\{ \frac{-2}{n + a - 1} \right\}. \tag{4.10.20}$$

Theorem 4.10.7 ((Saff & Varga, 1977)) *All the zeros of $y_n(z; a, b)$ lie in the cordioidal region $C(n, a)$.*

Theorem 4.10.7 sharpens an earlier result of Dočev which says that all the zeros of $y_n(z; a, 2)$ lie in the disc

$$D(n, a) := \{ z \in \mathbb{C} : |z| \le 2/(n + a - 1) \}. \tag{4.10.21}$$

Indeed, $C(n, a)$ is a proper subset of $D(n, a)$ except for the point $-2/(n + a - 1)$.

Theorem 4.10.8 ((Underhill, 1972), (Saff & Varga, 1977)) *For any integers a and $n \ge 1$, with $n + a \ge 2$, the zeros of $y_n(x; a, 2)$ satisfy*

$$|z| < \frac{2}{\mu(2n + a - 2)}, \tag{4.10.22}$$

where μ is the unique positive root of $\mu e^{\mu+1} = 1$.

Note that $\mu \approx 0.278465$.

It is more desirable to rescale the polynomials. Let $\mathcal{L}$ be the set of all zeros of the normalized polynomials

$$\left\{ y_n\left(\frac{2z}{n + a - 1}; a, 2\right) : n = N, a \in \mathbb{R}, n + a > 1 \right\}. \tag{4.10.23}$$

Under $z \to 2z/(n + a - 1)$ the cardioidal region (4.10.20) is mapped onto

$$C := \{ z = re^{i\theta} \in \mathbb{C} : 0 < r < (1 - \cos\theta)/2 \} \bigcup \{-1\}. \tag{4.10.24}$$

Theorem 4.10.9 *Each boundary point of C of (4.10.24) is an accumulation point of the set $\mathcal{L}$ of all zeros of the normalized polynomials in (4.10.23).*

Theorem 4.10.10 *For every $a \in \mathbb{R}$, there exists an integer $N = N(a)$ such that all the zeros of $y_n(z; a, 2)$ lie in $\{z : \operatorname{Re} z < 0\}$ for $n > N$. For $a < -2$, one can take $N = \lceil 2^{3-a} \rceil$.*

Theorem 4.10.10 was conjectured by Grosswald (Grosswald, 1978). de Bruin, Saff and Varga proved Theorems 4.10.9 and 4.10.10 in (de Bruin et al., 1981a), (de Bruin et al., 1981b).

In the above-mentioned papers of de Bruin, Saff and Varga, it is also proved that the zeros of $y_n(z; a, 2)$ lie in the annulus

$$A(n, a) := \left\{ z \in \mathbb{C} : \frac{2}{2n + a - 2/3} < |z| \le \frac{2}{n + a - 1} \right\}, \qquad (4.10.25)$$

which is stronger than (4.10.22) of Theorem 4.10.8.

Theorem 4.10.11 *Let $a \in \mathbb{R}$ and let ρ be the unique (negative) root of*

$$-\rho \exp\left(\sqrt{1 + \rho^2}\right) = 1 + \sqrt{1 + \rho^2} \qquad (\rho \approx -0.662743419), \qquad (4.10.26)$$

and let

$$K(\rho, a) := \frac{\rho\sqrt{1 + \rho^2} + (2 - a) \ln\left(\rho + \sqrt{1 + \rho^2}\right)}{\sqrt{1 + \rho^2}}.$$

Then for n odd, $\alpha_n(a)$, the unique negative zero of $y_n(z; a, 2)$ satisfies the asymptotic relationship

$$\frac{2}{\alpha_n(a)} = (2n + a - 2)\rho + K(\rho, a) + \mathcal{O}\left(\frac{1}{2n + a - 2}\right), \text{ as } n \to \infty. \quad (4.10.27)$$

Theorem 4.10.11 was proved in (de Bruin et al., 1981a) and (de Bruin et al., 1981b). Earlier, Luke and Grosswald conjectured (4.10.27) but only correctly predicted the main term, see (Luke, 1969a, p. 194) and (Grosswald, 1978, p. 93).

In §24.8 we shall state two conjectures on the irreducibility of the Bessel polynomials over $\mathbf{Q}$, the field of rational numbers.

Grosswald's book (Grosswald, 1978) contains broad applications of the Bessel polynomials, from proving the irrationality of π and e^r, for r rational, to probabilistic problems and electrical networks. A combinatorial model for the Bessel polynomials is in (Dulucq & Favreau, 1991).

Exercises

4.1   Prove that

$$2^n n! \lim_{\beta \to \infty} \beta^{-n} L_n^{(\beta^2/2)}\left(-\beta x + \beta^2/2\right) = H_n(x).$$

A combinatorial proof is in (Labelle & Yeh, 1989).

4.2   Show that

$$\lim_{\alpha \to \infty} \alpha^{-n} L_n^{(\alpha)}(\alpha x) = (1 - x)^n / n!$$

4.3   Derive the recursion relation

$$(x + 1) L_{n+1}^{(\alpha-1)}(x) = (\alpha - x) L_n^{(\alpha)}(x) - x L_{n-1}^{(\alpha+1)}(x).$$

4.4   Prove

$$\sum_{n=0}^{\infty} \frac{(-1)^n}{n + 1/2} \frac{\sin \sqrt{b^2 + \pi^2 (n + 1/2)^2}}{\sqrt{b^2 + \pi^2 (n + 1/2)^2}} = \frac{\pi}{2} \frac{\sin b}{b},$$

(Gosper et al., 1993).

Hint: Use Sonine's second integral (4.6.39) with $\mu = -1/2$ and Sonine's first integral, Exercise 1.3.

4.5   Generalize Exercise 4.4 to

$$\sum_{n=0}^{\infty} \frac{(-1)^n}{n + 1/2} \frac{J_\nu \left(\sqrt{b^2 + \pi^2 (n + 1/2)^2} \right)}{[b^2 + \pi^2 (n + 1/2)^2]^{\nu/2}}$$

$$= \frac{\pi}{2} b^{-\nu} J_\nu(b), \qquad b > 0, \quad \mathrm{Re}(\nu) > -\frac{1}{2},$$

(Gosper et al., 1993).

4.6   Carry out the details of the proof of the Kibble–Slepian formula outlined in Remark 4.7.2.

4.7   Prove the inverse relations

$$C_n^\nu(x) = \sum_{k=0}^{\lfloor n/2 \rfloor} {}_2F_0(-k, \nu + n - k; -; 1) \frac{(-1)^k (\nu)_{n-k}}{k! \, (n - 2k)!} H_{n-2k}(x),$$

$$\frac{H_n(x)}{n!} = \sum_{k=0}^{\lfloor n/2 \rfloor} \frac{(-1)^k (\nu + n - 2k)}{k! \, (\nu)_{n+1-2k}} C_{n-2k}^\nu(x).$$

4.8   Show that the Legendre polynomials $\{P_n(x)\}$ have the integral representations

$$P_n(x) = \frac{2}{n! \sqrt{\pi}} \int_0^\infty \exp \left(-t^2 \right) t^n H_n(xt) \, dt.$$

4.9   Establish the relationships

$$\int_0^t L_n(x(t - x)) \, dx = \frac{(-1)^n H_{2n+1}(t/2)}{2^{2n} (3/2)_n},$$

$$\int_0^t \frac{H_{2n} \left(\sqrt{x(t - x)} \right)}{\sqrt{x(t - x)}} = (-1)^n \pi 2^{2n} (1/2)_n L_n \left(t^2/4 \right),$$

where $L_n(x) = L_n^{(0)}(x)$.

4.10   (a) Prove that

$$x^\alpha e^{-x} \left[L_n^{(\alpha)}(x) \right]^2 = \int_0^\infty J_{2\alpha} \left(2\sqrt{xy} \right) e^{-y} y^\alpha \left[L_n^{(\alpha)}(y) \right]^2 dy,$$

holds for $\alpha > -1/2$. (**Hint**: Use the Poisson kernel.)

(b) Prove that

$$e^{-x/2} L_n^{(0)}(x) = \frac{2^{1-n}}{n!\sqrt{\pi}} \int_0^\infty e^{-t^2} H_n^2(t) \cos\left(\sqrt{2x}\, t \right) dt.$$

(c) Show that for $0 \le n \le p$, $p = 0, 1, \ldots$, we have

$$\int_{\mathbb{R}} e^{-x^2} H_p^2(x) H_{2n}(x)\, dx = \frac{2^{p+n} (2n)! \, (p!)^2 \sqrt{\pi}}{(p-n)! \, (n!)^2}.$$

4.11   Show that

$$\sum_{n_1,\ldots,n_k=0}^\infty \frac{t_1^{n_1} \cdots t_k^{n_k}}{n_1! \cdots n_k!} \int_{\mathbb{R}} \frac{e^{-x^2}}{\sqrt{\pi}} H_{n_1}(x) \cdots H_{n_k}(x)\, dx$$

$$= \exp\left(2 \sum_{1 \le i < j \le k} t_i t_j \right),$$

and use it to prove that

$$2^{-(n_1 + \cdots + n_k)/2} \int_{\mathbb{R}} \frac{e^{-x^2}}{\sqrt{\pi}} H_{n_1}(x) \cdots H_{n_k}(x)\, dx,$$

are nonnegative integers. A combinatorial interpretation is in (Azor et al., 1982), see also (Ismail et al., 1987).

4.12   Prove formula (4.6.43). Also prove that i^{-n}, $n = 0, 1, \ldots$, are the only eigenvalues of the Fourier transform.

4.13   For nonnegative integers m, show that

$$x^m = \frac{m!}{2^m} \sum_{n=0}^{\lfloor m/2 \rfloor} \frac{1}{n!} H_{m-2n}(x).$$

4.14   Show that

$$e^{ax} = e^{a^2/4} \sum_{n=0}^\infty \frac{a^n}{2^n \, n!} H_n(x).$$

4.15   Prove that

$$x^{\alpha/2} e^{-x} L_n^{(\alpha)}(x) = \frac{(-1)^n}{2} \int_0^\infty J_\alpha \left(\sqrt{xy} \right) y^{\alpha/2} e^{-y/2} L_n^{(\alpha)}(y)\, dy,$$

for $\alpha > -1$, $n = 0, 1, \ldots$.

4.16   Prove that

$$\int_{-1}^{1} x\left(1-x^2\right)^{\nu+1/2}\left[\frac{d}{dx}\,C_n^{\nu}(x)\right]\left[\frac{d}{dx}\,C_m^{\nu}(x)\right]dx$$

is zero unless $m-n=\pm 1$ and determine its value in these cases.

4.17   Prove the expansion

$$x^n=\sum_{k=0}^{n}\frac{(-1)^k\,n!\,(\alpha+1)_n}{(n-k)!\,(\alpha+1)_k}\,L_k^{(\alpha)}(x).$$

4.18   Prove that

$$\frac{\left(1-x^2\right)}{(\alpha+\beta+n+1)}\,\frac{d}{dx}\,P_n^{(\alpha,\beta)}(x)=\frac{2(\alpha+n)(\beta+n)}{(2n+\alpha+\beta)_2}\,P_{n-1}^{(\alpha,\beta)}(x)$$

$$+\frac{2n(\alpha-\beta)}{(2n+\alpha+\beta)(2n+\alpha+\beta+2)}\,P_n^{(\alpha,\beta)}(x)$$

$$+\frac{2n(n+1)}{(2n+\alpha+\beta+1)_2}\,P_{n+1}^{(\alpha,\beta)}(x).$$

4.19   Deduce

$$e^{-x^2}H_n(x)=\frac{(-1)^{\lfloor n/2\rfloor}}{\sqrt{\pi}}\,2^{n+1}\int_0^{\infty}e^{-t^2}t^n\cos(2xt)\,dt,\quad n\text{ even}$$

$$e^{-x^2}H_n(x)=\frac{(-1)^{\lfloor n/2\rfloor}}{\sqrt{\pi}}\,2^{n+1}\int_0^{\infty}e^{-t^2}t^n\sin(2xt)\,dt,\quad n\text{ odd}$$

from (4.6.41).

4.20   Establish the following relationship between Hermite and Laguerre polynomials

$$L_n^{(\alpha)}(x)=\frac{(-1)^n\Gamma(n+\alpha+1)}{\Gamma(\alpha+1/2)\sqrt{\pi}\,(2n)!}\int_{-1}^{1}\left(1-t^2\right)^{\alpha-1/2}H_{2n}\left(tx^{1/2}\right)dt,$$

for $\alpha>-1/2$.

4.21   Show that the function of the second kind associated with Legendre polynomials has the integral representation (Laplace integral):

$$Q_n(z)=\int_0^{\infty}\left\{z+\left(z^2-1\right)^{1/2}\cos\theta\right\}^{-n-1}d\theta,$$

$n=0,1,\ldots$. Find the corresponding integral representation for the ultraspherical (Gegenbauer) function of the second kind, where $z+\left(z^2-1\right)^{1/2}\cos\theta$ has its principal value when $\theta\neq 0$.

5

Some Inverse Problems

In this chapter we address the question of recovering the orthogonality measure of a set of polynomials from the knowledge of the recursion coefficients. We first treat the simple case of the ultraspherical polynomials $\{C_n^\nu\}$. This example illustrates the method without the technical details needed to treat the Pollaczek polynomials, for example.

5.1 Ultraspherical Polynomials

Recall the recurrence relation

$$2x(n+\nu)C_n^\nu(x) = (n+1)C_{n+1}^\nu(x) + (n+2\nu-1)C_{n-1}^\nu(x), n > 0, \quad (5.1.1)$$

and the initial conditions

$$C_0^\nu(x) = 1, \quad C_1^\nu(x) = 2x\nu. \quad (5.1.2)$$

Let $F(x,t)$ denote the formal power series $\sum_{n=0}^{\infty} C_n^\nu(x)t^n$. By multiplying (5.1.1) by t^n and add for all n, to turn the recursion (5.1.1) to the differential equation

$$2x\nu F(x,t) + 2xt\partial_t F(x,t) = \partial_t F(x,t) + t^2\partial_t F(x,t) + 2t\nu F(x,t), \quad (5.1.3)$$

after taking (5.1.2) into account. The differential equation (5.1.3) simplifies to

$$\partial_t F(x,t) = \frac{2\nu(x-t)}{1-2xt+t^2} F(x,t).$$

The solution of the above equation subject to $F(x,0) = 1$ is

$$F(x,t) = \sum_{n=0}^{\infty} C_n^\nu(x)t^n = \frac{1}{(1-2xt+t^2)^\nu}. \quad (5.1.4)$$

It is clear that we can reverse the above steps and start from (5.1.4) and derive (5.1.1)–(5.1.2), giving a rigorous justification to the derivation of (5.1.4). We follow the usual practice of defining $\sqrt{x^2-1}$ to be the branch of the square root for which $\sqrt{x^2-1}/x \to 1$ as $x \to \infty$ in the appropriate part of the complex x-plane. With this convention we let

$$e^{\pm i\theta} = x \pm \sqrt{x^2-1}. \quad (5.1.5)$$

133

Now let

$$1 - 2xt + t^2 = (1 - t/\rho_1)(1 - t/\rho_2) \quad \text{with } |\rho_1| \le |\rho_2|. \tag{5.1.6}$$

It is easy to see that $\left|e^{-i\theta}\right| \ne \left|e^{i\theta}\right|$ if and only if x is in the complex plane cut along $[-1, 1]$. Furthermore $\rho_1 = e^{-i\theta}$ for $\mathrm{Im}\, x > 0$ while $\rho_1 = e^{i\theta}$ for $\mathrm{Im}\, x < 0$.

Theorem 5.1.1 *The ultraspherical polynomials have the asymptotic property*

$$C_n^\nu(x) = \frac{n^{\nu-1}}{\Gamma(\nu)} \left[1 - \rho_1^2\right]^{-\nu} \rho_2^n [1 + o(1)], \quad \text{as } n \to \infty, \tag{5.1.7}$$

for $x \in \mathbb{C} \setminus [-1, 1]$.

Proof We first assume $0 < \nu < 1$. When $\mathrm{Im}\, x > 0$ we choose the comparison function

$$g(t) = \left[1 - \rho_1^2\right]^{-\nu} \left[1 - t/\rho_1\right]^{-\nu}, \tag{5.1.8}$$

in Theorem 1.2.3. The binomial theorem gives

$$g(t) = \left[1 - \rho_1^2\right]^{-\nu} \sum_{n=0}^\infty \frac{(\nu)_n}{n!} t^n.$$

Applying (1.3.7) and (1.4.7) we establish (5.1.7). For general ν it is easy to see that $g(t)$ in (5.1.8) is the dominant part in a comparison function, hence (5.1.7) follows. □

The monic polynomials associated with the ultraspherical polynomials are

$$P_n(x) = \frac{n!}{2^n(\nu)_n} C_n^\nu(x), \tag{5.1.9}$$

hence by defining $C_n^{*\nu}(x)$ as $2^n(\nu)_n P_n^*(x)/n!$ we see that $C_n^{*\nu}(x)$ satisfies the recursion (5.1.1) and the initial conditions

$$C_0^{*\nu}(x) = 0, \quad C_1^{*\nu}(x) = 2\nu. \tag{5.1.10}$$

Theorem 5.1.2 *We have*

$$C_n^{*\nu}(x) = 2\nu C_n^\nu(x) \left\{ \int_0^{\rho_1} \left[1 - 2xu + u^2\right]^{\nu-1} du \right\} [1 + o(1)], \tag{5.1.11}$$

for x in the complex plane cut along $[-1, 1]$.

Proof Let $F^*(x, t)$ denote the generating function $\sum_{n=0}^\infty C_n^{*\nu}(x)t^n$. In the case of $C_n^{*\nu}(x)$'s instead of the differential equation (5.1.3), the initial conditions (5.1.10) lead to the differential equation

$$2x\nu F^*(x,t) + 2xt\partial_t F^*(x,t) = \partial_t F^*(x,t) + t^2 \partial_t F^*(x,t) + 2t\nu F(x,t) - 2\nu.$$

Thus

$$F^*(x,t) = 2\nu \left[1 - 2xt + t^2\right]^{-\nu} \int_0^t \left[1 - 2xu + u^2\right]^{\nu-1} du. \qquad (5.1.12)$$

Clearly (5.1.12) implies the theorem.   □

In the notation of (2.6.1) the continued fraction associated with (5.1.1) corresponds to

$$A_n = 2(\nu + n)/(n+1), \quad B_n = 0, \quad C_n = (n + 2\nu - 1)/(n+1). \qquad (5.1.13)$$

Let

$$F(x) = \frac{A_0}{A_0 x-} \, \frac{C_1}{A_1 x-} \cdots, \qquad (5.1.14)$$

with A_n and C_n are defined by (5.1.13). Markov's theorem 2.6.2 implies

$$F(x) = \lim_{n\to\infty} \frac{\left(C_n^\nu(x)\right)^*}{C_n^\nu(x)} = 2\nu \int_0^{\rho_1} \left[1 - 2xu + u^2\right]^{\nu-1} du, \qquad (5.1.15)$$

for $x \notin [-1,1]$. The change of variable $u \to u\rho_1$ in (5.1.15) and the Euler integral representation (1.4.8) lead to

$$F(x) = 2\rho_1 \, {}_2F_1\left(1 - \nu, 1; \nu + 1; \rho_1^2\right), \quad x \notin [-1,1]. \qquad (5.1.16)$$

If we did not know the measure with respect to which the ultraspherical polynomials are orthogonal we can find it from (5.1.15) and the Perron–Stieltjes inversion formula (1.2.8)–(1.2.9). Since $F(x)$ has no poles and is single-valued across the real axis, it follows from the remarks following (1.2.8)–(1.2.9) that the orthogonality measure is absolutely continuous and is supported on $[-1,1]$. With $x = \cos\theta$, $0 < \theta < \pi$, we find

$$F\left(x - i0^+\right) - F\left(x + i0^+\right) = 2\nu \int_{e^{-i\theta}}^{e^{i\theta}} \left[1 - 2xu + u^2\right]^{\nu-1} du.$$

Letting $u = e^{i\theta} + \left[e^{-i\theta} - e^{i\theta}\right] v$ then deforming the contour of integration to $v \in [0,1]$ we get

$$\frac{F\left(x - i0^+\right) - F\left(x + i0^+\right)}{2\pi i} = \frac{\nu\Gamma^2(\nu)}{\pi\Gamma(2\nu)} \sin^{2\nu-1}\theta, \qquad (5.1.17)$$

and we obtain the normalized weight function

$$w_\nu(x) = 2^{2\nu-1}\frac{\Gamma(\nu+1)\Gamma(\nu)}{\pi\Gamma(2\nu)} \left(1 - x^2\right)^{\nu-1/2}. \qquad (5.1.18)$$

5.2 Birth and Death Processes

This section contains an application to birth and death processes of the method described in §5.1 to find the measure from the three term recurrence relation.

A birth and death process is a stationary Markov process whose states are labeled by nonnegative integers and whose transition probabilities

$$p_{m,n}(t) = \Pr\{X(t) = n \mid X(0) = m\} \tag{5.2.1}$$

satisfy the conditions

$$p_{mn}(t) = \begin{cases} \lambda_m t + o(t), & n = m + 1, \\ \mu_m t + o(t), & n = m - 1, \quad \text{as } t \to 0^+, \\ 1 - (\lambda_m + \mu_m)\, t + o(t), & n = m, \end{cases} \tag{5.2.2}$$

where $\lambda_m > 0$, $m = 0, 1, \dots, \mu_m > 0$, $m = 1, 2, \dots, \mu_0 \geq 0$. The λ_n's are the birth rates and the μ_n's are the death rates. The transition matrix P is

$$P(t) = (p_{m,n}(t)), \quad m, n = 0, 1, \dots. \tag{5.2.3}$$

The stationary requirement implies

$$P(s + t) = P(s)P(t).$$

We may consider birth and death processes with a finite state space, say $\{0, 1, \dots, N - 1\}$. In such case $\lambda_N = 0$ and we say that we have an absorbing barrier at state N. Unless we say otherwise the state space will be the nonnegative integers.

Theorem 5.2.1 *The transition probabilities $\{p_{m,n}(t) : m, n = 0, 1, \dots\}$ satisfy the Chapman–Kolomogorov equations*

$$\frac{d}{dt} p_{m,n}(t) = \lambda_{n-1} p_{m,n-1} + \mu_{n+1} p_{m,n+1} - (\lambda_n + \mu_n)\, p_{m,n}(t), \tag{5.2.4}$$

$$\frac{d}{dt} p_{m,n}(t) = \lambda_m p_{m+1,n} + \mu_m p_{m-1,n} - (\lambda_n + \mu_n)\, p_{m,n}(t). \tag{5.2.5}$$

Proof We compute $p_{m,n}(t + \delta t)$ in two different ways. The system can go from state m to state n in time increments of t and δt or in total time $t + \delta t$. From (5.2.3) it follows that $P(t)P(\delta t) = P(t + \delta t) = P(\delta t)P(t)$. Therefore

$$p_{m,n}(t + \delta t) = p_{m,n-1}(t)[\lambda_{n-1}\delta t] + p_{m,n+1}(t)\,[\mu_{n+1}\delta t]$$
$$+ p_{m,n}(t)\,[1 - (\lambda_n + \mu_n)\,\delta t] + o(t).$$

Subtract $p_{m,n}(t)$ from the above equation then divide by δt and let $\delta t \to 0$ we establish (5.2.4). Similarly (5.2.5) can be proved. $\qquad\square$

Let A be the tridiagonal matrix $\{a_{m,n} : m \geq 0, n \geq 0\}$

$$a_{n,n} = -\lambda_n - \mu_n, \quad a_{n,n+1} = \lambda_n, \quad a_{n,n-1} = \mu_n. \tag{5.2.6}$$

Birth and death processes have the properties

$$\begin{cases} \textbf{I} \quad \dot{P}(t) = P(t)A, & \textbf{II} \; \dot{P}(t) = AP(t), \\ \textbf{III} \; P(0) = I, & \textbf{IV} \; p_{m,n}(t) \geq 0, \\ \textbf{V} \; \sum_{n=0}^{\infty} p_{m,n}(t) \leq 1, m \geq 0, t \geq 0, & \textbf{VI} \; P(s+t) = P(s)P(t). \end{cases} \quad (5.2.7)$$

where I is the identity matrix.

The next step is to solve (5.2.4)–(5.2.5) using the method of separation of variables. The outline we give may not be rigorous but provides a good motivation for the result. We will also give a rigorous proof for the case of finitely many states.

Let

$$p_{m,n}(t) = f(t)Q_m F_n. \quad (5.2.8)$$

Since Q_m can not vanish identically then (5.2.4) yields

$$f'(t)/f(t) = [\lambda_{n-1}F_{n-1} + \mu_{n+1}F_{n+1} - (\lambda_n + \mu_n) F_n]/F_n = -x, \quad (5.2.9)$$

say, for some separation constant x. Therefore $f(t) = e^{-xt}$, up to a multiplicative constant. Thus the F_n's satisfy $F_{-1}(x) = 0$ and

$$-xF_n(x) = \lambda_{n-1}F_{n-1}(x) + \mu_{n+1}F_{n+1}(x) - (\lambda_n + \mu_n) F_n(x), \quad n > 0. \quad (5.2.10)$$

It is clear F_0 is arbitrary and up to a multiplicative constant we may take $F_0(x) = 1$. Now (5.2.5) and (5.2.8) show that the Q_n's, up to a multiplicative constant actor, are generated by

$$Q_0(x) = 1, \quad Q_1(x) = (\lambda_0 + \mu_0 - x)/\lambda_0, \quad (5.2.11)$$

$$\begin{aligned} &-xQ_n(x) \\ &= \lambda_n Q_{n+1}(x) + \mu_n Q_{n-1}(x) - (\lambda_n + \mu_n) Q_n(x), \quad n > 0. \end{aligned} \quad (5.2.12)$$

The relationships (5.2.10)–(5.2.12) show that

$$F_n(x) = \zeta_n Q_n(x), \quad (5.2.13)$$

with

$$\zeta_0 := 1, \quad \zeta_n = \prod_{j=1}^{n} \frac{\lambda_{j-1}}{\mu_j}. \quad (5.2.14)$$

Thus we have shown that the separation of variables gives a solution of the form

$$p_{m,n}(t) = \frac{1}{\zeta_m} \int_{\mathbb{R}} e^{-xt} F_m(x) F_n(x) \, d\mu(x), \quad (5.2.15)$$

for some measure μ which incorporates the separation constants. As $t \to 0$ we must have

$$\zeta_n \delta_{m,n} = \int_{\mathbb{R}} F_m(x) F_n(x) \, d\mu(x).$$

Hence the F_n's are orthogonal with respect to μ. What we have not proved but holds true is that any solution of the Chapman–Kolmogorov equations (5.2.4)–(5.2.5) has the form (5.2.15).

From (5.2.10) it is clear that the polynomials

$$P_n(x) := (-1)^n \mu_1 \cdots \mu_n F_n(x)$$

satisfy (2.2.1)–(2.2.2) with $\alpha_n = \lambda_n + \mu_n$, $\beta_n = \lambda_{n-1}\mu_n$, hence, by the spectral theorem, are orthogonal polynomials. In §7.2 we shall show that all the zeros of F_n, for all n, belong to $(0, \infty)$. Thus the support of any measure produced by the construction in the proof of the spectral theorem will be a subset of $[0, \infty)$.

Next we truncate the matrix A after N rows and columns and consider the resulting finite birth and death process. Let A_N and $P_N(t)$ be the $N \times N$ principal minors of A and P respectively. In this case the solution of **I–III** of (5.2.7) is

$$P_N(t) = \exp(tA_N) = \sum_{n=0}^{\infty} \frac{t^n}{n!} A_N^n \tag{5.2.16}$$

To diagonalize A_N, first note that the eigenvalues of A_N coincide with zeros of $F_N(x)$. Let $x_{N,1} > \cdots > x_{N,N}$, be the zeros of $F_N(x)$. Set

$$\mathbf{F_j} := \rho(x_{N,j})(F_0(x_{N,j}), \ldots, F_{N-1}(x_{N,j}), \tag{5.2.17}$$

$$\frac{1}{\rho(x_{N,j})} := \sum_{j=0}^{N-1} F_j^2(x_{N,j}) = -F_N'(x_{N,j}) F_{N-1}(x_{N,j})/\zeta_n, \tag{5.2.18}$$

and (2.2.4) was used in the last step. From (5.2.10) we see that $\mathbf{F}$ is a left eigenvector for A_N with the eigenvalue $x_{N,j}$. Let F be the matrix whose rows are formed by the vector $F_1, \ldots, F_N$. The Christoffel–Darboux formula (2.2.4) shows that the columns of F^{-1} are formed by the vectors $(F_0(x_{N,j})/\zeta_0, \cdots, F_{N-1}(x_{N,j})/\zeta_{N-1})$. Furthermore

$$F A_N F^{-1} = -(x_{N,j}\delta_{j,k}), \quad 1 \le j, k \le N.$$

Thus formula (5.2.16) becomes $P_N(t) = F^{-1}DF$, where D is the diagonal matrix $(\exp(-t\,x_{N,j})\,\delta_{j,k})$, $1 \le j, k \le N$. A calculation then yields the representation

$$p_{m,n}(t) = \frac{1}{\zeta_n} \sum_{j=1}^{N} \exp(-tx_{N,j}) F_m(x_{N,j}) F_n(x_{N,j}) \rho(x_{N,j}). \tag{5.2.19}$$

Note that the sum in (5.2.19) is $\int_{\mathbb{R}} e^{-tx} F_m(x) F_n(x) d\psi_N(x)$ where the measure ψ_N is as constructed in the spectral theorem, that is ψ_N has a mass $\rho(x_{N,j})$ at $x = x_{N,j}$. Indeed $F_0(x), \ldots, F_{N-1}$ are orthogonal with respect to ψ_N. By letting $N \to \infty$ we see that the F_n's are orthogonal with respect to the measure μ in (5.2.18). It must be emphasized that μ may not be unique.

If one only cares about the states of this process and not about the times of arrival then the appropriate process to consider is a random walk to which is associated a set of orthogonal polynomials defined by

$$R_{-1}(x) = 0, \, R_0(x) = 1,$$
$$xR_n(x) = m_n R_{n+1}(x) + \ell_n R_{n-1}(x), \qquad (5.2.20)$$
$$m_n = \lambda_n / (\lambda_n + \mu_n), \quad \ell_n = \mu_n / (\lambda_n + \mu_n),$$

see (Karlin & McGregor, 1958), (Karlin & McGregor, 1959). We shall refer to these polynomials as random walk polynomials. These polynomials are orthogonal on $[-1, 1]$ with respect to an even measure. The orthogonality relation is

$$\int_{-1}^{1} r_n(x) r_m(x) \, d\mu(x) = \delta_{m,n}/h_n, \qquad (5.2.21)$$

where

$$h_0 = 1, \quad h_n = \frac{\lambda_0 \lambda_1 \cdots \lambda_{n-1} (\lambda_n + \mu_n)}{\mu_1 \mu_2 \cdots \mu_n (\lambda_0 + \mu_0)}, \quad n > 0.$$

Note that the Laguerre polynomials are birth and death process polynomials with $\lambda_n = n + \alpha$, $\mu_n = n$. The Jacobi polynomials $P_n^{(\alpha,\beta)}(x + 1)$ correspond to a birth and death process but with rational birth and death rates. The Meixner and Charlier polynomials, §6.1, are also birth and death process polynomials.

We now outline a generating function method proved effective in determining measures of orthogonality of birth and death process polynomials when λ_n and μ_n are polynomials in n. Define $P_m(t, w)$ by

$$P_m(t, w) = \sum_{n=0}^{\infty} w^n p_{m,n}(t). \qquad (5.2.22)$$

The series defining $P_{m,n}(t, w)$ converges for $|w| \leq 1$ and all $t > 0$ since $\sum_{n=0}^{\infty} p_{m,n}(t)$ converges and $p_{m,n}(t) \geq 0$. The integral representation (5.2.15) gives

$$\zeta_m P_m(t, w) = \int_{0}^{\infty} e^{-tx} F_m(x) F(x, w) \, d\mu(x), \qquad (5.2.23)$$

with

$$F(x, w) := \sum_{n=0}^{\infty} w^n F_n(x). \qquad (5.2.24)$$

Now assume that λ_n and μ_{n+1} are polynomials in n, $n \geq 0$, and

$$\mu_0 = 0, \quad \tilde{\mu}_0 = \lim_{n \to 0} \mu_n. \qquad (5.2.25)$$

Multiply the forward Chapman–Kolmogorov equation (5.2.4) by w^n and add for $n \geq 0$, with $\lambda_{-1} p_{m,-1}(t) := 0$ we establish the partial differential equation

$$\frac{\partial}{\partial t} P_m(t, w)$$
$$= (1 - w) \left[w^{-1} \mu(\delta) - \lambda(\delta) \right] P_m(t, w) + \left[(1 - w^{-1}) \tilde{\mu}_0 - \mu_0 \right] p_{m,0}(t),$$

where

$$\delta := w\frac{\partial}{\partial w}, \quad \lambda(n) = \lambda_n, \quad \mu(n) = \mu_n. \tag{5.2.26}$$

Theorem 5.2.2 *As a formal power series, the generating function $F(x, w)$ satisfies the differential equation*

$$\left[(1-w)\{w^{-1}\mu(\delta) - \lambda(\delta)\} + x\right]F(x, w)$$
$$= \mu_0 - \tilde{\mu}_0(1 - w^{-1}). \tag{5.2.27}$$

If $F(x, w)$ converges in a neighborhood of $w = 0$, then F satisfies the additional boundary conditions

$$F(x, 0) = 1, \quad \int_{\mathbb{R}} F(x, w)\,d\mu(x) = 1. \tag{5.2.28}$$

All the classical polynomials are random walk polynomials or birth and death process polynomials, or limits of them, under some normalization. The choice $\lambda_n = n + 1$, $\mu_n = n + \alpha$ makes the birth and death process polynomials equal to Laguerre polynomials while $\lambda_n = n + \alpha + 1$, $\mu_n = n$ leads to multiples of Laguerre polynomials. With $\lambda_n = (n + 2\nu + 1)/[2(n + \nu)]$, $\mu_n = n/[2(n + \nu)]$, $r_n(x)$ is a multiple of $C_n^\nu(x)$, while $r_n = C_n^\nu(x)$ if $\lambda_n = (n + 1)/[2(n + \nu)]$, $\mu_n(n + 2\nu)/[2(n + \nu)]$. The interested reader may prove that $\left\{P_n^{(\alpha,\beta)}(x-1)\right\}$ are birth and death process polynomials corresponding to rational λ_n and μ_n.

Remark 5.2.1 *When $\mu_0 > 0$, there are two natural families of birth and death polynomials. The first is the family $\{Q_n(x)\}$ defined by (5.2.11)–(5.2.12). Another family is $\left\{\tilde{Q}_n(x)\right\}$ defined by*

$$\tilde{Q}_0(x) = 1, \quad \tilde{Q}_1 = (\lambda_0 - x)/\lambda_0 \tag{5.2.29}$$
$$-x\tilde{Q}_n(x) = \lambda_n\tilde{Q}_{n+1}(x) + \mu_n\tilde{Q}_{n-1}(x) - (\lambda_n + \mu_n)\tilde{Q}_n(x), \quad n > 0. \tag{5.2.30}$$

In effect, we redefine μ_0 to be zero. We do not see this phenomenon in the classical polynomials, but it starts to appear in the associated polynomials.

When the state space of a birth and death process consists of all integers and $\lambda_n\mu_n \neq 0$ for $n = 0, \pm 1, \ldots$, there is a similar theory which relates the transition probabilities of such processes to spectral measures of doubly infinite Jacobi matrices, see (Pruitt, 1962). The spectral theory of doubly infinite Jacobi matrices is available in (Berezans'kiĭ, 1968).

Queueing theory is a study of birth and death processes where the states of the system represent the number of customers in a queue. In the last twenty years, models were introduced in which the number of customers is now a continuous quantity. Such systems are referred to as fluid queues. These models have applications to fluid flows through reservoirs. Some of the works in this area are (Anick et al., 1982), (Mandjes & Ridder, 1995), (Scheinhardt, 1998), (Sericola, 1998), (Sericola, 2001), (Van Doorn & Scheinhardt, 1966). So far there is no theory connecting orthogonal polynomials and fluid queues, but there is probably a continuous analogue of orthogonal polynomials which will play the role played by orthogonal polynomials in birth and death processes.

The relations (4.6.5)–(4.6.6) between Hermite polynomials and $\left\{ L_n^{(\pm 1/2)}(x) \right\}$ carry over to general birth and death process polynomials. Let $\{F_n(x)\}$ be generated by (5.2.10) and

$$F_0(x) = 1, \quad F_1(x) = (\lambda_0 + \mu_0 - x) / \mu_1. \tag{5.2.31}$$

Let $\{\rho_n(x)\}$ be the corresponding monic polynomials, that is

$$\rho_n(x) = (-1)^n \left\{ \prod_{k=1}^{n} \mu_k \right\} F_n(x), \tag{5.2.32}$$

so that

$$x\rho_n(x) = \rho_{n+1}(x) + (\beta_{2n} + \beta_{2n+1}) \rho_n(x) + \beta_{2n}\beta_{2n-1}\rho_{n-1}(x), \tag{5.2.33}$$

where the β_n's are defined by

$$\lambda_n = \beta_{2n+1}, \ n \geq 0, \quad \mu_n = \beta_{2n}, \ n \geq 0. \tag{5.2.34}$$

Let $\{\sigma_n(x)\}$ be generated by $\sigma_0(x) = 1$, $\sigma_1(x) = x - \beta_1 - \beta_2$, and

$$x\sigma_n(x) = \sigma_{n+1}(x) + (\beta_{2n+1} + \beta_{2n+2}) \sigma_n(x) + \beta_{2n}\beta_{2n+1}\sigma_{n-1}(x). \tag{5.2.35}$$

Clearly $\{\sigma_n(x)\}$ is a second family of birth and death process polynomials. The polynomials $\{\rho_n(x)\}$ and $\{\sigma_n(x)\}$ play the role of $\left\{ L_n^{(\pm 1/2)}(x) \right\}$. Indeed, we can define a symmetric family of polynomials $\{\mathcal{F}_n(x)\}$ by

$$\mathcal{F}_0(x) = 1, \quad \mathcal{F}_1(x) = x, \tag{5.2.36}$$

$$\mathcal{F}_{n+1}(x) = x\mathcal{F}_n(x) - \beta_n\mathcal{F}_{n-1}(x), \tag{5.2.37}$$

which makes

$$\rho_n(x) = \mathcal{F}_{2n}\left(\sqrt{x}\right), \quad \sigma_n(x) = x^{-1/2}\mathcal{F}_{2n+1}\left(\sqrt{x}\right). \tag{5.2.38}$$

Moreover, given $\{\mathcal{F}_n\}$ one can define $\{\rho_n\}$ and $\{\sigma_n\}$ uniquely through (5.2.33) and (5.2.35), where $\{\lambda_n\}$ and $\{\mu_n\}$ are given by (5.2.34) with $\mu_0 = 0$.

We shall apply the above results in §21.1 and §21.9.

5.3 The Hadamard Integral

In this section we study some basic properties of the (simple) Hadamard integral (Hadamard, 1932). The Hadamard integrals will be used in §5.4 to determine the measure with respect to which the Pollaczek polynomials are orthogonal.

We say that an open subset Ω of the complex plane is a branched neighborhood of b if Ω contains a set of the form $D \setminus R_b$, where D is an open disc such that $b \in D$ and R_b is a half-line emanating at b and not bisecting D. We will usually assume that Ω is simply connected. Clearly, any open disc is a branched neighborhood of its boundary points. If D is the unit disc, $D - [0, \infty)$ is a branched neighborhood of 0.

Let Ω be a simply connected branched neighborhood of b and assume that ρ is a complex number which is not a negative integer. Assume further that $(t-b)^\rho$ is defined in Ω and that $g(t)$ is an analytic function having a power series expansion $\sum\limits_{n=0}^{\infty} a_n(b-t)^n$ around b which holds in a neighborhood of $\Omega \cup \{b\}$. We define the Hadamard integral

$$\int\limits_{z}^{b]} (b-t)^\rho g(t)\, dt, \quad z \in \Omega,$$

by the formula

$$\int\limits_{z}^{b]} (b-t)^\rho g(t)\, dt = \sum_{n=0}^{\infty} \frac{a_n}{\rho+n+1}\, (b-z)^{\rho+n+1}. \tag{5.3.1}$$

It is clear that when $\mathrm{Re}(\rho) > -1$, then

$$\int\limits_{z}^{b]} (b-t)^\rho g(t)\, dt = \int\limits_{z}^{b} (b-t)^\rho g(t)\, dt, \tag{5.3.2}$$

where the integral on the right side is over any path in Ω joining z and b.

More generally, assume that Ω' is a simply connected open set containing Ω and g is analytic in Ω' and has a power series expansion around b which holds in a neighborhood of $\Omega \cup \{b\}$. We define

$$\int\limits_{a}^{b]} (b-t)^\rho g(t)\, dt = \int\limits_{a}^{z} (b-t)^\rho g(t)\, dt + \int\limits_{z}^{b]} (b-t)^\rho g(t)\, dt, \quad a \in \Omega', \tag{5.3.3}$$

where $z \in \Omega$. Furthermore

$$\int\limits_{[b}^{a} (b-t)^\rho g(t)\, dt = - \int\limits_{a}^{b]} (b-t)^\rho g(t)\, dt. \tag{5.3.4}$$

If Ω is also a branched neighborhood of a and Ω' is a neighborhood of $\Omega \cup \{a\}$, and $g(t)$ is analytic in Ω' and $\rho, \sigma \neq -1, -2, \ldots$, then we define

$$\int\limits_{[a}^{b]} (t-a)^\sigma (b-t)^\rho g(t)\, dt = \int\limits_{[a}^{z} (t-a)^\sigma (b-t)^\rho g(t)\, dt$$

$$+ \int\limits_{z}^{b]} (t-a)^\sigma (b-t)^\rho g(t)\, dt, \tag{5.3.5}$$

where z is an point in Ω'.

The integral $\int\limits_{[a}^{b]} (t-a)^\sigma (b-t)^\rho g(t)\, dt$ is an extension of the integral $\int\limits_{a}^{b} (t-a)^\sigma (b-$

$t)^\rho g(t)\, dt$ from the proper cases $\text{Re}(\sigma) > -1$, $\text{Re}(\rho) > -1$ to the case when σ and $\rho \neq -1, -2, -3, \ldots$.

The definition of the Hadamard integral can be extended to a function $f(t)$ of the form

$$f(t) = \sum_{n=0}^{\infty} C_n (b-t)^{\rho+n}, \quad t \in \Omega. \tag{5.3.6}$$

Let g satisfy the same assumptions as in (5.3.3). The extended Hadamard integral is defined by

$$\int_a^{b]} f(t)g(t)\, dt = \sum_{n=0}^{N} C_n \int_a^{b]} (b-t)^{\rho+n} g(t)\, dt + \int_a^{b} h(t)g(t)\, dt, \tag{5.3.7}$$

where $h(t)$

$$h(t) = \sum_{n-N+1}^{\infty} C_n (b-t)^{\rho+n}$$

and $\text{Re}(\rho + n) > -1$ for $n > N$. Functions of the type in (5.3.6) are said to have an algebraic branch singularity at $t = b$. When f is given by (5.3.6), Ω is a branched neighborhood of a, and

$$g(t) = \sum_{n=0}^{\infty} a_n (t-a)^{\sigma+n} \tag{5.3.8}$$

with $\text{Re}(\sigma) \neq -1, -2, \ldots$, we define

$$\int_{[a}^{b]} f(t)g(t)\, dt = \int_{[a}^{z} f(t)g(t)\, dt + \int_z^{b]} f(t)g(t)\, dt, \quad z \in \Omega'. \tag{5.3.9}$$

It is not difficult to prove the following.

Theorem 5.3.1 *Let f be an analytic function in the simply connected branched neighborhood Ω of the point b, and assume that f has an algebraic branch singularity at b. Let $\{g_n\}$ be a sequence of analytic functions in a neighborhood Ω' of $\Omega \cup \{b\}$ converging uniformly to zero on compact subsets of Ω'. Then, for all $a \in \Omega'$ we have*

$$\lim_{n \to \infty} \int_a^{b]} f(t) g_n(t)\, dt = 0.$$

Corollary 5.3.2 *Let f, Ω, $\{g_n\}$ and Ω' be as in Theorem 5.3.1 but assume that $\{g_n\}$ converges to g on compact sets. Then*

$$\lim_{n \to \infty} \int_a^{b]} f(t) g_n(t)\, dt = \int_a^{b]} f(t)g(t)\, dt. \tag{5.3.10}$$

Corollary 5.3.3 *Let f, Ω, Ω' be as in the theorem, and assume that*

$$g(t) = \sum_{n=0}^{\infty} a_n (t-a)^n \qquad (5.3.11)$$

holds for $a \in \Omega$ and all $t \in \Omega'$. Then

$$\int_a^{b]} f(t)g(t)\, dt = \sum_{n=0}^{\infty} a_n \int_a^{b]} f(t)(t-a)^n\, dt. \qquad (5.3.12)$$

Since uniform convergence on compact subsets is sometimes difficult to check, the following corollary is often useful.

Corollary 5.3.4 *Let f, Ω, Ω', $\{g_n\}$ be as in Theorem 5.3.1, but assume only that $\{g_n\}$ is uniformly bounded on compact subsets of Ω' and that $\{g_n(t)\}$ converges to $g(t)$ for each t in a subset S of Ω' having a limit point in Ω'. Then*

$$\lim_{n \to \infty} \int_a^{b]} f(t)g_n(t)\, dt = \int_a^{b]} f(t)g(t)\, dt. \qquad (5.3.13)$$

We now study Hadamard integrals of functions that will arise in this work. These integrals are related to certain analytic functions in the cut plane $\mathbb{C} \setminus [-1, 1]$ that we will now introduce.

Let $\sqrt{z+1}$ be the branch of the square root of $z+1$ in $\mathbb{C} \setminus (-\infty, -1]$ that makes $\sqrt{z+1} > 0$ if $z > -1$, and $\sqrt{z-1}$ be the branch of the square root of $z-1$ in $\mathbb{C} \setminus (-\infty, 1]$ with $\sqrt{z-1} > 0$ for $z > 1$. Both $\sqrt{z+1}$ and $\sqrt{z-1}$ are single valued in the cut plane $\mathbb{C} \setminus (-\infty, 1]$. Let

$$\tau(z) = \sqrt{z+1}\,\sqrt{z-1}, \quad z \in \mathbb{C} \setminus (-\infty, 1]. \qquad (5.3.14)$$

Observe that when $x < -1$ we have

$$\lim_{\substack{y \to 0 \\ y > 0}} \sqrt{x+iy+1}\,\sqrt{x+iy-1} = i\sqrt{-x-1} \cdot i\sqrt{-x+1} = -\sqrt{x^2-1} \quad (5.3.15)$$

and

$$\lim_{\substack{y \to 0 \\ y < 0}} \sqrt{x+iy+1}\,\sqrt{x+iy-1} = \left(-i\sqrt{-x-1}\right) \cdot \left(-i\sqrt{-x+1}\right) = -\sqrt{x^2-1}. \qquad (5.3.16)$$

We now extend τ, by continuity, to the cut plane $\mathbb{C} \setminus [-1, 1]$. In order to do so we define

$$\tau(z) = -\sqrt{z^2-1}, \quad z < -1. \qquad (5.3.17)$$

Clearly, $\tau(z)$ is analytic in $\mathbb{C} - [-1, 1]$. In what follows we shall simply write

$$\tau(z) = \sqrt{z^2-1}. \qquad (5.3.18)$$

We now define the following analytic functions in $\mathbb{C} \setminus [-1, 1]$

$$\rho_2(z) = z + \tau(z) = z + \sqrt{z^2-1}, \quad \rho_1(z) = z - \tau(z) = z - \sqrt{z^2-1} \quad (5.3.19)$$

and

$$A(z) = -\lambda + \frac{az+b}{\tau(z)} = -\lambda + \frac{az+b}{\sqrt{z^2-1}},$$

$$B(z) = -\lambda - \frac{az+b}{\tau(z)} = -\lambda - \frac{az+b}{\sqrt{z^2-1}}. \qquad (5.3.20)$$

Here, a, b, λ are real numbers and

$$\lambda > -\frac{1}{2}, \qquad (5.3.21)$$

$$\alpha - \lambda \neq 0, 1, 2, \dots . \qquad (5.3.22)$$

We note that

$$\rho_2(x) = x + \sqrt{x^2-1}, \quad \rho_1(x) = x - \sqrt{x^2-1} \text{ if } x > 1, \qquad (5.3.23)$$

$$\rho_2(x) = x - \sqrt{x^2-1}, \quad \rho_1(x) = x + \sqrt{x^2+1} \text{ if } x < -1, \qquad (5.3.24)$$

$$A(x) = -\lambda \pm \frac{ax+b}{\sqrt{x^2-1}}, \quad B(x) = -\lambda \mp \frac{ax+b}{\sqrt{x^2-1}}, \quad \pm x > 1, \qquad (5.3.25)$$

$$\lim_{y\to 0\pm} \tau(x+iy) = \pm\sqrt{1-x^2}, \quad -1 \leq x \leq 1. \qquad (5.3.26)$$

The following functions are continuous on their domain of definition

$$\tau^+(x+iy) = \begin{cases} \tau(x+iy), & y > 0, \quad \tau(x), \quad |x| > 1, \quad y = 0, \\ i\sqrt{1-x^2}, & |x| \leq 1, \quad y = 0, \end{cases} \qquad (5.3.27)$$

$$\tau^-(x+iy) = \begin{cases} \tau(x+iy), & y < 0, \quad \tau(x), \quad |x| > 1, \quad y = 0, \\ -i\sqrt{1-x^2}, & |x| \leq 1, \quad y = 0, \end{cases} \qquad (5.3.28)$$

$$\rho_2^\pm(z) = z + \tau^\pm(z), \quad \rho_1^\pm(z) = z - \tau^\pm(z), \qquad (5.3.29)$$

$$A^\pm(z) = -\lambda + \frac{az+b}{\tau^\pm(z)}, \quad B^\pm(z) = -\lambda - \frac{az+b}{\tau^\pm(z)}. \qquad (5.3.30)$$

Observe that for $-1 \leq x \leq 1$ we have

$$\rho_2^-(x) = \overline{\rho_1^+(x)}, \quad \overline{\rho_1^-(x)} = \rho_2^-(x), \qquad (5.3.31)$$

and

$$A^-(x) = \overline{B^+(x)}, \quad \overline{B^-(x)} = A^+(x). \qquad (5.3.32)$$

To simplify the notation we will write when $-1 < x < 1$

$$\rho_2^+(x) = \rho_2(x), \quad \rho_1^+(x) = \rho_1(x); \quad A^+(x) = A(x), \quad B^+(x) = B(x). \qquad (5.3.33)$$

The following elementary result will be very useful.

Lemma 5.3.5 *For each z in $\mathbb{C}$, $\rho_2(z)$ and $\rho_1(z)$ are the solutions of the equation*

$$t^2 - 2zt + 1 = 0 \tag{5.3.34}$$

that satisfy

$$\rho_2(z) + \rho_1(z) = 2z, \quad \rho_2(z) - \rho_1(z) = 2\tau(z) = 2\sqrt{z^2 - 1}, \quad \rho_2(z)\rho_1(z) = 1. \tag{5.3.35}$$

Furthermore, $|\rho_1(z)| \leq |\rho_2(z)|$, with $|\rho_2(z)| = |\rho_1(z)|$ if and only if $-1 \leq z \leq 1$.

Now let

$$\Omega = \{z \notin [-1, 1]; B(z) \neq 0, 1, \dots\}. \tag{5.3.36}$$

Lemma 5.3.6 *For $z \in \Omega$ and all integers $n \geq 0$,*

$$\int_0^{1]} (1 - u)^{-B(z)-1} u^n \, du = \frac{n!}{(-B)_{n+1}}, \quad z \in \Omega. \tag{5.3.37}$$

The next theorem gives a series expansion for a Hadamard integral.

Theorem 5.3.7 *For every $z \in \Omega$, define $F(z)$ by*

$$F(z) = \int_0^{1]} \left(1 - \frac{\rho_1}{\rho_2} u\right)^{-A-1} (1 - u)^{-B-1} \, du. \tag{5.3.38}$$

Then the function $F(z)$ is analytic in Ω and is given by

$$F(z) = -\frac{1}{B} \sum_{n=0}^{\infty} \frac{(A+1)_n}{(-B-1)_n} \left(\frac{\rho_1}{\rho_2}\right)^n = -\frac{1}{B}\, {}_2F_1\left(\begin{array}{c} A+1, 1 \\ -B+2 \end{array}\middle| \frac{\rho_1}{\rho_2}\right). \tag{5.3.39}$$

The next theorem relates a Hadamard beta integral to an ordinary beta integral.

Theorem 5.3.8 *For $-1 < x < 1$, we have*

$$\int_{[0}^{1]} (1 - u)^{-B(x)-1} u^{-A(x)-1} \, du = \frac{\Gamma(-A(x))\Gamma(-B(x))}{\Gamma(2\lambda)}, \quad \lambda \neq 0, \tag{5.3.40}$$

and

$$\int_{[0}^{1]} (1 - u)^{-B(x)} u^{-A(x)-1} \, du = \frac{\Gamma(-B(x)+1)\Gamma(-A(x))}{\Gamma(2\lambda+1)}. \tag{5.3.41}$$

Proof Note in the first place that $-A - B = 2\lambda$. We shall only give a proof of (5.3.40) because (5.3.41) can be proved similarly. When $-1 < x < 1$, we have

$$A(x) = -\lambda - i\frac{ax+b}{\sqrt{1-x^2}}, \quad B(x) = -\lambda + i\frac{ax+b}{\sqrt{1-x^2}}, \tag{5.3.42}$$

so that $\mathrm{Re}\,(A(x)) = \mathrm{Re}(B(x)) = -\lambda$. If $\lambda > 0$, (5.3.40) and (5.3.41) are just the beta integral. Now, assume $-\frac{1}{2} < \lambda < 0$ and $0 < z < 1$. Clearly

$$\int_{[0}^{1]} (1-u)^{-B-1} u^{-A-1}\, du = \int_{[0}^{z} (1-u)^{-B-1} u^{-A-1}\, du + \int_{z}^{1]} (1-u)^{-B-1} u^{-A-1}\, du.$$

$$(5.3.43)$$

By the definition of the Hadamard integral,

$$\int_{[0}^{z} (1-u)^{-B-1} u^{-A-1}\, du = z^{-A} \sum_{n=0}^{\infty} \frac{(B+1)_n}{n!} \cdot \frac{z^n}{n-A}. \qquad (5.3.44)$$

For the time being we let λ be a complex number in the domain U given by $\mathrm{Re}(\lambda) > -\frac{1}{2}$, $\lambda \neq 0$. Then, the right side of (5.3.40) is an analytic function of λ in this domain, and an argument based on (5.3.44) shows that

$$f(\lambda) := \int_{0}^{z} (1-u)^{-B-1} u^{-A-1}\, du$$

is analytic in U. On the other hand, the function

$$g(\lambda) := \int_{z}^{1]} (1-u)^{-B-1} u^{-A-1}\, du = \int_{[0}^{1-z} u^{-B-1}(1-u)^{-A-1}\, du$$

is also analytic in U. Since, from (5.3.43),

$$f(\lambda) + g(\lambda) = \frac{\Gamma(-A)\Gamma(-B)}{\Gamma(2\lambda)}$$

for $\mathrm{Re}(\lambda) > 0$, the above equality also holds in U and, in particular, for $-\frac{1}{2} < \lambda < 0$. This completes the proof of the theorem. $\qquad \square$

5.4 Pollaczek Polynomials

The (general) Pollaczek polynomials $P_n^\lambda(x; a, b)$ satisfy the three term recurrence relation (Szegő, 1950b), (Chihara, 1978),

$$(n+1)P_{n+1}^\lambda(x; a, b) = 2[(n+\lambda+a)x + b]P_n^\lambda(x; a, b)$$
$$- (n+2\lambda-1)P_{n-1}^\lambda(x; a, b), \quad n > 0, \qquad (5.4.1)$$

and the initial conditions

$$P_0^\lambda(x; a, b) = 1, \quad P_1^\lambda(x; a, b) = 2(\lambda+a)x + 2b. \qquad (5.4.2)$$

Pollaczek (Pollaczek, 1949a) introduced these polynomials when $\lambda = 1/2$ and Szegő (Szegő, 1950b) generalized them by introducing the parameter λ. By comparing (5.4.1) and (5.1.1) we see that

$$C_n^{(\lambda)}(x) = P_n^\lambda(x; 0, 0). \qquad (5.4.3)$$

The monic polynomials associated with (5.4.1) and (5.4.2) are

$$Q_n^\lambda(x; a, b) := \frac{n!}{2^n (a + \lambda)_n} P_n^\lambda(x; a, b), \tag{5.4.4}$$

and the monic recurrence relation is

$$Q_0^\lambda(x; a, b) = 1, \quad Q_1^\lambda(x; a, b) = x + b/(\lambda + a)$$

$$Q_{n+1}^\lambda(x; a, b) = \left[x + \frac{b}{n + a + \lambda} \right] Q_n^\lambda(x; a, b)$$

$$- \frac{n(n + 2\lambda - 1)}{(a + \lambda + n - 1)_2} Q_n^\lambda(x; a, b). \tag{5.4.5}$$

It is easy to see from (5.4.1)–(5.4.2) that

$$P_n^\lambda(-x; a, b) = (-1)^n P_n^\lambda(x; a, -b), \tag{5.4.6}$$

hence there is no loss of generality in assuming $b \geq 0$.

Let

$$F(x, t) := \sum_{n=0}^{\infty} P_n^\lambda(x; a, b) \, t^n. \tag{5.4.7}$$

It is straightforward to use the technique of §5.1 to convert the recurrence relation (5.4.1) and (5.4.2) to the differential equation

$$\frac{\partial F}{\partial t} = \frac{2(\lambda + a)x + 2b - 2\lambda t}{1 - 2xt + t^2} F,$$

whose solution through a partial fraction decomposition is

$$\sum_{n=0}^{\infty} P_n^\lambda(x; a, b) \, t^n = \left(1 - te^{i\theta}\right)^{-\lambda + i\Phi(\theta)} \left(1 - te^{-i\theta}\right)^{-\lambda - i\Phi(\theta)}, \tag{5.4.8}$$

where

$$x = \cos\theta, \quad \text{and} \quad \Phi(\theta) := \frac{a \cos\theta + b}{\sin\theta}. \tag{5.4.9}$$

The generating function (5.4.8) leads to the explicit form

$$P_n^\lambda(\cos; a, b) = e^{in\theta} \frac{(\lambda - i\Phi(\theta))_n}{n!} \, {}_2F_1 \left(\begin{matrix} -n, \lambda + i\Phi(\theta) \\ -n - \lambda + i\Phi(\theta) \end{matrix} \middle| e^{-2i\theta} \right). \tag{5.4.10}$$

It is not clear that the right-hand side of (5.4.10) is a polynomial in $\cos\theta$. An interesting problem is to find an alternate representation for the above right-hand side which clearly exhibits its polynomial character.

The proof we give below of the orthogonality relation of the Pollaczek polynomials is due to Szegő and uses the following lemma

Lemma 5.4.1 *Let A and B be real and assume that $A > |B|$. Then*

$$\int_{-\pi}^{\pi} \exp\left(\frac{A \cos\theta + B}{i \sin\theta} \right) d\theta = 2\pi e^{-A}. \tag{5.4.11}$$

The above lemma was stated in (Szegő, 1950b) under the condition $A \geq |B|$. We do not believe the result is valid when $A = \pm B$.

The proof consists of putting $z = e^{i\theta}$ changing the integral to a contour integral over the unit circle with indentations at $z = \pm 1$, prove that the integration on the indentations goes to zero, then evaluate the integral by Cauchy's theorem. The only singularity inside the contour is at $z = 0$.

Theorem 5.4.2 *When $a > |b|$, $\lambda > 0$ then the Pollaczek polynomials satisfy the orthogonality relation*

$$\int_{-1}^{1} P_m^\lambda(x; a, b) \, P_n^\lambda(x; a, b) w^\lambda(x; a, b) \, dx$$

$$= \frac{2\pi \Gamma(n + 2\lambda) \, \delta_{m,n}}{2^{2\lambda} \, (n + \lambda + a) \, n!},$$

(5.4.12)

where

$$w^\lambda(x; a, b) = \left(1 - x^2\right)^{\lambda - 1/2} \exp\left(2\theta - \pi\right)\Phi(\theta)) \left|\Gamma(\lambda + i\Phi(\theta))\right|^2, \qquad (5.4.13)$$

for $x = \cos\theta \in (-1, 1)$.

Proof Let t_1 and t_2 be real, $|t_1| < 1$, $|t_2| < 1$. Define $H = H(\theta)$ by

$$H = \frac{(1 + t_1 t_2)\cos\theta - t_1 - t_2}{(1 - t_1 t_2)\sin\theta},$$

so that

$$\left(1 - t_1 e^{i\theta}\right)\left(1 - t_2 e^{i\theta}\right) = e^{i(\theta - \pi/2)}\left(1 - t_1 t_2\right)\sin\theta\,(1 + iH).$$

Since t_1 and t_2 are real and recalling (5.4.8) and (5.4.7) we find

$$F\left(\cos\theta, t_1\right) F\left(\cos\theta, t_2\right) w^\lambda(\cos\theta; a, b)\sin\theta$$
$$= \left[(1 - t_1 t_2)\sin\theta\right]^{-2\lambda} e^{(\pi - 2\theta)\Phi(\theta)}$$
$$\times (1 + iH)^{-\lambda + i\Phi(\theta)}(1 - iH)^{-\lambda - i\Phi(\theta)} w^\lambda(\cos\theta; a, b)\sin\theta.$$

Let I denote the integral of the above function on $(0, \pi)$ and use

$$\Gamma(\lambda \pm i\Phi(\theta))\left(1 \mp iH\right)^{-\lambda \mp i\Phi(\theta)} = \int_0^\infty e^{-(1 \mp iH)s} s^{\lambda \pm i\Phi(\theta) - 1} ds, \qquad (5.4.14)$$

to establish

$$I = (1 - t_1 t_2)^{-2\lambda} \int_0^\infty\int_0^\infty e^{-s_1 - s_2}\left(s_1 s_2\right)^{\lambda - 1}$$

$$\times \int_0^\pi \exp\left(-iH\left(s_1 - s_2\right) - i\Phi(\theta)\left(\log s_1 - \log s_2\right)\right) d\theta \, ds_1 ds_2.$$

Write $\int_0^\infty \int_0^\infty \cdots ds_1 ds_2$ as

$$\int_0^\infty \int_{s_2}^\infty \cdots ds_1 ds_2 + \int_0^\infty \int_{s_1}^\infty \cdots ds_2 ds_1,$$

then interchange s_1 and s_2 in the second integral. Interchanging s_1 and s_2 is equivalent to replacing θ by $-\theta$ in the integrand of the theta integral, hence the θ integral is now on $[-\pi, \pi]$. Thus the above relationship can be written in the form

$$I = (1 - t_1 t_2)^{-2\lambda} \int_0^\infty \int_{s_2}^\infty e^{-s_1 - s_2} (s_1 s_2)^{\lambda - 1}$$

$$\times \int_{-\pi}^{\pi} \exp\left(-iH(s_1 - s_2) - i\Phi(\theta)(\log s_1 - \log s_2)\right) d\theta \, ds_1 \, ds_2.$$

In the last equation use the substitution $s_1 = e^\sigma s_2$ in the inner integral. By Lemma 5.4.1 we obtain

$$\frac{I}{2\pi} = (1 - t_1 t_2)^{-2\lambda} \int_0^\infty \int_0^\infty s_2^{2\lambda - 1}$$

$$\times \exp\left(\sigma(\lambda - a) - s_2(1 + e^\sigma) + s_2 \frac{1 + t_1 t_2}{1 - t_1 t_2}(1 - e^\sigma)\right) ds_2 \, d\sigma$$

$$= (1 - t_1 t_2)^{-2\lambda} \Gamma(2\lambda) \int_0^\infty e^{\sigma(\lambda - a)} \left[1 + e^\sigma - \frac{1 + t_1 t_2}{1 - t_1 t_2}(1 - e^\sigma)\right]^{-2\lambda} d\sigma$$

$$= \frac{\Gamma(2\lambda)}{2^{2\lambda}} \int_0^\infty \sum_{n=0}^\infty \frac{(2\lambda)_n}{n!} (t_1 t_2)^n \, e^{-(\lambda + a + n)\sigma} \, d\sigma$$

$$= \frac{\Gamma(2\lambda)}{2^{2\lambda}} \sum_{n=0}^\infty \frac{(2\lambda)_n (t_1 t_2)^n}{n!(\lambda + a + n)},$$

after the application of the binomial theorem. The theorem now follows.   □

The proof of Theorem 5.4.2 given here is due to Szegő (Szegő, 1950b) who stated the result for $\lambda > -1$ and $a \geq |b|$. Upon the examination of the proof one can easily see that it is necessary that $\lambda > 0$ since (5.4.14) was used and $\lambda = \operatorname{Re} \alpha$. The measure of orthogonality when $a = \pm b$ may have discrete masses, as we shall see in the next section.

Let

$$1 > x_{n,1}(\lambda, a, b) > x_{n,2}(\lambda, a, b) > \cdots > x_{n,n}(\lambda, a, b) > -1, \qquad (5.4.15)$$

be the zeros of $P_n^\lambda(x; a, b)$ and let

$$x_{n,k}(\lambda, a, b) = \cos\left(\theta_{n,k}(\lambda, a, b)\right). \qquad (5.4.16)$$

Novikoff proved that

$$\lim_{n\to\infty} \sqrt{n}\,\theta_{n,k}(1/2;a,b) = \sqrt{2(a+b)}, \qquad (5.4.17)$$

(Novikoff, 1954). This should be contrasted with the case of ultraspherical polynomials where

$$\lim_{n\to\infty} \sqrt{n}\,\theta_{n,k}(\nu,0,0) = j_{\nu-1/2,k}.$$

Askey conjectured that (5.4.17) will continue to hold and guessed the form of error term. Askey's conjecture was proved in (Rui & Wong, 1996), and we now state it as a theorem.

Theorem 5.4.3 *We have*

$$\theta_{n,k}\left(\frac{1}{2};a,b\right) = \sqrt{\frac{a+b}{n}} + \frac{(a+b)^{1/6}}{2n^{5/6}}\,i_k + O\left(n^{-7/6}\right) \qquad (5.4.18)$$

where i_k is the kth positive zeros of the Airy function.

Rui and Wong proved an asymptotic formula for Pollaczek polynomials with $x = \cos\left(t/\sqrt{n}\right)$ which implies (5.4.18).

5.5 A Generalization

We now investigate the polynomials $\left\{P_n^\lambda(x;a,b)\right\}$ when the condition $a > |b|$ is violated. This section is based on (Askey & Ismail, 1984), and (Charris & Ismail, 1987). In order to study the asymptotics in the complex plane we follow the notation in (5.1.5)–(5.1.6). Recall that $\rho_1 = e^{-i\theta}$ if $\operatorname{Im} z > 0$ while $\rho_1 = e^{i\theta}$ if $\operatorname{Im} z < 0$. As in §5.2 we define a second solution to (5.4.1) with $P_0^{\lambda*}(x;a,b) = 0$, and $P_1^{\lambda*}(x;a,b) = 2(\lambda+a)$. With

$$F^*(x,t) := \sum_{n=0}^{\infty} P_n^{\lambda*}(x;a,b)\,t^n,$$

we convert the recurrence relation (5.4.1) through the new initial conditions to the differential equations

$$\frac{\partial F^*}{\partial t} - \frac{2(\lambda+a)x + 2b - 2\lambda t}{1 - 2xt + t^2}F^* = \frac{2(\lambda+a)}{1 - 2xt + t^2}$$

The appearance of the equations will be simplified if we use the notations

$$A = -\lambda + \frac{2b + a\left(\rho_1 + \rho_2\right)}{\rho_2 - \rho_1}, \qquad B = -\lambda + \frac{2b + a\left(\rho_1 + \rho_2\right)}{\rho_1 - \rho_2} \qquad (5.5.1)$$

Therefore

$$F^*(x,t) = 2(\lambda+a)\left(1 - t/\rho_2\right)^A \left(1 - t/\rho_1\right)^B$$

$$\times \int_0^t \left(1 - u/\rho_1\right)^{-B-1}\left(1 - u/\rho_2\right)^{-A-1} du, \qquad (5.5.2)$$

and we find

$$\lim_{n\to\infty} \frac{P_n^{\lambda*}(x;a,b)}{P_n^\lambda(x;a,b)} = 2(\lambda+a) \int_0^{\rho_1]} (1-u/\rho_1)^{-B-1} (1-u/\rho_2)^{-A-1}\, du, \quad (5.5.3)$$

for Im $x \neq 0$. In the present case the coefficients α_n and β_n in the monic form (5.4.5) are

$$\alpha_n = \frac{b}{n+\lambda+a}, \quad \beta_n = \frac{n(n+2\lambda-1)}{4(n+\lambda+a)(n+\lambda+a-1)}, \quad (5.5.4)$$

and are obviousely bounded. Thus the measure with respect to which the polynomials $\{P_n^\lambda(x;a,b)\}$ are orthogonal, say $\mu^\lambda(x;a,b)$ is compactly supported and Theorems 2.5.2 and 2.6.2 are applicable. Formula (5.5.3) implies

$$F^\lambda(z;a,b) := \int_\mathbb{R} \frac{d\mu^\lambda(y;a,b)}{z-y}$$

$$ \qquad\qquad\qquad\qquad\qquad (5.5.5)$$

$$= 2(\lambda+a) \int_0^{\rho_1]} (1-u/\rho_1)^{-B-1} (1-u/\rho_2)^{-A-1}\, du.$$

Using the Hadamard integral we write (5.5.5) in the more convenient form

$$F^\lambda(z;a,b) = -2\frac{(\lambda+a)}{B}\rho_1$$

$$\times \left[1 - (1-\rho_1^2)^{-A-1} \sum_{n=1}^\infty \frac{(A+1)_n}{n!} \left(\frac{\rho_1}{\rho_1-\rho_2}\right)^n \frac{n}{n-\rho_1} \right]. \qquad (5.5.6)$$

Before inverting the above Stieltjes transform to find μ^λ we determine the domains of the parameters λ, a, b. Recall from Theorems 2.5.2 and 2.2.1 that $\{P_n^\lambda(x;a,b)\}$ will orthogonal with respect to a positive measure if and only if α_n is real and $\beta_{n+1} > 0$, for all $n \geq 0$. Hence (5.5.4) implies that for orthogonality it is necessary and sufficient that

$$(n+2\lambda-1)(a+\lambda+n-1)(a+\lambda+n) > 0, \quad n = 1,2,\dots. \qquad (5.5.7)$$

It is easy to see that the inequalities (5.5.7) hold if and only if (i) or (ii) below hold,

(i) $\lambda > 0$, and $a+\lambda > 0$, (ii) $-1/2 < \lambda < 0$, and $-1 < a+\lambda < 0$. (5.5.8)

It is clear from (5.5.5)–(5.5.6) that the support of the absolutely continuous component of μ^λ is $[-1,1]$. Furthermore

$$\frac{d\mu^\lambda(x;a,b)}{dx} = \frac{F(x-i0^+) - F(x+i0^+)}{2\pi i}.$$

This establishes

$$\frac{d\mu^\lambda(x;a,b)}{dx} = \frac{(\lambda+a)}{\pi} \int_{e^{-i\theta}}^{e^{i\theta}} (1-ue^{i\theta})^{\lambda-1-i\Phi(\theta)} (1-ue^{-i\theta})^{-\lambda-1+i\Phi(\theta)}\, du.$$

The above integral is a beta integral when $\lambda > 0$. Theorem 5.3.8 gives

$$\frac{d\mu^\lambda(x; a, b)}{dx} = \frac{2^{2\lambda-1}(\lambda + a)}{\pi\Gamma(2\lambda)}\left(1 - x^2\right)^{\lambda - 1/2}$$
$$\times \exp\left((2\theta - \pi)\Phi(\theta)\right)\left|\Gamma(\lambda + i\Phi(\theta))\right|^2. \tag{5.5.9}$$

The measure μ^λ in (8.2.17) is normalized so that $\int_{\mathbb{R}} d\mu^\lambda(x; a, b) = 1$. This evaluates $d\mu^\lambda/dx$ in case (i). In case (ii) $-1/2 < \lambda < 0$ the integral giving μ^λ is now a Hadamard integral and one can argue that (8.2.17) continues to hold.

Let D be the set of poles of F^λ. Obviously, D coincides with the set of points supporting point masses for μ^λ. It is evident from (5.5.5) that the pole singularities of F^λ are at the solutions of

$$B(x) = n, \quad n = 0, 1, 2, \ldots. \tag{5.5.10}$$

Let

$$\Delta_n = (n + \lambda)^2 + b^2 - a^2,$$
$$x_n = \frac{-ab + (n + \lambda)\sqrt{\Delta_n}}{a^2 - (n + \lambda)^2}, \quad y_n = \frac{-ab - (n + \lambda)\sqrt{\Delta_n}}{a^2 - (n + \lambda)^2} \tag{5.5.11}$$

Using (5.3.20)–(5.3.25) and Lemma 5.3.6 one can prove the following theorems. The details are in Charris and Ismail (Charris & Ismail, 1987).

Theorem 5.5.1 *Let $a > |b|$. Then $D = \phi$ when $\lambda > 0$, but $D = \{x_0, y_0\}$, and $x_0 > 1$, $y_0 < -1$, if $\lambda < 0$.*

With the subdivision of the (λ, α) plane shown in Figure 1, one can prove the following theorem whose detailed proof follows from Theorem 4.25 in (Charris & Ismail, 1987); see also Theorem 6.2 in (Charris & Ismail, 1987).

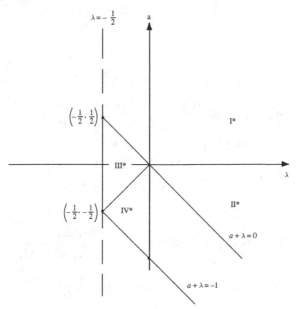

Theorem 5.5.2 *When $b \geq 0$ and $a \leq b$, the set D is as follows:*

Region I	(i)	$a < b.$ *Then* $D = \{x_n : n \geq 0\}.$
	(ii)	$a = b.$ *Then* $D = \emptyset.$
Region II	(i)	$-b \leq a < b.$ *Then* $D = \{x_n : n \geq 0\}.$
	(ii)	$a < -b.$ *Then* $D = \{x_n : n \geq 0\} \bigcup \{y_n : n \geq 0\}.$
Region III	(i)	$-b < a < b.$ *Then* $D = \{x_n : n \geq 0\},\ x_0 > 1.$
	(ii)	$a = -b \neq 0.$ *Then* $D = \{x_n : n \geq 1\}.$
	(iii)	$a < -b.$ *Then* $D = \{x_n : n > 1\} \bigcup \{y_n : n > 1\}.$
	(iv)	$a = b > 0 \ (= 0).$ *Then* $D = \{x_0\} \ (= \emptyset).$
Region IV	(i)	$-b < a.$ *Then* $D = \{x_n : n \geq 0\},\ x_0 > 1.$
	(ii)	$b = -a.$ *Then* $D = \{x_n : n \geq 1\}.$
	(iii)	$a < -b.$ *Then* $D = \{x_n : n \geq 1\} \bigcup \{y_n : n \geq 1\}.$

In all the regions $x_n < -1$ and $y_n > 1$ for $n \geq 1$. Also, $x_0 < -1$ and $y_0 > 1$ if $\lambda > 0$.

The symmetry relation

$$(-1)^n P_n^\lambda(x; a, -b) = P_n^\lambda(-x; a, b) \tag{5.5.12}$$

follows from (5.4.1) and (5.4.2). It shows that the case $a \leq -b$, $b \leq 0$ can be obtained from Theorem 5.5.2 interchanging x_n and y_n, $n \geq 0$.

We now determine the point masses located at the points in D. The point mass at $z = \zeta$ is the residue of $F^\lambda(z; a, b)$ at $z = \zeta$. The relationships (8.2.17) and (5.5.5) yield

$$\text{Res}\left\{F^\lambda(z; a, b) : z = \zeta\right\} = -2\frac{\lambda + a}{B'(\zeta)}\,\rho_1(\zeta) \quad \text{if } B(\zeta) = 0, \tag{5.5.13}$$

$$\text{Res}\left\{F^\lambda(z; a, b) : z = \zeta\right\} = -2(\lambda + a)\rho_1^{2n+1}\left(1 - \rho_1^2\right)^{2\lambda - 1}\frac{(2\lambda)_n}{n!}\frac{n}{B'(\zeta)}, \tag{5.5.14}$$

if $B(\zeta) = n \geq 1$.

Therefore

$$\text{Res}\left\{F^\lambda(z; a, b) : z = x_n\right\}$$
$$= (\lambda + a)\rho_1^{2n}\left(1 - \rho_1^2\right)^{2\lambda}\frac{(2\lambda)_n}{n!\,\sqrt{\Delta_n}}\frac{\left[a\sqrt{\Delta_n} - b(n + \lambda)\right]}{[a^2 - (n + \lambda)^2]},$$
$$\text{Res}\left\{F^\lambda(z; a, b) : z = y_n\right\} \tag{5.5.15}$$
$$= (\lambda + a)\rho_1^{2n}\left(1 - \rho_1^2\right)^{2\lambda}\frac{(2\lambda)_n}{n!\,\sqrt{\Delta_n}}\frac{\left[a\sqrt{\Delta_n} + b(n + \lambda)\right]}{[a^2 - (n + \lambda)^2]}.$$

Furthermore

$$\text{Res}\left\{F^\lambda(z; a, b) : z = x_0\right\} = -2(\lambda + a)\rho_1(x_0)\frac{\left[a\sqrt{\Delta_0} - b\lambda\right]^2}{\sqrt{\Delta_0}\,(a^2 - \lambda^2)^2}, \tag{5.5.16}$$

$$\text{Res}\left\{F^\lambda(z; a, b) : z = y_0\right\} = -2(\lambda + a)\rho_1(y_0)\frac{\left[a\sqrt{\Delta_0} + b\lambda\right]^2}{\sqrt{\Delta_0}\,(a^2 - \lambda^2)^2}. \tag{5.5.17}$$

With w^λ defined in (5.4.13) we have the orthogonality relation

$$
\int_{-1}^{1} w^\lambda(x;a,b) P_m^\lambda(x;a,b) P_n^\lambda(x;a,b)\, dx
$$

$$
+ \sum_{\zeta \in D} P_n^\lambda(\zeta;a,b) P_m^\lambda(\zeta;a,b) J_\zeta = \frac{2\pi\Gamma(n+2\lambda)}{2^{2\lambda}(n+\lambda+a)n!}\, \delta_{m,n},
$$

(5.5.18)

with

$$
J_\zeta = \frac{\pi\Gamma(2\lambda)}{\lambda+a}\, 2^{1-2\lambda}\, \mathrm{Res}\left\{ F^\lambda(z;a,b) : z = \zeta \right\}.
$$

(5.5.19)

The symmetric case $b = 0$ is in (Askey & Ismail, 1984). Their normalization was different because the Askey–Ismail polynomials arose as random walk polynomials, so their orthogonality measure is supported on $[-1,1]$. The Askey–Ismail normalization has the advantage of having the absolutely continuous part of μ supported on $[-\gamma,\gamma]$, for some γ, so we can let $\gamma \to 0$.

The random walk polynomials associated with

$$
\lambda_n = an + b, \quad \mu_n = n,
$$

(5.5.20)

were originally proposed by Karlin and McGregor, who only considered the case $a = 0$, (Karlin & McGregor, 1958). Surprisingly around the same time, Carlitz (independently and using a completely different approach) studied the same random walk polynomials ($\lambda_n = b, \mu_n = n$). We will include Carlitz' proof (Carlitz, 1958) at the end of this section.

We now present a summary of the results in (Askey & Ismail, 1984). Let

$$
G_n(x;a,b) = r_n(x)\, a^n \frac{(b/a)_n}{n!},
$$

(5.5.21)

with λ_n and μ_n as in (5.5.20). The recurrence relation satisfied by $\{G_n(x;a,b)\}$ is

$$
[b + n(a+1)]x G_n(x;a,b) = (n+1)G_{n+1}(x;a,b) + (an+b-a)G_{n-1}(x;a,b).
$$

(5.5.22)

Set

$$
\xi = \sqrt{(a+1)^2 x^2 - 4a},
$$

$$
\alpha = \frac{x(a+1)}{2a} + \frac{\xi}{2a}, \quad \beta = x(a+1)2a - \frac{\xi}{2a},
$$

(5.5.23)

and

$$
A = -\frac{b}{2a} - \frac{x(1-a)b}{2a\xi}, \quad B = -\frac{b}{2a} - \frac{x(1-a)b}{2a\xi}.
$$

(5.5.24)

Then

$$
\sum_{n=0}^{\infty} G_n(x;a,b) t^n = (1 - t/\alpha)^A (1 - t/\beta)^B,
$$

(5.5.25)

$$G_n(x; a, b) = \frac{(-B)_n}{n!} \beta^{-n} {}_2F_1\left(\begin{array}{c} -n, -A \\ -n + B + 1 \end{array}\bigg| \beta/\alpha\right)$$

$$= \frac{(b/a)_n}{n!} \alpha^{-n} {}_2F_1\left(\begin{array}{c} -n, -B \\ b/a \end{array}\bigg| - \xi\alpha\right).$$

(5.5.26)

Moreover,

$$\sum_{n=0}^{\infty} \frac{(\lambda)_n}{(b/a)_n} t^n G_n(x; a, b) = (1 - t/\alpha)^{-\lambda} {}_2F_1\left(\begin{array}{c} \lambda, -B \\ b/a \end{array}\bigg| \frac{t\xi}{1 - t/\alpha}\right). \quad (5.5.27)$$

To write down the orthogonality relation, we need the notation

$$x_k = (b + 2ak)[(b + k(a + 1))(b + ka(a + 1))]^{-1/2}$$

$$J_k = \frac{ba^k (b/a)_k}{2\, k!} \frac{[b(1 - a)]^{1+b/a}[b + k(a + 1)]^{k-1}}{[b + ka(a + 1)]^{k+1+b/a}}, \quad (5.5.28)$$

$$w(x; a, b) = \frac{b\, 2^{-1+b/a}}{\pi(a + 1)\Gamma(b/a)} (\sin\theta)^{-1+b/a}$$

$$\times \exp\left(\frac{b(a - 1)}{a(a + 1)} (\theta - \pi/2) \cot\theta\right) \left|\Gamma\left(\frac{b}{2a} + i\frac{b(1 - a)}{2a(a + 1)} \cot\theta\right)\right|^2,$$

(5.5.29)

$$x := \frac{2\sqrt{a}}{1 + a} \cos\theta, \quad 0 < \theta < \pi. \quad (5.5.30)$$

We have four parameter regions where $\{G_n\}$ are orthogonal with respect to a positive measure. In general, the orthogonality relation is

$$\int_{\frac{-2\sqrt{a}}{1+a}}^{\frac{2\sqrt{a}}{1+a}} G_m(x; a, b)\, G_n(x; a, b)\, w(x; a, b)\, dx$$

$$+ \sum_{k \in K} J_k \left\{G_m(x_k; a, b)\, G_n(x_k, a, b) + G_m(-x_k, a, b)\, G_m(-x_k, a, b)\right\}$$

$$= \frac{ba^n (b/a)_n}{n!\,[b + n(a + 1)]} \delta_{m,n}.$$

(5.5.31)

The polynomials $\{G_n\}$ are orthogonal with respect to a positive measure if and only if a and b belong to one of the following regions:

Region I $a > 1, b > 0.$ Here, K is empty.

Region II $0 \le a < 1, b > 0.$ Here, $K = \{0, 1, \dots\}$

Region III $a < 1, 0 < a + b < a.$ Here, $K = \{0\}$.

Region IV $a > 1, 0 < a + b < a.$ Here, $K = \{1, 2, \dots\}$.

When $a = 0$, the generating function becomes

$$\sum_{n=0}^{\infty} G_n(x; 0, b) t^n = e^{tb/x} (1 - xt)^{(1-x^2)b/x^2},$$ (5.5.32)

and the explicit form is

$$G_n(x; 0, b) = \sum_{k=0}^{n} \frac{b^{n-k}}{(n-k)!} x^{2k-n} \frac{(b(1 - 1/x^2))_k}{k!}.$$ (5.5.33)

Moreover,

$$(b + n) x \, G_n(x; 0, b) = (n + 1) \, G_{n+1}(x; 0, b) + b \, G_{n-1}(x; 0, a).$$ (5.5.34)

We now give Carlitz' proof of the orthogonality relation. He guessed the measure to have mass J_k at $\pm x_k$,

$$J_k = \frac{b(b+k)^{k-1}}{2\,(k!)} \exp(-k - b), \quad x_k = \sqrt{\frac{b}{b+k}},$$ (5.5.35)

$k = 0, 1, \ldots$. Let

$$I_n = \int_{\mathbb{R}} x^n G_n(x; 0, b) \, d\mu(x).$$

Since $G_n(-x; 0, b) = (-1)^n G_n(x, 0, b)$,

$$I_n = 2 \int_0^{\infty} x^n G_n(x; 0, b) \, d\mu(x)$$

$$= \sum_{k=0}^{\infty} \frac{b(b+k)^{k-1}}{k!} e^{-k-b} \sum_{j=0}^{n} \frac{b^{n-j}}{(n-j)!} \left(\frac{b}{b+k}\right)^j \frac{(-k)_j}{j!}$$

$$= b^{n+1} e^{-b} \sum_{j=0}^{n} \frac{(-1)^j e^{-j}}{j!\,(n-j)!} \sum_{k=0}^{\infty} \frac{(b+k+j)^{k-1}}{k!} e^{-k}.$$

Now (1.2.4) gives

$$I_n = \frac{b^{n+1} e^{-b}}{n!} \sum_{j=0}^{n} \frac{(-n)_j}{j!} \frac{e^b}{b+j} = \frac{b^n}{n!} \, {}_2F_1(-n, b; b + 1; 1).$$

Therefore, the Chu–Vandermonde sum leads to

$$I_n = b^n / (b + 1)_n.$$ (5.5.36)

Multiply (5.5.34) by x^{n-1} and integrate with respect to μ to get

$$(b + n) I_n = (n + 1) \int_{\mathbb{R}} x^{n-1} G_{n+1}(x; 0, b) \, d\mu(x) + b I_{n-1}.$$

Apply (5.5.36) and conclude that

$$\int_{\mathbb{R}} x^{n-1} G_{n+1}(x; 0, b) \, d\mu(x) = 0, \quad n > 0.$$

Since μ is symmetric around $x = 0$,

$$\int_{\mathbb{R}} x^{n-2k-1} G_n(x; 0, b) \, d\mu(x) = 0, \quad k = 0, 1, \ldots, \quad \lfloor (n-1)/2 \rfloor.$$

Moreover, (5.5.34) yields

$$(b+n) \int_{\mathbb{R}} x^{n-2k} G_n(x; 0, b) \, d\mu(x)$$

$$= (n+1) \int_{\mathbb{R}} x^{n-2k-1} G_{n+1}(x; 0, b) \, d\mu(x)$$

$$+ b \int_{\mathbb{R}} x^{n-2k-1} G_{n-1}(x; 0, b) \, d\mu(x).$$

From $k = 1$, we conclude that

$$\int_{\mathbb{R}} x^{n-4} G_n(x; 0, b) \, d\mu(x) = 0, \quad n \geq 4,$$

then, by induction, we prove that

$$\int_{\mathbb{R}} x^{n-2k} G_n(x; 0, b) \, d\mu(x) = 0, \quad k = 1, 2, \ldots \lfloor n/2 \rfloor,$$

and the orthogonality follows. Formula (1.2.4) implies $\int_{\mathbb{R}} d\mu(x) = 1$. Thus, the orthogonality relation is

$$\sum_{k=0}^{\infty} J_k \left\{ G_m(x_k; 0, b) \, G_n(x_k; 0, b) + G_m(-x_k; 0, b) \, G_n(-x_k; 0, b) \right\}$$

$$= \frac{b^{n+1}}{(n+1)! \, (b+n)} \delta_{m,n}.$$

(5.5.37)

Remark 5.5.1 *Carlitz' proof raises the question of finding a direct special function proof of the general orthogonality relation (5.5.18). It is unlikely that the integral and the sum in (5.5.18) can be evaluated separately, so what is needed is a version of the Lagrange expansion (1.2.4) or of Theorem 1.2.3, where one side is a sum plus an integral. A hint may possibly come from considering some special values of m ($= n$) in (5.5.18).*

Remark 5.5.2 *Carlitz' proof shows that (1.2.4) is what is behind the orthogonality of $\{G_n(x; 0, b)\}$. The more general (1.2.5) has not been used in orthogonal polynomials, and an interesting problem is to identify the orthogonal polynomials whose orthogonality relation uses (1.2.5).*

5.6 Associated Laguerre and Hermite Polynomials

The Laguerre polynomials are birth and death process polynomials with rates $\lambda_n = n + \alpha + 1$, $\mu_n = n$. According to Remark 5.2.1 we will have two birth and death process models arising from their associated polynomials. For these models we have

$$\text{Model I}: \lambda_n = n + c + \alpha + 1, \quad \mu_n = n + c, \ n \geq 0, \tag{5.6.1}$$

$$\text{Model II}: \lambda_n = n + c + \alpha + 1, \quad \mu_{n+1} = n + c, \ n \geq 0, \ \mu_0 = 0. \tag{5.6.2}$$

The treatment of associated Laguerre and Hermite polynomials presented here is from (Ismail et al., 1988).

Recall that the generating function satisfies the differential equation (5.2.27), which in this case becomes

$$w(1-w)\frac{\partial F}{\partial w} + [(1-w)\{c - (c+\alpha+1)w\} + xw]F = c(1-w)^\eta, \tag{5.6.3}$$

where

$$\eta := 0 \text{ in Model I}, \quad \eta := 1 \text{ in Model II}. \tag{5.6.4}$$

The general solution of (5.6.3) is

$$F(x,w) = w^{-c}(1-w)^{-\alpha-1} \exp\left(\frac{-x}{1-w}\right)$$
$$\times \left[C + c\int_a^w (1-u)^{\eta+\alpha-1} u^{c-1} \exp\left(\frac{-x}{1-u}\right) du \right], \tag{5.6.5}$$

for some constant C and a, $0 < a < 1$. When $c \geq 0$ the boundary condition $F(x,0) = 1$ implies the integral representation

$$F(x,w) = cw^{-c}(1-w)^{-\alpha-1} \exp\left(\frac{-x}{1-w}\right)$$
$$\times \int_0^w (1-u)^{\eta+\alpha-1} u^{c-1} \exp\left(\frac{-x}{1-u}\right) du.$$

In other words

$$F(x, z/(1+z)) = cz^{-c}(1+z)^{c+\alpha+1}$$
$$\times \int_0^z v^{c-1}(1+v)^{-\alpha-c-\eta} e^{x(v-z)}\, dv. \tag{5.6.6}$$

The second boundary condition in (5.2.28) gives

$$z^c(1+z)^{-c-\alpha-1}$$
$$= c\int_0^\infty \left\{ \int_0^z v^{c-1}(1+v)^{-\alpha-c-\eta} e^{-x(z-v)}\, dv \right\} d\mu(x). \tag{5.6.7}$$

The inner integral is a convolution of two functions, so we apply the Laplace transform to the above identity and obtain

$$\int_0^\infty \frac{d\mu(x)}{x+p} = \frac{\Psi(c+1, 1-\alpha; p)}{\Psi(c, 1-\alpha-\eta; p)}. \tag{5.6.8}$$

Recall that we require $\lambda_n > 0$, $\mu_{n+1} > 0$ for $n \geq 0$ and $\mu_0 \geq 0$. This forces

$$c \geq 0, \quad \text{and} \quad c + \alpha + 1 > 0, \quad \text{in Model I}$$
$$c > -1, \quad \text{and} \quad c + \alpha + 1 > 0, \quad \text{in Model II.} \tag{5.6.9}$$

If $0 > c > -1$ in Model II, the integral representation (5.6.6) is not valid so we go back to (5.6.5), and integrate by parts (by integrating cu^{c-1}) then apply the boundary condition (5.2.28). This establishes

$$\int_0^\infty \frac{d\mu(x)}{x + p} = \frac{\Psi(c+1, 2-\alpha; p) - \Psi(c+2, 2-\alpha; p)}{\alpha\Psi(c+1, 1-\alpha; p) + p\Psi(c+1, 2-\alpha; p)}. \tag{5.6.10}$$

Using the contiguous relations (6.6.6)–(6.6.7) of (Erdélyi et al., 1953b) we reduce the right-hand side of (5.6.10) to the right-hand side of (5.6.8). Thus (5.6.8) hold in all cases.

Theorem 5.6.1 *Let $\left\{ L_n^{(\alpha)}(x; c) \right\}$ and $\left\{ \mathcal{L}_n^{(\alpha)}(x; c) \right\}$ be the F_n's in Models I and II, respectively, and let μ_1 and μ_2 be their spectral measures. Then*

$$L_0^{(\alpha)}(x; c) = 1, L_1^{(\alpha)}(x; c) = \frac{2c + \alpha + 1 - x}{c + 1},$$
$$(2n + 2c + \alpha + 1 - x) L_n^{(\alpha)}(x; c) \tag{5.6.11}$$
$$= (n + c + 1) L_{n+1}^{(\alpha)}(x; c) + (n + c + \alpha) L_{n-1}^{(\alpha)}(x; c), \quad n > 0,$$

and

$$\mathcal{L}_0^{(\alpha)}(x; c) = 1, \mathcal{L}_1^{(\alpha)}(x; c) = \frac{c + \alpha + 1 - x}{c + 1},$$
$$(2n + 2c + \alpha + 1 - x) \mathcal{L}_n^{(\alpha)}(x; c) \tag{5.6.12}$$
$$= (n + c + 1) \mathcal{L}_{n+1}^{(\alpha)}(x; c) + (n + c + \alpha) \mathcal{L}_{n-1}^{(\alpha)}(x; c),$$

$$\int_0^\infty \frac{d\mu_j(x)}{x + p} = \frac{\Psi(c+1, 1-\alpha; p)}{\Psi(c, 2-\alpha-j; p)}, \quad j = 1, 2. \tag{5.6.13}$$

Moreover the measures μ_1 and μ_2 are absolutely continuous and

$$\mu_1'(x) = x^\alpha e^{-x} \frac{\left| \Psi(c, 1-\alpha; xe^{-i\pi}) \right|^{-2}}{\Gamma(c+1)\Gamma(1+c+\alpha)},$$
$$\mu_2'(x) = x^\alpha e^{-x} \frac{\left| \Psi(c, -\alpha, xe^{-i\pi}) \right|^{-2}}{\Gamma(c+1)\Gamma(1+c+\alpha)}. \tag{5.6.14}$$

Furthermore the polynomials $\left\{ L_n^{(\alpha)}(x; c) \right\}$ and $\left\{ \mathcal{L}_n^{(\alpha)}(x; c) \right\}$ have the orthogonality relations

$$\int_0^\infty p_{m,j}(x) p_{n,j}(x) \, d\mu_j(x) = \frac{(\alpha + c + 1)_n}{(c + 1)_n} \delta_{m,n}, \tag{5.6.15}$$

for $j = 1, 2$, where $p_{n,1} = L_n^{(\alpha)}(x; c)$ and $p_{n,2} = \mathcal{L}_n^{(\alpha)}(x; c)$.

Proof Equations (5.6.11)–(5.6.13) have already been proven. The orthogonality relations (5.6.15) follow from the three-term recurrence relations in (5.6.11)–(5.6.12). We will only evaluate μ'_2 of (5.6.14) because the evaluation of μ'_1 is similar. First apply $\Psi'(a, c; x) = -a\Psi(a + 1, c + 1; x)$ to write the right-hand side of (5.6.13) as

$$\frac{-1}{c} \frac{\Psi'(c, -\alpha; p)}{\Psi(c, -\alpha; p)} \qquad (5.6.16)$$

In our case we follow the notation in (Erdélyi et al., 1953a) and write

$$y_1 := \Phi(c, -\alpha; x), \quad y_2(x) := x^{1+\alpha}\Phi(c + \alpha + 1, 2 + \alpha; x)$$

for solutions of the confluent hypergeometric differential equation. In this case the Wronskian of y_1 and y_2 is

$$y_1(x)y'_2(x) - y_2(x)y'_1(x) = (1 + \alpha)x^{\alpha}e^x, \qquad (5.6.17)$$

(Erdélyi et al., 1953a, §6.3). The Perron–Stieltjes inversion formula (1.2.9), equations (5.6.16)–(5.6.17), and the relationships

$$y_1\left(xe^{i\pi}\right) = y_1\left(xe^{-i\pi}\right), \quad y'_1\left(xe^{i\pi}\right) = y'_1\left(xe^{-i\pi}\right),$$
$$y_2\left(xe^{i\pi}\right) = e^{2\pi i\alpha}y_2\left(xe^{-i\pi}\right), \quad y'_2\left(xe^{i\pi}\right) = e^{2\pi i\alpha}y'_2\left(xe^{-i\pi}\right),$$

establish the second equation in (5.6.14) after some lengthy calculations. $\qquad \square$

We now find explicit representations for the polynomials $\{L_n^{\alpha}(x; c)\}$ and $\{\mathcal{L}_n^{\alpha}(x; c)\}$. Expand $e^{x(v-z)}$ in (5.6.6) in power series and apply the integral representation (5.6.6) to obtain

$$F(x, x/(1+z))$$
$$= \Gamma(c + 1)(1 + z)^{c+\alpha+1} \sum_{m=0}^{\infty} \frac{(-xz)^m}{\Gamma(c + m + 1)} {}_2F_1\left(\begin{matrix} c, \alpha + \eta + c \\ m + c + 1 \end{matrix} \middle| -z\right),$$

where the beta integral evaluation was used. The Pfaff–Kummer transformation (1.4.9) and the binomial theorem lead to

$$F(x, w) = \sum_{m,j,k=0}^{\infty} \frac{(\alpha + 1 + m)_j(c)_k(m + 1 - \alpha - \eta)_k}{(c + 1)_m\, j!\, k!\, (m + c + 1)_k} (-x)^m\, w^{m+j+k}.$$

Upon equating coefficients we find

$$F_n(x) = F_n(x; \alpha, c, \eta) = \frac{(\alpha + 1)_n}{n!}$$
$$\times \sum_{m=0}^{n} \frac{(-n)_m\, x^m}{(c + 1)_m(\alpha + 1)_m} {}_3F_2\left(\begin{matrix} m - n, m + 1 - \alpha - \eta, c \\ -\alpha - n, \quad c + m + 1 \end{matrix} \middle| 1\right). \qquad (5.6.18)$$

Of course $F_n(x; \alpha, c, 0) = L_n^{\alpha}(x; c)$ and $F_n(x; \alpha, c, 1) = \mathcal{L}_n^{\alpha}(x; c)$.

In view of (4.6.5)–(4.6.6) we define associated Hermite polynomials by

$$H_{2n+1}(x; c) = 2x(-4)^n(1 + c/2)_n L_n^{1/2}\left(x^2; c/2\right)$$
$$H_{2n}(x; c) = (-4)^n(1 + c/2)_n \mathcal{L}_n^{-1/2}\left(x^2; c/2\right). \qquad (5.6.19)$$

Their orthogonality relations are

$$\int_{\mathcal{R}} \frac{H_m(x;c)H_n(x;c)}{\left|D_{-c}\left(xe^{i\pi/2}\sqrt{2}\right)\right|^2} \, dx = 2^n\sqrt{\pi}\,\Gamma(n+c+1)\,\delta_{m,n}. \qquad (5.6.20)$$

The function D_{-c} in (5.6.20) is a parabolic cylinder function

$$D_{2\nu}(2x) = 2^\nu e^{-x^2}\Psi\left(-\nu,1/2;2x^2\right). \qquad (5.6.21)$$

The polynomials $\left\{L_n^{(\alpha)}(x;c)\right\}$ and $\{H_n(x;c)\}$ were introduced in (Askey & Wimp, 1984), where their weight functions and explicit formulas were also found. The work (Ismail et al., 1988) realized that birth and death processes naturally give rise to two families of associated Laguerre polynomials and found an explicit representation and the weight function for the second family. They also observed that the second family manifested itself in the representation of $H_{2n}(x;c)$ in (5.6.19). The original representation in (Askey & Wimp, 1984) was different. The results on Model II are from (Ismail et al., 1988). It is then appropriate to call $\left\{L_n^{(\alpha)}(x;c)\right\}$ the Askey–Wimp polynomials and refer to $\left\{\mathcal{L}_n^{(\alpha)}(x;c)\right\}$ as the ILV polynomials, after the authors of (Ismail et al., 1988).

5.7 Associated Jacobi Polynomials

The techniques developed by Pollaczek in (Pollaczek, 1956) can be used to find orthogonality measures for several families of associated polynomials. In this section we not only find the orthogonality measures of two families of Jacobi polynomials, but we also present many of their algebraic properties.

A detailed study of the associated Jacobi polynomials is available in (Wimp, 1987) and (Ismail & Masson, 1991). The polynomials are denoted by $\left\{P_n^{(\alpha,\beta)}(x;c)\right\}$ and are generated by

$$P_{-1}^{(\alpha,\beta)}(x;c) = 0, \quad P_0^{(\alpha,\beta)}(x;c) = 1 \qquad (5.7.1)$$

and

$$\begin{aligned} 2(n&+c+1)(n+c+\gamma)(2n+2c+\gamma-1)p_{n+1} \\ &= (2n+2c+\gamma)\left[(2n+2c+\gamma-1)(2n+2c+\gamma+1)x\right.\\ &\quad \left.+(\alpha^2-\beta^2)\right]p_n - 2(n+c+\alpha) \\ &\quad \times(n+c+\beta)(2n+2c+\gamma+1)p_{n-1}, \end{aligned} \qquad (5.7.2)$$

where p_n stands for $P_n^{(\alpha,\beta)}(x;c)$ and

$$\gamma := \alpha + \beta + 1. \qquad (5.7.3)$$

We shall refer to $\left\{P_n^{(\alpha,\beta)}(x;c)\right\}$ as the Wimp polynomials. It is easy to see that $(-1)^n P_n^{(\beta,\alpha)}(-x;c)$ also satisfies (5.7.2) and has the same initial conditions as $P_n^{(\alpha,\beta)}(x)$. Thus,

$$P_n^{(\alpha,\beta)}(-x;c) = (-1)^n P_n^{(\beta,\alpha)}(x;c). \qquad (5.7.4)$$

Wimp proved the following theorem.

Theorem 5.7.1 *The associated Jacobi polynomials $P_n^{(\alpha,\beta)}(x;c)$ have the explicit form*

$$P_n^{(\alpha,\beta)}(x;c) = \frac{(\gamma+2c)_n(\alpha+c+1)_n}{(\gamma+c)_n\,n!}$$

$$\times \sum_{k=0}^{n} \frac{(-n)_k(n+\gamma+2c)_k}{(c+1)_k(\alpha+c+1)_k}\left(\frac{1-x}{2}\right)^k \tag{5.7.5}$$

$$\times {}_4F_3\left(\begin{array}{c} k-n, n+\gamma+k+2c, \alpha+c, c \\ \alpha+k+c+1, k+c+1, \gamma+2c-1 \end{array}\bigg| 1\right),$$

and satisfy the orthogonality relation

$$\int_{-1}^{1} P_m^{(\alpha,\beta)}(t;c)P_n^{(\alpha,\beta)}(t;c)\,w(t;c) = 0 \tag{5.7.6}$$

if $m \neq n$, where

$$w(t;c) := \frac{(1-t)^\alpha(1+t)^\beta}{|F(t)|^2} \tag{5.7.7}$$

and

$$F(t) := {}_2F_1\left(\begin{array}{c} c, 2-\gamma-c \\ 1-\beta \end{array}\bigg|\frac{1+t}{2}\right)$$

$$+K(c)(1+t)\,{}_2F_1\left(\begin{array}{c} \beta+c, 1-\alpha-c \\ 1+\beta \end{array}\bigg|\frac{1+t}{2}\right), \tag{5.7.8}$$

and

$$K(c) = e^{i\pi\beta}\frac{\Gamma(-\beta)\Gamma(c+\beta)\Gamma(c+\gamma-1)}{2\Gamma(\beta)\Gamma(c+\gamma-\beta-1)\Gamma(c)}. \tag{5.7.9}$$

Wimp also proved that $P_n^{(\alpha,\beta)}(x)$ satisfies the differential equation

$$A_0(x)\,y'''' + A_1(x)\,y''' + A_2(x)\,y'' + A_3(x)\,y' + A_4(x)\,y = 0, \tag{5.7.10}$$

with

$$A_0(x) = \left(1-x^2\right)^2$$
$$A_1(x) = -10x\left(1-x^2\right)$$
$$A_2(x) = -(1-x)^2\left(2K+2C+\gamma^2-25\right)$$
$$\qquad\quad + 2(1-x)\left(2k+2C+2\alpha\gamma\right)+2(\alpha+1)-26 \tag{5.7.11}$$
$$A_3(x) = 3(1-x)\left(2K+2C+\gamma^2-5\right)-6(K+C+\alpha\gamma+\beta-2)$$
$$A_4(x) = n(n+2)(n+\gamma+2c)(n+\gamma+2c-2),$$

where

$$K = (n+c)(n+\gamma+c), \quad C = (c-1)(c+\alpha+\beta). \tag{5.7.12}$$

Moreover, Wimp gave the representation

$$
P_n^{(\alpha,\beta)}(x;c) = \frac{\Gamma(c+1)\Gamma(\gamma+c)}{\alpha\Gamma(\alpha+c)\Gamma(\beta+c)(\gamma+2c-1)}
$$

$$
\times \left\{ \frac{\Gamma(c+\beta)\Gamma(n+\alpha+c+1)}{\Gamma(\gamma+c-1)\Gamma(n+c+1)} \; {}_2F_1\left(\begin{matrix} c, 2-\gamma-c \\ 1-\alpha \end{matrix} \middle| \frac{1-x}{2} \right) \right.
$$

$$
\times \; {}_2F_1\left(\begin{matrix} -n-c, n+\gamma+c \\ \alpha+1 \end{matrix} \middle| \frac{1-x}{2} \right)
$$

$$
- \frac{\Gamma(\alpha+c)}{\Gamma(c)} \; \frac{\Gamma(n+\beta+1+c)}{\Gamma(n+c+\gamma)}
$$

$$
\times \; {}_2F_1\left(\begin{matrix} 1-c, \gamma+c-1 \\ \alpha+1 \end{matrix} \middle| \frac{1-x}{2} \right)
$$

$$
\left. \times \; {}_2F_1\left(\begin{matrix} n+c+1, 1-n-\gamma-c \\ 1-\alpha \end{matrix} \middle| \frac{1-x}{2} \right) \right\}.
$$

$$(5.7.13)$$

G. N. Watson proved the following asymptotic formula

$$
{}_2F_1\left(\begin{matrix} a+n, b-n \\ c \end{matrix} \middle| \sin^2\theta \right) = \frac{n^{-c+1/2}\Gamma(c)}{\sqrt{\pi}} \frac{(\cos\theta)^{c-a-b-1/2}}{(\sin\theta)^{c-1/2}}
$$

$$
\times \cos[2n\theta + (a-b)\theta - \pi(c-1/2)/2],
$$

$$(5.7.14)$$

see (Luke, 1969a, (14), p. 187) or (Erdélyi et al., 1953b, (17), p. 77), Wimp used (5.7.14) to establish

$$
P_n^{(\alpha,\beta)}(x;c) \approx \frac{\Gamma(c+1)\Gamma(\gamma+c)(2n\pi)^{-1/2}2^{(\beta-\alpha)/2}}{(\gamma+2c-1)\Gamma(\alpha+c)\Gamma(\beta+c)(1-x^2)^{1/4}}
$$

$$
\times \left\{ 2^{\alpha} \frac{\Gamma(\beta+c)\Gamma(\alpha)}{\Gamma(\gamma+c-1)(1-x)^{\alpha/2}(1+x)^{\beta/2}} \right.
$$

$$
\times \; {}_2F_1\left(\begin{matrix} c, 2-\gamma-c \\ 1-\alpha \end{matrix} \middle| \frac{1-x}{2} \right)
$$

$$
\times \cos(n\theta + (c+\gamma/2)\theta - \pi(\alpha+1/2)/2)
$$

$$
+ \frac{\Gamma(\alpha+c)\Gamma(-\alpha)}{\Gamma(c)(1-x)^{-\alpha/2}(1+x)^{-\beta/2}} \; {}_2F_1\left(\begin{matrix} 1-c, \gamma+c-1 \\ \alpha+1 \end{matrix} \middle| \frac{1-x}{2} \right)
$$

$$
\left. \times \cos(n\theta + (c+\gamma/2)\theta + \pi(\alpha-1/2)/2) \right\},
$$

$$(5.7.15)$$

where $x = \cos\theta$, $0 < \theta < \pi$. By applying a result of (Flensted-Jensen & Koornwinder, 1975), Wimp discovered the generating function

$$\sum_{n=0}^{\infty} \frac{(c+\gamma)_n (c+1)_n}{n!\,(\gamma+2c+1)_n}\, t^n\, P_n^{(\alpha,\beta)}(x;c)$$

$$= \frac{1/\beta}{(\gamma+2c-1)} \left(\frac{2}{1+t+R}\right)^{\gamma+c} (\beta+c)(\gamma+c-1)$$

$$\times\, {}_2F_1\left(\begin{matrix} c, 2-\gamma-c \\ 1-\beta \end{matrix}\;\middle|\;\frac{1+x}{2}\right)\, {}_2F_1\left(\begin{matrix} -c, \gamma+c \\ \beta+1 \end{matrix}\;\middle|\;\frac{1+t-R}{2t}\right)$$

$$\times\, {}_2F_1\left(\begin{matrix} \alpha+c+1, \gamma+c \\ \gamma+2c+1 \end{matrix}\;\middle|\;\frac{2t}{1+t+R}\right) - c(\gamma+c-\beta-1) \qquad (5.7.16)$$

$$\times \left(\frac{1+t+R}{2}\right)^{\beta}\, {}_2F_1\left(\begin{matrix} c+\beta, 1-c-\alpha \\ \beta+1 \end{matrix}\;\middle|\;\frac{1+x}{2}\right)$$

$$\times\, {}_2F_1\left(\begin{matrix} c+\alpha+1, -c-\beta \\ 1-\beta \end{matrix}\;\middle|\;\frac{1+t-R}{2t}\right)$$

$$\times\, {}_2F_1\left(\begin{matrix} \gamma+c+\beta, \gamma+c \\ \gamma+2c+1 \end{matrix}\;\middle|\;\frac{2t}{1+t+R}\right),$$

where $R = \sqrt{1+t^2-2xt}$, as in (4.3.10). When $c=0$, (5.7.16) does not reduce to (4.3.11), but to the generating function

$$\sum_{n=0}^{\infty} \left(\frac{\alpha+\beta+1}{\alpha+\beta+1+n}\right) t^n\, P_n^{(\alpha,\beta)}(x)$$

$$= \left(\frac{2}{1+t+R}\right)^{\alpha+\beta+1} {}_2F_1\left(\begin{matrix} \alpha+1, \alpha+\beta+1 \\ \alpha+\beta+2 \end{matrix}\;\middle|\;\frac{2t}{1+t+R}\right), \qquad (5.7.17)$$

with R as in (4.3.10). It is easy to see that (4.3.11) follows if we multiply (5.7.17) by $t^{\alpha+\beta+1}$ then differentiate with respect to t.

The polynomials

$$Q_n(x) = \frac{(-1)^n (1+c)_n}{(\beta+c+1)_n}\, P_n^{(\alpha,\beta)}(x-1;c)$$

are birth and death process polynomials and satisfy (5.2.12) with

$$\lambda_n = \frac{2(n+c+\beta+1)(n+c+\alpha+\beta+1)}{(2n+2c+\alpha+\beta+1)_2}, \quad n \geq 0$$

$$\mu_n = \frac{2(n+c)(n+c+\alpha)}{(2n+2c+\alpha+\beta)_2}, \quad n \geq 0. \qquad (5.7.18)$$

For $c>0$, $\mu_0 \neq 0$. Remark 5.2.1 suggests that there is another family of birth and death process polynomials with birth and death rates as in (5.7.18), except that $\mu_0 = 0$. Ismail and Masson studied this family in (Ismail & Masson, 1991). Let $\left\{\mathcal{P}_n^{(\alpha,\beta)}(x;c)\right\}$ denote the orthogonal polynomials generated by (5.7.2) with the initial conditions

$$\mathcal{P}_0^{(\alpha,\beta)}(x;c) = 1, \quad \mathcal{P}_1^{(\alpha,\beta)}(x;c) = \frac{(1+\gamma)(\gamma+2c)_2}{2(c+1)(\gamma+c)} - \frac{\beta+c+1}{c+1}. \qquad (5.7.19)$$

We suggest calling $\left\{\mathcal{P}_n^{(\alpha,\beta)}(x;c)\right\}$ the Ismail–Masson polynomials. Ismail and Masson proved

$$\mathcal{P}_n^{(\alpha,\beta)}(x;c) = \frac{(-1)^n(\gamma+2c)_n(\beta+c+1)_n}{n!\,(\gamma+c)_n}\sum_{k=0}^{n}\frac{(-n)_k(\gamma+n+2c)_k}{(1+c)_k(c+1+\beta)_k}$$

$$\times\left(\frac{1+x}{2}\right)^k {}_4F_3\left(\begin{array}{c}k-n,n+\gamma+k+2c,c+\beta+1,c\\k+c+\beta+1,k+c+1,\gamma+2c\end{array}\bigg|\,1\right),$$

$$(5.7.20)$$

$$(-1)^n\mathcal{P}_n^{(\alpha,\beta)}(x;c) = \frac{(c+\beta+1)_n}{(c+1)_n}\,{}_2F_1\left(\begin{array}{c}-n-c,n+c+\gamma\\\beta+1\end{array}\bigg|\,\frac{1+x}{2}\right)$$

$$\times\,{}_2F_1\left(\begin{array}{c}c,1-c-\gamma\\-\beta\end{array}\bigg|\,\frac{1+x}{2}\right)-\frac{c(c+\alpha)_{n+1}(1+x)}{2\beta(\beta+1)(c+\gamma)_n}$$

$$\times\,{}_2F_1\left(\begin{array}{c}n+c+1,1-n-c-\gamma\\1-\beta\end{array}\bigg|\,\frac{1+x}{2}\right){}_2F_1\left(\begin{array}{c}1-c,c+\gamma\\2+\beta\end{array}\bigg|\,\frac{1+x}{2}\right).$$

$$(5.7.21)$$

Consequently

$$\mathcal{P}_n^{(\alpha,\beta)}(-1;c) = \frac{(-1)^n(c+\beta+1)_n}{(c+1)_n}.\qquad(5.7.22)$$

This also follows from (5.7.20) and the Pfaff–Saalschütz theorem. Applying Watson's asymptotic formula (5.7.14), Ismail and Masson proved

$$\mathcal{P}_n^{(\alpha,\beta)}(x;c) \approx \frac{\Gamma(\beta+1)\Gamma(c+1)}{\sqrt{n\pi}\,(c+\beta+1)}\left(\frac{1-x}{2}\right)^{-\alpha-1/2}\left(\frac{1+x}{2}\right)^{-\beta-1/2}$$

$$\times W(x)\cos[(n+c+\gamma/2)\theta+c+(2\gamma-1)/4-\eta],\qquad(5.7.23)$$

with

$$W(x) = \left|{}_2F_1\left(\begin{array}{c}c,-c-\beta-\alpha\\-\beta\end{array}\bigg|\,\frac{1+x}{2}\right)+\mathcal{K}(1+x)^{\beta+1}\right.$$

$$\left.\times\,{}_2F_1\left(\begin{array}{c}c+\beta+1,1-c-\alpha\\2+\beta\end{array}\bigg|\,\frac{1+x}{2}\right)\right|,\qquad(5.7.24)$$

$x = \cos\theta,\ \theta\in(0,\pi)$, and

$$\mathcal{K} = \frac{\Gamma(c+\gamma)\Gamma(c+\beta+1)}{\Gamma(c)\Gamma(c+\alpha)\Gamma(2+\beta)}\,2^{-\beta-1}e^{i\pi\beta}.$$

The phase shift η is derived from

$$W(x)\cos\eta = \left[{}_2F_1\left(\begin{array}{c}c,-c-\beta-\alpha\\-\beta\end{array}\bigg|\,\frac{1+x}{2}\right)+\mathcal{K}(1+x)^{\beta+1}\right.$$

$$\left.\times\,{}_2F_1\left(\begin{array}{c}c+\beta+1,1-c-\alpha\\2+\beta\end{array}\bigg|\,\frac{1+x}{2}\right)\right]\cos(\pi\beta/2),\qquad(5.7.25)$$

and

$$
W(x)\sin\eta = \left[\,{}_2F_1\left(\begin{array}{c}c,-c-\beta-\alpha\\-\beta\end{array}\middle|\frac{1+x}{2}\right) - \mathcal{K}(1+x)^{\beta+1}\right.
$$
$$
\left.\times\,{}_2F_1\left(\begin{array}{c}c+\beta+1,1-c-\alpha\\2+\beta\end{array}\middle|\frac{1+x}{2}\right)\right]\sin(\pi\beta/2).
$$

Ismail and Masson also gave the generating function

$$
\sum_{n=0}^{\infty}\frac{(\gamma+c)_n(c+1)_n t^n}{n!\,(\gamma+2c+1)_n}\,\mathcal{P}_n^{\alpha,\beta}(x;c)
$$
$$
=\left\{\frac{2}{1+t+R}\right\}^{c+\gamma}{}_2F_1\left(\begin{array}{c}c,1-c-\gamma\\-\beta\end{array}\middle|\frac{1+x}{2}\right)
$$
$$
\times\,{}_2F_1\left(\begin{array}{c}-c,c+\gamma\\1+\beta\end{array}\middle|\frac{1+t-R}{2t}\right)
$$
$$
\times\,{}_2F_1\left(\begin{array}{c}c+1+\alpha,\gamma\\\gamma+2c+1\end{array}\middle|\frac{2t}{1+t+R}\right) - \frac{c(c+\alpha)}{\beta(\beta+1)}
$$
$$
\times\left\{\frac{2}{1+t+R}\right\}^{c+1}\left(\frac{1+x}{2}\right){}_2F_1\left(\begin{array}{c}1-c,2+\gamma\\2+\beta\end{array}\middle|\frac{1+x}{2}\right)
$$
$$
\times\,{}_2F_1\left(\begin{array}{c}1-c-\gamma,c+1\\1-\beta\end{array}\middle|\frac{1+t-R}{2t}\right)
$$
$$
\times\,{}_2F_1\left(\begin{array}{c}\beta+c+1,c+1\\\gamma+2c+1\end{array}\middle|\frac{2t}{1+t+R}\right),
$$

where $R=\sqrt{1-2xt+t^2}$, as in (4.3.10). When $c=0$, the above generating function reduces to

$$
\sum_{n=0}^{\infty}\frac{\gamma}{\gamma+n}t^n P_n^{(\alpha,\beta)}(x) = \left(\frac{2}{1+t+R}\right)^{\gamma}{}_2F_1\left(\begin{array}{c}\alpha+1,\gamma\\\gamma+1\end{array}\middle|\frac{2t}{1+t+R}\right).\quad (5.7.26)
$$

One can prove (5.7.26) from (4.3.11).

Finally the orthogonality relation is

$$
\int_{-1}^{1} P_m^{(\alpha,\beta)}(x;c)P_n^{(\alpha,\beta)}(x;c)\frac{(1-x)^{\alpha}(1+x)^{\beta}}{W^2(x)}\,dx \qquad (5.7.27)
$$
$$
= h_n^{(\alpha,\beta)}(c)\delta_{m,n},
$$

where

$$
h_n^{(\alpha,\beta)}(c) = \frac{2^{\alpha+\beta+1}\Gamma(c+1)\Gamma^2(\beta+1)\Gamma(c+\alpha+n+1)\Gamma(c+\beta+n+1)}{(2n+2c+\gamma)\Gamma(c+\gamma+n)\Gamma^2(c+\beta+1)(c+1)_n}.
$$
$$
\qquad (5.7.28)
$$

Analogous to (4.10.10), one can define two families of associated Bessel polynomials by

$$y_n(x; a, b; c) = \lim_{\lambda \to \infty} \frac{n!}{(\lambda + 1)_n} P_n^{(\lambda, a - \lambda)}(1 + 2\lambda x/b; c) \qquad (5.7.29)$$

$$\mathcal{Y}_n(x; a, b; c) = \lim_{\lambda \to \infty} \frac{n!}{(\lambda + 1)_n} P_n^{(\lambda, a - \lambda)}(1 + 2\lambda x/b; c). \qquad (5.7.30)$$

Therefore $\gamma = a + 1$ and from (5.7.5) and (5.7.21) we find

$$
\begin{aligned}
y_n(x; a, b; c) &= \frac{(a + 1 + 2c)_n}{(a + 1 + c)_n} \sum_{k=0}^{n} \frac{(-n)_k(n + a + 1 + 2c)_k}{(c + 1)_k} \left(-\frac{x}{b}\right)^k \\
&\quad \times {}_3F_2 \left(\begin{matrix} k - n, n + a + 1 + 2c + k, c \\ k + c + 1, a + 2c \end{matrix} \middle| 1 \right), \\
\mathcal{Y}_n(x; a, b; c) &= \frac{(a + 1 + 2c)_n}{(a + c + 1)_n} \sum_{k=0}^{n} \frac{(-n)_k(a + 1 + n + 2c)_k}{(c + 1)_k(c + 1 + \beta)_k} \left(-\frac{x}{b}\right)^k \\
&\quad \times {}_3F_2 \left(\begin{matrix} k - n, n + a + 1 + k + 2c, c \\ k + c + 1, a + 1 + 2c \end{matrix} \middle| 1 \right).
\end{aligned}
\qquad (5.7.31)
$$

Generating functions and asymptotics can be established by taking limits of the corresponding formulas for associated Jacobi polynomials. We do not know of a weight function for either family of associated Bessel polynomials.

5.8 The J-Matrix Method

The J-Matrix method in physics leads naturally to orthogonal polynomials defined through three-term recurrence relations. The idea is to start with a Schrödinger operator T defined on $\mathbb{R}$, that is

$$T := -\frac{1}{2} \frac{d^2}{dx^2} + V(x). \qquad (5.8.1)$$

The operator T is densely defined on $L^2(\mathbb{R})$ and is symmetric. The idea is to find an orthonormal system $\{\varphi_n(x)\}$ which is complete in $L^2(\mathbb{R})$ such that φ_n is in the domain of T for every n and the matrix representation of T in $\{\varphi_n(x)\}$ is tridiagonal. In other words

$$\int_{\mathbb{R}} \overline{\varphi}_m T\varphi_n \, dx = 0, \text{ if } |m - n| > 1.$$

Next we diagonalize T, that is, set $T\psi_E = E\psi_E$ and assume

$$\psi_E(x) \sim \sum_{n=0}^{\infty} \varphi_n(x) \, p_n(E). \qquad (5.8.2)$$

Observe that

$$
\begin{aligned}
Ep_n(E) &= (E\psi_E, \varphi_n) = (T\psi_E, \varphi_n) \\
&= (T\varphi_{n-1}p_{n-1}(E) + T\varphi_n p_n(E) + T\varphi_{n+1}p_{n+1}(E), \varphi_n).
\end{aligned}
$$

Therefore,

$$
E\psi_n(E) = p_{n+1}(E)\,(T\varphi_{n+1}, \varphi_n) + p_n(E)\,(T\varphi_n, \varphi_n) \\
+ p_{n-1}(E)\,(T\varphi_{n-1}, \varphi_n)\,. \tag{5.8.3}
$$

If $(T\varphi_n, \varphi_{n+1}) \neq 0$ then (5.8.3) is a recursion relation for a sequence of orthogonal polynomials if and only if

$$
(T\varphi_n, \varphi_{n-1})\,(T\varphi_{n-1}, \varphi_n) > 0.
$$

The symmetry of T shows that the left-hand side of the above inequality is

$$
(\varphi_n, T\varphi_{n-1})\,(T\varphi_{n-1}, \varphi_n) = |(T\varphi_{n-1}, \varphi_n)|^2 > 0.
$$

The spectrum of T is now the support of the orthogonality measure of $\{p_n(e)\}$. This technique was developed in (Heller, 1975) and (Yamani & Fishman, 1975). See also (Broad, 1978), (Yamani & Reinhardt, 1975).

We first apply the above technique to the radial part of the Schrödinger operator for a free particle in 3 space. The operator now is

$$
H_0 = -\frac{1}{2}\frac{d^2}{dr^2} + \frac{\ell(\ell+1)}{2r^2}, \quad r > 0, \tag{5.8.4}
$$

where ℓ is an angular momentum number. The $\{\varphi_n\}$ basis is

$$
\varphi_n(r) = r^{\ell+1} e^{-r/2} L_n^{(2\ell+1)}(r), \quad n = 0, 1, \ldots\,. \tag{5.8.5}
$$

Using differential recurrence relations of Laguerre polynomials, we find that the matrix elements

$$
J_{m,n} = \int \varphi_m\,(H_0 - E)\,\varphi_n\,dx \tag{5.8.6}
$$

are given by

$$
J_{m,n} = \left(\frac{1}{8} + E\right)(n+1)\frac{\Gamma(2\ell+3+n)}{(n+1)!}\delta_{m,n+1} \\
+ \left(\frac{1}{8} - E\right)(2n+2\ell+2)\frac{\Gamma(n+2\ell+2)}{n!}\delta_{m,n} \tag{5.8.7} \\
+ \left(\frac{1}{8} + E\right)n\frac{\Gamma(n+2\ell+2)}{n!}\delta_{m,n-1}.
$$

Now $(H_0 - E)\psi_E = 0$ if and only if $JP = 0$, $J = (J_{m,n})$, $P = (u_0(E), u_1(E), \ldots)^T$. With

$$
x = \frac{E - 1/8}{E + 1/8}, \quad p_n(x) = \frac{\Gamma(n+2\ell+2)}{n!}u_n(E), \tag{5.8.8}
$$

we establish the following recurrence relation from (5.8.7)

$$
2x(n+\ell+1)p_n(x) = (n+1)\,p_{n+1}(x) + (n+2\ell+1)\,p_{n-1}(x). \tag{5.8.9}
$$

The recursion (5.8.9) is the three term recurrence relation for ultraspherical polynomials, see (4.5.3). Since the measure of $p_n(x)$ is absolutely continuous and is supported on $[-1, 1]$, we conclude that the spectrum of H_0 is continuous and is $[0, \infty)$

because $x \in [-1, 1]$ if and only if $E \in [0, \infty)$, as can be seen from (5.8.8). There are no bound states (discrete masses).

For the radial Coulomb problem, the Hamiltonian is

$$H = -\frac{1}{2} \frac{d^2}{dr^2} + \frac{\ell(\ell + 1)}{2r^2} + \frac{z}{r}. \qquad (5.8.10)$$

The Coulomb potential is attractive if $z < 0$ and repulsive if $z > 0$. When H_0 is replaced by H, the analogue of (5.8.9) is (Yamani & Reinhardt, 1975)

$$2[(n + \lambda + a)x - a] p_n(x) = (n + 1) p_{n+1}(x) + (n + 2\lambda - 1) p_{n-1}(x), \qquad (5.8.11)$$

where

$$x = \frac{E - 1/8}{E + 1/8}, \quad \lambda = \ell + 1, \quad a = 2z. \qquad (5.8.12)$$

In the above, $p_n(x)$ denotes $P_n^{(\lambda)}(x; a, -a)$. The recurrence relation (5.8.11) is the recurrence relation of Pollaczek polynomials. The measure is absolutely continuous when $z > 0$ (repulsive potential) and has infinite discrete part (bound states) when $z < 0$ (attractive potential). Indeed, $p_n(x) = P_n^{(\ell+1)}(x; 2z, -2z)$. It is important to note that the attractive Coulomb potential polynomials of (Bank & Ismail, 1985) have all the qualitative features of the more general Pollaczek polynomials treated in (Charris & Ismail, 1987) and, as such, deserve to be isolated and studied as a special polynomial system.

Indeed with

$$x_n = \left[a^2 + (\lambda + n)^2\right] / \left[a^2 - (\lambda + n)^2\right], \qquad (5.8.13)$$

$$J_k = 2^{4\lambda+1} \frac{(\lambda + a)}{k!} (2\lambda)_k \frac{\lambda + k + a}{\lambda + k - a} [-a(\lambda + k)]^{2\lambda}$$
$$\times (-a)(a + \lambda + k)^{2k-2} |\lambda + k - a|^{-2k-4\lambda}, \qquad (5.8.14)$$

$$w(x) = \frac{2^{2\lambda-1}(\lambda + a)}{\pi \, \Gamma(2\lambda)} (\sin \theta)^{2\lambda-1} |\Gamma(\lambda + ia \tan(\theta/2))|^2$$
$$\times \exp((2\theta - \pi)a \tan(\theta/2)), \qquad (5.8.15)$$

$x = \cos \theta$, the orthogonality relation becomes

$$\int_{-1}^{1} w(x) \, p_m(x) \, p_n(x) \, dx + \sum_{k \in K} p_m(x_k) \, p_n(x_k) \, J_k = \delta_{n,m},$$

where K is defined below and depends on the domain of the parameters:

Region I $\lambda \geq 0, a \geq 0,$ $K = $ empty.

Region II $\lambda > 0, 0 > a > -\lambda,$ $K = \{0, 1, 2, \dots\}.$

Region III $-1/2 < \lambda < 0, 0 < a < -\lambda,$ $K = \{0\}.$

Region IV $-1/2 < \lambda < 0, -1 - \lambda < a < 0,$ $K = \{1, 2, \dots\}.$

A discrete approximation to $Ty = \lambda y$ when T is given by (5.8.1) is

$$-\frac{1}{\delta^2}[y(x+\delta) - 2y(x) + y(x-\delta)] + \frac{1}{x}y(x) = \lambda y(x). \tag{5.8.16}$$

Aunola considered solutions of (5.8.16) of the form $y(x) = e^{\beta x}x\, g_n(x)$, where

$$g_n(x) = x^n + A_n x^{n-1} + \text{lower order terms.} \tag{5.8.17}$$

Substitute for y with g_n as in (5.8.17), in (5.8.16) then equate the coefficients of x^{n+1} and x^n. The result is that

$$\lambda = (1 - \cosh\beta\delta)/2, \quad \sinh\beta\delta = -\delta/(n+1).$$

Hence $\lambda = \lambda_n$ is given by

$$\delta^2\lambda_n = \sqrt{1 + \delta^2/(n+1)^2}. \tag{5.8.18}$$

Note that $\delta^2\lambda_n$ agree with the discrete spectrum in (5.5.11) with $\lambda = 1$, $a = 0$, $b = -\delta$, after a shift in the spectral variable. Indeed the discrete approximation $x = (m+1)\,\delta$ in (5.8.16) turns it to the recurrence relation (5.4.1) with $\lambda = 1$, $a = 0, b = -\delta, x = -\delta^2\lambda$, and $P_n^\lambda(x; a, b)$ replaced by $(-1)^n\, y_n$. The details are in (Aunola, 2005).

In the case of the quantum mechanical harmonic oscillator, the potential is $(C + 1/2)\, r^2$, $r > 0$, so the radial part of the Schrödinger wave equation is

$$-\frac{1}{2}\frac{d^2}{dr^2}\psi_E + \frac{\ell(\ell+1)}{r^2}\psi_E + (C+1/2)\, r^2\psi_E = E\psi_E.$$

It has a tridiagonal matrix representation in the basis

$$\chi_n(r) = r^{\ell+1}\exp\left(-r^2/2\right) L_n^{\ell+1/2}\left(r^2\right).$$

The coefficients in the expansion of $\psi_E(r)$ in $\{\chi_n(r)\}$ are multiples of the Meixner polynomials when $C > -1/2$, but when $C < -1/2$ the coefficients are multiples of the Meixner–Pollaczek polynomials.

Recent applications of the J-matrix method through the use of orthogonal polynomials to the helium atom and many body problems can be found in (Konovalov & McCarthy, 1994), (Konovalov & McCarthy, 1995), and (Kartono et al., 2005) and in their references. Applications to the spectral analysis of the three-dimensional Dirac equation for radial potential is in (Alhaidari, 2004c), while (Alhaidari, 2004a) treats the case of a Columb potential. The work (Alhaidari, 2004b) deals with the one-dimensional Dirac equation with Morse potential. Other examples are in (Alhaidari, 2005). In all these cases, the spectral analysis and expansion of wave functions in L^2 basis are done through the application of orthogonal polynomials.

5.9 The Meixner–Pollaczek Polynomials

These polynomials appeared first in (Meixner, 1934) as orthogonal polynomials of Sheffer A-type zero relative to $\dfrac{d}{dx}$, see Chapter 10 for definitions. Their recurrence relation is

$$(n+1)P_{n+1}^{(\lambda)}(x;\phi) - 2[x\sin\phi + (n+\lambda)\cos\phi]P_n^{(\lambda)}(x;\phi)$$
$$+(n+2\lambda-1)P_{n-1}^{(\lambda)}(x;\phi) = 0, \tag{5.9.1}$$

with the initial conditions

$$P_0^{(\lambda)}(x;\phi) = 1, \quad P_1^{(\lambda)}(x;\phi) = 2[x\sin\phi + \lambda\cos\phi]. \tag{5.9.2}$$

We shall assume $0 < \phi < \pi$ and $\lambda > 0$ to ensure orthogonality with respect to a positive measure.

One can turn (5.9.1)–(5.9.2) into a differential equation for the generating function $\sum_{n=0}^{\infty} P_n^{(\lambda)}(x;\phi)\,t^n$ and establish

$$\sum_{n=0}^{\infty} P_n^{(\lambda)}(x;\phi)\,t^n = \left(1 - te^{i\phi}\right)^{-\lambda+ix}\left(1 - te^{-i\phi}\right)^{-\lambda-ix}. \tag{5.9.3}$$

The generating function (5.9.3) leads to the explicit formula

$$P_n^{(\lambda)}(x;\phi) = \frac{(\lambda+ix)_n}{n!}\,e^{-in\phi}\,{}_2F_1\left(\begin{array}{c} -n, \lambda-ix \\ -n-\lambda-ix+1 \end{array}\middle|\, e^{2i\phi}\right). \tag{5.9.4}$$

By writing the right-hand side of (5.9.3) as

$$\left(\frac{1 - te^{-i\phi}}{1 - te^{i\phi}}\right)^{-\lambda-ix}\left(1 - te^{i\phi}\right)^{-2\lambda}$$

$$= \left\{1 + \frac{te^{i\phi}\left(1 - e^{-2i\phi}\right)}{1 - te^{i\phi}}\right\}^{-\lambda-ix}\left(1 - te^{i\phi}\right)^{-2\lambda}$$

$$= \sum_{k=0}^{\infty} \frac{(\lambda+ix)_k}{k!}\,t^k e^{ik\phi}\left(e^{-2i\phi} - 1\right)^k\left(1 - te^{i\phi}\right)^{-2\lambda-k}.$$

Expand $\left(1 - te^{i\phi}\right)^{-2\lambda-k}$ in powers of t and collect the coefficient of t^n. This leads to

$$P_n^{(\lambda)}(x;\phi) = \frac{(2\lambda)_n}{n!}\,e^{in\phi}\,{}_2F_1\left(\begin{array}{c} -n, \lambda+ix \\ 2\lambda \end{array}\middle|\, 1 - e^{-2i\phi}\right). \tag{5.9.5}$$

The t-singularities of the generating function (5.9.3) are $t = e^{\pm i\phi}$, and the application of Darboux's method leads to the asymptotoc formulas

$$P_n^{(\lambda)}(x;\phi) \approx \begin{cases} \frac{(\lambda-ix)_n}{n!}\left(1 - e^{-2i\phi}\right)^{-\lambda-ix}e^{in\phi}, & \operatorname{Im} x > 0, \\ \frac{(\lambda+ix)_n}{n!}\left(1 - e^{2i\phi}\right)^{-\lambda+ix}e^{-in\phi}, & \operatorname{Im} x < 0. \end{cases} \tag{5.9.6}$$

When x is real, then Darboux's method gives

$$P_n^{(\lambda)}(x;\phi) \approx \frac{(\lambda-ix)_n}{n!}\left(1 - e^{-2i\phi}\right)^{-\lambda-ix}e^{in\phi}$$

$$+ \frac{(\lambda+ix)_n}{n!}\left(1 - e^{2i\phi}\right)^{-\lambda+ix}e^{-in\phi}.$$

The orthonormal polynomials are $\left\{ P_n^{(\lambda)}(x; \varphi) \sqrt{n!/(2\lambda)_n} \right\}$. Hence, we have

$$\sqrt{\frac{n!}{(2\lambda)_n}} \, P_n^{(\lambda)}(x; \phi) \approx \frac{(2 \sin \phi)^{-\lambda}}{\sqrt{n}} \, \frac{e^{(\frac{\pi}{2} - \phi)x}}{\Gamma(\lambda - ix)} \, e^{i(n+\lambda)\phi - i\lambda\pi/2}$$

$$\times \exp(-ix \ln(2 \sin \varphi))$$

$$+ \text{complex conjugate}. \quad (5.9.7)$$

In analogy with the asymptotics of Hermite and Laguerre polynomials, (4.8.16) and (4.8.17), we expect the weight function to be

$$w_{MP}(x; \phi) = |\Gamma(\lambda - ix)|^2 \, e^{(2\phi - \pi)x}. \quad (5.9.8)$$

This can be confirmed by computing the asymptotics of the numerator polynomials and using properties of the moment problem, (Akhiezer, 1965).

Theorem 5.9.1 *The orthogonality relation for Meixner–Pollaczek polynomials is*

$$\int_{\mathbb{R}} w_{MP}(x; \phi) P_m^{(\lambda)}(x; \phi) P_n^{(\lambda)}(x; \phi) \, dx = \frac{\pi \Gamma(n + 2\lambda)}{(2 \sin \phi)^{2\lambda} n!} \, \delta_{m,n}. \quad (5.9.9)$$

The explicit formula (5.9.5) implies the generating functions

$$\sum_{n=0}^{\infty} P_n^{(\lambda)}(x; \phi) \frac{t^n}{(2\lambda)_n} = \exp\left(te^{i\phi}\right) {}_1F_1\left(\left. \begin{matrix} \lambda + ix \\ 2\lambda \end{matrix} \right| -2it \sin \varphi \right), \quad (5.9.10)$$

$$\sum_{n=0}^{\infty} \frac{(\gamma)_n}{(2\lambda)_n} \, P_n^{(\lambda)}(x; \phi) \frac{t^n}{e^{in\phi}} = (1 - t)^{-\gamma} {}_2F_1\left(\left. \begin{matrix} \gamma, \lambda + ix \\ 2\lambda \end{matrix} \right| \frac{1 - e^{-2i\phi}}{t - 1} t \right). \quad (5.9.11)$$

Exercises

5.1   Let $u_0(x) = 1$, $u_1(x) = ax + b$ and generate $u_n(x)$, $n > 1$ from

$$u_{n+1}(x) = 2xu_n(x) - u_{n-1}(x). \quad (E5.1)$$

(a) Show that $\{u_n(x)\}$ are orthogonal with respect to a positive measure μ, $d\mu = w dx + \mu_s$, w is supported on $[-1, 1]$, and μ_s has at most two masses and they are outside $[-1, 1]$. Evaluate w and μ_s explicitly.

(b) Express $u_n(x)$ as a sum of at most three Chebyshev polynomials.

(c) Generalize parts (a) and (b) by finding the measure of orthogonality for $\{u_n(x)\}$ if $u_n(x)$ solves (E5.1) for $n > m$ and

$$u_m(x) := \varphi(x), \quad u_{m+1}(x) := \psi(x),$$

Here φ, ψ have degrees m, $m + 1$, respectively, which have real simple and interlacing zeros. Show that

$$u_{n+m}(x) = \varphi(x) T_n(x) + [\psi(x) - x\varphi(x)] U_{n-1}(x).$$

6

Discrete Orthogonal Polynomials

In this chapter we treat the Meixner and Hahn polynomials and discuss their limiting cases. We also give a discrete analogue of the differential equations and discriminants of Chapter 3. It turned out that, in general, we do not have a closed form expression for the discriminants of Hahn and Meixner polynomials, but we have closed-form expressions for the discrete discriminants introduced in (Ismail, 2000a) and (Ismail et al., 2004).

6.1 Meixner Polynomials

The Meixner polynomials $\{M_n(x; \beta, c)\}$ are orthogonal with respect to a discrete measure whose distribution function has jumps $(\beta)_x c^x / x!$ at $x = 0, 1, \dots$. For integrability and positivity of the measure we need $c \in (0, 1)$. Let

$$w(x; \beta, c) = (\beta)_x c^x / x!, \quad x = 0, 1, \dots . \tag{6.1.1}$$

The attachment procedure explained at the beginning of §4.1 suggests letting

$$M_n(x; \beta, c) = \sum_{j=0}^{n} \frac{(-n)_j (-x)_j}{j!} c_{n,j}, \tag{6.1.2}$$

where $\{c_{n,j} : 0 \le j \le n\}$ are to be determined. This way the factor $(-x)_j$ is attached to the factor $1/x!$ in the weight function. The appropriate factor to attach to $(\beta)_x$ is $(\beta + x)_m$. We now evaluate the sum

$$\sum_{x=0}^{\infty} (\beta + x)_m \frac{(\beta)_x c^x}{x!} \sum_{j=0}^{n} \frac{(-n)_j (-x)_j}{j!} c_{n,j}$$

Since $(-x)_j = (-1)^j x(x-1) \cdots (x - j + 1)$, we see that the above sum is

$$\sum_{j=0}^{n} \frac{(-n)_j (-1)^j}{j!} c_{n,j} c^j \sum_{x=0}^{\infty} \frac{(\beta)_{x+j+m} c^x}{x!} c^x$$

$$= \sum_{j=0}^{n} \frac{(-n)_j (\beta)_{j+m} (-1)^j}{j!} c_{n,j} c^j (1-c)^{-\beta-j-m-1}$$

From here, as in §4.1, we see that the choice $c_{n,j} = (1 - 1/c)^j / (\beta)_j$ and the above quantity becomes

$$(1 - c)^{-\beta - m - 1}(\beta)_m \, {}_2F_1\left(\begin{matrix} -n, \beta + m \\ \beta \end{matrix} \,\middle|\, 1\right) = (1 - c)^{-\beta - m - 1} \frac{(\beta)_m (-m)_n}{(\beta)_n}.$$

Theorem 6.1.1 *The Meixner polynomials*

$$M_n(x; \beta, c) = {}_2F_1\left(\begin{matrix} -n, -x \\ \beta \end{matrix} \,\middle|\, 1 - \frac{1}{c}\right) \qquad (6.1.3)$$

satisfy the orthogonality relation

$$\sum_{x=0}^{\infty} M_n(x; \beta, c) M_m(x; \beta, c) \frac{(\beta)_x}{x!} c^x = \frac{n! \, (1 - c)^{-\beta}}{c^n \, (\beta)_n} \delta_{m,n}, \qquad (6.1.4)$$

for $\beta > 0, 0 < c < 1$. Their three-term recurrence relation is

$$- x M_n(x; \beta, c) = c(\beta + n)(1 - c)^{-1} M_{n+1}(x; \beta, c)$$
$$- [n + c(\beta + n)](1 - c)^{-1} M_n(x; \beta, c) + n(1 - c)^{-1} M_{n-1}(x; \beta, c). \qquad (6.1.5)$$

Proof From the analysis preceding the theorem, we see that (6.1.3) holds for $m < n$. The coefficient of x^n in the right-hand side of (6.1.3) is $(1 - 1/c)^n / (\beta)_n$. Therefore the left-hand side of (6.1.4) when $m = n$ is

$$\frac{(1 - 1/c)^n}{(\beta)_n} (1 - 1/c)^{-\beta - n - 1} (-n)_n = (1 - c)^{-\beta - 1} \frac{n!}{c^n (\beta)_n}, \qquad (6.1.6)$$

and (6.1.4) follows. The representation (6.1.3) implies

$$M_n(x, \beta, c) = \frac{(1 - 1/c)^n}{(\beta)_n} x^n$$

$$+ \frac{n(1 - 1/c)^n}{2c \, (\beta)_n} [c(2\beta + n - 1) + n - 1] x^{n-1} + \text{lower order terms.} \qquad (6.1.7)$$

Since we know that M_n must satisfy a three-term recurrence relation, we then use (6.1.7) to determine the coefficients if M_{n+1} and M_n from equating the coefficients of x^{n+1} and x^n on both sides. The coefficient of M_{n-1} can then be determined by setting $x = 0$ and noting that $M_n(0, \beta, c) = 1$ for all n. $\qquad \square$

We now derive the generating function

$$\sum_{n=0}^{\infty} \frac{(\beta)_n}{n!} M_n(x; \beta, c) t^n = (1 - t/c)^x (1 - t)^{-x - \beta}. \qquad (6.1.8)$$

To prove (6.1.8), multiply (6.1.3) by $(\beta)_n t^n / n!$ and use the fact $(-n)_k = (-1)^k n! / (n - k)!$. Similarly one can prove

$$\sum_{n=0}^{\infty} \frac{t^n}{n!} M_n(x; \beta, c) = e^t \, {}_1F_1\left(\begin{matrix} -x \\ \beta \end{matrix} \,\middle|\, \frac{1 - c}{c} t\right), \qquad (6.1.9)$$

$$\sum_{n=0}^{\infty} \frac{(\gamma)_n}{n!} M_n(x; \beta, c) t^n = (1 - t)^{-\gamma} \, {}_2F_1\left(\begin{matrix} \gamma, -x \\ \beta \end{matrix} \,\middle|\, \frac{(1 - c)t}{c(1 - t)}\right). \qquad (6.1.10)$$

Recall the finite difference operators

$$\Delta f(x) = (\Delta f)(x) := f(x+1) - f(x),$$
$$\nabla f(x) = (\nabla f)(x) := f(x) - f(x-1).$$

(6.1.11)

It is easy to see that

$$\Delta(-x)_j = -j(-x)_{j-1}, \quad \nabla(-x)_j = -j(-x+1)_{j-1}.$$

(6.1.12)

A direct calculation using (6.1.3) and (6.1.12) gives

$$\Delta M_n(x; \beta, c) = \frac{n(c-1)}{\beta c} M_{n-1}(x; \beta+1, c).$$

(6.1.13)

We now find the adjoint relation to (6.1.13). The proof is similar to the derivation of (4.2.3) from (4.2.2) using the orthogonality of Jacobi polynomials. We have

$$\frac{(n+1)(c-1)n!}{\beta c (1-c)^{\beta+1} c^n (\beta+1)_n} \delta_{m,n} = \sum_{x=0}^{\infty} \frac{(\beta+1)_x}{x!} c^x M_m(x; \beta, c) \Delta M_{n+1}(x; \beta, c).$$

Thus

$$-\frac{(n+1)!(1-c)^{-\beta}}{c^{n+1}(\beta)_{n+1}} \delta_{m,n}$$

$$= \sum_{x=0}^{\infty} M_{n+1}(x; \beta, c) \frac{(\beta)_x}{x!} c^x \left[\frac{x}{\beta c} M_m(x-1; \beta, c) - \frac{(\beta+x)}{\beta} M_m(x-1; \beta, c) \right].$$

Therefore the uniqueness of the orthogonal polynomials gives the following companion to (6.1.14)

$$c(\beta+x)M_n(x; \beta+1, c) - xM_n(x-1; \beta+1, c) = c\beta M_{n+1}(x; \beta, c). \quad (6.1.14)$$

Combining (6.1.14) and (6.1.13) we establish the second order difference equation

$$c(\beta+x)M_n(x+1; \beta, c) - [x + c(\beta+x)]M_n(x; \beta, c)$$
$$+ xM_n(x-1; \beta, c) = n(c-1)M_n(x; \beta, c).$$

(6.1.15)

It is important to note that the expression defining $M_n(x; \beta, c)$ in (6.1.3) is symmetric in x and n. Hence every formula we derive for $M_n(x; \beta, c)$ has a dual formula with x and n interchanged. Therefore we could have found (6.1.3) from (6.1.5).

Observe that (6.1.14) can be written in the form

$$\nabla \left[\frac{(\beta+1)_x}{x!} c^x M_n(x; \beta+1, c) \right] = \frac{(\beta)_x c^x}{x!} M_{n+1}(x; \beta, c).$$

Iterating the above form we get

$$\frac{(\beta)_x c^x}{x!} M_{n+k}(x; \beta, c) = \nabla^k \left[\frac{(\beta+k)_x}{x!} c^x M_n(x; \beta+k, c) \right]. \quad (6.1.16)$$

In particular we have the discrete Rodrigues formula

$$\frac{(\beta)_x c^x}{x!} M_n(x; \beta, c) = \nabla^n \left[\frac{(\beta+n)_x}{x!} c^x \right]. \quad (6.1.17)$$

The limiting relation

$$\lim_{c \to 1^-} M_n(x/(1-c); \alpha+1, c) = \frac{n!}{(\alpha+1)_n} L_n^{(\alpha)}(x), \qquad (6.1.18)$$

follows from (6.1.3) and (4.6.1). In the orthogonality relation (6.1.4) by writing $y = (1-c)x$,

$$\frac{(\beta)_x}{x!} = \frac{\Gamma(\beta + y/(1-c))}{\Gamma(\beta)\Gamma(1 + y/(1-c))} \approx \left(\frac{y}{1-c}\right)^{\beta-1} \frac{1}{\Gamma(\beta)},$$

as $c \to 1$, we see that as $c \to 1^-$, (6.1.4) is a Riemann sum approximation to (4.6.2) with the appropriate renormaliztion.

Another limiting case is

$$\lim_{\beta \to \infty} M_n(x; \beta, a/(\beta+a)) = C_n(x; a), \qquad (6.1.19)$$

where $C_n(x; a)$ are the Charlier polynomials

$$C_n(x; a) = {}_2F_0\left(\begin{matrix} -n, -x \\ - \end{matrix} \middle| -\frac{1}{a}\right). \qquad (6.1.20)$$

The orthogonality relation (6.1.4) and the generating function (6.1.8) imply

$$\sum_{x=0}^{\infty} C_m(x; a)C_n(x; a)\frac{a^x}{x!} = \frac{n!}{a^n} e^a \, \delta_{m,n}, \qquad (6.1.21)$$

$$\sum_{n=0}^{\infty} C_n(x; a)\frac{t^n}{n!} = (1 - t/a)^x e^t. \qquad (6.1.22)$$

On the other hand (6.1.14) and (6.1.13) establish the functional equations

$$\Delta C_n(x; a) = -\frac{n}{a}C_{n-1}(x; a), \qquad (6.1.23)$$

$$aC_n(x; a) - xC_{n-1}(x - 1; a) = aC_{n+1}(x; a). \qquad (6.1.24)$$

The following recurrence relation follows from (6.1.5) and (6.1.19)

$$-xC_n(x; a) = aC_{n+1}(x; a) - (n + a)C_n(x; a) + nC_{n-1}(x; a),$$
$$C_0(x; a) = 1, \quad C_1(x; a) = (a - x)/a. \qquad (6.1.25)$$

Rui and Wong derived uniform asymptotic developments for Charlier polynomials (Rui & Wong, 1994) which implies asymptotics of the kth largest zero of $C_n(x; a)$ as $n \to \infty$ and k is even allowed to depend on n.

6.2 Hahn, Dual Hahn, and Krawtchouk Polynomials

The Hahn polynomials are orthogonal with respect to a discrete whose mass at $x = k$ is $w(x; \alpha, \beta, N)$,

$$w(x; \alpha, \beta, N) := \frac{(\alpha + 1)_x}{x!} \frac{(\beta + 1)_{N-x}}{(N - x)!}, \quad x = 0, 1, \ldots, N. \qquad (6.2.1)$$

The attachment technique suggests we try a polynomial of the form

$$Q_n(x) = Q_n(x; \alpha, \beta, N) = \sum_{j=0}^{n} \frac{(-n)_j(-x)_j}{j!} c_{n,j}.$$

In order to find the coefficients $c_{n,j}$ we need to show that the sum

$$I_{m,n} := \sum_{x=0}^{N} (\alpha + x)_m Q_n(x; \alpha, \beta, N) w(x; \alpha, \beta, N), \qquad (6.2.2)$$

is zero for $m < n$. It is straightforward to see that

$$I_{m,n} = \sum_{j=0}^{n} \frac{(-n)_j}{j!} c_{n,j} \sum_{x=j}^{N} (\alpha + x)_m \frac{(-x)_j(\alpha + 1)_x}{x!} \frac{(\beta + 1)_{N-x}}{(N - x)!}$$

$$= \frac{(\beta + 1)_N}{N!} \sum_{j=0}^{n} \frac{(-n)_j}{j!} (-1)^j c_{n,j} \sum_{x=j}^{N} \frac{(\alpha + 1)_{x+m}}{(x - j)!} \frac{(-N)_x}{(-\beta - N)_x}$$

$$= \frac{(\beta + 1)_N}{N!} \sum_{j=0}^{n} \frac{(\alpha + 1)_{m+j}(-n)_j(-N)_j}{(-\beta - N)_j \, j!} (-1)^j c_{n,j}$$

$$\times \sum_{x=0}^{N-j} \frac{(\alpha + m + j)_x}{x!} \frac{(-N + j)_x}{(-\beta - N + j)_x}.$$

In the above steps we used (1.3.10). Now the last x sum is

$${}_2F_1(-N + j, \alpha + m + j; -\beta - N + j; 1) = \frac{(-\beta - N - \alpha - m)_{N-j}}{(-\beta - N + j)_{N-j}},$$

by (1.4.3). Thus

$$I_{m,n} = \frac{(\alpha + 1)_m(\beta + 1)_N}{N!}$$

$$\times \sum_{j=0}^{n} \frac{(\alpha + m + 1)_j(-n)_j(-N)_j(-b - \alpha - N - m)_{N-j}}{(-1)^j \, (-\beta - N)_j \, (-\beta - N + j)_{N-j} \, j!} c_{n,j}.$$

After some trials the reader can convince himself/herself that one needs to use the Pfaff–Saalschütz theorem (1.4.5) and that $c_{n,j}$ must be chosen as $(n + \alpha + \beta + 1)_j/[(\alpha + 1)_j(-N)_j]$. This establishes the following theorem.

Theorem 6.2.1 *The Hahn polynomials have the representation*

$$Q_n(x) = Q_n(x; \alpha, \beta, N)$$

$$= {}_3F_2\left(\begin{matrix} -n, \, n + \alpha + \beta + 1, \, -x \\ \alpha + 1, \, -N \end{matrix} \, \middle| \, 1 \right), \qquad n = 0, 1, \dots, N, \qquad (6.2.3)$$

and satisfy the orthogonality relation

$$\sum_{x=0}^{N} Q_m(x; \alpha, \beta, N) \, Q_n(x; \alpha, \beta, N) \, w(x; \alpha, \beta, N)$$

$$= \frac{n! \, (N - n)! \, (\beta + 1)_n (\alpha + \beta + n + 1)_{N+1}}{(N!)^2 \, (\alpha + \beta + 2n + 1) \, (\alpha + 1)_n} \delta_{m,n}. \qquad (6.2.4)$$

The relationship

$$\Delta Q_n(x; \alpha, \beta, N) = -\frac{n(n+\alpha+\beta+1)}{N(\alpha+1)} Q_{n-1}(x; \alpha+1, \beta+1, N-1) \quad (6.2.5)$$

follows from (6.1.12) and (6.2.3) in a straightforward manner. Moreover (6.2.3) and the Chu–Vandermonde sum (1.4.3) give the special point evaluations

$$Q_n(0; \alpha, \beta, N) = 1, \quad Q_n(0; \alpha, \beta, N) = Q_n(N; \alpha, \beta, N) = \frac{(-\beta)_n}{(\alpha+1)_n},$$

$$Q_n(-\alpha-1; \alpha, \beta, N) = (\alpha+\beta+N+2)_n \frac{(N-n)!}{N!}. \quad (6.2.6)$$

Furthermore (6.2.3) yields

$$Q_n(x; \alpha, \beta, N) = \frac{(\alpha+\beta+n+1)_n}{(\alpha+1)_n(-N)_n} x^n$$

$$+ \frac{n(\alpha+\beta+1)_{n-1}}{2(\alpha+1)_n(-N)_n} [(\alpha-\beta)(n-1) - 2N(n+\alpha)] x^{n-1} \quad (6.2.7)$$

$$+ \text{ lower order terms.}$$

From (6.2.7) and (6.2.6) we establish the three term recurrence relation, whose existence is guaranteed by the orthogonality,

$$-x Q_n(x; \alpha, \beta, N) = \lambda_n Q_{n+1}(x; \alpha, \beta, N) + \mu_n Q_{n-1}(x; \alpha, \beta, N)$$
$$- [\lambda_n + \mu_n] Q_n(x; \alpha, \beta, N), \quad (6.2.8)$$

with

$$\lambda_n = \frac{(\alpha+\beta+n+1)(\alpha+n+1)(N-n)}{(\alpha+\beta+2n+1)(\alpha+\beta+2n+2)},$$

$$\mu_n = \frac{n(n+\beta)(\alpha+\beta+n+N+1)}{(\alpha+\beta+2n)(\alpha+\beta+2n+1)}. \quad (6.2.9)$$

It readily follows from (6.2.3) and (6.1.3) that

$$\lim_{N\to\infty} Q_n(Nx; \alpha, \beta, N) = P_n^{(\alpha,\beta)}(1-2x) n! / (\alpha+1)_n, \quad (6.2.10)$$

$$\lim_{N\to\infty} Q_n(x; \alpha, N((1-c)/c), N) = M_n(x; \alpha, c), \quad (6.2.11)$$

that is, the Jacobi and Meixner polynomials are limiting cases of Hahn polynomials. The adjoint relation to (6.2.5) is

$$(x+\alpha)(N+1-x)Q_n(x; \alpha, \beta, N) - x(\beta+N+1-x)Q_n(x-1; \alpha, \beta, N)$$
$$= \alpha(N+1)Q_{n+1}(x; \alpha-1, \beta-1, N-1) \quad (6.2.12)$$

or, equivalently,

$$\nabla [w(x; \alpha, \beta, N) Q_n(x; \alpha, \beta, N)]$$
$$= \frac{N+1}{\beta} w(x; \alpha-1, \beta-1, N+1) Q_{n+1}(x; \alpha-1, \beta-1, N+1). \quad (6.2.13)$$

Combining the above relationships, we establish the second order difference equation

$$\frac{1}{w(x,\alpha,\beta,N)} \nabla(w(x;\alpha+1,\beta+1,N-1)\Delta Q_n(x;\alpha,\beta,N))$$
$$= -\frac{n(n+\alpha+\beta+1)}{(\alpha+1)(\beta+1)} Q_n(x;\alpha,\beta,N). \tag{6.2.14}$$

Equation (6.2.14), when expanded out, reads

$$(x-N)(\alpha+x+1)\nabla\Delta y_n(x) + [x(\alpha+\beta+2)-N(\alpha+1)]\nabla y_n(x)$$
$$= n(n+\alpha+\beta+1)y_n(x), \tag{6.2.15}$$

or, equivalently,

$$(x-N)(\alpha+x+1)y_n(x+1) - [(x-N)(\alpha+x+1)+x(x-\beta-N-1)]$$
$$\times y_n(x) + x(x-\beta-N-1)y_n(x-1) = n(n+\alpha+\beta+1)y_n(x). \tag{6.2.16}$$

where $y_n(x) = Q_n(x;\alpha,\beta,N)$.

Koornwinder showed that the orthogonality of the Hahn polynomials is equivalent to the orthogonality of the Clebsch–Gordon coefficients for $SU(2)$, or $3-j$ symbols, see (Koornwinder, 1981).

The Jacobi polynomials are limiting cases of the Hahn polynomials. Indeed

$$\lim_{N\to\infty} Q_n(Nx;\alpha,\beta,N) = P_n^{(\alpha,\beta)}(1-2x)/P_n^{(\alpha,\beta)}(1). \tag{6.2.17}$$

The Hahn polynomials provide an example of a finite set of orthogonal polynomials. This makes the matrix whose i,j entry is $\phi_i(j)$, $0 \le i,j \le N$ an orthogonal matrix and implies the dual orthogonality relation

$$\sum_{n=0}^{N} \phi_n(x)\phi_n(y)h_n = \delta_{x,y}/w(x), \tag{6.2.18}$$

for $x,y = 0,1,\ldots,N$.

We now introduce the dual Hahn polynomials. They arise when we interchange n and x in (6.2.3), and their orthogonality relation will follow from (6.2.4).

Definition 6.2.1 *The dual Hahn polynmials are*

$$R_n(\lambda(x);\gamma,\delta,N) = {}_3F_2\left(\begin{array}{c} -n,-x,x+\gamma+\delta+1 \\ \gamma+1,-N \end{array}\bigg| 1\right), \quad n = 0,1,2,\ldots,N, \tag{6.2.19}$$

where

$$\lambda(x) = x(x+\gamma+\delta+1).$$

When $\gamma > -1$, and $\delta > -1$ or for $\gamma < -N$ and $\delta < -N$, the orthogonality relation dual to (6.2.4) is

$$\sum_{x=0}^{N} \frac{(2x + \gamma + \delta + 1)(\gamma + 1)_x (-N)_x N!}{(-1)^2 (x + \gamma + \delta + 1)_{N+1} (\delta + 1)_x x!}$$

$$\times R_m(\lambda(x); \gamma, \delta, N) R_n(\lambda(x); \gamma, \delta, N) = \frac{\delta_{mn}}{\binom{\gamma + n}{n}\binom{\delta + N - n}{N - n}}. \qquad (6.2.20)$$

The three-term recurrence relation for the dual Hahn polynomials can be easily found to be

$$\lambda(x) R_n(\lambda(x)) = A_n R_{n+1}(\lambda(x)) - (A_n + C_n) R_n(\lambda(x)) + C_n R_{n-1}(\lambda(x)), \qquad (6.2.21)$$

where

$$R_n(\lambda(x)) := R_n(\lambda(x); \gamma, \delta, N)$$

and

$$A_n = (n + \gamma + 1)(n - N), \quad C_n = n(n - \delta - N - 1). \qquad (6.2.22)$$

The corresponding monic polynomials $P_n(x)$ satisfy the recurrence relation

$$x P_n(x) = P_{n+1}(x) - (A_n + C_n) P_n(x) + A_{n-1} C_n P_{n-1}(x), \qquad (6.2.23)$$

where

$$R_n(\lambda(x); \gamma, \delta, N) = \frac{1}{(\gamma + 1)_n (-N)_n} P_n(\lambda(x)).$$

The dual Hahn polynomials satisfy the difference equation

$$-n y(x) = B(x) y(x + 1) - [B(x) + D(x)] y(x) + D(x) y(x - 1),$$
$$y(x) = R_n(\lambda(x); \gamma, \delta, N), \qquad (6.2.24)$$

where

$$\begin{cases} B(x) = \dfrac{(x + \gamma + 1)(x + \gamma + \delta + 1)(N - x)}{(2x + \gamma + \delta + 1)(2x + \gamma + \delta + 2)} \\[3mm] D(x) = \dfrac{x(x + \gamma + \delta + N + 1)(x + \delta)}{(2x + \gamma + \delta)(2x + \gamma + \delta + 1)}. \end{cases}$$

The lowering operator formula is

$$R_n(\lambda(x + 1); \gamma, \delta, N) - R_n(\lambda(x); \gamma, \delta, N)$$
$$= -\frac{n(2x + \gamma + \delta + 2)}{(\gamma + 1)N} R_{n-1}(\lambda(x); \gamma + 1, \delta, N - 1) \qquad (6.2.25)$$

or, equivalently,

$$\frac{\Delta R_n(\lambda(x); \gamma, \delta, N)}{\Delta \lambda(x)} = -\frac{n}{(\gamma + 1)N} R_{n-1}(\lambda(x); \gamma + 1, \delta, N - 1). \qquad (6.2.26)$$

The raising operator formula is

$$(x + \gamma)(x + \gamma + \delta)(N + 1 - x)R_n(\lambda(x); \gamma, \delta, N)$$
$$- x(x + \gamma + \delta + N + 1)(x + \delta)R_n(\lambda(x - 1); \gamma, \delta, N)$$
$$= \gamma(N + 1)(2x + \gamma + \delta)R_{n+1}(\lambda(x); \gamma - 1, \delta, N + 1) \quad (6.2.27)$$

or, equivalently,

$$\frac{\nabla[\omega(x; \gamma, \delta, N)R_n(\lambda(x); \gamma, \delta, N)]}{\nabla\lambda(x)}$$
$$= \frac{1}{\gamma + \delta}\omega(x; \gamma - 1, \delta, N + 1)R_{n+1}(\lambda(x); \gamma - 1, \delta, N + 1), \quad (6.2.28)$$

where

$$\omega(x; \gamma, \delta, N) = \frac{(-1)^x(\gamma + 1)_x(\gamma + \delta + 1)_x(-N)_x}{(\gamma + \delta + N + 2)_x(\delta + 1)_x \, x!}.$$

Iterating (6.2.28), we derive the Rodrigues-type formula

$$\omega(x; \gamma, \delta, N)R_n(\lambda(x); \gamma, \delta, N)$$
$$= (\gamma + \delta + 1)_n \, (\nabla_\lambda)^n \, [\omega(x; \gamma + n, \delta, N - n)], \quad (6.2.29)$$

where

$$\nabla_\lambda := \frac{\nabla}{\nabla\lambda(x)}.$$

The following generating functions hold for $x = 0, 1, 2, \ldots, N$

$$(1 - t)^{N-x} \, {}_2F_1\left(\begin{matrix} -x, -x - \delta \\ \gamma + 1 \end{matrix}\Bigg| t\right) = \sum_{n=0}^{N} \frac{(-N)_n}{n!} R_n(\lambda(x); \gamma, \delta, N) \, t^n. \quad (6.2.30)$$

$$(1 - t)^x \, {}_2F_1\left(\begin{matrix} x - N, x + \gamma + 1 \\ -\delta - N \end{matrix}\Bigg| t\right)$$
$$= \sum_{n=0}^{N} \frac{(\gamma + 1)_n(-N)_n}{(-\delta - N)_n \, n!} R_n(\lambda(x); \gamma, \delta, N) \, t^n. \quad (6.2.31)$$

$$\left[e^t \, {}_2F_2\left(\begin{matrix} -x, x + \gamma + \delta + 1 \\ \gamma + 1, -N \end{matrix}\Bigg| -t\right)\right]_N = \sum_{n=0}^{N} \frac{R_n(\lambda(x); \gamma, \delta, N)}{n!} t^n. \quad (6.2.32)$$

$$\left[(1 - t)^{-a} \, {}_3F_2\left(\begin{matrix} a, -x, x + \gamma + \delta + 1 \\ \gamma + 1, -N \end{matrix}\Bigg| \frac{t}{t - 1}\right)\right]_N$$
$$= \sum_{n=0}^{N} \frac{(a)_n}{n!} R_n(\lambda(x); \gamma, \delta, N) \, t^n, \quad (6.2.33)$$

where a is an arbitrary parameter.

Definition 6.2.2 *The Krawtchouk polynomials are*

$$K_n(x; p, N) = {}_2F_1\left(\begin{matrix} -n, -x \\ -N \end{matrix} \middle| \frac{1}{p} \right), \quad n = 0, 1, 2, \ldots, N. \tag{6.2.34}$$

The limiting relation

$$\lim_{t \to \infty} Q_n(x; pt, (1-p)t, N) = K_n(x; p, N)$$

enables us to derive many results for the Krawtchouk polynomials from the corresponding results for the Hahn polynomials. In particular, we establish the orthogonality relation

$$\sum_{x=0}^{N} \binom{N}{x} p^x (1-p)^{N-x} K_m(x; p, N) K_n(x; p, N)$$
$$= \frac{(-1)^n n!}{(-N)_n} \left(\frac{1-p}{p} \right)^n \delta_{m,n}, \quad 0 < p < 1. \tag{6.2.35}$$

and the recurrence relation

$$- x K_n(x; p, N) = p(N-n) K_{n+1}(x; p, N)$$
$$- [p(N-n) + n(1-p)] K_n(x; p, N) + n(1-p) K_{n-1}(x; p, N). \tag{6.2.36}$$

The monic polynomials $\{P_n(x)\}$ satisfy the normalized recurrence relation

$$x P_n(x) = P_{n+1}(x) + [p(N-n) + n(1-p)] P_n(x) + np(1-p)(N+1-n) P_{n-1}(x), \tag{6.2.37}$$

where

$$K_n(x; p, N) = \frac{1}{(-N)_n p^n} P_n(x).$$

The corresponding difference equation is

$$-ny(x) = p(N-x)y(x+1) - [p(N-x) + x(1-p)]y(x) + x(1-p)y(x-1), \tag{6.2.38}$$

where

$$y(x) = K_n(x; p, N).$$

The lowering operator is

$$\Delta K_n(x; p, N) = -\frac{n}{Np} K_{n-1}(x; p, N-1). \tag{6.2.39}$$

On the other hand, the raising operator is

$$(N + 1 - x) K_n(x; p, N) - x \left(\frac{1-p}{p} \right) K_n(x-1; p, N)$$
$$= (N+1) K_{n+1}(x; p, N+1) \tag{6.2.40}$$

or, equivalently,

$$\nabla \left[\binom{N}{x} \left(\frac{p}{1-p} \right)^x K_n(x; p, N) \right] = \binom{N-n}{x} \left(\frac{p}{1-p} \right)^x K_{n+1}(x; p, N+1), \tag{6.2.41}$$

which leads to the Rodrigues-type formula

$$\binom{N}{x}\left(\frac{p}{1-p}\right)^x K_n(x;p,N) = \nabla^n\left[\binom{N-n}{x}\left(\frac{p}{1-p}\right)^x\right]. \quad (6.2.42)$$

The following generating functions hold for $x = 0, 1, 2, \ldots, N$

$$\left(1 - \frac{(1-p)}{p}t\right)^x (1+t)^{N-x} = \sum_{n=0}^{N}\binom{N}{n}K_n(x;p,N)\,t^n, \quad (6.2.43)$$

$$\left[e^t\,{}_1F_1\left(\begin{matrix}-x\\-N\end{matrix}\Big| -\frac{t}{p}\right)\right]_N = \sum_{n=0}^{N}\frac{K_n(x;p,N)}{n!}\,t^n, \quad (6.2.44)$$

and

$$\left[(1-t)^{-\gamma}\,{}_2F_1\left(\begin{matrix}\gamma,-x\\-N\end{matrix}\Big|\frac{t}{p(t-1)}\right)\right]_N$$

$$= \sum_{n=0}^{N}\frac{(\gamma)_n}{n!}K_n(x;p,N)\,t^n, \quad (6.2.45)$$

where γ is an arbitrary parameter.

The Krawtchouk polynomials are self-dual because they are symmetric in n and x. They are also the eigenmatrices of the Hamming scheme $H(n,q)$, (Bannai & Ito, 1984, Theorem 3.2.3). The orthogonality of the Krawtchouk polynomials is equivalent to the unitarity of unitary representations of $SU(2)$, (Koornwinder, 1982).

Krawtchouk polynomials have been applied to many areas of mathematics. We shall briefly discuss their role in coding theory. The Lloyd polynomials $L_n(x;p,N)$ are

$$L_n(x;p,N) = \sum_{m=0}^{n}\binom{N}{m}\frac{p^m}{(1-p)^m}K_m(x;p,N).$$

The important cases are when $1/(1-p)$ is an integer. It turns out that

$$L_n(x;p,N) = \frac{p^n}{(1-p)^n}\binom{N-1}{n}K_n(x;p,N-1),$$

so the zeros of L_n are related to the zeros of K_n. One question which arises in coding theory is to describe all integer zeros of K_n. In other words, for fixed p such that $1/(1-p)$ is an integer, describe all triples of positive integers (n,x,N) such that $K_n(x;p,N) = 0$, (Habsieger, 2001a). Habsieger and Stanton gave a complete list of solutions in the cases $N - 2n \in \{1,2,3,4,5,6\}$, $N - 2n = 8$, or x odd, see (Habsieger & Stanton, 1993). Let $N(n,N)$ be the number of integer zeros of $K_n(x;1/2,N)$. Two conjectures in this area are due to Krasikov and Litsyn, (Krasikov & Litsyn, 1996), (Habsieger, 2001a).

Conjecture 6.2.2 *For $2n - N < 0$, we have*

$$N(n,N) \leq \begin{cases} 3 & \textit{if n is odd} \\ 4 & \textit{if n is even.} \end{cases}$$

Conjecture 6.2.3 *Let* $n = \binom{m}{2}$. *Then the only integer zeros of* $K_n\left(x; 1/2, m^2\right)$ *are* 2, $m^2 - 2$ *and* $m^2/4$ *for* $m \equiv 2$ (mod 4).

Hong showed the existence of a noninteger zero for K_n when $1/p - 1$ is an integer greater than 2, see (Hong, 1986). For a survey of these results, see (Habsieger, 2001a). See also (Habsieger, 2001b).

The strong asymptotics of $K_n(x; p, N)$ when $n, N \to \infty$, $x > 0$ but n/N is fixed are in (Ismail & Simeonov, 1998), while a uniform asymptotic expansion is in (Li & Wong, 2000). Sharapudinov studied the asymptotic properties of $K_n(x; p, N)$ when $n, N \to \infty$ with $n = \mathcal{O}\left(N^{1/3}\right)$. He also studied the asymptotics of the zeros of $K_n(x; p, N)$ when $n = o\left(N^{1/4}\right)$. These results are in (Sharapudinov, 1988). More recently, Qiu and Wong gave an asymptotic expansion for the Krawtchouk polynomials and their zeros in (Qiu & Wong, 2004). The WKB technique has been applied in (Dominici, 2005) to the study of the asymptotics of $K_n(x; p, N)$.

Let $q = 1/(1-p)$ be a positive integer an denote the Hamming space $(\mathbb{Z}/q\mathbb{Z})^n$ by H, and O is the origin in H. For $X \subset H$, $X \neq \phi$ the radon transform T_X is defined on functions $f : H \to \mathbb{C}$ by

$$T_X(f)(u) = \sum_{v \in u + X} f(v),$$

for $u \in H$. For $\mathbf{x} = (x_1, \ldots, x_N)$, $\mathbf{y} = (y_1, \ldots, y_N)$ in H, the Hamming distance between $\mathbf{x}$ and $\mathbf{y}$ is

$$d(\mathbf{x}, \mathbf{y}) = |\{i : 1 \leq i \leq N \text{ and } x_i \neq y_i\}|.$$

Let

$$S(\mathbf{x}, n) = \{\mathbf{y} : \mathbf{y} \in H, d(\mathbf{x}, \mathbf{y}) = n\},$$
$$B(\mathbf{x}, n) = \{\mathbf{y} : \mathbf{y} \in H, d(\mathbf{x}, \mathbf{y}) \leq n\}.$$

Theorem 6.2.4 *The radon transform* $T_{S(O,n)}$ *is invertible if and only if the polynomial* $K_n(x, p, N)$, $q = 1/(1 - p)$, *has no integer roots. The radon transform* $T_{B(O,n)}$ *is invertible if and only if the (Lloyd) polynomial* $K_n(x, p, N - 1)$ *has no integer zeros.*

Theorem 6.2.4 is in (Diaconis & Graham, 1985) for $T_{S(O,n)}$, but Habsieger pointed out that their proof method works for $T_{B(O,n)}$, see (Habsieger, 2001a).

Another problem in graph theory whose solution involves zeros of Krawtchouk polynomials is a graph reconstruction problem. Let I be a subset of vertices of a graph G. Construct a new graph G_I by switching with respect to I. That is, if $u \in I$, $v \notin I$, then u and v are adjacent (nonadjacent) in G_I if and only if they are nonadjacent (adjacent) in G. Assume that G has N vertices. The n-switching deck is the multiset of unlabelled graphs $D_n(G) = \{G_I : |I| = n\}$. Stanley proved that G may be reconstructible from $D_n(G)$ if $K_n(x; 1/2, N)$ has no even zeros.

We have only mentioned samples of problems where an object has a certain property if the zeros of a Krawtchouk polynomial lie on the spectrum $\{0, 1, \ldots, N\}$.

6.3 Difference Equations

In this and the following section we extend most of the results of Chapter 3 to difference equations and discrete orthogonal polynomials. Let $\{p_n(x)\}$ be a family of polynomials orthogonal with respect to a discrete measure supported on $\{s, s + 1, \ldots, t\} \subset \mathbb{R}$, where s is finite but t is finite or infinite. Assume that the orthogonality relation is

$$\sum_{\ell=s}^{t} p_m(\ell)p_n(\ell)w(\ell) = \kappa_m \delta_{m,n}, \tag{6.3.1}$$

where w is a weight function normalized by

$$\sum_{\ell=s}^{t} w(\ell) = 1. \tag{6.3.2}$$

We make the assumption that w is not identically zero on $\mathbb{R} \setminus \{s, s + 1, \ldots, t\}$ and

$$w(s - 1) = 0, \quad w(t + 1) = 0. \tag{6.3.3}$$

Define $u(x)$ by

$$w(x + 1) - w(x) = -u(x + 1)w(x + 1). \tag{6.3.4}$$

The function $u(x)$ is the discrete analogue of the function $v(x)$ of §3.1. Although we require w and u to be defined only on the non-negative integers in $[s, t]$ we will make **the additional assumption** that u has an extension to a differentiable function on $[s + 1, t - 1]$.

In this notation the Christoffel–Darboux formula is

$$\sum_{\nu=0}^{n-1} \frac{p_\nu(x)p_\nu(y)}{\kappa_\nu} = \frac{\gamma_{n-1}}{\gamma_n \kappa_n} \frac{p_n(x)p_{n-1}(y) - p_n(y)p_{n-1}(x)}{x - y}. \tag{6.3.5}$$

In the sequel we will use the following property: If $q(x)$ is a polynomial of degree at most n and c is a constant, then

$$\sum_{\ell=s}^{t} \frac{p_n(\ell)q(\ell)}{\ell - c} w(\ell) = q(c) \sum_{\ell=s}^{t} \frac{p_n(\ell)}{\ell - c} w(\ell), \tag{6.3.6}$$

since $(q(\ell) - q(c))/(\ell - c)$ is a polynomial of ℓ of degree at most $n - 1$ and the orthogonality relation (6.3.1) holds.

Theorem 6.3.1 *Let*

$$p_n(x) = \gamma_n x^n + lower\ order\ terms,$$

satisfy (6.3.1). *Then,*

$$\Delta p_n(x) = A_n(x)p_{n-1}(x) - B_n(x)p_n(x), \tag{6.3.7}$$

where $A_n(x)$ and $B_n(x)$ are given by

$$A_n(x) = \frac{\gamma_{n-1}}{\gamma_n \kappa_{n-1}} \frac{p_n(t+1)p_n(t)}{(t-x)} w(t)$$

$$+ \frac{\gamma_{n-1}}{\gamma_n \kappa_{n-1}} \sum_{\ell=s}^{t} p_n(\ell)p_n(\ell-1) \frac{u(x+1)-u(\ell)}{(x+1-\ell)} w(\ell), \tag{6.3.8}$$

$$B_n(x) = \frac{\gamma_{n-1}}{\gamma_n \kappa_{n-1}} \frac{p_n(t+1)p_{n-1}(t)}{(t-x)} w(t)$$

$$+ \frac{\gamma_{n-1}}{\gamma_n \kappa_{n-1}} \sum_{\ell=s}^{t} p_n(\ell)p_{n-1}(\ell-1) \frac{u(x+1)-u(\ell)}{(x+1-\ell)} w(\ell). \tag{6.3.9}$$

A proof is in (Ismail et al., 2004) and is similar to the proof of Theorem 3.2.1 so it will be omitted. The proof uses the form (6.3.5).

It is clear that if $\{p_n(x)\}$ are orthonormal, that is $\kappa_n = 1$, they satisfy (3.1.6). In this case, since $\gamma_{n-1}/\gamma_n = a_n$, formulas (6.3.8) and (6.3.9) simplify to

$$A_n(x) = \frac{a_n p_n(t+1)p_n(t)}{(t-x)} w(t)$$

$$+ a_n \sum_{\ell=s}^{t} p_n(\ell)p_n(\ell-1) \frac{u(x+1)-u(\ell)}{(x+1-\ell)} w(\ell), \tag{6.3.10}$$

$$B_n(x) = \frac{a_n p_n(t+1)p_{n-1}(t)}{(t-x)} w(t)$$

$$+ a_n \sum_{\ell=s}^{t} p_n(\ell)p_{n-1}(\ell-1) \frac{u(x+1)-u(\ell)}{(x+1-\ell)} w(\ell), \tag{6.3.11}$$

respectively.

Relation (6.3.7) produces a lowering (annihilation) operator.

We now introduce the linear operator

$$L_{n,1} := \Delta + B_n(x). \tag{6.3.12}$$

By (6.3.7), $L_{n,1}p_n(x) = A_n(x)p_{n-1}(x)$, thus $L_{n,1}$ is a lowering operator. Solving (6.3.7) and (3.1.6) for p_{n-1} we get

$$\frac{1}{A_n(x)}[\Delta + B_n(x)]p_n(x) = \left(\frac{x-b_n}{a_n}\right) p_n(x) - \frac{a_{n+1}}{a_n} p_{n+1}(x).$$

Then, the operator $L_{n+1,2}$ defined by

$$L_{n+1,2} := -\Delta - B_n(x) + \frac{(x-b_n)}{a_n} A_n(x) \tag{6.3.13}$$

is a raising operator since $L_{n+1,2}p_n(x) = (a_{n+1}A_n(x)/a_n)p_{n+1}(x)$. These operators generate two second-order difference equations:

$$L_{n,2}\left(\frac{1}{A_n(x)}L_{n,1}\right) p_n(x) = \frac{a_n A_{n-1}(x)}{a_{n-1}} p_n(x), \tag{6.3.14}$$

$$L_{n+1,1}\left(\frac{a_n}{a_{n+1}A_n(x)}L_{n+1,2}\right) p_n(x) = A_{n+1}(x)p_n(x). \tag{6.3.15}$$

Using the formulas

$$\Delta(fg)(x) = \Delta f(x)\Delta g(x) + f(x)\Delta g(x) + g(x)\Delta f(x) \tag{6.3.16}$$

and

$$\Delta(1/f)(x) = -\Delta f(x)/(f(x)f(x+1)),$$

equation (6.3.14) can be written in the form

$$\Delta^2 p_n(x) + R_n(x)\Delta p_n(x) + S_n(x)p_n(x) = 0, \tag{6.3.17}$$

where

$$R_n(x) = -\frac{\Delta A_n(x)}{A_n(x)} + B_n(x+1) + \frac{B_{n-1}(x)A_n(x+1)}{A_n(x)}$$
$$- \frac{(x - b_{n-1})}{a_{n-1}}\frac{A_{n-1}(x)A_n(x+1)}{A_n(x)}, \tag{6.3.18}$$

$$S_n(x) = \left(B_{n-1}(x) - 1 - \frac{(x - b_{n-1})}{a_{n-1}}A_{n-1}(x)\right)\frac{B_n(x)A_n(x+1)}{A_n(x)}$$
$$+ B_n(x+1) + \frac{a_n A_{n-1}(x)A_n(x+1)}{a_{n-1}}. \tag{6.3.19}$$

For some applications it is convenient to have equation (6.3.17) written in terms of $y(x) = p_n(x)$, $y(x+1)$, and $y(x-1)$:

$$y(x+1) + (R_n(x-1) - 2)\,y(x)$$
$$+ [S_n(x-1) - R_n(x-1) + 1]\,y(x-1) = 0. \tag{6.3.20}$$

Similarly, equation (6.3.15) can be written in the form

$$\Delta^2 p_n(x) + \tilde{R}_n(x)\Delta p_n(x) + \tilde{S}_n(x)p_n(x) = 0, \tag{6.3.21}$$

where

$$\tilde{R}_n(x) = -\frac{\Delta A_n(x)}{A_n(x)} + B_n(x+1) + \frac{B_{n+1}(x)A_n(x+1)}{A_n(x)}$$
$$- \frac{(x + 1 - b_n)}{a_n}A_n(x+1), \tag{6.3.22}$$

$$\tilde{S}_n(x) = \left(B_n(x) - \frac{B_n(x)}{B_{n+1}(x)} - \frac{(x - b_n)}{a_n}A_n(x)\right)\frac{B_{n+1}(x)A_n(x+1)}{A_n(x)}$$
$$+ B_n(x+1) + \frac{a_{n+1}A_{n+1}(x)A_n(x+1)}{a_n} - \frac{A_n(x+1)}{a_n}. \tag{6.3.23}$$

Analogous to the functions of the second kind in §3.6 we define the function of the second kind $J_n(x)$ by

$$J_n(x) := \frac{1}{w(x)}\sum_{\ell=s}^{t}\frac{p_n(\ell)}{x - \ell}w(\ell), \qquad x \notin \{s, s+1, \ldots, t\}. \tag{6.3.24}$$

Indeed, $J_n(x)$ satisfies the three-term recurrence relation (3.1.6). The next theorem shows that it also satisfies the same finite difference equation

Theorem 6.3.2 *Assume that* (6.3.3) *holds and that the polynomials* $\{p_n\}$ *are orthonormal. Then, the function of the second kind* $J_n(x)$ *satisfies the first-order difference equation* (6.3.7).

For a proof see (Ismail et al., 2004).

From Theorem 6.3.2 it follows that the corresponding coefficients of (6.3.17) and (6.3.21) are equal. In particular, $R_n(x) = \tilde{R}_n(x)$ implies

$$B_{n+1}(x) - B_{n-1}(x)$$
$$= \frac{A_n(x)}{a_n} + \frac{(x - b_n)}{a_n} A_n(x) - \frac{(x - b_{n-1})}{a_{n-1}} A_{n-1}(x). \tag{6.3.25}$$

Adding these equations we obtain

$$\sum_{k=0}^{n-1} \frac{A_k(x)}{a_k} = B_n(x) + B_{n-1}(x) - \frac{(x - b_{n-1})}{a_{n-1}} A_{n-1}(x) + u(x+1) \tag{6.3.26}$$

Next, $S_n(x) = \tilde{S}_n(x)$ eventually leads to

$$\left(B_{n-1}(x) - \frac{(x - b_{n-1})}{a_{n-1}} A_{n-1}(x) \right) B_n(x)$$
$$- \left(B_n(x) - \frac{(x - b_n)}{a_n} A_n(x) \right) B_{n+1}(x) \tag{6.3.27}$$
$$= -\frac{A_n(x)}{a_n} + \frac{a_{n+1} A_n(x) A_{n+1}(x)}{a_n} - \frac{a_n A_{n-1}(x) A_n(x)}{a_{n-1}}.$$

In (6.3.27) we substitute for $B_{n-1}(x) - B_{n+1}(x)$ using (6.3.25) and simplify to obtain the identity

$$B_{n+1}(x) - \left(1 + \frac{1}{x - b_n} \right) B_n(x)$$
$$= \frac{a_{n+1} A_{n+1}(x)}{x - b_n} - \frac{a_n^2 A_{n-1}(x)}{a_{n-1}(x - b_n)} - \frac{1}{x - b_n}. \tag{6.3.28}$$

Theorem 6.3.3 *The functions* $A_n(x)$ *and* $B_n(x)$ *satisfy* (6.3.25), (6.3.26), *and* (6.3.28).

Theorem 6.3.4 *The functions* $A_n(x)$ *and* $B_n(x)$ *satisfy fifth-order nonhomogeneous recurrence relations.*

Proof Eliminating $B_{n+1}(x)$ from (6.3.25) and (6.3.28), and replacing n by $n+1$ we obtain

$$\left(1 + \frac{1}{x - b_{n+1}} \right) B_{n+1}(x) - B_n(x) = \frac{(x + 1 - b_{n+1})}{a_{n+1}} A_{n+1}(x)$$
$$- \frac{a_{n+2} A_{n+2}(x)}{x - b_{n+1}} - \frac{(x - b_n)}{a_n} A_n(x) + \frac{a_{n+1}^2 A_n(x)}{a_n(x - b_{n+1})} + \frac{1}{x - b_{n+1}}. \tag{6.3.29}$$

Solving the system formed by equations (6.3.28) and (6.3.29) for $B_n(x)$ and $B_{n+1}(x)$ and setting the solution for $B_n(x)$, with n replaced by $n+1$, equal to the solution for

$B_{n+1}(x)$ yields a fifth-order recurrence relation for $A_n(x)$. A fifth-order recurrence relation for $B_n(x)$ is obtained similarly.   □

6.4 Discrete Discriminants

Ismail (Ismail, 2000a) introduced the concept of a generalized discriminant associated with a degree-reducing operator T as

$$D(g;T) = (-1)^{\binom{n}{2}}\gamma^{-1}\operatorname{Res}\{g, Tg\}. \tag{6.4.1}$$

In other words

$$D(g;T) = (-1)^{n(n-1)/2}\gamma^{n-2}\prod_{j=1}^{n}(Tg)(x_j), \tag{6.4.2}$$

where g is as in (3.1.7). If $T = \dfrac{d}{dx}$ then formula (6.4.2) becomes (3.1.9). When $T = \Delta$ the generalized discriminant becomes the discrete discriminant

$$D(f_n;\Delta) = \gamma^{2n-2}\prod_{1\leq j<k\leq n}(x_j - x_k - 1)(x_j - x_k + 1). \tag{6.4.3}$$

Note that if in (6.4.2) we use the difference operator δ_h,

$$(\delta_h f)(x) = [f(x+h) - f(x)]/h$$

then the right-hand side of (6.4.3) becomes

$$h^{-n}\gamma^{2n-2}\prod_{1\leq j<k\leq n}(x_j - x_k - h)(x_j - x_k + h)$$

and it is clear that $h^n D(g;\Delta_h)$ reduces to the usual discriminant when $h \to 0$.

Ismail and Jing showed how the expression in (6.4.2) arises as a correlation function or expectation values in certain models involving vertex operators, for details see (Ismail & Jing, 2001).

Theorem 6.4.1 *Let* $\{p_n(x)\}$ *be a family of orthogonal polynomials generated by* (3.4.1) *and* (3.4.2). *Assume that* $\{p_n(x)\}$ *satisfy* (6.3.1). *Then,*

$$D(p_n;\Delta) = \prod_{j=1}^{n}A_n(x_{n,j})\prod_{k=1}^{n}\left(\xi_k^{2n-2k-1}\nu_k^{k-1}\right). \tag{6.4.4}$$

The proof is identical to the proof of Theorem 3.4.2 through Schur's theorem. Observe that if (6.3.7) is replaced by

$$\Delta p_n(x) = C_n(x)p_{n-1}(x) + D_n(x)p_n(x), \tag{6.4.5}$$

then (6.4.4) becomes

$$D(p_n;\Delta) = \prod_{j=1}^{n}C_n(x_{n,j})\prod_{k=1}^{n}\left(\xi_k^{2n-2k-1}\zeta_k^{k-1}\right). \tag{6.4.6}$$

We now compute the functions $A_n(x)$ and $B_n(x)$ for the Meixner and Hahn polynomials to illustrate the finite difference relation of §6.3. These will also be used to compute the discrete discriminants for the Meixner and Hahn polynomials.

Meixner Polynomials. From §6.1 we see that $x^n w(x; \beta, c) \to 0$ as $x \to \infty$, for every $n \geq 0$, where $w(x; \beta, c)$ is the weight function for the Meixner polynomials. From (6.3.4) and (6.1.1), it follows that

$$u(x) = -1 + \frac{w(x-1)}{w(x)} = -1 + \frac{x}{(\beta + x - 1)c}, \tag{6.4.7}$$

and clearly u has a differentiable extension to $[1, \infty)$. Furthermore

$$\frac{u(x+1) - u(\ell)}{x + 1 - \ell} = \frac{\beta - 1}{(\beta + x)(\beta + \ell - 1)c}.$$

Since $u(0) = -1$, $w(-1) = w(0)(1 + u(0)) = 0$. From (6.3.8), (6.3.6), and (6.1.1) we obtain

$$A_n(x) = \frac{\gamma_{n-1}}{\gamma_n \kappa_{n-1}} \frac{(\beta - 1)}{(\beta + x)c} \sum_{\ell=0}^{\infty} \frac{p_n(\ell)p_n(\ell - 1)w(\ell)}{\beta + \ell - 1}$$

$$= \frac{\gamma_{n-1}}{\gamma_n \kappa_{n-1}} \frac{(\beta - 1)}{(\beta + x)c} p_n(-\beta) \sum_{\ell=0}^{\infty} \frac{p_n(\ell)w(\ell)}{\beta + \ell - 1}$$

$$= \frac{\gamma_{n-1}}{\gamma_n \kappa_{n-1}} \frac{p_n(-\beta)(\beta - 1)}{(\beta + x)c} \sum_{\ell=0}^{\infty} \frac{(\beta)_\ell c^\ell}{\ell!(\beta + \ell - 1)} p_n(\ell)$$

$$= \frac{\gamma_{n-1}}{\gamma_n \kappa_{n-1}} \frac{p_n(-\beta)}{(\beta + x)c} \sum_{\ell=0}^{\infty} \frac{(\beta - 1)_\ell c^\ell}{\ell!} p_n(\ell).$$

From (6.1.3) and the above computation it follows that

$$A_n(x) = \frac{\gamma_{n-1}}{\gamma_n \kappa_{n-1}} \frac{p_n(-\beta)}{(\beta + x)c} \sum_{\ell=0}^{\infty} \frac{(\beta - 1)_\ell c^\ell}{\ell!} \sum_{k=0}^{n} \frac{(-n)_k(-\ell)_k}{(\beta)_k k!} \left(1 - \frac{1}{c}\right)^k$$

$$= \frac{\gamma_{n-1}}{\gamma_n \kappa_{n-1}} \frac{p_n(-\beta)}{(\beta + x)c} \sum_{k=0}^{n} \frac{(-n)_k}{k!(\beta)_k} \left(1 - \frac{1}{c}\right)^k (-1)^k \sum_{\ell=k}^{\infty} \frac{(\beta - 1)_\ell c^\ell}{(\ell - k)!}.$$

The binomial theorem sums the inner sum and we get

$$A_n(x) = \frac{\gamma_{n-1}}{\gamma_n \kappa_{n-1}} \frac{p_n(-\beta)}{(\beta + x)c} \sum_{k=0}^{n} \frac{(-n)_k(\beta - 1)_k(1 - c)^{1-\beta}}{k!(\beta)_k}$$

$$= \frac{\gamma_{n-1}}{\gamma_n \kappa_{n-1}} \frac{p_n(-\beta)(1 - c)^{1-\beta}}{(\beta + x)c} \, {}_2F_1(-n, \beta - 1; \beta; 1) \tag{6.4.8}$$

$$= \frac{\gamma_{n-1}}{\gamma_n \kappa_{n-1}} \frac{p_n(-\beta)(1 - c)^{1-\beta} n!}{(\beta + x)c(\beta)_n},$$

where we used the Chu–Vandermonde sum (1.4.3). The binomial theorem and (6.1.3), imply

$$p_n(-\beta) = {}_1F_0(-n; -; 1 - 1/c) = c^{-n}. \tag{6.4.9}$$

Furthermore, from (6.1.3) and (6.1.4) we get

$$\gamma_n = \frac{1}{(\beta)_n} \left(1 - \frac{1}{c} \right)^n, \qquad \kappa_n = \frac{c^{-n} n!}{(\beta)_n (1 - c)^\beta}. \qquad (6.4.10)$$

Substituting for $p_n(-\beta)$, γ_{n-1}/γ_n, and κ_{n-1} in (6.4.8) we obtain

$$A_n(x) = \frac{(\beta + n - 1)c}{(c - 1)} \frac{(\beta)_{n-1}(1 - c)^\beta c^{n-1}}{(n - 1)!} \frac{c^{-n}(1 - c)^{1-\beta} n!}{(\beta + x)c(\beta)_n}. \qquad (6.4.11)$$

The above equation simplifies to

$$A_n(x) = -\frac{n}{(\beta + x)c}. \qquad (6.4.12)$$

To find $B_n(x)$ we note that (6.3.8), (6.3.9), (6.1.1), (6.3.6), and (6.4.9) imply

$$B_n(x) = \frac{p_{n-1}(-\beta)}{p_n(-\beta)} A_n(x) = -\frac{n}{\beta + x}. \qquad (6.4.13)$$

Apply (6.4.12), (6.4.10), and (6.4.9) in the case of the recurrence relation (6.1.5) to get

$$\prod_{j=1}^{n} A_n(x_{n,j}) = \frac{(n/c)\gamma_n}{M_n(-\beta; \beta, c)} = \frac{n^n}{(\beta)_n} (1 - 1/c)^n. \qquad (6.4.14)$$

From Theorem 6.4.1, (6.1.5), and (6.4.14) we obtain

$$D\left(M_n(x; \beta, c); \Delta \right)$$
$$= \frac{n^n}{(\beta)_n} (1 - 1/c)^n \prod_{k=1}^{n} \left(\frac{c - 1}{(\beta + k - 1)c} \right)^{2n-2k-1} \prod_{k=2}^{n} \left(\frac{k - 1}{(\beta + k - 1)c} \right)^{k-1}.$$

Thus we have established the following theorem.

Theorem 6.4.2 *The Meixner polynomials satisfy*

$$\Delta M_n(x; \beta, c) = \frac{n}{\beta + x} M_n(x; \beta, c) - \frac{n}{(\beta + x)c} M_{n-1}(x; \beta, c) \qquad (6.4.15)$$

and their discrete discriminant is given by

$$D\left(M_n(x; \beta, c); \Delta \right) = \frac{(1 - 1/c)^{n^2-n}}{c^{n(n-1)/2}} \prod_{k=1}^{n} \frac{k^k}{(\beta + k - 1)^{2n-k-1}}. \qquad (6.4.16)$$

Hahn Polynomials. We can use an argument similar to what we used in the case of Meixner polynomials to evaluate $A_n(x)$ and $B_n(x)$. This approach is lengthy and the details are in (Ismail et al., 2004). The result is that $A_n(x)$ and $B_n(x)$ are given by

$$A_n(x) = \frac{n(\alpha + \beta + n + N + 1)(\beta + n)}{(\alpha + \beta + 2n)(x + \alpha + 1)(x - N)}, \qquad (6.4.17)$$

and

$$
\begin{aligned}
B_n(x) = &-\frac{n}{(\alpha + N + 1)(\alpha + \beta + 2n)} \\
&\times \left(\frac{(N - n + 1)(\beta + n)}{x + \alpha + 1} + \frac{(\alpha + n)(\alpha + \beta + n + N + 1)}{x - N} \right).
\end{aligned}
\tag{6.4.18}
$$

The three term recurrence relation for Hahn polynomials is (6.2.8). Hence the parameters ξ_n and ν_n in Schur's theorem, Lemma 3.4.1, for $n \geq 2$, are given by,

$$
\xi_n = -\frac{1}{b_{n-1}} = -\frac{(\alpha + \beta + 2n - 1)(\alpha + \beta + 2n)}{(\alpha + \beta + n)(\alpha + n)(N - n + 1)},
$$

and

$$
\nu_n = \frac{d_{n-1}}{b_{n-1}} = \frac{(n - 1)(\beta + n - 1)(\alpha + \beta + N + n)(\alpha + \beta + 2n)}{(\alpha + \beta + 2n - 2)(\alpha + \beta + n)(\alpha + n)(N - n + 1)}.
$$

From (6.2.6) and (6.4.17) it follows that

$$
\begin{aligned}
\gamma_n^{-2} \prod_{j=1}^{n} A_n(x_{n,j}) &= \frac{n^n(\alpha + \beta + N + n + 1)^n(\beta + n)^n}{(\alpha + \beta + 2n)^n p_n(-\alpha - 1)p_n(N)} \\
&= \frac{(-1)^n n^n(\alpha + \beta + N + n + 1)^n(\beta + n)^n(\alpha + 1)_n(N + 1 - n)_n}{(\alpha + \beta + 2n)^n(\beta + 1)_n(\alpha + \beta + N + 2)_n}.
\end{aligned}
\tag{6.4.19}
$$

Furthermore,

$$
\begin{aligned}
\gamma_n^2 \prod_{k=1}^{n} \left(\xi_k^{2n-2k-1} \zeta_k^{k-1} \right) &= \prod_{k=1}^{n} \left(\xi_k^{2n-2k+1} \zeta_k^{k-1} \right) \\
&= \prod_{k=1}^{n} \left(-\frac{(\alpha + \beta + 2k - 1)(\alpha + \beta + 2k)}{(\alpha + \beta + k)(\alpha + k)(N - k + 1)} \right)^{2n-2k+1} \prod_{j=1}^{n-1} j^j \\
&\times \prod_{k=1}^{n} \left(\frac{(\beta + k - 1)(\alpha + \beta + N + k)(\alpha + \beta + 2k)}{(\alpha + \beta + k)(\alpha + k)(\alpha + \beta + 2k - 2)(N - k + 1)} \right)^{k-1}.
\end{aligned}
\tag{6.4.20}
$$

Theorem 6.4.1, and (6.4.19), (6.4.20), and (6.4.3) yield

$$
\begin{aligned}
&D\left(Q_n(x; \alpha, \beta, N); \Delta\right) \\
&= \prod_{k=1}^{n} \left(\frac{k^k(\alpha + \beta + 2k - 1)^{2n-2k+1}(\alpha + \beta + 2k)^{2n-k}}{(\alpha + \beta + k)^{2n-k}(\alpha + k)^{2n-k-1}} \right. \\
&\qquad\qquad \left. \times \frac{(\beta + k)^{k-1}(\alpha + \beta + N + k + 1)^{k-1}}{(N - k + 1)^{2n-k-1}} \right).
\end{aligned}
$$

Thus we arrive at the explicit representation

$$
\begin{aligned}
&D\left(Q_n(x; \alpha, \beta, N); \Delta\right) \\
&= \prod_{k=1}^{n} \left(\frac{k^k(\alpha + \beta + n + k)^{n-k}(\alpha + \beta + N + k + 1)^{k-1}}{(\beta + k)^{1-k}(\alpha + k)^{2n-k-1}(N - k + 1)^{2n-k-1}} \right).
\end{aligned}
\tag{6.4.21}
$$

The discriminants of the Jacobi polynomials as in (3.4.16) can be obtained from the generalized discriminants of the Hahn polynomials through the limiting relation

(6.2.10) while the discrete discriminant for Meixner polynomials could have been obtained from (6.4.21) through the limiting process in (6.2.11).

6.5 Lommel Polynomials

The Lommel polynomials arise when we iterate (1.3.19). It is clear from iterating (1.3.19) that $J_{\nu+n}(z)$ is a linear combination of $J_\nu(z)$ and $J_{\nu-1}(z)$ with coefficients which are polynomials in $1/z$.

Theorem 6.5.1 *Define polynomials* $\{R_{n,\nu}\}$ *by*

$$R_{0,\nu}(z) = 1, \quad R_{1,\nu}(z) = 2\nu/z, \tag{6.5.1}$$

$$R_{n+1,\nu}(z) = \frac{2(n+\nu)}{z} R_{n,\nu}(z) - R_{n-1,\nu}(z). \tag{6.5.2}$$

Then

$$J_{\nu+n}(z) = R_{n,\nu}(z) J_\nu(z) - R_{n-1,\nu+1}(z) J_{\nu-1}(z). \tag{6.5.3}$$

Proof Use induction and formulas (6.5.1), (6.5.2), and (1.3.19). ☐

It is easy to see that both $J_{\nu+n}(z)$ and $(-1)^n J_{-\nu-n}(z)$ satisfy

$$\frac{2(\nu+n)}{z} f_n(z) = f_{n+1}(z) + f_{n-1}(z),$$

hence the proof of Theorem 6.5.1 also establishes

$$(-1)^n J_{-n-\nu}(z) = J_{-\nu}(z) R_{n,\nu}(z) + J_{1-\mu}(z) R_{n-1,\nu+1}(z).$$

Eliminating $R_{n-1,\nu+1}(z)$ beween the above identity and (6.5.3), we obtain

$$
\begin{aligned}
&J_{\nu+n}(z) J_{1-\nu}(z) + (-1)^n J_{-\nu-n}(z) J_{\nu-1}(z) \\
&= R_{n,\nu}(z) \left[J_\nu(z) J_{1-\nu}(z) + J_{-\nu}(z) J_{\nu-1}(z) \right],
\end{aligned}
\tag{6.5.4}
$$

for non-integer values of ν.

Lemma 6.5.2 *The Bessel functions satisfy the product formula*

$$J_\nu(z) J_\mu(z) = \sum_{r=0}^\infty \frac{(-1)^r (z/2)^{\mu+\nu+2r} (\mu+\nu+r+1)_r}{r! \, \Gamma(\mu+r+1) \, \Gamma(\nu+r+1)}. \tag{6.5.5}$$

Proof From (1.3.17) we see that the left-hand side of (6.5.5) is

$$
\frac{(z/2)^{\mu+\nu}}{\Gamma(\mu+1)\Gamma(\nu+1)} \sum_{m,n=0}^\infty \frac{(-1)^{m+n}(z/2)^{2m+2n}}{m! \, n! \, (\mu+1)_m (\nu+1)_n}
$$

$$
= \frac{(z/2)^{\mu+\nu}}{\Gamma(\mu+1)\Gamma(\nu+1)} \sum_{r=0}^\infty \frac{(-1)^r (z/2)^{2r}}{r! \, (\mu+1)_r} \sum_{n=0}^r \frac{r!(\mu+1)_r}{n! \, (r-n)! \, (\nu+1)_n (\mu+1)_{r-n}}.
$$

Applying (1.3.30) we see that the n sum is $_2F_1(-r, -r-\mu; \nu+1; 1)$ which is $(\mu+\nu+r+1)_r/(\nu+1)_r$, by the Chu–Vandermonde sum (1.4.3). The result then follows. ☐

Observe that formula (6.5.5) implies

$$(-1)^n J_{-\nu-n}(z) J_{\nu-1}(z) = \sum_{r=0}^{\infty} \frac{(-1)^{n+r} (z/2)^{2r-n-1} (-n+r)_r}{r! \, \Gamma(r+1-\nu-n) \, \Gamma(\nu+r)}$$

$$= \sum_{r=0}^{n} \frac{(-1)^{r+n} (z/2)^{2r-n-1} (-n+r)_r}{r! \, \Gamma(r+1-\nu-n) \, \Gamma(\nu+r)}$$

$$- \sum_{r=0}^{\infty} \frac{(-1)^r (z/2)^{2r+n+1} (1+r)_{r+n+1}}{(r+n+1)! \, \Gamma(r+2-\nu) \, \Gamma(\nu+n+r+1)}.$$

In the infinite sum use

$$\frac{(r+1)_{r+n+1}}{(r+n+1)!} = \frac{(2r+n+1)!}{r! \, (r+n+1)!} \frac{(r+n+2)_r}{r!},$$

to obtain

$$J_{\nu+n}(z) J_{1-\nu}(z) + (-1)^n J_{-\nu-n} J_{\nu-1}(z)$$

$$= \sum_{r=0}^{n} \frac{(-1)^{n-r} (z/2)^{2r-n-1} (-n+r)_r}{r! \, \Gamma(r+1-\nu-n) \, \Gamma(\nu+r)}. \tag{6.5.6}$$

In particular the case $n = 0$ is

$$J_{\nu}(z) J_{1-\nu}(z) + J_{-\nu} J_{\nu-1}(z) = \frac{2 \sin(\nu\pi)}{\pi z}, \tag{6.5.7}$$

where we used (1.3.9).

Theorem 6.5.3 *For all ν, the Lommel polynomials are given by*

$$R_{n,\nu}(z) = \sum_{r=0}^{\lfloor n/2 \rfloor} \frac{(-1)^r (n-r)! \, (\nu)_{n-r}}{r! \, (n-2r)! \, (\nu)_r} \left(\frac{2}{z}\right)^{n-2r}.$$

Proof When ν is not an integer apply (6.5.4) and (6.5.7) to see that

$$\frac{2 \sin(\nu\pi)}{\pi z} R_{n,\nu}(z) = \sum_{r=0}^{n} \frac{(-1)^{n-r} (z/2)^{2r-n-1} (-n+r)_r}{r! \, \Gamma(r+1-\nu-n) \, \Gamma(\nu+r)},$$

which simplifies to the equation in Theorem 6.5.3 after using (1.3.7). This also establishes the theorem for all ν since both sides of the equation in the theorem are rational functions of ν. $\qquad\square$

Following (Watson, 1944) we shall use the notation

$$h_{n,\nu}(z) := R_{n,\nu}(1/z) = \sum_{r=0}^{\lfloor n/2 \rfloor} \frac{(-1)^r (n-r)! \, (\nu)_{n-r}}{r! \, (n-2r)! \, (\nu)_r} \left(\frac{z}{2}\right)^{n-2r}. \tag{6.5.8}$$

The polynomials $\{h_{n,\nu}(x)\}$ are called the modified Lommel polynomials in (Watson, 1944). It is clear from (6.5.1) and (6.5.2) that that $\{h_{n,\nu}(x)\}$ is a system of orthogonal polynomials when $\nu > 0$. The large n behavior of $h_{n,\nu}(x)$ is needed in order to determine the measure with respect to which they are orthogonal.

Theorem 6.5.4 (Hurwitz) *The limit*

$$\lim_{n \to \infty} \frac{(z/2)^{n+\nu} R_{n,\nu+1}(z)}{\Gamma(n+\nu+1)} = J_\nu(z), \qquad (6.5.9)$$

holds uniformly on compact subsets of $\mathbb{C}$.

Proof From Theorem 6.5.3 we see that

$$\frac{(z/2)^n R_{n,\nu+1}(z)}{\Gamma(\nu+n+1)} = \sum_{r=0}^{\lfloor n/2 \rfloor} \frac{(-1)^r A(n,r)(z/2)^{2r}}{\Gamma(\nu+n+1)}, \qquad (6.5.10)$$

with

$$A(n,r) = \frac{(n-r)!\,\Gamma(\nu+n-r+1)}{(n-2r)!\,\Gamma(n+\nu+1)}.$$

Since $A(n,r) = (n-2r+1)_r/(n+\nu+1-r)_r$ then

$$|A(n,r)| \le \frac{(n-2r+1)_r}{(n+1-r-|\nu|)_r}, \qquad |\nu| < n/2, \ 0 \le r \le n/2.$$

This implies $|A(n,r)| \le 2$ and $A(n,k) \to 1$ as $n \to \infty$ for every fixed k. Therefore we can interchange the $n \to \infty$ limit and the sum in (6.5.10), through the use of the dominated convergence theorem for discrete measures (Tannery's theorem), (McDonald & Weiss, 1999). This establishes the theorem. □

From (6.5.2) and (6.5.1) and using the notation in Theorem 2.6.1 we see that

$$D_n(z) = R_{n,\nu}(z), \quad N_n(z) = 2\nu R_{n-1,\nu+1}(z).$$

Therefore Hurwitz's theorem establishes the validity of

$$\frac{J_\nu(z)}{J_{\nu-1}(z)} = \frac{1}{2\nu/z-} \frac{1}{2(\nu+1)/z-} \cdots \frac{1}{2(\nu+1)/z-} \cdots, \qquad (6.5.11)$$

for all finite z when $J_\nu(z) \ne 0$, and the continued fraction converges uniformly over all compact subsets of $\mathbb{C}$ not containing $z = 0$ or any zero of $z^{1-\nu} J_{\nu-1}(z)$. The case $\nu = 1/2$ of Theorem 6.5.5 was known to Lambert in 1761 who used it to pove the irationality of π because the continued fraction (6.5.11) becomes a continued fraction for $\tan z$, see (1.3.18). According to Wallisser (Wallisser, 2000), Lambert gave explicit formulas for the polynomials $R_{n,-1/2}(z)$ and $R_{n,1/2}(z)$ from which he established Hurwitz's theorem in the cases $\nu = 1/2, 3/2$ then proved (6.5.11) for $\nu = 1/2$. This is remarkable since Lambert had polynomials with no free parameters and parameter-dependent explicit formulas are much easier to prove.

Rewrite (6.5.1)–(6.5.2) in terms of $\{h_{n,\nu}(x)\}$ as

$$h_{0,\nu}(x) = 1, \quad h_{1,\nu}(x) = 2\nu x, \qquad (6.5.12)$$

$$2x(n+\nu)h_{n,\nu}(x) = h_{n+1,\nu}(x) + h_{n-1,\nu}(x). \qquad (6.5.13)$$

Theorem 6.5.5 For $\nu > 0$ the polynomials $\{h_{n,\nu}(x)\}$ are orthogonal with espect to a discrete measure α_ν normalized to have total mass 1, where

$$\int_{\mathbb{R}} \frac{d\alpha_\nu(t)}{z - t} = 2\nu \frac{J_\nu(1/z)}{J_{\nu-1}(1/z)}. \tag{6.5.14}$$

Proof The $h_{n,\nu}(x)$ polynomials are denominators of a continued fraction. The numerators $h_{n,\nu}^*(x)$ satisfy the initial conditions $h_{0,\nu}^*(x) = 0$, $h_{1,\nu}^*(x) = 2\nu$. Hence (6.5.13) gives $h_{n,\nu}^*(x) = 2\nu h_{n-1,\nu+1}(x)$. The monic form of (6.5.13) corresponds to

$$P_n(x) = 2^{-n} h_{n,\nu}(x)/(\nu)_n \tag{6.5.15}$$

with $\alpha_n = 0$, $\beta_n = [4(\nu + n)(\nu + n - 1)]^{-1}$. For $\nu > 0$, β_n is bounded and positive; hence, Theorems 2.5.4 and 2.5.5 guarantee the boundedness of the interval of orthogonality. Theorem 2.6.2 and (6.5.9) establish (6.5.14). $\square$

To invert the Stieltjes transform in (6.5.14), note that $J_\nu(1/z)/J_{\nu-1}(1/z)$ is a single-valued function with an essential singularity at $z = 0$ and pole singularities at $z = \pm 1/j_{\nu-1,n}$, $n = 1, 2, \ldots$, see (1.3.25). In view of (1.2.9), α_ν is a purely discrete measure supported on a compact set. Moreover

$$\alpha_\nu \left(\{1/j_{\nu-1,n}\}\right) = \mathrm{Res}\left\{2\nu J_\nu(1/z)/J_{\nu-1}(1/z) : z = 1/j_{\nu-1,n}\right\}$$

$$= \frac{-2\nu J_\nu(j_{\nu-1,n})}{j_{\nu-1,n}^2 J_{\nu-1}'(j_{\nu-1,n})}.$$

Thus (1.3.26) implies

$$\alpha_\nu \left(\{\pm 1/j_{\nu-1,n}\}\right) = \frac{2\nu}{j_{\nu-1,n}^2}. \tag{6.5.16}$$

It remains to verify whether $x = 0$ supports a mass. To verify this we use Theorem 2.5.6. Clearly (6.5.12)–(6.5.13) or (6.5.8) give

$$h_{2n+1,\nu}(0) = 0, \quad h_{2n}(0) = (-1)^n.$$

Therefore

$$\frac{P_{2n}^2(0)}{\zeta_{2n}} = \frac{4^{-2n}}{(\nu)_{2n}^2 \zeta_{2n}}.$$

Since $\zeta_n = \beta_1 \cdots \beta_n$, $\zeta_n = 4^{-n}/[(\nu)_n(\nu + 1)_n]$, and

$$\frac{P_{2n}^2(0)}{\zeta_{2n}} = \frac{(\nu)_n(\nu)_{n+1}}{(\nu)_{2n}^2} = \frac{\Gamma^2(\nu + n)(n + \nu)}{\Gamma^2(\nu + 2n)\Gamma(\nu)}.$$

Thus $\sum_{n=0}^{\infty} P_n^2(0)/\zeta_n$ diverges by Stirling's formula and $\alpha_\nu(\{0\}) = 0$. Thus we have proved the orthogonality relation

$$\sum_{k=1}^{\infty} \frac{1}{j_{\nu,k}^2} \{h_{n,\nu+1}(1/j_{\nu,k}) h_{m,\nu+1}(1/j_{\nu,k})$$

$$+ h_{n,\nu+1}(-1/j_{\nu,k}) h_{m,\nu+1}(-1/j_{\nu,k})\} = \frac{\delta_{m,n}}{2(\nu + n + 1)}. \tag{6.5.17}$$

H. M. Schwartz (Schwartz, 1940) gave a proof of (6.5.17) without justifying that $\alpha_\nu(\{0\}) = 0$. Later, Dickinson (Dickinson, 1954) rediscovered (6.5.17) but made a numerical error and did not justify $\alpha_\nu(\{0\}) = 0$. A more general class of polynomials was considered in (Dickinson et al., 1956), again without justifying that $x = 0$ does not support a mass. Goldberg corrected this slip and pointed out that in some cases of the class of polynomials considered by (Dickinson et al., 1956), $\mu(\{0\})$ may indeed be positive, see (Goldberg, 1965).

The Lommel polynomials can be used to settle a generalization of the Bourget hypothesis, (Bourget, 1866). Bourget conjectured that when ν is a nonnegative integer and m is a positive integer then $z^{-\nu} J_\nu(z)$ and $z^{-\nu-m} J_{\nu+m}(z)$ have no common zeros. Siegel proved that $J_\nu(z)$ is not an algebraic number when ν is a rational number and $z, z \neq 0$, is an algebraic number, (Siegel, 1929). If $J_\nu(z)$ and $J_{\nu+n}(z)$ have a common zero $z_0, z_0 \neq 0$, then (6.5.3) shows that $R_{n-1,\nu+1}(z_0) = 0$ since $z^{-\nu} J_\nu(z)$ and $z^{1-\nu} J_{\nu-1}(z)$ have no common zeros. Hence z_0 is an algebraic number. When ν is a rational number, this contradicts Siegel's theorem and then Bourget's conjecture follows not only for integer values of ν but also for any rational number ν.

Theorem 6.5.6 *We have*

$$
I_{n+1/2}(z) = \frac{1}{\sqrt{2\pi z}} \left[e^z \sum_{r=0}^{n} \frac{(-1)^r (n+r)! (2z)^{-r}}{r!(n-r)!} \right.
$$
$$
\left. + (-1)^{n+1} e^{-z} \sum_{r=0}^{n} \frac{(n+r)! (2z)^{-r}}{r!(n-r)!} \right],
$$
(6.5.18)

$$
I_{-n-1/2}(z) = \frac{1}{\sqrt{2\pi z}} \left[e^z \sum_{r=0}^{n} \frac{(-1)^r (n+r)! (2z)^{-r}}{r!(n-r)!} \right.
$$
$$
\left. + (-1)^n e^{-z} \sum_{r=0}^{n} \frac{(n+r)! (2z)^{-r}}{r!(n-r)!} \right].
$$
(6.5.19)

Proof Theorem 6.5.1, (1.3.18), and the definition of I_ν give

$$
I_{n+1/2}(z) = \sqrt{\frac{2}{\pi z}} \left[i^{-n} R_{n,1/2}(ix) \sinh x + i^{1-n} R_{n-1,3/2}(ix) \cosh x \right]
$$

and (6.5.7) yields (6.5.18), the result after some simplification. Formula (6.5.19) follows from (6.5.6) and (6.5.18). $\qquad\qquad\square$

Starting from (1.3.24) one can prove

$$
K_{n+\nu}(z) = i^n R_{n,\nu}(iz) K_\nu(z) + i^{n-1} R_{n-1,\nu+1}(iz) K_{\nu-1}(z).
$$
(6.5.20)

In particular we have

$$
K_{n+1/2}(z) = K_{1/2}(z) \left[i^n R_{n,1/2}(iz) + i^{n-1} R_{n-1,3/2}(iz) \right].
$$
(6.5.21)

Consequently

$$
y_n(x) = i^{-n} h_{n,1/2}(iz) + i^{1-n} h_{n-1,3/2}(iz).
$$
(6.5.22)

Wimp introduced a generalization of the Lommel polynomials in (Wimp, 1985). His polynomials arise when one iterates the three-term recurrence relation of the Coulomb wave functions (Abramowitz & Stegun, 1965) as in Theorem 6.5.1.

6.6 An Inverse Operator

Let $D = \dfrac{d}{dx}$ and $w_\nu(x)$ denote the weight function of the ultraspherical polynomials, see (4.5.4), that is

$$w_\nu(x) = \left(1 - x^2\right)^{\nu - 1/2}, \quad x \in (-1, 1). \tag{6.6.1}$$

Motivated by (4.5.5) we may define an inverse operator to D on $L^2\left[w_{\nu+1}(x)\right]$ by

$$(T_\nu\, g)\,(x) \sim \sum_{n=1}^{\infty} \frac{g_{n-1}}{2\nu} C_n^\nu(x) \quad \text{if } g(x) \sim \sum_{n=0}^{\infty} g_n C_n^{\nu+1}(x), \tag{6.6.2}$$

where $\sim$ means has the orthogonal expansion. In (6.6.2) it is tacitly assumed that $\nu > 0$ and

$$g \in L^2[w_{\nu+1}(x)] \quad \text{that is} \quad \sum_{n=0}^{\infty} |g_n|^2 \frac{(2\nu + 2)_n}{n!(\nu + n + 1)} < \infty. \tag{6.6.3}$$

Since the g_n's are the coefficients in the orthogonal ultraspherical expansion of g, we define D^{-1} on $L^2\left[w_{\nu+1}(x)\right]$ as the integral operator

$$(T_\nu\, g)\,(x) = \int_{-1}^{1} \left(1 - t^2\right)^{\nu + 1/2} K_\nu(x, t) g(t)\, dt, \tag{6.6.4}$$

where

$$K_\nu(x, t) = \frac{\Gamma(\nu)\, \pi^{-1/2}}{\Gamma(\nu + 1/2)} \sum_{n=1}^{\infty} \frac{(n-1)!\,(n+\nu)}{(2\nu + 1)_n} C_n^\nu(x) C_{n-1}^{\nu+1}(t). \tag{6.6.5}$$

The kernel $K_\nu(x, t)$ is a Hilbert–Schmidt kernel on $L^2[w_{\nu+1}(x)] \times L^2\left[w_{\nu+1}(x)\right]$ as can be seen from (4.5.4). Now (6.6.4) is the formal definition of T_ν.

We seek functions $g(x; \lambda)$ in $L^2\left[w_\nu(x)\right] \subset L^2\left[w_{\nu+1}(x)\right]$ such that

$$\lambda g(x; \lambda) = \int_{-1}^{1} \left(1 - t^2\right)^{\nu + 1/2} K_\nu(x, t) g(t; \lambda)\, dt, \tag{6.6.6}$$

with

$$g(x; \lambda) \sim \sum_{n=1}^{\infty} a_n(\lambda) C_n^\nu(x). \tag{6.6.7}$$

The coefficient of $C_n^\nu(x)$ on the left-hand side of (6.6.6) is $\lambda a_n(\lambda)$. The corresponding coefficient on the right-hand side is

$$\frac{\Gamma(\nu)(n-1)!(\nu+n)}{\Gamma(1/2)\Gamma(\nu+1/2)(2\nu+1)_n}$$

$$\times \int_{-1}^{1} \left(1-t^2\right) C_{n-1}^{\nu+1}(t) \left(1-t^2\right)^{\nu-1/2} g(t;\lambda)\, dt. \tag{6.6.8}$$

In view of (4.5.7) the integrand in above expression is

$$\left(1-t^2\right)^{\nu-1/2} g(t;\lambda) \left[\frac{(n+2\nu-1)_2}{2(\nu+n)} C_{n-1}^{\nu}(t) - \frac{(n)_2}{2(\nu+n)} C_{n+1}^{\nu}(t)\right].$$

Using (4.5.4) the expression in (6.6.8) becomes

$$\frac{a_{n-1}(\lambda)}{2(\nu+n-1)} - \frac{a_{n+1}(\lambda)}{2(\nu+n+1)}.$$

Therefore

$$\lambda a_n(\lambda) = \frac{a_{n-1}(\lambda)}{2(\nu+n-1)} - \frac{a_{n+1}(\lambda)}{2(\nu+n+1)}, \quad n > 1, \tag{6.6.9}$$

$$\lambda a_1(\lambda) = -\frac{a_2(\lambda)}{2(\nu+2)}. \tag{6.6.10}$$

Consider T_ν as a mapping

$$T_\nu : L^2\left(w_{\nu+1}(x)\right) \to L^2\left(w_\nu(x)\right) \subset L^2\left(w_{\nu+1}(x)\right).$$

Theorem 6.6.1 *Let R_ν be the closure of the span of $\{C_n^\nu(x) : n = 1, 2, \dots\}$ in $L^2\left[w_{\nu+1}(x)\right]$, then R_ν is an invariant subspace for T_ν in $L^2\left[w_{\nu+1}(x)\right]$, and*

$$L^2\left[w_{\nu+1}(x)\right] = R_\nu \oplus R_\nu^\perp$$

where

$$R_\nu^\perp = \text{span}\left\{\left(1-x^2\right)^{-1}\right\} \text{ for } \nu > 1/2 \text{ and } R_\nu^\perp = \{0\} \text{ for } 1/2 \geq \nu > 0.$$

Furthermore if we let $g(x;\lambda) \in R_\nu$ have the orthogonal expansion (6.6.7), then the eigenvalue equation (6.6.6) holds if and only if

$$\sum_{n=1}^{\infty} |a_n(\lambda)|^2 n^{2\nu-2} < \infty. \tag{6.6.11}$$

Proof Observe that $L^2\left[w_\nu(x)\right] \subset L^2\left[w_{\nu+1}(x)\right]$ and that T_ν maps $L^2\left[w_{\nu+1}(x)\right]$ into $L^2\left[w_\nu(x)\right]$. In fact T_ν is bounded and its norm is at most $1/\sqrt{2\nu+1}$. Therefore R_ν is an invariant subspace for T_ν in $L^2\left[w_{\nu+1}(x)\right]$ and $L^2\left[w_{\nu+1}(x)\right] = R_\nu \oplus R_\nu^\perp$.

Now, for every $f \in R_\nu^\perp$, we have

$$\int_{-1}^{1} w_{\nu+1}(x)|f(x)|^2 dx < \infty,$$

$$\int_{-1}^{1} f(x)C_n^\nu(x)w_{\nu+1}(x)\, dx = 0, n = 1, 2, \dots .$$

Since $\{C_n^\nu(x) : n = 0, 1, \dots\}$ is a complete orthogonal basis in $L^2 [w_\nu(x)]$ and $w_{\nu+1}(x) = (1 - x^2)\, w_\nu(x)$ we conclude that $R_\nu^\perp$ is the span of $1/(1 - x^2)$ if $\nu > 1/2$ but consists of the zero function if $0 < \nu \le 1/2$. If (6.6.6) holds then $g(x; \lambda) \in L^2 [w_\nu(x)]$, and this is equivalent to (6.6.11) since the right hand side of (4.5.4) $(m = n)$ is $O(n^{2\nu-2})$. On the other hand, if (6.6.11) and (6.6.6) hold, then we apply (4.5.10) and find

$$g(x; \lambda) = \sum_{n=1}^{\infty} \frac{\nu a_n(\lambda)}{n + \nu} C_n^{\nu+1}(x) - \sum_{n=1}^{\infty} \frac{\nu a_n(\lambda)}{n + \nu} C_{n-2}^{\nu+1}(x).$$

Thus $g(x; \lambda) \in L^2 [w_{\nu+1}(x)]$ and $g(x; \lambda)$ is indeed an eigenfunction of T_ν on R_ν.

$\square$

In order to verify (6.6.11) we need to renormalize the $a_n(\lambda)$'s. It is clear from (6.6.9) and (6.6.10) that $a_n(\lambda) = 0$ for all n if $a_1(\lambda) = 0$. Thus $a_1(\lambda)$ is a multiplicative constant and can be factored out. Set

$$a_n(\lambda) = i^{n-1}\frac{(\nu + n)}{(\nu + 1)} b_{n-1}(i\lambda)a_1(\lambda). \tag{6.6.12}$$

Therefore the b_n's are recusively generated by

$$b_{-1}(\lambda) := 0, \qquad b_0(\lambda) := 1,$$
$$2\lambda(\nu + n + 1)b_n(\lambda) = b_{n+1}(\lambda) + b_{n-1}(\lambda). \tag{6.6.13}$$

This recursive definition identifies $\{b_n(\lambda)\}$ as modified Lommel polynomials. In the notation of §6.5, we have

$$b_n(\lambda) = h_{n,\nu+1}(\lambda). \tag{6.6.14}$$

Theorem 6.6.2 *The convergence condition* (6.6.11) *holds if and only if λ is purely imaginary, $\lambda \neq 0$ and $J_\nu(i/\lambda) = 0$.*

Proof Clearly (6.6.12), (6.6.14) and (6.5.11) show that in order for (6.6.11) to hold it is necessary that $J_\nu(i/\lambda) = 0$ or possibly $\lambda = 0$. If $\lambda = 0$ then $b_{2n+1}(0) = 0$ and $b_{2n}(0) = (-1)^n$, as can be easily seen from (6.6.13). In this case (6.6.11) does not hold. It now remains to show that $J_\nu(i/\lambda) = 0$ is sufficient. From (6.5.3) and (6.5.10) we conclude that if $J_\nu(1/x) = 0$ then

$$J_{\nu-1}(1/x)h_{n-1,\nu+1}(x) = -J_{\nu+n}(1/x).$$

Therefore when $J_\nu(1/x) = 0$ we must have

$$J_{\nu-1}(1/x)h_{n-1,\nu+1}(x) \approx -(2x)^{-\nu-n}/\Gamma(\nu+n+1) \qquad (6.6.15)$$

as $n \to \infty$. Since $x^{-\nu}J_\nu(x)$ and $x^{-\nu-1}J_{\nu+1}(x)$ have no common zeros then (6.6.15) implies (6.6.11). □

Thus we have proved the following theorem.

Theorem 6.6.3 *Let the positive zeros of $J_\nu(x)$ be as in* (1.3.25). *Then the eigenvalues of the integral operator T_ν of* (6.6.4) *are* $\{\pm i/j_{\nu,k} : k = 1, 2, \dots\}$. *The eigenfunctions have the ultraspherical series expansion*

$$g\left(x;\ \pm i/j_{\nu,k}\right) \sim \sum_{n=1}^\infty (\mp i)^{n-1}\frac{(\nu+n)}{\nu+1}C_n^\nu(x)h_{n-1,\nu+1}\left(1/j_{\nu,k}\right). \qquad (6.6.16)$$

The eigenfunction $g\left(x, \pm i/j_{\nu,k}\right)$ is $e^{ixj_{\nu,k}}$. Theorem 6.6.3, formulas (6.5.3), and analytic continuation can be used to establish (4.8.3). A similar analysis using an L_2 space weighted with the weight function for Jacobi polynomials can be used to prove Theorem 4.8.3. The details are in (Ismail & Zhang, 1988).

Exercises

6.1   Show that

$$C_n(x+y;a) = \sum_{k=0}^n \binom{n}{k}(-y)_{n+k}a^{-n-k}C_k(x;a).$$

6.2   Prove that

$$(-a)^n C_n(x;a) = n!\,L_n^{(x-n)}(a).$$

7

Zeros and Inequalities

In this chapter, we study the monotonicity of zeros of parameter dependent orthogonal polynomials as functions of the parameter(s) involved. We also study bounds for the largest and smallest zeros of orthogonal polynomials.

Let $\{\phi_n(x;\tau)\}$ be a family of polynomials satisfying the initial conditions

$$\phi_0(x;\tau) = 1, \quad \phi_1(x;\tau) = (x - \alpha_0(\tau))/\xi_0(\tau), \tag{7.0.1}$$

and the recurrence relation

$$x\phi_n(x;\tau) = \xi_n(\tau)\phi_{n+1}(x;\tau) + \alpha_n(\tau)\phi_n(x;\tau) + \eta_n(\tau)\phi_{n-1}(x;\tau), \tag{7.0.2}$$

for $n > 0$. The corresponding monic polynomials are

$$P_n(x) = \left[\prod_{j=1}^{n} \xi_{j-1}(\tau) \right] \phi_n(x;\tau). \tag{7.0.3}$$

Furthermore in the notation of (2.2.1) $\beta_n = \eta_n\xi_{n-1}$. Hence, by the spectral theorem the polynomials, Theorem 2.5.1, $\{p_n(x;\tau)\}$ are orthogonal if and only if the positivity condition

$$\xi_{n-1}(\tau)\eta_n(\tau) > 0, \quad n = 1, 2, \ldots, \tag{7.0.4}$$

holds for all n. When the positivity condition (7.0.4) holds, then (7.0.3) implies orthogonality relation

$$\int_{\mathbb{R}} \phi_m(x;\tau)\phi_n(x;\tau)d\mu(x) = \zeta_n\delta_{mn}$$

$$\zeta_0 = 1 \quad \zeta_n = \prod_{j=1}^{n} \frac{\eta_j}{\xi_{j-1}}, \quad n > 0. \tag{7.0.5}$$

7.1 A Theorem of Markov

We now state and prove an extension of an extremely useful theorem of A. Markov (Szegő, 1975). We shall refer to Theorem 7.1.1 below as the generalized Markov's theorem.

Theorem 7.1.1 *Let $\{p_n(x;\tau)\}$ be orthogonal with respect to $d\alpha(x;\tau)$,*

$$d\alpha(x;\tau) = \rho(x;\tau)\,d\alpha(x), \qquad (7.1.1)$$

on an interval $I = (a,b)$ and assume that $\rho(x;\tau)$ is positive and has a continuous first derivative with respect to τ for $x \in I$, $\tau \in T = (\tau_1,\tau_2)$. Furthermore, assume that

$$\int_a^b x^j \rho_\tau(x;\tau)\,d\alpha(x), \qquad j=0,1,\dots,2n-1,$$

converge uniformly for τ in every compact subinterval of T. Then the zeros of $p_n(x;\tau)$ are increasing (decreasing) functions of τ, $\tau \in T$, if $\partial\{\ln\rho(x;\tau)\}/\partial\tau$ is an increasing (decreasing) function of x, $x \in I$.

Proof Let $x_1(\tau), x_2(\tau),\dots,x_n(\tau)$ be the zeros of $p_n(x;\tau)$. In this case, the mechanical quadrature formula (2.4.1)

$$\int_a^b p(x)\,d\alpha(x;\tau) = \sum_{i=1}^n \lambda_i(\tau)p\left(x_i(\tau)\right), \qquad (7.1.2)$$

holds for polynomials $p(x)$ of degree at most $2n-1$. We choose

$$p(x) = [p_n(x;\nu)]^2 / [x - x_k(\nu)], \qquad \nu \neq \tau,$$

then we differentiate (7.1.2) with respect to τ, use (7.1.1), then let $\nu \to \tau$. The result is

$$\int_a^b \frac{p_n^2(x;\tau)}{x - x_k(\tau)}\,\frac{\partial\rho(x;\tau)}{\partial\tau}\,d\alpha(x)$$

$$= \sum_{i=1}^n \left[p\left(x_i(\tau)\right)\lambda_i'(\tau) + \lambda_i(\tau)p'\left(x_i(\tau)\right)x_i'(\tau) \right]. \qquad (7.1.3)$$

The first term in the summand vanishes for all i while the second term vanishes when $i \neq k$. Therefore, (7.1.3) reduces to

$$\int_a^b \frac{p_n^2(x;\tau)}{x - x_k(\tau)}\,\frac{\rho_\tau(x;\tau)}{\rho(x;\tau)}\,d\alpha(x;\tau) = \lambda_k(\tau)\left\{p_n'\left(x_k(\tau);\tau\right)\right\}^2 x_k'(\tau). \qquad (7.1.4)$$

In view of the quadrature formula (7.1.2) the integral

$$\int_a^b \frac{p_n^2(x;\tau)}{x - x_k(\tau)}\,d\alpha(x;\tau)$$

vanishes, so we subtract $[\rho_\tau (x_k(\tau); \tau) / \rho (x_k(\tau); \tau)]$ times the above integral from the left-hand side of (7.1.4) and establish

$$
\int_a^b \frac{p_n^2(x;\tau)}{x - x_k(\tau)} \left\{ \frac{\rho_\tau(x;\tau)}{\rho(x;\tau)} - \frac{\rho_\tau (x_k(\tau); \tau)}{\rho (x_k(\tau); \tau)} \right\} d\alpha(x;\tau)
$$

$$
= \lambda_k(\tau) \{ p_n' (x_k(\tau); \tau) \}^2 x_k'(\tau).
$$

(7.1.5)

Theorem 7.1.1 now follows from (7.1.5) since the integrand has a constant sign on (a, b). $\qquad\square$

Markov's theorem is the case when $\alpha(x) = x$, (Szegő, 1975, §6.12). The above more general version is stated as Problem 15 in Chapter III of (Freud, 1971).

Theorem 7.1.2 *The zeros of a Jacobi polynomial $P_n^{(\alpha,\beta)}(x)$ or a Hahn polynomial $Q_n(x; \alpha, \beta, N)$ increase with β and decrease with α. The zeros of a Meixner polynomial $M_n(x; \beta, c)$ increase with β while the zeros of a Laguerre polynomial $L_n^{(\alpha)}(x)$ increase with α. In all these cases increasing (decreasing) means strictly increasing (decreasing) and the parameters are such that the polynomials are orthogonal.*

Proof For Jacobi polynomials $\rho(x; \alpha, \beta) = (1 - x)^\alpha (1 + x)^\beta$ and $\alpha(x) = x$, hence

$$
\frac{\partial \ln \rho(x; \alpha, \beta)}{\partial \beta} = \ln(1 + x),
$$

which increases with x. Similarly for the monotonicity in β. For the Hahn polynomials α is a step function with unit jumps at $0, 1, \ldots, N$.

$$
\rho(x; \alpha, \beta) = \frac{\Gamma(\alpha + 1 + x)}{\Gamma(\alpha + 1)} \frac{\Gamma(\beta + 1 + N - x)}{\Gamma(\beta + 1)}.
$$

Hence, by (1.3.5), we obtain

$$
\frac{\partial \ln \rho(x; \alpha, \beta)}{\partial \alpha} = \frac{\Gamma'(\alpha + 1 + x)}{\Gamma(\alpha + 1 + x)} - \frac{\Gamma'(\alpha + 1)}{\Gamma(\alpha + 1)}
$$

$$
= \sum_{n=0}^{\infty} \left[\frac{1}{\alpha + n + 1} - \frac{1}{\alpha + n + x + 1} \right],
$$

which obviously decreases with x. The remaining cases similarly follow. $\qquad\square$

7.2 Chain Sequences

Let A_N be a symmetric tridiagonal matrix with entries $a_{j,k}$, $0 \le j, k \le N - 1$,

$$
a_{j,j} = \alpha_j, \quad a_{j,j+1} = a_{j+1}, \quad 0 \le j < N. \tag{7.2.1}
$$

To determine its positive definiteness we apply Theorem 1.1.5. It is necessary that $\alpha_j > 0$, for all j. We also assume $a_j \neq 0$, otherwise A_N will be decomposed to two smaller matrices. The row operations: Row $i \to$ Row $i + c$ Row j, with $j < i$

will reduce the matrix to an upper triangular matrix without changing the principal minors. It is easy to see that the diagonal elements after the row reduction are

$$\alpha_0, \alpha_1 - \frac{a_1^2}{\alpha_0}, \alpha_2 - \frac{a_2^2}{\alpha_1 - a_1^2/\alpha_0}, \cdots.$$

The positivity of α_j and the above diagonal elements (called Pivots) is necessary and sufficient for the positive definiteness of A_N, as can be seen from Theorem 1.1.5. Now define $g_0 = 0$, and write $\alpha_1 - a_1^2/\alpha_0$ as $\alpha_1(1 - g_1)$. That is $g_1 = a_1^2/(\alpha_0\alpha_1)$, hence $g_1 \in (0, 1)$. The positivity of the remaining pivots is equivalent to

$$\frac{a_j^2}{\alpha_j\alpha_{j-1}} = g_j(1 - g_{j-1}), \quad 0 < j < N, \text{ and } 0 < g_j < 1. \qquad (7.2.2)$$

The above observations are from (Ismail & Muldoon, 1991) and motivate the following definition.

Definition 7.2.1 *A sequence $\{c_n : n = 1, 2, \ldots, N\}$, $N \le \infty$, is called a chain sequence if there exists another sequence $\{g_n : n = 0, 1, 2, \ldots, N\}$ such that*

$$c_n = g_n(1 - g_{n-1}), \, n > 0, \text{ with } 0 < g_n < 1, \, n > 0, \, 0 \le g_0 < 1.$$

If we need to specify whether N is finite or inifite we say $\{c_n\}$ is a finite (infinite) chain sequence, depending on whether N is finite or infinite. The sequence $\{g_n\}$ is called a parameter sequence for the sequence $\{c_n\}$.

Researchers in continued fractions allow g_n to take the values 0 or 1 for $n \ge 0$, but we shall adopt Chihara's terminology (Chihara, 1978) because it is the most suitable for the applications in this chapter.

Theorem 7.2.1 *A matrix A_N with entries as in (7.2.1) is positive definite if and only if*

(i) $\alpha_j > 0$, for $0 \le j < N$
(ii) *The sequence $\{a_j^2/(\alpha_j\alpha_{j-1}) : 0 < j < N\}$ is a chain sequence.*

Proof This follows from the argument preceding Definition 7.2.1.   □

Theorem 7.2.2 *Let $\{c_n\}_1^N$ be a chain sequence and assume that $0 < d_n \le c_n$ for all n, $1 \le n \le N$. Then $\{d_n\}_1^N$ is a chain sequence.*

Proof Define matrices A_N and B_N with $a_{j,j} = b_{j,j} = 1$ and $a_{j,j+1} = \sqrt{d_n}$, $b_{j,j+1} = \sqrt{c_n}$, $0 \le j < N$. Now A_N is positive definite if and only if $\{d_n\}_1^N$ is a chain sequence, that its pivots are positive. A simple calculation shows that the pivots of A_N are greater than or equal to the corresponding pivots of B_N, and the theorem follows.   □

Theorem 7.2.3 *Let A_N be as in (7.2.1). Then the eigenvalues of A_N belong to (a, b) if and only if*

(i) $\alpha_j \in (a, b)$, for $0 \le j < N$

(ii) *The sequence*

$$\frac{a_j^2}{(\alpha_j - x)(\alpha_{j-1} - x)}, \quad 0 < j < N,$$

is a chain sequence at $x = a$ *and at* $x = b$.

Proof The matrix $A_N - aI$ ($bI - A_N$, respectively) is positive definite if and only if all the eigenvalues of A_N are in (a, ∞) $(-\infty, b)$, respectively. The theorem now follows from Theorem 7.2.1. $\qquad\square$

Theorem 7.2.3 is due to Chihara in (Chihara, 1962). Chain sequences were used in (Wall & Wetzel, 1944) to study positive definite J-fractions.

Corollary 7.2.4 *Assume that* $\{P_n(x)\}$ *is a monic sequence of orthogonal polynomials recursively generated by (2.2.2) and (2.2.1). If*

$$u_n(t) = \frac{\beta_n}{(t - \alpha_n)(t - \alpha_{n-1})}, \quad n \geq 1, \qquad (7.2.3)$$

then the following are equivalent:

(i) *The true interval of orthogonality* $[\xi, \eta]$ *is contained in* (a, b),
(ii) $\alpha_n \in (a, b)$ *for all* $n \geq 1$, *and both* $\{u_n(a)\}$ *and* $\{u_n(b)\}$ *are chain sequences.*

Corollary 7.2.4 is a source of many examples of chain sequences because we know the true interval of orthogonality of many orthogonal polynomials.

Theorem 7.2.5 *The zeros of birth and death process polynomials belong to* $(0, \infty)$ *while the zeros of random walk polynomials belong to* $(-1, 1)$.

Proof If we write (5.2.12) in the symmetric form (7.2.1) then $\alpha_n = \lambda_n + \mu_n$ and $a_n = \sqrt{\lambda_{n-1}\mu_n}$. Clearly $\alpha_n > 0$ in this case. In order to apply Theorem 7.2.1, we consider the sequence

$$\frac{a_n^2}{\alpha_n \alpha_{n-1}} = \frac{\lambda_{n-1}\mu_n}{(\lambda_n + \mu_n)(\lambda_{n-1} + \mu_{n-1})}$$

$$= \frac{\mu_n}{(\lambda_n + \mu_n)}\left[1 - \frac{\mu_{n-1}}{\lambda_{n-1} + \mu_{n-1}}\right].$$

Thus the above sequence is a chain sequence and the zeros of all Q_n's are in $(0, \infty)$. For random walk polynomials $\alpha_n = 0$ and $a_n = \sqrt{m_{n-1}\ell_n}$, see (5.2.20). Again by Theorem 7.2.1 we need to verify that $m_{n-1}\ell_n/(\pm 1)^2$ is a chain sequence, which is obvious since $m_n + \ell_n = 1$. $\qquad\square$

We now treat the case of constant chain sequences.

Theorem 7.2.6 *A positive constant sequence* $\{c\}_1^{N-1}$, *is a chain sequence if and only if*

$$0 < c \leq \frac{1}{4\cos^2(\pi/(N+1))}. \qquad (7.2.4)$$

If $N = \infty$ the condition becomes $c \leq 1/4$.

Proof Let A_N be the symmetric tridiagonal matrix with $a_{j,j} = 1, 0 \leq j < N$, and $a_{j,j+1} = \sqrt{c}$. The characteristic polynomial of A_n, $n \leq N$, $n < \infty$ is $c^{n/2} U_{n-1} \left(\dfrac{x-1}{2\sqrt{c}} \right)$ whose zeros are $x = 1 + 2\sqrt{c} \cos \left(\dfrac{j\pi}{n+1} \right)$, $j = 1, \ldots, N$. Thus the smallest eigenvalue of A_N for $N < \infty$ is positive if and only if (7.2.4) holds. If $N = \infty$, then the spectrum of A_n is positive for all n if and only if (7.2.4) holds with $N = \infty$. $\qquad\square$

The next theorem gives upper and lower bounds for zeros of polynomials.

Theorem 7.2.7 *Let $\{P_n(x)\}$ be a sequence of monic polynomials satisfying (2.2.1), with $\beta_n > 0$, for $1 \leq n < N$ and let $\{c_n\}$ be a chain sequence. Set*

$$B := \max\{x_n : 0 < n < N\}, \quad and \quad A := \min\{y_n : 0 < n < N\}, \quad (7.2.5)$$

where x_n and y_n, $x_n \geq y_n$, are the roots of the equation

$$(x - \alpha_n)(x - \alpha_{n-1}) c_n = \beta_n, \qquad (7.2.6)$$

that is

$$x_n, y_n = \frac{1}{2}(\alpha_n + \alpha_{n-1}) \pm \frac{1}{2}\sqrt{(\alpha_n - \alpha_{n-1})^2 + 4\beta_n/c_n}. \qquad (7.2.7)$$

Then the zeros of $P_N(x)$ lie in (A, B).

Proof Let

$$f(x) := (x - \alpha_n)(x - \alpha_{n-1}) - \beta_n/c_n. \qquad (7.2.8)$$

It readily follows that f is positive at $x = \pm\infty$ and has two real zeros. Furthermore $f(\alpha_n) < 0$, hence $\alpha_n \in (A, B)$, $0 < n < N$. The second part in condition (ii) of Corollary 7.2.4 holds since $u_n(x) = c_n$ at $x = x_n, y_n$, and $u_n(A) \leq u_n(y_n)$, $u_n(B) \leq u_n(y_n)$. $\qquad\square$

Theorems 7.2.6–7.2.7 and the remaining results in this section are from (Ismail & Li, 1992).

Theorem 7.2.8 *Let $L(N, \alpha)$ and $S(N, \alpha)$ be the largest and smallest zeros of a Laguerre polynomial $L_N^{(\alpha)}(x)$. Then*

$$L(N, \alpha) < 2N + \alpha - 2 + \sqrt{1 + a(N-1)(N+\alpha-1)} \qquad (7.2.9)$$

for $\alpha > -1$, and

$$S(N, \alpha) > 2N + \alpha - 2 - \sqrt{1 + 4(N-1)(N+\alpha-1)} \qquad (7.2.10)$$

for $\alpha \geq 1$ where

$$a = 4\cos^2(\pi/(N+1)). \qquad (7.2.11)$$

Proof From (4.6.26) it follows that the monic Laguerre polynomials satisfy (2.2.1) with $\alpha_n = 2n + \alpha + 1$, $\beta_n = n(n + \alpha)$. Therefore

$$x_n, y_n = 2n + \alpha \pm \sqrt{1 + an(n + \alpha)}.$$

The result follows because x_n increases with n while y_n decreases with n. $\qquad\square$

For the associated Laguerre polynomials $\alpha_n = 2n + 2c + \alpha + 1$, $\beta_n = (n + c)(n + \alpha + c)$, see §2.9 and (4.6.26) and one can prove the following.

Theorem 7.2.9 *Let $L^{(c)}(N, \alpha)$ and $I^{(c)}(N, \alpha)$ be the largest and smallest zeros for an associated Laguerre polynomial of degree N and association parameter c. Then*

$$L^{(c)}(N, \alpha) < 2N + 2c + \alpha - 2 + \sqrt{1 + a(N + c - 1)(N + c + \alpha - 1)},$$
$$(7.2.12)$$

$$I^{(c)}(N, \alpha) > 2N + 2c + \alpha - 2 - \sqrt{1 + 4(N + c - 1)(N + c + \alpha - 1)},$$
$$(7.2.13)$$

where a is as in (7.2.10).

The associated Laguerre polynomials do not satisfy the second order differential equation, hence Sturmian's techniques of (Szegő, 1975) are not applicable.

For the Meixner polynomials of §6.1, we know that

$$\lim_{c \to 1} M_n \left(\frac{x\sqrt{c}}{1 - c}; \beta, c \right) = n! L_n^\beta(x).$$

The recursion coefficients α_n and β_n for $(-1)^n \beta_n M_n \left(\frac{x\sqrt{c}}{1-c}; \beta, c \right)$ are

$$\alpha_n = \sqrt{c}\beta + n(1 + c)/\sqrt{c}, \quad \beta_n = n(\beta + n - 1). \qquad (7.2.14)$$

Theorem 7.2.10 *Let $m_{N,1}$ be the largest zero of $M_N (x\sqrt{c}/(1 - c); \beta, c)$. Then, with α defined by (7.2.10) we have*

$$m_{N,1} \leq \sqrt{c}\beta + \left(N - \frac{1}{2} \right) \frac{1 + c}{\sqrt{c}} + \frac{1}{2\sqrt{c}} \sqrt{(1 + c)^2 + 4acN(N + \beta - 1)}.$$
$$(7.2.15)$$

The bound (7.2.15) is sharp in the sense

$$m_{N,1} = \frac{(1 + \sqrt{c})^2}{\sqrt{c}} N(1 + o(N)), \quad as \ N \to \infty. \qquad (7.2.16)$$

Proof In the present case x_n of (7.2.7) increases with n its maximum is when $n = N$ and we establish (7.2.16). Next consider the symmetric tridiagonal matrix associated with $\{M_n (x\sqrt{c}/(1 - c); \beta, c)\}$ for $n = 0, 1, \dots, N - 1$. Its diagonal entries are $\alpha_0, \dots, \alpha_{N-1}$ and the super diagonal entries are $\sqrt{\beta_1}, \dots, \sqrt{\beta_{N-1}}$. Let $e_1, \dots, e_N$ be the usual basis for $\mathbb{R}^n$ and for $k < N$, define $X = \left(\sum_{j=N-k+1}^{N} e_j \right) / \sqrt{k}$. Clearly

$\|\mathbf{X}\| = 1$, hence for fixed k and as $N \to \infty$, we get

$$\|A_N\|^2 \geq \|A_N \mathbf{X}\|^2$$
$$= \frac{1}{k}\left[\beta_{N-k} + 2\left(\beta_{N-1} + \alpha_{N-1}\right)^2 + (k-2)\left(\alpha_{N-1} + 2\sqrt{\beta_{N-1}}\right)^2\right]$$
$$\cdot [1 + o(1)].$$

Thus, as $N \to \infty$, we find

$$m_{N,1} = \|A_N\| \geq \|A_N \mathbf{X}\|$$
$$= N\left[\frac{1}{k} + \left(1 - \frac{2}{k}\right)\left(2 + \frac{1+c^2}{\sqrt{c}} + \frac{2}{k}\left(1 + \frac{1+c}{\sqrt{c}}\right)^2\right)\right]^{1/2} \quad [1 + o(1)],$$

which, by choosing k large, proves that $\liminf\limits_{N} m_{n,1}/N \geq (1 + \sqrt{c})^2/\sqrt{c}$ and (7.2.16) follows. $\square$

The work (Ismail & Li, 1992) also contains bounds on the largest and smallest zeros of Meixner–Pollaczek polynomials.

As an application of Theorem 7.2.1, we prove the following theorem whose proof is from (Szwarc, 1995).

Theorem 7.2.11 *Let $\{\varphi_n(x)\}$ be a sequence of polynomials orthogonal with respect to μ such that $\operatorname{supp}\mu \subset [0, \infty)$. If $\varphi_n(0) > 0$, then*

$$\int\limits_0^\infty e^{-tx}\varphi_m(x)\varphi_n(x)\, d\mu(x) \geq 0, \quad m, n \geq 0. \tag{7.2.17}$$

Proof Let $\psi_n(x) = \phi_n(a - x)$ and let μ_a be the measure with respect to which $\{\psi_n(x)\}$ is orthogonal. Since $\varphi_n(0) > 0$, we see that the leading term in $\psi_n(x)$ is positive. The integral in (7.2.17) is

$$\int\limits_0^a e^{-t(a-x)}\psi_m(x)\psi_n(x)\, d\mu_a(x) = e^{-ta}\sum_{k=0}^\infty \frac{t^k}{k!}\int\limits_0^a x^k\psi_m(x)\psi_n(x)\, d\mu_a(x), \tag{7.2.18}$$

where μ_a is a positive measure. The three term recurrence relation for ψ_n has the form

$$x\psi_n(x) = A_n\psi_{n+1}(x) + B_n\psi_n(x) + C_n\psi_{n-1}(x), \tag{7.2.19}$$

with $A_n > 0$, hence $C_n > 0$ follows from orthogonality. Theorem 7.2.1 implies $B_n > 0$, $n \geq 0$ hence $\int_0^a x\psi_n^2(x)\, d\mu(x) > 0$. The latter fact and induction establish the nonnegativity of $\int_0^a x^k\psi_m(x)\psi_n(x)\, d\mu_a$ and the theorem follows from (7.2.18). If μ is not compactly supported then we consider $\mu(x; N) = \chi_{[0,N]}\mu(x)$, construct polynomials $\varphi_n(x; N)$ orthogonal with respect to $\mu(x; N)$, then apply standard real analysis techniques to conclude that $\varphi_n(x; N) \to \varphi_n(x)$ as $N \to \infty$, and we establish (7.2.17) because the integral in (7.2.17) is a limit of nonnegative numbers. $\qquad\square$

7.3 The Hellmann–Feynman Theorem

Let S_ν be an inner product space with an inner product $\langle ., . \rangle_\nu$. The inner product may depend on a parameter ν which is assumed to vary continuously in an open interval $(a, b) = I$, say. If the inner product is ν-dependent, then we assume that there is a fixed set (independent of ν) which is dense in S_ν for all $\nu \in (a, b)$. The following version of the Hellmann–Feynman theorem was proved in (Ismail & Zhang, 1988).

Theorem 7.3.1 *Let H_v be a symmetric operator defined on S_ν and assume that ψ_ν is an eigenfunction of H_v corresponding to an eigenvalue λ_v. Furthermore assume that*

$$\lim_{\mu \to \nu} \langle \psi_\mu, \psi_\nu \rangle_\nu = \langle \psi_\nu, \psi_\nu \rangle_\nu, \tag{7.3.1}$$

holds and that the limit

$$\lim_{\mu \to \nu} \left\langle \frac{H_\mu - H_\nu}{\mu - \nu} \psi_\mu, \psi_\nu \right\rangle_\nu \quad \text{exists.} \tag{7.3.2}$$

If we define the action of $\partial H_\nu / \partial \nu$ on the eigenspaces by

$$\left\langle \frac{\partial H_\nu}{\partial \nu} \psi_\nu, \psi_v \right\rangle_\nu := \lim_{\mu \to \nu} \left\langle \frac{H_\mu - H_\nu}{\mu - \nu} \psi_\mu, \psi_\nu \right\rangle_\nu \tag{7.3.3}$$

then $d\lambda_\nu / d\nu$ exists for $\nu \in I$ and is given by

$$\frac{d\lambda_\nu}{d\nu} = \frac{\left\langle \frac{\partial H_\nu}{\partial \nu} \psi_\nu, \psi_\nu \right\rangle_\nu}{\langle \psi_\nu, \psi_\nu \rangle_\nu}. \tag{7.3.4}$$

Proof Clearly the eigenvalue equation $H_\mu \psi_\mu = \lambda_\mu \psi_\mu$ implies $\langle H_\mu \psi_\mu, \psi_\nu \rangle_\nu = \lambda_\mu \langle \psi_\mu, \psi_\nu \rangle_\nu$. Hence

$$(\lambda_\mu - \lambda_\nu) \langle \psi_\mu, \psi_\nu \rangle_\nu = \langle H_\mu \psi_\mu, \psi_\nu \rangle_\nu - \langle \psi_\mu, H_\nu \psi_\nu \rangle_\nu.$$

The symmetry of the operator H_ν implies

$$(\lambda_\mu - \lambda_\nu) \langle \psi_\mu, \psi_\nu \rangle_\nu = \langle (H_\mu - H_\nu) \psi_\mu, \psi_\nu \rangle_\nu. \tag{7.3.5}$$

We now divide by $\mu - \nu$ and then let $\mu \to \nu$ in (7.3.5). The limit of the right-hand side of (7.3.5) exists, for $\nu \in I$, and equals

$$\left\langle \frac{\partial H_\nu}{\partial \nu} \psi_\nu, \psi_\nu \right\rangle_\nu$$

while the second factor on the left-hand side tends to the positive number $\langle \psi_\nu, \psi_\nu \rangle_\nu$ as $\mu \to \nu$, $\nu \in I$. Thus, the limit of the remaining factor exists and (7.3.4) holds. This completes the proof. $\qquad\square$

In all the examples given here the eigenspaces are one-dimensional. In the cases when the geometric multiplicity of an eigenvalue λ_ν is larger than 1, the conditions (7.3.1) and (7.3.2) put restrictions on the geometric multiplicities of λ_ν when μ is near ν. Apparently this point was not clear in the physics literature and several papers with various assumptions on the dimensions of the eigenspaces have appeared recently; see (Alon & Cederbaum, 2003), (Balawender & Holas, 2004), (Fernandez, 2004), (Vatsaya, 2004), (Zhang & George, 2002), and (Zhang & George, 2004).

An immediate consequence of Theorem 7.3.1 is the following corollary.

Corollary 7.3.2 *If $\partial H_\nu / \partial \nu$ is positive (negative) definite then all the eigenvalues of H_ν increase (decrease) with ν.*

The advantage of the above formulation over its predecessors is the fact that $\partial H_\nu / \partial \nu$ need only to be defined on the eigenspaces. This is particularly useful in applications involving unbounded operators such as the Sturm–Liouville differential operators

$$\frac{d}{dx}\left(p(x)\frac{d}{dx}\right) + \nu^2 q(x),$$

see (Ismail & Zhang, 1989; Laforgia, 1985; Laforgia & Muldoon, 1986; Lewis & Muldoon, 1977). In this work, however, we shall deal mostly with finite dimensional spaces where it is easy to show that the derivative of a matrix operator is the matrix formed by the derivatives of the original matrix. At the end of this section, we shall briefly discuss the case of Sturm–Liouville differential operators.

Pupyshev's very informative article (Pupyshev, 2000) contains an historical survey of the physics literature on the Hellmann–Feynman theorem. A brief account of Hellmann's life (and his tragic death) is also included.

The spectral theorem for orthogonal polynomials asserts that positive measure $d\mu$ in (7.0.5) has infinite support and has moments of all orders. Furthermore recursion relations (7.0.1)–(7.0.2) generate a tridiagonal matrice $A_N = \{a_{ij}\}$, $N = 1, 2, \ldots$ or ∞, with

$$a_{m,n} = \xi_m(\tau)\delta_{m+1,n} + \alpha_m(\tau)\delta_{m,n} + \eta_m(\tau)\delta_{m,n+1},$$
$$m, n = 0, 1, \ldots, N-1. \tag{7.3.6}$$

Theorem 2.2.5, in a different normalization, shows that the characteristic polynomial of A_N, i.e., $\det(\lambda I - A_N)$, is a constant multiple of $p_N(\lambda; \tau)$, hence the eigenvalues of A_N are the zeros of $p_N(\lambda; \tau)$, say $\lambda_1, \lambda_2, \ldots, \lambda_N$. From Theorem 2.2.4, the eigenvalues are real and distinct. An eigenvector corresponding to the eigenvalue λ_j is $\mathbf{P}_j = (p_o(\lambda_j; \tau), \ldots, p_{N-1}(\lambda_j; \tau))^T$. It is easy to see that the matrix operator A_N is self-adjoint (Hermitian) on $\mathbb{R}^N$ equipped with the inner product

$$\langle \mathbf{U}, \mathbf{V} \rangle = \sum_{i=0}^{N-1} u_i v_i / \zeta_i, \text{ where } \mathbf{U} = (u_0, u_1, \ldots, u_{N-1}),$$
$$\mathbf{V} = (v_0, v_1, \ldots, v_{N-1}), \tag{7.3.7}$$

with

$$\zeta_0 = \zeta_0(\tau) = 1, \qquad \zeta_n = \zeta_n(\tau) = \prod_{j=0}^{n-1} \frac{\eta_{j+1}(\tau)}{\xi_j(\tau)}.$$

We now apply the Hellmann–Feynman theorem to the space S of finite sequences, $S = \{\mathbf{U} : \mathbf{U} = (u_0, u_1, \ldots, u_{N-1})\}$, with the inner product (7.3.7) and the matrix operator $H_\tau = A_N$. The conclusion, formula (7.2.4), is that if λ is a zero of $p_N(x; \tau)$ then

$$\left[\sum_{m=0}^{N-1} p_m^2(\lambda; \tau)/\zeta_m \right] \frac{d\lambda}{d\tau}$$

$$= \sum_{n=0}^{N-1} \frac{p_n(\lambda; \tau)}{\zeta_n} \left\{ \xi_n'(\tau) p_{n+1}(\lambda; \tau) + \alpha_n'(\tau) p_n(\lambda; \tau) + \eta_n'(\tau) p_{n-1}(\lambda; \tau) \right\}.$$

$$(7.3.8)$$

As an example, consider the associated Laguerre polynomials $\{L_n^\alpha(x; c)\}$ of Section 5.6.

Theorem 7.3.3 ((Ismail & Muldoon, 1991)) *The zeros of the associated Laguerre polynomials increase with α for $\alpha \geq 0$, and $c > -1$.*

Proof The corresponding orthonormal polynomials $\{p_n(x)\}$ are

$$p_n(x) = (-1)^n \sqrt{\frac{(c+1)_n}{(\alpha+c+1)_n}} \, L_n^{(\alpha)}(x; c). \qquad (7.3.9)$$

In the notation of (7.3.6) we have the recursion coefficients for $\{p_n(x)\}$

$$\xi_{n-1} = \eta_n = \sqrt{(n+c)(n+c+\alpha)}, \qquad \alpha_n = 2n + 2c + \alpha + 1.$$

Let A_n be the corresponding Jacobi matrix. The i, j entry of the derivative matrix $\partial A_N / \partial \alpha$ is

$$\frac{\sqrt{i+c+1}}{2\sqrt{i+c+\alpha+1}} \delta_{i,j-1} + \delta_{i,j} + \frac{\sqrt{i+c}}{2\sqrt{(i+c+\alpha)}} \delta_{i,j+1}.$$

Therefore the matrix $\partial A_N / \partial \alpha$ is real symmetric, diagonally dominant, with positive diagonal entries, hence is positive definite by Theorem 1.1.6. □

The special case $c = 0$, shows that the zeros of Laguerre polynomials increase with α, $\alpha \geq 0$. The stronger result that all the zeros of $\{L_n^\alpha(x)\}$ increase with α, $\alpha > -1$ follows from Markov's theorem, Theorem 7.1.2.

We remark that the weight function for the Askey–Wimp associated Laguerre polynomials is (see (Askey & Wimp, 1984)) $x^\alpha e^{-x} \left| \psi \left(c, 1 - \alpha, xe^{-i\pi} \right) \right|^{-2}$, ψ being the Tricomi ψ function (1.3.15) and we know of no way to express the derivative with respect to a parameter of the Tricomi ψ function in terms of simple special functions. Furthermore, if $c > -1$, $\alpha + c > -1$ but $1 + \alpha + 2c < 0$, the measure of orthogonality of the associated Laguerre polynomials has a discrete mass whose position depends on α, hence Theorem 7.1.1, is not applicable. The associated Laguerre polynomials do not satisfy a second-order differential equation of Sturm–Liouville type. They satisfy a fourth-order differential equation with polynomial coefficients (Askey & Wimp, 1984) which does not seem amenable to a Sturmian approach.

As another example, consider the Meixner polynomials. The corresponding Jacobi matrix $A_N = (a_{j,k})$ is

$$a_{j.k} = \frac{\sqrt{c(j+1)(j+\beta)}}{1-c} \delta_{j,j+1} + \frac{j+c(j+\beta)}{1-c} \delta_{j,k}$$
$$+ \frac{\sqrt{cj(j+\beta-1)}}{1-c} \delta_{j,j-1}. \tag{7.3.10}$$

One can apply Corollary 7.3.2 to see that the zeros of $M_n(x;\beta,c)$ increase with β when $\beta > 1$. The details are similar to our analysis of the associated Laguerre polynomials and will be omitted. The dependence of the zeros of the Meixner polynomials on the parameter c is interesting. It is more convenient to use the renormalization

$$p_n(x;\beta,c) := (-1)^n c^{n/2} \sqrt{\frac{(\beta)_n}{n!}} \, M_n\left(\frac{x\sqrt{c}}{1-c};\beta,c\right),$$

so that

$$p_{-1}(x;\beta,c) = 0, \quad p_0(x;\beta,c) = 1,$$
$$xp_n(x;\beta,c) = \sqrt{(n+1)(n+\beta)} \, p_{n+1}(x;\beta,c) \tag{7.3.11}$$
$$+ \left[\sqrt{c}\beta + n(1+c)/\sqrt{c}\right] p_n(x;\beta,c) + \sqrt{n(n+\beta-1)} \, p_{n-1}(x;\beta,c).$$

In view of (6.1.18), the zeros of $p_n(x;\beta,c)$ converge to the corresponding zeros of $L_n^{(\beta-1)}(x)$, as $c \to 1^-$. The next step is to estimate the rate at which the zeros of $p_n(x;\beta,c)$ tend to the corresponding zeros of $L_n^{(\beta-1)}(x)$, as $c \to 1^-$. Let

$$m_{n,1}(\beta,c) > \cdots > m_{n,n}(\beta,c), \text{ and } l_{n,1}(\alpha) > \cdots > l_{n,n}(\alpha) \tag{7.3.12}$$

be the zeros of $M_n(x;\beta,c)$ and $L_n^\alpha(x)$, respectively. We shall denote the zeros of $p_n(x;\beta,c)$ by $x_{n,j}(\beta,c)$, i.e.,

$$x_{n,j}(\beta,c) = \frac{1-c}{\sqrt{c}} m_{n,j}(\beta,c). \tag{7.3.13}$$

Theorem 7.3.4 ((Ismail & Muldoon, 1991)) *The quantities $x_{n,j}(\beta,c)$ increase with c on the interval $(n-1)/(\beta+n-1) < c < 1$ and converge to $l_{n,j}(\beta-1)$ as $c \to 1^-$.*

Proof Let A_n be the $n \times n$ truncation of the infinite tridiagonal matrix associated with (7.3.11) and apply Theorem 7.3.1 to get

$$\frac{\partial}{\partial c} x_{n,j}(\beta,c) = \sum_{k=0}^{n-1} \frac{\beta c + k(c-1)}{2c\sqrt{c}} \, p_k^2\left(x_{n,j}(\beta,c);\beta,c\right)$$
$$\times \left[\sum_{k=0}^{n-1} p_k^2\left(x_{n,j}(\beta,c);\beta,c\right)\right]^{-1}. \tag{7.3.14}$$

The coefficients are all positive for the given range of values of c and the theorem follows. $\qquad \square$

We now obtain two-sided inequalities for the zeros of the Meixner polynomials.

Theorem 7.3.5 ((Ismail & Muldoon, 1991)) *Let* $m_{n,j}(\beta, c)$ *and* $\ell_{n,j}(\alpha)$ *be as in* (7.3.12). *If* $0 < c < 1$, *then*

$$
\ell_{n,j}(\beta - 1) - \beta \left(1 - \sqrt{c}\right) < \frac{1 - c}{\sqrt{c}} m_{n,j}(\beta, c)
$$

$$
< \ell_{n,j}(\beta - 1) - \beta \left(1 - \sqrt{c}\right) + \frac{(n - 1)}{\sqrt{c}} \left(1 - \sqrt{c}\right)^2.
$$

(7.3.15)

Proof Observe that (7.3.14) holds for all $n > 0$ and all $c \in (0, 1)$. Since $\beta c \geq \beta c + k(c - 1) \geq \beta c + (n - 1)(c - 1)$, we get

$$
\frac{\beta}{2\sqrt{c}} > \frac{\partial M_{n,j}(\beta, c)}{\partial c} > \frac{\beta c + (n - 1)(c - 1)}{2c\sqrt{c}}.
$$

Integrating this inequality between c and 1 and using $x_{n,j}(\beta, 1) = \ell_{n,j}(\beta - 1)$ we get (7.3.15). □

Consider the class of polynomials $\{h_n(x)\}$ generated by

$$
h_0(x) = 1, \quad h_1(x) = x \, a_1(\tau),
$$
$$
x \, a_n(\tau) h_n(x) = h_{n+1}(x) + h_{n-1}(x),
$$

(7.3.16)

where $\{a_n(\tau)\}$ is a given sequence of positive numbers for all τ in a certain interval T. The polynomials $h_n(x)$ of odd (even) degrees are odd (even) functions.

Theorem 7.3.6 ((Ismail, 1987)) *The positive zeros of* $h_n(x)$ *are increasing (decreasing) differentiable functions of* τ, $\tau \in T$, *if* $a_n(\tau)$ *is a decreasing (increasing) differentiable function of* τ, $\tau \in T$, $0 \leq n < N$. *Moreover, if* λ *is a positive zero of* h_N *then*

$$
\frac{1}{\lambda} \frac{d\lambda}{d\tau} = - \frac{\sum\limits_{n=1}^{N-1} a_n'(\tau) h_n^2(\lambda)}{\sum\limits_{n=0}^{N-1} a_n(\tau) h_n^2(\lambda)}.
$$

(7.3.17)

Proof Let λ be a positive zero of $h_N(x)$. In this case $\zeta_n = a_0(\tau)/a_n(\tau)$ and (7.3.8) is

$$
\lambda' \sum_{n=0}^{N-1} a_n(\tau) h_n^2(\tau) = - \sum_{n=0}^{N-1} \frac{a_n'(\tau)}{a_n(\tau)} h_n(\lambda) \left[h_{n-1}(\lambda) + h_{n+1}(\lambda)\right].
$$

Using (7.3.16) we rewrite the above equation in the form

$$
\lambda' \sum_{n=0}^{N-1} a_n(\tau) h_n^2(\lambda) = -\lambda \sum_{n=0}^{N-1} a_n'(\tau) h_n^2(\lambda),
$$

which proves the theorem. □

The Lommel polynomials of §6.5 correspond to the case $a_n(\tau) = 2(n + \tau)$ while the q-Lommel polynomials (Ismail, 1982; Ismail & Muldoon, 1988) correspond to $a_n(\tau) = 2(1 - q^{\tau+n})$. Thus, the positive zeros of the Lommel and q-Lommel polynomials decrease with τ, $\tau \in (0, \infty)$. On the other hand, if λ is a positive zero of a Lommel polynomial then we apply Theorem 3.1 with $a_n(\tau) = 2(n + \tau)/\tau$ to see that $\lambda\tau$ increases with τ, $\tau > 0$. Similar results hold for the q-Lommel polynomials (Ismail & Muldoon, 1988). The class of polynomials when $a_n(\tau)$ is a function of $n + \tau$ was studied in Dickinson, Pollack, and Wannier (Dickinson et al., 1956) and later by Goldberg in (Goldberg, 1965). It is a simple exercise to extend the results of (Goldberg, 1965) to the more general case when $a_n(\tau)$ is not necessarily a function of $n + \tau$.

The case of the Lommel polynomials is interesting. Let

$$x_{N,1}(\nu) > x_{N,2}(\nu) > \cdots > x_{N,\lfloor N/2 \rfloor}(\nu) > 0$$

be the positive zeros of $h_{N,\nu}(x)$. Then (7.3.16) becomes

$$\frac{1}{x_{N,j}(\nu)} \frac{d}{d\nu} x_{N,j}(\nu) = -\frac{\displaystyle\sum_{k=0}^{N-1} h_{k,\nu}^2\left(x_{N,j}(\nu)\right)}{\displaystyle\sum_{k=0}^{N-1}(k + \nu) h_{k,\nu}^2\left(x_{N,j}(\nu)\right)}. \tag{7.3.18}$$

Since the polynomials $\{h_{n,\nu+1}(x)\}$ are orthogonal with respect to a probability measure with masses at $\pm 1/j_{\nu,n}$, $n = 1, 2, \ldots$, $x_{N,n}(\nu + 1) \to 1/j_{\nu,n}$ as $N \to \infty$. Moreover, the mass at $\pm 1/j_{\nu,n}$ is $2(\nu + 1)/j_{\nu,n}^2$ and the orthonormal polynomials are $\left\{\sqrt{1 + n/(\nu + 1)}\, h_{n,\nu}(x)\right\}$. Therefore, Theorem 21.1.8 implies

$$\sum_{k=0}^{\infty} \frac{k + \nu + 1}{\nu + 1} h_{k,\nu}^2\left(1/j_{\nu,n}\right) = \frac{j_{\nu,n}^2}{2(\nu + 1)}. \tag{7.3.19}$$

Hence, the limiting case $N \to \infty$ of (7.3.18) is

$$\frac{1}{j_{\nu,n}} \frac{dj_{\nu,n}}{d\nu} = 2\sum_{k=0}^{\infty} h_{k,\nu+1}^2\left(1/j_{\nu,n}\right). \tag{7.3.20}$$

Apply (6.5.3) to see that (7.3.20) is equivalent to

$$\frac{dj_{\nu,n}}{d\nu} = \frac{2}{j_{\nu,n} J_{\nu+1}^2(j_{\nu,n})} \sum_{k=0}^{\infty} J_{\nu+k+1}^2\left(j_{\nu,n}\right). \tag{7.3.21}$$

The relationships (7.3.20) and (7.3.21) were established in (Ismail & Muldoon, 1988). Ismail and Muldoon applied (7.3.20) to derive inequalities for zeros of Bessel functions, especially for $j_{\nu,1}$.

We now consider the case of differential operators. Let

$$H_\nu := -\frac{d}{dx}\left(p(x) \frac{d}{dx}\right) + \nu^2 q(x), \quad \nu \in (a, b) =: I, \tag{7.3.22}$$

with $p(x) \neq 0$, $p'(x)$ and $q(x)$ continuous on (c, d). Let

$$S = \left\{y : y \in L^2(c, d), y \in C^2(c, d), p(y)y(x)y'(x) = 0 \text{ at } x = c, d\right\}. \tag{7.3.23}$$

It is clear that H_ν is self-adjoint on S. Consider the eigenvalue problem

$$H_\nu y(x) = \lambda_\nu \phi(x) y(x), \quad y \in S. \tag{7.3.24}$$

Theorem 7.3.7 *Assume $\phi(x) \geq 0$ on (c, d), $\phi(x) \not\equiv 0$ on (c, d), and*

$$\lim_{\mu \to \nu} \int_c^d \phi(x) \psi_\mu(x) \psi_\nu(x)\, dx = \int_c^d \psi(x) \phi_\nu^2(x)\, dx,$$

$$\lim_{\mu \to \nu} \int_c^d q(x) \psi_\mu(x) \psi_\nu(x)\, dx = \int_c^d q(x) \psi_\nu^2(x)\, dx,$$

then $\dfrac{d\lambda_\nu}{d\nu}$ exists and

$$\frac{d\lambda_\nu}{d\nu} = 2\nu \left[\int_c^d q(x) \psi_\nu^2(x)\, dx \right] \Big/ \left[\int_c^d \phi(x) \psi_\nu^2(x)\, dx \right]. \tag{7.3.25}$$

If in addition $\int_c^d \phi(x) \psi_\nu^2(x)\, dx = 1$, then

$$\frac{d}{d\nu} \left(\frac{\lambda_\nu}{\nu} \right) = \int_c^d \left[2q(x) - \nu^{-2} \lambda_\nu \phi(x) \right] \psi_\nu^2(x)\, dx, \tag{7.3.26}$$

$$\frac{d}{d\nu} \left(\frac{\lambda_\nu}{\nu^2} \right) = -2\nu^{-3} \int_c^d p(x) \left[\psi_\nu'(x) \right]^2 dx. \tag{7.3.27}$$

Proof By definition

$$\left\langle \frac{\partial H}{\partial \nu} \psi_\nu, \psi_\nu \right\rangle = \lim_{\mu \to \nu} \int_c^d \frac{\mu^2 - \nu^2}{\mu - \nu} q(x) \psi_\mu(x) \psi_\nu(x)\, dx$$

$$= 2\nu \int_c^d q(x) \psi_\nu^2(x)\, dx,$$

hence (7.3.25) follows from the Hellmann–Feynman theorem. Formula (7.3.26) easily follows from (7.3.25).

To prove (7.3.27), note that

$$\lambda_\nu = -\int_c^d \left[\frac{d}{dx} p(x) \frac{d}{dx} \psi_\nu(x) \right] \psi_\nu(x)\, dx + \nu^2 \int_c^d q(x) \psi_\nu^2(x)\, dx$$

$$= \int_c^d p(x) \left[\psi_\nu'(x) \right]^2 dx + \nu^2 \int_c^d q(x) \psi_\nu^2(x)\, dx.$$

Hence (7.3.27) follows. $\qquad \square$

Theorem 7.3.8 *For $\nu > 0$, we have*

$$\frac{d}{d\nu} j_{\nu,k} = \frac{2\nu}{j_{\nu,k} J_{\nu+1}^2 (j_{\nu,k})} \int\limits_0^{j_{\nu,k}} J_\nu^2(t) \frac{dt}{t}. \qquad (7.3.28)$$

Moreover, for k fixed $j_{\nu,k}$ increases with ν while $j_{\nu,k}/\nu$ decreases with ν, for $\nu > 0$.

Proof Apply Theorem 7.3.7 with $p(x) = x$, $q(x) = 1/x$, $\phi(x) = x$, $a = 0$, $b = \infty$, $c = 0$, $d = 1$. The equation $H_\nu y = \lambda y$ is

$$-xy'' - y' + \frac{\nu^2}{x} = \lambda xy,$$

whose solutions are $J_\nu\left(\sqrt{\lambda}\, x\right)$, $Y_\nu\left(\sqrt{\lambda}\, x\right)$. The boundary conditions $y(0) = y(1) = 0$ imply $\lambda = j_{\nu,k}^2$ and $\psi_{\nu,k}(x) = CJ_\nu(j_{\nu,k}x)$. We evaluate C from $\int_0^1 x\left[\psi_{\nu,k}(x)\right]^2 dx = 1$. Set

$$A(a,b) = \int\limits_0^1 xJ_\nu(ax)J_\nu(bx)\, dx.$$

From (1.3.20) it follows that

$$\left(a^2 - b^2\right) x^2 J_\nu(ax)J_\nu(bx) = \left\{x^2 \frac{d^2}{dx^2} J_\nu(bx) + x \frac{d}{dx} J_\nu(bx)\right\} J_\nu(ax)$$
$$- \left\{x^2 \frac{d^2}{dx^2} J_\nu(ax) + x \frac{d}{dx} J_\nu(ax)\right\} J_\nu(bx).$$

Therefore

$$\left(a^2 - b^2\right) A(a,b) = \int\limits_0^1 \left\{J_\nu(ax) \frac{d}{dx} x \frac{d}{dx} J_\nu(bx) - J_\nu(bx) \frac{d}{dx} x \frac{d}{dx} J_\nu(ax)\right\} dx$$
$$= x\left\{J_\nu(ax)bJ_\nu'(bx) - J_\nu(bx)aJ_\nu'(ax)\right\}\big|_0^1$$
$$= bJ_\nu(a)J_\nu'(b) - aJ_\nu(b)J_\nu'(a).$$

Now take $b = j_{\nu,k}$ and let $a \to j_{\nu,k}$. The result is

$$A\left(j_{\nu,k}, j_{\nu,k}\right) = \frac{1}{2}\left\{J_\nu'\left(j_{\nu,k}\right)\right\}^2 = \frac{1}{2} J_{\nu+1}^2\left(j_{\nu,k}\right),$$

where we used (1.3.26). Thus (7.3.28) follows from (7.3.25). Indeed the normalized eigenfunction is

$$\frac{\sqrt{2j_{\nu,k}}}{J_{\nu+1}\left(j_{\nu,k}\right)} J_\nu\left(j_{\nu,k}x\right).$$

Now (7.3.28) follows from (7.3.25) and we conclude that λ_ν increases with ν for $\nu > 0$. The monotonicity of $j_{\nu,k}/\nu$ follows from (7.3.27). $\square$

Formula (7.3.28) is called Schläfli's formula, (Watson, 1944, §15.6).

7.4 Extreme Zeros of Orthogonal Polynomials

We now give theorems dealing with monotonicity properties of the largest or the smallest zeros of orthogonal polynomials. These results are particularly useful when the polynomials are defined through their recurrence relation (2.2.17). In many combinatorial applications, (Bannai & Ito, 1984), the positivity condition $A_{n-1}C_n > 0$ holds for $1 \leq n < N$ and does not hold for $n = N$, for some N. In such cases we have only a finite set of orthogonal polynomials $\{p_n(x; \tau) : n = 0, 1, \ldots, N - 1\}$ and one can prove that they are orthogonal with respect to a positive measure supported on the zeros of $p_N(x; \tau)$.

We now state the Perron–Frobenius theorem for tridiagonal matrices. We avoid stating the theorem in its full generality because we only need the special case stated below. The general version may be found in (Horn & Johnson, 1992).

Theorem 7.4.1 (Perron–Frobenius) *Let A and B be tridiagonal $n \times n$ matrices with positive off-diagonal elements and nonnegative diagonal elements. If the elements of $B - A$ are nonnegative then the largest eigenvalue of B is greater than the largest eigenvalue of A.*

In (5.2.11)–(5.2.12) we replace $Q_n(x)$ by $Q_n(x; \tau)$ and replace λ_n and μ_n by $\lambda_n(\tau)$ and $\mu_n(\tau)$; respectively. We also replace $R_n(x)$ by $R_n(x; \tau)$ in (5.2.20)–(5.2.21).

If the birth rates $\{\lambda_n(\tau)\}$ and death rates $\{\mu_n(\tau)\}$ are increasing (decreasing) functions of τ we apply the Perron–Frobenius theorem to $(-1)^n Q_n(x; \tau)$ and prove that the largest zero of $Q_n(x; \tau)$ is an increasing (decreasing) function of τ.

As we saw in §7.3, the true interval of orthogonality of birth and death process polynomials is a subset of $[0, \infty)$, while random walk polynomials have their true interval of orthogonality $\subset [-1, 1]$.

Theorem 7.4.2 ((Ismail, 1987)) *Let $\mu_0 = 0$ and assume that λ_n, $N > n \geq 0$, and λ_n/μ_n, $N > n > 0$, are differentiable monotone increasing (decreasing) functions of a parameter τ. Then the smallest zero of a birth and death process polynomial $Q_N(x; \tau)$ is also a differentiable monotone increasing (decreasing) function of the parameter τ.*

Proof Let λ be the smallest zero of $Q_N(x; \tau)$. Clearly all zeros of $Q_N(x; \tau)$ are differentiable functions of τ. Using (7.3.8) and (5.2.12), we get

$$\frac{d\lambda}{d\tau} \sum_{n=0}^{N-1} Q_n^2(\lambda; \tau)/\zeta_n$$

$$= \sum_{n=0}^{N-1} Q_n(\lambda; \tau) \left[-\lambda_n' Q_{n+1}(\lambda; \tau) - \mu_n^2 Q_{n-1}(\lambda; \tau) + (\lambda_n' + \mu_n') Q_n(\lambda; \tau) \right]$$

$$\tag{7.4.1}$$

where f' denotes differentiation with respect to τ and ζ_n is as in (5.2.14). It is easy to see that $\mu_0 = 0$ implies $Q_n(0; \tau) = 1$, $0 \leq n \leq N$. Therefore, $Q_n(\lambda; \tau) > 0$ since λ is to the left of the smallest zero of $Q_n(x; \tau)$. By (7.4.1) it remains to show that the quantity

$$-\lambda'_n Q_{n+1}(\lambda;\tau) - \mu'_n Q_{n-1}(\lambda;\tau) + (\lambda'_n + \mu'_n) Q_n(\lambda;\tau) \qquad (7.4.2)$$

which appears in the square bracket in (7.4.1) is positive. We use (5.2.12) to eliminate $Q_{n+1}(\lambda;\tau)$ from the expression (7.4.2). The result is that the expression (7.4.2) is a positive multiple of

$$\lambda\lambda'_n Q_n(\lambda;\tau) + \mu'_n \{Q_{n-1}(\lambda;\tau) - Q_n(\lambda;\tau)\} (\lambda_n/\mu_n)' .$$

The proof will be complete when we show that $g(\lambda) > 0$, where $g(x) = Q_{n-1}(x;\tau) - Q_n(x;\tau)$. The interlacing of the zeros of $Q_{n-1}(x;\tau)$ and $Q_n(x;\tau)$ causes the function to change sign in every open interval whose endpoints are consecutive zeros of $Q_n(x;\tau)$. Thus, $g(x)$ possesses $n-1$ zeros located between the zeros of $Q_n(x;\tau)$. Furthermore, $g(0) = 0$. This accounts for all zeros of $g(x)$ since $g(x)$ is a polynomial of degree n. Therefore, $g(x)$ does not vanish between $x = 0$ and the first zero of $Q_n(x;\tau)$. It is clear from (5.2.11) and (5.2.12) that the sign of the coefficient of x^n in $Q_n(x;\tau)$ is $(-1)^n$, hence the sign of the coefficient of x^n in $g(x)$ is $(-1)^{n-1}$. Thus $g(x) < 0$ on $(-\infty, 0)$ and $g(x)$ must be positive when $0 < x \le \lambda$. Therefore the expression in (7.4.2) is positive and (7.4.1) establishes the theorem. $\qquad\square$

Theorem 7.4.3 ((Ismail, 1987)) *Suppose that the m'_n's of (5.2.20) are differentiable monotone increasing (decreasing) functions of a parameter τ for $N > n \ge 0$ and $m_0(\tau) = 1$, i.e., $\mu_0(\tau) = 0$. Then the largest positive zero of $R_N(x;\tau)$ is a differentiable monotone decreasing (increasing) function of τ.*

Proof We denote the largest positive zero of $R_N(x;\tau)$ by Λ. The assumption $m_0(\tau) = 1$ and induction on n in (5.2.20) imply

$$R_N(1;\tau) = 1, \quad R_n(-x;\tau) = (-1)^n R_n(x;\tau). \qquad (7.4.3)$$

Let $x_{n,1} > x_{n,2} > \cdots > x_{n,n}$ be the zeros of $R_n(x;\tau)$. They lie in $(-1,1)$ and, in view of (7.4.3), are symmetric around the origin. In the present case (7.3.8) is

$$\Lambda' \sum_{n=0}^{N-1} R_n^2(\Lambda;\tau)/\zeta_n = \sum_{n=0}^{N-1} m'_n(\tau) R_n(\Lambda;\tau) \{R_{n+1}(\Lambda;\tau) - R_{n-1}(\Lambda;\tau)\} /\zeta_n.$$

$$(7.4.4)$$

The theorem will follow once we show that

$$R_n(\Lambda;\tau) \{R_{n+1}(\Lambda;\tau) - R_{n-1}(\Lambda;\tau)\} < 0, \quad 0 \le n < N. \qquad (7.4.5)$$

We now prove the claim (7.4.5). Define a function f by

$$f(x) = m_n(\tau) \{R_{n+1}(x;\tau) - R_{n-1}(x;\tau)\} .$$

Note that $f(x) = x R_n(x;\tau) - R_{n-1}(x;\tau)$ and $f(1) = f(-1) = 0$ follow from (7.4.3). Furthermore,

$$f(-x) = (-1)^{n+1} f(x).$$

We first consider the case of odd n. In this case $x_{n,(n+1)/2} = 0$ and f is an even polynomial function with $f(0) \ne 0$. Now (7.4.3) and the interlacing of zeros of

$R_i(x;\tau)$ and $R_{i-1}(x;\tau)$ give $(-1)^{j+1}R_{n-1}(x_{n,j},\tau) > 0$, $1 \leq j \leq n$. Thus, f has a zero in each interval $(x_{n,j}, x_{n,j+1})$, $1 \leq j < n$. But f is a polynomial of degree $n+1$ and vanishes at ± 1. Thus, f has only one zero in each interval $(x_{n,j}, x_{n,j+1})$, $1 \leq j < n$. This shows that f is negative on the interval $(x_{n,1}, 1)$ which contains $(\Lambda, 1)$. On the other hand, $R_n(x;\tau)$ is positive on $(\Lambda, 1)$; hence, (7.4.5) follows when n is odd. We now come to the case of even n. We similarly show that f has a zero in any interval $(x_{n,j}, x_{n,j+1})$, $j \neq n/2$. This accounts for $n-2$ zeros of f. The remaining zeros are $x = 0$, ± 1. This shows that f vanishes only once in each interval $(x_{n,j}, x_{n,j+1})$, $j \neq n/2$. Therefore, $f(x)$ is negative on $(\Lambda, 1)$. But $R_n(x;\tau)$ is positive on $(\Lambda, 1)$ and so we have proved (7.4.5) for even n, and the proof is complete. □

Theorem 7.4.4 *Let $\zeta(\nu)$ be a positive zero of an ultraspherical polynomial $C_n^\nu(x)$. Then $(1+\nu)^{1/2}\zeta(\nu)$ increases with ν, $\nu \geq -1/2$.*

Theorem 7.4.4 was stated in (Ismail, 1989) as a conjecture based on the application of the Perron–Frobeneius theorem and extensive numerical computations done by J. Letessier in an earlier version of this conjecture. The conjecture was proved in (Elbert & Siafarikas, 1999).

7.5 Concluding Remarks

Readers familiar with the literature on monotonicity of zeros of orthogonal polynomials will notice that we avoided discussing the very important and elegant Sturmian methods of differential equations. There are two reasons for this omission. The first is lack of space. The second is that excellent surveys on Sturm comparison method and related topics are readily available so we decided to concentrate on the relatively new discrete methods. The reader interested in Sturmian methods may consult the books of Szegő (Szegő, 1975) and Watson (Watson, 1944, pp. 517–521), and the research articles of Lorch (Lorch, 1977), Laforgia and Muldoon (Laforgia & Muldoon, 1986). For more recent results, see (Ahmed et al., 1982) and (Ahmed et al., 1986). The key results and methods of Makai (Makai, 1952) and Szegő and Turán (Szegő & Turán, 1961) are worth noting.

Szegő's book (Szegő, 1975) has an extensive bibliography covering a good part of the literature up to the early seventies. The interesting work (Laforgia & Muldoon, 1986) is a good source for some recent literature on the subject. Moreover, (Gatteschi, 1987) establishes new and rather complicated inequalities for zeros of Jacobi polynomials using Sturm comparison theorem. The bibliography in (Gatteschi, 1987) complements the above-mentioned references.

8

Polynomials Orthogonal on the Unit Circle

One way to generalize orthogonal polynomials on subsets of $\mathbb{R}$ is to consider orthogonality on curves in the complex plane. Among these generalizations, the most developed theory in the general theory of orthogonal polynomial on the unit circle. The basic sources for this chapter are (Grenander & Szegő, 1958), (Szegő, 1975), (Geronimus, 1962), (Geronimus, 1977), (Simon, 2004) and recent papers which will be cited at the appropriate places.

8.1 Elementary Properties

Let $\mu(\theta)$ be a probability measure supported on an infinite subset of $[-\pi, \pi]$.

$$\mu_n := \int_{-\pi}^{\pi} e^{-in\theta} \, d\mu(\theta), \quad n = 0, \pm 1, \pm 2, \ldots. \tag{8.1.1}$$

Let $\mathbf{T}_n$ be the Toeplitz matrix

$$\mathbf{T}_n = (c_{j-k}), \quad j, k = 0, 1, \ldots, n, \tag{8.1.2}$$

and D_n be its determinant

$$D_n := \det \mathbf{T}_n. \tag{8.1.3}$$

We associate with $\mathbf{T}_n$ the Hermitian form

$$\mathbf{H}_n := \sum_{j,k=0}^{n} c_{j-k} u_j \, \overline{u_k} = \int_{-\pi}^{\pi} \left| \sum_{j=0}^{n} u_j z^j \right|^2 \, d\mu(\theta), \tag{8.1.4}$$

where

$$z = e^{i\theta}. \tag{8.1.5}$$

Thus $D_n > 0$ for all $n \geq 0$. One can construct the polynomials orthonormal with respect to μ via a Gram–Schmidt procedure. Indeed these polynomials, which will be denoted by $\phi_n(z)$, are unique when the leading term is positive. The analogue of the orthonormal form of (2.1.6) is

$$
\phi_n(x) = \frac{1}{\sqrt{D_n D_{n-1}}}
\begin{vmatrix}
\mu_0 & \mu_{-1} & \cdots & \mu_{-n} \\
\mu_1 & \mu_0 & \cdots & \mu_{-n+1} \\
\vdots & \vdots & \cdots & \vdots \\
\mu_{n-1} & \mu_{n-2} & \cdots & \mu_{-1} \\
1 & z & \cdots & z^n
\end{vmatrix},
$$

$$
= \frac{1}{\sqrt{D_n D_{n-1}}}
\begin{vmatrix}
\mu_0 z - \mu_{-1} & \mu_{-1} z - \mu_{-2} & \cdots & \mu_{1-n} z - \mu_{-n} \\
\mu_1 z - \mu_0 & \mu_0 z - \mu_{-1} & \cdots & \mu_{-n+1} z - \mu_{-n+1} \\
\vdots & \vdots & \cdots & \vdots \\
\mu_{n-1} z - \mu_{n-2} & \mu_{n-2} z - \mu_{n-3} & \cdots & \mu_0 z - \mu_{-1}
\end{vmatrix},
$$

$$(8.1.6)$$

for $n > 0$ and can be similarly proved. Moreover $\phi_0(z) = 1$. Indeed

$$\phi_n(z) = \kappa_n z^n + \ell_n z^{n-1} + \text{lower order terms}, \qquad (8.1.7)$$

and

$$\kappa_n = \sqrt{D_{n-1}/D_n}. \qquad (8.1.8)$$

It is clear that

$$
\phi_n(0) = \frac{(-1)^n}{\sqrt{D_n D_{n-1}}}
\begin{vmatrix}
\mu_{-1} & \mu_{-2} & \cdots & \mu_{-n} \\
\mu_0 & \mu_{-1} & \cdots & \mu_{-n+1} \\
\vdots & \vdots & & \vdots \\
\mu_{n-2} & \mu_{n-3} & \cdots & \mu_{-1}
\end{vmatrix}. \qquad (8.1.9)
$$

If f is a polynomial of degree n then the reverse polynomial f^* is $z^n \overline{f}(1/z)$, that is

$$f^*(z) := \sum_{k=0}^{n} \overline{a_k}\, z^{n-k}, \quad \text{if } f(z) = \sum_{k=0}^{n} a_k z^k, \quad \text{and } a_n \neq 0, \qquad (8.1.10)$$

Theorem 8.1.1 *The minimum of the integral*

$$\int_{-\pi}^{\pi} |\pi(z)|^2 \, d\mu(\theta), \quad z = e^{i\theta} \qquad (8.1.11)$$

over all monic polynomials $\pi(z)$ of degree n is attained when polynomial $\pi(z) = \phi_n(z)/\kappa_n$. The minimum value of the integral is $1/\kappa_n^2$.

Proof Let $\pi(z) = \sum_{k=0}^{n} a_k \phi_k(z)$. Then $a_n = 1/\kappa_n$ and the integral is equal to $\kappa_n^{-2} + \sum_{k=0}^{n-1} |a_k|^2$, which proves the theorem. □

The kernel polynomials are

$$s_n(a, z) = \sum_{k=0}^{n} \overline{\phi_k(a)}\, \phi_k(z), \quad n = 0, 1, \ldots. \qquad (8.1.12)$$

Theorem 8.1.2 *Let a be a fixed complex constant and let $\pi(z)$ be a polynomial of degree n satisfying the constraint*

$$\int_{-\pi}^{\pi} |g(z)|^2 d\mu(\theta) = 1, \quad z = e^{i\theta}. \tag{8.1.13}$$

The maximum of $|\pi(a)|^2$ over $\pi(z)$ satisfying the constraint (8.1.12), is attained when

$$\pi(z) = \epsilon s_n(a, z) / [s_n(a, a)]^{1/2}, \tag{8.1.14}$$

where $|\epsilon| = 1$. The maximum value of $|\pi(a)|^2$ is $s_n(a, a)$.

Proof We set $\pi(z) = \sum_{k=0}^{n} a_k \phi_k(z)$, hence the constraint implies $\sum_{k=0}^{n} |a_k|^2 = 1$. Thus

$$|\pi(a)|^2 \leq \left[\sum_{k=0}^{n} |a_k|^2\right] \left[\sum_{k=0}^{n} |\phi_k(a)|^2\right] = s_n(a, a). \tag{8.1.15}$$

The equality is attained if and only if for some ϵ on the unit circle $a_k = \epsilon \phi_k(a)$, for all k, $0 \leq k \leq n$. $\qquad\square$

Theorem 8.1.3 *The kernel polynomials are the only polynomials having the reproducing property*

$$\int_{-\pi}^{\pi} s_n(a, z) \overline{\pi(z)} \, d\mu(\theta) = \overline{\pi(a)}, \quad z = e^{i\theta}, \tag{8.1.16}$$

for any polynomial $\pi(z)$ of degree at most n.

Proof To see that (8.1.16) holds, just expand $\pi(z)$ in $\{\phi_k(z)\}$. For the converse assume (8.1.16) holds with $s_n(a, z)$ replaced by $f(a, z)$. Applying (8.1.16) with $\pi(z) = \phi_k(z)$ the result readily follows. $\qquad\square$

Corollary 8.1.4 *We have*

$$s_n(a, z) = -\frac{1}{D_n} \begin{vmatrix} \mu_0 & \mu_{-1} & \cdots & \mu_{-n} & 1 \\ \mu_1 & \mu_0 & \cdots & \mu_{-n+1} & \overline{a} \\ \vdots & \vdots & \cdots & \vdots & \vdots \\ \mu_n & \mu_n & \cdots & \mu_0 & \overline{a}^n \\ 1 & z & \cdots & z^n & 0 \end{vmatrix}. \tag{8.1.17}$$

Proof Verify that the right-hand side of (8.1.16) has the reproducing property in Theorem 8.1.3, for $\pi(z) = z^k$, $k = 0, 1, \ldots, z^n$. $\qquad\square$

An immediate consequence of (8.1.5) is

$$s_n(a, z) = (\overline{a}z)^n \, s_n \left(1/\overline{z}, 1/\overline{a}\right), \tag{8.1.18}$$

and its limiting case, $a \to 0$,

$$s_n(0, z) = \kappa_n \phi_n^*(z). \tag{8.1.19}$$

Moreover

$$s_n(0, 0) = \sum_{k=0}^{n} |\phi_k(0)|^2 = \kappa_n^2 = D_{n-1}/D_n. \tag{8.1.20}$$

Consequently

$$|\phi_n(0)|^2 = \kappa_n^2 - \kappa_{n-1}^2. \tag{8.1.21}$$

In particular, this shows that κ_n does not decrease with n.

8.2 Recurrence Relations

Theorem 8.2.1 *The analogue of the Christoffel–Darboux identity is*

$$\begin{aligned} s_n(a, z) &= \sum_{k=0}^{n} \overline{\phi_k(a)}\, \phi_k(z) \\ &= \frac{\overline{\phi_{n+1}^*(a)}\, \phi_{n+1}^*(z) - \overline{\phi_{n+1}(a)}\, \phi_{n+1}(z)}{1 - \bar{a}z}. \end{aligned} \tag{8.2.1}$$

Moreover the polynomials $\{\phi_n(z)\}$ *satisfy the recurrence relations*

$$\kappa_n z \phi_n(z) = \kappa_{n+1} \phi_{n+1}(z) - \phi_{n+1}(0) \phi_{n+1}^*(z), \tag{8.2.2}$$

$$\kappa_n \phi_{n+1}(z) = \kappa_{n+1} z \phi_n(z) + \phi_{n+1}(0) \phi_n^*(z). \tag{8.2.3}$$

Proof Let $\pi(z)$ be a polynomial of degree at most n. Then, with $z = e^{i\theta}$ we find

$$\int_{-\pi}^{\pi} \frac{\overline{\phi_{n+1}^*(a)}\, \phi_{n+1}^*(z) - \overline{\phi_{n+1}(a)}\, \phi_{n+1}(z)}{1 - \bar{a}z}\, \overline{\pi(z)}\, d\mu(\theta)$$

$$= \overline{\pi(a)} \int_{-\pi}^{\pi} \frac{\overline{\phi_{n+1}^*(a)}\, \phi_{n+1}^*(z) - \overline{\phi_{n+1}(a)}\, \phi_{n+1}(z)}{1 - \bar{a}z}\, d\mu(\theta)$$

$$+ \int_{-\pi}^{\pi} \left[\overline{\phi_{n+1}^*(a)}\, \phi_{n+1}^*(z) - \overline{\phi_{n+1}(a)}\, \phi_{n+1}(z) \right] \frac{\overline{\pi(z)} - \overline{\pi(a)}}{1 - \bar{a}z}\, d\mu(\theta).$$

But $\pi(z) - \pi(a) = (z - a)g(z)$, and g has degree $\leq n - 1$, and with $z = e^{i\theta}$ we obtain

$$\int_{-\pi}^{\pi} \phi_{n+1}^*(z) \overline{zg(z)}\, d\mu(\theta) = \int_{-\pi}^{\pi} \overline{\phi_{n+1}(z)}\, z^n g(1/z)\, d\mu(\theta) = 0,$$

and

$$\int_{-\pi}^{\pi} \phi_{n+1}(z) \overline{zg(z)}\, d\mu(\theta) = 0.$$

Therefore

$$\frac{\overline{\phi_{n+1}^*(a)}\,\phi_{n+1}^*(z) - \overline{\phi_{n+1}(a)}\,\phi_{n+1}(z)}{1 - \overline{a}z} = c s_n(a, z)$$

where c is a constant. Interchanging z and a in the above equality and taking the complex conjugates we see that c does not depend on a. Let $z = a = 0$ and use (8.1.21) to see that $c = 1$ and (8.2.1) follows. Multiply (8.2.1) by $1 - \overline{a}z$ then equate the coefficients of $(\overline{a})^{n+1}$ to prove (8.2.2). By taking the reverse polynomial of both sides of (8.2.2) and eliminating $\phi_{n+1}^*(z)$ we establish (8.2.3). $\square$

The above proof is from (Szegő, 1975) and (Grenander & Szegő, 1958). One can prove (8.2.2)–(8.2.3) directly as follows; see (Akhiezer, 1965) and (Simon, 2004). The polynomial

$$\phi(z) := \kappa_n \phi_{n+1}(z) - \kappa_{n+1} z \phi_n(z).$$

has degree at most n. If $\phi(z) \equiv 0$, then (8.2.3) holds, otherwise for $1 \le k \le n$, $z = e^{i\theta}$ we have

$$\int_{-\pi}^{\pi} \phi(z)\,\overline{z^k}\,d\mu(\theta) = 0 - \kappa_{n+1} \int_{-\pi}^{\pi} \overline{z^{k-1}}\,\phi_n(z)\,d\mu(\theta) = 0.$$

In other words

$$0 = \int_{-\pi}^{\pi} z^k\,\overline{\phi(z)}\,d\mu(\theta) = \int_{-\pi}^{\pi} \overline{z^{n-k}}\,\phi^*(z)\,d\mu(\theta),$$

for $0 \le n - k < n$. Therefore $\phi^*(z)$ is a constant multiple of $\phi_n(z)$, that is $\phi(z) = c\phi_n^*(z)$, and c is found to $\phi(0)/\kappa_n$. This establishes (8.2.3). Similarly we prove (8.2.2) in the form

$$\kappa_n \phi_n^*(z) = \kappa_{n+1} \phi_{n+1}^*(z) - \overline{\phi_{n+1}(0)}\,\phi_{n+1}(z). \tag{8.2.4}$$

It is convenient to write the recurrence relations in terms of the monic polynomials

$$\Phi_n(z) = \phi_n(z)/\kappa_n. \tag{8.2.5}$$

Indeed we have

$$\Phi_{n+1}(z) = z\Phi_n(z) - \overline{\alpha_n}\,\Phi_n^*(z), \tag{8.2.6}$$

$$\Phi_{n+1}^*(z) = \Phi_n^*(z) - \alpha_n z \Phi_n(z), \tag{8.2.7}$$

where

$$\alpha_n = -\overline{\Phi_{n+1}(0)} = -\overline{\phi_{n+1}(0)}/\kappa_{n+1}. \tag{8.2.8}$$

The coefficients $\{\alpha_n\}$ are called the recursion coefficients or the Geronimus coefficients. In his recent book (Simon, 2004), Simon makes a strong case for calling them the Verblunsky coefficients. Note that (8.2.6)–(8.2.7) can be written as a system

$$\begin{pmatrix} \Phi_{n+1}(z) \\ \Phi_{n+1}^*(z) \end{pmatrix} = \begin{pmatrix} z & \overline{\alpha_n} \\ -\alpha_n z & 1 \end{pmatrix} \begin{pmatrix} \Phi_n(z) \\ \Phi_n^*(z) \end{pmatrix}. \tag{8.2.9}$$

If we eliminate ϕ_n^* between (8.2.2) and (8.2.3) we get the three-term recurrence relation in (Geronimus, 1977, XI.4, p. 91)

$$\kappa_n\phi_n(0)\phi_{n+1}(z) + \kappa_{n-1}\phi_{n+1}(0)z\phi_{n-1}(z)$$
$$= [\kappa_n\phi_{n+1}(0) + \kappa_{n+1}\phi_n(0)z]\,\phi_n(z),$$
(8.2.10)

see also (Geronimus, 1961). Note that the recursion coefficients in (8.2.2), (8.2.3) and (8.2.10) can be written in terms of determinants of the moments using (8.1.8) and (8.1.9). A treatment of polynomials orthogonal on the unit circle via maximum entropy was initiated in Henry Landau's very interesting article (Landau, 1987) and is followed in (Simon, 2004).

It is not difficult to use (8.2.6)–(8.2.7) to prove the following theorem, (Simon, 1998)

Theorem 8.2.2 (Verblunsky's formula) *Let μ be a probability measure on $[-\pi, \pi]$ with moments $\{\mu_n\}_0^\infty$. Let $\{\alpha_n\}_0^\infty$ be the recursion coefficients in (8.2.6)–(8.2.7) and $\mu_{-n} = \overline{\mu_n}$. Then*

(i) *The coefficients of Φ_n are polynomials in $\{\alpha_j : 0 \le j < n\}$ and $\{\overline{\alpha_j} : 0 \le j < n\}$ with integer coefficients.*

(ii) *For each n,*

$$\mu_{n+1} - \alpha_n \prod_{j=0}^{n-1}\left(1 - |\alpha_j|^2\right)$$

is a polynomial in $\{\alpha_j : 0 \le j < n\}$ and $\{\overline{\alpha_j} : 0 \le j < n\}$ with integer coefficients.

(iii) *The quantity $\alpha_n D_n$ is a polynomial in the variables $\{\mu_j : -n \le j \le n+1\}$.*

Theorem 8.2.3 *We have*

$$D_n = \frac{1}{(n+1)!}\int_{[-\pi,\pi]^n}\prod_{0\le j<k\le n}|e^{i\theta_j} - e^{i\theta_k}|^2 \prod_{j=0}^n d\mu(\theta_j),$$
(8.2.11)

$$\prod_{j=0}^{n-1}(1 - |\alpha_j^2|) = D_n/D_{n-1}.$$
(8.2.12)

Proof The proof of formula (8.2.11) is similar to the proof of Theorem 2.1.2. To prove (8.2.12), apply (8.2.6) with $\zeta_n = \int_{-\pi}^\pi |\Phi_n(z)|^2 d\mu(\theta)$ to get

$$\zeta_n = \int_{-\pi}^\pi |z\Phi_n(z)|^2 d\mu(\theta) = \zeta_{n+1} + |\alpha_n|^2\zeta_n,$$

so that

$$\zeta_n = \prod_{j=0}^{n-1}\left(1 - |\alpha_j|^2\right).$$

Now (8.2.12) follows from (8.1.20) and the fact that $\zeta_n = 1/\kappa_n^2$.   □

The positive definite continued J-fractions are related to the spectral function $\int_{\mathbb{R}} (z-t)^{-1} d\mu(t)$ via Markov's theorem, Theorem 2.6.2, when the recursion coefficients are bounded. In the case of the unit circle, let

$$f(z) = a_0 + \frac{1 - |a_0|^2 \, z}{\overline{a_0} \, z+} \, \frac{1}{a_1+} \, \frac{1 - |a_1|^2 \, z}{\overline{a_1} \, z+} \cdots, \tag{8.2.13}$$

then

$$\int_{-\pi}^{\pi} \frac{e^{i\theta} + z}{e^{i\theta} - z} \, d\mu(\theta) = \frac{1 + z f(z)}{1 - z f(z)}. \tag{8.2.14}$$

For details see (Khruschev, 2005) and (Jones & Thron, 1980).

Observe that (8.2.14) can be used to recover μ. Indeed if

$$F(z) = \int_{-\pi}^{\pi} \frac{e^{i\theta} + z}{e^{i\theta} - z} \, d\mu(\theta), \quad |z| < 1, \tag{8.2.15}$$

then the discrete part of μ produces poles in F and the isolated masses can be recovered from the residues of F at its isolated poles. On the other hand, with $z = re^{i\phi}$,

$$\mathrm{Re}\, F(z) = \int_{-\pi}^{\pi} \frac{(1 - r^2)}{1 + r^2 - 2r \cos(\theta - \varphi)} \, d\mu(\theta), \tag{8.2.16}$$

and the Poisson kernel yields

$$\mu'(\theta) = 2\pi \lim_{r \to 1^-} \mathrm{Re}\, F(z). \tag{8.2.17}$$

It is clear from (8.2.12) that $|\alpha_n| < 1$ and that

$$D_n = \prod_{j=0}^{n-1} \left(1 - |\alpha_j|^2\right)^{n-j}. \tag{8.2.18}$$

The next result shows how systems of orthogonal polynomials on $|\zeta| = 1$ is in one-to-one correspondence with pairs of special systems of polynomials orthogonal on $[-1, 1]$. The model is $\{z^n\}$ on $|z| = 1$ and the Chebyshev polynomials $\{\mathrm{Re}\, z^n\}$ and $\{\mathrm{Im}\, z^{n+1}/\mathrm{Im}\, z\}$ on $[-1, 1]$.

Theorem 8.2.4 *Let $d\mu(x)$ be a probability measure on $[-1, 1]$ and let ϕ_n be the polynomials orthonormal with respect to $d\mu(\cos\theta)$ on the unit circle. Assume further that $\{t_n(x)\}$ and $\{u_n(x)\}$ orthonormal sequences of polynomials whose measures of orthogonality are $d\mu(x)$ and $c_2 \left(1 - x^2\right) d\mu(x)$, respectively. With $x = (z + 1/z)/2$ we have*

$$\begin{aligned}
t_n(x) &= [1 + \phi_{2n}(0)/\kappa_{2n}]^{-1/2} \left[z^{-n}\phi_{2n}(z) + z^n\phi_{2n}(1/z)\right] \\
&= [1 - \phi_{2n}(0)/\kappa_{2n}]^{-1/2} \left[z^{-n+1}\phi_{2n-1}(z) + z^{n-1}\phi_{2n-1}(1/z)\right],
\end{aligned} \tag{8.2.19}$$

and

$$u_n(x) = \frac{z^{-n-1}\phi_{2n+2}(z) + z^{n+1}\phi_{2n+2}(1/z)}{\sqrt{1 - \phi_{2n+2}(0)/\kappa_{2n+2}}\,(z - 1/z)}$$

$$= \frac{z^{-n}\phi_{2n+1}(z) + z^n\phi_{2n+1}(1/z)}{\sqrt{1 + \phi_{2n+2}(0)/\kappa_{2n+2}}\,(z - 1/z)}.$$

(8.2.20)

Proof Observe that the polynomials ϕ_n have real coefficients because their measure is even in θ. To prove (8.2.19), and with $x = \cos\theta$ and $z = e^{i\theta}$ we first show that

$$\int_0^\pi \left[z^{-n}\phi_{2n}(z) + z^n\phi_{2n}(1/z)\right] T_k(\cos\theta)\,d\mu(\cos\theta) = 0,$$

for $0 \le k < n$. The above integral is

$$\frac{1}{2}\int_{-\pi}^\pi \phi_{2n}(z)\left[\overline{z^{n-k}} + \overline{z^{-n-k}}\right] d\mu(\cos\theta) = 0,$$

from the orthogonality of $\{\phi_k\}$. Since

$$z^{-n}\phi_{2n}(z) + z^n\phi_{2n}(1/z) = [\kappa_{2n} + \phi_{2n}(0)]\left[z^n + z^{-n} + \cdots\right]$$

and with $z = e^{i\theta}$, $x = \cos\theta$, we see that

$$\int_{-1}^1 \left|z^{-n}\phi_{2n}(z) + z^n\phi_{2n}(1/z)\right|^2 d\mu(x)$$

$$= [\kappa_{2n} + \phi_{2n}(0)]\int_{-1}^1 \left[z^{-n}\phi_{2n}(z) + z^n\phi_{2n}(1/z)\right][2T_n(x)]\,d\mu(x)$$

$$= [\kappa_{2n} + \phi_{2n}(0)]\int_{-\pi}^\pi \left[\phi_{2n}(z) + z^{2n}\overline{\phi_{2n}(z)}\right] d\mu(\cos\theta) = [\kappa_{2n} + \phi_{2n}(0)]/\kappa_{2n},$$

when $n > 0$. This establishes the first line in (8.2.19). Similarly

$$\int_0^\pi \left[z^{1-n}\phi_{2n-1}(z) + z^{n-1}\phi_{2n-1}(1/z)\right] T_k(\cos\theta)\,d\mu(\cos\theta) = 0,$$

for $0 \le k < n$. Hence the second line in (8.2.20) is a constant multiple of the first and the constant multiple can be determined by equating coefficients of x^n. The proof of (8.2.19) is similar and is left to the reader as an exercise. $\qquad\square$

Example 8.2.5 *The circular Jacobi orthogonal polynomials (CJ) are orthogonal with respect to the weight function*

$$w(\theta) = \frac{\Gamma^2(a+1)}{2\pi\Gamma(2a+1)}\left|1 - e^{i\theta}\right|^{2a}, \quad a > -1.$$

(8.2.21)

The polynomials orthogonal with respect to the above weight function arise in a class of random unitary matrix ensembles, the CUE, where the parameter a is related to the charge of an impurity fixed at $z = 1$ in a system of unit charges located on the unit circle at the complex values given by the eigenvalues of a member of this matrix ensemble (Witte & Forrester, 2000). From Theorem 8.2.3 and properties of the ultraspherical polynomials it follows that the orthonormal polynomials are

$$\phi_n(z) = \frac{(a)_n}{\sqrt{n!(2a+1)_n}} \, {}_2F_1(-n, a+1; -n+1-a; z),\qquad (8.2.22).$$

and the coefficients are

$$\kappa_n = \frac{(a+1)_n}{\sqrt{n!(2a+1)_n}} \quad n \geq 0, \qquad (8.2.23)$$

$$\ell_n = \frac{na}{n+a}\kappa_n \quad n \geq 1, \qquad (8.2.24)$$

$$\phi_n(0) = \frac{a}{n+a}\kappa_n \quad n \geq 0. \qquad (8.2.25)$$

The reciprocal polynomials are

$$\phi_n^*(z) = \frac{(a+1)_n}{\sqrt{n!(2a+1)_n}} \, {}_2F_1(-n, a; -n-a; z). \qquad (8.2.26)$$

The following theorem describes the location of zeros of $\phi_n(z)$ and $s_n(a, z)$.

Theorem 8.2.6 *For $|a| \lesssim 1$, the zeros of $s_n(a, z)$ lie in $|z| \gtrsim 1$ When $|a| = 1$ then all zeros of $s_n(a, z)$ lie on $|z| = 1$. The zeros of $\phi_n(z)$ are in $|z| < 1$. The zeros of $\phi_n^*(z)$ are in $|z| > 1$.*

Proof Let $s_n(a, \zeta) = 0$. and denote a generic polynomial of exact degree k by $\pi_k(z)$ for all k. With $z = e^{i\theta}$, it is clear that,

$$s_n(a, a) = \max \left\{ |\pi_n(a)|^2 : \int_{|z|=1} |\pi_n|^2 d\mu(\theta) = 1 \right\}$$

$$\geq \max \left\{ |\pi_1(a)s_n(z, a)/(z-\zeta)|^2 : \int_{|z|=1} |\pi_1(a)s_n(z, a)/(z-\zeta)|^2 \, d\mu(\theta) = 1 \right\}$$

$$\geq s_n(a, a),$$

when we take $\pi_1(z) = (z - \zeta)/\sqrt{s_n(a, a)}$. Thus all the inequalities in the above lines are equalities. Consider a probability measure ν, defined by

$$d\nu(\theta) = c \left| \frac{s_n(a, z)}{z - \zeta} \right|^2 d\mu(\theta), \quad z = e^{i\theta}, \qquad (8.2.27)$$

where c is a constant. The above shows that

$$s_n(a, a) = \max \left\{ |\pi_1(a)|^2 : \int_{|z|=1} |\pi_1|^2 \, d\nu(\theta) = 1 \right\},$$

and the maximum is attained when $\pi_1(z) = b(z - \zeta)$, b is a constant. There ζ is a zero of another kernel polynomial of degree 1. Let the moments of ν be denoted by ν_n. Thus ζ satisfies

$$\begin{vmatrix} \nu_0 & \nu_{-1} & 1 \\ \nu_1 & \nu_0 & \overline{a} \\ 1 & z & 0 \end{vmatrix} = 0,$$

that is $\zeta = (\nu_0 - \nu_1 \overline{a}) / (\nu_1 - \nu_0 \overline{a})$. This implies the assertion about the zeros of s_n since $|\nu_1| \le \nu_0 = 1$. The rest follows from (8.1.18). $\qquad\square$

8.3 Differential Equations

This section is based on (Ismail & Witte, 2001). We shall assume that μ is absolutely continuous, that is the orthogonality relation becomes

$$\int_{|\zeta|=1} \phi_m(\zeta) \overline{\phi_n(\zeta)} \, w(\zeta) \frac{d\zeta}{i\zeta} = \delta_{m,n}. \tag{8.3.1}$$

Thus $\kappa_n \; (> 0)$ can be found from the knowledge of $|\phi_k(0)|$. By equating coefficients of z^n in (8.2.10) and in view of (8.1.7) we find

$$\kappa_n \ell_{n+1} \phi_n(0) + \kappa_{n-1}^2 \phi_{n+1}(0) = \kappa_n^2 \phi_{n+1}(0) + \kappa_{n+1} \ell_n \phi_n(0).$$

Therefore

$$\kappa_n \ell_{n+1} = \kappa_{n+1} \ell_n + \overline{\phi_n(0)} \, \phi_{n+1}(0). \tag{8.3.2}$$

Formula (8.3.2) leads to

$$\ell_n = \kappa_n \sum_{j=0}^{n-1} \frac{\overline{\phi_j(0)} \, \phi_{j+1}(0)}{\kappa_j \kappa_{j+1}}. \tag{8.3.3}$$

Following the notation in Chapter 3, we set

$$w(z) = e^{-v(z)}. \tag{8.3.4}$$

Theorem 8.3.1 *Let $w(z)$ be differentiable in a neighborhood of the unit circle, has moments of all integral orders and assume that the integrals*

$$\int_{|\zeta|=1} \frac{v'(z) - v'(\zeta)}{z - \zeta} \zeta^n \, w(\zeta) \frac{d\zeta}{i\zeta}$$

exist for all integers n. Then the corresponding orthonormal polynomials satisfy the *differential relation*

$$\phi_n'(z) = n\frac{\kappa_{n-1}}{\kappa_n}\phi_{n-1}(z) - i\phi_n^*(z)\int\limits_{|\zeta|=1}\frac{v'(z) - v'(\zeta)}{z - \zeta}\phi_n(\zeta)\overline{\phi_n^*(\zeta)}\,w(\zeta)\,d\zeta$$

$$+ i\phi_n(z)\int\limits_{|\zeta|=1}\frac{v'(z) - v'(\zeta)}{z - \zeta}\phi_n(\zeta)\overline{\phi_n(\zeta)}\,w(\zeta)\,d\zeta.$$

$$\tag{8.3.5}$$

Proof One can derive

$$\phi_n'(z) = \sum_{k=0}^{n-1}\phi_k(z)\int\limits_{|\zeta|=1}\phi_n'(\zeta)\overline{\phi_k(\zeta)}\,w(\zeta)\,\frac{d\zeta}{i\zeta}$$

$$= \sum_{k=0}^{n-1}\phi_k(z)\int\limits_{|\zeta|=1}\left[v'(\zeta)\overline{\phi_k(\zeta)} + \overline{\zeta\phi_k(\zeta)} + \overline{\zeta^2\phi_k'(\zeta)}\right]\phi_n(\zeta)\,w(\zeta)\,\frac{d\zeta}{i\zeta},$$

through integration by parts, then rewriting the derivative of the conjugated polynomial in the following way

$$\frac{d}{d\zeta}\overline{\phi_n(\zeta)} = -\overline{\zeta^2\phi_n'(\zeta)},\tag{8.3.6}$$

since $\overline{\zeta} = 1/\zeta$ for $|\zeta| = 1$. The relationships (8.3.1) and (8.1.7) give

$$\phi_n'(z) = \int\limits_{|\zeta|=1}v'(\zeta)\,\phi_n(\zeta)\sum_{k=0}^{n-1}\overline{\phi_k(\zeta)}\,\phi_k(z)\,w(\zeta)\,\frac{d\zeta}{i\zeta}$$

$$+ \phi_{n-1}(z)\int\limits_{|\zeta|=1}\left[\overline{\zeta\phi_{n-1}(\zeta)} + \overline{\zeta^2\phi_{n-1}'(\zeta)}\right]\phi_n(\zeta)\,w(\zeta)\,\frac{d\zeta}{i\zeta}$$

$$= \int\limits_{|\zeta|=1}\frac{v'(z) - v'(\zeta)}{z - \zeta}\phi_n(\zeta)\left[\overline{\phi_n^*(\zeta)}\,\phi_n^*(z) - \overline{\phi_n(\zeta)}\,\phi_n(z)\right]w(\zeta)\,\frac{d\zeta}{i}$$

$$+ \phi_{n-1}(z)\left[\frac{\kappa_{n-1}}{\kappa_n} + (n-1)\frac{\kappa_{n-1}}{\kappa_n}\right].$$

This establishes (8.3.5). □

Use (8.2.2) to eliminate ϕ_n^* from (8.3.5), assuming $\phi_n(0) \neq 0$ to establish

$$\phi_n'(z) = n\frac{\kappa_{n-1}}{\kappa_n}\phi_{n-1}(z)$$

$$+ i\frac{\kappa_{n-1}}{\phi_n(0)}\,z\,\phi_{n-1}(z)\int\limits_{|\zeta|=1}\frac{v'(z) - v'(\zeta)}{z - \zeta}\phi_n(\zeta)\overline{\phi_n^*(\zeta)}\,w(\zeta)\,d\zeta$$

$$\tag{8.3.7}$$

$$+ i\phi_n(z)\int\limits_{|\zeta|=1}\frac{v'(z) - v'(\zeta)}{z - \zeta}\phi_n(\zeta)\left[\overline{\phi_n(\zeta)} - \frac{\kappa_n}{\phi_n(0)}\,\overline{\phi_n^*(\zeta)}\right]w(\zeta)\,d\zeta.$$

Observe that $\overline{\phi_n(\zeta)} - \dfrac{\kappa_n}{\phi_n(0)}\,\overline{\phi_n^*(\zeta)}$ is a polynomial of degree $n-1$. Let

$$A_n(z) = n\,\frac{\kappa_{n-1}}{\kappa_n}$$

$$+ i\,\frac{\kappa_{n-1}}{\phi_n(0)}\,z\int\limits_{|\zeta|=1}\frac{v'(z)-v'(\zeta)}{z-\zeta}\,\phi_n(\zeta)\,\overline{\phi_n^*(\zeta)}\,w(\zeta)\,d\zeta, \qquad (8.3.8)$$

$$B_n(z) = -i\int\limits_{|\zeta|=1}\frac{v'(z)-v'(\zeta)}{z-\zeta}\,\phi_n(\zeta)$$

$$\times\left[\overline{\phi_n(\zeta)} - \frac{\kappa_n}{\phi_n(0)}\,\overline{\phi_n^*(\zeta)}\right]w(\zeta)\,d\zeta. \quad (8.3.9)$$

For future reference we note that $A_0 = B_0 = 0$ and

$$A_1(z) = \kappa_1 - \phi_1(z)\,v'(z) - \frac{\phi_1^2(z)}{\phi_1(0)}\,M_1(z), \qquad (8.3.10)$$

$$B_1(z) = -v'(z) - \frac{\phi_1(z)}{\phi_1(0)}\,M_1(z), \qquad (8.3.11)$$

where M_1 is defined by

$$M_1(z) = \int\limits_{|\zeta|=1}\zeta\,\frac{v'(z)-v'(\zeta)}{z-\zeta}\,w(\zeta)\,\frac{d\zeta}{i\zeta}. \qquad (8.3.12)$$

Now rewrite (8.3.7) in the form

$$\phi_n'(z) = A_n(z)\,\phi_{n-1}(z) - B_n(z)\,\phi_n(z). \qquad (8.3.13)$$

Define differential operators $L_{n,1}$ and $L_{n,2}$ by

$$L_{n,1} = \frac{d}{dz} + B_n(z), \qquad (8.3.14)$$

and

$$L_{n,2} = -\frac{d}{dz} - B_{n-1}(z) + \frac{A_{n-1}(z)\kappa_{n-1}}{z\kappa_{n-2}}$$

$$+ \frac{A_{n-1}(z)\kappa_n\phi_{n-1}(0)}{\kappa_{n-2}\phi_n(0)}. \qquad (8.3.15)$$

After the elimination of ϕ_{n-1} between (8.3.7) and (8.2.10) we find that the operators $L_{n,1}$ and $L_{n,2}$ are annihilation and creation operators in the sense that they satisfy

$$L_{n,1}\,\phi_n(z) = A_n(z)\,\phi_{n-1}(z),$$

$$L_{n,2}\,\phi_{n-1}(z) = \frac{A_{n-1}(z)}{z}\,\frac{\phi_{n-1}(0)\kappa_{n-1}}{\phi_n(0)\kappa_{n-2}}\,\phi_n(z). \qquad (8.3.16)$$

Hence we have established the second-order differential equation

$$L_{n,2}\left(\frac{1}{A_n(z)}\,L_{n,1}\right)\phi_n(z) = \frac{A_{n-1}(z)}{z}\,\frac{\phi_{n-1}(0)\kappa_{n-1}}{\phi_n(0)\kappa_{n-2}}\,\phi_n(z), \qquad (8.3.17)$$

which will also be written in the following way

$$\phi_n'' + P(z)\phi_n' + Q(z)\phi_n = 0. \qquad (8.3.18)$$

Note that, unlike for polynomials orthogonal on the line, $L_{n,1}^*$ is not related to $L_{n,2}$. In fact if we let

$$(f,g) := \int\limits_{|\zeta|=1} f(\zeta)\,\overline{g(\zeta)}\,w(\zeta)\,\frac{d\zeta}{i\zeta}, \qquad (8.3.19)$$

then in the Hilbert space endowed with this inner product, the adjoint of $L_{n,1}$ is

$$\left(L_{n,1}^* f\right)(z) = z^2 f'(z) + z f(z) + \left[\overline{v(z) + B_n(z)}\right] f(z). \qquad (8.3.20)$$

To see this use integration by parts and the fact that for $|\zeta| = 1$, $\overline{g(\zeta)} = \bar{g}(1/\zeta)$.

Example 8.3.2 *The circular Jacobi polynomials have been defined already in Example 8.2.5. Using the differentiation formula and some contiguous relations for the hypergeometric functions, combined in the form*

$$(1-z)\frac{d}{dz}\,{}_2F_1(-n, a+1; 1-n-a; z)$$

$$= \frac{n(n+2a)}{n-1+a}\,{}_2F_1(1-n, a+1; 2-n-a; z)$$

$$- n\,{}_2F_1(-n, a+1; 1-n-a; z),$$

one finds the differential-recurrence relation

$$(1-z)\,\phi_n' = -n\,\phi_n + [n(n+2a)]^{1/2}\phi_{n-1}, \qquad (8.3.21)$$

and the coefficient functions

$$A_n(z) = \frac{\sqrt{n(n+2a)}}{1-z}, \qquad B_n(z) = \frac{n}{1-z}. \qquad (8.3.22)$$

The second order differential equation becomes

$$\phi_n'' + \phi_n'\left\{\frac{1-n-a}{z} - \frac{2a+1}{1-z}\right\} + \phi_n\,\frac{n(a+1)}{z(1-z)} = 0. \qquad (8.3.23)$$

Example 8.3.3 *Consider the following weight function which is a generalization of the weight function in the previous example*

$$w(z) = 2^{-1-2a-2b}\,\frac{\Gamma(a+b+1)}{\Gamma(a+1/2)\Gamma(b+1/2)}\,|1-z|^{2a}\,|1+z|^{2b}. \qquad (8.3.24)$$

Here $x = \cos\theta$. The corresponding orthogonal polynomials are known as Szegő polynomials (Szegő, 1975, §11.5). They can be expressed in terms of the Jacobi *polynomials via Theorem 8.2.3,*

$$z^{-n}\phi_{2n}(z) = \mathbb{A}P_n^{(a-1/2,b-1/2)}\left([z+z^{-1}]/2\right)$$
$$+\frac{1}{2}\mathbb{B}\left[z-z^{-1}\right]P_{n-1}^{(a+1/2,b+1/2)}\left([z+z^{-1}]/2\right), \tag{8.3.25}$$

$$z^{1-n}\phi_{2n-1}(z) = \mathbb{C}P_n^{(a-1/2,b-1/2)}\left([z+z^{-1}]/2\right)$$
$$+\frac{1}{2}\mathbb{D}\left[z-z^{-1}\right]P_{n-1}^{(a+1/2,b+1/2)}\left([z+z^{-1}]/2\right). \tag{8.3.26}$$

In their study of the equilibrium positions of charges confined to the unit circle subject to logarithmic repulsion Forrester and Rogers considered orthogonal polynomials defined on x which are just the first term of (8.3.25). Using the normalization amongst the even and odd sequences of polynomials, orthogonality between these two sequences and the requirement that the coefficient of z^{-n} on the right-hand side of (8.3.26) must vanish, one finds explicitly that the coefficients are

$$\mathbb{A} = \left\{\frac{n!(a+b+1)_n}{(a+1/2)_n(b+1/2)_n}\right\}^{1/2}, \qquad \mathbb{B} = \frac{1}{2}\mathbb{A},$$
$$\mathbb{C} = n\left\{\frac{(n-1)!(a+b+1)_{n-1}}{(a+1/2)_n(b+1/2)_n}\right\}^{1/2}, \qquad \mathbb{D} = \frac{n+a+b}{2n}\mathbb{C}. \tag{8.3.27}$$

Furthermore the following coefficients of the polynomials are found to be

$$\kappa_{2n} = 2^{-2n}\frac{(a+b+1)_{2n}}{\sqrt{n!(a+b+1)_n(a+1/2)_n(b+1/2)_n}},$$
$$\kappa_{2n-1} = 2^{1-2n}\frac{(a+b+1)_{2n-1}}{\sqrt{(n-1)!(a+b+1)_{n-1}(a+1/2)_n(b+1/2)_n}}, \tag{8.3.28}$$

$$\ell_{2n} = 2n\frac{a-b}{2n+a+b}\kappa_{2n},$$
$$\ell_{2n-1} = (2n-1)\frac{a-b}{2n+a+b-1}\kappa_{2n-1}, \tag{8.3.29}$$

$$\phi_{2n}(0) = \frac{a+b}{2n+a+b}\kappa_{2n},$$
$$\phi_{2n-1}(0) = \frac{a-b}{2n+a+b-1}\kappa_{2n-1}. \tag{8.3.30}$$

The three-term recurrences are then

$$2(a-b)\sqrt{n(n+a+b)}\,\phi_{2n}(z)$$
$$+2(a+b)\sqrt{(n+a-1/2)(n+b-1/2)}\,z\,\phi_{2n-2}(z) \tag{8.3.31}$$
$$= [(a+b)(2n+a+b-1)+(a-b)(2n+a+b)z]\,\phi_{2n-1}(z),$$

and

$$2(a+b)\sqrt{(n+a-1/2)(n+b-1/2)}\,\phi_{2n-1}(z)$$
$$+2(a-b)\sqrt{(n-1)(n+a+b-1)}\,z\,\phi_{2n-3}(z) \tag{8.3.32}$$
$$= [(a-b)(2n+a+b-2)+(a+b)(2n+a+b-1)z]\,\phi_{2n-2}(z),$$

when $a \neq b$ and both these degenerate to $\phi_{2n-1}(z) = z\,\phi_{2n-2}(z)$ when $a = b$.

Using the differential and recurrence relations for the Jacobi polynomials directly when $a \neq b$ one can establish

$$A_{2n-1}(z) = 2\sqrt{(n+a-1/2)(n+b-1/2)}\,\frac{a-b+(a+b)z}{(a-b)\,(1-z^2)}, \qquad (8.3.33)$$

$$B_{2n-1}(z) = \frac{4ab+(2n-1)\,[a+b+(a-b)z]}{(a-b)\,(1-z^2)}, \qquad (8.3.34)$$

$$A_{2n}(z) = 2\sqrt{n(n+a+b)}\,\frac{a+b+(a-b)z}{(a+b)\,(1-z^2)}, \qquad (8.3.35)$$

$$B_{2n}(z) = 2n\frac{a-b+(a+b)z}{(a+b)\,(1-z^2)}. \qquad (8.3.36)$$

Example 8.3.4 (Modified Bessel Polynomials) *Consider the weight function*

$$w(z) = \frac{1}{2\pi I_0(t)}\,\exp\left(\frac{1}{2}t\,[z+z^{-1}]\right), \qquad (8.3.37)$$

where I_ν is a modified Bessel function. This system of orthogonal polynomials has arisen from studies of the length of longest increasing subsequences of random words (Baik et al., 1999) and matrix models (Periwal & Shevitz, 1990), (Hisakado, 1996). The leading coefficient has the Toeplitz determinant form

$$\kappa_n^2(t) = I_0(t)\,\frac{\det\left(I_{j-k}(t)\right)_{0\leq j,k\leq n-1}}{\det\left(I_{j-k}(t)\right)_{0\leq j,k\leq n}}. \qquad (8.3.38)$$

The first few members of this sequence are

$$\kappa_1^2 = \frac{I_0^2(t)}{I_0^2(t)-I_1^2(t)}, \qquad (8.3.39)$$

$$\frac{\phi_1(0)}{\kappa_1} = -\frac{I_1(t)}{I_0(t)}, \qquad (8.3.40)$$

$$\kappa_2^2 = \frac{I_0(t)\left(I_0^2(t)-I_1^2(t)\right)}{(I_0(t)-I_2(t))\,[I_0(t)(I_0(t)+I_2(t))-2I_1^2(t)]}, \qquad (8.3.41)$$

$$\frac{\phi_2(0)}{\kappa_2} = \frac{I_0(t)I_2(t)-I_1^2(t)}{I_1^2(t)-I_0^2(t)}. \qquad (8.3.42)$$

One can think of the weight function in (8.3.37) as a Toda-type modification of the constant weight function $1/2\pi$. Therefore (8.3.37) is an example of a Schur flow; see (Ablowitz & Ladik, 1976, (2.6)), (Faybusovich & Gekhtman, 1999). Thanks to Leonid Golinskii for bringing this to my attention.

Gessel (Gessel, 1990) has found the exact power series expansions in t for the first three determinants which appear in the above coefficients. Some recurrence relations for the corresponding coefficients of the monic version of these orthogonal polynomials have been known (Periwal & Shevitz, 1990), (Hisakado, 1996), (Tracy & Widom, 1999) and we derive the equivalent results for κ_n, etc.

Lemma 8.3.5 ((Periwal & Shevitz, 1990)) *The reflection coefficient $r_n(t) \equiv \phi_n(0)/ \kappa_n$ for the modified Bessel polynomial system satisfies a form of the discrete Painléve II equation, namely the recurrence relation*

$$-2\frac{n}{t}\frac{r_n}{1-r_n^2} = r_{n+1} + r_{n-1}, \qquad (8.3.43)$$

for $n \geq 1$ and $r_0(t) = 1$, $r_1(t) = -I_1(t)/I_0(t)$.

Proof Firstly we make a slight redefinition of the external field $w(z) = \exp(-v(z + 1/z))$ for convenience. Employing integration by parts we evaluate

$$-\int v'(\zeta+1/\zeta)\left(1-1/\zeta^2\right)\phi_{n+1}(\zeta)\,\overline{\phi_n(\zeta)}\,w(\zeta)\,\frac{d\zeta}{i\zeta}$$

$$= \int \left[\phi_{n+1}(\zeta)\,\overline{\zeta^2\phi_n'(\zeta)} + \phi_{n+1}(\zeta)\,\overline{\zeta\phi_n(\zeta)} - \phi_{n+1}'(\zeta)\,\overline{\phi_n(\zeta)}\right]w(\zeta)\,\frac{d\zeta}{i\zeta}$$

$$= (n+1)\left[\frac{\kappa_n}{\kappa_{n+1}} - \frac{\kappa_{n+1}}{\kappa_n}\right], \qquad (8.3.44)$$

for general external fields $v(z)$ using (8.3.1) and (8.1.7) in a similar way to the proof of Theorem 8.3.1. However in this case $v'(\zeta + 1/\zeta) = -t/2$, a direct evaluation of the left-hand side yields

$$-\frac{1}{2}t\left(\frac{\ell_n}{\kappa_{n+1}} - \frac{\kappa_n \ell_{n+2}}{\kappa_{n+1}\kappa_{n+2}}\right),$$

and simplification of this equality in terms of the defined ratio and use of (8.3.3) gives the above result.   □

There is also a differential relation satisfied by these coefficient functions or equivalently a differential relation in t for the orthogonal polynomials themselves (Hisakado, 1996), (Tracy & Widom, 1999).

Lemma 8.3.6 *The modified Bessel polynomials satisfy the differential relation*

$$2\frac{d}{dt}\phi_n(z) = \left[\frac{I_1(t)}{I_0(t)} + \frac{\phi_{n+1}(0)}{\kappa_{n+1}}\frac{\kappa_n}{\phi_n(0)}\right]\phi_n(z)$$

$$- \frac{\kappa_{n-1}}{\kappa_n}\left[1 + \frac{\phi_{n+1}(0)}{\kappa_{n+1}}\frac{\kappa_n}{\phi_n(0)}z\right]\phi_{n-1}(z), \qquad (8.3.45)$$

for $n \geq 1$ and $\dfrac{d}{dt}\phi_0(z) = 0$. The differential equations for the coefficients are

$$\frac{2}{\kappa_n}\frac{d}{dt}\kappa_n = \frac{I_1(t)}{I_0(t)} + \frac{\phi_{n+1}(0)}{\kappa_{n+1}}\frac{\phi_n(0)}{\kappa_n}, \qquad (8.3.46)$$

$$\frac{2}{\phi_n(0)}\frac{d}{dt}\phi_n(0) = \frac{I_1(t)}{I_0(t)} + \frac{\phi_{n+1}(0)}{\kappa_{n+1}}\frac{\kappa_n}{\phi_n(0)} - \frac{\phi_{n-1}(0)}{\phi_n(0)}\frac{\kappa_{n-1}}{\kappa_n}, \qquad (8.3.47)$$

for $n \geq 1$.

Proof Differentiating the orthonormality relation (8.3.1) with respect to t one finds from the orthogonality principle for $m \leq n - 2$ that

$$\frac{d}{dt}\phi_n(z) + \frac{1}{2}z\phi_n(z) = a_n\,\phi_{n+1}(z) + b_n\,\phi_n(z) + c_n\,\phi_{n-1}(z) \qquad (8.3.48)$$

for some coefficients a_n, b_n, c_n. The first coefficient is immediately found to be $a_n = \frac{1}{2}\kappa_n/\kappa_{n+1}$. Consideration of the differentiated orthonormality relation for $m = n - 1$ sets another coefficient, $c_n = -\frac{1}{2}\kappa_{n-1}/\kappa_n$, while the case of $m = n$ leads to $b_n = \frac{1}{2}I_1(t)/I_0(t)$. Finally use of the three-term recurrence (8.2.10) allows one to eliminate $\phi_{n+1}(z)$ in favor of $\phi_n(z), \phi_{n-1}(z)$ and one arrives at (8.3.45). The differential equations for the coefficients $\kappa_n, \phi_n(0)$ in (8.3.46)–(8.3.47) follow from reading off the appropriate terms of (8.3.45).   □

Use of the recurrence relation and the differential relations will allow us to find a differential equation for the coefficients, and thus another characterization of the coefficients.

Lemma 8.3.7 *The reflection coefficient* $r_n(t)$ *satisfies the following second order differential equation*

$$\frac{d^2}{dt^2}r_n = \frac{1}{2}\left(\frac{1}{r_n+1} + \frac{1}{r_n-1}\right)\left(\frac{d}{dt}r_n\right)^2$$
$$-\frac{1}{t}\frac{d}{dt}r_n - r_n\left(1-r_n^2\right) + \frac{n^2}{t^2}\frac{r_n}{1-r_n^2}, \qquad (8.3.49)$$

with the boundary conditions determined by the expansion

$$r_n(t) \underset{t\to 0}{\sim} \frac{(-t/2)^n}{n!}\left\{1 + \left(\frac{n}{n+1} - \delta_{n,1}\right)\frac{t^2}{4} + O\left(t^4\right)\right\}, \qquad (8.3.50)$$

for $n \geq 1$. *The coefficient* r_n *is related by*

$$r_n(t) = \frac{z_n(t)+1}{z_n(t)-1}, \qquad (8.3.51)$$

to $z_n(t)$ *which satisfies the Painlevé transcendent P-V equation with the parameters*

$$\alpha = \beta = \frac{n^2}{8}, \quad \gamma = 0, \quad \delta = -2. \qquad (8.3.52)$$

Proof Subtracting the relations (8.3.46)–(8.3.47) leads to the simplified expression

$$r_{n+1} - r_{n-1} = \frac{2}{1-r_n^2}\frac{d}{dt}r_n, \qquad (8.3.53)$$

which should be compared to the recurrence relation, in a similar form

$$r_{n+1} + r_{n-1} = -\frac{2n}{t}\frac{r_n}{1-r_n^2}. \qquad (8.3.54)$$

The differential equation (8.3.49) is found by combining these latter two equations and the identification with the P-V can be easily verified.   □

In the unpublished manuscript (Golinskii, 2005) Golinskii studied the time evolution of $r_n(t)$. He proved that

$$\lim_{t \to \infty} r_n(t) = (-1)^n, \quad n = 0, 1, \ldots \tag{8.3.55}$$

and that

$$\lim_{t \to \infty} t \left(1 - r_n^2(t)\right) = n. \tag{8.3.56}$$

In fact, (8.3.56) follows from (8.3.55) and the recurrence relation (8.3.43).

As a consequence of the above we find that the coefficients for the modified Bessel polynomials can be determined by the Toeplitz determinant (8.3.38), by the recurrence relations (8.3.54) or by the differential equation (8.3.49). An example of the use of this last method we note

$$\kappa_n^2(t) = \frac{I_0(t)}{\sqrt{1 - r_n^2(t)}} \exp\left(-n \int_0^t \frac{ds}{s} \frac{r_n^2(s)}{1 - r_n^2(s)}\right). \tag{8.3.57}$$

We now indicate how to find the coefficients of the differential relations, $A_n(z)$, $B_n(z)$ and observe that

$$\frac{v'(z) - v'(\zeta)}{z - \zeta} = -\frac{t}{2} \left[\frac{1}{z\zeta^2} + \frac{1}{z^2\zeta}\right].$$

The above relationship and (8.3.7) yield

$$\phi_n'(z) = \frac{\kappa_{n-1}}{\kappa_n} \phi_{n-1}(z) \left[n + \frac{t}{2z}\right]$$

$$+ \frac{t}{2} \frac{\kappa_{n-1}}{\phi_n(0)} \phi_{n-1}(z) \int_{|\zeta|=1} \phi_n(\zeta) \overline{\zeta \phi_n^*(\zeta)} \, w(\zeta) \frac{d\zeta}{i\zeta} \tag{8.3.58}$$

$$+ \frac{t}{2z} \phi_n(z) \int_{|\zeta|=1} \phi_n(\zeta) \overline{\zeta} \left[\overline{\phi_n(\zeta)} - \frac{\kappa_n}{\phi_n(0)} \overline{\phi_n^*(\zeta)}\right] w(\zeta) \frac{d\zeta}{i\zeta}.$$

Easy calculations using (8.1.7) give

$$\zeta \left[\phi_n(\zeta) - \frac{\kappa_n}{\phi_n(0)} \phi_n^*(\zeta)\right] = -\frac{\kappa_{n-1}}{\kappa_n} \frac{\overline{\phi_{n-1}(0)}}{\phi_n(0)} \phi_n(\zeta) + \text{lower order terms.}$$

and

$$\zeta \frac{\phi_n^*(\zeta)}{\phi_n(0)} = \frac{\phi_{n+1}(\zeta)}{\kappa_{n+1}} + \left[\frac{\kappa_n \ell_n - \kappa_{n-1}\ell_{n-1}}{\kappa_n |\phi_n(0)|^2} - \frac{\ell_{n+1}}{\kappa_{n+1}\kappa_n}\right] \phi_n(\zeta)$$

$$+ \text{lower order terms.}$$

These identities together with (8.3.58) establish the differential-difference relation

$$\phi_n'(z) = \frac{\kappa_{n-1}}{\kappa_n} \left[n + \frac{t}{2z} + \frac{t}{2} \frac{\kappa_{n-1}}{\kappa_n} \frac{\overline{\phi_{n-1}(0)}}{\phi_n(0)} - \frac{t}{2} \frac{\overline{\phi_{n+1}(0)} \phi_n(0)}{\kappa_{n+1}\kappa_n}\right] \phi_{n-1}(z)$$

$$- \frac{t}{2z} \frac{\kappa_{n-1}}{\kappa_n} \frac{\overline{\phi_{n-1}(0)}}{\phi_n(0)} \phi_n(z).$$

8.4 Functional Equations and Zeros

In this section we continue the development of the previous discussion of the differential relations satisfied by orthogonal polynomials to find a functional equation and its relationship to the zeros of the polynomials. Expressing the second order differential equation (8.3.17) in terms of the coefficient functions $A_n(z)$ and $B_n(z)$ we have

$$
\phi_n'' + \left\{ B_n + B_{n-1} - A_n'/A_n - \frac{\kappa_{n-1}}{\kappa_{n-2}} \frac{A_{n-1}}{z} - \frac{\kappa_n}{\kappa_{n-2}} \frac{\phi_{n-1}(0)}{\phi_n(0)} A_{n-1} \right\} \phi_n'
$$
$$
+ \left\{ B_n' - B_n A_n'/A_n + B_n B_{n-1} - \frac{\kappa_{n-1}}{\kappa_{n-2}} \frac{A_{n-1} B_n}{z} \right.
$$
$$
\left. - \frac{\kappa_n}{\kappa_{n-2}} \frac{\phi_{n-1}(0)}{\phi_n(0)} A_{n-1} B_n + \frac{\kappa_{n-1}}{\kappa_{n-2}} \frac{\phi_{n-1}(0)}{\phi_n(0)} \frac{A_{n-1} A_n}{z} \right\} \phi_n = 0. \quad (8.4.1)
$$

Now by analogy with the orthogonal polynomials defined on the real line the coefficient of the ϕ_n' term above can be simplified.

Theorem 8.4.1 *Given that $v(z)$ is a meromorphic function in the unit disk then the following functional equation holds*

$$
B_n + B_{n-1} - \frac{\kappa_{n-1}}{\kappa_{n-2}} \frac{A_{n-1}}{z} - \frac{\kappa_n}{\kappa_{n-2}} \frac{\phi_{n-1}(0)}{\phi_n(0)} A_{n-1}
$$
$$
= \frac{1-n}{z} - v'(z). \quad (8.4.2)
$$

Proof From the definitions (8.3.8)–(8.3.9) we start with the following expression

$$
B_n + B_{n-1} - \frac{\kappa_{n-1}}{\kappa_{n-2}} \frac{A_{n-1}}{z} - \frac{\kappa_n}{\kappa_{n-2}} \frac{\phi_{n-1}(0)}{\phi_n(0)} A_{n-1}
$$
$$
= -(n-1) \left[\frac{1}{z} + \frac{\kappa_n}{\kappa_{n-1}} \frac{\phi_{n-1}(0)}{\phi_n(0)} \right]
$$
$$
+ i \int \frac{v'(z) - v'(\zeta)}{z - \zeta}
$$
$$
\times \left\{ -\phi_n \overline{\phi_n} + \frac{\kappa_n}{\phi_n(0)} \phi_n \overline{\phi_n^*} - \phi_{n-1} \overline{\phi_{n-1}} - \frac{\kappa_n}{\phi_n(0)} \zeta \phi_{n-1} \overline{\phi_{n-1}^*} \right\} w(\zeta) \, d\zeta
$$
$$
- i \frac{\kappa_n}{\phi_n(0)} \int [v'(z) - v'(\zeta)] \phi_{n-1} \overline{\phi_{n-1}^*} \, w(\zeta) d\zeta.
$$

Employing the recurrences (8.2.3)–(8.2.2), and the relation amongst coefficients (8.1.20) one can show that the factor in the first integral on the right-hand side above is

$$
-\phi_n \overline{\phi_n} + \frac{\kappa_n}{\phi_n(0)} \phi_n \overline{\phi_n^*} - \phi_{n-1} \overline{\phi_{n-1}} - \frac{\kappa_n}{\phi_n(0)} \zeta \phi_{n-1} \overline{\phi_{n-1}^*} = -\phi_n \overline{\phi_n} + \phi_n^* \overline{\phi_n^*}.
$$

Now since $|\zeta|^2 = 1$, one can show that the right-hand side of the above is zero from the Christoffel–Darboux sum (8.2.1). Consequently our right-hand side is now

$$
-(n-1)\left[\frac{1}{z} + \frac{\kappa_n}{\kappa_{n-1}}\frac{\phi_{n-1}(0)}{\phi_n(0)}\right]
$$

$$
- i\frac{\kappa_n}{\phi_n(0)}\left[v'(z)\int \phi_{n-1}\overline{\phi_{n-1}^*}\,w(\zeta)\,d\zeta - \int v'(\zeta)\phi_{n-1}\overline{\phi_{n-1}^*}\,w(\zeta)\,d\zeta\right].
$$

Taking the first integral in this expression and using the recurrence (8.2.3) and the decomposition $\zeta\phi_{n-1} = \kappa_{n-1}/\kappa_n\phi_n + \pi_{n-1}$ where $\pi_n \in \Pi_n$, Π_n being the space of polynomials of degree at most n, we find it reduces to $-i\phi_n(0)/\kappa_n$ from the normality of the orthogonal polynomials. Considering now the second integral above we integrate by parts and are left with

$$
\int \phi_{n-1}'\overline{\phi_{n-1}^*}\,w(\zeta)\,d\zeta + \int \phi_{n-1}\overline{\phi_{n-1}^*}'\,w(\zeta)\,d\zeta,
$$

and the first term here must vanish as ϕ_{n-1}^* can be expressed in terms of ϕ_{n-1},ϕ_n from (8.2.3) but $\phi_{n-1}' \in \Pi_{n-2}$. The remaining integral, the second one above, can be treated in the following way. First, express the conjugate polynomial in terms of the polynomial itself via (8.2.2) and employ the relation for its derivative (8.3.6). Further noting that $\zeta\phi_{n-1}' = (n-1)\phi_{n-1} + \pi_{n-2}$, $\zeta\phi_{n-2} = \kappa_{n-2}/\kappa_{n-1}\phi_{n-1} + \pi_{n-2}$, and $\zeta^2\phi_{n-2}' = (n-2)\kappa_{n-2}/\kappa_{n-1}\phi_{n-1} + \pi_{n-2}$ along with the orthonormality relation, the final integral is nothing but $-i(n-1)\phi_{n-1}(0)/\kappa_{n-1}$. Combining all this, the final result is (8.4.2). $\qquad\square$

Remark 8.4.1 *The zeros of the polynomial* $\phi_n(z)$ *will be denoted by* $\{z_j\}_{1\le j\le n}$ *and are confined within the unit circle* $|z| < 1$. *One can construct a real function* $|T(z_1,\ldots,z_n)|$ *from*

$$
T(z_1,\ldots,z_n) = \prod_{j=1}^n z_j^{-n+1}\frac{e^{-v(z_j)}}{A_n(z_j)}\prod_{1\le j<k\le n}(z_j - z_k)^2, \tag{8.4.3}
$$

such that the zeros are given by the stationary points of this function.

This function has the interpretation of being the total energy function for n mobile unit charges in the unit disk interacting with a one-body confining potential, $v(z) + \ln A_n(z)$, an attractive logarithmic potential with a charge $n-1$ at the origin, $(n-1)\ln z$, and repulsive logarithmic two-body potentials, $-\ln(z_i - z_j)$, between pairs of charges. However all the stationary points are saddle-points, a natural consequence of analyticity in the unit disk. Following §3.5, one can show that the conditions for the stationary points of function $T(z_1,\ldots,z_n)$ above lead to a system of equations

$$
-v'(z_j) - \frac{A_n'(z_j)}{A_n(z_j)} - \frac{n-1}{z_j} + 2\sum_{1\le k\le n, k\ne j}\frac{1}{z_j - z_k} = 0, \tag{8.4.4}
$$

for $j = 1,\ldots,n$. Then, as in §3.5, we have the n conditions expressed as

$$
f''(z_j) + \left\{\frac{1-n}{z_j} - v'(z_j) - \frac{A_n'(z_j)}{A_n(z_j)}\right\}f'(z_j) = 0, \tag{8.4.5}
$$

for $j = 1, \cdots, n$. The result then follows.

Remark 8.4.2 *The functional equation (8.4.2) actually implies a very general recurrence relation on the orthogonal system coefficients $\kappa_n, \phi_n(0)$. In general if it is possible to relate the differential recurrence coefficients A_n, B_n to these polynomial coefficients, then the functional equation dictates that equality holds for all z, and thus for independent terms in z. For rational functions this can be applied to the coefficients of monomials in z.*

Remark 8.4.3 *Equation (8.3.17) is one way of expressing the second order differential equation for the orthogonal polynomials; however, one can perform the elimination in the opposite order and find*

$$L_{n+1,1}\left(\frac{z}{A_n(z)}L_{n+1,2}\right)\phi_n(z)$$
$$= \frac{\kappa_n\phi_n(0)}{\kappa_{n-1}\phi_{n+1}(0)}A_{n+1}(z)\,\phi_n(z). \qquad (8.4.6)$$

Requiring the coefficients of ϕ_n' in (8.4.6) and (8.3.16) to be equal is equivalent to (8.4.2).

To see this, write (8.4.6) in full, that is

$$\phi_n'' + \left\{B_{n+1} + B_n - A_n'/A_n - \frac{\kappa_n}{\kappa_{n-1}}\frac{A_n}{z} - \frac{\kappa_{n+1}}{\kappa_{n-1}}\right.$$
$$\left.\frac{\phi_n(0)}{\phi_{n+1}(0)}A_n + \frac{1}{z}\right\}\phi_n'$$
$$+ \left\{B_n' - B_nA_n'/A_n + B_{n+1}B_n - \frac{\kappa_n}{\kappa_{n-1}}\frac{A_nB_{n+1}}{z}\right. \qquad (8.4.7)$$
$$-\frac{\kappa_{n+1}}{\kappa_{n-1}}\frac{\phi_n(0)}{\phi_{n+1}(0)}A_nB_{n+1} + \frac{\kappa_n}{\kappa_{n-1}}\frac{\phi_n(0)}{\phi_{n+1}(0)}\frac{A_nA_{n+1}}{z}$$
$$\left.+\frac{B_n}{z} - \frac{\kappa_{n+1}}{\kappa_{n-1}}\frac{\phi_n(0)}{\phi_{n+1}(0)}\frac{A_n}{z}\right\}\phi_n = 0.$$

Equating the coefficients of $\phi_n'(z)$ in (8.4.1) and (8.4.7) we derive an inhomogeneous first order difference equation, whose solution is

$$B_n + B_{n-1} - \frac{\kappa_{n-1}}{\kappa_{n-2}}\frac{A_{n-1}}{z} - \frac{\kappa_n}{\kappa_{n-2}}\frac{\phi_{n-1}(0)}{\phi_n(0)}A_{n-1}$$
$$= \frac{1-n}{z} + \text{function of } z \text{ only}. \qquad (8.4.8)$$

This function can be simply evaluated by setting $n = 1$, evaluating the integrals after noting $B_0 = 0$ and the cancellations, and yields the result $-v'(z)$.

In Example 8.2.5, we can verify that the general form for the T-function is correct in the case of the circular Jacobi polynomials by a direct evaluation

$$T(z_1, \ldots, z_n)$$
$$= \prod_{j=1}^{n} z_j^{1-n-a} (1 - z_j)^{a+1} (z_j - 1)^a \prod_{1 \le j < k \le n} (z_j - z_k)^2, \tag{8.4.9}$$

where we have used the identity

$$|1 - z|^{2a} = (1 - z)^a (1 - 1/z)^a = z^{-a} (1 - z)^a (z - 1)^a, \tag{8.4.10}$$

on $|z| = 1$ to suitably construct a locally analytic weight function. One can show that the stationary points for this problem are the solution to the set of equations

$$\frac{1 - n - a}{z_j} - \frac{2a + 1}{1 - z_j} + 2 \sum_{j \ne k} \frac{1}{z_j - z_k} = 0, \quad 1 \le j \le n, \tag{8.4.11}$$

so that the polynomial $f(z) = \prod_{j=1}^{n} (z - z_j)$ satisfies the relations

$$f''(z_j) + f'(z_j) \left\{ \frac{1 - n - a}{z_j} - \frac{2a + 1}{1 - z_j} \right\} = 0. \tag{8.4.12}$$

Consequently we find that

$$z(1 - z)f''(z) + f'(z)\left\{(1 - n - a)(1 - z) - (2a + 1)z\right\} + Qf(z) = 0, \tag{8.4.13}$$

for some constant Q independent of z, but possibly dependent on n and a and is identical to the second order differential equations (8.3.23).

In Example 8.3.3, using the expressions (8.3.33)–(8.3.36) one can verify that the identity (8.4.2) holds and in particular becomes

$$B_n + B_{n-1} - \frac{\kappa_{n-1}}{\kappa_{n-2}} \frac{A_{n-1}}{z} - \frac{\kappa_n}{\kappa_{n-2}} \frac{\phi_{n-1}(0)}{\phi_n(0)} A_{n-1}$$
$$= -\frac{n-1}{z} - \frac{a+b}{z} - \frac{2a}{1-z} + \frac{2b}{1+z}, \tag{8.4.14}$$

for both the odd and even sequences. Consequently the coefficients in the second order differential equation are

$$P_n(z) = -\frac{n + a + b - 1}{z} - \frac{2a + 1}{1 - z} + \frac{2b + 1}{1 + z} - \frac{a \pm b}{a \mp b + (a \pm b)z}, \tag{8.4.15}$$

$$Q_{2n}(z) = 2n \frac{a(a+1)(1+z)^2 - b(b+1)(1-z)^2}{z(1-z^2)[a + b + (a - b)z]}, \tag{8.4.16}$$

and

$$Q_{2n-1}(z) = \frac{(2n-1)\left[a(a+1)(1+z)^2\right]}{z(1-z^2)[a - b + (a + b)z]}$$
$$+ \frac{(2n-1)\left[b(b+1)(1-z)^2 - 2ab(1-z^2)\right] + 4ab}{z(1-z^2)[a - b + (a + b)z]}. \tag{8.4.17}$$

Similarly we can verify that the general form for the T-function is correct in the case of the Szegő polynomials by using the identity

$$|1 - z|^{2a}|1 + z|^{2b} = z^{-a-b}(1 - z)^a(z - 1)^a(1 + z)^{2b}, \qquad (8.4.18)$$

to suitably analytically continue the weight function. The stationary points for this problem are the solution to the following system of nonlinear equations

$$
\begin{aligned}
\frac{1 - n - a - b}{z_j} - \frac{2a + 1}{1 - z_j} &+ \frac{2b + 1}{1 + z_j} \\
- \frac{a \mp b}{a \pm b + (a \mp b)z_j} &+ 2\sum_{j \neq k} \frac{1}{z_j - z_k} = 0,
\end{aligned}
\qquad (8.4.19)
$$

for $1 \leq j \leq n$, such that the polynomial $f(z) = \prod_{j=1}^{n}(z - z_j)$ satisfies the relations

$$
f''(z_j) + f'(z_j) \left\{ \frac{1 - n - a - b}{z_j} \right.
$$
$$
\left. - \frac{2a + 1}{1 - z_j} + \frac{2b + 1}{1 + z_j} - \frac{a \mp b}{a \pm b + (a \mp b)z_j} \right\} = 0.
\qquad (8.4.20)
$$

Finally we find that

$$
f''(z) + Q(z)f(z)
$$
$$
+ f'(z) \left\{ \frac{1 - n - a - b}{z} - \frac{2a + 1}{1 - z} + \frac{2b + 1}{1 + z} - \frac{a \mp b}{a \pm b + (a \mp b)z} \right\} = 0,
\qquad (8.4.21)
$$

for some constant Q independent of z, but possibly dependent on n and a. The coefficient of the first derivative term is identical to the expression for $P(z)$ in (8.4.15).

Example 8.4.2 *One can also verify the functional relation (8.4.2) for the modified Bessel polynomials. Forming the left-hand side of this identity we find this reduces to*

$$
B_n + B_{n-1} - \frac{\kappa_{n-1}}{\kappa_{n-2}} \frac{A_{n-1}}{z} - \frac{\kappa_n}{\kappa_{n-2}} \frac{\phi_{n-1}(0)}{\phi_n(0)} A_{n-1}
$$
$$
= -\frac{n-1}{z} - \frac{t}{2z^2} - (n-1) \frac{\kappa_n}{\kappa_{n-1}} \frac{\phi_{n-1}(0)}{\phi_n(0)}
$$
$$
- \frac{t}{2} \frac{\kappa_n \kappa_{n-2}}{\kappa_{n-1}^2} \frac{\phi_{n-2}(0)}{\phi_n(0)} + \frac{t}{2} \frac{\phi_{n-1}^2(0)}{\kappa_{n-1}^2}. \qquad (8.4.22)
$$

Now the last three terms on the right-hand side of the above equation simplify to $t/2$ using the recurrence relation (8.3.43), showing that the general functional relation holds. In fact, as remarked earlier, this relation implies the recurrence relation itself.

8.5 Limit Theorems

In this section we mention several important limit theorems of orthogonal polynomials and Toeplitz determinants. The results will be stated without proof because detailed proofs are lengthy and are already available in Simon's recent book (Simon, 2004).

We write

$$d\mu(\theta) = w(\theta) \frac{d\theta}{2\pi} + d\mu_s(\theta),$$

where μ_s is singular with respect to $d\theta/2\pi$, and $w(\theta)$ is the Radon–Nikadym derivative of μ with respect to $d\theta/2\pi$. It is assumed that $w(\theta) > 0$ almost everywhere in θ.

Theorem 8.5.1 *Let $\mathbf{H}_n$ ba as in (8.1.4). For ζ in the open unit disc, let $m_n(\zeta)$ be the minimum of $\mathbf{H}_n$ under the side condition $\sum\limits_{k=0}^{n} u_k \zeta^k = 1$. There, for $\zeta = re^{i\phi}$,*

$$\lim_{n\to\infty} m_n(\zeta) = \left(1 - |\zeta|^2\right) \exp\left(\frac{1}{2\pi} \int\limits_{-\pi}^{\pi} \frac{\left(1 - r^2\right) \ln w(\theta)\, d\theta}{1 - 2r\cos(\phi - \theta) + r^2} \right). \qquad (8.5.1)$$

If $\ln w$ is not integrable on $[-\pi, \pi]$, the integral on the right-hand side is defined as $-\infty$.

Theorem 8.5.2 *((Simon, 2004, Theorem 1.5.7)) The following are equivalent:*

(i) $\sum\limits_{k=0}^{\infty} |\alpha_k|^2 < \infty$

(ii) $\lim\limits_{n\to\infty} \kappa_n < \infty$

(iii) *The polynomials in z are not dense in $L^2[C, \mu]$, C is the unit circle.*

The following strong version of the Szegő limit theorem is in (Ibragimov, 1968). It is also stated and proved in (Simon, 2004).

Theorem 8.5.3 *Let $d\mu = w\, d\theta/2\pi$ and assume that*

$$\int\limits_{-\pi}^{\pi} \ln(w(\theta)) \frac{d\theta}{2\pi} > -\infty. \qquad (8.5.2)$$

Define $\left\{\hat{L}_n\right\}$ by

$$\ln(w(\theta)) = \sum_{n=-\infty}^{\infty} \hat{L}_n e^{in\theta},$$

and the Szegő function $D(z)$ by

$$D(z) = \exp\left(\frac{1}{2}\hat{L}_0 + \sum_{n=0}^{\infty} \hat{L}_n z^n \right). \qquad (8.5.3)$$

The the four quantities below are equal (all may be infinite):

(i) $\lim\limits_{n\to\infty} D_n \exp\left(-\frac{(n+1)}{2\pi} \int\limits_{-\pi}^{\pi} \ln(w(\theta))\, d\theta \right);$

(ii) $\prod_{j=0}^{\infty} \left(1 - |\alpha_j|^2\right)^{-j-1}$;

(iii) $\exp\left(\sum_{n=1}^{\infty} n \left|\hat{L}_n\right|^2\right)$;

(iv) $\exp\left(\dfrac{1}{\pi} \int_{|z| \leq 1} \left|\dfrac{dD(z)}{dz}\right|^2 |D(z)|^{-2} d^2 z\right)$.

Assume that w satisfies (8.5.2). We form an analytic function $h(z)$ whose real part is the harmonic function

$$\frac{1}{2n} \int_{-\pi}^{\pi} \ln(w(\theta)) \frac{(1 - r^2)\, d\theta}{1 - 2r\cos(\phi - \theta) + r^2},$$

$r < 1$, $z = re^{i\phi}$. We further assume $h(0) = 0$ and define a function $g(z)$ via

$$g(z) = \exp(h(z)/2), \quad |z| < 1.$$

Theorem 8.5.4 *Let $d\mu = w(\theta) \dfrac{d\theta}{2\pi} + d\mu_s$, μ_s is singular and assume that (8.5.2) holds. Then the following limiting relations hold:*

(i) $\displaystyle\lim_{n\to\infty} s_n(z, z) = \dfrac{1}{1 - |z|^2} \dfrac{1}{|g(z)|^2}, \quad |z| < 1;$

(ii) $\displaystyle\lim_{n\to\infty} s_n(\zeta, z) = \dfrac{1}{1 - \bar{\zeta}z} \dfrac{1}{\overline{g(\zeta)}\, g(z)}, \quad for\ |z| < 1,\ |\zeta| < 1;$

(iii) $\displaystyle\lim_{n\to\infty} z^{-n}\phi_n(z) = 1/\bar{g}\left(z^{-1}\right), \quad |z| > 1;$

(iv) $\displaystyle\lim_{n\to\infty} \phi_n(z) = 0, \quad |z| < 1.$

For a proof, see §3.4 in (Grenander & Szegő, 1958).

8.6 Modifications of Measures

In this section, we state the analogues of §2.7 for polynomials orthogonal on the unit circle. We start with the analogue of the Christoffel formula.

Theorem 8.6.1 *Let $\{\phi_n(z)\}$ be orthonormal with respect to a probability measure μ and let $G_{2m}(z)$ be a polynomial of precise degree $2m$ such that*

$$z^{-m} G_{2m}(z) = |G_{2m}(z)|, \quad |z| = 1.$$

Define polynomials $\{\psi_n(z)\}$ by

$$G_{2m}(z)\,\psi_n(z) = \begin{vmatrix} \phi^*(z) & z\phi^*(z) & \cdots & z^{m-1}\phi^*(z) \\ \phi^*(\alpha_1) & \alpha_1\phi^*(\alpha_1) & \cdots & \alpha_1^{m-1}\phi^*(\alpha_1) \\ \phi^*(\alpha_2) & \alpha_2\phi^*(\alpha_2) & \cdots & \alpha_2^{m-1}\phi^*(\alpha_2) \\ \vdots & \vdots & \cdots & \vdots \\ \phi^*(\alpha_{2m}) & \alpha_{2m}\phi^*(\alpha_{2m}) & \cdots & \alpha_{2m}^{m-1}\phi^*(\alpha_{2m}) \\ \phi(z) & z\phi(z) & \cdots & z^m\phi(z) \\ \phi(\alpha_1) & \alpha_1\phi(\alpha_1) & \cdots & \alpha_1^m\phi(\alpha_1) \\ \phi(\alpha_2) & \alpha_2\phi(\alpha_2) & \cdots & \alpha_2^m\phi(\alpha_2) \\ \vdots & \vdots & \cdots & \vdots \\ \phi(\alpha_{2m}) & \alpha_{2m}\phi(\alpha_{2m}) & \cdots & \alpha_{2m}^m\phi^*(\alpha_{2m}) \end{vmatrix}$$

(8.6.1)

where $\alpha_1, \alpha_2, \ldots, \alpha_{2m}$ are the zeros of $G_{2m}(z)$ and ϕ stands for ϕ_{n+m}.

For zeros of multiplicity r, $r > 1$, replace the corresponding rows in (8.6.1) by the derivatives of order $0, 1, \ldots, r-1$ of the polynomials in the first row evaluated at that zero.

Then $\{\psi_n(z)\}$ are orthogonal with respect to $\left|G_{2m}\left(e^{i\theta}\right)\right| d\mu(\theta)$ on the unit circle.

The proof of Theorem 8.6.1 uses two lemmas, which we will state and prove first.

Lemma 8.6.2 *Each polynomial in the first row of (8.6.1), when divided by z^m, is orthogonal to any polynomial of degree at most $n-1$ with respect to μ.*

Proof Let $\pi_{n-1}(z)$ be a polynomial of degree at most $n-1$. Then, for the polynomials $z^\ell\phi_{n+m}(z)$, $0 \le \ell \le m$, and $z = e^{i\theta}$, we have

$$\int_{-\pi}^{\pi} \frac{z^\ell\phi_{n+m}(z)}{z^m}\,\overline{\pi_{n-1}(z)}\,d\mu(\theta) = \int_{-\pi}^{\pi} \phi_{n+m}(z)\,\overline{z^{m-\ell}\pi_{n-1}(z)}\,d\mu(\theta) = 0.$$

On the other hand, for the polynomials $z^\ell\phi_{n+m}^*(z)$, $0 \le \ell < m$, we have

$$\int_{-\pi}^{\pi} \overline{z^{\ell-m}\phi_{n+m}^*(z)}\,P_{n-1}(z)\,d\mu(\theta)$$

$$= \int_{-\pi}^{\pi} z^{m-\ell}\,\overline{\phi_{n+m}^*(z)}\,P_{n-1}(z)\,d\mu(\theta)$$

$$= \int_{-\pi}^{\pi} z^{m-\ell}\,z^{-n-m}\phi_{n+m}(z)\,P_{n-1}(z)\,d\mu(\theta)$$

$$= \int_{-\pi}^{\pi} \phi_{n+m}(z)\,\overline{z^{\ell+n}P_{n-1}(1/z)}\,d\mu(\theta) = 0,$$

since $0 \le \ell < m$. □

Lemma 8.6.3 *The determinant in* (8.6.1) *is a polynomial of precise degree* $2m + n$.

Proof Assume the coefficient of $z^m \phi_{n+m}(z)$ is zero; i.e., the determinant we get from crossing out the first row and last column of our original matrix is zero. Then there exist constants, not all zero, $\lambda_0, \lambda_1, \ldots, \lambda_{m-1}$ and $\gamma_0, \gamma_1, \ldots, \gamma_{m-1}$, such that the polynomials $g(z)$ defined by

$$
\begin{aligned}
g(z) := & \left(\lambda_0 + \lambda_1 z + \cdots + \lambda_{m-1} z^{m-1} \right) \phi_{n+m}(z) \\
& + \left(\gamma_0 + \gamma_1 z + \cdots + \gamma_{m-1} z^{m-1} \right) \phi_{n+m}^*(z)
\end{aligned}
$$

vanishes for $z = \alpha_1, \alpha_2, \ldots, \alpha_{2m}$. This shows that $g(z)$ has the form $g(z) = G_{2m}(z) \pi_{n-1}(z)$ for some $\pi_{n-1}(z)$. We know that $g(z)$ is not identically zero as the zeros of $\phi(z)$ lie in $|z| < 1$ and the zeros of $\phi^*(z)$ lie in $|z| > 1$. From Lemma 8.6.2 we know $g(z)/z^m$ is orthogonal to any polynomial of degree less than n. Thus,

$$
0 = \int_{-\pi}^{\pi} \frac{g(z)}{z^m} \, \overline{\pi_{n-1}(z)} \, d\nu(\theta) = \int_{-\pi}^{\pi} \frac{G_{2m}(z) \pi_{n-1}(z)}{z^m} \, \overline{\pi_{n-1}(z)} \, d\nu(\theta)
$$

$$
= \int_{-\pi}^{\pi} |\pi_{n-1}(z)|^2 \, |G_{2m}(z)| \, d\nu(\theta)
$$

which implies $\rho_{n-1}(z) \equiv 0$ and, consequently, $g(z) \equiv 0$.   □

Proof of Theorem 8.6.1   From Lemma 8.6.3 and the form of the determinant in (8.6.1), each $\psi_n(z)$ is a polynomial of degree n. From Lemma 8.6.2 we see that for any $\pi_{n-1}(z)$

$$
\int_{-\pi}^{\pi} \frac{G_{2m}(z) \psi_n(z)}{z^m} \, \overline{\pi_{n-1}(z)} \, d\nu(\theta) = 0;
$$

that is,

$$
\int_{-\pi}^{\pi} \psi_n(z) \, \overline{\pi_{n-1}(z)} \, |G_{2m}(z)| \, d\nu(\theta) = 0.
$$

Thus, the polynomials $\{\psi_n(z)\}$ are constant multiples of the polynomials orthonormal with respect to $|G_{2m}(z)| \, d\nu(\theta)$.   □

This form of Theorem 8.6.1 is from (Ismail & Ruedemann, 1992). A different version of Theorem 8.6.1 containing both the ϕ_n's and their kernel polynomials is in (Godoy & Marcellán, 1991).

To prove Uvarov's type theorem for polynomials orthogonal on the unit circle, we proceed in two steps. First, we modify the measure by dividing μ by $|G_{2m}(z)|$. In Step 2, we combine Step 1 with Theorem 8.6.1.

Theorem 8.6.4 *Let* $\{\phi_n(z)\}$ *be orthonormal with respect to a probability measure* $\mu(\theta)$ *on* $z = e^{i\theta}$ *and let* $G_{2m}(z)$ *be a polynomial of precise degree* $2m$ *such that*

$$
z^{-m} G_{2m}(z) = |G_{2m}(z)| > 0, \qquad z = e^{i\theta}.
$$

Define a new system of polynomials $\{\psi_n(z)\}$, $n = 2m, 2m + 1, \ldots,$ by

$$
\psi_n(z) = \begin{vmatrix}
\phi^*(z) & z\phi^*(z) & \cdots & z^{m-1}\phi^*(z) \\
L_{\beta_1}(\phi^*) & L_{\beta_1}(z\phi^*) & \cdots & L_{\beta_1}(z^{m-1}\phi^*) \\
L_{\beta_2}(\phi^*) & L_{\beta_2}(z\phi^*) & \cdots & L_{\beta_2}(z^{m-1}\phi^*) \\
\vdots & \vdots & & \vdots \\
L_{\beta_{2m}}(\phi^*) & L_{\beta_{2m}}(z\phi^*) & \cdots & L_{\beta_{2m}}(z^{m-1}\phi^*) \\
\phi(z) & z\phi(z) & \cdots & z^m\phi(z) \\
L_{\beta_1}(\phi) & L_{\beta_1}(z\phi) & \cdots & L_{\beta_1}(z^m\phi) \\
L_{\beta_2}(\phi) & L_{\beta_2}(z\phi) & \cdots & L_{\beta_2}(z^m\phi) \\
\vdots & \vdots & & \vdots \\
L_{\beta_{2m}}(\phi) & L_{\beta_{2m}}(z\phi) & \cdots & L_{\beta_{2m}}(z^m\phi)
\end{vmatrix}
\qquad (8.6.2)
$$

where the zeros of $G_{2m}(z)$ are $\{\beta_1, \beta_2, \ldots, \beta_{2m}\}$, $\phi(z)$ denotes $\phi(z)$, and where we define

$$
L_\beta(p) := \int_{-\pi}^{\pi} p(\xi) \overline{\left(\frac{\xi^m}{\xi - \beta}\right)} \, d\nu(\theta), \quad \xi = e^{i\theta}.
$$

For zeros of multiplicity h, $h > 1$, we replace the corresponding rows in the determinant (8.6.2) by

$$
L_\beta^k(p) := \int_{-\pi}^{\pi} p(\xi) \overline{\left(\frac{\xi^m}{(\xi - \beta)^k}\right)} \, d\nu(\theta), \quad \xi = e^{i\theta},
$$

$k = 1, 2, \ldots, h$ *acting on the first row.*

Under the above assumptions $\{\psi_n(z)\}$ are the orthonormal polynomials associated with the distribution $(1/|G_{2m}(z)|) \, d\nu(\theta)$ on the unit circle, $z = e^{i\theta}$, up to multiplicative constants, for $n \geq 2m$.

Proof Assume for the moment that the zeros of $G_{2m}(z)$ are pairwise distinct.

Now, if $k \geq 2m$ and $\rho_k(z)$ is of precise degree k we have

$$
\rho_k(z) = G_{2m}(z) \, q(z) + r(z)
$$

with the degree of $r(z)$ less than $2m$. Thus define

$$
q_{k-2m}(z) = \frac{\rho_k(z)}{G_{2m}(z)} - \frac{r(z)}{G_{2m}(z)},
$$

where in case $k < 2m$ we set $r(z) \equiv \rho_k(z)$ and $q_{k-2m}(z) \equiv 0$. In either case, $q_{k-2m}(z)$ has degree at most $k - 2m$.

Now we decompose $r(z)/G_{2m}(z)$ via partial fractions, i.e.,

$$
\frac{r(z)}{G_{2m}(z)} = \sum_{i=1}^{2m} \frac{A_i(\rho_k)}{z - \beta_i},
$$

where the $\{A_i\,(\rho_k)\}$ are constants depending on ρ_k. Assuming $k \leq n - 1$ we have for every

$$\gamma(z) \in \text{Span}\left\{\phi(z), \phi^*(z), z\phi(z), z\phi^*(z), \ldots, z^{m-1}\phi(z), z^{m-1}\phi^*(z), z^m\phi(z)\right\},$$

where ϕ denotes ϕ_{n-m}, that

$$\int_{-\pi}^{\pi} \gamma(z)\,\overline{z^m q_{k-2m}(z)}\,d\mu(\theta) = 0$$

and thus

$$\int_{-\pi}^{\pi} \gamma(z)\,\overline{\rho_k(z)}\,\frac{1}{|G_{2m}(z)|}\,d\mu(\theta) = \sum_{i=1}^{2m}\left[\overline{A_i\,(\rho_k)}\int_{-\pi}^{\pi}\gamma(z)\overline{\left(\frac{z^m}{z-\beta_i}\right)}\,d\mu(\theta)\right]$$

for $k \leq n - 1$.

Hence if we let $\psi_n(z)$ be defined as in Theorem 8.6.4 above, we get

$$\int_{-\pi}^{\pi}\psi_n(z)\,\overline{\rho_k(z)}\,\frac{1}{|G_{2m}(z)|}\,d\mu(\theta) = 0, \quad k \leq n - 1$$

by linearity as under integration the first row in the determinant will be a linear combination of the lower rows. (If $G_{2m}(z)$ has multiple zeros we simply change the form of the partial fraction decomposition.) However, we still must show that $\phi_n(z)$ is of precise degree n. For that we will require $n \geq 2m$. Thus we are missing the first $2m$ polynomials in our representation.

Assume the coefficient of $z^m\phi_{n-m}(z)$ is zero; i.e., the determinant we get from crossing out the first row and last column of our matrix is zero. Then there exist constants $\lambda_0, \lambda_1, \ldots, \lambda_{m-1}$ and $\mu_0, \mu_1, \ldots, \mu_{m-1}$, not all zero, such that if we let $\gamma(z)$ be defined by

$$\begin{aligned}\gamma(z) &:= \left(\lambda_0 + \lambda_1 + \cdots + \lambda_{m-1}z^{m-1}\right)\phi_{n-m}(z) \\ &\quad + \left(\mu_0 + \mu_1 z + \cdots + \mu_{m-1}z^{m-1}\right)\phi_{n-m}^*(z)\end{aligned}$$

we have $L_{\beta_1}(\gamma) = 0$ for every i.

This means

$$\int_{-\pi}^{\pi}\gamma(z)\,\overline{\rho_k(z)}\,\frac{1}{|G_{2m}(z)|}\,d\mu(\theta) = 0$$

for every polynomial $\rho_k(z)$ of degree $k \leq n - 1$ and, in particular, for $\gamma(z)$ as well. Thus

$$\int_{-\pi}^{\pi}|\gamma(z)|^2\,\frac{1}{|G_{2m}(z)|}\,d\mu(\theta) = 0$$

which implies that $\gamma(z) \equiv 0$. However, if $n \geq 2m$ then $\gamma(z)$ cannot be identically zero. Thus the polynomials $\{\psi(z)\}$ are constant multiples of the polynomials orthonormal with respect to $(1/|G_{2m}(z)|)\,d\nu(\theta)$. $\square$

We may combine Theorems 8.6.1 and 8.6.4 and establish the following theorem which covers the modification by a rational function.

Theorem 8.6.5 *Let $\{\phi_n(z)\}$ and μ be as in Theorem 8.6.1 and let $G_{2m}(z)$ and $H_{2k}(z)$ be polynomials of precise degrees $2m$ and $2k$, respectively, such that*

$$z^{-m}G_{2m}(z) = |G_{2m}(z)|, \quad z^{-k}H_{2k}(z) = |H_{2k}(z)| > 0, \quad |z| = 1.$$

Assume the zeros of $G_{2m}(z)$ are $\{\alpha_1, \alpha_2, \ldots, \alpha_{2m}\}$ and the zeros of $H_{2k}(z)$ are $\{\beta_1, \beta_2, \ldots, \beta_{2k}\}$. Let $\phi(z)$ denote $\phi_{n+m-k}(z)$ and $s = m + k$. For $n \geq 2k$ define $\psi_n(z)$ by

$$G_{2m}(z)\psi_n(z) = \begin{vmatrix}
\phi^*(z) & z\phi^*(z) & \cdots & z^{s-1}\phi^*(z) \\
\phi^*(\alpha_1) & \alpha_1\phi^*(\alpha_1) & \cdots & \alpha_1^{s-1}\phi^*(\alpha_1) \\
\phi^*(\alpha_2) & \alpha_2\phi^*(\alpha_2) & \cdots & \alpha_2^{s-1}\phi^*(\alpha_2) \\
\vdots & \vdots & & \vdots \\
\phi^*(\alpha_{2m}) & \alpha_{2m}\phi^*(\alpha_{2m}) & \cdots & \alpha_{2m}^{s-1}\phi^*(\alpha_{2m}) \\
L_{\beta_1}(\phi^*) & L_{\beta_1}(z\phi^*) & \cdots & L_{\beta_1}(z^{s-1}\phi^*) \\
L_{\beta_2}(\phi^*) & L_{\beta_2}(z\phi^*) & \cdots & L_{\beta_2}(z^{s-1}\phi^*) \\
\vdots & \vdots & & \vdots \\
L_{\beta_{2k}}(\phi^*) & L_{\beta_{2k}}(z\phi^*) & \cdots & L_{\beta_{2k}}(z^{s-1}\phi^*) \\
\phi(z) & z\phi(z) & \cdots & z^s\phi(z) \\
\phi(\alpha_1) & \alpha_1\phi(\alpha_1) & \cdots & \alpha_1^s\phi(\alpha_1) \\
\phi(\alpha_2) & \alpha_2\phi(\alpha_2) & \cdots & \alpha_2^s\phi(\alpha_2) \\
\vdots & \vdots & & \vdots \\
\phi(\alpha_{2m}) & \alpha_{2m}\phi(\alpha_{2m}) & \cdots & \alpha_{2m}^s\phi(\alpha_{2m}) \\
L_{\beta_1}(\phi) & L_{\beta_1}(z\phi) & \cdots & L_{\beta_1}(z^s\phi) \\
L_{\beta_2}(\phi) & L_{\beta_2}(z\phi) & \cdots & L_{\beta_2}(z^s\phi) \\
\vdots & \vdots & & \vdots \\
L_{\beta_{2k}}(\phi) & L_{\beta_{2k}}(z\phi) & \cdots & L_{\beta_{2k}}(z^s\phi)
\end{vmatrix},$$

$$(8.6.3)$$

where we define

$$L_\beta(p) := \int_{-\pi}^{\pi} p(\xi)\,\overline{(\xi^s/(\xi - \beta))}\,d\nu(\theta), \quad \xi = e^{i\theta}.$$

For zeros of $H_{2k}(z)$ of multiplicity h, $h > 1$, we replace the corresponding rows in the determinant by

$$L_\beta^r(p) := \int_{-\pi}^{\pi} p(\xi)\,\overline{(\xi^s/(\xi - \beta)^r)}\,d\nu(\theta), \quad \xi = e^{i\theta},$$

$r = 1, 2, \ldots, h$ *acting on the first row.*

For zeros of $G_{2m}(z)$ of multiplicity h, $h > 1$, we replace the corresponding row in the determinant by the derivatives of order $0, 1, 2, \ldots, h - 1$ of the polynomials *in the first row, evaluated at that zero. (As usual, $\rho_r^*(z) = z^r \bar{\rho}_r(z^{-1})$, for $\rho_r(z)$ a polynomial of degree r.)*

Then $\{\psi_n(z)\}$ are constant multiples of the polynomials orthonormal with respect to $|G_{2m}/H_{2k}(z)| \, d\nu(\theta)$ on the unit circle $z = e^{i\theta}$.

The paper (Ismail & Ruedemann, 1992) contains applications of Theorems 8.6.1, 8.6.4–8.6.5 to derive explicit formulas for certain polynomials.

Exercises

8.1   Assume that $\phi_n(0) \neq 0$, $n = 0, 1, \dots$ and that

$$\kappa_n \phi_{n+1}(0) = c \kappa_{n+1} \phi_n(0),$$

for some constant c. Prove that there is only one polynomial sequence with this property; see the Rogers–Szegő polynomials of Chapter 17.

8.2   Let $\Phi(z)$ be a monic polynomial of degree m which has all its zeros in the open unit disc. then there is a system of monic orthogonal polynomials $\{\Phi_n(z)\}$ such that $\Phi_m(z) = \Phi(z)$. Let $\{\alpha_n\}$ be the recursion coefficients of $\{\Phi_n(z)\}$. Then α_n are uniquely determined if $0 \leq n < m$. Moreover the moments $\{\mu_j : 0 \leq j \leq n\}$ are uniquely determined.

Hint: This result is in (Geronimus, 1946) and is the unit circle analogue of Wendroff's theorem, Theorem 2.10.1.

8.3   Prove that $h(z) = D(z)/D(0)$.

8.4   Let

$$M_k(z) = \begin{pmatrix} z & \overline{\alpha_k} \\ -\alpha_k z & 1 \end{pmatrix}.$$

Show that

$$\begin{pmatrix} \Phi_n(z) \\ \Phi_n^*(z) \end{pmatrix} = M_{n-1}(z) \cdots M_0(z) \begin{pmatrix} 1 \\ 1 \end{pmatrix}.$$

8.5   Fill in the details of the following proof of (8.2.6). Start with

$$\frac{\Phi_n(z) - z^n}{z^{n-1}} = \sum_{k=0}^{n-1} c_{n,k} \overline{\Phi}_k(1/z), \tag{E8.1}$$

so that $c_{n,k} \int_{-\pi}^{\pi} |\Phi_k(z)|^2 \, d\mu(\theta) = 0 - \int_{-\pi}^{\pi} z \Phi_k(z) \, d\mu(\theta)$, hence $c_{n,k}$ does not depend on n. Conclude that

$$\Phi_{n+1}(z) - z\Phi_n(z) = c_{n,n} \Phi_n^*(z),$$

and evaluate $c_{n,n}$, (Akhiezer, 1965, §5.2).

8.6   Prove that Theorem 8.1.1 holds if $\pi(x)$ is assumed to have degree at most n.

8.7   Prove Theorem 8.2.2.

9

Linearization, Connections and Integral Representations

In this chapter, we study connection coefficients of several orthogonal polynomials and the coefficients in the linearization coefficients of products of two or more polynomials. Interesting combinatorial and positivity questions arise in this context and some of them are treated in this chapter. Continuous analogues of these are integral representations and product formulas. These are also treated.

Given a system of orthogonal polynomials $\{P_n(x; \mathbf{a})\}$ depending on parameters $\alpha_1, \ldots, \alpha_s$, an interesting question is to find the connection coefficients $c_{n,k}(\mathbf{a}, \mathbf{b})$ in the expansion

$$P_n(x; \mathbf{b}) = \sum_{k=0}^{n} c_{n,k}(\mathbf{a}, \mathbf{b}) P_k(x; \mathbf{a}). \tag{9.0.1}$$

We use the vector notation

$$\mathbf{a} = (a_1, \ldots, a_s), \quad \mathbf{b} = (b_1, \ldots, b_s). \tag{9.0.2}$$

Another problem is to say something about the linearization coefficients $c_{m,n,k}$ in

$$P_m(x; \mathbf{a}) P_n(x; \mathbf{a}) = \sum_{k=0}^{m+n} c_{m,n,k}(\mathbf{a}) P_k(x; \mathbf{a}). \tag{9.0.3}$$

When we cannot find the coefficients explicitly, one usually tries to find sign patterns, or unimodality conditions satisfied by the coefficients. Evaluating the linearization coefficients in (9.0.3) amounts to evaluating the integrals

$$c_{m,n,k}(\mathbf{a}) \zeta_k(\mathbf{a}) = \int_{\mathbb{R}} P_m(x; \mathbf{a}) P_n(x; \mathbf{a}) P_k(x; \mathbf{a}) \, d\mu(x; \mathbf{a}), \tag{9.0.4}$$

where

$$\int_{\mathbb{R}} P_m(x; \mathbf{a}) P_n(x; \mathbf{a}) \, d\mu(x; \mathbf{a}) = \zeta_n(\mathbf{a}) \delta_{m,n}. \tag{9.0.5}$$

This raises the question of studying sign behavior of the integrals

$$I(n_1, \ldots, n_k) := \int_{\mathbb{R}} P_{n_1}(x; \mathbf{a}) \cdots P_{n_k}(x; \mathbf{a}) \, d\mu(x, \mathbf{a}). \tag{9.0.6}$$

Observe that (9.0.1) is a finite dimensional problem. Assume $d\mu(x; \mathbf{a}) = w(x; \mathbf{a}) dx$.

When we can evaluate $c_{n,k}(\mathbf{a}, \mathbf{b})$ then we have solved the infinite dimensional expansion problem of expanding $w(x; \mathbf{a})P_k(x; \mathbf{b})/w(x; \mathbf{a})$ in $\{P_n(x; \mathbf{a})\}$. In fact,

$$w(x; \mathbf{a})P_k(x; \mathbf{b}) \sim \sum_{n=k}^{\infty} c_{n,k}(\mathbf{a}, \mathbf{b})P_n(x; \mathbf{a})w(x; \mathbf{a}). \tag{9.0.7}$$

Polynomials with nonnegative linearization coefficients usually have very special properties. One such property is that they lead to convolution structures as we shall see later in this chapter. One special property is stated next as a theorem.

Theorem 9.0.1 *Let $\{p_n(x)\}$ be orthonormal with respect to μ and assume μ is supported on a subset of $(-\infty, \xi]$. Also assume that*

$$p_n^N(x) = \sum_k c(k, N, n)p_k(x), \quad c(k, N, n) \geq 0. \tag{9.0.8}$$

Then

$$|p_n(x)| \leq p_n(\xi), \quad \mu\text{-almost everywhere.} \tag{9.0.9}$$

Proof Using the fact that the zeros of p_n lie in $(-\infty, \xi)$ for all n we find

$$\int_{\mathbb{R}} p_n^{2N}(x)\, d\mu(x) = c(0, 2N, n) \leq \sum_{k \geq 0} c(k, 2N, n)p_k(\xi) = p_n^{2N}(\xi).$$

Therefore

$$\left| \int_{\mathbb{R}} p_n^{2N}(x)\, d\mu(x) \right|^{1/(2N)} \leq p_n(\xi).$$

By letting $N \to \infty$ and using the fact that the L_∞ norm is the limit of the L_p norm as $p \to \infty$, we establish (9.0.9). $\qquad\square$

One approach to evaluate connection coefficients is to think of (9.0.1) as a polynomial expansion problem. A general formula for expanding a hypergeometric function in hypergeometric polynomials was established in (Fields & Wimp, 1961). This was generalized in (Verma, 1972) to

$$\sum_{m=0}^{\infty} a_m b_m \frac{(zw)^m}{m!} = \sum_{n=0}^{\infty} \frac{(-z)^n}{n!\,(\gamma + n)_n} \left(\sum_{r=0}^{\infty} \frac{b_{n+r}\, z^r}{r!\,(\gamma + 2n + 1)_r} \right)$$
$$\times \left[\sum_{s=0}^{n} \frac{(-n)_s (n + \gamma)_s}{s!} a_s w^s \right]. \tag{9.0.10}$$

When z, w are replaced by $z\gamma$ and w/γ, respectively, we establish the companion formula

$$\sum_{m=0}^{\infty} a_m b_m (zw)^m = \sum_{n=0}^{\infty} \frac{(-z)^n}{n!} \left[\sum_{j=0}^{\infty} \frac{b_{n+j}}{j!} z^j \right] \left[\sum_{k=0}^{n} (-n)_k a_k w^k \right], \tag{9.0.11}$$

by letting $\gamma \to \infty$. Fields and Ismail (Fields & Ismail, 1975) showed how to derive (9.0.10) and other identities from generating functions of Boas and Buck type. This essentially uses Lagrange inversion formulas. q-analogues are in (Gessel & Stanton, 1983) and (Gessel & Stanton, 1986).

Proof of (9.0.10) Let

$$P_n(w) = \frac{(\gamma)_n}{n!} \sum_{s=0}^{n} \frac{(-n)_s (n+\gamma)_s}{s! \, (\gamma)_{2s}} a_s (4w)^s. \tag{9.0.12}$$

Therefore

$$\sum_{n=0}^{\infty} P_n(w) \, t^n = \sum_{n,s=0}^{\infty} (-4tw)^s \frac{a_s}{s!} \sum_{n=0}^{\infty} \frac{(\gamma+2s)_n}{n!} \, t^n,$$

and we have established the generating function

$$\sum_{n=0}^{\infty} P_n(w) \, t^n = (1-t)^{-\gamma} \sum_{s=0}^{\infty} \frac{(-4tw)^s}{(1-t)^{2s}} \frac{a_s}{s!}. \tag{9.0.13}$$

Set $u = -4t/(1-t)^2$, and choose the branch $t = -u \left[1 + \sqrt{1-u} \right]^{-2}$, and $u \in (-1,1)$. This makes $1 - t = 2 \left(1 + \sqrt{1-u} \right)^{-1}$. Hence (1.4.14) leads to

$$\sum \frac{a_s}{s!} w^s u^s = \sum_{n=0}^{\infty} \frac{(-u)^n}{4^n} \left[\frac{1 + \sqrt{1-u}}{2} \right]^{-\gamma-2n} P_n(w)$$

$$= \sum_{n=0}^{\infty} \frac{(-u)^n}{4^n} \, {}_2F_1 \left(\begin{matrix} n+\gamma/2, \, n+(\gamma+1)/2 \\ 2n+\gamma+1 \end{matrix} \middle| \, u \right) P_n(w).$$

Consequently, we find

$$\frac{a_m w^m}{m!} = \sum_{s+n=m} \frac{(-1)^n (2n+\gamma)_{2s}}{4^n s! \, (2n+\gamma+1)_s} P_n(w). \tag{9.0.14}$$

Therefore, the left-hand side of (9.0.10) is

$$\sum_{n,s=0}^{\infty} \frac{(-1)^n (2n+\gamma)_{2s}}{4^n s! \, (2n+\gamma+1)_s} b_{n+s} \, z^{n+s} P_n(w),$$

and (9.0.10) follows if we replace a_m and b_m by $(\gamma)_{2m} a_m/4^m$, and $4^m b_m/(\gamma)_{2m}$, respectively. $\qquad \square$

Observe that the whole proof rested on the inverse relations (9.0.12) and (9.0.14). This works when $\{P_n(w)\}$ has a generating function of Boas and Buck type, see (Fields & Ismail, 1975).

9.1 Connection Coefficients

We now solve the connection coefficient problem of expressing a Jacobi polynomial as a series in Jacobi polynomials with different parameters.

Theorem 9.1.1 *The connection relation for Jacobi polynomials is*

$$P_n^{(\gamma,\delta)}(x) = \sum_{k=0}^{n} c_{n,k}(\gamma,\delta;\alpha,\beta)\, P_k^{(\alpha,\beta)}(x),$$

$$c_{n,k}(\gamma,\delta;\alpha,\beta) = \frac{(\gamma+k+1)_{n-k}(n+\gamma+\delta+1)_k}{(n-k)!\,\Gamma(\alpha+\beta+2k+1)}\,\Gamma(\alpha+\beta+k+1) \quad (9.1.1)$$

$$\times {}_3F_2\left(\begin{array}{c} -n+k,\, n+k+\gamma+\delta+1,\, \alpha+k+1 \\ \gamma+k+1,\, \alpha+\beta+2k+2 \end{array}\middle|\, 1\right).$$

In particular,

$$C_n^{\gamma}(x) = \sum_{k=0}^{\lfloor n/2 \rfloor} \frac{(\gamma-\beta)_k (\gamma)_{n-k}}{k!\,(\beta+1)_{n-k}} \left(\frac{\beta+n-2k}{\beta}\right) C_{n-2k}^{\beta}(x). \qquad (9.1.2)$$

Proof From the orthogonality relation (4.1.2) and the Rodrigues formula (4.2.8), integration by parts and the use of (4.2.2) and (4.1.1), we find that $h_k^{(\alpha,\beta)}\, c_{n,k}(\gamma,\delta;\alpha,\beta)$ is given by

$$\int_{-1}^{1} P_n^{(\gamma,\delta)}(x) P_k^{(\alpha,\beta)}(x)(1-x)^{\alpha}(1+x)^{\beta}\, dx$$

$$= \frac{(-1)^k}{2^k k!} \int_{-1}^{1} P_n^{(\gamma,\delta)}(x) \frac{d^k}{dx^k}\left[(1-x)^{\alpha+k}(1+x)^{\beta+k}\right] dx$$

$$= \frac{1}{2^k k!} \int_{-1}^{1} (1-x)^{\alpha+k}(1+x)^{\beta+k} \frac{d^k}{dx^k} P_n^{(\gamma,\delta)}(x)\, dx$$

$$= \frac{(n+\gamma+\delta+1)_k}{4^k k!} \int_{-1}^{1} P_{n-k}^{(\gamma+k,\delta+k)}(x)(1-x)^{\alpha+k}(1+x)^{\beta+k}\, dx.$$

The above expression is

$$= \frac{(\gamma+k+1)_{n-k}(n+\gamma+\delta+1)_k}{4^k (n-k)!\, k!} \sum_{j=0}^{n-k} \frac{(k-n)_j (\gamma+\delta+1+n+k)_j}{2^j\,(\gamma+k+1)_j\, j!}$$

$$\times \int_{-1}^{1} (1-x)^{\alpha+k+j}(1+x)^{\beta+k}\, dx$$

$$= \frac{(\gamma+k+1)_{n-k}(n+\gamma+\delta+1)_k}{2^{-\alpha-\beta-1}(n-k)!\, k!} \frac{\Gamma(\alpha+k+1)\Gamma(\beta+k+1)}{\Gamma(\alpha+\beta+2k+2)}$$

$$\times {}_3F_2\left(\begin{array}{c} -n+k,\, n+k+\gamma+\delta+1,\, \alpha+k+1 \\ \gamma+k+1,\, \alpha+\beta+2k+2 \end{array}\middle|\, 1\right).$$

The theorem now follows from (4.1.3). $\qquad\qquad\square$

Observe that the symmetry relation (4.2.4) and (9.1.1) imply the following transformation between $_3F_2$ functions

$$(-1)^n (\gamma + k + 1)_{n-k} \, _3F_2 \left(\begin{matrix} -n+k, n+\gamma+\delta+1, \alpha+k+1 \\ \gamma+k+1, \alpha+\beta+2k+1 \end{matrix} \middle| 1 \right)$$

$$= (-1)^k (\delta + k + 1)_{n-k} \, _3F_2 \left(\begin{matrix} -n+k, n+\gamma+\delta+1, \beta+k+1 \\ \delta+k+1, \alpha+\beta+2k+1 \end{matrix} \middle| 1 \right). \tag{9.1.3}$$

Theorem 9.1.1 also follows from (9.0.10) with the parameter identification

$$\gamma = \alpha + \beta + 1, \quad a_s = 1/(\alpha+1)_s, \quad w = (1-x)/2, \quad z = 1,$$
$$b_s = (-N)_s (N + \gamma + \delta + 1)_s (\alpha+1)_s / (\gamma+1)_s.$$

Corollary 9.1.2 *We have the connection relation*

$$P_n^{(\alpha,\delta)}(x) = \sum_{k=0}^{n} d_{n,k} P_k^{(\alpha,\beta)}(x) \tag{9.1.4}$$

with

$$d_{n,k} = \frac{(\alpha+k+1)_{n-k}(n+\alpha+\delta+1)_k \Gamma(\alpha+\beta+k+1)}{(n-k)! \, \Gamma(\alpha+\beta+n+k+1)}$$
$$\times (\beta - \delta + 2k - n)_{n-k}. \tag{9.1.5}$$

Let $c_{n,k}(\gamma, \delta; \alpha, \beta)$ be as in (9.1.2). By interating (9.1.1) we discover the orthogonality relation

$$\delta_{n,j} = \sum_{k=j}^{n} c_{n,k}(\gamma, \delta; \alpha, \beta) c_{k,j}(\alpha, \beta; \gamma, \delta). \tag{9.1.6}$$

In other words

$$\delta_{n,j} = \frac{(\gamma+1)_n (\alpha+\beta+1)_j}{(\alpha+1)_j (\gamma+\delta+j+1)_j}$$

$$\times \sum_{k=j}^{n} \frac{(n+\gamma+\delta+1)_k (\alpha+1)_k (\alpha+\beta+j+1)_k}{(\gamma+1)_k (\alpha+\beta+1)_{2k} (n-k)! (k-j)!}$$

$$\times \, _3F_2 \left(\begin{matrix} k-n, k+n+\gamma+\delta+1, k+1+\alpha \\ k+\gamma+1, 2k+\alpha+\beta+1 \end{matrix} \middle| 1 \right)$$

$$\times \, _3F_2 \left(\begin{matrix} -k+j, k+j+\alpha+\beta+1, \gamma+j+1 \\ \alpha+j+1, \gamma+\delta+2j+1 \end{matrix} \middle| 1 \right).$$

Moreover, we also have

$$\sum_{k=j}^{n} c_{n,k}(\gamma, \delta, \alpha, \beta) c_{k,j}(\alpha, \beta, \rho, \sigma) = c_{n,j}(\gamma, \delta, \rho, \sigma). \tag{9.1.7}$$

Indeed, (9.1.6) corresponds to $\alpha = \rho$ and $\beta = \sigma$.

The Wilson polynomials are

$$W_n(x; a, b, c, d) = \frac{(a+b)_n(a+c)_n(a+d)_n}{a^n}$$

$$\times {}_4F_3 \left(\begin{array}{c} -n, n+a+b+c+d-1, a+i\sqrt{x}, a-i\sqrt{x} \\ a+b, a+c, a+d \end{array} \middle| 1 \right). \quad (9.1.8)$$

They were introduced by James Wilson in (Wilson, 1980). Applying (9.0.10) with

$$\gamma = a+b+c+d-1,$$

$$a_s = \frac{(a+i\sqrt{x})_s (a-i\sqrt{x})_s}{(a+b)_s(a+c)_s(a+d)_s},$$

$$b_s = \frac{(-N)_s(a+b)_s(a+c)_s(a+d)_s}{(a+b')_s (a+c')_s (a+d')_s} (a+b'+c'+d'+N-1),$$

to prove that

$$W_n (x; a, b', c', d') = \sum_{k=0}^{n} c_k W_k(x; a, b, c, d), \quad (9.1.9)$$

with

$$c_k = (a+b'+c'+d'+N-1)_k$$

$$\times \frac{n! (a+b'+k)_{n-k} (a+c'+k)_{n-k} (a+d'+k)_{n-k}}{a^{n-k} k! (a+b+c+d+k-1)_k (n-k)!}$$

$$\times {}_5F_4 \left(\begin{array}{c} -n+k, a+b+k, a+c+k, a+d+k, a+b'+c'+d'+N+k \\ a+b'+k, a+c'+k, a+d'+k, a+b+c+d+2k-1 \end{array} \middle| 1 \right). \quad (9.1.10)$$

We next state and prove two general theorems on the nonnegativity of the connection coefficients.

Theorem 9.1.3 ((Wilson, 1970)) *Let $\{p_n(x)\}$ and $\{s_n(x)\}$ be polynomial sequences with positive leading terms and assume that $\{p_n(x)\}$ is orthonormal with respect to μ. If*

$$\int_{\mathbb{R}} s_m(x)s_n(x) \, d\mu(x) \le 0, \quad n \ne m,$$

then

$$p_n(x) = \sum_{k=0}^{n} c_{n,k} s_k(x), \quad \text{with } a_{n,k} \ge 0. \quad (9.1.11)$$

Proof Clearly $a_{n,n} > 0$. Let

$$I_{jk} := \int_{\mathbb{R}} s_j(x) s_k(x) \, d\mu(x).$$

By orthogonality, for $j < n$,

$$0 = \int_{\mathbb{R}} p_n(x) s_j(x) \, d\mu(x) = \sum_{k=0}^{n} c_{n,k} I_{j,k}.$$

For fixed n, let $\mathbf{X} = (c_{n,0}, c_{n,1}, \ldots, c_{n,n})^T$ and denote $\int_{\mathbb{R}} p_n(x) s_n(x)\, d\mu(x)$ by β.
Thus

$$\beta = \int_{\mathbb{R}} p_n(x)\,(u_n x^n + \cdots)\, d\mu(x) = \frac{u_n}{\gamma_n} \int_{\mathbb{R}} P_n^2(x)\, d\mu(x) = \frac{u_n}{\gamma_n} > 0,$$

where $p_n(x) = \gamma_n x^n + \cdots$. We now choose $\mathbf{Y} = (0, 0, \ldots, 0, 1)^T \in \mathbb{R}^{n+1}$,
$A = (I_{jk} : 0 \leq j, k \leq n)$, so that $A\mathbf{X} = \mathbf{Y}$. Since A is symmetric, positive definite
and all its off-diagonal elements are negative, it is a Stieltjes matrix (see (Varga,
2000, p. 85). By Corollary 3, on page 85 of (Varga, 2000), all elements of A^{-1} are
nonnegative, hence $\mathbf{X}$ has nonnegative components and the theorem follows. $\qquad\square$

Theorem 9.1.4 ((Askey, 1971)) *Let $\{P_n(x)\}$ and $\{Q_n(x)\}$ be monic orthogonal
polynomials satisfying*

$$\begin{aligned}
xQ_n(x) &= Q_{n+1}(x) + A_n Q_n(x) + B_n Q_{n-1}(x), \\
xP_n(x) &= P_{n+1}(x) + \alpha_n P_n(x) + \beta_n P_{n-1}(x).
\end{aligned} \tag{9.1.12}$$

If

$$A_k \leq \alpha_n, \quad B_{k+1} \leq \beta_{n+1}, \quad 0 \leq k \leq n,\ n \geq 0, \tag{9.1.13}$$

then

$$Q_n(x) = \sum_{k=0}^{n} c_{n,k} P_k(x), \tag{9.1.14}$$

with $c_{n,k} \geq 0$, $0 \leq k \leq n$, $n \geq 0$.

Proof It is clear that $c_{n,n} = 1$. Assume (9.1.14) and use (9.1.12), with the convention
$c_{n,n+1} = c_{n,-1} = P_{k-1} := 0$, to get

$$\begin{aligned}
Q_{n+1}(x) &= xQ_n(x) - A_n Q_n(x) - B_n Q_{n-1}(x) \\
&= (x - A_n) \sum_{k=0}^{n} c_{n,k} P_k(x) - B_n \sum_{k=0}^{n-1} c_{n-1,k} P_k(x) \\
&= \sum_{k=0}^{n} c_{n,k} \left[P_{k+1}(x) + \beta_k P_{k-1}(x) + (\alpha_k - A_n) P_k(x) \right] \\
&\quad - B_n \sum_{k=0}^{n-1} c_{n-1,k} P_k(x) \\
&= P_{n+1}(x) + \sum_{k=0}^{n-1} \left[c_{n,k-1} + c_{n,k+1} \beta_{k+1} \right. \\
&\quad \left. + (\alpha_k - A_n)\, c_{n,k} - B_n c_{n-1,k} \right] P_k(x).
\end{aligned}$$

Therefore

$$c_{n+1,n} = c_{n,n-1} + (\alpha_n - A_n), \tag{9.1.15}$$

$$c_{n+1,k} = c_{n,k-1} + c_{n,k+1}\beta_{k+1} + (a_k - A_n)\, c_{n,k} - B_n c_{n-1,k}, 0 < k < n, \tag{9.1.16}$$

$$c_{n+1,0} = c_{n,1}\beta_1 + (a_0 - A_n)\, c_{n,0} - B_n c_{n-1,0},$$

hence

$$c_{n+1,0} = (\alpha_0 - A_n)\, c_{n,0} + (\beta_1 - B_n)\, c_{n,1} + B_n\,(c_{n,1} - c_{n-1,0}). \tag{9.1.17}$$

We proceed by induction on n to establish $c_{n,k} \geq 0$. Assume $c_{m,k} \geq 0$ has been proved for all $k \leq m$, $m \leq n$ and consider $c_{n+1,k}$. Clearly, $c_{n+1,n+1} = 1 > 0$. If $k = n$, then $c_{n+1,n} \geq 0$ follows from (9.1.15). If $0 \leq k < n$, rewrite (9.1.16) as

$$c_{n+1,k} - c_{n,k-1} = (\alpha_k - A_n)\, c_{n,k} + (\beta_{k+1} - B_n)$$
$$\times\, c_{n,k+1} + B_n\,(c_{n,k+1} - c_{n-1,k}).$$

Since the first two terms on the above right-hand side are ≥ 0, we obtain

$$c_{n+1,k} - c_{n,k-1} \geq B_n\,[c_{n,k+1} - c_{n-1,k}],$$

and by iteration

$$c_{n+1,k} - c_{n,k-1} \geq B_n B_{n-1} \cdots B_{n-j}\,[c_{n-j,k+j+1} - c_{n-j-1,k+1}]. \tag{9.1.18}$$

Choosing $j = \lfloor(n-k-1)/2\rfloor$ we either have $k+j+1 = n-j$ or $k+j+2 = n-j$. In the first case, the quantity in $[\,]$ in (9.1.18) vanishes and $c_{n+k,k\geq0}$ follows. In the second case, we see that the quantity in $[\,]$ is

$$c_{(n+k+2)/2,(n+k)/2} - c_{(n+k)/2,(n+k-2)/2} = \alpha_{(n+k)/2} - A_{(n+k)/2} \geq 0,$$

where we applied (9.1.15) in the last step.   □

Szwarc observed that although the nonnegativity of the linearization and connection coefficients is invariant under $P_n(x) \rightarrow c_n P_n(\lambda x)$, $c_n > 0$, sometimes a renormalization simplifies the study of the linearization or connection coefficients. In general, we write the three term recurrence relation as

$$x\varphi_n(x) = A_n\varphi_{n+1}(x) + B_n\varphi_n(x) + C_n\varphi_{n-1}(x), \quad n \geq 0, \tag{9.1.19}$$

with $C_0\varphi_{-1}(x) := 0$, and $A_n > 0$, $C_{n+1} > 0$, $n = 0, 1, \ldots$.

Theorem 9.1.5 ((Szwarc, 1992a)) *Let $\{r_n(x)\}$ and $\{s_n(x)\}$ satisfy $r_0(x) = s_0(x) = 1$, and*

$$xr_n(x) = A_n r_{n+1}(x) + B_n r_n(x) + C_n r_{n-1}(x),$$
$$xs_n(x) = A'_n s_{n+1}(x) + B'_n s_n(x) + C'_n s_{n-1}(x), \tag{9.1.20}$$

for $n \geq 0$ with $C_0 r_{-1}(x) = C'_0 s_{-1}(x) := 0$. Assume that

(i) $C'_m \geq C_n$ *for* $m \leq n$,
(ii) $B'_m \geq B_m$ *for* $m \leq n$,

(iii) $C'_m + A'_m \geq C_n + A_n$ for $m \leq n$,

(iv) $A'_m \geq C_n$ for $m < n$.

Then the connection coefficients $c(n, k)$ *in*

$$r_n(x) = \sum_{k=0}^{n} c(n, k) s_k(x), \tag{9.1.21}$$

are nonnegative.

The proof in (Szwarc, 1992a) uses discrete boundary value problems. Szwarc also used the same technique to give a proof of Askey's theorem, Theorem 9.1.4.

Corollary 9.1.6 ((Szwarc, 1992a)) *Assume that* $\{r_n(x)\}$ *are generated by* (9.1.20), *for* $n \geq 0$, *and* $r_0(x) = 1$, $r_{-1}(x) = 0$. *If* $C_n \leq 1/2$, $A_n + C_n \leq 1$, $B_n \leq 0$. *Then* $r_n(x)$ *can be represented as a linear combination of the Chebyshev polynomials of the first and second kinds.*

Corollary 9.1.7 ((Szwarc, 1992a)) *Let* E *denote the closure of the area enclosed by the ellipse whose foci are* ± 1. *Under the assumptions of Corollary 9.1.6, the max of* $|\varphi_n(z)|$ *for* $z \in E$, *is attained at the right endpoint of the major axis.*

Theorem 9.1.8 ((Szwarc, 1992a)) *Let* $\{r_n(x)\}$ *and* $\{s_n(x)\}$ *be as in Theorem 9.1.5 with* $B_n = B'_n = 0$, $n \geq 0$. *Assume that*

(i) $C'_{2m} \geq C_{2n}$ *and* $C'_{2m+1} \geq C_{2n+1}$, *for* $0 < m \leq n$,

(ii) $A'_{2m} + C'_{2m} \geq A_{2n} + C_{2n}$, *and* $A'_{2m+1} + C'_{2m+1} \geq A_{2n+1} + C_{2n+1}$, *for* $m \leq n$,

(iii) $A'_{2m} > A_{2n}$ *and* $A'_{2n+1} \geq A_{2m+1}$ *for* $m < n$.

Then the connection coefficients in (9.1.21) *are nonnegative. The same conclusion holds if* (i)–(iii) *are replaced by*

(a) $C_1 \geq C'_1 \geq C_2 \geq C'_2 \geq \cdots$, $B_0 \geq B'_0 \geq B_1 \geq B'_1 \geq \cdots$,

(b) $A_0 + C_0 \geq A'_0 + C'_0 \geq A_1 + C_1 \geq A'_1 + C'_1 \geq \cdots$,

(c) $A'_m \geq C_n$ *for* $m < n$.

9.2 The Ultraspherical Polynomials and Watson's Theorem

We evaluate the connection coefficients for the ultraspercial polynomials in two different ways and by equating the answers we discover the terminating version of Watson's theorem (1.4.11).

Theorem 9.2.1 *The ultraspherical polynomials satisfy the connection relation*

$$C_n^\lambda(x) = \sum_{k=0}^{n} a_{n,k}(\lambda, \nu) C_{n-2k}^\nu(x), \tag{9.2.1}$$

where

$$a_{n,k}(\lambda, \nu) = \frac{(n + \nu - 2k)\Gamma(\nu)\,\Gamma(\lambda + k - \nu)\,\Gamma(n - k + \lambda)}{\Gamma(\lambda)\Gamma(\lambda - \nu)\Gamma(n - k + \nu + 1)\,k!}. \tag{9.2.2}$$

Proof From (4.5.4) we get

$$\frac{(2\nu)_{n-2k}\Gamma(1/2)\Gamma(\nu+1/2)}{(n-2k)!\,(\nu+n-2k)\Gamma(\nu)}a_{n,k}(\lambda,\nu)$$

$$=\int_{-1}^{1}C_n^\lambda(x)C_{n-2k}^\nu(x)\left(1-x^2\right)^{\nu-1/2}\,dx. \tag{9.2.3}$$

First assume that $\nu>1/2$. Apply (4.5.12) with n replaced by $n-2k$, integrate by parts $n-2k$ times then apply (4.5.5) to see that the right-hand side of equation (9.2.3) is

$$\frac{2^{2k-n}\,(2\nu)_{n-2k}}{(n-2k)!\,(\nu+1/2)_{n-2k}}\int_{-1}^{1}\left(1-x^2\right)^{\nu+n-2k-1/2}\frac{d^{n-2k}}{dx^{n-2k}}C_n^\lambda(x)\,dx$$

$$=\frac{(\lambda)_{n-2k}\,(2\nu)_{n-2k}}{(n-2k)!\,(\nu+1/2)_{n-2k}}\int_{-1}^{1}\left(1-x^2\right)^{\nu+n-2k-1/2}C_{2k}^{\lambda+n-2k}(x)\,dx$$

Insert the representation (4.5.16) for $C_{2k}^{\lambda+n-2k}(x)$ to see that the above expression is

$$\frac{(\lambda)_{n-2k}\,(2\nu)_{n-2k}(2\lambda+2n-4k)_{2k}}{(n-2k)!\,(\nu+1/2)_{n-2k}}\sum_{j=0}^{k}\frac{(-1)^j2^{-2j}/j!}{(\lambda+n-2k+1/2)_j(2k-2j)!}$$

$$\times2\int_{0}^{1}\left(1-x^2\right)^{\nu+n-2k+j-1/2}x^{2k-2j}dx$$

$$=\frac{(\lambda)_{n-2k}\,(2\nu)_{n-2k}(2\lambda+2n-4k)_{2k}}{(n-2k)!\,(\nu+1/2)_{n-2k}}\frac{\Gamma(1/2)\Gamma(\nu+n-2k+1/2)}{2^{2k}k!\,\Gamma(\nu+n-k+1)}$$

$$\times{}_2F_1(-k,n-2k+\nu+1/2;\lambda+n-2k+1/2;1)$$

$$=\frac{(\lambda)_{n-2k}\,(2\nu)_{n-2k}(\lambda+n-2k)_k}{(n-2k)!\,(\nu+1/2)_{n-2k}}\frac{\Gamma(1/2)\Gamma(\nu+n-2k+1/2)(\lambda-\nu)_k}{k!\,\Gamma(\nu+n-k+1)}$$

$$=\frac{(\lambda)_{n-k}\,(2\nu)_{n-2k}\Gamma(1/2)\Gamma(\nu+1/2)}{(n-2k)!\,k!\,\Gamma(\nu+n-k+1)}(\lambda-\nu)_k.$$

Equating the above expression with the left-hand side of (9.2.3) we establish (9.2.2) for $\nu>1/2$. This restriction can then be removed by analytic continuation and the proof is complete. $\square$

We now come to Watson's theorem. Using (9.1.1) and (4.5.1) we see that $c_{n,n-k}(\lambda-1/2,\lambda-1/2;\nu-1/2,\nu-1/2)=0$ if k is odd. That is

$${}_3F_2\left(\begin{array}{c}-2k-1,2n-2k-1+2\lambda,n-2k+\nu-1/2\\n-2k+\lambda-1/2,2\nu+2n-4k-1\end{array}\bigg|1\right)=0. \tag{9.2.4}$$

On the other hand when k is even

$$c_{n,n-2k}(\lambda-1/2,\lambda-1/2;\nu-1/2,\nu-1/2)$$

$$=\frac{(2\nu)_{n-2k}(\lambda+1/2)_n}{(\nu+1/2)_{n-2k}(2\lambda)_n}a_{n,k}(\lambda,\nu).$$

Therefore

$$
{}_3F_2\left(\begin{array}{c} -2k, 2n-2k+2\lambda, n-2k+\nu+1/2 \\ n-2k+\lambda+1/2, 2\nu+2n-4k+1 \end{array} \middle| 1\right)
$$
$$
= \frac{(2\nu)_{n-2k}(\lambda+1/2)_n}{(\nu+1/2)_{n-2k}(2\lambda)_n} \frac{(n+\nu-2k)\Gamma(\nu)(\lambda-\nu)_k(\lambda)_{n-k}}{k!\Gamma(\nu+n-k+1)}.
$$

(9.2.5)

Formulas (9.2.4) and (9.2.5) establish the terminating form of Watson's theorem (1.4.12).

9.3 Linearization and Power Series Coefficients

As already mentioned in §9.0, it is desirable to study sign behavior of integrals of the type (9.0.5). Through generating functions we can transform this problem to investigating coefficients in power series expansions of certain functions. One such generating function arose in the early 1930s when K. O. Friedrichs and H. Lewy studied the discretization of the time dependent wave equation in two dimensions. To prove the convergence of the finite diference scheme to a solution of the wave equation, they needed the nonegativity of the coefficients $A(k, m, n)$ in the expansion

$$
\frac{1}{(1-r)(1-s)+(1-r)(1-t)+(1-s)(1-t)}
$$
$$
= \sum_{k,m,n=0}^{\infty} A(k,m,n) r^k s^m t^n.
$$

(9.3.1)

G. Szegő (Szegő, 1933) solved this problem using the Sonine integrals for Bessel functions and observed that

$$
A(k,m,n) = \int_0^\infty e^{-3x} L_k(x) L_m(x) L_n(x)\,dx,
$$

(9.3.2)

where

$$
L_n(x) = L_n^{(0)}(x).
$$

(9.3.3)

Therefore, in view of (4.6.2), the Friedrichs–Lewy problem is equivalent to showing that the linearization coefficients in

$$
e^{-2x} L_m(x)\, e^{-2x} L_n(x) = \sum_{k=0}^\infty A(k,m,n) e^{-2x} L_k(x).
$$

(9.3.4)

Szegő raised the question of proving the nonnegativity of $A(k, m, n)$ directly from (9.3.4) (Szegő, 1933). Askey and Gasper (Askey & Gasper, 1972) observed that that the nonnegativity of $c_{m,n,k}(a)$ implies the nonnegativity of $c_{m,n,k}(b)$ for $b > a$, where

$$
e^{-ax} L_n^{(\alpha)}(x) e^{-ax} L_n^{(\alpha)}(x) e^{-ax} = \sum_{k=0}^\infty c_{m,n,k}(a) e^{-ax} L_k^{(\alpha)}(x).
$$

(9.3.5)

To see this, observe that $c_{m,n,k}(a)$ is a positive multiple of

$$\int_0^\infty x^\alpha e^{-(a+1)x} L_k^{(\alpha)}(x) L_m^{(\alpha)}(x) L_n^{(\alpha)}(x)\, dx$$

$$= (a+1)^{-\alpha-1} \int_0^\infty x^\alpha e^{-x} L_k^{(\alpha)}\left(\frac{x}{a+1}\right) L_m^{(\alpha)}\left(\frac{x}{a+1}\right) L_n^{(\alpha)}\left(\frac{x}{a+1}\right)\, dx.$$

The Askey–Gasper observation now follows from Theorem 4.6.5. Formulas like (9.3.5) suggest that we consider the numbers

$$A^{(\alpha)}(n_1,\ldots,n_k;\mu) := \int_0^\infty \frac{x^\alpha e^{-\mu x}}{\Gamma(\alpha+1)} L_{n_1}^{(\alpha)}(x) \cdots L_{n_k}^{(\alpha)}(x)\, dx, \qquad (9.3.6)$$

with $\alpha > -1$. The generating function (4.6.4) and the Gamma function integral establish the generating function

$$\sum_{j=1}^k \sum_{n_j=0}^\infty A^{(\alpha)}(n_1,\ldots,n_k;\mu)\, t_1^{n_1} \cdots t_k^{n_k}$$

$$= \prod_{j=1}^k (1-t_j)^{-\alpha-1} \left[\mu + \sum_{j=1}^k t_j/(1-t_j)\right]^{-\alpha-1} \qquad (9.3.7)$$

$$= \left[\mu + \sum_{j=1}^k (-1)^j (\mu-j)\sigma_j\right]^{-\alpha-1},$$

where σ_j is the jth elementary symmetric function of $t_1,\ldots,t_k$.

Askey and Gasper also raised the question of finding the smallest μ for which $A^{(\alpha)}(k,m,n;\mu) \geq 0$, for $\alpha \geq 0$. In §9.4 we shall see that

$$(-1)^{k+m+n} A^{(\alpha)}(k,m,n;1) \geq 0, \qquad \alpha \geq 0. \qquad (9.3.8)$$

On the other hand Askey and Gasper proved $A^\alpha(k,m,n;2) \geq 0$, for $\alpha \geq 0$ (Askey & Gasper, 1977), so the smallest $\mu \in (1,2]$.

We next prove a result of (Gillis et al., 1983) which is useful in establishing inequalities of power series coefficients.

Theorem 9.3.1 *Assume that $F(x_1,\ldots,x_{n-1})$ and $G(x_1,\ldots,x_{n-1})$ are polynomials. Assume further that*

$$\text{(i)} \qquad \frac{1}{F(x_1,\ldots,x_{n-1}) - x_n G(x_1,\ldots,x_{n-1})},$$

$$\text{(ii)} \qquad [F(x_1,\ldots,x_{n-1})]^{-\alpha}$$

have nonnegative power series coefficients, for all $\alpha > 0$. Then

$$[F(x_1,\ldots,x_{n-1}) - x_n G(x_1,\ldots,x_{n-1})]^{-\beta} \qquad (9.3.9)$$

has nonnegative power series coefficients for $\beta \geq 1$.

Proof The power series expansion of the function in (i) is

$$\sum_{k=0}^{\infty} \frac{(G(x_1,\ldots,x_{n-1}))^k}{(F(x_1,\ldots,x_{n-1}))^{k+1}} x_n^k,$$

and (i) implies the nonnegativity of the power series coefficients in G^k/F^{k+1}. Therefore

$$[F - x_n G]^{-\beta} = F^{1-\beta} \sum_{k=0}^{\infty} \frac{(\beta)_k}{k!} \frac{G^k}{F^{k+1}} x_n^k,$$

has nonnegative power series cofficients. $\square$

Theorem 9.3.2 ((Gillis et al., 1983)) *If $[A(x,y) - zB(x,y)]^{-\alpha}$ and $[C(x,y) - zD(x,y)]^{-\alpha}$ have nonnegative power series coefficients for $\alpha > 0$ so also does $[A(x,y)C(z,u) - B(x,y)D(z,u)]^{-\alpha}$.*

Proof The power series of

$$[A(x,y) - zB(x,y)]^{-\alpha} = [A(x,y)]^{1-\alpha} \sum_{n=0}^{\infty} \frac{(\alpha)_n}{n!} \frac{[B(x,y)]^n}{[A(x,y)]^{n+1}} z^n,$$

$$[C(x,y) - zD(x,y)]^{-\alpha} = [C(x,y)]^{1-\alpha} \sum_{n=0}^{\infty} \frac{(\alpha)_n}{n!} \frac{[D(x,y)]^n}{[C(x,y)]^{n+1}} z^n$$

have nonnegative coefficients. Hence for every n the power series expansions of both

$$[A(x,y)]^{1-\alpha} \frac{[B(x,y)]^n}{[A(x,y)]^{n+1}} \text{ and } [C(x,y)]^{1-\alpha} \frac{[D(x,y)]^n}{[C(x,y)]^{n+1}}$$

have nonnegative coefficients. Thus

$$[A(x,y)C(x,y)]^{1-\alpha} \frac{[B(x,y)D(x,y)]^n}{[A(x,y)C(x,y)]^{n+1}}$$

has nonnegative power series coefficients and the result follows. $\square$

Theorem 9.3.3 ((Askey & Gasper, 1977)) *For the inequalities $A^{(\alpha)}(k,m,n;2) \geq 0$ to hold it is necessary and sufficient that $\alpha \geq \left(\sqrt{17} - 5\right)/2$.*

Proof (Gillis et al., 1983) In (9.3.7) set

$$\mu = 2, \quad k = 3, \quad B_n^{(\alpha)}(k,m,n) = 2^{k+m+n+\alpha+1} A^{(\alpha)}(k,m,n;2)$$

and let $R = 1 - x - y - z + 4xyz$. The generating function

$$\sum_{k,m,n=0}^{\infty} B^{(\alpha)}(k,m,n)x^k y^m z^n = R^{-\alpha-1} \tag{9.3.10}$$

follows from (9.3.7). It is easy to see that

$$\partial_x R^{-\alpha-1} = 2(y\partial_y - z\partial_z)R^{-\alpha-1}$$
$$+(1+2z)[x\partial_x - y\partial_y + z\partial_z + \alpha + 1]R^{-\alpha-1}. \tag{9.3.11}$$

From (9.3.11) and upon equating the coefficients of $x^k y^m z^n$ we derive the recursion relation

$$(k+1)B^{(\alpha)}(k+1,m,n) = (\alpha+1+k+m+n)B^{(\alpha)}(k,m,n)$$
$$+2(\alpha+k-m+n)B^{(\alpha)}(k,m,n-1). \tag{9.3.12}$$

By symmetry it suffices to prove the result for $k \geq m \geq n$. The coefficients in the recurrence relation (9.3.12) are positive if $n \geq 1$ and the result will then follow by induction from $B^{(\alpha)}(k,k,1) \geq 0$, $k \geq 1$, which we now prove. Observe that

$$[1-x-y-z+4xyz]^{-\alpha-1} = \frac{[1-z(1-4xy)/(1-x-y)]^{-\alpha-1}}{[1-x-y]^{\alpha+1}},$$

which yields

$$\sum_{k,m=0}^{\infty} B^{(\alpha)}(k,m,1)x^k y^m = \frac{(\alpha+1)(1-4xy)}{(1-x-y)^{\alpha+2}}$$

$$= (\alpha+1)(1-4xy)\sum_{j=0}^{\infty} \frac{(\alpha+2)_j}{j!}(x+y)^j.$$

Equating coefficients of $x^k y^k$ and noting that j must be even we establish

$$B^{(\alpha)}(k,k,1) = \frac{(\alpha+1)_{2k+1}}{(k!)^2} - 4\frac{(\alpha+1)_{2k-1}}{((k-1)!)^2}$$

$$= \frac{(\alpha+1)_{2k-1}}{(k!)^2}\left[(\alpha+2k)(\alpha+2k+1) - 4k^2\right].$$

Thus $B^{(\alpha)}(k,k,1) \geq 0$ for all $k \geq 1$ if and only if $\alpha^2 + \alpha(4k+1) + 2k \geq 0$ for $k \geq 1$. From the cases of k large we conclude that $\alpha \geq -1/2$, and by taking $k=1$ we see that $\alpha \geq \left(-5+\sqrt{17}\right)/2$. It is clear that this condition is also sufficient. $\quad\square$

Corollary 9.3.4 *The Friedrichs–Lewy numbers in (9.3.2) are nonnegative.*

Proof See the Askey–Gasper observation above (9.3.5). $\quad\square$

Theorems 9.3.1 and 9.3.2 will be used in §9.4 to establish inequalities for linearization coefficients.

In the rest of this section, we state and prove some general results concerning the nonnegativity of linearization coefficients.

Theorem 9.3.5 ((Askey, 1970a)) *Let $P_0(x) = 1$, $P_1(x) = x + c$ and*

$$P_1(x)P_n(x) = P_{n+1}(x) + \alpha_n P_n(x) + \beta_n P_{n-1}(x). \tag{9.3.13}$$

Then if $\alpha_n \geq 0$, $\beta_{n+1} > 0$, $\alpha_{n+1} \geq \alpha_n$, $\beta_{n+2} \geq \beta_{n+1}$, $n = 0,1,\ldots$, we have

$$P_m(x)P_n(x) = \sum_{k=|n-m|}^{m+n} C(m,n,k)P_k(x).$$

with $C(m,n,k) \geq 0$.

Proof By symmetry, assume $m \leq n$ and that $C(j, k, \ell) \geq 0, j = 0, 1, \ldots, m, j < \ell$. Then

$$
\begin{aligned}
P_{m+1}(x)P_n(x) &= \left[P_1(x)P_m(x) - \alpha_m P_m(x) - \beta_m P_{m-1}(x) \right] P_n(x) \\
&= P_m(x) \left[P_{n+1}(x) + \alpha_n P_n(x) + \beta_n P_{n-1}(x) \right] \\
&\quad - \alpha_m P_m(x)P_n(x) - \beta_m P_{m-1}(x)P_n(x),
\end{aligned}
$$

hence

$$
\begin{aligned}
P_{m+1}(x)P_n(x) &= P_m(x)P_{n+1}(x) + \left(\alpha_n - \alpha_m \right) P_m(x)P_n(x) \\
&\quad + \left(\beta_n - \beta_m \right) P_{m-1}(x)P_n(x) \qquad\qquad (9.3.14) \\
&\quad + \beta_m \left[P_m(x)P_{n-1}(x) - P_{m-1}(x)P_n(x) \right].
\end{aligned}
$$

The first three terms on the right-hand side have nonnegative linearization coefficients, so we only need to prove that

$$
P_m(x)P_{n-1}(x) - P_{m-1}(x)P_n(x)
$$

has nonnegative linearization coefficient. Indeed, (9.3.14) shows that the quantity $\Delta_{m,n}(x) := P_{m+1}(x)P_n(x) - P_m(x)P_{n+1}(x)$ has nonnegative linearization coefficients if $\Delta_{m-1,n-1}(x)$ has the same property. Thus $\Delta_{m,n}(x)$ has nonnegative linearization coefficients if $\Delta_{0,n-m}(x)$ has the same property. But,

$$
\begin{aligned}
\Delta_{0,n-m}(x) &= P_1(x)P_{n-m}(x) - P_0(x)P_{n-m+1}(x) \\
&= \alpha_{n-m} P_{n-m}(x) + \beta_{n-m} P_{n-m-1}(x)
\end{aligned}
$$

which has nonnegative coefficients, by the induction hypothesis, and the theorem follows from (9.3.14). □

Askey noted that the proof of Theorem 9.3.5 also establishes monotonicity of the linearization coefficients, see (Askey, 1970b) and, as such, it is not sharp. It is true, however, that Theorem 9.3.5 covers most of the cases when the nonnegativity of the linearization coefficients is known; for example, for Jacobi polynomials.

We now use the general normalization in (9.1.19).

Theorem 9.3.6 *If B_n, C_n, $A_n + C_n$ are nondecreasing and $C_n \leq A_n$ for all n, then the linearization coefficients of $\{\varphi_n(x)\}$ are nonnegative.*

Theorem 9.3.7 *Assume that $B_n = 0$ and C_{2n}, C_{2n+1}, $A_{2n} + C_{2n}$, $A_{2n+1} + C_{2n+1}$ are nondecreasing. If, in addition, $C_n \leq A_n$ for all $n \geq 0$, then $\{\varphi_n(x)\}$ have nonnegative linearization coefficients.*

Theorems 9.3.6 and 9.3.7 are due to R. Szwarc, who proved them using discrete boundary value problem techniques. The proofs are in (Szwarc, 1992c) and (Szwarc, 1992d). Szwarc noted that the conditions in Theorems 9.3.6–9.3.7 are invariant under $n \to n+c$, hence if they are satisfied for a polynomial sequence they will be satisfied for the corresponding associated polynomials.

In order to state a theorem of Koornwinder, we first explain its set-up. Let X and Y be compact Hausdorff spaces with Borel measures μ and ν, respectively, such that $\mu(E_1)$ and $\nu(E_2)$ are positive and finite for every open nonempty sets E_1, E_2,

$E_1 \subset X$, $E_2 \subset Y$. Let $\{p_n(x)\}$ and $\{r_n(x)\}$ be families of orthogonal continuous functions on X and Y with $r_0(x) = 0$. Set

$$\int_X p_m(x)\,\overline{p_n(x)}\,d\mu(x) = \frac{\delta_{m,n}}{\pi_n},$$

$$\int_Y r_m(y)\overline{r_n(y)}\,d\nu(n) = \frac{\delta_{m,n}}{\rho_n},$$

(9.3.15)

where $0 < \pi_n \rho_n < \infty$.

Theorem 9.3.8 ((Koornwinder, 1978)) *Assume that*

$$p_n(x)\,p_n(y) = \sum_\ell a_{mn\ell} \pi_\ell\,\overline{p_\ell(x)},$$

(9.3.16)

with only finitely many nonzero terms. Suppose that Λ is a continuous mapping from $X \times X \times Y$ to X such that for each n there is an addition formula of the form

$$p_n(\Lambda(x,y,t)) = \sum_k c_{n,k}\, p_n^{(k)}(x)\,\overline{p_n^{(k)}(y)}\, r_k(t)$$

(9.3.17)

where $p_n^{(k)}$ is continuous on X, $p_n^{(0)} = p_n$, $c_{n,k} \geq 0$ but $c_{n,0} > 0$. Assume further that for every fixed n the set $\{c_{n,k} : k = 0,1,\dots\}$ is finite. Then the coefficients $a_{mn\ell}$ in (9.3.16) are nonnegative.

Koornwinder showed that (9.3.17) holds for the disc polynomials and, through a limiting procedure, proved that the coefficients in the expansion of $L_m^{(\alpha)}(\lambda x)\,L_n^{(\alpha)}$ $((1-\lambda)x)$ in $L_n^{(\alpha)}(x)$ are nonnegative for $\lambda \in [0,1]$.

9.4 Linearization of Products and Enumeration

In this section we state combinatorial interpretations of linearization coefficients for certain polynomials. The key is MacMahon's Master Theorem of partitions stated below. A proof is in volume 2 of (MacMahon, 1916). A more modern proof is in (Cartier & Foata, 1969).

Theorem 9.4.1 (MacMahon's Master Theorem) *Let $a\,(n_1, n_2, \dots, a_k)$ be the coefficient of $x_1^{n_1} x_2^{n_2} \cdots x_k^{n_k}$ in*

$$\left(\sum_{j=1}^k a_{1,j} x_j\right)^{n_1} \left(\sum_{j=1}^k a_{2,j} x_j\right)^{n_2} \cdots \left(\sum_{j=1}^k a_{k,j} x_j\right)^{n_k}.$$

(9.4.1)

Then

$$\sum_{n_1,\dots,n_k=0}^\infty a\,(n_1, n_2, \dots, a_k)\, t_1^{n_1} t_2^{n_2} \cdots t_k^{n_k} = 1/\det V,$$

(9.4.2)

and V is the matrix $\{v_{i,j}\}$,

$$v_{i,i} = 1 - a_{i,i} t_i, \quad v_{i,j} = -a_{i,j} t_i, \text{ for } i \neq j.$$

(9.4.3)

The entries $a_{i,j}$ in the Master theorem form a matrix which we shall refer to as the A matrix. Now consider the **Derangement Problem** where we have k boxes with box number j full to capacity with indistinguishable objects (balls) of type j. We then redistribute the objects in the boxes in such a way that no object stays in the box it originally occupied. We assume that box number j has capacity n_j.

Theorem 9.4.2 ((Even & Gillis, 1976)) *Let $D\,(n_1, n_2, \ldots, n_k)$ be the number of derangements. Then*

$$D\,(n_1, n_2, \ldots, n_k) = (-1)^{n_1 + \cdots + n_k} \int_0^\infty e^{-x} L_{n_1}(x) \cdots L_{n_k}(x)\,dx. \qquad (9.4.4)$$

Proof It is easy to see that the A matrix of the derangement problem is given by $a_{i,j} = 1 - \delta_{i,j}$. An exercise in determinants shows that the determinant of the corresponding V matrix is

$$\det V = 1 - \sigma_2 - 2\sigma_3 - \cdots - (k-1)\sigma_k \qquad (9.4.5)$$

with $\sigma_1, \ldots, \sigma_k$ denoting the elementary symmetric functions of $t_1, t_2, \ldots, t_k$. Let $E\,(n_1, n_2, \ldots, n_k)$ denote the right-hand side of (9.4.4). Therefore

$$\sum_{n_1, \ldots, n_k = 0}^\infty E\,(n_1, n_2, \ldots, n_k)\, t_1^{n_1} t_2^{n_2} \cdots t_k^{n_k}$$

$$= \sum_{n_1, \ldots, n_k = 0}^\infty (-1)^{n_1 + \cdots + n_k} A^{(0)}\,(n_1, \ldots, n_k; 1)\, t_1^{n_1} \cdots t_k^{n_k}$$

$$= \left[1 - \sum_{j=2}^k (j-1)\sigma_j \right]^{-1},$$

by (9.3.7). Thus, (9.4.5) shows that the above expression is $1/\det V$ and the proof is complete. $\qquad\square$

For $k > 2$ we define

$$C^{(\alpha)}\,(n_1, \ldots, n_k, b_1, \ldots, b_k) = \int_0^\infty \frac{x^\alpha e^{-x}}{\Gamma(\alpha + 1)} \prod_{j=1}^k L_{n_j}^{(\alpha)}(b_j x)\,dx. \qquad (9.4.6)$$

Koornwinder studied the case $k = 3$, $b_1 = 1$, $b_2 + b_3 = 1$, $b_1 \geq 0$, $b_2 \geq 0$. The more general case treated here is from (Askey et al., 1978).

Theorem 9.4.3 *We have the generating function*

$$G^{(\alpha)}(b_1, \ldots, b_k; t_1 \cdots t_k) :=$$

$$\sum_{n_1, \ldots, n_k = 0}^{\infty} C^{(\alpha)}(n_1, \ldots, n_k; b_1, \ldots, b_k) t_1^{n_1} \cdots t_k^{n_k} \tag{9.4.7}$$

$$= \left[\prod_{j=1}^{k}(1 - t_j) + \sum_{l=1}^{k} b_l \prod_{j=1, j \neq l}^{k}(1 - t_j) \right]^{-\alpha - 1}.$$

Proof The generating function (6.4.5) shows that the left-hand side of (9.4.6) is

$$\prod_{j=1}^{k}(1 - t_j)^{-\alpha - 1} \int_0^{\infty} \frac{x^{\alpha} e^{-x}}{\Gamma(\alpha + 1)} \exp\left[\sum_{l=1}^{k}(-b_l t_l x / (1 - t_l)) \right] dx$$

which simplifies to the right-hand side of (9.4.6). □

Theorem 9.4.4 *The generating function $G^{(\alpha)}$ satisfies*

$$1/G^{(0)}(b_1, \ldots, b_k; t_1, \ldots, t_k)$$

$$= \prod_{j=1}^{k}(1 - t_j) + \sum_{l=1}^{k} b_l \prod_{j=1, j \neq l}^{k}(1 - t_j) \tag{9.4.8}$$

$$= \det(\delta_{i,j} - a_{ij} t_j),$$

where

$$a_{ii} = 1 - b_i, \quad a_{ij} = -\sqrt{b_i b_j}, \quad i \neq j. \tag{9.4.9}$$

Proof The first equality is from Theorem 9.4.2 so we prove the second. We shall use induction over k. Assume the theorem holds for k and consider the case of $k + 1$ variables. We may assume that $t_j \neq 0$ for all j because otherwise the theorem trivially follows. If $\det(a_{ij})$ is expanded in a power series all the coefficients are determined except for the coefficient of $t_1 t_2 \cdots t_{k+1}$. And by induction they all satisfy the second equality in (9.4.8). So it only remains to show that

$$\det\left(\delta_{ij} - \sqrt{b_i b_j}\right) = 1 - \sum_{j=1}^{k+1} b_j$$

Again this is proved by induction. Expand the left-hand side of the above equation in a power series of b_j's. If $b_i = 0$ then the left-hand side of the above equation is the same determinant with the ith rows and columns deleted, and both sides are equal by the induction hypothesis. Doing this for for $i = 1, 2, \ldots, k + 1$ gives all the coefficents except the coefficient of $b_1 b_2 \cdots b_{k+1}$. This is $\det\left(-\sqrt{b_i b_j}\right)$ which is clearly zero. This completes the proof of (9.4.8) and thus of Theorem 9.4.3. □

Theorem 9.4.5 *Let a_{ij} be as in (9.4.9). The coefficient of $t_1^{n_1} t_2^{n_2} \cdots t_k^{n_k}$ in the expansion*

$$\prod_{i=1}^{k} \left(\sum_{j=1}^{k} a_{ij} t_j \right)^{n_i} \tag{9.4.10}$$

is $C^{(0)}(n_1, n_2, \cdots, n_k; b_1, b_2, \cdots, n_k)$.

Proof Apply the MacMahon Master Theorem and Theorems 9.4.2–9.4.3, with $A = (a_{ij})$. □

Theorem 9.4.6 *The inequality*

$$C^{(\alpha)}(\ell, m, n; \lambda, 1 - \lambda, 1) \geq 0, \tag{9.4.11}$$

holds for $\alpha \geq 0$, $\lambda \in [0, 1]$, with strict inequality if $\ell = 0$ and $\lambda \in (0, 1)$.

Proof First consider the case $\alpha = 0$ and let A be the matrix a_{ij}. From Theorem 9.4.4 we see that $C^{(\alpha)}(\ell, m, n; \lambda, 1 - \lambda, 1)$ is the coefficient of $r^\ell s^m t^n$ in

$$\left[-\sqrt{\lambda} r - \sqrt{1 - \lambda} s \right]^n \left[-\sqrt{\lambda(1 - \lambda)} r + \lambda s - \sqrt{1 - \lambda} t \right]^m$$
$$\left[(1 - \lambda) r - \sqrt{\lambda(1 - \lambda)} s - \sqrt{\lambda} t \right]^\ell.$$

Expand the above expression as a power series as

$$(-1)^n \sum_{i,j} \binom{\ell}{i} \left[(1 - \lambda) r - \sqrt{\lambda(1 - \lambda)} s \right]^i \left(-\sqrt{\lambda} t \right)^{\ell - i}$$
$$\times \binom{m}{j} \left[-\sqrt{\lambda(1 - \lambda)} r + \lambda s \right]^j \left(-\sqrt{1 - \lambda} t \right)^{m-j}$$
$$\times \left[\sqrt{\lambda} r + \sqrt{1 - \lambda} s \right]^n$$
$$= (-1)^{\ell + m + n} \sum_{i,j} (-1)^i \binom{\ell}{i} \binom{m}{j} \left[\sqrt{(1 - \lambda)} r - \sqrt{\lambda} s \right]^{i+j}$$
$$\times t^{\ell + m - i - j} \lambda^{(\ell + j - i)/2} (1 - \lambda)^{(m + i - j)/2}$$
$$= (-1)^{\ell + m + n} \sum_{i,j,p,q} (-1)^{j+p} \binom{\ell}{i} \binom{m}{j} \binom{i+j}{p} \binom{n}{q} r^{p+q}$$
$$\times s^{i+j+n-p-q} t^{\ell+m-i-j} \lambda^{(q+2j+\ell-p)/2} (1 - \lambda)^{(p+n-q+m+i-j)/2}.$$

The term $r^\ell s^m t^n$ arises if and only if $p + q = \ell$, $i + j + n - p - q = m$, and $k + m - i - j = n$. Therefore for $\ell + m \geq n$ we eliminate j and q and find that

$C^{(0)}(\ell, m, n; \lambda, 1 - \lambda, 1)$ is given by

$$\sum_{i,p} \binom{\ell}{i}\binom{m}{\ell + m - n - i}\binom{\ell + m - n}{p}\binom{n}{\ell - p}$$

$$\times \lambda^{2\ell + m - n - i - p}(1 - \lambda)^{n - \ell + p + i}$$

$$= \lambda^{2\ell + m - n}(1 - \lambda)^{n - k}\frac{(\ell + m - n)! \, n!}{k! \, m!}$$

$$\times \left[\sum_i (-1)^i [(1 - \lambda)/\lambda]^i \binom{\ell}{i}\binom{m}{n - \ell + i}\right]^2,$$

which is clearly nonegative but $C^{(0)}(0, m, n; \lambda, 1 - \lambda, 1) > 0$ or $\lambda \in (0, 1)$. This proves the theorem for $\alpha = 0$. Now the generating function (9.4.7)

$$G^{\alpha + 1}(\lambda, 1 - \lambda, 1, r, s, t)$$

$$= [1 - (1 - \lambda)r - \lambda s - \lambda rt - (1 - \lambda)st + rst]^{-\alpha - 1}$$

Apply Theorem 9.3.1 with

$$F(r, s) = 1 - (1 - \lambda)r - \lambda s, \quad G(r, s) = \lambda r + (1 - \lambda)s - rs,$$

for $\lambda \in [0, 1]$ to complete the proof.   □

Ismail and Tamhankar proved Theorem 9.4.6 when $\alpha = 0$ in (Ismail & Tamhankar, 1979). In the same paper, they also proved the positivity of the numbers $A^{(0)}(k, m, n; z)$ of Theorem 9.3.3.

It is important to note that we have proved that when $\lambda \in (0, 1)$, $C^{(0)}(\ell, m, n; \lambda, 1 - \lambda, 1)$ is a positive multiple of the square of

$$\sum_i (-1)^i [(1 - \lambda)/\lambda]^i \binom{\ell}{i}\binom{m}{n - \ell + i} \tag{9.4.12}$$

But the expression in (9.4.12) is

$$\frac{m!}{(n - \ell)!(m + \ell - n)!} \, {}_2F_1(-\ell, n - \ell - m; n - \ell + 1; (1 - \lambda)/\lambda),$$

which can vanish for certain λ in $(0, 1)$.

From §9.0 it follows that

$$L_m^{(\alpha)}(\lambda x)L_n^{(\alpha)}((1 - \lambda)x)$$

$$= \sum_{\ell = 0}^{m + n} \frac{\ell!}{\Gamma(\ell + \alpha + 1)} C^{(\alpha)}(n, m, \ell; \lambda, 1 - \lambda, 1)L_\ell^{(\alpha)}(x) \tag{9.4.13}$$

Thus the linearization coefficients in the above formula are nonnegative for $\lambda \in [0, 1]$. Itertate formula (9.4.13) to see that

$$L_{n_1}^{(\alpha)}(a_1 x) \, L_{n_2}^{(\alpha)}(a_2 x) \cdots L_{n_k}^{(\alpha)}(a_k x) = \sum_{\ell = 0}^{n_1 + \cdots + n_k} c_\ell L_\ell^{(\alpha)}(x),$$

and $c_\ell \geq 0$ provided that $\alpha \geq 0$, $a_j \geq 0$, $1 \leq j \leq k$, and $a_1 + \cdots + a_k = 1$.

The Meixner polynomials are discrete generalizations of Laguerre polynomials as can be seen from (6.1.18). This suggests generalizing the numbers $A^{(\alpha)}$ $(n_1, \ldots, n_k; \mu)$ to

$$M^{(\beta)}(n_1, \ldots, n_k; \mu) := (-1)^{n_1 + \cdots + n_k} (1 - c^\mu)^\beta$$

$$\times \sum_{x=0}^{\infty} \frac{(\beta)_x c^{x\mu}}{x!} M_{n_1}(x; \beta, c) \ldots M_{n_k}(x; \beta, c). \tag{9.4.14}$$

Therefore using (6.1.8) we get

$$\sum_{n_1, \ldots, n_k = 0}^{\infty} \left\{ \prod_{j=1}^{k} \frac{(\beta)_{n_j}}{n_j!} t_j^{n_j} \right\} M^{(\beta)}(n_1, \ldots, n_k; \mu)$$

$$= \left[1 + \frac{1 - c^{\mu-1}}{1 - c^\mu} \sigma_1 + \frac{1 - c^{\mu-2}}{1 - c^\mu} \sigma_2 + \cdots + \frac{1 - c^{\mu-k}}{1 - c^\mu} \sigma_k \right]^{-\beta}, \tag{9.4.15}$$

(Askey & Ismail, 1976). In the special case $\mu = \beta = 1$, $M^{(1)}(n_1, \ldots, n_k, 1)$ have the following combinatorial interpretation. Consider k boxes, where the box number j contains n_j indistinguishable objects of type j. The types are different. We redistribute these objects in such a way that each box ends up with the same number of objects it originally contained and no object remains in its original container. We then assign weights to the derangements we created. A derangement has the weight c^{-a} where a is the number of objects that ended up in a box of lower index than the box it originally occupied; that is, a is the number of objects that "retreated." Theorem 9.4.1 and (9.4.15) prove that $M^{(1)}(n_1, \ldots, n_k; \mu)$ is the sum of these weighted derangements, (Askey & Ismail, 1976).

In (Zeng, 1992), Zeng extended these weighted derangement interpretations to the linearization coefficients of all orthogonal polynomials which are of Sheffer A-type zero relative to $\frac{d}{dx}$; see Chapter 10.

Two conjectures involving the positivity of coefficients in formal power series will be stated in 24.3.

9.5 Representations for Jacobi Polynomials

In this section we consider representations of Jacobi polynomials as integrals involving the nth power of a function. These are similar in structure to the Laplace integral (4.5.17) but are double integrals.

Theorem 9.5.1 ((Braaksma & Meulenbeld, 1971)) *The integral representation*

$$P_n^{(\alpha,\beta)}(x) = \frac{2^n \, \Gamma(\alpha + n + 1) \, \Gamma(\beta + n + 1)}{\pi \, \Gamma(\alpha + 1/2) \, \Gamma(\beta + 1/2) \, (2n)!}$$

$$\times \int_0^\pi \int_0^\pi \left[i\sqrt{1-x} \cos\phi + \sqrt{1+x} \cos\psi \right]^{2n} \tag{9.5.1}$$

$$\times (\sin\phi)^{2\alpha} (\sin\psi)^{2\beta} \, d\phi \, d\psi,$$

holds for $\mathrm{Re}\,\alpha > -1/2$, $\mathrm{Re}\,\beta > -1/2$.

Proof The right-hand side of (4.3.3) is

$$\frac{\Gamma(\alpha+1)\,\Gamma(\beta+1)\,e^{-i\pi\beta/2}}{(t(1-x)/2)^{\alpha/2}(t(1+x)/2)^{\beta/2}}\,J_\alpha\left(\sqrt{2t(1-x)}\right)J_\alpha\left(i\sqrt{2t(1+x)}\right).$$

Apply (4.8.5) with $n=0$ to write the above as

$$\frac{\Gamma(\alpha+1)\,\Gamma(\beta+1)}{\pi\,\Gamma(\alpha+1/2)\Gamma(\beta+1/2)}$$

$$\times\int_0^\pi\!\!\int_0^\pi\left\{\cos\left[\sqrt{2t(1-x)}\cos\phi\right]\cos\left[i\sqrt{2t(1+x)}\cos\psi\right]\right\} \tag{9.5.2}$$

$$\times(\sin\phi)^{2\alpha}\,(\sin\psi)^{2\beta}\,d\phi\,d\psi.$$

The addition of

$$\sin\left[\sqrt{2t(1-x)}\cos\phi\right]\sin\left[i\sqrt{2t(1+x)}\cos\psi\right]$$

to the term in {} does not change the value of the integral and replaces the term in {} by

$$\cos\left[\sqrt{2t(1-x)}\cos\phi-i\sqrt{2t(1+x)}\cos\psi\right]. \tag{9.5.3}$$

The coefficient of t^n in (9.5.3) is

$$\frac{(-2)^n}{(2n)!}\left[\sqrt{1-x}\cos\phi-i\sqrt{1+x}\cos\psi\right]^{2n}. \tag{9.5.4}$$

Thus the coefficient of t^n in the left-hand side of (4.3.3) is

$$\frac{\Gamma(\alpha+1)\,\Gamma(\beta+1)}{\pi\,2^n\,\Gamma(\alpha+1/2)\Gamma(\beta+1/2)}$$

$$\times\int_0^\pi\!\!\int_0^\pi\left[i\sqrt{1-x}\cos\phi+\sqrt{1+x}\cos\psi\right]^{2n} \tag{9.5.5}$$

$$\times(\sin\phi)^{2\alpha}\,(\sin\psi)^{2\beta}\,d\phi\,d\psi,$$

and the proof is complete. □

Theorem 9.5.2 *We have the integral representation*

$$\frac{P_n^{(\alpha,\beta)}(x)}{P_n^{(\alpha,\beta)}(1)}=\frac{2\Gamma(\alpha+1)\,n!}{\Gamma(1+(\alpha+\beta)/2)\,\Gamma((\alpha-\beta)/2)\,(\alpha+\beta+1)_n}$$

$$\times\int_0^1 u^{\alpha+\beta+1}\left(1-u^2\right)^{-1+(\alpha-\beta)/2}C_n^{(\alpha+\beta+1)/2}\left(1+u^2(x-1)\right)\,du, \tag{9.5.6}$$

valid for $\mathrm{Re}(\alpha)>\mathrm{Re}(\beta)$, $\mathrm{Re}(\alpha+\beta)>-2$.

Proof Use the Euler integral representation (1.4.8) to see that the right-hand side of (4.3.1) is

$$\frac{\Gamma(\alpha+1)}{\Gamma(1+(\alpha+\beta)/2)\,\Gamma((\alpha-\beta)/2)}$$

$$\times \int_0^1 u^{(\alpha+\beta)/2}\,(1-u)^{-1+(\alpha-\beta)/2}\left[(1-t)^2 - 2tu(x-1)\right]^{-(\alpha+\beta+1)/2}du$$

The coefficient of t^n in

$$\left[1+t^2 - 2t(1+u(x-1))\right]^{-(\alpha+\beta+1)/2}$$

is $C_n^{(\alpha+\beta+1)/2}(1+u(x-1))$, hence (9.5.6) holds. $\qquad\square$

The integral representations in (9.5.6) are important because every integral representation for C_n^ν will lead to a double integral representation for Jacobi polynomials. Indeed, the Laplace first integral, (4.5.17), implies

$$\frac{P_n^{(\alpha,\beta)}(x)}{P_n^{(\alpha,\beta)}(1)} = \frac{2\Gamma(\alpha+1)}{\Gamma(1/2)\,\Gamma((\alpha-\beta)/2)}$$

$$\times \int_0^1\int_0^\pi r^{\alpha+\beta+1}\,(1-r^2)^{-1+(\alpha-\beta)/2}\,(\sin\phi)^{\alpha+\beta} \qquad (9.5.7)$$

$$\times \left[1 - r^2(1-x) + ir\cos\phi\,\sqrt{(1-x)\,(2-r^2(1-x))}\right]^n d\phi\,du,$$

for $\operatorname{Re}\alpha > \operatorname{Re}\beta$, and $\operatorname{Re}(\alpha+\beta) > -2$.

Another Laplace-type integral is

$$\frac{P_n^{(\alpha,\beta)}(x)}{P_n^{(\alpha,\beta)}(1)} = \int_0^\pi\int_0^1\left[\frac{1+x-(1-x)r^2}{2} + i\sqrt{1-x^2}\,r\cos\varphi\right]^n d\mu_{\alpha,\beta}(r,\varphi),$$

$$(9.5.8)$$

where

$$d\mu_{\alpha,\beta} = (r,\varphi) := c_{\alpha,\beta}\,(1-r^2)^{\alpha-\beta-1}\,r^{2\beta+1}(\sin\varphi)^{2\beta}dr\,d\varphi,$$
$$c_{\alpha,\beta} := 2\Gamma(\alpha+1)/\left[\sqrt{\pi}\,\Gamma(\alpha-\beta)\Gamma(\beta+1/2)\right], \qquad (9.5.9)$$

which holds for $\alpha > \beta > -1/2$.

Proof of (9.5.8) Expand the integrand in (9.5.8) to see that the right-hand side of (9.5.8) equals

$$
\int_0^\pi \int_0^1 \sum_{k=0}^{\lfloor n/2 \rfloor} \binom{n}{2k} 2^{2k-n} \left[1 + x - (1-x)r^2 \right]^{n-2k}
$$

$$
\times (-1)^k \left(1 - x^2 \right)^k r^{2k} (\cos \varphi)^{2k} d\mu_{\alpha,\beta}(r,\varphi)
$$

$$
= \frac{\Gamma(\alpha+1)2^{-n}}{\sqrt{\pi}\,\Gamma(\alpha-\beta)\Gamma(\beta+1/2)} \sum_{k=0}^{\lfloor n/2 \rfloor} \frac{n!\,(-1)^k (1-x)^k (1+x)^{n-k}}{k!\,(1/2)_k (n-2k)!}
$$

$$
\times \int_0^\pi \int_0^1 r^{k+\beta}(1-\nu)^{\alpha-\beta-1}(\sin \varphi)^{2/3}(\cos \varphi)^{2k} \left[1 - \frac{(1-x)}{1+x} \nu \right]^{n-2k} d\nu \, d\varphi.
$$

The ν integral is evaluated by (1.4.8), while the φ integral is a beta integral. Thus, the above is

$$
\frac{\Gamma(\alpha+1)2^{-n}}{\Gamma(\alpha-\beta)\Gamma(\beta+1/2)} \sum_{k=0}^{\lfloor n/2 \rfloor} \frac{n!\,(-1)^k (1-x)^k (1+x)^{n-k}}{k!\,\Gamma(k+1/2)(n-2k)!}
$$

$$
\times \frac{\Gamma(\beta+1/2)\Gamma(k+1/2)}{\Gamma(\beta+k+1)} \frac{\Gamma(\beta+k+1)\Gamma(\alpha-\beta)}{\Gamma(\alpha+k+1)}
$$

$$
\times {}_2F_1 \left(\begin{array}{c} 2k-n, \beta+k+1 \\ \alpha+k+1 \end{array} \middle| \frac{1-x}{1+x} \right).
$$

By expanding the ${}_2F_1$ as a j sum then let $m = k + j$ and write the sums as sums over m and k the above becomes

$$
2^{-n} \sum_{m=0}^n \frac{n!\,(x-1)^m (x+1)^{n-m}}{(\alpha+1)_m} (\beta+1)_m \sum_{k=0}^{m \wedge (n-m)} \frac{1/(\beta+1)_k}{k!\,(m-k)!\,(n-m-k)!}.
$$

The k-sum is

$$
\frac{1}{m!\,(n-m)!} {}_2F_1 \left(\begin{array}{c} -m, n-m \\ \beta+1 \end{array} \middle| 1 \right) = \frac{(\beta+m+1)_{n-m}}{m!\,(n-m)!\,(\beta+1)_{n-m}},
$$

by the Chu–Vandermonde sum. Formula (9.5.8) now follows from (4.3.6). □

9.6 Addition and Product Formulas

Theorem 9.6.1 *The Jacobi polynomials have the product formula*

$$
\frac{P_n^{(\alpha,\beta)}(x)P_n^{(\alpha,\beta)}(y)}{P_n^{(\alpha,\beta)}(1)} = \int_0^\pi \int_0^1 P_n^{(\alpha,\beta)} \left[\frac{1}{2}(1+x)(1+y) \right. \tag{9.6.1}
$$

$$
\left. + \frac{1}{2}(1-x)(1-y)r^2 + \sqrt{(1-x^2)(1-y^2)}\, r \cos \varphi - 1 \right] d\mu_{\alpha,\beta}(r,\varphi),
$$

where $\mu_{\alpha,\beta}$ is defined in (9.5.9).

Proof Use (4.3.19) to express the left-hand side of (9.6.1) as a finite sum, then represent $P_k^{(\alpha,\beta)}$ as an integral using (9.5.8). The result is that the left-hand side of (9.6.1) is

$$\int_0^\pi \int_0^1 \frac{(-1)^n(\beta+1)}{n!} \sum_{k=0}^n \frac{(-n)_k(\alpha+\beta+n+1)_k}{k!\,(\beta+1)_k 2^k}$$

$$\times \left[\frac{1}{2}(1+x)(1+y) + (1-x)(1-y)r^2 + 2r\cos\varphi\sqrt{(1-x^2)\,(1-y^2)}\right]^k$$

$$\times d\mu_{\alpha,\beta}(r,\varphi),$$

and the result follows from (4.1.4). $\qquad\qquad\qquad\qquad\qquad\qquad\square$

In the case of ultraspherical polynomials, the representation (9.6.1) reduces to a single integral because the Laplace first integral for the ultraspherical polynomial is a single integral, see (4.5.17). After applying (4.5.15) we establish

$$\frac{C_n^\nu(x)C_n^\nu(y)}{C_n^\nu(1)} = \frac{\Gamma(\nu+1/2)}{\sqrt{\pi}\,\Gamma(\nu)}$$

$$\times \int_0^\pi C_n^\nu\left(xy + \sqrt{(1-x^2)\,(1-y^2)}\,\cos\varphi\right)(\sin\varphi)^{2\nu-1}d\varphi. \qquad (9.6.2)$$

Next, we state the Gegenbauer addition theorem.

Theorem 9.6.2 *The ultraspherical polynomials satisfy the addition theorem*

$$C_n^\nu(\cos\theta\cos\varphi + \sin\theta\sin\varphi\cos\psi)$$

$$= \sum_{k=0}^n a_{k,n}^\nu(\sin\theta)^k C_{n-k}^{\nu+k}(\cos\theta)(\sin\varphi)^k C_{n-k}^{\nu+k}(\cos\varphi)C_k^{\nu-1/2}(\cos\psi), \qquad (9.6.3)$$

with

$$a_{k,n}^\nu = \frac{\Gamma(\nu-1/2)(\nu)_k}{\Gamma(2\nu+n+k)}\,(n-k)!\,\Gamma(2\nu+2k). \qquad (9.6.4)$$

Proof Expand $C_n^\nu\left(xy + \sqrt{(1-x^2)\,(1-y^2)}\,z\right)$ in $\left\{C_k^{\nu-1/2}(z)\right\}$. The coefficient of $C_k^{\nu-1/2}(z)$ is

$$\frac{k!\,(\nu+k-1/2)\Gamma(\nu-1/2)}{(2\nu-1)_k\sqrt{\pi}\,\Gamma(\nu)}\int_{-1}^1 \left(1-z^2\right)^{\nu-1}$$

$$\times C_k^{\nu-1/2}(z)C_n^\nu\left(xy + \sqrt{(1-x^2)\,(1-y^2)}\,z\right)dz$$

$$= \frac{(-1)^k(\nu+k-1/2)\Gamma(\nu-1/2)}{2^k\sqrt{\pi}\,\Gamma(\nu+k)}$$

$$\times \int_{-1}^1 C_n^\nu\left(xy + \sqrt{(1-x^2)\,(1-y^2)}\,z\right)\left[\frac{d^k}{dz^k}\left(1-z^2\right)^{\nu+k-1}\right]dz,$$

where we used the Rodrigues formula (4.5.12). Apply (4.5.5) and integration by parts to reduce the above to

$$\left(1-x^2\right)^{k/2}\left(1-y^2\right)^{k/2}\frac{(\nu+k-1/2)\Gamma(\nu-1/2)(\nu)_k}{\sqrt{\pi}\,\Gamma(\nu+k)}\int_{-1}^{1}\left(1-z^2\right)^{\nu+k-1}$$

$$\times C_{n-k}^{\nu+k}\left(xy+\sqrt{\left(1-x^2\right)\left(1-y^2\right)}\,z\right)dz$$

and the result follows from (9.6.2). $\square$

Theorem 9.6.3 (Koornwinder) *The addition theorem for Jacobi polynomials is*

$$P_n^{(\alpha,\beta)}\left(2\cos^2\theta\cos^2\tau+2\sin^2\theta\sin^2\tau r^2+\sin 2\theta\sin 2\tau r\cos\phi-1\right)$$

$$=\sum_{k=0}^{n}\sum_{\ell=0}^{k}c_{n,k,\ell}^{(\alpha,\beta)}(\sin\theta)^{2k-\ell}(\cos\theta)^{\ell}P_{n-k}^{(\alpha+2k-\ell,\beta+\ell)}(\cos 2\theta)$$

$$\times\sin(\tau)^{2k-\ell}(\cos\tau)^{\ell}P_{n-k}^{(\alpha+2k-\ell,\beta+\ell)}(\cos 2\tau)$$

$$\times r^{\ell}P_{k-\ell}^{(\alpha-\beta-1,\beta+\ell)}\left(2r^2-1\right)P_{\ell}^{(\beta-1/2,\beta-1/2)}(\cos\phi),$$

(9.6.5)

where

$$c_{n,k,\ell}^{(\alpha,\beta)}=\frac{(\alpha+2k-\ell)(\beta+\ell)(n+\alpha+\beta+1)_k}{(\alpha+k)\left(\beta+\frac{1}{2}\ell\right)(\beta+1)_k\left(\beta+\frac{1}{2}\right)_\ell}$$

$$\times\frac{(\beta+n-k+\ell+1)_{k-\ell}}{(\alpha+k+1)_{n-\ell}}(2\beta+1)_\ell(n-k)!$$

(9.6.6)

There are several proofs of Theorem 9.6.3, but the one we give below is from (Koornwinder, 1977). For information and proofs, the reader may consult (Koornwinder, 1972), (Koornwinder, 1973), (Koornwinder, 1974), and (Koornwinder, 1975).

One can think of (9.6.5) as an expansion of its left-hand side in orthogonal polynomials in the variables $x=\cos^2\tau$, $y=r^2\sin^2\tau$, $z=2^{-1/2}r\sin(2\tau)\cos\phi$. We first assume $\alpha>\beta>-1/2$. Let $S\subset\mathbb{R}^3$ be

$$S=\left\{(x,y,z):0<x+y<1,z^2<2xy\right\}.$$

Let $\mathcal{H}_n$ be the class of orthogonal polynomials of degree n on S with respect to the weight function

$$w(x,y,z)=(1-x-y)^{\alpha-\beta-1}\left(2xy-z^2\right)^{\beta-1/2}.$$

(9.6.7)

Lemma 9.6.4 *The polynomials*

$$p_{n,k,\ell}(x,y,z)=P_{n-k}^{(\alpha+2k-\ell,\beta+\ell)}(2x-1)(1-x)^{k-\ell}$$

$$\times P_{k-\ell}^{(\alpha-\beta-1,\beta+1)}((x+2y-1)/(1-x))(xy)^{\ell/2}P_{\ell}^{(\beta-1/2,\beta-1/2)}\left(z/\sqrt{2xy}\right),$$

(9.6.8)

for $n\geq k\geq\ell\geq 0$ form an orthogonal basis of $\mathcal{H}_n$, which is obtained by orthogonalizing the linear independent polynomials

$$1,x,y,z,x^2,xy,xz,y^2,yz,z^2,x^3,\ldots.$$

Proof Clearly, the function $p_{n,k,\ell}(x, y, z)$ is a polynomial of degree n in x, y, z of degree k in y, z and of degree ℓ in z. Hence, $p_{n,k,\ell}(x, y, z)$ is a linear combination of monomials $x^{m_1-m_2}y^{m_2-m_3}z^{m_3}$ with "highest" term const. $x^{n-k}y^{k-\ell}z^{\ell}$. Let $u = 2x - 1, v = (x + 2y - 1)/(1 - x), w = z(2xy)^{-1/2}$. The mapping $(x, y, z) \rightarrow (u, v, w)$ is a diffeomorphism from R onto the cubic region $\{(u, v, w) : -1 - u < 1, -1 < v < 1, -1 < w < 1\}$. By making this substitution and by using the orthogonality properties of Jacobi polynomials it follows that

$$\iiint\limits_{R} p_{n,k,\ell}(x, y, z)p_{n',k',\ell'}(x, y, z)\, w(x, y, z)\, dx\, dy\, dz$$

$$= \delta_{n,n'}\delta_{k,k'}\delta_{\ell,\ell'} 2^{-2\alpha-2k-\ell-1} h_{n-k}^{(\alpha+2k-\ell,\beta+\ell)} h_{k-\ell}^{(\alpha-\beta-\ell,\beta+\ell)} h_{\ell}^{(\beta-1/2,\beta-1/2)}.$$

$\square$

Let S be a bounded subset of $\mathbb{R}^m$ and let $w = w(x_1, \ldots, x_m)$ be a positive continuous integrable function on S. We denote by $\mathcal{H}_n$ the class of all polynomials $p(\mathbf{x})$ which has the property

$$\int\limits_{S} p(\mathbf{x})\, q(\mathbf{x})\, w(\mathbf{x})\, d\mathbf{x} = 0, \quad \mathbf{x} = (x_1, \ldots, x_m)$$

if $q(\mathbf{x})$ is a polynomial of degree less than n. $\mathcal{H}_n$ can be chosen in infinitely many ways, but one way is to apply the Gram–Schmidt orthogonalization process to

$$x_1^{n_1-n_2} x_2^{n_2-x_3} \cdots x_{m-1}^{n_{m-1}-n_m} x_m^{n_m}, \quad n_1 \ge n_2 \ge \cdots \ge n_m \ge 0,$$

which are arranged by the lexicographic ordering of the k-tuples $(n_1, n_2, \ldots, n_k)$.

Let $\{p_s(\mathbf{x}) : 0 \le s \le N\}$ be an orthogonal basis of $\mathcal{H}_n$ and let $\zeta_s = \int_S p_s^2(\mathbf{x}) w(\mathbf{x})\, d\mathbf{x}$. The kernel

$$K(\mathbf{x}, \mathbf{y}) = \sum_{s=0}^{N} p_s(\mathbf{x})\, p_s(\mathbf{y})/\zeta_s, \quad \mathbf{x}, \mathbf{y} \in S$$

is the kernel polynomial or reproducing kernel of $\mathcal{H}_n$. Note that $K(\mathbf{x}, \mathbf{y})$ is independent of the choice of the orthogonal basis. In particular, if T is an isometry maping $\mathbb{R}^m$ on to $\mathbb{R}^m$ such that $T(S) = S$ and $w(T\mathbf{x}) = w(\mathbf{x})$, then

$$K(T\mathbf{x}, T\mathbf{y}) = K(\mathbf{x}, \mathbf{y}). \tag{9.6.9}$$

Proof of Theorem 9.6.3 Let

$$K((x, y, z), (x', y', z')) = \sum_{k=0}^{n}\sum_{\ell=0}^{k} \|p_{n,k,\ell}\|^{-2} p_{n,k,\ell}(x, y, z)\, p_{n,k,\ell}(x', y', z').$$

$$\tag{9.6.10}$$

It follows from (9.6.8) that $p_{n,k,\ell}(1, 0, 0) = 0$ if $(n, k, \ell) \neq (n, 0, 0)$. Hence

$$K((x, y, z), (1, 0, 0)) = \|p_{n,0,0}\|^2 P_n^{(\alpha,\beta)}(1)P_n^{(\alpha,\beta)}(2x - 1). \tag{9.6.11}$$

Any rotation around the axis $\{(x, y, z) \mid x = y, z = 0\}$ of the cone maps the region S onto itself and leaves the weight function $w(x, y, z)$ invariant. In particular, consider a rotation of this type over an angle -2θ. It maps point $\left(\cos^2\theta, \sin^2\theta, 2^{-1/2}\sin 2\theta\right)$ onto $(1, 0, 0)$ and point (x, y, z) onto a point (ξ, η, ζ) where $\xi = x\cos^2\theta + y\sin^2\theta + 2^{-1/2}z\sin 2\theta$. Hence, by (9.6.9), (9.6.8), (9.6.10) and (9.6.11) we have

$$
\|p_{n,0,0}\|^{-2} P_n^{(\alpha,\beta)}(1) P_n^{(\alpha,\beta)}\left(2\left(x\cos^2\theta + y\sin^2\theta + 2^{-1/2}z\sin 2\theta\right) - 1\right)
$$

$$
= K\left((x, y, z), \left(\cos^2\theta, \sin^2\theta, 2^{-1/2}\sin 2\theta\right)\right)
$$

$$
= \sum_{k=0}^{n}\sum_{\ell=0}^{k} \|p_{n,k,\ell}\|^{-2} P_{k-1}^{(\alpha-\beta-1,\beta+\ell)}(1) P_\ell^{(\beta-1/2,\beta-1/2)}(1)
$$

$$
\times (\sin\theta)^{2k-1}(\cos\theta)^\ell P_{n-k}^{(\alpha+2k-\ell,\beta+\ell)}(\cos 2\theta)
$$

$$
\times P_{n-k}^{(\alpha+2k-\ell,\beta+\ell)}(2x-1)(1-x)^{k-\ell} P_{k-\ell}^{(\alpha-\beta-1,\beta+\ell)}((x+2y-1)/(1-x))
$$

$$
\times (xy)^{1/2} P_\ell^{(\beta-1/2,\beta-1/2)}\left((2xy)^{-1/2}z\right).
$$

$$(9.6.12)$$

The substitution of $x = \cos^2\tau$, $y = r\sin^2\tau$, $z = 2^{-1/2}r\sin 2\tau\cos\phi$ gives (9.6.5) with

$$
c_{n,k,\ell}^{(\alpha,\beta)} = \frac{\|p_{n,0,0}\|^2 P_{k-\ell}^{(\alpha-\beta-1,\beta+\ell)}(1) P_\ell^{(\beta-1/2,\beta-1/2)}(1)}{\|p_{n,k,\ell}\|^2 P_n^{(\alpha,\beta)}(1)}
$$

Using the expression for $\|p_{n,k,\ell}\|^2$ at the end of the proof of Lemma 9.6.4 we see that the coefficients $c_{n,k,\ell}$ are given by (9.6.6). $\qquad\square$

9.7 The Askey–Gasper Inequality

The main result of this section is the inequality (9.7.1), which was a key ingredient in de Branges' proof of the Bieberbach conjecture, (de Branges, 1985). de Branges needed the case $\alpha \geq 2$.

Theorem 9.7.1 ((Askey & Gasper, 1976)) *The inequality*

$$
\frac{(\alpha+2)_n}{n!}\, {}_3F_2\left(\begin{array}{c} -n, n+\alpha+2, (\alpha+1)/2 \\ \alpha+1, (\alpha+3)/2 \end{array}\middle|\, x\right) \geq 0, \tag{9.7.1}
$$

for $0 < x < 1$, $\alpha > -2$.

Proof Use the integral representation

$$
{}_3F_2\left(\begin{array}{c} a_1, a_2, a_3 \\ b_1, b_2 \end{array}\middle|\, x\right) = \frac{\Gamma(b_1)}{\Gamma(a_3)\Gamma(b_1-a_3)} \int_0^1 t^{a_3-1}(1-t)^{b_1-a_3-1}
$$

$$
\times {}_2F_1\left(\begin{array}{c} a_1, a_2 \\ b_2 \end{array}\middle|\, xt\right) dt \tag{9.7.2}
$$

to obtain

$$
g(x, \lambda) = \frac{\Gamma(2\lambda-1)}{\Gamma^2(\lambda-1/2)} \int_0^1 \{t(1-t)\}^{\lambda-3/2} C_n^\lambda(1-2xt)\, dt, \tag{9.7.3}
$$

where

$$g(x, \lambda) = \frac{(2\lambda)_n}{n!} \, {}_3F_2 \left(\begin{matrix} -n, n + 2\lambda, \lambda - 1/2 \\ 2\lambda - 1, \lambda + 1/2 \end{matrix} \, \middle| \, x \right). \tag{9.7.4}$$

In (9.7.3), let $t \to (1+t)/2$ to get

$$g(x, \lambda) = \frac{\Gamma(2\lambda - 1)}{\Gamma^2(\lambda - 1/2)} \, 2^{4-2\lambda} \int\limits_{-1}^{1} \left(1 - t^2\right)^{\lambda - 3/2} C_n^\lambda(1 - x - xt) \, dt. \tag{9.7.5}$$

Let

$$m = \lfloor \lambda \rfloor - 2, \quad \lambda = (\alpha + 2)/2, \quad \lambda \geq 2. \tag{9.7.6}$$

The differential recurrence relation (4.5.5) transforms (9.7.5) to

$$g(x, \lambda) = \frac{(-x)^{-m} 2^{4-2\lambda-m} \Gamma(2\lambda - 1)}{\Gamma^2(\lambda - 1/2)(\lambda - m)_m}$$

$$\times \int\limits_{-1}^{1} \left(1 - t^2\right)^{\lambda - 3/2} \frac{\partial^m}{\partial t^m} \left\{ C_{n+m}^{\lambda - m}(1 - x - xt) \right\} dt.$$

Now, integrate by parts m times to get

$$g(x, \lambda) = \frac{x^{-m} 2^{4-2\lambda-m} \Gamma(2\lambda - 1)}{\Gamma^2(\lambda - 1/2)(\lambda - m)_m}$$

$$\times \int\limits_{-1}^{1} C_{n+m}^{\lambda - m}(1 - x - xt) \frac{d^m}{dt^m} \left(1 - t^2\right)^{\lambda - 3/2} dt$$

$$= \frac{(-x)^m 2^{4-2\lambda} \Gamma(2\lambda - 1) m! \, (\lambda - m - 1/2)_m}{\Gamma^2(\lambda - 1/2)(\lambda - m)_m (2\lambda - 2m - 2)_m}$$

$$\times \int\limits_{-1}^{1} C_{n+m}^{\lambda - m}(1 - x - xt) \left(1 - t^2\right)^{\lambda - m - 3/2} C_m^{\lambda - m - 1}(t) \, dt.$$

The Rodrigues formula (4.5.12) was used in the last step. Therefore, $g(x, \lambda)(-1)^m$ is a positive multiple of the coefficient of $C_m^{\lambda-m-1}(t)$ in the expansion of $C_{n+m}^{\lambda-m}(1 - x - xt)$ in $\left\{ C_j^{\lambda-m-1}(x) \right\}_{j=0}^{m+n}$.

Apply (9.6.3) with $t = \cos \psi$, $\sin \theta = \sqrt{x}$, $\varphi = -\theta$ to see that $C_n^\nu(1 - x - xt)$ can be expanded in terms of $(-1)^m C_m^{\nu-1/2}(t)$ with positive coefficients. On the other hand, (9.1.2) proves that $(-1)^n C_n^{\nu-1/2}(x)$ can be expanded in $(-1)^n C_{n-2k}^\mu(x)$, if $\mu < \nu - 1/2$. Therefore, the coefficient of $C_m^{\lambda-m-1}(t)$ in the expansion of $C_{n+m}^{\lambda-m}(1 - x - xt)$ has the sign $(-1)^m$ and (9.7.1) follows. $\qquad \square$

Exercises

9.1   Prove the equivalence of Conjectures 24.3.1 and 24.3.2.

10

The Sheffer Classification

In this chapter we briefly outline ideas from the Sheffer classification (Sheffer, 1939) and umbral calculus initiated by Rota and developed in a series of papers by Rota and his collaborators as well as by other individuals. In particular we single out the work (Rota et al., 1973). Our treatment, however, is more general than the published work in the sense that we assume nothing about the starting operator T. The existing treatments assume T is a special operator.

10.1 Preliminaries

Let T be a linear operator defined on polynomials. We say that a polynomial sequence $\{\phi_n(x)\}$ belongs to T if T reduces the degree of a polynomial by one and

$$T\phi_n(x) = \phi_{n-1}(x), \quad n > 0. \tag{10.1.1}$$

Theorem 10.1.1 *Let a polynomial sequence $\{f_n(x)\}$ belong to T. The polynomial sequence $\{g_n(x)\}$ also belongs to T if and only if there exists a sequence of constants $\{a_n\}$, with $a_0 \neq 0$ such that*

$$g_n(x) = \sum_{k=0}^{n} a_{n-k} f_k(x), \quad n \geq 0. \tag{10.1.2}$$

Proof Both $\{f_n(x)\}$ and $\{g_n(x)\}$ are bases for the space of polynomials over $\mathbb{C}$, hence the connection coefficients in

$$g_n(x) = \sum_{k=0}^{n} c_{n,k} f_k(x), \tag{10.1.3}$$

exist, with $c_{n,n} \neq 0$. Apply T to (10.1.3) to see that

$$g_{n-1}(x) = \sum_{k=1}^{n} c_{n,k} f_{k-1}(x),$$

and the uniqueness of the connection coefficients implies $c_{n,k} = c_{n-1,k-1}$. Therefore, by iteration we conclude that $c_{n,k} = c_{n-k,0}$ and (10.1.2) follows. Conversely

282

if (10.1.2) hold then

$$Tg_n(x) = \sum_{k=1}^{n} a_{n-k} f_{k-1}(x) = \sum_{k=0}^{n-1} a_{n-k-1} f_k(x) = g_{n-1}(x),$$

and the theorem follows. $\qquad\qquad\square$

The series in this chapter should be treated as formal power series. An equivalent form of Theorem 10.1.1 is the following corollary.

Corollary 10.1.2 *If* $\{f_n(x)\}$ *belong to* T, *then* $\{g_n(x)\}$ *belongs to* T *if and only only if the generating functions relationship*

$$\sum_{n=0}^{\infty} g_n(x)t^n = A(t) \sum_{n=0}^{\infty} f_n(x)t^n \qquad (10.1.4)$$

holds, where $A(t) = \sum_{n=0}^{\infty} a_n t^n$ *with* $a_0 \neq 0$.

Theorem 10.1.3 *Let a linear operator* T *defined on polynomials and* Tx^n *has precise degree* $n - 1$. *Given a polynomial sequence* $\{f_n(x)\}$ *there exists an operator* J

$$J = J(x, T) = \sum_{k=0}^{\infty} a_k(x) T^{k+1}, \qquad (10.1.5)$$

with $a_k(x)$ *of degree at most* k *such that* $\{f_n(x)\}$ *belongs to* J.

Proof Define a_0 by $a_0 = f_0(x)/[Tf_1(x)]$. Then define the polynomials $\{a_k(x)\}$ by induction through

$$a_n(x) T^{n+1} f_{n+1}(x) = -\sum_{k=0}^{n-1} a_k(x) T^{k+1} f_{n+1}(x), n > 0. \qquad (10.1.6)$$

Clearly a_k has degree at most k and the operator J of (10.1.5) with the above a_k's makes $Jf_n(x) = f_{n-1}(x)$. $\qquad\qquad\square$

Definition 10.1.1 *A polynomial sequence* $\{f_n(x)\}$ *is called of Sheffer A-type* m *relative to* T *if the polynomials* $a_k(x)$ *in the operator* J *to which* $\{f_n(x)\}$ *belongs, have degrees at most* m *and one of them has precise degree* m. *We say that* $\{f_n(x)\}$ *is an Appell set relative to* T *if* $J = T$, *that is* $TP_n(x) = P_{n-1}(x)$.

Sheffer (Sheffer, 1939) introduced the above classification for $T = \dfrac{d}{dx}$. He also observed that an earlier result of Meixner (Meixner, 1934) can be interpreted as characterizing all orthogonal polynomials of Sheffer A-type zero relative to $\dfrac{d}{dx}$ and gave another proof of Meixner's result. The totality of the measures that the polynomials are orthogonal with respect to turned out to be precisely the probability measures of the exponential distributions.

The function e^{xt} satisfies $\dfrac{d}{dx}e^{xt} = te^{xt}$. For general T, we assume that there is a function $\mathcal{E}$ of two variables such that

$$\mathcal{E}(0;t) = 1 \quad \text{and} \quad T_x\mathcal{E}(x;t) = t\mathcal{E}(x;t), \tag{10.1.7}$$

where T_x means T acts on the x variable. We shall assume the existence of $\mathcal{E}$ and that the expansion

$$\mathcal{E}(x;t) = \sum_{n=0}^{\infty} u_n(x)t^n, \tag{10.1.8}$$

holds. Analytically, this requires the analyticity of $\mathcal{E}$ as a function of t in a neighborhood of $t = 0$. Clearly

$$u_0(x) = 1, \quad \text{and} \quad Tu_n(x) = u_{n-1}(x)$$

follow from (10.1.7). By induction, we see that we can find $u_n(x)$ of exact degree n such that (10.1.7)–(10.1.8) hold.

Theorem 10.1.4 *Let $\{f_n(x)\}$ be of Sheffer A-type zero relative to T and belongs to J. Then there is $A(t)$ with $A(0) \neq 0$ and*

$$\sum_{n=0}^{\infty} f_n(x)t^n = A(t)\mathcal{E}(x;H(t)), \tag{10.1.9}$$

where $H(t)$ is the inverse function to $J(t)$, that is

$$J(t) = \sum_{n=0}^{\infty} a_k t^{k+1}, \quad J(H(t)) = H(J(t)) = t. \tag{10.1.10}$$

Conversely, if (10.1.9)–(10.1.10) hold then $\{f_n(x)\}$ is of Sheffer A-type zero relation to T and belongs to $\sum_{k=0}^{\infty} a_k T^{k+1}$.

Proof Assume (10.1.9)–(10.1.10). Then

$$\sum_{n=0}^{\infty} Jf_n t^n = J\sum_{n=0}^{\infty} f_n(x)t^n$$

$$= A(t)J\mathcal{E}(x;H(t)) = A(t)\sum_{k=0}^{\infty} a_k(H(t))^{k+1}\mathcal{E}(x;H(t))$$

$$= tA(t)\mathcal{E}(x;H(t)).$$

Therefore, $Jf_n(x) = f_{n-1}(x)$. Next assume $\{f_n(x)\}$ belongs to J, $J = \sum_{0}^{\infty} a_k T^{k+1}$, and the a_k's are constants. Let H be as in (10.1.10) and $\{u_n(x)\}$ be as in (10.1.8). Clearly, $\mathcal{E}(x, H(t)) = \sum_{n=0}^{\infty} g_n(x)t^n$. Thus, in view of (10.1.7),

$$\sum_{n=0}^{\infty} J g_n(x)t^n = J\mathcal{E}(x; H(t)) = J(H(t))\mathcal{E}(x; H(t))$$

$$= t \sum_{n=0}^{\infty} g_n(x)t^n.$$

Therefore $\{g_n(x)\}$ belongs to J and (10.1.9) follows. □

Corollary 10.1.5 *A polynomial sequence* $\{f_n(x)\}$ *is an Appell sequence relative to* T *if and only if there is a power series* $A(t) = \sum_{n=0}^{\infty} a_n t^n$, $a_0 \neq 0$ *and* $\sum_{n=0}^{\infty} f_n(x)t^n = A(t)\mathcal{E}(x; t)$.

Corollary 10.1.6 *A polynomial sequence* $\{f_n(x)\}$ *is of Sheffer A-type zero if and only if it has the generating function*

$$\sum_{n=0}^{\infty} f_n(x)\,t^n = A(t)\exp(xH(t)), \tag{10.1.11}$$

where $A(t)$ *and* $H(t)$ *are as in* (10.1.10).

Jiang Zeng gave combinatorial interpretations for the linearization coefficients for the Meixner and Meixner–Pollaczek polynomials and noted that his techniques give combinatorial interpretations for the linearization coefficients of all orthogonal polynomials of Sheffer A-type zero relative to $\dfrac{d}{dx}$. The interested reader may consult (Zeng, 1992). The recent work (Anshelevich, 2004) gives free probability interpretations of the class of orthogonal polynomials which are also of Sheffer A-type zero relative to $\dfrac{d}{dx}$.

10.2 Delta Operators

In this section we introduce delta operators and study their properties.

Definition 10.2.1 *A linear operator* T *is said to be shift invariant if* $(TE^y f)(x) = (E^y T f)(x)$ *for all polynomials* f, *where* E^y *is the shift by* y, *that is*

$$(E^y f)(x) = f(x + y). \tag{10.2.1}$$

Definition 10.2.2 *A linear operator* $\mathcal{Q}$ *acting on polynomials is called a delta operator if* $\mathcal{Q}$ *is a shift-invariant operator and* $\mathcal{Q}\,x$ *is a nonzero constant.*

Theorem 10.2.1 *Let* $\mathcal{Q}$ *be a delta operator. Then*
  (i) $\mathcal{Q}\,a = 0$ *for any constant* a.
  (ii) *If* $f_n(x)$ *is a polynomials of degree* n *in* x *then* $\mathcal{Q}\,f_n(x)$ *is a polynomial of exact degree* $n - 1$.

Proof Let $Qx = c \neq 0$. Thus

$$QE^a x = Qx + Qa = c + Qa.$$

On the other hand $QE^a x = E^a Qx = E^a c = c$. Thus $Qa = 0$ and (i) follows. To prove (ii), let $Qx^n = \sum_j b_{n,j} x^j$. Then $E^y Qx^n = QE^y x^n$ implies

$$\sum_j b_{n,j}(x+y)^j = E^y Qx^n = QE^y x^n = Q(x+y)^n. \tag{10.2.2}$$

Since $(x+y)^n - y^n$ is a polynomial in y of degree $n-1$, $Q(x+y)^n$ is a polynomial in y of degree at most $n-1$. Thus $b_{n,j} = 0$ for $j \geq n$. Equating coefficients of y^{n-1} on both sides of (10.2.2) we find $b_{n,n-1} = nQx$. Thus $b_{n,n-1} \neq 0$ and Qx^n is of exact degree $n-1$. This proves (ii). □

Definition 10.2.3 *A polynomial sequence $\{f_n(x)\}$ is called of binomial type if the f_n's satisfy the addition theorem*

$$E^y f_n(x) = f_n(x+y) = \sum_{k=0}^{n} \binom{n}{k} f_k(x) f_{n-k}(y). \tag{10.2.3}$$

The model polynomials of binomial type are the monomials $\{x^n\}$.

Definition 10.2.4 *Let Q be a delta operator. A polynomial sequence $\{f_n(x)\}$ is called the sequence of basic polynomials for Q if*

(i) $f_0(x) = 1$
(ii) $f_n(0) = 0$, *for all* $n > 0$.
(iii) $Qf_n(x) = nf_{n-1}(x)$, $n \geq 0$, *with* $f_{-1}(x) := 0$.

Theorem 10.2.2 *Every delta operator has a unique sequence of basic polynomials.*

Proof We take $f_0(x) = 1$, and construct the polynomials recursively from (iii), and determine the constant term from (ii). □

Theorem 10.2.3 *A polynomial sequence is of binomial type if and only if it is a basic sequence of some delta operator.*

Proof Let $\{f_n(x)\}$ be a basic sequence of a delta operator Q. Thus $Q^k f_n(x)\big|_{x=0} = n!\,\delta_{n,k}$. Therefore

$$f_n(x) = \sum_{k=0}^{\infty} \frac{f_k(x)}{k!} \, Q^k f_n(y)\big|_{y=0}$$

hence any polynomial $p(x)$ satisfies

$$p(x) = \sum_{k=0}^{\infty} \frac{f_k(x)}{k!} \, Q^k p(y)\big|_{y=0}. \tag{10.2.4}$$

In (10.2.4) take $p(x) = E^c f_n(x)$. Thus

$$Q^k E^c p(y) = E^c Q^k f_n(y) = \frac{n!}{(n-k)!} f_{n-k}(y+c),$$

and (10.2.4) implies $\{f_n(x)\}$ is of binomial type. Conversely let $\{f_n(x)\}$ be of binomial type and define an operator Q by $Q f_n(x) = n f_{n-1}(x)$ with $f_{-1} := 0$. To prove that Q is shift invariant, first note

$$E^y f_n(x) = f_n(x+y) = \sum_{k=0}^{n} \frac{f_k(y)}{k!} Q^k f_n(x). \tag{10.2.5}$$

Extend (10.2.5) to all polynomials so that

$$E^y p(x) = \sum_{k=0}^{n} \frac{f_k(y)}{k!} Q^k p(x).$$

With $p \to Qp$ we find

$$(E^y Q) p(x) = \sum_{k=0}^{n} \frac{g_k(y)}{k!} Q^{k+1} p(x) = Q \left(\sum_{k=0}^{n} \frac{g_k(y)}{k!} Q^k p(x) \right)$$

$$= Q E^y p(x),$$

hence Q is shift invariant, so Q is a delta operator. $\qquad\square$

Theorem 10.2.4 (Expansion Theorem) *Let $\{f_n(x)\}$ be a basic sequence of a delta operator Q and let T be a shift invariant operator. Then*

$$T = \sum_{k=0}^{\infty} \frac{a_k}{k!} Q^k, \qquad a_k := T f_k(y)|_{y=0}. \tag{10.2.6}$$

Proof Extend (10.2.5) to all polynomials via

$$p(x+y) = \sum_{k=0}^{n} \frac{f_k(y)}{k!} Q^k p(x). \tag{10.2.7}$$

Apply T to (10.2.7) then set $y = 0$ after writing $T E^y$ as $E^y T$. This establishes (10.2.6). $\qquad\square$

Corollary 10.2.5 *Any two shift invariant operators commute.*

10.3 Algebraic Theory

In (Joni & Rota, 1982) and (Ihrig & Ismail, 1981) it was pointed out that a product of functionals on the vector space of polynomials can be defined through

$$\langle LM \mid p(x) \rangle = \langle L \otimes M \mid \Delta p(x) \rangle, \tag{10.3.1}$$

where Δ is a comultiplication on the bialgebra $K[x]$, of polynomials over a field K.

Definition 10.3.1 *Let V_1 and V_2 be two modules over K. The tensor product of the linear functional L_1 and L_2 maps $V_1 \otimes V_2$ into K via*

$$\langle L_1 \otimes L_2 \,|\, v_1 \otimes v_2 \rangle = \langle L_1 \,|\, v_1 \rangle \langle L_2 \,|\, v_2 \rangle .$$ (10.3.2)

We want to characterize all polynomial sequences $\{p_n(x)\}$ which can be treated as if they were x^n. To do so we introduce a new multiplication "$*$" on $K[x]$.

In this section, $\{p_n(x)\}$ will no longer denote orthonormal polynomials but will denote a polynomial sequence.

Definition 10.3.2 *Let $\{p_n(x)\}$ be a given polynomial sequence. Then $_*K[x]$ will denote the algebra of polynomials over K with the usual addition and multiplication by scalars, but the product is*

$$p_m * p_n = p_{m+n}$$ (10.3.3)

and is extended to $_*K[x]$ by linearity.

The map $\Delta : K[x] \to K[x] \otimes K[x]$ defined by

$$\Delta(x) = x \otimes 1 + 1 \otimes x$$ (10.3.4)

and extended to arbitrary polynomials by

$$\Delta(p(x)) = p(\Delta(x)), \quad \text{for all } p \in K[x],$$ (10.3.5)

is an algebra homomorphism.

Definition 10.3.3 *Let L and M be linear functionals on $K[x]$. The product and $*$ product of L and M are defined by*

$$\langle LM \,|\, p(x) \rangle = \langle L \otimes M \,|\, \Delta p(x) \rangle,$$ (10.3.6)
$$\langle L * M \,|\, p(x) \rangle = \langle L \otimes M \,|\, \Delta^* p(x) \rangle,$$ (10.3.7)

where Δ^* is the comultipication on $_*K[x]$ defined by $\Delta^*(x) = x \otimes 1 + 1 \otimes x$ and extended as an algebra homomorphism using the $*$ product (10.3.3).

Since our model will be $\{x^n\}$, it is natural to assume that $\{p_n(x)\}$ in (10.3.3) satisfy

$$p_0(x) = 1, \qquad p_1(0) = 0.$$ (10.3.8)

Theorem 10.3.1 *Assume that $\{p_n(x)\}$ is a polynomial sequence satisfying (10.3.8) and defining a star product. The comultiplications Δ and Δ^* are equal if and only if $\{p_n(x)\}$ is of binomial type.*

Proof Since $\Delta(x) = \Delta^*(x)$, and $p_1(x) = p_1(a)x$, we find

$$\Delta^*(p_1(x)) = 1 \otimes p_1(x) + p_1(x) \otimes 1.$$

Therefore, with p^{n*} meaning the $*$ product of p n times we have

$$\Delta^* \left(p_n(x) \right) = \Delta^* \left(p_1(x)^{n*} \right) = \left(\Delta^* \left(p_1(x) \right) \right)^{n*}$$

$$= \left(p_1(x) \otimes 1 + 1 \otimes p_1(x) \right)^{n*} = \sum_{k=0}^{n} \binom{n}{k} \left(p_1(x) \otimes 1 \right)^{k*} \left(1 \otimes p_1(x) \right)^{(n-k)*}$$

$$= \sum_{k=0}^{n} \binom{n}{k} \left(p_1(x) \right)^{k*} \otimes \left(p_1(x) \right)^{(n-k)*} = \sum_{k=0}^{n} \binom{n}{k} p_k(x) \otimes p_{n-k}(x).$$

On the other hand $\Delta \left(p_n(x) \right) = p_n \left(x \otimes 1 + 1 \otimes x \right)$. Thus $\Delta = \Delta^*$ if and only if

$$p_n \left(x \otimes 1 + 1 \otimes x \right) = \sum_{k=0}^{n} \binom{n}{k} p_k(x) \otimes p_{n-k}(x)$$

$$= \sum_{k=0}^{n} \binom{n}{k} p_k(x \otimes 1) \otimes p_{n-k}(1 \otimes x).$$

Hence $\Delta = \Delta^*$ if and only if $\{ p_n(x) \}$ is of binomial type. $\qquad \square$

If $\{ p_n(x) \}$ is of binomial type then the product of functional in (10.3.1) has the property

$$\langle LM \,|\, p_n \rangle = \sum_{k=0}^{n} \binom{n}{k} \langle L \,|\, p_k \rangle \langle M \,|\, p_{n-k} \rangle. \tag{10.3.9}$$

Using this product of functionals one can establish several properties of polynomials of binomial type and how they relate to functionals. In particular we record the following results whose proofs can be found in (Ihrig & Ismail, 1981). By a degree reducing operator T we mean an operator whose action reduces the degree of a polynomial by 1.

Theorem 10.3.2 *Let $p_n(x)$ be a polynomial sequence of binomial type. Then any polynomial p has the expansion*

$$p(x) = \sum_{j=0}^{\infty} \frac{1}{j!} \left\langle \tilde{L}^j \,|\, p(x) \right\rangle p_j(x), \tag{10.3.10}$$

where $\tilde{L}$ is the functional $\tilde{L} p_n(x) = \delta_{n,0}$. Moreover there exists a degree reducing operator Q and a functional L such that

$$p(x) = \sum_{n=0}^{\infty} \langle L \,|\, Q^n \, p(x) \rangle \frac{p_n(x)}{n!} \quad p(x) \in K[x]. \tag{10.3.11}$$

The expansions given in this section provide alternatives to orthogonal expansions when the polynomials under consideration are not necessarily orthogonal.

Exercises

10.1   Let

$$\sum_{n=0}^{\infty} P_n(w) t^n = A(t) \phi(w H(t)),$$

where $A(t)$, $H(t)$ and $\phi(t)$ are formal power series with $H(t) = \sum_{n=1}^{\infty} h_n t^n$,

$A(t) = \sum_{n=0}^{\infty} a_n t^n$, $\phi(t) = \sum_{n=0}^{\infty} \phi_n t^n$ with $\phi_0 h_1 a_0 \neq 0$.

(a) Prove that $P_n(w)$ is a polynomial in w of degree n and find its leading term.

(b) Set $u = H(t)$ so that $t = t(u)$ and set

$$\{t(u)\}^n / A(t) = \sum_{j=0}^{\infty} \lambda_{n,j} u^{n+j}.$$

Show that

$$\phi_m w^m = \sum_{n=0}^{m} \lambda_{n,m-n} P_n(w).$$

(c) Conclude that

$$\sum_{n=0}^{\infty} \phi_m b_m (zw)^m = \sum_{n=0}^{\infty} z^n R_n(z) P_n(w),$$

where

$$R_n(z) = \sum_{m=0}^{\infty} b_{n+m} \lambda_{n,m} z^m.$$

(d) With

$$A(t)\{H(t)\}^n = \sum_{j=0}^{\infty} \mu_{n,j} t^{n+j},$$

show that the inverse relation to (b) is

$$P_m(w) = \sum_{j=0}^{n} \mu_{j,n-j} \phi_j w^j.$$

(e) Write down the inverse relations in (b) and (d) for parts (i)–(iii) below.

(i) $H(t) = -4t(1-t)^{-2}$, $A(t) = (1-t)^{-c}$. Show that the expansion formula in (c) becomes

$$\sum_{m=0}^{\infty} \phi_m b_m (zw)^m$$

$$= \sum_{n=0}^{\infty} \frac{(c)_{2n}(-z)^n}{n!\,(c)_{n+1}} \left[\sum_{j=0}^{\infty} \frac{(c+2n)_j}{j!} b_{n+j} z^j \right]$$

$$\times \sum_{k=0}^{n} \frac{(-n)_k (c+2k)}{(c+n+1)_k} \phi_k w^k,$$

which generalizes a formula in (Fields & Wimp, 1961) and is due to (Verma, 1972).

(ii) Repeat part (i) for $H(t) = -t/(1-t)$, $A(t) = (1-t)^{-c}$ to prove

$$\sum_{n=0}^{\infty} \phi_m b_m (zw)^m = \sum_{n=0}^{\infty} \frac{(c)_n}{n!} (-z)^n \left[\sum_{j=0}^{\infty} \frac{(n+c)_j}{j!} b_{n+j} z^j \right]$$

$$\times \sum_{k=0}^{n} \frac{(-n)_k}{(c)_k} \phi_k w^k,$$

which generalizes another result of (Fields & Wimp, 1961).

(iii) Repeat part (i) for $A(t) = (1+t^2)^{-\nu}$, $H(t) = 2t/(1+t^2)$ and establish

$$\sum_{n=0}^{\infty} \phi_m b_m (zw)^m$$

$$= \sum_{n=0}^{\infty} \frac{\nu+n}{2n} z^n \left[\sum_{j=0}^{\infty} \frac{(\nu+n+j)_j (z/2)^j}{j! (n+\nu+j)} b_{n+2j} \right]$$

$$\times \sum_{k=0}^{\lfloor n/2 \rfloor} \binom{2k-n-\nu}{k} \phi_{n-2k} (2w)^{n-2k},$$

see (Fields & Ismail, 1975).

(f) By interchanging $\{\phi_m\}$ and $\{b_m\}$ in part (c), one can derive a dual expansion because the right-hand side is not necessarily symmetric in $\{\phi_m\}$ and $\{b_m\}$.

This exercise is based on the approach given in (Fields & Ismail, 1975). For a careful convergence analysis of special cases, see Chapter 9 of (Luke, 1969b).

10.2 Use Exercise 10.1 to expand the Cauchy kernel $1/(z-w)$ in a series of ultraspherical polynomials $\{C_n^\nu(x)\}$.

10.3 Use Exercise 10.1 to give another proof of (4.8.2).
Hint: Set $x+1 = w$ in (4.8.2).

10.4 Let $\{\phi_n(x)\}$ be of Sheffer A-type zero relative to $\dfrac{d}{dx}$.

(a) Prove that

$$g_n(m, x) := \frac{d^m}{dx^m} \phi_{n+m}(x)$$

is also of Sheffer A-type zero relative to $\dfrac{d}{dx}$ and belongs to the same operator as does $\{\phi_n(x)\}$.

(b) Show that

$$\psi_n(x) = \varphi_n(x) / \prod_{j=1}^{m} (1+\rho_j)_n$$

is of Sheffer A-type m relative to $\dfrac{d}{dx}$, where $\rho_1, \ldots, \rho_m$ are constants, none of which equals -1.

10.5   If $\{P_n(x)\}$ are the Legendre polynomials, show that $\{\phi_n(x)\}$,

$$\phi_n(x) := \frac{(1+x^2)^{n/2}}{n!} P_n\left(\frac{x}{\sqrt{1+x^2}}\right),$$

is of Sheffer A-type zero relative to $\dfrac{d}{dx}$, while $\{\psi_n(x)\}$,

$$\psi_n(x) := \frac{(x-1)^n}{(n!)^2} P_n\left(\frac{x+1}{x-1}\right)$$

is of Sheffer A-type zero relative to $\dfrac{d}{dx} x \dfrac{d}{dx}$.

11

q-Series Preliminaries

11.1 Introduction

Most of the second half of this monograph is a brief introduction to the theory of
q-orthogonal polynomials. We have used a novel approach to the development of
those parts needed from the theory of basic hypergeometric functions. This chapter contains preliminary analytic results needed in the later chapters. One important
difference between our approach to basic hypergeometric functions and other approaches, for example those of Andrews, Askey and Roy (Andrews et al., 1999),
Gasper and Rahman (Gasper & Rahman, 1990), or of Bailey (Bailey, 1935) and
Slater (Slater, 1964) is our use of the divided difference operators of Askey and Wilson, the q-difference operator, and the identity theorem for analytic functions.

The identity theorem for analytic functions can be stated as follows.

Theorem 11.1.1 *Let $f(z)$ and $g(z)$ be analytic in a domain Ω and assume that*
$f(z_n) = g(z_n)$ for a sequence $\{z_n\}$ converging to an interior point of Ω. Then
$f(z) = g(z)$ at all points of Ω.

A proof of Theorem 11.1.1 is in most elementary books on complex analysis, see
for example, (Hille, 1959, p. 199), (Knopp, 1945).

In Chapter 12 we develop those parts of the theory of basic hypergeometric functions that we shall use in later chapters. Sometimes studying orthogonal polynomials
leads to other results in special functions. For example the Askey–Wilson polynomials of Chapter 15 lead directly to the Sears transformation, so the Sears transformation (12.4.3) is stated and proved in Chapter 12 but another proof is given in Chapter
15.

11.2 Orthogonal Polynomials

As we saw in Example 2.5.3 the measure with respect to which a polynomial sequence is orthogonal may not be unique.

An important criterion for the nonuniqueness of μ is stated as the following theorem (Akhiezer, 1965), (Shohat & Tamarkin, 1950).

Theorem 11.2.1 *The measure μ is not unique if and only if the series*

$$\sum_{n=0}^{\infty} |p_n(z)|^2 / \zeta_n, \tag{11.2.1}$$

converges for all z. For uniqueness it is sufficient that it diverges for one $z \notin \mathbf{R}$. If μ is unique and the series in (11.2.1) converges at $z = x_0 \in \mathbf{R}$ then μ, normalized to have total mass 1, has a mass at x_0 and the mass is

$$\left[\sum_{n=0}^{\infty} |p_n(z)|^2 / \zeta_n \right]^{-1}. \tag{11.2.2}$$

A very useful theorem to recover the absolutely continuous component of the orthogonality measure from the asymptotic behavior of the polynomials is the following theorem of (Nevai, 1979), see Corollary 40, page 140.

Theorem 11.2.2 (Nevai) *If in addition to the assumptions of the spectral theorem we assume*

$$\sum_{n=0}^{\infty} \left[\left| \sqrt{\beta_n} - \frac{1}{2} \right| + |\alpha_n| \right] < \infty, \tag{11.2.3}$$

then μ has an absolutely continuous component μ' supported on $[-1, 1]$. Furthermore if μ has a discrete part, then it will lie outside $(-1, 1)$. In addition the limiting relation

$$\limsup_{n \to \infty} \left[\sqrt{1-x^2} \frac{P_n(x)}{\sqrt{\xi_n}} - \sqrt{\frac{2\sqrt{1-x^2}}{\pi\mu'(x)}} \sin((n+1)\theta - \varphi(\theta)) \right] = 0, \tag{11.2.4}$$

holds, with $x = \cos\theta \in (-1, 1)$. In (11.2.4) $\varphi(\theta)$ does not depend on n.

The orthonormal Chebyshev polynomials of the first kind are $T_n(x)\sqrt{2/\pi}$, and their weight function is $1/\sqrt{1-x^2}$. Nevai's theorem then relates the asymptotics of general polynomials to those of the Chebyshev polynomials.

11.3 The Bootstrap Method

In the subsequent chapters we shall often make use of a procedure we shall call the "bootstrap method" where we may obtain new orthogonal functions from old ones. Assume that we know a generating function for a sequence of orthogonal polynomials $\{P_n(x)\}$ satisfying (2.1.5). Let such a generating function be

$$\sum_{n=0}^{\infty} P_n(x) t^n / c_n = G(x, t), \tag{11.3.1}$$

with $\{c_n\}$ a suitable numerical sequence of nonzero elements. Thus the orthogonality relation (2.1.5) is equivalent to

$$\int_{-\infty}^{\infty} G(x, t_1) G(x, t_2) \, d\mu(x) = \sum_{n=0}^{\infty} \zeta_n \frac{(t_1 t_2)^n}{c_n^2}, \tag{11.3.2}$$

provided that we can justify the interchange of integration and sums.

The idea is to use

$$G(x, t_1) G(x, t_2) \, d\mu(x)$$

as a new measure, the total mass of which is given by (11.3.1), and then look for a system of functions (preferably polynomials) orthogonal or biorthogonal with respect to this new measure. If such a system is found one can then repeat the process. It it clear that we cannot indefinitely continue this process. The functions involved will become too complicated at a certain level, and the process will then terminate.

If μ has compact support it will often be the case that (11.3.1) converges uniformly for x in the support and $|t|$ sufficiently small. In this case the justification of interchanging sums and integrals is obvious.

We wish to formulate a general result with no assumptions about the support of μ. For $0 < \rho \leq \infty$ we denote by $D(0, \rho)$ the set of $z \in \mathbf{C}$ with $|z| < \rho$. Recall that if the measure μ in (2.1.5) is not unique then the moment problem associated with $\{P_n(x)\}$ is called indeterminate.

Theorem 11.3.1 *Assume that* (2.1.5) *holds and that the power series*

$$\sum_{n=0}^{\infty} \frac{\sqrt{\zeta_n}}{c_n} z^n \qquad (11.3.3)$$

has a radius of convergence ρ with $0 < \rho \leq \infty$.

1. *Then there is a μ-null set $N \subseteq \mathbf{R}$ such that* (11.3.1) *converges absolutely for $|t| < \rho$, $x \in \mathbf{R} \setminus N$. Furthermore* (11.3.1) *converges in $L^2(\mu)$ for $|t| < \rho$, and* (11.3.2) *holds for $|t_1|, |t_2| < \rho$.*
2. *If μ is indeterminate then* (11.3.1) *converges absolutely and uniformly on compact subsets of $\Omega = \mathbf{C} \times D(0, \rho)$, and G is holomorphic in Ω.*

Proof For $0 < r_0 < r < \rho$ there exists $C > 0$ such that $\left(\sqrt{\zeta_n} / |c_n| \right) r^n \leq C$ for $n \geq 0$, and we find

$$\left\| \sum_{n=0}^{N} |p_n(x)| \, r_0^n / |c_n| \right\|_{L^2(\mu)} \leq \sum_{n=0}^{N} \frac{\sqrt{\zeta_n}}{|c_n|} r^n \, (r_0/r)^n \leq C \sum_{n=0}^{\infty} \left(\frac{r_0}{r} \right)^n < \infty,$$

which by the monotone convergence theorem implies that

$$\sum_{n=0}^{\infty} |p_n(x)| \frac{r_0^n}{|c_n|} \in L^2(\mu),$$

and in particular the sum is finite for μ-almost all x. This implies that there is a μ-null set $N \subseteq \mathbf{R}$ such that $\sum_{n=0}^{\infty} p_n(x) \, (t^n/c_n)$ is absolutely convergent for $|t| < \rho$ and $x \in \mathbf{R} \setminus N$.

The series (11.3.1) can be considered as a power series with values in $L^2(\mu)$, and by assumption its radius of convergence is ρ. It follows that (11.3.1) converges to $G(x, t)$ in $L^2(\mu)$ for $|t| < \rho$, and the validity of (11.3.2) is a consequence of Parseval's formula.

If μ is indeterminate it is well known that $\sum_{n=0}^{\infty} |p_n(x)|^2 / \zeta_n$ converges uniformly on compact subsets of $\mathbb{C}$, cf. (Akhiezer, 1965), (Shohat & Tamarkin, 1950), and the assertion follows.   □

11.4 *q*-Differences

A discrete analogue of the derivatives is the q-difference operator

$$(D_q f)(x) = (D_{q,x} f)(x) = \frac{f(x) - f(qx)}{(1 - q)x}. \tag{11.4.1}$$

It is clear that

$$D_{q,x} x^n = \frac{1 - q^n}{1 - q} x^{n-1}, \tag{11.4.2}$$

and for differentiable functions

$$\lim_{q \to 1^-} (D_q f)(x) = f'(x).$$

Some of the arguments in the coming chapters will become more transparent if we keep in mind the concept of q-differentiation and q-integration. The reason is that we can relate the q-results to the case $q = 1$ of classical special functions.

For finite a and b the q-integral is

$$\int_0^a f(x)\, d_q x := \sum_{n=0}^{\infty} \left[aq^n - aq^{n+1} \right] f(aq^n), \tag{11.4.3}$$

$$\int_a^b f(x)\, d_q x := \int_0^b f(x)\, d_q x - \int_0^a f(x)\, d_q x. \tag{11.4.4}$$

It is clear from (11.4.3)–(11.4.4) that the q-integral is an infinite Riemann sum with the division points in a geometric progression. We would then expect $\int_a^b f(x)\, d_q x \to \int_a^b f(x)\, dx$ as $q \to 1$ for continuous functions. The q-integral over $[0, \infty)$ uses the division points $\{ q^n : -\infty < n < \infty \}$ and is

$$\int_0^\infty f(x)\, d_q x := (1 - q) \sum_{n=-\infty}^{\infty} q^n f(q^n). \tag{11.4.5}$$

The relationship

$$\int_a^b f(x)g(qx)\, d_q x = q^{-1} \int_a^b g(x)f(x/q)\, d_q x$$
$$+ q^{-1}(1 - q)[ag(a)f(a/q) - bg(b)f(b/q)] \tag{11.4.6}$$

follows from series rearrangements. The proof is straightforward and will be omitted.

Consider the weighted inner product

$$\langle f, g \rangle_q := \int_a^b f(t)\, \overline{g(t)}\, w(t)\, d_q t$$

$$= (1-q) \sum_{k=0}^{\infty} f(y_k)\, \overline{g(y_k)}\, y_k w(y_k) \qquad (11.4.7)$$

$$- (1-q) \sum_{k=0}^{\infty} f(x_k)\, \overline{g(x_k)}\, x_k w(x_k),$$

where

$$x_k := aq^k, \qquad y_k := bq^k, \qquad (11.4.8)$$

and $w(x_k) > 0$ and $w(y_k) > 0$ for $k = 0, 1, \ldots$. We will take $a \le 0 \le b$.

Theorem 11.4.1 *An analogue of integration by parts for $D_{q,x}$ is*

$$\langle D_{q,x} f, g \rangle_q = -f(x_0)\, \overline{g(x_{-1})}\, w(x_{-1}) + f(y_0)\, \overline{g(y_{-1})}\, w(y_{-1})$$

$$- q^{-1} \left\langle f, \frac{1}{w(x)} D_{q^{-1},x}(g(x)w(x)) \right\rangle_q, \qquad (11.4.9)$$

provided that the series on both sides of (11.4.7) converge absolutely and

$$\lim_{n \to \infty} w(x_n) f(x_{n+1}) \overline{g(x_n)} = \lim_{n \to \infty} w(y_n) f(y_{n+1}) \overline{g(y_n)} = 0. \qquad (11.4.10)$$

Proof We have

$$-\frac{\langle D_{q,x} f, g \rangle_q}{1-q} = \lim_{n \to \infty} \sum_{k=0}^{n} \frac{f(x_k) - f(x_{k+1})}{x_k - x_{k+1}} \overline{g(x_k)}\, x_k w(x_k)$$

$$- \lim_{n \to \infty} \sum_{k=0}^{n} \frac{f(y_k) - f(y_{k+1})}{y_k - y_{k+1}} \overline{g(y_k)}\, y_k\, w(y_k)$$

$$= \lim_{n \to \infty} \sum_{k=0}^{n} f(x_k) \left\{ \frac{\overline{g(x_k)}\, x_k\, w(x_k)}{x_k - x_{k+1}} - \frac{\overline{g(x_{k-1})}\, x_{k-1} w(x_{k-1})}{x_{k-1} - x_k} \right\}$$

$$- \lim_{n \to \infty} \sum_{k=0}^{n} f(y_k) \left\{ \frac{\overline{g(y_k)}\, y_k w(y_k)}{y_k - y_{k+1}} - \frac{\overline{g(y_{k-1})}\, y_{k-1} w(y_{k-1})}{y_{k-1} - y_k} \right\}$$

$$+ f(x_0) \frac{\overline{g(x_{-1})}\, x_{-1} w(x_{-1})}{x_{-1} - x_0} - f(y_0) \frac{\overline{g(y_{-1})}\, y_{-1} w(y_{-1})}{y_{-1} - y_0}$$

$$= f(x_0) \frac{\overline{g(x_{-1})}\, x_{-1} w(x_{-1})}{x_{-1} - x_0} - f(y_0) \frac{\overline{g(y_{-1})}\, y_{-1} w(y_{-1})}{y_{-1} - y_0}$$

$$+ \frac{q^{-1}}{1-q} \left\langle f, \frac{1}{w(x)} D_{q^{-1},x}(g(x)w(x)) \right\rangle_q.$$

The result now follows since x_k and y_k are given by (11.4.8), so that $x_{k\pm1} = q^{\pm1} x_k$, $y_{k\pm1} = q^{\pm1} y_k$. $\qquad \square$

We will need an inner product corresponding to $q > 1$. To this end for $0 < q < 1$, we set

$$
\langle f, g \rangle_{q^{-1}} := -\frac{(1-q)}{q} \sum_{n=0}^{\infty} f(r_n) \overline{g(r_n)} r_n w(r_n)
$$

$$
+ \frac{(1-q)}{q} \sum_{n=0}^{\infty} f(s_n) \overline{g(s_n)} s_n w(s_n), \tag{11.4.11}
$$

where

$$
r_n := \alpha q^{-n}, \quad s_n = \beta q^{-n}, \tag{11.4.12}
$$

and w is a function positive at r_n and s_n. The quantity $\langle ., . \rangle_{q^{-1}}$ is the definition of the weighted inner product in this case. A proof similar to that of Theorem 11.4.1 establishes the following analogue of integration by parts:

$$
\langle D_{q^{-1},x} f, g \rangle_{q^{-1}} = -f(r_0) \frac{\overline{g(r_{-1})} r_{-1} w(r_{-1})}{r_{-1} - r_0} + f(s_0) \frac{\overline{g(s_{-1})} s_{-1} w(s_{-1})}{s_{-1} - s_0}
$$

$$
- q \left\langle f, \frac{x}{w(x)} D_{q,x}(g(x)w(x)) \right\rangle_{q^{-1}}, \tag{11.4.13}
$$

provided that both sides are well-defined and

$$
\lim_{n \to \infty} \left[-w(r_n) r_n f(r_{n+1}) \overline{g(r_n)} + w(s_n) f(s_{n+1}) \overline{g(s_n)} \right] = 0. \tag{11.4.14}
$$

The product rule for D_q is

$$
(D_q f g)(x) = f(x)(D_q g)(x) + g(qx)(D_q f)(x). \tag{11.4.15}
$$

12

q-Summation Theorems

Before we can state the summation theorems needed in the development of q-orthogonal polynomials we wish to introduce some standard notation.

12.1 Basic Definitions

The q-shifted factorials are

$$(a;q)_0 := 1, \quad (a;q)_n := \prod_{k=1}^{n} \left(1 - aq^{k-1}\right), \quad n = 1, 2, \ldots, \text{ or } \infty, \qquad (12.1.1)$$

and the multiple q-shifted factorials are defined by

$$(a_1, a_2, \ldots, a_k; q)_n := \prod_{j=1}^{k} (a_j; q)_n. \qquad (12.1.2)$$

We shall also use

$$(a;q)_\alpha = (a;q)_\infty / (aq^\alpha; q)_\infty, \qquad (12.1.3)$$

which agrees with (12.1.1) when $\alpha = 0, 1, 2, \ldots$ but holds for general α when $aq^\alpha \neq q^{-n}$ for a nonnegative integer n. The q-binomial coefficient is

$$\begin{bmatrix} n \\ k \end{bmatrix}_q := \frac{(q;q)_n}{(q;q)_k (q;q)_{n-k}}. \qquad (12.1.4)$$

Unless we say otherwise we shall always assume that

$$0 < q < 1. \qquad (12.1.5)$$

A basic hypergeometric series is

$$\begin{aligned}
{}_r\phi_s \left(\begin{matrix} a_1, \ldots, a_r \\ b_1, \ldots, b_s \end{matrix} \middle| q, z \right) &= {}_r\phi_s \left(a_1, \ldots, a_r; b_1, \ldots, b_s; q, z \right) \\
&= \sum_{n=0}^{\infty} \frac{(a_1, \ldots, a_r; q)_n}{(q, b_1, \ldots, b_s; q)_n} z^n \left(-q^{(n-1)/2} \right)^{n(s+1-r)}.
\end{aligned}$$

$$(12.1.6)$$

Note that $\left(q^{-k}; q\right)_n = 0$ for $n = k+1, k+2, \ldots$. To avoid trivial singularities or indeterminancies in (12.1.6) we shall always assume, unless otherwise stated, that none of the denominator parameters $b_1, \ldots, b_s$ in (12.1.6) has the form q^{-k}, $k = 0, 1, \ldots$. If one of the numerator parameters is of the form q^{-k} then the sum on the right-hand side of (12.1.6) is a finite sum and we say that the series in (12.1.6) is **terminating**. A series that does not terminate is called **nonterminating**.

The radius of convergence of the series in (12.1.6) is 1, 0 or ∞ accordingly as $r = s+1$, $r > s+1$ or $r < s+1$, as can be seen from the ratio test.

These notions extend the notions of shifted and multishifted factorials and the generalized hypergeometric functions introduced in §1.3. It is clear that

$$\lim_{q \to 1^-} \frac{(q^a; q)_n}{(1-q)^n} = (a)_n, \tag{12.1.7}$$

hence

$$\lim_{q \to 1^-} {}_r\phi_s \left(\begin{matrix} q^{a_1}, \ldots, q^{a_r} \\ q^{b_1}, \ldots, q^{b_s} \end{matrix} \,\middle|\, q, z(1-q)^{s+1-r} \right)$$

$$= {}_rF_s \left(\begin{matrix} a_1, \ldots, a_r \\ b_1, \ldots, b_s \end{matrix} \,\middle|\, (-1)^{s+1-r} z \right), \quad r \le s+1. \tag{12.1.8}$$

There are two key operators used in our analysis of q functions. The first is the q-difference operator D_q defined in (11.4.1). The second is the Askey–Wilson operator $\mathcal{D}_q$, which will be defined below. Given a polynomial f we set $\breve{f}\left(e^{i\theta}\right) := f(x)$, $x = \cos\theta$, that is

$$\breve{f}(z) = f((z+1/z)/2), \quad z = e^{\pm i\theta}. \tag{12.1.9}$$

In other words we think of $f(\cos\theta)$ as a function of $e^{i\theta}$ or $e^{-i\theta}$. In this notation the Askey–Wilson divided difference operator $\mathcal{D}_q$ is defined by

$$(\mathcal{D}_q f)(x) := \frac{\breve{f}\left(q^{1/2}e^{i\theta}\right) - \breve{f}\left(q^{-1/2}e^{i\theta}\right)}{\breve{e}\left(q^{1/2}e^{i\theta}\right) - \breve{e}\left(q^{-1/2}e^{i\theta}\right)}, \quad x = \cos\theta, \tag{12.1.10}$$

with

$$e(x) = x. \tag{12.1.11}$$

A calculation reduces (12.1.10) to

$$(\mathcal{D}_q f)(x) = \frac{\breve{f}\left(q^{1/2}e^{i\theta}\right) - \breve{f}\left(q^{-1/2}e^{i\theta}\right)}{\left(q^{1/2} - q^{-1/2}\right)(z-1/z)/2}, \quad x = (z+1/z)/2. \tag{12.1.12}$$

It is important to note that although we use $x = \cos\theta$, θ is not necessarily real. In fact z and z^{-1} are defined as

$$z, z^{-1} = x \pm \sqrt{x^2 - 1}, \quad |z| \ge 1. \tag{12.1.13}$$

The branch of the square root is taken such that $\sqrt{x+1} > 0$, $x > -1$. This makes $z = e^{\pm i\theta}$ if $\operatorname{Im} x \lessgtr 0$.

As an example, let us apply $\mathcal{D}_q$ to a Chebyshev polynomial. Recall that the Chebyshev polynomials of the first kind and second kinds, $T_n(x)$ and $U_n(x)$, respectively, are

$$T_n(\cos\theta) := \cos(n\theta), \tag{12.1.14}$$

$$U_n(\cos\theta) := \frac{\sin((n+1)\theta)}{\sin\theta}. \tag{12.1.15}$$

Both T_n and U_n have degree n. Thus

$$\check{T}_n(z) = \left(z^n + z^{-n}\right)/2. \tag{12.1.16}$$

A calculation gives

$$\mathcal{D}_q T_n(x) = \frac{q^{n/2} - q^{-n/2}}{q^{1/2} - q^{-1/2}} U_{n-1}(x). \tag{12.1.17}$$

Therefore

$$\lim_{q\to 1} \left(\mathcal{D}_q f\right)(x) = f'(x), \tag{12.1.18}$$

holds for $f = T_n$, hence for all polynomials, since $\{T_n(x)\}_0^\infty$ is a basis for the vector space of all polynomials and $\mathcal{D}_q$ is a linear operator. In Chapter 16 we will extend the definition of $\mathcal{D}_q$ to q-differentiable functions and show how to obtain the Wilson operator (Wilson, 1982) as a limiting case of $\mathcal{D}_q$.

In defining $\mathcal{D}_q$ we implicitly used the q-shifts

$$\left(\eta_q \check{f}\right)(z) = \check{f}\left(q^{1/2}z\right), \quad \left(\eta_q^{-1}\check{f}\right)(z) = \check{f}\left(q^{-1/2}z\right). \tag{12.1.19}$$

The product rule for $\mathcal{D}_q$ is

$$\mathcal{D}_q(fg) = \eta_q f \mathcal{D}_q g + \eta_q^{-1} g \mathcal{D}_q f. \tag{12.1.20}$$

The averaging operator $\mathcal{A}_q$

$$\left(\mathcal{A}_q f\right)(x) = \frac{1}{2}\left[\eta_q \check{f}(z) + \eta_q^{-1}\check{f}(z)\right], \tag{12.1.21}$$

enables us to rewrite (12.1.20) in the more symmetric form

$$\mathcal{D}_q(fg) = (\mathcal{A}_q f)(\mathcal{D}_q g) + (\mathcal{A}_q g)(\mathcal{D}_q f). \tag{12.1.22}$$

An induction argument using (12.1.20) implies (Cooper, 1996)

$$\mathcal{D}_q^n f(x) = \frac{(2z)^n q^{n(3-n)/4}}{(q-1)^n}$$
$$\times \sum_{k=0}^n \begin{bmatrix} n \\ k \end{bmatrix}_q \frac{q^{k(n-k)} z^{-2k}\eta^{2k-n}\check{f}(z)}{(q^{n-2k+1}z^{-2};q)_k (z^2 q^{2k+1-n};q)_{n-k}}. \tag{12.1.23}$$

We will use the Askey–Wilson operator to derive some of the summation theorems needed in our treatment but before we do so we need to introduce a q-analogue of the gamma function. The q-gamma function is

$$\Gamma_q(z) := \frac{(q;q)_\infty}{(1-q)^{z-1}(q^z;q)_\infty}. \tag{12.1.24}$$

It satisfies the functional equation

$$\Gamma_q(z+1) = \frac{1-q^z}{1-q}\,\Gamma_q(z), \tag{12.1.25}$$

and extends the shifted factorial in the sense $\Gamma_q(n) = (q;q)_n/(1-q)^n$. The q-analogue of the Bohr–Mollerup theorem asserts that the only log convex solution to

$$y(x+1) = (1-q^x)\,y(x)/(1-q), \quad y(1) = 1$$

is $y(x) = \Gamma_q(x)$ (Andrews et al., 1999). A very elegant proof of

$$\lim_{q\to 1^-} \Gamma_q(z) = \Gamma(z), \tag{12.1.26}$$

is due to R. W. Gosper. We include here the version in Andrews' wonderful monograph (Andrews, 1986), for completeness.

Proof of (12.1.26)

$$\begin{aligned}
\Gamma_q(z+1) &= \left(\prod_{n=0}^{\infty} \frac{(1-q^{n+1})}{(1-q^{n+z+1})}\right)(1-q)^{-z} \\
&= \left(\prod_{n=1}^{\infty} \frac{(1-q^n)}{(1-q^{n+z})}\right)(1-q)^{-z} \\
&= \prod_{n=1}^{\infty} \frac{(1-q^n)\,(1-q^{n+1})^z}{(1-q^{n+z})\,(1-q^n)^z}.
\end{aligned}$$

The last step follows from the fact that

$$\prod_{n=1}^{m} \frac{(1-q^{n+1})^z}{(1-q^n)^z} = \frac{(1-q^{m+1})^z}{(1-q)^z}$$

which tends to $(1-q)^{-z}$ as $m\to\infty$. Therefore

$$\begin{aligned}
\lim_{q\to 1^-} \Gamma_q(z+1) &= \prod_{n=1}^{\infty} \frac{n}{n+z}\left(\frac{n+1}{n}\right)^z \\
&= \prod_{n=1}^{\infty} \frac{n}{n+z}\left(1+\frac{1}{n}\right)^z = \Gamma(z+1),
\end{aligned}$$

where the last statement is from (Rainville, 1960). $\qquad\qquad\square$

Formula (12.1.26) will be useful in formulating the limiting results $q \to 1$ of what is covered in this work.

12.2 Expansion Theorems

In the calculus of the Askey–Wilson operator the basis $\{\phi_n(x;a) : 0 \le n < \infty\}$

$$\phi_n(x;a) := \left(ae^{i\theta}, ae^{-i\theta}; q\right)_n = \prod_{k=0}^{n-1}\left[1 - 2axq^k + a^2q^{2k}\right], \tag{12.2.1}$$

plays the role played by the monomials $\{x^n : 0 \le n < \infty\}$ in the differential and integral calculus.

Theorem 12.2.1 *We have*

$$\mathcal{D}_q \left(ae^{i\theta}, ae^{-i\theta}; q\right)_n = -\frac{2a\left(1 - q^n\right)}{1 - q} \left(aq^{1/2}e^{i\theta}, aq^{1/2}e^{-i\theta}; q\right)_{n-1}. \qquad (12.2.2)$$

Proof Here we take $f(x) = \left(ae^{i\theta}, ae^{-i\theta}; q\right)_n$, hence $\breve{f}(z) = (az, a/z; q)_n$. The rest is an easy calculation. $\qquad \square$

Theorem 12.2.1 shows that the Askey–Wilson operator $\mathcal{D}_q$ acts nicely on the polynomials $\left(ae^{i\theta}, ae^{-i\theta}; q\right)_n$. Therefore it is natural to use

$$\left\{\left(ae^{i\theta}, ae^{-i\theta}; q\right)_n : n = 0, 1, \dots\right\}$$

as a basis for polynomials when we deal with the Askey–Wilson operator. Our next theorem provides an expansion formula for polynomials in terms of the basis

$$\left\{\left(ae^{i\theta}, ae^{-i\theta}; q\right)_n : n = 0, 1, \dots\right\}.$$

Theorem 12.2.2 (Expansion Theorem) *Let f be a polynomial of degree n, then*

$$f(x) = \sum_{k=0}^n f_k \left(ae^{i\theta}, ae^{-i\theta}; q\right)_k, \qquad (12.2.3)$$

where

$$f_k = \frac{(q-1)^k}{(2a)^k (q;q)_k} q^{-k(k-1)/4} \left(\mathcal{D}_q^k f\right)(x_k) \qquad (12.2.4)$$

with

$$x_k := \frac{1}{2}\left(aq^{k/2} + q^{-k/2}/a\right). \qquad (12.2.5)$$

Proof It is clear that the expansion (12.2.3) exists, so we now compute the f_k's. Formula (12.2.2) yields

$$\mathcal{D}_q^k \left(ae^{i\theta}, ae^{-i\theta}; q\right)_n \Big|_{x=x_k} \qquad (12.2.6)$$

$$= (2a)^k \frac{q^{(0+1+\cdots+k-1)/2}(q;q)_n}{(q-1)^k(q;q)_{n-k}} \left(aq^{k/2}e^{i\theta}, aq^{k/2}e^{-i\theta}; q\right)_{n-k} \Big|_{e^{i\theta}=aq^{k/2}}$$

$$= \frac{(q;q)_k}{(q-1)^k} (2a)^k q^{k(k-1)/4} \delta_{k,n}.$$

The theorem now follows by applying $\mathcal{D}_q^j$ to both sides of (12.2.3) then setting $x = x_j$. $\qquad \square$

We need some elementary properties of the q-shifted factorials. It is clear from (12.1.1) that

$$(a; q)_n = (a; q)_\infty / (aq^n; q)_\infty, \quad n = 0, 1, \dots.$$

This suggests the following definition for q-shifted factorials of negative order

$$(a;q)_n := (a;q)_\infty / (aq^n;q)_\infty = 1/(aq^n;q)_{-n}, \quad n = -1,-2,\ldots . \quad (12.2.7)$$

It is easy to see that

$$(a;q)_m \, (aq^m;q)_n = (a;q)_{m+n}, \quad m,n = 0,\pm1,\pm2,\ldots . \quad (12.2.8)$$

Some useful identities involving q-shifted factorials are

$$\left(aq^{-n};q\right)_k = \frac{(a;q)_k(q/a;q)_n}{(q^{1-k}/a;q)_n} q^{-nk}, \quad (12.2.9)$$

$$\left(aq^{-n};q\right)_n = (q/a;q)_n(-a)^n q^{-n(n+1)/2}, \quad (12.2.10)$$

$$(a;q)_{n-k} = \frac{(a;q)_n}{(q^{1-n}/a;q)_k}(-a)^{-k} q^{\frac{1}{2}k(k+1)-nk}, \quad (12.2.11)$$

$$\frac{(a;q)_{n-k}}{(b;q)_{n-k}} = \frac{(a;q)_n \left(q^{1-n}/b;q\right)_k}{(b;q)_n \left(q^{1-n}/a;q\right)_k} \left(\frac{b}{a}\right)^k, \quad (12.2.12)$$

$$\left(a;q^{-1}\right)_n = (1/a;q)_n(-a)^n q^{-n(n-1)/2}. \quad (12.2.13)$$

The identities (12.2.9)–(12.2.13) follow from the definitions (12.1.1) and (12.2.8).

We are now in a position to prove the q-analogue of the Pfaff–Saalschütz theorem. Recall that a basic hypergeometric function (12.1.6) is called **balanced** if

$$r = s+1 \quad \text{and} \quad qa_1a_2\cdots a_{s+1} = b_1b_2\cdots b_s. \quad (12.2.14)$$

Theorem 12.2.3 (q-Pfaff–Saalschütz) *The sum of a terminating balanced $3\phi_2$ is given by*

$$3\phi_2 \left(\begin{matrix} q^{-n}, a, b \\ c, d \end{matrix} \, \middle| \, q, q \right) = \frac{(d/a, d/b;q)_n}{(d, d/ab;q)_n}, \quad (12.2.15)$$

with $cd = abq^{1-n}$.

Proof Apply Theorem 12.2.2 to the function

$$f(\cos\theta) = \left(be^{i\theta}, be^{-i\theta};q\right)_n.$$

Using (12.2.2), (12.2.3) and (12.2.4) we obtain

$$f_k = \frac{(q;q)_n \, (b/a)^k}{(q;q)_k(q;q)_{n-k}} \left(bq^{k/2}e^{i\theta}, bq^{k/2}e^{-i\theta};q\right)_{n-k} \Bigg|_{e^{i\theta}=aq^{k/2}}$$

$$= \frac{(q;q)_n \, (b/a)^k}{(q;q)_k(q;q)_{n-k}} \left(abq^k, b/a;q\right)_{n-k}.$$

Therefore (12.2.3) becomes

$$\frac{\left(be^{i\theta}, be^{-i\theta};q\right)_n}{(q;q)_n} = \sum_{k=0}^{n} \frac{b^k \left(ae^{i\theta}, ae^{-i\theta};q\right)_k \left(abq^k, b/a;q\right)_{n-k}}{a^k(q;q)_k(q;q)_{n-k}},$$

that is

$$\frac{\left(be^{i\theta}, be^{-i\theta};q\right)_n}{(q, ab;q)_n} = \sum_{k=0}^{n} \frac{\left(ae^{i\theta}, ae^{-i\theta};q\right)_k}{(q, ab;q)_k} \left(\frac{b}{a}\right)^k \frac{(b/a;q)_{n-k}}{(q;q)_{n-k}}. \quad (12.2.16)$$

Using (12.2.12) we can rewrite the above equation in the form

$$\frac{(be^{i\theta}, be^{-i\theta}; q)_n}{(ab, b/a; q)_n} = {}_3\phi_2\left(\begin{array}{c} q^{-n}, ae^{i\theta}, ae^{-i\theta} \\ ab, q^{1-n}a/b \end{array}\bigg| q, q\right),$$

which is equivalent to (12.2.15). □

Our next result gives a q-analogue of the Chu–Vandermonde sum and Gauss's theorem for hypergeometric functions stated in §1.4. For proofs, we refer the interested reader to (Andrews et al., 1999) and (Slater, 1964).

Theorem 12.2.4 *We have the q-analogue of the Chu–Vandermonde sum*

$$ {}_2\phi_1\left(q^{-n}, a; c; q, q\right) = \frac{(c/a; q)_n}{(c; q)_n} a^n, \tag{12.2.17}$$

and the q-analogue of Gauss's theorem

$$ {}_2\phi_1(a, b; c; q, c/ab) = \frac{(c/a, c/b; q)_\infty}{(c, c/ab; q)_\infty}, \quad |c/ab| < 1. \tag{12.2.18}$$

Proof Let $n \to \infty$ in (12.2.16). Taking the limit inside the sum is justified since $(a, b; q)_k/(q, c; q)_k$ is bounded. The result is (12.2.18). When $b = q^{-n}$ then (12.2.18) becomes

$$ {}_2\phi_1\left(q^{-n}, a; c; q, cq^n/a\right) = \frac{(c/a; q)_n}{(c; q)_n}. \tag{12.2.19}$$

To prove (12.2.17) we express the left-hand side of (12.2.19) as a sum, over k say, replace k by $n - k$, then apply (12.2.12) and arrive at formula (12.2.17) after some simplifications and substitutions. This completes the proof. □

The approach presented so far is from the author's paper (Ismail, 1995).

When we replace a, b, c by q^a, q^b, q^c, respectively, in (12.2.18), then apply (12.1.7), (12.1.17) and (12.1.18), we see that (12.2.18) reduces to Gauss's theorem (Rainville, 1960)

$$ {}_2F_1(a, b; c; 1) = \frac{\Gamma(c)\Gamma(c - a - b)}{\Gamma(c - a)\Gamma(c - b)}, \quad \text{Re}(c - a - b) > 0. \tag{12.2.20}$$

Remark 12.2.1 *Our proof of Theorem 12.2.4 shows that the terminating q-Gauss sum (12.2.19) is equivalent to the terminating q-Chu–Vandermonde sum (12.2.17). It is not true however that the nonterminating versions of (12.2.19) and (12.2.17) are equivalent. The nonterminating version of (12.2.19) is (12.2.18) but the nonterminating version of (12.2.17) is*

$$\frac{(aq/c, bq/c; q)_\infty}{(q/c; q)_\infty} {}_2\phi_1\left(\begin{array}{c} a, b \\ c \end{array}\bigg| q, q\right)$$

$$+ \frac{(a, b; q)_\infty}{(c/q; q)_\infty} {}_2\phi_1\left(\begin{array}{c} aq/c, bq/c \\ q^2/c \end{array}\bigg| q, q\right) \tag{12.2.21}$$

$$= (abq/c; q)_\infty.$$

We shall give a proof of (12.2.21) in Chapter 18 when we discuss the Al-Salam–Carlitz polynomials. This is one place where orthogonal polynomials provide an insight into the theory of basic hypergeometric functions.

Theorem 12.2.5 *If $|z| < 1$ or $a = q^{-n}$ then*

$$_1\phi_0(a; -; q, z) = \frac{(az; q)_\infty}{(z; q)_\infty}. \tag{12.2.22}$$

Proof Let $c = abz$ in (12.2.18) then let $b \to 0$. The result is (12.2.22). □

Note that as $q \to 1^-$ the left-hand side of (12.2.22), with a replaced by q^a, tends to $\sum_{n=0}^\infty (a)_n z^n / n!$, hence, by the binomial theorem the right-hand side must tend to $(1 - z)^{-a}$ and we have

$$\lim_{q \to 1^-} \frac{(q^a z; q)_\infty}{(z; q)_\infty} = (1 - z)^{-a}. \tag{12.2.23}$$

Theorem 12.2.6 (Euler) *We have*

$$e_q(z) := \sum_{n=0}^\infty \frac{z^n}{(q; q)_n} = \frac{1}{(z; q)_\infty}, \quad |z| < 1, \tag{12.2.24}$$

and

$$E_q(z) := \sum_{n=0}^\infty \frac{z^n}{(q; q)_n} q^{n(n-1)/2} = (-z; q)_\infty. \tag{12.2.25}$$

Proof Formula (12.2.24) is the special case $a = 0$ of (12.2.22). To get (12.2.25), we replace z by $-z/a$ in (12.2.22) and let $a \to \infty$. This and (12.1.1) establish (12.2.25) and the proof is complete. □

The left-hand sides of (12.2.24) and (12.2.25) are q-analogues of the exponential function. It readily follows that $e_q((1 - q)x) \to e^x$, and $E_q((1 - q)x) \to e^x$ as $q \to 1^-$.

The terminating version of the q-binomial theorem is

$$_1\phi_0\left(q^{-n}; -; q, z\right) = \left(q^{-n} z; q\right)_n = (-z)^n q^{-n(n+1)/2}(q/z; q)_n, \tag{12.2.26}$$

which follows from (12.2.22). The above identity may be written as

$$(z; q)_n = \sum_{k=0}^n \begin{bmatrix} n \\ k \end{bmatrix}_q q^{\binom{k}{2}}(-z)^k. \tag{12.2.27}$$

The $_6\phi_5$ summation theorem

$$_6\phi_5\left(\begin{array}{c} a, q\sqrt{a}, -q\sqrt{a}, b, c, d \\ \sqrt{a}, -\sqrt{a}, aq/b, aq/c, aq/d \end{array} \middle| q, \frac{aq}{bcd}\right)$$
$$= \frac{(aq, aq/bc, aq/bd, aq/cd; q)_\infty}{(aq/b, aq/c, aq/d, aq/bcd; q)_\infty}, \tag{12.2.28}$$

evaluates the sum of a very well-poised $_6\phi_5$ and was first proved by Rogers. The definition of a very well-poised series is given in (12.5.12). When $d = q^{2n}$, (12.2.28) follows from applying Cooper's formula (12.1.23) to the function

$$f(\cos\theta) = \left(\alpha e^{i\theta}, \alpha e^{-i\theta}; q\right)_\infty / \left(\beta e^{i\theta}, \beta e^{-i\theta}; q\right)_\infty. \qquad (12.2.29)$$

Indeed

$$\mathcal{D}_q f(\cos\theta) = \frac{2(\beta - \alpha)}{1 - q} \frac{\left(\alpha q^{1/2} e^{i\theta}, \alpha q^{1/2} e^{-i\theta}; q\right)_\infty}{\left(\beta q^{-1/2} e^{i\theta}, \beta q^{-1/2} e^{-i\theta}; q\right)_\infty}$$

gives

$$\mathcal{D}_q^n f(\cos\theta) = \frac{2^n \beta^n (\alpha/\beta; q)_n}{(1 - q)^n} q^{n(1-n)/4} \frac{\left(\alpha q^{n/2} e^{i\theta}, \alpha q^{n/2} e^{-i\theta}; q\right)_\infty}{\left(\beta q^{-n/2} e^{i\theta}, \beta q^{-n/2} e^{-i\theta}; q\right)_\infty}.$$

Replace n by $2n$, then substitute in (12.1.23) to obtain, with $z = e^{i\theta}$,

$$\frac{\beta^{2n} (\alpha/\beta; q)_{2n} (\alpha q^n z, \alpha q^n/z; q)_\infty}{(\beta q^{-2n} z, \beta q^{-n}(z; q))}$$

$$= z^{2n} q^n \sum_{k=0}^{2n} \frac{\left(q^{-2n}, \beta q^{-n} z, z^2 q^{-2n}; q\right)_k}{(q, \alpha q^{-n} z; q)_k} \frac{\left(\alpha q^{n-k}/z; q\right)_k}{\left(\beta q^{n-k}/z; q\right)_k}$$

$$\times \frac{\left(1 - z^2 q^{2k-2n}\right)}{(z^2 q^{-2n}; q)_{k+1+2n}} \frac{(\alpha q^n/z, \alpha q^{-n} z; q)_\infty}{(\beta q^n/z, \beta q^{-n} z; q)_\infty} q^{2nk}.$$

After replacing z by zq^n and simplification we find

$$_6\phi_5 \left(\begin{matrix} z^2, qz, -qz, \beta z, qz/\alpha, q^{-2n} \\ z, -z, qz/\beta, \alpha z, q^{2n+1} z^2 \end{matrix} \middle| q, q^{2n} \frac{\alpha}{\beta} \right)$$

$$= \frac{(\alpha/\beta, qz^2; q)_{2n}}{(\alpha z, qz/\beta; q)_{2n}}. \qquad (12.2.30)$$

Write the right-hand side of the above as

$$\left(\alpha/\beta, qz^2, \alpha z q^{2n}, q^{2n+1} z/\beta; q\right)_\infty / \left(\alpha z, qz/\beta, \alpha q^{2n}/\beta, q^{2n+1} z^2; q\right)_\infty$$

then observe that both sides of (12.2.30) are analytic functions in q^{2n} in a neighborhood of the origin. The identity theorem now establishes (12.2.28).

The terminating version of (12.2.28) is

$$_6\phi_5 \left(\begin{matrix} a, q\sqrt{a}, -q\sqrt{a}, b, c, q^{-n} \\ \sqrt{a}, -\sqrt{a}, qa/b, qa/c, q^{n+1}a \end{matrix} \middle| q, \frac{aq^{n+1}}{bc} \right)$$

$$= \frac{(aq, aq/bc; q)_n}{(aq/b, aq/c; q)_n}. \qquad (12.2.31)$$

12.3 Bilateral Series

Recall that $(a; q)_n$ for $n < 0$ has been defined in (12.2.7). A bilateral basic hypergeometric function is

$$_m\psi_m \left(\begin{matrix} a_1, \ldots, a_m \\ b_1, \ldots, b_m \end{matrix} \middle| q, z \right) = \sum_{-\infty}^{\infty} \frac{(a_1, \ldots, a_m; q)_n}{(b_1, \ldots, b_m; q)_n} z^n. \qquad (12.3.1)$$

It is easy to see that the series in (12.3.1) converges if

$$\left| \frac{b_1 b_2 \cdots b_m}{a_1 a_2 \cdots a_m} \right| < |z| < 1. \tag{12.3.2}$$

Our next result is the Ramanujan ${}_1\psi_1$ sum.

Theorem 12.3.1 *The following holds for* $|b/a| < |z| < 1$

$${}_1\psi_1(a; b; q, z) = \frac{(b/a, q, q/az, az; q)_\infty}{(b, b/az, q/a, z; q)_\infty}. \tag{12.3.3}$$

Proof (Ismail, 1977b) Observe that both sides of (12.3.3) are analytic function of b for $|b| < |az|$ and, by (12.2.7), we have

$${}_1\psi_1(a; b; q, z) = \sum_{n=1}^{\infty} \frac{(a; q)_n}{(b; q)_n} z^n + \sum_{n=0}^{\infty} \frac{(q/b; q)_n}{(q/a; q)_n} \left(\frac{b}{az} \right)^n.$$

Furthermore when $b = q^{m+1}$, m a positive integer, then $1/(b; q)_n = (bq^n; q)_{-n} = 0$ for $n < -m$, see (12.2.7). Therefore

$${}_1\psi_1\left(a; q^{m+1}; q, z\right) = \sum_{n=-m}^{\infty} \frac{(a; q)_n}{(q^{m+1}; q)_n} z^n$$

$$= z^{-m} \frac{(a; q)_{-m}}{(q^{m+1}; q)_{-m}} \sum_{n=0}^{\infty} \frac{(aq^{-m}; q)_n}{(q; q)_n} z^n$$

$$= z^{-m} \frac{(a; q)_{-m}}{(q^{m+1}; q)_{-m}} \frac{(azq^{-m}; q)_\infty}{(z; q)_\infty}$$

$$= \frac{z^{-m}(az; q)_\infty (q, azq^{-m}; q)_m}{(z; q)_\infty (aq^{-m}; q)_m}.$$

Using (12.2.8) and (12.2.10) we simplify the above formula to

$${}_1\psi_1(a; b; q, z) = \frac{\left(q^{m+1}/a, q, q/az, az; q\right)_\infty}{\left(q^{m+1}, q^{m+1}/az, q/a, z; q\right)_\infty},$$

which is (12.3.3) with $b = q^{m+1}$. The identity theorem for analytic functions then establishes the theorem.   □

Another proof of Theorem 12.3.1 using functional equations is in (Andrews & Askey, 1978). A probabilistic proof is in (Kadell, 1987). Combinatorial proofs are in (Kadell, 2005) and (Yee, 2004). Recently, Schlosser showed that the ${}_1\psi_1$ sum follows from the Pfaff–Saalschütz theorem, (Schlosser, 2005). For other proofs, see the references in (Andrews & Askey, 1978), (Gasper & Rahman, 1990), (Ismail, 1977b). The combinatorics of the ${}_1\psi_1$ sum have been studied in (Corteel & Lovejoy, 2002).

Theorem 12.3.2 (Jacobi Triple Product Identity) *We have*

$$\sum_{-\infty}^{\infty} q^{n^2} z^n = \left(q^2, -qz, -q/z; q^2\right)_\infty. \tag{12.3.4}$$

Proof Formula (12.3.3) implies

$$
\sum_{-\infty}^{\infty} q^{n^2} z^n = \lim_{c\to 0} {}_1\psi_1\left(-1/c; 0; q^2, qzc\right) = \lim_{c\to 0} \frac{\left(q^2, -qz, -q/z; q^2\right)_\infty}{\left(-q^2 c, qcz; q^2\right)_\infty}
$$

$$
= \left(q^2, -qz, -q/z; q^2\right)_\infty,
$$

which is (12.3.4). $\square$

As we proved the ${}_1\psi_1$ sum from the q-binomial theorem, one can use (12.2.28) to prove

$$
{}_6\psi_6\left(\begin{array}{c} q\sqrt{a}, -q\sqrt{a}, b, c, d, e \\ \sqrt{a}, -\sqrt{a}, aq/b, aq/c, aq/d, aq/e \end{array}\Bigg| q, \frac{qa^2}{bcde}\right)
$$
$$
= \frac{(aq, aq/bc, aq/bd, aq/be, q/cd, q/ce, q/de, q, q/a; q)_\infty}{(aq/b, aq/c, aq/d, aq/e, q/b, q/c, q/d, q/e, aq^2/bcde; q)_\infty}.
$$
                                                                           (12.3.5)

Theorem 12.3.3 *The Ramanujan q-beta integral*

$$
\int_0^\infty t^{c-1} \frac{(-tb, -qa/t; q)_\infty}{(-t, -q/t; q)_\infty}\, dt = \frac{\pi}{\sin(\pi c)} \frac{(q^c, q^{1-c}, ab; q)_\infty}{(aq^c, bq^{-c}, q; q)_\infty}
$$
                                                                           (12.3.6)

holds for $|q^c a| < 1$, $|q^{-c} b| < 1$.

Proof Write $\int_0^\infty$ as $\sum_{n=-\infty}^{\infty} \int_{q^{n+1}}^{q^n}$, then replace t by tq^n to see that the left-hand side of (12.3.6) is

$$
\int_q^1 t^{c-1} \sum_{n=-\infty}^{\infty} \frac{(-tbq^n, -q^{1-n}a/t; q)_\infty}{(-tq^n, -q^{1-n}/t; q)_\infty}\, q^{nc}\, dt.
$$

The above sum is

$$
\frac{(-tb, -qa/t; q)_\infty}{(-t, -q/t; q)_\infty}\, {}_1\psi_1\left(-t/a; -bt; q, aq^c\right)
$$
$$
= \frac{(q, ab; q)_\infty}{(aq^c, bq^{-c}; q)_\infty}\, \frac{(-q^c t, -q^{1-c}/t; q)_\infty}{(-t, -q/t; q)_\infty}.
$$

Therefore, the left-hand side of (12.3.6) is

$$
\frac{(q, ab; q)_\infty}{(aq^c, bq^{-c}; q)_\infty} \int_q^1 \frac{(-q^c t, -q^{1-c}/t; q)_\infty}{(-t, -q/t; q)_\infty}\, t^{c-1} dt.
$$
                                                                           (12.3.7)

The integral in (12.3.7) depends only on c, so we denote it by $f(c)$. The special case $a = 1$, $b = q$ gives

$$
\frac{(q, q; q)_\infty}{(q^c, q^{1-c}; q)_\infty}\, f(c) = \int_0^\infty \frac{t^{c-1}}{1+t}\, dt = \frac{\pi}{\sin(\pi c)},
$$

for $1 > \operatorname{Re} c > 0$. This evaluates $f(c)$ and (12.3.6) follows. $\square$

One can rewrite (12.3.6) in terms of gamma and q-gamma functions in the form

$$\int_0^\infty t^{c-1} \frac{(-tq^b, -q^{1+a}/t; q)_\infty}{(-t, -q/t; q)_\infty} \, dt = \frac{\Gamma(c)\,\Gamma(1-c)\,\Gamma_q(a+c)\,\Gamma_q(b-c)}{\Gamma_q(c)\,\Gamma_q(1-c)\,\Gamma_q(a+b)}. \quad (12.3.8)$$

The proof of Theorem 12.3.3 given here is new.

12.4 Transformations

A very important transformation in the theory of basic hypergeometric functions is the Sears transformation, (Gasper & Rahman, 1990, (III.15)). It can be stated as

$$
{}_4\phi_3\left(\begin{matrix} q^{-n}, a, b, c \\ d, \ e, \ f \end{matrix} \middle| q, q \right)
$$

$$
= \left(\frac{bc}{d}\right)^n \frac{(de/bc, df/bc; q)_n}{(e, f; q)_n} \, {}_4\phi_3\left(\begin{matrix} q^{-n}, a, d/b, d/c \\ d, \ de/bc, \ df/bc \end{matrix} \middle| q, q \right), \quad (12.4.1)
$$

where $abc = defq^{n-1}$. We feel that this transformation can be better motivated if expressed in terms of the Askey–Wilson polynomials

$$
\omega_n(x; a, b, c, d \,|\, q) := {}_4\phi_3\left(\begin{matrix} q^{-n}, abcdq^{n-1}, ae^{i\theta}, ae^{-i\theta} \\ ab, \ ac, \ ad \end{matrix} \middle| q, q \right). \quad (12.4.2)
$$

Theorem 12.4.1 *We have*

$$
\omega_n(x; a, b, c, d \,|\, q) = \frac{a^n (bc, bd; q)_n}{b^n (ac, ad; q)_n} \, \omega_n(x; b, a, c, d \,|\, q). \quad (12.4.3)
$$

It is clear that (12.4.1) and (12.4.3) are equivalent.

Proof of Theorem 12.4.1 Using (12.2.2) we see that

$$
\mathcal{D}_q\omega_n(x; a, b, c, d \,|\, q) = -\frac{2aq\,(1-q^{-n})\,(1-abcdq^{n-1})}{(1-q)(1-ab)(1-ac)(1-ad)} \\ \times \omega_{n-1}\left(x; aq^{1/2}, bq^{1/2}, cq^{1/2}, dq^{1/2} \,|\, q \right). \quad (12.4.4)
$$

On the other hand we can expand $\omega_n(x; a, b, c, d \,|\, q)$ in $\{\phi_k(x; b)\}$ and get

$$
\omega_n(x; a, b, c, d \,|\, q) = \sum_{k=0}^n f_k \left(be^{i\theta}, be^{-i\theta}; q\right)_k,
$$

and (12.2.4) and (12.4.4) yield

$$
f_k = \frac{a^k q^k \left(q^{-n}, abcdq^{n-1}; q\right)_k}{b^k (q, ab, ac, ad; q)_k} \\ \times \omega_{n-k}\left(\left(bq^{k/2} + q^{-k/2}/b\right)/2; aq^{k/2}, bq^{k/2}, cq^{k/2}, dq^{k/2} \,|\, q\right) \\ = q^k \frac{a^k \left(q^{-n}, abcdq^{n-1}; q\right)_k}{b^k (q, ab, ac, ad; q)_k} \, {}_3\phi_2\left(\begin{matrix} q^{k-n}, abcdq^{n+k-1}, a/b \\ acq^k, \ adq^k \end{matrix} \middle| q, q \right).
$$

Now (12.2.15) sums the $3\phi_2$ function and we find

$$f_k = \frac{a^k\,(q^{-n}, abcdq^{n-1}; q)_k\,(q^{1-n}/bd; q)_{n-k}\,(bc; q)_n}{b^k(q, ab, bc, ad; q)_k\,(q^{1-n}/ad; q)_{n-k}\,(ac; q)_n}\,q^k$$

and we obtain (12.4.3) after some manipulations. This completes the proof. $\square$

We will also obtain the Sears transformation as a consequence of orthogonal polynomials in Chapter 15, see the argument following Theorem 15.2.1. The proof given in Chapter 15 also uses Theorem 12.2.3.

A limiting case of the Sears transformation is the useful transformation of Theorem 12.4.2.

Theorem 12.4.2 *The following $3\phi_2$ transformation holds*

$$\,_3\phi_2\left(\begin{matrix} q^{-n}, a, b \\ c, d \end{matrix}\,\middle|\, q, q\right) = \frac{b^n\,(d/b; q)_n}{(d; q)_n}\,_3\phi_2\left(\begin{matrix} q^{-n}, b, c/a \\ c, q^{1-n}b/d \end{matrix}\,\middle|\, q, aq/d\right). \qquad (12.4.5)$$

Proof In (12.4.1) set $f = abcq^{1-n}/de$ then let $c \to 0$ so that $f \to 0$ while all the other parameters remain constant. The result is

$$\,_3\phi_2\left(\begin{matrix} q^{-n}, a, b \\ d, e \end{matrix}\,\middle|\, q, q\right)$$

$$= \frac{(-e)^n q^{n(n-1)/2}\,(aq^{1-n}/e; q)_n}{(e; q)_n}\,_3\phi_2\left(\begin{matrix} q^{-n}, a, d/b \\ d, q^{1-n}a/e \end{matrix}\,\middle|\, q, bq/e\right).$$

The result now follows from (12.2.10). $\square$

An interesting application of (12.4.5) follows by letting b and d tend to ∞ in such a way that b/d remains bounded. Let $b = \lambda d$ and let $d \to \infty$ in (12.4.5). The result is

$$\,_2\phi_1\left(\begin{matrix} q^{-n}, a \\ c \end{matrix}\,\middle|\, q, q\lambda\right) = (\lambda q^{1-n}; q)_n \sum_{j=0}^{n} \frac{(q^{-n}, c/a; q)_j\,q^{j(j-1)/2}}{(q, c, \lambda q^{1-n}; q)_j}\,(-\lambda aq)^j.$$

Now replace λ by λq^{n-1} and observe that the above identity becomes the special case $\gamma = q^n$ of

$$\,_2\phi_1\left(\begin{matrix} a, 1/\gamma \\ c \end{matrix}\,\middle|\, q, \gamma\lambda\right) = \frac{(\lambda; q)_\infty}{(\lambda\gamma; q)_\infty} \sum_{j=0}^{\infty} \frac{(1/\gamma, c/a; q)_j q^{j(j-1)/2}}{(q, c, \lambda; q)_j}\,(-\lambda a\gamma)^j. \quad (12.4.6)$$

Since both sides of the relationship (12.4.6) are analytic functions of γ when $|\gamma| < 1$ and they are equal when $\gamma = q^n$ then they must be identical for all γ if $|\gamma| < 1$.

It is more convenient to write the identity (12.4.6) in the form

$$\sum_{n=0}^{\infty} \frac{(A, C/B; q)_n}{(q, C, Az; q)_n}\,q^{n(n-1)/2}(-Bz)^n = \frac{(z; q)_\infty}{(Az; q)_\infty}\,_2\phi_1(A, B; C; q, z). \quad (12.4.7)$$

In terms of basic hypergeometric functions (12.4.7) takes the form

$$\,_2\phi_2(A, C/B; C, Az; q, Bz) = \frac{(z; q)_\infty}{(Az; q)_\infty}\,_2\phi_1(A, B; C; q, z). \qquad (12.4.8)$$

Observe that (12.4.7) or (12.4.8) is the q-analogue of the Pfaff–Kummer transformation, (Rainville, 1960), (Slater, 1964),

$$_2F_1(a, b; c; z) = (1 - z)^{-a} \, _2F_1(a, c - b; c; z/(z - 1)),\qquad(12.4.9)$$

which holds when $|z| < 1$ and $|z/(z - 1)| < 1$.

In Chapters 14 and 15 we will encounter the following $_3\phi_2$ transformation

$$_3\phi_2\left(\begin{matrix}q^{-n}, a, b\\ c, 0\end{matrix}\,\middle|\, q, q\right) = \frac{(b; q)_n a^n}{(c; q)_n}\, _2\phi_1\left(\begin{matrix}q^{-n}, c/b\\ q^{1-n}/b\end{matrix}\,\middle|\, q, q/a\right).\qquad(12.4.10)$$

Proof of (12.4.10) Let $c \to 0$ in (12.4.5) to get

$$_3\phi_2\left(\begin{matrix}q^{-n}, a, b\\ d, 0\end{matrix}\,\middle|\, q, q\right) = \frac{(d/b; q)_n b^n}{(d; q)_n}\, _2\phi_1\left(\begin{matrix}q^{-n}, b\\ q^{1-n}b/d\end{matrix}\,\middle|\, q, qa/d\right).$$

On the $_2\phi_1$ side replace the summation index, say k by $n - k$, then apply (12.2.12) to obtain a result equivalent to (12.4.10).   □

The transformation (12.4.10) has an interesting application. Since the left-hand side is symmetric in a, b then

$$(b; q)_n a^n \, _2\phi_1\left(\begin{matrix}q^{-n}, c/b\\ q^{1-n}/b\end{matrix}\,\middle|\, q, q/a\right) = (a; q)_n b^n \, _2\phi_1\left(\begin{matrix}q^{-n}, c/a\\ q^{1-n}/a\end{matrix}\,\middle|\, q, q/b\right).$$
$$(12.4.11)$$

Now replace a and b by q^{1-n}/a and q^{1-n}/b, respectively, and use (12.2.8) to get

$$_2\phi_1\left(\begin{matrix}q^{-n}, q^{n-1}bc\\ b\end{matrix}\,\middle|\, q, q^n a\right)$$
$$= \frac{(q^{1-n}/a; q)_n \, a^n}{(q^{1-n}/b; q)_n \, b^n}\, _2\phi_1\left(\begin{matrix}q^{-n}, q^{n-1}ac\\ a\end{matrix}\,\middle|\, q, q^n b\right)$$
$$= \frac{(a, bq^n; q)_\infty}{(b, aq^n; q)_\infty}\, _2\phi_1\left(\begin{matrix}q^{-n}, q^{n-1}ac\\ a\end{matrix}\,\middle|\, q, q^n b\right).$$

Now observe that the above equation, with c replaced by cq, is the case $\gamma = q^n$ of the transformation

$$_2\phi_1\left(\begin{matrix}1/\gamma, bc\gamma\\ b\end{matrix}\,\middle|\, q, \gamma a\right) = \frac{(a, b\gamma; q)_\infty}{(b, a\gamma; q)_\infty}\, _2\phi_1\left(\begin{matrix}1/\gamma, ac\gamma\\ a\end{matrix}\,\middle|\, q, \gamma b\right).\qquad(12.4.12)$$

Since $\gamma = 0$ is a removable singularity, both sides of the above identity are analytic functions of γ in the open unit disc. Hence, the validity of (12.4.12) for the sequence $\gamma = q^n$ implies its validity for $|\gamma| < 1$.

It is more convenient to cast (12.4.12) in the form

$$_2\phi_1\left(\begin{matrix}a, b\\ c\end{matrix}\,\middle|\, q, z\right) = \frac{(az, c/a; q)_\infty}{(c, z; q)_\infty}\, _2\phi_1\left(\begin{matrix}a, abz/c\\ az\end{matrix}\,\middle|\, q, c/a\right).\qquad(12.4.13)$$

The transformation (12.4.13) is an iterate of the Heine transformation (12.5.2).

12.5 Additional Transformations

A basic hypergeometric series (12.1.6) when $a_1 = b_1 q^k$, $k = 0, 1, \ldots$, is reducible to a sum of lower functions. The reason is

$$\frac{(q^k b; q)_n}{(b; q)_n} = \frac{(b; q)_{n+k}}{(b; q)_k (b; q)_n} = \frac{(bq^n; q)_k}{(b; q)_k}$$

$$= \frac{1}{(b; q)_k} \,_1\phi_0 \left(q^{-k}; -; q, q^{k+n} \, b \right),$$

by (12.2.26), so that

$$\sum_{n=0}^{\infty} \lambda_n \frac{(bq^k; q)_n}{(b; q)_n} = \frac{1}{(b; q)_k} \sum_{s=0}^{k} \frac{(q^{-k}; q)_s}{(q; q)_s} b^s q^{ks} \sum_{n=0}^{\infty} \lambda_n q^{ns}. \tag{12.5.1}$$

One useful application of this idea leads to the Heine transformation.

Theorem 12.5.1 *The Heine transformation*

$$\,_2\phi_1 \left(\begin{matrix} a, b \\ c \end{matrix} \,\middle|\, q, z \right) = \frac{(b, az; q)_\infty}{(c, z; q)_\infty} \,_2\phi_1 \left(\begin{matrix} c/b, z \\ az \end{matrix} \,\middle|\, q, b \right), \tag{12.5.2}$$

holds for $|z| < 1$, $|b| < 1$.

Proof When $b = cq^k$, the left-hand side of (12.5.2) is

$$\frac{1}{(c; q)_k} \sum_{s=0}^{k} \frac{(q^{-k}; q)_s}{(q; q)_s} c^s q^{ks} \,_1\phi_0 \left(a; -; q, q^s z \right)$$

$$= \frac{1}{(c; q)_k} \sum_{s=0}^{k} \frac{(q^{-k}; q)_s}{(q; q)_s} c^s q^{ks} \frac{(azq^s; q)_\infty}{(q^s z; q)_\infty}$$

$$= \frac{(az; q)_\infty}{(z; q)_\infty (c; q)_k} \,_2\phi_1 \left(q^{-k}, z; az; q, cq^k \right).$$

Thus (12.5.2) holds on the sequence $b = cq^k$ and the rest follows from the identity theorem for analytic functions. $\square$

Corollary 12.5.2 *The following transformation*

$$\,_2\phi_1 \left(\begin{matrix} a, b \\ c \end{matrix} \,\middle|\, q, z \right) = \frac{(abz/c; q)_\infty}{(z; q)_\infty} \,_2\phi_1 \left(\begin{matrix} c/a, c/b \\ c \end{matrix} \,\middle|\, q, abz/c \right). \tag{12.5.3}$$

holds, subject to $|z| < 1$, $|abz| < |c|$.

Proof Apply (12.4.13) to the right-hand side of (12.5.2). $\square$

It is clear that (12.5.3) is the analogue of the Euler transformation

$$\,_2F_1(a, b; c; z) = (1 - z)^{c-a-b} \,_2F_1(c - a, c - b; c; z). \tag{12.5.4}$$

Corollary 12.5.3 (Bailey–Daum sum) *For* $|q| < |b|$ *we have*

$$2\phi_1(a, b; aq/b; q, -q/b) = \frac{(-q; q)_\infty \, (aq, aq^2/b^2; q^2)_\infty}{(-q/b, aq/b; q)_\infty}. \tag{12.5.5}$$

Proof Apply (12.5.2) to $2\phi_1(b, a; aq/b; q, -q/b)$ to see that it is

$$\frac{(a, -q; q)_\infty}{(aq/b, -q/b; q)_\infty} \, 2\phi_1\left(\begin{array}{c} q/b, -q/b \\ -q \end{array}\middle| q, a\right)$$

$$= \frac{(a, -q; q)_\infty}{(aq/b, -q/b; q)_\infty} \, 1\phi_0\left(q^2/b^2; -; q^2, a\right).$$

The result follows from (12.2.22).   □

The next set of transformations will be used in the text. At this time we neither know how to motivate them in terms of what we have done so far, nor are we able to give proofs simpler than those in (Gasper & Rahman, 1990). The transformations in question are listed below:

$$2\phi_1\left(\begin{array}{c} A, B \\ C \end{array}\middle| q, Z\right)$$

$$+ \frac{(B, q/C, C/A, AZ/q, q^2/AZ; q)_\infty}{(C/q, Bq/C, q/A, AZ/C, Cq/AZ; q)_\infty} \, 2\phi_1\left(\begin{array}{c} Aq/C, Bq/C \\ q^2/C \end{array}\middle| q, Z\right)$$

$$= \frac{(ABZ/C, q/C; q)_\infty}{(AZ/C, q/A; q)_\infty} \, 2\phi_1\left(\begin{array}{c} C/A, Cq/ABZ \\ Cq/AZ \end{array}\middle| q, Bq/C\right), \tag{12.5.6}$$

$$2\phi_1\left(\begin{array}{c} A, B \\ C \end{array}\middle| q, Z\right) = \frac{(B, C/A, AZ, q/AZ; q)_\infty}{(C, B/A, Z, q/Z; q)_\infty}$$

$$\times \, 2\phi_1\left(\begin{array}{c} A, Aq/C \\ Aq/B \end{array}\middle| q, \frac{Cq}{ABZ}\right) + \frac{(A, C/B, BZ, q/BZ; q)_\infty}{(C, A/B, Z, q/Z; q)_\infty} \tag{12.5.7}$$

$$\times \, 2\phi_1\left(\begin{array}{c} B, Bq/C \\ Bq/A \end{array}\middle| \frac{Cq}{ABZ}\right),$$

$$3\phi_2\left(\begin{array}{c} A, B, C \\ D, E \end{array}\middle| q, \frac{DE}{ABC}\right) = \frac{(E/B, E/C; q)_\infty}{(E, E/BC; q)_\infty} \, 3\phi_2\left(\begin{array}{c} D/A, B, C \\ D, BCq/E \end{array}\middle| q, q\right)$$

$$+ \frac{(D/A, B, C, DE/BC; q)_\infty}{(D, E, BC/E, DE/ABC; q)_\infty} \, 3\phi_2\left(\begin{array}{c} E/B, E/C, DE/ABC \\ DE/BC, qE/BC \end{array}\middle| q, q\right). \tag{12.5.8}$$

The Singh quadratic transformation is

$$4\phi_3\left(\begin{array}{c} A^2, B^2, C, D \\ AB\sqrt{q}, -AB\sqrt{q}, -CD \end{array}\middle| q, q\right)$$

$$= 4\phi_3\left(\begin{array}{c} A^2, B^2, C^2, D^2 \\ A^2B^2q, -CD, -CDq \end{array}\middle| q^2, q^2\right). \tag{12.5.9}$$

A series of the type

$$3+r\phi_{2+r}(a_1, a_2, \ldots, a_{r+3}; b_1, \ldots, b_{r+2}; q, z) \tag{12.5.10}$$

is called **very well-poised** if

$$a_2 = -a_3 = q\sqrt{a_1}, \; b_2 = -b_3 = \sqrt{a_1}, \; a_{j+3}b_{j+2} = qa_1, \; 1 \le j \le r. \quad (12.5.11)$$

Bailey used the notation

$$_{3+r}W_{r+2}\,(\alpha; a_1, \ldots, a_r; z)$$
$$= {}_{r+3}\phi_r \left(\begin{matrix} \alpha, q\sqrt{\alpha}, -q\sqrt{\alpha}, a_1, \ldots, a_r \\ \sqrt{a}, -\sqrt{a}, \alpha q/a_1, \ldots, \alpha q/a_r \end{matrix} \middle| q, z \right) \quad (12.5.12)$$

to denote a very well-poised series.

A transformation due to Bailey (Bailey, 1935) relates a very well-poised $_8\phi_7$ to a sum of two balanced $_4\phi_3$'s. It is

$$_8\phi_7 \left(\begin{matrix} a, q\sqrt{a}, -q\sqrt{a}, b, c, d, e, f \\ \sqrt{a}, -\sqrt{a}, aq/b, aq/c, aq/d, aq/e, aq/f \end{matrix} \middle| q, \frac{a^2q^2}{bcdef} \right)$$
$$= \frac{(aq, aq/de, aq/df, aq/ef; q)_\infty}{(aq/d, aq/e, aq/f, aq/def; q)_\infty} {}_4\phi_3 \left(\begin{matrix} d, e, f, aq/bc \\ aq/b, aq/c, def/a \end{matrix} \middle| q, q \right)$$
$$+ \frac{\left(aq, d, e, f, a^2q^2/bdef, a^2q^2/cdef; q \right)_\infty}{(aq/b, aq/c, , aq/d, , aq/f, aq/bcdef, def/aq; q)_\infty}$$
$$\times {}_4\phi_3 \left(\begin{matrix} aq/de, aq/df, aq/ef, a^2q^2/bcdef \\ aq^2/def, a^2q^2/bdef, a^2q^2/cdef \end{matrix} \middle| q, q \right). \quad (12.5.13)$$

In particular we have the Watson transformation

$$_8\phi_7 \left(\begin{matrix} a, q\sqrt{a}, -q\sqrt{a}, b, c, d, e, q^{-n} \\ \sqrt{a}, -\sqrt{a}, aq/b, aq/c, aq/d, aq/e, aq^{n+1} \end{matrix} \middle| q, \frac{a^2q^{n+2}}{bcde} \right)$$
$$= \frac{(aq, aq/de; q)_n}{(aq/d, aq/e; q)_n} {}_4\phi_3 \left(\begin{matrix} aq/bc, d, e, q^{-n} \\ aq/b, aq/c, deq^{-n}/a \end{matrix} \middle| q, q \right). \quad (12.5.14)$$

A useful $_8\phi_7$ to $_8\phi_7$ transformation is

$$_8W_7\,(a; b, c, d, e, f; a^2q^2/bcdef)$$
$$= \frac{(aq, aq/ef, \lambda q/e, \lambda q/f; q)_\infty}{(aq/e, aq/f, \lambda q, \lambda q/ef; q)_\infty}$$
$$_8W_7(\lambda; \lambda b/a, \lambda c/a, \lambda d/a, e, f; aq/ef), \quad (12.5.15)$$

where $\lambda = qa^2/bcd$.

12.6 Theta Functions

We need to use identities among theta functions, so below we say a few words about theta functions. We follow the notation in Whittaker and Watson (Whittaker & Watson, 1927, Chapter 21). The four theta functions have the infinite product representations (Whittaker & Watson, 1927, §21.3),

$$\vartheta_1(z,q) = 2q^{1/4} \sin z \left(q^2, q^2 e^{2iz}, q^2 e^{-2iz}; q^2\right)_\infty, \tag{12.6.1}$$

$$\vartheta_3(z,q) = \left(q^2, -q e^{2iz}, -q e^{-2iz}; q^2\right)_\infty, \tag{12.6.2}$$

$$\vartheta_2(z,q) = 2q^{1/4} \cos z \left(q^2, -q^2 e^{2iz}, -q^2 e^{-2iz}; q^2\right)_\infty, \tag{12.6.3}$$

$$\vartheta_4(z,q) = \left(q^2, q e^{2iz}, q e^{-2iz}; q^2\right)_\infty. \tag{12.6.4}$$

We shall follow the notation in Whittaker and Watson and drop q when there is no ambiguity. In (Whittaker & Watson, 1927, Exercise 3, p. 488) we find

$$\vartheta_1(y \pm z)\vartheta_4(y \mp z)\vartheta_2(0)\vartheta_3(0)$$
$$= \vartheta_1(y)\vartheta_4(y)\vartheta_2(z)\vartheta_3(z) \pm \vartheta_1(z)\vartheta_4(z)\vartheta_2(y)\vartheta_3(y). \tag{12.6.5}$$

Moreover

$$\frac{d}{dz}\left(\frac{\vartheta_1(z)}{\vartheta_4(z)}\right) = \vartheta_4^2(0)\frac{\vartheta_2(z)}{\vartheta_4(z)}\frac{\vartheta_3(z)}{\vartheta_4(z)}, \tag{12.6.6}$$

is stated on page 478 of (Whittaker & Watson, 1927).

The Jacobi triple product identity gives the following trigonometric representations

$$\vartheta_1(z,q) = q^{1/4}\sum_{-\infty}^{\infty}(-1)^n q^{n^2+n}\sin(2n+1)z, \tag{12.6.7}$$

$$\vartheta_3(z,q) = 2\sum_{-\infty}^{\infty} q^{n^2}\cos(2nz), \tag{12.6.8}$$

$$\vartheta_2(z,q) = q^{1/4}\sum_{-\infty}^{\infty} q^{n^2+n}\cos(2n+1)z, \tag{12.6.9}$$

$$\vartheta_4(z,q) = 2\sum_{-\infty}^{\infty}(-1)^n q^{n^2}\cos(2nz). \tag{12.6.10}$$

Exercises

12.1   Let $(E_q f)(x) = f(qx)$. Prove the Leibniz rule

$$(D_q^n fg)(x) = \sum_{k=0}^{n} \begin{bmatrix} n \\ k \end{bmatrix}_q (D_q^k f)(x)(E_q^k D_q^{n-k} g)(x).$$

12.2   (a) Prove the operational formula

$$e_q(\lambda D_q)e_q(ax) = e_q(\lambda a)e_q(ax).$$

(b) Show that

$$e_q(\lambda D_q) f(x)g(x) = \sum_{k=0}^{\infty}\frac{\lambda^k}{(q;q)_k}(D_q^k f)(x)(E_q^k e_q(\lambda D_q) g)(x).$$

12.3   Let

$$h_n(z) = \sum_{k=0}^{n} \begin{bmatrix} n \\ k \end{bmatrix}_q z^k.$$

(a) Prove that

$$e_q(D_q) x^n = h_n(x).$$

(b) Show that

$$\sum_{n=0}^{\infty} \frac{h_n(z)}{(q;q)_n} t^n = \frac{1}{(t, tz; q)_\infty}.$$

(c) Use 12.2(b) and 12.3(b) to show that

$$\sum_{n=0}^{\infty} \frac{h_n(z) h_n(\zeta)}{(q;q)_n} t^n = \frac{(t^2 z\zeta; q)_\infty}{(t, tz, t\zeta, tz\zeta; q)_\infty}.$$

(d) Show that (Carlitz, 1972)

$$\sum_{n=0}^{\infty} \frac{h_n(z) h_{n+k}(\zeta)}{(q;q)_n} t^n = \frac{(t^2 z\zeta; q)_\infty}{(t, tz, t\zeta, tz\zeta; q)_\infty}$$

$$\times \sum_{r=0}^{k} \frac{(q, t\zeta, tz\zeta; q)_r}{(q, t^2 z\zeta; q)_r} \frac{\zeta^{k-r}}{(q;q)_{k-r}}.$$

Note: The polynomials $\{h_n(z)\}$ are related to the q-Hermite polynomials of Chapter 13 by $H_n(\cos\theta \,|\, q) = e^{in\theta} H_n(e^{-2i\theta})$. In fact, (c) gives another derivation of the Poisson kernel of $\{H_n(x \,|\, q)\}$ while (d) generalizes the Poisson kernel.

12.4   Evaluate $D_q^n (1 - z)^{-1}$ and use the result to prove Theorem 12.2.5 when $a = q^n$. Use the identity theorem for analytic functions to prove Theorem 12.2.5 for all a.

13

Some q-Orthogonal Polynomials

In this chapter we study the continuous q-ultraspherical, continuous q-Hermite polynomials and q-Pollaczek polynomials. The first two first appeared in Rogers' work on the Rogers–Ramanujan identities in 1893–95 (Askey & Ismail, 1983) while the q-Pollaczek polynomials are of a very recent vintage, (Charris & Ismail, 1987). In addition, the Al-Salam–Ismail polynomials are mentioned in conjunction with the Rogers–Ramanujan identities. Several special systems of orthogonal polynomials are treated in the later sections, including the q-Pollaczek polynomials and some q-polynomials from (Ismail & Mulla, 1987) and (Al-Salam & Ismail, 1983).

Fejer generalized the Legendre polynomials to polynomials $\{p_n(x)\}$ having generating functions

$$\sum_{n=0}^{\infty} \phi_n(\cos\theta)t^n = \left| F\left(re^{i\theta}\right) \right|^2, \qquad (13.0.1)$$

where $F(z)$ is analytic in a neighborhood of $z = 0$. The Legendre polynomials correspond to the case $F(z) = (1 - z)^{-1/2}$. Fejer proved that the zeros of the generalized Legendre polynomials share many of the properties of the zeros of the Legendre and ultraspherical polynomials. For an account of these results see (Szegő, 1933). Feldheim (Feldheim, 1941b) and Lanzewizky (Lanzewizky, 1941) independently proved that the only orthogonal generalized Legendre polynomials are either the ultraspherical polynomials or the q-ultraspherical polynomials or special cases of them. They proved that F has to be F_1 or F_2, or some limiting cases of them, where

$$F_1(z) = (1 - z)^{-\nu}, \quad F_2(z) = \frac{(\beta z; q)_\infty}{(z; q)_\infty}. \qquad (13.0.2)$$

For a proof of this characterization, see (Andrews et al., 1999). The weight function for the q-Hermite polynomials was found by Allaway (Allaway, 1972) and Al-Salam and Chihara (Al-Salam & Chihara, 1976) while the weight function for the more general q-ultraspherical polynomials was found by Askey and Ismail (Askey & Ismail, 1980), and Askey and Wilson (Askey & Wilson, 1985) using different methods. Allaway's result was published in (Allaway, 1980).

13.1 q-Hermite Polynomials

The continuous q-Hermite polynomials $\{H_n(x\,|\,q)\}$ are generated by the recursion relation

$$2xH_n(x\,|\,q) = H_{n+1}(x\,|\,q) + (1 - q^n)\,H_{n-1}(x\,|\,q), \qquad (13.1.1)$$

and the initial conditions

$$H_0(x\,|\,q) = 1, \quad H_1(x\,|\,q) = 2x. \qquad (13.1.2)$$

Our first task is to derive a generating function for $\{H_n(x\,|\,q)\}$. Let

$$H(x,t) := \sum_{n=0}^{\infty} H_n(x\,|\,q)\frac{t^n}{(q;q)_n}. \qquad (13.1.3)$$

Multiply (13.1.1) by $t^n/(q;q)_n$, add for $n = 1, 2, \ldots$, and take into account the initial conditions (13.1.2). We obtain the functional equation

$$H(x,t) - H(x,qt) = 2xtH(x,t) - t^2 H(x,t).$$

Therefore

$$H(x,t) = \frac{H(x,qt)}{1 - 2xt + t^2} = \frac{H(x,qt)}{(1 - te^{i\theta})\,(1 - te^{-i\theta})}, \quad x = \cos\theta. \qquad (13.1.4)$$

This suggests iterating the functional equation (13.1.4) to get

$$H(\cos\theta, t) = \frac{H\,(\cos\theta, q^n t)}{(te^{i\theta}, te^{-i\theta}; q)_n}.$$

As $n \to \infty$, $H\,(x, q^n t) \to H(x,0) = 1$. This motivates the next theorem.

Theorem 13.1.1 *The continuous q-Hermite polynomials have the generating function*

$$\sum_{n=0}^{\infty} H_n(\cos\theta\,|\,q)\frac{t^n}{(q;q)_n} = \frac{1}{(te^{i\theta}, te^{-i\theta}; q)_\infty}. \qquad (13.1.5)$$

Proof It is straightforward to see that the left-hand side of (13.1.5) satisfies the functional equation

$$\left(1 - 2xt + t^2\right) F(x,t) = F(x,qt). \qquad (13.1.6)$$

Since the right-hand side of (13.1.5) is analytic in t in a neighborhood of $t = 0$ then it can be expanded in a power series in t and by substituting the expansion $F(x,t) = \sum_{n=0}^{\infty} f_n(x)t^n/(q;q)_n$ into (14.1.6) and equating coefficients of t^n, we find that the f_n's satisfy the three-term recurrence relation (14.1.1) and agree with $H_n(x\,|\,q)$ when $n = 0$, $n = 1$. Thus $f_n = H_n(x\,|\,q)$ for all n and the proof is complete. □

We indicated a rigorous proof of Theorem 13.1.1 in order to show how to justify the formal argument leading to it. In future results of similar nature, we will only give the formal proof and the more rigorously inclined reader can easily fill in the details.

To obtain an explicit formula for the H_n's we expand $1/\left(te^{\pm i\theta};q\right)_\infty$ by (12.2.24), then multiply the resulting series. This gives

$$H_n(\cos\theta\,|\,q) = \sum_{k=0}^{n} \frac{(q;q)_n}{(q;q)_k(q;q)_{n-k}}\, e^{i(n-2k)\theta}. \tag{13.1.7}$$

Since $H_n(x\,|\,q)$ is a real polynomial one can use (13.1.7) to get

$$\begin{aligned}
H_n(\cos\theta\,|\,q) &= \sum_{k=0}^{n} \frac{(q;q)_n}{(q;q)_k(q;q)_{n-k}}\, \cos(n-2k)\theta \\
&= \sum_{k=0}^{n} \frac{(q;q)_n}{(q;q)_k(q;q)_{n-k}}\, \cos(|n-2k|)\theta.
\end{aligned} \tag{13.1.8}$$

The representation (13.1.8) reflects the polynomial character of $H_n(x\,|\,q)$ since

$$\cos((n-2k)\theta) = \cos(|(n-2k)|\theta)$$

which is a polynomial in $\cos\theta$ of degree $|n-2k|$.

Theorem 13.1.2 *The continuous q-Hermite polynomials have the following properties*

$$H_n(-x\,|\,q) = (-1)^n H_n(x\,|\,q), \tag{13.1.9}$$

and

$$\max\{|H_n(x\,|\,q)| : -1 \le x \le 1\} = H_n(1\,|\,q) = (-1)^n H_n(-1\,|\,q), \tag{13.1.10}$$

and the maximum is attained only at $x = \pm 1$.

Proof Replace θ by $\pi - \theta$ in (13.1.8) to get (13.1.9). The rest follows from (13.1.7) and the triangular inequality.   □

An immediate consequence of (13.1.10) is that the series on the left-hand side of (13.1.5) converges uniformly in x for $x \in [-1,1]$ for every fixed t provided that $|t| < 1$.

Theorem 13.1.3 *The continuous q-Hermite polynomials satisfy the orthogonality relation*

$$\int_{-1}^{1} H_m(x\,|\,q)H_n(x\,|\,q)w(x\,|\,q)\,dx = \frac{2\pi(q;q)_n}{(q;q)_\infty}\,\delta_{m,n}, \tag{13.1.11}$$

where

$$w(x\,|\,q) = \frac{\left(e^{2i\theta},e^{-2i\theta};q\right)_\infty}{\sqrt{1-x^2}}, \quad x = \cos\theta,\ 0 \le \theta \le \pi. \tag{13.1.12}$$

The proof of Theorem 13.1.3 is based on the following Lemma:

Lemma 13.1.4 *We have the following evaluation*

$$\int_0^\pi e^{2ij\theta} \left(e^{2i\theta}, e^{-2i\theta}; q\right)_\infty d\theta = \frac{\pi(-1)^j}{(q;q)_\infty} \left(1 + q^j\right) q^{j(j-1)/2}. \qquad (13.1.13)$$

Proof Let I_j denote the left side of (13.1.13). The Jacobi triple product identity (12.3.4) gives

$$I_j = \int_0^\pi e^{2ij\theta} \left(1 - e^{2i\theta}\right) \left(qe^{2i\theta}, e^{-2i\theta}; q\right)_\infty d\theta$$

$$= \int_0^\pi \frac{e^{2ij\theta} \left(1 - e^{2i\theta}\right)}{(q;q)_\infty} \sum_{n=-\infty}^{\infty} (-1)^n q^{n(n+1)/2} e^{2in\theta} d\theta$$

$$= \sum_{n=-\infty}^{\infty} \frac{(-1)^n q^{n(n+1)/2}}{2(q;q)_\infty} \int_{-\pi}^{\pi} \left(1 - e^{i\theta}\right) e^{i(j+n)\theta} d\theta.$$

The result now follows from the orthogonality of the trigonometric functions on $[-\pi, \pi]$. $\qquad \square$

Proof of Theorem 13.1.3 Since the weight function $w(x \mid q)$ is an even function of x, it follows that (13.1.11) trivially holds if $|m - n|$ is odd. Thus there is no loss of generality in assuming $m \leq n$ and $n - m$ is even. It is clear that we can replace $n - 2k$ by $|n - 2k|$ in (13.1.8). Therefore it suffices to evaluate the following integrals for $0 \leq j \leq n/2$.

$$\int_0^\pi e^{i(n-2j)\theta} H_n(\cos\theta \mid q) \left(e^{2i\theta}, e^{-2i\theta}; q\right)_\infty d\theta$$

$$= \sum_{k=0}^n \frac{(q;q)_n}{(q;q)_k (q;q)_{n-k}} \int_0^\pi e^{2i(n-j-k)\theta} \left(e^{2i\theta}, e^{-2i\theta}; q\right)_\infty d\theta$$

$$= \frac{\pi}{(q;q)_\infty} \sum_{k=0}^n \frac{(-1)^{j+k+n}(q;q)_n}{(q;q)_k (q;q)_{n-k}} \left(1 + q^{n-j-k}\right) q^{(n-j-k)(n-j-k-1)/2}$$

$$= \frac{(-1)^{n+j}\pi}{(q;q)_\infty} q^{(n-j)(n-j-1)/2} \left[{}_1\phi_0\left(q^{-n}; -; q, q^{j+1}\right) + q^{n-j}{}_1\phi_0\left(q^{-n}; -; q, q^j\right)\right].$$

By (12.2.26) we evaluate the ${}_1\phi_0$ and after some simplification we obtain

$$\int_0^\pi e^{i(n-2j)\theta} H_n(\cos\theta \mid q)(e^{2i\theta}, e^{-2i\theta}; q)_\infty d\theta$$

$$= \frac{(-1)^{n+j}\pi}{(q;q)_\infty} q^{(n-j)(n-j-1)/2}[(q^{-n+j+1}; q)_n + q^{n-j}(q^{-n+j}; q)_n]. \qquad (13.1.14)$$

For $0 < j < n$ it is clear that the right-hand side of (13.1.14) vanishes. When $j = 0$, the right-hand side of (13.1.14) is

$$\frac{\pi}{(q;q)_\infty} q^{n(n-1)/2}(-1)^n q^n \left(q^{-n};q\right)_n.$$

Thus

$$\int_0^\pi e^{i(n-2j)\theta} H_n(\cos\theta\,|\,q) \left(e^{2i\theta}, e^{-2i\theta}; q\right)_\infty d\theta = \frac{\pi(q;q)_n}{(q;q)_\infty}\delta_{j,0}, \quad 0 \le j < n.$$

$$(13.1.15)$$

This calculation establishes (13.1.11) when $m < n$. When $m = n$ we use (13.1.7) and (13.1.15) to obtain

$$\int_0^\pi H_m(\cos\theta\,|\,q)H_n(\cos\theta\,|\,q)\left(e^{2i\theta}, e^{-2i\theta};q\right)_\infty d\theta$$

$$= 2\int_0^\pi e^{i(n)\theta}H_n(\cos\theta\,|\,q)\left(e^{2i\theta}, e^{-2i\theta};q\right)_\infty d\theta$$

$$= \frac{2\pi(q;q)_n}{(q;q)_\infty}.$$

It is worth noting that

$$H_n(x\,|\,q) = (2x)^n + \text{lower order terms}, \qquad (13.1.16)$$

which follows from (13.1.1) and (13.1.2). □

Theorem 13.1.5 *The linearization of products of continuous q-Hermite polynomials is given by*

$$H_m(x\,|\,q)H_n(x\,|\,q) = \sum_{k=0}^{m\wedge n} \frac{(q;q)_m(q;q)_n}{(q;q)_k(q;q)_{m-k}(q;q)_{n-k}} H_{m+n-2k}(x\,|\,q), \quad (13.1.17)$$

where

$$m \wedge n := \min\{m, n\}. \qquad (13.1.18)$$

Proof It is clear from (13.1.9) that $H_m(x\,|\,q)H_n(x\,|\,q)$ has the same parity as $H_{m+n}(x\,|\,q)$. Therefore there exists a sequence $\{a_{m,n,k} : 0 \le k \le m \wedge n\}$ such that

$$H_m(x\,|\,q)H_n(x\,|\,q) = \sum_{k=0}^{m\wedge n} a_{m,n,k}H_{m+n-2k}(x\,|\,q) \qquad (13.1.19)$$

and $a_{m,n,k}$ is symmetric in m and n. Furthermore

$$a_{m,0,k} = a_{0,n,k} = \delta_{k,0} \qquad (13.1.20)$$

holds and (13.1.16) implies

$$a_{m,n,0} = 1. \qquad (13.1.21)$$

Multiply (13.1.19) by $2x$ and use the three-term recurrence relation (13.1.1) to obtain

$$\sum_{k=0}^{(m+1)\wedge n} a_{m+1,n,k} H_{m+n+1-2k}(x \mid q)$$

$$+ (1 - q^m) \sum_{k=0}^{(m-1)\wedge n} a_{m-1,n,k} H_{m+n-1-2k}(x \mid q)$$

$$= \sum_{k=0}^{m\wedge n} a_{m,n,k} \left[H_{m+n+1-2k}(x \mid q) + \left(1 - q^{m+n-2k}\right) H_{m+n-1-2k}(x \mid q) \right],$$

with $H_{-1}(x \mid q) := 0$. This leads us to the system of difference equations

$$a_{m+1,n,k+1} - a_{m,n,k+1} = \left(1 - q^{m+n-2k}\right) a_{m,n,k} - (1 - q^m) a_{m-1,n,k}, \quad (13.1.22)$$

subject to the initial conditions (13.1.20) and (13.1.21). When $k = 0$ equations (13.1.22) and (13.1.21) imply

$$a_{m+1,n,1} = a_{m,n,1} + q^m (1 - q^n),$$

which leads to

$$a_{m,n,1} = (1 - q^n) \sum_{k=0}^{m-1} q^k = \frac{(1 - q^m)(1 - q^n)}{1 - q}. \qquad (13.1.23)$$

Setting $k = 1$ in (13.1.22) and applying (13.1.23) we find

$$a_{m+1,n,2} = a_{m,n,2} + q^{m-1} (1 - q^m)(1 - q^n)\left(1 - q^{n-1}\right)/(1 - q),$$

whose solution is

$$a_{m,n,2} = \frac{(1 - q^n)\left(1 - q^{n-1}\right)(1 - q^m)\left(1 - q^{m-1}\right)}{(1 - q)(1 - q^2)}.$$

From here we suspect the pattern

$$a_{m,n,k} = \frac{(q;q)_m (q;q)_n}{(q;q)_{m-k}(q;q)_{n-k}(q;q)_k},$$

which can be proved from (13.1.22) by a straightforward induction. $\qquad\qquad\square$

Let $V_n(q)$ denote an n-dimensional vector space over a field with q-elements. The q-binomial coefficient $\begin{bmatrix} n \\ k \end{bmatrix}_q$ counts the number of $V_k(q)$ such that $V_k(q)$ is a subspace of a fixed $V_n(q)$. One can view $H_n(\cos\vartheta \mid q)$ as a generating function for $\begin{bmatrix} n \\ k \end{bmatrix}_q$, $k = 0, 1, \ldots$, since

$$z^{-n} H_n(\cos\theta \mid q) = \sum_{k=0}^{n} \begin{bmatrix} n \\ k \end{bmatrix}_q z^{-2k}, \quad z = e^{i\theta}.$$

Using this interpretation, one can prove (13.1.17) by classifying the subspaces of a $V_{n+m}(q)$ according to the dimensions of their intersections with $V_n(q)$ and $V_m(q)$. For details, see (Ismail et al., 1987).

Our next result is the computation of the Poisson kernel of the continuous q-Hermite polynomials.

Theorem 13.1.6 *The Poisson kernel of the H_n's is*

$$\sum_{n=0}^{\infty} \frac{H_n(\cos\theta \mid q) H_n(\cos\phi \mid q)}{(q;q)_n} t^n$$

$$= \frac{(t^2;q)_\infty}{\left(te^{i(\theta+\phi)}, te^{i(\theta-\phi)}, te^{-i(\theta+\phi)}, te^{-i(\theta-\phi)}; q\right)_\infty}. \qquad (13.1.24)$$

Moreover, the evaluation of the Poisson kernel is equivalent to the linearization formula (13.1.17).

Proof Multiply (13.1.17) by $t_1^m t_2^n/(q;q)_m(q;q)_n$ and add for $m, n = 0, 1, \ldots$. The generating function (13.1.5) implies

$$\frac{1}{(t_1 e^{i\theta}, t_1 e^{-i\theta}, t_2 e^{i\theta}, t_2 e^{-i\theta}; q)_\infty} = \sum_{m\geq k, n\geq k, k\geq 0} \frac{H_{m+n-2k}(\cos\theta \mid q) t_1^m t_2^n}{(q;q)_{m-k}(q;q)_{n-k}(q;q)_k}$$

$$= \sum_{k,m,n=0}^{\infty} \frac{H_{m+n}(\cos\theta \mid q) t_1^m t_2^n (t_1 t_2)^k}{(q;q)_m(q;q)_n(q;q)_k}$$

$$= \frac{1}{(t_1 t_2; q)_\infty} \sum_{m,n=0}^{\infty} \frac{H_{m+n}(\cos\theta \mid q) t_1^m t_2^n}{(q;q)_m(q;q)_n},$$

where we used (12.2.24). In the last sum replace $m + n$ by s then replace t_1 and t_2 by $t_1 e^{i\phi}$ and $t_1 e^{-i\phi}$, respectively. Therefore

$$\frac{(t_1^2; q)_\infty}{\left(t_1 e^{i(\theta+\phi)}, t_1 e^{i(\phi-\theta)}, t_1 e^{i(\theta-\phi)}, t_1 e^{-i(\theta+\phi)}; q\right)_\infty}$$

$$= \sum_{s=0}^{\infty} \frac{H_s(\cos\theta \mid q) t_1^s}{(q;q)_s} \sum_{n=0}^{s} \frac{(q;q)_s e^{i(s-2n)\phi}}{(q;q)_n(q;q)_{s-n}}.$$

In view of (13.1.7) the n sum is $H_s(\cos\phi \mid q)$ and (13.1.24) follows. The above steps can be reversed and, starting with (13.1.24), we equate coefficients of $t_1^m t_2^n$ and establish (13.1.17). $\qquad\qquad\square$

In the notation of Exercise 12.3, $H_n(\cos\theta \mid q) = e^{in\theta} h_n\left(e^{-2i\theta}\right)$, hence Exercise 12.3(c) is equivalent to (13.1.24).

The linearization formula (13.1.17) has an inverse which will be our next theorem.

Theorem 13.1.7 *The inverse to (13.1.17) is*

$$\frac{H_{n+m}(x \mid q)}{(q;q)_m(q;q)_n} = \sum_{k=0}^{m\wedge n} \frac{(-1)^k q^{k(k-1)/2}}{(q;q)_k} \frac{H_{n-k}(x \mid q)}{(q;q)_{n-k}} \frac{H_{m-k}(x \mid q)}{(q;q)_{m-k}}. \qquad (13.1.25)$$

Proof As in the proof of Theorem 13.1.6 we have

$$\frac{(t_1 t_2; q)_\infty}{(t_1 e^{i\theta}, t_1 e^{-i\theta}, t_2 e^{i\theta}, t_2 e^{-i\theta}; q)_\infty} = \sum_{m,n=0}^{\infty} \frac{H_{n+m}(x \mid q)}{(q; q)_m (q; q)_n} t_1^m t_2^n. \qquad (13.1.26)$$

Now expand $(t_1 t_2; q)_\infty$ by (12.2.25) and use (13.1.5) to expand the rest of the left-hand side of (13.1.26) then equate coefficients of $t_1^m \, t_2^n$. The result is (13.1.25). $\qquad\square$

The value of $H_n(x \mid q)$ can be found in closed form at three special points, $x = 0$, $x = \pm \left(q^{1/4} + q^{-1/4} \right) /2$ through the generating function (13.1.5). Indeed

$$\sum_{n=0}^{\infty} \frac{H_n(0 \mid q)}{(q; q)_n} t^n = \frac{1}{(it, -it; q)_\infty} = \frac{1}{(-t^2; q^2)_\infty} = \sum_{0}^{\infty} \frac{(-1)^n t^{2n}}{(q^2; q^2)_n}.$$

Hence

$$H_{2n+1}(0 \mid q) = 0, \quad \text{and} \quad H_{2n}(0 \mid q) = (-1)^n (-q; q)_n. \qquad (13.1.27)$$

Moreover with $\xi = \left(q^{1/4} + q^{-1/4} \right) /2$, (13.1.5) yields

$$\sum_{n=0}^{\infty} \frac{H_n(\xi \mid q)}{(q; q)_n} t^n = \frac{1}{\left(tq^{1/4}, tq^{-1/4}; q \right)_\infty} = \frac{1}{\left(tq^{-1/4}; q^{1/2} \right)_\infty}$$

$$= \sum_{n=0}^{\infty} \frac{t^n q^{-n/4}}{\left(q^{1/2}; q^{1/2} \right)_n}.$$

Therefore

$$H_n \left(\left(q^{1/4} + q^{-1/4} \right) /2 \mid q \right) = q^{-n/4} \left(-q^{1/2}; q^{1/2} \right)_n. \qquad (13.1.28)$$

Of course $H_n(-\xi \mid q) = (-1)^n H_n(\xi \mid q)$.

The Askey–Wilson operator acts on $H_n(x \mid q)$ in a natural way.

Theorem 13.1.8 *The polynomials* $\{ H_n(x \mid q) \}$ *have the ladder operators*

$$\mathcal{D}_q H_n(x \mid q) = \frac{2(1 - q^n)}{1 - q} q^{(1-n)/2} H_{n-1}(x \mid q) \qquad (13.1.29)$$

and

$$\frac{1}{w(x \mid q)} \mathcal{D}_q \left\{ w(x \mid q) H_n(x \mid q) \right\} = -\frac{2 q^{-n/2}}{1 - q} H_{n+1}(x \mid q), \qquad (13.1.30)$$

where $w(x \mid q)$ *is as defined in (13.1.12).*

Proof Apply $\mathcal{D}_q$ to (13.1.5) and get

$$\sum_{n=0}^{\infty} \frac{t^n}{(q; q)_n} \mathcal{D}_q H_n(x \mid q) = \frac{2t/(1 - q)}{\left(tq^{-1/2} e^{i\theta}, tq^{-1/2} e^{-i\theta}; q \right)_\infty}.$$

The above and (13.1.5) imply (13.1.27). $\qquad\square$

Since $\left(q^{-1};q^{-1}\right)_n = (-1)^n(q;q)_n q^{-n(n+1)/2}$, we derive

$$H_n\left(\cos\theta\,|\,q^{-1}\right) = \sum_{k=0}^{n} \frac{(q;q)_n}{(q;q)_k(q;q)_{n-k}} q^{k(k-n)} e^{i(n-2k)\theta} \qquad (13.1.31)$$

from (13.1.7).

Theorem 13.1.9 *The polynomials* $\left\{H_n\left(x\,|\,q^{-1}\right)\right\}$ *have the generating function*

$$\sum_{n=0}^{\infty} \frac{H_n\left(\cos\theta\,|\,q^{-1}\right)}{(q;q)_n} (-1)^n t^n q^{\binom{n}{2}} = \left(te^{i\theta}, te^{-i\theta};q\right)_\infty. \qquad (13.1.32)$$

Proof Insert $H_n\left(\cos\theta\,|\,q^{-1}\right)$ from (13.1.31) into the left-hand side of (13.1.32) to see that

$$\sum_{n=0}^{\infty} \frac{H_n\left(\cos\theta\,|\,q^{-1}\right)}{(q;q)_n} (-1)^n t^n q^{\binom{n}{2}}$$

$$= \sum_{n\geq k\geq 0} \frac{q^{k(k-n)+n(n-1)/2}}{(q;q)_k(q;q)_{n-k}} (-t)^n e^{i(n-2k)\theta}$$

$$= \sum_{k=0}^{\infty} \frac{(-t)^k}{(q;q)_k} q^{k(k-1)/2} e^{-ik\theta} \sum_{n=0}^{\infty} \frac{q^{n(n-1)/2}(-t)^n}{(q;q)_n} e^{in\theta}$$

and the result follows from Euler's theorem (12.2.25). □

The q-Hermite polynomials are q-analogues of the Hermite polynomials. Indeed

$$\lim_{q\to 1^-} \left(\frac{2}{1-q}\right)^{n/2} H_n\left(x\sqrt{2/(1-q)}\,|\,q\right) = H_n(x), \qquad (13.1.33)$$

which can be verified using (13.1.1) and (4.5.27). It is an interesting exercise to see how the orthogonality relation for $\{H_n(x\,|\,q)\}$ tends to the orthogonality relation for $\{H_n(x)\}$.

13.2 q-Ultraspherical Polynomials

The continuous q-ultraspherical polynomials is a one-parameter family which generalizes the q-Hermite polynomials. They are defined by

$$C_n(\cos\theta;\beta\,|\,q) = \sum_{k=0}^{n} \frac{(\beta;q)_k(\beta;q)_{n-k}}{(q;q)_k(q;q)_{n-k}} e^{i(n-2k)\theta}. \qquad (13.2.1)$$

It is clear that

$$\begin{aligned} C_n(x;0\,|\,q) &= H_n(x\,|\,q)/(q;q)_n, \\ C_n(-x;\beta\,|\,q) &= (-1)^n C_n(x;\beta\,|\,q), \\ C_n(x;\beta\,|\,q) &= \frac{2^n(\beta;q)_n}{(q;q)_n} x^n + \text{lower order terms}. \end{aligned} \qquad (13.2.2)$$

Although the C_n's are special cases of the Askey–Wilson polynomials of Chapter 15 we, nevertheless, give an independent proof of their orthogonality. The proof given here is from (Askey & Ismail, 1983).

The representation (13.3.1) is equivalent to the $_2\phi_1$ representation

$$C_n(\cos\theta;\beta\,|\,q) = \frac{(\beta;q)_n\,e^{in\theta}}{(q;q)_n}\, {}_2\phi_1\left(\begin{matrix} q^{-n},\beta \\ q^{1-n}/\beta \end{matrix}\,\bigg|\, q, qe^{-2i\theta}/\beta\right). \tag{13.2.3}$$

Theorem 13.2.1 *The orthogonality relation*

$$\int_{-1}^{1} C_m(x;\beta\,|\,q)C_n(x;\beta\,|\,q)w(x\,|\,\beta)\,dx \tag{13.2.4}$$

$$= \frac{2\pi(\beta,q\beta;q)_\infty}{(q,\beta^2;q)_\infty}\frac{(1-\beta)\,(\beta^2;q)_n}{(1-\beta q^n)\,(q;q)_n}\,\delta_{m,n}$$

holds for $|\beta| < 1$, *with*

$$w(\cos\theta\,|\,\beta) = \frac{(e^{2i\theta},e^{-2i\theta};q)_\infty}{(\beta e^{2i\theta},\beta e^{-2i\theta};q)_\infty}(\sin\theta)^{-1}. \tag{13.2.5}$$

Proof We shall first assume that $\beta \neq q^k$, $k = 1,2,\dots$ then extend the result by analytic continuation to $|\beta| < 1$. Since the weight function is even, it follows from (13.2.2) that (13.2.4) holds trivially if $n - m$ is odd. Thus it suffices to evaluate

$$I_{m,n} := \int_0^\pi e^{i(n-2m)\theta}C_n(\cos\theta;\beta\,|\,q)\frac{(e^{2i\theta},e^{-2i\theta};q)_\infty}{(\beta e^{2i\theta},\beta e^{-2i\theta};q)_\infty}\,d\theta, \tag{13.2.6}$$

for $0 \le m \le n$. From (13.2.2) and the $_1\psi_1$ sum (12.3.3) we find

$$\frac{1}{\pi}I_{m,n} = \frac{1}{\pi}\sum_{k=0}^{n}\frac{(\beta;q)_k(\beta;q)_{n-k}}{(q;q)_k(q;q)_{n-k}}$$

$$\times \int_0^\pi \frac{(e^{2i\theta},qe^{-2i\theta};q)_\infty}{(\beta e^{2i\theta},\beta e^{-2i\theta};q)_\infty}\,e^{(2n-2m-2k)i\theta}\left(1-e^{-2i\theta}\right)d\theta$$

$$= \frac{(\beta,\beta q;q)_\infty}{(q,\beta^2;q)_\infty}\sum_{k=0}^{n}\frac{(\beta;q)_k(\beta;q)_{n-k}}{(q;q)_k(q;q)_{n-k}}$$

$$\times \sum_{j=-\infty}^{\infty}\frac{\beta^j(1/\beta;q)_j}{(\beta;q)_j}\left[\delta_{j,k+m-n}-\delta_{j,k+m-n+1}\right]$$

$$= \frac{(\beta;q)_n\,(\beta,\beta q;q)_\infty}{(q;q)_n\,(q,\beta^2;q)_\infty}$$

$$\times \left[\frac{\beta^{m-n}(1/\beta;q)_{m-n}}{(\beta;q)_{m-n}}\,{}_3\phi_2\left(\begin{matrix} q^{-n},\beta,q^{m-n}/\beta \\ q^{1-n}/\beta,q^{m-n}\beta \end{matrix}\,\bigg|\,q,q\right)\right.$$

$$\left. -\frac{\beta^{m-n+1}(1/\beta;q)_{m-n+1}}{(\beta;q)_{m-n+1}}\,{}_3\phi_2\left(\begin{matrix} q^{-n},\beta,q^{m-n+1}/\beta \\ q^{1-n}/\beta,q^{m-n+1}\beta \end{matrix}\,\bigg|\,q,q\right)\right]$$

Thus

$$I_{m,n} = \pi \frac{(\beta;q)_n (\beta, \beta q; q)_\infty}{(q;q)_n (q, \beta^2; q)_\infty} \left[\frac{\beta^{m-n}(1/\beta;q)_{m-n} \left(q^{m-n}, \beta^2; q\right)_n}{(\beta;q)_{m-n} \left(q^{m-n}\beta, \beta; q\right)_n} \right.$$
$$\left. - \frac{\beta^{m-n+1}(1/\beta;q)_{m-n+1} \left(q^{m-n+1}, \beta^2; q\right)_n}{(\beta;q)_{m-n+1} \left(q^{m-n+1}\beta, \beta; q\right)_n} \right]. \tag{13.2.7}$$

We used (12.2.15) in the last step. The factor $(q^{m-n};q)_n$ vanishes for $m \le n$ and causes the first term in [ ] in (13.2.7) to vanish for $m \le n$ while the factor $\left(q^{m-n+1};q\right)_n$ annihilates the second term in [ ] for $m < n$. If $m = n$ then

$$I_{n,n} = \frac{\pi(\beta, q\beta; q)_\infty}{(q, \beta^2; q)_\infty} \frac{(\beta^2;q)_n}{(q\beta;q)_n}.$$

Thus (13.2.4) holds for $m < n$ and if $m = n$ its left-hand side is $2(\beta;q)_n I_{n,n}/(q;q)_n$. This completes the proof. □

It is straightforward to see that (13.2.1) implies the generating function

$$\sum_{n=0}^\infty C_n(\cos\theta; \beta \,|\, q) t^n = \frac{(t\beta e^{i\theta}, t\beta e^{-i\theta}; q)_\infty}{(t e^{i\theta}, t e^{-i\theta}; q)_\infty}. \tag{13.2.8}$$

We now derive a second generating function for the q-ultraspherical polynomials. Apply the Pfaff–Kummer transformation (12.4.7) to the representation (13.2.3) to get

$$C_n(\cos\theta; \beta \,|\, q) = \frac{(\beta, q^{1-n}e^{-2i\theta}/\beta; q)_n}{(q;q)_n} e^{in\theta}$$
$$\times \sum_{k=0}^n \frac{(q^{-n}, q^{1-n}/\beta^2; q)_k}{(q, q^{1-n}/\beta, q^{1-n}e^{-2i\theta}/\beta; q)_k} q^{\binom{k}{2}} (-q)^k e^{-2ik\theta}.$$

Replace k by $n - k$ and simplify to arrive at the representation

$$C_n(\cos\theta; \beta \,|\, q) = \frac{(\beta^2;q)_n e^{-in\theta}}{(q;q)_n \beta^n} \, {}_3\phi_2\left(\begin{matrix} q^{-n}, \beta, \beta e^{2i\theta} \\ \beta^2, 0 \end{matrix} \,\middle|\, q, q \right). \tag{13.2.9}$$

One immediate consequence of (15.5.1) is the generating function

$$\sum_{n=1}^\infty \frac{C_n(\cos\theta; \beta \,|\, q)}{(\beta^2;q)_n} q^{\binom{n}{2}} t^n = (-t e^{-i\theta}; q)_\infty \, {}_2\phi_1\left(\begin{matrix} \beta, \beta e^{2i\theta} \\ \beta^2 \end{matrix} \,\middle|\, q, -t e^{-i\theta} \right). \tag{13.2.10}$$

The ${}_2\phi_1$ on the right-hand side of (15.5.2) is an analogue of a modified Bessel function I_ν, with $\beta = q^\nu$. Another representation for $C_n(x; \beta \,|\, q)$ follows from comparing the weight function $w(x \,|\, \beta)$ with the weight function $w(x; t \,|\, q)$ in (15.2.4). The result is (Askey & Ismail, 1983)

$$C_n(\cos\theta; \beta \,|\, q) = \frac{(\beta^2;q)_n}{\beta^{n/2}(q;q)_n} \, {}_4\phi_3\left(\begin{matrix} q^{-n}, q^n\beta^2, \sqrt{\beta}\,e^{i\theta}, \sqrt{\beta}\,e^{-i\theta} \\ \beta q^{1/2}, -\beta q^{1/2}, -\beta \end{matrix} \,\middle|\, q, q \right). \tag{13.2.11}$$

As $q \to 1$, the representation (13.2.11) with $\beta = q^\nu$ reduces to the second line in (4.5.1).

From (13.2.8) it follows that

$$[1 - 2xt + t^2] \sum_{n=0}^{\infty} C_n(x; \beta \,|\, q) t^n = [1 - 2x\beta t + \beta^2 t^2] \sum_{n=0}^{\infty} C_n(x; \beta \,|\, q)(qt)^n$$

and upon equating like powers of t we establish the recurrence relation

$$2x \left(1 - \beta q^n\right) C_n(x; \beta \,|\, q) = \left(1 - q^{n+1}\right) C_{n+1}(x; \beta \,|\, q) \\ + \left(1 - \beta^2 q^{n-1}\right) C_{n-1}(x; \beta \,|\, q), \tag{13.2.12}$$

for $n > 0$. The initial values of the C_n's are

$$C_0(x; \beta \,|\, q) = 1, \quad C_1(x; \beta \,|\, q) = 2x(1 - \beta)/(1 - q). \tag{13.2.13}$$

The special cases $\beta = q$ of (13.2.1) or (13.2.12)–(13.2.13) give

$$C_n(x; q \,|\, q) = U_n(x), \quad n \geq 0. \tag{13.2.14}$$

On the other hand

$$\lim_{\beta \to 1} \frac{(1 - \beta^2 q^n)}{(1 - \beta^2)} C_n(x; \beta \,|\, q) = T_n(x), \quad \lim_{q \to 1} C_n(x; q^\nu \,|\, q) = C_n^\nu(x), \tag{13.2.15}$$

for $n \geq 0$, $\{C_n^\nu(x)\}$ being the ultraspherical polynomials of §4.5.

It is clear from (13.2.1) that

$$\max \{C_n(x; \beta \,|\, q) : -1 \leq x \leq 1\} = C_n(1; \beta \,|\, q). \tag{13.2.16}$$

Unlike the ultraspherical polynomials $C_n(1; \beta \,|\, q)$, for general β, does not have a closed form expression. However

$$C_n \left(\left(\beta^{1/2} + \beta^{-1/2} \right) /2; \beta \,|\, q \right) = (-1)^n C_n \left(- \left(\beta^{1/2} + \beta^{-1/2} \right) /2; \beta \,|\, q \right) \\ = \beta^{-n/2} \frac{(\beta^2; q)_n}{(q; q)_n}, \tag{13.2.17}$$

since in this case $e^{i\theta} = \beta^{1/2}$ and the left-hand side of (13.2.1) is

$$\beta^{n/2} \frac{(\beta; q)_n}{(q; q)_n} \, {}_2\phi_1 \left(q^{-n}, \beta; q^{1-n}/\beta; q, q/\beta^2 \right),$$

which can be summed by (12.2.19). The answer simplifies via (12.2.10) to (13.2.17). Furthermore

$$\max \left\{ C_n(x; \beta \,|\, q) : |x| \leq \left(\beta^{1/2} + \beta^{-1/2} \right) /2, x \text{ real} \right\} = \beta^{-n/2} \frac{(\beta^2; q)_n}{(q; q)_n}. \tag{13.2.18}$$

The value of $C_n(0; \beta \,|\, q)$ can also be found in closed form. This evaluation follows from (13.2.8) and the answer is

$$C_{2n+1}(0; \beta \,|\, q) = 0, \quad C_{2n}(0; \beta \,|\, q) = \frac{(-1)^n (\beta^2; q^2)_n}{(q^2; q^2)_n}. \tag{13.2.19}$$

In particular formula (13.1.27) is the special case $\beta = 0$ of (13.2.19).

An important special case of the C_n's is

$$\lim_{\beta \to \infty} \beta^{-n} C_n(x; \beta \mid q) = \frac{q^{n(n-1)/2}(-1)^n}{(q; q)_n} H_n \left(x \mid q^{-1}\right) \qquad (13.2.20)$$

where $H_n\left(x \mid q^{-1}\right)$ is as in (13.1.31).

Theorem 13.2.2 *The orthogonality relation* (13.2.4) *is equivalent to the evaluation of the integral*

$$\int_0^\pi \frac{\left(t_1\beta e^{i\theta}, t_1\beta e^{-i\theta}, t_2\beta e^{i\theta}, t_2\beta e^{-i\theta}, e^{2i\theta}, e^{-2i\theta}; q\right)_\infty}{\left(t_1 e^{i\theta}, t_1 e^{-i\theta}, t_2 e^{i\theta}, t_2 e^{-i\theta}, \beta e^{2i\theta}, \beta e^{-2i\theta}; q\right)_\infty} d\theta$$

$$= \frac{(\beta, q\beta; q)_\infty}{(q, \beta^2; q)_\infty} \, {}_2\phi_1 \left(\beta^2, \beta; q\beta; q, t_1 t_2\right), |t_1| < 1, |t_2| < 1. \qquad (13.2.21)$$

Proof We will see in §13.4 that

$$C_n(1; \beta \mid q)/n \to (\beta; q)_\infty^2 / (q; q)_\infty^2. \qquad (13.2.22)$$

The Lebesgue bounded convergence theorem allows us to expand the integrand in (13.2.20) and interchange the summation and integration and establish the desired equivalence. □

The analogue of (13.1.29) is

$$\mathcal{D}_q C_n(x; \beta \mid q) = \frac{2(1-\beta)}{1-q} \, q^{(1-n)/2} \, C_{n-1}(x; q\beta \mid q), \qquad (13.2.23)$$

and can be proved similarly using the generating function (13.2.8). The above is a lowering operator for $\{C_n(x; \beta \mid q)\}$ and the raising operator can be found by using the generating function (13.2.8). The result is

$$\mathcal{D}_q[w(x \mid \beta) C_n(x; \beta \mid q)]$$

$$= -\frac{2q^{-\frac{1}{2}n} \left(1 - q^{n+1}\right) \left(1 - \beta^2 q^{n-1}\right)}{(1-q)\left(1 - \beta q^{-1}\right)} \, w\left(x \mid \beta q^{-1}\right) C_{n+1}\left(x; \beta q^{-1} \mid q\right).$$

$$(13.2.24)$$

13.3 Linearization and Connection Coefficients

Rogers' connection coefficient formula for the continuous q-ultraspherical polynomials is (Rogers, 1894)

$$C_n(x; \gamma \mid q) = \sum_{k=0}^{\lfloor n/2 \rfloor} \frac{\beta^k (\gamma/\beta; q)_k (\gamma; q)_{n-k}}{(q; q)_k (q\beta; q)_{n-k}} \frac{(1 - \beta q^{n-2k})}{(1 - \beta)} C_{n-2k}(x; \beta \mid q).$$

$$(13.3.1)$$

Two important special cases are, cf. (13.2.2),

$$C_n(x;\gamma\,|\,q) = \sum_{k=0}^{\lfloor n/2 \rfloor} \frac{(-\gamma)^k(\gamma;q)_{n-k}}{(q;q)_k} q^{\binom{k}{2}} \frac{H_{n-2k}(x\,|\,q)}{(q;q)_{n-2k}}, \tag{13.3.2}$$

$$\frac{H_n(x\,|\,q)}{(q;q)_n} = \sum_{k=0}^{\lfloor n/2 \rfloor} \frac{\beta^k}{(q;q)_k(q\beta;q)_{n-k}} \frac{\left(1-\beta q^{n-2k}\right)}{(1-\beta)} C_{n-2k}(x;\beta\,|\,q). \tag{13.3.3}$$

The proof of (13.3.1) will be in three steps. We first prove (13.3.1) for $\beta = q$ and general γ then use (13.2.23) to extend to $\beta = q^j$. A pattern for the coefficients in (13.3.1) will then emerge. The fact that both sides are rational functions of β and we have proved it for enough values of β will establish the result. Another proof which uses integration by parts will be given in §16.4 and is new. This proof mirrors the proof of the case $q = 1$ (Theorem 9.2.1).

Theorem 13.3.1 *The connection relation* (13.3.1) *holds.*

Proof Let $\beta = q$. We use the orthogonality relation (13.2.4), so we need to evaluate the integral

$$I_{n,k} := \int_{-1}^{1} C_n(x;\gamma\,|\,q) U_{n-2k}(x)\sqrt{1-x^2}\,dx. \tag{13.3.4}$$

Now (13.2.1) implies

$$I_{n,k} = \frac{1}{2} \sum_{j=0}^{n} \frac{(\gamma;q)_j(\gamma;q)_{n-j}}{(q;q)_j(q;q)_{n-j}}$$

$$\times \int_0^{\pi} \cos(n-2j)\theta\,[\cos((n-2k)\theta) - \cos((n-2k+2)\theta)]\,d\theta$$

$$= \frac{\pi}{4} \sum_{j=0}^{n} \frac{(\gamma;q)_j(\gamma;q)_{n-j}}{(q;q)_j(q;q)_{n-j}} [\delta_{j,k} + \delta_{j,n-k} - \delta_{j,k-1} + \delta_{j,n-k+1}]$$

$$= \frac{\pi}{2} \left[\frac{(\gamma;q)_k(\gamma;q)_{n-k}}{(q;q)_k(q;q)_{n-k}} - \frac{(\gamma;q)_{k-1}(\gamma;q)_{n-k+1}}{(q;q)_{k-1}(q;q)_{n-k+1}} \right],$$

which gives (13.3.1) for $\beta = q$. Apply $\mathcal{D}_q^j$ to (13.3.1) with $\beta = q$ and use n replace by $n+j$. In view of (13.2.23) we get

$$(\gamma;q)_j q^{j(1-2n-j)/4} C_n\left(x;q^j\gamma\,|\,q\right)$$

$$= \sum_{k=0}^{\lfloor n/2 \rfloor} \frac{q^k(\gamma/q;q)_k(\gamma;q)_{n-k+j}}{(q;q)_k(q^2;q)_{n-k+j}} \frac{\left(1-q^{j+n-2k+1}\right)}{(1-q)}$$

$$\times (q;q)_j q^{j(1-2n-j+4k)/4} C_{n-2k}\left(x;q^{j+1}\,|\,q\right).$$

The result now follows since

$$\frac{(q;q)_j(\gamma;q)_{n-k+j}}{(1-q)(q^2;q)_{n-k+j}(\gamma;q)_j} = \frac{(\gamma q^j;q)_{n-k}}{(1-q^{j+1})(q^{j+2};q)_{n-k}}.$$

This proves Theorem 13.3.1. □

The limiting case $\beta \to \infty$ of (13.3.3) is

$$\frac{H_n(x\,|\,q)}{(q;q)_n} = \sum_{k=0}^{\lfloor n/2 \rfloor} \frac{(-1)^k q^{k(3k-2n-1)/2}}{(q;q)_k (q;q)_{n-2k}} H_{n-2k}\left(x\,|\,q^{-1}\right), \qquad (13.3.5)$$

whose inverse is

$$H_n\left(x\,|\,q^{-1}\right) = \sum_{s=0}^{\lfloor n/2 \rfloor} \frac{q^{-s(n-s)}(q;q)_n}{(q;q)_s (q;q)_{n-2s}} H_{n-2s}(x\,|\,q), \qquad (13.3.6)$$

and is also a limiting case of (13.3.3).

In view of the orthogonality relation (13.2.1) the connection coefficient formula (13.3.1) is equivalent to the integral evaluation

$$\int_0^\pi \frac{\left(t\gamma e^{i\theta}, t\gamma e^{-i\theta}, e^{2i\theta}, e^{-2i\theta}; q\right)_\infty}{\left(te^{i\theta}, te^{-i\theta}, \beta e^{2i\theta}, \beta e^{-2i\theta}; q\right)_\infty} C_m(\cos\theta; \beta\,|\,q)\, d\theta$$

$$= \frac{(\beta, q\beta; q)_\infty (\gamma; q)_m t^m}{(q, \beta^2; q)_\infty (q\beta; q)_m} {}_2\phi_1\left(\gamma/\beta, \gamma q^m; q^{m+1}\beta; q, \beta t^2\right). \quad (13.3.7)$$

Since $C_n(x; q\,|\,q) = U_n(x)$ is independent of q we can then we can use (13.3.2) and (13.3.3) to establish the change of basis formula (Bressoud, 1981)

$$\frac{H_n(x\,|\,q)}{(q;q)_n} = \sum_{k=0}^{\lfloor n/2 \rfloor} \frac{q^k \left(1 - q^{n-2k+1}\right)}{(q;q)_k (q;q)_{n-k+1}}$$
$$\times \sum_{j=0}^{\lfloor n/2 \rfloor - k} \frac{(-1)^j p^{\binom{j+1}{2}} (p;p)_{n-2k-j}}{(p;p)_j} \frac{H_{n-2k-2j}(x\,|\,p)}{(p;p)_{n-2k-2j}}. \qquad (13.3.8)$$

Similarly we get the more general connection formula

$$C_n(x;\gamma\,|\,q) = \sum_{k=0}^{\lfloor n/2 \rfloor} \frac{q^k (\gamma/q; q)_k (\gamma; q)_{n-k} \left(1 - q^{n-2k+1}\right)}{(q;q)_k (q^2; q)_{n-k}(1-q)}$$
$$\times \sum_{j=0}^{\lfloor n/2 \rfloor - k} \frac{(\beta)^j (p/\beta; q)_j (p;p)_{n-2k-j}}{(p;p)_j (p\beta; q)_{n-2k-j}}$$
$$\times \frac{\left(1 - \beta p^{n-2k-2j}\right)}{(1-\beta)} C_{n-2k-2j}(x; \beta\,|\,p). \qquad (13.3.9)$$

Theorem 13.3.2 ((Rogers, 1894)) *We have the linearization formula*

$$C_m(x;\beta\,|\,q)C_n(x;\beta\,|\,q)$$
$$= \sum_{k=0}^{m\wedge n} \frac{(q;q)_{m+n-2k}(\beta;q)_{m-k}(\beta;q)_{n-k}(\beta;q)_k \left(\beta^2;q\right)_{m+n-k}}{(\beta^2;q)_{m+n-2k}(q;q)_{m-k}(q;q)_{n-k}(q;q)_k(\beta q;q)_{m+n-k}} \qquad (13.3.10)$$
$$\times \frac{\left(1 - \beta q^{m+n-2k}\right)}{(1-\beta)} C_{m+n-2k}(x;\beta\,|\,q).$$

Proof In view of the first equation in (13.2.2) and the linearization formula for $\{H_n(x \mid q)\}$ (13.1.7), we expect a linearization formula of the type

$$C_m(x; \beta \mid q) \, C_n(x; \beta \mid q)$$

$$= \sum_{k=0}^{m \wedge n} \frac{(q)_{m+n-2k} a_{m,n,k}}{(q;q)_k (q;q)_{m-k} (q;q)_{n-k}} C_{m+n-2k}(x; \beta \mid q).$$

As in the proof of Theorem 13.1.5, we set up a difference equation for $a_{m,n,k}$ with n fixed by using the three-term recurrence relation (13.2.12). Solving the resulting difference equation establishes (13.3.10). $\qquad\square$

Remark 13.3.1 *Computer algebra packages are extremely useful in determing connection coefficients and linearization coefficients. Indeed, one can guess the linearization coefficients in (13.2.14) by finding them for few small values of m and n. Once the correct pattern is detected, one can easily prove it by induction.*

The linearization formula (13.3.10) leads to an interesting integral operator. Multiply (13.3.10) by $r^m s^n$, sum over $m, n \geq 0$, then replace m, n by $m + k, n + k$, respectively, to find

$$\frac{\left(\beta r e^{i\theta}, \beta r e^{-i\theta}, \beta s e^{i\theta}, \beta s e^{-i\theta}; q\right)_\infty}{(r e^{i\theta}, r e^{-i\theta}, s e^{i\theta}, s e^{-i\theta}; q)_\infty}$$

$$= \sum_{m,n=0}^{\infty} \frac{(q, \beta^2; q)_{m+n} \, (\beta; q)_m (\beta; q)_n}{(\beta^2, \beta q; q)_{m+n} \, (q; q)_m (q; q)_n} \frac{1 - \beta q^{m+n}}{1 - \beta}$$

$$\times C_{m+n}(\cos\theta; \beta \mid q) \, r^m s^n \, {}_2\phi_1 \left(\begin{array}{c} \beta, \beta^2 q^{m+n} \\ \beta q^{m+n+1} \end{array} \bigg| \, q, rs \right).$$

Theorem 13.3.3 ((Ismail & Stanton, 1988)) *We have the bilinear generating function*

$$\frac{\left(\beta t e^{i(\theta+\phi)}, \beta t e^{i(\theta-\phi)}, \beta t e^{i(\phi-\theta)}, \beta t e^{-i(\theta+\phi)}; q\right)_\infty}{\left(t e^{i(\theta+\phi)}, t e^{i(\theta-\phi)}, t e^{i(\phi-\theta)}, t e^{-i(\theta+\phi)}; q\right)_\infty}$$

$$= \sum_{n=0}^{\infty} \frac{(q, \beta^2; q)_n}{(\beta^2, \beta q; q)_n} \frac{1 - \beta q^n}{1 - \beta} t^n \, {}_2\phi_1 \left(\begin{array}{c} \beta, \beta^2 q^n \\ \beta q^{n+1} \end{array} \bigg| \, q, t^2 \right)$$

$$\times C_n(\cos\theta; \beta \mid q) C_n(\cos\phi; \beta \mid q). \quad (13.3.11)$$

Proof Put $r = t e^{i\phi}$, $s = t e^{-i\phi}$ in the formula preceeding this theorem. $\qquad\square$

Theorem 13.3.3 implies that the left-hand side of (13.3.11) is a symmetric Hilbert–Schmidt kernel and (13.3.11) is the expansion guaranteed by Mercer's theorem, (Tricomi, 1957), for $\beta \in (-1, 1)$.

13.4 Asymptotics

For x in the complex plane set

$$x = \cos\theta, \quad \text{and} \quad e^{\pm i\theta} = x \pm \sqrt{x^2 - 1}. \tag{13.4.1}$$

We choose the branch of the square root that makes $\sqrt{x^2 - 1}/x \to 1$ as $x \to \infty$. This makes

$$\left| e^{-i\theta} \right| \le \left| e^{i\theta} \right|, \tag{13.4.2}$$

with strict inequality if and only if $x \notin [-1, 1]$.

The t-singularities of the generating function (13.2.8) are at $t = e^{\pm i\theta} q^{-n}$, $n = 0, 1, \dots$. Therefore when $x \notin [-1, 1]$, the singularity with smallest absolute value is $t = e^{i\theta}$ and a comparison function is

$$\frac{(\beta, \beta e^{-2i\theta}; q)_\infty}{(1 - te^{-i\theta})(q, e^{-2i\theta}; q)_\infty}.$$

The coefficient of t^n in the comparison function is

$$e^{-ni\theta}(\beta, \beta e^{-2i\theta}; q)_\infty / (q, e^{-2i\theta}; q)_\infty.$$

Thus as $n \to \infty$

$$C_n(x; \beta \,|\, q) = \frac{e^{-in\theta}(\beta, \beta e^{-2i\theta}; q)_\infty}{(q, e^{-2i\theta}; q)_\infty}[1 + o(1)], \quad x = \cos\theta \in \mathbf{C} \setminus [-1, 1].$$
$$\tag{13.4.3}$$

For $x \in (-1, 1)$ both $e^{\pm i\theta}$ have the same modulus and a comparison function will be

$$\frac{(\beta, \beta e^{-2i\theta}; q)_\infty}{(1 - te^{-i\theta})(q, e^{-2i\theta}; q)_\infty} + \frac{(\beta, \beta e^{-2i\theta}; q)_\infty}{(1 - te^{-i\theta})(q, e^{-2i\theta}; q)_\infty},$$

and we established

$$C_n(\cos\theta; \beta \,|\, q) = 2\sqrt{\frac{(\beta, \beta, \beta e^{2i\theta}, \beta e^{-2i\theta}; q)_\infty}{(q, q, e^{2i\theta}, e^{-2i\theta}; q)_\infty}} \cos(n\theta + \phi)[1 + o(1)],$$
$$\tag{13.4.4}$$
$$x = \cos\theta \in (-1, 1), \quad \text{as } n \to \infty$$

and

$$\phi = \arg\left[\frac{(\beta e^{2i\theta}; q)_\infty}{(e^{2i\theta}; q)_\infty}\right]. \tag{13.4.5}$$

Finally at $x = 1$ a comparison function is of the form

$$\frac{(\beta; q)_\infty^2}{(q; q)_\infty^2 (1 - t)^2} + \frac{\text{constant}}{1 - t}$$

and we get

$$(-1)^n C_n(-1; \beta \,|\, q) = C_n(1; \beta \,|\, q) = n \frac{(\beta; q)_\infty^2}{(q; q)_\infty^2}[1 + o(1)], \quad \text{as } n \to \infty. \tag{13.4.6}$$

In the normalization in Theorem 11.2.2,

$$\alpha_n = 0, \quad \beta_n = \frac{\left(1 - \beta^2 q^{n-1}\right)\left(1 - q^n\right)}{4\left(1 - \beta q^n\right)\left(1 - \beta q^{n-1}\right)}, \quad \zeta_n = \frac{1 - \beta}{1 - \beta q^n}\frac{\left(\beta^2; q\right)_n}{(q; q)_n}$$

$$\frac{P_n(x)}{\sqrt{\zeta_n}} = \frac{C_n(x; \beta \mid q)}{\sqrt{u_n}}, \quad u_n = \frac{1 - \beta}{1 - \beta q^n}\frac{\left(\beta^2; q\right)_n}{(q; q)_n}.$$

Therefore

$$\frac{C_n(\cos\theta; \beta \mid q)}{\sqrt{\zeta_n}} = 2\sqrt{\frac{\left(q\beta, \beta, \beta e^{2i\theta}, \beta e^{-2i\theta}; q\right)_\infty}{\left(q, \beta^2, e^{2i\theta}, e^{-2i\theta}; q\right)_\infty}}\cos(n\theta + \phi)[1 + o(1)],$$

$$x = \cos\theta \in (-1, 1), \quad \text{as } n \to \infty$$

$$(13.4.7)$$

It is interesting to note that the asymptotic formulas (13.4.3)–(13.4.6) and Theorem 12.2.1 show that the C_n's are orthogonal with respect to a measure with no discrete part. In addition Nevai's theorem, Theorem 12.2.4 and (13.4.4) predict that the C_n's are orthogonal with respect to a weight function

$$\frac{\left(q, \beta^2, e^{2i\theta}, e^{-2i\theta}; q\right)_\infty}{2\pi\sin\theta\left(q\beta, \beta, \beta e^{2i\theta}, \beta e^{-2i\theta}; q\right)_\infty}, \quad x = \cos\theta$$

whose total x-mass is 1. This is equivalent to (13.2.4).

13.5 Application: The Rogers–Ramanujan Identities

The Rogers–Ramanujan identities are

$$\sum_{n=0}^\infty \frac{q^{n^2}}{(q; q)_n} = \frac{1}{\left(q, q^4; q^5\right)_\infty}, \qquad (13.5.1)$$

$$\sum_{n=0}^\infty \frac{q^{n^2 + n}}{(q; q)_n} = \frac{1}{\left(q^2, q^3; q^5\right)_\infty}. \qquad (13.5.2)$$

The Rogers–Ramanujan identities and their generalizations play a central role in the theory of partition (Andrews, 1986) and (Andrews, 1976b). MacMahon's interpretations will be established at the end of this section but in the meantime we will concentrate on the analytic identities. Let

$$I(t) = \frac{(q; q)_\infty}{2\pi}\int_0^\pi \left(te^{i\theta}, te^{-i\theta}, e^{2i\theta}, e^{-2i\theta}; q\right)_\infty d\theta. \qquad (13.5.3)$$

Proof of (13.5.1) From the generating function (13.1.32), the connection coefficient formula (13.3.6), and the orthogonality relation (13.1.11), we have

$$I(t) = \frac{(q; q)_\infty}{2\pi}\sum_{l=0}^\infty (-t)^l q^{\binom{l}{2}}\sum_{s=0}^{[l/2]}\frac{q^{s(s-l)}}{(q; q)_s (q; q)_{l-2s}}\delta_{l-2s,0}$$

$$= \sum_{n=0}^\infty \frac{q^{n^2 - n}}{(q; q)_n}t^{2n}. \qquad (13.5.4)$$

For the product side choose $t = \sqrt{q}$, and expand the infinite products by the Jacobi triple product identity (12.3.4) using

$$\left(q, e^{2i\theta}, e^{-2i\theta}; q\right)_\infty = \sum_{j=-\infty}^\infty (-1)^j q^{\binom{j+1}{2}} e^{2ij\theta} \left(1 - e^{2i\theta}\right). \qquad (13.5.5)$$

Since the integrand in $I(t)$ is an even function of θ, we integrate on $[-\pi, \pi]$, and use the exponential orthonormality of $\{e^{in\theta}/\sqrt{2\pi} : -\infty < n < \infty\}$ to find

$$I(\sqrt{q}) = \frac{1}{4\pi(q;q)_\infty} \sum_{n,j=-\infty}^\infty (-1)^j q^{\binom{j+1}{2}} (-1)^n q^{n^2/2} \int_{-\pi}^\pi e^{i\theta(2j-n)} \left(1 - e^{2i\theta}\right) d\theta$$

$$= \frac{1}{2(q;q)_\infty} \sum_{-\infty}^\infty (-1)^j q^{\binom{j+1}{2}} \left[q^{2j^2} - q^{2(j+1)^2}\right].$$

In the second sum replace j by $-1 - j$ and establish

$$I\left(\sqrt{q}\right) = \frac{1}{(q;q)_\infty} \sum_{j=-\infty}^\infty (-1)^j q^{(5j^2+j)/2} = \frac{\left(q^5, q^3, q^2; q^5\right)_\infty}{(q;q)_\infty} = \frac{1}{\left(q, q^4; q^5\right)_\infty}, \qquad (13.5.6)$$

where the Jacobi triple product identity was used again in the last step. Now (13.5.4) and (13.5.6) establish (13.5.1). $\qquad \square$

Proof of (13.5.2) The other Rogers–Ramanujan identity is proven by choosing $t = q$ and writing the integrand as $\left(e^{i\theta}, qe^{-i\theta}, qe^{2i\theta}, e^{-2i\theta}; q\right)_\infty \times \left(1 + e^{i\theta}\right)$. The rest of the proof is similar to the proof of (13.5.1) and is omitted. $\qquad \square$

We now give a generalization of the Rogers–Ramanujan identities.

Theorem 13.5.1 *The following identity holds for $m = 0, 1, \ldots$*

$$\sum_{n=0}^\infty \frac{q^{n^2+mn}}{(q;q)_n} = \frac{1}{(q;q)_\infty} \sum_{s=0}^m \begin{bmatrix} m \\ s \end{bmatrix}_q q^{2s(s-m)} \left(q^5, q^{3+4s-2m}, q^{2-4s+2m}; q^5\right)_\infty. \qquad (13.5.7)$$

Proof Consider the integral

$$I_m(t) = \frac{(q;q)_\infty}{2\pi} \int_0^\pi H_m(\cos\theta \,|\, q) \left(te^{i\theta}, te^{-i\theta}, e^{2i\theta}, e^{-2i\theta}; q\right)_\infty d\theta. \qquad (13.5.8)$$

Expand $\left(te^{i\theta}, te^{-i\theta}; q\right)_\infty$ in $\{H_j\left(x \,|\, q^{-1} |\, q\right)\}$ by (3.3.9) then expand the $H_j\left(x \,|\, q^{-1}\right)$'s in terms of $H_j(x \,|\, q)$'s using (3.3.8), and apply the q-Hermite orthogonality. The result is that $I_m(t)$ is given by

$$I_m(t) = \sum_{\ell=0}^\infty (-t)^\ell q^{\binom{\ell}{2}} \sum_{s=0}^{[\ell/2]} \frac{q^{s(s-\ell)}}{(q;q)_s} \delta_{\ell-2s,m}.$$

Hence $I_m(t)$ has the series representation

$$I_m(t) = (-t)^m q^{\binom{m}{2}} \sum_{n=0}^{\infty} \frac{q^{n^2-n}}{(q;q)_n} \left(t^2 q^m\right)^n. \tag{13.5.9}$$

As in the proof of (13.5.1) we choose $t = \sqrt{q}$, then apply (12.3.4) twice to obtain

$$I_m\left(\sqrt{q}\right) = \frac{(-1)^m q^{m^2/2}}{(q;q)_\infty} \sum_{s=0}^{m} \begin{bmatrix} m \\ s \end{bmatrix}_q q^{2s(s-m)} \left(q^5, q^{3+4s-2m}, q^{2-4s+2m}; q^5\right)_\infty. \tag{13.5.10}$$

Now (13.5.9) and (13.5.10) establish the desired result. □

Note that the terms $4s - 2m \equiv 1 \pmod{5}$ in (13.5.10) vanish. On the other hand if $4s - 2m \equiv 0, 4 \pmod{5}$ in (13.5.10), the infinite products may be rewritten as a multiple of the Rogers–Ramanujan product $1/\left(q, q^4; q^5\right)_\infty$, while $4s - 2m \equiv 1, 3 \pmod{5}$ leads to a multiple of $1/\left(q^2, q^3; q^5\right)_\infty$. A short calculation reveals that (13.5.7) is

$$\sum_{n=0}^{\infty} \frac{q^{n^2+mn}}{(q;q)_n} = \frac{(-1)^m q^{-\binom{m}{2}} a_m(q)}{(q, q^4; q^5)_\infty} + \frac{(-1)^{m+1} q^{-\binom{m}{2}} b_m(q)}{(q^2, q^3; q^5)_\infty} \tag{13.5.11}$$

where

$$a_m(q) = \sum_\lambda (-1)^\lambda q^{\lambda(5\lambda-3)/2} \begin{bmatrix} m-1 \\ \lfloor \frac{m+1-5\lambda}{2} \rfloor \end{bmatrix}_q,$$

$$b_m(q) = \sum_\lambda (-1)^\lambda q^{\lambda(5\lambda+1)/2} \begin{bmatrix} m-1 \\ \lfloor \frac{m+1-5\lambda}{2} \rfloor \end{bmatrix}_q. \tag{13.5.12}$$

The polynomials $a_m(q)$ and $b_m(q)$ were considered by Schur in conjunction with his proof of the Rogers–Ramanujan identities. See (Andrews, 1976b) and (Garrett et al., 1999) for details. We shall refer to $a_m(q)$ and $b_m(q)$ as the Schur polynomials.

Our next result is an inverse relation to (13.5.7).

Theorem 13.5.2 *The following identity holds*

$$\frac{\left(q^{3-2k}, q^{2+2k}; q^5\right)_\infty}{\left(q, q^2, q^3, q^4; q^5\right)_\infty} = \sum_{j=0}^{[k/2]} (-1)^j q^{2j(j-k)+j(j+1)/2} \begin{bmatrix} k-j \\ j \end{bmatrix}_q \sum_{s=0}^{\infty} \frac{q^{s^2+s(k-2j)}}{(q;q)_s}. \tag{13.5.13}$$

Observe that (13.5.13) provides an infinite family of extensions to both Rogers–Ramanujan identities. This is so since the cases $k = 0, 1$ of (13.5.13) yield (13.5.1) and (13.5.2) respectively. Furthermore the relationships (13.5.7) and (13.5.13) are inverse relations.

Proof of Theorem 13.5.2 Define $J(t)$ by

$$J(k) := \frac{(q;q)_\infty}{2\pi} \int_0^\pi \left(\sqrt{q}e^{i\theta}, \sqrt{q}e^{-i\theta}, e^{2i\theta}, e^{-2i\theta}; q\right)_\infty U_k(\cos\theta) \, d\theta. \tag{13.5.14}$$

Thus (12.3.2) yields

$$J(k) = \frac{1}{2(q;q)_\infty} \sum_{m,n=-\infty}^{\infty} q^{(m^2+N^2+n)/2}(-1)^{m+n}$$

$$\times \int_{-\pi}^{\pi} e^{i(m-2n)\theta} \left[e^{ik\theta} - e^{-i(k+2)\theta} \right] d\theta.$$

$$= \frac{1}{2(q;q)_\infty} \sum_{n=-\infty}^{\infty} \left[q^{[(2n-k)^2+n(n+1)]/2} (-1)^{n+k} \right.$$

$$\left. -(-1)^{n+k} q^{[(2n+k+2)^2+n(n+1)]/2} \right].$$

In the second sum replace n by $-n-1$ to see that the two sums are equal and we get

$$J(k) = \frac{(-1)^k q^{k^2/2}}{(q;q)_\infty} \sum_{n=-\infty}^{\infty} (-1)^n q^{5n^2/2} q^{n(1-4k)/2}$$

$$= \frac{(-1)^k q^{k^2/2}}{(q;q)_\infty} \left(q^5, q^{3-2k}, q^{2+2k}; q^5 \right)_\infty.$$

Therefore

$$J(k) = (-1)^k q^{k^2/2} \frac{\left(q^{3-2k}, q^{2+2k}; q^5 \right)_\infty}{(q, q^2, q^3, q^4; q^5)_\infty}. \tag{13.5.15}$$

We now evaluate the integral in a different way. We have

$$J(k) = \frac{(q;q)_\infty}{2\pi} \sum_{n=0}^{\infty} \frac{(-1)^n q^{n^2/2}}{(q;q)_n} \int_0^\pi H_n \left(\cos\theta \,|\, q^{-1} \right)$$

$$\times U_k(\cos\theta) \left(e^{2i\theta}, e^{-2i\theta}; q \right)_\infty d\theta.$$

Since $U_k(x) = C_k(x; q \,|\, q)$ we use the connection coefficient formula (3.3.1) to get

$$J(k) = \sum_{\infty > n \geq 2s \geq 0} \frac{(-1)^n q^{s(s-n)+n^2/2}}{(q;q)_s (q;q)_{n-2s}}$$

$$\times \sum_{j=0}^{[k/2]} \frac{(q;q)_{k-j}(-1)^j}{(q;q)_j} q^{j(j+1)/2} \delta_{n-2s, k-2j}$$

Therefore

$$J(k) = \sum_{s \geq 0, 0 \leq 2j \leq k} \frac{(-1)^{k+j} q^{s^2+s(k-2j)+(k-2j)^2/2}}{(q;q)_s (q;q)_{k-2j}} \frac{(q;q)_{k-j}}{(q;q)_j} q^{j(j+1)/2}$$

$$= (-1)^k q^{k^2/2} \sum_{s=0}^{\infty} \sum_{j=0}^{[k/2]} \frac{(-1)^j q^{s^2+s(k-2j)+2j^2-2kj+j(j+1)/2}(q;q)_{k-j}}{(q;q)_s (q;q)_{k-2j}(q;q)_j}.$$

This completes the proof. $\square$

The results and proofs presented so far are from (Garrett et al., 1999).

We now come to the number theoretic interpretations of the Rogers–Ramanujan identities. A **partition** of n, $n = 1, 2, \ldots$, is a finite sequence $(n_1, n_2, \ldots n_k)$, with $n_i \leq n_{i+1}$, so that $n = \sum_{i=1}^{k} n_i$. For example $(1, 1, 1, 3)$, $(1, 2, 3)$ and $(1, 1, 2, 2)$ are partitions of 6. The number of parts in a partition $(n_1, n_2, \ldots, n_k)$ is k and its parts are $n_1, n_2, \ldots, n_k$.

Let $p(n)$ be the number of partitions of n, with $p(0) := 1$. Euler proved

$$\sum_{n=0}^{\infty} p(n)q^n = \frac{1}{(q; q)_\infty}. \tag{13.5.16}$$

Proof Expand the infinite product in (13.5.16) as

$$\prod_{n=1}^{\infty} \left[\sum_{s=0}^{\infty} q^{ns} \right].$$

Thus the coefficient of q^m on the right-hand side of (13.5.16) is the number of ways of writing m as $\sum_j n_j s_j$. In other words s_j is the number of parts each of which is n_j and (13.5.16) follows. $\qquad\square$

From the idea behind Euler's theorem it follows that if

$$\sum_{n=0}^{\infty} p(n; 1, 4)q^n = 1/ \left(q, q^4; q^5 \right)_\infty,$$

$$\tag{13.5.17}$$

$$\sum_{n=0}^{\infty} p(n; 2, 3)q^n = 1/ \left(q^2, q^3; q^5 \right)_\infty,$$

then $p(0; 1, 4) = p(0; 2, 3) = 1$ and $p(n; 1, 4)$ is the number of partitions of n into parts congruent to 1, or 4 modulo 5, while $p(n; 2, 3)$ is the number of partitions of n into parts congruent to 2, or 2 modulo 5. This provides a partition theoretic interpretation of the right-hand sides of (13.5.1) and (13.5.2). In order to interpret the left-hand sides of these identities we need to develop more machinery.

A partition $(n_1, n_2, \ldots, n_k)$ can be represented graphically by putting the parts of the partition on different levels and each part n_i is represented by n_i dots. The number of parts at a level is greater than or equal the number of parts at any level below it. For example the partition $(1, 1, 3, 4, 6, 8)$ will be represented graphically as

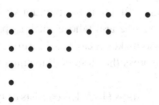

Interchanging rows and columns of a partition $(n_1, n_2, \ldots, n_k)$ gives its conjugate partition. For instance the partition conjugate to $(1, 1, 3, 4, 6, 8)$ is $(1, 1, 2, 2, 3, 4, 4, 6)$.

It is clear that $1/(q; q)_k$ is the generating function of partitions into parts each of which is at most k. Since conjugation is a bijection on partitions then $1/(q; q)_k$ is also the generating function of partitions into at most k-parts. Note that the identity $k^2 = \sum_{j=1}^{k} (2j - 1)$ is a special partition, so for example 4^2 gives the partition

Therefore $q^{k^2}/(q;q)_k$ is the generating function of partitions of n into k parts differing by at least 2. Similarly $k^2 + k = \sum_{j=1}^{k} (2j)$ shows that $q^{k^2+k}/(q;q)_k$ is the generating function of partitions of n into k parts differing by at least 2, and each part is at least 2. This establishes the following theorem.

Theorem 13.5.3 *The number of partitions of n into parts congruent to 1 or 4 modulo 5 equals the number of partitions of n into parts which differ by at least 2. Furthermore the partitions of n into parts congruent to 2 or 3 modulo 5 are equinumerous as the partitions of n into parts which differ by at least 2 and the smallest part is at least 2.*

13.6 Related Orthogonal Polynomials

We return to the analysis of (13.5.11). The left side of (13.5.11) is the generating function for partitions with difference at least two whose smallest part is at least $m + 1$. Andrews (Andrews, 1970) gave a polynomial generalization of the Rogers–Ramanujan identities by showing that

$$a_m(q) = \sum_j q^{j^2+j} \begin{bmatrix} m - j - 2 \\ j \end{bmatrix}_q,$$

$$b_m(q) = \sum_j q^{j^2} \begin{bmatrix} m - j - 1 \\ j \end{bmatrix}_q. \tag{13.6.1}$$

The polynomials $\{a_m\}$ and $\{b_m\}$ have the following combinatorial interpretations: $a_m(q)$ $(b_m(q))$ is the generating function for partitions with difference at least 2 whose largest part is at most $m - 2$ and whose smallest part is at least 2 (1). The representations in (13.5.13) also makes it easy to determine the large m asymptotics of $a_m(q)$ and $b_m(q)$, hence express the Rogers–Ramanujan continued fraction as a quotient of two infinite series.

Andrews' proof of the relationships (13.5.13) consists of first showing that the left-hand side ℓ_m of (13.5.12) satisfy the recurrence relation $\ell_m - \ell_{m+1} = q^{m+1}\ell_{m+2}$. This implies that $\{a_m(q)\}$ and $\{b_m(q)\}$ are solutions of the three term recurrence relation

$$y_{m+2} = y_{m+1} + q^m y_m, \tag{13.6.2}$$

with the initial conditions

$$a_0(q) = 1, a_1(q) = 0, \quad \text{and} \quad b_0(q) = 0, b_1(q) = 1. \tag{13.6.3}$$

This implies that $\{a_m(q)\}$ and $\{b_m(q)\}$ form a basis of solutions to (13.5.15).

The above observations lead to another proof of (13.5.11) from the knowledge of the Rogers–Ramanujan identities. The proof is as follows. Denote the left-hand side of (13.5.11) by $F_m(q)$. It is straightforward to establish

$$F_m(q) - F_{m+1}(q) = q^{m+1} F_{m+2}(q). \tag{13.6.4}$$

Now (13.6.4) shows that $(-1)^m q^{\binom{m}{2}} F_m(q)$ satisfies (13.6.2) hence must be of the form $A a_m(q) + B b_m(q)$ and A and B can be found from the initial conditions (13.6.3). This proves (13.5.11). More importantly (13.6.2) allows us to define y_m for $m < 0$ from y_0 and y_1. If we set

$$y_{-m} = (-1)^m z_{m+1} q^{-m(m+1)/2}, \tag{13.6.5}$$

then we find that z_m satisfies (13.6.2). Applying the initial conditions (13.6.3) we see that

$$\begin{aligned} b_{1-m}(q) &= (-1)^m q^{-\binom{m}{2}} a_m(q), \quad m \geq 1, \\ a_{1-m}(q) &= (-1)^{m+1} q^{-\binom{m}{2}} b_m(q), \quad m \geq 1. \end{aligned} \tag{13.6.6}$$

Theorem 13.6.1 *The generalized Rogers–Ramanujan identities* (13.5.11) *hold for all integers* m, *where* $a_m(q)$ *and* $b_m(q)$ *are given by* (13.6.1) *for* $m \geq 0$ *and when* $m < 0$ *we use* (13.6.6) *to find closed form expressions for* $a_m(q)$ *and* $b_m(q)$.

Carlitz proved the case $m \leq 0$ of Theorem 13.6.1 in (Carlitz, 1959).

Theorem 13.6.2 *The following quintic transformations*

$$\begin{aligned}
\sum_{n=0}^{\infty} \frac{q^{n^2}(qf)^{2n}}{(q;q)_n} &= \frac{(f^4 q^5; q)_\infty}{(f^4 q^5, f^6 q^{10}; q^5)_\infty (f^2 q^3; q)_\infty} \\
\times \sum_{n=0}^{\infty} \frac{1 - f^6 q^{10n+5}}{1 - f^6 q^5} & \frac{(f^6 q^5, f^4 q^{10}; q^5)_n}{(q^5, f^2; q^5)_n} \frac{(f^2; q)_{5n}}{(f^4 q^6; q)_{5n}} q^{5\binom{n}{2}} (-f^4 q^{10})^n \\
&= \frac{(f^4 q^9, f^2 q^5, f^4 q^6; q^5)_\infty}{(f^2 q^3; q)_\infty} \, {}_3\phi_2\left(\begin{matrix} f^2 q^2, f^2 q^3, f^2 q^5 \\ f^4 q^9, f^4 q^6 \end{matrix} \middle| q^5, f^2 q^5 \right) \\
&= \frac{(f^4 q^8, f^2 q^6, f^4 q^6; q^5)_\infty}{(f^2 q^3; q)_\infty} \, {}_3\phi_2\left(\begin{matrix} f^2 q, f^2 q^3, f^2 q^4 \\ f^4 q^8, f^4 q^6 \end{matrix} \middle| q^5, f^2 q^6 \right),
\end{aligned} \tag{13.6.7}$$

hold.

Observe that the Rogers–Ramanujan identities (13.5.1) and (13.5.2) correspond to the special cases $f = q^{-1}$ and $f = q^{-1/2}$ in the last two forms of (13.6.7).

Our proof of Theorem 13.6.2 relies on the connection coefficient formula

$$
\frac{H_n(x\mid q)}{(q;q)_n} = \sum_{k=0}^{[n/2]} \frac{q^k\left(1-q^{n-2k+1}\right)}{(q;q)_k(q;q)_{n-k+1}}
$$
$$
\times \sum_{j=0}^{[n/2]-k} \frac{(-1)^j p^{\binom{j+1}{2}}(p;p)_{n-2k-j}}{(p;p)_j} \frac{H_{n-2k-2j}(x\mid p)}{(p;p)_{n-2k-2j}},
$$

(13.6.8)

The details are in (Garrett et al., 1999).

Al-Salam and Ismail found a common generalization of $\{a_n(q)\}$ and $\{b_n(q)\}$. They introduced and studied the polynomials,

$$
U_0(x;a,b) := 1, \qquad U_1(x;a,b) := x(1+a),
$$
$$
x\left(1+aq^n\right)U_n(x;a,b) = U_{n+1}(x;a,b) + bq^{n-1}U_{n-1}(x;a,b),
$$

(13.6.9)

for $q \in (0,1)$, $b > 0$, $a > -1$. Set

$$
F(x;a) := \sum_{k=0}^{\infty} \frac{(-1)^k x^k q^{k(k-1)}}{(q,-a;q)_k},
$$

(13.6.10)

(Al-Salam & Ismail, 1983).

Theorem 13.6.3 *We have*

$$
U_n^*(x;a,b) = (1+a)U_{n-1}(x;qa,qb),
$$

(13.6.11)

$$
\sum_{n=0}^{\infty} U_n(x;a,b)t^n = \sum_{m=0}^{\infty} \frac{(bt/(ax);q)_m}{(xt;q)_{m+1}} (axt)^m q^{m(m-1)/2},
$$

(13.6.12)

$$
U_n(x;a,b) = \sum_{k=0}^{\lfloor n/2 \rfloor} \frac{(-a,q;q)_{n-k}(-b)^k x^{n-2k}}{(-a,q;q)_k(q;q)_{n-2k}} q^{k(k-1)}.
$$

(13.6.13)

Moreover

$$
\lim_{n\to\infty} x^{-n}U_n(x;a,b) = (-a;q)_\infty F\left(b/x^2;a\right),
$$

(13.6.14)

Proof Formula (13.6.11) follows from (13.6.9). Next, multiply the recursion in (13.6.9) by t^n and add for all n and take into account the initial conditions in (13.6.9) to derive a functional equation for the generating function, which can then be solved and leads to (13.6.12). Equating coefficients of t^n in (13.6.12) establishes (13.6.13). Finally (13.6.14) follows from (13.6.13) and Tannery's theorem. $\square$

Through straightforward manipulations one can prove the following lemma

Lemma 13.6.4 *The functions $F(z;a)$ satisfy*

$$
(z-a)F(zq;a) + (a-q)F(z;a) + qF(z/q;a) = 0,
$$
$$
[(1+a)qF(z;a) - F(z/q;a)] = zF(qz;qa).
$$

(13.6.15)

Theorem 13.6.5 *The functions $F(z;a)$ and $F(qz;qa)$ have no common zeros.*

Proof Assume both functions have a common zero $z = \xi$. Then $F(\xi/q; a) = 0$, by (13.6.16), and (13.6.15) implies $F(q\xi; a) = 0$. Applying (13.6.15) repeatedly we prove that $F(q^n\xi; a) = 0$ for $n = 0, 1, \ldots$. This contradicts the identity theorem for analytic functions because $F(z; a)$ is an entire function of z. $\qquad\square$

Theorem 13.6.6 *Let $\mu^{(a)}$ be the normalized orthogonality measure of $\{U_n(x; a, b)\}$. Then*

$$\int_{\mathbb{R}} \frac{d\mu^{(a)}(y)}{z - y} = \frac{F\left(qbz^{-2}; qa\right)}{zF\left(bz^{-2}; a\right)}, \qquad (13.6.16)$$

where $zF\left(bz^{-2}; a\right) \neq 0$.

Proof Apply Markov's theorem and formulas (13.6.11) and (13.6.14). $\qquad\square$

Theorem 13.6.7 *For $a > -1$, $q \in (0, 1)$, the function $F(z; a)$ has only positive simple zeros. The zeros of $F(z; a)$ and $F(zq, aq)$ interlace. The measure $\mu^{(a)}$ is supported at $\left\{\pm\sqrt{b/x_n(a)}\right\}$ where $\{x_n(a)\}$ are the zeros of $F(z; a)$ arranged in increasing order.*

Proof The singularities of $z^{-1}F\left(qbz^{-2}; qa\right)/F\left(bz^{-2}; a\right)$ agree with the singularities of $\int_{\mathbb{R}} d\mu^{(a)}(y)/(z-y)$, hence all are real. These singularities must be all the zeros of $F\left(bz^{-2}; a\right)$ plus possibly $z = 0$, since $F(qz; qa)$ and $F(z; a)$ have no common zeros. The fact that the right-hand side of (13.6.16) is single-valued proves that $\mu^{(a)}$ is discrete. The positivity of $\mu^{(a)}$ implies the positivity of the residue of the right-hand side of (13.6.16) at its poles, hence the interlacing property holds. To show that $x = 0$ supports no positive mass, we show that $\sum_0^\infty \left[\tilde{U}_n(0; a, b)\right]^2 = \infty$, $\tilde{U}_n$ being the orthonormal polynomials. From (13.6.9) we have

$$\left[\tilde{U}_n(0; a, b)\right]^2 = U_n^2(0; a, b) \frac{1 + aq^n}{1 + a} b^{-n} q^{-n(n-1)/2}.$$

The divergence of $\sum_{n=0}^\infty \left[\tilde{U}_n(0; a, b)\right]^2$ now follows from (13.6.11). The rest follows from the Perron–Stieltjes inversion formula. $\qquad\square$

Observe that the case $a = 0$, $x = 1$, $b = 1$ of the recursion in (13.6.9) becomes (13.6.3). Indeed,

$$a_{n+2}(q) = U_n\left(1; 0, -q^2\right), \qquad b_{n+2} = U_{n+1}(1; 0, -q). \qquad (13.6.17)$$

The q-Lommel polynomials of §14.4 are $\{U_n\left(2x; -q^\nu, q^\nu\right)\}$.

Remark 13.6.1 *Note that applying Darboux's method to (13.6.12) shows that*

$$\lim_{n\to\infty} x^{-n} U_n(x, a, b) = \sum_{m=0}^\infty \left(\frac{b}{ax^2}; q\right)_m \frac{a^m q^{\binom{m}{2}}}{(q; q)_m}. \qquad (13.6.18)$$

Set

$$G(x; a) = \sum_{m=0}^{\infty} \frac{(x; q)_m}{(q; q)_m} a^m q^{\binom{m}{2}}.$$ (13.6.19)

Thus, (13.6.18) *and* (13.6.14) *imply*

$$(-a; q)_\infty F(x; a) = G(x/a; a),$$ (13.6.20)

which can be proved independently.

The continued fraction associated with (13.6.9) is

$$\frac{1+a}{x(1+a)-} \frac{b}{x(1+aq)-} \cdots \frac{bq^{n-2}}{x(1+aq^n)} \cdots$$
$$= \frac{F\left(bq/x^2; aq\right)}{F\left(b/x^2; a\right)} = (1+a) \frac{G\left(b/ax^2; aq\right)}{G\left(b/ax^2; a\right)}.$$ (13.6.21)

The continued fraction evaluation (13.6.21) appeared in Ramanujan's lost notebook.
George Andrews gave a proof of it in (Andrews, 1981) without identifying its partial
numerators or denominators.

13.7 Three Systems of q-Orthogonal Polynomials

Charris and Ismail introduced and extensively studied a q-analogue of the Pollaczek
polynomials in (Charris & Ismail, 1987). The same polynomials appeared later when
W. A. Al-Salam and T. S. Chihara found all families of orthogonal polynomials having generating functions of the form

$$A(t) \prod_{n=0}^{\infty} \frac{1 - \delta x H(q^m t)}{1 - \theta x K(q^m t)} = \sum_{n=0}^{\infty} P_n(x) t^n; \quad m \geq 0, \ n \geq 0.$$

For $\delta\theta \neq 0$ they showed that all solutions are given by the q-Pollaczek polynomials
plus two exceptional cases, see (Al-Salam & Chihara, 1987).

We shall follow the notation of (Charris & Ismail, 1987) and denote the polynomials by $\{F_n(x; U, \Delta, V)\}$, or $\{F_n(x)\}$ for short. The polynomials are generated
by

$$F_0(x) = 1, \quad F_{-1}(x) = 0,$$ (13.7.1)

and

$$2\left[(1 - U\Delta q^n) x + V q^n\right] F_n(x) = \left(1 - q^{n+1}\right) F_{n+1}(x)$$
$$+ \left(1 - \Delta^2 q^{n-1}\right) F_{n-1}(x), \quad n > 0.$$ (13.7.2)

The polynomials $\{F_n(x)\}$ have the generating function

$$\sum_{n=0}^{\infty} F_n(\cos \theta) t^n = \frac{(t/\xi, t/\eta; q)_\infty}{(te^{i\theta}, te^{-i\theta}; q)_\infty},$$ (13.7.3)

where

$$1 + 2q(V - x\Delta U)\Delta^{-2} y + q^2 \Delta^{-2} y^2 = (1 - q\xi y)(1 - q\eta y),$$ (13.7.4)

and ξ and η depend on x, and satisfy

$$\xi\eta = \Delta^{-2}. \tag{13.7.5}$$

The numerators $\{F_n^*(x)\}$ have the generating function

$$\sum_{n=0}^{\infty} F_n^*(\cos\theta)t^n = 2t(1 - U\Delta)\sum_{n=0}^{\infty} \frac{(t/\xi, t/\eta; q)_n \, q^n}{(te^{i\theta}, te^{-i\theta}; q)_{n+1}}. \tag{13.7.6}$$

The generating function (13.7.3) implies the explicit representation

$$F_n(\cos\theta) = e^{in\theta}\frac{(e^{-i\theta}/\xi; q)_n}{(q; q)_n} \, {}_2\phi_1\left(\begin{matrix} q^{-n}, e^{i\theta}/\eta \\ q^{1-n}e^{i\theta}\xi \end{matrix} \,\middle|\, q, qe^{-i\theta}\xi\right). \tag{13.7.7}$$

It was shown in (Charris & Ismail, 1987) that the orthogonality relation of the F_n's is

$$\int_0^{\pi} \frac{(e^{2i\theta}, e^{-2i\theta}; q)_\infty}{(e^{i\theta}/\xi, e^{-i\theta}/\xi, e^{i\theta}/\eta, e^{-i\theta}/\eta; q)_\infty}$$

$$\times F_m(\cos\theta; U, \Delta, V)F_n(\cos\theta; U, \Delta, V)d\theta \tag{13.7.8}$$

$$= \frac{2\pi}{(q, \Delta^2; q)_\infty}\frac{(\Delta^2; q)_n}{(1 - U\Delta q^n)(q; q)_n}\delta_{m,n},$$

valid for $q, U, \Delta \in [0, 1)$ and $1 - U^2 \pm 2V > 0$. No direct special function proof of (13.7.8) is known and finding such proof is a very interesting problem. The proof in (Charris & Ismail, 1987) uses Darboux's asymptotic method and Markov's theorem (Theorem 2.6.2).

The associated continuous q-ultraspherical polynomials $\left\{C_n^{(\alpha)}(x; \beta \,|\, q)\right\}$ (Bustoz & Ismail, 1982) satisfy the three-term recurrence relation

$$2x\left(1 - \alpha\beta q^n\right)C_n^{(\alpha)}(x; \beta \,|\, q) = \left(1 - \alpha q^{n+1}\right)C_{n+1}^{(\alpha)}(x; \beta \,|\, q)$$

$$+ \left(1 - \alpha\beta^2 q^{n-1}\right)C_{n-1}^{(\alpha)}(x; \beta \,|\, q), \ n > 0, \tag{13.7.9}$$

and the initial conditions

$$C_0^{(\alpha)}(x; \beta \,|\, q) = 1, \quad C_1^{(\alpha)}(x; \beta \,|\, q) = \frac{2(1 - \alpha\beta)}{(1 - \alpha q)}x. \tag{13.7.10}$$

A generating function is (Bustoz & Ismail, 1982)

$$\sum_{n=0}^{\infty} C_n^{(\alpha)}(x; \beta \,|\, q)t^n = \frac{1 - \alpha}{1 - 2xt + t^2} \, {}_2\phi_1\left(\begin{matrix} q, \beta te^{i\theta}, \beta te^{-i\theta} \\ qte^{i\theta}, qte^{-i\theta} \end{matrix} \,\middle|\, q, \alpha\right). \tag{13.7.11}$$

Let $\mu(.; \alpha, \beta)$ be the orthogonality measure of $\left\{C_n^{(\alpha)}(x; \beta \,|\, q)\right\}$. Then

$$\int_{\mathbb{R}} \frac{d\mu(t; \alpha, \beta)}{x - t} = \frac{2(1 - \alpha\beta) \, {}_2\phi_1(\beta, \beta\rho_1^2; q\rho_1^2; q, q, \alpha)}{(1 - \alpha)\rho_2 \, {}_2\phi_1(\beta, \beta\rho_1^2; q\rho_1^2; q, q\alpha)}, \tag{13.7.12}$$

for $x \notin \mathbb{R}$, where ρ_1 and ρ_2 are defined in (5.3.19).

The large n asymptotics of $C_n^{(\alpha)}(x; \beta \mid q)$ are given by

$$C_n^{(\alpha)}(\cos \theta; \beta \mid q) \approx \frac{(1 - \alpha)i}{2 \sin \theta} e^{-i(n+1)\theta} {}_2\phi_1 \left(\begin{matrix} \beta e^{2i\theta}, \beta \\ q e^{2i\theta} \end{matrix} \Bigg| q, \alpha \right)$$

$$+ \text{ a similar term with } \theta \text{ replaced by } -\theta, \qquad (13.7.13)$$

$0 < \theta < \pi$, which follows from Darboux's method. The orthonormal polynomials are

$$p_n(x) = C_n^{(\alpha)}(x; \beta \mid q) \sqrt{\frac{(1 - \alpha\beta q^n)(\alpha q; q)_n}{(1 - \alpha\beta)(\alpha\beta^2; q)_n}} \qquad (13.7.14)$$

Thus Nevai's theorem, Theorem 11.2.2, implies

$$w(\cos \theta; \alpha, \beta) = \frac{2}{\pi} \frac{(1 - \alpha\beta)(\alpha\beta^2; q)_\infty}{(1 - \alpha)(\alpha; q)_\infty} \left| {}_2\phi_1 \left(\begin{matrix} \beta e^{2i\theta}, \beta \\ q e^{2i\theta} \end{matrix} \Bigg| q, \alpha \right) \right|^{-2}, \qquad (13.7.15)$$

and the orthogonality relation is

$$\int_0^\pi C_m^{(\alpha)}(\cos \theta; \beta \mid q) C_n^{(\alpha)}(\cos \theta; \beta \mid q) w(\cos \theta; \alpha, \beta) \sin \theta \, d\theta$$

$$\qquad (13.7.16)$$

$$= \frac{(1 - \alpha\beta)(a\beta^2; q)_n}{(1 - \alpha\beta q^n)(\alpha q; q)_n} \delta_{m,n},$$

when the measure is purely abosolutely continuous. The orthogonality measure has no singular part if the denominator in (13.7.13) has no zeros. Bustoz and Ismail proved the orthogonality measure is absolutely continuous if $0 < q < 1$ and

$$0 < \beta < 1, \quad 0 < \alpha < 1, \quad \text{or} \quad q^2 \le \beta < 1, \quad -1 < \alpha < 0. \qquad (13.7.17)$$

The condition in (13.1.17) plus $0 < q < 1$ are sufficient, but are far from being necessary. For details, see (Bustoz & Ismail, 1982).

The continued fraction associated with (13.7.9) is

$$\frac{1}{x-} \frac{\beta_1}{x-} \frac{\beta_2}{\cdots} \qquad (13.7.18)$$

where

$$\beta_n = \frac{1}{4} \frac{(1 - \alpha\beta^2 q^{n-1})(1 - \alpha q^n)}{(1 - \alpha\beta q^n)(1 - \alpha\beta q^{n-1})}. \qquad (13.7.19)$$

We now treat an interesting example of orthogonal polynomials from (Ismail & Mulla, 1987). Let

$$\theta_0^{(a)}(x; q) = 1, \quad \theta_1^{(a)}(x; q) = 2x - a, \qquad (13.7.20)$$

$$2x \, \theta_n^{(a)}(x; q) = \theta_{n+1}^{(a)}(x; q) + a q^n \, \theta_n^{(a)}(x; q) + \theta_{n-1}^{(a)}(x; q). \qquad (13.7.21)$$

It is routine to establish the generating function

$$\sum_{n=0}^\infty \theta_n^{(a)}(x; q) \, t^n = \sum_{n=0}^\infty \frac{(-at)^k q^{\binom{k}{2}}}{(t/\rho_2(x), t/\rho_1(x); q)_{k+1}}, \qquad (13.7.22)$$

where $\rho_1(x)$ and $\rho_2(x)$ are as in Lemma 5.3.5. The numerator polynomials are

$$\left(\theta_n^{(a)}(x;q)\right)^* = 2\,\theta_{n-1}^{(a)}(x;q), \quad n \geq 0, \tag{13.7.23}$$

since $\theta_{-1}^{(a)}(x;q)$ can be interpreted as zero from (13.7.21). Let

$$M(x;a,q) := \sum_{k=0}^{\infty} \frac{(-a\rho_1(x))^k\, q^{\binom{k}{2}}}{(q;q)_k\, (\rho_1^2(x);q)_{k+1}}. \tag{13.7.24}$$

Applying Darboux's method to (13.7.22) and making use of Markov's thorem, Theorem 2.6.2, we see that

$$\int_{\mathbb{R}} \frac{d\psi(t;a,q)}{x-t} = \frac{2\rho_1(x)M(x;aq,q)}{M(x;a,q)}, \quad \operatorname{Im} x \neq 0, \tag{13.7.25}$$

where

$$\int_{\mathbb{R}} \theta_m^{(a)}(x;q)\,\theta_n^{(a)}(x;q)\, d\psi(x;a,q) = \delta_{m,n}. \tag{13.7.26}$$

Moreover for $x = \cos\theta$, $0 < \theta < \pi$, Darboux's method yields

$$\theta_n^{(a)}(\cos\theta;q) \approx 2\left|\sum_{k=0}^{\infty} \frac{(-ae^{i\theta})^k\, q^{\binom{k}{2}}}{(q;q)_k\, (e^{2i\theta};q)_{k+1}}\right| \cos(n\theta + \varphi), \tag{13.7.27}$$

where

$$\varphi = \arg\left(\sum_{k=0}^{\infty} \frac{(-ae^{i\theta})^k\, q^{\binom{k}{2}}}{(q;q)_k\, (e^{2i\theta};q)_{k+1}}\right). \tag{13.7.28}$$

Nevai's theorem, Theorem 11.2.2, implies

$$\frac{d\psi(x;a,q)}{dx} = \frac{2}{\pi}\sqrt{1-x^2}\left|\sum_{k=0}^{\infty} \frac{(-ae^{i\theta})^k\, q^{\binom{k}{2}}}{(q,qe^{2i\theta};q)_k}\right|^{-2}. \tag{13.7.29}$$

One can prove that ψ is absolutely continuous if $q \in (0,1)$ and

$$|a|\,q < 1 + q - \sqrt{1+q^2}, \quad \text{or } |a| \leq (1-q)^2, \tag{13.7.30}$$

see (Ismail & Mulla, 1987).

The continued J-fraction associated with $\left\{Q_n^{(a)}(x;q)\right\}$ is

$$\frac{2}{2x-a-} \frac{1}{2x-aq-} \frac{1}{2x-aq^2-} \cdots \frac{1}{2x-aq^n-} \cdots$$
$$= 2\rho_1(x)\frac{M(x;aq,q)}{M(x;aq)}. \tag{13.7.31}$$

Darboux' method also shows that

$$\theta_n^{(a)}(1;q) \approx (n+1)\sum_{k=0}^{\infty} \frac{(-a)^k q^{\binom{k}{2}}}{(q;q)_\infty^2}. \tag{13.7.32}$$

Moreover (13.7.20) and (13.7.21) show that

$$\theta_n^{(a)}(-x;q) = (-1)^n \theta_n^{(-a)}(x;q).$$

It follows from Theorem 11.2.1 and (13.7.32) that $x = \pm 1$ do not support any discrete masses for any $a \in \mathbb{R}$.

It turned out that special cases of the continued fraction (13.7.31) are related to continued fractions in Ramanujan's notes which became known as "the lost notebook." For details see (Ismail & Stanton, 2005). One special case is when $x = 1/2 = \cos(\pi/3)$. At this point the continued fraction does not converge, but convergents of order $3k + s$, $s = -1, 0, 1$ converge. This follows from (13.7.27). The result is

$$
\begin{aligned}
&\lim_{k\to\infty} \left(\frac{1}{1-} \frac{1}{1+q-} \frac{1}{1+q^2-} \cdots \frac{1}{1+q^{3k+s}} \right) \\
&= -\omega^2 \frac{(q^2;q^3)_\infty}{(q;q^3)_\infty} \frac{\omega^{s+1} - (\omega^2 q;q)_\infty / (\omega q;q)_\infty}{\omega^{s-1} - (\omega^2 q;q)_\infty / (\omega q;q)_\infty},
\end{aligned}
\tag{13.7.33}
$$

where $\omega = e^{2\pi i/3}$. This was proved in (Andrews et al., 2003) and (Andrews et al., 2005). A proof using the polynomials $\theta_n^{(a)}(x;q)$ is in (Ismail & Stanton, 2005). Ismail and Stanton also extended (13.7.33) to any kth root of unity by letting $x = \cos(\pi/k)$. Related results on continued fractions which become transparent through the use of orthogonal polynomials are in (Andrews, 1990) and (Berndt & Sohn, 2002).

Exercises

13.1   Let $w(x \mid \beta)$ be the weight function for $\{C_n(x; \beta \mid q)\}$. Prove that

$$\int_{-1}^{1} x w(x \mid \beta q) \{\mathcal{D}_q C_m(x; \beta \mid q)\} \{\mathcal{D}_q C_n(x; \beta \mid q)\} \, dx$$

is zero unless $m - n = \pm 1$ and determine its value in these cases.

13.2   Evaluate the coefficients $\{d_{n,k}\}$ in

$$\frac{(\beta e^{2i\theta}, \beta e^{-2i\theta}; q)_\infty}{(\gamma e^{2i\theta}, \gamma e^{-2i\theta}; q)_\infty} C_n(\cos\theta; \gamma \mid q)$$

$$= \sum_{k=0}^{\infty} d_{k,n} C_n(\cos\theta; \beta \mid q),$$

(Askey & Ismail, 1983). In particular, show that

$$\left[(1+\beta)^2 - 4\beta x^2 \right] \mathcal{D}_q C_n(x; \beta \mid q)$$

$$= \sum_{k=-1}^{1} c_{n,k} C_{n+k}(x; \beta \mid q),$$

for some constants $c_{n,0}, c_{n,\pm 1}$. Evaluate $c_{n,0}, c_{n,\pm 1}$.

Hint: Use Exercise 2.9.

13.3   Carry out the details of the proof of Theorem 13.3.2.

13.4   Prove that (13.2.1) is equivalent to the hypergeometric representation (13.2.3).

13.5   Fill in the details of deriving (13.7.12) and (13.7.15)–(13.7.16).

13.6   Consider the convergents of (13.7.18), $\{C_n\}$. Prove that when $x = 1/2$, $C_{3n+\epsilon}$ converges for $\epsilon = 0, \pm 1$ and find its limits, (Ismail & Stanton, 2005).

13.7   Repeat Exercise 13.5 for the continued fraction associated with (13.7.20)–(13.7.21), (Andrews et al., 2005) and (Ismail & Stanton, 2005).
        Note: This was stated in Ramanujan's lost notebook.

13.8   Use Nevai's theorem, Theorem 11.2.2, to generalize Exercises 13.5 and 13.6 to general moduli.

13.9   Use (16.4.2) to evaluate the quantized discriminants
        (a) $D\left(H_n(x\,|\,q); \mathcal{D}_q\right)$,
        (b) $D\left(C_n(x; \beta\,|\,q); \mathcal{D}_q\right)$

        Hint: For (b), rewrite (13.2.23) in the form

        $$\left(1 - 2\left(2x^2 - 1\right)\beta + \beta^2\right) \mathcal{D}_q C_n(x; \beta\,|\,q)$$
        $$= A_n(x)C_{n-1}(x; \beta\,|\,q) - B_n(x)C_n(x; \beta\,|\,q),$$

        and evaluate $A_n(x)$ and $B_n(x)$. See Exercise 13.2.

13.10   Let

        $$H_n(x\,|\,p) = \sum_{k=0}^{\lfloor n/2 \rfloor} c_{n,n-2k}(p, q) H_{n-2k}(x\,|\,q).$$

        (a) Prove that

        $$c_{2n,0}(p, q) = \sum_{j=-n}^{n} (-1)^j q^{n-j} q^{j(j+1)/2} \begin{bmatrix} 2n \\ n-1 \end{bmatrix}_p.$$

        (b) Show that

        $$c_{2n,0}\left(q^2, q\right) = (-1)^n q^{n^2} \left(q; q^2\right)_n,$$
        $$c_{2n,0}(-q, q) = (-q)^n \left(-1; q^2\right)_n$$
        $$c_{2n,0}\left(q^{1/2}, q\right) = q^{n/2} \left(q^{1/2}; q\right)_n,$$
        $$c_{2n,0}\left(q^{1/3}, q\right) = q^{n/3} \left(q^{2n/3}; q^{-1/3}\right)_n$$
        $$c_{2n,0}\left(q^{2/3}, q\right) = q^{2n/3} \left(q^{1/3}; q^{2/3}\right)_n.$$

        This material is from (Ismail & Stanton, 2003b) and leads to Rogers–Ramanujan type identities.

13.11   Show that the continuous q-ultraspherical polynomials have the following properties.

        (a) Prove that

        $$\frac{\left(\gamma t e^{i\theta}, \gamma t e^{-i\theta}; q\right)_\infty}{\left(t e^{i\theta}, t e^{-i\theta}; q\right)_\infty} = \sum_{n=0}^{\infty} \frac{1 - \beta q^n}{1 - \beta} C_n(\cos\theta; \beta\,|\,q) F_n(t),$$

where

$$F_n(t) = \frac{t^n (\gamma; q)_n}{(q\beta; q)_n} \, {}_2\phi_1 \left(\begin{array}{c} \gamma/\beta, \gamma q^n \\ \gamma q^{n+1} \end{array} \middle| q, t^2 \right).$$

(b) Deduce, (Koornwinder, 2005b, (2.20))

$$\left(-isq^{\frac{1}{2}} e^{i\theta}, -isq^{\frac{1}{2}} e^{-i\theta}; q \right)_\infty = \frac{q^{\frac{1}{2}\alpha^2}}{s^\alpha} \frac{(q; q)_\infty}{(q^{\alpha+1}; q)_\infty}$$

$$\times \sum_{k=0}^{\infty} i^k q^{\frac{1}{2}k^2 + \frac{1}{2}k\alpha} \frac{1 - q^{\alpha+k}}{1 - q^\alpha}$$

$$\times J_{\alpha+k}^{(2)} \left(2sq^{-\frac{1}{2}\alpha}; q \right) C_k$$

$$\times \left(\cos\theta; q^\alpha \,|\, q \right).$$

13.12   Let $\{a_n(q)\}$ and $\{b_n(q)\}$ be as in §13.6. Let

$$A(t) = \sum_{n=0}^{\infty} a_n(q) \, t^n, \quad B(t) = \sum_{n=0}^{\infty} b_n(q) \, t^n.$$

(a) Show that

$$A(t) = \sum_{n=0}^{\infty} \frac{t^{2n} q^{n(n-1)}}{(t; q)_n}, \quad B(t) = \sum_{n=0}^{\infty} \frac{t^{2n+1} q^{n^2}}{(t; q)_{n+1}}.$$

(b) Deduce the representations in (13.6.1) from part (a).

14

Exponential and q-Bessel Functions

In this chapter we explore properties of the functions e_q and E_q and introduce a third exponential function $\mathcal{E}_q$ which is closely related to the Askey–Wilson operators. We prove two addition theorems for $\mathcal{E}_q$ which correspond to $e^{xy}e^{xz} = e^{x(y+z)}$ and $e^{xy}e^{zy} = e^{(x+z)y}$. We also introduce Jackson's q-Bessel functions and derive some of their properties. Several results involving the q-exponential and q-Bessel functions will also be derived including an analogue of the expansion of a plane wave in spherical harmonics.

14.1 Definitions

A consequence of Theorem 12.2.6 is that the functions e_q and E_q satisfy

$$e_q(x)E_q(-x) = 1. \tag{14.1.1}$$

There is no addition theorem like $e^{x+y} = e^x e^y$ for the functions e_q and E_q. H. S. A. Potter (Potter, 1950) and Schützenberger (Schützenberger, 1953) proved that if A and B satisfy the commutation relation

$$BA = qAB, \tag{14.1.2}$$

then

$$(A + B)^n = \sum_{k=0}^{n} \begin{bmatrix} n \\ k \end{bmatrix}_q A^k B^{n-k}. \tag{14.1.3}$$

This is easy to prove by induction and implies

$$e_q(A + B) = e_q(A)e_q(B). \tag{14.1.4}$$

The functions e_q and E_q are the exponential functions associated with D_q in the sense

$$D_q e_q(xy) = \frac{y}{1-q} e_q(xy), \quad D_{q^{-1}} E_q(xy) = \frac{y}{1-q} E_q(xy). \tag{14.1.5}$$

351

The q-exponential function

$$
\mathcal{E}_q(\cos\theta,\cos\phi;\alpha) := \frac{(\alpha^2;q^2)_\infty}{(q\alpha^2;q^2)_\infty} \sum_{n=0}^\infty \frac{(\alpha e^{-i\phi})^n}{(q;q)_n} q^{n^2/4}
$$
$$
\times \left(-e^{i(\phi+\theta)}q^{(1-n)/2}, -e^{i(\phi-\theta)}q^{(1-n)/2};q\right)_n
$$
(14.1.6)

was introduced in (Ismail & Zhang, 1994). In view of (12.2.2), formula (14.1.6) implies

$$
\mathcal{D}_q\mathcal{E}_q(x,y;\alpha) = \frac{2\alpha q^{1/4}}{1-q}\mathcal{E}_q(x,y;\alpha).
$$
(14.1.7)

Furthermore we define

$$
\mathcal{E}_q(x;\alpha) = \mathcal{E}_q(x,0;\alpha).
$$
(14.1.8)

In other words

$$
\mathcal{E}_q(\cos\theta;\alpha) := \frac{(\alpha^2;q^2)_\infty}{(q\alpha^2;q^2)_\infty}
$$
$$
\times \sum_{n=0}^\infty \left(-ie^{i\theta}q^{(1-n)/2}, -ie^{-i\theta}q^{(1-n)/2};q\right)_n
$$
(14.1.9)
$$
\times \frac{(-i\alpha)^n}{(q;q)_n} q^{n^2/4}
$$

Define $u_n(x,y)$ by

$$
u_n(\cos\theta,\cos\phi) = e^{-in\phi}\left(-e^{i(\phi+\theta)}q^{(1-n)/2}, -e^{i(\phi-\theta)}q^{(1-n)/2};q\right)_n.
$$
(14.1.10)

It is easy to see that $u_n(x,y) \to 2^n(x+y)^n$ as $q \to 1$. Hence

$$
\lim_{q\to 1}\mathcal{E}_q(x;(1-q)t/2) = \exp(tx).
$$

Lemma 14.1.1 *We have*

$$
\mathcal{E}_q(0;\alpha) = 1.
$$
(14.1.11)

Proof Using (14.1.9) we see that

$$
\frac{(q\alpha^2;q^2)_\infty}{(\alpha^2;q^2)_\infty}\mathcal{E}_q(0;\alpha) = \sum_{n=0}^\infty \frac{(q^{(1-n)/2},-q^{(1-n)/2};q)_n}{(q;q)_n} q^{n^2/4}(-i\alpha)^n
$$
$$
= \sum_{n=0}^\infty \frac{(q^{1-n};q^2)_n}{(q;q)_n} q^{n^2/4}(-i\alpha)^n.
$$

When n is odd $\left(q^{1-n}; q^2\right)_n = 0$. Therefore

$$\frac{\left(q\alpha^2; q^2\right)_\infty}{\left(\alpha^2; q^2\right)_\infty} \mathcal{E}_q(0; \alpha) = \sum_{n=0}^\infty \frac{\left(q^{1-2n}; q^2\right)_{2n}}{(q;q)_{2n}} q^{n^2} (-1)^n \alpha^{2n}$$

$$= \sum_{n=0}^\infty \frac{\left(q^{1-2n}; q^2\right)_n (q; q^2)_n}{(q, q^2; q^2)_n} q^{n^2} (-1)^n \alpha^{2n}$$

$$= \sum_{n=0}^\infty \frac{(q; q^2)_n}{(q^2; q^2)_n} (-1)^n \alpha^{2n} = \frac{\left(q\alpha^2; q^2\right)_\infty}{\left(\alpha^2; q^2\right)_\infty},$$

by the q-binomial theorem (12.2.22) and the proof is complete. $\qquad\square$

Observe that (14.1.7), (14.1.11) show that $\mathcal{E}_1\left(x; -q^{-1/4}t(1-q)/2\right)$ is $\mathcal{E}(x; t)$ for $T = \mathcal{D}_q$, see (10.1.7)–(10.1.8).

Theorem 14.1.2 *The function $\mathcal{E}_q$ has the q-hypergeometric representation*

$$\mathcal{E}_q(\cos\theta; t) = \frac{\left(-t; q^{1/2}\right)_\infty}{\left(qt^2; q^2\right)_\infty} {}_2\phi_1\left(\begin{array}{c} q^{1/4}e^{i\theta}, q^{1/4}e^{-i\theta} \\ -q^{1/2} \end{array} \middle| q^{1/2}, -t\right). \qquad (14.1.12)$$

Proof Set

$$\phi_n(\cos\theta) := \left(q^{1/2}e^{i\theta}, q^{1/4}e^{-i\theta}; q^{1/2}\right)_n. \qquad (14.1.13)$$

It is easy to see that

$$\mathcal{D}_q\phi_n(x) = 2q^{1/4}\frac{(1-q^n)}{q-1}\phi_{n-1}(x). \qquad (14.1.14)$$

Therefore

$$\frac{(q-1)^n q^{-n/4}}{2^n(q;q)_n} \phi_n(x), \quad \text{and} \quad \frac{(1-q)^n q^{n(n-1)/4}}{2^n(q;q)_n} u_n(x,0)$$

belong to $\mathcal{D}_q$. Thus Corollary 10.1.2 implies

$$\mathcal{E}_q(\cos\theta; t) = A(t) \, {}_2\phi_1\left(\begin{array}{c} q^{1/4}e^{i\theta}, q^{1/4}e^{-i\theta} \\ -q^{1/2} \end{array} \middle| q^{1/2}, -t\right),$$

hence (14.1.11) gives

$$\frac{1}{A(t)} = {}_2\phi_1\left(\begin{array}{c} q^{1/4}i, -q^{1/4}i \\ -q^{1/2} \end{array} \middle| q^{1/2}, -t\right) = \sum_0^\infty \frac{\left(-q^{1/2}; q\right)_n}{(q;q)_n} (-t)^n$$

$$= \frac{\left(tq^{1/2}; q\right)_\infty}{(-t; q)_\infty},$$

by the q-binomial theorem. $\qquad\square$

One can prove that for real θ and t, we have

$$\text{Re}\,\mathcal{E}_q(\cos\theta; it) = {}_2\phi_1\left(\begin{array}{c} -e^{2i\theta}, -e^{-2i\theta} \\ q \end{array} \middle| q^2, qt^2\right),$$

$$\text{Im}\,\mathcal{E}_q(\cos\theta; it) = \frac{2tq^{1/4}\cos\theta}{1-q} {}_2\phi_1\left(\begin{array}{c} -qe^{2i\theta}, -qe^{-2i\theta} \\ q^3 \end{array} \middle| q^2, qt^2\right).$$

$$(14.1.15)$$

The functions on the right-hand sides of (14.1.15) are q-analogues of the cosine and sine functions, respectively.

Jackson introduced the q-Bessel functions

$$J_\nu^{(1)}(z;q) = \frac{(q^{\nu+1};q)_\infty}{(q;q)_\infty} \sum_{n=0}^\infty \frac{(-1)^n (z/2)^{\nu+2n}}{(q, q^{\nu+1};q)_n}, \qquad (14.1.16)$$

$$J_\nu^{(2)}(z;q) = \frac{(q^{\nu+1};q)_\infty}{(q;q)_\infty} \sum_{n=0}^\infty \frac{(-1)^n (z/2)^{\nu+2n}}{(q, q^{\nu+1};q)_n} q^{n(\nu+n)}. \qquad (14.1.17)$$

This notation is from (Ismail, 1982) and is different from Jackson's original notation. It is easy to see that $J_\nu^{(k)}((1-q)z;q) \to J_\nu(z)$ as $q \to 1^-$. F. H. Jackson (Jackson, 1903; Jackson, 1904; Jackson, 1905) studied the cases of integer ν which are normally referred to as Bessel coefficients. An algebraic setting for q-Bessel functions and their generalization is in (Floreanini & Vinet, 1994).

It is clear that $z^{-\nu} J_\nu^{(2)}(z;q)$ is entire but $z^{-\nu} J_\nu^{(1)}(z;q)$ is analytic in $|z| < 2$.

Theorem 14.1.3 *The identity*

$$J_\nu^{(1)}(z;q) = \frac{J_\nu^{(2)}(z;q)}{(-z^2/4;q)_\infty}, \qquad (14.1.18)$$

holds for $z < 2$ and analytically continues $z^{-\nu} J_\nu^{(1)}(z;q)$ to a meromorphic function outside $|z| \le 2$. Furthermore we have

$$q^\nu J_{\nu+1}^{(k)}(z;q) = \frac{2(1-q^\nu)}{z} J_\nu^{(k)}(z;q) - J_{\nu-1}^{(k)}(z;q) \qquad (14.1.19)$$

$$J_\nu^{(1)}(z\sqrt{q};q) = q^{\nu/2} \left[J_\nu^{(1)}(z;q) + \frac{z}{2} J_{\nu+1}^{(1)}(z;q) \right] \qquad (14.1.20)$$

$$J_\nu^{(1)}(z\sqrt{q};q) = q^{-\nu/2} \left[J_\nu^{(1)}(z;q) - \frac{z}{2} J_{\nu-1}^{(1)}(z;q) \right]. \qquad (14.1.21)$$

Proof The function $(z/2)^{-\nu} \left(-z^2/4;q \right)_\infty J_\nu^{(1)}(z;q)$ is an even analytic function in a neighborhood of the origin and the coefficient of $(z/2)^{2n}$ in its Taylor expansion is

$$= \frac{(q^{\nu+1};q)_\infty}{(q;q)_n (q;q)_\infty} \sum_{k=0}^n \frac{q^{(n-k)(n-k-1)/2}(-1)^k}{(q, q^{\nu+1};q)_k (q;q)_{n-k}}$$

$$= \frac{q^{n(n-1)/2}(q^{\nu+1};q)_\infty}{(q;q)_n (q;q)_\infty} \lim_{b\to 0} {}_2\phi_1\left(q^{-n}, b; q^{\nu+1}; q, q \right)$$

$$= \frac{q^{n(n-1)/2}(q^{\nu+1};q)_\infty}{(q;q)_n (q;q)_\infty} \lim_{b\to 0} \frac{b^n (q^{\nu+1}/b;q)_n}{(q^{\nu+1};q)_n},$$

which easily simplifies to the coefficient of $(z/2)^{2n}$ in $(z/2)^{-\nu} J_\nu^{(2)}(z;q)$. The proofs of (14.1.18)–(14.1.20) also follow by equating coefficients of like powers of z. $\qquad\Box$

It readily follows from (14.1.16)–(14.1.17) that

$$\lim_{q\to 1^-} J_\nu^{(k)}(x(1-q);q) = J_\nu(x). \qquad (14.1.22)$$

It is not difficult to establish the q-difference equations

$$\left(1 + qx^2/4\right) J_\nu^{(2)}(qx; q) + J_\nu^{(2)}(x; q)$$
$$= \left(q^{\nu/2} + q^{-\nu/2}\right) J_\nu^{(2)}\left(\sqrt{q}\,x; q\right),$$

(14.1.23)

$$J_\nu^{(1)}(qx; q) + \left(1 + x^2/4\right) J_\nu^{(1)}(x; q)$$
$$= \left(q^{\nu/2} + q^{-\nu/2}\right) J_\nu^{(1)}\left(\sqrt{q}\,x; q\right),$$

(14.1.24)

directly follow from (14.1.16)–(14.1.17).

The functions $I_\nu(z; q)$ and $K_\nu(z; q)$ can be defined in a way similar to $I_\nu(z)$ and $K_\nu(z)$ of (1.3.17) and (1.3.23). Indeed, see (Ismail, 1981)

$$I_\nu^{(k)}(z; q) = e^{-i\pi\nu/2} J_\nu^{(k)}\left(ze^{i\pi/2}; q\right), \quad k = 1, 2,$$

(14.1.25)

$$K_\nu^{(k)}(z; q) = \frac{\pi}{2}\frac{I_{-\nu}^{(k)}(z; q) - I_\nu^{(k)}(z; q)}{\sin(\pi\nu)}, \quad k = 1, 2,$$

(14.1.26)

with $K_n^{(k)}(z; q) = \lim_{\nu\to n} K_\nu^{(k)}(z; q)$, $n = 0, \pm1, \pm2, \dots$. Observe that $K_\nu^{(k)}(z; q)$ is an even function of ν. The functions $K_\nu^{(j)}$, and $I_\nu^{(j)}$ satisfy

$$2\frac{1 - q^\nu}{z} K_\nu^{(j)}(z; q) = q^\nu K_{\nu+1}^{(j)}(z; q) - K_{\nu-1}^{(j)}(z; q),$$
$$2\frac{1 - q^\nu}{z} I_\nu^{(j)}(z; q) = I_{\nu-1}^{(j)}(z; q) - q^\nu I_{\nu+1}^{(j)}(z; q),$$

(14.1.27)

for $j = 1, 2$.

Some of the recent literature considered the function

$$E^{(\alpha)}(x; q) := \sum_{n=0}^\infty \frac{x^n}{(q; q)_n} q^{\alpha n^2}, \quad 0 < \alpha, \ 0 < q < 1,$$

(14.1.28)

as a q-analogue of the exponential function (Atakishiyev, 1996). Below we shall show that $E^{(\alpha)}(x; q)$ and $\mathcal{E}_q(x; t)$ are entire functions of order zero hence have infinitely many zeros. The asymptotics and graphs of the large zeros of $E^{1/4}(z; q)$ have been studied in detail in (Nelson & Gartley, 1994). q-analogues of the logarithmic function are in (Nelson & Gartley, 1996).

Lemma 14.1.4 *Let $\{f_n\}$ be a bounded sequence with infinitely many nonzero terms and let*

$$f(z) = \sum_{n=0}^\infty f_n p^{n^2} z^n, \quad 0 < p < 1.$$

(14.1.29)

Then $\rho(f) = 0$ and $f(z)$ has infinitely many zeros.

Proof By Theorem 1.2.5 it suffices to show that $\rho(f) = 0$. With $|f_n| \le C$ and $|z| \le r$, we have

$$M(r, f) \le C \sum_{n=0}^\infty p^{n^2} r^n < C \sum_{n=-\infty}^\infty p^{n^2} r^2 = C\left(p^2, -pr, -p/r; p^2\right)_\infty.$$

Set $r = p^{-2(N+\epsilon)}$, for $-\frac{1}{2} \le \epsilon < \frac{1}{2}$ and $N = 0, 1, 2, \ldots$. Clearly

$$\left(-pr; p^2\right)_\infty = \left(-p^{2N+1-2\epsilon}; p^2\right)_N \left(-p^{1-2\epsilon}; p^2\right)_\infty$$

$$= p^{-\left(N^2+2N\epsilon\right)} \left(-p^{1-2\epsilon}; p^2\right)_N \left(-p; p^2\right)_\infty.$$

Hence for fixed p

$$\ln M(r, f) \le -(N+\epsilon)^2 \ln p + O(1) = -\frac{(\ln r)^2}{4 \ln p} + O(1),$$

which implies

$$\limsup_{r \to \infty} \frac{\ln M(r, f)}{\ln^2 r} \le \frac{1}{4 \ln p^{-1}}, \tag{14.1.30}$$

and $\rho(f) = 0$. $\square$

Note that (14.1.30) is stronger than $\rho(f) = 0$.

Corollary 14.1.5 *The function $E^{(\alpha)}(x; q)$ has infinitely many zeros.*

By a slight modification of the proof of Lemma 14.1.4 it follows that

$$\lim_{r \to \infty} \frac{\ln M\left(r, E^\alpha(\cdot, q)\right)}{\ln^2 r} = \frac{1}{4\alpha \ln q^{-1}}. \tag{14.1.31}$$

Theorem 14.1.6 ((Ismail & Stanton, 2003b)) *The maximum modulus of $\mathcal{E}_q(x, t)$, for fixed t, $|t| < 1$, and $0 < q < 1$, satisfies*

$$\lim_{r \to \infty} \frac{\ln M\left(r, \mathcal{E}_q(\cdot, t)\right)}{\ln^2 r} = \frac{1}{\ln q^{-1}}.$$

The proof uses (14.1.12) and is similar to the proof of (14.1.31).

14.2 Generating Functions

We first prove an analogue of the generating function (4.6.28).

Theorem 14.2.1 *We have*

$$\left(qt^2; q^2\right)_\infty \mathcal{E}_q(x; t) = \sum_{n=0}^\infty \frac{q^{n^2/4} t^n}{(q; q)_n} H_n(x \mid q). \tag{14.2.1}$$

Proof Recall

$$u_n(x, y) = e^{-in\varphi} \left(-q^{(1-n)/2} e^{i(\varphi+\theta)}, -q^{(1-n)/2} e^{i(\varphi-\theta)}; q\right)_n. \tag{14.2.2}$$

Formula (12.2.2) implies

$$D_{q,x} u_n(x, y) = 2q^{(1-n)/2} \frac{(1 - q^n)}{(1 - q)} u_{n-1}(x, y). \tag{14.2.3}$$

Therefore (14.2.3) and (13.1.29) show that $c_n H_n(x \mid q)$ and $c_n u_n(x, y)$ belong to $\mathcal{D}_q$ with

$$c_n = \frac{(1-q)^n}{(q;q)_n} 2^{-n} q^{n(n-1)/4}. \tag{14.2.4}$$

Thus, by Corollary 10.1.2, there is a power series $A(t) = \sum\limits_{n=0}^{\infty} a_n t^n$, with $a_0 \neq 0$ so that

$$\sum_{n=0}^{\infty} \frac{q^{n^2/4}}{(q;q)_n} t^n H_n(x \mid q) = A(t)\mathcal{E}_q(x;t). \tag{14.2.5}$$

But $\mathcal{E}_q(x;t)$ is entire in x, hence the series side in (14.2.5) is an entire function of x. Set $x = 0$ in (14.2.5) and apply (13.1.27). The result then follows from (12.2.24). $\qquad\square$

An important classical expansion formula is the expansion (4.8.3) of the plane wave in spherical harmonics. Ismail and Zhang (Ismail & Zhang, 1994) gave a q-analogue of this expansion. Their formula is

$$\mathcal{E}_q(x; i\alpha/2) = \frac{(2/\alpha)^\nu (q;q)_\infty}{(-q\alpha^2/4; q^2)_\infty (q^{\nu+1}; q)_\infty} \sum_{n=0}^{\infty} \frac{(1 - q^{n+\nu})}{(1 - q^\nu)} q^{n^2/4} i^n$$
$$\times J_{\nu+n}^{(2)}(\alpha; q) \, C_n(x; q^\nu \mid q). \tag{14.2.6}$$

We shall refer to (14.2.6) as the q-plane wave expansion. Different proofs of (14.2.6) were given in (Floreanini & Vinet, 1995a), (Floreanini & Vinet, 1995b), (Ismail et al., 1996), and (Ismail et al., 1999). The proof by Floreanini and Vinet (Floreanini & Vinet, 1995a) is group theoretic and is of independent interest. For a proof of the plane wave expansion and its connections to the addition theorem of Bessel functions see (Watson, 1944, Chapter 11).

Proof of (14.2.6) Use Theorem 14.2.1, and expand the q-Hermite polynomials in terms of the q-ultraspherical polynomials through (13.3.3). The result is

$$\left(q\alpha^2; q^2\right)_\infty \mathcal{E}_q(x; \alpha) = \sum_{n=0}^{\infty} \frac{(1 - \beta q^n)}{(1 - \beta)} \alpha^n q^{n^2/4} C_n(x; \beta \mid q)$$
$$\times \sum_{k=0}^{\infty} \frac{\alpha^{2k} \beta^k}{(q;q)_k (\beta q; q)_{n+k}} q^{k(n+k)}. \tag{14.2.7}$$

With $\alpha \to i\alpha/2$, $\beta = q^\nu$, the k-sum contributes the q-Bessel function and the infinite products to (14.1.5). $\qquad\square$

Observe that the formal interchange of q and q^{-1} amounts to interchanging the formal series expansions in z of $(z;q)_\infty$ and $1/\left(zq^{-1};q^{-1}\right)_\infty$. Furthermore for $|q| \neq 1$, it readily follows from (14.1.6) and (14.1.8) that

$$\mathcal{E}_q(x; \alpha) = \mathcal{E}_{q^{-1}}\left(x; -\alpha\sqrt{q}\right). \tag{14.2.8}$$

Thus, we would expect Theorem 14.2.1 to be equivalent to the following corollary.

Corollary 14.2.2 *We have*

$$\frac{1}{(\alpha^2; q^2)_\infty} \mathcal{E}_q(x; \alpha) = \sum_{n=0}^{\infty} \frac{q^{n^2/4} \alpha^n}{(q; q)_n} H_n(x \,|\, q^{-1}). \tag{14.2.9}$$

Proof Substitute for $H_n(x \,|\, q)/(q; q)_n$ from (13.3.5) in the right-hand side of (14.2.1), replace n by $n + 2k$, then evaluate the k sum by (12.2.25). The result is (14.2.9).   □

In view of the orthogonality relation (13.2.4) formula (14.2.6) yields

$$\frac{2\pi i^n \left(q^{2\nu}; q\right)_n (q^\nu; q)_\infty q^{n^2/4}}{\alpha^\nu (q^{2\nu}; q)_\infty (q; q)_n (-q\alpha^2; q^2)_\infty} J_{n+\nu}^{(2)}(2\alpha; q)$$

$$= \int_0^\pi \mathcal{E}_q(\cos\theta; i\alpha) C_n (\cos\theta; q^\nu \,|\, q) \frac{(e^{2i\theta}, e^{-2i\theta}; q)_\infty}{(\beta e^{2i\theta}, \beta e^{-2i\theta}; q)_\infty} d\theta. \tag{14.2.10}$$

14.3 Addition Formulas

The function $\mathcal{E}_q(x; \alpha)$ is a q-analogue of $e^{\alpha x}$, but unlike $e^{\alpha x}$, $\mathcal{E}_q(x; \alpha)$ is not symmetric in x and α. The variables x and α seem to play different roles, so one would expect the function $\mathcal{E}_q(x; \alpha)$ to have two different addition theorems. This expectation will be confirmed in this section. We shall prove two addition theorems for the $\mathcal{E}_q$ functions. They are commutative q-analogues of

$$\exp(\alpha(x + y)) = \exp(\alpha x) \exp(\alpha y),$$

$$e^{\alpha x} e^{\beta x} = \sum_{n=0}^{\infty} \frac{\alpha^n x^n (1 + \beta/\alpha)^n}{n!}. \tag{14.3.1}$$

Theorem 14.3.1 *The $\mathcal{E}_q$ function have the addition theorems*

$$\mathcal{E}_q(x, y; \alpha) = \mathcal{E}_q(x; \alpha) \mathcal{E}_q(y; \alpha), \tag{14.3.2}$$

and

$$(q\alpha^2, q\beta^2; q^2)_\infty \mathcal{E}_q(x; \alpha) \mathcal{E}_q(x; \beta)$$

$$= \sum_{n=0}^{\infty} q^{n^2/4} \alpha^n H_n(x \,|\, q) \left(-\alpha\beta q^{(n+1)/2}; q\right)_\infty \frac{(-q^{(1-n)/2}\beta/\alpha; q)_n}{(q; q)_n}. \tag{14.3.3}$$

Proof Formula (14.2.3) implies that $c_n u_n(x, y)$ and $c_n u_n(x, 0)$ belong to $\mathcal{D}_q$ with c_n given by (14.2.4). Corollary 10.1.2 now implies

$$\mathcal{E}_q(x, y; \alpha) = \mathcal{E}_q(x; \alpha) A(\alpha, y), \tag{14.3.4}$$

where A is independent of x. With $x = 0$ in (14.3.4) and applying (14.1.8) and (14.1.11), we find $A(\alpha, y) = \mathcal{E}_q(y; \alpha)$. This establishes (14.3.2). From Theorem

14.2.1 and (13.1.19) we get

$$\left(q\alpha^2, q\beta^2; q^2\right)_\infty \mathcal{E}_q(x;\alpha)\mathcal{E}_q(x;\beta)$$

$$= \sum_{m,n=0}^{\infty} q^{(m^2+n^2)/4}\alpha^m\beta^n \sum_{k=0}^{\min(m,n)} \frac{H_{m+n-2k}(x\mid q)}{(q;q)_k(q;q)_{m-k}(q;q)_{n-k}}$$

$$= \sum_{m,n=0}^{\infty} q^{(m^2+n^2)/4}\alpha^m\beta^n \frac{H_{m+n}(x\mid q)}{(q;q)_m(q;q)_n} \sum_{k=0}^{\infty} \frac{\alpha^k\beta^k}{(q;q)_k} q^{k(k+m+n)/2}$$

Euler's formula (12.2.25) shows that the above is

$$= \sum_{m,n=0}^{\infty} q^{(m^2+n^2)/4}\alpha^m\beta^n \left(-\alpha\beta q^{(m+n+1)/2}; q\right)_\infty \frac{H_{m+n}(x\mid q)}{(q;q)_m(q;q)_n}$$

$$= \sum_{N=0}^{\infty} \left(-\alpha\beta q^{(N+1)/2}; q\right)_\infty \frac{\alpha^N H_N(x\mid q)}{(q;q)_N} q^{N^2/4}$$

$$\times {}_1\phi_0\left(q^{-N}; -; q, -q^{(N+1)/2}\beta/\alpha\right),$$

which simplifies to (14.3.3), after applying (12.2.22). This completes the proof. □

Clearly (14.3.2) and (14.3.3) are q-analogues of the first and second formulas in (14.3.1), respectively.

The addition theorem (14.3.2) is due to Suslov (Suslov, 1997), while (14.3.3) is due to Ismail and Stanton (Ismail & Stanton, 2000). The proof of (14.3.2) given here is from Ismail and Zhang (Ismail & Zhang, 2005), while a proof of (14.3.2) in (Suslov, 1997) is wrong.

14.4 q-Analogues of Lommel and Bessel Polynomials

This section is based on our work (Ismail, 1982). As in §6.5, we iterate (14.1.8) and establish

$$q^{n\nu+n(n-1)/2} J^{(k)}_{\nu+n}(x;q) = R_{n,\nu}(x;q)J^{(k)}_{\nu}(x;q) \tag{14.4.1}$$
$$- R_{n-1,\nu+1}(x;q)J^{(k)}_{\nu-1}(x;q),$$

where $k = 1, 2$, and $R_{n,\nu}(x;q)$ is a polynomial in $1/x$ of degree n. With

$$h_{n,\nu}(x;q) = R_{n,\nu}(1/x;q), \tag{14.4.2}$$

we have

$$2x\left(1 - q^{n+\nu}\right)h_{n,\nu}(x;q) = h_{n+1,\nu}(x;q) + q^{n+\nu-1}h_{n-1,\nu}(x;q), \tag{14.4.3}$$

$$h_{0,\nu}(x;q) = 1, \quad h_{1,\nu}(x;q) = 2\left(1 - q^{n+\nu}\right)x. \tag{14.4.4}$$

Theorem 14.4.1 *The polynomials $\{h_{n,\nu}(x;q)\}$ have the explicit form*

$$h_n(x;q) = \sum_{j=0}^{\lfloor n/2 \rfloor} \frac{(-1)^j \left(q^\nu, q; q\right)_{n-j}}{(q, q^\nu; q)_j (q;q)_{n-2j}} (2x)^{n-2j} q^{j(j+\nu-1)}. \tag{14.4.5}$$

and the generating function

$$\sum_{n=0}^{\infty} h_{n,\nu}(x;q)t^n = \sum_{j=0}^{\infty} \frac{(-2xtq^{\nu})^j \left(-\frac{1}{2}t/x;q\right)_j}{(2xt;q)_{j+1}} q^{j(j-1)/2}. \qquad (14.4.6)$$

Proof Let $G(x,t) = \sum_{n=0}^{\infty} h_{n,\nu}(x;q)t^n$. Multiply (14.4.3) by t^{n+1} and add for $n \geq 1$ to derive the q-difference equation

$$(1 - 2xt)G(x,t) = \left[1 - 2xtq^{\nu}\left(1 + \frac{1}{2}t/x\right)\right] G(x,qt).$$

We also used (14.4.4). Through repeated applications of $t \to qt$ we arrive at (14.4.6).

$\square$

Lemma 14.4.2 *The functions* $z^{-\nu}J_{\nu}^{(2)}(z;q)$ *and* $z^{-\nu-1}J_{\nu+1}^{(2)}(z;q)$ *have no common zeros for ν real.*

Proof A calculation using the definition of $J_{\nu}^{(2)}$ gives

$$J_{\nu}^{(2)}\left(\sqrt{q}\,z;q\right) - q^{\nu/2}J_{\nu}^{(2)}(z;q) = \frac{x}{2}q^{\nu+1}J_{\nu+1}^{(2)}\left(\sqrt{q}\,z;q\right). \qquad (14.4.7)$$

If u is a common zero of of $J_{\nu}^{(2)}(z;q)$ and $J_{\nu+1}^{(2)}(z;q)$ then, by (14.4.7), $J_{\nu}^{(2)}(u/q;q)$ must also vanish. It is clear that u can not be purely imaginary. The q-difference equation (14.1.22) will show that $\sqrt{q}\,u$ is also a zero of $J_{\nu}^{(2)}(z;q)$. Hence

$$\left(uq^{n/2}\right)^{-\nu} J_{\nu}^{(2)}\left(q^{n/2}\,u;q\right) = 0,$$

for $n = 0, 1, \ldots$, which contradicts the fact that $\lim_{n\to\infty} \left(zq^{n/2}\right)^{-\nu} J_{\nu}^{(2)}\left(q^{n/2}\,z;q\right) \neq 0$, for any $z \neq 0$.

$\square$

Theorem 14.4.3 *The q-Lommel polynomials satisfy*

$$\lim_{n\to\infty} \frac{R_{n,\nu+1}(x;q)}{(x/2)^{n+\nu}\,(q;q)_{\infty}} = J_{\nu}^{(2)}(x;q), \qquad (14.4.8)$$

$$h_{n,\nu}^*(x;q) = 2\left(1 - q^{\nu}\right)h_{n-1,\nu+1}(x;q). \qquad (14.4.9)$$

Moreover for $\nu > 0$, $\{h_{n,\nu}(x;q)\}$ are orthogonal with respect to a purely discrete measure α_{ν}, with

$$\int_{\mathbb{R}} \frac{d\alpha_{\nu}(t;q)}{z - t} = 2\left(1 - q^{\nu}\right) \frac{J_{\nu}^{(2)}(1/z;q)}{J_{\nu-1}^{(2)}(1/z;q)}, \qquad z \notin \text{supp}\,\mu. \qquad (14.4.10)$$

Furthermore for $\nu > -1$, $z^{-\nu}J_{\nu}^{(2)}(z;q)$ has only real and simple zeros. Let

$$0 < j_{\nu,1}(q) < j_{\nu,2}(q) \cdots < j_{\nu,n}(q) < \cdots, \qquad (14.4.11)$$

be the positive zeros of $J_\nu^{(2)}(z;q)$. Then $\{h_{n,\nu}(x;q)\}$ satisfy the orthogonality relation

$$\int_{\mathcal{R}} h_{m,\nu}(x;q)h_{n,\nu}(x;q)d\alpha_\nu(x) = q^{n(2\nu+n-1)/2}\frac{1-q^\nu}{1-q^{\nu+n}}\delta_{m,n}, \qquad (14.4.12)$$

where α_ν is supported on $\{\pm 1/j_{\nu,n}(q) : n = 1, 2, \ldots\} \cup \{0\}$, but $x = 0$ does not support a positive mass.

Proof Formula (14.4.9) follows from the definition of $h_{n,\nu}^*$, while (14.4.8) follows from (14.4.5) and the bounded convergence theorem. Markov's theorem and (14.4.8)–(14.4.9) establish (14.4.10). Since the right-hand side of (14.4.10) is single-valued across $\mathbb{R}$ then α_ν is discrete and has masses at the singularities of the function $J_\nu^{(2)}(1/z;q)/J_{\nu-1}^{(2)}(1/z;q)$. This establishes (14.4.12). The essential singularity $z = 0$ supports no mass as can be seen from applying Theorem 2.5.6 and using $h_{2n+1,\nu}(0;q) = 0$, and $h_{2n,\nu}(0;q) = (-1)^n q^{n(n+\nu-1)}$, and (14.4.12). By Theorem 14.4.1, the pole singularities of the right-hand side of (14.4.10) are the zeros of $J_\nu^{(2)}(1/z;q)$. □

Let

$$\alpha_\nu\{\pm 1/j_{n,\nu-1}\} = (1-q^\nu)\,A_n(\nu)/j_{n,\nu-1}^2(q). \qquad (14.4.13)$$

It readily follows that (14.4.10) is the Mittag–Leffler expansion

$$\sum_{k=1}^\infty A_k(\nu+1)\left[\frac{1}{z - j_{\nu,k}(q)} + \frac{1}{z + j_{\nu,k}(q)}\right] = -2\frac{J_{\nu+1}^{(2)}(z;q)}{J_\nu^{(2)}(z;q)}.$$

The coefficients A_n of (14.4.13) satisfy

$$\begin{aligned}
\frac{d}{dz}\,J_\nu^{(2)}(z;q)\Big|_{z=j_{\nu,n}(q)} &= -2\frac{J_{\nu+1}^{(2)}(j_{\nu,n}(q);q)}{A_n(\nu+1)} \\
&= 2q^{-\nu}\frac{J_{\nu-1}^{(2)}(j_{\nu,n}(q);q)}{A_n(\nu+1)}.
\end{aligned} \qquad (14.4.14)$$

The second equality follows from (14.1.18). We then express (14.4.12) in the form

$$\begin{aligned}
&\sum_{k=1}^\infty \frac{A_k(\nu+1)}{j_{\nu,k}^2(q)}\, h_{n,\nu+1}\left(1/j_{\nu,k}(q);q\right) h_{m,\nu+1}\left(1/j_{\nu,k}(q);q\right) \\
&+ \sum_{k=1}^\infty \frac{A_k(\nu+1)}{j_{\nu,k}^2(q)}\, h_{n,\nu+1}\left(-1/j_{\nu,k}(q);q\right) h_{m,\nu+1}\left(-1/j_{\nu,k}(q);q\right) \qquad (14.4.15) \\
&\qquad = \frac{q^{\nu n+n(n+1)/2}}{1-q^{n+\nu+1}}\delta_{m,n},
\end{aligned}$$

for $\nu > -1$. When $q \to 1$ we use (1.3.26) to find $\lim_{q\to 1} A_n(\nu+1) = 2$ and with some analysis one can show that (14.4.15) tends to (6.5.17).

We now come to the Bessel polynomials. Motivated by (4.10.6), Abdi (Abdi, 1966) defined q-Bessel polynomials by

$$Y_n(x,a) = \frac{(q^{a-1};q)_n}{(q;q)_n} \, {}_2\phi_1\left(q^{-n}, q^{n+a-1}; 0, q, x/2\right).$$ (14.4.16)

There is an alternate approach to discover the q-analogue of the Bessel polynomials. They should arise from the q-Bessel functions in the same way as the Bessel polynomials came from Bessel functions.

This was done in (Ismail, 1981) and leads to a different polynomial sequence. Iterate (14.1.6) and get

$$q^{\nu n + n(n-1)/2} K^{(j)}_{\nu+n}(z;q)$$
$$= i^n R_{n,\nu}(ix;q) K^{(j)}_\nu(z;q) + i^{n-1} R_{n-1,\nu+1}(ix;q) K^{(j)}_{\nu-1}(z;q),$$ (14.4.17)

which implies

$$q^{n^2/2} K^{(j)}_{n+1/2}(z;q)$$
$$= \left[i^n R_{n,1/2}(ix) + i^{n-1} R_{n-1,3/2}(ix)\right] K^{(j)}_{1/2}(z;q).$$ (14.4.18)

In analogy with (6.5.20)–(6.5.22) we define

$$y_n(x \mid q) = i^{-n} h_{n,1/2}(ix) + i^{1-n} h_{n-1,3/2}(ix).$$ (14.4.19)

By considering the cases of odd and even n in (14.4.19) and applying (14.4.5) we derive the explicit representation

$$y_n\left(x \mid q^2\right) = q^{n(n-1)/2} \, {}_2\phi_1\left(q^{-n}, q^{n+1}, -q; q, -2qx\right).$$ (14.4.20)

The analogue of $y_n(x;a) \, (= y_n(x;a,2))$ is

$$y_n\left(x;a \mid q^2\right) = q^{n(n-1)/2} \, {}_2\phi_1\left(q^{-n}, q^{n+a-1}, -q; q, -2qx\right)$$ (14.4.21)

Clearly $y_n\left(x/\left(1-q^2\right), a \mid q^2\right) \to {}_2F_0(-n, n+a-1, -; -x/2) = y_n(x;a)$, as $q \to 1$.

Theorem 14.4.4 *Set*

$$w_{QB}(z;a) = \sum_{n=0}^{\infty} \frac{(-1;q)_n}{(q^{a-1};q)_n} \, (-2z)^{-n}.$$ (14.4.22)

For $r > 1/2$ the polynomials $y_n(z;a \mid q)$ satisfy the orthogonality relation

$$\frac{1}{2\pi i} \oint_{|z|=r} y_n\left(z;a \mid q^2\right) y_n\left(z;a \mid q^2\right) w_{QB}(z;a) dz$$
$$= \frac{(-1)^{n+1} q^{n^2} \left(-q^{a-1}, q; q\right)_n}{\left(-q, q^{a-1}; q\right)_n \left(1 - q^{2n+a-1}\right)} \delta_{m,n}.$$ (14.4.23)

Proof Clearly for $m \leq n$ we have

$$q^{-\binom{n}{2}} \frac{1}{2\pi i} \oint_{|z|=r} z^m y_n \left(z; a \,|\, q^2\right) y_n \left(z; a \,|\, q^2\right) w_{QB}(z; a) dz$$

$$= \sum_{k=0}^{n} \sum_{j=0}^{\infty} \frac{\left(q^{-n}, q^{n+a-1}; q\right)_k}{(q, -q; q)_k \left(q^{a-1}; q\right)_j} \frac{(-1; q)_j}{(-2)^j} \frac{(-2q)^k}{2\pi i} \frac{1}{\oint_{|z|=r} z^{m+k-j} dz}$$

$$= \sum_{k=0}^{n} \frac{\left(q^{-n}, q^{n+a-1}; q\right)_k}{(q, -q; q)_k \left(q^{a-1}; q\right)_{k+m+1}} \frac{(-1; q)_{k+m+1}}{(-2)^{m+1}} q^k$$

$$= \frac{(-1; q)_{m+1}}{\left(q^{a-1}; q\right)_{m+1}(-2)^{m+1}} \, {}_3\phi_2 \left(\begin{matrix} q^{-n}, q^{n+a-1}, -q^{m+1} \\ -q, q^{a+m} \end{matrix} \,\middle|\, q, q \right)$$

$$= -\frac{(-q; q)_m \left(-q^{a-1}, q^{m+1-n}; q\right)_n}{\left(q^{a-1}; q\right)_{m+1}(-2)^m \left(q^{a+m}, -q^{-n}; q\right)_n},$$

and the ${}_3\phi_2$ was summed by Theorem 12.2.3. It is clear that the last term above vanishes for $m < n$. Thus the left-hand side of (14.4.23) is

$$-\frac{q^{\binom{n+1}{2}} \left(q^{-n}, q^{a+n-1}; q\right)_n}{(q, -q)_n} \frac{q^{\binom{n}{2}} \left(-q, -q^{a-1}, q; q\right)_n}{\left(q^{a-1}; q\right)_{2n+1} \left(-q^{-n}; q\right)_n},$$

which simplifies to the right-hand side of (14.4.23), and the proof is complete. $\square$

14.5 A Class of Orthogonal Functions

Consider functions of the form

$$\mathcal{F}_k(x) = \sum_{n=0}^{\infty} u_n \, r_n(x_k) \, p_n(x), \tag{14.5.1}$$

where $\{p_n(x)\}$ is complete orthonormal system and $\{r_n(x)\}$ is a complete system, orthonormal with respect to a discrete measure. Let the orthogonality relations of the real polynomials $\{p_n(x)\}$ and $\{r_n(x)\}$ be

$$\int_E p_m(x)p_n(x)d\mu(x) = \delta_{m,n}, \quad \sum_{k=1}^{\infty} \rho(x_k) \, r_m(x_k)r_n(x_k) = \delta_{m,n}, \tag{14.5.2}$$

respectively. The set E may or may not be bounded.

Theorem 14.5.1 ((Ismail, 2001b)) *Assume that $\{u_n\}$ in (14.5.1) is a sequence of points lying on the unit circle and that $\{p_n(x)\}$ and $\{r_n(x)\}$ are orthogonal with respect to unique positive measures. Then the system $\left\{ \sqrt{\rho(x_k)} \, \mathcal{F}_k(x) \right\}$ is a complete orthonormal system in $L^2(\mathbb{R}, \mu)$.*

Proof First $\mathcal{F}$ is well defined since $\{u_n r_n(x_k)\} \in \ell^2$. Parseval's formula gives

$$\int_E \mathcal{F}_k(x) \overline{\mathcal{F}_j(x)} \, d\mu(x) = \sum_{n=0}^{\infty} r_n(x_k) \, r_n(x_j) = \frac{1}{\rho(x_k)} \delta_{j,k},$$

where we used the dual orthogonality (Theorem 2.11.1) in the last step. To prove the completeness assume that $f \in L^2[\mu]$ and

$$\int_E f(x) \overline{\mathcal{F}_k(x)} \, d\mu(x) = 0$$

for $k = 1, 2, \ldots$. Thus f has an orthogonal expansion $\sum_{n=0}^{\infty} f_n p_n(x)$. Moreover $\{f_n\} \in \ell^2$ and

$$0 = \int_E f(x) \overline{\mathcal{F}_k(x)} \, d\mu(x) = \sum_{n=0}^{\infty} f_n \, \overline{u_n} \, r_n(x_k), \quad k = 1, 2, \ldots.$$

The sequence $\{f_n \overline{u_n}\} \in \ell^2$, hence the completeness of $\{r_n(x)\}$ implies that $f_n = 0$ for all n. $\square$

Let $f(x)$ be a polynomial. Following (Ismail, 2001b) expand $f(x)$ as $\sum_{k=0}^{m} f_k p_k(x)$. From the definition of the orthogonal coefficients write

$$p_k(x) = \sum_{j=1}^{\infty} \mathcal{F}_j(x) r_k(x_j) \rho(x_j),$$

$$r_n(x_j) = \int_E \mathcal{F}_j(x) p_n(x) d\mu(x).$$

Thus we find

$$f(x) = \sum_{k=0}^{m} f_k p_k(x) = \sum_{k=0}^{m} f_k \sum_{j=1}^{\infty} \mathcal{F}_j(x) r_k(x_j) \rho(x_j)$$

$$= \sum_{j=1}^{\infty} \mathcal{F}_j(x) \rho(x_j) \sum_{k=0}^{m} f_k r_k(x_j)$$

$$= \sum_{j=1}^{\infty} \mathcal{F}_j(x) \rho(x_j) \sum_{k=0}^{m} f_k \int_E \mathcal{F}_j(x) p_k(x) d\mu(x).$$

Example 14.5.2 *Let us consider the example of q-Lommel polynomials. In this case*

$$p_n(x) = \sqrt{\frac{(1 - q^{n+\nu})(q; q)_n}{(1 - q^\nu)(q^{2\nu}; q)_n}} \, C_n(x; q^\nu \,|\, q),$$

$$\tag{14.5.3}$$

$$r_n(x) = \sqrt{\frac{(1 - q^{n+\nu})}{(1 - q^\nu)}} \, q^{-n\nu/2 - n(n-1)/4} \, h_{n,\nu}(x; q).$$

When $x = 1/j_{\nu-1,k}(q)$, *then*

$$h_{n,\nu}(x; q) = q^{n\nu + n(n-1)/2} J_{\nu+n}^{(2)}(1/x; q) / J_\nu^{(2)}(1/x; q)$$

and we see that the functions

$$
\mathcal{F}_k(x) = \sum_{n=0}^{\infty} u_n q^{n(2\nu+n-1)/4} \frac{(1-q^{n+\nu})}{(1-q^\nu)}
$$

$$
\times \sqrt{\frac{(q;q)_n}{(q^{2\nu};q)_n}} \frac{J^{(2)}_{\nu+n}(\alpha_k;q)}{J^{(2)}_{\nu}(\alpha_k;q)} C_n\left(x; q^\nu \mid q\right),
\tag{14.5.4}
$$

with $x_k := \pm 1/j_{\nu-1,k}$ form a complete orthonormal system in $L^2[-1,1]$ weighted by the normalized weight function

$$
w(x;\nu) := \frac{\left(q, q^{2\nu}; q\right)_\infty}{2\pi \left(q^\nu, q^{\nu+1}; q\right)_\infty} \frac{\left(e^{2i\theta}, e^{-2i\theta}; q\right)_\infty}{\left(q^\nu e^{2i\theta}, q^\nu e^{-2i\theta}; q\right)_\infty} \frac{1}{\sqrt{1-x^2}},
\tag{14.5.5}
$$

and $x = \cos\theta$.

The orthogonality and completeness of the special case $\nu = 1/2$ of the system $\{\mathcal{F}_k(x)\}$ is the main result of (Bustoz & Suslov, 1998), where technical q-series theory was used resulting in lengthy proofs which will not extend to the general functions $\{\mathcal{F}_k(x)\}$.

Another interesting example is to take $\{r_n(x)\}$ to be multiples of the q-analogue of Wimp's polynomials (Wimp, 1985) and take p_n to be continuous q-Jacobi polynomials. The q-Wimp polynomials are defined and studied in (Ismail et al., 1996) where a q-plane wave expansion is given.

Some open problems will be mentioned in §24.2.

14.6 An Operator Calculus

It is clear that all constants are annihilated by $\mathcal{D}_q$. On the other hand functions f for which $\breve{f}$ satisfies $\breve{f}\left(q^{1/2}z\right) = \breve{f}\left(q^{-1/2}z\right)$ are also annihilated by $\mathcal{D}_q$. One example is

$$
f(\cos\theta) = \frac{(\cos\theta - \cos\phi)\left(qe^{i(\theta+\phi)}, qe^{-i(\theta+\phi)}qe^{i(\theta-\phi)}, qe^{i(\phi-\theta)}; q\right)_\infty}{\left(q^{1/2}e^{i(\theta+\phi)}, q^{1/2}e^{-i(\theta+\phi)}, q^{1/2}e^{i(\theta-\phi)}, q^{1/2}e^{i(\phi-\theta)}; q\right)_\infty},
$$

$$
= (\cos\theta - \cos\phi) \prod_{n=0}^{\infty} \frac{1 - 2\cos\theta q^{n+1}e^{i\phi} + q^{2n+2}e^{2i\phi}}{1 - 2\cos\theta q^{n+1/2}e^{i\phi} + q^{2n+1}e^{2i\phi}}
$$

$$
\times \prod_{k=0}^{\infty} \frac{1 - 2\cos\theta q^{k+1}e^{-i\phi} + q^{2k+2}e^{-2i\phi}}{1 - 2\cos\theta q^{k+1/2}e^{-i\phi} + q^{2k+1}e^{-2i\phi}},
$$

for a fixed ϕ. This motivated the following definition.

Definition 14.6.1 *A q-constant is a function annihilated by $\mathcal{D}_q$.*

Most of the series considered here are formal series, so we will assume that our series are formal series unless we state otherwise.

We will define the q-translation operator through its action on the continuous q-Hermite polynomials.

Define polynomials $\{g_n(x)\}$ by

$$\frac{g_n(x)}{(q;q)_n} = \sum_{k=0}^{[n/2]} \frac{q^k}{(q^2;q^2)_k} \frac{H_{n-2k}(x\,|\,q)}{(q;q)_{n-2k}} q^{(n-2k)^2/4}. \tag{14.6.1}$$

We will prove that

$$g_n(\cos\theta) = q^{n^2/4}\left(1 + e^{2i\theta}\right) e^{-in\theta} \left(-q^{2-n}e^{2i\theta};q^2\right)_{n-1}, \tag{14.6.2}$$

for $n > 0$ and $g_0(x) = 1$. It readily follows from (13.1.29) that

$$\mathcal{D}_q g_n(x) = 2q^{1/4}\frac{1-q^n}{1-q}\,g_{n-1}(x). \tag{14.6.3}$$

Since $H_n(-x\,|\,q) = (-1)^n H_n(x\,|\,q)$, (14.6.1) gives $g_n(-x) = (-1)^n g_n(x)$.

We define the action of operator of translation by y, E_q^y on $H_n(x\,|\,q)$ to be

$$E_q^y H_n(x\,|\,q) = H_n\left(x \overset{\circ}{+} y\,|\,q\right)$$
$$:= \sum_{m=0}^{n} \begin{bmatrix} n \\ m \end{bmatrix}_q H_m(x\,|\,q)g_{n-m}(y)q^{(m^2-n^2)/4}. \tag{14.6.4}$$

In other words

$$q^{n^2/4}\frac{H_n\left(x \overset{\circ}{+} y\,|\,q\right)}{(q;q)_n}$$
$$= \sum_{0\le m,j,m+2j\le n} \frac{q^{j+(m^2+(n-m-2j)^2)/4}}{(q^2;q^2)_j} \frac{H_m(y\,|\,q)H_{n-m-2j}(x\,|\,q)}{(q;q)_m(q;q)_{n-m-2j}}. \tag{14.6.5}$$

We then extend E_q^y to the space of all polynomials by linearity. Since both $H_n(x\,|\,q)$ and $g_n(x)$ tend to $(2x)^n$ as $q \to 1$, (14.6.4) or (14.6.5) shows that E_q^y, tends to the usual translation by y as $q \to 1$, hence $\overset{\circ}{+}$ becomes $+$ as $q \to 1$.

Theorem 14.6.1 *We have*

$$E_q^0 = identity, \quad and \quad g_n(0) = \delta_{n,0}. \tag{14.6.6}$$

Proof First note that (14.6.1) gives $g_{2n+1}(0) = 0$, and

$$g_{2n}(0) = \sum_{k=0}^{n} \frac{(-1)^{n-k}\left(q;q^2\right)_{n-k}(q;q)_{2n}}{(q^2;q^2)_k\,(q;q)_{2n-2k}} q^{k+(n-k)^2}$$

$$= \left(q;q^2\right)_n \sum_{k=0}^{n} \frac{(-1)^{n-k}\left(q^2;q^2\right)_n}{(q;q^2)_k\,(q^2;q^2)_{n-k}} q^{k+(n-k)^2}$$

$$= q^{n^2}\left(q;q^2\right)_n \sum_{k=0}^{n} \frac{\left(q^{-2n};q^2\right)_k}{(q^2;q^2)_k} q^{2k} = 0,$$

for $n > 0$, where we applied the q-binomial theorem in the last step. Thus $g_n(0) = \delta_{n,0}$ and the theorem follows. $\qquad\square$

Recall that E^y, the operator of translation by y satisfies $(E^y f)(x) = (E^x f)(y)$ and both $= f(x + y)$. This property is also shared by E_q^y, since in (14.6.5) we may replace m by $n - 2j - m$ which transforms the right-hand side of (14.6.5) to the same expression with x and y interchanged. Hence $E_q^y p(x) = E_q^x p(y)$ for all polynomials p. Therefore $\overset{\circ}{+}$ is a commutative operation.

Theorem 14.6.2 *The q-translation E_q^y commutes with the Askey–Wilson operator $\mathcal{D}_{q,x}$ on the vector space of polynomials over the field of complex numbers.*

Proof Apply (13.1.29) and (14.6.4). □

From (14.2.1) and Euler's theorem we find

$$\mathcal{E}_q(x; \alpha) = \sum_{n=0}^{\infty} \frac{g_n(x)}{(q; q)_n} \alpha^n. \tag{14.6.7}$$

Proof of (14.6.2) Denote the right-hand side of (14.6.2) by $u_n(x)$, then show that $\mathcal{D}_q u_n = 2q^{1/4} (1 - q^n) u_{n-1}(x)/(1 - q)$. Thus, Corollary 10.1.2 and (14.6.7) imply the existence of a formal power series $A(t)$ such that

$$\sum_{n=0}^{\infty} \frac{u_n(x) t^n}{(q; q)_n} = A(t) \mathcal{E}_q(x; t).$$

Since $u_n(0) = \delta_{n,0}$ and $\mathcal{E}_q(0; t) = 1$, we conclude that $A(t) = 1$ and (14.6.2) follows. □

Proof of (14.1.27) In (14.6.7) replace $g_n(x)$ by the expression in (14.6.2) then take real and imaginary parts. □

Theorem 14.6.3 *The q-translations commute, that is $E_q^y E_q^z = E_q^z E_q^y$. Furthermore*

$$(E_q^y f)(x) = (E_q^x f)(y). \tag{14.6.8}$$

Proof Formula (14.6.4) implies that

$$\sum_{n=0}^{\infty} \frac{\alpha^n}{(q; q)_n} q^{n^2/4} E_q^y H_n(x \mid q) = \sum_{n=0}^{\infty} \frac{g_n(y)}{(q; q)_n} \alpha^n \sum_{m=0}^{\infty} \frac{H_m(x \mid q)}{(q; q)_m} \alpha^m q^{m^2/4}$$
$$= (q\alpha^2; q^2)_{\infty} \mathcal{E}_q(x; \alpha) \mathcal{E}_q(y; \alpha).$$

Thus

$$\sum_{n=0}^{\infty} \frac{\alpha^n}{(q; q)_n} q^{n^2/4} E_q^z E_q^y H_n(x \mid q)$$

$$= (q\alpha^2; q^2)_{\infty} \mathcal{E}_q(x; \alpha) \mathcal{E}_q(y; \alpha) \mathcal{E}_q(z; \alpha). \tag{14.6.9}$$

The right-hand side of the above equation is symmetric in y and z hence $E_q^y E_q^z = E_q^z E_q^y$ on polynomials and the first part of the theorem follows. The rest follows from $g_n(0) = \delta_{n,0}$. □

Observe that (14.6.9) with $z = 0$ is

$$E_q^y \mathcal{E}_q(x; \alpha) = \mathcal{E}_q \left(x \overset{\circ}{+} y; \alpha \right) = \mathcal{E}_q(x; \alpha) \mathcal{E}_q(y; \alpha). \qquad (14.6.10)$$

Note that (14.6.7) and (14.6.10) yield

$$g_n \left(x \overset{\circ}{+} y \right) = \sum_{k=0}^{n} \begin{bmatrix} n \\ k \end{bmatrix}_q g_k(x) g_{n-k}(y), \qquad (14.6.11)$$

a reminiscent of the functional equation of polynomials of binomial type, see Definition 10.2.3. Note further that (14.6.10) shows that $\mathcal{E}_q(x; \alpha)$ solves the functional equation

$$f \left(x \overset{\circ}{+} y \right) = f(x) f(y). \qquad (14.6.12)$$

The above functional equation is an analogue of the Cauchy functional equation $f(x+y) = f(x)f(y)$ whose only measurable solutions are of the form $\exp(\alpha x)$. The question of characterizing all solutions to (14.6.12), or finding minimal assumptions under which the solution to (14.6.12) is unique, seems to be a difficult problem. However, a partial answer, where the assumptions are far from being minimal, is given next.

Theorem 14.6.4 *Let* $f(x) = \sum\limits_{n=0}^{\infty} f_n g_n(x)/(q; q)_n$ *for all x in a domain Ω in the complex plane, and assume that the series converges absolutely and uniformly on compact subsets of Ω. Assume further that* $\sum\limits_{n=0}^{\infty} f_n g_n \left(x \overset{\circ}{+} y \right) /(q; q)_n$ *also converges absolutely and uniformly for all x and y in any compact subset of Ω and define* $f \left(x \overset{\circ}{+} y \right)$ *by the latter series. If f satisfies (14.6.12) then $f(x) = \mathcal{E}_q(x; \alpha)$ on Ω for some α.*

Proof Substitute for f in (14.6.12) and use the functional relationship (14.4.11) to get

$$\sum_{m,n=0}^{\infty} f_{m+n} \frac{g_m(x)}{(q; q)_m} \frac{g_n(y)}{(q; q)_n} = \sum_{m,n=0}^{\infty} f_m f_n \frac{g_m(x)}{(q; q)_m} \frac{g_n(y)}{(q; q)_n}.$$

Thus $f_{m+n} = f_m f_n$ which implies $f_n = \alpha^n$ for some α and we find $f(x) = \mathcal{E}_q(x; \alpha)$. $\square$

A q-shift-invariant operator is any linear operator mapping polynomials to polynomials which commutes with E_q^y for all complex numbers y. A q-delta operator $\mathcal{Q}$ is a q-shift-invariant operator for which $\mathcal{Q}x$ is a nonzero constant.

Theorem 14.6.5 *Let $\mathcal{Q}$ be a q-delta operator. Then*

(i) $\mathcal{Q} a = 0$ *for any q-constant a.*

(ii) *If $p_n(x)$ is a polynomials of degree n in x then $\mathcal{Q} p_n(x)$ is a polynomial of degree $n - 1$.*

The proof is similar to the proof of Theorem 10.2.1 and will be omitted.

We next study q-infinitesimal generators. Recall that polynomials of binomial type are those polynomial sequences $\{p_n(x)\}$ which satisfy the addition theorem (10.2.3). The model polynomials of this kind are the monomials $\{x^n\}$ and (10.2.3) is the binomial theorem. In the q-case the model polynomials are the continuous q-Hermite polynomials and (14.6.4) is indeed a q-analogue of the binomial theorem and as $q \to 1$ it tends to the binomial theorem.

We now derive an operational representation for E_q^y in terms of $\mathcal{D}$. Formula (13.1.29) implies

$$\mathcal{D}_{q,x}^k q^{n^2/4} \frac{H_n(x \mid q)}{(q;q)_n} = \left(\frac{2q^{1/4}}{1-q}\right)^k q^{(n-k)^2/4} \frac{H_{n-k}(x \mid q)}{(q;q)_{n-k}}. \tag{14.6.13}$$

Clearly (14.6.4) and (14.6.13) yield

$$\begin{aligned} & E_q^y \left(q^{n^2/4} \frac{H_n(x \mid q)}{(q;q)_n} \right) \\ = & \sum_{j,m\geq 0} \frac{q^{j+m^2/4} H_m(y \mid q)}{(q^2;q^2)_j (q;q)_m} \left(\frac{1-q}{2q^{1/4}} \mathcal{D}_{q,x}\right)^{m+2j} q^{n^2/4} \frac{H_n(x \mid q)}{(q;q)_n}. \end{aligned} \tag{14.6.14}$$

Thus by extending (14.6.14) to all polynomials we have established the following theorem.

Theorem 14.6.6 *The q-translation has the operational representation*

$$\begin{aligned} \left(E_q^y f\right)(x) &= f\left(x \overset{\circ}{+} y\right) \\ &= \sum_{j,m\geq 0} \frac{q^{j+m^2/4} H_m(y \mid q)}{(q^2;q^2)_j (q;q)_m} \left(\frac{1-q}{2q^{1/4}}\right)^{m+2j} \mathcal{D}_{a,x}^{m+2j} f(x), \end{aligned} \tag{14.6.15}$$

for polynomials f.

It must be noted that the right-hand side of (14.6.15) is a finite sum. Moreover Theorem 14.6.6 is an q-analogue of the Taylor series. Indeed as $q \to 1$, $(1-q)^m/(q;q)_m \to 1/m!$, $\mathcal{D}_{q,x} \to \frac{d}{dx}$, $(1-q)^j/(q^2;q^2)_j \to 2^{-j}/j!$, so the remaining $(1-q)^j$ tends to zero unless $j = 0$, and (14.6.15) becomes the Taylor series for f.

Motivated by Theorem 14.6.6 we use the operator notation

$$E_q^y = \sum_{j,m\geq 0} \frac{q^{j+m^2/4} H_m(y)}{(q^2;q^2)_j (q;q)_m} \left(\frac{1-q}{2q^{1/4}}\right)^{m+2j} \mathcal{D}_{q,x}^{m+2j}. \tag{14.6.16}$$

In other words

$$\begin{aligned} E_q^y = & \sum_{m=0}^{\infty} \frac{q^{m^2/4}}{(q;q)_m} H_m(y \mid q) \left(\frac{1-q}{2q^{1/4}}\right)^m \mathcal{D}_{q,x}^m \\ & \times \left[\left(q \frac{(1-q)^2}{4\sqrt{q}} \mathcal{D}_{q,x}^2; q^2\right)_{\infty}\right]^{-1}. \end{aligned} \tag{14.6.17}$$

Therefore with

$$\mathcal{B}_{q,x} := \frac{(1-q)}{2q^{1/4}} \mathcal{D}_{q,x},$$ (14.6.18)

we have established the operational representation

$$E_q^y = \mathcal{E}_q\left(y; \mathcal{B}_{q,x}\right).$$ (14.6.19)

operators.

Theorem 14.6.7 *The composition of translation operators satisfies*

$$E_q^y E_q^z = E_q^w, \quad \text{where} \quad w = y \overset{\circ}{+} z.$$ (14.6.20)

Proof The right-hand side of (14.6.20) is

$$\mathcal{E}_q\left(y \overset{\circ}{+} z; \mathcal{B}_{q,x}\right) = \frac{1}{\left(q\mathcal{B}_{q,x}^2; q^2\right)_\infty} \sum_{n=0}^\infty H_n\left(y \overset{\circ}{+} z \,|\, q\right) \frac{q^{n^2/4}}{(q;q)_n} \mathcal{B}_{q,x}^n$$

$$= \frac{1}{\left(q\mathcal{B}_{q,x}^2; q^2\right)_\infty} \sum_{n=0}^\infty \sum_{m=0}^n \frac{H_m(y\,|\,q)}{(q;q)_m} \frac{g_{n-m}(z)}{(q;q)_{n-m}} q^{m^2/4} \mathcal{B}_{q,x}^n$$

$$= \frac{1}{\left(q\mathcal{B}_{q,x}^2; q^2\right)_\infty} \sum_{m=0}^\infty \frac{H_m(y\,|\,q)}{(q;q)_m} q^{m^2/4} \mathcal{B}_{q,x}^m \sum_{n=0}^\infty \frac{g_n(z)}{(q;q)_n} \mathcal{B}_{q,x}^n$$

$$= \mathcal{E}_q\left(y; \mathcal{B}_{q,x}\right) \mathcal{E}_q\left(z; \mathcal{B}_{q,x}\right) = E_q^y E_q^z.$$

Hence (14.6.20) holds.   □

Recall that the infinitesimal generator of a semigroup $T(t)$ is the limit, in the strong operator topology, as $t \to 0$ of $[T(t) - T(0)]/t$. We also have that $T(0)$ is the identity operator. A standard example is the shift operator being $\exp\left(t\left(\dfrac{d}{dx}\right)\right)$. In this case the infinitesimal generator is $\dfrac{d}{dx}$. This example has a q-analogue. Consider the one parameter family of operators E_q^y, that is

$$T(y) = \mathcal{E}_q\left(y; \mathcal{B}_{q,x}\right).$$ (14.6.21)

Thus $T(0) = I$. It readily follows that $\mathcal{D}_{q,y} T(y)$ at $y = 0$ is $\mathcal{D}_{q,x}$.

Another application of the q-translation is to inroduce a q-analogue of the Gauss–Weierstrass transform, (Hirschman & Widder, 1955). For polynomials f define

$$F_W(x) = \frac{(q;q)_\infty}{2\pi} \int_{-1}^1 f\left(x \overset{\circ}{+} y\right) w(y\,|\,q)\, dy,$$ (14.6.22)

where w is the weight function of the q-Hermite polynomials defined by (13.1.12).

Theorem 14.6.8 *The transform (14.6.22) has the inversion formula*

$$f(x) = \left(\frac{1}{4} q^{1/2}(1-q)^2 \mathcal{D}_{q,x}^2; q^2\right)_\infty F_W(x).$$ (14.6.23)

Proof Clearly (14.2.1) implies

$$\frac{1}{(qt^2; q^2)_\infty} = \frac{(q; q)_\infty}{2\pi} \int_{-1}^{1} \mathcal{E}_q(y; t)\, w(y \mid q)\, dy.$$

For polynomials f we have

$$\frac{1}{\left(\frac{1}{4}q^{1/2}(1-q)^2 \mathcal{D}_{q,x}^2; q^2\right)_\infty} f(x) = \int_{-1}^{1} \mathcal{E}_q\left(y; \frac{1}{2}q^{-1/4}(1-q)\mathcal{D}_{q,x}\right) f(x)\, w(y)\, dy$$

$$= \int_{-1}^{1} f(x \overset{\circ}{+} y)\, w(y)\, dy,$$

and the theorem follows. $\qquad\qquad\qquad\qquad\qquad\qquad\qquad\qquad\qquad\square$

Formula (14.6.23) is the exact analogue of the classical real inversion formula, as in (Hirschman & Widder, 1955).

14.7 Polynomials of q-Binomial Type

The continuous q-Hermite polynomials are the model for what we will call polynomials of q-binomial type and the functional relationship (14.6.5) will be used to define the class of polynomials of q-binomial type. As in §10.3, $\{p_n(x)\}$ denotes a polynomial sequence which is not necessarily orthonormal.

Definition 14.7.1 *A sequence of q-polynomials $\{p_n(x) : n = 0, 1, \dots\}$ is called a sequence of q-binomial type if:*

(i) *For all n, $p_n(x)$ is of exact degree n,*
(ii) *The identities*

$$p_n(x \overset{\circ}{+} y) = \sum_{m,j \geq 0} \begin{bmatrix} n \\ m \end{bmatrix}_q \begin{bmatrix} n-m \\ 2j \end{bmatrix}_q (q; q^2)_j\, q^j\, p_m(y)\, p_{n-m-2j}(x),$$

(14.7.1)

hold for all n.

Thus $q^{n^2/4} H_n(x \mid q)$ is of q-binomial type. It is also clear that as $q \to 1$ (14.7.1) tends to (10.2.3) since the limit of $(q; q^2)_j$ as $q \to 1$ is $\delta_{j,0}$.

Recall that for polynomials of binomial type the sequence of basic polynomials was required to satisfy $p_n(0) = \delta_{n,0}$. By inspecting (14.7.1), (14.6.4) and (14.6.5) we observe that we made no assumptions on $H_n(x \mid q)$ at any special point but we demanded $g_n(0) = \delta_{n,0}$. This motivates the following definition.

Definition 14.7.2 *Assume that $\mathcal{Q}$ is a q-delta operator. A polynomial sequence $\{p_n(x)\}$ is called the sequence of basic polynomials for $\mathcal{Q}$ if*

(i) $p_0(x) = 1$

(ii) $\tilde{g}_n(0) = 0$, *for all $n > 0$, where*

$$\tilde{g}_n(x) := \sum_{k=0}^{[n/2]} \frac{(q;q)_n\, q^k}{(q^2;q^2)_k} \frac{p_{n-2k}(x)}{(q;q)_{n-2k}}. \tag{14.7.2}$$

(iii) $\mathcal{Q}p_n(x) = (1-q^n)\, p_{n-1}(x)$.

Theorem 14.7.1 *Every q-delta operator has a unique sequence of basic polynomials.*

Proof We take $p_0(x) = 1$, and construct the polynomials recursively from (iii), by applying Theorem 14.6.5, and determine the constant term from (ii).   □

Note that (14.7.2) shows that $\tilde{g}_n(x)$ is a polynomial sequence. It will be useful to rewrite (14.7.1) as

$$E_q^y p_n(x) = p_n(x \overset{\circ}{+} y) = \sum_{m=0}^{n} \begin{bmatrix} n \\ m \end{bmatrix}_q p_m(x)\tilde{g}_{n-m}(y). \tag{14.7.3}$$

Theorem 14.7.2 *A polynomial sequence is of q-binomial type if and only if it is a basic sequence for some q-delta operator.*

Proof Let $\{p_n(x)\}$ be a basic sequence of a q-delta operator $\mathcal{Q}$. From the above definition we see that $\mathcal{Q}\,\tilde{g}_n(x) = (1-q^n)\,\tilde{g}_{n-1}(x)$, hence $\mathcal{Q}^k\tilde{g}_n(x)\big|_{x=0} = (q;q)_n\delta_{k,n}$. Therefore

$$\tilde{g}_n(x) = \sum_{k=0}^{\infty} \frac{\tilde{g}_k(x)}{(q;q)_k}\, \mathcal{Q}^k\tilde{g}_n(y)\big|_{y=0}$$

hence any polynomial p satisfies

$$p(x) = \sum_{k=0}^{\infty} \frac{\tilde{g}_k(x)}{(q;q)_k}\, \mathcal{Q}^k p(y)\big|_{y=0}. \tag{14.7.4}$$

In (14.7.4) take $p(x) = E_q^z p_n(x)$. Thus

$$\mathcal{Q}^k p(y)\big|_{y=0} = E_q^z \mathcal{Q}^k p_n(y)\big|_{y=0} = \frac{(q;q)_n}{(q;q)_{n-k}}\, p_{n-k}(z),$$

and (14.7.4) proves that $\{p_n(x)\}$ is of q-binomial type. Conversely let $\{p_n(x)\}$ be of q-binomial type and define a linear operator $\mathcal{Q}$ on all polynomials by $\mathcal{Q}p_n(x) = (1-q^n)\, p_{n-1}(x)$, with $p_{-1}(x) := 0$. Define q-translations by (14.7.3). Now (14.7.3) with $y = 0$ and the linear independence of $\{p_n(x)\}$ imply $g_n(0) = \delta_{n,0}$, so we only need to show that the operator $\mathcal{Q}$ we constructed commutes with q-translations. Define $\tilde{g}_n$ by (14.7.2). Write (14.7.3) in the form

$$E_q^y p_n(x) = \sum_{k=0}^{n} \frac{\tilde{g}_k(y)}{(q;q)_k}\, \mathcal{Q}^k p_n(x)$$

which can be extended to

$$E_q^y p(x) = \sum_{k=0}^{n} \frac{\tilde{g}_k(y)}{(q;q)_k}\, \mathcal{Q}^k p(x).$$

Replace p by $\mathcal{Q}p$ to get

$$\left(E_q^y \mathcal{Q}\right) p(x) = \sum_{k=0}^{n} \frac{\tilde{g}_k(y)}{(q;q)_k} \mathcal{Q}^{k+1} p(x) = \mathcal{Q} \sum_{k=0}^{n} \frac{\tilde{g}_k(y)}{(q;q)_k} \mathcal{Q}^k p(x) = \mathcal{Q} E_q^y p(x).$$

Hence $\mathcal{Q}$ is a q-delta operator. $\qquad\qquad\qquad\qquad\qquad\qquad\square$

It is important to note that (14.7.2) is equivalent to the following functional relationship between generating functions of $\{p_n(x)\}$ and $\{\tilde{g}_n(x)\}$

$$\sum_{n=0}^{\infty} p_n(x) \frac{t^n}{(q;q)_n} = (qt^2;q^2)_{\infty} \sum_{n=0}^{\infty} \tilde{g}_n(x) \frac{t^n}{(q;q)_n}. \qquad (14.7.5)$$

Theorem 14.7.3 (Expansion Theorem) *Let $\{p_n(x)\}$ be a basic sequence of a q-delta operator $\mathcal{Q}$ and let T be a q-shift-invariant operator. Then*

$$T = \sum_{k=0}^{\infty} \frac{a_k}{(q;q)_k} \mathcal{Q}^k, \quad a_k := T\tilde{g}_k(y)\big|_{y=0}. \qquad (14.7.6)$$

Proof Again (14.7.3) can be extended to all polynomials via

$$p(x \overset{\circ}{+} y) = \sum_{k=0}^{n} \frac{\tilde{g}_k(y)}{(q;q)_k} \mathcal{Q}^k p(x). \qquad (14.7.7)$$

Apply T to (14.7.7) then set $y = 0$ after writing $T\, E_q^y$ as $E_q^y\, T$ to establish (14.7.6). $\qquad\qquad\qquad\qquad\qquad\qquad\qquad\qquad\qquad\qquad\qquad\square$

Theorem 14.7.4 *Let $\mathbf{F}$ and Σ be the rings (over the complex numbers) of formal power series in the variable t and q-shift-invariant operators, respectively. Assume that $\mathcal{Q}$ be a q-delta operator. Then the mapping ϕ from $\mathbf{F}$ onto Σ, defined by*

$$\phi(f) = T, \quad f(t) = \sum_{k=0}^{\infty} \frac{a_k}{(q;q)_k} t^k, \quad T = \sum_{k=0}^{\infty} \frac{a_k}{(q;q)_k} \mathcal{Q}^k, \qquad (14.7.8)$$

is an isomorphism.

The proof is similar to the proof of Theorem 10.2.4.

Corollary 14.7.5 *A q-shift-invariant operator T is invertible if and only if $T1 \neq 0$. A q-delta operator P is invertible if and only if $p(t) = \phi^{-1}(P)$, satisfies $p(0) = 0$ and $p'(0) \neq 0$.*

The next result is a characterization of basic polynomials of q-delta operators in terms of their generating functions.

Theorem 14.7.6 *Let $\{p_n(x)\}$ be a basic sequence of polynomials of a q-delta operator $\mathcal{Q}$ and let $\mathcal{Q} = f(\mathcal{Q}_{q,x})$ where $\phi(f) = \mathcal{Q}$. Then*

$$\sum_{n=0}^{\infty} \frac{\tilde{g}_n(x)}{(q;q)_n} t^n = \mathcal{E}_q\left(x; \mathbf{c} f^{-1}(t)\right), \quad \mathbf{c} := \frac{(1-q)}{2q^{1/4}}, \tag{14.7.9}$$

$$\sum_{n=0}^{\infty} \frac{p_n(x)}{(q;q)_n} t^n = \left(qt^2; q^2\right)_\infty \mathcal{E}_q\left(x; \mathbf{c} f^{-1}(t)\right), \tag{14.7.10}$$

where $\tilde{g}_n$ is as in (14.7.2).

Proof From (14.7.2) it follows that $\mathcal{Q}\tilde{g}_n = (1 - q^n)\tilde{g}_{n-1}$. Expand E_q^a in a formal power series in $\mathcal{Q}$ using (14.7.6). Thus

$$E_q^a = \sum_{n=0}^{\infty} \frac{\tilde{g}_n(a)}{(q;q)_n} \mathcal{Q}^n. \tag{14.7.11}$$

With

$$\mathcal{Q} = f(\mathcal{D}_{q,x})$$

we obtain from (4.6.20) that

$$\mathcal{E}_q(a; \mathcal{A}_{q,x}) = \sum_{n=0}^{\infty} \frac{\tilde{g}_n(a)}{(q;q)_n} [f(\mathcal{D}_{q,x})]^n.$$

The theorem now follows from (14.7.5) and Theorem 14.7.6. $\qquad\square$

Corollary 14.7.7 *Any two q-shift-invariant operators commute.*

Observe that the inverse relation to (14.7.2) is

$$p_n(x) = \sum_{k=0}^{n} \frac{(q;q)_n}{(q^2;q^2)_k} (-1)^k q^{k^2} \frac{\tilde{g}_{n-2k}(x)}{(q;q)_{n-2k}}, \tag{14.7.12}$$

which follows from (14.7.9)–(14.7.10).

It is clear that (14.7.1) and (14.7.9) imply the binomial type relation

$$\tilde{g}_n(x \overset{\circ}{+} y) = \sum_{k=0}^{n} \begin{bmatrix} n \\ k \end{bmatrix}_q \tilde{g}_k(x) \tilde{g}_{n-k}(y). \tag{14.7.13}$$

One can study the Sheffer classification relative to $\mathcal{D}_q$ using results from Chapter 10. In particular we have the following.

Theorem 14.7.8 *A polynomial sequence $\{p_n(x)\}$ is of Sheffer-A type zero relative to $\mathcal{D}_q$ if and only if*

$$\sum_{n=0}^{\infty} \frac{p_n(x)t^n}{(q;q)_n} = A(t)\,\mathcal{E}_q(x; H(t)), \tag{14.7.14}$$

where

$$H(t) = \sum_{n\geq 1} h_n t^n, \quad A(t) = \sum_{n=0}^{\infty} a_n t^n, \quad a_0 h_1 \neq 0. \tag{14.7.15}$$

The class of polynomials of q-A type zero relative to $\mathcal{D}_q$ when $H(t) = J(t) = t$ will be called q-Appell polynomials. In view of (13.1.29) the polynomial sequence $\left\{ q^{n^2/2} H_n(x \mid q) \right\}$ is q-Appell. Waleed Al-Salam (Al-Salam, 1995) has proved that the only orthogonal q-Appell polynomial sequence is a sequence of constant multiples of continuous q-Hermite polynomials. The problem of characterizing all orthogonal polynomials which are q-A type zero relative to $\mathcal{D}_q$ remains open.

14.8 Another q-Umbral Calculus

We briefly outline another q-analogue of polynomials of binomial type.

Definition 14.8.1 *A polynomial sequence $\{p_n(x)\}$ is called an Eulerian family if its members satisfy the functional equation*

$$p_n(xy) = \sum_{k=0}^{n} \begin{bmatrix} n \\ k \end{bmatrix}_q p_k(x) y^k p_{n-k}, \quad n = 0, 1, \dots \tag{14.8.1}$$

The model for Eulerian families of polynomials is $\{\theta_n(x)\}$,

$$\theta_0(x) := 1, \quad \theta_n(x) = \prod_{j=0}^{n-1} \left(x - q^j \right), \quad n > 0. \tag{14.8.2}$$

In this case we use

$$\Delta(x) = x \otimes x, \tag{14.8.3}$$

instead of the Δ in (10.3.4). This map is not grade-preserving, but is an algebra map. The product of functionals L and M is defined on any polynomial by

$$\langle LM \mid p(x) \rangle = \langle L \otimes M \mid \Delta p(x) \rangle = \langle L \otimes M \mid p(\Delta x) \rangle. \tag{14.8.4}$$

Theorem 14.8.1 *A polynomial sequence $\{p_n(x)\}$ is an Eulerian family of polynomials if and only if*

$$\langle LM \mid p_n(x) \rangle = \sum_{k=0}^{n} \begin{bmatrix} n \\ k \end{bmatrix}_q \langle L \mid p_k(x) \rangle \langle M \mid x^k p_{n-k}(x) \rangle. \tag{14.8.5}$$

The proof is straightforward, see (Ihrig & Ismail, 1981).

Theorem 14.8.2 *A polynomial sequence $\{p_n(x)\}$ with $p_0(x) = 1$ is an Eulerian family if and only if it has a generating function of the form*

$$\sum_{n=0}^{\infty} p_n(x) \frac{t^n}{(q;q)_n} = \frac{f(xt)}{f(t)}, \tag{14.8.6}$$

where

$$f(t) = \sum_{n=0}^{\infty} \gamma_n t^n / (q;q)_n, \quad \gamma_0 = 1, \gamma_n \neq 0, n = 1, 2, \dots. \tag{14.8.7}$$

Proofs are in (Andrews, 1971) and (Ihrig & Ismail, 1981). Note that the coefficient of x^n in $p_n(x)$ is γ_n.

The polynomials $\{\theta_n(x, y)\}$,

$$\theta_0(x, y) := 1, \quad \theta_n(x, y) := \prod_{k=0}^{n-1} (x - q^k y), \tag{14.8.8}$$

appeared in (Hahn, 1949a), (Goldman & Rota, 1970) and (Andrews, 1971). There series expansion is

$$\theta_n(x, y) = \sum_{k=0}^{n} \begin{bmatrix} n \\ k \end{bmatrix}_q (-1)^k q^{k(k-1)/2} x^{n-k} y^k. \tag{14.8.9}$$

Definition 14.8.2 *The q-translation $\mathcal{E}^y$ is defined on monomials by*

$$\mathcal{E}^y x^n := \theta_n(x, -y) = x^n (-y/x; q)_n,$$

and extended to all polynomials as a linear operator.

Thus

$$\mathcal{E}^y \left(\sum_{n=0}^{m} f_n x^n \right) = \sum_{n=0}^{m} f_n \theta_n(x, -y), \quad m = 0, 1, \ldots . \tag{14.8.10}$$

It readily follows from (14.8.9) and (14.8.10) that

$$\mathcal{E}^y = \sum_{k=0}^{\infty} \frac{1}{(q; q)_k} q^{k(k-1)/2} y^k (1 - q)^k D_{q,x}^k,$$

that is

$$\mathcal{E}^y p(x) = (y(q - 1) D_{q,x}; q)_\infty p(x), \tag{14.8.11}$$

for polynomials p.

One can define q-constants as those functions defined for all x and are annihilated by D_q. If a q-constant g is continuous at $x = 0$, then $D_q g(x) = 0$ implies $g(x) = g(qx)$, hence $g(x) = g(xq^n)$, $n = 1, 2, \ldots$, and by letting $n \to \infty$, it follows that g is a constant. Since we will require q-constants to be continuous at $x = 0$, we will not distinguish between constants and q-constants.

We define q-shift invariant operators as those linear operators T whose domain contains all polynomials and T commutes with $\mathcal{E}^y$. It can be proved that T is q-shift invariant if and only if there is a sequence of constants $\{a_n\}$ such that

$$T = \sum_{n=0}^{\infty} a_n D_q^n.$$

15

The Askey–Wilson Polynomials

In this chapter we shall build the theory of the Askey–Wilson polynomials through a method of attachment. This method combines generating functions and summation theorems in what seems to be a simple but powerful technique to get new orthogonal or biorthogonal functions from old ones. Sections 15.1 and 15.2 are mostly based on (Berg & Ismail, 1996). An intermediate step is the Al-Salam–Chihara polynomials, whose properties resemble those of Laguerre polynomials. The Askey–Wilson polynomials are q-analogues of the Wigner 6-j symbols, (Biedenharn & Louck, 1981). Their $q \to 1$ limit gives the Wilson polynomials (Wilson, 1980).

15.1 The Al-Salam–Chihara Polynomials

The Al-Salam–Chihara polynomials arose as part of a characterization theorem in (Al-Salam & Chihara, 1976). The characterization problems will be stated in §20.4. Al-Salam and Chihara recorded the three-term recurrence relation and a generating function. The weight function was found by Askey and Ismail, who also named the polynomials after the ones who first identified the polynomials, see (Askey & Ismail, 1984).

In this section, we derive the orthogonality relation of the Al-Salam–Chihara polynomials by starting from the continuous q-Hermite polynomials. The orthogonality of the continuous q-Hermite polynomials special case $\beta = 0$ of (13.2.21)

$$\int_0^\pi \frac{\left(e^{2i\theta}, e^{-2i\theta}; q\right)_\infty}{\left(t_1 e^{i\theta}, t_1 e^{-i\theta}, t_2 e^{i\theta}, t_2 e^{-i\theta}; q\right)_\infty} \, d\theta = \frac{2\pi}{(q, t_1 t_2; q)_\infty}, \quad |t_1|, |t_2| < 1. \quad (15.1.1)$$

The next step is to find polynomials $\{p_n(x; t_1, t_2 \,|\, q)\}$ orthogonal with respect to the weight function

$$w_1(x; t_1, t_2 \,|\, q) := \frac{\left(e^{2i\theta}, e^{-2i\theta}; q\right)_\infty}{\left(t_1 e^{i\theta}, t_1 e^{-i\theta}, t_2 e^{i\theta}, t_2 e^{-i\theta}; q\right)_\infty} \frac{1}{\sqrt{1 - x^2}}, \quad x = \cos\theta,$$

$$(15.1.2)$$

which is positive for $t_1, t_2 \in (-1, 1)$ and its total mass is given by (15.1.1). Here we follow a clever technique of attachment which was used by Andrews and Askey (Andrews & Askey, 1985), and by Askey and Wilson in (Askey & Wilson, 1985).

Write $\{p_n(x; t_1, t_2 \mid q)\}$ in the form

$$p_n(x; t_1, t_2 \mid q) = \sum_{k=0}^{n} \frac{(q^{-n}, t_1 e^{i\theta}, t_1 e^{-i\theta}; q)_k}{(q; q)_k} a_{n,k}, \qquad (15.1.3)$$

then compute $a_{n,k}$ from the fact that $p_n(x; t_1, t_2 \mid q)$ is orthogonal to $(t_2 e^{i\theta}, t_2 e^{-i\theta}; q)_j$, $j = 0, 1, \ldots, n-1$. As we saw in (13.2.1) $(a e^{i\theta}, a e^{-i\theta}; q)_k$ is a polynomial in x of degree k. The reason for choosing the bases $\{(t_1 e^{i\theta}, t_1 e^{-i\theta}; q)_k\}$ and $\{(t_2 e^{i\theta}, t_2 e^{-i\theta}; q)_j\}$ is that they attach nicely to the weight function in (15.1.1), and (15.1.2) enables us to integrate products of their elements against the weight function $w_1(x; t_1, t_2 \mid q)$. Indeed

$$(t_1 e^{i\theta}, t_1 e^{-i\theta}; q)_k (t_2 e^{i\theta}, t_2 e^{-i\theta}; q)_j w_1(x; t_1, t_2 \mid q) = w_1(x; t_1 q^k, t_2 q^j \mid q).$$

Therefore we have

$$\int_{-1}^{1} (t_2 e^{i\theta}, t_2 e^{-i\theta}; q)_j p_n(x; t_1, t_2 \mid q) w_1(x; t_1, t_2 \mid q) \, dx$$

$$= \sum_{k=0}^{n} \frac{(q^{-n}; q)_k}{(q; q)_k} a_{n,k} \int_{0}^{\pi} \frac{(e^{2i\theta}, e^{-2i\theta}; q)_\infty \, d\theta}{(t_1 q^k e^{i\theta}, t_1 q^k e^{-i\theta}, t_2 q^j e^{i\theta}, t_2 q^j e^{-i\theta}; q)_\infty}$$

$$= \frac{2\pi}{(q; q)_\infty} \sum_{k=0}^{n} \frac{(q^{-n}; q)_k \, a_{n,k}}{(q; q)_k (t_1 t_2 q^{k+j}; q)_\infty}$$

$$= \frac{2\pi}{(q, t_1 t_2 q^j; q)_\infty} \sum_{k=0}^{n} \frac{(q^{-n}, t_1 t_2 q^j; q)_k}{(q; q)_k} a_{n,k}.$$

At this stage we look for $a_{n,k}$ as a quotient of products of q-shifted factorials in order to make the above sum vanish for $0 \le j < n$. The q-Chu–Vandermonde sum (12.2.17) suggests

$$a_{n,k} = q^k / (t_1 t_2; q)_k.$$

Therefore

$$\int_{-1}^{1} (t_2 e^{i\theta}, t_2 e^{-i\theta}; q)_j p_n(x) w_1(x; t_1, t_2 \mid q) \, dx$$

$$= \frac{2\pi}{(q, t_1 t_2 q^j; q)_\infty} \, {}_2\phi_1\left(q^{-n}, t_1 t_2 q^j; t_1 t_2; q, q\right)$$

$$= \frac{2\pi (q^{-j}; q)_n}{(q, t_1 t_2 q^j; q)_\infty (t_1 t_2; q)_n} \left(t_1 t_2 q^j\right)^n.$$

It follows from (15.1.3) and (12.2.1) that the coefficient of x^n in $p_n(x; t_1, t_2 \mid q)$ is

$$(-2t_1)^n q^{n(n+1)/2} (q^{-n}; q)_n / (q, t_1 t_2; q)_n = (2t_1)^n / (t_1 t_2; q)_n. \qquad (15.1.4)$$

This leads to the orthogonality relation

$$\int_{-1}^{1} p_m\left(x; t_1, t_2 \mid q\right) p_n\left(x; t_1, t_2 \mid q\right) w_1\left(x; t_1, t_2 \mid q\right) dx$$

(15.1.5)

$$= \frac{2\pi (q; q)_n t_1^{2n}}{(q, t_1 t_2; q)_\infty (t_1 t_2; q)_n} \delta_{m,n}.$$

Furthermore the polynomials are given by

$$p_n\left(x; t_1, t_2 \mid q\right) = {}_3\phi_2\left(\begin{array}{c} q^{-n}, t_1 e^{i\theta}, t_1 e^{-i\theta} \\ t_1 t_2, 0 \end{array} \middle| q, q\right).$$

(15.1.6)

The polynomials we have just found are the Al-Salam–Chihara polynomials and were first identified by W. Al-Salam and T. Chihara (Al-Salam & Chihara, 1976). Their weight function was given in (Askey & Ismail, 1984) and (Askey & Wilson, 1985).

Observe that the orthogonality relation (15.1.5) and the uniqueness of the polynomials orthogonal with respect to a positive measure show that $t_1^{-n} p_n(x)$ is symmetric in t_1 and t_2. This gives the transformation formula

$$
{}_3\phi_2\left(\begin{array}{c} q^{-n}, t_1 e^{i\theta}, t_1 e^{-i\theta} \\ t_1 t_2, 0 \end{array} \middle| q, q\right)
$$

(15.1.7)

$$
= (t_1/t_2)^n \, {}_3\phi_2\left(\begin{array}{c} q^{-n}, t_2 e^{i\theta}, t_2 e^{-i\theta} \\ t_1 t_2, 0 \end{array} \middle| q, q\right),
$$

as a byproduct of our analysis.

Our next task is to repeat the process with the Al-Salam–Chihara polynomials as our starting point. The representation (15.1.6) needs to be transformed to a form more amenable to generating functions. This can be done in two different ways. One way is to derive a three-term recurrence relation.

Theorem 2.2.1 shows that there exists constants A_n, B_n, C_n such that

$$2x p_n\left(x; t_1, t_2 \mid q\right) = A_n p_{n+1}\left(x; t_1, t_2 \mid q\right)$$
$$+ B_n p_n\left(x; t_1, t_2 \mid q\right) + C_n p_{n-1}\left(x; t_1, t_2 \mid q\right).$$

Since the coefficient of x^n in p_n is given by (15.1.4), then $A_n = \left(1 - t_1 t_2 q^n\right)/t_1$. Moreover, the choices $e^{i\theta} = t_1, t_2$ give

$$p_n\left((t_1 + 1/t_1)/2; t_1, t_2 \mid q\right) = 1,$$
$$p_n\left((t_2 + 1/t_2)/2; t_1, t_2 \mid q\right) = {}_2\phi_1\left(q^{-n}, t_1/t_2; 0; q, q\right) = (t_1/t_2)^n,$$

by (12.2.17). Therefore

$$B_n + C_n = t_1 + t_2 q^n,$$
$$B_n + C_n t_2/t_1 = t_2 + t_1 q^n,$$

and we establish the three-term recurrence relation

$$\left[2x - (t_1 + t_2) q^n\right] t_1 \, p_n\left(x; t_1, t_2 \mid q\right)$$
$$= \left(1 - t_1 t_2 q^n\right) p_{n+1}\left(x; t_1, t_2 \mid q\right) + t_1^2 \left(1 - q^n\right) p_{n-1}\left(x; t_1, t_2 \mid q\right).$$

(15.1.8)

The initial conditions are

$$p_0\left(x; t_1, t_2 \,|\, q\right) = 1, \quad p_1\left(x; t_1, t_2 \,|\, q\right) = t_1\left(2x - t_1 - t_2\right). \tag{15.1.9}$$

Set

$$F(x, t) = \sum_{n=0}^{\infty} \frac{(t_1 t_2; q)_n}{(q; q)_n} \frac{t^n}{t_1^n} \, p_n\left(\cos\theta; t_1, t_2 \,|\, q\right).$$

Multiplying (15.1.8) by $(t_1 t_2; q)_n \, t_1^{-n-1} t^{n+1}/(q; q)_n$ and adding for $n = 1, 2, \ldots$, and taking (5.1.9) into account we establish the functional equation

$$F(x, t) = \frac{1 - t\left(t_1 + t_2\right) + tt_1 t_2}{1 - 2xt + t^2} \, F(x, qt),$$

which implies

$$\sum_{n=0}^{\infty} \frac{(t_1 t_2; q)_n}{(q; q)_n} \, p_n\left(\cos\theta; t_1, t_2 \,|\, q\right) (t/t_1)^n = \frac{(tt_1, tt_2; q)_\infty}{(te^{-i\theta}, te^{i\theta}; q)_\infty}. \tag{15.1.10}$$

Expand the right-hand side of (15.1.10) by the binomial theorem and find the alternate representation

$$p_n\left(x; t_1, t_2 \,|\, q\right) = \frac{(t_1 e^{-i\theta}; q)_n \, t_1^n e^{in\theta}}{(t_1 t_2; q)_n} \, {}_2\phi_1\left(\begin{matrix} q^{-n}, t_2 e^{i\theta} \\ q^{1-n} e^{i\theta}/t_1 \end{matrix} \,\middle|\, q, q e^{-i\theta}/t_1\right). \tag{15.1.11}$$

Another way to derive (15.1.10) is to write the ${}_3\phi_2$ in (15.1.6) as a sum over k then replace k by $n - k$. Applying (12.2.11) and (12.2.12) we obtain

$$p_n\left(x; t_1, t_2 \,|\, q\right) = \frac{(t_1 e^{i\theta}, t_1 e^{-i\theta}; q)_n}{(t_1 t_2; q)_n} \, q^{-n(n-1)/2}(-1)^n$$

$$\times \sum_{k=0}^{n} \frac{(-t_2/t_1)^k \left(q^{-n}, q^{1-n}/t_1 t_2; q\right)_k}{(q, q^{1-n} e^{i\theta}/t_1, q^{1-n} e^{-i\theta}/t_1; q)_k} \, q^{k(k+1)/2}.$$

Then apply the q-analogue of Pfaff–Kummer transformation (12.4.7) with

$$A = q^{-n}, \ B = t_2 e^{i\theta}, \ C = q^{1-n} e^{i\theta}/t_1, \ z = q e^{-i\theta}/t_1$$

to replace the right-hand side of the above equation by a ${}_2\phi_1$ series. This gives the alternate ${}_2\phi_1$ representation in (15.1.11).

Using (12.2.12) we express a multiple of p_n as a Cauchy product of two sequences. The result is

$$p_n\left(\cos\theta; t_1, t_2 \,|\, q\right) = \frac{(q; q)_n t_1^n}{(t_1 t_2; q)_n} \sum_{k=0}^{n} \frac{(t_2 e^{i\theta}; q)_k}{(q; q)_k} \, e^{-ik\theta} \frac{(t_1 e^{-i\theta}; q)_{n-k}}{(q; q)_{n-k}} \, e^{i(n-k)\theta}. \tag{15.1.12}$$

When $x \notin [-1, 1]$ and with $\left|e^{-i\theta}\right| < \left|e^{i\theta}\right|$, formula (15.1.12) leads to the asymptotic formula

$$\lim_{n \to \infty} \frac{p_n\left(\cos\theta; t_1, t_2 \,|\, q\right)}{t_1^n \, e^{-in\theta}} = \frac{(t_1 e^{-i\theta}, t_2 e^{-i\theta}; q)_\infty}{(t_1 t_2, e^{-2i\theta}; q)_\infty}. \tag{15.1.13}$$

It readily follows from (15.1.12) that the p_n's have the generating function (15.1.10) and satisfy the three term recurrence relation (15.1.8). Another consequence of (15.1.12) is

$$\max\left\{|p_n\left(x;t_1,t_2\,|\,q\right)|:-1\leq x\leq 1\right\}=|p_n\left(1;t_1,t_2\,|\,q\right)|\leq Cn\left|t_1\right|^n,\quad(15.1.14)$$

for some constant C which depends only on t_1 and t_2.

As in the proof of (14.1.27) we establish the difference recursion relation

$$\mathcal{D}_qp_n\left(x;t_1,t_2\,|\,q\right)=\frac{\left(1-q^n\right)t_1q^{n-1}}{\left(1-t_1t_2\right)\left(1-q\right)}\,p_{n-1}\left(x;q^{1/2}t_1,q^{1/2}t_2\,|\,q\right).\quad(15.1.15)$$

When $q>1$, we can replace q by $1/q$ and realize that the polynomials involve two new parameters t_1 and t_2, and (15.1.8) can be normalized to become

$$\left[2xq^n+t_1+t_2\right]r_n\left(x;t_1,t_2\right)$$
$$=\left(t_1t_2+q^n\right)r_{n+1}\left(x;t_1,t_2\right)+\left(1-q^n\right)r_{n-1}\left(x;t_1,t_2\right).\quad(15.1.16)$$

We also assume

$$r_0\left(x;t_1,t_2\right):=1,\quad r_1\left(x;t_1,t_2\right)=\frac{\left(2x+t_1+t_2\right)}{\left(1+t_1t_2\right)}.\quad(15.1.17)$$

Similar to (15.1.10) we derive

$$\sum_{n=0}^{\infty}r_n\left(\sinh\xi;t_1,t_2\right)\frac{\left(1/t_1t_2;q\right)_n}{\left(q;q\right)_n}\left(t_1t_2t\right)^n=\frac{\left(-te^{\xi},te^{-\xi};q\right)_{\infty}}{\left(tt_1,tt_2;q\right)_{\infty}}.\quad(15.1.18)$$

From (15.1.18) we derive the explicit formula

$$r_n\left(\sinh\xi;t_1,t_2\right)=\frac{\left(q;q\right)_n}{t_1^n\left(1/t_1t_2;q\right)_n}\sum_{k=0}^{n}\frac{\left(e^{-\xi}/t_2;q\right)_k}{\left(q;q\right)_k}\frac{\left(-e^{\xi}/t_1;q\right)_{n-k}}{\left(q;q\right)_{n-k}}\left(\frac{t_1}{t_2}\right)^k.$$
$$(15.1.19)$$

It must be emphasized that the Al-Salam–Chihara polynomials are q-analogues of Laguerre polynomials; see Exercise 15.2.

15.2 The Askey–Wilson Polynomials

The orthogonality relation (15.1.5), the bound (15.1.14), and the generating function (15.1.11) imply the Askey–Wilson q-beta integral, (Askey & Wilson, 1985), (Gasper & Rahman, 1990), (Gasper & Rahman, 2004)

$$\int_0^{\pi}\frac{\left(e^{2i\theta},e^{-2i\theta};q\right)_{\infty}}{\prod_{j=1}^{4}\left(t_je^{i\theta},t_je^{-i\theta};q\right)_{\infty}}\,d\theta=\frac{2\pi\left(t_1t_2t_3t_4;q\right)_{\infty}}{\left(q;q\right)_{\infty}\prod_{1\leq j<k\leq4}\left(t_jt_k;q\right)_{\infty}},\quad\left|t_1\right|,\left|t_2\right|<1.$$
$$(15.2.1)$$

Other proofs are in (Rahman, 1984) and (Askey, 1983).

The polynomials orthogonal with respect to the weight function whose total mass is given by (15.2.1) are the Askey–Wilson polynomials. To save space we shall use the vector notation $\mathbf{t}$ to denote the ordered tupple (t_1,t_2,t_3,t_4). Their weight function is

$$w\left(x; t_1, t_2, t_3, t_4 \mid q\right) = \frac{\left(e^{2i\theta}, e^{-2i\theta}; q\right)_\infty}{\displaystyle\prod_{j=1}^{4}\left(t_j e^{i\theta}, t_j e^{-i\theta}; q\right)_\infty} \frac{1}{\sqrt{1-x^2}}, \qquad (15.2.2)$$

$x = \cos\theta$. We now find their explicit representation and establish their orthogonality relation. We use the bases $\left\{\left(t_1 e^{i\theta}, t_1 e^{-i\theta}; q\right)_k\right\}$ and $\left\{\left(t_2 e^{i\theta}, t_2 e^{-i\theta}; q\right)_k\right\}$ because they can be easily attached to the weight function. Let

$$p_n(x; \mathbf{t} \mid q) = \sum_{k=0}^{n} \frac{\left(q^{-n}, t_1 e^{i\theta}, t_1 e^{-i\theta}; q\right)_k}{(q; q)_k} a_{n,k}, \qquad (15.2.3)$$

where the $a_{n,k}$'s are to be determined. Therefore

$$\int_0^\pi \left(t_2 e^{i\theta}, t_2 e^{-i\theta}; q\right)_j p_n(\cos\theta; \mathbf{t} \mid q) w(\cos\theta; \mathbf{t} \mid q) \sin\theta \, d\theta$$

$$= \sum_{k=0}^{n} \frac{(q^{-n}; q)_k}{(q; q)_k} a_{n,k} \int_0^\pi w\left(\cos\theta; t_1 q^k, t_2 q^j, t_3, t_4 \mid q\right) \sin\theta \, d\theta$$

$$= \sum_{k=0}^{n} \frac{(q^{-n}; q)_k}{(q; q)_k} a_{n,k} \frac{2\pi \left(q^{j+k} t_1 t_2 t_3 t_4; q\right)_\infty /(q; q)_\infty}{\left(q^{j+k} t_1 t_2, q^k t_1 t_3, q^k t_1 t_4, q^j t_2 t_3, q^j t_2 t_4, t_3 t_4; q\right)_\infty}$$

$$= \sum_{k=0}^{n} \frac{(q^{-n}; q)_k}{(q; q)_k} a_{n,k} \frac{\left(q^j t_1 t_2, t_1 t_3, t_1 t_4; q\right)_k}{\left(q^j t_1 t_2 t_3 t_4; q\right)_k}$$

$$\times \frac{2\pi \left(q^j t_1 t_2 t_3 t_4; q\right)_\infty}{\left(q, q^j t_1 t_2, t_1 t_3, t_1 t_4, q^j t_2 t_3, q^j t_2 t_4, t_3 t_4; q\right)_\infty}.$$

In order to use the $_3\phi_2$ sum (13.2.19) we choose

$$a_{n,k} = q^k \frac{\left(t_1 t_2 t_3 t_4 q^{n-1}; q\right)_k}{\left(t_1 t_2, t_1 t_3, t_1 t_4; q\right)_k} a_{n,0}.$$

Therefore

$$\int_0^\pi \left(t_2 e^{i\theta}, t_2 e^{-i\theta}; q\right)_j p_n(\cos\theta; \mathbf{t} \mid q) \, w(\cos\theta; \mathbf{t} \mid q) \sin\theta \, d\theta$$

$$= \frac{2\pi \left(q^j t_1 t_2 t_3 t_4; q\right)_\infty a_{n,0}}{\left(q, q^j t_1 t_2, t_1 t_3, t_1 t_4, q^j t_2 t_3, q^j t_2 t_4, t_3 t_4; q\right)_\infty} \frac{\left(q^{-j}, q^{1-n}/t_3 t_4; q\right)_n}{\left(t_1 t_2, q^{1-j-n}/t_1 t_2 t_3 t_4; q\right)_n}.$$

For $j \leq n$ we use (13.2.13) and (13.2.16) to see that the right-hand side of the above equation is

$$\frac{2\pi \left(q^j t_1 t_2 t_3 t_4; q\right)_\infty a_{n,0}}{\left(q, q^j t_1 t_2, t_1 t_3, t_1 t_4, q^j t_2 t_3, q^j t_2 t_4, t_3 t_4; q\right)_\infty}$$

$$\times \frac{(q, t_3 t_4; q)_n}{\left(t_1 t_2, q^j t_1 t_2 t_3 t_4; q\right)_n} (-t_1 t_2)^n q^{n(n-1)/2} \delta_{j,n}.$$

Since

$$\left(t_1 e^{i\theta}, t_1 e^{-i\theta}; q\right)_n = (t_1/t_2)^n \left(t_2 e^{i\theta}, t_2 e^{-i\theta}; q\right)_n + \text{lower order terms},$$

we have

$$\int_0^\pi \left(t_1 e^{i\theta}, t_1 e^{-i\theta}; q\right)_j p_n(\cos\theta; \mathbf{t} \mid q) w(\cos\theta; \mathbf{t} \mid q) \sin\theta \, d\theta$$

$$= \frac{2\pi \left(q^j t_1 t_2 t_3 t_4; q\right)_\infty a_{n,0}}{(q, q^j t_1 t_2, t_1 t_3, t_1 t_4, q^j t_2 t_3, q^j t_2 t_4, t_3 t_4; q)_\infty}$$

$$\times \frac{(q, t_3 t_4; q)_n \left(-t_1^2\right)^n}{(t_1 t_2, q^j t_1 t_2 t_3 t_4; q)_n} q^{n(n-1)/2} \delta_{j,n}.$$

Hence if $m \le n$ then

$$\int_{-1}^1 p_m(x; \mathbf{t} \mid q) p_n(x; \mathbf{t} \mid q) w(x; \mathbf{t}) \, dx$$

$$= \frac{\left(q^{-m}, t_1 t_2 t_3 t_4 q^{m-1}; q\right)_m}{(q, t_1 t_2, t_1 t_3, t_1 t_4; q)_m} \frac{(q, t_3 t_4; q)_n}{(t_1 t_2, q^m t_1 t_2 t_3 t_4; q)_n}$$

$$\times \frac{2\pi (-1)^n \left(q^m t_1 t_2 t_3 t_4; q\right)_\infty a_{n,0}^2}{(q, q^m t_1 t_2, t_1 t_3, t_1 t_4, q^m t_2 t_3, q^m t_2 t_4, t_3 t_4; q)_\infty} (t_1)^{2n} q^{n(n+1)/2} \delta_{m,n}.$$

With the choice

$$a_{n,0} := t_1^{-n} (t_1 t_2, t_1 t_3, t_1 t_4; q)_n,$$

we have established the following result.

Theorem 15.2.1 *The Askey–Wilson polynomials satisfy the orthogonality relation*

$$\int_{-1}^1 p_m(x; \mathbf{t} \mid q)\, p_n(x; \mathbf{t} \mid q)\, w(x; \mathbf{t} \mid q)\, dx$$

$$= \frac{2\pi \left(t_1 t_2 t_3 t_4 q^{2n}; q\right)_\infty \left(t_1 t_2 t_3 t_4 q^{n-1}; q\right)_n}{(q^{n+1}; q)_\infty \prod\limits_{1 \le j < k \le 4} (t_j t_k q^n; q)_\infty} \delta_{m,n}, \quad (15.2.4)$$

for $\max\{|t_1|, |t_2|, |t_3|, |t_4|\} < 1$.

The analysis preceding Theorem 15.2.1 shows that the polynomials under consideration have the basic hypergeometric representation

$$p_n(x; \mathbf{t} \mid q) = t_1^{-n} (t_1 t_2, t_1 t_3, t_1 t_4; q)_n$$

$$\times {}_4\phi_3 \left(\begin{array}{c} q^{-n}, t_1 t_2 t_3 t_4 q^{n-1}, t_1 e^{i\theta}, t_1 e^{-i\theta} \\ t_1 t_2, \; t_1 t_3, \; t_1 t_4 \end{array} \;\middle|\; q, q \right). \quad (15.2.5)$$

Observe that the weight function in (15.2.2) and the right-hand side of (15.2.4) are symmetric functions of t_1, t_2, t_3, t_4. The weight function in (15.2.2) and (15.2.4) is positive when $\max\{|t_1|, |t_2|, |t_3|, |t_4|\} < 1$ and the uniqueness of the polynomials orthogonal with respect to a positive measure shows that the Askey–Wilson polynomials are symmetric in the four parameters t_1, t_2, t_3, t_4. This symmetry is the Sears transformation in the form

$$
t_1^{-n} (t_1 t_2, t_1 t_3, t_1 t_4; q)_n \, {}_4\phi_3 \left(\begin{matrix} q^{-n}, t_1 t_2 t_3 t_4 q^{n-1}, t_1 e^{i\theta}, t_1 e^{-i\theta} \\ t_1 t_2, \ t_1 t_3, \ t_1 t_4 \end{matrix} \, \middle| \, q, q \right)
$$
$$
= t_2^{-n} (t_2 t_1, t_2 t_3, t_2 t_4; q)_n \, {}_4\phi_3 \left(\begin{matrix} q^{-n}, t_1 t_2 t_3 t_4 q^{n-1}, t_2 e^{i\theta}, t_2 e^{-i\theta} \\ t_2 t_1, \ t_2 t_3, \ t_2 t_4 \end{matrix} \, \middle| \, q, q \right),
$$

which we saw in Chapter 12, see (12.4.1) and Theorem 12.4.1. The case $q = 1$ is the Whipple transformation. The Whipple transformation gives all the symmetries of the Wigner 6-j symbols, (Biedenharn & Louck, 1981). These symmetries were discovered independently by physicists.

We now establish a generating function for the Askey–Wilson polynomials following a technique due to (Ismail & Wilson, 1982).

Theorem 15.2.2 *The Askey–Wilson polynomials have the generating function*

$$
\sum_{n=0}^{\infty} \frac{p_n(\cos\theta; \mathbf{t} \mid q)}{(q, t_1 t_2, t_3 t_4; q)_n} \, t^n
$$
$$
= {}_2\phi_1 \left(\begin{matrix} t_1 e^{i\theta}, t_2 e^{i\theta} \\ t_1 t_2 \end{matrix} \, \middle| \, q, t e^{-i\theta} \right) {}_2\phi_1 \left(\begin{matrix} t_3 e^{-i\theta}, t_4 e^{-i\theta} \\ t_3 t_4 \end{matrix} \, \middle| \, q, t e^{i\theta} \right). \quad (15.2.6)
$$

Proof Apply (12.4.1) with

$$
a = t_1 e^{i\theta}, \ b = t_1 e^{-i\theta}, \ c = t_1 t_2 t_3 t_4 q^{n-1}, \ d = t_1 t_2, \ e = t_1 t_3, \ f = t_1 t_4,
$$

to obtain

$$
p_n(x; \mathbf{t} \mid q) = \left(t_1 t_2, q^{1-n} e^{i\theta}/t_3, q^{1-n} e^{i\theta}/t_4; q \right)_n \left(t_3 t_4 q^{n-1} e^{-i\theta} \right)^n
$$
$$
\times {}_4\phi_3 \left(\begin{matrix} q^{-n}, t_1 e^{i\theta}, t_2 e^{i\theta}, q^{1-n}/t_3 t_4 \\ t_1 t_2, \ q^{1-n} e^{i\theta}/t_3, \ q^{1-n} e^{i\theta}/t_4 \end{matrix} \, \middle| \, q, q \right). \quad (15.2.7)
$$

Using (12.2.10) we write

$$
\left(q^{1-n} e^{i\theta}/t_3, q^{1-n} e^{i\theta}/t_4; q \right)_n = e^{2in\theta} q^{-n(n-1)} (t_3 t_4)^{-n} \left(t_3 e^{-i\theta}, t_4 e^{-i\theta}; q \right)_n.
$$

Furthermore, if the summation index of the ${}_4\phi_3$ in (15.2.7) is k then we may use (12.2.12) to get

$$
\frac{\left(q^{-n}, q^{1-n}/t_3 t_4; q \right)_k}{\left(q^{1-n} e^{i\theta}/t_3, q^{1-n} e^{i\theta}/t_4; q \right)_k}
$$
$$
= \frac{(q, t_3 t_4; q)_n}{(q, t_3 t_4; q)_{n-k}} \frac{\left(t_3 e^{-i\theta}, t_4 e^{-i\theta}; q \right)_{n-k}}{\left(t_3 e^{-i\theta}, t_4 e^{-i\theta}; q \right)_n} \left(q e^{2i\theta} \right)^{-k}.
$$

Therefore

$$
\begin{aligned}
&\frac{p_n\left(x ; t_1, t_2, t_3, t_4 \mid q\right)}{\left(q, t_1 t_2, t_3 t_4 ; q\right)_n} \\
&= \sum_{k=0}^n \frac{\left(t_1 e^{i\theta}, t_2 e^{i\theta} ; q\right)_k}{\left(q, t_1 t_2 ; q\right)_k} e^{-i k \theta} \frac{\left(t_3 e^{-i\theta}, t_4 e^{-i\theta} ; q\right)_{n-k}}{\left(q, t_3 t_4 ; q\right)_{n-k}} e^{i(n-k)\theta} .
\end{aligned}
\tag{15.2.8}
$$

It is clear that (15.2.8) implies (15.2.6) and the proof is complete. $\qquad \square$

We now write the orthogonality relation (15.2.4) in terms of the generating function (15.2.6). Multiply (15.2.4) by

$$
\frac{t_5^m t_6^n}{\left(q, t_1 t_2, t_3 t_4 ; q\right)_m\left(q, t_1 t_2, t_3 t_4 ; q\right)_n}
$$

and add for all $m, n \geq 0$. This leads to the evaluation of the following integral

$$
\begin{aligned}
& \int_0^\pi \prod_{j=5}^6 {}_2\phi_1\left(\begin{array}{c} t_1 e^{i\theta}, t_2 e^{i\theta} \\ t_1 t_2 \end{array} \middle| q, t_j e^{-i\theta}\right) {}_2\phi_1\left(\begin{array}{c} t_3 e^{-i\theta}, t_4 e^{-i\theta} \\ t_3 t_4 \end{array} \middle| q, t_j e^{i\theta}\right) \\
& \qquad\qquad \times \frac{\left(e^{2i\theta}, e^{-2i\theta} ; q\right)_\infty}{\prod_{j=1}^4\left(t_j e^{i\theta}, t_j e^{-i\theta} ; q\right)_\infty} d\theta \\
& = \frac{2\pi\left(t_1 t_2 t_3 t_4 ; q\right)_\infty}{(q ; q)_\infty \prod_{1 \leq j<k \leq 4}\left(t_j t_k ; q\right)_\infty} \\
& \times {}_6\phi_5\left(\begin{array}{c} \sqrt{t_1 t_2 t_3 t_4 / q},\, -\sqrt{t_1 t_2 t_3 t_4 / q},\, t_1 t_3,\, t_1 t_4,\, t_2 t_3,\, t_2 t_4 \\ \sqrt{t_1 t_2 t_3 t_4 q},\, -\sqrt{t_1 t_2 t_3 t_4 q},\, t_1 t_2,\, t_3 t_4,\, t_1 t_2 t_3 t_4 / q \end{array} \middle| q, t_5 t_6\right),
\end{aligned}
\tag{15.2.9}
$$

valid for $\max \left\{\left|t_1\right|,\left|t_2\right|,\left|t_3\right|,\left|t_4\right|,\left|t_5\right|,\left|t_6\right|\right\}<1$.

Formula (15.2.9) provides an integral representation for a ${}_6\phi_5$ function.

The Askey–Wilson polynomials are orthogonal polynomials, hence they satisfy a three-term recurrence relation of the form

$$
2 x p_n(x ; \mathbf{t} \mid q)=A_n p_{n+1}(x ; \mathbf{t} \mid q)+B_n p_n(x ; \mathbf{t} \mid q)+C_n p_{n-1}(x ; \mathbf{t} \mid q) . \tag{15.2.10}
$$

The coefficient of x^n in p_n is $2^n\left(t_1, t_2, t_3, t_4 q^{n-1} ; q\right)_n$. Equating the coefficients of x^{n+1} on both sides of (15.2.10) we get

$$
A_n=\frac{1-t_1 t_2 t_3 t_4 q^{n-1}}{\left(1-t_1 t_2 t_3 t_4 q^{2n-1}\right)\left(1-t_1 t_2 t_3 t_4 q^{2n}\right)} . \tag{15.2.11}
$$

We next choose the special values $e^{-i\theta}=t_1, t_2$ in (15.2.6) and obtain

$$
\begin{aligned}
p_n\left(\left(t_1+1/t_1\right) / 2 ; \mathbf{t} \mid q\right) &=\left(t_1 t_2, t_1 t_3, t_1 t_4 ; q\right)_n t_1^{-n}, \\
p_n\left(\left(t_2+1/t_2\right) / 2 ; \mathbf{t} \mid q\right) &=\left(t_2 t_1, t_2 t_3, t_2 t_4 ; q\right)_n t_2^{-n}.
\end{aligned}
$$

With A_n given by (15.2.11), we substitute $x=\left(t_j+1/t_j\right) / 2, j=1,2$ in (15.2.10) then solve for B_n and C_n. The result is

$$C_n = \frac{(1 - q^n) \prod_{1 \leq j < k \leq 4} \left(1 - t_j t_k q^{n-1}\right)}{\left(1 - t_1 t_2 t_3 t_4 q^{2n-2}\right)\left(1 - t_1 t_2 t_3 t_4 q^{2n-1}\right)}, \tag{15.2.12}$$

$$B_n = t_1 + t_1^{-1} - A_n t_1^{-1} \prod_{j=2}^{4} (1 - t_1 t_j q^n)$$

$$- \frac{t_1 C_n}{\prod_{2 \leq k \leq 4} \left(1 - t_1 t_k q^{n-1}\right)}. \tag{15.2.13}$$

Rahman and Verma proved the following addition theorem which reduces the Gegenbauer addition theorem as $q \to 1$.

Theorem 15.2.3 ((Rahman & Verma, 1986a)) *We have*

$$p_n\left(z; a, aq^{1/2}, -a, -aq^{1/2} \mid q\right)$$

$$= \sum_{k=0}^{n} \frac{(q;q)_n \left(a^4 q^n, a^4/q, a^2 q^{1/2}, -a^2 q^{1/2}, -a^2; q\right)_k a^{n-k}}{(q;q)_k (q;q)_{n-k} \left(a^4/q;q\right)_{2k} \left(a^2 q^{1/2}, -a^2 q^{1/2}, -a^2; q\right)_n}$$

$$\times p_{n-k}\left(x; aq^{k/2}, aq^{(k+1)/2}, -aq^{k/2}, -aq^{(k+1)/2} \mid q\right) \tag{15.2.14}$$

$$\times p_{n-k}\left(y; aq^{k/2}, aq^{(k+1)/2}, -aq^{k/2}, -aq^{(k+1)/2} \mid q\right)$$

$$\times p_k\left(z; ae^{i(\theta+\phi)}, ae^{-i(\theta+\phi)}, ae^{i(\theta-\phi)}, ae^{i(\phi-\theta)} \mid q\right),$$

where $x = \cos\theta$, $y = \cos\phi$.

The q-ultraspherical polynomials are constant multiples of a special case of the Askey–Wilson polynomials as we saw in (13.2.11)

$$p_n\left(x; \sqrt{\beta}, -\sqrt{\beta}, \sqrt{\beta q}, -\sqrt{\beta q} \mid q\right) = (q, -\beta; q)_n \frac{(q\beta^2; q^2)_n}{(\beta^2; q)_n} C_n(x; \beta \mid q). \tag{15.2.15}$$

It is a simple exercise to use (15.2.15) to let $q \to 1$ in (15.2.14) and see that it reduces to the Gegenbauer addition theorem, Theorem 9.6.2.

15.3 Remarks

The continuous q-ultraspherical polynomials correspond to the case

$$t_1 = -t_2 = \sqrt{\beta}, \quad \text{and} \quad t_3 = -t_4 = \sqrt{q\beta}$$

as can be seen from comparing the weight functions (13.2.5) and (15.2.2). Thus $C_n(x; \beta \mid q)$ must be a constant multiple of an Askey–Wilson polynomial of degree

n and the above parameters. Therefore

$$
{}_2\phi_1\left(\begin{array}{c} q^{-n}, \beta \\ q^{1-n}/\beta \end{array} \middle| q, \frac{qe^{-2i\theta}}{\beta}\right)
$$
$$
= \frac{(\beta^2; q)_n\, e^{-in\theta}}{\beta^{n/2}(\beta; q)_n}\, {}_4\phi_3\left(\begin{array}{c} q^{-n}, q^n\beta^2, \sqrt{\beta}e^{i\theta}, \sqrt{\beta}e^{-i\theta} \\ -\beta, \beta\sqrt{q}, -\beta\sqrt{q} \end{array} \middle| q, q\right). \tag{15.3.1}
$$

The constant multiple was computed from the fact that the leading coefficient in $C_n(x; \beta \,|\, q)$ is $2^n(\beta; q)_n/(q; q)_n$. When we replace β by q^b in (15.3.1) and let $q \to 1$ we obtain the quadratic transformation

$$
{}_2F_1\left(\begin{array}{c} -n, b \\ 1-n-b \end{array} \middle| q, x^2\right) = \frac{x^n(2b; q)_n}{(b; q)_n}\, {}_2F_1\left(\begin{array}{c} -n, n+2b \\ b+1/2 \end{array} \middle| -\frac{(1-x)^2}{4x}\right). \tag{15.3.2}
$$

The success in evaluating the Askey–Wilson integral (15.2.1) raises the question of evaluating the general integral

$$
I(t_1, t_2, \ldots, t_k) := \frac{(q; q)_\infty}{2\pi} \int_0^\pi \frac{\left(e^{2i\theta}, e^{-2i\theta}; q\right)_\infty}{\displaystyle\prod_{j=1}^k \left(t_j e^{i\theta}, t_j e^{-i\theta}; q\right)_\infty}\, d\theta. \tag{15.3.3}
$$

The evaluation of this integral is stated below, but its known proof uses combinatorial ideas that are outside the scope of this book.

Theorem 15.3.1 *Let*

$$
I(t_1, t_2, \ldots, t_k) = \sum_{n_1, \ldots n_k = 0}^\infty g(n_1, n_2, \ldots, n_k) \prod_{j=1}^k \frac{t_j^{n_j}}{(q; q)_{n_j}}. \tag{15.3.4}
$$

Then

$$
g(n_1, n_2, \ldots, n_k) = \sum_{n_{ij}} \prod_{\ell=1}^k \begin{bmatrix} n_\ell \\ n_{\ell 1}, \ldots n_{\ell k} \end{bmatrix}_q \prod_{1 \le i < j \le k} (q; q)_{n_{ij}}\, q^B, \tag{15.3.5}
$$

where the summation is over all non-negative integral symmetric matrices (n_{ij}) such that $n_{ii} = 0$ and $\sum_{i=1}^k n_{ij} = n_j$ for $1 \le j \le k$. Furthermore

$$
B = \sum_{1 \le i < j < m < \ell \le k} n_{im} n_{j\ell}. \tag{15.3.6}
$$

Theorem 15.3.1 is in (Ismail et al., 1987). The q-multinomial coefficient in (15.3.5) is

$$
\begin{bmatrix} n \\ n_1, \ldots n_k \end{bmatrix}_q := (q; q)_n \Big/ \prod_{j=1}^k (q; q)_{n_j}, \quad \sum_{j=1}^k n_j = n. \tag{15.3.7}
$$

Observe that when $k = 4$ then $B = n_{13}n_{24}$ and

$$I(t_1,\ldots,t_4)$$
$$= \sum_{n_{ij},1\le i,j\le 4} \frac{(t_1t_2)^{n_{12}}(t_1t_3)^{n_{13}}(t_1t_4)^{n_{14}}(t_2t_3)^{n_{23}}(t_2t_4)^{n_{24}}(t_3t_4)^{n_{34}}}{(q;q)_{n_{12}}(q;q)_{n_{13}}(q;q)_{n_{14}}(q;q)_{n_{23}}(q;q)_{n_{24}}(q;q)_{n_{34}}} q^{n_{13}n_{24}}.$$

The sums over $n_{12}, n_{14}, n_{23}, n_{13}$, and n_{34} are evaluable by (12.2.24) and we find

$$I(t_1,\ldots,t_4) = \frac{1}{(t_1t_2,t_1t_4,t_2t_3,t_3t_4;q)_\infty} \sum_{n_{24}=0}^{\infty} \frac{(t_2t_4)^{n_{24}}}{(q;q)_{n_{24}}(t_1t_3q^{n_{24}};q)_\infty}$$

$$= \frac{1}{(t_1t_2,t_1t_3,t_1t_4,t_2t_3,t_3t_4;q)_\infty} \sum_{n_{24}=0}^{\infty} \frac{(t_1t_3;q)_{n_{24}}}{(q;q)_{n_{24}}}(t_2t_4)^{n_{24}},$$

and the evaluation of $I(t_1,t_2,t_3,t_4)$ follows from (12.2.22).

In addition to Theorem 15.3.1, (Ismail et al., 1987) contains combinatorial integrations of the moments, and the polynomials as generating functions of certain statistics with generating function variables, q; and x and q; respectively.

When $k = 5$, $I(t_1,\ldots,t_5)$ is a multiple of $_3\phi_2$. The invariance of $I(t_1,\ldots,t_5)$ under $t_j \leftrightarrow t_j$ gives all the known transformation formulas for $_3\phi_2$'s. Details are in Exercise 15.3.

Koornwinder established a second addition theorem for the continuous q-ultraspherical polynomials in (Koornwinder, 2005b). His result is

$$p_n\left(\cos\theta, aq^{1/2}s/t, aq^{1/2}t/s, -aq^{1/2}st, -aq^{1/2}/st \mid q\right)$$
$$= (-1)^n q^{n^2/2} \sum_{k=0}^{n} a^{n-k}(a^2q^{k+1};q)_{n-k} q^{k/2}$$
$$\times \frac{(q^{-n},a^4q^{n+1};q)_k}{(q,a^4q^k;q)_k}\frac{(-a^2s^2q^{k+1},-a^2q^{k+1}/t^2;q)_{n-k}}{(s/t)^{n-k}}$$
$$\times {}_2\phi_2\left(\begin{matrix}q^{k-n},a^4q^{n+k+1}\\a^2q^{k+1},-a^2s^2q^{k+1}\end{matrix}\middle| q,-s^2q\right)$$
$$\times {}_2\phi_2\left(\begin{matrix}q^{k-n},a^4q^{n+k+1}\\a^2q^{k+1},-a^2q^{k+1}/t^2\end{matrix}\middle| q,-q/t^2\right)$$
$$p_k\left(\cos\theta;a,-a,aq^{1/2},-aq^{1/2}\mid q\right).$$

(15.3.8)

The special case $a = 1$ is in (Koornwinder, 1993).

15.4 Asymptotics

In this section we derive a series representation for the Askey–Wilson polynomials which implies a complete asymptotic expansion for them.

Theorem 15.4.1 *With $z = e^{i\theta}$ and $x = \cos\theta$ we have*

$$\frac{p_n(\cos\theta; \mathbf{t} \mid q)}{(q, t_1 t_2, t_3 t_4; q)_n}$$

$$= z^n \frac{(t_1/z, t_3/z; q)_\infty}{(z^{-2}, q; q)_\infty} \sum_{m=0}^{\infty} \frac{(qz/t_1, qz/t_3; q)_m}{(q, qz^2; q)_m}$$

$$\times {}_2\phi_2\left(\begin{array}{c} t_1 z, t_1/z \\ t_1 t_2, t_1 q^{-m}/z \end{array} \middle| q, t_2 q^{-m}/z\right) {}_2\phi_2\left(\begin{array}{c} t_3 z, t_3/z \\ t_3 t_4, t_3 q^{-m}/z \end{array} \middle| q, t_4 q^{-m}/z\right)$$

$$+ \text{ a similar term with } z \text{ and } 1/z \text{ interchanged.} \quad (15.4.1)$$

Proof Let $F(x,t)$ denote the right-hand side of (15.2.6). Apply the q-analogue of the Pfaff–Kummer transformation, (12.4.7) to the ${}_2\phi_1$'s in $F(x,t)$. Thus

$$F(x,t) = \frac{(tt_1, tt_3; q)_\infty}{(tz, t/z; q)_\infty} \sum_{k,j=0}^{\infty} \frac{(t_1 z, t_1/z; q)_k (t_3 z, t_3/z; q)_j}{(q, t_1 t_2, tt_1; q)_k (q, t_3 t_4, tt_3; q)_j}$$

$$\times q^{\binom{j}{2}+\binom{k}{2}} (-tt_4)^j (-tt_1)^k.$$

Cauchy's theorem shows that

$$\frac{p_n(\cos\theta; \mathbf{t} \mid q)}{(q, t_1 t_2, t_3 t_4; q)_n} = \frac{1}{2\pi i} \int_C F(x,t) t^{-n-1}\, dt, \quad (15.4.2)$$

where C is the contour $\{t : |t| = r\}$, with $r < |e^{-i\theta}| = 1/|z|$. Now think of the contour C as a contour around the point $t = \infty$ with the wrong orientation, so it encloses all the poles of $F(x,t)$. Therefore the right-hand side of (15.4.2) is $-\sum$ Residues. Now $t = 0$ is outside the contour and the singularities of F inside are $t = q^{-m}z^{\pm 1}$, $m = 0, 1, \ldots$. It is straightforward to see that

$$\text{Res}\{F(x,t) : t = zq^{-m}\}$$

$$= -\frac{(q^{-m}zt_1, q^{-m}zt_3; q)_\infty}{(q^{-m}; q)_m (q, q^{-m}z^2; q)_\infty} \left(zq^{-m}\right)^{-n}$$

$$\times \sum_{k,j=0}^{\infty} \frac{(t_1 z, t_1/z; q)_k (t_3 z, t_3/z; q)_j}{(q, t_1 t_2, t_1 zq^{-m}; q)_k (q, t_3 t_4, t_3 zq^{-m}; q)_j}$$

$$\times q^{\binom{j}{2}+\binom{k}{2}} (-t_4)^j (-t_1)^k \left(q^{-m}z\right)^{j+k}$$

$$= -z^{-n} \frac{(zt_1, zt_3; q)_\infty (q/zt_1, q/zt_3; q)_m}{(q; q)_m (q, z^2; q)_\infty (q, q/z^2; q)_m} (t_1 t_3 q^n)^m$$

$$\times {}_2\phi_2\left(\begin{array}{c} t_1 z, t_1/z \\ t_1 t_2, t_1 zq^{-m} \end{array} \middle| q, t_2 zq^{-m}\right) {}_2\phi_2\left(\begin{array}{c} t_3 z, t_3/z \\ t_3 t_4, t_3 zq^{-m} \end{array} \middle| q, t_4 zq^{-m}\right).$$

For the residue at $t = q^{-m}/z$ replace z by $1/z$, and we establish Theorem 15.4.1. $\quad\square$

Observe that the series (15.4.1) is both an explicit formula and an asymptotic series.

Theorem 15.4.1 is from (Ismail, 1986). The special case of the q-ultraspherical polynomials was proved earlier using a different method by Rahman and Verma (Rahman & Verma, 1986b) and takes the form

$$
C_n(\cos\theta;\beta\,|\,q) = \frac{(\beta,\beta e^{2i\theta};q)_\infty}{(q,e^{2i\theta};q)_\infty}\, e^{-in\theta}\, {}_2\phi_1\left(\begin{array}{c} q/\beta, qe^{-2i\theta}/\beta \\ qe^{-2i\theta} \end{array}\middle|\, q,\beta^2 q^n\right) \tag{15.4.3}
$$
$$
+ \text{ a similar term with } \theta \text{ with } -\theta.
$$

It is convenient to rewrite (15.4.3) as a q-integral in the form

$$
C_n(\cos\theta;\beta\,|\,q) = \frac{2i\sin\theta\,\left(\beta,\beta,\beta e^{2i\theta},\beta e^{-2i\theta};q\right)_\infty (\beta^2;q)_n}{(1-q)\,(q,\beta^2,e^{2i\theta},e^{-2i\theta};q)_\infty (q;q)_n}
$$
$$
\times \int_{e^{i\theta}}^{e^{-i\theta}} u^n\, \frac{(que^{i\theta},que^{-i\theta};q)_\infty}{(\beta ue^{i\theta},\beta ue^{-i\theta};q)_\infty}\, d_q u. \tag{15.4.4}
$$

It is clear that (15.4.4) is a moment representation for $C_n(x;\beta\,|\,q)$. We shall return to moment representations in §15.7.

Rahman and Verma observed that (15.4.3) has several interesting applications. First, if we multiply (15.4.3) by $(\lambda;q)_n t^n/(\beta^2;q)_n$ and sum over n we get

$$
\sum_{n=1}^{\infty} C_n(\cos\theta;\beta\,|\,q)\frac{(\lambda;q)_n t^n}{(\beta^2;q)_n} = \frac{2i\sin\theta\,\left(\beta,\beta,\beta e^{2i\theta},\beta e^{-2i\theta};q\right)_\infty}{(1-q)\,(q,\beta^2,e^{2i\theta},e^{-2i\theta};q)_\infty}
$$
$$
\times \int_{e^{i\theta}}^{e^{-i\theta}} \frac{(que^{i\theta},que^{-i\theta},\lambda ut;q)_\infty}{(\beta ue^{i\theta},\beta ue^{-i\theta},ut;q)_\infty}\, d_q u. \tag{15.4.5}
$$

15.5 Continuous q-Jacobi Polynomials and Discriminants

Definition 15.5.1 *We take* $t_1 = q^{(2\alpha+1)/4}$, $t_2 = q^{(2\alpha+3)/4}$, $t_3 = -q^{(2\beta+1)/4}$, $t_4 = -q^{(2\beta+3)/4}$ *in the definition of the Askey–Wilson polynomials in* (15.2.5), *and let*

$$
P_n^{(\alpha,\beta)}(\cos\theta\,|\,q) = \frac{(q^{\alpha+1};q)_n}{(q;q)_n}
$$
$$
\times {}_4\phi_3\left(\begin{array}{c} q^{-n}, q^{n+\alpha+\beta+1}, q^{(2\alpha+1)/4}e^{i\theta}, q^{(2\alpha+1)/4}e^{-i\theta} \\ q^{\alpha+1}, -q^{(\alpha+\beta+1)/2}, -q^{(\alpha+\beta+2)/2} \end{array}\middle|\, q;q\right). \tag{15.5.1}
$$

For $\alpha > -1/2$ and $\beta > -1/2$ we have

$$\frac{1}{2\pi} \int_{-1}^{1} w\left(x \,|\, q^\alpha, q^\beta\right) P_m^{(\alpha,\beta)}(x \,|\, q) \, P_n^{(\alpha,\beta)}(x \,|\, q) \, dx$$

$$= \frac{\left(q^{(\alpha+\beta+2)/2}, q^{(\alpha+\beta+3)/2}; q\right)_\infty}{\left(q, q^{\alpha+1}, q^{\beta+1}, -q^{(\alpha+\beta+1)/2}, -q^{(\alpha+\beta+2)/2}; q\right)_\infty}$$

$$\times \frac{\left(1 - q^{\alpha+\beta+1}\right)\left(q^{\alpha+1}, q^{\beta+1}, -q^{(\alpha+\beta+3)/2}; q\right)_n}{\left(1 - q^{2n+\alpha+\beta+1}\right)\left(q, q^{\alpha+\beta+1}, -q^{(\alpha+\beta+1)/2}; q\right)_n} \, q^{(2\alpha+1)n/2} \delta_{mn}, \quad (15.5.2)$$

where

$$\sin\theta \, w\left(\cos\theta \,|\, q^\alpha, q^\beta\right)$$

$$= \left| \frac{\left(e^{2i\theta}; q\right)_\infty}{\left(q^{(2\alpha+1)/4}e^{i\theta}, q^{(2\alpha+3)/4}e^{i\theta}, -q^{(2\beta+1)/4}e^{i\theta}, -q^{(2\beta+3)/4}e^{i\theta}; q\right)_\infty} \right|^2 \quad (15.5.3)$$

$$= \left| \frac{\left(e^{i\theta}, -e^{i\theta}; q^{1/2}\right)_\infty}{\left(q^{(2\alpha+1)/4}e^{i\theta}, -q^{(2\beta+1)/4}e^{i\theta}; q^{1/2}\right)_\infty} \right|^2 .$$

The recurrence relation for the polynomials $\{\phi_n(x)\}$,

$$\phi_n(x \,|\, q) := \frac{(q; q)_n}{(q^{\alpha+1}; q)_n} \, P_n^{(\alpha,\beta)}(x \,|\, q) \quad (15.5.4)$$

is

$$2x\phi_n(x \,|\, q) = A_n \phi_{n+1}(x \,|\, q)$$
$$+ \left[q^{(2\alpha+1)/4} + q^{-(2\alpha-1)/4} - (A_n + C_n) \right] \phi_n(x \,|\, q) + C_n \phi_{n-1}(x \,|\, q), \quad (15.5.5)$$

where

$$A_n = \frac{\left(1 - q^{n+\alpha+1}\right)\left(1 - q^{n+\alpha+\beta+1}\right)\left(1 + q^{n+(\alpha+\beta+1)/2}\right)\left(1 + q^{n+(\alpha+\beta+2)/2}\right)}{q^{(2\alpha+1)/4}\left(1 - q^{2n+\alpha+\beta+1}\right)\left(1 - q^{2n+\alpha+\beta+2}\right)},$$

$$C_n = \frac{q^{(2\alpha+1)/4}\left(1 - q^n\right)\left(1 - q^{n+\beta}\right)\left(1 + q^{n+(\alpha+\beta)/2}\right)\left(1 + q^{n+(\alpha+\beta+1)/2}\right)}{\left(1 - q^{2n+\alpha+\beta}\right)\left(1 - q^{2n+\alpha+\beta+1}\right)}.$$

The monic polynomials satisfy the recurrence relation

$$xP_n(x) = P_{n+1}(x) + \frac{1}{2}\left[q^{(2\alpha+1)/4} + q^{-(2\alpha-1)/4} - (A_n + C_n) \right] P_n(x)$$
$$+ \frac{1}{4} A_{n-1} C_n P_{n-1}(x), \quad (15.5.6)$$

where

$$P_n^{(\alpha,\beta)}(x \,|\, q) = \frac{2^n q^{(2\alpha+1)n/4}\left(q^{n+\alpha+\beta+1}; q\right)_n}{\left(q, -q^{(\alpha+\beta+1)/2}, -q^{(\alpha+\beta+2)/2}; q\right)_n} \, P_n(x).$$

The lowering operator is

$$\mathcal{D}_q P_n^{(\alpha,\beta)}(x \,|\, q) = \frac{2q^{-n+(2\alpha+5)/4}\left(1 - q^{n+\alpha+\beta+1}\right)}{(1 - q)\left(1 + q^{(\alpha+\beta+1)/2}\right)\left(1 + q^{(\alpha+\beta+2)/2}\right)} \, P_{n-1}^{(\alpha+1,\beta+1)}(x \,|\, q)$$

$$(15.5.7)$$

while the raising operator is

$$\mathcal{D}_q \left[w\left(x \mid q^\alpha, q^\beta\right) P_n^{(\alpha,\beta)}(x \mid q) \right]$$

$$= -2q^{-(2\alpha+1)/4} \frac{\left(1 - q^{n+1}\right)\left(1 + q^{(\alpha+\beta-1)/2}\right)\left(1 + q^{(\alpha+\beta)/2}\right)}{1-q}$$

$$\times w\left(x \mid q^{\alpha-1}, q^{\beta-1}\right) P_{n+1}^{(\alpha-1,\beta-1)}(x \mid q). \quad (15.5.8)$$

The following Rodrigues-type formula follows from iterating (15.5.8)

$$w\left(x \mid q^\alpha, q^\beta\right) P_n^{(\alpha,\beta)}(x \mid q)$$

$$= \left(\frac{q-1}{2}\right)^n \frac{q^{n(n+2\alpha)/4}}{\left(q, -q^{(\alpha+\beta+1)/2}, -q^{(\alpha+\beta+2)/2}; q\right)_n} \mathcal{D}_q^n \left[w\left(x \mid q^{\alpha+n}, q^{\beta+n}\right)\right].$$

$$(15.5.9)$$

The generating function

$$_2\phi_1 \left(\begin{matrix} q^{(2\alpha+1)/4} e^{i\theta}, q^{(2\alpha+3)/4} e^{i\theta} \\ q^{\alpha+1} \end{matrix} \middle| q; e^{-i\theta} t \right)$$

$$\times {}_2\phi_1 \left(\begin{matrix} -q^{(2\beta+1)/4} e^{-i\theta}, -q^{(2\beta+3)/4} e^{-i\theta} \\ q^{\beta+1} \end{matrix} \middle| q; e^{i\theta} t \right) \quad (15.5.10)$$

$$= \sum_{n=0}^{\infty} \frac{\left(-q^{(\alpha+\beta+1)/2}, -q^{(\alpha+\beta+2)}; q\right)_n}{\left(q^{\alpha+1}, q^{\beta+1}; q\right)_n} \frac{P_n^{(\alpha,\beta)}(x \mid q)}{q^{(2\alpha+1)n/4}} t^n$$

follows from (15.2.6). Moreover, one can establish the generating functions

$$_2\phi_1 \left(\begin{matrix} q^{(2\alpha+1)/4} e^{i\theta}, -q^{(2\beta+1)/4} e^{i\theta} \\ -q^{(\alpha+\beta+1)/2} \end{matrix} \middle| q; e^{-i\theta} t \right)$$

$$\times {}_2\phi_1 \left(\begin{matrix} q^{(2\alpha+3)/4} e^{-i\theta}, -q^{(2\beta+3)/4} e^{-i\theta} \\ -q^{(\alpha+\beta+3)/2} \end{matrix} \middle| q; e^{i\theta} t \right) \quad (15.5.11)$$

$$= \sum_{n=0}^{\infty} \frac{\left(-q^{(\alpha+\beta+2)/2}; q\right)_n}{\left(-q^{(\alpha+\beta+3)/2}; q\right)_n} \frac{P_n^{(\alpha,\beta)}(x \mid q)}{q^{(2\alpha+1)n/4}} t^n,$$

$$_2\phi_1 \left(\begin{matrix} q^{(2\alpha+1)/4} e^{i\theta}, -q^{(2\beta+3)/4} e^{i\theta} \\ -q^{(\alpha+\beta+2)/2} \end{matrix} \middle| q; e^{-i\theta} t \right)$$

$$\times {}_2\phi_1 \left(\begin{matrix} q^{(2\alpha+3)/4} e^{-i\theta}, -q^{(2\beta+1)/4} e^{-i\theta} \\ -q^{(\alpha+\beta+2)/2} \end{matrix} \middle| q; e^{i\theta} t \right) \quad (15.5.12)$$

$$= \sum_{n=0}^{\infty} \frac{\left(-q^{(\alpha+\beta+1)/2}; q\right)_n}{\left(-q^{(\alpha+\beta+2)/2}; q\right)_n} \frac{P_n^{(\alpha,\beta)}(x \mid q)}{q^{((2\alpha+1)/4)n}} t^n$$

Remark 15.5.1 *In (Rahman, 1981), M. Rahman takes $t_1 = q^{1/2}$, $t_2 = q^{\alpha+1/2}$, $t_3 = -q^{\beta+1/2}$ and $t_4 = -q^{1/2}$ in the definition of the Askey–Wilson polynomials to obtain after renormalizing*

$$P_n^{(\alpha,\beta)}(x;q) = \frac{\left(q^{\alpha+1}, -q^{\beta+1}; q\right)_n}{(q, -q; q)_n} \; _4\phi_3 \left(\begin{matrix} q^{-n}, q^{n+\alpha+\beta+1}, q^{1/2} e^{i\theta}, q^{1/2} e^{-i\theta} \\ q^{\alpha+1}, -q^{\beta+1}, -q \end{matrix} \middle| q; q \right).$$

$$(15.5.13)$$

Theorem 15.5.1 *These two q-analogues of the Jacobi polynomials are connected by*

$$P_n^{(\alpha,\beta)}\left(x \mid q^2\right) = \frac{(-q;q)_n}{(-q^{\alpha+\beta+1};q)_n} q^{n\alpha} P_n^{(\alpha,\beta)}(x;q). \qquad (15.5.14)$$

Proof The weight function for $P_n^{(\alpha,\beta)}\left(x \mid q^2\right)$ is

$$\frac{\left(e^{2i\theta}, e^{-2i\theta}; q^2\right)_\infty / \sin\theta}{\left(q^{\alpha+1/2}e^{i\theta}, -q^{\beta+1/2}e^{i\theta}, q^{\alpha+1/2}e^{-i\theta}, -q^{\beta+1/2}e^{-i\theta}; q\right)_\infty}$$

$$= \frac{\left(e^{2i\theta}, e^{-2i\theta}; q\right)_\infty / \sin\theta}{\left|\left(q^{\alpha+1/2}e^{i\theta}, -q^{\beta+1/2}e^{i\theta}, q^{1/2}e^{i\theta}, -q^{1/2}e^{i\theta}; q\right)_\infty\right|^2},$$

when α and β are real. Therefore both sides of (15.5.14) are orthogonal with respect to the same weight function, hence they must be constant multiples of each other. The constants can be evaluated by finding the leading terms in both $P_n^{(\alpha,\beta)}\left(x \mid q^2\right)$ and $P_n^{(\alpha,\beta)}(x;q)$. $\qquad\square$

The continuous q-Jacobi polynomials can be evaluated at

$$x_1 = \frac{1}{2}\left(q^{(2\alpha+1)/4} + q^{-(2\alpha+1)/4}\right), \quad x_2 = -\frac{1}{2}\left(q^{(2\beta+1)/4} + q^{-(2\beta+1)/4}\right). \qquad (15.5.15)$$

Indeed

$$P_n^{(\alpha,\beta)}\left(x_1 \mid q\right) = \frac{(q^{\alpha+1};q)_n}{(q;q)_n}, \quad P_n^{(\alpha,\beta)}\left(x_2 \mid q\right) = \frac{(q^{\beta+1})_n}{(q;q)_n}(-1)^n q^{(\alpha-\beta)n/2}. \qquad (15.5.16)$$

The evaluation at x_2 follows from the Pfaff–Saaschütz theorem.

The continuous q-Jacobi polynomials given by (15.5.11) and the continuous q-ultraspherical polynomials are connected by the quadratic transformations

$$C_{2n}\left(x; q^\lambda \mid q\right) = \frac{(q^\lambda, -q; q)_n}{(q^{1/2}, -q^{1/2})_n} q^{-\frac{1}{2}n} P_n^{(\lambda-\frac{1}{2}, -\frac{1}{2})}\left(2x^2 - 1; q\right),$$

$$C_{2n+1}\left(x; q^\lambda \mid q\right) = \frac{(q^\lambda, -q; q)_{n+1}}{(q^{1/2}, -q^{1/2})_{n+1}} q^{-n/2} x P_n^{(\lambda-\frac{1}{2}, \frac{1}{2})}\left(2x^2 - 1; q\right). \qquad (15.5.17)$$

The continuous q-Jacobi polynomials are essentially invariant under $q \to q^{-1}$. Indeed

$$P_n^{(\alpha,\beta)}\left(x \mid q^{-1}\right) = q^{-n\alpha} P_n^{(\alpha,\beta)}(x \mid q),$$
$$P_n^{(\alpha,\beta)}\left(x; q^{-1}\right) = q^{-n(\alpha+\beta)} P_n^{(\alpha,\beta)}(x; q). \qquad (15.5.18)$$

Theorem 15.5.2 *The continuous q-Jacobi polynomials have the property*

$$\left(1 - 2xq^{(2\alpha+1)/4} + q^{\alpha+1/2}\right)\left(1 + 2xq^{(2\beta+1)/4} + q^{\beta+1/2}\right)$$
$$\times D_q P_n^{(\alpha,\beta)}(x \mid q) = A_n(x) P_{n-1}^{(\alpha,\beta)}(x \mid q) + B_n(x) P_n^{(\alpha,\beta)}(x \mid q), \qquad (15.5.19)$$

where

$$A_n(x) = 2\frac{\left(1 - q^{\alpha+n}\right)\left(1 - q^{\beta+n}\right)}{(1-q)\left(1 - q^{n+(\alpha+\beta)/2}\right)}\left(1 + q^{(\alpha+\beta+1)/2}\right)q^{(\alpha-n)/2+3/4},$$

$$B_n(x) = 2\frac{\left(1 - q^n\right)\left(1 - q^{(\alpha-\beta)/2}\right)}{(1-q)\left(1 - q^{n+(\alpha+\beta)/2}\right)}\left(1 + q^{n+\alpha+\beta+1/2}\right)q^{(\beta-n)/2+3/4}$$

$$-4q^{(\alpha+\beta+2-n)/2}\frac{(1-q^n)}{1-q}x$$

$$(15.5.20)$$

Proof Clearly (15.5.3) implies

$$\frac{w\left(x \mid q^{\alpha+1}, q^{\beta+1}\right)}{w\left(x \mid q^{\alpha}, q^{\beta}\right)} = \left(1 - 2xq^{(2\alpha+1)/4} + q^{\alpha+1/2}\right)$$
$$\times \left(1 + 2xq^{(2\beta+1)/4} + q^{\beta+1/2}\right).$$

Therefore, Theorem 2.7.1 shows that there are constants a, b, c such that

$$\left(1 - 2xq^{(2\alpha+1)/4} + q^{\alpha+1/2}\right)\left(1 + 2xq^{(2\beta+1)/4} + q^{\beta+1/2}\right)\mathcal{D}_q P_n^{(\alpha,\beta)}(x \mid q)$$
$$= (ax + b)\, P_n^{(\alpha,\beta)}(x \mid q) + c P_{n-1}^{(\alpha,\beta)}(x \mid q).$$

$$(15.5.21)$$

By equating coefficients of x^{n+1} in (15.5.21) we find

$$a = -4\frac{(1-q^n)}{1-q}q^{(\alpha+\beta+2-n)/2}.$$

Applying (15.5.16) we solve the equations

$$-ax_j = b + c P_{n-1}^{(\alpha,\beta)}(x_j \mid q)/P_n^{(\alpha,\beta)}(x_j \mid q), \quad j = 1, 2,$$

and evaluate b and c. The result now follows from (15.5.7). ∎

Note that as $q \to 1$, formula (15.5.19) reduces to (3.3.16).

The relevant discriminant of the continuous q-Jacobi, in the notation of (6.4.1), is $D\left(P_n^{(\alpha,\beta)}, \mathcal{D}_q\right)$.

Theorem 15.5.3 *The quantized discriminant $D\left(f, \mathcal{D}_q\right)$ for continuous q-Jacobi polynomials is given by*

$$D\left(P_n^{(\alpha,\beta)}(x \mid q); \mathcal{D}_q\right) = \frac{2^{n^2-2n}q^{n(n-1)\alpha}}{(1-q)^n}\prod_{j=1}^{n}\left(1 - q^{\alpha+\beta+n+j}\right)^{n-j}$$

$$\times \prod_{j=1}^{n}\left(1-q^j\right)^{j-2n+2}\left(1-q^{\alpha+j}\right)^{j-1}\left(1-q^{\beta+j}\right)^{j-1}$$

$$\times \prod_{j=1}^{2n}\left(1 + q^{(\alpha+\beta+j)/2}\right)^{j-2n}.$$

Proof The result follows from the definition of $D\left(P_n^{(\alpha,\beta)}(x \mid q), \mathcal{D}_q\right)$, (15.5.19), and Lemma 3.4.1. $\qquad\square$

Theorem 15.5.3 is not in the literature, but we felt it was important enough to be recorded in this volume.

15.6 q-Racah Polynomials

The q-Racah polynomials were introduced in (Askey & Wilson, 1979). They are defined by

$$R_n(\mu(x); \alpha, \beta; \gamma, \delta) := {}_4\phi_3\left(\begin{array}{c} q^{-n}, \alpha\beta q^{n+1}, q^{-x}, \gamma\delta q^{x+1} \\ \alpha q, \beta\delta q, \gamma q \end{array} \middle| q, q\right), \qquad (15.6.1)$$

where

$$\mu(x) = q^{-x} + \gamma\delta q^{x+1}, \qquad (15.6.2)$$

and $\alpha q = q^{-N}$, for some positive integer N. Clearly, $\left(q^{-x}, \gamma\delta q^{x+1}; q\right)_m$ is a polynomial of degree m in $\mu(x)$, hence $R_n(\mu(x))$ is a polynomial of exact degree n in $\mu(x)$. Indeed

$$R_n(\mu(x); \alpha, \beta, \gamma, \delta)$$
$$= \frac{\left(\alpha\beta a^{n+1}; q\right)_n (\mu(x))^n}{(\alpha q, \beta\delta q, \gamma q; q)_n} + \text{lower order terms.} \qquad (15.6.3)$$

It is important to note that $\mu(x)$ depends on γ and δ. The case $q \to 1$ gives the Racah polynomials of the Racah–Wigner algebra of quantum mechanics, (Biedenharn & Louck, 1981). See also (Askey & Wilson, 1982).

Let

$$w(x; \alpha, \beta, \gamma, \delta) := \frac{(\alpha q, \beta\delta q, \gamma q, \gamma\delta q; q)_x}{(q, \gamma\delta q/\alpha, \gamma q/\beta, \delta q; q)_x} \frac{\left(1 - \gamma\delta q^{2x+1}\right)}{(\alpha\beta q)^x (1 - \gamma\delta q)}. \qquad (15.6.4)$$

Theorem 15.6.1 *A discrete analogue of the Askey–Wilson integral is*

$$\sum_{x=0}^{N} w(x; \alpha, \beta, \gamma, \delta) = \frac{\left(\gamma/\alpha\beta, \delta/\alpha, 1/\beta, \gamma\delta q^2; q\right)_\infty}{(1/\alpha\beta q, \gamma\delta q/\alpha, \gamma q/\beta, \delta q; q)_\infty}. \qquad (15.6.5)$$

Proof The left-hand side of (15.6.5) is

$${}_6\phi_5\left(\begin{array}{c} \gamma\delta q, \sqrt{\gamma\delta q}\, q, -\sqrt{\gamma\delta q}\, q, \alpha q, \beta\delta q, \gamma q \\ \sqrt{\gamma\delta q}, -\sqrt{\gamma\delta q}, \gamma\delta q/\alpha, \gamma q/\beta, \delta q \end{array} \middle| q, \frac{1}{\alpha\beta q}\right).$$

Since $\alpha q = q^{-N}$, the ${}_6\phi_5$ terminates and (12.2.31) shows that its sum equals the right-hand side of (15.6.5). $\qquad\square$

With above choice of α, (15.6.5) is

$$\sum_{x=0}^{N} w\left(x; q^{-N-1}, \beta, \gamma, \delta\right) = \frac{\left(1/\beta, q^2\gamma\delta; q\right)_N}{(q\delta; q\gamma/\beta; q)_N}. \qquad (15.6.6)$$

Theorem 15.6.2 *The q-Racah polynomials satisfy the orthogonality relation*

$$\sum_{x=0}^{N} w(x;\alpha,\beta,\gamma,\delta)R_m(\mu(x);\alpha,\beta;\gamma,\delta)R_n(\mu(x);\alpha,\beta;\gamma,\delta) \tag{15.6.7}$$

$$= h_n\delta_{m,n},$$

where

$$h_n = \frac{(\gamma/\alpha\beta,\delta/\alpha,1/\beta,\gamma\delta q^2;q)_\infty}{(1/\alpha\beta q,\gamma\delta q/\alpha,\gamma q/\beta,\delta q;q)_\infty} \tag{15.6.8}$$

$$\times\frac{(1-\alpha\beta q)(\gamma\delta q)^n}{(1-\alpha\beta q^{2n+1})}\frac{(q,\alpha\beta q/\gamma,\alpha q/\delta,\beta q;q)_n}{(\alpha q,\alpha\beta q,\beta\delta q,\gamma q;q)_n}.$$

Proof It is clear that $\left(\gamma q^{x+1},q^{-x}/\delta;q\right)_s$ is a polynomial in $\mu(x)$ of exact degree s. Consider the sums

$$I_{j,k} = \sum_{x=0}^{N}\left(\gamma q^{x+1},q^{-x}/\delta;q\right)_j\left(q^{-x},\gamma\delta q^{x+1};q\right)_k w(x;\alpha,\beta,\gamma,\delta). \tag{15.6.9}$$

It is easy to see that

$$\left(q^{-x},\gamma\delta q^{x+1};q\right)_k = \frac{(-1)^k(q;q)_x(\gamma\delta q;q)_{x+k}}{(q;q)_{x-k}(\gamma\delta q;q)_x}q^{-kx+k(k-1)/2}$$

$$\left(q^{-x}/\delta,\gamma q^{x+1};q\right)_j = \frac{(-1)^j(\gamma q;q)_{x+j}}{\delta^j(\gamma q;q)_x}q^{-xj+j(j-1)/2}\frac{(\delta q;q)_x}{(\delta q;q)_{x-j}}.$$

Therefore $I_{j,k}$ is given by

$$\frac{(-1)^{j+k}q^{\binom{k}{2}+\binom{j}{2}}}{\delta^j\left(q^{-N}\beta\right)^k q^{k(k+j)}}\frac{(\gamma q;q)_{j+k}(\gamma\delta q;q)_{2k}}{(\gamma q;q)_{k-j}}\frac{(1-\gamma\delta q^{2k+1})}{1-\gamma\delta q}$$

$$\times\frac{(\beta\delta q,q^{-N};q)_k}{(\gamma q/\beta,\gamma\delta q^{2+N};q)_k}\sum_{x=0}^{N-k}w\left(\alpha q^k,\beta q^j,\gamma q^{j+k},\delta q^{k-j}\right).$$

We apply (15.6.6) to see that the above expression is

$$\frac{(-1)^{j+k}q^{\binom{j}{2}-\binom{k+1}{2}-kj+kN}}{\delta^j\beta^k}\frac{(\gamma\delta q^2;q)_{N+k}(\gamma q;q)_{j+k}(q^{-j}/\beta;q)_{N-k}}{(\delta q;q)_{k-j}(q^{1+k-j}\delta,\gamma q^{k+1}/\beta;q)_{N-k}}$$

$$\times\frac{(q^{-N},\beta\delta q;q)_k}{(q\gamma/\beta,\gamma\delta q^{2+N};q)_k},$$

which simplifies to

$$\frac{(-1)^j}{\delta^j}q^{\binom{j}{2}}\frac{(\delta q^{j+1},q^{-N},\beta\delta q;q)_k}{(q^{1-N+j}\beta,\delta\gamma q^{2+N};q)_k}$$

$$\times\frac{(\gamma\delta q^2;q)_{N+k}}{(\gamma q/\beta;q)_N}\frac{(\gamma q;q)_j}{(\delta q;q)_{N-j}}\left(q^{-j}/\beta;q\right)_N.$$

Thus for $j \leq n$ we find

$$\sum_{x=0}^{N} w(x; \alpha, \beta, \gamma, \delta) \left(q^{-x}/\delta, \gamma q^{x+1}; q\right)_j p_n(\mu(x); \alpha, \beta, \gamma, \delta)$$

$$= \frac{(-1)^j q^{\binom{j}{2}}}{\delta^j} \frac{\left(\gamma \delta q^2; q\right)_N (\gamma q; q)_j \left(q^{-j}/\beta; q\right)_N}{(\gamma q/\beta; q)_N (q\delta; q)_{N-j}}$$

$$\times {}_3\phi_2 \left(\begin{matrix} q^{-n}, \gamma q^{j+1}, \beta q^{n-N} \\ q\gamma, \beta q^{1-N+j} \end{matrix} \; ; q, q \right).$$

The ${}_3\phi_2$ can be summed by (12.2.15) and the above expression is

$$\frac{(-1)^j q^{\binom{j}{2}}}{\delta^j} \frac{\left(\gamma \delta q^2, q^{-j}/\beta; q\right)_n (\gamma q; q)_j}{(\gamma q/\beta; q)_N (q\delta; q)_{N-j}} \frac{\left(q^{-j}, q\gamma q^{N-n}/\beta; q\right)_n}{(q\gamma, q^{N-n-j}/\beta; q)_n}, \tag{15.6.10}$$

which clearly vanishes for $j < n$. Since

$$\delta^n \left(q^{-x}/\delta, \gamma q^{x+1}; q\right)_n - \left(q^{-x}, \gamma \delta q^{x+1}; q\right)_n$$

is a polynomial in $q^{-x} + \gamma \delta q^{x+1}$ of degree less than n, then the left-hand side of (15.6.7) is zero when $m < n$. Moreover it is

$$\delta^n q^n \frac{\left(q^{-n}, \beta q^{n-N}; q\right)_n}{(q, q^{-N}, \beta\delta q, \gamma q; q)_n}$$

times the expression in (15.6.10) with $j = n$. Thus

$$h_n = \frac{(-1)^n q^{\binom{n}{2}}}{(\gamma q/\beta; q)_N} \frac{\left(\gamma \delta q^2, q^{-n}/\beta, q^{-n}, q\gamma q^{N-n}/\beta, q^{-n}, \beta q^{n-N}; q\right)_n}{(q\delta; q)_{N-n} (q\gamma, q^{N-2n}/\beta, q^{-N}, \beta\delta q, q; q)_n},$$

which simplifies to the expression in (15.6.7). $\qquad\square$

By reparameterizing the parameters of the Askey–Wilson we can prove that the R_n's satisfy

$$\left(q^{-x} - 1\right) \left(1 - \gamma\delta q^{x+1}\right) R_n(\mu(x))$$
$$= A_n R_{n+1}(\mu(x)) - (A_n + C_n) R_n(\mu(x)) + C_n R_{n-1}(\mu(x)), \tag{15.6.11}$$

where

$$A_n = \frac{\left(1 - \alpha q^{n+1}\right)\left(1 - \alpha\beta q^{n+1}\right)\left(1 - \beta\delta q^{n+1}\right)\left(1 - \gamma q^{n+1}\right)}{(1 - \alpha\beta q^{2n+1})(1 - \alpha\beta q^{2n+2})},$$
$$C_n = \frac{q\left(1 - q^n\right)\left(1 - \beta q^n\right)\left(\gamma - \alpha\beta q^n\right)\left(\gamma - \alpha q^n\right)}{(1 - \alpha\beta q^{2n})(1 - \alpha\beta q^{2n+1})}, \tag{15.6.12}$$

with $R_0(\mu(x)) = 1$, $R_{-1}(\mu(x)) = 0$. Here we used $R_n(\mu(x))$ for $R_n(\mu(x); \alpha, \beta, \gamma, \delta)$. It is clear from (15.6.1) that $R_n(\mu(x))$ is symmetric under $x \leftrightarrow n$. Hence, (15.6.11) shows that $R_n(\mu(n))$ solves the difference equation

$$\left(q^{-n} - 1\right)\left(1 - \gamma\delta q^{n+1}\right) y(x)$$
$$= A_x y(x+1) - (A_x + C_x) y(x) + C_x y(x-1). \tag{15.6.13}$$

In fact, (15.6.13) can be factored as a product of two first-order operators. Indeed

$$\frac{\Delta R_n(u(x); \alpha, \beta, \gamma, \delta)}{\Delta \mu(x)}$$

$$= \frac{q^{1-n} (1 - q^n) (1 - \alpha\beta q^{n+1})}{(1 - q) (1 - \alpha q) (1 - \beta\delta q) (1 - \gamma q)} R_{n-1}(\mu(x); \alpha q, \beta q, \gamma q, \delta),$$ (15.6.14)

$$\frac{\nabla (\bar{w}(x; \alpha, \beta, \gamma, \delta) R_n(\mu(x); \alpha, \beta, \gamma, \delta))}{\nabla \mu(x)}$$

$$= \frac{\bar{w}(x; \alpha/q, \beta/q, \gamma/q, \delta)}{(1 - q) (1 - \gamma\delta)} R_{n+1}(\mu(x); \alpha/q, \beta/q, \gamma/q, \delta),$$ (15.6.15)

where

$$\bar{w}(x; \alpha, \beta, \gamma, \delta) = \frac{(\alpha q, \beta\delta q, \gamma q, \gamma\delta q; q)_x}{(q, q\gamma\delta/\alpha, \gamma q/\beta, \delta q; q)_x} (\alpha\beta)^{-x}.$$ (15.6.16)

Repeated applications of (15.6.15) gives the Rodrigues formula

$$\bar{w}(x; \alpha, \beta, \gamma, \delta) R_n(x; \alpha, \beta, \gamma, \delta)$$

$$= (1 - q)^n \left(\frac{\nabla}{\nabla \mu(x)} \right)^n \bar{w}(x; \alpha q^n, \beta q^n, \gamma q^n, \delta).$$ (15.6.17)

One can prove the following generating functions using the Sears transformation. For $x = 0, 1, 2, \ldots, N$ we have

$${}_2\phi_1 \left(\begin{matrix} q^{-x}, \alpha\gamma^{-1}\delta^{-1}q^{-x} \\ \alpha q \end{matrix} \middle| q; \gamma\delta q^{x+1}t \right) {}_2\phi_1 \left(\begin{matrix} \beta\delta q^{x+1}, \gamma q^{x+1} \\ \beta q \end{matrix} \middle| q; q^{-x}t \right)$$

$$= \sum_{n=0}^{N} \frac{(\beta\delta q, \gamma q; q)_n}{(\beta q, q; q)_n} R_n(\mu(x); \alpha, \beta, \gamma, \delta \,|\, q) \, t^n, \quad (15.6.18)$$

if $\beta\delta q = q^{-N}$ or $\gamma q = q^{-N}$,

$${}_2\phi_1 \left(\begin{matrix} q^{-x}, \beta\gamma^{-1}q^{-x} \\ \beta\delta q \end{matrix} \middle| q; \gamma\delta q^{x+1}t \right) {}_2\phi_1 \left(\begin{matrix} \alpha q, \gamma q^{x+1} \\ \alpha\delta^{-1}q \end{matrix} \middle| q; q^{-x}t \right)$$

$$= \sum_{n=0}^{N} \frac{(\alpha q, \gamma q; q)_n}{(\alpha\delta^{-1}q, q; q)_n} R_n(\mu(x); \alpha, \beta, \gamma, \delta \,|\, q) \, t^n, \quad (15.6.19)$$

if $\alpha q = q^{-N}$ or $\gamma q = q^{-N}$,

$${}_2\phi_1 \left(\begin{matrix} q^{-x}, \delta^{-1}q^{-x} \\ \gamma q \end{matrix} \middle| q; \gamma\delta q^{x+1}t \right) {}_2\phi_1 \left(\begin{matrix} \alpha q^{x+1}, \beta\delta q^{x+1} \\ \alpha\beta\gamma^{-1}q \end{matrix} \middle| q; q^{-x}t \right)$$

$$= \sum_{n=0}^{N} \frac{(\alpha q, \beta\delta q; q)_n}{(\alpha\beta\gamma^{-1}q, q; q)_n} R_n(\mu(x); \alpha, \beta, \gamma, \delta \,|\, q) \, t^n, \quad (15.6.20)$$

if $\alpha q = q^{-N}$ or $\beta\delta q = q^{-N}$.

15.7 q-Integral Representations

In this section we derive representations for some q-orthogonal polynomials as moments. Let $p_n(x) = \int_E u^n d\mu(u, x)$. This has two applications. One is that it that we can start with a power series identity $f(t) = \sum_{n=0}^{\infty} a_n t^n$ then replace t by ut and apply the moment functional to u. This replaces u^n by $p_n(x)$ through the moment representation and leads to generating functions. This enables us to derive bilinear generating functions from linear generating functions. This is the spirit of the umbral calculus (Roman & Rota, 1978) and the symbolic method of (Kaplansky, 1944). Another application is to evaluate determinants of orthogonal polynomials because the moment representation identifies such determinants as Hankel determinants of some other measure. The applications will be presented in §15.8. The contents of Sections 15.7, 15.8 and 15.9 are based on our joint work (Ismail & Stanton, 1997) and (Ismail & Stanton, 2002).

The Al-Salam–Chihara polynomials may be renormalized in two ways so that the three-term recurrence relation is linear in q^n. Specifically if,

$$\hat{p}_n(x; t_1, t_2) := p_n(x; t_1, t_2) / t_1^n$$

$$c_n(x; t_1, t_2) := \frac{(t_1 t_2; q)_n}{(q; q)_n t_1^n} p_n(x; t_1, t_2) \tag{15.7.1}$$

then the recurrence relation (15.1.8) becomes

$$2x\hat{p}_n(x; t_1, t_2)$$
$$= (1 - t_1 t_2 q^n) \hat{p}_{n+1}(x; t_1, t_2) + (1 - q^n) \hat{p}_{n-1}(x; t_1, t_2) \tag{15.7.2}$$
$$+ (t_1 + t_2) q^n \hat{p}_n(x; t_1, t_2), \quad n > 0,$$

$$2x c_n(x; t_1, t_2)$$
$$= (1 - q^{n+1}) c_{n+1}(x; t_1, t_2) + (1 - t_1 t_2 q^{n-1}) c_{n-1}(x; t_1, t_2) \tag{15.7.3}$$
$$+ (t_1 + t_2) q^n c_n(x; t_1, t_2), \quad n > 0,$$

with the initial conditions, see (15.1.9),

$$\hat{p}_0(x; t_1, t_2) = 1 = c_0(x; t_1, t_2),$$
$$(1 - t_1 t_2) \hat{p}_1(x; t_1, t_2) / (1 - q) = (2x - t_1 - t_2) / (1 - q) = c_1(x; t_1, t_2).$$

The following theorem is from (Ismail & Stanton, 1997) and (Ismail & Stanton, 1998).

Theorem 15.7.1 *The Al-Salam–Chihara polynomials have the q-integral representations*

$$\frac{p_n(\cos\theta; t_1, t_2)}{t_1^n} = \frac{\left(t_1 e^{i\theta}, t_1 e^{-i\theta}, t_2 e^{i\theta}, t_2 e^{-i\theta}; q\right)_\infty}{(1 - q)e^{i\theta} \left(q, t_1 t_2, q e^{2i\theta}, e^{-2i\theta}; q\right)_\infty}$$

$$\times \int_{e^{-i\theta}}^{e^{i\theta}} y^n \frac{\left(q y e^{i\theta}, q y e^{-i\theta}; q\right)_\infty}{(t_1 y, t_2 y; q)_\infty} d_q y, \tag{15.7.4}$$

$$\frac{(t_1 t_2; q)_n}{(q; q)_n} \frac{p_n (\cos \theta; t_1, t_2)}{t_1^n} = \frac{(t_1 e^{i\theta}, t_1 e^{-i\theta}, q e^{i\theta}/t_1, q e^{-i\theta}/t_1; q)_\infty}{2(1-q) i \sin \theta \, (q, q, q e^{2i\theta}, q e^{-2i\theta}; q)_\infty}$$

$$\times \int_{e^{-i\theta}}^{e^{i\theta}} y^n \frac{(q y e^{i\theta}, q y e^{-i\theta}, t_2/y; q)_\infty}{(q y/t_1, t_1 y, q/(y t_1); q)_\infty} \, d_q y. \tag{15.7.5}$$

Proof We seek an integral representation

$$c_n (x; t_1, t_2) = \int_a^b y^n f(y) \, d_q y, \tag{15.7.6}$$

with f satisfying the boundary conditions

$$f(a/q) = f(b/q) = 0. \tag{15.7.7}$$

We demand that a and b are finite, hence the moment problem is determinate. Substitute the representation (15.7.6) for the c's in (15.7.3), then equate the coefficients of y^n. The result, after applying (11.4.9), is that f must satisfy the functional equation

$$f(y) = \frac{q}{t_1 t_2} \frac{(1 - 2xyq + q^2 y^2)}{(1 - qy/t_1)(1 - qy/t_2)} f(qy). \tag{15.7.8}$$

It is clear that

$$u(y) = \frac{(\lambda y, q/(\lambda y); q)_\infty}{(\mu y, q/(\mu y); q)_\infty} \quad \text{implies} \quad \frac{u(u)}{u(qy)} = \frac{\lambda}{\mu}. \tag{15.7.9}$$

Thus a solution to (15.7.8) which satisfies the boundary conditions (15.7.7) is given by

$$f(y) = \frac{(q y e^{i\theta}, q y e^{-i\theta}, \lambda y, q/(\lambda y); q)_\infty}{(q y/t_1, q y/t_2, y\mu, q/(y\mu); q)_\infty}, \quad \text{with} \quad q\mu = t_1 t_2 \lambda \tag{15.7.10}$$

with $x = \cos \theta$, $a = e^{-i\theta}$ and $b = e^{i\theta}$. Since a and b are finite, if f exists it will be unique. We then choose $\mu = t_1$ and $\lambda = q/t_2$ so that

$$g(\cos \theta) c_n (\cos \theta; t_1, t_2) = \frac{1}{1-q} \int_{e^{-i\theta}}^{e^{i\theta}} y^n \frac{(q y e^{i\theta}, q y e^{-i\theta}, t_2/y; q)_\infty}{(q y/t_1, t_1 y, q/(y t_1); q)_\infty} \, d_q y, \tag{15.7.11}$$

for some function $g(\cos \theta)$, independent of n. Next we determine the function g. The recurrence relation (15.7.3) has two linear independent polynomial solutions c_n and c_n^*, with $c_{-1} = 0$ but $c_{-1}^* \neq 0$. Actually we only proved that the q-integral in (15.7.11) solves (15.7.3), so in order to prove that it is a multiple of c_n it suffices to show that it vanishes when $n = -1$ and is not zero when $n = 0$. For general n write the right-hand side of (15.7.11) as

$$e^{i\theta} \sum_{m=0}^{\infty} \frac{(q^{m+1} e^{2i\theta}, q^{m+1}, q^{-m} t_2 e^{-i\theta}; q)_\infty}{(q^{m+1} e^{i\theta}/t_1, q^m t_1 e^{i\theta}, q^{1-m} e^{-i\theta}/t_1; q)_\infty} e^{in\theta} q^{m(n+1)}$$

$$- \text{ a similar term with } \theta \text{ replaced by } - \theta,$$

which simplifies to

$$e^{i(n+1)\theta} \frac{\left(qe^{2i\theta}, q, e^{-i\theta}t_2; q\right)_\infty}{\left(qe^{i\theta}/t_1, t_1 e^{i\theta}, qe^{-i\theta}/t_1; q\right)_\infty}$$

$$\times \left[{}_2\phi_1 \left(\begin{matrix} qe^{i\theta}/t_1, qe^{i\theta}/t_2 \\ qe^{2i\theta} \end{matrix} \middle| q, t_1 t_2 q^n \right) + e^{-2in\theta} \frac{\left(t_2 e^{i\theta}, e^{-2i\theta}, t_1 e^{i\theta}; q\right)_\infty}{\left(t_2 e^{-i\theta}, e^{2i\theta}, t_1 e^{-i\theta}; q\right)_\infty} \right.$$

$$\left. \times {}_2\phi_1 \left(\begin{matrix} qe^{-i\theta}/t_1, qe^{-i\theta}/t_2 \\ qe^{-2i\theta} \end{matrix} \middle| q, t_1 t_2 q^n \right) \right].$$

$$(15.7.12)$$

When $n = -1$ the ${}_2\phi_1$ functions are summed by the q-Gauss theorem and the above expression vanishes. When $n = 0$ apply (12.5.8) with the parameter identification

$$A = qe^{i\theta}/t_1, \quad B = qe^{i\theta}/t_2, \quad C = qe^{2i\theta}, \quad Z = t_1 t_2. \qquad (15.7.13)$$

The result is that the choice

$$g(\cos\theta) = \frac{e^{i\theta} \left(qe^{2i\theta}, q, t_2 e^{-i\theta}; q\right)_\infty}{\left(qe^{i\theta}/t_1, t_1 e^{i\theta}, qe^{-i\theta}/t_1; q\right)_\infty} \frac{\left(q, e^{-2i\theta}; q\right)_\infty}{\left(t_2 e^{-i\theta}, t_1 e^{-i\theta}; q\right)_\infty}, \qquad (15.7.14)$$

makes $c_0 = 1$ and (15.7.5) follows. Similarly one can prove (15.7.4). $\qquad \square$

We now consider the q-Pollaczek polynomials. From (13.7.6) and (15.1.11) it follows that

$$F_n(x; U, \Delta, V) = \frac{(1/(\xi\eta); q)_n}{(q; q)_n} \eta^n p_n\left(x; 1/\eta, 1/\xi\right), \qquad (15.7.15)$$

and we can use Theorem 15.7.1 to state similar results for the q-Pollaczek polynomials.

Theorem 15.7.2 *The q-Pollaczek polynomials have the q-integral representations*

$$\frac{(q; q)_n}{(\Delta^2; q)_n} F_n(x; U, \Delta, V) = \frac{\left(e^{i\theta}/\eta, e^{-i\theta}/\eta, e^{i\theta}/\xi, e^{-i\theta}/\xi; q\right)_\infty}{(1-q)e^{i\theta} \left(q, qe^{2i\theta}, qe^{-2i\theta}; q\right)_\infty}$$

$$\times \int_{e^{-i\theta}}^{e^{i\theta}} y^n \frac{\left(qye^{i\theta}, qye^{-i\theta}; q\right)_\infty}{(y/\eta, y/\xi; q)_\infty} \, d_q y, \qquad (15.7.16)$$

$$F_n(x; U, \Delta, V) = \frac{\left(\eta e^{i\theta}, \eta e^{-i\theta}, e^{i\theta}/\eta, e^{-i\theta}/\eta; q\right)_\infty}{2(1-q)i\sin\theta \left(q, q, qe^{2i\theta}, qe^{-2i\theta}; q\right)_\infty}$$

$$\times \int_{e^{-i\theta}}^{e^{i\theta}} y^n \frac{\left(qye^{i\theta}, qye^{-i\theta}, 1/(\xi y); q\right)_\infty}{(qy\eta, y/\eta, q\eta/y; q)_\infty} \, d_q y, \qquad (15.7.17)$$

We now consider the continuous q-Hermite polynomials. Clearly $H_n(x \mid q) = \hat{p}_n(x; 0, 0)$, so that Theorem 15.7.1 gives integral representations for the q-Hermite polynomials. For (15.7.4) this is immediate, while it is not clear how to let $t_1 = t_2 = 0$ in (15.7.5). Now we carry out this limit, and we also give two additional q-integral representations for q-Hermite polynomials.

Theorem 15.7.3 *The continuous q-Hermite polynomials have the q-integral representations*

$$H_n(\cos\theta \,|\, q) = \frac{1}{(1-q)e^{i\theta}\,(q, qe^{2i\theta}, e^{-2i\theta}; q)_\infty}$$

$$\times \int\limits_{e^{-i\theta}}^{e^{i\theta}} y^n \left(qye^{i\theta}, qye^{-i\theta}; q\right)_\infty d_q y, \tag{15.7.18}$$

$$\frac{H_n(\cos\theta \,|\, q)}{(q;q)_n} = \frac{(\lambda e^{i\theta}, \lambda e^{-i\theta}, qe^{i\theta}/\lambda, qe^{-i\theta}/\lambda; q)_\infty}{2(1-q)i\sin\theta\,(q, q, qe^{2i\theta}, qe^{-2i\theta}; q)_\infty}$$

$$\times \int\limits_{e^{-i\theta}}^{e^{i\theta}} y^n \frac{(qye^{i\theta}, qye^{-i\theta}; q)_\infty}{(\lambda y, qy/\lambda, \lambda/y, q/(\lambda y); q)_\infty} \, d_q y, \tag{15.7.19}$$

$$\frac{H_n\left(\cos\theta \,|\, q^2\right)}{(q;q)_n} = \frac{(\sqrt{q}\,e^{i\theta}, \sqrt{q}\,e^{i\theta}, \sqrt{q}\,e^{-i\theta}, \sqrt{q}\,e^{-i\theta}; q)_\infty}{2(1-q)i\sin\theta\,(q, q, qe^{2i\theta}, qe^{-2i\theta}; q)_\infty}$$

$$\times \int\limits_{e^{-i\theta}}^{e^{i\theta}} y^n \frac{(qye^{i\theta}, qye^{-i\theta}, -\sqrt{q}/y; q)_\infty}{(\sqrt{q}\,y, \sqrt{q}\,y, \sqrt{q}/y; q)_\infty} \, d_q y. \tag{15.7.20}$$

$$\frac{H_n(\cos\theta \,|\, q^2)}{(-q;q)_n} = \frac{(qe^{2i\theta}, qe^{-2i\theta}; q^2)_\infty}{2(1-q)i\sin\theta\,(q, -q, qe^{2i\theta}, qe^{-2i\theta}; q)_\infty}$$

$$\times \int\limits_{e^{-i\theta}}^{e^{i\theta}} y^n \frac{(qye^{i\theta}, qye^{-i\theta}; q)_\infty}{(qy^2; q^2)_\infty} \, d_q y. \tag{15.7.21}$$

Note that the right-hand side of (15.7.19) is independent of λ.

Proof Formula (15.7.18) is the special case $t_1 = t_2 = 0$ of Theorem 15.7.1.

Next we motivate the integral in (15.7.19). In terms of the polynomials $\hat{H}_n(x \,|\, q) = H_n(x \,|\, q)/(q;q)_n$, the three-term recurrence relation for q-Hermite polynomials becomes

$$2x\hat{H}_n(x \,|\, q) = \left(1 - q^{n+1}\right)\hat{H}_{n+1}(x \,|\, q) + \hat{H}_{n-1}(x \,|\, q). \tag{15.7.22}$$

Here again we see that writing $\hat{H}_n(x \,|\, q) = \int_a^b y^n f(y)\, d_q y$ requires f to satisfy

$$f(y) = \left(1 - qye^{i\theta}\right)\left(1 - qye^{-i\theta}\right)\left(qy^2\right)^{-1} f(qy).$$

Solving the above functional equation gives rise to the two solutions

$$\int\limits_0^{e^{\pm i\theta}} y^n \frac{(qye^{i\theta}, qye^{-i\theta}; q)_\infty}{(\lambda y, \lambda/y, q/(\lambda y), qy/\lambda; q)_\infty} \, d_q y$$

and the integral in (15.7.19) is a linear combination of these two solutions.

We next show that (15.7.19) is implied by (15.7.18). The right-hand side of (15.7.19) is

$$
\lim_{h \to 0} \left[\frac{e^{in\theta}}{(q, e^{-2i\theta}; q)_\infty} \, {}_2\phi_1 \left(\begin{matrix} 1/h, 1/h \\ qe^{2i\theta} \end{matrix} \,\middle|\, q, h^2 e^{2i\theta} q^{n+2} \right) \right.
$$
$$
\left. + \frac{e^{-in\theta}}{(q, e^{2i\theta}; q)_\infty} \, {}_2\phi_1 \left(\begin{matrix} 1/h, 1/h \\ qe^{-2i\theta} \end{matrix} \,\middle|\, q, h^2 e^{2i\theta} q^{n+2} \right) \right].
$$

Apply the Heine transformation (12.5.3) to the above ${}_2\phi_1$s we reduce the above combination to

$$
\frac{e^{in\theta}}{(q; q)_n \, (e^{-2i\theta}; q)_\infty} \, {}_2\phi_1 \left(\begin{matrix} 0, 0 \\ qe^{2i\theta} \end{matrix} \,\middle|\, q, q^{n+1} \right)
$$
$$
+ \frac{e^{-in\theta}}{(q; q)_n \, (e^{2i\theta}; q)_\infty} \, {}_2\phi_1 \left(\begin{matrix} 0, 0 \\ qe^{-2i\theta} \end{matrix} \,\middle|\, q, q^{n+1} \right),
$$

which by direct evaluation is $1/(q; q)_n$ times the right-hand side of (15.7.18).

To prove (15.7.20) we set $p_n(x \,|\, q) = H_n \left(x \,|\, q^2 \right) / (q; q)_n$, then observe that the three-term recurrence relation for $H_n(x \,|\, q)$ gives the recursion

$$
2x p_n(x \,|\, q) = \left(1 - q^{n+1} \right) p_{n+1}(x \,|\, q) + (1 + q^n) p_{n-1}(x \,|\, q). \tag{15.7.23}
$$

Comparing (15.7.23) and (15.7.3) we conclude that $p_n(x \,|\, q) = c_n \left(x; \sqrt{q}, -\sqrt{q} \right)$, so that (15.7.20) is a special case of (15.7.5). Finally (15.7.21) follows from Theorem 15.7.1 in a similar way. $\qquad\qquad\square$

Comparing (15.1.10) and (13.2.8) we find

$$
C_n(\cos\theta; \beta \,|\, q) = \frac{(\beta^2; q)_n}{(q; q)_n} \frac{p_n \left(\cos\theta; \beta e^{i\theta}, \beta e^{-i\theta} \,|\, q \right)}{\beta^n e^{in\theta}}, \tag{15.7.24}
$$

hence Theorem 15.7.1 is transformed to

$$
C_n(\cos\theta; \beta \,|\, q) = \frac{\left(\beta e^{2i\theta}, \beta e^{-2i\theta}, \beta, \beta; q \right)_\infty}{(1 - q)e^{i\theta} \left(q, \beta^2, qe^{2i\theta}, qe^{-2i\theta}; q \right)_\infty}
$$
$$
\times \frac{(\beta^2; q)_n}{(q; q)_n} \int_{e^{-i\theta}}^{e^{i\theta}} y^n \frac{\left(qy e^{i\theta}, qy e^{-i\theta}; q \right)_\infty}{\left(\beta e^{i\theta} y, \beta e^{-i\theta} y; q \right)_\infty} \, d_q y, \tag{15.7.25}
$$

and

$$
C_n(\cos\theta; \beta \,|\, q) = \frac{\left(\beta e^{2i\theta}, qe^{-2i\theta}/\beta, \beta, q/\beta; q \right)_\infty}{2(1 - q)i\sin\theta \, (q, q, qe^{2i\theta}, qe^{-2i\theta}; q)_\infty}
$$
$$
\times \int_{e^{-i\theta}}^{e^{i\theta}} y^n \frac{\left(qy e^{i\theta}, qy e^{-i\theta}, \beta e^{-i\theta}/y; q \right)_\infty}{\left(qy e^{-i\theta}/\beta, \beta y e^{i\theta}, qe^{-i\theta}/(\beta y); q \right)_\infty} \, d_q y. \tag{15.7.26}
$$

Formula (15.7.25) already appeared as (15.4.4).

The q-integral representations as moments presented in this section can be used to evaluate Hankel determinants of orthogonal polynomials. The details are in (Ismail, 2005b).

15.8 Linear and Multilinear Generating Functions

In this section we apply the q-integral representations of the previous section to derive a variety of generating functions. Some of the generating functions can also be derived by other techniques. The work (Koelink & Van der Jeugt, 1999) uses the positive discrete series representations of the quantized universal enveloping algebra $\mathcal{U}_q(\mathrm{su}(1,1))$. The idea is that in the tensor products of two such representations, two sets of eigenfunctions of a certain operator arise. The eigenfunctions turn out to be Al-Salam–Chihara and Askey–Wilson polynomials and bilinear generating functions arise in this natural way.

In view of (15.7.24), every linear or multilinear generating function for the polynomials $\{p_n(x; t_1, t_2 \mid q)\}$ leads to a similar result for $\{C_n(x; \beta \mid q)\}$. We shall not record these results here.

Theorem 15.8.1 *We have the linear generating function*

$$
\sum_{n=0}^{\infty} \frac{(t_1 t_2, \lambda/\mu; q)_n}{(q, q; q)_n} p_n \left(\cos\theta; t_1, t_2\right) \mu^n
$$
$$
= \frac{e^{i\theta} \left(t_2 e^{-i\theta}, t_1 e^{-i\theta}, t_1 \lambda e^{i\theta}; q\right)_{\infty}}{2i \sin\theta \, (q, q e^{-2i\theta}, t_1 \mu e^{i\theta}; q)_{\infty}}
$$
$$
\times {}_3\phi_2 \left(\begin{matrix} q e^{i\theta}/t_1, q e^{i\theta}/t_2, t_1 \mu e^{i\theta} \\ q e^{2i\theta}, t_1 \lambda e^{i\theta} \end{matrix} \;\middle|\; q, t_1 t_2 \right)
$$

$$
- \text{ a similar term with } \theta \text{ replaced by} - \theta.
$$

(15.8.1)

Proof Formula (15.7.5) and the q-binomial theorem show that the left-hand side of (15.8.1) is

$$
\frac{\left(t_1 e^{i\theta}, t_1 e^{-i\theta}, q e^{i\theta}/t_1, q e^{-i\theta}/t_1; q\right)_{\infty}}{2(1-q)i \sin\theta \, (q, q, q e^{2i\theta}, q e^{-2i\theta}; q)_{\infty}}
$$
$$
\times \int_{e^{-i\theta}}^{e^{i\theta}} \frac{\left(q y e^{i\theta}, q y e^{-i\theta}, t_2/y, \lambda t_1 y; q\right)_{\infty}}{(q y/t_1, t_1 y, q/(y t_1), \mu y t_1; q)_{\infty}} \, d_q y.
$$

(15.8.2)

It is easy to see that

$$
\int_{e^{-i\theta}}^{e^{i\theta}} \frac{\left(qye^{i\theta}, qye^{-i\theta}, t_2/y, \lambda t_1 y; q\right)_\infty}{\left(qy/t_1, t_1 y, q/\left(yt_1\right), \mu y t_1; q\right)_\infty} \frac{d_q y}{1-q}
$$

$$
= \sum_{m=0}^{\infty} \frac{\left(q^{m+1}e^{2i\theta}, q^{m+1}, q^{-m}t_2 e^{-i\theta}, t_1 \lambda e^{i\theta} q^m; q\right)_\infty}{\left(q^{m+1}e^{i\theta}/t_1, q^m t_1 e^{i\theta}, q^{1-m}e^{-i\theta}/t_1, t_1 \mu e^{i\theta} q^m; q\right)_\infty} e^{i\theta} q^m
$$

$$
\quad - \text{ a similar term with } \theta \to -\theta
$$

$$
= \frac{e^{i\theta}\left(qe^{2i\theta}, q, t_2 e^{-i\theta}, t_1 \lambda e^{i\theta}; q\right)_\infty}{\left(qe^{i\theta}/t_1, t_1 e^{i\theta}, qe^{-i\theta}/t_1, t_1 \mu e^{i\theta}; q\right)_\infty}
$$

$$
\times \;{}_3\phi_2 \left(\begin{array}{c} qe^{i\theta}/t_1, qe^{i\theta}/t_2, t_1 \mu e^{i\theta} \\ qe^{2i\theta}, t_1 \lambda e^{i\theta} \end{array} \middle|\; q, t_1 t_2 \right)
$$

$$
\quad - \text{ a similar term with } \theta \text{ replaced by } -\theta.
$$

Therefore (15.8.2) and the above calculation indicate that the left-hand side of (15.8.1) is

$$
\frac{e^{i\theta}\left(t_2 e^{-i\theta}, t_1 e^{-i\theta}, t_1 \lambda e^{i\theta}; q\right)_\infty}{2i\sin\theta \left(q, qe^{-2i\theta}, t_1 \mu e^{i\theta}; q\right)_\infty}\;{}_3\phi_2 \left(\begin{array}{c} qe^{i\theta}/t_1, qe^{i\theta}/t_2, t_1 \mu e^{i\theta} \\ qe^{2i\theta}, t_1 \lambda e^{i\theta} \end{array} \middle|\; q, t_1 t_2 \right)
$$

$$
\quad - \text{ a similar term with } \theta \text{ replaced by } -\theta,
$$

and the theorem follows. $\qquad\qquad\square$

Theorem 15.8.2 *We have the following bilinear generating functions for the Al-Salam–Chihara polynomials*

$$
\sum_{n=0}^{\infty} \frac{(t_1 t_2, s_1 s_2; q)_n}{(q, q; q)_n}\, p_n\left(\cos\theta; t_1, t_2\right) p_n\left(\cos\phi; s_1, s_2\right) \left(\frac{t}{t_1 s_1}\right)^n
$$

$$
= \frac{\left(t_1 e^{-i\theta}, t_2 e^{-i\theta}, t s_1 e^{i\theta}, t s_2 e^{i\theta}; q\right)_\infty}{\left(q, e^{-2i\theta}, te^{i(\theta+\phi)}, te^{i(\theta-\phi)}; q\right)_\infty} \tag{15.8.3}
$$

$$
\times \;{}_4\phi_3 \left(\begin{array}{c} te^{i(\theta+\phi)}, te^{i(\theta-\phi)}, qe^{i\theta}/t_1, qe^{i\theta}/t_2 \\ t s_1 e^{i\theta}, t s_2 e^{i\theta}, qe^{2i\theta} \end{array} \middle|\; q, t_1 t_2 \right)
$$

$$
\quad + \text{ a similar term with } \theta \text{ replaced by } -\theta,
$$

and

$$\sum_{n=0}^{\infty} \frac{(t_1 t_2; q)_n}{(q;q)_n t_1^n} t^n p_n(\cos\theta; t_1, t_2) p_n(\cos\phi; s_1, s_2)$$

$$= \frac{(s_1 e^{-i\phi}, s_2 e^{-i\phi}, t t_1 s_1 e^{i\phi}, t t_2 s_1 e^{i\phi}; q)_\infty}{(s_1 s_2, e^{-2i\phi}, t s_1 e^{i(\theta+\phi)}, t s_1 e^{i(\phi-\theta)}; q)_\infty}$$

$$\times {}_4\phi_3\left(\begin{array}{c} s_1 e^{i\phi}, s_2 e^{i\phi}, t s_1 e^{i(\theta+\phi)}, t s_1 e^{i(\phi-\theta)} \\ q e^{2i\phi}, t t_1 s_1 e^{i\phi}, t t_2 s_1 e^{i\phi} \end{array} \middle| q, q \right) \qquad (15.8.4)$$

$$+ \frac{(s_1 e^{i\phi}, s_2 e^{i\phi}, t t_1 s_1 e^{-i\phi}, t t_2 s_1 e^{-i\phi}; q)_\infty}{(s_1 s_2, e^{2i\phi}, t s_1 e^{i(\theta-\phi)}, t s_1 e^{-i(\theta+\phi)}; q)_\infty}$$

$$\times {}_4\phi_3\left(\begin{array}{c} s_1 e^{-i\phi}, s_2 e^{-i\phi}, t s_1 e^{-i(\theta+\phi)}, t s_1 e^{i(\theta-\phi)} \\ q e^{-2i\phi}, t t_1 s_1 e^{-i\phi}, t t_2 s_1 e^{-i\phi} \end{array} \middle| q, q \right).$$

Proof Replace $p_n(\cos\theta; t_1, t_2)$ by its integral representation in (15.7.5), then use (15.1.10) to see that the left-hand side of (15.8.3) is

$$\frac{\left(t_1 e^{i\theta}, t_1 e^{-i\theta}, q e^{i\theta}/t_1, q e^{-i\theta}/t_1; q\right)_\infty}{2(1-q)i\sin\theta\, (q, q, q e^{2i\theta}, q e^{-2i\theta}; q)_\infty}$$

$$\times \int_{e^{-i\theta}}^{e^{i\theta}} \frac{\left(qy e^{i\theta}, qy e^{-i\theta}, t_2/y, t s_1 y, t s_2 y; q\right)_\infty}{(qy/t_1, t_1 y, q/(yt_1), t y e^{i\phi}, t y e^{-i\phi}; q)_\infty} d_q y.$$

This expression simplifies to the right-hand side of (15.8.3). To prove (15.8.4) start with (15.1.10), replace t by ty, then multiply by

$$\frac{\left(qy e^{i\phi}, qy e^{-i\phi}; q\right)_\infty}{(s_1 y, s_2 y; q)_\infty}$$

and q-integrate over y between $e^{-i\phi}$ and $e^{i\phi}$. The result follows from (15.7.4). □

An unexpected transformation formula results from Theorem 15.8.2, namely the fact that its right-hand side is invariant under the interchanges

$$(\theta, \phi, t_1, t_2, s_1, s_2) \to (\phi, \theta, s_1, s_2, t_1, t_2).$$

This establishes the next corollary.

Corollary 15.8.3 *The combination*

$$\frac{\left(t_1 e^{-i\theta}, t_2 e^{-i\theta}, t s_1 e^{i\theta}, t s_2 e^{i\theta}; q\right)_\infty}{(q, e^{-2i\theta}, t e^{i(\theta+\phi)}, t e^{i(\theta-\phi)}; q)_\infty}$$

$$\times {}_4\phi_3\left(\begin{array}{c} t e^{i(\theta+\phi)}, t e^{i(\theta-\phi)}, q e^{i\theta}/t_1, q e^{i\theta}/t_2 \\ t s_1 e^{i\theta}, t s_2 e^{i\theta}, q e^{2i\theta} \end{array} \middle| q, t_1 t_2 \right) \qquad (15.8.5)$$

$$+ \text{ a similar term with } \theta \text{ replaced by } -\theta,$$

is invariant under the permutation $(\theta, \phi, t_1, t_2, s_1, s_2) \to (\phi, \theta, s_1, s_2, t_1, t_2)$.

It is important to emphasize that the $_4\phi_3$'s appearing in the transformation Corollary 15.8.3 are not balanced and most of the known transformations of this type involve balanced series. We do not know of an alternate proof of Corollary 15.8.3.

Theorem 15.8.4 *If $s_1 s_2 = t_1 t_2$ then*

$$\sum_{n=0}^{\infty} \frac{(t_1 t_2; q)_n}{(q; q)_n t_1^n} t^n p_n(\cos\theta; t_1, t_2)\, p_n(\cos\phi; s_1, s_2)$$

$$= \frac{(tt_1 s_1 e^{i\phi}, tt_2 s_1 e^{i\phi}, ts_1 s_2 e^{i\theta}, ts_1^2 e^{i\theta}; q)_\infty}{(tt_1 t_2 s_1 e^{i(\theta+\phi)}, ts_1 e^{i(\theta-\phi)}, ts_1 e^{i(\theta+\phi)}, ts_1 e^{i(\phi-\theta)}; q)_\infty}$$

$$\times {}_8\phi_7 \left(\begin{matrix} s_1^2 ts_2 e^{i(\theta+\phi)}/q, s_1\sqrt{qts_2}e^{i(\theta+\phi)/2}, -s_1\sqrt{qts_2}e^{i(\theta+\phi)/2}, t_2 e^{i\theta}, t_1 e^{i\theta}, \\ s_1\sqrt{ts_2/q}e^{i(\theta+\phi)/2}, -s_1\sqrt{ts_2/q}e^{i(\theta+\phi)/2}, tt_1 s_1 e^{i\phi}, tt_2 s_1 e^{i\phi}, \\ s_1 e^{i\phi}, s_2 e^{i\phi}, ts_1 e^{i(\theta+\phi)} \\ ts_1 s_2 e^{i\theta}, s_1^2 te^{i\theta}, s_1 s_2 \end{matrix} \middle| q, ts_1 e^{-i(\theta+\phi)} \right).$$

$$(15.8.6)$$

Proof Apply (12.5.13) to (15.8.4). □

In view of (15.7.24), Theorem 15.8.4 leads to the following result.

Theorem 15.8.5 *We have the Poisson-type kernel*

$$\sum_{n=0}^{\infty} \frac{(q; q)_n}{(\beta^2; q)_n} C_n(\cos\theta; \beta \,|\, q)\, C_n(\cos\phi; \beta \,|\, q)\, t^n$$

$$= \frac{(\beta t e^{i(\theta+\phi)}, \beta t e^{i(\phi-\theta)}, \beta t e^{i(\theta-\phi)}, \beta t e^{i(\theta+\phi)}; q)_\infty}{(\beta^2 t e^{i(\theta+\phi)}, t e^{i(\theta-\phi)}, t e^{i(\theta+\phi)}, t e^{i(\phi-\theta)}; q)_\infty} \qquad (15.8.7)$$

$$\times {}_8W_7 \left(\beta^2 t e^{i(\theta+\phi)}/q; \beta, \beta e^{2i\theta}, \beta e^{2i\phi}, \beta, e^{i(\theta+\phi)}; t e^{-i(\theta+\phi)} \right).$$

An application of (12.5.15) recasts (15.8.7) in the more symmetric form, due to (Gasper & Rahman, 1983),

$$\sum_{n=0}^{\infty} \frac{(q; q)_n}{(\beta^2; q)_n} C_n(\cos\theta; \beta \,|\, q)\, C_n(\cos\phi; \beta \,|\, q)\, t^n$$

$$= \frac{(\beta, t^2, \beta t e^{i(\theta+\phi)}, \beta t e^{i(\theta-\phi)}, \beta t e^{i(\phi-\theta)}, \beta t e^{-i(\theta+\phi)}; q)_\infty}{(\beta^2, \beta t^2, t e^{i(\theta+\phi)}, t e^{i(\theta-\phi)}, t e^{i(\phi-\theta)}, t e^{-i(\theta+\phi)}; q)_\infty} \qquad (15.8.8)$$

$$\times {}_8W_7 \left(\beta t^2/q; t e^{i(\theta+\phi)}, t e^{i(\theta-\phi)}, t e^{i(\phi-\theta)}, t e^{-i(\theta+\phi)}, \beta; \beta \right).$$

Theorem 15.8.4 first appeared in (Askey et al., 1996). The case $t_1 = s_1$, $t_2 = s_2$ of (15.8.6) is the Poisson kernel for $\{p_n(x; t_1, t_2)\}$.

The moment representations not only give an integral representation for the Al-Salam–Chihara polynomials but also they give q-integral representations for other solutions to the same three term recurrence relation. For details we refer the interested reader to (Ismail & Stanton, 2002).

We next state a bibasic version of Theorem 15.8.2.

Theorem 15.8.6 *Let* $p_n\left(x;t_1,t_2\mid q\right)$ *denote the Al-Salam–Chihara polynomials with base* q. *Then*

$$\sum_{n=0}^{\infty} p_n\left(\cos\theta;t_1,t_2\mid q\right) p_n\left(\cos\phi;s_1,s_2\mid p\right) \frac{(t_1t_2;q)_n\,(s_1s_2;p)_n}{(q;q)_n(p;p)_n}\left(\frac{t}{t_1s_1}\right)^n$$

$$= \frac{\left(t_1e^{-i\theta},t_2e^{-i\theta};q\right)_{\infty}}{(q,e^{-2i\theta};q)_{\infty}} \sum_{k=0}^{\infty} \frac{\left(qe^{i\theta}/t_1,t_1e^{i\theta},qe^{i\theta}/t_2;q\right)_k}{(q,qe^{2i\theta},t_1e^{i\theta};q)_k}\,(t_1t_2)^k$$

$$\times\ \frac{\left(ts_1q^ke^{i\theta},ts_2q^ke^{i\theta};p\right)_{\infty}}{\left(tq^ke^{i(\theta+\phi)},tq^ke^{i(\theta-\phi)};p\right)_{\infty}}$$

$$+\ a\ similar\ term\ with\ \theta\ replaced\ by\ -\theta.$$

$$(15.8.9)$$

The proof is similar to the proof of Theorem 15.8.2 and will be omitted. An immediate consequence of Theorem 15.8.6 is the following bibasic version of Corollary 15.8.3. Theorem 15.8.6 is also in (Van der Jeugt & Jagannathan, 1998) with a quantum group derivation.

Corollary 15.8.7 *The right-hand side of* (15.8.9) *is symmetric under interchanging*

$$(t_1,t_2,s_1,s_2,\theta,\phi,p,q)\ \ with\ \ (s_1,s_2,t_1,t_2,\phi,\theta,q,p).$$

We now give two generating functions which follow from (15.7.19) and one which follows from (15.7.21)

$$\sum_{n=0}^{\infty} \frac{H_{n+k}(\cos\theta\mid q)}{(q;q)_{n+k}}\,t^n = \frac{e^{ik\theta}}{(1-te^{i\theta})(e^{-2i\theta};q)_{\infty}}\,{}_1\phi_2\left(\begin{array}{c}te^{i\theta}\\qe^{2i\theta},qte^{i\theta}\end{array}\middle|\,q,q^{k+2}e^{2i\theta}\right)$$

$$+\ a\ similar\ term\ with\ \theta\ replaced\ by\ -\theta.$$

$$(15.8.10)$$

In fact one can get the more general result

$$\sum_{n=0}^{\infty} \frac{H_{n+k}(\cos\theta\mid q)}{(q;q)_{n+k}}\,\frac{(\lambda;q)_n\,t^n}{(q;q)_n}$$

$$= \frac{\left(\lambda te^{i\theta};q\right)_{\infty}e^{ik\theta}}{(te^{i\theta},e^{-2i\theta};q)_{\infty}}\,{}_1\phi_2\left(\begin{array}{c}te^{i\theta}\\qe^{2i\theta},\lambda te^{i\theta}\end{array}\middle|\,q,q^{k+2}e^{2i\theta}\right) \qquad (15.8.11)$$

$$+\ a\ similar\ term\ with\ \theta\ replaced\ by\ -\theta.$$

Theorem 15.8.8 *A generating function for the* q-*Hermite polynomials is*

$$\sum_{n=0}^{\infty} \frac{(\lambda;q)_n}{(q^2;q^2)_n}\,H_n\left(x\mid q^2\right) t^n$$

$$= \frac{\left(\lambda te^{i\theta};q\right)_{\infty}}{(te^{i\theta};q)_{\infty}}\,{}_3\phi_2\left(\begin{array}{c}\lambda,\sqrt{q}\,e^{i\theta},-\sqrt{q}\,e^{i\theta}\\\lambda te^{i\theta},-q\end{array}\middle|\,q,te^{-i\theta}\right). \qquad (15.8.12)$$

Proof Multiply both sides of equation (15.7.21) by $(\lambda;q)_n t^n/(q;q)_n$, sum on n and use the q-binomial theorem. The right-hand side becomes

$$\frac{\left(\sqrt{q}\,e^{-i\theta}, -\sqrt{q}\,e^{-i\theta}, \lambda t e^{i\theta}; q\right)_\infty}{(-q, e^{-2i\theta}, t e^{i\theta}; q)_\infty}\, {}_3\phi_2\left(\begin{array}{c} \sqrt{q}\,e^{i\theta}, -\sqrt{q}\,e^{i\theta} t e^{i\theta} \\ q e^{2i\theta}, \lambda t e^{i\theta} \end{array} \middle| q, q\right)$$

$$+\frac{\left(\sqrt{q}\,e^{i\theta}, -\sqrt{q}\,e^{i\theta}, \lambda t e^{-i\theta}; q\right)_\infty}{(-q, e^{2i\theta}, t e^{-i\theta}; q)_\infty}\, {}_3\phi_2\left(\begin{array}{c} \sqrt{q}\,e^{-i\theta}, -\sqrt{q}\,e^{-i\theta} t e^{-i\theta} \\ q e^{-2i\theta}, \lambda t e^{-i\theta} \end{array} \middle| q, q\right).$$

Apply the transformation (12.5.8) with the parameter identification

$$A = \lambda, \quad B = -C = -\sqrt{q}\,e^{i\theta}, \quad D = \lambda t e^{i\theta}, \quad E = -q,$$

to reduce the combination of the ${}_3\phi_2$ functions to the right-hand side of (15.8.12).

□

Another proof of (15.8.12) is in (Ismail & Stanton, 2002).

Observe that the ${}_3\phi_2$ in Theorem 15.8.8 is essentially bibasic on base q and q^2. If $\lambda = 0$ or $\lambda = -q$ the ${}_3\phi_2$ may be summed to infinite products and we recover standard generating functions.

Using (15.7.18) and (13.1.24) we can derive a trilinear generating function for the continuous q-Hermite polynomials.

Theorem 15.8.9 *We have the following trilinear generating functions*

$$\sum_{n=0}^\infty H_{n+k}(\cos\psi\,|\,q) H_n(\cos\theta\,|\,q) H_n(\cos\phi\,|\,q)\frac{t^n}{(q;q)_n}$$

$$=\frac{e^{ik\psi}\left(t^2 e^{2i\psi}; q\right)_\infty}{\left(e^{-2i\psi}, t e^{i(\psi+\theta+\phi)}, t e^{i(\psi+\theta-\phi)}, t e^{i(\psi+\phi-\theta)}, t e^{i(\psi-\theta-\phi)}; q\right)_\infty}$$

$$\times {}_6\phi_5\left(\begin{array}{c} t e^{i(\psi+\theta+\phi)}, t e^{i(\psi+\theta-\phi)}, t e^{i(\psi+\phi-\theta)}, t e^{i(\psi-\theta-\phi)}, 0, 0 \\ q e^{2i\psi}, t e^{i\psi}, -t e^{i\psi}, \sqrt{q}\, t e^{i\psi}, -\sqrt{q}\, t e^{i\psi} \end{array} \middle| q, q^{k+1} e^{i\psi}\right)$$

$$+ \text{ a similar term with } \psi \text{ replaced by } -\psi.$$

(15.8.13)

and

$$\sum_{n=0}^\infty \frac{H_{n+k}(\cos\psi\,|\,q)}{(q;q)_{n+k}(q;q)_n} H_n(\cos\theta\,|\,q) H_n(\cos\phi\,|\,q)\, t^n$$

$$=\frac{e^{ik\psi}\left(t^2 e^{2i\psi}; q\right)_\infty}{\left(e^{-2i\psi}, t e^{i(\psi+\theta+\phi)}, t e^{i(\psi+\theta-\phi)}, t e^{i(\psi+\phi-\theta)}, t e^{i(\psi-\theta-\phi)}; q\right)_\infty}$$

$$\times {}_4\phi_5\left(\begin{array}{c} t e^{i(\psi+\theta+\phi)}, t e^{i(\psi+\theta-\phi)}, t e^{i(\psi+\phi-\theta)}, t e^{i(\psi-\theta-\phi)} \\ q e^{2i\psi}, t e^{i\psi}, -t e^{i\psi}, \sqrt{q}\, t e^{i\psi}, -\sqrt{q}\, t e^{i\psi} \end{array} \middle| q, q^{k+2} e^{3i\psi}\right)$$

$$+ \text{ a similar term with } \psi \text{ replaced by } -\psi.$$

(15.8.14)

Furthermore, the right-hand sides of (15.8.13) *and* (15.8.14) *are symmetric under any permutation of* θ, ϕ, *and* ψ.

Proof Replace t by ty in (13.1.24), then multiply by y^k, and then use (15.7.18) to establish (15.8.13). Similarly using (15.7.8) and (13.1.24) we establish (15.8.14).

When $k = 0$ the left-hand sides of (15.8.13) and (15.8.14) are clearly symmetric in θ, ψ, and ψ. Demanding that the the right-hand sides are also symmetric proves the last assertion in the theorem.   □

The trilinear generating function (15.8.13) contains two important product formulas for the continuous q-Hermite polynomials which will be stated in the next theorem.

Theorem 15.8.10 *With $K(\cos\theta, \cos\phi, \cos\psi)$ denoting the right-hand side of* (15.8.13), *we have the product formulas*

$$H_n(\cos\theta \,|\, q)H_n(\cos\phi|q) = \frac{(q;q)_\infty (q;q)_n}{2\pi t^n (q;q)_{n+k}} \int_0^\pi K(\cos\theta, \cos\phi, \cos\psi) \tag{15.8.15}$$
$$\times\, H_{n+k}(\cos\psi \,|\, q)\left(e^{2i\psi}, e^{-2i\psi};q\right)_\infty d\psi,$$

and

$$H_n(\cos\theta \,|\, q)H_{n+k}(\cos\psi \,|\, q) = \frac{(q;q)_\infty}{2\pi t^n} \int_0^\pi K(\cos\theta, \cos\phi, \cos\psi) \tag{15.8.16}$$
$$\times\, H_n(\cos\phi \,|\, q)\left(e^{2i\phi}, e^{-2i\phi};q\right)_\infty d\phi.$$

The special case $\lambda = 0$ of (15.8.12) yields

$$\frac{H_n(\cos\theta \,|\, q^2)}{(q^2;q^2)_n} = \sum_{k=0}^n \frac{(qe^{2i\theta};q^2)_k}{(q^2;q^2)_k\,(q;q)_{n-k}} e^{i(n-2k)\theta}. \tag{15.8.17}$$

Note that (13.1.1), (13.1.2), (15.7.3), and the initial conditions of $c_n(x;t_1,t_2)$ imply

$$c_n(\cos 2\theta; -1, -q \,|\, q^2) = \frac{H_{2n}(\cos\theta \,|\, q)}{(q^2;q^2)_n},$$
$$2\cos\theta c_n\left(\cos 2\theta; -q^2, -q \,|\, q^2\right) = \frac{H_{2n+1}(\cos\theta \,|\, q)}{(q^2;q^2)_n}.$$

Thus Theorem 15.7.1 gives q-integral moment representations for the following functions:

$$\frac{H_{2n}(x \,|\, q)}{(q^2;q^2)_n}, \quad \frac{H_{2n+1}(x \,|\, q)}{(-q;q^2)_n}, \quad \frac{H_{2n+1}(x \,|\, q)}{(q^2;q^2)_n}, \quad \frac{H_{2n+1}(x \,|\, q)}{(-q^3;q^2)_n}.$$

One can also derive several generating functions involving $H_{2n}(x \,|\, q)$ and $H_{2n+1}(x \,|\, q)$ from the corresponding results in §15.7.

15.9 Associated q-Ultraspherical Polynomials

In this section we give moment representations for the associated continuous q-ultraspherical polynomials which lead to three generating functions. This section is based on (Ismail & Stanton, 2002).

Set

$$C_n^{(\alpha)}(x;\beta\,|\,q) = \int_a^b y^n f(y) d_q y$$

then find from (13.7.9) that f satisfies

$$f(y) = \frac{q}{\alpha\beta^2} \frac{\left(1 - qye^{i\theta}\right)\left(1 - qye^{-i\theta}\right)}{\left(1 - qye^{i\theta}/\beta\right)\left(1 - qye^{-i\theta}/\beta\right)} f(qy).$$

This suggests that we consider the functions

$$\int_{e^{-i\theta}}^{e^{i\theta}} \frac{y^n}{1-q} \frac{(qye^{i\theta}, qye^{-i\theta}, \lambda y, q/(\lambda y); q)_\infty}{(\mu y, q/(\mu y), qye^{i\theta}/\beta, qye^{-i\theta}/\beta; q)_\infty} d_q y,$$

with

$$q\mu = \lambda\alpha\beta^2. \tag{15.9.1}$$

We choose $\lambda = qe^{i\theta}/\beta$, $\mu = \alpha\beta e^{i\theta}$ and consider the functions

$$\Phi_n(\theta;\beta,\alpha) = \int_{e^{-i\theta}}^{e^{i\theta}} \frac{y^n}{1-q} \frac{(qye^{i\theta}, qye^{-i\theta}, \beta e^{-i\theta}/y; q)_\infty}{(\alpha\beta e^{i\theta}y, qe^{-i\theta}/(\alpha\beta y), qye^{-i\theta}/\beta; q)_\infty} d_q y. \tag{15.9.2}$$

Theorem 15.9.1 *The functions $\Phi_n(\theta,\beta,\alpha)$ have the hypergeometric representation*

$$\Phi_n(\theta,\beta,\alpha) = e^{i(n+1)\theta} \frac{(q, \alpha q^{n+1}, qe^{2i\theta}, e^{-2i\theta}; q)_\infty}{(q/\beta, \alpha\beta q^n, \alpha\beta e^{2i\theta}, qe^{2i\theta}/(\alpha\beta); q)_\infty}$$

$$\times {}_2\phi_1\left(\begin{array}{c} q^{-n}/\alpha, \beta \\ q^{1-n}/(\alpha\beta) \end{array}\bigg|\, q, \frac{q}{\beta} e^{-2i\theta}\right), \quad n \geq 0. \tag{15.9.3}$$

Proof The right-hand side of (15.9.2) is

$$e^{i(n+1)\theta} \frac{(q, qe^{2i\theta}, \beta e^{-2i\theta}; q)_\infty}{(q/\beta, \alpha\beta e^{2i\theta}, qe^{-2i\theta}/(\alpha\beta); q)_\infty} {}_2\phi_1\left(\begin{array}{c} q/\beta, qe^{2i\theta}/\beta \\ qe^{2i\theta} \end{array}\bigg|\, q, \alpha\beta^2 q^n\right)$$

$$-\frac{e^{-i(n+1)\theta} (q, qe^{-2i\theta}, \beta; q)_\infty}{(\alpha\beta, q/(\alpha\beta), qe^{-2i\theta}/\beta; q)_\infty} {}_2\phi_1\left(\begin{array}{c} q/\beta, qe^{-2i\theta}/\beta \\ qe^{-2i\theta} \end{array}\bigg|\, q, \alpha\beta^2 q^n\right)$$

$$=\frac{e^{i(n+1)\theta} (q, qe^{2i\theta}, \beta e^{-2i\theta}; q)_\infty}{(q/\beta, \alpha\beta e^{2i\theta}, qe^{-2i\theta}/(\alpha\beta); q)_\infty} \left[{}_2\phi_1\left(\begin{array}{c} q/\beta, qe^{2i\theta}/\beta \\ qe^{2i\theta} \end{array}\bigg|\, q, \alpha\beta^2 q^n\right)\right.$$

$$-e^{-2i(n+1)\theta} \frac{(qe^{-2i\theta}, \beta, q/\beta, \alpha\beta e^{2i\theta}, qe^{-2i\theta}/(\alpha\beta); q)_\infty}{(qe^{2i\theta}, \beta e^{-2i\theta}, \alpha\beta, q/(\alpha\beta), qe^{-2i\theta}/\beta; q)_\infty}$$

$$\left.\times {}_2\phi_1\left(\begin{array}{c} q/\beta, qe^{-2i\theta}/\beta \\ qe^{-2i\theta} \end{array}\bigg|\, q, \alpha\beta^2 q^n\right)\right].$$

Apply the transformation (12.5.6) with $A = qe^{2i\theta}/\beta$, $B = q/\beta$, $C = qe^{2i\theta}$, $Z = \alpha\beta^2 q^n$ to simplify the right-hand side of the above equation to the right-hand side of (15.9.3). $\quad\square$

Corollary 15.9.2 *The function* $v_n(\theta; \beta, \alpha)$ *defined by*

$$v_n(\theta; \beta, \alpha) = \frac{\Phi_n(\theta; \beta, \alpha)}{\Phi_0(\theta; \beta, \alpha)}$$

$$= e^{in\theta} \frac{(\alpha\beta; q)_n}{(q\alpha; q)_n} \, {}_2\phi_1 \left(\begin{array}{c} q^{-n}/\alpha, \beta \\ q^{1-n}/(\alpha\beta) \end{array} \Big| \, q, \frac{q}{\beta} e^{-2i\theta} \right) \qquad (15.9.4)$$

satisfies the three-term recurrence relation (13.7.9).

When $\alpha = 1$ the extreme right-hand side of (15.9.4) reduces to the q-ultraspherical polynomial $C_n(\cos\theta; \beta \,|\, q)$. For $\alpha \neq 1$ it may not be a polynomial but, nevertheless, is a solution to (13.7.9).

The solution of (13.7.9) given in (15.9.4) has a restricted β domain. We give two other solutions of (13.7.9) which are defined for a wider domain of β. Unlike Φ_n, constructing these two solutions will not require the application of transformations of basic hypergeometric series. However we will need to verify the three-term recurrence relation for $n = 0$.

Let $y^n f(y, \theta)$ be the integrand in (15.9.2). We have already indicated that both $\int_0^{e^{\pm i\theta}} y^n f(y, \theta) d_q y$, for $n > 0$ satisfy the recurrence (13.7.9). Define $v_n^{\pm}(\theta; \alpha, \beta)$ by

$$v_n^{\pm}(\theta; \alpha, \beta) := e^{\pm(n+1)i\theta} \, {}_2\phi_1 \left(\begin{array}{c} q/\beta, qe^{\pm 2i\theta}/\beta \\ qe^{\pm 2i\theta} \end{array} \Big| \, q, \alpha\beta^2 q^n \right). \qquad (15.9.5)$$

This comes from the q-integral (15.9.2) on $[0, e^{\pm i\theta}]$. Both v_n^+ and v_n^- satisfy (13.7.9) for $n > 0$ and we will see later that they are linearly independent functions of n for $\theta \neq k\pi$, $k = 0, \pm 1, \dots$.

We now verify that v_n^+ and v_n^- satisfy (13.7.9) if $n = 0$. To do so assume

$$-1 < \alpha\beta^2/q < 1, \qquad (15.9.6)$$

so that $v_{-1}^{\pm}$ is well-defined. Upon reexamining the analysis preceding Theorem 15.9.1 we see that when $a = 0$, the boundary term in equation (11.4.6) will vanish if $ug(u)f(u/q) \to 0$ as $u \to 0$ if u is of the form ζq^m for fixed ζ and $m \to \infty$. In our case it suffices to prove that

$$\lim_{m \to \infty} \left(\beta e^{-i\theta} q^{-m}/\zeta; q\right)_\infty \big/ \left(qe^{-i\theta} q^{-m}/(\zeta\alpha\beta); q\right)_\infty = 0.$$

It is clear that the above limit is a bounded function times

$$\lim_{m \to \infty} \frac{\left(\beta e^{-i\theta} q^{-m}/\zeta; q\right)_m}{\left(qe^{-i\theta} q^{-m}/(\zeta\alpha\beta); q\right)_m} = \lim_{m \to \infty} \left(\alpha\beta^2/q\right)^m \frac{\left(q\zeta e^{i\theta}/\beta; q\right)_m}{\left(\alpha\beta\zeta e^{i\theta}/q; q\right)_m} = 0.$$

Note that (13.7.9)–(13.7.2) and (15.9.6) imply $C_{-1}^{(\alpha)}(x; \beta \,|\, q) = 0$.

One can directly verify that $v_n^{\pm}$ satisfy (13.7.9) by substituting the right-hand side of (15.9.5) in (13.7.9) and equating coefficients of various powers of α. In fact this shows that $v_n^{\pm}$ satisfies (13.7.9) for all n for which $\left|\alpha\beta q^{n-1}\right| < 1$. To relax this restriction we need to analytically continue the ${}_2\phi_1$ in (15.9.5) using the transformations (12.4.13), and (12.5.2)–(12.5.3).

We show that v_n^+ and v_n^- are linearly independent functions of n by showing that the Casorati determinant

$$\Delta_n = v_{n+1}^+(\theta; \beta, \alpha)v_n^-(\theta; \beta, \alpha) - v_n^+(\theta; \beta, \alpha)v_{n+1}^-(\theta; \beta, \alpha),$$

does not vanish. Recall that the Casorati determinant

$$\Delta_n := u_{n+1}v_n - v_{n+1}u_n$$

of two solutions u_n and v_n of a difference equation

$$a_n y_n = b_n y_{n+1} + c_n y_{n-1}, \tag{15.9.7}$$

has the property

$$\Delta_n = \Delta_m \prod_{k=m}^{n} [c_k/b_k]. \tag{15.9.8}$$

Equation (15.9.8) implies

$$\Delta_n = \frac{(q\alpha\beta^2; q)_{n-1}}{(q^3\alpha; q)_{n-1}} \Delta_1,$$

and since $e^{\mp i(n+1)\theta}v_n^\pm \to 1$ as $n \to \infty$, then we have $\Delta_n \to 2i\sin\theta$ as $n \to \infty$. Hence

$$\Delta_n = \frac{(\alpha q^{n+2}; q)_\infty}{(\alpha\beta^2 q^n; q)_\infty} 2i\sin\theta. \tag{15.9.9}$$

This confirms the linear independence of $v_n^\pm$ when $\theta \neq k\pi$. Note also that

$$\Delta_{-1} = \frac{(\alpha q; q)_\infty}{(\alpha\beta^2/q; q)_\infty} 2i\sin\theta. \tag{15.9.10}$$

Since both $v_n^\pm$ satisfy (13.7.9), there exists $A(\theta)$ and $B(\theta)$ such that

$$C_n^{(\alpha)}(\cos\theta; \beta \,|\, q) = A(\theta)v_n^+(\theta; \beta, \alpha) + B(\theta)v_n^-(\theta; \beta, \alpha).$$

To determine A and B use the initial conditions $C_{-1}^{(\alpha)} = 0$, and $C_0^{(\alpha)} = 1$ and (15.9.10). The result is

$$C_n^{(\alpha)}(\cos\theta; \beta \,|\, q) = \frac{(\alpha\beta^2/q; q)_\infty}{2i\sin\theta(\alpha q; q)_\infty}$$
$$\times \left[v_{-1}^-(\theta; \beta, \alpha)v_n^+(\theta; \beta, \alpha) - v_{-1}^+(\theta; \beta, \alpha)v_n^-(\theta; \alpha, \beta) \right]. \tag{15.9.11}$$

Formula (15.9.11) is Rahman and Tariq's result (Rahman & Tariq, 1997, (3.4)) but this proof is from (Ismail & Stanton, 2002).

Theorem 15.9.3 ((Ismail & Stanton, 2002)) *We have*

$$\sum_{n=0}^{\infty} \frac{(\lambda;q)_n}{(q;q)_n} C_{n+k}^{(\alpha)}(\cos\theta; \beta \mid q) t^n$$

$$= e^{ik\theta} \frac{(\lambda t e^{i\theta}, \alpha\beta^2/q; q)_\infty}{(1-e^{2i\theta})(te^{i\theta}, \alpha q; q)_\infty} \, {}_2\phi_1 \left(\begin{array}{c} q/\beta, qe^{-2i\theta}/\beta \\ qe^{-2i\theta} \end{array} \middle| q, \frac{\alpha\beta^2}{q} \right) \qquad (15.9.12)$$

$$\times {}_3\phi_2 \left(\begin{array}{c} q/\beta, qe^{2i\theta}/\beta, te^{i\theta} \\ qe^{2i\theta}, \lambda t e^{i\theta} \end{array} \middle| q, \alpha\beta^2 q^k \right)$$

+ *a similar term with θ replaced by* $-\theta$.

Proof This is a direct consequence of (15.9.11). $\qquad\square$

The cases $\lambda = q$ or $k = 0$ of Theorem 15.9.3 are in (Rahman & Tariq, 1997).

Theorem 15.9.4 ((Ismail & Stanton, 2002)) *We have the bilinear generating function*

$$\sum_{n=0}^{\infty} C_n (\cos\phi; \beta_1 \mid q) \, C_n^{(\alpha)}(\cos\theta; \beta \mid q) \, t^n$$

$$= \frac{(\alpha\beta^2/q, \beta_1 te^{i(\theta+\phi)}, \beta_1 te^{i(\theta-\phi)}; q)_\infty}{(1-e^{-2i\theta})(\alpha q, te^{i(\theta+\phi)}, te^{i(\theta-\phi)}; q)_\infty}$$

$$\times {}_2\phi_1 \left(\begin{array}{c} q/\beta, qe^{-2i\theta}/\beta \\ qe^{-2i\theta} \end{array} \middle| q, \frac{\alpha\beta^2}{q} \right) \qquad (15.9.13)$$

$$\times {}_4\phi_3 \left(\begin{array}{c} q/\beta, qe^{2i\theta}/\beta, te^{i(\theta+\phi)}, te^{i(\theta-\phi)} \\ qe^{2i\theta}, \beta_1 te^{i(\theta+\phi)}, \beta_1 te^{i(\theta-\phi)} \end{array} \middle| q, \alpha\beta^2 \right)$$

+ *a similar term with θ replaced by* $-\theta$.

Proof Multiply (15.9.11) by $C_n (\cos\phi; \beta_1 \mid q) t^n$ and add, then use the generating function (13.2.8). $\qquad\square$

We now give a Poisson-type kernel for the polynomials under consideration.

Theorem 15.9.5 ((Ismail & Stanton, 2002)) *A bilinear generating function for the associated continuous q-ultraspherical polynomials is given by*

$$\sum_{n=0}^{\infty} C_n^{(\alpha_1)} (\cos\phi; \beta_1 \mid q) \, C_n^{(\alpha)}(\cos\theta; \beta \mid q) \, t^n$$

$$= \frac{(1-\alpha_1)(\alpha\beta^2/q; q)_\infty}{(1-e^{-2i\theta})(\alpha q; q)_\infty} \, {}_2\phi_1 \left(\begin{array}{c} q/\beta, qe^{-2i\theta}/\beta \\ qe^{-2i\theta} \end{array} \middle| q, \frac{\alpha\beta^2}{q} \right)$$

$$\times \sum_{k=0}^{\infty} \frac{(q/\beta, qe^{2i\theta}/\beta; q)_k \, \alpha^k \beta^{2k}}{[1 - 2\cos\phi \, te^{i\theta} q^k + t^2 q^{2k} e^{2i\theta}](q, qe^{2i\theta}; q)_k} \qquad (15.9.14)$$

$$\times {}_3\phi_2 \left(\begin{array}{c} q, \beta_1 tq^k e^{i(\theta+\phi)} \beta_1 tq^k e^{i(\theta-\phi)} \\ q^{k+1} te^{i(\theta+\phi)}, q^{k+1} te^{i(\theta-\phi)} \end{array} \middle| q, \alpha_1 \right)$$

+ *a similar term with θ replaced by* $-\theta$.

The case $\alpha = \alpha_1$ of (15.9.13) is in (Rahman & Tariq, 1997).

15.10 Two Systems of Orthogonal Polynomials

We will study two systems of orthogonal polynomials introduced by Ismail and Rahman and arise when we introduce an association parameter in the recurrence relation of the Askey–Wilson polynomials. All the results in this section are from (Ismail & Rahman, 1991).

Consider the three-term recurrence relation

$$2xy_\alpha(x) = A_\alpha y_{\alpha+1}(x) + B_\alpha y_\alpha(x) + C_\alpha y_{\alpha-1}(x), \qquad (15.10.1)$$

where

$$A_\alpha = \frac{(1 - t_1 t_2 t_3 t_4 q^{\alpha-1}) \prod_{k=2}^{4} (1 - t_1 t_k q^\alpha)}{t_1 (1 - t_1 t_2 t_3 t_4 q^{2\alpha-1}) (1 - t_1 t_2 t_3 t_4 q^{2\alpha})}, \qquad (15.10.2)$$

$$C_\alpha = \frac{t_1 (1 - q^\alpha) \prod_{2 \le j < k \le 4} (1 - t_j t_k q^{\alpha-1})}{(1 - t_1 t_2 t_3 t_4 q^{2\alpha-2}) (1 - t_1 t_2 t_3 t_4 q^{2\alpha-1})}, \qquad (15.10.3)$$

$$B_\alpha = t_1 + t_1^{-1} - A_\alpha - C_\alpha. \qquad (15.10.4)$$

In this notation when $x \to -x - t_1 - 1/t_1$ and $\alpha \to n + \alpha$ (15.10.1) becomes a three-term recurrence relation for birth and death process polynomials.

Throughout this section we choose z such that

$$x := \frac{1}{2} [z + 1/z], \quad |z| \le 1. \qquad (15.10.5)$$

Theorem 15.10.1 *Let $x \in \mathbb{C} \setminus [-1, 1]$. Then the functions*

$$r_\alpha(z) := \frac{(t_2 t_3 t_4 q^\alpha / z; q)_\infty \prod_{k=2}^{4} (t_1 t_k q^\alpha; q)_\infty}{(t_1 z q^\alpha; q)_\infty \prod_{2 \le j < k \le 4} (t_j t_k q^\alpha; q)_\infty} \left(\frac{a}{z}\right)^\alpha \qquad (15.10.6)$$

$$\times \, _8W_7 \left(t_2 t_3 t_4 / q z; t_2/z, t_3/z, t_4/z, t_1 t_2 t_3 t_4 q^{\alpha-1}, q^{-\alpha}; q, q z/t_1\right),$$

and

$$s_\alpha(z) := \frac{(t_1 t_2 t_3 t_4 q^{2\alpha}, t_2 t_3 t_4 z q^\alpha; q)_\infty \prod_{k=2}^{4} (t_k z q^{\alpha+1}; q)_\infty}{(t_2 t_2 t_4 z q^{2\alpha+1}, q^{\alpha+1}; q)_\infty \prod_{2 \le j < k \le 4} (t_j t_k q^\alpha; q)_\infty} (az)^\alpha \qquad (15.10.7)$$

$$\times \, _8W_7 \left(t_2 t_3 t_4 z q^{2\alpha}; t_2 t_3 q^\alpha, t_2 t_4 q^\alpha, t_3 t_4 q^\alpha, q^{\alpha+1}, z q/t_1; q, t_1 z\right),$$

are linear independent solutions of (15.10.1).

Proof We only outline the idea of the proof. The long details are in (Ismail & Rahman, 1991). To prove that r_α and s_α satisfy (15.10.1), one derives contiguous relations for both functions then show that (15.10.1) is a common contiguous relation.

In view of $|z| < 1$ then

$$\lim_{\alpha \to +\infty} (z/t_1)^\alpha \, r_\alpha(z) = {}_6W_5 \left(t_2 t_3 t_4/qz; t_2/z, t_3/z, t_4/z; q, z^2 \right)$$

$$= \frac{(t_2 t_3 t_4/z; t_2 z, t_3 z, t_4 z; q)_\infty}{(t_2 t_3, t_2 t_4, t_3 t_4; q)_\infty}, \tag{15.10.8}$$

and

$$\lim_{\alpha \to +\infty} (z t_1)^{-\alpha} \, s_\alpha(z) = {}_1\phi_0 \left(za/t_1; q, t_1 z \right) = \frac{(z^2; q)_\infty}{(t_1 z; q)_\infty}, \tag{15.10.9}$$

provided that $|t_1 z| < 1$. The linear independence of r_α and s_α now follows since the functions have different asymptotics as $\alpha \to +\infty$. $\qquad\square$

The functions r_α and s_α have become known as the Ismail–Rahman functions.

The difference equation (15.10.1) with $\alpha \to n + \alpha$ becomes a three-term recurrence relation for a family of birth and death process polynomials. Let

$$z_n^\alpha(x) = t_1^{-n} \prod_{k=2}^{4} (1 - t_1 t_k q^\alpha)_n \, y_{n+\alpha}(x). \tag{15.10.10}$$

The functional relation (15.10.1) becomes

$$2x z_n^{(\alpha)}(x) = A_n^{(\alpha)} \, z_{n+1}^{(\alpha)}(x) + B_n^{(\alpha)} \, z_n^{(\alpha)}(x) + C_n^{(\alpha)} \, z_{n-1}^{(\alpha)}(x), \tag{15.10.11}$$

with

$$A_n^{(\alpha)} = \frac{1 - t_1 t_2 t_3 t_4 q^{n+\alpha-1}}{\left(1 - t_1 t_2 t_3 t_4 q^{2n+2\alpha-1}\right)\left(1 - t_1 t_2 t_3 t_4 q^{2n+2\alpha}\right)}, \tag{15.10.12}$$

$$C_n^{(\alpha)} = \frac{(1 - q^{n+\alpha}) \prod_{1 \le j < k \le 4} \left(1 - t_j t_k q^{n+\alpha-1}\right)}{\left(1 - t_1 t_2 t_3 t_4 q^{2n+2\alpha-2}\right)\left(1 - t_1 t_2 t_3 t_4 q^{2n+2\alpha-1}\right)}, \tag{15.10.13}$$

and

$$B_n^{(\alpha)} = t_1 + t_1^{-1} - A_n^{(\alpha)} t_1^{-1} \prod_{j=2}^{4} \left(1 - t_1 t_j q^{n+\alpha}\right)$$

$$- \frac{t_1 C_n^{(\alpha)}}{\prod_{2 \le j < k \le 4} \left(1 - t_j t_k q^{n+\alpha-1}\right)}. \tag{15.10.14}$$

Remark 5.2.1 suggests two polynomial solutions, say $\left\{ p_n^{(\alpha)} \right\}$ and $\left\{ q_n^{(\alpha)} \right\}$ of (15.10.11) having the initial conditions

$$p_0^{(\alpha)}(x; \mathbf{t} \,|\, q) = 1, \quad p_1^{(\alpha)}(x; \mathbf{t} \,|\, q) = \left[2x - B_0^{(\alpha)} \right] / A_0^{(\alpha)},$$

$$q_0^{(\alpha)}(x; \mathbf{t} \,|\, q) = 1, \quad q_1^{(\alpha)}(x; \mathbf{t} \,|\, q) = \left[2x - \tilde{B}_0^{(\alpha)} \right] / A_0^{(\alpha)} \tag{15.10.15}$$

with

$$\tilde{B}_0^{(\alpha)} := t_1 + t_1^{-1} - A_0^{(\alpha)} t_1^{-1} \prod_{j=2}^{4} \left(1 - t_1 t_j q^\alpha\right). \tag{15.10.16}$$

To simplify the writing we used the simplified notation $p_n^{(\alpha)}(x; \mathbf{t} \mid q)$ and $q_n^{(\alpha)}(x; \mathbf{t} \mid q)$ unless there is a need to exhibit the dependence on the parameters. Ismail and Rahman established the representations

$$p_n^{(\alpha)}(x; \mathbf{t} \mid q) = z t_1^{1-2\alpha-n} \frac{\prod\limits_{2 \leq j < k \leq 4} (t_j t_k; q)_\infty}{\prod\limits_{k=2}^{4} (t_j t_k q^{\alpha+n}, t_k z; q)_\infty}$$
$$\times \frac{(q^\alpha, t_1 z; q)_\infty (t_2 t_4 q^{\alpha-1}, t_3 t_4 q^{\alpha-1}; q)_\infty}{(1 - t_1 t_2 t_3 t_4 q^{2\alpha-2})(t_2 t_3 t_4/z; q)_\infty}$$
$$\times \{s_{\alpha-1}(z) r_{n+\alpha}(z) - r_{\alpha-1}(z) s_{n+\alpha}(z)\} \tag{15.10.17}$$

and

$$q_n^{(\alpha)}(x; \mathbf{t} \mid q) = z t_1^{1-2\alpha-n} \frac{\prod\limits_{2 \leq j < k \leq 4} (t_j t_k; q)_\infty}{\prod\limits_{k=2}^{4} (t_j t_k q^{\alpha+n}, t_k z; q)_\infty}$$
$$\times \frac{(q^\alpha, t_1 z; q)_\infty (t_2 t_4 q^{\alpha-1}, t_3 t_4 q^{\alpha-1}; q)_\infty}{(1 - t_1 t_2 t_3 t_4 q^{2\alpha-2})(t_2 t_3 t_4/z; q)_\infty}$$
$$\times \{(s_{\alpha-1}(z) - s_\alpha(z)) r_{n+\alpha}(z) - (r_{\alpha-1}(z) - r_\alpha(z)) s_{n+\alpha}(z)\} \tag{15.10.18}$$

Using transformation theory of basic hypergeometric series, Ismail and Rahman found the following compact representations

$$p_n^{(\alpha)}(\cos\theta, \mathbf{t} \mid q) = (t_1 t_2 q^\alpha, t_1 t_3 q^\alpha, t_1 t_4 q^\alpha; q)_n \, t_1^{-n}$$
$$\times \sum_{k=0}^{n} \frac{(q^{-n}, t_1 t_2 t_3 t_4 q^{2\alpha+n-1}, t_1 t_2 t_3 t_4 q^{2\alpha-1}, t_1 e^{i\theta}, t_2 e^{-i\theta}; q)_k}{(q, t_1 t_2 q^\alpha, t_1 t_3 q^\alpha, t_1 t_4 q^\alpha, t_1 t_2 t_3 t_4 q^{\alpha-1}; q)_\infty} q^k \tag{15.10.19}$$
$$\times {}_8 W_7 \left(t_1 t_2 t_3 t_4 q^{2\alpha+k-2}; q^\alpha, t_2 t_3 q^{\alpha-1}, t_2 t_4 q^{\alpha-1}, t_3 t_4 q^{\alpha-1}, \right.$$
$$\left. q^{k+1}, q^{k-n}, t_1 t_2 t_3 t_4 q^{2\alpha+n+k-1}; q, t_1^2\right),$$

and

$$q_n^{(\alpha)}(\cos\theta, \mathbf{t} \mid q) = (t_1 t_2 q^\alpha, t_1 t_3 q^\alpha, t_1 t_4 q^\alpha; q)_n \, t_1^{-n}$$
$$\times \sum_{k=0}^{n} \frac{(q^{-n}, t_1 t_2 t_3 t_4 q^{2\alpha+n-1}, t_1 t_2 t_3 t_4 q^{2\alpha-1}, t_1 e^{i\theta}, t_2 e^{-i\theta}; q)_k}{(q, t_1 t_2 q^\alpha, t_1 t_3 q^\alpha, t_1 t_4 q^\alpha, t_1 t_2 t_3 t_4 q^{\alpha-1}; q)_\infty} q^k \tag{15.10.20}$$
$$\times {}_8 W_7 \left(t_1 t_2 t_3 t_4 q^{2\alpha+k-2}; q^\alpha, t_2 t_3 q^{\alpha-1}, t_2 t_4 q^{\alpha-1}, t_3 t_4 q^{\alpha-1}, \right.$$
$$\left. q^k, q^{k-n}, t_1 t_2 t_3 t_4 q^{2\alpha+n+k-1}; q, q t_1^2\right).$$

Formulas (15.10.7)–(15.10.8) and a Wronskian-type formula can be used to establish the limiting relations

$$\lim_{n \to \infty} z^n p_n^{(\alpha)}(x; \mathbf{t} \mid q) = (t_1 z)^{1-\alpha} s_{\alpha-1}(z)$$
$$\times \frac{(q^\alpha, t_1 z; q)_\infty \prod\limits_{2 \leq i < j \leq 4} (t_i t_j q^{\alpha-1}; q)_\infty}{(1 - t_1 t_2 t_3 t_4 q^{2\alpha-2})(t_1 t_2 t_3 t_4 q^{\alpha-1}, z^2; q)_\infty}, \tag{15.10.21}$$

and

$$\lim_{n\to\infty} z^n q_n^{(\alpha)}(x;\mathbf{t}\,|\,q) = (t_1 z)^{1-\alpha}\left[s_{\alpha-1}(z) - s_\alpha(z)\right]$$

$$\times \frac{(q^\alpha, t_1 z; q)_\infty \prod\limits_{2\le i<j\le 4}\left(t_i t_j q^{\alpha-1}; q\right)_\infty}{(1 - t_1 t_2 t_3 t_4 q^{2\alpha-2})\left(t_1 t_2 t_3 t_4 q^{\alpha-1}, z^2; q\right)_\infty}, \tag{15.10.22}$$

for $|t_1 z| < 1$.

Theorem 15.10.2 *Let $\mu^{(1)}(x;\mathbf{t},\alpha)$ and $\mu^{(2)}(x;\mathbf{t},\alpha)$ be the probability measuress with respect to which $p_n^{(\alpha)}$ and $p_n^{(\alpha)}$ are orthogonal. Then*

$$\int_{\mathbb{R}} \frac{d\mu^{(1)}(y;\mathbf{t},\alpha)}{x - y}$$

$$= \frac{2z\left(t_2 t_3 t_4 z q^{2\alpha-1}; q\right)_2}{(1 - t_2 t_3 t_4 z q^{\alpha-1})\prod\limits_{k=2}^{4}(1 - t_k z q^\alpha)}$$

$$\times \frac{{}_8W_7\left(t_2 t_3 t_4 z q^{2\alpha}; t_2 t_3 q^\alpha, t_2 t_4 q^\alpha, t_3 t_4 q^\alpha, q^{\alpha+1}, zq/t_1; q, t_1 z\right)}{{}_8W_7\left(t_2 t_3 t_4 z q^{2\alpha-2}; t_2 t_3 q^{\alpha-1}, t_2 t_4 q^{\alpha-1}, t_3 t_4 q^{\alpha-1}, q^\alpha, qz/t_1; q, t_1 z\right)}, \tag{15.10.23}$$

and

$$\int_{\mathbb{R}} \frac{d\mu^{(2)}(y;\mathbf{t},\alpha)}{x - y}$$

$$= \frac{2z\left(t_2 t_3 t_4 z q^{2\alpha-1}; q\right)_2}{(1 - t_2 t_3 t_4 z q^{\alpha-1})\prod\limits_{k=2}^{4}(1 - t_k z q^\alpha)}$$

$$\times \frac{{}_8W_7\left(t_2 t_3 t_4 z q^{2\alpha}; t_2 t_3 q^\alpha, t_2 t_4 q^\alpha, t_3 t_4 q^\alpha, q^{\alpha+1}, zq/t_1; q, t_1 z\right)}{{}_8W_7\left(t_2 t_3 t_4 z q^{2\alpha-2}; t_2 t_3 q^{\alpha-1}, t_2 t_4 q^{\alpha-1}, t_3 t_4 q^{\alpha-1}, q^\alpha, z/t_1; q, qt_1 z\right)}, \tag{15.10.24}$$

which are valid in the complex x-plane cut along $[-1, 1]$.

Proof It readily follows that

$$\left[p_n^{(\alpha)}(x;\mathbf{t}\,|\,q)\right]^* = \left[p_n^{(\alpha)}(x;\mathbf{t}\,|\,q)\right]^* = \frac{2}{A_0^{(\alpha)}}\, p_{n-1}^{(\alpha+1)}(x;\mathbf{t}\,|\,q). \tag{15.10.25}$$

The asymptotic formulas (15.10.8)–(15.10.9) and Markov's theorem, Theorem 2.6.2, establish the theorem from the fact that

$$(t_1 z)^{1-\alpha}\left[s_{\alpha-1}(z) - s_\alpha(z)\right]$$

$$= \frac{\left(t_1 t_2 t_3 t_4 q^{2\alpha-2}\right)\left(1 - t_2 t_3 t_4 z q^{\alpha-1}\right)\prod\limits_{k=2}^{4}\left(t_k z q^\alpha; q\right)_\infty}{(q^\alpha, t_2 t_3 t_4 z q^{2\alpha-1}; q)_\infty \prod\limits_{2\le i<j\le 4}\left(t_i t_j q^{\alpha-1}; q\right)_\infty}$$

$$\times {}_8W_7\left(t_2 t_3 t_4 z q^{2\alpha-2}; t_2 t_3 q^{\alpha-1}, t_2 t_4 q^{\alpha-1}, t_3 t_4 q^{\alpha-1}, q^\alpha, z/t_1; q, qt_1 z\right). \tag{15.10.26}$$

The proof of (15.10.26) involves technical q-series transformations. The details are in (Ismail & Rahman, 1991). $\square$

Theorem 15.10.3 *The absolutely continuous components of $\mu^{(1)}$ and $\mu^{(2)}$ are given by*

$$
\frac{d\mu^{(1)}(\cos\theta; \mathbf{t}, \alpha)}{d\theta} = \left(1 - t_1 t_2 t_3 t_4 q^{2\alpha-2}\right) \left(t_1 t_2 t_3 t_4 q^{2\alpha-2}; q\right)_\infty
$$

$$
\times \frac{\left(q^{\alpha+1}; q\right)_\infty \displaystyle\prod_{1 \leq j < k \leq 4} (t_j t_k q^\alpha; q)_\infty}{2\pi \left(1 - t_1 t_2 t_3 t_4 q^{\alpha-2}\right) \left(t_1 t_2 t_3 t_4 q^{\alpha-2}, t_1 t_2 t_3 t_4 q^{2\alpha}; q\right)_\infty}
$$

$$
\times \frac{\left(e^{2i\theta}, e^{-2i\theta}, q^{\alpha+1} e^{2i\theta}, q^{\alpha+1} e^{-2i\theta}; q\right)_\infty}{(q e^{2i\theta}, q e^{-2i\theta}; q)_\infty \displaystyle\prod_{k=1}^{4} (t_k e^{i\theta}, t_k e^{-i\theta}; q)_\infty}
$$

$$
\times \left| {}_8 W_7(q^\alpha e^{2i\theta}; q e^{i\theta}/t_1, q e^{i\theta}/t_2, q e^{i\theta}/t_3, q e^{i\theta}/t_4, q^\alpha; q, t_1 t_2 t_3 t_4 q^{\alpha-2}) \right|^2,
$$

$$\tag{15.10.27}$$

and

$$
\frac{d\mu^{(2)}(\cos\theta; \mathbf{t}, \alpha)}{d\theta} = \frac{\left(q^{\alpha+1}; q\right)_\infty \displaystyle\prod_{1 \leq j < k \leq 4} (t_j t_k q^\alpha; q)_\infty}{2\pi \left(t_1 t_2 t_3 t_4 q^{2\alpha}; q\right)_\infty}
$$

$$
\times \frac{\left(t_1 t_2 t_3 t_4 q^{2\alpha-1}; q\right)_\infty}{\left(t_1 t_2 t_3 t_4 q^{\alpha-1}; q\right)_\infty} \frac{1 - 2 t_1 x q^\alpha + t_1^2 q^{2\alpha}}{1 - 2 t_1 x + t_1^2}
$$

$$
\times \frac{\left(e^{2i\theta}, e^{-2i\theta}, q^{\alpha+1} e^{2i\theta}, q^{\alpha+1} e^{-2i\theta}; q\right)_\infty}{(q e^{2i\theta}, q e^{-2i\theta}; q)_\infty \displaystyle\prod_{k=1}^{4} (t_k e^{i\theta}, t_k e^{-i\theta}; q)_\infty}
$$

$$
\times \left| {}_8 W_7(q^\alpha e^{2i\theta}; e^{i\theta}/t_1, q e^{i\theta}/t_2, q e^{i\theta}/t_3, q e^{i\theta}/t_4, q^\alpha; q, t_1 t_2 t_3 t_4 q^{\alpha-1}) \right|^2,
$$

$$\tag{15.10.28}$$

respectively.

Two proofs are in (Ismail & Rahman, 1991). One uses the Perron–Stieltjes inversion formula (1.2.8)–(1.2.9); the other develops the large n asymptotics on $(-1, 1)$, then applies Theorem 11.2.2. The technical details are too long to be reproduced here.

Exercises

15.1   Consider the integral (Ismail & Stanton, 1988)

$$
\mathcal{J}(t_1, t_2, t_3, t_4) := \frac{\left(q, \beta^2; q\right)_\infty}{2\pi(\beta, \beta; q)_\infty}
$$

$$
\times \int_0^\pi \prod_{j=1}^{4} \frac{\left(\beta t_j e^{i\theta}, \beta t_j e^{-i\theta}; q\right)_\infty}{\left(t_j e^{i\theta}, t_j e^{-i\theta}; q\right)_\infty} \frac{\left(e^{2i\theta}, e^{-2i\theta}; q\right)_\infty}{\left(\beta e^{2i\theta}, \beta e^{-2i\theta}; q\right)_\infty} \, d\theta.
$$

(a) Prove that

$$\mathcal{J}\left(\rho e^{i\phi}, \rho e^{-i\phi}, \sigma e^{i\psi}, \sigma e^{-i\psi}\right)$$

$$= \sum_{n=0}^{\infty} \frac{(q, \beta^2; q)_n}{(\beta; q)_{n+1}(\beta; q)_n} (\rho\sigma)^n C_n(\cos\phi; \beta \,|\, q) C_n(\cos\psi; \beta \,|\, q)$$

$$\times {}_2\phi_1\left(\begin{matrix} \beta, \beta^2 q^n \\ \beta q^{n+1} \end{matrix}\,\middle|\, q, \rho^2\right) {}_2\phi_1\left(\begin{matrix} \beta, \beta^2 q^n \\ \beta q^{n+1} \end{matrix}\,\middle|\, q, \sigma^2\right),$$

for $|\rho| < 1$, $|\sigma| < 1$, $\beta \in (-1, 1)$, $|t_j| < 1$, $1 \le j \le 4$.

Note: Integral $\mathcal{J}$ generalizes the Askey–Wilson integral and reduces it when $\beta = 0$.

(b) Find the eigenvalues and eigenfunctions of the integral operator

$$(Tf)(x) = \int_0^{\pi} \mathcal{J}\left(\rho e^{i\theta}, \rho e^{-i\theta}, \sigma e^{i\phi}, \sigma e^{-i\phi}\right)$$

$$\times w(\cos\phi \,|\, \beta) f(\cos\phi) \, d\phi,$$

$x = \cos\theta \in (-1, 1)$, $f \in L^2(w(x \,|\, \beta), -1, 1)$, and

$$w(\cos\phi \,|\, \beta) = \frac{\left(e^{2i\phi}, e^{-2i\phi}; q\right)_\infty}{\left(\beta e^{2i\phi}, \beta e^{-2i\phi}; q\right)_\infty},$$

by first proving that $\mathcal{J}$ is a Hilbert–Schmidt kernel (Tricomi, 1957). The details are in (Ismail & Stanton, 1988).

(c) Prove that the identity in (a) is the expansion in Mercer's theorem, (Tricomi, 1957).

15.2 Let $[n] = (1 - q^n)/(1 - q)$. A version of the q-Charlier polynomials may be defined by the three-term recurrence

$$C_{n+1}(x; a, q) = (x - aq^n - [n]_q) C_n(x; a, q) - a[n]_q C_{n-1}(x; a, q),$$
$$C_0(x; a, q) = 1, \quad C_{-1}(x; a, q) = 0.$$

(a) Establish the generating function

$$\sum_{n=0}^{\infty} C_n(x; a, q) \frac{t^n}{(q; q)_n} = \frac{(t(1-a)/(1-q); q)_\infty}{(bte^{i\theta}, bte^{-i\theta}; q)_\infty},$$

where $b = \sqrt{a/(1-q)}$ and $x = (1-q)^{-1} + 2b\cos\theta$.

(b) Show that

$$C_n(x; a, q) = \sum_{k=0}^{n} \begin{bmatrix} n \\ k \end{bmatrix}_q (-a)^{n-k} q^{\binom{n-k}{2}} \prod_{i=0}^{k-1} (x - [i]_q).$$

(c) Recall that the unsigned Stirling numbers of the first kind $|s(n, k)|$

count the number of permutations in S_n with exactly k cycles, and

$$x(x-1)\cdots(x-n+1) = \sum_{k=1}^{n} |s(n,k)|x^k(-1)^{n-k}$$

$$= \sum_{\sigma \in S_n} x^{\# \text{ cycles}(\sigma)}(-1)^{n-\# \text{ cycles}}.$$

(E15.1)

By defining an appropriate q-version of (E15.1) give a combinatorial interpretation for $C_n(x; a, q)$.

(d) The moments of the Charlier polynomials are $\mu_n = \sum_{k=1}^{n} S(n,k)a^k$. Show that the moments of the q-Charlier polynomials are

$$\mu_n(q) = \sum_{k=1}^{n} S_q(n,k)a^k,$$

for an appropriately defined set of q-Stirling numbers of the second kind which satisfy

$$S_q(n,k) = S_q(n-1, k-1) + [k]_q S_q(n-1, k).$$

(**Possible Hint:** Find an orthogonality relation for q-Stirling numbers of the first kind (which you have in (c)) and the second kind (defined above) which implies orthogonality of $C_n(x; a, q)$ to $C_0(x; a, q)$. Or use part (e).)

(e) Recall that an RG-function is a word w such that if $i+1$ occurs in w, then i must occur to the left $i+1$ in w. For example, 112321341 is an RG-function, but 1122423 is not. Also recall that there is a bijection between RG-functions of length n whose entries are exactly $1, 2, \ldots, k$ denoted $RG(n, k)$ and set partitions of $[n]$ using weighted Motzkin paths (Viennot, 1983), show that

$$\mu_n(q) = \sum_{k=1}^{n} \sum_{w \in RG(n,k)} a^k q^{rs(w)} = \sum_{k=1}^{n} S_q(n,k)a^k,$$

where $rs(w) = $ "$right - smaller(w)$." $rs(w)$ is computed in the following way: for each entry $w_i \in w$ find the cardinality of $\{j : j < w_i, j$ occurs to the right of $w_i\}$. Then add all values to find $rs(w)$. For example, if $w = 1213114221$, then $rs(w) = 0 + 1 + 0 + 2 + 0 + 0 + 3 + 1 + 1$.

(f) Show that the RG-statistic "lb = left-bigger" is equidistributed with rs.

(g) Write down the continued fraction which is the moment generating function.

(h) By considering the Al-Salam–Carlitz polynomials which appear in Chapter 18, find an explicit representing measure for $C_n(x; a, q)$ when $a > 0, 0 < q < 1$. What happens if $q \to 1$?

(i) What q-version of no-crossing set partitions gives a nice q-Catalan as moments?

15.3   Another q-analogue of Charlier polynomials may be recursively defined by

$$p_0(x) = 1, \quad p_{-1}(x) = 0,$$

$$p_{n+1}(x) = (x - (a + [n]_q))\, p_n(x) - a[n]_q\, p_{n-1}(x).$$

(a) Show that $p_n(x)$ is the Al-Salam–Chihara polynomial

$$p_n(x) = \left(\frac{a}{1-q}\right)^{n/2}$$

$$\times Q_n\left(\frac{1}{2}\sqrt{\frac{1-q}{a}}\,(x - a - 1/(1-q))\,;\, \frac{-1}{\sqrt{a(1-q)}}, 0\,\Big|\, q\right).$$

(b) Establish the generating function

$$\sum_{n=0}^{\infty} p_n(x)\,\frac{t^n}{n!_q} = \frac{(-t;q)_\infty}{\left(\sqrt{a(1-q)}\, te^{i\theta},\, \sqrt{a(1-q)}\, te^{-i\theta}; q\right)_\infty},$$

where

$$\cos\theta = \frac{1}{2}\sqrt{\frac{1-q}{a}}\,(x - a - 1/(1-q)).$$

(c) Using (b), or otherwise, derive the explicit representation

$$p_n(x) = \sum_{k=0}^{n}\begin{bmatrix} n \\ k \end{bmatrix}_q (-a)^{n-k} q^{k(k-n)} \prod_{i=0}^{k-1}\{x - [i]_q + (aq^{-i} - 1)\}.$$

(d) Using the Askey–Wilson integral, show that

$$p_{n_1}(x)\, p_{n_2}(x) = \sum_{n_3=0}^{n_1+n_2} c\,(n_1, n_2, n_3)\, p_{n_3}(x),$$

where

$$c\,(n_1, n_2, n_3) = \sum_{\ell=\max(0, n_1-n_3, n_2-n_3)}^{(n_1+n_2-n_3)/2}$$

$$\frac{n_1!_q n_2!_q a^\ell q^{\left(\frac{n_1+n_2-n_3-2\ell}{2}\right)}/\ell!_q}{(n_3 - n_1 + \ell)!_q\,(n_3 - n_2 + \ell)!_q\,(n_1 + n_2 - n_3 - 2\ell)!_q},$$

and $k!_q = (q;q)_k/(1-q)^k$.

(e) Conclude that the linearization coefficient $c\,(n_1, n_2, n_3)$ is polynomial in a and q with positive integer coefficients. Anshelevich gave a combinatorial interpretation for these coefficients in (Anshelevich, 2005).

(f) Let μ be the orthogonality measure of $\{p_n(x)\}$. Show that $\mu = \mu_{ac} + \mu_s$, where μ_{ac} is absolutely continuous and supported on

$$\left[a + \frac{1}{1-q} - 2\sqrt{\frac{a}{1-q}},\, a + \frac{1}{1-q} + 2\sqrt{\frac{a}{1-q}}\right],$$

and μ_s is a discrete measure.

(g) If $a(1-q) \le 1$, then μ_s has a finite discrete part whose support is $\{x_n : n = 0, 1, \ldots, m\}$, where

$$m = \max\left\{n : a(1-q) \le q^{2n}\right\},$$

$$x_n = \frac{1}{2}\left[\sqrt{a(1-q)}\, q^{-n} + q^n / \sqrt{a(1-q)}\right].$$

Determine all the masses in this case. Also prove that $\mu_s = 0$ if $a(1-q) > 1$.

(h) Show that μ converges in the weak star topology to the normalized orthogonality measure of Charlier polynomials as in (6.1.21), as $q \to 1^-$.

15.4   Prove that (Ismail et al., 1987)

$$\frac{(q;q)_\infty}{2\pi} \int_0^\pi \frac{\left(e^{2i\theta}, e^{-2i\theta}; q\right)_\infty\, d\theta}{\prod\limits_{j=1}^{5}\left(t_j e^{i\theta}, t_j e^{-i\theta}; q\right)_\infty}$$

$$= \frac{(t_1 t_2 t_3 t_5, t_2 t_3 t_4 t_5, t_1 t_4; q)_\infty}{\prod\limits_{1 \le j < k \le 5}(t_j t_k; q)_\infty}\, {}_3\phi_2\left(\begin{matrix} t_2 t_3, t_2 t_5, t_3 t_5 \\ t_1 t_2 t_3 t_5, t_2 t_3 t_4 t_5 \end{matrix}\ \middle|\ q, t_1 t_4\right).$$

Note that the left-hand side is invariant under $t_i \leftrightarrow t_j$, but the form of the right-hand side is only invariant under $t_1 \leftrightarrow t_4$, $t_2 \leftrightarrow t_3$, $t_2 \leftrightarrow t_5$. The invariance under $t_1 \leftrightarrow t_2$, for example, leads to a transformation formula.

15.5   Show that (15.2.14) tends to (9.6.3) as $q \to 1$.

15.6   Evaluate $P_n(x; \mathbf{t}\,|\,q)$ at the x values $(t_j + 1/t_j)/2$, $j = 1, 2, 3, 4$.

15.7   Let $w(x; t_1, t_2\,|\,q)$ be the weight function of the Al-Salam–Chihara polynomials as in (15.1.2). Define a probability measure μ by

$$d\mu(x, t_1, t_2) = \frac{(t_1 t_2, q; q)_\infty}{2\pi}\, w(x; t_1, t_2\,|\,q)\, dx.$$

(a) Prove that

$$\int_{-1}^{1} H_n(x\,|\,q)\, d\mu\left(x; te^{i\phi}, te^{-i\phi}\right) = t^n H_n(\cos\phi\,|\,q).$$

(b) Prove that if part (a) holds for a probability measure μ and $t > 0$, $\phi \in [-\pi, \pi]$, then μ must be as given above, (Bryc et al., 2005).

(c) Find all the eigenvalues and eigenfunctions of the integral operator

$$(Tf)(y) = \int_{-1}^{1} f(x)\, d\mu\left(x; te^{i\phi}, te^{-i\phi}\right),$$

with $y = \cos\phi$.

15.8   Let

$$M_n(x \mid q) = \begin{vmatrix} H_0(x \mid q) & H_1(x \mid q) & \cdots & H_{n-1}(x \mid q) \\ H_1(x \mid q) & H_2(x \mid q) & \cdots & H_n(x \mid q) \\ \vdots & \vdots & & \vdots \\ H_{n-1}(x \mid q) & H_n(x \mid q) & \cdots & H_{2n-2}(x \mid q) \end{vmatrix}.$$

Prove that $M_{n+1} = \left[(-1)^n q^{n(n-1)/2} (q;q)_n / (1-q^n) \right] M_n$, hence evaluate M_n, (Bryc et al., 2005).

15.9   Prove that $q^{-(2\alpha+1)n/4} P_n^{(\alpha,\beta)}(x \mid q)$ is symmetric in α and β. Write this as a ${}_4\phi_3$ transformation.

15.10   Find all values of x for which $P_n^{(\alpha,\beta)}(x \mid q)$ can be evaluated in closed form.

15.11   Prove that (Ismail & Stanton, 2003b)

$$\sum_{n=0}^{\infty} \frac{H_{2n}(\cos\theta \mid q)}{(q^2;q^2)_n} t^n = \frac{(-t;q)_\infty}{(te^{2i\theta}, te^{-2i\theta}; q^2)_\infty},$$

$$\sum_{n=0}^{\infty} \frac{H_n(\cos\theta \mid q^2)}{(q;q)_n} t^n = \frac{(qt^2;q^2)_\infty}{(te^{i\theta}, te^{-i\theta}; q)_\infty}.$$

Hint: Expand the right-hand sides in series of q-Hermite polynomials using orthogonality.

15.12   Show that

$$\lim_{q \to 1^-} p_n \left(2 + x(1-q); q^{(\alpha+1)/2}, q^{(\alpha+1)/2} \right)$$

is a multiple of a Laguerre polynomial $L_n^{(\alpha)}(x)$.

16

The Askey–Wilson Operators

In this chapter we develop a calculus for the Askey–Wilson operator $\mathcal{D}_q$. In addition to an inner product, two basic ingredients are needed. They are an analogue of integration by parts, and a concept of an indefinite integral or a right inverse to $\mathcal{D}_q$. These will be developed along with an analogue of a Sturm–Liouville theory of second-order Askey–Wilson operators.

16.1 Basic Results

We shall use the inner product associated with the Chebyshev weight $\left(1 - x^2\right)^{-1/2}$ on $(-1, 1)$, namely

$$\langle f, g \rangle := \int_{-1}^{1} f(x) \, \overline{g(x)} \, \frac{dx}{\sqrt{1 - x^2}}. \tag{16.1.1}$$

Recall the definitions of $\check{f}$ and the Askey–Wilson operator $\mathcal{D}_q$ as per (12.1.9), (12.1.10) and (12.1.12). Observe that the definition (12.1.9) requires $\check{f}(z)$ to be defined for $\left|q^{\pm 1/2} z\right| = 1$ as well as for $|z| = 1$. In particular $\mathcal{D}_q$ is well-defined on $H_{1/2}$, where

$$H_\nu := \left\{ f : f((z + 1/z)/2) \text{ is analytic for } q^\nu \le |z| \le q^{-\nu} \right\}. \tag{16.1.2}$$

The operator $\mathcal{D}_q$ is well-defined on polynomials and we shall see that on the Askey–Wilson polynomials it plays the role played by $\frac{d}{dx}$ on the classical polynomials of Jacobi, Hermite and Laguerre.

Theorem 16.1.1 (Integration by parts) *The Askey–Wilson operator $\mathcal{D}_q$ satisfies*

$$\langle \mathcal{D}_q f, g \rangle = \frac{\pi \sqrt{q}}{1 - q} \left[f\left(\frac{1}{2}\left(q^{1/2} + q^{-1/2}\right)\right) \overline{g(1)} - f\left(-\frac{1}{2}\left(q^{1/2} + q^{-1/2}\right)\right) \overline{g(-1)} \right]$$
$$- \left\langle f, \sqrt{1 - x^2} \, \mathcal{D}_q \left(g(x) \left(1 - x^2\right)^{-1/2}\right) \right\rangle, \tag{16.1.3}$$

for $f, g \in H_{1/2}$.

Proof It is clear that (16.1.1) and (12.1.12) imply

$$\langle \mathcal{D}_q f, g \rangle = \int_0^\pi \frac{\check{f}\left(q^{1/2}e^{i\theta}\right) - \check{f}\left(q^{-1/2}e^{i\theta}\right)}{\left(q^{1/2} - q^{-1/2}\right) i \sin\theta} \, g(\cos\theta) \, d\theta. \tag{16.1.4}$$

The integrand in (16.1.4) is singular at $\theta = 0, \pi$, so the integral in (16.1.4) is defined as a Cauchy principal value and we need to consider the integrals

$$I_\epsilon := \int_\epsilon^{\pi-\epsilon} \frac{\check{f}\left(q^{1/2}e^{i\theta}\right) - \check{f}\left(q^{-1/2}e^{i\theta}\right)}{\left(q^{1/2} - q^{-1/2}\right) i \sin\theta} \, g(\cos\theta) \, d\theta. \tag{16.1.5}$$

Since

$$\check{f}\left(q^{-1/2}e^{-i\theta}\right) = f\left(\left(q^{-1/2}e^{-i\theta} + q^{1/2}e^{i\theta}\right)/2\right) = \check{f}\left(q^{1/2}e^{i\theta}\right),$$

we are led to

$$I_\epsilon = \int_\epsilon^{\pi-\epsilon} \frac{\check{f}\left(q^{1/2}e^{i\theta}\right)}{i\left(q^{1/2} - q^{-1/2}\right)} \frac{g(\cos\theta)}{\sin\theta} \, d\theta - \int_{-\epsilon}^{\epsilon-\pi} \frac{\check{f}\left(q^{-1/2}e^{-i\theta}\right)}{i\left(q^{1/2} - q^{-1/2}\right)} \frac{g(\cos\theta)}{\sin\theta} \, d\theta$$

$$= \left(\int_{-\pi+\epsilon}^{-\epsilon} + \int_\epsilon^{\pi-\epsilon}\right) \frac{\check{f}\left(q^{1/2}e^{i\theta}\right)}{i\left(q^{1/2} - q^{-1/2}\right)} \frac{g(\cos\theta)}{\sin\theta} \, d\theta.$$

Therefore I_ϵ has the alternate representation

$$I_\epsilon = \left(\int_C - \int_{C_\epsilon} - \int_{C'_\epsilon}\right) \frac{\check{f}\left(q^{1/2}z\right)}{\left(q^{1/2} - q^{-1/2}\right)} \frac{\bar{g}(z)}{(z - 1/z)/2} \frac{dz}{iz},$$

where C is the unit circle indented at ± 1 with circular arcs centered at ± 1 and radii equal to ϵ; and C_ϵ and C'_ϵ are the circular arcs

$$C_\epsilon = \left\{z : z = 1 + \epsilon e^{i\theta}, \ -(\pi - \epsilon)/2 \leq \theta \leq (\pi - \epsilon)/2\right\}$$

and

$$C'_\epsilon = \left\{z : z = -1 + \epsilon e^{i\theta}, \ -(\pi - \epsilon)/2 \leq \theta \leq (\pi - \epsilon)/2\right\}.$$

Both C_ϵ and C'_ϵ are positively oriented.

Define a function ϕ by

$$\phi(z) := 2 \frac{\check{f}\left(q^{1/2}z\right)}{\left(q^{1/2} - q^{-1/2}\right)} \frac{\bar{g}(z)}{iz(z - 1/z)}. \tag{16.1.6}$$

The residues of ϕ at $z = \pm 1$ are

$$\frac{f\left(\left(q^{1/2} + q^{-1/2}\right)/2\right) \overline{g(1)}}{i\left(q^{1/2} - q^{-1/2}\right)}, \quad -\frac{f\left(-\left(q^{1/2} + q^{-1/2}\right)/2\right) \overline{g(-1)}}{i\left(q^{1/2} - q^{-1/2}\right)},$$

respectively. Thus the limits as $\epsilon \to 0^+$ of the contour integrals over C_ϵ and C'_ϵ are

q-Sturm-Liouville problems

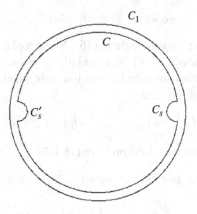

Fig. 16.1.

given by

$$\lim_{\epsilon \to 0^+} \int_{C_\epsilon} \phi(z)\, dz = -\pi i \ \text{Res}\{\phi(z) : z = 1\}$$

and

$$\lim_{\epsilon \to 0^+} \int_{C_\epsilon'} \phi(z)\, dz = -\pi i \ \text{Res}\{\phi(z) : z = -1\}.$$

Let C_1 be the circle $|z| = q^{-1/2}$. It is clear that

$$\int_C \phi(z)\, dz = \int_{C_1} \phi(z)\, dz - 2\pi i \left[\text{Res}\{\phi(z) : z = 1\} + \text{Res}\{\phi(z) : z = -1\} \right].$$

Thus we have established the following representation for the left-hand side of (16.1.4)

$$
\begin{aligned}
\langle \mathcal{D}_q f, g \rangle &= \lim_{\epsilon \to 0^+} I_\epsilon \\
&= \frac{\sqrt{q}\pi}{1-q} \left[f\left(\left(q^{1/2} + q^{-1/2}\right)/2 \right) \overline{g(1)} - f\left(-\left(q^{1/2} + q^{-1/2}\right)/2 \right) \overline{g(-1)} \right] \\
&\quad + 2 \int_{C_1} \frac{\breve{f}\left(q^{1/2} z\right)}{q^{1/2} - q^{-1/2}} \frac{\overline{\breve{g}(z)}}{z - 1/z} \frac{dz}{iz}.
\end{aligned}
$$

$$(16.1.7)$$

In the last integral we replace z by $zq^{-1/2}$ to change the integral over C_1 to an integral over the unit circle. The result is

$$2 \int_{C_1} \frac{\breve{f}\left(q^{1/2} z\right)}{q^{1/2} - q^{-1/2}} \frac{\overline{\breve{g}(z)}}{z - 1/z} \frac{dz}{iz} = \int_{|z|=1} \frac{\breve{f}(z)}{i} \frac{\overline{\breve{h}\left(q^{-1/2} z\right)}}{q^{-1/2} - q^{1/2}} \frac{dz}{z},$$

$$(16.1.8)$$

where

$$h(\cos\theta) = g(\cos\theta)/\sin\theta.$$

Finally in the integral over the unit circle in (16.1.8) we replace z by $e^{i\theta}$ then write the integral as a sum of integrals over $[-\pi, 0]$ and $[0, \pi]$. In the integral over the range $[-\pi, 0]$ replace θ by $-\theta$ then combine the two integrals which are now over $[0, \pi]$. Combining this with the observation

$$\breve{h}\left(q^{-1/2}z\right)|_{z=e^{-i\theta}} = -\breve{h}\left(q^{1/2}e^{i\theta}\right)$$

we obtain (16.1.3). This completes the proof of (16.1.3).   □

Theorem 16.1.1 indicates that $\mathcal{D}_q^*$, the adjoint of $\mathcal{D}_q$ is given by

$$\mathcal{D}_q^* f = -\sqrt{1-x^2}\,\mathcal{D}_q\left(\left(1-x^2\right)^{-1/2} f\right).$$

We will see in §6.3 that $\mathcal{D}_q$ and its adjoint play the roles of lowering and raising operators on the Askey–Wilson polynomials and these operators provide an Infield–Hull factorization of a second order q-Sturm–Liouville operator whose eigenfunctions are the Askey–Wilson polynomials.

We now come to the analogue of the indefinite integral. The Chebyshev polynomials $\{T_n(x)\}$ and $\{U_n(x)\}$ have been defined in (4.5.18) and their orthogonality relations are (4.5.19)–(4.5.20). As we saw in (12.1.15) $\mathcal{D}_q$ maps $T_n(x)$ to a multiple of $U_{n-1}(x)$. We now give another definition of $\mathcal{D}_q$. In order to study the action of $\mathcal{D}_q$ on functions defined on $[-1, 1]$ without having to extend them to the complex plane. To rectify this difficulty we propose to define $\mathcal{D}_q$ on a dense subset of $L^2\left[\left(1-x^2\right)^{-1/2}, [-1, 1]\right]$.

Definition 16.1.1 *A function* $f \in L^2\left[\left(1-x^2\right)^{-1/2}, [-1, 1]\right]$, *is called q-differentiable if f has a Fourier–Chebyshev expansion*

$$f(x) \sim \sum_{n=0}^{\infty} f_n T_n(x), \tag{16.1.9}$$

with the Fourier–Chebyshev coefficients $\{f_n\}$ satisfying

$$\sum_{n=0}^{\infty} \left|(1-q^n)\,q^{-n/2}\,f_n\right|^2 < \infty. \tag{16.1.10}$$

Furthermore if f is q-differentiable then $\mathcal{D}_q f$ is defined as the unique (almost everywhere) function whose Fourier–Chebyshev expansion is

$$(\mathcal{D}_q f)(x) \sim \sum_{n=1}^{\infty} \frac{q^{n/2} - q^{-n/2}}{q^{1/2} - q^{-1/2}}\, f_n\, U_{n-1}(x). \tag{16.1.11}$$

Obviously (16.1.10) holds on a dense subset S of $L^2\left[\left(1-x^2\right)^{-1/2}, [-1, 1]\right]$. Moreover, $\mathcal{D}_q$ maps S into a dense subset of $L^2\left[\left(1-x^2\right)^{1/2}, [-1, 1]\right]$.

Our next objective is to define a right inverse to $\mathcal{D}_q$. In other words we seek an operator $\mathcal{D}_q^{-1}$ so that

$$\mathcal{D}_q \mathcal{D}_q^{-1} = I. \qquad (16.1.12)$$

Let $\mathcal{D}_q f = g$ so that $f(x) \sim \sum\limits_{n=1}^{\infty} f_n T_n(x)$, $g(x) \sim \sum\limits_{n=0}^{\infty} g_n U_{n-1}(x)$ and

$$f_n = g_n \frac{\left(q^{1/2} - q^{-1/2}\right)}{\left(q^{n/2} - q^{-n/2}\right)}, \qquad n > 0, \qquad (16.1.13)$$

To recover f from the knowledge of g, we expect to have

$$\sum_{n=0}^{\infty} f_n T_n(x) = \sum_{n=0}^{\infty} g_n \frac{\left(q^{1/2} - q^{-1/2}\right)}{\left(q^{n/2} - q^{-n/2}\right)} T_n(x)$$

$$= \frac{2}{\pi} \sum_{n=0}^{\infty} \left(\int_{-1}^{1} g(y) U_{n-1}(y) \sqrt{1 - y^2} \, dy \right) \frac{\left(q^{1/2} - q^{-1/2}\right)}{\left(q^{n/2} - q^{-n/2}\right)} T_n(x)$$

$$= \frac{2}{\pi} (1 - q) q^{-1/2} \int_{-1}^{1} g(y) \left[\sum_{n=0}^{\infty} \frac{T_n(x) U_{n-1}(y)}{1 - q^n} q^{n/2} \right] \sqrt{1 - y^2} \, dy.$$

This formal procedure hints at defining $\mathcal{D}_q^{-1}$ as an integral operator whose kernel is given by

$$F(x, y) := \frac{2(1 - q)}{\pi \sqrt{q}} \sum_{n=0}^{\infty} \frac{T_n(x) U_{n-1}(y)}{1 - q^n} q^{n/2}. \qquad (16.1.14)$$

It is more convenient to use the new variables θ, ϕ,

$$x = \cos \theta, \qquad y = \cos \phi. \qquad (16.1.15)$$

The kernel F of (16.1.14) takes the form

$$F(\cos \theta, \cos \phi) = \frac{(1 - q) q^{-1/2}}{\pi \sin \phi} \sum_{n=0}^{\infty} \frac{2 \cos(n\theta) \sin(n\phi)}{1 - q^n} q^{n/2}$$

$$= \frac{(1 - q) q^{-1/2}}{\pi \sin \phi} \sum_{n=0}^{\infty} \frac{q^{n/2}}{1 - q^n} [\sin(n(\theta + \phi)) - \sin(n(\theta - \phi))].$$

$$(16.1.16)$$

Observe that

$$\int_{-1}^{1} F(x, y) g(y) \sqrt{1 - y^2} \, dy = \int_{0}^{\pi} F(\cos \theta, \cos \phi) g(\cos \phi) \sin^2 \phi \, d\phi$$

$$= \int_{-\pi}^{\pi} G(\cos \theta, \cos \phi) g(\cos \phi) \sin \phi \, d\phi,$$

with

$$G(\cos \theta, \cos \phi) = \frac{(1 - q)}{\pi \sqrt{q}} \sum_{n=0}^{\infty} \frac{q^{n/2}}{1 - q^n} \sin(n(\theta + \phi)). \qquad (16.1.17)$$

The logarithmic derivative of $\vartheta_4(z, q)$ has the Fourier series expansion

$$\frac{\vartheta_4'(z, q)}{\vartheta_4(z, q)} = 2i \sum_{k=0}^{\infty} \left[\frac{e^{-2iz}q^{2k+1}}{1 - e^{-2iz}q^{2k+1}} - \frac{e^{2iz}q^{2k+1}}{1 - e^{2iz}q^{2k+1}} \right],$$

as can be seen from (12.6.1). By expanding $1/\left[1 - e^{\pm 2iz}q^{2k+1}\right]$ in a geometric series then evaluating the k sum we establish

$$\frac{\vartheta_4'(z, q)}{\vartheta_4(z, q)} = 4 \sum_{n=0}^{\infty} \frac{q^n}{1 - q^{2n}} \sin(2nz). \qquad (16.1.18)$$

This is Exercise 11, p. 489 in (Whittaker & Watson, 1927). Thus (16.1.14), (16.1.16), (16.1.17) and (16.1.18) motivate our our next definition

Definition 16.1.2 *The inverse operator $\mathcal{D}_q^{-1}$ is defined as the integral operator*

$$\left(\mathcal{D}_q^{-1}g\right)(\cos\theta) = \frac{1-q}{4\pi\sqrt{q}} \int_{-\pi}^{\pi} \frac{\vartheta_4'\left((\theta+\phi)/2, \sqrt{q}\right)}{\vartheta_4\left((\theta+\phi)/2, \sqrt{q}\right)} g(\cos\phi) \sin\phi \, d\phi \qquad (16.1.19)$$

on the space $L^2\left[\left(1-x^2\right)^{1/2}, [-1, 1]\right]$.

Observe that the kernel of the integral operator (16.1.19) is uniformly bounded when (x, y) $(= (\cos\theta, \cos\phi)) \in [-1, 1] \times [-1, 1]$. Thus the operator $\mathcal{D}_q^{-1}$ is well-defined and bounded on $L^2\left[\left(1-x^2\right)^{1/2}, [-1, 1]\right]$. Furthermore $\mathcal{D}_q^{-1}$ is a one-to-one mapping from $L^2\left[\left(1-x^2\right)^{1/2}, [-1, 1]\right]$ into $L^2\left[\left(1-x^2\right)^{-1/2}, [-1, 1]\right]$.

Theorem 16.1.2 *On the space $L^2\left[\left(1-x^2\right)^{1/2}, [-1, 1]\right]$, the operator $\mathcal{D}_q\mathcal{D}_q^{-1}$ is the identity operator.*

Proof Replace ϑ_4'/ϑ_4 in (16.1.19) by the expansion (16.1.18) then apply Parseval's formula. In view of the uniform convergence of the series in (16.1.18) we may reverse the order of integration and summation in (16.1.19). Thus it follows that the steps leading to Definition 16.1.2 can be reversed and we obtain the desired result. □

Note that the kernel G of (16.1.17) is defined and bounded for all $\phi \in [-\pi, \pi]$ and all θ for which $\left|q^{1/2}e^{i\theta}\right| \leq 1$. This allows us to extend the definition of $\mathcal{D}_q^{-1}$ to the interior of the ellipse $\left|z + \sqrt{z^2 - 1}\right| = q^{-1/2}$ in the complex z-plane. This ellipse has foci at ± 1 and its major and minor axes are $q^{-1/2} \pm q^{1/2}$, respectively. Its equation in the xy-plane is

$$\frac{x^2}{a^2} + \frac{y^2}{b^2} = 1, \quad a = \frac{1}{2}\left(q^{-1/2} + q^{1/2}\right), \quad b = \frac{1}{2}\left(q^{-1/2} - q^{1/2}\right). \qquad (16.1.20)$$

It is worth mentioning that provided we exercise some care the definition (16.1.11) of the Askey–Wilson operator combined with (16.1.10) yields the result $\mathcal{D}_q\mathcal{D}_q^{-1} = I$. One reason for being particularly careful is that $\mathcal{D}_q^{-1}g$ may not be in the domain of $\mathcal{D}_q$ because in order to use (16.1.11) we need to assume that f has an analytic extension to a domain in the complex plane containing $\left|z \pm \sqrt{z^2 - 1}\right| \leq q^{-1/2}$. Let $f(\cos \theta)$ denote the right-hand side of (16.1.19).

Theorem 16.1.3 *Let $g(x)$ be a continuous function on $[-1, 1]$ except for finitely many jumps. Then with $\mathcal{D}_q^{-1}$ defined as in (16.1.19), the limiting relation*

$$\lim_{p \to q^+} \left(\mathcal{D}_p \mathcal{D}_q^{-1} g\right)(x) = g(x)$$

holds at the points of continuity of g.

Proof It is easy to see that $\mathcal{D}_p f(x)$ is well-defined provided that $1 > p > q$ and

$$(\mathcal{D}_p f)(\cos \theta) = \int_{-\pi}^{\pi} \mathcal{D}_p G(\cos \theta, \cos \phi) g(\cos \phi) \sin \phi \, d\phi$$

$$= \frac{(1 - q)p^{1/2}}{\pi(1 - p)q^{1/2} \sin \theta}$$

$$\times \int_{-\pi}^{\pi} \sum_{n=0}^{\infty} \frac{(1 - p^n)}{(1 - q^n)} (q/p)^{n/2} \cos(n(\theta + \phi)) g(\cos \phi) \sin \phi \, d\phi.$$

By writing $\dfrac{1 - p^n}{1 - q^n}$ as $1 + \dfrac{q^n - p^n}{1 - q^n}$, and denoting q/p by r^2 we see that

$$\lim_{p \to q^+} \left(\mathcal{D}_p \mathcal{D}_q^{-1} g\right)(\cos \theta) = \frac{1}{\pi \sin \theta} \lim_{r \to 1^-} \int_{-\pi}^{\pi} \left[\sum_{n=0}^{\infty} r^n \cos(n(\theta + \phi)) \right]$$

$$\times g(\cos \phi) \sin \phi \, d\phi + 0$$

$$= \frac{1}{\pi \sin \theta} \lim_{r \to 1^-} \int_{-\pi}^{\pi} \left[\frac{1}{2} + \sum_{n=0}^{\infty} r^n \cos(n(\theta + \phi)) \right]$$

$$\times g(\cos \phi) \sin \phi \, d\phi,$$

since $g(\cos \phi) \sin \phi$ is an odd function. Therefore

$$\lim_{p \to q^+} \left(\mathcal{D}_p \mathcal{D}_q^{-1} g\right)(\cos \theta) = \lim_{r \to 1^-} \frac{1}{2\pi \sin \theta} \int_{-\pi}^{\pi} \frac{(1 - r^2) g(\cos \phi) \sin \phi}{1 - 2r \cos(\theta + \phi) + r^2} \, d\phi.$$

The above limit exists and equals $g(\cos \theta)$ at the points of continuity of g if $g(\cos \theta)$ is continuous on $[-\pi, \pi]$ except for finitely many jumps, (Nehari, 1961, p. 147). □

The kernel ϑ_4'/ϑ_4 of (16.1.18) and (16.1.19) has appeared earlier in conformal mappings. Let ζ be a fixed point in the interior of the ellipse (16.1.20) in the complex plane and let $f(z, \zeta)$ be the Riemann mapping function that maps the interior of the ellipse (16.1.20) conformally onto the open unit disc and satisfies $f(\zeta, \zeta) = 0$ and $f'(\zeta, \zeta) > 0$. It is known, (Nehari, 1952, p. 260), that

$$f(z, \zeta) = g(z, \zeta) - g(\zeta, \zeta), \tag{16.1.21}$$

where

$$g(z, \zeta) = \sqrt{\frac{\pi}{K(\zeta, \zeta)}} \sum_{n=0}^{\infty} \frac{T_n(z)\, \overline{U_n(\zeta)}}{\rho^n - \rho^{-n}}, \tag{16.1.22}$$

and the Bergman kernel $K(z, \zeta)$ is

$$K(z, \zeta) := \frac{4}{\pi} \sum_{n=1}^{\infty} \frac{(n+1)\, U_n(z)\, \overline{U_n(\zeta)}}{\rho^{n+1} - \rho^{-n-1}}, \quad \rho := (a+b)^2 = \left(b + \sqrt{b^2+1}\right)^2. \tag{16.1.23}$$

In fact the Bergman kernel of the ellipse is a constant multiple of $f'(z, \zeta)$. It is clear that $g(z, \zeta)$ is a constant multiple of our kernel G of (16.1.17) with $\rho = q^{-1/2}$, so that $q = (b + \sqrt{b^2+1})^{-4} = e^{-4u}$ if $b = \sinh u$. Szegő gave a nice treatment of the above facts in (Szegő, 1950a).

The connection between the Riemann mapping function $f(z, \zeta)$ of the ellipse (16.1.20) and our kernel may seem very surprising at a first glance. However this may not be a complete surprise because if f is real analytic in $(-1, 1)$ it will have an extension which is analytic in the open unit disc and (16.1.11) will be meaningful if $|q^{1/2}e^{i\theta}| < 1$; which is the interior of the ellipse (16.1.20). Furthermore the Chebyshev polynomials $\{U_n(z)\}$ are orthogonal on the unit disc with respect to the Lebesgue measure in the plane.

16.2 A q-Sturm–Liouville Operator

Let

$$(f, g) := \int_{-1}^{1} f(x)\, \overline{g(x)}\, dx. \tag{16.2.1}$$

It is evident that (16.1.3) implies that

$$(\mathcal{D}_q f, g) = -(f, \mathcal{D}_q g), \tag{16.2.2}$$

holds for all $f, g \in H_{1/2}$.

Lemma 16.2.1 *For all $f, g \in H_{1/2}$ we have*

$$\int_{q}^{q^{1/2}} \left(\left[\breve{f}, \breve{g}\right](x) - \left[\breve{f}, \breve{g}\right](-x) \right) dx = 0, \tag{16.2.3}$$

where

$$\left[\breve{f}, \breve{g}\right](x) = \breve{f}(x)\breve{g}\left(q^{-1/2}x\right) - \breve{f}\left(q^{-1/2}x\right)\breve{g}(x). \tag{16.2.4}$$

Proof We first observe that

$$(\mathcal{D}_q f, g) = \frac{iq^{1/2}}{1-q} \int_0^\pi \left\{ \check{f}\left(q^{1/2} e^{i\theta}\right) - \check{f}\left(q^{-1/2} e^{i\theta}\right) \right\} \bar{\check{g}}\left(e^{i\theta}\right) d\theta$$

$$= \frac{iq^{1/2}}{1-q} \left\{ \int_0^\pi \check{f}\left(q^{1/2} e^{i\theta}\right) \bar{\check{g}}\left(e^{i\theta}\right) d\theta - \int_{-\pi}^0 \check{f}\left(q^{1/2} e^{i\theta}\right) \bar{\check{g}}\left(e^{i\theta}\right) d\theta \right\}.$$

If C_2^+ denotes $\{z = q^{1/2} e^{i\theta} : 0 < \theta < \pi\}$, then

$$\int_0^\pi \check{f}\left(q^{1/2} e^{i\theta}\right) \bar{\check{g}}\left(e^{i\theta}\right) d\theta = \int_{C_2^+} \check{f}\left(q^{1/2} z\right) \bar{\check{g}}(z) \frac{dz}{iz}$$

$$- \left[\int_{-1}^{-q^{1/2}} + \int_{q^{1/2}}^1 \right] \check{f}(q^{1/2} x) \bar{\check{g}}(x) \frac{dx}{ix}$$

$$= - \int_0^{-\pi} \check{f}\left(e^{i\theta}\right) \bar{\check{g}}\left(q^{1/2} e^{i\theta}\right) d\theta + \left[\int_{-1}^{-q^{1/2}} + \int_{q^{1/2}}^1 \right] \check{f}(x) \bar{\check{g}}\left(q^{-1/2} x\right) \frac{dx}{ix}$$

since $\bar{\check{g}}\left(q^{-1/2} e^{-\theta}\right) = \bar{\check{g}}\left(q^{1/2} e^{\theta}\right)$. It follows that

$$(\mathcal{D}_q f, g) - (f, \mathcal{D}_q g) = -\frac{iq^{1/2}}{q-1} \left\{ \int_0^\pi \left[\check{f}\left(q^{1/2} e^{i\theta}\right) \bar{\check{g}}\left(e^{i\theta}\right) + \bar{\check{g}}\left(q^{1/2} e^{i\theta}\right) \check{f}\left(e^{i\theta}\right) \right] d\theta \right.$$

$$\left. - \int_{-\pi}^0 \left[\check{f}\left(q^{1/2} e^{i\theta}\right) \bar{\check{g}}\left(e^{i\theta}\right) + \bar{\check{g}}\left(q^{1/2} e^{i\theta}\right) \check{f}\left(e^{i\theta}\right) \right] d\theta \right\}$$

$$= -\frac{q^{1/2}}{q-1} \left\{ \int_{-q^{-1/2}}^{-q} + \int_q^{q-1/2} \right\} \left[\check{f}\bar{\check{g}} \right] (x) \frac{dx}{x}$$

$$= \frac{q^{1/2}}{1-q} \int_q^{q^{1/2}} \left(\left[\check{f}\bar{\check{g}} \right] (x) - \left[\check{f}\bar{\check{g}} \right] (-x) \right) \frac{dx}{x}$$

and (16.2.3) follows. $\qquad\qquad\qquad\qquad\qquad\qquad\qquad\qquad\qquad\qquad\square$

Let $\mathcal{H}_w$ denote the weighted space $L^2(-1, 1; w(x)dx)$ with inner product

$$(f, g)_w := \int_{-1}^1 f(x) \overline{g(x)} \, w(x) \, dx, \quad \|f\|_w := (f, f)_w^{1/2} \qquad (16.2.5)$$

and let T be defined by

$$Tf(x) := Mf(x) \qquad\qquad\qquad\qquad\qquad (16.2.6)$$

for f in H_1, where

$$(Mf)(x) = -\frac{1}{w(x)} \mathcal{D}_q \left(p \mathcal{D}_q f \right)(x). \tag{16.2.7}$$

We shall assume that p and w are positive on $(-1, 1)$ and also satisfy

(i) $\quad p(x)/\sqrt{1-x^2} \in H_{1/2}$, $1/p \in L(-1, 1)$,

(ii) $\quad w(x) \in L(-1, 1)$, $1/w \in L\left(-1, 1; \dfrac{dx}{(1-x^2)}\right)$. $\tag{16.2.8}$

The expression Mf is therefore defined for $f \in H_1$, and the operator T acts in $\mathcal{H}_w$. Furthermore, the domain H_1 of T is dense in $\mathcal{H}_w$ since it contains all polynomials.

Theorem 16.2.2 *The operator T is symmetric in $\mathcal{H}_w$ and positive.*

Proof We infer from (16.2.2) that for all $f, g \in H_1$,

$$(Tf, f)_w = -(\mathcal{D}_q [p\mathcal{D}_q f], f) = (p\mathcal{D}_q f, \mathcal{D}_q f)$$

$$= \int_{-1}^{1} p(x) |\mathcal{D}_q f(x)|^2 \, dx, \tag{16.2.9}$$

hence T is positive. Another application of (16.2.2) implies

$$(Tf, f)_w = (f, Tf)_w,$$

and the symmetry of T follows. $\qquad\square$

Theorem 16.2.3 *Let $y_1, y_2 \in H_1$ be solutions to*

$$Ty = \lambda y, \tag{16.2.10}$$

with $\lambda = \lambda_1$ and λ_2, respectively. Then the λ's are real and

$$\int_{-1}^{1} y_1(x) \overline{y_2(x)} \, w(x) \, dx = 0. \tag{16.2.11}$$

if $\lambda_1 \neq \lambda_2$.

Proof It follows from (12.2.8) that

$$(\lambda_1 - \overline{\lambda_2}) \int_{-1}^{1} w(x) y_1(x) \overline{y_2(x)} \, dx = (\lambda_1 y_1, y_2) - (y_1, \lambda_2 y_2)$$

$$= (Ty_1, y_2) - (y_1, Ty_2) = 0,$$

by the symmetry of T. Taking $\lambda_1 = \lambda_2$ then λ_1 shows that the eigenvalues must be real. If $\lambda_1 \neq \lambda_2$ then

$$(\lambda_1 - \lambda_2) \int_{-1}^{1} w(x) |y_1(x)|^2 \, dx = 0, \tag{16.2.12}$$

and (16.2.11) follows. □

Let $\mathcal{Q}(T)$ denote the form domain of T and $\tilde{T}$ its Friedrichs extension. Recall that $\mathcal{Q}(T)$ is the completion of H_1 with respect to $\|.\|_{\mathcal{Q}}$, where

$$\|f\|_{\mathcal{Q}}^2 := \int_{-1}^{1} p(x) |\mathcal{D}_q f|^2 \, dx + \|f\|_w^2, \tag{16.2.13}$$

and if $(.,.)_{\mathcal{Q}}$ denotes the inner product on $\mathcal{Q}(T)$, then for all $f \in \mathcal{D}\left(\tilde{T}\right)$ and $g \in \mathcal{Q}(T)$,

$$(f, g)_{\mathcal{Q}} = \left(\left[\tilde{T} + I\right] f, g\right)_w. \tag{16.2.14}$$

where I is the identity on $\mathcal{H}_w$. We have that $f \in \mathcal{Q}(T)$ if and only if there exists a sequence $\{f_n\} \subset H_1$ such that $\|f - f_n\|_{\mathcal{Q}} \to 0$; hence $\|f - f_n\|_w \to 0$ and $\{\mathcal{D}_q f_n\}$ is a Cauchy sequence in $L^2(-1, 1; p(x)dx)$, with limit F say. From (16.2.7) and (16.2.2) it follows that for $\phi \in H_{1/2}$,

$$\int_{-1}^{1} F(x) \phi(x) \, dx = \lim_{n \to \infty} \int_{-1}^{1} (\mathcal{D}_q f_n)(x) \, \phi(x) \, dx = - \lim_{n \to \infty} \int_{-1}^{1} f_n(x) \, \mathcal{D}_q \phi(x) dx,$$

$$= - \int_{-1}^{1} f(x) \, \mathcal{D}_q \phi(x) \, dx + \mathcal{O}\left(\|f - f_n\|_w \left[\int_{-1}^{1} |\mathcal{D}_q \phi(x)|^2 \frac{dx}{w(x)}\right]^{1/2}\right)$$

$$= - \int_{-1}^{1} f(x) \, \mathcal{D}_q \phi(x) \, dx.$$

Thus, in analogy with distributional derivatives, we shall say that $F = \mathcal{D}_q f$ in the generalized sense. Note that this proves that F is unique up to a function that vanishes almost everywhere, so different Cauchy sequences f_n give the same $\mathcal{D}_q f$.

We conclude that the norm on $\mathcal{Q}(T)$ is defined by (16.2.13) with $\mathcal{D}_q f$ now understood in the generalized sense. Also, it follows in a standard way that

$$\mathcal{D}\left(T^*\right) = \{f : f, Mf \in \mathcal{H}_w\}, \quad T^* f = Mf, \tag{16.2.15}$$

$$\mathcal{D}\left(\tilde{T}\right) = \mathcal{Q}(T) \cap \mathcal{D}\left(T^*\right)$$
$$= \left\{f : p^{1/2} \mathcal{D}_q f \in L^2(-1, 1), \ Mf \in \mathcal{H}_w\right\}. \tag{16.2.16}$$

If T is the operator in the Askey–Wilson case, that is

$$w(x) = w(x; \mathbf{t}), \quad p(x) = w(x; q^{1/2}\mathbf{t}),$$

then the Askey–Wilson polynomials satisfy (16.3.6), that is

$$T p_n(x; \mathbf{t}) = -\lambda_n p_n(x; \mathbf{t}).$$

Since $p_n(x; \mathbf{t})$ is of degree n, $(n = 0, 1, \dots)$ and the polynomials are dense in $\mathcal{H}_w$, it follows that $\{p_n(x; \mathbf{t})\}$ forms a basis for $\mathcal{H}_w$. Hence T has a selfadjoint closure $\overline{T} = \tilde{T}$ and $\tilde{T}$ has a discrete spectrum consisting of the eigenvalues λ_n in (16.3.7), $n = 0, 1, \dots$. If $f \in \mathcal{Q}(T)$ then setting $p_n(x) \equiv p_n(x; \mathbf{t})$, we have

$$(f, p_n)_Q = (f, [T + I]p_n)_w = (\lambda_n + 1)(f, p_n)_w$$

and, in particular, $\zeta_n \equiv \zeta_n(\mathbf{t})$ denoting the right-hand side of (15.2.4) we get

$$(p_m, p_n)_Q = (\lambda_n + 1)\zeta_n \delta_{m,n}.$$

It follows that $e_n = p_n / \sqrt{\zeta_n (\lambda_n + 1)}$, $(n = 0, 1, \dots)$ is an orthonormal basis for $\mathcal{Q}(T)$. Thus $f \in \mathcal{Q}(T)$ if and only if

$$\sum_{n=0}^{\infty} |(f, e_n)_Q|^2 = \sum_{n=0}^{\infty} \frac{\lambda_n + 1}{\zeta_n} |(f, p_n)_w|^2 < \infty, \tag{16.2.17}$$

$f \in \mathcal{D}\left(\tilde{T}\right)$ if and only if

$$\sum_{n=0}^{\infty} \left| \left([\tilde{T} + I] f, \frac{p_n}{\zeta_n^{1/2}} \right)_w \right|^2 = \sum_{n=0}^{\infty} \frac{(\lambda_n + 1)^2}{\zeta_n} |(f, p_n)_w|^2 < \infty \tag{16.2.18}$$

and, for $f \in \mathcal{D}\left(\tilde{T}\right)$

$$\tilde{T}f = \sum_{n=0}^{\infty} \left(\tilde{T}f, \frac{p_n}{\zeta_n^{1/2}} \right)_w \frac{p_n}{\zeta_n^{1/2}} = \sum_{n=0}^{\infty} \frac{\lambda_n}{\zeta_n} (f, p_n)_w p_n. \tag{16.2.19}$$

This section is based on (Brown et al., 1996).

16.3 The Askey–Wilson Polynomials

We now return to the Askey–Wilson polynomials and apply Theorem 16.2.2 to give a new proof of the orthogonality of the Askey–Wilson polynomials. Recall from (12.2.2) that

$$\mathcal{D}_q \left(ae^{i\theta}, ae^{-i\theta}; q \right)_k = -2a \frac{1 - q^k}{1 - q} \left(aq^{1/2} e^{i\theta}, aq^{1/2} e^{-i\theta}; q \right)_{k-1}. \tag{16.3.1}$$

Apply (16.3.1) to the explicit form (15.2.5) of the Askey–Wilson polynomials to obtain

$$\mathcal{D}_q p_n(x; \mathbf{t} \mid q)$$

$$= 2 \frac{(1 - q^n)(1 - t_1 t_2 t_3 t_4 q^{n-1})}{(1 - q) q^{(n-1)/2}} p_{n-1}\left(x; q^{1/2} \mathbf{t} \mid q \right). \tag{16.3.2}$$

Thus $\mathcal{D}_q$ is a lowering operator for the p_n's since it lowers their degrees by 1. Our next result gives the raising operator.

Theorem 16.3.1 *The Askey–Wilson polynomials satisfy*

$$\frac{2q^{(1-n)/2}}{q-1}\, w(x;\mathbf{t}\,|\,q)\, p_n(x;\mathbf{t}\,|\,q)$$
$$= \mathcal{D}_q \left(w\left(x;q^{1/2}\mathbf{t}\,|\,q\right) p_{n-1}\left(x;q^{1/2}\mathbf{t}\,|\,q\right)\right), \quad (16.3.3)$$

where $w(x;\mathbf{t}\,|\,q)$ *is defined in* (15.2.2).

Proof It is easy to express $w(x;\mathbf{t}\,|\,q)$ in the form

$$w(\cos\theta;\mathbf{t}\,|\,q)$$
$$= \frac{2i\, e^{-i\theta}\left(e^{2i\theta}, qe^{-2i\theta}; q\right)_\infty}{\left(t_1 e^{i\theta}, t_1 e^{-i\theta}, t_2 e^{i\theta}, t_2 e^{-i\theta}, t_3 e^{i\theta}, t_3 e^{-i\theta}, t_4 e^{i\theta}, t_4 e^{-i\theta}; q\right)_\infty}, \quad (16.3.4)$$

and an easy calculation using (12.1.10) gives

$$\frac{\mathcal{D}_q w\left(x;q^{1/2}\mathbf{t}\,|\,q\right)}{w(x;\mathbf{t}\,|\,q)} = \frac{2}{q-1}\left[2\left(1-\sigma_4\right)\cos\theta - \sigma_1 + \sigma_3\right]. \quad (16.3.5)$$

where σ_j is the jth elementary symmetric function of $t_1,\ldots,t_4$. This leads to

$$\frac{\mathcal{D}_q \left[w\left(x;q^{1/2}\mathbf{t}\,|\,q\right) p_{n-1}\left(x;q^{1/2}\mathbf{t}\,|\,q\right)\right]}{w(x;\mathbf{t}\,|\,q)}$$

$$= \frac{(t_1t_2q, t_1t_3q, t_1t_4q; q)_{n-1}}{(t_1\sqrt{q})^{n-1} w\left(x;t_1,t_2,t_3,t_4\,|\,q\right)} \sum_{k=0}^{n-1} \frac{(q^{1-n}, \sigma_4q^n; q)_k\, q^k}{(q, t_1t_2q, t_1t_3q, t_1t_4q; q)_k}$$
$$\times \mathcal{D}_q w\left(x; t_1 q^{k+1/2}, t_2 q^{1/2}, t_3 q^{1/2}, t_4 q^{1/2}\,|\,q\right)$$

$$= \frac{2\left(t_1t_2, t_1t_3, t_1t_4; q\right)_n}{(q-1)t_1^n q^{(n-1)/2}} \sum_{k=0}^{n-1} \frac{(q^{1-n}, \sigma_4q^n; q)_k}{(q;q)_k\,(t_1t_2, t_1t_3, t_1t_4; q)_{k+1}} \frac{w\left(x; t_1 q^k, t_2, t_3, t_4\,|\,q\right)}{w\left(x; t_1, t_2, t_3, t_4\,|\,q\right)}$$
$$\times \left[2t_1 q^k \cos\theta\left(1 - t_1t_2t_3t_4q^k\right)\right.$$
$$\left. -t_1 q^k\left(t_2 + t_3 + t_4\right) + t_1^2 q^{2k}\left(t_2t_3 + t_2t_4 + t_3t_4 - 1\right) + t_1t_2t_3t_4q^k\right].$$

The term in square brackets on the right-hand side can be written as

$$\left(1 - t_1t_2q^k\right)\left(1 - t_1t_3q^k\right)\left(1 - t_1t_4q^k\right)$$
$$- \left(1 - t_1 q^k e^{i\theta}\right)\left(1 - t_1 q^k e^{-i\theta}\right)\left(1 - \sigma_4 q^k\right).$$

438   *The Askey–Wilson Operators*

Therefore

$$\frac{\mathcal{D}_q\left[w\left(x;q^{1/2}\mathbf{t}\mid q\right)p_{n-1}\left(x;q^{1/2}\mathbf{t}\mid q\right)\right]}{w(x;\mathbf{t}\mid q)}$$

$$= \frac{2\left(t_1t_2,t_1t_3,t_1t_4;q\right)_n}{(q-1)t_1^n q^{(n-1)/2}}\left[\sum_{k=0}^{n-1}\frac{\left(q^{1-n},\sigma_4 q^n,t_1e^{i\theta},t_1e^{-i\theta};q\right)_k}{(q,t_1t_2,t_1t_3,t_1t_4;q)_k}\right.$$

$$\left. - \sum_{k=1}^{n}\frac{\left(q^{1-n},\sigma_4 q^n;q\right)_{k-1}\left(1-q^k\right)\left(t_1e^{i\theta},t_1e^{-i\theta};q\right)_k}{(q,t_1t_2,t_1t_3,t_1t_4;q)_k}\left(1-\sigma_4 q^{k-1}\right)\right]$$

$$= \frac{2\left(t_1t_2,t_1t_3,t_1t_4;q\right)_n}{(q-1)t_1^n q^{(n-1)/2}}\sum_{k=1}^{n}\frac{\left(q^{1-n},\sigma_4 q^n;q\right)_{k-1}\left(t_1e^{i\theta},t_1e^{-i\theta};q\right)_k}{(q,t_1t_2,t_1t_3,t_1t_4;q)_k}$$

$$\times \left[\left(1-q^{(k-n)}\right)\left(1-\sigma_4 q^{n+k-1}\right)-\left(1-q^k\right)\left(1-\sigma_4 q^{k-1}\right)\right].$$

Putting all this together establishes (16.3.3) since the term in square brackets is

$$q^k\left(1-q^{-n}\right)\left(1-\sigma_4 q^{n-1}\right).\quad\square$$

Remark 16.3.1 *The proof we gave of Theorem* 16.3.1 *assumes only results derived in this chapter in order to provide an alternate approach to the theory of the Askey–Wilson polynomials. If we were to assume the orthogonality relation* (15.2.4) *then we would use the following argument. Denote the coefficient of* $\delta_{m,n}$ *on the right-hand side of* (15.2.4) *by* $\zeta_n\left(t_1,t_2,t_3,t_4\right)$. *Thus*

$$\zeta_{n-1}\left(q^{1/2}\mathbf{t}\right)\delta_{m,n}$$

$$= \left\langle p_{m-1}\left(x;q^{1/2}\mathbf{t}\mid q\right),\left(1-x^2\right)^{1/2}w\left(x;q^{1/2}\mathbf{t}\mid q\right)p_{n-1}\left(x;q^{1/2}\mathbf{t}\mid q\right)\right\rangle$$

$$= \frac{(1-q)q^{(n-1)/2}/2}{(1-q^m)(1-\sigma_4 q^{m-1})}$$

$$\times \left\langle \mathcal{D}_q p_m(x;\mathbf{t}\mid q),\left(1-x^2\right)^{1/2}w\left(x;q^{1/2}\mathbf{t}\mid q\right)p_{n-1}\left(x;q^{1/2}\mathbf{t}\mid q\right)\right\rangle$$

$$= \frac{(1-q)q^{(n-1)/2}/2}{(1-q^m)(1-\sigma_4 q^{m-1})}$$

$$\times \left\langle p_m(x;\mathbf{t}\mid q),\left(1-x^2\right)^{1/2}\mathcal{D}_q\left(w(x;q^{1/2}\mathbf{t}\mid q)p_{n-1}\left(x;q^{1/2}\mathbf{t}\right)\mid q\right)\right\rangle.$$

The set of all polynomials is dense in $C([-1,1])$ *which is dense in the Hilbert space* $L^2\left[w\left(x;t_1,t_2,t_3,t_4\mid q\right);-1,1\right]$ *since* $w\left(x;t_1,t_2,t_3,t_4\mid q\right)$ *is continuous when* $|t_j|<1$ *for* $1\leq j\leq 4$. *Thus the Askey–Wilson polynomials are complete in the space* $L^2\left[w\left(x;t_1,t_2,t_3,t_4\mid q\right);-1,1\right]$. *But we have shown that*

$$\frac{\mathcal{D}_q\left(w\left(x;q^{1/2}\mathbf{t}\mid q\right)p_{n-1}\left(x;q^{1/2}\mathbf{t}\mid q\right)\right)}{w(x;\mathbf{t}\mid q)}$$

is orthogonal to $p_m(x; \mathbf{t} \mid q)$ *for all* $n \neq m$. *Therefore there exists a constant* C_n
such that

$$
\frac{\mathcal{D}_q \left(w \left(x; q^{1/2} \mathbf{t} \mid q \right) p_{n-1} \left(x; q^{1/2} \mathbf{t} \mid q \right) \right)}{w(x; \mathbf{t} \mid q)}
$$
$$
= C_n p_n(x; \mathbf{t} \mid q).
$$

The constant C_n *can be computed from*

$$
\zeta_{n-1} \left(q^{1/2} \mathbf{t} \right) = \frac{(1-q) q^{(n-1)/2} C_n}{2 \left(1 - q^n \right) \left(1 - \sigma_4 q^{n-1} \right)} \zeta_n(\mathbf{t}),
$$

and Theorem 16.3.1 *follows.*

Theorem 16.3.2 *The Askey–Wilson polynomials satisfy the eigenvalue problem*

$$
\frac{1}{w(x; \mathbf{t})} \mathcal{D}_q \left(w \left(x; q^{1/2} \mathbf{t} \mid q \right) \mathcal{D}_q \, p_n(x; \mathbf{t}) \right) \tag{16.3.6}
$$
$$
= \lambda_n \, p_n(x; \mathbf{t}),
$$

whose eigenvalues $\{\lambda_n\}$ *are*

$$
\lambda_n := \frac{4q}{(1-q)^2} \left(1 - q^{-n} \right) \left(1 - \sigma_4 q^{n-1} \right), \quad n = 1, 2, \dots . \tag{16.3.7}
$$

Proof Replace $\mathcal{D}_q p_n(x; \mathbf{t} \mid q)$ in (16.3.6) by $p_{n-1} \left(x; q^{1/2} \mathbf{t} \mid q \right)$ then apply (16.3.3).
Simple manipulations will establish (16.3.6)–(16.3.7). $\qquad \square$

By iterating (16.3.3) one can derive the Rodrigues formula

$$
w(x; \mathbf{t} \mid q) \, p_n(x; \mathbf{t} \mid q) = \left(\frac{q-1}{2} \right)^n q^{n(n-1)/4} \mathcal{D}_q^n \left[w(x; q^{n/2} \mathbf{t} \mid q) \right]. \tag{16.3.8}
$$

We now come to the orthogonality relation of the Askey–Wilson polynomials.

Theorem 16.3.3 *The orthogonality relation* (15.2.4) *is implied by the eigenvalue
problem* (16.3.6).

Proof The function

$$
\frac{1-x}{x(1-ux)}
$$

strictly decreases on $(0, 1)$ if $0 < u < 1$. Thus when $0 < t_1 t_2 t_3 t_4 < q$ the eigenvalues λ_n of (16.3.8) are distinct and the Askey–Wilson polynomials are orthogonal
with respect to $w(x; t_1, t_2, t_3, t_4 \mid q)$. When $m \neq n$, Theorem 16.2.3 establishes
(15.2.4) when $0 < t_1 t_2 t_3 t_4 < q$. Now Theorem 11.1.1 enables us to extend the
validity of (15.2.4) when $m \neq n$ to $t_j \in (0, 1)$ for $1 \leq j \leq 4$. Thus it remains only

to consider the case $m = n$. We have

$$\left(\frac{2}{q-1}\right)^n q^{n(1-n)/4} \int_{-1}^{1} p_n^2(x;\mathbf{t}\,|\,q) w(x;\mathbf{t}\,|\,q)\,dx$$

$$= \left\langle \mathcal{D}_q^n \left[w\left(x, q^{n/2}\mathbf{t}\,|\,q\right) \right], \sqrt{1-x^2}\,p_n(x;\mathbf{t}\,|\,q) \right\rangle$$

$$= (-1)^n \left\langle w\left(x, q^{n/2}\mathbf{t}\,|\,q\right), \sqrt{1-x^2}\,\mathcal{D}_q^n p_n(x;\mathbf{t}\,|\,q) \right\rangle,$$

where we have used (16.3.8) and (16.3.3). Let $\zeta_n(\mathbf{t})$ denote the left-hand side of (15.2.4) with $m = n$. On applying (16.3.2) we obtain

$$\zeta_n(\mathbf{t}) = \left(q, \sigma_4 q^{n-1}; q\right)_n \zeta_0\left(q^{n/2}\mathbf{t}\right).$$

But the three term recurrence relation (15.2.10) and the explicit coefficients (15.2.11)–(15.2.13) when combined with (2.2.16)–(2.2.18) yield

$$\zeta_n(\mathbf{t}) = \frac{(q, t_1 t_2, t_1 t_3, t_1 t_4, t_2 t_3, t_2 t_4, t_3 t_4; q)_n\,(\sigma_4/q; q)_{2n}}{(t_1 t_2 t_3 t_4/q; q)_n\,(t_1 t_2 t_3 t_4; q)_{2n}}\,\zeta_0(\mathbf{t}). \qquad (16.3.9)$$

Thus we have established the functional equation

$$\zeta_0(\mathbf{t})/\zeta_0\left(q^{n/2}\mathbf{t}\right)$$

$$= \frac{(t_1 t_2 t_3 t_4/q, t_1 t_2 t_3 t_4 q^{n-1}; q)_n\,(\sigma_4; q)_{2n}}{(t_1 t_2, t_1 t_3, t_1 t_4, t_2 t_3, t_2 t_4, t_3 t_4; q)_n\,(t_1 t_2 t_3 t_4/q; q)_{2n}}. \qquad (16.3.10)$$

As $n \to \infty$, (16.3.10) becomes

$$\zeta_0(\mathbf{t}) = \frac{(t_1 t_2 t_3 t_4; q)_\infty}{(t_1 t_2, t_1 t_3, t_1 t_4, t_2 t_3, t_2 t_4, t_3 t_4; q)_\infty}\,\zeta_0(0,0,0,0). \qquad (16.3.11)$$

Now $\zeta_0(0,0,0,0)$ has already appeared as the case $m = n$ of (13.1.11) which was proved using the Jacobi triple product identity (12.2.25). The result is

$$\zeta_0(0,0,0,0) = 2\pi/(q;q)_\infty.$$

Therefore $\zeta_0(0,0,0,0)$ satisfies

$$\zeta_0\left(t_1, t_2, t_3, t_4\right) = \int_{-1}^{1} w\left(x, t_1, t_2, t_3, t_4\,|\,q\right)\,dx$$

$$= \frac{2\pi\,(\sigma_4; q)_\infty}{(q, t_1 t_2, t_1 t_3, t_1 t_4, t_2 t_3, t_2 t_4, t_3 t_4; q)_\infty}, \qquad (16.3.12)$$

and (16.3.9) and (16.3.12) show that $\zeta_n\left(t_1, t_2, t_3, t_4\right)$ equals the right-hand side of (15.2.4). This completes the proof of Theorem 16.3.3. $\square$

Motivated by Theorem 16.3.2 we consider the polynomials solutions of

$$\frac{1}{w(x;\mathbf{t})}\,\mathcal{D}_q\left[w\left(x; q^{1/2}\mathbf{t}\right)\mathcal{D}_q\,f(x)\right] = \lambda\,f(x), \qquad (16.3.13)$$

Theorem 16.3.4 *If f is a polynomial solution of* (16.3.13) *of degree n, then $\lambda = \lambda_n$ and f is a constant multiple of $p_n(x;\mathbf{t}\,|\,q)$.*

Proof Let

$$f(x) = \sum_{k=0}^{n} a_k \left(t_1 e^{i\theta}, t_1 e^{-i\theta}; q\right)_k . \tag{16.3.14}$$

Substitute for f from (16.3.14) into (16.3.13) and use (16.3.1) and (16.3.5) to get

$$\lambda\, w(x; \mathbf{t}\,|\,q)\, f(x)$$

$$= \frac{2t_1}{q-1} \sum_{k=1}^{n} (1 - q^k)\, a_k \mathcal{D}_q \left[\left(t_1 q^{1/2} e^{i\theta}, t_1 q^{1/2} e^{-i\theta}; q\right)_{k-1} w\left(x; q^{1/2}\mathbf{t}\,|\,q\right) \right]$$

$$= \frac{2t_1}{q-1} \sum_{k=0}^{n-1} (1 - q^{k+1})\, a_{k+1} \mathcal{D}_q \left[w\left(x; t_1 q^{k+1/2}, t_2 q^{1/2}, t_3 q^{1/2}, t_4 q^{1/2}\,|\,q\right) \right]$$

$$= \frac{4t_1}{(q-1)^2} \sum_{k=0}^{n-1} (1 - q^{k+1})\, a_{k+1} w\left(x; t_1 q^k, t_2, t_3, t_4\,|\,q\right)$$

$$\times \left[2x\left(1 - t_1 t_2 t_3 t_4 q^k\right) - t_2 - t_3 - t_4 + t_1 q^k\left(t_2 t_3 + t_2 t_4 + t_3 t_4 - 1\right) + t_2 t_3 t_4 \right].$$

Since

$$w\left(x; t_1 q^k, t_2, t_3, t_4\,|\,q\right) = w\left(x; t_1, t_2, t_3, t_4\,|\,q\right)\left(t_1 e^{i\theta}, t_1 e^{-i\theta}; q\right)_k,$$

we get

$$\lambda f(x) = \frac{4t_1}{(q-1)^2} \sum_{k=0}^{n-1} (1 - q^{k+1})\, a_{k+1} \left(t_1 e^{i\theta}, t_1 e^{-i\theta}; q\right)_k \left[2x\left(1 - t_1 t_2 t_3 t_4 q^k\right)\right.$$

$$\left. - t_2 - t_3 - t_4 + t_1 q^k\left(t_2 t_3 + t_2 t_4 + t_3 t_4 - 1\right) + t_2 t_3 t_4 \right]. \tag{16.3.15}$$

As in the proof of Theorem 16.3.1 we write

$$t_1 q^k \left[2x\left(1 - \sigma_4 q^k\right) - t_2 - t_3 - t_4 + t_1 q^k\left(t_2 t_3 + t_2 t_4 + t_3 t_4 - 1\right) + t_2 t_3 t_4\right]$$

$$= \left(1 - t_1 t_2 q^k\right)\left(1 - t_1 t_3 q^k\right)\left(1 - t_1 t_4 q^k\right)$$

$$- \left(1 - \sigma_4 q^k\right)\left(1 - t_1 q^k e^{i\theta}\right)\left(1 - t_1 q^k e^{-i\theta}\right). \tag{16.3.16}$$

We now use the relationships (16.3.14)–(16.3.16), and upon equating coefficients of $\left(t_1 e^{i\theta}, t_1 e^{-i\theta}; q\right)_k$ on both sides of (16.3.13) we obtain

$$\frac{(q-1)^2 \lambda}{4}\, a_k = q^{-k}\left(1 - q^{k+1}\right)\left(1 - t_1 t_2 q^k\right)\left(1 - t_1 t_3 q^k\right)\left(1 - t_1 t_4 q^k\right) a_{k+1}$$

$$- q^{1-k}\left(1 - \sigma_4 q^{k-1}\right)\left(1 - q^k\right) a_k.$$

Thus

$$a_{k+1} = \frac{\left[q\left(1 - \sigma_4 q^{k-1}\right)\left(1 - q^k\right) + q^k (q-1)^2 \lambda/4\right]}{\left(1 - q^{k+1}\right)\left(1 - t_1 t_2 q^k\right)\left(1 - t_1 t_3 q^k\right)\left(1 - t_1 t_4 q^k\right)}\, a_k. \tag{16.3.17}$$

This shows that $a_{n+1} = 0$ but $a_n \neq 0$ if and only if $\lambda = \lambda_n$. When $\lambda = \lambda_n$ it is straightforward to see that (16.3.17) shows that a_k is given by

$$a_k = \frac{q^k\left(q^{-n}, t_1 t_2 t_3 t_4 q^{n-1}; q\right)_k}{(q, t_1 t_2, t_1 t_3, t_1 t_4; q)_k}\, a_0. \tag{16.3.18}$$

Thus we have proved Theorem 16.3.4.   □

It is useful to write (16.3.6) in expanded form. Using (12.1.21)–(12.1.22) and the facts

$$\pi_2(x) := \frac{\mathcal{D}_q w(x, q^{1/2}\mathbf{t})}{w(x, \mathbf{t})} \tag{16.3.19}$$
$$= -q^{-1/2} \left[2\left(1 + \sigma_4\right) x^2 - \left(\sigma_1 + \sigma_3\right) x - 1 + \sigma_2 - \sigma_4 \right],$$

$$\pi_1(x) := \frac{\mathcal{A}_q w(x, q^{1/2}\mathbf{t})}{w(x, \mathbf{t})} = \frac{2\left[2\left(\sigma_4 - 1\right) x + \sigma_1 - \sigma_3\right]}{(1 - q)} \tag{16.3.20}$$

transforms (16.3.6) to

$$\pi_2(x)\mathcal{D}_q^2 y(x) + \pi_1(x)\mathcal{A}_q \mathcal{D}_q y(x) = \lambda_n y(x),$$

and λ_n is given by (16.3.7).

16.4 Connection Coefficients

We will establish the Nassrallah–Rahman integral evaluation

$$\int_0^\pi \frac{\left(e^{2i\theta}, e^{-2i\theta}; q\right)_\infty \left(\alpha e^{i\theta}, \alpha e^{-i\theta}; q\right)_\infty d\theta}{\prod_{j=1}^5 \left(a_j e^{i\theta}, a_j e^{-i\theta}; q\right)_\infty}$$

$$= \frac{2\pi \left(\alpha/a_4, \alpha a_4, a_1 a_2 a_3 a_4, a_1 a_3 a_4 a_5, a_2 a_3 a_4 a_5; q\right)_\infty}{(q, a_1 a_2 a_3 a_4^2 a_5; q)_\infty \prod_{1 \le j < k \le 5} \left(a_j a_k; q\right)_\infty}$$

$$\times {}_8W_7 \left(a_1 a_2 a_3 a_4^2 \beta/q; a_1 a_4, a_2 a_4, a_3 a_4, a_4 a_5, a_1 a_2 a_3 a_4 a_5/\beta; q, \alpha/a_4\right), \tag{16.4.1}$$

when $|a_j| < 1$; $1 \le j \le 5$. The proof depends on the following lemma.

Lemma 16.4.1 *We have the evaluation*

$$\int_0^\pi \frac{\left(e^{2i\theta}, e^{-2i\theta}; q\right)_\infty \left(\alpha e^{i\theta}, \alpha e^{-i\theta}; q\right)_n d\theta}{\prod_{j=1}^4 \left(a_j e^{i\theta}, a_j e^{-i\theta}; q\right)_\infty}$$

$$= \frac{2\pi \left(\alpha/a_4, \alpha a_4; q\right)_n \left(a_1 a_2 a_3 a_4; q\right)_\infty}{(q; q)_\infty \prod_{1 \le j < k \le 4} \left(a_j a_k; q\right)_\infty}$$

$$\times {}_4\phi_3 \left(\begin{matrix} q^{-n}, a_1 a_4, a_2 a_4, a_3 a_4 \\ \alpha a_4, a_1 a_2 a_3 a_4, q^{1-n} a_4/\alpha \end{matrix} \middle| \; q, q \right) \tag{16.4.2}$$

$$= \frac{2\pi \left(a_1 a_2 a_3 a_4; q\right)_\infty \left(\alpha a_4; q\right)_n}{(q; q)_\infty \prod_{1 \le j < k \le 4} \left(a_j a_k; q\right)_\infty}$$

$$\times \sum_{k=0}^n \frac{(q; q)_n}{(q; q)_k} \frac{\left(a_1 a_4, a_2 a_4, a_3 a_4; q\right)_k \left(\alpha/a_4; q\right)_{n-k}}{\left(\alpha a_4, a_1 a_2 a_3 a_4; q\right)_k (q; q)_{n-k}} \left(\frac{\alpha}{a_4}\right)^k,$$

where $|a_j| < 1$, $1 \le j \le 5$.

Proof Apply (12.2.16) with $b = \alpha$ and $a = a_4$ to see that the left-hand side of (16.4.2) is

$$\sum_{k=0}^{n} \frac{(q, \alpha a_4; q)_n \, (\alpha/a_4)^k}{(\alpha a_4, q; q)_k \, (q; q)_{n-k}} \, (\alpha/a_4; q)_{n-k} \, I\left(a_1, a_2, a_3, q^k a_4\right)$$

$$= \frac{2\pi \, (q, \alpha a_4; q)_n \, (a_1 a_2 a_3 a_4; q)_\infty}{(q; q)_\infty \displaystyle\prod_{1 \le j < k \le 4} (a_j a_k; q)_\infty}$$

$$\times \sum_{k=0}^{n} \frac{(a_1 a_4, a_2 a_4, a_3 a_4; q)_k \, (\alpha/a_4; q)_{n-k}}{(q, \alpha a_4, a_1 a_2 a_3 a_4; q)_k \, (q; q)_{n-k}} \, (\alpha/a_4)^k \,,$$

and the results follow. $\qquad\square$

Proof of (16.4.1) Let $\alpha = a_5 q^n$ and apply Lemma 16.4.1. Next apply (12.5.14) to the $_4\phi_3$ in Lemma 16.4.1 with the choices:

$$a = a_1 a_4, \quad b = a_2 a_4, \quad c = a_3 a_4, \quad e = a_1 a_2 a_3 a_4,$$
$$d = \alpha a_4, \quad f = q^{1-n} a_4/\alpha, \quad \mu = a_1 a_2 a_3 a_4^2 \alpha q^{n-1}.$$

This establishes the theorem when $\alpha = a_5 q^n$. Since both sides of (16.4.2) are analytic functions of α the identity theorem for analytic functions establishes the result. $\qquad\square$

We are now in a position to evaluate the connection coefficients in the expansion of an Askey–Wilson polynomial in similar polynomials.

Theorem 16.4.2 *The Askey–Wilson polynomials have the connection relation*

$$p_n(x; \mathbf{b}) = \sum_{k=o}^{n} c_{n,k}(\mathbf{a}, \mathbf{b}) \, p_k(x; \mathbf{a}), \tag{16.4.3}$$

where

$$c_{n,k}(\mathbf{b}, \mathbf{a}) = \frac{(q; q)_n \, b_4^{k-n} \, (b_1 b_2 b_3 b_4 q^{n-1}; q)_k \, (b_1 b_4, b_2 b_4, b_3 b_4; q)_n}{(q; q)_{n-k} \, (q, a_1 a_2 a_3 a_4 q^{k-1}; q)_k \, (b_1 b_4, b_2 b_4, b_3 b_4; q)_k}$$

$$\times q^{k(k-n)} \sum_{j,l \ge 0} \frac{(q^{k-n}, b_1 b_2 b_3 b_4 q^{n+k-1}, a_4 b_4 q^k; q)_{j+l} \, q^{j+l}}{(b_1 b_4 q^k, b_2 b_4 q^k, b_3 b_4 q^k; q)_{j+l} \, (q; q)_j (q; q)_l} \tag{16.4.4}$$

$$\times \frac{(a_1 a_4 q^k, a_2 a_4 q^k, a_3 a_4 q^k; q)_l \, (b_4/a_4; q)_j}{(a_4 b_4 q^k, a_1 a_2 a_3 a_4 q^{2k}; q)_l} \left(\frac{b_4}{a_4}\right)^l.$$

Proof Clearly the coefficients $c_{n,k}$ exist and are given by

$$h_k(\mathbf{a}) c_{n,k} = \left\langle \sqrt{1 - x^2} \, p_n(x; \mathbf{b}), w(x; \mathbf{a}) p_k(x; \mathbf{a}) \right\rangle, \tag{16.4.5}$$

where $\langle f, g \rangle$ is the inner product $\int_{-1}^{1} f\bar{g} \left(1 - x^2\right)^{-1/2} dx$, see (16.1.1). We use the integration by parts formula (16.1.3) and the Rodrigues formula (16.3.8) to find

$$
h_k(\mathbf{a}) c_{n,k} = \left[\frac{q-1}{2}\right]^k q^{k(k-1)/4} \left\langle \sqrt{1 - x^2}\, p_n(x; \mathbf{b}), \mathcal{D}_q^k w\left(x; q^{k/2}\mathbf{a}\right) \right\rangle
$$

$$
= \left[\frac{1-q}{2}\right]^k q^{k(k-1)/4} \left\langle \mathcal{D}_q^k p_n(x; \mathbf{b}), \sqrt{1 - x^2}\, w\left(x; q^{k/2}\mathbf{a}\right) \right\rangle
$$

$$
= q^{k(k-n)/2} \left(b_1 b_2 b_3 b_4 q^{n-1}; q\right)_k \frac{(q; q)_n}{(q; q)_{n-k}}
$$

$$
\times \left\langle p_{n-k}\left(x; q^{k/2}\mathbf{b}\right), \sqrt{1 - x^2}\, w\left(x; q^{k/2}\mathbf{a}\right) \right\rangle
$$

$$
= b_4^{k-n} \left(b_1 b_2 b_3 b_4 q^{n-1}; q\right)_k \left(b_1 b_4 q^k, b_2 b_4 q^k, b_3 b_4 q^k; q\right)_{n-k}
$$

$$
\times q^{k(k-n)} \frac{(q; q)_n}{(q; q)_{n-k}} \sum_{j=0}^{n-k} \frac{\left(q^{k-n}, b_1 b_2 b_3 b_4 q^{n+k-1}; q\right)_j}{\left(q, b_1 b_4 q^k, b_2 b_4 q^k, b_3 b_4 q^k; q\right)_j} q^j
$$

$$
\times \left\langle \phi_j\left(x; b_4 q^{k/2}\right), \sqrt{1 - x^2}\, w\left(x; q^{k/2}\mathbf{a}\right) \right\rangle.
$$

Using Lemma 16.4.1 we see that the j-sum is

$$
\frac{2\pi \left(a_1 a_2 a_3 a_4 q^{2k}; q\right)_\infty}{(q; q)_\infty \prod_{1 \le r < s \le 4} (a_r a_s q^k; q)_\infty}
$$

$$
\times \sum_{j=0}^{n-k} \frac{\left(q^{k-n}, b_1 b_2 b_3 b_4 q^{n+k-1}, a_4 b_4 q^k; q\right)_j}{\left(b_1 b_4 q^k, b_2 b_4 q^k, b_3 b_4 q^k; q\right)_j}
$$

$$
\times \sum_{l=0}^{j} \frac{\left(a_1 a_4 q^k, a_2 a_4 q^k, a_3 a_4 q^k; q\right)_l}{\left(q, a_4 b_4 q^k, a_1 a_2 a_3 a_4 q^{2k}; q\right)_l} \frac{(b_4/a_4; q)_{j-l}}{(q; q)_{j-l}} \left(\frac{b_4}{a_4}\right)^l,
$$

and some manipulations and the use of (12.2.12) one completes the proof. □

Theorem 16.4.2 is due to (Ismail & Zhang, 2005). Although Askey and Wilson (Askey & Wilson, 1985) only considered the case when $a_4 = b_4$, they were aware that the connection coefficients are double sums, as Askey kindly pointed out in a private conversation. To get the Askey–Wilson result set $a_4 = b_4$ in (16.4.4) to obtain

$$
c_{n,k}\left(b_1, b_2, b_3, a_4; a_1, a_2, a_3, a_4\right)
$$

$$
= \left(b_1 b_2 b_3 a_4 q^{n-1}; q\right)_k \frac{q^{k(k-n)}(q; q)_n \left(b_1 a_4 q^k, b_2 a_4 q^k, b_3 a_4 q^k; q\right)_{n-k}}{a_4^{n-k}(q; q)_{n-k}\left(q, a_1 a_2 a_3 a_4 q^{k-1}; q\right)_k} \tag{16.4.6}
$$

$$
\times {}_5\phi_4\left(\begin{matrix} q^{k-n}, b_1 b_2 b_3 a_4 q^{n+k-1}, a_1 a_4 q^k, a_2 a_4 q^k, a_3 a_4 q^k \\ b_1 a_4 q^k, \quad b_2 a_4 q^k, \quad b_3 a_4 q^k, \quad a_1 a_2 a_3 a_4 q^{2k} \end{matrix} \;\middle|\; q, q\right).
$$

Askey and Wilson also pointed out that if in addition to $a_4 = b_4$, we also have $b_j = a_j$ for $j = 2, 3$ then the ${}_5\phi_4$ becomes a ${}_3\phi_2$ which can be summed by the q-analogue of the Pfaff–Saalschütz theorem. This is evident from (16.4.6).

Now define an $(N + 1) \times (N + 1)$ lower triangular matrix $C(\mathbf{a}, \mathbf{b})$ whose n, k element is $c_{n,k}(\mathbf{a}, \mathbf{b})$, with $0 \le k \le n \le N$. Thus (16.4.3) is

$$X(\mathbf{b}) = C(\mathbf{b}, \mathbf{a}) X(\mathbf{a}), \qquad (16.4.7)$$

where $X(\mathbf{a})$ is a column vector whose jth component is $p_j(x; \mathbf{a})$, $0 \le j \le N$. Therefore the family of matrices $C(\mathbf{a}, \mathbf{b})$ has the property

$$C(\mathbf{c}, \mathbf{b})C(\mathbf{b}, \mathbf{a}) = C(\mathbf{c}, \mathbf{a}). \qquad (16.4.8)$$

Furthermore

$$[C(\mathbf{b}, \mathbf{a})]^{-1} = C(\mathbf{a}, \mathbf{b}). \qquad (16.4.9)$$

The implications of the orthogonality relation $C(\mathbf{b}, \mathbf{a}) C(\mathbf{a}, \mathbf{b}) = I$, I being the identity matrix are still under investigation.

We now prove the Nassrallah–Rahman integral, (16.4.1).

Theorem 16.4.3 *We have for* $|a_j| < 1$; $1 \le j \le 5$, *the integral evaluation* (16.4.1) *holds.*

Proof Let $a_5 = \alpha q^n$ and apply Lemma 16.4.1. Next apply (12.5.14) to the $_4\phi_3$ in Lemma 16.4.1 with the choices:

$$a = a_1 a_4, \quad b = a_2 a_4, \quad c = a_3 a_4, \quad e = a_1 a_2 a_3 a_4,$$
$$d = \alpha a_4, \quad f = q^{1-n} a_4/\alpha, \quad \mu = a_1 a_2 a_3 a_4^2 \alpha q^{n-1}.$$

This establishes the theorem when $a_5 = \alpha q^n$. Since both sides of (16.4.1) are analytic functions of α, the identity theorem for analytic functions establishes the result.
$\square$

16.5 Bethe Ansatz Equations of XXZ Model

In this section we show how to solve a generalization of the Bethe Ansatz equations for XXZ model using the Askey–Wilson operators. This section is based on (Ismail et al., 2005).

The one-dimensional $U_q\left(sl_2(\mathbb{C})\right)$-invariant XXZ model of spin $1/2$ of a size $2N$ with the open (Dirichlet) boundary condition described by the Hamiltonian (Sklyanin, 1988), (Kulish & Sklyanin, 1991),

$$H_{\text{XXZ}}^{(o)} = -\sum_{j=1}^{2N-1} \left(\sigma_j^1 \sigma_{j+1}^1 + \sigma_j^2 \sigma_{j+1}^2 + \triangle \sigma_j^3 \sigma_{j+1}^3\right) - \frac{q - q^{-1}}{2}\left(\sigma_1^3 - \sigma_{2N}^3\right),$$

$$(16.5.1)$$

where $\triangle := \left(q + q^{-1}\right)/2$ and σ_n^j are the Pauli matrices acting on the jth site:

$$\sigma^1 = \begin{pmatrix} 0 & 1 \\ 1 & 0 \end{pmatrix}, \quad \sigma^2 = \begin{pmatrix} 0 & -i \\ i & 0 \end{pmatrix}, \quad \sigma^3 = \begin{pmatrix} 1 & 0 \\ 0 & -1 \end{pmatrix}.$$

The diagonalization problem of the Hamiltonian has been investigated by means of solutions of the following Bethe Ansatz equations,

$$\left(\frac{\sin\left(\lambda_k + \frac{1}{2}\eta\right)}{\sin\left(\lambda_k - \frac{1}{2}\eta\right)}\right)^{2N} = \prod_{j\neq k, j=1}^{n} \frac{\sin\left(\lambda_k + \lambda_j + \eta\right)\sin\left(\lambda_k - \lambda_j + \eta\right)}{\sin\left(\lambda_k + \lambda_j - \eta\right)\sin\left(\lambda_k - \lambda_j - \eta\right)}, \quad 1 \le k \le n.$$

To solve this probem Ismail, Lin, and Roan considered a more general problem, namely the system of equations

$$\prod_{\ell=1}^{2N} \frac{\sin\left(\lambda_k + s_\ell\eta\right)}{\sin\left(\lambda_k - s_\ell\eta\right)} = \prod_{j\neq k, j=1}^{n} \frac{\sin\left(\lambda_k + \lambda_j + \eta\right)\sin\left(\lambda_k - \lambda_j + \eta\right)}{\sin\left(\lambda_k + \lambda_j - \eta\right)\sin\left(\lambda_k - \lambda_j - \eta\right)}, \quad 1 \le k \le n,$$

$$(16.5.2)$$

where s_ℓ's are $2N$ complex numbers. As we saw in the electrostatic equilibrium problems, the solution of (16.5.2) will determine the quantities $\{\cos\left(2\lambda_j\right)\}$ which are the roots of a polynomial solution of a Sturm–Liouville-type equation involving the Askey–Wilson operator. For $N = 2$, the system of equations (16.5.2) is solved by the zeros of the Askey–Wilson polynomials.

In the rest of this section we shall use the following convention

$$q = e^{2i\eta}, \quad \theta = 2\lambda, \quad (\text{hence } x = \cos 2\lambda). \tag{16.5.3}$$

For given functions $w(x)$, $p(x)$, $r(x)$, we consider the following q-Sturm–Liouville equation

$$\frac{1}{w(x)}\mathcal{D}_q\left((p(x)\mathcal{D}_q)\, y\right)(x) = r(x)y(x). \tag{16.5.4}$$

By (12.1.22), one can rewrite the equation (16.5.2) in the form

$$\Pi(x)\mathcal{D}_q^2 f(x) + \Phi(x)\left(\mathcal{A}_q\mathcal{D}_q f\right)(x) = r(x)f(x), \tag{16.5.5}$$

where the functions Π, Φ are defined by

$$\Pi(x) = \frac{1}{w(x)}\mathcal{A}_q p(x), \quad \Phi(x) = \frac{1}{w(x)}\mathcal{D}_q p(x). \tag{16.5.6}$$

The form (16.5.4) is the symmetric form of (16.5.6).

It readily follows that

$$\left(\mathcal{A}_q\mathcal{D}_q f\right)(x) = \frac{qe^{i\theta}}{(q-1)\left(qe^{2i\theta}-1\right)\left(e^{2i\theta}-q\right)}$$
$$\times \left\{\left(e^{2i\theta}-q\right)\eta_{q^2} - \left(qe^{2i\theta}-1\right)\eta_{q^{-2}} + (q-1)\left(e^{2i\theta}+1\right)\right\} f(x);$$

$$\mathcal{D}_q^2 f(x) = \frac{2q^{3/2}e^{i\theta}}{i(1-q)^2\sin\theta\left(qe^{2i\theta}-1\right)\left(e^{2i\theta}-q\right)}$$
$$\times \left\{\left(e^{2i\theta}-q\right)\eta_{q^2} + \left(qe^{2i\theta}-1\right)\eta_{q^{-2}} - (q+1)\left(e^{2i\theta}-1\right)\right\} f(x).$$

Thus a root $x_0 = \cos 2\lambda_0$ of a polynomial $f(x)$, that is $f(x_0) = 0$, necessarily satisfies the equation

$$\left\{\left(e^{4i\lambda_0}-q\right)\left(\Pi(x_0) - \Phi(x_0)\sin\eta\sin 2\lambda_0\right)\eta_{q^2}\right\} f(x_0)$$
$$+ \left\{\left(qe^{4i\lambda_0}-1\right)\left(\Pi(x_0) + \Phi(x_0)\sin\eta\sin 2\lambda_0\right)\eta_{q^{-2}}\right\} f(x_0) = 0$$

or, equivalently,

$$\left(\frac{\eta_{q^2} f}{\eta_{q^{-2}} f}\right)(x_0) = \frac{-\sin(2\lambda_0 + \eta)(\Pi(x_0) + \Phi(x_0)\sin\eta\sin 2\lambda_0)}{\sin(2\lambda_0 - \eta)(\Pi(x_0) - \Phi(x_0)\sin\eta\sin 2\lambda_0)}. \quad (16.5.7)$$

For a polynomial $f(x)$ of degree n with distinct simple roots $x_1, \ldots, x_n$, one writes

$$f(x) = \gamma \prod_{j=1}^{n}(x - x_j) = \gamma \prod_{j=1}^{n}(\cos 2\lambda - \cos 2\lambda_j) \quad \gamma \neq 0.$$

It is straightforward to see that

$$\eta_{q^2} f(x_k) = \frac{\gamma}{2^n} \prod_{j=1}^{n}\left(q e^{2i\lambda_k} - e^{2i\lambda_j} + q^{-1}e^{-2i\lambda_k} - e^{-2i\lambda_j}\right),$$

$$= (-1)^n \gamma \prod_{j=1}^{n} \sin(\lambda_k + \lambda_j + \eta)\sin(\lambda_k - \lambda_j + \eta).$$

Similarly

$$\eta_{q^{-2}} f(x_k) = (-1)^n \gamma \prod_{j=1}^{n} \sin(\lambda_k + \lambda_j - \eta)\sin(\lambda_k - \lambda_j - \eta).$$

Now observe that (16.5.7) indicates that the roots $x_1, \ldots, x_n$ of $f(x)$ satisfy the system of equations,

$$\frac{-\sin(2\lambda_k + \eta)[\Pi(x_k) + \Phi(x_k)\sin\eta\sin 2\lambda_k]}{\sin(2\lambda_k - \eta)[\Pi(x_k) - \Phi(x_k)\sin\eta\sin 2\lambda_k]}$$
$$= \prod_{j=1}^{n} \frac{\sin(\lambda_k + \lambda_j + \eta)\sin(\lambda_k - \lambda_j + \eta)}{\sin(\lambda_k + \lambda_j - \eta)\sin(\lambda_k - \lambda_j - \eta)}, \quad 1 \leq k \leq n.$$

In other words we arrive at the system of nonlinear equations

$$\frac{\Pi(x_k) + \Phi(x_k)\sin\eta\sin 2\lambda_k}{\Pi(x_k) - \Phi(x_k)\sin\eta\sin 2\lambda_k} = \prod_{j\neq k, j=1}^{n} \frac{\sin(\lambda_k + \lambda_j + \eta)\sin(\lambda_k - \lambda_j + \eta)}{\sin(\lambda_k + \lambda_j - \eta)\sin(\lambda_k - \lambda_j - \eta)},$$

$$(16.5.8)$$

for $1 \leq k \leq n$, which we call the Bethe Ansatz equations associated with $\Pi(x), \Phi(x)$. As in the problem of Heine and Stieltjes (which will be described in Chapter 20), we shall only consider those q-Sturm–Liouville problem where the coefficients $\Pi(x)$, $\Phi(x)$, $r(x)$ of (16.5.5) are polynomials in x with the degrees

$$\deg \Pi = 1 + \deg \Phi = 2 + \deg r = N \geq 2. \quad (16.5.9)$$

For $2N$ complex numbers t_j, $1 \leq j \leq 2N$, we denote

$$\mathbf{t} = (t_1, \ldots, t_{2N}),$$

and σ_j the j-th elementary symmetric function of t_i's for $0 \leq j \leq 2N$, with $\sigma_0 := 1$. We define the possibly signed weight function $w(x, \mathbf{t})$,

$$w(x, \mathbf{t}) := \frac{\left(e^{iN\theta}, e^{-iN\theta}; q^{N/2}\right)_{\infty}}{\sin(N\theta/2) \prod_{j=1}^{2N}\left(t_j e^{i\theta}, t_j e^{-i\theta}; q\right)_{\infty}}, \quad (16.5.10)$$

which can also be written in the following form,

$$
w(x, \mathbf{t}) = \frac{2ie^{-iN\theta/2} \left(e^{iN\theta}, q^{N/2}e^{-iN\theta}; q^{N/2} \right)_\infty}{\prod\limits_{j=1}^{2N} \left(t_j e^{i\theta}, t_j e^{-i\theta}; q \right)_\infty}
$$

$$
= \frac{-2ie^{iN\theta/2} \left(q^{N/2}e^{iN\theta}, e^{-iN\theta}; q^{N/2} \right)_\infty}{\prod_{j=1}^{2N} \left(t_j e^{i\theta}, t_j e^{-i\theta}; q \right)_\infty}.
$$

(16.5.11)

With $w(x) = w(x, \mathbf{t})$, $p(x) = w\left(x, q^{1/2}\mathbf{t}\right)$ in (16.5.2), we shall consider the following equations,

$$
\frac{1}{w(x, \mathbf{t})} \mathcal{D}_q \left(\left(w\left(x, q^{1/2}\mathbf{t}\right) \mathcal{D}_q \right) y \right)(x) = r(x)y(x),
$$

(16.5.12)

where $r(x)$ is a polynomial of degree at most $N - 2$. With the notation

$$
\Pi(x; \mathbf{t}) = \frac{1}{w(x, \mathbf{t})} \mathcal{A}_q w\left(x, q^{1/2}\mathbf{t}\right), \quad \Phi(x; \mathbf{t}) = \frac{1}{w(x, \mathbf{t})} \mathcal{D}_q w\left(x, q^{1/2}\mathbf{t}\right),
$$

(16.5.13)

equation (16.5.12) becomes

$$
\Pi(z; \mathbf{t})\mathcal{D}_q^2 y + \Phi(z; \mathbf{t}) \mathcal{A}_q \mathcal{D}_q y = r(x)y.
$$

(16.5.14)

This generalizes the Askey–Wilson equation since when $N = 2$, $r(x) \equiv r \in \mathbb{R}$ and $t_j \in (-1, 1)$, $1 \leq j \leq 4$ and $0 < q < 1$, the weight function $w_\mathbf{t}(x)$ is a positive function on $(-1, 1)$, and the solutions of the equation (16.5.12) are the Askey–Wilson polynomials.

Theorem 16.5.1 *The functions* $\Pi(x; \mathbf{t})$, $\Phi(x; \mathbf{t})$ *are polynomials of* x *of degree* N, $N - 1$ *respectively, and have the following explicit forms,*

$$
\Pi(x; \mathbf{t}) = -q^{-N/4} \left\{ (-1)^N \sigma_N + \sum_{\ell=0}^{N-1} (-1)^\ell \left(\sigma_\ell + \sigma_{2N-\ell} \right) T_{N-\ell}(x) \right\},
$$

$$
\Phi(x; \mathbf{t}) = \frac{2q^{-N/4}}{q^{1/2} - q^{-1/2}} \sum_{\ell=0}^{N-1} (-1)^\ell \left(\sigma_\ell - \sigma_{2N-\ell} \right) U_{N-\ell-1}(x).
$$

(16.5.15)

Conversely, for given polynomials Π *and* Φ *of degrees* N *and* $N - 1$ *respectively, there is a unique* $2N$-*element set* $\{t_j : 1 \leq j \leq 2N\}$ *such that* $\Pi(x) = \Pi(x, \mathbf{t})$ *and* $\Phi(x) = \Phi(x; \mathbf{t})$ *and* (16.5.13) *holds.*

Proof Assume that w is given by (16.5.10). A calculation using (16.5.13) shows that

$\Pi(x;\mathbf{t})$ is given by

$$\frac{iq^{N/4}\sin(N\theta/2)\displaystyle\prod_{j=1}^{2N}\left(t_je^{i\theta},t_je^{-i\theta};q\right)_\infty}{\left(e^{iN\theta},e^{-iN\theta};q^{N/2}\right)_\infty}$$

$$\times\left[\frac{e^{-iN\theta/2}\left(q^{N/2}e^{iN\theta},e^{-iN\theta};q^{N/2}\right)_\infty}{\displaystyle\prod_{j=1}^{2N}\left(t_jqe^{i\theta},t_je^{-i\theta};q\right)_\infty}-\frac{e^{iN\theta/2}\left(e^{iN\theta},q^{N/2}e^{-iN\theta};q^{N/2}\right)_\infty}{\displaystyle\prod_{j=1}^{2N}\left(t_je^{i\theta},t_jqe^{-i\theta};q\right)_\infty}\right],$$

which indicates that $\Pi(x;\mathbf{t})$ is of the form

$$= iq^{N/4}\sin(N\theta/2)\left[\frac{e^{-iN\theta/2}}{1-e^{iN\theta}}\prod_{j=1}^{2N}\left(1-t_je^{i\theta}\right)-\frac{e^{iN\theta/2}}{1-e^{-iN\theta}}\prod_{j=1}^{2N}\left(1-t_je^{-i\theta}\right)\right]$$

$$= \frac{-q^{-N/4}}{2}\left[e^{-iN\theta}\prod_{j=1}^{2N}\left(1-t_je^{i\theta}\right)+e^{iN\theta}\prod_{j=1}^{2N}\left(1-t_je^{-i\theta}\right)\right]$$

$$= \frac{-q^{-N/4}}{2}\sum_{\ell=0}^{2N}(-1)^\ell\sigma_\ell\left(e^{i(\ell-N)\theta}+e^{i(N-\ell)\theta}\right)=-q^{-N/4}\sum_{\ell=0}^{2N}(-1)^\ell\sigma_\ell\cos(N-\ell)\theta$$

$$= -q^{-N/4}\left\{(-1)^N\sigma_N+\sum_{l=0}^{N-1}(-1)^\ell\left(\sigma_\ell+\sigma_{2N-\ell}\right)T_{N-\ell}(x)\right\}.$$

Similarly we prove that $\Phi(x,\mathbf{t})$ is as in (16.5.15). To see the converse statement, given Π and Φ we expand them in Chebyshev polynomials of the first and second kinds, respectively, then define σ_0 by $\sigma_0=1$ and σ_N by $(-1)^{N+1}q^{N/4}$ times the constant term in the expansion (16.5.14) in terms of Chebyshev polynomials. Then define the remaining σ's through finding $\sigma_\ell\pm\sigma_{2N-\ell}$ from the coefficients in Π and Φ in (16.5.15). $\square$

It is important to note that Theorem 16.5.1 gives a constructive way of identifying the parameters $t_1,\ldots,t_{2N}$ and $w(x;\mathbf{t})$ from the functional equation (16.5.14).

Theorem 16.5.2 *Let $t_\ell=q^{-s_\ell}=e^{-2i\eta s_\ell}$ for $1\le\ell\le 2N$. The Bethe Ansatz equations (16.5.8) associated with the polynomials $\Pi(x;\mathbf{t})$, $\Phi(x;\mathbf{t})$ have the form (16.5.2).*

Proof From the proof of Theorem 16.5.1, we conclude that

$$\Pi(x;\mathbf{t})=\frac{-q^{-N/4}}{2}\left[e^{-2iN\lambda_k}\prod_{j=1}^{2N}\left(1-t_je^{2i\lambda_k}\right)+e^{2iN\lambda_k}\prod_{j=1}^{2N}\left(1-t_je^{-2i\lambda_k}\right)\right],$$

$$\Phi(x_k;\mathbf{t})\sin\eta\sin 2\lambda_k=\frac{q^{-N/4}}{2}\left[e^{-2iN\lambda_k}\prod_{j=1}^{2N}\left(1-t_je^{2i\lambda_k}\right)\right.$$

$$\left.-e^{2iN\lambda_k}\prod_{j=1}^{2N}\left(1-t_je^{-2i\lambda_k}\right)\right],$$

hence

$$\Pi\left(x_k;\mathbf{t}\right)+\Phi\left(x_k;\mathbf{t}\right)\sin\eta\sin 2\lambda_k=-q^{-N/4}e^{2iN\lambda_k}\prod_{j=1}^{2N}\left(1-t_je^{-2i\lambda_k}\right)$$

$$=-q^{-N/4}e^{-2iN\lambda_k}\prod_{j=1}^{2N}\left(e^{2i\lambda_k}-t_j\right),$$

$$\Pi\left(x_k;\mathbf{t}\right)-\Phi\left(x_k;\mathbf{t}\right)\sin\eta\sin 2\lambda_k=-q^{-N/4}e^{-2iN\lambda_k}\prod_{j=1}^{2N}\left(1-t_je^{2i\lambda_k}\right).$$

By substituting $t_j=q^{-s_j}$ in the above formula, the result of this theorem follows from (16.5.8). $\qquad\square$

If $|q|>1$ replace q by $1/p$, use the invariance of $\mathcal{D}_q$ and $\mathcal{A}_q$ under $q\to q^{-1}$ to rederive 16.5.2 with q replaced by $1/p$. This covers the case $|q|>1$.

As in the electrostatic equilibrium problem we have transformed the problem of solving a system of nonlinear equations, the Bethe Ansatz equations (16.5.8), to the problem of finding a polynomial solution of a functional equation, (16.5.14) whose zeros x_j, $1\le j\le n$, solve the system (16.5.8).

Theorem 16.5.3 *Let $N=2$ and $\eta=i\zeta$, $\zeta>0$. Then for all n the system (16.5.2) has a unique solution provided that $s_j<0$, $1\le j\le 4$. Furthermore all the λ's are in $(0,\pi/2)$.*

Proof Let y be a polynomial of degree n with zeros $\cos\left(2\lambda_j\right)$, $1\le j\le n$. We know that (16.5.2) implies the validity of (16.5.5) for $x=\cos\left(2\lambda_j\right)$. Here Π_1 and Φ of degrees 2 and 1, respectively. Thus Theorem 16.3.3 shows that y must be a multiple of an Askey–Wilson polynomial. Since the Askey–Wilson polynomials are orthogonal on $[-1,1]$, all their zeros are in $(-1,1)$. $\qquad\square$

Baxter introduced his T-Q equation in (Baxter, 1982) and used it to diagonalize the XYZ Hamiltonian. For the XYZ model, the T-Q equation takes the form

$$a(z)Q(qz)+b(z)Q(z)+c(z)Q(z/q)=0.$$

The difference between (16.5.14) and the above is that (16.5.14) is written in the polynomial variable x. For additional information on Baxter's Q-operator, see (Derkachov et al., 2003). The algebraic Bethe Ansatz for integrable systems is explained in (Faddeev, 1998). The interested reader may also consult (Kulish & Sklyanin, 1982), (Takhtadzhan, 1982), and (Takhtadzhan & Faddeev, 1979).

Van Diejen observed that Theorem 16.5.3 identifies the equilibrium configurartions of the BC-type Ruijsenaars–Schneider system with the zeros of Askey-Wilson polynomials, (Van Diejen, 2005).

The Bethe Ansatz equations of the Heisenberg XXX spin chain can be handled in a similar fashion. The Hamiltonian is

$$H_{XXX}=-J\sum_{j=1}^{L}\left(\sigma_j^1\sigma_{j+1}^1+\sigma_j^2\sigma_{j+1}^2+\sigma_j^3+\sigma_{j+1}^3-1\right),\quad J<0.$$

This was proposed in (Heisenberg, 1928) and solved in (Bethe, 1931). We write the variable z in the form,

$$z = e^{i\theta} = q^{-iy},$$

and again the parameters, $t_j = q^{-s_j}$. A calculation shows that as $q \to 1$, the transformation $\check{f}(z)$ and the operators $\eta_{q^{\pm 1}}$, $\mathcal{D}_q$, $\mathcal{A}_q$ become $\check{f}(y)$, $\eta_{\pm}$, W, A, respectively, where

$$\check{f}(y) := f(x) \quad \text{with } x = y^2;$$

$$(\eta_{\pm} f)(x) := \check{f}\left(y \pm \frac{i}{2}\right)$$

$$(Wf)(x) := \frac{1}{2yi}\left(\eta_+ f - \eta_- f\right)(x) \qquad (16.5.16)$$

$$(Af)(x) := \frac{1}{2}\left(\eta_+ f + \eta_- f\right)(x).$$

The above divided difference operator W is called the Wilson operator (Wilson, 1980). We have the relation

$$W(fg) = (Wf)(Ag) + (Af)(Wg).$$

Analogous to the q-Sturm–Liouville problem (16.5.4), we consider the following difference equation

$$\frac{1}{w(x)} W\left(p(x)Wf\right)(x) = r(x)f(x)$$

which is equivalent to the Sturm–Liouville problem in the form,

$$\Pi(x)W^2 f(x) + \Phi(x)(AWf)(x) = r(x)f(x), \qquad (16.5.17)$$

where Π, Φ are the functions defined by

$$\Pi(x) = \frac{1}{w(x)} Ap(x), \quad \Phi(x) = \frac{1}{w(x)} Wp(x). \qquad (16.5.18)$$

We seek polynomial solutions $f(x)$ to (16.5.17) under the assumptions that $\Pi(x)$, $\Phi(x)$, and $r(x)$ are polynomials of degrees N, $N-1$ and $N-2$, respectively.

From the relations

$$AWf(x) = \frac{1}{i(4y^2 + 1)}\left\{(y - i/2)\eta_+^2 - (y + i/2)\eta_-^2 + i\right\}f(x);$$

$$W^2 f(x) = \frac{-1}{y(4y^2 + 1)}\left\{(y - i/2)\eta_+^2 + (y + i/2)\eta_-^2 - 2y\right\}f(x),$$

it follows that if $f(x_0) = 0$ for $x_0 = y_0^2$, then

$$\left(\frac{\eta_+^2 f}{\eta_-^2 f}\right)(x_0) = \frac{-(y_0 + i/2)(\Pi(x_0) - iy_0\Phi(x_0))}{(y_0 - i/2)(\Pi(x_0) + iy_0\Phi(x_0))}. \qquad (16.5.19)$$

Let $f(x)$ be a polynomial of degree n with zeros x_j for $1 \le j \le n$, and with $x = y^2$, $x_j = y_j^2$, we see that

$$f(x) = \gamma \prod_{j=1}^{n} (x - x_j) = \gamma \prod_{j=1}^{n} (y^2 - y_j^2), \quad \gamma \neq 0.$$

Thus

$$\eta_\pm^2 f\left(x_k\right) = \gamma \prod_{j=1}^{n} \left(y_k - y_j \pm i\right)\left(y_k + y_j \pm i\right).$$

By (16.5.19), y_j's satisfy the following system of equations,

$$\frac{\Pi\left(x_k\right) - y_k i\Phi\left(x_k\right)}{\Pi\left(x_k\right) + y_k i\Phi\left(x_k\right)} = \prod_{j\neq k,j=1}^{n} \frac{\left(y_k - y_j + i\right)\left(y_k + y_j + i\right)}{\left(y_k - y_j - i\right)\left(y_k + y_j - i\right)}, \quad 1 \le k \le n.$$

$$(16.5.20)$$

In this case the analogue of the function w in (16.5.10) is

$$\omega(x,\mathbf{s}) := \frac{\prod_{\ell=1}^{2N} \Gamma\left(-s_\ell + iy\right)\Gamma\left(-s_\ell - iy\right)}{\Gamma(iNy)\Gamma(-iNy)}, \qquad (16.5.21)$$

where $\mathbf{s} = (s_1, \ldots, s_{2N})$. Consequently

$$p(x) = \omega\left(x, \mathbf{s} - \frac{1}{2}\right), \qquad \frac{1}{2} := \left(\frac{1}{2}, \ldots, \frac{1}{2}\right).$$

Therefore the polynomials Π and Φ are

$$\Pi(x;\mathbf{s}) = \frac{1}{\omega(x,\mathbf{s})} A\omega\left(x, \mathbf{s} - \frac{1}{2}\right), \qquad \Phi(x;\mathbf{s}) = \frac{1}{\omega(x,\mathbf{s})} W\omega\left(x, \mathbf{s} - \frac{1}{2}\right).$$

Theorem 16.5.4 *For a given* $\mathbf{s} = (s_1, \ldots, s_{2N})$ *with an even* N, *denote* ς_j *the* j-*th elementary symmetric function of* s_i's *for* $0 \le j \le 2N$, $(\varsigma_0 := 1)$. *Then* $\Pi(x;\mathbf{s})$, $x\Phi(x;\mathbf{s})$ *are the polynomials of* x *of degree at most* N *with following expressions,*

$$\Pi(x;\mathbf{t}) = (-1)^{\frac{N}{2}} \left\{ x^N + \sum_{j=0}^{N-1} (-1)^j \left(\frac{1}{2}\varsigma_{2j+1} - \varsigma_{2j+2}\right) x^{N-j-1} \right\},$$

$$x\Phi(x;\mathbf{t}) = (-1)^{\frac{N}{2}} \left\{ \sum_{j=0}^{N-1} (-1)^j \left(\frac{1}{2}\varsigma_{2j} - \varsigma_{2j+1}\right) x^{N-l\ell} + \frac{1}{2}\varsigma_{2N} \right\}.$$

The roots, $x_k = y_k^2$, $k = 1, \ldots, n$, *of a degree* n *polynomial solution* $f(x)$ *of the Sturm–Liouville problem* (16.5.17) *satisfy the following Bethe Ansatz type equations,*

$$\frac{y_k - i/2}{y_k + i/2} \prod_{\ell=1}^{2N} \frac{y_k + s_\ell i}{y_k - s_\ell i} = \prod_{j\neq k,j=1}^{n} \frac{\left(y_k - y_j + i\right)\left(y_k + y_j + i\right)}{\left(y_k - y_j - i\right)\left(y_k + y_j - i\right)}, \quad 1 \le k \le n.$$

$$(16.5.22)$$

The proof is similar to the proofs of Theorems 16.5.1–16.5.2 and will be omitted

Remark 16.5.1 *For the Bethe Ansatz equations* (16.5.22) *with* N *odd, one can reduce the problem to the above theorem for some even* N' *by adding certain zero-value* s_j's. *By the similar method, one enables to apply the above theorem to Bethe Ansatz*

problem of the following type with $s_1, \ldots, s_M \in \mathbb{C}$ and a positive integer M,

$$\prod_{\ell=1}^{M} \frac{y_k + s_\ell i}{y_k - s_\ell i} = \prod_{\substack{j \neq k, j=1}}^{n} \frac{(y_k - y_j + i)(y_k + y_j + i)}{(y_k - y_j - i)(y_k + y_j - i)}, \quad 1 \leq k \leq n, \quad (16.5.23)$$

There is extensive literature on the XXX model and we refer the interested reader to (Babujian, 1983) and (Takhtadzhan, 1982).

Exercises

16.1   Show that if $f \in H_{1/2}$ then

$$\left(\mathcal{D}_q^{-1}\mathcal{D}_q - \mathcal{D}_q\mathcal{D}_q^{-1}\right) f(x) = \int_0^\pi f(\cos\theta)\, d\theta.$$

16.2   Show that the zeros of a continuous q-ultraspherical polynomial $C_n(x; \beta \mid q)$ solve the nonlinear system

$$\frac{\sin(2\lambda_k + s)\sin(2\lambda_k + s + \eta)}{\sin(2\lambda_k - s)\sin(2\lambda_k - s - \eta)} = \prod_{\substack{j \neq k, j=1}}^{n} \frac{\sin(\lambda_k + \lambda_j - \eta)\sin(\lambda_k - \lambda_j - \eta)}{\sin(\lambda_k + \lambda_j + \eta)\sin(\lambda_k - \lambda_j + \eta)},$$

for $1 \leq k \leq n$. Identify β and q in terms of s and η.

16.3   Let $\mathbf{t} = (t_1, t_2, t_3, t_4)$ and let $h_n(\mathbf{t})$ denote the coefficient of $\delta_{m,n}$ in the right-hand side of (15.2.4). Assume that f has the expansion

$$f(x) \sim \sum_{n=0}^{\infty} c_n p_n(x, \mathbf{t}).$$

(a) Show that

$$c_n = \frac{(1-q)^n q^{n(n-1)/4}}{2^n h_n(\mathbf{t})} \int_{-1}^{1} w\left(x, q^{n/2}\mathbf{t}\right) \mathcal{D}_q^n f(x)\, dx.$$

(b) Which class of functions can be expanded as above where c_n is given by (a)?

17

q-Hermite Polynomials on the Unit Circle

In this chapter we study unit circle analogues of the q-Hermite polynomials and their generalizations. We present a four-parameter family of biorthogonal rational functions.

17.1 The Rogers–Szegő Polynomials

We have already investigated the polynomials orthogonal on $[-1, 1]$ with respect to the weight function $\left(e^{2i\theta}, e^{-2i\theta}; q\right)_\infty / \sqrt{1 - x^2}$, $x = \cos\theta$. The key idea is to expand the weight function using the Jacobi triple product identity. We wish to construct the polynomials orthogonal on the unit circle with respect to the weight function

$$w_c(z \mid q) := \left(q^{1/2}z, q^{1/2}/z; q\right)_\infty. \qquad (17.1.1)$$

We used the subscript c for "circle." It is clear that $w_c(z \mid q)$ is positive on the unit circle. Assume that the polynomials orthogonal on $|z| = 1$ with respect to the measure $w_c(z \mid q)dz/(iz)$ are $\{\mathcal{H}_n(z \mid q)\}$ and

$$\mathcal{H}_n(z \mid q) = \sum_{k=0}^{n} \frac{(q^{-n}; q)_k}{(q; q)_k} a_{n,k} z^k. \qquad (17.1.2)$$

Theorem 17.1.1 *The polynomials $\{\mathcal{H}_n(z \mid q)\}$ are given by*

$$\mathcal{H}_n(z \mid q) = \sum_{k=0}^{n} \frac{(q; q)_n}{(q; q)_k (q; q)_{n-k}} \left(q^{-1/2}z\right)^k, \qquad (17.1.3)$$

and satisfy the orthogonality relation

$$\frac{1}{2\pi i} \int_{|z|=1} \overline{\mathcal{H}_m(z \mid q)} \, \mathcal{H}_n(z \mid q) w_c(z \mid q) \, \frac{dz}{z} = q^{-n} \frac{(q; q)_n}{(q; q)_\infty} \delta_{m,n}. \qquad (17.1.4)$$

Proof It suffices to compute the integrals

$$I_{n,j} := \frac{1}{2\pi i} \int_{|z|=1} \bar{z}^j \, \mathcal{H}_n(z \mid q) w_c(z \mid q) \, \frac{dz}{z}, \qquad (17.1.5)$$

for $0 \le j \le n$. Therefore, by the Jacobi triple product identity

$$
I_{n,j} = \sum_{k=0}^{n} \frac{(q^{-n};q)_k \, a_{n,k}}{(q;q)_k(q;q)_\infty} \sum_{-\infty < r < \infty} (-1)^r q^{r^2/2} \frac{1}{2\pi} \int_0^{2\pi} e^{i(r-j+k)\theta} \, d\theta
$$

$$
= \sum_{k=0}^{n} \frac{(q^{-n};q)_k \, a_{n,k}}{(q;q)_k(q;q)_\infty} (-1)^{k-j} q^{(k-j)^2/2}.
$$

In order to sum the above series we need to get rid of the factor $q^{k^2/2}$. With a little experimentation we make the choice

$$
a_{n,k} = (-1)^k q^{nk} q^{-k^2/2}. \tag{17.1.6}
$$

With the above choice we get

$$
I_{n,j} = \frac{q^{j^2/2}(-1)^j}{(q;q)_\infty} {}_1\phi_0\left(q^{-n};-;q,q^{n-j}\right) = \frac{q^{j^2/2}(-1)^j}{(q;q)_\infty} \left(q^{-j};q\right)_n,
$$

where we used (12.2.22). After some simplification we establish

$$
I_{n,j} = \frac{(q;q)_n}{(q;q)_\infty} q^{-n/2} \delta_{j,n}, \quad 0 \le j \le n.
$$

The choice (17.1.6) puts $\mathcal{H}_n(z \mid q)$ of (17.1.2) in the form (17.1.3). To prove (17.1.4) observe that its left-hand side is

$$
\frac{(q;q)_n}{(q;q)_0(q;q)_n} q^{-n/2} I_{n,n} \delta_{m,n},
$$

which is the right-hand side of (17.1.4) and the proof is complete. $\qquad\square$

The polynomials $\{\mathcal{H}_n(z \mid q)\}$ are called the Rogers–Szegő polynomials because they resemble the continuous q-Hermite polynomials of Rogers whose work inspired Szegő (Szegő, 1926) to consider them. Szegő (Szegő, 1926) was motivated to consider them by his desire to give a nontrivial example of his theory of orthogonal polynomials on the unit circle. An exposition of the Szegő theory is available in (Grenander & Szegő, 1958) and (Simon, 2004).

Our next goal is to find a generating function for the polynomials $\{\mathcal{H}_n(z \mid q)\}$ and record the integral implied by the orthogonality relation (17.1.4).

Multiply (17.1.3) by $t^n/(q;q)_n$ and add for $n = 0, 1, \ldots$, then apply (12.2.22). This establishes the generating function

$$
\sum_{n=1}^{\infty} \mathcal{H}_n(z \mid q) \frac{t^n}{(q;q)_n} = \frac{1}{\left(t, q^{-1/2} tz; q\right)_\infty}. \tag{17.1.7}
$$

From (17.1.3) it follows that

$$
\max\left\{\mathcal{H}_n(z \mid q) : |z| < r\right\} = \mathcal{H}_n(r \mid q) \tag{17.1.8}
$$

Theorem 17.1.2 *The Ramanujan q-beta integral*

$$
\frac{1}{2\pi i} \int_{|z|=1} \frac{\left(q^{1/2}z, q^{1/2}/z; q\right)_\infty}{\left(q^{-1/2}t_1 z, q^{-1/2}t_2/z; q\right)_\infty} \frac{dz}{z} = \frac{(t_1, t_2; q)_\infty}{(q, t_1 t_2/q; q)_\infty}. \tag{17.1.9}
$$

is equivalent to the orthogonality relation (17.1.4).

Proof Divide the left-hand side of (17.1.9) by $(t_1, t_2; q)_\infty$ then expand

$$\frac{(q^{1/2}z, q^{1/2}/z; q)_\infty}{(t_1, t_2, q^{-1/2}t_1 z, q^{-1/2}t_2/z; q)_\infty}$$

in powers of t_1 and t_2 using the generating function (17.1.7). We can the interchange the sums with integration since $|\mathcal{H}_n(z \,|\, q)| \leq \mathcal{H}_n(1 \,|\, q)$ for $|z| \leq 1$, by (17.1.8), and $\mathcal{H}_n(1 \,|\, q) = q^{-n/2}(q; q)_n / (\sqrt{q}; \sqrt{q})_n$.

From here it readily follows that (17.1.9) is equivalent to (17.1.4) and the proof is complete. $\square$

The operator $\mathcal{D}_{q,z}$ of (12.1.12) acts nicely on the $\mathcal{H}_n$'s. It is straightforward to derive

$$\mathcal{D}_{q,z}\mathcal{H}_n(z \,|\, q) = \frac{q^{-1/2}(1 - q^n)}{1 - q} \mathcal{H}_{n-1}(z \,|\, q). \qquad (17.1.10)$$

This shows that $\mathcal{D}_{q,z}$ acts as a lowering operator on the $\mathcal{H}_n$'s. In order to find a raising operator we need to compute the adjoint of $\mathcal{D}_{q,z}$ with respect to a suitable inner product.

Consider the inner product

$$\langle f, g \rangle_c := \frac{1}{2\pi i} \int\limits_{|z|=1} f(z) \overline{g(z)} \frac{dz}{z}, \qquad (17.1.11)$$

defined for functions analytic in a domain containing the closed unit disc. If f is analytic in $r_1 < |z| < r_2$ then $\overline{f}$ will denote the function whose Laurent coefficients are the complex conjugates of the corresponding Laurent coefficients of f. Thus $\overline{f}$ is also analytic in $r_1 < |z| < r_2$. Let

$$\mathcal{F}_\nu := \left\{ f : f(z) \text{ is analytic for } q^\nu \leq |z| \leq q^{-\nu} \right\}. \qquad (17.1.12)$$

The space $\mathcal{F}_\nu$ is an inner product space with the inner product (17.1.11). It is clear that if $f \in \mathcal{F}_\nu$ then $\overline{f} \in \mathcal{F}_\nu$.

Theorem 17.1.3 *The adjoint of the q-difference operator $\mathcal{D}_{q,z}$ on $\mathcal{F}_1$ is $T_{q,z}$,*

$$(T_{q,z}f)(z) = \frac{z[f(z) - qf(qz)]}{1 - q}, \qquad (17.1.13)$$

that is

$$\langle \mathcal{D}_{q,z}f, g \rangle_c = \langle f, T_{q,z}g \rangle_c, \quad \text{for } f, g \in \mathcal{F}_1. \qquad (17.1.14)$$

Proof We have

$$
\langle \mathcal{D}_{q,z} f, g \rangle_c = \frac{1}{2\pi i} \int\limits_{|z|=1} \frac{f(z) - f(qz)}{(1-q)z} \, \overline{g}(z) \, \frac{dz}{z}
$$

$$
= \frac{1}{2\pi i} \int\limits_{|z|=1} \frac{f(z)}{(1-q)z} \, \overline{g}\left(z^{-1}\right) \frac{dz}{z}
$$

$$
- \frac{1}{2\pi i} \int\limits_{|z|=q^{-1}} \frac{f(qz)}{(1-q)z} \, \overline{g}\left(z^{-1}\right) \frac{dz}{z},
$$

since $f(z)$ and $\overline{g}\left(z^{-1}\right)$ are analytic in $1 \le |z| \le q^{-1}$. We replace z by $q^{-1}z$ in the last integral. This gives

$$
\langle \mathcal{D}_{q,z} f, g \rangle_c = \frac{1}{2\pi i} \int\limits_{|z|=1} f(z) \frac{\overline{g}\left(z^{-1}\right) - q\overline{g}\left(qz^{-1}\right)}{(1-q)z} \frac{dz}{z}
$$

$$
= \frac{1}{2\pi i} \int\limits_{|z|=1} f(z) \frac{\overline{g(z) - qg(qz)}}{(1-q)z} \frac{dz}{z},
$$

$$(17.1.15)$$

and the proof is complete. □

Observe that $T_{q,z}$ can be written in the form

$$
(T_{q,z} f)(z) = qz^2 (\mathcal{D}_{q,z} f)(z) + z f(z). \tag{17.1.16}
$$

We now show that $T_{q,z}$ is a raising operator for the polynomials $\mathcal{H}_n(x \mid q)$. Write the orthogonality relation (17.1.4) as

$$
q^{-n} \frac{(q;q)_n}{(q;q)_\infty} \delta_{m,n} = \langle \mathcal{H}_m(z \mid q), w_c(z \mid q)\mathcal{H}_n(z \mid q)\rangle_c
$$

$$
= \frac{q^{1/2}(1-q)}{1 - q^{m+1}} \langle \mathcal{D}_{q,z}\mathcal{H}_{m+1}(z \mid q), w_c(z \mid q)\mathcal{H}_n(z \mid q)\rangle_c
$$

$$
= \frac{q^{1/2}(1-q)}{1 - q^{m+1}} \langle \mathcal{H}_{m+1}(z \mid q), T_{q,z} w_c(z \mid q)\mathcal{H}_n(z \mid q)\rangle_c.
$$

Therefore the function

$$
\frac{1}{w_c(z \mid q)} T_{q,z} \left(w_c(z \mid q) \, \mathcal{H}_n(z \mid q)\right)
$$

is orthogonal to $\mathcal{H}_m(z \mid q)$ for all $n \ne m + 1$ and is in L^2 of the unit circle weighted by the continuous weight function $w_c(z \mid q)$. The $\mathcal{H}_n$'s are complete in the aforementioned L^2 space and we deduce that

$$
\frac{1}{w_c(z \mid q)} T_{q,z} \left(w_c(z \mid q) \, \mathcal{H}_n(z \mid q)\right)
$$

must be a constant multiple of $\mathcal{H}_{n+1}(z \mid q)$.

Theorem 17.1.4 *The raising operator for* $\{\mathcal{H}_n\}$ *is* $T_{q,z}$ *in the sense*

$$\frac{1}{w_c(z\,|\,q)}\,T_{q,z}\,(w_c(z\,|\,q)\,\mathcal{H}_n(z\,|\,q)) = \frac{\sqrt{q}}{1-q}\,\mathcal{H}_{n+1}(z\,|\,q). \qquad (17.1.17)$$

Theorem 17.1.4 follows by direct evaluation of the left-hand side of (17.1.17). When we combine (17.1.10) and (17.1.17) we arrive at the following theorem.

Theorem 17.1.5 *The polynomials* $\{\mathcal{H}_n(z\,|\,q)\}$ *satisfy the q-Sturm–Liouville equation*

$$\frac{1}{w_c(z\,|\,q)}\,T_{q,z}\,(w_c(z\,|\,q)\,\mathcal{D}_{q,z}\mathcal{H}_n(z\,|\,q)) = \lambda_n\mathcal{H}_n(z\,|\,q), \qquad (17.1.18)$$

where

$$\lambda_n = \frac{(1-q^n)}{(1-q)^2}. \qquad (17.1.19)$$

Observe that the eigenvalues (17.1.19) are distinct and positive.

Equation (17.1.18) suggests that we consider the more general symmetric operator

$$(Mf)(z) := \frac{1}{\omega(z)}\,(T_{q,z}\,(p(z)\,\mathcal{D}_{q,z}f))\,(z), \qquad (17.1.20)$$

where $\omega(z)$ is real on the unit circle with some restrictions on p and ω to follow. Let $\mathcal{H}_\omega$ denote the inner product space L^2 of the unit circle equipped with the inner product

$$(f,g)_\omega := \frac{1}{2\pi i}\int\limits_{|z|=1} f(z)\,\overline{g(z)}\,\omega(z)\,\frac{dz}{z}, \qquad (17.1.21)$$

and let

$$T := M_{|\mathcal{F}_2}\text{ in }\mathcal{H}_w.$$

We shall assume that p and ω satisfy

1. $p(z) > 0$ a.e. on $|z| = 1$, $p \in \mathcal{F}_1$, $1/p \in L(\{z : |z| = 1\})$; $\qquad (17.1.22)$
2. On the unit circle $\omega(z) > 0$ a.e. and both ω and $1/\omega$ are integrable.

The expression Mf is therefore defined for $f \in \mathcal{F}_2$, and the operator T acts in $\mathcal{H}_w$. Furthermore, the domain $\mathcal{F}_2$ of T is dense in $\mathcal{H}_\omega$ since it contains all Laurent polynomials.

Theorem 17.1.6 *The operator* T *is symmetric in* $\mathcal{H}_\omega$ *and* T *is a positive operator.*

Proof For all $f, g \in \mathcal{F}_2$ it follows that $(Tf, g)_w = (f, Tg)_w$, hence the operator T is symmetric. If $f \in \mathcal{F}_2$ then

$$(f, Tf)_\omega = \langle f, T_{q,z}\,(p(z)\mathcal{D}_{q,z}f)\rangle = \langle \mathcal{D}_{q,z}f, p(z)\mathcal{D}_{q,z}f\rangle$$
$$= \frac{1}{2\pi i}\int\limits_{|z|=1} p(z)\,|\mathcal{D}_{q,z}f(z)|^2\,\frac{dz}{z}, \qquad (17.1.23)$$

which proves that $T \geq 0$ and completes the proof of our theorem. $\qquad\square$

Corollary 17.1.7 *Let $y_1, y_2 \in \mathcal{F}_2$ be solutions to*

$$\frac{1}{w(z)} \left(T_{q,z} \left(p(z) \mathcal{D}_{q,z} f \right) \right)(z) = \lambda f, \qquad (17.1.24)$$

with $\lambda = \lambda_1$ and $\lambda = \lambda_2$, respectively and assume $\lambda_1 \neq \lambda_2$. Then y_1 and y_2 are orthogonal in the sense

$$\frac{1}{2\pi i} \int\limits_{|z|=1} w(z) y_1(z) \overline{y_2(z)} \frac{dz}{z} = 0. \qquad (17.1.25)$$

Furthermore the eigenvalues of (17.1.24) are all real.

Corollary 17.1.7 follows from Theorem 17.1.6 and the fact that the eigenvalues of symmetric operators are real and the eigenspaces are mutually orthogonal.

Note that Corollary 17.1.7 and (17.1.18) show that the polynomials $\mathcal{H}_n(z \,|\, q)$ are orthogonal with respect to $w_c(z \,|\, q)$ of (17.1.1). One can also evaluate the integrals

$$\zeta_n := \frac{1}{2\pi i} \int\limits_{|z|=1} |\mathcal{H}_n(z \,|\, q)|^2 \, w_c(z \,|\, q) \frac{dz}{z}$$

as follows

$$
\begin{aligned}
\zeta_n &= \langle \mathcal{H}_n(z \,|\, q), w_c(z \,|\, q) \mathcal{H}_n(z \,|\, q) \rangle_c \\
&= \frac{q^{1/2}(1-q)}{1-q^{n+1}} \langle \mathcal{D}_{q,z} \mathcal{H}_{n+1}(z \,|\, q), w_c(z \,|\, q) \mathcal{H}_n(z \,|\, q) \rangle_c \\
&= \frac{q^{1/2}(1-q)}{1-q^{n+1}} \langle \mathcal{H}_{n+1}(z \,|\, q), T_{q,z} \left(w_c(z \,|\, q) \mathcal{H}_n(z \,|\, q) \right) \rangle_c \\
&= \frac{q}{1-q^{n+1}} \langle \mathcal{H}_{n+1}(z \,|\, q), w_c(z \,|\, q) \mathcal{H}_{n+1}(z \,|\, q) \rangle_c = \frac{q}{1-q^{n+1}} \zeta_{n+1}.
\end{aligned}
$$

Hence $\zeta_n = q^{-n}(q;q)_n \zeta_0$. But the Jacobi triple product identity (12.3.4) gives $\zeta_0 = 1/(q;q)_\infty$. This analysis gives an alternate derivation of the orthogonality relation (17.1.4).

Theorem 17.1.8 *Let $f(z)$ be a nontrivial solution of (17.1.24) and assume that f is analytic in a neighborhood of the origin. Then λ must be of the form (17.1.19), for an $n = 1, 2, \ldots$ and f is a constant multiple of $\mathcal{H}_n(z \,|\, q)$.*

Proof Let $f(z) = \sum\limits_{n=1}^{\infty} f_k z^k / (q;q)_k$ in (17.1.24). Equating coefficients of z^n in the resulting equation yields

$$f_{k+1} = q^{-(k+1)/2} \left[(1-q)^2 \lambda - 1 + q^k \right] f_k. \qquad \square$$

17.2 Generalizations

We now wish to explore the polynomials orthogonal with respect to the integrand in the q-beta integral (17.1.9). Unlike the weight functions we have encountered so far the weight function we are now interested in is no longer real on the unit circle. It is real on $|z| = 1$ only when $t_2 = \overline{t_1}$. Set

$$w_c\left(z; t_1, t_2 \mid q\right) := \frac{\left(q^{1/2}z, q^{1/2}/z; q\right)_\infty}{\left(q^{-1/2}t_1 z, q^{-1/2}t_2/z; q\right)_\infty}. \tag{17.2.1}$$

It is clear from (17.2.1) that a candidate for the polynomials orthogonal with respect to $w_c(z; t_1, t_2 \mid q)$ may be of the form

$$p_n^c\left(z; t_1, t_2 \mid q\right) = \sum_{k=0}^{n} \frac{\left(q^{-n}, q^{-1/2}t_1 z; q\right)_k}{(q; q)_k} a_{n,k}. \tag{17.2.2}$$

Now consider the integrals

$$I_{n,j} := \frac{1}{2\pi i} \int\limits_{|z|=1} w_c\left(z; t_1, t_2 \mid q\right) \overline{\left(q^{-1/2}\overline{t_2}z; q\right)_j}\, p_n^c\left(z; t_1, t_2 \mid q\right) \frac{dz}{z}. \tag{17.2.3}$$

Substitute for $p_n^c\left(z; t_1, t_2 \mid q\right)$ from (17.2.2) into the above equation (17.2.3) to get

$$
\begin{aligned}
I_{n,j} &= \sum_{k=0}^{n} \frac{\left(q^{-n}; q\right)_k}{(q; q)_k} a_{n,k} \frac{1}{2\pi i} \int\limits_{|z|=1} w_c\left(z; t_1 q^k, t_2 q^j \mid q\right) \frac{dz}{z} \\
&= \sum_{k=0}^{n} \frac{\left(q^{-n}; q\right)_k a_{n,k}}{(q; q)_k (q; q)_\infty} \frac{\left(q^k t_1, q^j t_2; q\right)_\infty}{\left(q^{k+j-1}t_1 t_2; q\right)_\infty} \\
&= \frac{\left(t_1, q^j t_2; q\right)_\infty}{\left(q, q^{j-1}t_1 t_2; q\right)_\infty} \sum_{k=0}^{n} \frac{\left(q^{-n}, q^{j-1}t_1 t_2; q\right)_k a_{n,k}}{(q, t_1; q)_k},
\end{aligned}
$$

after making use of (17.1.9). Obviously the choice

$$a_{n,k} = \frac{\left(t_1; q\right)_k}{\left(t_1 t_2/q; q\right)_k} q^k \tag{17.2.4}$$

works because we can use the q-analogue of the Chu–Vandermonde sum, (12.2.17). The result is that

$$I_{n,j} = \frac{\left(t_1, t_2 q^j; q\right)_\infty \left(q^{-j}; q\right)_n}{\left(q, q^{j-1}t_1 t_2; q\right)_\infty \left(q^{-1}t_1 t_2; q\right)_n} \left(t_1 t_2 q^{j-1}\right)^n.$$

This analysis proves the following theorem.

Theorem 17.2.1 *The p_n^c's have the representation*

$$p_n^c\left(z; t_1, t_2 \mid q\right) = {}_3\phi_2 \left(\begin{matrix} q^{-n}, q^{-1/2}zt_1, t_1 \\ q^{-1}t_1 t_2, 0 \end{matrix} \;\middle|\; q, q \right), \tag{17.2.5}$$

and satisfy the biorthogonality relation

$$
\begin{aligned}
&\frac{1}{2\pi i} \int\limits_{|z|=1} \overline{p_m^c\left(z; \overline{t_2}, \overline{t_1} \mid q\right)}\, p_n^c\left(z; t_1, t_2 \mid q\right) w_c\left(z; t_1, t_2 \mid q\right) \frac{dz}{z} \\
&\qquad = \frac{\left(t_1, t_2; q\right)_\infty (q; q)_n}{\left(q, t_1 t_2/q; q\right)_\infty \left(t_1 t_2/q; q\right)_n} \left(t_1 t_2/q\right)^n \delta_{m,n}.
\end{aligned}
\tag{17.2.6}
$$

The transformation (12.4.10) shows that

$$p_n^c\left(z;t_1,t_2\,|\,q\right) = \frac{(q;q)_n\,t_1^n}{(t_1t_2/q;q)_n}\,\pi_n^c\left(z;t_1,t_2\,|\,q\right),\qquad(17.2.7)$$

where

$$\pi_n^c\left(z;t_1,t_2\,|\,q\right) = \sum_{k=0}^{n}\frac{(t_1;q)_k\,(t_2/q;q)_{n-k}}{(q;q)_k\,(q;q)_{n-k}}\left(q^{-1/2}z\right)^k.\qquad(17.2.8)$$

In terms of the π_n's the orthogonality relation (17.2.6) becomes

$$\frac{1}{2\pi i}\int_{|z|=1}\overline{\pi_m^c\left(z;\overline{t_2},\overline{t_1}\,|\,q\right)}\,\pi_n^c\left(z;t_1,t_2\,|\,q\right)w_c\left(z;t_1,t_2\,|\,q\right)\frac{dz}{z}$$

$$= \frac{(t_1,t_2;q)_\infty\,(t_1t_2/q;q)_n}{(q,t_1t_2/q;q)_\infty\,(q;q)_n}\,q^{-n}\delta_{m,n}.\qquad(17.2.9)$$

The form (17.2.8) makes it immediate to obtain the generating function

$$\sum_{n=0}^{\infty}\pi_n^c\left(z;t_1,t_2\,|\,q\right)t^n = \frac{\left(tt_1q^{-1/2}z,tt_2/q;q\right)_\infty}{\left(t,tq^{-1/2}z;q\right)_\infty}.\qquad(17.2.10)$$

The polynomials $\{\pi_n^c\left(x;t_1,t_2\,|\,q\right)\}$ were introduced in (Pastro, 1985) and are called the Pastro polynomials.

Theorem 17.2.2 ((Al-Salam & Ismail, 1994)) *The q-beta integral*

$$\frac{1}{2\pi i}\int_{|z|=1}w_c\left(z;t_1,t_2,t_3,t_4\,|\,q\right)\frac{dz}{z}$$

$$= \frac{\left(t_1,t_2,t_3,t_4,t_1t_2t_3t_4q^{-2};q\right)_\infty}{\left(q,t_1t_2/q,t_1t_4/q,t_2t_3/q,t_3t_4/q;q\right)_\infty},\qquad(17.2.11)$$

with

$$w_c\left(z;t_1,t_2,t_3,t_4\,|\,q\right) := \frac{\left(q^{1/2}z,q^{1/2}/z,t_1t_3q^{-1/2}z,t_2t_4q^{-1/2}/z;q\right)_\infty}{\left(t_1q^{-1/2}z,t_2q^{-1/2}/z,t_3q^{-1/2}z,t_4q^{-1/2}/z;q\right)_\infty},$$

$$(17.2.12)$$

holds when $|t_j| < q^{1/2}$ *for* $1 \le j \le 4$.

Proof For $t_j \in \left(-q^{1/2},q^{1/2}\right)$, $j = 1, 2, 3, 4$ the theorem follows from (7.2.14) and (7.2.13), since (7.2.11) implies

$$\left|\pi^c\left(z;t_1,t_2\,|\,q\right)\right| \le \pi^c\left(1;t_1,t_2\,|\,q\right)\qquad(17.2.13)$$

with equality if and only if $z = 1$, and this allows us to interchange summation and integration. We then extend (17.2.11) to complex t_j's by analytic continuation. This completes our proof. $\qquad\square$

We now look for functions biorthogonal with respect to the weight function in (17.2.12). Let

$$\phi_n\left(z; t_1, t_2, t_3, t_4 \,|\, q\right) = \sum_{k=0}^{n} \frac{\left(q^{-n}, q^{-1/2}t_1 z; q\right)_k}{\left(q, q^{-1/2}t_1 t_3 z; q\right)_k} \, b_{n,k}. \tag{17.2.14}$$

We then consider the integrals

$$J_{n,j} := \frac{1}{2\pi i} \int_{|z|=1} \frac{\left(q^{-1/2}\overline{t_2}z; q\right)_j}{\left(q^{-1/2}\overline{t_2 t_4}z; q\right)_j} \tag{17.2.15}$$

$$\times \phi_n\left(z; t_1, t_2, t_3, t_4 \,|\, q\right) w_c\left(z; t_1, t_2, t_3, t_4 \,|\, q\right) \frac{dz}{z}.$$

Therefore

$$J_{n,j} = \sum_{k=0}^{n} \frac{\left(q^{-n}; q\right)_k}{(q; q)_k} \, b_{n,k} \, \frac{\left(t_1 q^k, t_2 q^j, t_3, t_4, t_1 t_2 t_3 t_4 q^{k+j-2}; q\right)_\infty}{\left(q, t_1 t_2 q^{k+j-1}, t_1 t_4 q^{k-1}, t_2 t_3 q^{j-1}, t_3 t_4/q; q\right)_\infty},$$

$$= \frac{\left(t_1, t_2 q^j, t_3, t_4, t_1 t_2 t_3 t_4 q^{j-2}; q\right)_\infty}{\left(q, t_1 t_2 q^{j-1}, t_1 t_4/q, t_2 t_3 q^{j-1}, t_3 t_4/q; q\right)_\infty}$$

$$\times \sum_{k=0}^{n} \frac{\left(q^{-n}, t_1 t_2 q^{j-1}, t_1 t_2/q; q\right)_k}{\left(q, t_1, t_1 t_2 t_3 t_4 q^{j-2}; q\right)_k} \, b_{n,k}.$$

This leads to the choice

$$b_{n,k} = \frac{\left(t_1, t_1 t_2 t_3 t_4 q^{n-2}; q\right)_k}{\left(t_1 t_2/q, t_1 t_4/q; q\right)_k} \, q^k,$$

and the integrals $J_{n,j}$ become

$$J_{n,j} = \frac{\left(t_1, t_2 q^j, t_3, t_4, t_1 t_2 t_3 t_4 q^{j-2}; q\right)_\infty}{\left(q, t_1 t_2 q^{j-1}, t_1 t_4/q, t_2 t_3 q^{j-1}, t_3 t_4/q; q\right)_\infty},$$

$$\times {}_3\phi_2 \left(\begin{matrix} q^{-n}, t_1 t_2 q^{j-1}, t_1 t_2 t_3 t_4 q^{n-3} \\ t_1 t_2/q, t_1 t_2 t_3 t_4 q^{j-2} \end{matrix} \,\bigg|\, q, q \right)$$

$$= \frac{\left(t_1, t_2 q^j, t_3, t_4, t_1 t_2 t_3 t_4 q^{j-2}; q\right)_\infty}{\left(q, t_1 t_2 q^{j-1}, t_1 t_4/q, t_2 t_3 q^{j-1}, t_3 t_4/q; q\right)_\infty}$$

$$\times \frac{\left(t_3 t_4/q, q^{j+1-n}; q\right)_n}{\left(t_1 t_2 t_3 t_4 q^{j-2}, q^{2-n}/t_1 t_2; q\right)_n},$$

which clearly vanishes for $j < n$. Thus for $j \le n$ we have

$$J_{n,j} = \frac{\left(t_1, t_2 q^n, t_3, t_4, t_1 t_2 t_3 t_4 q^{2n-2}; q\right)_\infty}{\left(q^{n+1}, t_1 t_2 q^{n-1}, t_1 t_4 q^{n-1}, t_2 t_3 q^{n-1}, t_3 t_4/q; q\right)_\infty} \, \delta_{n,j}. \tag{17.2.16}$$

We state what we have done so far about the ϕ_n's as a theorem.

Theorem 17.2.3 *The rational functions*

$$\phi_n\left(z; t_1, t_2, t_3, t_4 \,|\, q\right) := {}_4\phi_3 \left(\begin{matrix} q^{-n}, t_1 q^{-1/2} z, t_1, t_1 t_2 t_3 t_4 q^{n-3} \\ t_1 t_2/q, t_1 t_4/q, t_1 t_3 q^{-1/2} z \end{matrix} \,\bigg|\, q, q \right) \tag{17.2.17}$$

satisfy the biorthogonality relation

$$\frac{1}{2\pi i} \int_{|z|=1} \overline{\phi_m \left(z; \overline{t_2}, \overline{t_1}, \overline{t_4}, \overline{t_3} \mid q\right)} \phi_n \left(z; t_1, t_2, t_3, t_4 \mid q\right)$$

$$\times w_c \left(z; t_1, t_2, t_3, t_4 \mid q\right) \frac{dz}{z}$$

$$= \frac{\left(t_1, t_2, t_3, t_4, t_1 t_2 t_3 t_4 q^{-2}; q\right)_\infty}{\left(q^{n+1}, t_1 t_2/q, t_2 t_3/q, t_1 t_4/q, t_3 t_4/q; q\right)_\infty} \frac{\left(t_1 t_2 t_3 t_4 q^{n-3}; q\right)_n}{\left(t_1 t_2 t_3 t_4 q^{-2}; q\right)_{2n}} \left(\frac{t_1 t_2}{q}\right)^n \delta_{m,n}.$$

$$\tag{17.2.18}$$

The biorthogonal rational functions $\{\phi_n\}$ are the exact analogue of the Askey–Wilson polynomials. They were found in (Al-Salam & Ismail, 1994) and we will refer to them as the Al-Salam–Ismail biorthogonal functions.

17.3 q-Difference Equations

This section is based on (Ismail & Witte, 2001).

Motivated by the convention $w = e^{-v}$ of Chapter 3, we set

$$\left(D_q w\right)(z) = -u(qz)w(qz), \tag{17.3.1}$$

where D_q is the q-difference operator defined in (11.4.1). In other words

$$w(z) = w(qz)[1 - (1 - q)zu(qz)], \quad |z| = 1. \tag{17.3.2}$$

First we give a q-analogue of Theorem 8.3.1.

Theorem 17.3.1 *If $w(z)$ is analytic in the ring $q < |z| < 1$ and is continuous on its boundary then*

$$\left(D_q \phi_n\right)(z) = \frac{\kappa_{n-1}}{\kappa_n} \frac{1 - q^n}{1 - q} \phi_{n-1}(z)$$

$$- i\phi_n^*(z) \int_{|\zeta|=1} \frac{u(\zeta) - u(qz)}{\zeta - qz} \phi_n(\zeta) \overline{\phi_n^*(q\zeta)} \, w(\zeta) \, d\zeta \tag{17.3.3}$$

$$+ i\phi_n(z) \int_{|\zeta|=1} \frac{u(\zeta) - u(qz)}{\zeta - qz} \phi_n(\zeta) \overline{\phi_n(q\zeta)} \, w(\zeta) \, d\zeta.$$

Proof Expand $D_q \phi_n(z)$ in a series of the ϕ_n's to see that

$$(1 - q)\left(D_q \phi_n\right)(z) = \int_{|\zeta|=1} \sum_{k=0}^{n-1} \phi_k(z) \overline{\phi_k(\zeta)} \left[\phi_n(\zeta) - \phi_n(q\zeta)\right] w(\zeta) \frac{d\zeta}{i\zeta^2}.$$

Write the above integral as a difference of two integrals involving $\phi_n(\zeta)$ and $\phi_n(q\zeta)$, then in the second integral replace ζ by ζ/q. Under such transformation $\overline{\phi_k(\zeta)}$ is transformed to $\overline{\phi_k(q\zeta)}$, since $|\zeta| = 1$. Furthermore (17.3.1) gives

$$w(\zeta/q) = [1 + \zeta u(\zeta)(1 - 1/q)] w(\zeta). \tag{17.3.4}$$

Therefore

$$(1-q)\,(D_q\phi_n)\,(z) = \int\limits_{|\zeta|=1} \sum_{k=0}^{n-1} \phi_k(z)\,\overline{\zeta\phi_k(\zeta)}\,\phi_n(\zeta)w(\zeta)\,\frac{d\zeta}{i\zeta}$$

$$+ \int\limits_{|\zeta|=1} \sum_{k=0}^{n-1} \phi_k(z)\left[-q\overline{\zeta\phi_k(q\zeta)} + u(\zeta)(1-q)\,\overline{\phi_k(q\zeta)}\right]\phi_n(\zeta)w(\zeta)\,\frac{d\zeta}{i\zeta}$$

$$= \frac{\kappa_{n-1}}{\kappa_n}\phi_{n-1}(z) - q^n\,\frac{\kappa_{n-1}}{\kappa_n}\phi_{n-1}(z)$$

$$+(1-q)\int\limits_{|\zeta|=1}\phi_n(\zeta)u(\zeta)\sum_{k=0}^{n-1}\phi_k(z)\,\overline{\phi_k(q\zeta)}\,w(\zeta)\,\frac{d\zeta}{i\zeta}.$$

The result now follows from (8.2.1).   □

We next substitute for $\phi_n^*(z)$ in (17.3.3) from (8.2.2), if $\phi_n(0) \neq 0$, and establish

$$(D_q\phi_n)\,(z) = A_n(z)\phi_{n-1}(z) - B_n(z)\phi_n(z), \tag{17.3.5}$$

with

$$A_n(z) = \frac{\kappa_{n-1}}{\kappa_n}\frac{1-q^n}{1-q}$$

$$+ i\,\frac{\kappa_{n-1}}{\phi_n(0)}\,z\int\limits_{|\zeta|=1}\frac{u(\zeta)-u(qz)}{\zeta-qz}\,\phi_n(\zeta)\,\overline{\phi_n^*(q\zeta)}\,w(\zeta)\,d\zeta \tag{17.3.6}$$

and

$$B_n(z) = -i\int\limits_{|\zeta|=1}\frac{u(\zeta)-u(qz)}{\zeta-qz}\,\phi_n(\zeta)$$

$$\times\left[\overline{\phi_n(q\zeta)} - \frac{\kappa_n}{\phi_n(0)}\,\overline{\phi_n^*(q\zeta)}\right]w(\zeta)\,d\zeta. \tag{17.3.7}$$

These are the q-analogues of (8.3.8), (8.3.9) and (8.3.13). Here again we set

$$L_{n,1} = D_q + B_n(z), \tag{17.3.8}$$

$$L_{n,2} = -D_q - B_{n-1}(z) + \frac{A_{n-1}(z)\kappa_{n-1}}{z\kappa_{n-2}} + \frac{A_{n-1}(z)\kappa_n\phi_{n-1}(0)}{\kappa_{n-2}\phi_n(0)}. \tag{17.3.9}$$

The ladder operators are

$$L_{n,1}\phi_n(z) = A_n(z)\phi_{n-1}(z),$$

$$L_{n,2}\phi_{n-1}(z) = \frac{\phi_{n-1}(0)\kappa_{n-1}}{\phi_n(0)\kappa_{n-2}}\frac{A_{n-1}(z)}{z}\,\phi_n(z). \tag{17.3.10}$$

This results in the q-difference equation

$$L_{n,2}\left(\frac{1}{A_n(z)}L_{n,1}\right)\phi_n(z) = \frac{A_{n-1}(z)}{z}\frac{\phi_{n-1}(0)\kappa_{n-1}}{\phi_n(0)\kappa_{n-2}}\phi_n(z). \tag{17.3.11}$$

The following theorem gives a q-analogue of the functional equation (8.4.2).

Theorem 17.3.2 *If $u(z)$ is analytic in the annular region $q < |z| < 1$ then the following functional equation for the coefficients $A_n(z)$, $B_n(z)$ holds*

$$B_n + B_{n-1} - \frac{\kappa_{n-1}}{\kappa_{n-2}} \frac{A_{n-1}}{z} - \frac{\kappa_n}{\kappa_{n-2}} \frac{\phi_{n-1}(0)}{\phi_n(0)} A_{n-1}$$

$$= -\frac{n-1}{qz} - \frac{u(qz)}{q} - \frac{1-q}{q} \sum_{j=0}^{n-1} \left[B_{j+1} - \frac{\kappa_j}{\kappa_{j-1}} \frac{A_j}{z} \right]. \quad (17.3.12)$$

Proof Two alternative forms of the second order q-difference equation are possible, namely (17.3.11) and the following,

$$L_{n+1,1} \left(\frac{z}{A_n(z)} L_{n+1,2} \right) \phi_n(z) = \frac{\kappa_n \phi_n(0)}{\kappa_{n-1} \phi_{n+1}(0)} A_{n+1}(z) \phi_n(z). \quad (17.3.13)$$

These two equations, written out in full are, respectively,

$$D_q^2 \phi_n(z) + \left\{ B_n(qz) + \frac{A_n(qz)}{A_n(z)} B_{n-1}(z) - \frac{D_q A_n(z)}{A_n(z)} \right.$$

$$\left. - \frac{\kappa_{n-1}}{\kappa_{n-2}} \frac{A_n(qz) A_{n-1}(z)}{A_n(z) z} - \frac{\kappa_n}{\kappa_{n-2}} \frac{\phi_{n-1}(0)}{\phi_n(0)} \frac{A_n(qz) A_{n-1}(z)}{A_n(z)} \right\} D_q \phi_n(z)$$

$$+ \left\{ D_q B_n(z) - \frac{B_n(z)}{A_n(z)} D_q A_n(z) + \frac{A_n(qz)}{A_n(z)} B_n(z) B_{n-1}(z) \right.$$

$$- \frac{\kappa_{n-1}}{\kappa_{n-2}} \frac{A_n(qz)}{A_n(z)} \frac{A_{n-1}(z) B_n(z)}{z}$$

$$- \frac{\kappa_n}{\kappa_{n-2}} \frac{\phi_{n-1}(0)}{\phi_n(0)} \frac{A_n(qz)}{A_n(z)} A_{n-1}(z) B_n(z)$$

$$\left. + \frac{\kappa_{n-1}}{\kappa_{n-2}} \frac{\phi_{n-1}(0)}{\phi_n(0)} \frac{A_n(qz) A_{n-1}(z)}{z} \right\} \phi_n(z) = 0, \quad (17.3.14)$$

and

$$D_q^2 \phi_n(z) + \left\{ \frac{A_n(qz)}{qA_n(z)} B_{n+1}(z) + B_n(qz) - \frac{D_q A_n(z)}{qA_n(z)} \right.$$

$$\left. - \frac{\kappa_n}{\kappa_{n-1}} \frac{A_n(qz)}{qz} - \frac{\kappa_{n+1}}{\kappa_{n-1}} \frac{\phi_n(0)}{\phi_{n+1}(0)} A_n(qz) + \frac{1}{qz} \right\} D_q \phi_n(z)$$

$$+ \left\{ D_q B_n(z) - \frac{B_n(z)}{qA_n(z)} D_q A_n(z) + \frac{A_n(qz)}{qA_n(z)} B_{n+1}(z) B_n(z) \right.$$

$$- \frac{\kappa_n}{\kappa_{n-1}} \frac{A_n(qz) B_{n+1}(z)}{qz} - \frac{\kappa_{n+1}}{\kappa_{n-1}} \frac{\phi_n(0)}{\phi_{n+1}(0)} \frac{A_n(qz) B_{n+1}(z)}{q}$$

$$+ \frac{\kappa_n}{\kappa_{n-1}} \frac{\phi_n(0)}{\phi_{n+1}(0)} \frac{A_n(qz) A_{n+1}(z)}{qz}$$

$$\left. + \frac{B_n(z)}{qz} - \frac{\kappa_{n+1}}{\kappa_{n-1}} \frac{\phi_n(0)}{\phi_{n+1}(0)} \frac{A_n(qz)}{qz} \right\} \phi_n(z) = 0. \quad (17.3.15)$$

A comparison of the coefficients of the first q-difference terms leads to the difference equation

$$\frac{1}{q} B_{n+1}(z) - B_{n-1}(z) - \frac{\kappa_n}{\kappa_{n-1}} \frac{A_n(z)}{qz} + \frac{\kappa_{n-1}}{\kappa_{n-2}} \frac{A_{n-1}(z)}{z}$$
$$- \frac{\kappa_{n+1}}{\kappa_{n-1}} \frac{\phi_n(0)}{\phi_{n+1}(0)} A_n(z) + \frac{\kappa_n}{\kappa_{n-2}} \frac{\phi_{n-1}(0)}{\phi_n(0)} A_{n-1}(z) = -\frac{1}{qz}. \quad (17.3.16)$$

Using the results for the first coefficients

$$B_1(z) = -u(qz) - \frac{\phi_1(qz)}{\phi_1(0)} M_1(z), \quad (17.3.17)$$

$$\frac{A_0(z)}{\kappa_{-1}} = -z M_1(z), \quad (17.3.18)$$

with

$$M_1(z) \equiv \int\limits_{|\zeta|=1} \zeta \frac{u(\zeta) - u(qz)}{\zeta - qz} w(\zeta) \frac{d\zeta}{i\zeta}, \quad (17.3.19)$$

this difference equation can be summed to yield the result in (17.3.12). □

Define the inner product

$$(f, g) = \int\limits_{|z|=1} f(\zeta) \overline{g(\zeta)} w(\zeta) \frac{d\zeta}{i\zeta}. \quad (17.3.20)$$

With respect to this inner product the adjoint of $L_{n,1}$ is

$$\left(L_{n,1}^* f\right)(z) = z^2 \left[q - (1-q)z\overline{u(z)} \right] D_q f(z) + zf(z) + \left[\overline{B_n(z)} + \overline{u(z)} \right] f(z), \quad (17.3.21)$$

provided that $w(z)$ is analytic in $q < |z| < 1$ and is continuous on $|z| = 1$ and $|z| = q$. The proof follows from the definition of D_q and the fact $\overline{g(\zeta)} = \bar{g}(1/\zeta)$, when $|\zeta| = 1$. Observe that as $q \to 1$, the right-hand side of (17.3.21) tends to the right-hand side of (8.3.20), as expected.

Example 17.3.3 *Consider the Rogers–Szegő polynomials* $\{\mathcal{H}_n(z \,|\, q)\}$, *where*

$$w(z) = \frac{(q^{1/2}z, q^{1/2}/z; q)_\infty}{2\pi(q; q)_\infty}, \quad \mathcal{H}_n(z \,|\, q) = \sum_{k=0}^{n} \frac{(q; q)_n q^{-k/2} z^k}{(q; q)_k (q; q)_{n-k}}. \quad (17.3.22)$$

In this case

$$\phi_n(z) = \frac{q^{n/2}}{\sqrt{(q; q)_n}} \mathcal{H}_n(z \,|\, q), \quad \phi_n(0) = \frac{q^{n/2}}{\sqrt{(q; q)_n}}, \quad \kappa_n = \frac{1}{\sqrt{(q; q)_n}}. \quad (17.3.23)$$

It is easy to see that

$$u(z) = \frac{\sqrt{q}}{1-q} + \frac{qz^{-1}}{1-q} \quad (17.3.24)$$

Thus $[u(\zeta) - u(qz)]/(\zeta - qz)$ is $-1/[(1 - q)\zeta z]$. A simple calculation gives

$$(D_q \phi_n)(z) = \frac{(1 - q^n)^{3/2}}{1 - q} \phi_{n-1}(z) + \frac{\kappa_{n-1}\phi_{n-1}(z)}{\phi_n(0)(1 - q)} \int\limits_{|\zeta|=1} \phi_n(\zeta) \overline{\phi_n^*(q\zeta)} \, w(\zeta) \frac{d\zeta}{i\zeta},$$

which simplifies to

$$(D_q \phi_n)(z) = \frac{\sqrt{1 - q^n}}{1 - q} \phi_{n-1}(z), \tag{17.3.25}$$

since $\kappa_n \overline{\phi_n^*(q\zeta)} - \phi_n(0)\overline{\phi_n(q\zeta)}$ is a polynomial of degree $n - 1$. The functional equation (17.3.25) can be verified independently by direct computation.

Exercises

17.1   Determine the large n asymptotics of the orthonormal Rogers–Szegő polynomials.

17.2   Evaluate the Szegő function $g(z)$ for the Rogers–Szegő polynomials using Theorem 8.5.4 and Exercise 17.1.

17.3   Prove that the zeros of the Rogers–Szegő polynomials lie on $\{z : |z| = q^{1/2}\}$, (Mazel et al., 1990).
        Hint: Prove that $z^{-n}\mathcal{H}_n\left(q^{1/2}z^2 \mid q\right)$ is a family of orthogonal polynomials on $[-1, 1]$ in $x = \left(z + z^{-1}\right)/2$.

17.4   Let $t_n(x) = H_n(x \mid q)/\sqrt{(q; q)_n}$ in Theorem 8.2.3. Find explicit representation for $u_n(x)$ of Theorem 8.2.3, hence find explicit formulas for the corresponding orthonormal polynomials $\{\phi_n(z)\}$.

17.5   Evaluate the function $h(z)$ and $D(z)$ of §8.5 for the Rogers–Szegő polynomials.

17.6   Find a generating function for the Al-Salam–Ismail biorthogonal functions $\{\phi_n\left(x; t_1, t_2, t_3, t_4 \mid q\right)\}$.
        Hint: Mimic the proof of Theorem 15.2.2.

17.7   Using the generating function in Exercise 17.6, find a unit circle analogue of (15.2.9).

17.8   Establish a Rodrigues formula for the Rogers–Szegő polynomials.

18

Discrete q-Orthogonal Polynomials

In this chapter we use two different approaches to develop the theory of explicitly defined discrete orthogonal polynomials. One method is similar to what we did in Chapter 5 where we start with the Al-Salam–Carlitz polynomials and work our way up to the big q-Jacobi polynomials. The second approach uses discrete q-Sturm–Liouville problems.

18.1 Discrete Sturm–Liouville Problems

We now study the q-Sturm–Liouville problems

$$\frac{1}{w(x)} D_{q^{-1},x} \left(p(x) D_{q,x} Y(x,\lambda) \right) = \lambda\, Y(x,\lambda) \tag{18.1.1}$$

$$\frac{1}{W(x)} D_{q,x} \left(P(x) D_{q^{-1},x} Z(x,\lambda) \right) = \Lambda\, Z(x,\lambda). \tag{18.1.2}$$

We assume that

$$w(x) > 0, \text{ and } p(x) > 0, \text{ for } x = x_k,\, x = y_k, \tag{18.1.3}$$

$$W(x) > 0, \text{ and } P(x) > 0, \text{ for } x = r_k,\, x = s_k, \tag{18.1.4}$$

where $\{x_k\}$ and $\{y_k\}$ are as in (11.4.8), while $\{r_k\}$ and $\{s_k\}$ are as in (11.4.12).

As usual, the values of λ for which $Y(x,\lambda)$ satisfies (18.1.1) and $\langle Y(\cdot,\lambda), Y(\cdot,\lambda)\rangle$ is finite, are called eigenfunctions. The eigenfunctions are assumed to take finite values at x_{-1} and y_{-1}. Moreover, we assume $w\,(x_{-1}) - w\,(y_{-1}) = 0$. The eigenfunctions of (18.1.2) are similarly defined.

Theorem 18.1.1 *Under the above assumptions the operator*

$$T = \frac{1}{w} D_{q^{-1},x}\, p D_q$$

is symmetric, hence it has real eigenvalues, and the eigenfunctions corresponding to distinct eigenvalues are orthogonal.

Proof From Theorem 11.4.1, we find

$$\langle f, Tg \rangle = -q \left\langle D_{q,x}f, \frac{p}{w} D_{q,x}\, g \right\rangle = -q \overline{\left\langle D_{q,x}\, g, \frac{p}{w} D_{q,x}\, f \right\rangle}$$

$$= \overline{\left\langle g, \frac{1}{w} D_{q^{-1}x}\, p\, D_{q,x}\, f \right\rangle} = \langle Tf, g \rangle,$$

hence, T is symmetric. $\qquad\square$

A rigorous theory of q-Sturm is now available in (Annaby & Mansour, 2005b). This paper corrects many of the results in (Exton, 1983). The author of (Exton, 1983) also used inconsistent notation.

18.2 The Al-Salam–Carlitz Polynomials

The Al-Salam–Carlitz q-polynomials provide a one parameter family of discrete orthogonal polynomials and in the theory of discrete orthogonal polynomials will play the role played by the continuous q-Hermite polynomials in the previous chapters. The Al-Salam–Carlitz polynomials $\left\{ U_n^{(a)}(x; q) \right\}$ have the generating function (Al-Salam & Carlitz, 1965), (Chihara, 1978)

$$G(x; t) := \sum_{n=0}^{\infty} U_n^{(a)}(x; q) \frac{t^n}{(q; q)_n} = \frac{(t, at; q)_\infty}{(tx; q)_\infty}, \qquad (18.2.1)$$

and satisfy the the three-term recurrence relation

$$U_{n+1}^{(a)}(x; q) = [x - (1+a)q^n]\, U_n^{(a)}(x; q)$$
$$+ aq^{n-1}\,(1 - q^n)\, U_{n-1}^{(a)}(x; q), \qquad n > 0, \qquad (18.2.2)$$

and the initial conditions

$$U_0^{(a)}(x; q) := 1, \qquad U_1^{(a)}(x; q) := x - (1+a). \qquad (18.2.3)$$

Following the same procedure used to derive the generating function (13.1.3) we can easily show that the generating function (18.2.1) is equivalent to (18.2.2) and (18.2.3). Note that $\{U_n^a(x; q)\}$ are essentially birth and death process polynomials with rates $\lambda_n = aq^n$ and $\mu_n = 1 - q^n$.

Our first objective is to establish the orthogonality relation

$$\int_{\mathbb{R}} U_m^{(a)}(x; q) U_n^{(a)}(x; q)\, d\mu^{(a)}(x) = (-a)^n q^{n(n-1)/2}(q; q)_n \delta_{m,n}, \qquad a < 0,$$

$$(18.2.4)$$

with $\mu^{(a)}$ a discrete probability measure on $[a, 1]$ given by

$$\mu^{(a)} = \sum_{n=0}^{\infty} \left[\frac{q^n \varepsilon_{q^n}}{(q, q/a; q)_n (a; q)_\infty} + \frac{q^n \varepsilon_{aq^n}}{(q, aq; q)_n (1/a; q)_\infty} \right]. \qquad (18.2.5)$$

In (18.2.5) ε_y denotes a unit mass supported at y. The form of the orthogonality relation (18.2.4)–(18.2.5) given in (Al-Salam & Carlitz, 1965) and (Chihara, 1978) contained a complicated looking form of a normalization constant. The value of the constant was simplified in (Ismail, 1985).

The inner product associated with the Al-Salam–Carlitz polynomials correspond to $b = 1$ in (11.4.8). The orthogonality relation (18.2.4) can be written in the form

$$\int_a^1 \frac{(qx, qx/a; q)_\infty U_m^{(a)}(x; q)\, U_n^{(a)}(x; q)}{(q, a, q/a; q)_\infty (1 - q)}\, d_q x = (-a)^n q^{n(n-1)/2} (q; q)_n \delta_{m,n}.$$

(18.2.6)

Theorem 18.2.1 *The polynomials $U_n^{(a)}(x; q)$ are given by*

$$U_n^{(a)}(x; q) = \sum_{k=0}^n \frac{(q; q)_n (-a)^{n-k}}{(q; q)_k (q; q)_{n-k}} q^{(n-k)(n-k-1)/2} x^k (1/x; q)_k.$$

(18.2.7)

Proof In the right-hand side of (18.2.1), expand $(at; q)_\infty$ by (12.2.25) and expand $(t; q)_\infty/(tx; q)_\infty$ by the q-binomial theorem (12.2.22). This leads to (18.2.6) upon equating like powers of t. □

It is worth noting that (18.2.6) is equivalent to the hypergeometric representation

$$U_n^{(a)}(x; q) = (-a)^n q^{n(n-1)/2} {}_2\phi_1\left(q^{-n}, 1/x; 0; q, qx/a\right).$$

(18.2.8)

Theorem 18.2.2 *The $U_n^{(a)}$'s have the lowering (annihilation) and raising (creation) operators*

$$D_{q,x} U_n^{(a)}(x; q) = \frac{1 - q^n}{1 - q} U_{n-1}^{(a)}(x; q),$$

(18.2.9)

and

$$\frac{1}{(qx, qx/a; q)_\infty} D_{q^{-1},x}\left((qx, qx/a; q)_\infty U_n^{(a)}(x; q)\right) = \frac{q^{1-n}}{a(1 - q)} U_{n+1}^{(a)}(x; q),$$

(18.2.10)

respectively.

Proof The relationship (18.2.9) follows by applying $D_{q,x}$ to both sides of (18.2.8). To prove (18.2.10) first note that

$$x^k (1/x; q)_k = \prod_{j=0}^{k-1} (x - q^j) = (-1)^k q^{k(k-1)/2} \left(q^{1-k}x; q\right)_k.$$

Now

$$D_{q^{-1},x}\left((qx, qx/a; q)_\infty U_n^{(a)}(x; q)\right) = \sum_{k=0}^n \frac{(q; q)_n a^{n-k}}{(q; q)_k (q; q)_{n-k}}$$
$$\times q^{(n-k)(n-k-1)/2} q^{k(k-1)/2}(-1)^n D_{q^{-1},x}\left(q^{1-k}x, qx/a; q\right)_\infty.$$

On the other hand

$$D_{q^{-1},x}\left(q^{1-k}x, qx/a; q\right)_\infty = \frac{q/a}{(1-q)}\left(q^{1-k}x, qx/a; q\right)_\infty \left[xq^{-k} - aq^{-k} - 1\right].$$

Therefore the left-hand side of (18.2.10) is

$$\frac{q/a}{1-q}\sum_{k=0}^{n}\frac{(q;q)_n(-a)^{n-k}}{(q;q)_k(q;q)_{n-k}}q^{-k+(n-k)(n-k-1)/2}$$
$$\times\left[x^{k+1}(1/x;q)_{k+1}-ax^k(1/x;q)_k\right].$$

We then express the above expression as a single sum and calculate the coefficient of $x^k(1/x;q)_k$ and find that it equals the coefficient of $x^k(1/x;q)_k$ on the right-hand side of (18.2.10). □

Corollary 18.2.3 *The polynomials $\left\{U_n^{(a)}(x;q)\right\}$ satisfy the q-Sturm–Liouville equation*

$$\frac{1}{(qx,qx/a;q)_\infty}D_{q^{-1},x}\left((qx,qx/a;q)_\infty D_{q,x}U_n^{(a)}(x;q)\right)$$
$$=\frac{(1-q^n)q^{2-n}}{a(1-q)^2}U_n^{(a)}(x;q).$$
(18.2.11)

This corollary follows from (18.2.9) and (18.2.10). An equivalent form of (18.2.11) is

$$\left[a+x^2-x(1+a)\right]D_{q^{-1},x}D_{q,x}U_n^{(a)}(x;q)$$
$$+\frac{q(x-1-a)}{1-q}D_qU_n^{(a)}(x;q)=\frac{(1-q^n)q^{2-n}}{(1-q)^2}U_n^{(a)}(x;q).$$
(18.2.12)

Theorem 18.2.4 *The orthogonality relations* (18.2.4) *or* (18.2.6) *hold.*

Proof In the present case

$$w(x)=p(x)=(qx,qx/a;q)_\infty.$$
(18.2.13)

The polynomial $U_n^{(a)}(x;q)$ is an eigenfunction of (18.2.11) with the eigenvalue $\lambda_n=(1-q^n)q^{2-n}/\left[-a(1-q)^2\right]$. It is clear that these eigenvalues are distinct. We now apply Theorem (18.1.1) with $b=1$ since with w as in (18.2.13), $w(x_{-1})=w(y_{-1})=0$. The eigenvalues $\{\lambda_n\}$ are distinct then $\left\langle U_n^{(a)},U_m^{(a)}\right\rangle_q=0$ for $m\neq n$ and the polynomials $\left\{U_n^{(a)}(x;q)\right\}$ are orthogonal with respect to $w(x)$. It only remains to compute the normalizing constants $\zeta_n(a)$, $\zeta_n(a)=\left\langle U_n^{(a)},U_n^{(a)}\right\rangle$. Clearly

$$\zeta_{n-1}(a)\left(\frac{1-q^n}{1-q}\right)^2=\left(\frac{1-q^n}{1-q}\right)^2\left\langle U_{n-1}^{(a)},U_{n-1}^{(a)}\right\rangle_q$$
$$=\left\langle D_{q,x}U_n^{(a)},D_{q,x}U_n^{(a)}\right\rangle_q=-q^{-1}\left\langle U_n^{(a)},\frac{1}{w}D_{q^{-1},x}wD_{q,x}U_n^{(a)}\right\rangle_q$$
$$=-\frac{q^{1-n}(1-q^n)}{a(1-q)^2}\left\langle U_n^{(a)},U_n^{(a)}\right\rangle_q=-\frac{q^{1-n}(1-q^n)}{a(1-q)^2}\zeta_n(a).$$

Thus $\zeta_n(a)=-aq^{n-1}(1-q^n)\zeta_{n-1}(a)$ and we find

$$\zeta_n(a)=(-a)^n(q;q)_nq^{n(n-1)/2}\zeta_0(a).$$
(18.2.14)

We now evaluate $\zeta_0(a)$. Observe that

$$\zeta_0(a) = \sum_{n=0}^{\infty} q^n (q^{n+1}, q^{n+1}/a; q)_\infty - a \sum_{n=0}^{\infty} q^n \left(aq^{n+1}, q^{n+1}; q\right)_\infty. \qquad (18.2.15)$$

Thus

$$\zeta_0(a) = (q; q)_\infty \sum_{n=0}^{\infty} \frac{q^n}{(q; q)_n} \left[(q^{n+1}/a; q)_\infty - a \left(q^{n+1}a; q\right)_\infty \right]$$

$$= (q; q)_\infty \sum_{n=0}^{\infty} \frac{q^n}{(q; q)_n} \left[\sum_{k=0}^{\infty} \frac{(-1)^k q^{k(k-1)/2}}{(q; q)_k a^k} q^{k(n+1)} \right.$$

$$\left. - \sum_{k=0}^{\infty} \frac{(-1)^k q^{k(k-1)/2}}{(q; q)_k} a^{k+1} q^{k(n+1)} \right],$$

where we used Euler's formula (12.2.25). Interchanging the k sum with the n sum and using (12.2.24) to evaluate the n sum we find

$$\zeta_0(a) = (q; q)_\infty \sum_{k=0}^{\infty} \frac{(-1)^k q^{k(k+1)/2}}{(q; q)_k (q^{k+1}; q)_\infty} \left[a^{-k} - a^{k+1} \right]$$

$$= \sum_{k=0}^{\infty} (-1)^k q^{k(k+1)/2} \left[a^{-k} - a^{k+1} \right] = \sum_{n=-\infty}^{\infty} q^{k^2/2} \left(-\sqrt{q}/a \right)^k.$$

Hence the Jacobi triple product identity yields

$$\zeta_0(a) = (q, a, q/a; q)_\infty. \qquad (18.2.16)$$

In evaluating the last sum we used the Jacobi triple product identity (12.3.4). $\qquad \Box$

In (Al-Salam & Carlitz, 1965), (18.2.6) was established without the evaluation of $\zeta_0(a)$. In fact, $\zeta_0(a)$ was left as in (18.2.5). The evaluation of $\zeta_0(a)$ appears in (Ismail, 1985).

The next result readily follows from (18.2.10).

Theorem 18.2.5 *The $U_n^{(a)}$'s have the Rodrigues-type formula*

$$U_n^{(a)}(x; q) = \frac{(1-q)^n a^n q^{n(n-3)/2}}{(qx, qx/a; q)_\infty} D_{q^{-1}, x}^n \left\{ (qx, qx/a; q)_\infty \right\}. \qquad (18.2.17)$$

We now consider the second family of Al-Salam–Carlitz polynomials, the polynomials $\left\{ V_n^{(a)}(x; q) \right\}$ which are generated by

$$V_0^{(a)}(x; q) = 1, \quad V_1^{(a)}(x; q) = x - 1 - a, \qquad (18.2.18)$$

and

$$V_{n+1}^{(a)}(x; q) = \left[x - (1+a)q^{-n} \right] V_n^{(a)}(x; q)$$
$$- aq^{1-2n} (1 - q^n) V_{n-1}^{(a)}(x; q), \quad n > 0. \qquad (18.2.19)$$

These polynomials correspond to formally replacing q by $1/q$ in $U_n^{(a)}(x; q)$. The

$V_n^{(a)}$'s are orthogonal with respect to a positive measure if and only if $0 < aq$, $-1 < q < 1$.

Theorem 18.2.6 *The polynomials have the generating function*

$$V^{(a)}(x;t) := \sum_{n=0}^{\infty} V_n^{(a)}(x;q) \frac{q^{\binom{n}{2}}}{(q;q)_n} (-t)^n$$

$$= \frac{(xt;q)_\infty}{(t,at;q)_\infty}, \quad |t| < \min(1, 1/a).$$

(18.2.20)

The proof consists of multiplying (18.2.19) by $q^{\binom{n+1}{2}} t^{n+1}/(q;q)_{n+1}$ then derive a functional equation for $V(x,t)$ whose solution gives (18.2.20). The details are omitted.

Corollary 18.2.7 *The $V_n^{(a)}$'s have the explicit form*

$$V_n^{(a)}(x;q) = (-1)^n q^{-n(n-1)/2} \sum_{k=0}^{n} \frac{(q;q)_n a^{n-k}}{(q;q)_k (q;q)_{n-k}} (x;q)_k.$$

(18.2.21)

Proof Expand $(xt;q)_\infty/(t;q)_\infty$ and $1/(at;q)_\infty$ by the q-binomial theorem (12.2.17) and Euler's formula (12.2.19), respectively. Equate the coefficients of the like powers of t in (18.2.21) and after some manipulations we obtain (18.2.22). One can also derive (18.2.21) by replacing q by $1/q$ in (18.2.7). $\qquad\square$

Theorem 18.2.8 *The lowering and raising operators for the $V_n^{(a)}$'s are*

$$D_{q^{-1},x} V_n^{(a)}(x;q) = q^{1-n} \frac{(1-q^n)}{(1-q)} V_{n-1}^{(a)}(x;q),$$

(18.2.22)

and

$$(x, x/a; q)_\infty D_{q,x} \frac{V_n^{(a)}(x;q)}{(x, x/a; q)_\infty} = \frac{q^n}{a(q-1)} V_{n+1}^{(a)}(x;q),$$

(18.2.23)

respectively.

Proof Replacing q by $1/q$ in (18.2.9) establishes (18.2.22). Formula (18.2.22) also follows by applying $D_{q^{-1},x}$ to (18.2.21). The relationship

$$D_{q,x} \frac{V_n^{(a)}(x;q)}{(x, x/a; q)_\infty} = (-1)^n q^{-\binom{n}{2}} \sum_{k=0}^{n} \frac{(q;q)_n a^{n-k}}{(q;q)_k (q;q)_{n-k}} D_{q,x} \frac{1}{(xq^k, x/a; q)_\infty}$$

$$= (-1)^n q^{-\binom{n}{2}} \sum_{k=0}^{n} \frac{(q;q)_n a^{n-k}}{(q;q)_k (q;q)_{n-k}} \frac{1}{(xq^k, x/a; q)_\infty} \frac{q^k + (1 - xq^k)/a}{1-q}$$

implies that the left-hand side of (18.2.23) is given by

$$\frac{(-1)^n q^{-\binom{n}{2}}}{(1-q)} \sum_{k=0}^{n} \frac{(q;q)_n a^{n-k}}{(q;q)_k (q;q)_{n-k}} \left[q^k (x;q)_k + a^{-1}(x;q)_{k+1} \right].$$

Formula (18.2.23) now follows from rearranging the terms in the above expression. $\qquad\square$

Theorem 18.2.9 *The $V_n^{(a)}$'s satisfy the q-Sturm–Liouville equation*

$$(x, x/a; q)_\infty D_{q,x} \frac{1}{(x, x/a; q)_\infty} D_{q^{-1},x} V_n^{(a)}(x; q) = -\frac{1 - q^n}{a(1 - q)^2} V_n^{(a)}(x; q).$$

$$(18.2.24)$$

Proof Combine (18.2.22) and (18.2.23). $\qquad\square$

The q-difference equation (18.2.23) is

$$\left[a - x(1 + a) + x^2\right] D_{q,x} D_{q^{-1},x} V_n$$
$$+ \frac{1 + a - x}{1 - q} D_{q^{-1},x} V_n + \frac{1 - q^n}{(1 - q)^2} V_n = 0.$$

$$(18.2.25)$$

Observe that the coefficients in (18.2.25) correspond to replacing q by $1/q$ in (18.2.12).

We shall see in §21.9 that the orthogonality measure for $\left\{V_n^{(a)}(x; q)\right\}$ is not unique.

Theorem 18.2.10 *An orthogonality relation for $\left\{V_n^{(a)}(x; q)\right\}$ is*

$$\sum_{k=0}^\infty \frac{a^k q^{k^2}}{(q, aq; q)_k} V_m^{(a)}\left(q^{-k}; q\right) V_n^{(a)}\left(q^{-k}; q\right) = \frac{(q; q)_n a^n}{(qa; q)_n q^{n^2}} \delta_{m,n}.$$

$$(18.2.26)$$

Proof Integrate $(x/a; q)_m V_n^{(a)}(x; q)$ with respect to the measure in (18.2.26) and denote this integral by $I_{m,n}$. Clearly for $m \le n$, we have

$$(-1)^n q^{\binom{n}{2}} I_{m,n} = \sum_{k=0}^\infty \frac{q^{k^2} a^k \left(q^{-k}/a; q\right)_m}{(q, aq; q)_k} \sum_{j=0}^n \frac{(q; q)_n a^{n-j}}{(q; q)_j (q; q)_{n-j}} \left(q^{-k}; q\right)_j$$

$$= \sum_{j=0}^n \frac{(q; q)_n a^{n-m} (-1)^{j+m}}{(q; q)_j (q; q)_{n-j}} q^{\binom{j}{2}+\binom{m}{2}} \sum_{k=j}^\infty \frac{q^{k^2} a^{k-j} q^{-k(m+j)}}{(q; q)_{k-j} (aq; q)_{k-m}}.$$

The inner sum is

$$\frac{q^{-jm}}{(aq; q)_{j-m}} \lim_{\lambda \to \infty} {}_2\phi_1\left(\begin{array}{c} \lambda, \lambda \\ aq^{j-m+1} \end{array} \middle| \frac{a}{\lambda^2} q^{j-m+1}\right)$$

$$= \frac{q^{-jm}}{(aq; q)_{j-m}} \lim_{s \to \infty} {}_2\phi_1\left(\begin{array}{c} q^{-s}, q^{-s} \\ aq^{j-m+1} \end{array} \middle| aq^{j-m+1+2s}\right)$$

$$= \frac{q^{-jm}}{(aq; q)_{j-m}} \frac{1}{(aq^{j-m+1}; q)_\infty} = \frac{q^{-jm}}{(aq; q)_\infty}.$$

Hence

$$(-1)^n q^{\binom{n}{2}} I_{m,n} = \frac{(-1)^m q^{\binom{m}{2}} a^{n-m}}{(aq;q)_\infty} \, {}_1\phi_0\left(q^{-n}; -; q, q^{n-m}\right)$$

$$= \frac{(-1)^m q^{\binom{m}{2}} a^{n-m}}{(aq;q)_\infty} \left(q^{-m}; q\right)_n.$$

Thus $I_{m,n} = 0$ for $m < n$ and $I_{n,n} = (-1)^n q^{-n(n+1)/2} (q;q)_n / (aq;q)_\infty$. The left-hand side of (18.2.26) when $n = m$ is $a^n (-1)^n q^{-n(n-1)/2} I_{n,n}$ and the proof is complete. $\qquad\square$

18.3 The Al-Salam–Carlitz Moment Problem

We give another illustration of the technique of Chapter 5 by recovering the measure of orthogonality for the $U_n^{(a)}$'s from the three-term recurrence relation (18.2.2). Let

$$U^*(x,t) = \sum_{n=1}^\infty \frac{U_n^{(a)*}(x;q)}{(q;q)_n} \, t^n. \tag{18.3.1}$$

The U_n^*'s are generated by (18.2.2) and the initial values 0, 1. If we multiply the recurrence relation by $t^{n+1}/(q;q)_n$ and add for $n = 1, 2, \ldots$, we see that the generating function (18.3.1) satisfies the functional equation

$$U^*(x,t) = \frac{t}{1-xt} + \frac{(1-t)(1-at)}{1-xt} U^*(x, qt),$$

so that

$$U^*(x,t) = \sum_{n=1}^\infty \frac{U_n^{(a)*}(x;q)}{(q;q)_n} \, t^n = t \sum_{n=1}^\infty \frac{q^n (t, at; q)_n}{(xt;q)_{n+1}}. \tag{18.3.2}$$

Both (18.2.1) and (18.3.1) are meromorphic functions of t and the pole nearest to the origin is at $t = 1/x$. Thus

$$\frac{(1/x, a/x; q)_\infty}{(1 - xt)(q;q)_\infty} \quad \text{and} \quad \frac{1}{x(1 - xt)} \sum_{n=1}^\infty \frac{q^n (1/x, q/x; q)_n}{(q;q)_n}, \tag{18.3.3}$$

are comparison functions for (18.2.1) and (18.3.1), respectively. Therefore as $n \to \infty$ we have, for $x \neq 0$,

$$U_n^{(a)}(x;q) = (1/x, a/x; q)_\infty \, x^n [1 + o(1)], \tag{18.3.4}$$

$$U_n^{(a)*}(x;q) = (q;q)_\infty x^{n-1} \sum_{k=0}^\infty \frac{q^k (1/x, a/x; q)_k}{(q;q)_k} [1 + o(1)]. \tag{18.3.5}$$

Hence, for non-real z we have

$$F(z) := \lim_{n\to\infty} \frac{U_n^{(a)*}(z;q)}{U_n^{(a)}(z;q)} = \frac{(q;q)_\infty}{z(1/z, a/z; q)_\infty} \, {}_2\phi_1(1/z, a/z; 0; q, q). \tag{18.3.6}$$

Since the coefficients in the three-term recurrence relation (18.2.2) are bounded, $F(z)$ in (18.3.6) is $\int_{\mathbb{R}} d\mu(t)/(z - t)$, see (2.6.2). Hence, the U_n's satisfy the orthogonality relation (18.2.4). From (18.3.6) it is clear that $F(x)$ is meromorphic with poles at $x = q^n$, aq^n, $n = 0, 1, \dots$ and $x = 0$ is an essential singularity. The Perron–Stieltjes inversion formula shows that μ is a discrete measure supported on $\{aq^n, q^n : n = 0, 1, \dots\}$ and $x = 0$ may support a point mass. The mass at an isolated mass point t is the residue of F at $x = t$. Now (18.3.6) implies

$$\text{Res}\,\{F(x) : x = q^n\} = \frac{1}{(q^{-n};q)_n\,(aq^{-n};q)_\infty}\,{}_2\phi_1\left(q^{-n}, aq^{-n}; 0; q, q\right)$$

$$= \frac{q^n}{(q, q/a; q)_n (a; q)_\infty},$$

$$(18.3.7)$$

where the Chu–Vandermonde sum was used. Similarly

$$\text{Res}\,\{F(x) : x = aq^n\} = \frac{q^n}{(q, qa; q)_n (1/a; q)_\infty}.$$

$$(18.3.8)$$

Let $\mu(0)$ be the possible mass at $x = 0$. Since μ is normalized to have a unit total mass then

$$1 - \mu(0) = \sum_{n=1}^{\infty} \frac{q^n}{(q, q/a; q)_n (a; q)_\infty} + \sum_{n=1}^{\infty} \frac{q^n}{(q, qa; q)_n (1/a; q)_\infty}.$$

$$(18.3.9)$$

The right-hand side in (18.3.9) is $(aq, 1/a; q)_\infty$ times $\zeta_0(a)$ of (18.2.16), so by (18.2.15) we see that the series on the right-hand side of (18.3.9) sums to 1 implying $\mu(0) = 0$. Thus we have given an alternate proof of (18.2.4).

18.4 q-Jacobi Polynomials

We proceed and apply the bootstrap method and the attachment procedure to the identity implied by the orthogonality of the U_ns. First observe that the radius of convergence of (18.2.1) is $\rho = \infty$, so we get

$$\int_{\mathbb{R}} G\left(x; t_1\right) G\left(x; t_2\right) d\mu^{(a)}(x) = \sum_{n=1}^{\infty} \frac{(-at_1 t_2)^n}{(q; q)_n} q^{n(n-1)/2} = (at_1 t_2; q)_\infty,$$

$$(18.4.1)$$

for $t_1, t_2 \in \mathbb{C}$. The second equality follows by Euler's formula (12.2.20). This establishes the integral

$$\int_{\mathbb{R}} \frac{d\mu^{(a)}(x)}{(xt_1, xt_2; q)_\infty} = \frac{(at_1 t_2; q)_\infty}{(t_1, t_2, at_1, at_2; q)_\infty}.$$

$$(18.4.2)$$

When we substitute for $\mu^{(a)}$ from (18.2.5) in (18.4.2) we obtain

$$\frac{(at_1 t_2; q)_\infty}{(t_1, t_2, at_1, at_2; q)_\infty}$$

$$= \sum_{n=0}^{\infty} \frac{q^n / (q, q/a; q)_n}{(a, t_1 q^n, t_2 q^n; q)_\infty} + \sum_{n=0}^{\infty} \frac{q^n / (q, qa; q)_n}{(1/a, at_1 q^n, at_2 q^n; q)_\infty},$$

which when expressed in terms of basic hypergeometric functions is the nonterminating analogue of the Chu–Vandermonde theorem stated in (12.2.16).

We now restrict our attention to the case $t_1, t_2 \in \left(a^{-1}, 1\right)$ which ensures that

$1/\left(xt_1, xt_2; q\right)_\infty$ is a positive weight function on $[a, 1]$. The next step is to find polynomials orthogonal with respect to $d\mu^{(a)}(x)/\left(xt_1, xt_2; q\right)_\infty$. Define $P_n(x)$ by

$$P_n(x) = \sum_{k=0}^{n} \frac{\left(q^{-n}, xt_1; q\right)_k}{(q; q)_k} q^k a_{n,k} \qquad (18.4.3)$$

where $a_{n,k}$ will be chosen later. Using (18.4.2) it is easy to see that

$$\int_{\mathbb{R}} P_n(x)\, (xt_2; q)_m\, \frac{d\mu^{(a)}(x)}{\left(xt_1, xt_2; q\right)_\infty}$$

$$= \sum_{k=0}^{n} \frac{\left(q^{-n}; q\right)_k}{(q; q)_k} q^k a_{n,k} \frac{\left(at_1 t_2 q^{k+m}; q\right)_\infty}{\left(t_1 q^k, at_1 q^k, t_2 q^m, at_2 q^m; q\right)_\infty}$$

$$= \frac{\left(at_1 t_2 q^m; q\right)_\infty}{\left(t_1, at_1, t_2 q^m, at_2 q^m; q\right)_\infty} \sum_{k=0}^{n} \frac{\left(q^{-n}, t_1, at_1; q\right)_k}{\left(q, at_1 t_2 q^m; q\right)_k} a_{n,k} q^k.$$

The choice $a_{n,k} = (\lambda; q)_k / \left(t_1, at_1; q\right)_k$ allows us to apply the q-Chu–Vandermonde sum (12.2.13). The choice $\lambda = at_1 t_2 q^{n-1}$ leads to

$$\int_{\mathbb{R}} P_n(x)\, (xt_2; q)_m\, \frac{d\mu^{(a)}(x)}{\left(xt_1, xt_2; q\right)_\infty}$$

$$= \frac{\left(at_1 t_2 q^m; q\right)_\infty \left(q^{m+1-n}; q\right)_n \left(at_1 t_2 q^{n-1}\right)^n}{\left(t_1, at_1, t_2 q^m, at_2 q^m; q\right)_\infty \left(at_1 t_2 q^m; q\right)_n}. \qquad (18.4.4)$$

Obviously, the right-hand side of (18.4.4) vanishes for $0 \le m < n$. The coefficient of x^n in $P_n(x)$ is

$$\frac{\left(q^{-n}, at_1 t_2 q^{n-1}; q\right)_n}{\left(q, t_1, at_1; q\right)_n} (-t_1)^n\, q^{n(n+1)/2} = \frac{\left(at_1 t_2 q^{n-1}; q\right)_n}{\left(t_1, at_1; q\right)_n} t_1^n.$$

Therefore

$$P_n(x) = \varphi_n\left(x; a, t_1, t_2\right) = {}_3\phi_2 \left(\begin{matrix} q^{-n}, at_1 t_2 q^{n-1}, xt_1 \\ t_1, at_1 \end{matrix} \,\middle|\, q, q \right), \qquad (18.4.5)$$

satisfies the orthogonality relation

$$\int_{a}^{1} \varphi_m\left(x; a, t_1, t_2\right) \frac{\varphi_n\left(x; a, t_1, t_2\right)}{(q, a, q/a; q)_\infty (1-q)} \frac{(qx, qx/a; q)_\infty}{\left(xt_1, xt_2; q\right)_\infty}\, d_q x$$

$$\qquad (18.4.6)$$

$$= \frac{\left(q, t_2, at_2, at_1 t_2 q^{n-1}; q\right)_n \left(at_1 t_2 q^{2n}; q\right)_\infty}{\left(t_1, at_1, t_2, at_2; q\right)_\infty \left(t_1, at_1; q\right)_n} \left(-at_1^2\right)^n q^{n(n-1)} \delta_{m,n}.$$

The polynomials $\{\varphi_n\left(x; a, t_1, t_2\right)\}$ are the big q-Jacobi polynomials of Andrews and Askey (Andrews & Askey, 1985) in a different normalization. The Andrews–Askey normalization is

$$P_n(x; \alpha, \beta, \gamma : q) = {}_3\phi_2 \left(\begin{matrix} q^{-n}, \alpha\beta q^{n+1}, x \\ \alpha q, \gamma q \end{matrix} \,\middle|\, q, q \right). \qquad (18.4.7)$$

Note that we may rewrite the orthogonality relation (18.4.6) in the form

$$
\int_a^1 \frac{(qx, qx/a; q)_\infty}{(xt_1, xt_2; q)_\infty} t_1^{-m} (t_1, at_1; q)_m \frac{\varphi_m (x; a, t_1, t_2)}{(1 - q)(q, a, q/a; q)_\infty}
$$

$$
\times t_1^{-n} (t_1, at_1; q)_n \, \varphi_n (x; a, t_1, t_2) \, d_q x \qquad (18.4.8)
$$

$$
= \frac{(q, t_1, at_1, t_2, at_2, at_1 t_2 q^{n-1}; q)_n}{(t_1, at_1, t_2, at_2; q)_\infty}
$$

$$
\times (at_1 t_2 q^{2n}; q)_\infty (-a)^n \, q^{n(n-1)/2} \, \delta_{m,n}.
$$

Since $(qx, qx/a; q)_\infty / (xt_1, xt_2; q)_\infty$ and the right-hand side of (18.4.8) are symmetric in t_1 and t_2 then

$$
t_1^{-n} (t_1, at_1; q)_n \, \varphi_n (x; a, t_1, t_2)
$$

must be symmetric in t_1 and t_2. This gives the $3\phi_2$ transformation

$$
{}_3\phi_2 \left(\begin{array}{c} q^{-n}, at_1 t_2 q^{n-1}, xt_1 \\ t_1, at_1 \end{array} \middle| \, q, q \right)
$$

$$
= \frac{t_1^n (t_2, at_2; q)_n}{t_2^n (t_1, at_1; q)_n} \, {}_3\phi_2 \left(\begin{array}{c} q^{-n}, at_1 t_2 q^{n-1}, xt_2 \\ t_2, at_2 \end{array} \middle| \, q, q \right). \qquad (18.4.9)
$$

The big q-Jacobi polynomials were introduced by Hahn and developed by George Andrews and Richard Askey, who coined the name and used the notation in (18.4.7); see (Andrews & Askey, 1985) and (Andrews et al., 1999). The big q-Jacobi polynomials generalize the Jacobi polynomials. The parameter identification between the Andrews–Askey notation and our own notation is

$$
x \to t_1 x, \quad \alpha = t_1/q, \quad \beta = at_2/q, \quad \gamma = at_1/q. \qquad (18.4.10)
$$

The proofs given here are from (Berg & Ismail, 1996). The little q-Jacobi are

$$
p_n(x; \alpha, \beta) = \lim_{a \to \infty} \varphi_n (ax; a, \alpha q, \alpha q/a)
$$

$$
= {}_2\phi_1 \left(q^{-n}, \alpha\beta q^{n+1}; q\alpha; q, qx \right). \qquad (18.4.11)
$$

Moreover

$$
\lim_{q \to 1} p_n \left(x; q^\alpha, q^\beta \right) = {}_2F_1(-n, n + \alpha + \beta + 1; \alpha + 1; x),
$$

hence

$$
\lim_{q \to 1} p_n \left(x; q^\alpha, q^\beta \right) = \frac{n!}{(\alpha + 1)_n} P_n^{(\alpha,\beta)} (1 - 2x). \qquad (18.4.12)
$$

Their orthogonality relation is

$$
\sum_{k=0}^\infty \frac{(\beta q; q)_k}{(q; q)_k} (aq)^k p_m \left(q^k; \alpha, \beta \right) p_n \left(q^k; \alpha, \beta \right)
$$

$$
= \frac{(\alpha\beta q^2)_\infty}{(\alpha q; q)_\infty} \frac{(1 - \alpha\beta q)}{(1 - \alpha\beta q^{2n+1})} \frac{(q; \beta q; q)_n}{(\alpha q; \alpha\beta q; q)_n} \delta_{m,n}. \qquad (18.4.13)
$$

This follows from (18.4.6) and (18.4.11). We prefer our notation over the notation in (18.4.7) because the lowering operator in our notation is much simpler; compare (18.4.22) and (Koekoek & Swarttouw, 1998, (3.5.7)). For convenience, we record (18.4.5) in the notation of (18.4.7),

$$
\int_{\gamma q}^{\alpha q} \frac{(x/\alpha, x/\gamma; q)_\infty}{(x, \beta x/\gamma; q)_\infty} P_m(x; \alpha, \beta, \gamma : q) P_n(x; \alpha, \beta, \gamma; q) d_q x
$$

$$
= \frac{\alpha q(1-q) \left(q, \alpha\beta q^2, \gamma/\alpha, \alpha/\gamma; q\right)_\infty}{(\alpha q, \beta q, \gamma q, \alpha\beta q/\gamma; q)_\infty} \frac{(1-\alpha\beta q)}{(1-\alpha\beta q^{2n+1})}
$$

$$
\times \frac{(q, \beta q, q\alpha\beta/\gamma; q)_n}{(\alpha q, \alpha\beta q, \gamma q; q)_n} (-\alpha\gamma)^n q^{n(n+3)/2} \delta_{m,n}. \quad (18.4.14)
$$

To find the three term recurrence relation satisfied by φ_n, we set

$$
(xt_1 - 1)\,\varphi_n\,(x; a, t_1, t_2) = A_n \varphi_{n+1}\,(x; a, t_1, t_2)
$$
$$
+ B_n \varphi_n\,(x; a, t_1, t_2) + C_n \varphi_{n+1}\,(x; a, t_1, t_2). \quad (18.4.15)
$$

Since $\varphi_n\,(1/t_1; a, t_1, t_2) = 1$, we find from (18.4.15) that

$$
B_n = -A_n - C_n. \quad (18.4.16)
$$

It readily follows from (18.4.5) that

$$
\varphi_n\,(x; a, t_1, t_2) = \frac{\left(at_1 t_2 q^{n-1}; q\right)_n}{(t_1, at_1; q)_n} t_1^n x^n + \text{lower order terms}, \quad (18.4.17)
$$

which implies

$$
A_n = \frac{(1 - t_1 q^n)\,(1 - at_1 q^n)\,\left(1 - at_1 t_2 q^{n-1}\right)}{\left(1 - at_1 t_2 q^{2n}\right)\left(1 - at_1 t_2 q^{2n-1}\right)}. \quad (18.4.18)
$$

Apply the Chu–Vandermonde theorem to obtain

$$
\varphi_n\,(1, a, t_1, t_2) = {}_2\phi_1\left(q^{-n}, at_1 t_2 q^{n-1}; at_1; q, q\right) = \left(q^{1-n}/t_2; q\right)_n (at_1)^n
$$
$$
= (-1)^n (t_2; q)_n (at_1/t_2)^n q^{n(n-1)/2}.
$$

Use the above evaluation at $x = 1$ in (18.4.15) and use (14.4.15) to obtain

$$
C_n = -\frac{at_1^2 q^n\,(1 - q^n)\,\left(1 - t_2 q^{n-1}\right)\left(1 - at_1 q^{n-1}\right)}{\left(1 - at_1 t_2 q^{2n-1}\right)\left(1 - at_1 t_2 q^{2n-2}\right)}. \quad (18.4.19)
$$

Theorem 18.4.1 *The big q-Jacobi polynomials satisfy*

$$
(xt_1 - 1)\,y_n(x) = A_n y_{n+1}(x) - (A_n + C_n)\,y_n(x) + C_n y_{n-1}(x). \quad (18.4.20)
$$

In particular, after the scaling $x \to (x+1)/t_1$, $\{\varphi_n\}$ becomes a family of birth and death process polynomials with birth rates $\{A_n\}$ and death rates $\{C_n\}$.

It is easy to see that

$$
D_{q,x}(tx; q)_k = -\frac{1 - q^k}{1 - q} t(qtx; q)_{k-1}. \quad (18.4.21)
$$

Therefore

$$D_{q,x}\varphi_n\left(x;a,t_1,t_2\right) = \frac{t_1 q^{1-n}\left(1-q^n\right)\left(1-at_1 t_2 q^{n-1}\right)}{(1-q)(1-t_1)(1-at_1)}\,\varphi_{n-1}\left(x;a,qt_1,qt_2\right).$$

$$(18.4.22)$$

This shows that $D_{q,x}$ is a lowering operator for $\{\varphi_n\}$. We now proceed to find a raising operator.

Theorem 18.4.2 *The polynomials* $\{\varphi_n\left(x;a,t_1,t_2\right)\}$ *satisfy*

$$\frac{(t_1 x/q, t_2 x/q; q)_\infty}{(qx, qx/a; q)_\infty}\,D_{q^{-1},x}\left(\frac{(qx, qx/a; q)_\infty}{(t_1 x, t_2 x; q)_\infty}\varphi_n\left(x;a,t_1,t_2\right)\right)$$

$$= -\frac{(1-t_1/q)(1-at_1/q)\,q^2}{at_1(1-q)}\,\varphi_{n+1}\left(x;a,t_1/q,t_2/q\right).$$

$$(18.4.23)$$

Proof In view of (18.4.5) and the fact

$$(xt_1; q)_k / (xt_1; q)_\infty = 1/\left(xt_1 q^k; q\right)_\infty$$

the left-hand side of (18.4.23) is

$$-q\sum_{k=0}^{n}\frac{\left(q^{-n}, at_1 t_2 q^{n-1}, xt_1/q; q\right)_k q^k}{(1-q)(q, t_1, at_1; q)_k}$$

$$\times\left[1 + \frac{1}{a} - t_1 q^{k-1} - \frac{t_2}{q} + x\left(t_1 t_2 q^{k-2} - 1/a\right)\right].$$

The quantity in square brackets can be put in the form

$$\left(1 - t_1 q^{k-1}\right)\left(1 - q^{1-k}/at_1\right) + \left(1 - xt_1 q^{k-1}\right)\left(1 - at_1 t_2 q^{k-2}\right)q^{1-k}/at_1.$$

After some manipulations we establish (18.4.23) and the proof is complete. $\qquad\square$

An immediate consequence of Theorem 18.4.2 is the Rodrigues-type formula

$$\varphi_n\left(x;a,t_1,t_2\right) = \frac{a^n t_1^n (1-q)^n}{(t_1, at_1; q)_n\,q^n}\,\frac{(t_1 x, t_2 x; q)_\infty}{(qx, qx/a; q)_\infty}$$

$$\times D_{q^{-1},x}^n\left(\frac{(qx, qx/a; q)_\infty}{(q^n t_1 x, q^n t_2 x; q)_\infty}\right).$$

$$(18.4.24)$$

Combining (18.4.12) and (18.4.14) we see that the big q-Jacobi polynomials are solutions to the q-Sturm–Liouville problem

$$\frac{(t_1 x, t_2 x; q)_\infty}{(qx, qx/a; q)_\infty}\,D_{q^{-1},x}\left(\frac{(qx, qx/a; q)_\infty}{(qt_1 x, qt_2 x; q)_\infty}D_{q,x}\varphi_n\left(x;a,t_1,t_2\right)\right)$$

$$= \frac{(1-q^n)\left(1-at_1 t_2 q^{n-1}\right)q^{2-n}}{a(1-q)^2}\,\varphi_n\left(x;a,t_1,t_2\right).$$

$$(18.4.25)$$

The q-difference equation (18.4.25) when expanded out becomes

$$\left[x^2 - x(1+a) + a \right] D_{q^{-1},x} D_{q,x} y$$
$$+ q \, \frac{(1 - at_1 t_2)\, x + a\, (t_1 + t_2) - a - 1}{(1-q)} \, D_{q,x} y$$
$$= \frac{(1 - q^n)\, (1 - at_1 t_2 q^{n-1})}{(1-q)^2} \, q^{2-n} \, y. \quad (18.4.26)$$

Thus $y = \varphi_n\,(x; a, t_1, t_2)$ is a solution to (18.4.26). Thus, the little q-Jacobi polynomials satisfy

$$x(x-1) D_{q^{-1},x} D_{q,x} y + q\, \frac{\left(1 - q^2 \alpha \beta \right) x + q\alpha - 1}{1-q} \, D_{q,x} y$$
$$= \frac{(1 - q^n)\, (1 - \alpha\beta q^{n+1})}{(1-q)^2} \, y. \quad (18.4.27)$$

Theorem 18.4.3 *The q-Sturm–Liouville property* (18.4.15) *implies the orthogonality relation* (18.4.6).

The proof is similar to our proof of Theorem 18.2.4. We use Theorem 18.1.1 to show that the right-hand side of (18.4.6) vanishes when $m \neq n$. To compute ζ_n ($=$ the left-hand side of (18.4.6) when $m = n$) we follow our proof of Theorem 18.2.4 and relate ζ_n to ζ_0. Finally the value of ζ_0 is found from (18.4.2).

The large n asymptotics of the big and little q-Jacobi polynomials were developed in (Ismail & Wilson, 1982) through the application of Darboux's method to generating functions.

In (12.4.1), set $f = (abc/de)q^{1-n}$ and let $a \to \infty$. This leads to

$$\varphi_n\,(x; a, t_1, t_2) = \frac{(q^{1-n}/t_2 x; q)_n}{(at_1; q)_n} \, \left(at_1 t_2 x q^{n-1} \right)^n$$
$$\times\, {}_3\phi_2 \left(\begin{array}{c} q^{-n}, q^{1-n}/at_2, 1/x \\ t_1, q^{1-n}/t_2 x \end{array} \bigg|\, q, q \right). \quad (18.4.28)$$

Write the ${}_3\phi_2$ as $\sum\limits_{k=0}^{n}$ then replace k by $n - k$ to obtain

$$\varphi_n\,(x; t_1, t_2, a) = \frac{(q, at_2; q)_n}{(at_1; q)_n} \sum_{k=0}^{n} \frac{(1/x; q)_{n-k}\, (t_1 x)^{n-k}}{(q, t_1; q)_{n-k}}$$
$$\times \frac{(-at_1)^k\, (t_2 x; q)_k\, q^{k(k-1)/2}}{(q, at_2; q)_k}, \quad (18.4.29)$$

after applying (12.2.11)–(12.2.12). Therefore, we have established the generating function (Ismail & Wilson, 1982)

$$\sum_{n=0}^{\infty} \varphi_n\,(x; t_1, t_2, a) \, \frac{(at_1; q)_n}{(q, at_2; q)_n} \, t^n$$
$$= {}_2\phi_1 \left(1/x, 0; t_1; q, t_1 x \right) {}_1\phi_1 \left(t_2 x; at_2; q, at_1 t \right). \quad (18.4.30)$$

Theorem 18.4.4 *We have the asymptotic formulas*

$$\lim_{n\to\infty} (t_1 x)^{-n} \varphi_n (x; a, t_1, t_2) = \frac{(1/x, a/x; q)_\infty}{(t_1, at_1; q)_\infty}, \tag{18.4.31}$$

for $x \neq 0$, q^m, aq^m, $m = 0, 1, \ldots$; and

$$\lim_{n\to\infty} \frac{q^{nm-\binom{n}{2}}}{(-at_1)^n} \varphi_n (q^m; a, t_1, t_2) = \frac{(t_2; q)_\infty \, q^{m^2}}{(at; q)_\infty (t_1, t_2; q)_m},$$

$$\lim_{n\to\infty} \frac{q^{nm-\binom{n}{2}}}{(-t_1)^n} \varphi_n (aq^m; a, t_1, t_2) = \frac{(at_2; q)_\infty}{(t_1; q)_\infty} \frac{1}{(t_1; q)_m}. \tag{18.4.32}$$

Proof From (18.4.29) we see that the left-hand side of (18.4.31) is

$$\frac{(1/x, at_2; q)_\infty}{(t_1, at_1; q)_\infty} \sum_{k=0}^{\infty} \frac{(t_2 x; q)_k}{(q, at_2; q)_k} (a/x)^k \lim_{c\to\infty} \frac{(c; q)_k}{c^k}$$

$$= \frac{(1/x, at_2; q)_\infty}{(t_1, at_1; q)_\infty} \lim_{c\to\infty} {}_2\phi_1 (c, t_2 x; at_2; q, a/xc)$$

$$= \frac{(1/x, at_2; q)_\infty}{(t_1, at_1; q)_\infty} \frac{(a/x; q)_\infty}{(at_2; q)_\infty},$$

where we used (12.2.18). This proves (18.4.31). If $x = q^m$, use (18.4.28) to prove the first equation in (18.4.32) after some manipulations. To prove the second part of (18.4.32), apply (12.4.5) to the representation (18.4.5) to get

$$\varphi_n (x; a, t_1, t_2) = \left(at_1 t_2 q^{n-1} \right)^n \frac{\left(q^{1-n}/at_2; q \right)_n}{(t_1; q)_n}$$

$$\times {}_3\phi_2 \left(\begin{array}{c} q^{-n}, at_1 t_2 q^{n-1}, a/x \\ at_1, at_2 \end{array} \bigg| q, q \right). \tag{18.4.33}$$

When $x = aq^m$ we can let $n \to \infty$ in (18.4.32) and establish the second part of (18.4.32). $\qquad\square$

The little and big q-Jacobi functions can be defined as

$$\frac{1}{w(x)} \int_\alpha^\beta \frac{w(t) u_n(t)}{x - t} w(t) \, dt,$$

where $u_n(x)$ is a litttle (big) q-Jacobi polynomial, $(\alpha, \beta) = (0, 1)$ $((\alpha, \beta) = (a, 1))$, respectively, and w is the corresponding weight function. Kadell introduced a different type of little q-Jacobi functions in (Kadell, 2005) and used it to give new derivations of several summation theorems for q-series.

18.5 q-Hahn Polynomials

The q-Hahn polynomials are

$$Q_n(x; \alpha, \beta, N) = Q_n(x; \alpha, \beta, N; q)$$

$$= {}_3\phi_2 \left(\begin{matrix} q^{-n}, \alpha\beta q^{n+1}, x \\ \alpha q, q^{-N} \end{matrix} \middle| q, q \right), \tag{18.5.1}$$

$n = 0, 1, \ldots, N$. Their orthogonality relation is

$$\sum_{j=0}^{N} \frac{(\alpha q, q^{-N}; q)_j}{(q, q^{-N}/\beta; q)_j} (\alpha\beta q)^{-j} Q_m\left(q^{-j}; \alpha, \beta, N\right) Q_n\left(q^{-j}; \alpha, \beta, N\right)$$

$$= \frac{(\alpha\beta q^2; q)_N}{(\beta q; q)_N (\alpha q)^N} \frac{(q, \alpha\beta q^{N+2}, \beta q; q)_n}{(\alpha q, \alpha\beta q, q^{-N}; q)_n} \frac{(1 - \alpha\beta q)(-\alpha q)^n}{(1 - \alpha\beta q^{2n+1})} q^{\binom{n}{2} - Nn} \delta_{m,n}. \tag{18.5.2}$$

Proof of (18.5.2) Set

$$I_{m,n} = \sum_{j=0}^{N} \frac{(\alpha q, q^{-N}; q)_j}{(q, q^{-N}/\beta; q)_j} (\alpha\beta q)^{-j} \left(\beta q^{N+1-j}; q\right)_m Q_n\left(q^{-j}; \alpha, \beta, N\right),$$

and assume $m \leq n$. Thus, $I_{m,n}$ equals

$$\sum_{k=0}^{n} \frac{(q^{-n}, \alpha\beta q^{n+1}; q)_k}{(q, \alpha q, q^{-N}; q)_k} q^k \sum_{j=k}^{N} \frac{(\alpha q, q^{-N}; q)_j}{(q, q^{-N}/\beta; q)_j} \frac{(-1)^k q^{\binom{k}{2}-jk}}{(\alpha\beta q)^j} \frac{(q; q)_j}{(q; q)_{j-k}}$$

$$\times (-1)^m \left(\beta q^{N+1}\right)^m q^{\binom{m}{2}-jm} \frac{(q^{-N-m}/\beta; q)_{m+j}}{(q^{-m-N}/\beta; q)_j}$$

$$= (-\beta)^m q^{\binom{m+1}{2}+mN} \left(q^{-N-m}/\beta; q\right)_m \sum_{k=0}^{n} \frac{(q^{-n}, \alpha\beta q^{n+1}; q)_k}{(q, q^{-m-N}/\beta; q)_k} (-1)^k$$

$$\times \frac{q^{-\binom{k+1}{2}}}{(\alpha\beta)^k} {}_2\phi_1 \left(\begin{matrix} \alpha q^{k+1}, q^{k-N} \\ q^{-m-N+k}/\beta \end{matrix} \middle| q, \frac{q^{-m-k-1}}{\alpha\beta} \right)$$

$$= (-\beta)^m q^{\binom{m+1}{2}+mN} \frac{(q^{-m-N}/\beta; q)_m}{(q^{-m-N}/\beta; q)_N}$$

$$\times \sum_{k=0}^{n} \frac{(q^{-n}, \alpha\beta q^{n+1}; q)_k (q^{-m-N-1}/\alpha\beta; q)_{N-k} q^{-\binom{k+1}{2}}}{(q; q)_k (\alpha\beta)^k} (-1)^k$$

$$= \frac{(-\beta)^m}{(q^{-m-N}/\beta; q)_N} q^{\binom{m+1}{2}+mN} \left(q^{-N-m}/\beta; q\right)_m \left(q^{-N-m-1}/\alpha\beta; q\right)_N$$

$$\times {}_2\phi_1 \left(\begin{matrix} q^{-n}, \alpha\beta q^{n+1} \\ \alpha\beta q^{m+2} \end{matrix} \middle| q, q^{m+1} \right).$$

The ${}_2\phi_1$ is summed by formula (12.2.19) and equals $(q^{m+1-n}; q)_n / (\alpha\beta q^{m+1}; q)_n$. Therefore for $m < n$, the left-hand side of (18.5.2) is zero. When $m = n$, the left-hand side of (18.5.2) is

$$\frac{\left(q^{-n}, \alpha\beta q^{n+1}; q\right)_n}{\left(q, \alpha q, q^{-N}; q\right)_n} \frac{q^n I_{n,n}}{\left(\beta q^{N+1}\right)^n},$$

and some lengthy calculations reduce it to the right-hand side of (18.5.2).   □

The three-term recurrence relation is

$$(1 - x)Q_n(x; \alpha, \beta, N) = A_n Q_{n+1}(x; \alpha, \beta, N)$$
$$- (A_n + C_n) Q_n(x; \alpha, \beta, N) + C_n Q_{n-1}(x; \alpha, \beta, N), \tag{18.5.3}$$

where

$$A_n = -\frac{\left(1 - q^{n-N}\right)\left(1 - \alpha q^{n+1}\right)\left(1 - \alpha\beta q^{n+1}\right)}{\left(1 - \alpha\beta q^{2n+1}\right)\left(1 - \alpha\beta q^{2n+2}\right)}$$
$$C_n = \frac{\alpha q^{n-N}\left(1 - q^n\right)\left(1 - \alpha\beta q^{n+N+1}\right)\left(1 - \beta q^n\right)}{\left(1 - \alpha\beta q^{2n}\right)\left(1 - \alpha\beta q^{2n+1}\right)}. \tag{18.5.4}$$

The lowering operator for Q_n is

$$D_{q^{-1}} Q_n(x; \alpha, \beta, N) = \frac{q^{1-n}\left(1 - q^n\right)\left(1 - \alpha\beta q^{n+1}\right)}{(1 - q)(1 - \alpha q)\left(1 - q^{-N}\right)} Q_{n-1}(x; \alpha q, \beta q, N - 1). \tag{18.5.5}$$

With

$$w(x, \alpha, \beta, N) = \frac{\left(\alpha q, q^{-N}; q\right)_u}{\left(q, q^{-N}/\beta; q\right)_u} \frac{1}{(\alpha\beta)^u}, \tag{18.5.6}$$

where $x = q^{-u}$, $u = 0, 1, \ldots, N$. The raising operator is

$$D_q \left(w(x; \alpha q, \beta q, N - 1)Q_n(x; \alpha q, \beta q, N - 1)\right)$$
$$= \frac{w(x, \alpha, \beta, N)}{1 - q} Q_{n+1}(x, \alpha, \beta, N). \tag{18.5.7}$$

The Rodrigues formula is

$$Q_n(x; \alpha, \beta, N) = \frac{(1 - q)^k}{w(x; \alpha, \beta, N)}$$
$$\times D_q^k \left(w\left(x; \alpha q^k, \beta q^k, N - k\right) Q_{n-k}\left(x; \alpha q^k, \beta q^k, N - k\right)\right). \tag{18.5.8}$$

In particular,

$$Q_n(x; \alpha, \beta, N) = \frac{(1 - q)^n}{w(x; \alpha, \beta, N)} D_q^n \left(w\left(x, \alpha q^n, \beta q^n, N - n\right)\right). \tag{18.5.9}$$

The second-order operator equation is

$$\frac{1}{w(x; \alpha, \beta, N)} D_q \left(w(x; \alpha q, \beta q, N - 1) D_{q^{-1}} Q_n(x; \alpha, \beta, N)\right)$$
$$= \frac{q^{1-n}\left(1 - q^n\right)\left(1 - \alpha\beta q^{n+1}\right)}{(1 - q)^2(1 - \alpha q)\left(1 - q^{-N}\right)} Q_n(x; \alpha, \beta, N).$$

The generating functions

$$\sum_{n=0}^{N} \frac{(q^{-N};q)_n}{(q,\beta q;q)_n} Q_n(x;\alpha,\beta,N)\, t^n$$

$$= {}_1\phi_1\left(\begin{array}{c} x \\ \alpha q \end{array}\Big| q, \alpha q t\right) {}_2\phi_1\left(\begin{array}{c} xq^{-N}, 0 \\ \beta q \end{array}\Big| q, xt\right) \tag{18.5.10}$$

and

$$\sum_{n=0}^{N} \frac{(\alpha q, q^{-N};q)_n}{(q;q)_n} q^{-\binom{n}{2}} Q_n(x;\alpha,\beta,N)\, t^n$$

$$= {}_2\phi_1\left(\begin{array}{c} x, \beta q^{N+1} x \\ 0 \end{array}\Big| q, -\alpha t q^{1-N}/x\right) {}_2\phi_0\left(\begin{array}{c} q^{-N}/x, \alpha q/x \\ \underline{\hspace{1cm}} \end{array}\Big| q, -tx\right) \tag{18.5.11}$$

hold when $x = 1, q^{-1}, \ldots, q^{-N}$.

18.6 *q*-Differences and Quantized Discriminants

In this section we develop a theory of q-difference equations for general discrete q-orthogonal polynomials and compute their q-discriminants. This section is based on (Ismail, 2003a).

Let $\{p_n(x)\}$ satisfy the orthogonality relation

$$\int_a^b p_m(x)p_n(x)w(x)\, d_q x = \delta_{m,n}. \tag{18.6.1}$$

One can mimic the proofs in §3.2 and establish the following theorem.

Theorem 18.6.1 *Let $\{p_n(x)\}$ be a sequence of discrete q-orthonormal polynomials. Then they have a lowering (annihilation) operator of the form*

$$D_q p_n(x) = A_n(x)p_{n-1}(x) - B_n(x)p_n(x),$$

where $A_n(x)$ and $B_n(x)$ are given by

$$A_n(x) = a_n \left. \frac{w(y/q)p_n(y)p_n(y/q)}{x - y/q} \right]_a^b$$

$$+ a_n \int_a^b \frac{u(qx) - u(y)}{qx - y} p_n(y)p_n(y/q)w(y)\, d_q y, \tag{18.6.2}$$

$$B_n(x) = a_n \left. \frac{w(y/q)p_n(y)p_{n-1}(y/q)}{x - y/q} \right]_a^b$$

$$+ a_n \int_a^b \frac{u(qx) - u(y)}{qx - y} p_n(y)p_{n-1}(y/q)w(y)\, d_q y, \tag{18.6.3}$$

where u is defined by

$$D_q w(x) = -u(qx)w(qx), \tag{18.6.4}$$

and $\{a_n\}$ are the recursion coefficients.

As in §3.2, we set

$$L_{1,n} := B_n + D_q,$$

$$L_{2,n} := \frac{x - b_n - 1}{a_{n-1}} A_{n-1}(x) - B_{n-1}(x) - D_q.$$

One can prove the following lowering and raising relations

$$L_{1,n} p_n(x) = A_n(x) p_{n-1}(x) \tag{18.6.5}$$

$$L_{2,n} p_{n-1}(x) = \frac{a_n}{a_{n-1}} A_{n-1}(x) p_n(x), \tag{18.6.6}$$

and apply them to derive the second-order q-difference equations

$$D_q^2 p_n(x) + R_n(x) D_q p_n(x) + S_n(x) p_n(x) = 0 \tag{18.6.7}$$

with

$$R_n(x) = B_n(qx) - \frac{D_q A_n(x)}{A_n(x)} + \frac{A_n(qx)}{A_n(x)} \left[B_{n-1}(x) - \frac{(x - b_n - 1) A_{n-1}(x)}{a_{n-1}} \right], \tag{18.6.8}$$

$$S_n(x) = \frac{a_n}{a_{n-1}} A_n(qx) a_{n-1}(x) + D_q B_n(x) - \frac{B_n(x)}{A_n(x)} D_q A_n(x)$$

$$+ B_n(x) \frac{A_n(qx)}{A_n(x)} \left[B_{n-1}(x) - \frac{(x - b_n - 1) A_{n-1}(x)}{a_{n-1}} \right]. \tag{18.6.9}$$

A more symmetric form of (18.6.7) is

$$p_n(qx) - \left[1 + q + (1-q)x R_n(x/q) \right] p_n(x)$$
$$+ \left[q + x^2 (1-q)^2 q^{-1} S_n(x/q) + (1-q)x R_n(x/q) \right] p_n(x/q) = 0. \tag{18.6.10}$$

Recall that the generalized discriminant associated with a degree reducing linear operator T is defined by (6.4.2). When $T = D_q$ we find

$$D(f; D_q) = \gamma^{2n-2} q^{\binom{n}{2}} \prod_{1 \le i < j \le n} \left(q^{-\frac{1}{2}} x_i - q^{\frac{1}{2}} x_j \right) \left(q^{\frac{1}{2}} x_i - q^{-\frac{1}{2}} x_j \right). \tag{18.6.11}$$

We shall call this the q-discriminant, (Ismail, 2003a), and denote it by $D(f; q)$. In other words

$$D(f; q) = \gamma^{2n-2} q^{\binom{n}{2}} \prod_{1 \le i < j \le n} \left(x_i^2 + x_j^2 - x_i x_j \left(q + q^{-1} \right) \right). \tag{18.6.12}$$

Theorem 18.6.2 *Let $\{p_n\}$ satisfy (18.6.1). The q-discriminant of p_n is given by*

$$D(p_n; q) = \left[\prod_{j=1}^{n} \frac{A_n(x_{nj})}{a_n} \right] \left[\prod_{k=1}^{n} a_k^{2k-2n+2} \right],$$

where $\{a_n\}$ are the recursion coefficients of $\{x_{nj} : 1 \leq j \leq n\}$ are the zeros of p_n.

The proof follows from Theorem 18.6.1 and Schur's theorem, Lemma 3.4.1.

In the case of the little q-Jacobi polynomials as defined in (18.4.11), the recursion coefficients $\{a_n\}$ are

$$a_n = aq^{n-1/2} \sqrt{\frac{(1-q^n)(1-\alpha q^n)(1-\beta q^n)(1-\alpha\beta q^n)}{(1-\alpha\beta q^{2n-1})(1-\alpha\beta q^{2n})(1-\alpha\beta q^{2n+1})}}.$$

We next compute the q-discriminant of $p_n(x, \alpha, \beta)$ but we will use a normalization which tends to $P_n^{(\alpha,\beta)}(1-2x)$ as $q \to 1$.

Theorem 18.6.3 ((Ismail, 2003a)) *The q-discriminant $\Delta_n(a, b)$ of the little q-Jacobi polynomials*

$$\frac{(aq; q)_n}{(q; q)_n} \, {}_2\phi_1\left(\begin{array}{c} q^{-n}, abq^{n+1} \\ aq \end{array} \Bigg| q, qx\right)$$

is given by

$$\Delta_n(a, b) = a^{n(n-1)/2} q^{-n(n-1)(n+1)/3} \prod_{j=1}^{n} \left(\frac{1-q^j}{1-q}\right)^{j+2-2n}$$

$$\times \prod_{k=1}^{n} \left(\frac{1-aq^{k-1}}{1-q}\right)^{k-1} \left(\frac{1-bq^k}{1-q}\right)^{k-1} \left(\frac{1-abq^{n+k}}{1-q}\right)^{n-k}.$$

18.7 A Family of Biorthogonal Rational Functions

As we shall see in §21.9, the moment problem associated with $\left\{V_n^{(a)}(x; q)\right\}$ is determinate if and only if $0 < a \leq q$ or $1/q \leq a$. In the first case the unique solution is

$$m^{(a)} = (aq; q)_\infty \sum_{n=0}^{\infty} \frac{a^n q^{n^2}}{(q, aq; q)_n} \varepsilon_{q^{-n}}, \tag{18.7.1}$$

and in the second case it is

$$\sigma^{(a)} = (q/a; q)_\infty \sum_{n=0}^{\infty} \frac{a^{-n} q^{n^2}}{(q, q/a; q)_n} \varepsilon_{aq^{-n}}, \tag{18.7.2}$$

cf. (Berg & Valent, 1994). The total mass of these measures was evaluated to 1 in (Ismail, 1985). Recall that ϵ_b is a a unit measure supported at $x = b$.

If $q < a < 1/q$ the problem is indeterminate and both measures are solutions. In (Berg & Valent, 1994) the following one-parameter family of solutions with an analytic density was found

$$\nu(x; a, q, \gamma) = \frac{\gamma |a-1|(q, aq, q/a; q)_\infty}{\pi a\left[(x/a; q)_\infty^2 + \gamma^2 (x; q)_\infty^2\right]}, \qquad \gamma > 0. \tag{18.7.3}$$

In the above, $a = 1$ has to be excluded. For a similar formula when $a = 1$, see (Berg & Valent, 1994).

If μ is one of the solutions of the moment problem we have the orthogonality relation

$$\int_{\mathbb{R}} V_m^{(a)}(x;q) V_n^{(a)}(x;q)\, d\mu(x) = a^n q^{-n^2}(q;q)_n \delta_{m,n}. \tag{18.7.4}$$

The power series (11.3.3) has the radius of convergence $\sqrt{q/a}$, and therefore (11.3.2) becomes

$$\int_{\mathbb{R}} \frac{(xt_1, xt_2;q)_\infty\, d\mu(x)}{(t_1, at_1, t_2, at_2;q)_\infty} = \int_{\mathbb{R}} V^{(a)}(x,t_1)\, V^{(a)}(x,t_2)\, d\mu(x)$$

$$= \sum_{n=1}^\infty \frac{(at_1 t_2/q)^n}{(q;q)_n}$$

$$= \frac{1}{(at_1 t_2/q;q)_\infty}, \quad |t_1|, |t_2| < \sqrt{q/a}.$$

This identity with $\mu = m^{(a)}$ or $\mu = \sigma^{(a)}$ is nothing but the q-analogue of the Gauss theorem, (12.2.18).

Specializing to the density (18.4.18) we get

$$\int_{\mathbb{R}} \frac{(xt_1, xt_2;q)_\infty\, dx}{(x/a;q)_\infty^2 + \gamma^2(x;q)_\infty^2} = \frac{\pi a\,(t_1, at_1, t_2, at_2;q)_\infty}{|a-1|\gamma\,(q, aq, q/a, at_1 t_2/q;q)_\infty}, \tag{18.7.5}$$

valid for $q < a < 1/q$, $a \neq 1$, $\gamma > 0$.

We now seek polynomials or rational functions that are orthogonal with respect to a measure ν defined by

$$d\nu(x) = (xt_1, xt_2;q)_\infty\, d\mu(x), \tag{18.7.6}$$

where the measure μ satisfies (18.7.4). Next we integrate $1/\left[(xt_1;q)_k (xt_2;q)_j\right]$ with respect to the measure ν, which is positive if $-\sqrt{q/a} < t_1, t_2 < 0$. Set

$$\psi_n(x;a,t_1,t_2) := \sum_{k=0}^n \frac{(q^{-n};q)_k}{(q;q)_k} \frac{q^k a_{n,k}}{(xt_1;q)_k}. \tag{18.7.7}$$

The rest of the analysis is similar to our treatment of the U_n's. We get

$$\int_{\mathbb{R}} \frac{\psi_n(x;a,t_1,t_2)}{(xt_2;q)_m}\, d\nu(x)$$

$$= \sum_{k=0}^n \frac{(q^{-n};q)_k}{(q;q)_k} q^k a_{n,k} \int_{\mathbb{R}} (xt_1 q^k, xt_2 q^m;q)_\infty\, d\mu(x),$$

and if we choose $a_{n,k} = (t_1, at_1;q)_k / (at_1 t_2/q;q)_k$ the above expression is equal to

$$\frac{(t_1, at_1, t_2 q^m, at_2 q^m;q)_\infty}{(at_1 t_2 q^{m-1};q)_\infty}\, {}_2\phi_1\left(\begin{array}{c} q^{-n}, at_1 t_2 q^{m-1} \\ at_1 t_2/q \end{array} \middle|\, q,q\right),$$

$$= \frac{(t_1, at_1, t_2 q^m, at_2 q^m;q)_\infty (q^{-m};q)_n}{(at_1 t_2 q^{m-1};q)_\infty (at_1 t_2/q;q)_n}\, (at_1 t_2 q^{m-1})^n,$$

which is 0 for $m < n$. We have used the Chu–Vandermonde sum (12.2.17). Since ν is symmetric in t_1, t_2, this leads to the biorthogonality relation

$$
\int_{\mathbb{R}} \psi_m (x; a, t_2, t_1) \, \psi_n (x; a, t_1, t_2) \, d\nu(x)
$$

$$
= \frac{(t_1, at_1, t_2, at_2; q)_{\infty} (q; q)_n}{(at_1t_2/q; q)_{\infty} (at_1t_2/q; q)_n} (at_1t_2/q)^n \, \delta_{m,n}.
$$

(18.7.8)

The ψ_n's are given by

$$
\psi_n (x; a, t_1, t_2) = {}_3\phi_2 \left(\begin{array}{c} q^{-n}, t_1, at_2 \\ xt_1, at_1t_2 \end{array} \bigg| q, q \right).
$$

(18.7.9)

They are essentially the rational functions studied by Al-Salam and Verma in (Al-Salam & Verma, 1983). Al-Salam and Verma used the notation

$$
R_n(x; \alpha, \beta, \gamma, \delta; q) = {}_3\phi_2 \left(\begin{array}{c} \beta, \alpha\gamma/\delta, q^{-n} \\ \beta\gamma/q, \alpha qx \end{array} \bigg| q, q \right).
$$

(18.7.10)

The translation between the two notations is

$$
\psi_n (x; a, t_1, t_2) = R_n \left(\beta xq^{-1}/\alpha; \alpha, \beta, \gamma, \delta \right),
$$

(18.7.11)

with

$$
t_1 = \beta, \quad t_2 = \beta\delta/q\alpha, \quad a = \alpha\gamma/\beta\delta.
$$

(18.7.12)

Note that R_n has only three free variables since one of the parameters $\alpha, \beta, \gamma, \delta$ can be absorbed by scaling the independent variable.

Exercises

18.1   Use the q-integral representation (15.7.18) to evaluate the determinant $M_n(x \,|\, q)$,

$$
M_n(x \,|\, q) = \begin{vmatrix} H_0(x \,|\, q) & H_1(x \,|\, q) & \cdots & H_{n-1}(x \,|\, q) \\ H_1(x \,|\, q) & H_2(x \,|\, q) & & H_n(x \,|\, q) \\ \vdots & \vdots & & \vdots \\ H_{n-1}(x \,|\, q) & H_n(x \,|\, q) & \cdots & H_{2n-2}(x \,|\, q) \end{vmatrix},
$$

see Exercise 15.7.

18.2   Evaluate the determinants whose i, j entries are

   (a) $H_{i+j}(x \,|\, q)/(q; q)_{i+j}, \quad 0 \le i, j \le n - 1$,
   (b) $H_{i+j} \left(x \,|\, q^2 \right) /(-q; q)_{i+j}, \quad 0 \le i, j \le n - 1$,
   (c) $C_{i+j}(x; \beta \,|\, q), \quad 0 \le i, j \le n - 1$.

For related results, see (Ismail, 2005b).

19

Fractional and q-Fractional Calculus

In this chapter, we define operators of fractional calculus and their q-analogues and mention their applications to orthogonal polynomials. These operators have many other applications which we do not treat. For example, they can be used to solve dual integral and series equations which arise in crack problems in elasticity, see (Sneddon, 1966). Applications to special functions via Liebniz rule for fractional calculus are in (Osler, 1970; Osler, 1972; Osler, 1973). A theory of fractional difference operators has also been developed (Diaz & Osler, 1974). One important property of fractional integrals is that certain multiples of them map some orthogonal polynomials to orthogonal polynomials. We will also present other operators which preserve orthogonality.

In (Balakrishnan, 1960), A. V. Balakrishnan introduced a method of constructing fractional powers of a wide class of closed linear operators including the infinitesimal generators of semigroups. One can use this approach to define fractional powers of $\dfrac{d}{dx}$, Δ, and D_q. Westphal gave a new definition of fractional powers of infinitesimal generators in (Westphal, 1974). In the same paper, Westphal applied her results to fractional powers of $\dfrac{d}{dx}$ and Δ. The reader may consult (Westphal, 1974) for references, especially to the older literature.

19.1 The Riemann–Liouville Operators

The fractional integral and differential operators arose from an attempt to interpret repeated integration or differentiations noninteger number of times. For example it easily follows by induction that

$$
\int_a^x \int_a^{x_n} \cdots \int_a^{x_2} f(x_1)\, dx_1 \cdots dx_n
$$
$$
= \int_a^x \frac{(x-x_1)^{n-1}}{(n-1)!} f(x_1)\, dx_1. \tag{19.1.1}
$$

When n is not necessarily a positive integer (19.1.1) leads directly to the Riemann–Liouville fractional integral operator defined by

$$(I_a^\alpha f)(x) = \int_a^x \frac{(x-t)^{\alpha-1}}{\Gamma(\alpha)} f(t)\, dt, \qquad (19.1.2)$$

where f is a locally integrable function, and $\operatorname{Re}\alpha > 0$. Note that $I_a^\alpha f$ is a convolution of f and g, $g(u) := u^{\alpha-1}/\Gamma(\alpha)$ for $u > 0$, hence $I_a^\alpha f$ is integrable for integrable f.

Theorem 19.1.1 *The following index law holds good*

$$I_a^\alpha I_a^\beta = I_a^{\alpha+\beta}. \qquad (19.1.3)$$

Proof It is clear that

$$(I_a^\alpha (I_a^\beta f))(x) = \int_a^x \frac{(x-t)^{\alpha-1}}{\Gamma(\alpha)} \int_a^t \frac{(t-y)^{\beta-1}}{\Gamma(\beta)} f(y)\, dy\, dt$$

$$= \int_a^x \frac{f(y)}{\Gamma(\alpha)\Gamma(\beta)} \int_y^x (x-t)^{\alpha-1}(t-y)^{\beta-1}\, dt\, dy.$$

Make the substitution $t = y + v(x - y)$ in the t integration to obtain

$$(I_a^\alpha (I_a^\beta f))(x) = \int_a^x \frac{(x-y)^{\alpha+\beta-1}}{\Gamma(\alpha)\Gamma(\beta)} f(y) \int_0^1 (1-v)^{\alpha-1} v^{\beta-1}\, dv\, dy, \qquad (19.1.4)$$

and the theorem follows from the beta integral evaluation (1.3.3). □

Let $\mathcal{L}$ denote the Laplace transform

$$(\mathcal{L}f)(s) = \int_0^\infty e^{-st} f(t)\, dt. \qquad (19.1.5)$$

The convolution associated with the Laplace transform is

$$(f * g)(x) = \int_0^x f(t)g(x-t)\, dt. \qquad (19.1.6)$$

This means $\mathcal{L}(f * g) = [(\mathcal{L}f][(\mathcal{L}g]$. Therefore

$$(\mathcal{L}(I_0^\alpha f))(s) = \left[\int_0^\infty e^{-st} f(t)\, dt\right]\left[\int_0^\infty \frac{y^{\alpha-1}}{\Gamma(\alpha)} e^{-sy}\, dy\right],$$

and we have established the following theorem.

Theorem 19.1.2 $s^{-\alpha}$ *is a multiplier for the Laplace transform, for* $\operatorname{Re}\alpha > 0$. *Indeed*

$$(\mathcal{L}(I_0^\alpha f))(s) = s^{-\alpha}(\mathcal{L}f)(s), \qquad \operatorname{Re}\alpha > 0. \qquad (19.1.7)$$

Let D or D_x denote $\dfrac{d}{dx}$. It readily follows from (19.1.2) that

$$D^n I_a^\alpha = I_a^{\alpha-n}, \quad \operatorname{Re}\alpha > n. \tag{19.1.8}$$

Indeed, we can define D^α to be $D^n I_a^{n-\alpha}$, $\operatorname{Re}\alpha > 0$ with $n = \lceil \alpha \rceil$. An easy exercise is to show that

$$D^\alpha D^\beta = D^{\alpha+\beta}, \quad \operatorname{Re}\alpha > 0, \ \operatorname{Re}\beta > 0,$$
$$D^\alpha = D^n I_a^{n-\alpha} \text{ for } \operatorname{Re}\alpha < n, \quad n = 0, 1, \ldots . \tag{19.1.9}$$

The fractional integral operators provide operators whose actions change parameters in Jacobi and Laguerre polynomials. An application of the beta integral evaluation gives

$$x^{-\lambda-\alpha} I_0^\lambda \left(x^\alpha P_n^{(\alpha,\beta)}(1-2x) \right)$$
$$= \frac{\Gamma(\alpha+n+1)}{\Gamma(\alpha+\lambda+n+1)} P_n^{(\alpha+\lambda,\beta-\lambda)}(1-2x). \tag{19.1.10}$$

Similarly, we can prove the more general result

$$x^{1-\lambda-\mu} I_0^\lambda \left(x^{\mu-1} P_n^{(\alpha,\beta)}(1-2x) \right)$$
$$= \frac{(\alpha+1)_n \Gamma(\mu)}{n!\,\Gamma(\lambda+\mu)} \, {}_3F_2 \left(\begin{matrix} -n, n+\alpha+\beta+1, \mu \\ \alpha+1, \lambda+\mu \end{matrix} \, \middle| \, x \right). \tag{19.1.11}$$

Moreover, one can show

$$x^{1-\lambda-\mu} I_0^\lambda \left(x^{\mu-1} L_n^{(\alpha)}(x) \right) = \frac{(\alpha+1)_n}{n!\,\Gamma(\lambda+\mu)} \, {}_2F_2 \left(\begin{matrix} -n, \mu \\ \alpha+1, \lambda+\mu \end{matrix} \, \middle| \, x \right). \tag{19.1.12}$$

In particular

$$x^{-\lambda-\alpha} I_0^\lambda \left(x^\alpha L_n^{(a)}(x) \right) = \frac{(\alpha+1)_n}{\Gamma(\lambda+\alpha+n+1)} L_n^{(\lambda+\alpha)}(x). \tag{19.1.13}$$

We now determine the adjoint of I_0^α under the inner product

$$(f, g) = \int_0^\infty f(x) g(x) \, dx, \tag{19.1.14}$$

defined on real functions in $L_1(\mathbb{R}) \cap L_2(\mathbb{R})$. A simple calculation shows that the adjoint of I_0^α is W^α,

$$(W^\alpha f)(x) = \int_x^\infty \frac{(t-x)^{\alpha-1}}{\Gamma(\alpha)} f(t) \, dt. \tag{19.1.15}$$

A useful formula is

$$W^\lambda \left(e^{-x} L_n^{(\alpha)}(x) \right) = e^{-x} L_n^{(\alpha+1-\lambda)}(x), \tag{19.1.16}$$

which can be proved as follows. Substitute for $L_n^{(\alpha)}(x)$ from (4.6.1) in the left-hand side of (19.1.16), then replace t by $t + x$ to see that

$$W^\lambda \left(e^{-x} L_n^{(\alpha)}(x) \right) = \frac{e^{-x}(\alpha+1)_n}{\Gamma(\lambda)\,n!} \sum_{k=0}^{n} \frac{(-n)_k}{k!\,(\alpha+1)_k} \int_0^\infty e^{-t}(x+t)^{k+\lambda-1}\,dt$$

$$= \frac{e^{-x}(\alpha+1)_n}{\Gamma(\lambda)\,n!} \sum_{k=0}^{n}\sum_{j=0}^{k} \frac{(-n)_k\,x^j}{(\alpha+1)_k\,j!\,(k-j)!}\,\Gamma(\lambda+k-j).$$

After replacing k by $k + j$ and interchanging sums, the k sum is a terminating $_2F_1$ which is summed by the Chu–Vandermonde sum and (19.1.16) follows.

The operator W^α is called the Weyl fractional integral and originated in the theory of Fourier series. For $f \in L^2(-\pi, \pi)$ we write

$$f(x) \sim \sum_{-\infty}^{\infty} f_n e^{inx}, \; x \in (-\pi, \pi), \text{ if } f_n = \frac{1}{2\pi} \int_{-\pi}^{\pi} f(x)\,e^{-inx}\,dx. \qquad (19.1.17)$$

We will normalize f by $f_0 = 0$, that is, replace f by $f - \frac{1}{2\pi}\int_{-\pi}^{\pi} f(t)\,dt$. Weyl's original idea was to use $\frac{d}{dx} e^{inx} = in\,e^{inx}$ to define the fractional integral of a 2π periodic function f, $f \sim \sum f_n e^{inx}$, by

$$(\mathbb{W}_\alpha f)(x) \sim \sum_{n\neq 0} \frac{f_n}{(in)^\alpha} e^{inx}, \; \text{Re}\,\alpha > 0, \, x \in [-\pi, \pi]. \qquad (19.1.18)$$

The series in (19.1.18) represents a function in $L^2(-\pi, \pi)$ since $\{f_n\} \in \ell^2$. It is clear that $\mathbb{W}_\alpha f$ is smoother than f for $\alpha > 0$. Chapter 12 of (Zygmund, 1968) contains a detailed analysis of the mapping properties of $\mathbb{W}_\alpha$. One can rewrite (19.1.18) as

$$(\mathbb{W}_\alpha f)(x) = \frac{1}{2\pi} \int_0^{2\pi} f(t)\,\Psi_\alpha(x-t)\,dt, \qquad (19.1.19)$$

where

$$\Psi_\alpha(t) = \sum_{n\neq 0} \frac{e^{int}}{(in)^\alpha}. \qquad (19.1.20)$$

For $0 < \alpha < 1$, $0 < x < 2\pi$ one can apply the Poisson summation formula and prove that

$$(\mathbb{W}_\alpha f)(x) = \frac{1}{\Gamma(\alpha)} \int_{-\infty}^{x} f(t)\,(x-t)^{\alpha-1}\,dt.$$

The details are in §12.8 of (Zygmund, 1968).

It is clear that $\mathbb{W}_\alpha \mathbb{W}_\beta = \mathbb{W}_{\alpha+\beta}$. We define $\mathbb{W}_0$ to be the identity operator.

A variant of W^α when f is defined on $[0, 1]$ is

$$(S_{\lambda,\mu} f)(x) = \frac{(1-x)^{-\lambda-\mu} \Gamma(\lambda+\mu+1)}{\Gamma(\lambda+1)\Gamma(\mu)} \int\limits_x^1 (t-x)^{\mu-1} (1-t)^\lambda f(1-t)\, dt,$$

(19.1.21)

for $\operatorname{Re}\lambda > -1$, $\operatorname{Re}\mu > 0$. It is easy to prove

$$(S_{\lambda,\mu} t^n)(x) = \frac{(\lambda+1)_n}{(\lambda+\mu+1)_n}(1-x)^n,$$

hence

$$S_{\beta,\mu} P_n^{(\alpha,\beta)}(2x-1) = \frac{(\beta+1)_n}{(\beta+\mu)_n} P_n^{(\alpha-\mu,\beta+\mu)}(1-2x).$$

(19.1.22)

There are more general operators called the Erdélyi–Kober operators which have found applications in Elasticity. Their properties can be found in (Sneddon, 1966). Fractional integrals and derivatives are also useful in solving dual integral and dual series solutions. The dual integral and series equations arise in the solution of various types of crack problems in elastic media. Recently, several authors also considered linear differential equations of fractional order. The interested reader may consult the journal "Fractional Calculus and Their Applications" for some of the current research in this area.

An important class of fractional integral operators has been introduced in (Butzer et al., 2002a), (Butzer et al., 2002b) and (Butzer et al., 2002c). The new operators are

$$(\mathcal{J}_{0+,\mu}^\alpha f)(x) = \frac{1}{\Gamma(\alpha)} \int\limits_0^x \left(\ln\frac{x}{u}\right)^{\alpha-1} \left(\frac{x}{u}\right)^\mu f(u)\,\frac{du}{u}, \quad x > 0,\ \alpha > 0, \quad (19.1.23)$$

$$(\mathcal{J}_{-,\mu}^\alpha f)(x) = \frac{1}{\Gamma(\alpha)} \int\limits_x^\infty \left(\ln\frac{u}{x}\right)^{\alpha-1} \left(\frac{x}{u}\right)^\mu \frac{du}{u}, \quad x > 0. \quad (19.1.24)$$

The semigroup property

$$\mathcal{J}_{0+,\mu}^\alpha \mathcal{J}_{0+,\mu}^\beta f = \mathcal{J}_{0+,\mu}^{\alpha+\beta} f, \quad \alpha > 0,\ \beta > 0,$$
$$\mathcal{J}_{-,\mu}^\alpha \mathcal{J}_{-,\mu}^\beta f = \mathcal{J}_{-,\mu}^{\alpha+\beta} f, \quad \alpha > 0,\ \beta > 0$$

(19.1.25)

holds for these operators. These operators act as multipliers for the Mellin transform. It will be interesting to apply the operators $\mathcal{J}_{0+,\mu}^\alpha$ and $\mathcal{J}_{-,\mu}^\alpha$ to special functions and orthogonal polynomials.

19.2 Bilinear Formulas

The eigenvalue problem

$$\lambda_n \varphi_n(x) = \int\limits_a^b K(x,t)\,\varphi_n(t)\, dt, \quad b > a > 0, \quad (19.2.1)$$

is a basic problem of applied mathematics. The kernel $K(x,t)$ is assumed to belong to $L_1\left(E^2\right)\cap L_2\left(E^2\right)$, $E=(a,b)$ and b is finite or infinite. When K is symmetric, that is $K(x,y)=K(y,x)$, the integral operator in (19.2.1) is symmetric and $\{\varphi_n(x)\}$ forms a complete sequence of complete orthogonal functions in $L_2(E)$, see (Tricomi, 1957). Similarly, if $K(x,t)$ is symmetric in $L_2\left(E^2\right)$ and w is a weight function, then the eigenfunctions $\{\varphi_n(x)\}$ of the eigenvalue problem

$$\lambda_n\varphi_n(x)=\int_a^b K(x,t)\sqrt{w(t)/w(x)}\,\varphi_n(t)\,dt \qquad (19.2.2)$$

will be complete and orthogonal in $L_2(a,b,w)$. If the kernel $K(x,t)$ in (19.2.2) is continuous, symmetric, square integrable and has positive eigenvalues, then by Mercer's theorem (Tricomi, 1957, p. 25) it will have the representation

$$K(x,t)=\sqrt{w(t)w(x)}\sum_{n=0}^{\infty}\frac{\lambda_n}{\zeta_n}\varphi_n(x)\,\varphi_n(t), \qquad (19.2.3)$$

where

$$\zeta_n=\int_a^b w(x)\left[\varphi_n(x)\right]^2 dx. \qquad (19.2.4)$$

Assume that the kernel K is nice enough to justify exchanging integration over x and t. Using (19.2.3) again we get

$$\lambda_n^2\varphi_n(z)\sqrt{w(z)}=\int_a^b K^{(2)}(x,t)\varphi_n(t)\sqrt{w(t)}\,dt, \qquad (19.2.5)$$

where

$$K^{(2)}(z,t)=\int_a^b K(z,x)\,K(x,t)\,dx. \qquad (19.2.6)$$

If K is continuous, squarte integrable on $[a,b]\times[a,b]$, symmetric kernel with positive eigenvalues, then $K^{(2)}$ will inherit the same properties. This leads to

$$K^{(2)}(x,t)=\sqrt{w(x)w(t)}\sum_{n=0}^{\infty}\frac{\lambda_n^2}{\zeta_n}\varphi_n(x)\varphi_n(t). \qquad (19.2.7)$$

One can reverse the problem and start with a complete system of functions orthogonal with respect to $w(x)$ on $[a,b]$. If one can construct a continuous square integrable kernel $K(x,t)$ with positive eigenvalues such that (19.2.2) holds, then (19.2.3) will hold. We will give several examples of this technique in §19.3.

19.3 Examples

We illustrate the technique in §19.2 by considering the examples of Laguerre and Jacobi polynomials. We will also treat the Hahn polynomials as an example of a polynomial sequence orthogonal on a finite set.

Example 19.3.1 (Laguerre Polynomials) *A special case of (4.6.38) is the fractional integral representation*

$$\frac{x^{\nu} L_n^{(\nu)}(x)}{\Gamma(n+\nu+1)} = \frac{1}{\Gamma(\nu-\alpha)} \int_0^x (x-t)^{\nu-\alpha-1} \frac{t^{\alpha} L_n^{(\alpha)}(t)}{\Gamma(n+\alpha+1)}\, dt, \quad \nu > \alpha. \quad (19.3.1)$$

Clearly, (19.3.1) can be written in terms of $I_0^{\nu-\alpha}$. The W^{α} version is

$$e^{-x} L_n^{(\alpha)}(x) = \frac{1}{\Gamma(\nu-\alpha)} \int_x^{\infty} L_n^{(\nu)}(u)\,(u-x)^{\nu-\alpha-1} e^{-u}\, du, \quad (19.3.2)$$

which folows from the series representation for L_n^{ν} and the Chu–Vandermonde sum. The orthogonality relation (4.6.2) implies

$$\frac{\Gamma^2(\alpha+n+1)\Gamma^2(\nu-\alpha)}{n!\,\Gamma(\nu+n+1)}\, \delta_{m,n}$$

$$= \int_0^{\infty} x^{-\nu} e^{-x} \left\{ \int_0^{\infty} (x-t)^{\nu-\alpha-1} t^{\alpha} L_n^{(\alpha)}(t)\, dt \right\}$$

$$\times \left\{ \int_0^{\infty} (x-u)^{\nu-\alpha-1} u^{\alpha} L_n^{(\alpha)}(u)\, du \right\} dx$$

$$= \int_0^{\infty}\int_0^{\infty} t^{\alpha} e^{-t} L_n^{(\alpha)}(t) u^{\alpha} L_n^{(\alpha)}(u)$$

$$\times \int_{\max\{u,t\}}^{\infty} x^{-\nu}(x-t)^{\nu-\alpha-1}(x-u)^{\nu-\alpha-1} e^{t-x}\, dx\, du\, dt.$$

The completeness of $\left\{ L_n^{(\alpha)}(x) \right\}$ in $L_2\,[0,\infty, x^{\alpha} e^{-x}]$ establishes

$$\lambda_n^{(1)} L_n^{(\alpha)}(x) = \int_0^{\infty} K_1(x,t) L_n^{(\alpha)}(t)\, dt, \quad (19.3.3)$$

where

$$\lambda_n^{(1)} = \frac{\Gamma(\alpha+n+1)\Gamma^2(\nu-\alpha)}{\Gamma(n+\nu+1)}, \quad (19.3.4)$$

and

$$K_1(x,t) = t^{\alpha} e^{x} \int_{\max\{x,t\}}^{\infty} w^{-\nu}(w-x)^{\nu-\alpha-1}(w-t)^{\nu-\alpha-1} e^{-w}\, dw. \quad (19.3.5)$$

The functions $\left\{ x^{\alpha/2} e^{-x/2} L_n^{(\alpha)}(x) \right\}$ form a complete orthogonal basis for $L_2[(0,\infty)]$ and the kernel $e^{(t-x)/2}(x/t)^{\alpha/2} K_1(x,t)$ is positive and symmetric, see (Tricomi,

1957), for terminology. Therefore $\left\{L_n^{(\alpha)}(x)\right\}$ *are all the eigenfunctions of*

$$\lambda y(x) = \int_0^\infty K_1(x,t)y(t)\,dt.$$

Similarly, (19.3.2) proves that the eigenfunctions of

$$\lambda y(x) = \int_0^\infty K_2(x,t)y(t)\,dt, \tag{19.3.6}$$

with

$$K_2(x,t) = x^{-\nu}e^{-t}\int_0^{\min\{x,t\}}(x-w)^{\nu-\alpha-1}(t-w)^{\nu-\alpha-1}w^\alpha e^w\,dw, \tag{19.3.7}$$

are $\left\{L_n^{(\nu)}(t)\right\}$ *and the corresponding eigenvalues are also* $\left\{\lambda_n^{(1)}\right\}$.
The spectral resolutions of K_1 *and* K_2 *are*

$$e^t t^{-\alpha} K_1(x,t) = \frac{\Gamma^2(\nu-\alpha)}{\Gamma(\nu+1)}\sum_{n=0}^\infty \frac{n!}{(\nu+1)_n}L_n^{(\alpha)}(x)L_n^{(\alpha)}(t), \tag{19.3.8}$$

$$e^t t^{-\nu} K_2(x,t) = \frac{\Gamma^2(\nu-\alpha)}{\Gamma(\alpha+1)\Gamma^2(\nu+1)}\sum_{n=0}^\infty \frac{n!\,(\alpha+1)_n}{(\nu+1)_n^2}L_n^{(\nu)}(x)L_n^{(\nu)}(t), \tag{19.3.9}$$

for $x,t > 0$ *and* $\nu > \alpha > -1$.

Example 19.3.2 (Jacobi Polynomials) *To put the orthogonality of Jacobi polynomials on* $[0,1]$, *set*

$$J_n(x;\alpha,\beta) := {}_2F_1(-n, n+\alpha+\beta+1; \beta+1; x). \tag{19.3.10}$$

The operators which raise and lower the parameters α *and* β *are*

$$(T_{\lambda,\mu}f)(x) = \frac{x^{-\lambda-\mu}\Gamma(\lambda+\mu+1)}{\Gamma(\lambda+1)\Gamma(\mu)}\int_0^x t^\lambda(x-t)^{\mu-1}f(t)\,dt, \tag{19.3.11}$$

and

$$(S_{\lambda,\mu}f)(x) = \frac{(1-x)^{-\lambda-\mu}\Gamma(\lambda+\mu+1)}{\Gamma(\lambda+1)\Gamma(\mu)}\int_x^1 (t-x)^{\mu-1}(1-t)^\lambda f(1-t)\,dt,$$

$$\tag{19.3.12}$$

where $\lambda > -1$, $\mu > 0$ *in both cases. The beta integral yields*

$$T_{\lambda,\mu}x^n = \frac{(\lambda+1)_n}{(\lambda+\mu+1)_n}x^n, \quad S_{\lambda,\mu}x^n = \frac{(\lambda+1)_n}{(\lambda+\mu+1)_n}(1-x)^n.$$

Therefore, we have

$$(T_{\beta,\mu}J_n(\cdot;\alpha,\beta))(x) = J_n(x;\alpha-\mu,\beta+\mu) \tag{19.3.13}$$

and

$$(S_{\beta,\mu} J_n(\cdot;\alpha,\beta))(x) = J_n(1-x;\alpha-\mu,\beta+\mu). \qquad (19.3.14)$$

A calculation analogous to the Laguerre case establishes

$$\lambda_n^{(3)} J_n(x;\alpha,\beta) = \int_0^1 K_3(x,t) J(t;\alpha,\beta)\, dt,$$

$$\lambda_n^{(4)} J_n(x;\alpha,\beta) = \int_0^1 K_4(x,t) J(t;\alpha,\beta)\, dt, \qquad (19.3.15)$$

where

$$\lambda_n^{(3)} = \frac{\Gamma(\beta+n+1)\Gamma(\alpha-\mu+n+1)\Gamma^2(\mu)}{\Gamma(\mu+\beta+n+1)\Gamma(\alpha+n+1)},$$

$$\lambda_n^{(4)} = \frac{\Gamma(\alpha+n+1)\Gamma(\beta+\mu+n+1)\Gamma^2(\mu)}{\Gamma(\alpha+\mu+n+1)\Gamma(\beta+n+1)}. \qquad (19.3.16)$$

The kernels K_3 and K_4 are defined by

$$K_3(x,t) = (1-x)^{-\alpha} t^{\beta} \int_{\max\{x,t\}}^1 w^{-\beta-\mu}(1-w)^{\alpha-\mu}(w-x)^{\mu-1}(w-t)^{\mu-1}\, dw,$$

$$K_4(x,t) = (1-t)^{\alpha} t^{-\beta} \int_0^{\min\{x,t\}} (x-w)^{\mu-1}(t-w)^{\mu-1} w^{\alpha-\mu}(1-w)^{-\mu-\beta}\, dw.$$

$$(19.3.17)$$

In the above we assumed that

$$\alpha > \mu - 1 > -1, \text{ and } \beta > -1.$$

The approach outline in §19.2 establishes

$$\frac{K_3(x,t)}{t^{\beta}(1-t)^{\alpha}} = \sum_{n=0}^{\infty} \frac{\lambda_n^{(3)}(\alpha+\beta+2n)\Gamma(\mu+\beta+n+1)\Gamma(\alpha+\beta+n+1)}{n!\,\Gamma^2(\mu)\Gamma^2(\beta+1)\Gamma(\alpha-\mu+n+1)}$$
$$\times J_n(x;\alpha,\beta)\, J_n(t;\alpha,\beta)$$

$$\frac{K_4(x,t)}{t^{\beta}(1-t)^{\alpha}} = \sum_{n=0}^{\infty} \frac{\lambda_n^{(4)}(\alpha+\beta+2n)\Gamma(\mu+\beta+n+1)\Gamma(\alpha+\beta+n+1)}{n!\,\Gamma^2(\mu)\Gamma^2(\beta+1)\Gamma(\alpha-\mu+n+1)}$$
$$\times J_n(x;\alpha,\beta)\, J_n(t;\alpha,\beta).$$

$$(19.3.18)$$

To derive reproducing kernels for Hahn polynomials, we need a sequence to function transform which maps the Hahn polynomials to orthogonal polynomials. Let

$$\phi_N(x) = \sum_{k=0}^N \alpha_k x^k, \text{ with } \alpha_k \neq 0, \quad 0 \le k \le N.$$

Define a transform S_N on finite sequences $\{f(n) : n = 0, \ldots, N\}$ by

$$S_N\left[f; \phi_N; x\right] = \sum_{m=0}^{N} \frac{(-x)^n}{m!} \phi_N^{(n)}(x)\, f(n).$$ (19.3.19)

It is easy to derive

$$S_N\left[\binom{n}{j}; \phi_N; x\right] = (-1)^j \alpha_j x^j, \quad j = 0, 1, \ldots, N,$$ (19.3.20)

from the Taylor series.

The transform (19.3.19) with $\phi_N(x) = (1-x)^N$, has the property

$$S_N\left[Q_j(n; \alpha, \beta, N); (1-x)^N; x\right] = J_j(x; \beta, \alpha),$$ (19.3.21)

which follows from (19.3.20). By rewriting the orthogonality of $\left\{P_n^{(\alpha,\beta)}(x)\right\}$ in terms of $\{J_n(x; \beta, \alpha)\}$ then replace $J_n(x; \beta, \alpha)$ by the left-hand side of (19.3.21) we find

$$\sum_{r,s=0}^{N} Q_n(r; \alpha, \beta, N)\, Q_m(s; \alpha, \beta, N) \binom{N}{r}\binom{N}{s} \frac{\Gamma(\beta + 2N - r - s + 1)}{\Gamma(\alpha + \beta + 2n - r - s + 2)}$$
$$= \frac{\Gamma(\alpha+1)\Gamma(\beta+n+1)\, n!}{(\alpha+\beta+2n+1)\Gamma(\alpha+n+1)\Gamma(\alpha+\beta+n+1)} \delta_{m,n}.$$

Comparing the above relationship with the orthogonality relation of $\{Q_n\}$ as given in (6.2.4), we conclude that $Q_n(x; \alpha, \beta, N)$ solves the discrete integral equation

$$\lambda\, y(x) = \sum_{s=0}^{N} \xi(x, s)\, y(s),$$ (19.3.22)

where

$$\xi(x, s) = \frac{(-N)_s (\beta + N - x + 1)_{N-s}\, x!^2 (N-x)!}{(-N)_x (\alpha+1)_x\, s!\, (\alpha+\beta+2)_{2N-x-s}},$$
$$\lambda = \lambda_n = \frac{(N-1)!}{\Gamma(\alpha+\beta+n+N+1)} \binom{N}{n}.$$ (19.3.23)

Thus we proved

$$\sum_{n=0}^{N} \frac{\lambda_n}{h_n} Q_n(x; \alpha, \beta, N)\, Q_n(y; \alpha, \beta, N) = \frac{\xi(x, y)}{w(y; \alpha, \beta, N)},$$

where w is as in (6.2.1) and h_n is the coefficient of $\delta_{m,n}$ in (6.2.4).

The approach outlined in §19.2, as well as the results of §19.3, are from the author's paper (Ismail, 1977a).

T. Osler derived a Leibniz rule for fractional derivatives and applied it to derive identities for special functions. He also showed the Leibniz rule is related to Parseval's formula. The interested reader may consult (Osler, 1970), (Osler, 1972), (Osler, 1973).

19.4 q-Fractional Calculus

A q-analogue of the Riemann–Liouville fractional integral is

$$(I^\alpha f)(x;q) := \frac{x^\alpha}{\Gamma_q(\alpha)} \int_0^x (qt/x;q)_{\alpha-1} f(t) \, d_q t, \quad \alpha \neq -1, -2, \cdots, \quad (19.4.1)$$

where $(u;q)_\alpha$ is as in (12.1.3). A simple calculation yields

$$(I^\alpha f)(x;q) = (1-q)^\alpha x^\alpha \sum_{n=0}^\infty \frac{(q^\alpha;q)_n}{(q;q)_n} q^n f(xq^n). \quad (19.4.2)$$

We shall always assume $0 < q < 1$. We shall use (19.4.2) as the definition because it is defined in a wider domain of α, so we will apply I^α to functions for which (19.4.2) exisits. The function f is assumed to be q-integrable in the sense $\sum_{n=0}^\infty q^n |f(xq^n)| < \infty$, for all $x \geq 0$. The formula

$$(I^\alpha t^\beta)(x;q) = \frac{\Gamma_q(\beta+1) x^{\alpha+\beta}}{\Gamma_q(\alpha+\beta+1)} \quad (19.4.3)$$

follows from the q-binomial theorem.

Al-Salam proved the following q-analogue of (19.1.1)

$$\int_a^x \int_a^{x_n} \cdots \int_a^{x_2} f(x_1) \, d_q x_1 \, d_q x_2 \cdots d_q x_n$$

$$= \frac{(1-q)^{n-1} x^{n-1}}{(q;q)_{n-1}} \int_a^x (qt/x;q)_{n-1} f(t) \, dt, \quad (19.4.4)$$

see (Al-Salam, 1966b).

Theorem 19.4.1 *We have*

$$I^\alpha I^\beta = I^{\alpha+\beta} \quad (19.4.5)$$

and

$$D_q I^\alpha = I^{\alpha-1} \quad (19.4.6)$$

Proof Formula (19.4.3) follows from the q-Chu–Vandermonde sum (12.2.17) while (19.4.5) follows from the definitions of I^α and D_q. $\qquad\square$

The operators I^α were introduced for positive integers a in (Al-Salam, 1966b) and announced for general α in (Al-Salam, 1966a). They also appeared in (Agarwal, 1969), where the operators of q-fractional integration are defined by

$$D_q^\alpha f(x) := I^{-\alpha} f(x) = (1-q)^{-\alpha} x^{-\alpha} \sum_{n=0}^\infty \frac{(q^{-\alpha};q)_n}{(q;q)_n} q^n f(xq^n). \quad (19.4.7)$$

A q-analogue of W^α, see (19.1.15), is

$$(K^\alpha f)(x; q) := \frac{q^{\alpha(\alpha-1)/2}}{\Gamma_q(\alpha)} \int_x^\infty t^{\alpha-1}(x/t; q)_{\alpha-1} f(tq^{1-\alpha}) \, d_q t. \qquad (19.4.8)$$

A calculation gives

$$(K^\alpha f)(x) = q^{-\alpha(\alpha+1)/2} x^\alpha (1-q)^\alpha \sum_{k=0}^\infty \frac{(q^\alpha; q)_k}{(q; q)_k} q^{-k\alpha} f(xq^{-a-k}). \qquad (19.4.9)$$

$$\int_0^\infty f(x) (K^\alpha g)(x) = \int_0^\infty g(xq^{-\alpha}) (I^\alpha f)(x) \, d_q x. \qquad (19.4.10)$$

The q-analogue of the Laplace transform is

$$L_q(f; s) = \frac{1}{(1-q)} \int_0^{1/s} (qsx; q)_\infty f(x) \, d_q x, \qquad (19.4.11)$$

see (Hahn, 1949a, §9) and (Abdi, 1960), (Abdi, 1964).

Lemma 19.4.2 *The following q-Laplace transform formula holds*

$$L_q(x^\alpha; s) = \frac{(q; q)_\infty}{(q^{\alpha+1}; q)_\infty} s^{-\alpha-1}, \qquad (19.4.12)$$

$\alpha \neq -1, -2, \ldots$.

Proof It is clear that

$$L_q(x^\alpha; s) = \frac{1}{s} \sum_{n=0}^\infty (q^{n+1}; q)_\infty \left(\frac{q^n}{s}\right)^\alpha q^n$$

$$= \frac{(q; q)_\infty}{s^{\alpha+1}} \sum_{n=0}^\infty \frac{q^{(\alpha+1)n}}{(q; q)_n}$$

and (19.4.12) follows from Euler's formula.   □

In particular

$$L_q(x^n; s) = \frac{(q; q)_n}{s^{n+1}}. \qquad (19.4.13)$$

In the notation of §14.8, we define the convolution of two functions f and g by

$$(f * g)(x) = \frac{q^{-1}}{1-q} \int_0^x f(t) \mathcal{E}^{-qt} g(x) \, d_q t, \qquad (19.4.14)$$

where $\mathcal{E}^t$ is defined in (14.8.11). It is assumed that the q-integral in (19.4.14) exists.

Example 19.4.3 *A calculation using the q-binomial theorem leads to*

$$x^\alpha * x^\beta = x^{\alpha+\beta+1} \frac{(q, q^{\alpha+\beta+2}; q)_\infty}{(q^{\alpha+1}, q^{\beta+1}; q)_\infty}. \tag{19.4.15}$$

It convenient to rewrite (19.4.15) *in the form*

$$x^\alpha * x^\beta = x^{\alpha+\beta+1} \frac{\Gamma_q(\alpha+1)\Gamma_q(\beta+1)}{\Gamma_q(\alpha+\beta+2)}, \tag{19.4.16}$$

which makes the limit as $q \to 1^-$ *transparent. Clearly* (19.4.15) *and* (19.4.16) *show that the convolution* $*$ *is commutative.*

We now define L_q on functions of the form

$$f(x) = \sum_{k=0}^\infty f_k x^{\lambda_k}, \tag{19.4.17}$$

where $\{\lambda_k\}$ is a sequence of complex numbers and x^{λ_k} is defined on $\mathbb{C}$ cut along a ray eminating from 0 to ∞. If λ_k is an integer for all k, then no cut is needed.

Definition 19.4.1 *For* f *of the form* (19.4.17), *we define*

$$L_q(f; s) = \sum_{k=0}^\infty f_k \frac{(q; q)_\infty}{(q^{1+\lambda_k}; q)_\infty} s^{-\lambda_k-1} \tag{19.4.18}$$

provided that the series converges absolutely and uniformly in a sector, $\theta_1 < \arg s < \theta_2$.

When $\lambda_k = \alpha + k$ (19.4.18) becomes

$$L_q\left(\sum_{k=0}^\infty f_k x^{\alpha+k}; s\right) = \frac{(q; q)_\infty}{(q^{\alpha+1}; q)_\infty} \sum_{k=0}^\infty f_k \frac{(q^{\alpha+1}; q)_k}{s^{\alpha+k+1}}. \tag{19.4.19}$$

In view of (19.4.16) the convolution following theorem holds.

Theorem 19.4.4 *We have*

$$L_q(f * g; s) = L_q(f; s) L_q(g; s), \tag{19.4.20}$$

for functions f *and* g *of the form* (19.4.17).

Theorem 19.4.5 *The operator* I^α *is a multiplier for* L_q *on the set of functions of the type* (19.4.17). *Specifically,*

$$L_q(I^\alpha f; s) = \frac{(1-q)^\alpha}{s^{\alpha+1}} L_q(f; s). \tag{19.4.21}$$

Proof Prove (19.4.21) when $f = x^\beta$ then extend it by linearity. $\square$

Al-Salam and Ismail introduced a q-analogue of a special Erdelyi–Kober fractional integral operator (Sneddon, 1966) as

$$\left(I^{(\alpha,\eta)}f\right)(x) = \frac{(q^\alpha, q^\eta; q)_\infty}{(q, q^{\alpha+\eta}; q)_\infty} \int_0^1 t^{\alpha-1} \frac{(qt; q)_\infty}{(tq^\eta; q)_\infty} \frac{f(xt)}{(1-q)} \, d_q t. \qquad (19.4.22)$$

An easy exercise is to prove that

$$\left(I^{(\alpha,\eta)}x^\beta\right)(x) = \frac{(q^\alpha, q^{\alpha+\beta+\eta}; q)_\infty}{(q^{\alpha+\eta}, q^{\alpha+\beta}; q)_\infty} x^\beta. \qquad (19.4.23)$$

The little q-Jacobi $p_n(x; a, b)$ are defined by (18.4.11). We have, (Al-Salam & Ismail, 1977),

$$I^{(\alpha,\eta)}p_n\left(x; q^\gamma, q^\delta\right) = \sum_{k=0}^n \frac{(q^{-n}, q^{\gamma+\delta+n+1}, q^\alpha; q)_k}{(q, q^{\gamma+1}, q^{\alpha+\eta}; q)_k} q^k x^k.$$

Therefore

$$I^{(\alpha+1,\eta)}p_n\left(x; q^\alpha, q^\delta\right) = p_n\left(x; q^{\alpha+\eta}, q^{\delta-\eta}\right). \qquad (19.4.24)$$

Using the procedure in §19.2, one can establish reproducing kernels and bilinear formulas involving the little q-Jacobi polynomials. The detailed results are in (Al-Salam & Ismail, 1977).

Al-Salam and Verma found a Liebniz rule of q-fractional derivatives and applied it to derive functional relations among q-special functions. They extended some results of Osler, (Osler, 1970), (Osler, 1972), (Osler, 1973) to q-fractional derivatives. The details are in (Al-Salam & Verma, 1975a), (Al-Salam & Verma, 1975b). Annaby and Mansour gave a detailed treatment of q-fractional integrals in (Annaby & Mansour, 2005a).

19.5 Some Integral Operators

Consider the family of operators

$$(S_r f)(\cos\theta) := \frac{\left(q, t^2; q\right)_\infty}{2\pi}$$

$$\times \int_0^\pi \frac{\left(e^{2i\phi}, e^{-2i\phi}; q\right)_\infty f(\cos\phi)}{\left(re^{i(\theta+\phi)}, re^{i(\theta-\phi)}, re^{i(\phi-\theta)}, re^{-i(\theta+\phi)}; q\right)_\infty} \, d\phi, \quad t \in (0,1). \qquad (19.5.1)$$

The operators S_r have the semigroup property

$$S_r S_s = S_{r+s}, \quad \text{for} \quad r, s, r+s \in (-1, 1). \qquad (19.5.2)$$

Theorem 19.5.1 *The Al-Salam–Chihara polynomials have the connection formula*

$$S_r\left(\frac{p_n(\cos\phi; t_1, t_2)}{(t_1 e^{i\phi}, t_1 e^{-i\phi})_\infty}\right)(\cos\theta) = \frac{p_n(\cos\theta; t_1 r, t_2/r)}{(t_1 re^{i\theta}, t_1 re^{-i\theta})_\infty}. \qquad (19.5.3)$$

Proof The left-hand side of (19.5.3) is

$$\frac{\left(q, r^2; q\right)_\infty}{2\pi} \int\limits_0^\pi \frac{\left(e^{2i\phi}, e^{-2i\phi}; q\right)_\infty}{\left(re^{i(\theta+\phi)}, re^{i(\theta-\phi)}, re^{i(\phi-\theta)}, re^{-i(\theta+\phi)}; q\right)_\infty}$$

$$\times \sum_{k=0}^n \frac{(q^{-n}; q)_k \ q^k \ d\phi}{(q, t_1 t_2; q)_k \ (t_1 q^k e^{i\phi}, t_1 q^k e^{-i\phi}; q)_\infty}$$

$$= \frac{1}{(rt_1 e^{i\theta}, rt_1 e^{-i\theta}; q)_\infty} {}_3\phi_2 \left(\begin{matrix} q^{-n}, rt_1 e^{i\theta}, rt_1 e^{-i\theta} \\ t_1 t_2, 0 \end{matrix} \ \middle| \ q, q \right),$$

where the Askey–Wilson integral was used in the last step. The result now follows from the above equation. $\qquad\square$

Since the Al-Salam–Chihara polynomials are q-analogues of Laguerre polynomials, the operators S_r may be thought of as some q-fractional integrals. A similar proof establishes the following extension to the Askey–Wilson polynomials.

Theorem 19.5.2 *When* $\max \{|r|, |t_1|, |t_2|, |t_3/r|, |t_4/r|\} < 1$ *then*

$$S_r \left(\frac{p_n (\cos \phi; t_1, t_2, t_3, t_4)}{(t_1 e^{i\phi}, t_1 e^{-i\phi}, t_2 e^{i\phi}, t_2 e^{-i\phi}; q)_\infty} \right)$$

$$= r^n \frac{(t_1 t_2 r^2 q^n; q)_\infty \ p_n (\cos \theta; t_1 r, t_2 r, t_3/r, t_4/r)}{(t_1 t_2 q^n, t_1 r e^{i\theta}, t_1 r e^{-i\theta}, t_2 e^{i\theta}/r, t_2 e^{-i\theta}/r; q)_\infty}. \tag{19.5.4}$$

Theorem 19.5.2 is from (Nassrallah & Rahman, 1985).

We next apply (19.5.3) to derive a bilinear generating function for the Al-Salam–Chihara polynomials. The idea is to use the orthogonality relations (15.1.5) and (19.5.3) to get

$$\frac{2\pi (q; q)_n t_1^{2n} r^{2n}}{(q, t_1 t_2; q)_\infty (t_1 t_2; q)_n} \delta_{m,n}$$

$$= \int\limits_0^\pi S_r \left(\frac{p_m (\cos \phi; t_1, t_2)}{(t_1 e^{i\phi}, t_1 e^{-i\phi})_\infty} \right) (\cos \theta) S_r \left(\frac{p_n (\cos \xi; t_1, t_2)}{(t_1 e^{i\xi}, t_1 e^{-i\xi})_\infty} \right) (\cos \theta)$$

$$\times \frac{(rt_1 e^{i\theta}, rt_1 e^{-i\theta}, e^{2i\theta}, e^{-2i\theta}; q)_\infty}{(t_2 e^{i\theta}/r, t_2 e^{-i\theta}/r; q)_\infty} \ d\theta$$

$$= \frac{(q, q; q)_\infty}{4\pi^2} \int\limits_0^\pi \int\limits_0^\pi \int\limits_0^\pi P_H(\cos \theta, \cos \phi, r) P_H(\cos \theta, \cos \xi, r)$$

$$\times p_m (\cos \phi; t_1, t_2) \ p_n (\cos \xi; t_1, t_2)$$

$$\times \frac{(t_1 r e^{i\theta}, t_1 r e^{-i\theta}, e^{2i\theta}, e^{-2i\theta}; q)_\infty}{(t_2 e^{i\theta}/r, t_2 e^{-i\theta}/r; q)_\infty} \ d\phi \, d\xi \, d\theta.$$

The θ integral can be evaluated by the Nassrallah–Rahman integral (16.4.3) and we find

$$\frac{2\pi(q;q)_n t_1^{2n} r^{2n}}{(q,t_1t_2;q)_\infty (t_1t_2;q)_n} \delta_{m,n} = \frac{(q,t_1t_2;q)_\infty}{2\pi} \int_0^\pi P_m \left(\cos\phi; t_1, t_2\right)$$

$$\times \int_0^\pi K\left(\cos\phi, \cos\xi, t_1, t_2, r^2\right) P_n \left(\cos\xi; t_1, t_2\right) d\xi \, d\phi,$$

where the kernel K is

$$K(\cos\phi, \cos\xi, t_1, t_2, r)$$

$$= \frac{(t_1 re^{i\phi}, t_1 re^{-i\phi}, t_2 re^{i\xi}, t_2 re^{-i\xi}; q)_\infty / (t_1 t_2 r; q)_\infty}{(t_2 e^{i\phi}, t_2 e^{-i\phi}, t_2 e^{i\xi}, t_2 e^{-i\xi}, re^{i(\phi+\xi)}, re^{i(\phi-\xi)}, re^{i(\xi-\phi)} re^{-i(\phi+\xi)}; q)_\infty}$$

$$\times {}_8W_7(t_1 t_2 r/q; t_2 e^{i\phi}, t_2 e^{-i\phi}, r, t_1 e^{i\xi}, t_1 e^{-i\xi}; q, r).$$

Recall the weight function $w(x, t_1, t_2)$ defined in (15.1.2). Set

$$W(\cos\theta; t_1, t_2) := \frac{(e^{2i\theta}, e^{-2i\theta}; q)_\infty}{(t_1 e^{i\theta}, t_1 e^{-i\theta}, t_2 e^{i\theta}, t_1 e^{-i\theta}; q)_\infty} = \frac{w(\cos\theta; t_1, t_2)}{\sin\theta}.$$

$$(19.5.5)$$

Since the Al-Salam–Chihara polynomials, $|t_1| < 1$, $|t_2| < 1$ are orthogonal with respect to $w(x, t_1, t_2)$ on $[-1, 1]$ and the weight function is continuous on $[-1, 1]$ then $\{p_n(x; t_1, t_2)\}$ are complete in $L_2[-1, 1, w(x; t_1, t_2)]$. Therefore

$$\frac{(q, t_1t_2; q)_\infty}{2\pi} \int_0^\pi K(\cos\phi, \cos\xi, t_1, t_2, r) p_n(\cos\xi; t_1, t_2) d\xi$$

$$= r^n w(\cos\phi, t_1, t_2) p_n(\cos\phi, t_1, t_2).$$

Thus the functions

$$t_1^{-n} \sqrt{\frac{(q, t_1t_2; q)_\infty (t_1t_2; q)_n}{2\pi(q;q)_n}} \sqrt{w(\cos\theta; t_1, t_2)} \, p_n(\cos\theta; t_1, t_2)$$

are orthonormal eigenfunctions of an integral operator with a positive symmetric kernel,

$$\frac{(q, t_1t_2; q)_\infty K(\cos\theta, \cos\phi, t_1, t_2, r)}{2\pi \sqrt{w(\cos\theta; t_1, t_2) w(\cos\phi; t_1, t_2)}}.$$

Since these eigenfunctions are complete in $L_2[0, \pi]$, then they constitute all the eigenfunctions. Finally, Mercer's theorem (Tricomi, 1957) implies the bilinear formula

$$\sum_{n=0}^\infty p_n(x; t_1, t_2) p_n(y; t_1, t_2) r^n t_1^{-2n} \frac{(t_1t_2; q)_n}{(q;q)_n} = K(x, y, t_1, t_2, r). \quad (19.5.6)$$

Observe that (19.5.6) is the Poisson kernel for the Al-Salam–Chihara polynomials. Our derivation assumes $|t_1 r| < 1$ and $|t_2| < |r| < 1$ because the orthogonality relation (15.1.5) holds for $|t_1|, |t_2| < 1$.

Theorem 19.5.3 *The Poisson kernel* (19.5.6) *holds for*
$\max\{|t_1|, |t_2|, |r|\} < 1$.

Proof From (15.1.13) it follows that for constant C, we have

$$\left|\sqrt{w\left(x; t_1, t_2\right)}\, p_n\left(x; t_1, t_2\right)\right| \le C\, n\, t_1^n, \quad \text{for } -1 \le x \le 1. \tag{19.5.7}$$

Hence the left-hand side of (19.5.6) is an analytic function of r in the open unit disc if $|t_1|, |t_2| \in (-1, 1)$ and $x, y \in [-1, 1]$. On the other hand the right-hand side of (19.5.6) is also analytic in r for $|r| < 1$ under the same restrictions. Hence our theorem follows from the identity theorem for analytic functions. $\qquad\square$

After proving Theorem 19.5.2, Nassrallah and Rahman used it to prove that the Askey–Wilson polynomials are eigenfunctions of an integral equation with a symmetric kernel. They established the integral equation

$$\int_{-1}^{1} K_r(x, y \,|\, q) p_n\left(y; t_1, t_2, t_3, t_4\right) = \lambda_n p_n\left(x; t_1, t_2, t_3, t_4\right), \tag{19.5.8}$$

where

$$K_r(x, y \,|\, q) = \frac{\left(t_1 t_2, t_3 t_4 r^{-2}, q, q, r^2, r^2, t_3 e^{i\theta}, t_3 e^{-i\theta}, t_4 e^{i\theta}, t_4 e^{-i\theta}; q\right)_\infty}{4\pi^2 \left(t_3 t_4, t_1 t_2 r^2; q\right)_\infty}$$

$$\times \int_0^\pi w\left(\cos\phi; t_3/r, t_4/r, re^{i\theta}, re^{-i\theta} \,|\, q\right) w\left(y; t_1, t_2, re^{i\phi}, re^{-i\phi} \,|\, q\right)$$

$$\times \left(t_1 re^{i\phi}, t_1 re^{-i\phi}, t_2 re^{i\phi}, t_2 re^{-i\phi}; q\right)_\infty \sin\phi\, d\phi, \tag{19.5.9}$$

provided that $\max\{|t_1|, |t_2|, |t_3/r|, |t_4/r|, |r|\} < 1$. The eigenvalues $\{\lambda_n\}$ are

$$\lambda_n = \frac{\left(t_1 t_2, t_3 t_4/r^2; q\right)_n}{\left(t_3 t_4, t_1 t_2 r^2; q\right)_n}\, r^{2n}.$$

Exercises

19.1   Let $D = \dfrac{d}{dx}$ and define the action of $(I - D)^\alpha$ on polynomials by its Taylor series. Show that

$$(I - D)^{\alpha+n}\, \frac{x^n}{n!} = (-1)^n L_n^{(\alpha)}(x).$$

19.2   Define $(I - D)^{-1} g(x)$ to be the function f which solves $(I - D) f(x) = g(x)$ and $f(0) = 0$.

(a) Show that

$$(I - D)^{-n} g(x) = (-1)^n e^x \int_0^x \frac{(x - t)^n}{n!}\, e^{-t} g(t)\, dt.$$

(b) Formulate a definition for $(I - D)^{-\nu}$ when $\nu > 0$, but not an integer. Prove an index law for such operators.

(c) By changing variables, define $(I - cD)^{-\nu}$ for a constant c.

20

Polynomial Solutions to Functional Equations

In this chapter we study polynomial solutions to equations of the type

$$f(x)Ty(x) + g(x)Sy(x) + h(x)y(x) = \lambda_n y(x), \qquad (20.0.1)$$

where S and T are linear operators which map a polynomial of precise degree n to a polynomial of exact degree $n-1$ and $n-2$, respectively. Moreover, f, g, h are polynomials and $\{\lambda_n\}$ is a sequence of constants. We require f, g, h to be independent of n and demand for every n equation (20.0.1) has a polynomial solution of exact degree n. It is tacitly assumed that S annihilates constants and T annihilates polynomials of degree 1. We describe the solutions when S and T involve $\frac{d}{dx}$, Δ, D_q and $\mathcal{D}_q$. In §20.4 we state Leonard's theorem, which characterizes orthogonal polynomials whose duals are also orthogonal. We also describe characterization theorems for classes of orthogonal polynomials.

20.1 Bochner's Theorem

S. Bochner (Bochner, 1929) considered polynomial solutions to (20.0.1) when $S = \dfrac{d}{dx}$ and $T = S^2$. W. Brenke considered the same problem with the added assumption that $\{y_n(x)\}$ are orthogonal (Brenke, 1930). In this section, we prove their results and give our generalization of Bochner's theorem.

Lemma 20.1.1 *Let S, T be as above. If (20.0.1) has a polynomial solution of exact degree n for $n = 0, 1, \ldots, N$, $N > 2$, then f and g have degrees at most 2 and 1, respectively, and we may take $\lambda_0 = 0$ and $h \equiv 0$. We may take $N = \infty$.*

Proof By adding $-\lambda_0 y(x)$ to both sides of (20.0.1) we may assume that $\lambda_0 = 0$. Let $y_n(x) = x^n +$ lower order terms, be a solution of (20.0.1). The result follows from substituting $y_0(x) = 1$, $y_1(x) = x + a$, $y_2(x) = x^2 + bx + c$ in (20.0.1). $\qquad\square$

We shall denote the exact degree of a polynomial p by $\deg(p)$.

Theorem 20.1.2 ((Bochner, 1929)) *Let $S = \dfrac{d}{dx}$, $T = S^2$. Then λ_n and a solution*
y_n are given by:

(i) $f(x) = 1 - x^2$, $\quad g(x) = \beta - \alpha - x(\alpha + \beta + 2)$,
$$\lambda_n = -n(n + \alpha + \beta + 1), \quad y_n = P_n^{(\alpha,\beta)}(x),$$

(ii) $f(x) = x^2$, $\quad g(x) = ax + 1$,
$$\lambda_n = n(n + a - 1), \quad y_n(x) = y_n(x; a, 1),$$

(iii) $f(x) = x^2$, $\quad g(x) = ax$,
$$\lambda_n = n(n + a - 1), \quad y_n(x) = x^n$$

(iv) $f(x) = x$, $\quad g(x) = 1 + \alpha - x$,
$$\lambda_n = -n, \quad y_n(x) = L_n^{(\alpha)}(x).$$

(v) $f(x) = 1$, $\quad g(x) = -2x$
$$\lambda_n = -2n, \quad y_n(x) = H_n(x).$$

Proof First assume that $\deg(f) = 2$, $\deg(g) \leq 1$. If f has two distinct roots then
the scaling $x \to ax + b$ allows us to take $f(x) = 1 - x^2$. Define α and β by
$g(x) = \beta - \alpha - x(\alpha + \beta + 2)$. This makes $\lambda_n = -n(n + \alpha + \beta + 1)$ and we see
that $y(x)$ is a contants multiple of $P_n^{(\alpha,\beta)}(x)$. The cases when α is a negative integer
are limiting cases, which clearly exist as can be seen from (4.6.1). If $\alpha + \beta + 2 = 0$
then g is a constant and y is still a Jacobi polynomial. When f has a double root the
scaling $x \to \alpha x + \beta$ makes f as in (ii) and $g(x) = ax + 1$ or $g(x) = ax$. In the
first case it is easy to verify that $\lambda_n = n(n + a - 1)$ and y must be a genaralized
Bessel polynomial (4.10.6). If $g(x) = ax$ then the solution must be as in case (iii).
Next assume $\deg(f) = \deg(g) = 1$. Again, through $x \to ax + b$, we may assume
$f(x) = x$ and $g(x) = 1 + \alpha - x$ or $g(x) = \alpha$. The first option leads to case (iv),
but the second option makes $\lambda_n = 0$ and we do not get polynomials solutions for all
n. If $\deg(f) = 1$ and $\deg(g) = 0$, then after rescaling we get case (iv). If f is a
constant and $\deg(g) = 1$ then a rescaling makes $f(x) = 1$ and $g(x) = -2x$ and we
find λ_n and y_n as in case (v). $\qquad\square$

E. J. Routh (Routh, 1884) proved the relevant cases of Theorem 20.1.2 under one
of the following additional assumptions:

(A) y satisfies a Rodrigues formula
(B) y satisfies a three-term recurrence of the type

$$A_n y_{n+1} + (B_n + C_n x)\, y_n + D_n y_{n-1} = 0,$$

with $A_n C_n D_{n+1} \neq 0$, $n = 0, 1, \ldots$, where we assume $D_0 y_{-1} := 0$.

He concluded that (A) is equivalent (B) when y is assumed to satisfy (20.0.1) with
$S = \dfrac{d}{dx}$ and $T = S^2$. In particular, he noted one case of orthogonal polynomials
contained in case (i) is $\alpha = a + ib$, $\beta = a - ib$, $x \to ix$. In this case,

$$P_n(x; a, b) = (-i)^n P_n^{(a+ib, a-ib)}(ix). \tag{20.1.1}$$

The Schrödinger form of the differential equation is

$$\frac{\exp(-2b\arctan x)}{(1+x^2)^a} \frac{d}{dx}\left[(1+x^2)^{a+1}\exp(2b\arctan x)y_n'\right]$$

$$= n(n+2a+1)y_n. \tag{20.1.2}$$

Equation (20.1.2) is of Sturm–Liouville type and the polynomials $\{p_n(x;a,b)\}$ are orthogonal with respect to the weight function

$$w(x) = (1+x^2)^a \exp(2b\arctan x), \tag{20.1.3}$$

provided that $\int_{\mathbb{R}} w\,dx$ is finite, i.e., $a < -1$. We will have only a finite number of orthogonal polynomials because w does not have moments of all orders. The same system of polynomials has been studied in (Askey, 1989a), where the orthoganality relation was proved by direct evaluation of integrals. It is clear that when $b = 0$, $w(x)$ reduces to the probability density function of the student t-distribution, so for general b, w is the probability density function of a generalization of the student t-distribution.

W. Brenke considered polynomial solutions to (20.0.1), but he focused on orthogonal polynomials, see (Brenke, 1930). He missed the Bessel polynomials because he did not consider the limiting case of (i) in Theorem 20.1.2 when you let $\alpha, -\beta \to \infty$ with $\alpha + \beta$ fixed, after scaling x. He also missed the orthoghonal polynomial system found by Routh because he considered only infinite systems of orthogonal polynomials.

The following motivation explains Theorem 20.1.2 and the generalizations discussed below. We seek a polynomial solution of degree n to the differential equation

$$f(x)y''(x) + g(x)y'(x) = \lambda y(x). \tag{20.1.4}$$

We know that one of the coefficients in f or g is not zero, hence there is no loss of generality in choosing it equal to 1. Thus f and g contain four free parameters. The scaling $x \to ax + b$ of the independent variable absorbs two of the four parameters. The eigenvalue parameter λ is then uniquely determined by equating coefficients of x^n in (20.1.4) since y has degree n. This reduces (20.1.4), in general, to a Jacobi differential equation whose polynomial solution, in general, is a Jacobi polynomial. The other cases are special or limiting cases of Jacobi polynomials. This approach also explains what happens if (20.0.1) involves D_q or $\mathcal{D}_q$. In the case of D_q the scaling $x \to ax$ is allowed so only one parameter can be absorbed by scaling. On the other hand no scaling is allowed if (20.0.1) contains $\mathcal{D}_q$. This means that the general polynomial solutions of (20.0.1) will contain three parameters or four parameters if (20.0.1) contains D_q, or $\mathcal{D}_q$, respectively.

Remark 20.1.1 *The operators $\mathcal{D}_q$ and $\mathcal{A}_q$ are invariant under $q \to q^{-1}$. Moreover*

$$(a;1/q)_n = (-a)^n q^{n(1-n)/2}(1/a;q)_n. \tag{20.1.5}$$

Therefore

$$q^{3n(n-1)/2} p_n(x;t_1,t_2,t_3,t_4 \mid 1/q)$$

$$= (-t_1t_2t_3t_4)^n p_n(x;1/t_1,1/t_2,1/t_3,1/t_4 \mid q). \tag{20.1.6}$$

It must be emphasized that Bochner's theorem classifies second order differential equations of Sturm–Liouville type with polynomial solutions. We next prove the corresponding theorem when $T = \mathcal{D}_q^2$, $S = \mathcal{A}_q \mathcal{D}_q$, that is we consider

$$f(x)\mathcal{D}_q^2 y_n(x) + g(x)\mathcal{A}_q \mathcal{D}_q y_n(x) + h(x) = \lambda_n y_n(x). \tag{20.1.7}$$

Recall that the Askey–Wilson polynomials satisfy

$$\pi_2(x)\mathcal{D}_q^2 y(x) + \pi_1(x)\mathcal{A}_q \mathcal{D}_q y(x) = \lambda_n y(x), \tag{20.1.8}$$

with π_1 and π_2 given by (16.3.20) and (16.3.19), respectively. Clearly, Lemma 20.1.1 implies that $h \equiv 0$, $\deg(f) \leq 2$ and $\deg(g) \leq 1$, and $\lambda_0 = 0$. To match (20.1.7) with (20.1.8), let

$$f(x) = f_0 x^2 + f_1 x + f_2, \qquad g(x) = g_0 x + g_1 \tag{20.1.9}$$

If $2q^{1/2} f_0 + (1 - q)g_0 \neq 0$ then through a suitable multiplier we can assume that $2q^{1/2} f_0 + (1 - q)g_0 = -8$ and then determine the σs uniquely, hence we determine the parameters t_1, t_2, t_3, t_4 up to permutations. Theorem 16.3.4 then proves that λ_n is given by (16.3.7), and (20.1.7) has only one polynomial solution, a constant multiple of an Askey–Wilson polynomial $p_n(x, \mathbf{t})$. If $2q^{1/2} f_0 + (1 - q)g_0 = 0$ but $|f_0| + |g_0| \neq 0$ then we let $q = 1/p$, and apply (20.1.6) and Lemma 20.1.1 to see that (20.1.8) is transformed to a similar equation where the σs are elementary symmetric functions of $1/t_1, 1/t_2, 1/t_3, 1/t_4$ and q is replaced by $1/q$. Finally, if $f_0 = g_0 = 0$, then $\lambda_n = 0$ for all n and with $u = \mathcal{D}_q y$, we see that

$$(f_1 x + f_2)\,\mathcal{D}_q u(x) + g_1 \mathcal{A}_q u(x) = 0. \tag{20.1.10}$$

Substituting $u(x) = \sum_{k=0}^{n} u_k \phi_k(x; a)$ in (20.1.10) and equating coefficients of ϕ_k for all k, we see that it is impossible to find polynomial solutions to (20.1.8) of all degrees. This establishes the following theorem.

Theorem 20.1.3 ((Ismail, 2003b)) *Given an equation of the form (20.1.7) has a polynomial solution $y_n(x)$ of degree n for every n, $n = 0, 1, \ldots$ if and only if $y_n(x)$ is a multiple of $p_n(x; t_1, t_2, t_3, t_4 \,|\, q)$ for some parameters t_1, t_2, t_3, t_4, including limiting cases as one or more of the paramaters tends to ∞. In all these cases (20.1.7) can always be reduced to (20.1.8), or a special or limiting case of it.*

Recently, Grünbaum and Haine (Grünbaum & Haine, 1996), (Grünbaum & Haine, 1997) have studied the bispectral problem of finding simultaneous solutions to the eigenvalue problem $L p_n(x) = \lambda_n p_n(x)$ and $M p_n(x) = x p_n(x)$, where L is a second-order Askey–Wilson operator and M is a second-order difference equation in n.

Remark 20.1.2 *It is important to note that solutions to (20.1.8) may not satisfy the orthogonality relation for the Askey–Wilson polynomials of Theorem 15.2.1. For example, formulas (15.2.10)–(15.2.13) show that the polynomials $r_n(x) = \lim_{t_4 \to \infty} p_n(x; \mathbf{t})$ satisfy*

$$2x r_n(x) = A_n r_{n+1}(x) + C_n r_{n-1}(x) + \left[t_1 + t_1^{-1} - A_n - C_n\right] r_n(x), \tag{20.1.11}$$

with

$$A_n = \frac{\left(1 - t_1 t_2 q^n\right)\left(1 - t_1 t_3 q^n\right)}{t_1 t_2 t_3\, q^{2n}}, \quad C_n = \frac{\left(1 - q^n\right)\left(1 - t_2 t_3 q^{n-1}\right)}{t_1 t_2 t_3\, q^{2n-1}}. \quad (20.1.12)$$

For orthogonality it is necessary and sufficient that $A_{n-1}C_n > 0$ for all $n > 0$, a condition which may or may not be satisfied. In fact the corresponding moment problem is indeterminate for $q \in (0,1)$, and t_1, t_2, t_3 are such that $A_{n-1}C_n > 0$, $n > 0$, (Akhiezer, 1965), (Shohat & Tamarkin, 1950). On the other hand if $q > 1$, the moment problem is determinate when $A_{n-1}C_n > 0$ for all $n > 0$. In fact, the latter polynomials are special Askey–Wilson polynomials, as can be seen from (20.1.6).

One possible generalization of Bochner's theorem is to consider polynomial solutions to

$$f(x)\, y'' + g(x)\, y' + h(x)\, y = 0. \quad (20.1.13)$$

More precisely, Heine considered the following problem, (Szegő, 1975, §6.8).

Problem. Given polynomials f and g of degrees at most $p+1$ and p, respectively, and a positive integer n, find all polynomials h such that (20.1.13) has a polynomial solution y of exact degree n.

Heine proved that, in general, there are exactly

$$\sigma(n,p) = \binom{n+p-1}{n} \quad (20.1.14)$$

choices of h which make (20.1.13) have a polynomial solution. Indeed, $\sigma(n,p)$ is always an upper bound and is attained in many cases.

Later, Stieltjes proved the following, (Szegő, 1975, §6.8).

Theorem 20.1.4 *Let f and g have precise degrees $p+1$ and p, respectively, and assume that f and g have positive leading terms. Assume further that the zeros of f and g are real, simple, and interlaced. Then there are exactly $\sigma(n,p)$ polynomials h of degree $p-1$ such that (20.1.13) has a polynomial solution of exact degree n. Moreover, for every such h, (20.1.13) has a unique polynomial solution, up to a multiplicative constant.*

Heine's idea is to first observe that we are searching for $n + p - 1$ unknowns, n of them are the coefficients in y and $p-1$ of them are the coefficients of h because we can always assume that y is monic and take one of the nonzero coefficients of h to be equal to 1. Heine then observes that, in general, we can prescribe any $p-1$ of these unknowns and find the remaining n unknowns by equating the coefficients of all powers of x in (20.1.13) to zero. Stieltjes makes this argument more precise by characterizing all $\sigma(n,p)$ solutions $y(x)$ in the following way: The n zeros of any solution are distributed in all possible ways in the p intevals defined by the $p+1$ zeros of $f(x)$. For details of Stieltjes' treatment, see (Szegő, 1975, §6.8).

Note that the polynomial solution of (20.1.13) alluded to in Theorem 20.1.4 is unique, up to a multiplicative constant.

H. L. Krall (Krall, 1938) considered orthogonal polynomial solutions to

$$\sum_{s=0}^{N} \pi_s(x)\, y^{(s)}(x) = \lambda y(x), \qquad (20.1.15)$$

where $\pi_s(x)$, $1 \le s \le N$ are real valued smooth functions on the real line, $\pi_N(x) \not\equiv 0$, and λ is a real parameter. One can prove that in order for (20.1.15) to have polynomial solutions of degree n, $n = 0, 1, \ldots, N$, then (20.1.15) must have the form

$$(L_n y)(x) := \sum_{s=0}^{N} \pi_s(x)\, y^{(s)}(x) = \lambda_n y(x), \quad n = 0, 1, \ldots, N, \qquad (20.1.16)$$

with π_s a polynomial of degree at most s, $0 \le s < N$ and π_N of exact degree N. Moreover, with $\pi_s(x) = \sum_{j=0}^{s} \pi_{ss}\, x^s$, the eigenvalues $\{\lambda_n\}$ are given by

$$\lambda_n = \sum_{s=0}^{n} \pi_{ss} \frac{n!}{(n-s)!}.$$

Recall the definition of signed orthogonal polynomials in Remark 2.1.2.

Theorem 20.1.5 *A differential equation of the type* (20.1.16) *has signed orthogonal polynomial solutions of degree n, $n = 0, 1, \ldots$, if and only if*

$$\text{(i) } D_n := \det |\mu_{i+j}|_{i,j=0}^{n} \ne 0, \quad n = 0, 1, \ldots,$$

and

$$\text{(ii) } S_k(m) := \sum_{s=2k+1}^{N} \sum_{j=0}^{s} \binom{s-k-1}{k} U(m - 2k - 1, s - 2k - 1)\, \pi_{s,s-j}\, \mu_{m-j},$$

for $k = 0, 1, \ldots, \lfloor (N-1)/2 \rfloor$ and $m = 2k+1, 2k+2, \ldots$, and

$$U(0, k) := 0, \quad U(n, k) = (-1)^k (-n)_k, \quad n > 0.$$

Related results are in (Krall, 1936a) and (Krall, 1936b).

For more recent literature on this problem, see (Kwon et al., 1994), (Kwon et al., 1993), (Kwon & Littlejohn, 1997), and (Yoo, 1993).

20.2 Difference and q-Difference Equations

We now consider the difference equations

$$f(x)\nabla\Delta y_n(x) + g(x)\nabla y_n(x) = \lambda_n y_n(x),$$
$$f(x)\nabla\Delta y_n(x) + g(x)\Delta y_n(x) = \lambda_n y_n(x).$$

Since

$$\nabla\Delta y_n(x) = y_n(x+1) - 2y_n(x) + y_n(x-1),$$

it follows that

$$f(x)\nabla\Delta y_n(x) + g(x)\nabla y_n(x) = (f(x) + g(x))\nabla\Delta y_n(x) + g(x)\nabla y_n(x).$$

The degrees of f and g are at most 2 and one, respectively. Thus there is no loss of generality in considering

$$f(x)\nabla\Delta y_n(x) + g(x)\nabla y_n(x) = \lambda_n y_n(x). \tag{20.2.1}$$

Theorem 20.2.1 *The difference equation* (20.2.1) *has a polynomial solution of degree n for $n = 0, 1, \ldots, M$, $M > 2$, up to scaling the x variable, if and only if*

(i) $f(x) = (x + \alpha + 1)(x - N)$, $g(x) = x(\alpha + \beta + 2) - N(\alpha + 1) \neq 0$,
$\lambda_n = n(n + \alpha + \beta + 1)$, $y_n = Q_n(x; \alpha, \beta, N)$,

(ii) $f(x) = x(x - N)$, $g(x) \equiv 0$,
$$\lambda_n = n(n-1), \quad y_n = x\,{}_3F_2\left(\begin{array}{c} -n+1, n, 1-x \\ 2, 1-N \end{array}\middle| 1\right),$$

(iii) $f(x) = c(x + \beta)$, $\beta \neq 0$, $g(x) = (c-1)x + c\beta$,
$\lambda_n = n(c-1)$, $y_n = M_n(x; \beta, c)$,

(iv) $f(x) = cx$, $g(x) = (c-1)x$,
$\lambda_n = n(c-1)$, $y_n = x\,{}_2F_1(1-n, 1-x; 2; 1 - 1/c)$.

Proof When f has precise degrees 2 and $g \not\equiv 0$, rescaling we may assume $f(0) = g(0)$ and $f(x) = x^2 + \cdots$. It is clear that there is no loss of generality in taking f and g as in (i), but N may or may not be a positive integer. It is true, however, that $Q_n(x; \alpha, \beta, N)$ is well-defined whether for all N, and we have the restriction $n < N$ only when N is a positive integer. Case (ii) corresponds to the choice $\beta = -1$, and the limiting case $y_n(x) = \lim_{\alpha \to -1^+} (\alpha + 1)Q_n(x; \alpha, \beta, N)$.

We next consider the case when f has degree 1 or 0. Then $g(x)$ must have precise degree 1. If f has exact degree 1, then there is no loss of generality in assuming $f(0) = g(0)$ and we may take f and g as in (iii) and we know that $y_n = M_n(x; \beta, c)$. Case (iv) corresponds to $\lim_{\beta \to 0} \beta M_n(x; \beta, c)$.   □

O. H. Lancaster (Lancaster, 1941) analyzed self-adjoint second and higher order difference equations. Since self-adjoint operators have orthogonal eigenfunctions, he characterized all orthogonal polynomials solutions to (20.2.1). We have seen neither Theorem 20.2.1 nor its proof in the literature, but it is very likely to exist in the literature.

We now come to equations involving D_q. Consider the case

$$T = D_{q^{-1},x}D_{q,x}, \qquad S = D_{q,x}. \tag{20.2.2}$$

As in the case of (20.2.1), the choices of S and T in (20.2.2) are equivalent to the choice $T = D_{q^{-1},x}D_{q,x}$, $S = D_{q^{-1},x}$. In this case, we treat the following subcases separately

(i) f has two distinct roots, neither of them is $x = 0$.

(ii) f has two distinct roots, one of which is $x = 0$.

(iii) f has a double root which $\neq 0$.

(iv) f has $x = 0$ as a double zero.

(v) $f(x) = x + c$, $c \neq 0$.

(vi) $f(x) = x$.

We first consider the cases when g has precise degree 1. In case (i) we scale x as $x \to cx$ to make $f(x) = 0$ at $x = 1, a$. Thus, $f(x) = (x-1)(x-a)$ and we can find parameters t_1 and t_2 such that

$$g(x) = q\left[(1 - at_1t_2)\,x + a\,(t_1 + t_2) - a - 1\right]/(1-q)$$

and we identify (20.0.1) with (18.4.26), so a polynomial solution is $y = \varphi_n\,(x; a, t_1, t_2)$. In case (ii) we take $f(x) = x(x-1)$, after scaling x (if necessary) and identify (20.0.1) with (18.4.27), so a solution is $p_n(x; \alpha, \beta)$. In case (iii) we scale x to make $f(x) = (x-1)^2$, choose $a = 1$ in (18.4.26) and find t_1 and t_2 from the knowledge of g, hence a polynomial solution is $y = \varphi_n\,(x; 1, t_1, t_2)$, which do not form a system of orthogonal polynomials.

20.3 Equations in the Askey–Wilson Operators

In §16.5 we showed that solving the Bethe Ansatz equations (16.5.2) was equivalent to finding polynommial solutions to (16.5.14). Observe that this is the exact problem raised by Heine, but $\dfrac{d}{dx}$ is replaced by $\mathcal{D}_q$. In work in progress we proved that, in general, there are $\sigma(n, N-1)$ choices of $r(x)$ in equation (16.5.14) in order for (16.5.14) to have a polynomial solution. Here, σ is as in (20.1.14). This raises the question of finding solutions to (16.5.14). In the absence of a concept of regular singular points of (16.5.14), we offer a method to find formal solutions. This section is based on (Ismail et al., 2005).

Recall the definition of $\phi_n(x; a)$ in (12.2.1). This can be generalized from polynomials to functions by

$$\phi_\alpha(x; a) = \frac{\left(ae^{i\theta}, ae^{-i\theta}; q\right)_\infty}{\left(aq^\alpha e^{i\theta}, aq^\alpha e^{-i\theta}; q\right)_\infty}. \tag{20.3.1}$$

It readily follows that

$$\mathcal{D}\phi_\alpha(x; a) = \frac{(1 - q^\alpha)}{2a(q-1)}\,\phi_{\alpha-1}\left(x; aq^{1/2}\right),$$

$$\mathcal{A}_q\phi_\alpha(x; a) = \phi_{\alpha-1}\left(x; aq^{1/2}\right)\left[1 - aq^{-1/2}\left(1 + q^\alpha\right)x + a^2q^{\alpha-1}\right]. \tag{20.3.2}$$

The second formula in (20.3.2) holds when $\alpha = 0$. Furthermore we have

$$2\mathcal{A}_q\phi_\alpha(x; a) = \left(1 + q^{-\alpha}\right)\phi_\alpha\left(x; aq^{1/2}\right)$$

$$+ \left(1 - q^{-\alpha}\right)\left(1 + a^2q^{2\alpha-1}\right)\phi_{\alpha-1}\left(x; aq^{1/2}\right). \tag{20.3.3}$$

The concept of singularities of differential equations is related to the analytic properties of the solutions in a neighborhood of the singularities. We have no knowledge of a geometric way to describe the corresponding situation for equations like (16.5.14). In the present setup, the analogue of a function analytic in a neighborhood of a point $(a + a^{-1})/2$ is a function which has a convergent series expansion of the form $\sum\limits_{n=0}^\infty c_n\phi_n(x; a)$. We have no other characterization of these q-analytic functions.

It is easy to see that when a is not among the $2N$ parameters $\{\zeta_1, \ldots, \zeta_{2N}\}$, where

$$\zeta_j := \left(t_j + t_j^{-1}\right)/2$$

then one can formally expand a solution y as $\sum_{n=0}^{\infty} y_n \phi_n(x; a)$, substitute the series expansion in (16.5.14) and recursively compute the coefficients y_n. This means that the only singular points are $\zeta_1, \ldots, \zeta_{2N}$ and possibly 0 and ∞. Expanding around $x = \zeta_j$ boils down to taking $a = t_j$ and using an expansion of the form

$$y(x) = \sum_{n=0}^{\infty} y_n \, \phi_{n+\alpha}(x; t_j). \tag{20.3.4}$$

There is no loss of generality in taking $j = 1$. Observe that $r(x)\phi_{n+\alpha}(x; t_1)$ is a linear combination of $\{\phi_{m+\alpha}(x; t_1) : n \leq m \leq n + N - 2\}$. Furthermore we note that (16.5.13) implies

$$\frac{1}{w(x; \mathbf{t})} \mathcal{D}_q \left(w\left(x; q^{1/2}\mathbf{t}\right) \mathcal{D}_q \phi_{n+\alpha}(x; t_1) \right)$$

$$= \frac{1 - q^{n+\alpha}}{2t_1(q-1)w(x; \mathbf{t})} \mathcal{D}_q \left(w\left(x; q^{1/2}\vec{a}\right) \phi_{n+\alpha-1}\left(x; q^{1/2}t_1\right) \right)$$

$$= \frac{1 - q^{n+\alpha}}{2t_1(q-1)w(x; \mathbf{t})} \mathcal{D}_q \left(w\left(x; t_1 q^{n+\alpha-1/2}, t_2, \ldots, t_{2N}\right) \right)$$

$$= \frac{(1 - q^{n+\alpha}) w\left(x; t_1 q^{n+\alpha-1}, t_2, \ldots, t_{2N}\right)}{2t_1(q-1)w(x; \mathbf{t})} \Phi\left(x; t_1 q^{n+\alpha-1}, t_2, \ldots, t_{2N}\right)$$

$$= \frac{1 - q^{n+\alpha}}{2t_1(q-1)} \Phi\left(x; t_1 q^{n+\alpha-1}, t_2, \ldots, t_{2N}\right) \phi_{n+\alpha-1}(x; t_1).$$

We substitute the expansion (20.3.4) for y in (16.5.14), and reduce the left-hand side of (16.5.14) to

$$\sum_{n=0}^{\infty} \frac{1 - q^{n+\alpha}}{2t_1(q-1)} \Phi\left(x; t_1 q^{n+\alpha-1}, t_2, \ldots, t_{2N}\right) \phi_{n+\alpha-1}(x; t_1) \, y_n. \tag{20.3.5}$$

The smallest subscript of a ϕ in $r(x)y(x)$ on the right-hand side of (16.5.14) is α. On the other hand, (20.3.5) implies that $\phi_{\alpha-1}$ appears on the left-hand side of (16.5.14). Thus the coefficient of $\phi_{\alpha-1}(x; t_1)$ must be zero. To determine this coefficient we set

$$\Phi\left(x; q^{\alpha-1}t_1, t_2, \ldots, t_{2N}\right) = \sum_{j=0}^{N-1} d_j\left(q^\alpha\right) \phi_j\left(x; t_1 q^{\alpha-1}\right), \tag{20.3.6}$$

and after making use of $\phi_n\left((a + a^{-1})/2; a\right) = \delta_{n,0}$ we find that

$$d_0\left(q^\alpha\right) = \Phi\left(\left(t_1 q^{\alpha-1} + t_1^{-1}q^{1-\alpha}\right)/2; q^{\alpha-1}t_1, t_2, \ldots, t_{2N}\right).$$

Thus the vanishing of the coefficient of $\phi_{\alpha-1}(x; t_1)$ on the left-hand side of (16.5.14) implies the vanishing of $(1 - q^\alpha) d_0(q^\alpha)$, that is

$$(1 - q^\alpha) \Phi\left(\left(t_1 q^{\alpha-1} + t_1^{-1}q^{1-\alpha}\right)/2; q^{\alpha-1}t_1, t_2, \ldots, t_{2N}\right) = 0. \tag{20.3.7}$$

Theorem 20.3.1 *Assume* $|t_j| \leq 1$, *for all* j. *Then the only solution(s) of* (20.3.7) *are given by* $q^\alpha = 1$, *or* $q^\alpha = q/(t_1 t_j)$, $j = 2, \ldots, 2N$.

Proof From (20.3.7) it is clear that $q^\alpha = 1$ is a solution. With $x = \left(t_1 q^{\alpha-1} + t_1^{-1} q^{1-\alpha} \right)/2$ as in (20.3.7) we find $e^{i\theta} = t_1 q^{\alpha-1}$, or $t_1^{-1} q^{1-\alpha}$. In the former case, $2i \sin \theta = t_1 q^{\alpha-1} - t_1^{-1} q^{1-\alpha}$, hence (20.3.7) and (16.5.15) imply

$$\frac{2i \left(1 - t_1^2 q^{2\alpha-2} \right)}{t_1 q^{\alpha-1} - q^{1-\alpha}/t_1} \prod_{j=2}^{2N} \left(1 - t_1 t_j q^{\alpha-1} \right) = 0,$$

which gives the result. On the other hand if $e^{i\theta} = q^{1-\alpha}/t_1$, then we reach the same solutions. $\qquad\square$

Ismail and Stanton (Ismail & Stanton, 2003b) used two bases in addition to $\{\phi_n(x; a)\}$ for polynomial expansions. Their bases are

$$\rho_n(\cos \theta) := \left(1 + e^{2i\theta} \right) \left(-q^{2-n} e^{2i\theta}; q^2 \right)_{n-1} e^{-in\theta}, \tag{20.3.8}$$

$$\phi_n(\cos \theta) := \left(q^{1/4} e^{i\theta}, q^{1/4} e^{-i\theta}; q^{1/2} \right)_n. \tag{20.3.9}$$

They satisfy

$$\mathcal{D}_q \rho_n(x) = 2q^{(1-n)/2} \frac{1 - q^n}{1 - q} \rho_{n-1}(x), \tag{20.3.10}$$

$$\mathcal{D}_q \phi_n(x) = -2q^{1/4} \frac{1 - q^n}{1 - q} \phi_{n-1}(x). \tag{20.3.11}$$

There is no theory known for expanding solutions of Askey–Wilson operator equations in terms of such bases.

20.4 Leonard Pairs and the q-Racah Polynomials

The material in this section is based on (Terwilliger, 2001), (Terwilliger, 2002), and (Terwilliger, 2004). The goal is to characterize the q-Racah polynomials and their special and limiting cases through an algebraic property. The result is called Leonard's theorem, which first appeared in a different form from what is given here in the work (Leonard, 1982). Originally, the problem arose in the context of association schemes and the P and Q polynomials in (Bannai & Ito, 1984).

Definition 20.4.1 *Let* V *denote a vector space over a field* $\mathbb{K}$ *with finite positive dimension. By a Leonard pair on* V, *we mean an ordered pair of linear transformations* $A : V \to V$ *and* $A^* : V \to V$ *which satisfy both* (i) *and* (ii) *below.*

(i) *There exists a basis for* V *with respect to which the matrix representing* A *is irreducible tridiagonal and the matrix representing* A^* *is diagonal.*

(ii) *There exists a basis for* V *with respect to which the matrix representing* A *is diagonal and the matrix representing* A^* *is irreducible tridiagonal.*

Usually A^* denotes the conjugate-transpose of a linear transformation A. We emphasize that this convention is not used in Definition 20.4.1. In a Leonard pair A, A^*, the linear transformations A and A^* are arbitrary subject to (i) and (ii) above.

A closely-related object is a Leonard system which will be defined after we make an observation about Leonard pairs.

Lemma 20.4.1 *Let V denote a vector space over $\mathbb{K}$ with finite positive dimension and let A, A^* deonte a Leonard pair on V. Then the eigenvalues of A are mutually distinct and contained in $\mathbb{K}$. Moreover, the eigenvalues of A^* are mutually distinct and contained in $\mathbb{K}$.*

To prepare for the definition of a Leonard system, we recall a few concepts from linear algebra. Let d denote a nonnegative integer and let $\mathrm{Mat}_{d+1}(\mathbb{K})$ denote the $\mathbb{K}$-algebra consisting of all $d + 1$ by $d + 1$ matrices which have entries in $\mathbb{K}$. We index the rows and columns by $0, 1, \ldots, d$. Let $\mathbb{K}^{d+1}$ denote the $\mathbb{K}$-vector space consisting of all $d + 1$ by 1 matrices which have entries in $\mathbb{K}$. Now we index the rows by $0, 1, \ldots, d$. We view $\mathbb{K}^{d+1}$ as a left module for $\mathrm{Mat}_{d+1}(\mathbb{K})$. Observe that this module is irreducible. For the rest of this section, $\mathcal{A}$ will denote a $\mathbb{K}$-algebra isomorphic to $\mathrm{Mat}_{d+1}(\mathbb{K})$. By an $\mathcal{A}$-module we mean a left $\mathcal{A}$-module. Let V denote an irreducible $\mathcal{A}$-module. Note that V is unique up to isomorphism of $\mathcal{A}$-modules, and that V has dimension $d + 1$. Let $v_0, v_1, \ldots, v_d$ denote a basis for V. For $X \in \mathcal{A}$ and $Y \in \mathrm{Mat}_{d+1}(\mathbb{K})$, we say Y represents X with respect to $v_0, v_1, \ldots, v_d$ whenever $Xv_j = \sum_{i=0}^{d} Y_{ij} v_i$ for $0 \leq j \leq d$. Let A denote an element of $\mathcal{A}$. A is called multiplicity-free whenever it has $d + 1$ mutually distinct eigenvalues in $\mathbb{K}$. Let A denote a multiplicity-free element of $\mathcal{A}$. Let $\theta_0, \theta_1, \ldots, \theta_d$ denote an ordering of the eigenvalues of A, and for $0 \leq i \leq d$ we set

$$E_i = \prod_{\substack{0 \leq j \leq d \\ j \neq i}} \frac{A - \theta_j I}{\theta_i - \theta_j}, \qquad (20.4.1)$$

where I denotes the identity of $\mathcal{A}$. Observe that:

(i) $AE_i = \theta_i E_i$ $(0 \leq i \leq d)$;

(ii) $E_i E_j = \delta_{ij} E_i$ $(0 \leq i, j \leq d)$;

(iii) $\sum_{i=0}^{d} E_i = I$;

(iv) $A = \sum_{i=0}^{d} \theta_i E_i$.

Let $\mathcal{D}$ denote the subalgebra of $\mathcal{A}$ generated by A. Using (i)–(iv), we find the sequence $E_0, E_1, \ldots, E_d$ is a basis for the $\mathbb{K}$-vector space $\mathcal{D}$. We call E_i the primitive idempotent of A associated with θ_i. It is helpful to think of these primitive idempotents as follows. Observe that $V = \bigoplus_{j=0}^{d} E_j V$, $E_i V$. Moreover, for $0 \leq i \leq d$, $E_i V$ is the (one-dimensional) eigenspace of A in V associated with the eigenvalue θ_i, and E_i acts on V as the projection onto this eigenspace. Furthermore, $\{A^i \mid 0 \leq i \leq d\}$

is a basis for the $\mathbb{K}$-vector space $\mathcal{D}$ and that $\prod_{i=0}^{d} (A - \theta_i I) = 0$. By a Leonard pair in $\mathcal{A}$, we mean an ordered pair of elements taken from $\mathcal{A}$ which act on V as a Leonard pair in the sense of Definition 20.4.1. We call $\mathcal{A}$ the ambient algebra of the pair and say the pair is over $\mathbb{K}$. We refer to d as the diameter of the pair. We now define a Leonard system.

Definition 20.4.2 *By a Leonard system in $\mathcal{A}$ we mean a sequence*

$$\Phi := \left(A; A^*; \{E_i\}_{i=0}^{d} ; \{E_i^*\}_{i=0}^{d} \right)$$

which satisfies (i)–(v) *below.*

(i) *Each of A, A^* is a multiplicity-free element in $\mathcal{A}$.*

(ii) *$E_0, E_1, \ldots, E_d$ is an ordering of the primitive idempotents of A.*

(iii) *$E_0^*, E_1^*, \ldots, E_d^*$ is an ordering of the primitive idempotents of A^*.*

(iv) $E_i A^* E_j = \begin{cases} 0, & \text{if } |i - j| > 1; \\ \neq 0, & \text{if } |i - j| = 1 \end{cases}$ $(0 \le i, j \le d)$.

(v) $E_i^* A E_j^* = \begin{cases} 0, & \text{if } |i - j| > 1; \\ \neq 0, & \text{if } |i - j| = 1 \end{cases}$ $(0 \le i, j \le d)$.

The number d is called the diameter of Φ. We call $\mathcal{A}$ the ambient algebra of Φ.

We comment on how Leonard pairs and Leonard systems are related. In what follows, V denotes an irreducible $\mathcal{A}$-module. Let $\left(A; A^*; \{E_i\}_{i=0}^{d} ; \{E_i^*\}_{i=0}^{d} \right)$ denote a Leonard system in $\mathcal{A}$. For $0 \le i \le d$, let v_i denote a nonzero vector in $E_i V$. Then the sequence $v_0, v_1, \ldots, v_d$ is a basis for V which satisfies Definition 20.4.1(ii). For $0 \le i \le d$, let v_i^* denote a nonzero vector in $E_i^* V$. Then the sequence $v_0^*, v_1^*, \ldots, v_d^*$ is a basis for V which satisfies Definition 20.4.1(i). By these comments the pair A, A^* is a Leonard pair in $\mathcal{A}$. Conversely, let A, A^* denote a Leonard pair in $\mathcal{A}$. Then each of A, A^* is multiplicity-free by Lemma 20.4.2. Let $v_0, v_1, \ldots, v_d$ denote a basis for V which satisfies Definition 20.4.1(ii). For $0 \le i \le d$, the vector v_i is an eigenvector for A; let E_i denote the corresponding primitive idempotent. Let $v_0^*, v_1^*, \ldots, v_d^*$ denote a basis for V which satisfies Definition 20.4.1(i). For $0 \le i \le d$ the vector v_i is an eigenvector for A^*; let E_i denote the corresponding primitive idempotent. Then $\left(A; A^*; \{E_i\}_{i=0}^{d} ; \{E_i^*\}_{i=0}^{d} \right)$ is a Leonard system in $\mathcal{A}$. In summary, we have the following.

Lemma 20.4.2 *Let A and A^* denote elements of $\mathcal{A}$. Then the pair A, A^* is a Leonard pair in $\mathcal{A}$ if and only if the following* (i) *and* (ii) *hold.*

(i) *Each of A, A^* is multiplicity-free.*

(ii) *There exists an ordering $E_0, E_1, \ldots, E_d$ of the primitive idempotents of A and there exists an ordering $E_0^*, E_1^*, \ldots, E_d^*$ of the primitive idempotents of A^* such that $\left(A; A^*; \{E_i\}_{i=0}^{d} ; \{E_i^*\}_{i=0}^{d} \right)$ is a Leonard system in $\mathcal{A}$.*

Recall the notion of isomorphism for Leonard pairs and Leonard systems. Let A, A^* denote a Leonard pair in $\mathcal{A}$ and let $\sigma : \mathcal{A} \to \mathcal{A}'$ denote an isomorphism of $\mathbb{K}$-algebras. Note that the pair $A^\sigma, A^{*\sigma}$ is a Leonard pair in $\mathcal{A}'$.

Definition 20.4.3 *Let* A, A^* *and* B, B^* *denote Leonard pairs over* $\mathbb{K}$. *By an isomorphism of Leonard pairs from* A, A^* *to* B, B^* *we mean an isomorphism of* $\mathbb{K}$-*algebras from the ambient algebra of* A, A^* *to the ambient algebra of* B, B^* *which sends* A *to* B *and* A^* *to* B^*. *The Leonard pairs* A, A^* *and* B, B^* *are said to be isomorphic whenever there exists an isomorphism of Leonard pairs from* A, A^* *to* B, B^*.

Let Φ denote the Leonard system from Definition 20.4.2 and let $\sigma : \mathcal{A} \to \mathcal{A}'$ denote an isomorphism of $\mathbb{K}$-algebras. We write

$$\Phi^\sigma := \left(A^\sigma; A^{*\sigma}; \{E_i^\sigma\}_{i=0}^d ; \{E_i^{*\sigma}\}_{i=0}^d \right)$$

and observe Φ^σ is a Leonard system in $\mathcal{A}'$.

Definition 20.4.4 *Let* Φ *and* Φ' *denote Leonard systems over* $\mathbb{K}$. *By an isomorphism of Leonard systems from* Φ *to* Φ' *we mean an isomorphism of* $\mathbb{K}$ *algebras* σ *from the ambient algebra of* Φ *to the ambient algebra of* Φ' *such that* $\Phi^\sigma = \Phi'$. *The Leonard systems* Φ, Φ' *are said to be isomorphic whenever there exists an isomorphism of Leonard systems from* Φ *to* Φ'.

A given Leonard system can be modified in several ways to get a new Leonard system. For instance, let Φ denote the Leonard system from Definition 20.4.2. Then each of the following three sequences is a Leonard system in $\mathcal{A}$.

$$\Phi^* := \left(A^*; A; \{E_i^*\}_{i=0}^d ; \{E_i\}_{i=0}^d \right),$$

$$\Phi^\downarrow := \left(A; A^*; \{E_i\}_{i=0}^d ; \{E_{d-i}^*\}_{i=0}^d \right),$$

$$\Phi^\Downarrow := \left(A; A^*; \{E_{d-i}\}_{i=0}^d ; \{E_i^*\}_{i=0}^d \right).$$

Viewing $*, \downarrow, \Downarrow$ as permutations on the set of all Leonard systems,

$$*^2 = \downarrow^2 = \Downarrow^2 = 1, \tag{20.4.2}$$

$$\Downarrow * = * \downarrow, \quad \downarrow * = * \Downarrow, \quad \downarrow\Downarrow = \Downarrow\downarrow . \tag{20.4.3}$$

The group generated by symbols $*, \downarrow, \Downarrow$ subject to the relations (20.4.2), (20.3.3) is the dihedral group D_4. Recall that D_4 is the group of symmetries of a square, and has 8 elements. It is clear that $*, \downarrow, \Downarrow$ induce an action of D_4 on the set of all Leonard systems. Two Leonard systems will be called *relatives* whenever they are in the same orbit of this D_4 action. The relatives of Φ are as follows:

name	relative
Φ	$\left(A; A^*; \{E_i\}_{i=0}^d ; \{E_i^*\}_{i=0}^d\right)$
$\Phi^\downarrow$	$\left(A; A^*; \{E_i\}_{i=0}^d ; \{E_{d-i}^*\}_{i=0}^d\right)$
$\Phi^{\downarrow\downarrow}$	$\left(A; A^*; \{E_{d-i}\}_{i=0}^d ; \{E_{d-i}^*\}_{i=0}^d\right)$
Φ^*	$\left(A^*; A; \{E_i^*\}_{i=0}^d ; \{E_i\}_{i=0}^d\right)$
$\Phi^{\downarrow *}$	$\left(A^*; A; \{E_{d-i}^*\}_{i=0}^d ; \{E_i\}_{i=0}^d\right)$
$\Phi^{\downarrow\downarrow *}$	$\left(A^*; A; \{E_i^*\}_{i=0}^d ; \{E_{d-i}\}_{i=0}^d\right)$
$\Phi^{\downarrow\downarrow *}$	$\left(A^*; A; \{E_{d-i}^*\}_{i=0}^d ; \{E_{d-i}\}_{i=0}^d\right)$

We remark there may be some isomorphisms among the above Leonard systems.

We now define the parameter array of a Leonard system. This array consists of four sequences of scalars: the eigenvalue sequence, the dual eigenvalue sequence, the first split sequence and the second split sequence. The eigenvalue sequence and dual eigenvalue sequence are defined as follows.

Definition 20.4.5 *Let Φ denote the Leonard system from Definition 20.4.2. For $0 \le i \le d$, we let θ_i (resp. θ_i^*) denote the eigenvalue of A (resp. A^*) associated with E_i (resp. E_i^*). We refer to $\theta_0, \theta_1, \ldots, \theta_d$ as the eigenvalue sequence of Φ. We refer to $\theta_0^*, \theta_1^*, \ldots, \theta_d^*$ as the dual eigenvalue sequence of Φ. We observe $\theta_0, \theta_1, \ldots, \theta_d$ are mutually distinct and contained in $\mathbb{K}$. Similarly, $\theta_0^*, \theta_1^*, \ldots, \theta_d^*$ are mutually distinct and contained in $\mathbb{K}$.*

We now define the first split sequence and the second split sequence. Let Φ denote the Leonard system from Definition 20.4.2. In (Terwilliger, 2001), it was shown that there exists scalars $\varphi_1, \varphi_2, \ldots, \varphi_d$ in $\mathbb{K}$ and there exists an isomorphism of $\mathbb{K}$-algebras $\sharp : \mathcal{A} \to \mathrm{Mat}_{d+1}(\mathbb{K})$ such that

$$
A^\sharp = \begin{pmatrix}
\theta_0 & 0 & & & & & \\
1 & \theta_1 & 0 & & & & \\
0 & 1 & \theta_2 & & & & \\
& & & \ddots & \ddots & & \\
& & & & \ddots & \ddots & \\
& & & & & \ddots & \ddots & \theta_{d-1} & 0 \\
& & & & & & 0 & 1 & \theta_d
\end{pmatrix},
$$

$$
A^{*\sharp} = \begin{pmatrix}
\theta_0^* & \varphi_1 & 0 & & & & \\
0 & \theta_1^* & \varphi_2 & & & & \\
0 & 0 & \theta_3^* & & & & \\
& & & \ddots & \ddots & & \\
& & & & \ddots & \ddots & \\
& & & & & \ddots & \ddots & \theta_{d-1}^* & \varphi_d \\
& & & & & & 0 & 0 & \theta_d^*
\end{pmatrix}
$$

$$(20.4.4)$$

where the θ_i, θ_i^* are from Definition 20.4.5. The sequence $\natural, \varphi_1, \varphi_2, \ldots, \varphi_d$ is uniquely determined by Φ. We call the sequence $\varphi_1, \varphi_2, \ldots, \varphi_d$ the first split sequence of Φ. We let $\phi_1, \phi_2, \ldots, \phi_d$ denote the first split sequence of $\Phi^{\Downarrow}$ and call this the second split sequence of Φ. For notational convenience, we define $\varphi_0 = 0$, $\varphi_{d+1} = 0$, $\phi_0 = 0$, $\phi_{d+1} = 0$.

Definition 20.4.6 *Let Φ denote the Leonard system from Definition 20.4.2. By the parameter array of Φ we mean the sequence $(\theta_i, \theta_i^*, i = 0, \ldots, d; \varphi_j, \phi_j, j = 1, \ldots, d)$, where $\theta_0, \theta_1, \ldots, \theta_d$ (resp. $\theta_0^*, \theta_1^*, \ldots, \theta_d^*$) is the eigenvalue sequence (resp. dual eigenvalue sequence) of Φ and $\varphi_1, \varphi_2, \ldots, \varphi_d$ (resp. $\phi_1, \phi_2, \ldots, \phi_d$) is the first split sequence (resp. second split sequence) of Φ.*

The following theorem characterizes Leonard systems in terms of the parameter array.

Theorem 20.4.3 *Let d denote a nonnegative integer and let*

$$\theta_0, \theta_1, \ldots, \theta_d; \qquad\qquad \theta_0^*, \theta_1^*, \ldots \theta_d^*;$$
$$\varphi_1, \varphi_2, \ldots, \varphi_d; \qquad\qquad \phi_1, \phi_2, \ldots, \phi_d$$

denote scalars in $\mathbb{K}$. Then there exists a Leonard system Φ over $\mathbb{K}$ with parameter array $(\theta_i, \theta_i^*, i = 0, \ldots, d; \varphi_j, \phi_j, j = 1, \ldots, d)$ if and only if (i)–(v) hold below.

(i) $\varphi_i \neq 0, \qquad \phi_i \neq 0$ $\hfill (1 \leq i \leq d),$

(ii) $\theta_i \neq \theta_j, \qquad \theta_i^* \neq \theta_j^* \quad$ *if* $i \neq j$, $\hfill (0 \leq i, j \leq d),$

(iii) $\varphi_i = \phi_1 \displaystyle\sum_{h=0}^{i-1} \frac{\theta_h - \theta_{d-h}}{\theta_0 - \theta_d} + (\theta_i^* - \theta_0^*)(\theta_{i-1} - \theta_d)$ $\hfill (1 \leq i \leq d),$

(iv) $\phi_i = \varphi_1 \displaystyle\sum_{h=0}^{i-1} \frac{\theta_h - \theta_{d-h}}{\theta_0 - \theta_d} + (\theta_i^* - \theta_0^*)(\theta_{d-i+1} - \theta_0)$ $\hfill (1 \leq i \leq d),$

(v) *The expressions*

$$\frac{\theta_{i-2} - \theta_{i+1}}{\theta_{i-1} - \theta_i}, \qquad \frac{\theta_{i-2}^* - \theta_{i+1}^*}{\theta_{i-1}^* - \theta_i^*}$$

are equal and independent of i for $2 \leq i \leq d - 1$.

Moreover, if (i)–(v) hold above then Φ is unique up to isomorphism of Leonard systems.

One nice feature of the parameter array is that it is modified in a simple way as one passes from a given Leonard system to a relative.

Theorem 20.4.4 *Let Φ denote a Leonard system with parameter array $(\theta_i, \theta_i^*, i = 0, \ldots, d; \varphi_j, \phi_j, j = 1, \ldots, d)$. Then (i)–(iii) hold below.*

(i) *The parameter array of Φ^* is*
$(\theta_i^*, \theta_i, i = 0, \ldots, d; \varphi_j, \phi_{d-j+1}, j = 1, \ldots, d)$.

(ii) *The parameter array of $\Phi^{\downarrow}$ is*
$\left(\theta_i, \theta_{d-i}^*, i = 0, \ldots, d; \phi_{d-j+1}, \varphi_{d-j+1}, j = 1, \ldots, d\right)$.

(iii) *The parameter array of $\Phi^{\Downarrow}$ is*

$$(\theta_{d-i}, \theta_i^*, i = 0, \ldots, d; \phi_j, \varphi_j, j = 1, \ldots, d).$$

Definition 20.4.7 *Let Φ be as in Definition 20.4.2. Set*

$$a_i = \operatorname{tr}(E_i^* A), \; 0 \le i \le d, \quad x_i = \operatorname{tr}(E_i^* A E_{i-1}^* A), \; 1 \le i \le d. \qquad (20.4.5)$$

For convenience, we take $x_0 = 0$.

Definition 20.4.8 *Let Φ, a_i and x_i be as above. Define a sequence of polynomials $\{P_k(\lambda) : 0 \le k \le d+1\}$, via*

$$P_{-1}(\lambda) = 0, \quad P_0(\lambda) = 1, \qquad (20.4.6)$$

$$\lambda P_i(\lambda) = P_{i+1}(\lambda) + a_i P_i(\lambda) + x_i P_{i-1}(\lambda), \quad 0 \le i \le d. \qquad (20.4.7)$$

It is clear that P_{d+1} is the characteristic polynomial of the Jacobi matrix associated with (20.4.4)–(20.4.5). Indeed

$$P_i(A) E_0^* V = E_i^* V, \quad 0 \le i \le d, \qquad (20.4.8)$$

for any irreducible A-module V. Moreover

$$P_i(A) E_0^* = E_i^* A^i E_0^*, \quad 0 \le i \le d. \qquad (20.4.9)$$

It turns out that

$$P_{d+1}(\lambda) = \prod_{i=0}^{d} (\lambda - \theta_i). \qquad (20.4.10)$$

Analogous to Definition 20.4.8, we define parameters a_i^*, x_i^* by

$$a_i^* = \operatorname{tr}(E_i A^*), \; 0 \le i \le d, \quad x_i^* = \operatorname{tr}(E_i A^* E_{i-1} A^*), \; 1 \le i \le d. \qquad (20.4.11)$$

It turns out that $x_i \ne 0$, $x_i^* \ne 0$, $1 \le i \le d$, and we follow the convention of taking x_0^* as 0. One then defines another system of monic polynomials $\{P_k^*(\lambda) : 0 \le k \le d+1\}$ by

$$P_{-1}^*(\lambda) = 0, \quad P_0^*(\lambda) = 1, \qquad (20.4.12)$$

$$\lambda P_i^*(\lambda) = P_{i+1}^*(\lambda) + a_i^* P_i^*(\lambda) + x_i^* P_{i-1}^*(\lambda), \quad 0 \le i \le d. \qquad (20.4.13)$$

The reader should not confuse the star notation here with the notation for the numerator polynomials of the polynomials in (20.4.6)–(20.4.7). As expected

$$P_{d+1}^*(\lambda) = \prod_{j=0}^{d} (\lambda - \theta_j^*). \qquad (20.4.14)$$

It can be shown that $P_i(\theta_0) \ne 0$, $P_i^*(\theta_0^*) \ne 0$, $0 \le i \le d$. One can prove the following theorem (Terwilliger, 2004).

Theorem 20.4.5 *Let Φ denote a Leonard system. Then*

$$\frac{P_i(\theta_j)}{P_i(\theta_0)} = \frac{P_j^*(\theta_i^*)}{P_j^*(\theta_0^*)}, \quad 0 \le i, j \le d. \qquad (20.4.15)$$

The equations (20.4.15) are called the Askey–Wilson duality.

We now state Leonard's theorem (Leonard, 1982) without a proof. For a proof, see (Terwilliger, 2004).

Theorem 20.4.6 *Let d be a nonnegative integer and assume that we are given monic polynomials $\{P_k : 0 \le k \le d+1\}$, $\{P_k^* : 0 \le k \le d+1\}$ in $\mathbb{K}[\lambda]$ satisfying (20.4.4)–(20.4.5) and (20.4.10)–(20.4.11). Given scalars $\{\theta_j : 0 \le j \le d\}$, $\{\theta_j^* : 0 \le j \le d\}$, satisfying*

$$\theta_j \neq \theta_k, \quad \theta_j^* \neq \theta_k^*, \quad if \ j \neq k, \quad 0 \le j, k \le d,$$
$$P_i(\theta_0) \neq 0, \ 0 \le i \le d, \quad P_i^*(\theta_0^*) \neq 0, \ 0 \le i \le d,$$

and the θ's and θ^*'s are related to P_{d+1} and P_{d+1}^* through (20.4.10) and (20.4.14). If (20.4.15) holds then there exists a Leonard system Φ over $\mathbb{K}$ which has the monic polynomials $\{P_k : 0 \le k \le d+1\}$, dual monic polynomials $\{P_k^* : 0 \le k \le d+1\}$, the eigensequences $\{\theta_j : 0 \le j \le d\}$, and the dual eigensequences $\{\theta_j^* : 0 \le j \le d\}$. The system Φ is unique up to isomorphism of Leonard systems.

20.5 Characterization Theorems

This section describes characterizations of orthogonal polynomials in certain classes of polynomial sequences.

Theorem 20.5.1 ((Meixner, 1934), (Sheffer, 1939)) *Let $\{f_n(x)\}$ be of Sheffer A-type zero relative to $\dfrac{d}{dx}$. Then $\{f_n(x)\}$ is orthogonal if and only if we have one of the following cases:*

(i) $A(t) = e^{-t^2}$, $H(t) = 2t$

(ii) $A(t) = (1 - t)^{-\alpha - 1}$, $H(t) = -t/(1 - t)$

(iii) $A(t) = (1 - t)^{-\beta}$, $H(t) = \ln((1 - t/c)/(1 - t))$

(iv) $A(t) = \left\{ \left(1 - te^{i\phi}\right)\left(1 - te^{-i\phi}\right) \right\}^{-\lambda}$,
     $H(t) = i\operatorname{Log}\left(1 - te^{i\phi}\right) - i\operatorname{Log}\left(1 - te^{-i\phi}\right)$

(v) $A(t) = (1 + t)^N$, $H(t) = \operatorname{Log}\left(\frac{1 - (1-p)t/p}{1+t}\right)$, $N = 1, 2, \ldots$

(vi) $A(t) = e^t$, $H(t) = \ln(1 - t/a)$.

The orthogonal polynomials in cases (i), (ii) and (iii) are Hermite, Laguerre, and Meixner polynomials, respectively. Cases (iv) and (v) correspond to the Meixner–Pollaczek and Krawtchouk polynomials, respectively. Case (vi) corresponds to the Charlier polynomials.

The way to prove Theorem 20.5.1 is to express the coefficients of x^n, x^{n-1}, x^{n-2} in $f_n(x)$ in terms of coefficients of the power series expansions of $H(t)$ and $A(t)$. Then substitute for f_n, f_{n+1}, f_{n-1} in a three-term recurrence relation and obtain necessary conditions for the recursion coefficients. After some lengthy algebraic calculations one finds a set of necessary conditions for the recursion coefficients.

After verifying that the conditions are also sufficient one obtains a complete description of the orthogonal polynomials $\{f_n(x)\}$. This method of proof is typical in the characterization theorems described in this section.

We next state a characterization theorem due to Feldheim and Lanzewizky, (Feldheim, 1941b), (Lanzewizky, 1941).

Theorem 20.5.2 *The only orthogonal polynomials $\{\phi_n(x)\}$ which have a generating function of the type* (13.0.1), *where $F(z)$ is analytic in a neighborhood of $z = 0$ are:*

1. *The ultraspherical polynomials when $F(z) = (1-z)^{-\nu}$,*
2. *The q-ultraspherical polynomials when $F(z) = (\beta z; q)_\infty / (z; q)_\infty$,*

or special cases of them.

Observe that if $\{\phi_n(x)\}$ of Sheffer A-type zero relative to $\dfrac{d}{dx}$, then Theorem 10.1.4 implies

$$\sum_{n=0}^{\infty} \phi_n(x) t^n = A(t) \exp(xH(t)),$$

so that

$$\phi_n(x+y) = \sum_{k=0}^{n} r_k(x) s_{n-k}(y), \tag{20.5.1}$$

with

$$\sum_{k=0}^{\infty} r_k(x) t^k = A_1(t) \exp(xH(t)), \quad \sum_{k=0}^{\infty} s_k(x) t^k = A_2(t) \exp(xH(t)),$$

and $A_1(t) A_2(t) = A(t)$. This led Al-Salam and Chihara to characterize orthogonal polynomials in terms of a functional equation involving a Cauchy convolution as in (20.5.1).

Theorem 20.5.3 ((Al-Salam & Chihara, 1976)) *Assume that $\{r_n(x)\}$ and $\{s_n(x)\}$ are orthogonal polynomials and consider the polynomials $\{\phi_n(x,y)\}$ defined by*

$$\phi_n(x,y) = \sum_{k=0}^{n} r_k(x) s_{n-k}(y).$$

Then $\{\phi_n(x,y)\}$ is a sequence of orthogonal polynomials in x for infinitely many values of y if and only if $\{r_n(x)\}$ and $\{s_n(x)\}$ are Sheffer A-type zero and $\phi_n(x,y) = \phi_n(x+y)$, or $\{r_n(x)\}$, $\{s_n(x)\}$ and $\{\phi_n(x,y)\}$ are Al-Salam–Chihara polynomials.

Al-Salam and Chihara considered the class of polynomials $\{Q_n(x)\}$ with generating functions

$$A(t) \prod_{k=0}^{\infty} \frac{1 - axH\left(tq^k\right)}{1 - bxK\left(tq^k\right)} = \sum_{n=0}^{\infty} Q_n(x) t^n, \tag{20.5.2}$$

with $H(t) = \sum\limits_{n=1}^{\infty} h_n t^n$, $K(t) = \sum\limits_{n=1}^{\infty} k_n t^n$, $h_1 k_1 \neq 0$, and $|a| + |b| \neq 0$.

Theorem 20.5.4 ((Al-Salam & Chihara, 1987)) *The only orthogonal sequences* $\{Q_n(x)\}$ *with generating functions of the type* (20.5.2) *are*

    (i) *The Al-Salam–Chihara polynomials if* $ab = 0$.

    (ii) *The q-Pollaczek polynomials if* $ab \neq 0$.

Al-Salam and Ismail characterized the orthogonal polynomials $\{\phi_n(x)\}$ for which $\{\phi_n(q^n x)\}$ are also orthogonal. Their result is given in the following theorem.

Theorem 20.5.5 ((Al-Salam & Ismail, 1983)) *Let* $\{P_n(x)\}$ *be sequence of symmetric orthogonal polynomials satisfying the three term recurrence relation*

$$xP_n(x) = P_{n+1}(x) + \beta_n P_{n-1}$$
$$P_0(x) = 1, \qquad P_1(x) = cx. \tag{20.5.3}$$

A necessary and sufficient condition for a $\{P_n(q^n x)\}$ *to be also a sequence of orthogonal polynomials is that* $\beta_n = q^{2n-2}$ *and* β_1 *is arbitrary.*

It is clear that the polynomials in Theorem 20.5.4 generalize the Schur polynomials of §13.6.

Two noteworthy characterization theorems will be mentioned in §24.7. They are the Geronimus problem in Problem 24.7.3 and Chihara's classification of all orthogonal Brenke-type polynomials, (Chihara, 1968), (Chihara, 1971).

It is easy to see that the Jacobi, Hermite, and Laguerre polynomials have the property

$$\pi(x)P_n'(x) = \sum_{k=-1}^{1} c_{n,k} P_{n+k}(x), \tag{20.5.4}$$

where $\pi(x)$ is a polynomial of degree at most 2 which does not depend on n.

Theorem 20.5.6 ((Al-Salam & Chihara, 1972)) *The only orthogonal polynomials having the property* (20.5.4) *are the Jacobi, Hermite, and Laguerre polynomials or special or limiting cases of them.*

Askey raised the question of characterizing all orthogonal polynomials satisfying

$$\pi(x)P_n'(x) = \sum_{k=-r}^{s} c_{n,k} P_{n+k}(x), \tag{20.5.5}$$

where $\pi(x)$ is a polynomial independent of n. This problem was solved by Maroni in (Maroni, 1985), (Maroni, 1987), (Bonan et al., 1987).

A q-analogue of (20.5.4) was proved in (Datta & Griffin, 2005). It states that the only orthogonal polynomials satisfying

$$\pi(x)D_q P_n(x) = \sum_{k=-1}^{1} c_{n,k} P_{n+k}(x), \tag{20.5.6}$$

where $\pi(x)$ is a polynomial of degree at most 2, are the big q-Jacobi polynomials or one of its special or limiting cases.

Theorem 20.5.7 *Let $\{\phi_n(x)\}$ be a sequence of orthogonal polynomials. Then the following are equivalent.*

(i) *The polynomials $\{\phi_n(x)\}$ are Jacobi, Hermite, and Laguerre polynomials or special cases of them.*

(ii) *$\{\phi_n(x)\}$ possesses a Rodrigues-type formula*

$$\phi_n(x) = c_n \frac{1}{w} \frac{d^n}{dx^n} \{w(x)\pi^n(x)\},$$

where w is nonnegative on an interval and $\pi(x)$ is a polynomial independent of n.

(iii) *The polynomial sequence $\{\phi_n(x)\}$ satisfies a nonlinear equation of the form*

$$\frac{d}{dx}\{\phi_n(x)\phi_{n-1}(x)\}$$
$$= \{b_n x + c_n\}\phi_n(x)\phi_{n-1}(x) + d_n\phi_n^2(x) + f_n\phi_{n-1}^2(x),$$

$n > 0$, where $\{b_n\}$, $\{c_n\}$, $\{d_n\}$ and $\{f_n\}$ are sequences of constants.

(iv) *Both $\{\phi_n(x)\}$ and $\{\phi'_{n+1}(x)\}$ are orthogonal polynomial sequences.*

From Chapter 4, we know that (i) implies (ii)–(iv). McCarthy proved that (iv) is equivalent to (i). In the western literature the fact that (iv) implies (i) is usually attributed to Hahn, but Geronimus in his work (Geronimus, 1977) attributes this result to (Sonine, 1887). Routh proved that (ii) implies (i); see (Routh, 1884).

Theorem 20.5.8 *(Hahn, 1937) The only orthogonal polynomials whose derivatives are also orthogonal are Jacobi, Laguerre and Hermite polynomials and special cases of them.*

Krall and Sheffer generalized Theorem 20.5.8 by characterizing all orthogonal polynomials $\{P_n(x)\}$ for which the polynomials $\{Q_n(x)\}$,

$$Q_n(x) := \sum_{j=0}^{m} a_j(x) \frac{d^j}{dx^j} P_n(x),$$

are orthogonal. It is assumed that m is independent of n and $a_j(x)$ is a polynomial of degree at most j. They also answered the same question if $Q_n(x)$ is

$$Q_n(x) = \sum_{j=0}^{m} a_j(x) \frac{d^{j+1}}{dx^{j+1}} P_{n+1}(x),$$

under the same assumptions on m and $a_j(x)$. These results are in (Krall & Sheffer, 1965).

A discrete analogue of Theorem 20.5.8 was recently proved in (Kwon et al., 1997). This result is the following

Theorem 20.5.9 *Let $\{\phi_n(x)\}$ and $\{\nabla^r\phi_{n+r}(x)\}$ be orthogonal polynomials. Then $\{\phi_n(x)\}$ are the Hahn polynomials, or special limiting cases of them.*

Polynomial Solutions to Functional Equations

The book (Lesky, 2005) reached me shortly before this book went to press. It is devoted to characterization theorems for classical continuous, discrete, and q-orthogonal polynomials.

21

Some Indeterminate Moment Problems

After a brief introduction to the Hamburger Moment Problem, we study several systems of orthogonal polynomials whose measure of orthogonality is not unique. This includes the continuous q-Hermite polynomials when $q > 1$, the Stieltjes–Wigert polynomials, and the q-Laguerre polynomials. We also introduce a system of biorthogonal rational functions.

21.1 The Hamburger Moment Problem

The moment problem is the problem of finding a probability distribution from its moments. In other words, given a sequence of real numbers $\{\mu_n\}$ the problem is to find a positive measure μ with infinite support such that $\mu_n = \int_{\mathbb{R}} t^n d\mu(t)$. This is called the Hamburger moment problem if there is no restriction imposed on the support of μ. The moment problem is a Stieltjes moment problem if the support of μ is restricted to being a subset of $[0, \infty)$. The Hausdorff moment problem requires μ to be supported in $[0, 1]$. Our principal references on the moment problem are (Akhiezer, 1965), (Shohat & Tamarkin, 1950) and (Stone, 1932). Most of the results in this chapter are from the papers (Ismail & Masson, 1994) and (Ismail, 1993).

When μ is unique the moment problem is called determinate, otherwise it is called indeterminate. Theorem 11.2.1 gives useful criteria for determinacy and indeterminacy of Hamburger moment problems.

Consider the polynomials

$$A_n(z) = z \sum_{k=0}^{n-1} P_k^*(0) P_k^*(z)/\zeta_k, \qquad (21.1.1)$$

$$B_n(z) = -1 + z \sum_{k=0}^{n-1} P_k^*(0) P_k(z)/\zeta_n, \qquad (21.1.2)$$

$$C_n(z) = 1 + z \sum_{k=0}^{n-1} P_k(0) P_k^*(z)/\zeta_n, \qquad (21.1.3)$$

$$D_n(z) = z \sum_{k=0}^{n-1} P_k(0) P_k(z)/\zeta_n. \qquad (21.1.4)$$

The Christofffel–Darboux formula (2.2.4) implies

$$A_{n+1}(z) = \left[P_{n+1}^*(z)P_n^*(0) - P_{n+1}^*(0)P_n^*(z)\right]/\zeta_n, \qquad (21.1.5)$$

$$B_{n+1}(z) = \left[P_{n+1}(z)P_n^*(0) - P_{n+1}^*(0)P_n(z)\right]/\zeta_n, \qquad (21.1.6)$$

$$C_{n+1}(z) = \left[P_{n+1}^*(z)P_n(0) - P_{n+1}(0)P_n^*(z)\right]/\zeta_n, \qquad (21.1.7)$$

$$D_{n+1}(z) = \left[P_{n+1}(z)P_n(0) - P_{n+1}(0)P_n(z)\right]/\zeta_n. \qquad (21.1.8)$$

The above equations and the Casorati determinant (Wronskian) evaluation imply

$$A_n(z)D_n(z) - B_n(z)C_n(z) = 1, \qquad (21.1.9)$$

and letting $n \to \infty$ we get

$$A(z)D(z) - B(z)C(z) = 1. \qquad (21.1.10)$$

Theorem 21.1.1 *In an indeterminate moment problem the polynomials $A_n(z)$, $B_n(z)$, $C_n(z)$, $D_n(z)$ converge uniformly to entire functions $A(z)$, $B(z)$, $C(z)$, $D(z)$, respectively.*

Theorem 21.1.1 follows from Theorem 11.2.1 and the Cauchy–Schwartz inequality.

The Nevanlinna matrix is

$$\begin{pmatrix} A(z) & C(z) \\ B(z) & D(z) \end{pmatrix} \qquad (21.1.11)$$

and its determinant is 1.

Theorem 21.1.2 *Let N denote the class of functions $\{\sigma\}$, which are analytic in the open upper half plane and map it into the lower half plane, and satisfy $\sigma(\bar{z}) = \overline{\sigma(z)}$. Then the formula*

$$\int_{\mathbb{R}} \frac{d\mu(t;\sigma)}{z - t} = \frac{A(z) - \sigma(z)\,C(z)}{B(z) - \sigma(z)\,D(z)}, \quad z \notin \mathcal{R} \qquad (21.1.12)$$

establishes a one-to-one correspondence between the solutions μ of the moment problem and functions σ in the class N, augmented by the constant ∞.

A solution of the moment problem is called N-extremal if σ is a real constant including $\pm\infty$. It is clear from (21.1.12) that all the N-extremal measures are discrete.

When a moment problem is indeterminate then the matrix operator defined by the action of the Jacobi matrix on l^2 has deficiency indices $(1, 1)$. The spectral measures of the selfadjoint extensions of this operator are in one-to-one correspondence with the N-extremal solutions of an indeterminate moment problem. The details are in (Akhiezer, 1965, Chapter 4). For an up-to-date account, the reader will be well-advised to consult Simon's recent article (Simon, 1998).

The following theorem is Theorem 2.3.3 in (Akhiezer, 1965).

Theorem 21.1.3 *Let μ be a solution of an indeterminate moment problem. Then the corresponding orthonormal polynomials form a complete system in $L^2(\mathbb{R}, \mu)$ if and only μ is an N-extremal solution.*

The solutions of an indeterminate moment problem form a convex set whose extreme points are precisely the measures μ which make the polynomials dense in $L^1(\mathbb{R}, \mu)$, see (Akhiezer, 1965, Theorem 2.3.4) after correcting L^2_w to L^1_w, as can be seen from the proof.

Theorem 21.1.4 ((Gabardo, 1992)) *Let* $z = x + iy$, $y > 0$ *and* X *be the class of absolutely continuous solutions to an indeterminate moment problem. Then the entropy integral*

$$\frac{1}{\pi} \int_{\mathbb{R}} \frac{y \ln \mu'(t)}{(x - t)^2 + y^2} \, dt$$

attains its maximum on X *when* μ *satisfies* (21.1.13) *with* $\sigma(z) = \beta - i\gamma$, $\operatorname{Im} z > 0$, $\gamma > 0$.

In general the functions A and C are harder to find than the functions B and D, so it is desirable to find ways of determining measures from (21.1.12) without the knowledge of A and C. The following two theorems achieve this goal.

Theorem 21.1.5 *Let* σ *in* (21.1.12) *be analytic in* $\operatorname{Im} z > 0$, *and assume* σ *maps* $\operatorname{Im} z > 0$ *into* $\operatorname{Im} \sigma(z) < 0$. *If* $\mu(x, \sigma)$ *does not have a jump at* x *and* $\sigma(x \pm i0)$ *exist then*

$$\frac{d\mu(x; \sigma)}{dx} = \frac{\sigma\left(x - i0^+\right) - \sigma\left(x + i0^+\right)}{2\pi i \, |B(x) - \sigma\left(x - i0^+\right) D(x)|^2}. \tag{21.1.13}$$

Proof The inversion formula (1.2.8)–(1.2.9) implies

$$\frac{d\mu(x; \sigma)}{dx} = \frac{1}{2\pi i} \left[\frac{A(x) - \sigma\left(x - i0^+\right) C(x)}{B(x) - \sigma\left(x - i0^+\right) D(x)} - \frac{A(x) - \sigma\left(x + i0^+\right) C(x)}{B(x) - \sigma\left(x + i0^+\right) D(x)} \right]$$

which equals the right-hand side of (21.1.13), after the application of the identity (21.1.10). ☐

Theorem 21.1.5 is due to (Berg & Christensen, 1981) and (Ismail & Masson, 1994).

Corollary 21.1.6 ((Berg & Christensen, 1981)) *Let* $\gamma > 0$. *The indeterminate moment has a solution* μ *with*

$$\mu'(x) = \frac{\gamma/\pi}{\gamma^2 B^2(x) + D^2(x)}. \tag{21.1.14}$$

Proof In Theorem 21.1.5 choose σ as $\sigma(z) = -i\gamma$ for $\operatorname{Im} z > 0$, and $\sigma(\bar{z}) = \overline{\sigma(z)}$. ☐

Theorem 21.1.7 ((Ismail & Masson, 1994)) *Let* $F(z)$ *denote either side of* (21.1.12). *If* F *has an isolated pole singularity at* $z = u$ *then*

$$\operatorname{Res}[F(z) \text{ at } z = u] = \operatorname{Res}\left[\frac{1}{B(z)\,[B(z) - \sigma(z)D(z)]} \text{ at } z = u \right]. \tag{21.1.15}$$

Proof At a pole $z = u$ of $F(z)$, $\sigma(u) = B(u)/D(u)$, so that

$$A(u) - \sigma(u)C(u) = \frac{A(u)D(u) - B(u)C(u)}{B(u)} = \frac{1}{B(u)},$$

and the theorem follows. $\qquad\qquad\qquad\qquad\qquad\qquad\qquad\qquad\qquad\square$

Theorem 21.1.8 *Assume that an N-extremal measure μ has a point mass at $x = u$. Then*

$$\mu(u) = \left[\sum_{n=0}^{\infty} P_n^2(u)/\zeta_n\right]^{-1}, \qquad (21.1.16)$$

For a proof the reader may consult (Akhiezer, 1965), (Shohat & Tamarkin, 1950).

Theorem 21.1.9 ((Berg & Pedersen, 1994)) *The entire functions A, B, C, D have the same order, type and Phragmén–Lindelöf indicator.*

An example of a moment problem where the orders of A, B, C, D are finite and positive is in (Berg & Valent, 1994). Many of the examples we will study in this chapter have entire functions of order zero. Indeed, the entire functions have the property

$$M(f, r) = \exp\left(c(\ln r)^2\right). \qquad (21.1.17)$$

We propose the following definition.

Definition 21.1.1 *An entire function of order zero is called of q-order ρ if*

$$\rho = \varlimsup_{r \to +\infty} \frac{\ln\ln M(r, f)}{\ln\ln r}. \qquad (21.1.18)$$

If f has q-order ρ, $\rho < \infty$, its q-type is σ, where

$$\sigma = \inf\left\{K : M(f, r) < \exp\left(K(\ln r)^\rho\right)\right\}. \qquad (21.1.19)$$

Moreover, the q-Phragmén–Lindelöf indicator is

$$h(\theta) = \varlimsup_{r \to +\infty} \frac{\ln\left|f\left(re^{i\theta}\right)\right|}{(\ln r)^\rho}, \qquad (21.1.20)$$

if f has q-order ρ, $\rho < \infty$.

A conjecture regarding q-orders, q-types and q-Phragmén–Lindelöf indicators will be formulated in Chapter 24 as Conjecture 24.4.4.

Ramis studied growth of entire function solutions to linear q-difference equations in (Ramis, 1992). He also observed that the property (21.1.18) holds for the functions he encounered but he called ρ, q-type because in the context of difference equations it corresponds to the type of entire function solutions.

At the end of §5.2, we explained how a monic symmetric family of orthogonal polynomials $\{\mathcal{F}_n(x)\}$ gives rise to two families of monic birth and death process polynomials, $\{\rho_n(x)\}, \{\sigma_n(x)\}$. See (5.2.31)–(5.2.38).

Theorem 21.1.10 ((Chihara, 1982)) *Assume that the Hamburger moment problem associated with $\{F_n(x)\}$ is indeterminate. Then the Hamburger moment problem associated with $\{\mathcal{F}_n(x)\}$ is also indeterminate and the Nevanlinna polynomials A_n, B_n, C_n, D_n associated with $\{\mathcal{F}_n(x)\}$ satisfy*

$$A_{2n+1}(z) = -F_n^*\left(z^2\right)/\pi_n, \tag{21.1.21}$$

$$B_{2n+1}(z) = -F_n\left(z^2\right)/\pi_n, \tag{21.1.22}$$

$$C_{2n+1}(z) = \lambda_n \pi_n \left[\frac{F_n^*\left(z^2\right)}{\pi_n} - \frac{F_{n+1}^*\left(z^2\right)}{\pi_{n+1}}\right], \tag{21.1.23}$$

$$D_{2n+1}(z) = \frac{\lambda_n \pi_n}{z} \left[\frac{F_n\left(z^2\right)}{\pi_n} - \frac{F_{n+1}\left(z^2\right)}{\pi_{n+1}}\right]. \tag{21.1.24}$$

In the above formula

$$\pi_n := \prod_{j=0}^{n-1} \frac{\lambda_j}{\mu_{j+1}}, \quad n > 0, \quad \pi_0 := 1.$$

21.2 A System of Orthogonal Polynomials

Recall that the continuous q-Hermite polynomials $\{H_n(x\,|\,q)\}$ satisfy the three term recurrence relation (13.1.1) and the initial conditions (13.1.2). When $q > 1$ the H_n's are orthogonal on the imaginary axis, so we need to renormalize the polynomials in order to make them orthogonal on the real axis. The proper normalization is

$$h_n(x\,|\,q) = i^{-n} H_n(ix\,|\,1/q), \tag{21.2.1}$$

which gives

$$h_0(x\,|\,q) = 1, \qquad h_1(x\,|\,q) = 2x, \tag{21.2.2}$$

$$h_{n+1}(x\,|\,q) = 2x h_n(x\,|\,q) - q^{-n}(1-q^n)\,h_{n-1}(x\,|\,q), \quad n > 0, \tag{21.2.3}$$

and now we assume $0 < q < 1$. Ismail and Masson (Ismail & Masson, 1994) referred to the polynomials $h_n(x\,|\,q)$ as the continuous q^{-1}-Hermite polynomials, or the q^{-1}-Hermite polynomials. Askey was the first to study these polynomials in (Askey, 1989b), where he found a measure of orthogonality for these polynomials. This was shortly followed by the detailed study of Ismail and Masson in (Ismail & Masson, 1994).

The formulas in the rest of this chapter will greatly simplify if we use the change of variable

$$x = \sinh \xi. \tag{21.2.4}$$

Theorem 21.2.1 *The polynomials $\{h_n(x\,|\,q)\}$ have the closed form*

$$h_n(\sinh \xi\,|\,q) = \sum_{k=0}^{n} \frac{(q;q)_n}{(q;q)_k(q;q)_{n-k}} (-1)^k q^{k(k-n)} e^{(n-2k)\xi}. \tag{21.2.5}$$

and the generating function

$$\sum_{n=0}^{\infty} \frac{t^n q^{n(n-1)/2}}{(q;q)_n} h_n(\sinh \xi \,|\, q) = (-te^\xi, te^{-\xi}; q)_\infty. \tag{21.2.6}$$

Proof Substitute from (21.2.1) into (13.1.7) to obtain the explicit representation (21.2.5). Next, multiply (21.2.5) by $t^n q^{n^2/2}/(q;q)_n$ and add for $n \geq 0$. The result after replacing n by $n + k$ is

$$\sum_{n=0}^{\infty} \frac{t^n q^{n^2/2}}{(q;q)_n} h_n(\sinh \xi \,|\, q) = \sum_{n,k=0}^{\infty} \frac{t^{n+k} q^{(n^2+k^2)/2}(-1)^k}{(q;q)_k(q;q)_n} e^{(n-k)\xi}.$$

The right-hand side can now be summed by Euler's formula (12.2.25), and (21.2.6) has been established. □

Corollary 21.2.2 *The polynomials $\{h_n\}$ have the property*

$$h_{2n+1}(0\,|\,q) = 0, \quad \text{and} \quad h_{2n}(0\,|\,q) = (-1)^n q^{-n^2} \left(q;q^2\right)_n. \tag{21.2.7}$$

Corollary 21.2.2 also follows directly from (21.2.2)–(21.2.3).

The result of replacing x by ix and q by $1/q$ in (13.1.17) is the linearization formula

$$q^{\binom{m}{2}} \frac{h_m(x\,|\,q)}{(q;q)_m} q^{\binom{n}{2}} \frac{h_n(x\,|\,q)}{(q;q)_n}$$
$$= \sum_{k=0}^{m\wedge n} \frac{q^{-n-m+\binom{k+1}{2}+\binom{m-k+1}{2}+\binom{n-k+1}{2}}}{(q;q)_k(q;q)_{m-k}(q;q)_{n-k}} h_{m+n-2k}(x\,|\,q). \tag{21.2.8}$$

Theorem 21.2.3 *The Poisson kernel (or the q-Mehler formula) for the polynomials $\{h_n(x\,|\,q)\}$ is*

$$\sum_{n=0}^{\infty} h_n(\sinh \xi \,|\, q) h_n(\sinh \eta \,|\, q) \frac{q^{n(n-1)/2}}{(q;q)_n} t^n \tag{21.2.9}$$
$$= (-te^{\xi+\eta}, -te^{-\xi-\eta}, te^{\xi-\eta}, te^{\eta-\xi}; q)_\infty \,/\, (t^2/q;q)_\infty.$$

Proof Multiply (21.2.8) by $s^m t^n$ and sum over m and n for $m \geq 0, n \geq 0$. Using (21.2.6) we see that the left side sums to

$$(-se^\xi, se^{-\xi}, -te^\xi, te^{-\xi}; q)_\infty.$$

On the other hand, the right-hand side (after interchanging the m and n sums with the k sum, then replacing m and n by $m + k$ and $n + k$, respectively), becomes

$$\sum_{m,n=0}^{\infty} \frac{s^m t^n}{(q;q)_m(q;q)_n} q^{\binom{m}{2}+\binom{n}{2}} h_{m+n}(x\,|\,q) \sum_{k=0}^{\infty} \frac{(st)^k q^{\binom{k}{2}-k}}{(q;q)_k}.$$

Now Euler's formula (12.2.25) and rearrangement of series reduce the above expression to

$$(-st/q;q)_\infty \sum_{j=0}^\infty h_j(x\,|\,q) \sum_{n=0}^j \frac{s^{j-n}t^n}{(q;q)_n(q;q)_{j-n}} q^{\binom{n}{2}+\binom{j-n}{2}}.$$

At this stage we found it more convenient to set

$$s = te^\eta, \quad t = -te^{-\eta}, \quad \text{and} \quad x = \sinh\xi, \quad y = \sinh\eta. \tag{21.2.10}$$

The above calculations lead to

$$(-te^{\xi+\eta}, te^{\eta-\xi}, te^{\xi-\eta}, -te^{-\xi-\eta};q)_\infty \,/\, (t^2/q;q)_\infty$$
$$= \sum_{j=0}^\infty h_j(\sinh\eta\,|\,q)\, t^j q^{j(j-1)/2} \sum_{n=0}^j \frac{q^{n(n-j)}(-1)^n e^{(j-2n)\xi}}{(q;q)_n(q;q)_{j-n}}.$$

This and (21.2.5) establish the theorem. $\qquad\square$

Technically, the Poisson kernel is the left hand side of (21.2.9) with t replaced by qt.

It is clear from (21.2.2) and (21.2.3) that the corresponding orthonormal polynomials are

$$p_n(x) = q^{n(n+1)/4}h_n(x\,|\,q)/\sqrt{(q;q)_n}. \tag{21.2.11}$$

Theorem 21.2.4 *The moment problem associated with $\{h_n(x\,|\,q)\}$ is indeterminate.*

Proof In view of Theorem 11.2.1, it suffices to show that the corresponding orthonormal polynomials are square summable for a non-real z, that is the series $\sum_{n=0}^\infty |p_n(z)|^2 < \infty$. The series in question is the left-hand side of (21.2.9) with $t = q$ and $\eta = \bar\xi$. $\qquad\square$

The bilinear generating function (21.2.9) can be used to determine the large n asymptotics of $h_n(x\,|\,q)$. To see this let $\xi = \eta$ and apply Darboux's asymptotic method. The result is

$$h_n^2(\sinh\xi\,|\,q)\, q^{n^2/2}$$
$$= [(-1)^n \left(\sqrt{q}\,e^{2\xi}, \sqrt{q}\,e^{-2\xi}, -\sqrt{q}, -\sqrt{q};q\right)_\infty \tag{21.2.12}$$
$$+ \left(-\sqrt{q}\,e^{2\xi}, -\sqrt{q}\,e^{-2\xi}, \sqrt{q}, \sqrt{q};q\right)_\infty][1 + o(1)],$$

as $n \to \infty$. Thus

$$h_{2n}^2(\sinh\xi\,|\,q) = \frac{q^{-2n^2}}{2} [\left(-\sqrt{q}\,e^{2\xi}, -\sqrt{q}\,e^{-2\xi}, \sqrt{q}, \sqrt{q};q\right)_\infty \tag{21.2.13}$$
$$+ \left(\sqrt{q}\,e^{2\xi}, \sqrt{q}\,e^{-2\xi}, -\sqrt{q}, -\sqrt{q};q\right)_\infty][1 + o(1)], \quad n \to \infty,$$

and

$$h_{2n+1}^2(\sinh\xi\,|\,q) = \frac{q^{-(2n+1)^2/2}}{2} [\left(-\sqrt{q}\,e^{2\xi}, -\sqrt{q}\,e^{-2\xi}, \sqrt{q}, \sqrt{q};q\right)_\infty \tag{21.2.14}$$
$$- \left(\sqrt{q}\,e^{2\xi}, \sqrt{q}\,e^{-2\xi}, -\sqrt{q}, -\sqrt{q};q\right)_\infty][1 + o(1)], \quad n \to \infty.$$

We shall comment on (21.2.13) and (21.2.14) in the next section.

21.3 Generating Functions

In this section we derive two additional generating functions which are crucial in computing the strong asymptotics of $\{h_n(x \mid q)\}$ using Darboux's method. The application of the method of Darboux requires a generating function having singularities in the finite complex plane. The generating function (21.2.6) is entire so we need to find a generating function suitable for the application of Darboux's method.

Set

$$h_n(x \mid q) = q^{-n^2/4} \left(\sqrt{q}; \sqrt{q}\right)_n s_n(x). \tag{21.3.1}$$

In terms of the s_n's, the recurrence relation (21.2.3) becomes

$$\left(1 - q^{(n+1)/2}\right) s_{n+1}(x) = 2xq^{(n+1/2)/2} s_n(x) - \left(1 + q^{n/2}\right) s_{n-1}(x). \tag{21.3.2}$$

Therefore, the generating function

$$G(x,t) := \sum_{n=0}^{\infty} s_n(x) \, t^n$$

transforms (21.3.2) to the q-difference equation

$$G(x,t) = \frac{1 + 2xq^{1/4}t - t^2 q^{1/2}}{1 + t^2} G\left(x, \sqrt{q}\, t\right).$$

By iterating the above functional equation we find

$$G(x,t) = \frac{\left(t\alpha, t\beta; \sqrt{q}\right)_n}{\left(-t^2; q\right)_n} G\left(x, q^{n/2} t\right).$$

Since $G(x,t) \to 1$ as $t \to 0$ we let $n \to \infty$ in the above functional equation. This establishes the following theorem.

Theorem 21.3.1 *The generating function*

$$\sum_{n=0}^{\infty} s_n(x) t^n = \sum_{n=0}^{\infty} \frac{q^{n^2/4} t^n}{\left(\sqrt{q}; \sqrt{q}\right)_n} h_n(x \mid q) = \frac{\left(t\alpha, t\beta; \sqrt{q}\right)_\infty}{\left(-t^2; q\right)_\infty}, \tag{21.3.3}$$

holds, where

$$\alpha = -\left(x + \sqrt{x^2 + 1}\right) q^{1/4} = -q^{1/4} e^{\xi},$$
$$\beta = \left(\sqrt{x^2 + 1} - x\right) q^{1/4} = q^{1/4} e^{-\xi}. \tag{21.3.4}$$

The t singularities with smallest absolute value of the right side of (21.3.3) are $t = \pm i$. Thus Darboux's method gives

$$s_n(x) = \left[\frac{\left(i\alpha, i\beta; \sqrt{q}\right)_\infty}{2(q; q)_\infty}(-i)^n + \frac{\left(-i\alpha, -i\beta; \sqrt{q}\right)_\infty}{2(q; q)_\infty} i^n\right][1 + o(1)], \tag{21.3.5}$$

as $n \to \infty$.

Theorem 21.3.2 *The large n behavior of the h_n's is described by*

$$h_{2n}(x \mid q) = q^{-n^2} \frac{\left(\sqrt{q}; \sqrt{q}\right)_\infty}{2(q;q)_\infty} (-1)^n$$

$$\times \left[(i\alpha, i\beta; \sqrt{q})_\infty + (-i\alpha, -i\beta; \sqrt{q})_\infty\right] [1 + o(1)],$$

(21.3.6)

$$h_{2n+1}(x \mid q) = iq^{-n^2-n-1/4} \frac{\left(\sqrt{q}; \sqrt{q}\right)_\infty}{2(q;q)_\infty} (-1)^{n+1}$$

$$\times \left[(i\alpha, i\beta; \sqrt{q})_\infty - (-i\alpha, -i\beta; \sqrt{q})_\infty\right] [1 + o(1)].$$

(21.3.7)

We next derive a different generating function which leads to a single term asymptotic term instead of the two term asymptotics in (21.3.6)–(21.3.7). This is achieved because both $h_{2n}(x \mid q)$ and $h_{2n+1}(x \mid q)/x$ are polynomials in x^2 of degree n. The recurrence relation (21.2.3) implies

$$4x^2 h_n(x \mid q) = h_{n+2}(x \mid q) + \left[q^{-n-1}(1+q) - 2\right] h_n(x \mid q)$$

$$+ q^{1-2n} \left(1 - q^n\right) \left(1 - q^{n-1}\right) h_{n-2}(x \mid q).$$

(21.3.8)

From here and the initial conditions (21.2.2) it readily follows that both $\{h_{2n}(\sqrt{x} \mid q)\}$ and $h_{2n+1}(\sqrt{x} \mid q)/\sqrt{x}$ are constant multiples of special Al-Salam–Chihara polynomials, see §15.1, with q replaced by $1/q$. For $0 < p < 1$, Askey and Ismail (Askey & Ismail, 1984) used the normalization

$$v_0(x) = 1, \quad v_1(x) = (a-x)/(1-p),$$

(21.3.9)

$$\left(1 - p^{n+1}\right) v_{n+1}(x) = (a - xp^n) v_n(x) - \left(b - cp^{n-1}\right) v_{n-1}(x), n > 0.$$

(21.3.10)

We used $v_n(x)$ to mean $\{v_n(x; p; a, b, c)\}$. An easy exercise recasts (21.3.9)–(21.3.10) in the form of the generating function (Askey & Ismail, 1984)

$$\sum_{n=0}^{\infty} v_n(x; p; a, b, c)\, t^n = \prod_{n=0}^{\infty} \left[\frac{1 - xtp^n + ct^2 p^{2n}}{1 - atp^n + bt^2 p^{2n}}\right].$$

(21.3.11)

Comparing (21.3.9)–(21.3.10) with (21.2.2)–(21.2.3) we find that

$$v_n \left(\left(4x^2 + 2\right)\sqrt{q}; q^2; q^{1/2} + q^{-1/2}, 1, q\right)$$

$$= \frac{q^{n(n-1/2)}(-1)^n}{(q^2; q^2)_n} h_{2n}(x \mid q).$$

(21.3.12)

Similarly we establish

$$v_n \left(q^{3/2}\left(4x^2 + 2\right); q^2; q^{1/2} + q^{-1/2}, 1, q^3\right)$$

$$= \frac{(-1)^n q^{n(n+1/2)}}{(2x)\,(q^2; q^2)_n} h_{2n+1}(x \mid q).$$

(21.3.13)

Theorem 21.3.3 *We have the generating functions*

$$\sum_{n=0}^{\infty} \frac{h_{2n}(x\,|\,q)}{(q^2;q^2)_n}\, t^n q^{n(n-1/2)}$$

$$= \prod_{n=0}^{\infty} \left[\frac{1 + \left(4x^2 + 2\right) tq^{2n+1/2} + t^2 q^{4n+1}}{1 + (1+q)\, tq^{2n-1/2} + t^2 q^{4n}} \right],$$ (21.3.14)

and

$$\sum_{n=0}^{\infty} \frac{h_{2n+1}(x\,|\,q)}{(q^2;q^2)_n}\, t^n q^{n(n+1/2)}$$

$$= 2x \prod_{n=0}^{\infty} \frac{1 + \left(4x^2 + 2\right) tq^{2n+3/2} + t^2 q^{4n+3}}{1 + (1+q)tq^{2n-1/2} + t^2 q^{4n}}.$$ (21.3.15)

Proof The identifications (21.3.12) and (21.3.13) and the generating function (21.3.11) establish the theorem.   □

The right-hand sides of (21.3.14) and (21.3.15) have only one singularity of small-est absolute value, hence applying Darboux's method leads to a single main term in the asymptotic expansion of h_{2n} and h_{2n+1}. Indeed it is straightforward to derive the following result.

Theorem 21.3.4 *The large n asymptotics of h_{2n} and h_{2n+1} are given by*

$$h_{2n}(x\,|\,q) = \frac{(-1)^n q^{-n^2}}{(q;q^2)_\infty}$$

$$\times \prod_{k=0}^{\infty} \left[1 - \left(4x^2 + 2\right) q^{2k+1} + q^{4k+2}\right] [1 + o(1)], \quad n \to \infty$$ (21.3.16)

and

$$h_{2n+1}(x\,|\,q) = \frac{(-1)^n q^{-n(n+1)}}{(q;q^2)_\infty}\, 2x$$

$$\times \prod_{k=0}^{\infty} \left[1 - \left(4x^2 + 2\right) q^{2k+2} + q^{4k+4}\right] [1 + o(1)], \quad n \to \infty.$$ (21.3.17)

The corresponding orthonormal polynomials are $\left\{ h_n(x\,|\,q)q^{n(n+1)/4} / \sqrt{(q;q)_n} \right\}$. From (21.3.16)–(21.3.17) it is now clear that the sum of squares of absolute values of the orthonormal h_n's converge for every x in the complex plane. This confirms the indeterminacy of the moment problem via Theorem 11.2.1.

We next turn to the numerator polynomials $\{h_n^*(x\,|\,q)\}$. They satisfy (21.2.3) and the initial conditions

$$h_0^*(x\,|\,q) = 0, \quad h_1^*(x\,|\,q) = 2.$$ (21.3.18)

We then have

$$P_n^*(x) = 2^{-n} h_n^*(x\,|\,q).$$ (21.3.19)

Following the renormalization (21.3.1) we let

$$h_n^*(x \mid q) = q^{-n^2/4} \left(\sqrt{q}; \sqrt{q}\right)_n s_n^*(x).$$
(21.3.20)

The s_n^*'s also satisfy (21.3.2), but $s_0^*(x) = 0$, $s_1^*(x) = 2q^{1/4}/\left(1 - \sqrt{q}\right)$.

Theorem 21.3.5 *The polynomials $\{h_n^*(x \mid q)\}$ have the generating function*

$$\sum_{n=0}^{\infty} s_n^*(x) t^n = \sum_{n=0}^{\infty} \frac{q^{-n^2/4} t^n}{\left(\sqrt{q}; \sqrt{q}\right)_n} h_n^*(x \mid q) = 2q^{1/4} t \sum_{n=0}^{\infty} \frac{(t\alpha, t\beta; \sqrt{q})_n}{(-t^2; q)_{n+1}} q^{n/2}.$$
(21.3.21)

Proof The generating function

$$G^*(x, t) = \sum_{n=0}^{\infty} s_n^*(x) t^n$$

transforms the recurrence relation (21.3.2) to

$$G^*(x, t) = \frac{1 + 2xq^{1/4}t - t^2 q^{1/2}}{1 + t^2} G^* \left(x, \sqrt{q}\, t\right) + \frac{2q^{1/4}t}{1 + t^2}.$$

The solution to the above q-difference equation with the initial conditions

$$G^*(x, 0) = 0, \qquad \frac{\partial G^*}{\partial t}(x, t) \bigg|_{t=0} = s_1^*(x)$$

is given by (21.3.21). □

Now Darboux's method gives, as $n \to \infty$,

$$\begin{aligned}
h_n^*(x \mid q) = {} &-q^{(1-n^2)/4} \left(\sqrt{q}; \sqrt{q}\right)_{\infty} i^{n+1} \\
&\times \left[(-1)^{n+1} \, _2\phi_1 \left(i\alpha\sqrt{q}, i\beta\sqrt{q}; -\sqrt{q}; \sqrt{q}, \sqrt{q}\right) \right. \\
&\left. + \, _2\phi_1 \left(-i\alpha\sqrt{q}, -i\beta\sqrt{q}; -\sqrt{q}; \sqrt{q}, \sqrt{q}\right)\right] [1 + o(1)].
\end{aligned}$$
(21.3.22)

In order to simplify the right side of (21.3.22) we need to go back to the recurrence relation in (21.2.3) and obtain separate generating functions for $\{h_{2n}^*(x \mid q)\}$ and $\{h_{2n+1}^*(x \mid q)\}$ as we did for the h_n's. In other words, we need a generating function for v_n^*. Using the recursion (21.3.10) and the initial conditions

$$v_0^*(x; p; a, b, c) = 0, \qquad v_1^*(x; p; a, b, c) = 1/(p - 1),$$
(21.3.23)

we derive the generating function

$$\begin{aligned}
&\sum_{n=0}^{\infty} v_n^*(x; p; a, b, c)\, t^n \\
&= \frac{-t}{1 - at + bt^2} \sum_{n=0}^{\infty} p^n \prod_{j=0}^{n-1} \frac{1 - xtp^j + ct^2 p^{2j}}{1 - atp^{j+1} + bt^2 p^{2j+2}}.
\end{aligned}$$
(21.3.24)

Therefore, for $n \to \infty$, we have

$$v_n^* \left(x; q^2; q^{1/2} + q^{-1/2}, 1, c \right) = \frac{-q^{(n+1)/2}}{1 - q}$$

$$\times \sum_{k=0}^{\infty} \frac{q^{2k}}{(q^2, q^3; q^2)_k} \prod_{j=0}^{k-1} \left[1 - xq^{2j+1/2} + cq^{4j+1} \right] [1 + o(1)].$$

(21.3.25)

It is a routine task to verify that

$$h_{2n}^*(x \,|\, q) = 4x \left(q^2; q^2 \right)_n q^{1/2} (-1)^n$$

$$\times q^{-n(n-1/2)} v_n^* \left(y; q^2, q^{1/2} + q^{-1/2}, 1, q \right).$$

(21.3.26)

Thus (21.3.24) implies

$$h_{2n}^*(x \,|\, q) = 4x(-1)^{n+1} q^{-n^2} \frac{q}{1 - q} \left(q^2; q^2 \right)_{\infty}$$

$$\times {}_2\phi_1 \left(qe^{2\xi}, qe^{-2\xi}; q^3; q^2, q^2 \right) [1 + o(1)].$$

(21.3.27)

To determine the asymptotics of $h_{2n+1}^*(x \,|\, q)$, we set

$$w_n \left(q^{3/2} \left(2 + 4x^2 \right) \right) = (-1)^n q^{n(n+1/2)} h_{2n+1}^*(x \,|\, q) / \left(q^3; q^2 \right)_n.$$

Thus

$$w_0(y) = 2, \quad w_1(y) = 2 \left[q^{-1/2} \left(1 + q^2 \right) - y \right] / \left(1 - q^3 \right),$$

with

$$y := q^{3/2} \left(2 + 4x^2 \right).$$

The recurrence relation (21.3.10) implies

$$\left(1 - q^{2n+3} \right) w_{n+1}(y) + \left[q^{2n} y - (1 + q) / \sqrt{q} \right] w_n(y)$$

$$+ \left(1 - q^{2n} \right) w_{n-1}(y) = 0, \quad n > 0.$$

Consider the generating function

$$W(y, t) := \sum_{n=0}^{\infty} w_n(y) \, t^n.$$

The defining equations of the w_n's lead to the q difference equation

$$W(y, t) \left\{ 1 - t(1 + q) / \sqrt{q} + t^2 \right\} - \left\{ q - ty + q^2 t^2 \right\} W \left(y, q^2 t \right)$$

$$= (1 - q) w_0(y) + \left(1 - q^3 \right) tw_1(y) + t \left[y - (1 + q) / \sqrt{q} \right] w_0(y).$$

Therefore, with $x = \sinh \xi$, y becomes $2q^{3/2} \cosh 2\xi$ and we have

$$W(y, t) = \frac{q \left(1 - t\sqrt{q} e^{2\xi} \right) \left(1 - t\sqrt{q} e^{-2\xi} \right)}{\left(1 - t\sqrt{q} \right) \left(1 - t/\sqrt{q} \right)} W \left(y, q^2 t \right) + \frac{2(1 - q)}{\left(1 - t/\sqrt{q} \right)}.$$

By iteration we obtain

$$\sum_{n=0}^{\infty} w_n(y) \, t^n = \frac{2(1 - q)}{1 - t/\sqrt{q}} \, {}_3\phi_2 \left(te^{2\xi} \sqrt{q}, te^{-2\xi} \sqrt{q}, q^2; tq^{1/2}, tq^{3/2}; q^2, q \right).$$

This establishes, via Darboux's method, the limiting relation

$$w_n(y) = 2(1-q)q^{-n/2} \, {}_2\phi_1\left(qe^{2\xi}, qe^{-2\xi}; q; q^2, q\right)[1 + o(1)],$$

which implies

$$\begin{aligned}
h_{2n+1}^*(x\,|\,q) &= 2\left(q; q^2\right)_\infty (-1)^n q^{-n(n+1)} \\
&\times {}_2\phi_1\left(qe^{2\xi}, qe^{-2\xi}; q; q^2, q\right)[1 + o(1)].
\end{aligned} \tag{21.3.28}$$

21.4 The Nevanlinna Matrix

The monic polynomials are

$$P_n(x) = 2^{-n} h_n(x\,|\,q), \quad P_n^*(x) = 2^{-n} h_n^*(x\,|\,q), \tag{21.4.1}$$

and the coefficients of the monic form of (21.2.3) has the coefficients

$$\alpha_n = 0, \quad \beta_n = \frac{1}{4} q^{-n}\left(1 - q^n\right), \quad n > 0. \tag{21.4.2}$$

hence

$$\zeta_n = 4^{-n}(q; q)_n \, q^{-n(n+1)/2} \tag{21.4.3}$$

Furthermore

$$P_{2n+1}(0) = 0, \quad P_{2n}^*(0) = 0,$$
$$P_{2n}(0) = (-1)^n \beta_1 \beta_3 \cdots \beta_{2n-1}, \quad P_{2n+1}^*(0) = (-1)^n \beta_2 \beta_4 \cdots \beta_{2n}. \tag{21.4.4}$$

Hence

$$\begin{aligned}
P_{2n}(0) &= (-1/4)^n q^{-n^2}\left(q; q^2\right)_n, \\
P_{2n+1}^*(0) &= (-1/4)^n q^{-n(n+1)}\left(q^2; q^2\right)_n
\end{aligned} \tag{21.4.5}$$

Now (21.1.5)–(21.1.8) yield

$$\begin{aligned}
A_{2n+1}(z) &= A_{2n}(z), \quad B_{2n+1}(z) = B_{2n}(z) \\
C_{2n+1}(z) &= C_{2n+2}(z), \quad D_{2n+1}(z) = D_{2n+2}(z),
\end{aligned}$$

and

$$\begin{aligned}
A_{2n}(z) &= \frac{(-1)^{n-1} P_{2n}^*(z)}{\beta_1 \beta_3 \cdots \beta_{2n-1}} = \frac{(-1)^{n-1} q^{n^2}}{(q; q^2)_n} h_{2n}^*(z), \\
B_{2n}(z) &= \frac{(-1)^{n-1} P_{2n}(z)}{\beta_1 \beta_3 \cdots \beta_{2n-1}} = \frac{(-1)^{n-1} q^{n^2}}{(q; q^2)_n} h_{2n}(z\,|\,q), \\
C_{2n+2}(z) &= \frac{(-1)^n P_{2n+1}^*(z)}{\beta_2 \beta_4 \cdots \beta_{2n}} = \frac{(-1)^n q^{n(n+1)}}{2\left(q^2; q^2\right)_n} h_{2n+1}^*(z), \\
D_{2n+2}(z) &= \frac{(-1)^n P_{2n+1}(z)}{\beta_2 \beta_4 \cdots \beta_{2n}} = \frac{(-1)^n q^{n(n+1)}}{2\left(q^2; q^2\right)_n} h_{2n+1}(z\,|\,q).
\end{aligned} \tag{21.4.6}$$

Theorem 21.4.1 *The entire functions A, C, B, and D are given by*

$$A(\sinh \xi) = \frac{4xq\,(q^2;q^2)_\infty}{(1-q)\,(q;q^2)_\infty}\, {}_2\phi_1\left(qe^{2\xi}, qe^{-2\xi}; q^3; q^2, q^2\right) \qquad (21.4.7)$$

$$C(\sinh \xi) = \frac{(q;q^2)_\infty}{(q^2;q^2)_\infty}\, {}_2\phi_1\left(qe^{2\xi}, qe^{-2\xi}; q; q^2, q\right), \qquad (21.4.8)$$

$$B(\sinh \xi) = -\,(q;q^2)_\infty^{-2}\left(qe^{2\xi}, qe^{-2\xi}; q^2\right)_\infty$$
$$= \frac{\vartheta_1(i\xi)}{2iq^{1/4}\,(q;q)_\infty\,(q^2;q^2)_\infty}, \qquad (21.4.9)$$

and

$$D(\sinh \xi) = \frac{x}{(q;q)_\infty}\left(q^2 e^{2\xi}, q^2 e^{-2\xi}; q^2\right)_\infty$$
$$= -\vartheta_4(i\xi)/(q;q)_\infty\,(q;q^2)_\infty, \qquad (21.4.10)$$

respectively.

Proof Apply (21.4.6), (21.3.27)–(21.3.28) and (21.3.16)–(21.3.17). □

Observe that (21.3.16)–(21.3.17) and (21.3.6)–(21.3.7) lead to the identities

$$\left(iq^{1/4}e^\xi, -iq^{1/4}e^{-\xi}; \sqrt{q}\right)_\infty + \left(-iq^{1/4}e^\xi, iq^{1/4}e^{-\xi}; \sqrt{q}\right)_\infty$$
$$= \frac{2\left(qe^{2\xi}, qe^{-2\xi}; q^2\right)_\infty}{\left(\sqrt{q};q\right)_\infty (q;q^2)}. \qquad (21.4.11)$$

and

$$\left(iq^{1/4}e^\xi, -iq^{1/4}e^{-\xi}; \sqrt{q}\right)_\infty - \left(-iq^{1/4}e^\xi, iq^{1/4}e^{-\xi}; \sqrt{q}\right)_\infty$$
$$= -\frac{4iq^{1/4}\sinh\xi\left(q^2 e^{2\xi}, q^2 e^{-2\xi}; q^2\right)_\infty}{\left(\sqrt{q};q\right)_\infty (q;q^2)_\infty}. \qquad (21.4.12)$$

The identities (21.4.11) and (21.4.12) give infinite product representations of the real and imaginary parts of $\left(iq^{1/4}e^\xi, -iq^{1/4}e^{-\xi}; \sqrt{q}\right)_\infty$ and are instances of quartic transformations. When (21.4.11) and (21.4.12) are expressed in terms of theta functions, they give the formulas in (Whittaker & Watson, 1927, Example 1, p. 464).

Similarly comparing (21.3.27)–(21.3.28) with (21.3.22) we discover the quartic transformations

$${}_2\phi_1\left(iq^{1/4}e^\xi, -iq^{1/4}e^{-\xi}; -q^{1/2}; q^{1/2}, q^{1/2}\right)$$
$$-{}_2\phi_1\left(iq^{1/4}e^{-\xi}, -iq^{1/4}e^\xi; -q^{1/2}; q^{1/2}, q^{1/2}\right)$$
$$= \frac{4ixq^{3/4}\,(q^2;q^2)_\infty}{(q-1)\,(q^{1/2};q^{1/2})_\infty}\, {}_2\phi_1\left(qe^{2\xi}, qe^{-2\xi}; q^3; q^2, q^2\right), \qquad (21.4.13)$$

and

$$2\phi_1\left(iq^{1/4}e^\xi, -iq^{1/4}e^{-\xi}; -q^{1/2}; q^{1/2}, q^{1/2}\right)$$
$$+ 2\phi_1\left(iq^{1/4}e^{-\xi}, -iq^{1/4}e^\xi; -q^{1/2}; q^{1/2}, q^{1/2}\right) \tag{21.4.14}$$
$$= \frac{2\,(q; q^2)_\infty}{(q^{1/2}; q^{1/2})_\infty}\, 2\phi_1\left(qe^{2\xi}, qe^{-2\xi}; q; q^2, q\right).$$

The quartic transformations (21.4.13)–(21.4.14) first appeared in (Ismail & Masson, 1994).

21.5 Some Orthogonality Measures

We now discuss the N-extremal measures. Recall that the N-extremal measures are discrete and are supported at the zeros of $B(x) - \sigma D(x)$, σ being a constant in $[-\infty, +\infty]$. These zeros are all real and simple. It is interesting to note that the h_n's are symmetric, that is $h_n(-x) = (-1)^n h_n(x)$, but the masses of the extremal measures are symmetric about the origin only when $\sigma = 0, \pm\infty$. This is so because the Stieltjes transform $\int_{\mathbb{R}} \frac{d\mu(t)}{x - t}$ of a normalized symmetric measure $(d\mu(-t) = d\mu(t))$ is always an odd function of x but it is clear from (21.4.7)–(21.4.10) that $A(x)$ and $D(x)$ are odd functions but $B(x)$ and $C(x)$ are even functions.

Let $\{x_n(\sigma)\}_{-\infty}^\infty$ be the zeros of $B(x) - \sigma D(x)$ arranged in increasing order

$$\cdots < x_{-n}(\sigma) < x_{-n+1}(\sigma) < \cdots < x_n(\sigma) < x_{n+1}(\sigma) < \cdots . \tag{21.5.1}$$

The zeros of $D(x)$ are $\{x_n(-\infty)\}_{-\infty}^\infty$ and are labeled as

$$\cdots < x_{-2}(-\infty) < x_{-1}(-\infty) < x_0(-\infty) = 0 < x_1(-\infty) < \cdots .$$

In general $x_n(\sigma)$ is a real analytic strictly increasing function of σ and increases from $x_n(-\infty)$ to $x_{n+1}(-\infty)$ as σ increases from $-\infty$ to $+\infty$. Furthermore the sequences $\{x_n(\sigma_1)\}$ and $\{x_n(\sigma_2)\}$ interlace when $\sigma_1 \neq \sigma_2$. This is part of Theorem 2.13, page 60 in (Shohat & Tamarkin, 1950). A proof is in (Shohat & Tamarkin, 1950), see Theorem 10.41, pp. 584–589.

Lemma 21.5.1 *The function $B(z)/D(z)$ is increasing on any open interval whose end points are consecutive zeros of $D(z)$.*

Proof This readily follows from (12.6.6) and (21.4.9)–(21.4.10). □

The graph of $B(x)/D(x)$ resembles the graph of the cotangent function, so for $\sigma \in (-\infty, \infty)$ define $\eta = \eta(\sigma)$ as the unique solution of

$$\sigma = B(\sinh \eta)/D(\sinh \eta), \quad 0 = x_0(-\infty) < \sinh \eta < x_1(-\infty). \tag{21.5.2}$$

We define $\eta(\pm\infty)$ by

$$\eta(-\infty) = 0, \quad \sinh(\eta(\infty)) = x_1(-\infty) = x_0(\infty) = \left(q^{-1} - q\right)/2.$$

With the above choice of σ

$$\int_{\mathbb{R}} \frac{d\mu(y)}{\sinh \xi - y} = \frac{A(\sinh \xi)D(\sinh \eta) - B(\sinh \eta)C(\sinh \xi)}{B(\sinh \xi)D(\sinh \eta) - B(\sinh \eta)D(\sinh \xi)}, \qquad (21.5.3)$$

for $\xi \notin \mathbb{R}$.

Theorem 21.5.2 *We have the infinite product representation*

$$B(\sinh \xi)D(\sinh \eta) - B(\sinh \eta)D(\sinh \xi)$$
$$= \frac{-1}{2a(q;q)_\infty} \left(ae^\xi, -ae^{-\xi}, -qe^\xi/a, qe^{-\xi}/a; q\right)_\infty, \qquad (21.5.4)$$

where η and σ are related by (21.5.2) and

$$a = e^{-\eta}. \qquad (21.5.5)$$

Proof Apply (21.4.9)–(21.4.10) to see that the above cross product is

$$\left[2iq^{1/4}(q;q)_\infty^2 \left(q, q^2; q^2\right)_\infty\right]^{-1} \left[\vartheta_1(i\xi)\vartheta_4(i\eta) - \vartheta_1(i\eta)\vartheta_4(i\xi)\right].$$

The product formula, (12.6.5) reduces the left-hand side of (21.5.4) to

$$\frac{\vartheta_2(i(\xi + \eta)/2)\vartheta_3(i(\xi + \eta)/2)\vartheta_1(i(\xi - \eta)/2)\vartheta_4(i(\xi\eta)/2)}{2i\sqrt{q}\,(q;q)_\infty^3 \,(q^2, -q, -q^2; q^2)_\infty^2}.$$

The infinite product representations (12.6.1) and (12.6.5) simplify the above expression to the desired result. $\qquad \square$

The orthogonality relation for the h_n's is

$$\int_{\mathbb{R}} h_m(x \mid q)h_n(x \mid q)\,d\mu(x) = q^{-n(n+1)/2}\,\delta_{m,n}. \qquad (21.5.6)$$

Theorem 21.5.3 *The N-extremal measures are parametrized by a parameter a, such that*

$$q < a < 1, \quad \sinh \eta = \frac{1}{2}\,(1/a - a)\,. \qquad (21.5.7)$$

Every such a determines a unique parameter σ given by (21.5.2), and the N-extremal measure is supported on $\{x_n(a) : n = 0, \pm 1, \ldots\}$, with

$$x_n(a) = \frac{1}{2}\left[q^{-n}/a - aq^n\right], \quad n = 0, \pm 1, \pm 2, \ldots. \qquad (21.5.8)$$

At $x_n(a)$, $n = 0, \pm 1, \pm 2, \ldots$, the N-extremal measure has mass

$$\mu\left(\{x_n(a)\}\right) = \frac{a^{4n}q^{n(2n-1)}\left(1 + a^2 q^{2n}\right)}{(-a^2, -q/a^2, q; q)_\infty}. \qquad (21.5.9)$$

Proof To determine the poles of the left-hand side of (21.5.3), use (21.2.3) and Theorem 21.5.2. The poles are precisely the sequence $\{x_n(a) : n = 0, \pm 1, \ldots,\}$ given by (21.5.8). To find the mass at $x_n(a)$ let $t = q$, $e^\eta = e^\xi = q^{-n}/a$ in Theorem 21.2.3

and use Theorem 21.1.8. Another way is to apply Theorem 21.1.7 and compute the residue of

$$\frac{D(\sinh\eta)}{B(x)D(\sinh\eta) - D(x)B(\sinh\eta)}$$

at $x = x_n(a)$ directly through the application of Theorem 21.5.2. □

Note that there is no loss of generality in assuming $q \leq a < 1$ in (21.5.7) since the set $\{x_n(a)\}$ is invariant under replacing a by aq^j, for any integer j.

Recall that the measures are normalized to have a total mass equal to unity. Therefore

$$\left(-qa^2, -qa^{-2}, q; q\right)_\infty$$
$$= \sum_{n=0}^\infty a^{4n} q^{n(2n-1)} \frac{\left(1 + a^2 q^{2n}\right)}{1 + a^2} + \sum_{n=1}^\infty a^{-4n} q^{n(2n-1)} \frac{\left(a^2 + q^{2n}\right)}{1 + a^2}$$

$$= \frac{1}{1 + a^2} \sum_{n=0}^\infty a^{4n} q^{n(2n-1)} + \sum_{n=1}^\infty a^{4n+2} q^{n(2n+1)}$$

$$+ \frac{1}{1 + a^2} \sum_{n=-1}^{-\infty} a^{4n+2} q^{n(2n+1)} + \sum_{n=-1}^{-\infty} a^{4n} q^{n(2n-1)}$$

$$= \frac{1}{1 + a^2} \left[\sum_{n=-\infty}^\infty a^{4n} q^{n(2n-1)} + \sum_{n=-\infty}^\infty a^{4n+2} q^{n(2n+1)} \right].$$

Observe that the first and second sums above are the even and odd parts of the series $\sum_{n=-\infty}^\infty \left(a^2/\sqrt{q}\right)^n q^{n^2/2}$, respectively. Therefore $\int_{\mathbb{R}} d\mu(x, \sigma) = 1$ is equivalent to the Jacobi triple product identity

$$\sum_{n=-\infty}^\infty z^n p^{n^2} = \left(p^2, -pz, -p/z; p^2\right)_\infty. \tag{21.5.10}$$

It is important to note that we have not used (21.5.10) in any computations leading to (21.5.9) and, as such, we obtain the Jacobi triple product identity as a by-product of our analysis. This enforces the point that many of the summation theorems for special functions arise from problems in orthogonal polynomials. To illustrate this point further we rewrite the orthogonality relation (21.5.6) in terms of generating functions, cf. (21.2.6). The result is

$$\int_{\mathbb{R}} \prod_{j=1}^2 \left(-t_j e^\xi, t_j e^{-\xi}; q\right)_\infty d\mu(x; \sigma) = \left(-t_1 t_2/q; q\right)_\infty. \tag{21.5.11}$$

Another interesting orthogonality measure correspond to $\sigma(z)$ being a nonreal constant in the upper and lower half planes with $\sigma(\bar{z}) = \overline{\sigma(z)}$. For ζ in the open upper half plane define

$$\sigma(z) = -B(\zeta)/D(\zeta), \quad \text{Im } z > 0, \quad \sigma(\bar{z}) = \overline{\sigma(z)}.$$

The mapping $z = \sinh \eta$ is a one-to-one mapping of the strip

$$D := \{\eta : 0 < \operatorname{Im} \eta < \pi/2\} \cup \{\eta : \operatorname{Im} \eta = \pi/2, \operatorname{Re} \eta \le 0\}, \qquad (21.5.12)$$

onto the half plane $\operatorname{Im} z > 0$. Taking into account that $B(z)$ is an even function and $D(z)$ is an odd function we then rewrite $\sigma(z)$ as

$$\sigma(z) := B(-\sinh \eta)/D(-\sinh \eta), \operatorname{Im} z > 0, \qquad \sigma(\overline{z}) = \overline{\sigma(z)}. \qquad (21.5.13)$$

If we denote the corresponding spectral measure by $d\mu(t; \eta)$ then

$$\frac{A(z)D(\sinh \eta) + C(z)B(\sinh \eta)}{B(z)D(\sinh \eta) + D(z)B(\sinh \eta)} = \int_{\mathbb{R}} \frac{d\mu(t; \eta)}{z - t}, \qquad \eta \in D. \qquad (21.5.14)$$

To find μ first observe the left-hand side of (21.5.14) has no poles as can be seen from Theorem 21.5.2. Thus μ is absolutely continuous and Theorem 21.1.5 yields (Ismail & Masson, 1994)

$$\frac{d\mu(x; \eta)}{dx} = \frac{B(\sinh \eta)D(\sinh \overline{\eta}) - B(\sinh \overline{\eta})D(\sinh \eta)}{2\pi i \, |B(x)D(-\sinh \eta) - D(x)B(-\sinh \eta)|^2}.$$

After applying (21.5.4) and some simplifications we obtain

$$\frac{d\mu(x; \eta)}{dx} = \frac{e^{2\eta_1} \sin \eta_2 \cosh \eta_1}{\pi}$$
$$\times \frac{(q, -qe^{2\eta_1}, -qe^{-2\eta_1}; q)_\infty \left|(qe^{2i\eta_2}; q)_\infty\right|^2}{\left|(e^{\xi+\eta}, -e^{\eta-\xi}, -qe^{\xi-\eta}, qe^{-\xi-\eta}; q)_\infty\right|^2}, \qquad (21.5.15)$$

with $x = \sinh \xi$, $\eta = \eta_1 + i\eta_2$. The orthogonality relation (21.5.6) establishes the q-beta integral (Ismail & Masson, 1994)

$$\int_{\mathbb{R}} \frac{\prod_{j=1}^{2} \left(-t_j e^\xi, t_j e^{-\xi}; q\right)_\infty}{\left|(e^{\xi+\eta}, -e^{\eta-\xi}, -qe^{\xi-\eta}, qe^{-\xi-\eta}; q)_\infty\right|^2} \cosh \xi \, d\xi \qquad (21.5.16)$$
$$= \frac{\pi e^{-2\eta_1} \left(-t_1 t_2/q; q\right)_\infty}{\sin \eta_2 \cosh \eta_1 \left(q, -qe^{2\eta_1}, -qe^{-2\eta_1}\right)_\infty \left|(qe^{2i\eta_2})_\infty\right|^2},$$

with $\eta_1, \eta_2 \in \mathbb{R}$.

21.6 Ladder Operators

The parameterization here is $x = \sinh \xi$, so we define

$$\check{f}(z) = f\left((z - z^{-1})/2\right), \qquad x = (z - z^{-1})/2, \qquad (21.6.1)$$

and the analogues of the Askey–Wilson operator $\mathcal{D}_q$ and the averaging operator $\mathcal{A}_q$ are

$$(\mathcal{D}_q f)(x) = \frac{\check{f}(q^{1/2}z) - \check{f}(q^{-1/2}z)}{(q^{1/2} - q^{-1/2})\left[(z + z^{-1})/2\right]}, \qquad (21.6.2)$$

$$(\mathcal{A}_q f)(x) = \frac{1}{2}\left[\check{f}\left(q^{1/2}z\right) + \check{f}\left(q^{-1/2}z\right)\right], \qquad (21.6.3)$$

respectively, with $x = \left(z - z^{-1}\right)/2$. So, we may think of z as e^ξ. The product rule for $\mathcal{D}_q$ is

$$\mathcal{D}_q f \, g = \mathcal{A}_q f \, \mathcal{D}_q g + \mathcal{A}_q g \, \mathcal{D}_q f \qquad (21.6.4)$$

The analogue of the inner product (16.1.1) is

$$\langle f, g \rangle = \int_{\mathcal{R}} f(x) \, \overline{g(x)} \frac{d\,x}{\sqrt{1+x^2}}. \qquad (21.6.5)$$

Theorem 21.6.1 *Let* $f, g \in L^2 \left(\mathbb{R}, \left(1 + x^2\right)^{-1/2}\right)$ *then*

$$\langle \mathcal{D}_q f, g \rangle = - \left\langle f, \sqrt{1+x^2} \, \mathcal{D}_q \left(g(x) \left(1 + x^2\right)^{-1/2}\right) \right\rangle. \qquad (21.6.6)$$

Proof We have

$$\left(q^{1/2} - q^{-1/2}\right) \langle \mathcal{D}_q f, g \rangle = \int_0^\infty \frac{\check{f}\left(q^{1/2}u\right) - \check{f}\left(q^{-1/2}u\right)}{\left(u^2 + 1\right)/2} \, \check{g}(u) \, du$$

$$= \int_0^\infty \frac{\check{f}(u) \, \check{g}\left(q^{-1/2}u\right)}{\left(q^{-1/2}u^2 + q^{1/2}\right)/2} \, du - \int_0^\infty \frac{\check{f}(u) \, \check{g}\left(q^{1/2}u\right)}{\left(q^{1/2}u^2 + q^{-1/2}\right)/2} \, du,$$

which implies the result. $\qquad\qquad\square$

Applying $\mathcal{D}_q$ and $\mathcal{A}_q$ to the generating function (21.2.6) we obtain

$$\mathcal{D}_q \, h_n(x \,|\, q) = 2\frac{1 - q^n}{1 - q} \, q^{(1-n)/2} h_{n-1}(x \,|\, q), \qquad (21.6.7)$$

$$\mathcal{A}_q \, h_n(x \,|\, q) = q^{n/2} h_n(x \,|\, q) + x(1 - q^n) \, q^{-n/2} \, h_{n-1}(x \,|\, q). \qquad (21.6.8)$$

It can be verified from (21.6.7)–(21.6.8) and the defining relations (21.2.2)–(21.2.3) that $y = h_n(x \,|\, q)$ solves

$$q^{1/2} \left(1 + 2x^2\right) \mathcal{D}_q^2 y + \frac{4q}{q-1} \, x \mathcal{A}_q \mathcal{D}_q y = \lambda y, \qquad (21.6.9)$$

with $\lambda = \lambda_n$,

$$\lambda_n = -\frac{4q(1 - q^n)}{(1 - q)^2}. \qquad (21.6.10)$$

Theorem 21.6.2 *Assume that a polynomial* p_n *of degree* n *satisfies* (21.6.9). *Then* $\lambda = \lambda_n$ *and* p_n *is a constant multiple of* h_n.

Proof Let $p_n(x) = \sum_{k=0}^n c_k \, h_k(x \,|\, q)$ and substitute in (21.6.9) then equate coefficients of $h_k(x \,|\, q)$ to find that $c_k \left(\lambda_k - \lambda\right) = 0$. Since $c_n \neq 0$, $\lambda = \lambda_n$. The monotonicity of the λ's proves that $c_k = 0$, $0 \leq k < n$ and the theorem follows. $\qquad\square$

Theorem 21.6.3 *Consider the eigenvalue problem:*

$$\frac{1}{w(x)} \mathcal{D}_q \left(p(x)\mathcal{D}_q y \right) = \lambda y, \tag{21.6.11}$$

$$y, p\mathcal{D}_q y \in L^2 \left(\left(1 + x^2\right)^{-1/2} \right), \tag{21.6.12}$$

for $p(x) \geq 0$, $w(x) > 0$ for all $x \in \mathcal{R}$. The eigenvalues of this eigenvalue problem are real. The eigenfunctions corresponding to distinct eigenvalues are orthogonal with respect to w.

Proof If λ is a complex eigenvalue with eigenfunction y then $\bar{y}$ is an eigenfunction with eigenvalue $\bar{\lambda}$. Moreover

$$(\lambda - \bar{\lambda}) \int_{\mathcal{R}} y(x)\overline{y(x)}\, w(x)\, dx$$

$$= \left\langle \mathcal{D}_q \left(p\mathcal{D}_q y \right), \sqrt{1 + x^2}\, y \right\rangle - \left\langle y\sqrt{1 + x^2}, \mathcal{D}_q \left(p\mathcal{D}_q y \right) \right\rangle$$

$$= - \left\langle p\mathcal{D}_q y, \sqrt{1 + x^2}\, \mathcal{D}_q y \right\rangle + \left\langle \mathcal{D}_q y \sqrt{1 + x^2}, p\mathcal{D}_q y \right\rangle = 0,$$

hence λ is real. If y_1 and y_2 are eigenfunctions with eigenvalues λ_1 and λ_2 then it follows that

$$(\lambda_1 - \lambda_2) \int_{\mathcal{R}} y_1(x)\overline{y_2(x)}\, w(x)\, dx$$

$$= \left\langle \mathcal{D}_q \left(p\mathcal{D}_q y_1 \right), \sqrt{1 + x^2}\, y_2 \right\rangle - \left\langle y_1 \sqrt{1 + x^2}, \mathcal{D}_q \left(p\mathcal{D}_q y_2 \right) \right\rangle$$

$$= - \left\langle p\mathcal{D}_q y_1, \sqrt{1 + x^2}\, \mathcal{D}_q y_2 \right\rangle + \left\langle \mathcal{D}_q y_1 \sqrt{1 + x^2}, p\mathcal{D}_q y_2 \right\rangle = 0,$$

and the theorem follows. $\qquad\qquad\square$

Theorem 21.6.4 *The h_n's are orthogonal with respect to the weight functions*

$$w_1(x) = \frac{\left(1 + x^2\right)^{-1/2}}{(-qe^{2\xi}, -qe^{-2\xi}; q)_\infty}, \tag{21.6.13}$$

$$w_2(x) = \exp\left(\frac{2}{\ln q} \left[\ln\left(x + \sqrt{x^2 + 1} \right) \right]^2 \right), \tag{21.6.14}$$

and

$$w_3(x; a) := \frac{1}{(ae^\xi, \bar{a}e^\xi, -qe^\xi/a, -qe^\xi/\bar{a}; q)_\infty}$$

$$\times \frac{1}{(-ae^{-\xi}, -\bar{a}e^{-\xi}, qe^{-\xi}/a, qe^{-\xi}/\bar{a}; q)_\infty}, \quad \operatorname{Im} a \neq 0. \tag{21.6.15}$$

Proof It suffices to prove that the h_n's satisfy (21.6.11) with $\lambda = \lambda_n$, as given by (21.6.10) and $p = w = w_1$, or $p = w = w_2$. Equivalently this is equivalent to showing that

$$\frac{1}{w_j(x)} \mathcal{D}_q w_j(x) = \frac{4qx}{q - 1}, \quad \frac{1}{w_j(x)} \mathcal{A}_q w_j(x) = q^{1/2} \left(2x^2 + 1 \right), \tag{21.6.16}$$

$j = 1, 2, 3$, which follow by direct calculations. $\qquad\square$

It is important to observe that the indeterminacy of the moment problem is manifested in the fact that the equations

$$\frac{1}{w(x)} \mathcal{D}_q w(x) = \frac{4\,qx}{q-1}, \quad \frac{1}{w(x)} \mathcal{A}_q w(x) = q^{1/2}\left(2x^2 + 1\right) \qquad (21.6.17)$$

and the supplementary condition $w(x) > 0$ on $\mathbb{R}$ do not determine the weight function w uniquely.

The weight function w_1 was found by Askey in (Askey, 1989b) while w_3 is the weight function in (21.5.15) and is due to Ismail and Masson, (Ismail & Masson, 1994). The weight function w_2 is in (Atakishiyev, 1994).

21.7 Zeros

We now give estimates on the largest zero of h_n and derive a normalized asymptotic formula.

Theorem 21.7.1 *All the zeros of $h_N(x \mid q)$ belong to*

$$\left(-\sqrt{q^{-N} - 1}, \ \sqrt{q^{-N} - 1}\right).$$

Proof In view of (21.2.3) we may apply Theorem 7.2.3 and compare $\left(q^{-N} - 1\right) / \left(4x^2\right)$ with the chain sequence $1/4$ and establish the theorem. $\qquad\square$

To see that $q^{-n/2}$ is the correct order of magnitude of the largest zero we consider the polynomials $\left\{h_n\left(q^{-n/2}y \mid q\right)\right\}$.

Theorems 21.7.2–21.7.4 are from (Ismail, 2005a).

Theorem 21.7.2 *Let*

$$x = \sinh \xi_n, \quad with \quad e^{\xi_n} = yq^{-n/2}. \qquad (21.7.1)$$

Then

$$\lim_{n \to \infty} \frac{q^{n^2/2}}{y^n} h_n(x \mid q) = \sum_{k=0}^{\infty} \frac{(-1)^k q^{k^2}}{(q;q)_k} \left(\frac{1}{y^2}\right)^k. \qquad (21.7.2)$$

Proof Substitute $e^{\xi_n} = yq^{-n/2}$ in (21.2.5) and justify interchanging the summation and limiting processes. $\qquad\square$

The special case $a = 0$ of Theorem 13.6.7 shows that the function

$$A_q(z) = \sum_{k=0}^{\infty} \frac{(-1)^k q^{k^2}}{(q;q)_k} z^k \qquad (21.7.3)$$

has only positive simple zeros whose only accumulation point is $z = +\infty$. The function $A_q(z)$ plays the role played by the Airy function in the asymptotics of Hermite and Laguerre polynomials. Let

$$x_{n,1}^{(h)} > x_{n,2}^{(h)} > \cdots > x_{n,n}^{(h)} = -x_{n,1}^{(h)}$$

be the zeros of $h_n(x \mid q)$ and let

$$0 < i_1(q) < i_2(q) < \cdots$$

be the zeros of $A_q(x)$.

Theorem 21.7.3 *For fixed j, we have*

$$\lim_{n \to \infty} q^{n/2} x_{n,j}^{(h)} = \frac{1}{\sqrt{i_j(q)}}. \tag{21.7.4}$$

Theorem 21.7.4 *A complete asymptotic expansion of $\{h_n (\sinh \xi_n \mid q)\}$ is given by*

$$\frac{q^{n^2/2}}{y^n} h_n (\sinh \xi_n \mid q) = \sum_{j=0}^{\infty} \frac{q^{j(n+1)}}{(q;q)_j y^{2j}} A_q \left(q^j / y^2 \right), \quad n \to \infty. \tag{21.7.5}$$

Proof Put $e^{\xi_n} = y q^{-n/2}$ in (21.2.5) and write $(q;q)_n/(q;q)_{n-k}$ as $\left(q^{n-k+1};q \right)_\infty / \left(q^{n+1};q \right)_\infty$ then expand it by the q-binomial theorem as

$$\sum_{j=0}^{k} \left(q^{-k};q \right)_j q^{(n+1)j}/(q;q)_j.$$

Interchanging the k and j sums proves the theorem. $\qquad\qquad \square$

It will be of interest to apply Riemann–Hilbert problem techniques to the polynomials $\{h_n(x \mid q)\}$. Deriving uniform asymptotic expansions is an interesting open problem, and its solution will be very useful.

If one wishes to prove the orthogonality of $\{h_n(x \mid q)\}$ with respect to w_1, w_2, w_3 via Theorem 21.6.4, one needs only to evaluate the integrals $\int_{\mathbb{R}} w_j(x)\,dx, j = 1, 2, 3$. For completeness, we include direct proofs of the orthogonality. In view of (21.2.6), it suffices to prove that

$$I_j := \int_{\mathbb{R}} w_j(\sinh \xi) \left(-t e^{\xi}, t e^{-\xi}, -s e^{\xi}, s e^{-\xi}; q \right)_\infty \cosh \xi \, d\xi$$

is a function of st. For I_1 or I_3, let $u = e^{\xi}$. Hence

$$I_1 = \int_0^{\infty} \frac{(-tu, t/u, -su, s/u; q)_\infty}{(-qu^2, -qu^2; q)_\infty} \frac{du}{u}.$$

Write $\int_0^{\infty}$ as $\displaystyle\sum_{n=-\infty}^{\infty} \int_{q^{n+1}}^{q^n}$, then let $u = q^n v$ to get

$$I_1 = \int_q^1 \frac{(-tv, t/v, -sv, s/v; q)_\infty}{(-qv^2, -q/v^2; q)_\infty}$$

$$\times \left[\sum_{-\infty}^{\infty} \frac{(tq^{-n}/v, sq^{-n}/v; q)_n \left(-qv^2; q \right)_{2n}}{(-tv, -sv; q)_n \left(-q^{1-2n}/v^2; q \right)_{2n}} \right] \frac{dv}{v}.$$

The sum in the square brackets can be written in the form

$$\lim_{\epsilon \to 0} {}_6\psi_6 \left(\begin{array}{c} ivq, -ivq, qv/t, qv/s, 1/\epsilon, 1/\epsilon \\ iv, -iv - tv, -sv, -qv^2\epsilon, -qv^2\epsilon \end{array} \middle| q, \frac{st}{q} v^2\epsilon^2 \right).$$

Evaluate the ${}_6\psi_6$ from (12.3.5) and conclude that

$$I_1 = (-st/q, q; q)_\infty \int\limits_q^1 \frac{dv}{v} = \left(\ln q^{-1} \right) (q, -st/q; q)_\infty,$$

which proves that

$$\int\limits_{\mathbb{R}} w_1(x) h_m(x \mid q) h_n(x \mid q)\, dx = (q; q)_\infty \ln q^{-1} (q; q)_n q^{-\binom{n+1}{2}} \delta_{m,n}. \qquad (21.7.6)$$

The evaluation of I_3 is similar and will be omitted.

We use a different idea to demonstrate that w_2 is a weight function for $\{h_n(x \mid q)\}$. The proof given here is new. One can argue by induction that because of the three term recurrence relation (21.2.3) it suffices to prove that $\int_{\mathbb{R}} w_2(x)\, h_n(x \mid q)\, dx = 0$, if $n > 0$. Moreover, because $h_n(-x \mid q) = (-1)^n h_n(x \mid q)$, we only need to evaluate the integral

$$J_n := \int\limits_{\mathbb{R}} w_2(x) h_{2n}(x \mid q)\, dx.$$

Clearly

$$\int\limits_{\mathbb{R}} w_2(\sinh \xi) e^{m\xi} \cosh \xi\, d\xi = \frac{1}{2} \int\limits_{\mathbb{R}} \exp\left(\frac{2}{\ln q} \xi^2 \right) \left[e^{(m+1)\xi} + e^{(m-1)\xi} \right] d\xi$$

$$= \frac{1}{2} \int\limits_{\mathbb{R}} \left\{ \exp\left(\frac{2}{\ln q} \left[\left(\xi + \frac{(m+1)}{4} \ln q \right)^2 - \frac{(m+1)^2}{8} \ln q \right] \right) \right.$$

$$\left. + \exp\left(\frac{2}{\ln q} \left[\left(\xi + \frac{(m-1)}{4} \ln q \right)^2 - \frac{(m-1)^2}{8} \ln q \right] \right) \right\}$$

$$= \frac{1}{2} \sqrt{\frac{\pi}{2} \ln q^{-1}} \left[q^{-(m+1)^2/8} + q^{-(m-1)^2/8} \right].$$

Therefore (21.2.5) yields

$$
\begin{aligned}
J_n &= \sum_{k=0}^{2n} \frac{(q;q)_{2n}}{(q;q)_k (q;q)_{2n-k}} (-1)^k q^{k(k-2n)} \int_{\mathbb{R}} w_2(\sinh \xi) e^{(2n-2k)\xi} \cosh \xi \, d\xi \\
&= \frac{1}{2}\sqrt{\frac{\pi}{2}\ln q^{-1}} \sum_{k=0}^{2n} \frac{(q;q)_{2n} q^{k(k-2n)}}{(q;q)_k (q;q)_{2n-k}} (-1)^k \left[q^{-(n-k+1/2)^2/2} + q^{-(n-k-1/2)^2/2} \right] \\
&= \frac{1}{2}\sqrt{\frac{\pi}{2}\ln q^{-1}} \sum_{k=0}^{2n} \frac{(q^{-2n};q)_k q^{nk+k/2}}{(q;q)_k} \left[q^{(k-n)/2} + q^{(n-k)/2} \right] q^{-\frac{n^2}{2}-\frac{1}{8}} \\
&= \frac{q^{\frac{n^2}{2}-\frac{1}{8}}}{2}\sqrt{\frac{\pi}{2}\ln q^{-1}} \{ q^{-\frac{n}{2}} {}_1\phi_0\left(q^{-2n};\text{---};q,q^{n+1}\right) + q^{\frac{n}{2}} {}_1\phi_0\left(q^{-2n};\text{---};q,q^{n}\right) \} \\
&= q^{-\frac{n^2}{2}-\frac{1}{8}}\sqrt{\frac{\pi}{2}\ln q^{-1}} \, \delta_{n,0}.
\end{aligned}
$$

Finally, (21.2.5) and the above calculations establish

$$
\int_{\mathbb{R}} w_2(x) \, h_m(x \mid q) \, h_n(x \mid q) \, dx = q^{-\binom{n+1}{2}}(q;q)_n \, q^{-1/8}\sqrt{\frac{\pi}{2}\ln q^{-1}} \, \delta_{m,n}.
$$

$$(21.7.7)$$

21.8 The q-Laguerre Moment Problem

The q-Laguerre polynomials are

$$
L_n^{(\alpha)}(x;q) = \frac{(q^{\alpha+1};q)_n}{(q;q)_n} \sum_{k=0}^{n} \begin{bmatrix} n \\ k \end{bmatrix}_q q^{\alpha k + k^2} \frac{(-x)^k}{(q^{\alpha+1};q)_k}, \tag{21.8.1}
$$

which can be written as

$$
L_n^{(\alpha)}(x;q) = \frac{(q^{\alpha+1};q)_n}{(q;q)_n} \sum_{k=0}^{n} \frac{(q^{-n};q)_k}{(q;q)_k} q^{\binom{k+1}{2}} \frac{x^k q^{(\alpha+n)k}}{(q^{\alpha+1};q)_k}. \tag{21.8.2}
$$

The Stieltjes–Wigert polynomials are

$$
S_n(x;q) = \lim_{\alpha \to \infty} L_n^{(\alpha)}\left(xq^{-\alpha};q\right) = \frac{1}{(q;q)_n} \sum_{k=0}^{n} \frac{(q^{-n};q)_k}{(q;q)_k} q^{\binom{k+1}{2}} x^k q^{nk}. \tag{21.8.3}
$$

Theorem 21.8.1 *The q-Laguerre polynomials satisfy the orthogonality relation*

$$
\int_0^{\infty} L_m^{(\alpha)}(x;q) L_n^{(\alpha)}(x;q) \frac{x^\alpha dx}{(-x;q)_\infty} = -\frac{\pi}{\sin(\pi\alpha)} \frac{(q^{-\alpha};q)_\infty}{(q;q)_\infty} \frac{(q^{\alpha+1};q)_n}{q^n (q;q)_n} \delta_{m,n}.
$$

$$(21.8.4)$$

If $\alpha = k$, $k = 0, 1, \ldots,$ the right-hand side of (21.8.4) is interpreted as

$$
\left(\ln q^{-1}\right) q^{-\binom{k+1}{2}-n} \left(q^{n+1};q\right)_k \delta_{m,n}.
$$

Proof For $m \le n$ we find

$$\frac{(q;q)_n}{(q^{\alpha+1};q)_n} \int_0^\infty x^m L_n^{(\alpha)}(x;q) \frac{x^\alpha dx}{(-x;q)_\infty}$$

$$= \sum_{k=0}^n \frac{(q^{-n};q)_k}{(q;q)_k} q^{\binom{k+1}{2}} \frac{q^{k(\alpha+n)}}{(q^{\alpha+1};q)_k} \int_0^\infty \frac{x^{\alpha+k+m}}{(-x;q)_\infty} dx$$

$$= \sum_{k=0}^n \frac{(q^{-n};q)_k}{(q;q)_k} q^{\binom{k+1}{2}} \frac{q^{k(\alpha+n)}}{(q^{\alpha+1};q)_k} \frac{(-1)^{k+m+1}\pi}{\sin(\pi\alpha)} \frac{(q^{-\alpha-m-k};q)_\infty}{(q;q)_\infty}$$

$$= \frac{\pi(-1)^{m+1}(q^{-\alpha-m};q)_\infty}{\sin(\pi\alpha)(q;q)_\infty} {}_2\phi_1\left(\begin{matrix} q^{-n}, q^{\alpha+m+1} \\ q^{\alpha+1} \end{matrix} \,\middle|\, q, q^{n-m} \right),$$

where we used (12.3.6) to evaluate the integral on the second line. The ${}_2\phi_1$ can be evaluated by the Chu–Vandermonde sum. The result is that

$$\int_0^\infty x^m L_n^{(\alpha)}(x;q) \frac{x^\alpha dx}{(-x;q)_\infty} = \frac{\pi(-1)^{m+n+1}(q^{-\alpha-m};q)_\infty}{\sin(\pi\alpha)(q;q)_\infty(q;q)_{n-m}} q^{-mn+\binom{n}{2}}, \quad (21.8.5)$$

hence the integral vanishes for $m < n$. Thus the left-hand side of (21.8.4) is

$$\delta_{mn} q^{\alpha n+n^2} \frac{(-1)^n}{(q;q)_n} \int_0^\infty x^n L_n^{(\alpha)}(x;q) \frac{x^\alpha dx}{(-x;q)_\infty},$$

which simplifies to the right-hand side of (21.8.4) after using (21.8.5). □

The solutions to the moment problem are normalized to have total mass 1. Let w be the normalized weight function

$$w_{QL}(x;\alpha) = -\frac{x^\alpha}{(-x;q)_\infty} \frac{(q;q)_\infty}{(q^{-\alpha};q)_\infty} \frac{\sin(\pi\alpha)}{\pi}, \quad x \in (0,\infty). \quad (21.8.6)$$

and set

$$\mu_n(\alpha) := \int_0^\infty x^n w_{QL}(x;\alpha)\, dx. \quad (21.8.7)$$

A calculation using (12.3.6) gives

$$\mu_n(\alpha) = q^{-\alpha n-\binom{n+1}{2}} \left(q^{\alpha+1};q\right)_n. \quad (21.8.8)$$

It is clear that $w_{QL}(x;\alpha)$ satisfies the functional equation

$$f(qx) = q^\alpha (1+x) f(x), \quad x > 0. \quad (21.8.9)$$

Theorem 21.8.2 ((Christiansen, 2003a)) *Let f be a positive and measurable function on $(0,\infty)$ so that $\int_0^\infty f(x)\, dx = 1$. If f satisfies (21.8.9), then f is the density of an absolutely continuous measure whose moments are given by (21.8.8).*

Proof We claim that all the moments

$$\mu_n := \int_0^\infty x^n f(x)\, dx$$

are finite. To prove this for $n = 1$, note that

$$1 = \mu_0 = \int_0^\infty f(x)\, dx = q \int_0^\infty f(qx)\, dx = q^{\alpha+1} \int_0^\infty (1+x)\, f(x)\, dx.$$

hence μ_1, being equal to $\int_0^\infty (1+x)\, f(x)\, dx - \int_0^\infty f(x)\, dx$, is finite. By induction using

$$\mu_n = q^{n+1} \int_0^\infty x^n f(qx)\, dx = q^{\alpha+n+1} \int_0^\infty x^n (1+x)\, f(x)\, dx,$$

we conclude that μ_{n+1} is finite if μ_n is finite. Further, this gives

$$\mu_{n+1} = \left(q^{-\alpha-n-1} - 1\right) \mu_n,$$

hence μ_n is given by (21.8.8). □

The rest of this section is based on (Ismail & Rahman, 1998).

Theorem 21.8.3 *The q-Laguerre polynomials are orthogonal with respect to the weight function*

$$w_{QL}(x; \alpha, c, \lambda) = \frac{x^{\alpha-c}}{C} \frac{(-\lambda x, -q/\lambda x; q)_\infty}{(-x, -\lambda q^c x, -q^{1-c}/\lambda x; q)_\infty}, \quad x \in (0, \infty), \quad (21.8.10)$$

$\alpha > 0$, $\lambda > 0$, *where C is chosen to make $\int_0^\infty w_{QL}(x; \alpha, c, \lambda)\, dx = 1$.*

Proof Apply (21.8.9) with $f = w_{QL}(x; \alpha, c, \lambda)$. □

The q-Laguerre polynomials satisfy the three-term recurrence relation

$$-xq^{2n+\alpha+1} L_n^{(\alpha)}(x; q) = \left(1 - q^{n+1}\right) L_{n+1}^{(\alpha)}(x; q)$$
$$+ \left(1 - q^{n+\alpha}\right) q L_{n-1}^{(\alpha)}(x; q) - \left[1 + q - q^{n+1} - q^{n+\alpha+1}\right] L_n^{(\alpha)}(x; q). \quad (21.8.11)$$

The monic polynomials are

$$P_n(x) := (-1)^n (q; q)_n q^{-n(n+\alpha)} L_n^{(\alpha)}(x; q). \quad (21.8.12)$$

Theorem 21.8.4 *For $\alpha \neq$ a negative integer and as $n \to \infty$, we have*

$$L_n^{(\alpha)}(x; q) = \sum_{j=0}^m \frac{(-1)^j q^{jn}}{(q; q)_j}\, q^{j(j+1)/2} L_\infty^{(\alpha-j)}(x; q) + O\left(q^{(m+1)n}\right), \quad (21.8.13)$$

where

$$L_\infty^{(\alpha)}(x;q) = \frac{(q^{\alpha+1};q)_\infty}{(q;q)_\infty} \sum_{k=0}^\infty \frac{q^{k(k+\alpha)}(-x)^k}{(q,q^{\alpha+1};q)_k} \qquad (21.8.14)$$

$$= x^{-\alpha/2} J_\alpha^{(2)}\left(2\sqrt{x};q\right).$$

Proof Use (21.8.1) to see that the left-hand side of (21.8.13) is

$$\frac{(q^{\alpha+1};q)_\infty}{(q;q)_\infty} \sum_{k=0}^n \frac{(-x)^k q^{k(k+\alpha)}}{(q,q^{\alpha+1};q)_k} \frac{(q^{n-k+1};q)_\infty}{(q^{\alpha+n+1};q)_\infty}$$

$$= \frac{(q^{\alpha+1};q)_\infty}{(q;q)_\infty} \sum_{j=0}^\infty \frac{q^{j(\alpha+n+1)}}{(q;q)_j} \sum_{k=0}^n \frac{(-x)^k q^{k(k+\alpha)}\left(q^{-k-\alpha};q\right)_j}{(q,q^{\alpha+1};q)_k}.$$

Since

$$\left(q^{-\alpha-k};q\right)_j = \frac{(q^{-\alpha};q)_j\left(q^{\alpha+1};q\right)_k}{(q^{\alpha+1-j};q)_k} q^{-jk},$$

it follows that

$$L_n^{(\alpha)}(x;q) = \frac{(q^{\alpha+1};q)_\infty}{(q;q)_\infty} \sum_{j=0}^\infty \frac{q^{j(\alpha+n+1)}\left(q^{-\alpha};q\right)_j}{(q;q)_j} \sum_{k=0}^n \frac{(-x)^k q^{k(k+\alpha)}}{(q,q^{\alpha-j+1};q)_k} q^{-kj}$$

Replace the k-sum by $\sum_{k\geq0} - \sum_{k\geq n+1}$ and observe that the second sum is $\mathcal{O}\left(q^{n^2}\right)$. After some simplification the theorem follows.   □

An immediate consequence of (21.8.13) is

$$L_n^{(\alpha)}(x;q) = L_\infty^{(\alpha)}(x;q) - \frac{q^{n+1}}{1-q} L_\infty^{(\alpha-1)}(x;q) + \mathcal{O}\left(q^{2n}\right), \qquad (21.8.15)$$

as $n \to \infty$, hence

$$P_n(x) = (-1)^n (q;q)_\infty q^{-n(n+\alpha)}$$

$$\times \left\{ L_\infty^{(\alpha)}(x;q) + \frac{q^{n+1}}{1-q} \left[L_\infty^{(\alpha)}(x;q) - L_\infty^{(\alpha-1)}(x;q) \right] + \mathcal{O}\left(q^{2n}\right) \right\},$$

$$(21.8.16)$$

as $n \to \infty$.

It can be proved that

$$P_n(0) = \left(q^{\alpha+1};q\right)_n (-1)^n q^{-n(n+\alpha)},$$

$$P_n^*(0) = \frac{(-1)^{n-1}}{1-q^\alpha} q^{-n(n+\alpha)+\alpha} \left\{ \left(q^{\alpha+1};q\right)_n - (q;q)_n \right\}.$$

Theorem 21.8.5 *The numerator polynomials* $\left\{ L_n^{*(\alpha)}(x;q) \right\}$ *have the generating function*

$$\sum_{n=0}^{\infty} L_n^{*(\alpha)}(x;q)\, t^n = -\sum_{n=0}^{\infty} \frac{(-x)^n}{(t, q^{-\alpha};q)_{n+1}}$$

$$+\frac{(tq^{\alpha+1}, -t;q)_\infty}{(t;q)_\infty} \sum_{n=0}^{\infty} \frac{(-x)^n}{(t, q^{-\alpha};q)_{n+1}} \sum_{m=0}^{\infty} \frac{(-x)^m}{(t, tq^{\alpha+1};q)_m}. \tag{21.8.17}$$

Proof Denote the left-hand side of (21.8.17) by $P(x,t)$. Since $L_0^{*(\alpha)}(x;q) = 0$, $L_1^{*(\alpha)}(x;q) = -q^{\alpha+1}/(1-q)$, and $\left\{ L_n^{*(\alpha)}(x;q) \right\}$ satisfies (21.8.11) we conclude from (21.8.11) that $P(x,t)$ satisfies

$$\left[1 - t(1+q) + qt^2\right] P(x,t) - \left[1 - t\left(q + q^{\alpha+1}\right) + q^{\alpha+2}t^2\right] P(x, qt)$$
$$+xtq^{\alpha+1} P\left(x, q^2 t\right) = -q^{\alpha+1}t.$$

We set $P(x,t) = f(x,t)/(t;q)_\infty$ so that f satisfies

$$f(x,t) - f(x,qt)\left(1 - tq^{\alpha+1}\right) + xtq^{\alpha+1} f\left(x, q^2 t\right) = -q^{\alpha+1}t\left(q^2 t; q\right)_\infty. \tag{21.8.18}$$

To find f, let $f(x,t) = \sum_{n=0}^{\infty} f_n(x)\, t^n$, and (21.8.18) implies the following recursive property of $\{f_n(x)\}$,

$$f_0(x) = 0,$$

$$f_n(x) = -q^{n+\alpha} \frac{\left(1 + xq^{n-1}\right)}{1 - q^n} f_{n-1}(x) + \frac{(-1)^n q^{\alpha+1}}{(q;q)_n} q^{(n-1)(n+2)/2}, \tag{21.8.19}$$

$n > 0$. The change of variables

$$g_n(x) := \frac{(-1)^n (q;q)_n}{(-x;q)_n q^{\alpha n}} q^{-\binom{n+1}{2}} f_n(x),$$

transforms (21.8.19) to

$$g_n(x) := g_{n-1}(x) + \frac{q^{\alpha - \alpha n}}{(-x;q)_n},$$

which by telescopy yields

$$f_n(x) = \frac{(-qx;q)_{n-1}}{(q;q)_n} (-1)^n q^{\alpha n + \binom{n+1}{2}} \sum_{k=0}^{n-1} \frac{q^{-\alpha k}}{(-qx;q)_k}, \tag{21.8.20}$$

for $n > 0$. Therefore, $f(x,t) = \sum_{n=0}^{\infty} f_{n+1}(x)\, t^{n+1}$ and we have

$$f(x,t) = -q^\alpha t \sum_{n=0}^{\infty} \sum_{k=0}^{n} \frac{t^n (-qx;q)_n (-1)^n}{(q;q)_{n+1}(-qx;q)_k} q^{\alpha(n-k)+\binom{n+2}{2}}. \tag{21.8.21}$$

Using

$$\sum_{k=0}^{n} \frac{z^k}{(a;q)_k} = \lim_{b \to 0} \left[{}_2\phi_1\left(\begin{matrix} q, b \\ a \end{matrix} \middle| q, z \right) - \frac{(b;q)_{n+1}}{(a;q)_{n+1}} z^{n+1} {}_2\phi_1\left(\begin{matrix} bq^{n+1}, b \\ aq^{n+1} \end{matrix} \middle| q, z \right) \right],$$

and the Heine transformation (12.5.2) we find that

$$\sum_{k=0}^{n} \frac{z^k}{(a;q)_k} = \frac{1}{(a;q)_\infty} \left[\sum_{k=0}^{\infty} \frac{(-a)^k q^{k(k-1)/2}}{(q;q)_k (1-zq^k)} - z^{n+1} \sum_{k=0}^{\infty} \frac{(-aq^{n+1})^k q^{k(k-1)/2}}{(q;q)_k (1-zq^k)} \right].$$

By (21.8.21) we express $f(x,t)$ in the form

$$\begin{aligned}
f(x,t) = -q^\alpha t \sum_{n=0}^{\infty} &\frac{(-tq^\alpha)^n q^{\binom{n+2}{2}}}{(q;q)_{n+1} (-xq^{n+1};q)_\infty} \\
\times &\left[\sum_{k=0}^{\infty} \frac{x^k q^{k(k+1)/2}}{(q;q)_k (1-q^{k-\alpha})} - q^{-\alpha(n+1)} \sum_{k=0}^{\infty} \frac{x^k q^{k(n+1)+\binom{k+1}{2}}}{(q;q)_k (1-q^{k-\alpha})} \right].
\end{aligned} \tag{21.8.22}$$

We now evaluate the n sum. The first series is

$$\begin{aligned}
-q^\alpha t \sum_{n=0}^{\infty} \frac{(-tq^\alpha)^n q^{\binom{n+2}{2}}}{(q;q)_{n+1} (-xq^{n+1};q)_\infty} &= \sum_{n=0}^{\infty} \frac{(-tq^\alpha)^{n+1}}{(q;q)_{n+1}} q^{\binom{n+2}{2}} \sum_{j=0}^{\infty} \frac{(-xq^{n+1})^j}{(q;q)_j} \\
&= \sum_{j=0}^{\infty} \frac{(-x)^j}{(q;q)_j} \sum_{n=1}^{\infty} \frac{(-tq^\alpha)^n}{(q;q)_n} q^{\binom{n+1}{2}+nj} \\
&= \sum_{j=0}^{\infty} \frac{(-x)^j}{(q;q)_j} \left[\sum_{n=0}^{\infty} \frac{(-tq^\alpha)^n}{(q;q)_n} q^{\binom{n+1}{2}+nj} - 1 \right] \\
&= \sum_{j=0}^{\infty} \frac{(-x)^j}{(q;q)_j} \left[(tq^{\alpha+j+1};q)_\infty - 1 \right] \\
&= (tq^{\alpha+1};q)_\infty \sum_{j=0}^{\infty} \frac{(-x)^j}{(q,tq^{\alpha+1};q)_j} - \frac{1}{(-x;q)_\infty}.
\end{aligned}$$

The second n-sum in (21.8.22) is

$$\sum_{n=0}^{\infty} \frac{(-t)^{n+1} q^{k(n+1)+\binom{n+2}{2}}}{(q;q)_{n+1} (-xq^{n+1};q)_\infty},$$

which is the same as the first series, but $t \to tq^{k-\alpha}$. Therefore

$$\sum_{n=0}^{\infty} \frac{(-t)^{n+1} q^{k(n+1)+\binom{n+2}{2}}}{(q;q)_{n+1} (-xq^{n+1};q)_\infty} = (qt;q)_\infty \sum_{j=0}^{\infty} \frac{(-x)^j}{(q;q)_j (qt;q)_{j+k}} - \frac{1}{(-x;q)_\infty}.$$

Replacing the n sum in (21.8.22) by the expressions derived above, we get

$$\begin{aligned}
f(x,t) = \sum_{j,k=0}^{\infty} &\frac{(-1)^j x^{j+k} (tq^{\alpha+j+1};q)_\infty q^{\binom{k+1}{2}}}{(q;q)_j (q;q)_k (1-q^{k-\alpha})} \\
- \sum_{j,k=0}^{\infty} &\frac{(-1)^j x^{j+k} (tq^{k+j+1};q)_\infty q^{\binom{k+1}{2}}}{(q;q)_j (q;q)_k (1-q^{k-\alpha})}.
\end{aligned}$$

Upon setting $n = j + k$ and replacing j by $n - k$ in the second double series above we find that

$$\sum_{j,k=0} \frac{(-1)^j x^{j+k} \left(tq^{k+j+1};q\right)_\infty q^{\binom{k+1}{2}}}{(q;q)_j (q;q)_k (1 - q^{k-\alpha})}$$

$$= \sum_{n=0}^\infty \frac{(-x)^n \left(tq^{n+1};q\right)_\infty}{(q;q)_n (1 - q^{-\alpha})} {}_2\phi_1 \left(\begin{array}{c} q^{-n}, q^{-\alpha} \\ q^{1-\alpha} \end{array} \middle| q, q^{n+1} \right)$$

$$= \sum_{n=0}^\infty \frac{(-x)^n \left(tq^{n+1};q\right)_\infty}{(q^{-\alpha};q)_{n+1}} = (t;q)_\infty \sum_{n=0}^\infty \frac{(-x)^n}{(t, q^{-\alpha};q)_{n+1}},$$

where the q-Gauss theorem was used. Therefore

$$f(x,t) = \left(tq^{\alpha+1};q\right)_\infty \sum_{j=0}^\infty \frac{(-x)^j}{(q, tq^{\alpha+1};q)_j} \sum_{k=0}^\infty \frac{x^k q^{\binom{k+1}{2}}}{(q;q)_k (1 - q^{k-\alpha})}$$

$$- (t;q)_\infty \sum_{n=0}^\infty \frac{(-x)^n}{(t, q^{-\alpha};q)_{n+1}}.$$

On the other hand

$$\sum_{k=0}^\infty \frac{x^k q^{\binom{k+1}{2}}}{(q;q)_k (1 - q^{k-\alpha})} = \frac{1}{1 - q^{-\alpha}} \lim_{b\to\infty} {}_2\phi_1 \left(\begin{array}{c} q^{-\alpha}, b \\ q^{1-\alpha} \end{array} \middle| q, -\frac{xq}{b} \right)$$

$$= \frac{1}{1 - q^{-\alpha}} \lim_{b\to\infty} \frac{(-x;q)_\infty}{(-xq/b;q)_\infty} {}_2\phi_1 \left(\begin{array}{c} q, q^{1-\alpha}/b \\ q^{1-\alpha} \end{array} \middle| q, -x \right)$$

$$= (-x;q)_\infty \sum_{n=0}^\infty \frac{(-x)^n}{(q^{-\alpha};q)_{n+1}},$$

where we used (12.5.3). This proves that

$$f(x,t) = \left(tq^{\alpha+1}, -x;q\right)_\infty \sum_{n=0}^\infty \frac{(-x)^n}{(q^{-\alpha};q)_{n+1}} \sum_{m=0}^\infty \frac{(-x)^m}{(q, tq^{\alpha+1};q)_m}$$

$$- \sum_{m=0}^\infty \frac{\left(tq^{m+1};q\right)_\infty (-x)^m}{(q^{-\alpha};q)_{m+1}} \tag{21.8.23}$$

and (21.8.17) follows. $\qquad\qquad\square$

Since $f(x,t)$ is an entire function of t, we derive the partial fraction decomposition

$$\frac{1}{(t;q)_m} = \sum_{j=0}^{m-1} \frac{(-1)^j q^{j(j+1)/2}}{(q;q)_j (q;q)_{m-1-j}} \frac{1}{1 - tq^j},$$

and apply Darboux's method to establish the complete asymptotic expansion

$$L_n^{*(\alpha)}(x;q) = \frac{1}{(q;q)_\infty}$$

$$\times \left\{ \sum_{j=0}^m \frac{q^{jn} (-1)^j q^{j(j+1)/2}}{(q;q)_j} f\left(y, q^{-j}\right) + O\left(q^{n(m+1)}\right) \right\}. \tag{21.8.24}$$

Clearly (21.8.23) implies

$$f(0,t) = \frac{q^\alpha}{1-q^\alpha}\left[(qt;q)_\infty - \left(tq^{\alpha+1};q\right)_\infty\right].\qquad(21.8.25)$$

It is now straightforward to prove that

$$D(x) = xL_\infty^{(\alpha+1)}(x;q),$$
$$B(x) = \frac{q^\alpha}{1-q^\alpha}\left[\frac{(q;q)_\infty}{(q^{\alpha+1};q)_\infty} - 1\right]xL_\infty^{(\alpha+1)}(x;q) - \frac{(q;q)_\infty}{(q^{\alpha+1};q)_\infty}L_\infty^{(\alpha)}(x;q).$$
$$(21.8.26)$$

Moreover

$$A(x) = \frac{1-q}{(q^\alpha,-x;q)_\infty}\left[\sum_{n=0}^\infty \frac{x^n q^{n(n+1)/2}}{(q;q)_n(1-q^{n-\alpha})}\right.$$
$$\times\left\{(q^\alpha;q)_\infty L_\infty^{(\alpha)}(x;q) + ((q;q)_\infty - (q^{\alpha+1};q)_\infty)L_\infty^{(\alpha-1)}(x;q)\right\}$$
$$-\frac{(q^\alpha,q)_\infty}{(q^{-\alpha};q)_\infty}L_\infty^{(-\alpha)}(x;q)$$
$$\left.- ((q;q)_\infty - (q^{\alpha+1};q)_\infty)\frac{(q;q)_\infty}{(q^{-\alpha};q)_\infty}L_\infty^{(1-\alpha)}(x;q)\right]$$
$$(21.8.27)$$

and

$$C(x) = \frac{x}{(-x;q)_\infty}L_\infty^{(\alpha+1)}(x;q)\sum_{n=0}^\infty \frac{x^n q^{n(n+1)/2}}{(q;q)_n(1-q^{n-\alpha})}$$
$$+\frac{(q;q)_\infty}{(-y,q^{-\alpha};q)}L_\infty^{(\alpha-1)}(x;q).$$
$$(21.8.28)$$

Corollary 21.1.6 gives the orthogonality relation

$$\int_{\mathbb{R}}\frac{L_m^{(\alpha)}(x;q)L_n^{(\alpha)}(x;q)\,dx}{\left[(q;q)_\infty L_\infty^{(\alpha)}(x;q)(q^{\alpha+1};q)_\infty\right]^2 + b^2x^2\left[L_\infty^{(\alpha+1)}(x;q)\right]^2}$$
$$= \frac{\pi(q^{\alpha+1};q)_n}{q^n b(q;q)_n}\delta_{m,n}.$$
$$(21.8.29)$$

In particular, (21.8.8) implies the unusual integrals

$$\int_{\mathbb{R}}\frac{x^n\,dx}{\left[(q;q)_\infty L_\infty^{(\alpha)}(x;q)/(q^{\alpha+1};q)_\infty\right]^2 + b^2x^2\left[L_\infty^{(\alpha+1)}(x;q)\right]^2}$$
$$= \frac{\pi}{b}(q^{\alpha+1};q)_n q^{-\alpha n - n(n+1)/2},$$
$$(21.8.30)$$

$n = 0, 1, \ldots$.

It is clear from (21.8.11) that the q-Laguerre polynomials are birth and death process polynomials, hence their zeros are positive. We will arrange the zeros as

$$x_{n,1}^{(L)}(q,\alpha) > x_{n,2}^{(L)}(q,\alpha) > \cdots > x_{n,n}^{(L)}(q,\alpha).\qquad(21.8.31)$$

The monic recursion coefficients are

$$\alpha_n = q^{-2n-\alpha-1}\left[\left(1-q^{n+1}\right)+q\left(1-q^{n+\alpha}\right)\right], \quad n \geq 0,$$
$$\beta_n = q^{-4n-2\alpha+1}\left(1-q^n\right)\left(1-q^{n+\alpha}\right), \quad n > 0. \tag{21.8.32}$$

Clearly $\alpha_k + \alpha_{k-1}$ and β_k increase with k. Moreover, $\alpha_k - \alpha_{k-1}$ also increases with k. Hence, Theorem 7.2.6 gives

$$x_{n,1}^{(L)}(q,\alpha) \leq q^{-2n-\alpha-1}g(q), \tag{21.8.33}$$

where

$$g(q) = \frac{1}{2}\left[(1+q)\left(1+q^2\right) + \sqrt{\left(1+q-q^2-q^3\right)^2 + 16q^3}\right]. \tag{21.8.34}$$

Theorem 21.8.6 (Plancherel–Rotach asymptotics (Ismail, 2005a)) *The limiting relation*

$$\lim_{n\to\infty} \frac{q^{n^2} L_n^{(\alpha)}\left(x_n(t);q\right)}{(-t)^n} = \frac{1}{(q;q)_\infty} A_q(1/t). \tag{21.8.35}$$

holds uniformly on compact subsets of the complex t-plane, not containing $t = 0$. Moreover

$$x_{n,k}^{(L)}(q,\alpha) = \frac{q^{-2n-\alpha}}{i_k(q)}\{1+o(1)\}, \tag{21.8.36}$$

holds for fixed k.

The asymptotics of the zeros in (21.8.36) is a consequence of (21.8.35).

Proof of Theorem 21.8.6 It is convenient to rewrite (21.8.11) in the form

$$L_n^{(\alpha)}(x;q) = \left(q^{\alpha+1};q\right)_n (-x)^n \sum_{k=0}^n \frac{q^{(n-k)(\alpha+n-k)}(-x)^{-k}}{(q;q)_k\left(q,q^{\alpha+1};q\right)_{n-k}}. \tag{21.8.37}$$

From (21.8.31) we see that the normalization around the largest zero is

$$x = x_n(t) := q^{-2n-\alpha}t, \tag{21.8.38}$$

hence

$$L_n^{(\alpha)}\left(x_n(t);q\right) = (-t)^n \left(q^{\alpha+1};q\right)_n q^{-n^2} \sum_{k=0}^n \frac{q^{k^2}(-t)^{-k}}{(q;q)_k\left(q,q^{\alpha+1};q\right)_{n-k}}, \tag{21.8.39}$$

and the theorem follows from the discrete bounded convergence theorem. $\qquad\square$

We briefly mention some properties of the Stieltjes–Wigert polynomials $\{S_n(x;q)\}$. They are defined in (21.8.3). Many results of the Stieltjes–Wigert polynomials follow from the corresponding results for q-Laguerre polynomials by applying the limit in (21.8.3). The recurrence relation is

$$-q^{2n+1}xS_n(x;q) = \left(1-q^{n+1}\right)S_{n+1}(x;q) + qS_{n-1}(x;q)$$
$$- \left[1+q-q^{n+1}\right]S_n(x;q). \tag{21.8.40}$$

One can derive the generating functions

$$\sum_{n=0}^{\infty} S_n(x;q)t^n = \frac{1}{(t;q)_\infty}\, {}_0\phi_1\left(\begin{array}{c} - \\ 0 \end{array}\bigg| q; -qxt\right),$$

(21.8.41)

$$\sum_{n=1}^{\infty} q^{\binom{n}{2}} S_n(x;q)t^n = (-t;q)_\infty\, {}_0\phi_2\left(\begin{array}{c} - \\ 0,-t \end{array}\bigg| qxt\right).$$

(21.8.42)

A common generalization of (21.8.41)–(21.8.42) is

$$\sum_{n=0}^{\infty} (\gamma;q)_n S_n(x;q)t^n = \frac{(\gamma t;q)_\infty}{(t;q)_\infty}\, {}_1\phi_2\left(\begin{array}{c} \gamma \\ 0,\gamma t \end{array}\bigg| q, -qxt\right).$$

(21.8.43)

With

$$w(x;q) = 1/(-x,-q/x;q)_\infty$$

we have

$$D_q S_n(x;q) = \frac{q}{q-1}\, S_{n-1}\left(q^2 x; q\right),$$

(21.8.44)

$$\frac{1}{w\left(q^{-1}x;q\right)} D_q\left[w(x;q)S_n(x;q)\right] = \frac{1-q^n}{q(1-q)}\, S_{n+1}\left(q^{-1}x;q\right).$$

(21.8.45)

Two orthogonality relations are

$$\int_0^{\infty} S_m(x;q)S_n(x;q)w(x;q)\,dx = -\frac{\ln q}{q^n}\frac{(q;q)_\infty}{(q;q)_n}\delta_{m,n},$$

$$\int_0^{\infty} S_m(x;q)S_n(x;q)\exp\left(-\gamma^2\ln^2 x\right)dx = \frac{\sqrt{\pi}\,q^{-n-1/2}}{\gamma(q;q)_n}\delta_{m,n}$$

(21.8.46)

with $\gamma^2 = -1/(2\ln q)$. The polynomial $S_n(x;q)$ solves the q-difference equation

$$-x\left(1-q^n\right)y(x) = xy(qx) - (x+1)y(x) + y(x/q).$$

(21.8.47)

The Hamburger moment problem is obviously indeterminate.

Theorem 21.8.7 ((Ismail, 2005a)) *The Stieltjes–Wigert polynomials have the explicit representation*

$$q^{n^2}(-t)^{-n}S_n\left(tq^{-2n};q\right) = \frac{1}{(q;q)_\infty}\sum_{s=0}^{\infty}\frac{(-1)^s}{(q;q)_s} q^{s(s+1)/2}q^{sn}A_q\left(q^{-s}/t\right).$$

(21.8.48)

Formula (21.8.48) is an explicit formula whose right-hand side is a complete asymptotic expansion of its left-hand side. Wang and Wong obtained a different expansion using Riemann–Hilbert problem techniques in (Wang & Wong, 2005c).

Let $x_{n,1}^{(SW)} > x_{n,2}^{(SW)} > \cdots > x_{n,n}^{(SW)}$ be the zeros of $S_n(x;q)$. Theorem 21.8.7 implies that

$$x_{n,k}^{(SW)} = \frac{q^{-2n}}{i_k(q)}\{1+o(1)\}, \quad \text{as } n \to \infty,$$

(21.8.49)

for fixed k. The Wang–Wong expansion is uniform and their analogue of (21.8.49) holds even when k depends on n. Let

$$F(x) := \frac{1}{2} \int_0^x \frac{(x - \tau)\, d\tau}{\sqrt{e^\tau - 1}}, \qquad (21.8.50)$$

and

$$x_{n,k}^{(SW)} = q^{1/2}\, 4^{t_{n,k}}\, q^{-(n+1/2)(1+t_{n,k})}. \qquad (21.8.51)$$

Theorem 21.8.8 ((Wang & Wong, 2005c)) *With the above notation we have*

$$F\left(a\left(1 - t_{n,k}\right)\right) = \frac{2}{3} \left(i_k\right)^{3/2} \ln q^{-1} + \mathcal{O}\left((n + 1/2)^{-1/3}\right), \qquad (21.8.52)$$

where $a = \ln\left(4q^{-(n+1/2)}\right)$.

We do not know how to show the equivalence of (21.8.49) and the asymptotics implied by (21.8.52). The derivation of the Wang–Wong formula involves the equilibrium measure of the Stieltjes–Wigert polynomials.

21.9 Other Indeterminate Moment Problems

Borozov, Damashinski and Kulish studied the polynomials $\{u_n(x)\}$ generated by

$$u_0(x) = 1, \quad u_1(x) = 2x,$$

$$u_{n+1}(x) = 2x u_n(x) - \frac{(q^{-n} - q^n)}{(q^{-1} - q)}\, u_{n-1}(x), \quad n > 0, \qquad (21.9.1)$$

with $0 < q < 1$. The corresponding moment problem is indeterminate. The polynomials $\{u_n(x)\}$ give another q-analogue of Hermite polynomials since

$$\lim_{q \to 1^-} u_n(x) = H_n(x).$$

The u_n's do arise in a q-deformed model of the harmonic oscillator. These results are from (Borzov et al., 2000). No simple explicit representations for $u_n(x)$, $A(x)$, $B(x)$, $C(x)$ or $D(x)$ is available. No concrete orthogonality measures are known.

In a private conversation, Dennis Stanton proved

$$u_m(x) u_n(x) = \sum_{k=0}^{m \wedge n} \frac{(q;q)_m (q;q)_n \left(-q^{m+n+1-2k}; q\right)_k}{(q;q)_k (q;q)_{m-k} (q;q)_{n-k}} (-1)^k$$

$$\times q^{-k(2m+2n+1-3k)/2}\, u_{m+n-2k}(x), \qquad (21.9.2)$$

using a combinatorial argument. Set

$$x_n(t) = \frac{1}{2} \frac{q^{-n/2} t}{\sqrt{q^{-1} - q}}, \quad u_n\left(x_n(t)\right) = \frac{q^{-n^2/2} t^n}{(q^{-1} - q)^{n/2}}\, v_n(t) \qquad (21.9.3)$$

and observe that $x_n(t) = x_{n\pm1}\left(tq^{\pm1/2}\right)$. Therefore (21.9.1) becomes

$$v_{n+1}(qt) = v_n(t) - t^{-2} \left(1 - q^{2n}\right) v_{n-1}(t/q). \qquad (21.9.4)$$

If $\lim\limits_{n\to\infty} v_n(t)$ exists, which we strongly believe to be the case, then the Plancherel–Rotach type asymptotic formula

$$\lim_{n\to\infty} v_n(t) = A_q\left(1/t^2\right), \tag{21.9.5}$$

follows from (21.9.4) since $A_q(t)$ satisfies

$$A_q\left(t/\sqrt{q}\right) = A_q(t) - tA_q\left(t/\sqrt{q}\right). \tag{21.9.6}$$

The functional equation (21.9.6) follows from the definition of A_q in (21.7.3).

Chen and Ismail introduced the following variation on $\{h_n(x\,|\,q)\}$ in (Chen & Ismail, 1998b)

$$P_0(x\,|\,q) = 1, \quad P_1(x\,|\,q) = qx,$$
$$xP_n(x\,|\,q) = q^{-n-1}P_{n+1}(x\,|\,q) + q^{-n}P_{n-1}(x\,|\,q), \quad n > 0. \tag{21.9.7}$$

They gave the generating function

$$\sum_{n=0}^{\infty} P_n(x\,|\,q)t^n = \sum_{m=0}^{\infty} \frac{q^{m(m+1)/2}(xt)^m}{(-qt^2;q^2)_{m+1}}, \tag{21.9.8}$$

and established the explicit representation

$$P_n(x\,|\,q) = q^{n(n+1)/2} \sum_{j=0}^{\lfloor n/2 \rfloor} \frac{(q^2;q^2)_{n-j}\,(-1)^j q^{2j(j-n)}}{(q^2;q^2)_j\,(q^2;q^2)_{n-2j}} x^{n-2j}. \tag{21.9.9}$$

Theorem 21.9.1 ((Chen & Ismail, 1998b)) *The elements of the Nevanlinna matrix of $\{P_n(x\,|\,q)\}$ are given by*

$$B(z) = -\sum_{m=0}^{\infty} \frac{(-1)^m z^{2m} q^{2m^2}}{(q^2;q^2)_{2m}}, \quad C(z) = -B(qz),$$
$$D(z) = \sum_{m=0}^{\infty} \frac{(-1)^m z^{2m+1} q^{2m(m+1)}}{(q^2;q^2)_{2m+1}}, \quad A(z) = qD(qz). \tag{21.9.10}$$

In Corollary 21.1.6, choose $\gamma = q^{-1/2}$ to establish the orthogonality relation

$$\int_{\mathbb{R}} \frac{P_m(x\,|\,q)P_n(x\,|\,q)}{|E(ix;q)|^2}\, dx = \pi q^{-1/2}\delta_{m,n}, \tag{21.9.11}$$

where

$$E(z;q) = \sum_{n=0}^{\infty} \frac{z^n q^{n^2/2}}{(q^2;q^2)_n}. \tag{21.9.12}$$

In the notation of (14.1.28), $E(z;q) = E^{(1/4)}\left(z;q^2\right)$.

One can use (21.9.9) to prove the Plancherel–Rotach type asymptotics

$$\lim_{n\to\infty} q^{\binom{n}{2}}t^{-n}P_n\left(tq^{-n}\,|\,q\right) = A_{q^2}\left(1/t^2\right). \tag{21.9.13}$$

Lemma 21.9.2 *With*

$$\beta := \frac{\ln x}{\ln q} - \left\lfloor \frac{\ln x}{\ln q} \right\rfloor,$$

the function $|E(ix;q)|$ *has the asymptotic behavior*

$$|E(ix;q)| \approx \frac{q^{\beta^2/2}}{(-q;q)_\infty} \exp\left(\frac{(\ln x)^2}{2\ln q^{-1}}\right) \sqrt{(q^{2\beta+1}, q^{1-2\beta}; q^2)_\infty}, \qquad (21.9.14)$$

as $x \to +\infty$.

Proof Write $(q^2; q^2)_n$ in (21.9.12) as $(q, -q; q)_n$, then expand

$$1/(-q;q)_n = (-q^{n+1};q)_\infty / (-q;q)_\infty$$

by Euler's theorem. Thus

$$E(z;q) = \frac{1}{(-q;q)_\infty} \sum_{n,j=0}^\infty \frac{z^n q^{n^2/2} q^{j(j-1)/2}}{(q;q)_n (q;q)_j} q^{(n+1)j}$$

$$= \frac{1}{(-q;q)_\infty} \sum_{j=0}^\infty \frac{q^{j(j+1)/2}}{(q;q)_j} \left(-zq^{j+1/2};q\right)_\infty$$

$$= \frac{(-zq^{1/2};q)_\infty}{(-q;q)_\infty} \sum_{j=0}^\infty \frac{q^{j(j+1)/2}}{(q, -q^{1/2}z;q)_j}.$$

Thus for real x, and as $x \to +\infty$, we find

$$|E(ix;q)|^2 \approx \left(ixq^{1/2}, -ixq^{1/2};q\right)_\infty / (-q, -q;q)_\infty. \qquad (21.9.15)$$

Set $x = q^{\beta-n}$, with $0 \le \beta < 1$ and n a positive integer. With this x we get

$$(-q;q)_\infty |E(ix;q)| \approx \left|\left(iq^{\beta-n+1/2};q\right)_\infty\right| = \left|\left(iq^{\beta-n+1/2};q\right)_n \left(iq^{\beta+1/2};q\right)_\infty\right|$$

$$= q^{n(2\beta-n)/2} \left|\left(iq^{\beta+1/2};q\right)_\infty \left(iq^{-\beta+1/2};q\right)_n\right|.$$

Since $n(n - 2/\beta) = (\ln x/\ln q)^2 - \beta^2$, the result follows (21.9.15) and the above calculations. $\qquad \square$

Lemma 21.9.2 explains the term $t^{-n}q^{n^2/2}$ on the left-hand side of (21.9.13). It is basically the square root of the weight function as expected. Indeed, (21.9.14) shows that

$$\left|E\left(iq^{-n}t;q\right)\right| = \frac{q^{\gamma^2/2}\sqrt{(q^{2\gamma+1}, q^{1-2\gamma}; q^2)_\infty}}{(-q;q)_\infty} \exp\left(\frac{\ln^2 t}{2\ln q^{-1}}\right) t^n q^{-n^2/2},$$

where $\gamma (= \beta)$ is the fractional part of $\ln t/\ln q$.

Theorem 21.9.3 *With* $f(z) = A(z)$, $B(z)$, $C(z)$ *or* $D(z)$ *where* A, B, C, D *are as in* (21.9.10) *we have*

$$\lim_{r \to +\infty} \frac{\ln M(r, f)}{\ln^2 r} = \frac{1}{2\ln q^{-1}}.$$

The proof is left to the reader, see (Chen & Ismail, 1998b).

Berg and Valent introduced orthogonal polynomials associated with a birth and death process with rates

$$\lambda_n = (4n+1)(4n+2)^2(4n+3), \quad \mu_n = (4n-1)(4n)^2(4n+1), \quad n \ge 0, \quad (21.9.16)$$

in (Berg & Valent, 1994). In their analysis of the Nevanlinna functions, Berg and Valent used the notations

$$\delta_j(z) = \sum_{n=0}^{\infty} \frac{(-1)^n z^{4n+j}}{(4n+j)!}, \quad j = 0, 1, 2, 3, \quad (21.9.17)$$

$$K_0 = \sqrt{2} \int_0^1 \frac{du}{\sqrt{1-u^4}} = \frac{[\Gamma(1/4)]^2}{4\sqrt{\pi}}, \quad (21.9.18)$$

$$\Delta_j(z) = \frac{K_0}{\sqrt{2}} \int_0^1 \delta_j(uz)\, \mathrm{cn}\,(K_0 u)\, du, \quad j = 0, 1, 2, 3. \quad (21.9.19)$$

Berg and Valent proved that

$$A(z) = \frac{1}{\sqrt{z}} \Delta_2(\zeta) - \frac{4\xi}{\pi} \Delta_0(\zeta), \quad C(z) = \frac{4}{\pi} \Delta_0(\zeta),$$

$$B(z) = -\delta_0(\zeta) - \frac{4}{\pi}\xi\sqrt{z}\delta_2(\zeta), \quad D(z) = \frac{4}{\pi}\sqrt{z}\delta_2(\zeta), \quad (21.9.20)$$

where

$$\zeta = z^{1/4} K_0/\sqrt{2}, \quad \xi = \frac{1}{4\sqrt{2}} \int_0^{K_0} u^2\, \mathrm{cn}\, u\, du. \quad (21.9.21)$$

They also proved that A, B, C, D have common order $1/4$, type $K_0/\sqrt{2}$ and also have the common Phragmén–Londelöff indicator

$$h(\theta) = \frac{K_0}{2} \left(|\cos(\theta/4)| + |\sin(\theta/4)| \right), \quad \theta \in \mathbb{R}. \quad (21.9.22)$$

Chen and Ismail observed that the Berg–Valent polynomials can be considered as the polynomials $\{P_n(x)\}$ in the notation of (5.2.32)–(5.2.33). They then considered the corresponding polynomials $\{\mathcal{F}_n(x)\}$ and applied Theorem 21.1.9 to compute the functions A, B, C, D.

Theorem 21.9.4 ((Chen & Ismail, 1998b)) *Let $\{\mathcal{F}_n(x)\}$ be generated by*

$$\mathcal{F}_0(x) = 1, \quad \mathcal{F}_1(x) = x, \quad (21.9.23)$$

$$\mathcal{F}_{n+1}(x) = x\mathcal{F}_n(x) - 4n^2\left(4n^2 - 1\right)\mathcal{F}_{n-1}(x). \quad (21.9.24)$$

Then the corresponding Hamburger moment problem is indeterminate and the elements of the Nevanlinna matrix are given by

$$A(z) = \Delta_2\left(K_0\sqrt{z/2}\right), \quad B(z) = -\delta_0\left(K_0\sqrt{z/2}\right),$$
$$C(z) = \frac{4}{\pi}\Delta_0\left(K_0\sqrt{z/2}\right), \quad D(z) = \frac{4}{\pi}\delta_2\left(K_0\sqrt{z/2}\right). \tag{21.9.25}$$

In the later work (Ismail & Valent, 1998), the polynomials $\{\mathcal{F}_n(x)\}$ were generalized to the polynomials $\{G_n(x;a)\}$ generated by

$$G_0(x;a) = 1, \quad G_1(x;a) = a - x/2, \tag{21.9.26}$$

$$-xG_n(x;a) = 2(n+1)(2n+1)G_{n+1}(x;a)$$
$$+2n(2n+1)G_{n-1}(x;a) - 2a(2n+1)^2 G_n(x;a), \quad n > 0. \tag{21.9.27}$$

Moreover, Ismail and Valent established the generating functions

$$\sum_{n=0}^{\infty} G_n(x;a)\frac{t^n}{2n+1} = \frac{\sin\left(\sqrt{x}\,g(t)\right)}{\sqrt{xt}}, \tag{21.9.28}$$

$$\sum_{n=0}^{\infty} G_n(x;a)t^n = \left(1 - 2at + t^2\right)^{-1/2}\cos\left(\sqrt{x}\,g(t)\right), \tag{21.9.29}$$

where

$$g(t) = \frac{1}{2}\int_0^t u^{-1/2}\left(1 - 2au + u^2\right)^{-1/2} du. \tag{21.9.30}$$

Since

$$G_n(x;a) = (-1)^n G_n(-x;-a),$$

we will only treat the cases $a > 0$.

Theorem 21.9.5 ((Ismail & Valent, 1998)) *The Hamburger moment problem associated with $\{G_n(x;a)\}$ is determinate if and only if $a \in (1,\infty) \cup (-\infty,-1)$. Moreover when $a > 1$, the continued J-fraction associated with the G_n's converges to $J_1(x;a)$,*

$$J_1(x;a) := -\int_0^{e^{-\phi/2}} \frac{\sin\left(\sqrt{x}\left(g\left(e^{-\phi}\right) - g\left(u^2\right)\right)\right)}{\sqrt{x}\cos\left(\sqrt{x}\,g\left(e^{-\phi}\right)\right)} du, \tag{21.9.31}$$

where $a = \cosh\phi$, $\phi > 0$. When $a > 1$, $\{G_n\}$ satisfy

$$\int_{\mathbb{R}} G_m(x;a)G_n(x;a)\,d\mu(x) = (2n+1)\,\delta_{m,n},$$

where μ is a discrete measure with masses at

$$x_n = \frac{(n+1/2)^2\pi^2}{k\mathbf{K}^2(k)}, \quad n = 0,1,\ldots, \tag{21.9.32}$$

and

$$\mu\left(\{x_n\}\right) = \frac{\pi^2}{k\mathbf{K}^2(k)} \frac{(2n+1)q^{n+1/2}}{1 - q^{2n+1}}, \quad n = 0, 1, \dots . \tag{21.9.33}$$

In (21.9.32) and (21.9.33), $k = e^{-\phi}$ and $\mathbf{K}(k)$ is the complete elliptic integral of the first kind and

$$q = \exp\left(-\pi \mathbf{K}'(k)/\mathbf{K}(k)\right).$$

Theorem 21.9.6 ((Ismail & Valent, 1998)) *Let $a \in (-1, 1)$, so we may assume $a = \cos\phi$, $\phi \in [0, \pi/2]$ and set*

$$\mathbf{K} = \mathbf{K}\left(\cos(\phi/2)\right), \quad \mathbf{K}' = \mathbf{K}\left(\sin(\phi/2)\right). \tag{21.9.34}$$

Then

$$B(x) = \frac{2}{\pi} \ln(\cot(\phi/2)) \sin\left(\sqrt{x}\,\mathbf{K}/2\right) \sinh\left(\sqrt{x}\,\mathbf{K}'/2\right)$$
$$+ \cos\left(\sqrt{x}\,\mathbf{K}/2\right) \cosh\left(\sqrt{x}\,\mathbf{K}'/2\right), \tag{21.9.35}$$
$$D(x) = -\frac{4}{\pi} \sin\left(\sqrt{x}\,\mathbf{K}/2\right) \sinh\left(\sqrt{x}\,\mathbf{K}'/2\right).$$

Corollary 21.9.7 *The function*

$$w(x) = \frac{1}{2} \left[\cos\left(\sqrt{x}\,\mathbf{K}\right) + \cosh\left(\sqrt{x}\,\mathbf{K}'\right)\right]^{-1}, \tag{21.9.36}$$

is a normalized weight function for $\{G_n(x; a)\}$.

It will be interesting to prove

$$\int_{\mathbb{R}} w(x)\, dx = 1, \tag{21.9.37}$$

directly from (21.9.36).

Motivated by the generating functions (21.9.28) and (21.9.29) Ismail, Valent, and Yoon characterized orthogonal polynomials which have a generating fucntion of the type

$$\sum_{n=0}^{\infty} C_n(x; \alpha, \beta)t^n = (1 - At)^\alpha (1 - Bt)^\beta \cos(\sqrt{x}g(t)), \tag{21.9.38}$$

or

$$\sum_{n=0}^{\infty} S_n(x; \alpha, \beta)t^n = (1 - At)^\alpha (1 - Bt)^\beta \frac{\sin\left(\sqrt{x}\,g(t)\right)}{\sqrt{xt}}, \tag{21.9.39}$$

where g is as in (21.9.30).

Theorem 21.9.8 ((Ismail et al., 2001)) *Let $\{C_n(x; \alpha, \beta)\}$ be orthogonal polynomials having the generating function (21.9.38). Then it is necessary that $AB \neq 0$, and $\alpha = \beta = 0$; $\alpha = \beta = -1/2$; $\alpha = \beta + 1/2 = 0$; or $\alpha = \beta - 1/2 = 0$.*

Since $AB \neq 0$, we may rescale t, so there is no loss of generality in assuming $AB = 1$. The case $\alpha = \beta = -1/2$ gives the polynomials $\{G_n(x;a)\}$. When $\alpha = \beta = 0$, it turned out that $C_0 = 0$, and the polynomials $\{C_n(x;0,0)/x\}$ are orthogonal. Indeed these polynomials are related to the polynomials $\{\psi_n(x)\}$ in equation (5.3) of (Carlitz, 1960) via

$$v_n(x;a) = k^{-n}\psi_n(-kx), \quad \text{with } A + B = k + 1/k \qquad (21.9.40)$$

where

$$v_n(x;a) = \frac{2(n+1)}{x} C_{n+1}(x;0,0), \quad a := (A+B)/2. \qquad (21.9.41)$$

In this case we have the generating function

$$\sum_{n=0}^{\infty} v_n(x;a)t^n = \frac{\sin(\sqrt{x}\,g(t))}{\sqrt{xt\,(1-2at+t^2)}}, \qquad (21.9.42)$$

and $\{v_n(x;a)\}$ satisfies the recurrence relation

$$-xv_n(x;a) = 2(n+1)(2n+3)v_{n+1}(x;a) - 8a(n+1)^2 v_n(x;a)$$
$$+2(n+1)(2n+1)v_{n-1}(x;a), \qquad (21.9.43)$$

with the initial conditions $v_0(x;a) = 1$, $v_{-1}(x;a) = 0$. It is clear that

$$v_n(x;a) = S_n(x;-1/2,-1/2), \quad \text{and } v_n(-x;-a) = (-1)^n v_n(x,a). \qquad (21.9.44)$$

Theorem 21.9.9 *The moment problem associated with $\{v_n(x)\}$ is determinate if $a \in (-\infty,-1) \cup (1,\infty)$. If $a > 1$ then the v_n's are orthogonal with respect to a measure μ whose Stieltjes transform is*

$$\int_0^{\infty} \frac{d\mu(t)}{x-t} = \frac{1}{2} \int_0^{e^{-\phi}} \frac{\sin\left(\sqrt{x}\,\left(g(u) - g\left(e^{-\phi}\right)\right)\right)}{\sin\left(\sqrt{x}\,g\left(e^{-\phi}\right)\right)}\,du, \qquad (21.9.45)$$

where $a = \cosh\phi$, $\phi > 0$. Moreover the continued J-fraction

$$\frac{-1/6}{a_0 x + b_0 -} \frac{c_1}{a_1 x + b_1 -} \cdots,$$
$$a_n := \frac{-1/2}{(n+1)(2n+3)}, \quad b_n := \frac{4a(n+1)}{2n+3}, \quad c_n := \frac{2n+1}{2n+3}, \qquad (21.9.46)$$

converges uniformly to the right-hand side of (21.9.45) on compact subsets of $\mathbb{C}$ not containing zeros of $\sin\left(\sqrt{x}\,g\left(e^{-\phi}\right)\right)$. The orthogonality relation of the v_n's is

$$\frac{\pi^4}{k^2 K^4(k)} \sum_{j=1}^{\infty} \frac{j^3 q^j}{1-q^{2j}}\, v_m(x_j;a)\, v_n(x_j;a) = (n+1)\delta_{m,n}. \qquad (21.9.47)$$

In the above, $k = e^{-\phi}$ and $x_n = n^2\pi^2/\left[kK^2(k)\right]$.

The contiued fraction result in Theorem 21.9.9 is from (Rogers, 1907) and has been stated as (94.20) in (Wall, 1948). The orthogonality relation (21.9.47) was stated in (Carlitz, 1960) through direct computation without proving that the moment problem is determinate. Theorem 21.9.9 was proved in (Ismail et al., 2001) and the proofs of the orthogonality relation and the contiued fraction evaluation in (Ismail et al., 2001) are new. The continued fraction (21.9.32) is also in Wall's book (Wall, 1948). Recently, Milne gave interpretations of continued fractions which are Laplace transforms of functions related to elliptic functions, see (Milne, 2002). Moreover, the generating functions (21.9.28) and (21.9.29) satisfy Lamé equation and raises interesting questions about the role of the classes of orthogonal polynomials inves-tigated here and solutions of Lamé equation. David and Gregory Chudnovsky seem to have been aware of this connection, and the interested reader should consult their work (Chudnovsky & Chudnovsky, 1989). A class of polynomials related to elliptic functions has been extensively studied in (Lomont & Brillhart, 2001). The Lomont–Brillhart theory has its origins in Carlitz's papers (Carlitz, 1960) and (Carlitz, 1961), and also in (Al-Salam, 1965). Lomont and Brilhart pointed out in (Lomont & Brill-hart, 2001) that Al-Salam's characterization result in (Al-Salam, 1990) is incorrect because he missed several cases and this is what led to the mammoth work (Lomont & Brillhart, 2001).

A. C. Dixon introduced a family of elliptic functiions arising from the cubic curve

$$x^3 + y^3 - 3\alpha xy = 1.$$

His work (Dixon, 1890) is a detailed study of these functions. He defined the func-tions $\mathrm{sm}(u, \alpha)$ and $\mathrm{cm}(u, \alpha)$ as the solutions to the coupled system

$$s'(u) = c^2(u) - \alpha s(u), \quad c'(u) = -s^2(u) + \alpha c(u), \tag{21.9.48}$$

subject to the initial conditions

$$s(0) = 0, \quad c(0) = 1.$$

In this notation, $s = \mathrm{sm}$, $c = \mathrm{cm}$. In his doctoral dissertation, Eric Conrad estab-lished the continued fraction representations (Conrad, 2002)

$$\frac{1}{x} \int_0^\infty e^{-ux} \, \mathrm{sm}(u, 0) \, du = \frac{1}{x^3 + b_0 -} \frac{a_1}{x^3 + b_1 -} \cdots \frac{a_n}{x^3 + b_n -}, \tag{21.9.49}$$

with

$$a_n := (3n - 2)(3n - 1)^2(3n)^2(3n + 1), \quad b_n := 2(3n + 1)\left[(3n + 1)^2 + 1\right], \tag{21.9.50}$$

$$\frac{1}{2} \int_0^\infty e^{-ux} \, \mathrm{sm}^2(u, 0) \, du = \frac{1}{x^3 + b_0 -} \frac{a_1}{x^3 - b_1 -} \cdots \frac{a_n}{x^3 + b_n -}, \tag{21.9.51}$$

with

$$a_n := (3n - 1)(3n)^2(3n + 1)^2(3n + 2), \quad b_n := 2(3n + 2)\left[(3n + 2)^2 + 1\right], \tag{21.9.52}$$

$$\frac{1}{6x} \int_0^\infty e^{-ux} \, \mathrm{sm}^3(u, 0) \, du = \frac{1}{x^3 + b_0 -} \frac{a_1}{x^3 + b_1 -} \cdots \frac{a_n}{x^3 + b_n -} \cdots, \tag{21.9.53}$$

with

$$a_n := 3n(3n+1)^2(3n+2)^2(3n+3), \quad b_n := 2(3n+3)\left[(3n+3)^2+1\right].$$
(21.9.54)

Another group of continued fractions from (Conrad, 2002) is

$$\frac{1}{x^2}\int_0^\infty e^{-ux}(\mathrm{cm}(u,0))\,du = \frac{1}{x^3+b_0-}\;\frac{a_1}{x^3+b_1-}\cdots\frac{a_n}{x^2+b_n-}\cdots \quad (21.9.55)$$

where

$$a_n := (3n-2)^2(3n-1)^2(3n)^2, \quad b_n := (3n-1)(3n)^2+(3n+1)^2(3n+2),$$
(21.9.56)

$$\frac{1}{x}\int_0^\infty e^{-ux}\,\mathrm{sm}(u,0)\,\mathrm{cm}(u,0)\,du = \frac{1}{x^3+b_0-}\;\frac{a_1}{x^3+b_1-}\cdots\frac{a_n}{x^3+b_n-}\cdots,$$

(21.9.57)

where

$$a_n := (3n-1)^3(3n)^2(3n+1)^2, \quad b_n = 3n(3n+1)^2+(3n+2)^2(3n+3), \quad (21.9.58)$$

and

$$\frac{1}{2}\int_0^\infty e^{-ux}\,\mathrm{sm}^2(u,0)\,\mathrm{cm}(u,0)e^{-ux}\,du = \frac{1}{x^3+b_0-}\;\frac{a_1}{x^3+b_1-}\cdots\frac{a_n}{x^3+b_n-}\cdots,$$

(21.9.59)

where

$$a_n := (3n)^2(3n+1)^2(3n+2)^2, \quad b_n = (3n+1)(3n+2)^2+(3n+3)^2(3n+4).$$
(21.9.60)

For details, see (Conrad & Flajolet, 2005).

The spectral properties of the orthogonal polynomials associated with the J-fractions (21.9.49)–(21.9.60), with $x^3 \to x$ should be very interesting. It is easy to see that all such polynomials come from birth and death processes with cubic rates so that the corresponding orthogonal polynomials are birth and death process polynomials. Indeed the continued J-fractions (21.9.49)–(21.9.54) arise from birth and death processes with

(i) $\lambda_n = (3n+1)(3n+2)^2, \quad \mu_n = (3n)^2(3n+1)$

(ii) $\lambda_n = (3n+2)(3n+3)^2, \quad \mu_n = (3n+1)^2(3n+2)$

(iii) $\lambda_n = (3n+3)(3n+4)^2, \quad \mu_n = (3n+2)^2(3n+3),$

respectively, while the continued fractions in (21.9.55)–(21.9.60) correspond to

(iv) $\lambda_n = (3n+1)^2(3n+2), \quad \mu_n = (3n)^2(3n-1)$

(v) $\lambda_n = (3n+2)^2(3n+3), \quad \mu_n = 3n(3n+1)^2$

(vi) $\lambda_n = (3n+3)^2(3n+4), \quad \mu_n = (3n+1)(3n+2)^2,$

respectively. Gilewicz, Leopold, and Valent solved the Hamburger moment problem associated with cases (i), (iv) and (v) in (Gilewicz et al., 2005). It is clear that (i)–(iii) are cases of the infinite family

$$\lambda_n = (3n + c + 1)(3n + c + 2)^2, \quad \mu_n = (3n + c)^2(3n + c + 1). \quad (21.9.61)$$

On the other hand (iv)–(vi) are contained in the infinite family

$$\lambda_n = (3n + c + 1)^2(3n + c + 2), \quad \mu_n = (3n + c - 1)^2(3n + c)^2. \quad (21.9.62)$$

We now come to the moment problem associated with the polynomials $\left\{V_n^{(a)}(x; q)\right\}$. The positivity condition (12.4.4) implies $a > 0$. Theorem 18.2.6 shows that the generating function $V^{(a)}(x, t)$ is analytic for $|t| < \min\{1, 1/a\}$. Thus a comparison function is

$$\begin{cases} (1 - t)^{-1}(x; q)_\infty/(q, a; q)_\infty, & \text{if } a > 1, \\ (1 - at)^{-1}(x/a; q)_\infty/(q, 1/a; q)_\infty, & \text{if } a < 1, \\ (1 - t)^{-2}(x; q)_\infty/(q; q)_\infty^2 & \text{if } a = 1. \end{cases}$$

Therefore

$$V_n^{(a)}(x; q) = (-1)^n q^{-\binom{n}{2}}[1 + o(1)]$$

$$\times \begin{cases} (x; q)_\infty/(a; q)_\infty, & \text{if } a > 1, \\ a^n(x/a; q)_\infty/(1/a; q)_\infty, & \text{if } a < 1, \\ n(x; q)_\infty/(q; q)_\infty & \text{if } a = 1 \end{cases} \quad (21.9.63)$$

as $n \to \infty$. To do the asymptotics for $V_n^{(a)*}(x; q)$ we derive a generating function by multiplying the three term recurrence relation (18.2.19) by $(-t)^{n+1}q^{\binom{n+1}{2}}/(q; q)_{n+1}$ and add the results for $n > 0$ taking into account the initial conditions $V_0^{(a)*}(x; q) = 0$, $V_1^{(a)*}(x; q) = 1$. This gives the generating function

$$\sum_{n=1}^\infty \frac{(-t)^n q^{\binom{n}{2}}}{(q; q)_n} V_n^{(a)*}(x; q) = -t \sum_{n=1}^\infty \frac{q^n(xt; q)_n}{(t, at; q)_{n+1}}. \quad (21.9.64)$$

By introducing suitable comparison functions we find

$$V_n^{(a)*}(x; q) = \frac{(-1)^n(q; q)_\infty}{q^{n(n-1)/2}}[1 + o(1)]$$

$$\times \begin{cases} a^n(a - 1)^{-1}{}_2\phi_1(x.0; aq; q, q) & \text{if } a > 1, \\ a^n(x/a; q)_\infty/(1/a; q)_\infty, & \text{if } a < 1, \\ n(x; q)_\infty/(q; q)_\infty & \text{if } a = 1. \end{cases} \quad (21.9.65)$$

The entire functions A, B, C, D can be computed from (21.9.7)–(21.9.9) and (21.1.5)–(21.1.8).

21.10 Some Biorthogonal Rational Functions

As we saw in Chapter 15, the orthogonality of the Askey–Wilson polynomials follows from the evaluation of the Askey–Wilson integral (15.2.1). Since the left-hand side of (15.2.1) is the product of four q-Hermite polynomials integrated against their orthogonality measure, one is led to consider the integral

$$I = I(t_1, t_2, t_3, t_4)$$

$$:= \int_{\mathbb{R}} \prod_{j=1}^{4} \left(-t_j \left(x + \sqrt{x^2 + 1} \right), t_j \left(\sqrt{x^2 + 1} - x \right); q \right)_{\infty} d\mu(x), \tag{21.10.1}$$

where μ is any solution of the moment problem associated with $\{h_n(x \mid q)\}$. It is assumed that the integral in (21.10.1) exists.

Theorem 21.10.1 *The integral in* (21.10.1) *is given by*

$$I(t_1, t_2, t_3, t_4) = \frac{\prod_{1 \le j < k \le 4} (-t_j t_k / q; q)_{\infty}}{(t_1 t_2 t_3 t_4 q^{-3}; q)_{\infty}}. \tag{21.10.2}$$

Proof An iterate of (21.2.8) is

$$q^{\binom{s}{2}+\binom{m}{2}+\binom{n}{2}} \frac{h_m(x \mid q) h_n(x \mid q) h_s(x \mid q)}{(q;q)_m (q;q)_n (q;q)_s}$$

$$= \sum_{k=0}^{m \wedge n} \frac{q^{-k + \binom{k}{2} + \binom{m-k}{2} + \binom{n-k}{2} - \binom{m+n-2k}{2}}}{(q;q)_k (q;q)_{m-k} (q;q)_{n-k}} (q;q)_{m+n-2k}$$

$$\times \sum_{j=0}^{s \wedge (m-n-2k)} \frac{q^{-j + \binom{j}{2} + \binom{s-j}{2} + \binom{m+n-2k-j}{2}}}{(q;q)_j (q;q)_{s-j} (q;q)_{m+n-2k-j}} h_{m+n+s-2k-2j}(x \mid q).$$

Multiply both sides by $t_1^r t_2^s t_3^m t_4^n q^{\binom{r}{2}} h_r(x)/(q)_r$, integrate with respect to μ, then add the results for all $m, n, r, s \ge 0$. The orthogonality relation (21.5.6) forces $r = m + n + s - 2k - 2j$. Thus $j = k + (m + n + s r)/2$. The sums of integrals of the left sides is $I(t_1, t_2, t_3, t_4)$ as can be seen from (21.2.6) and the Lebesgue convergence theorem. Therefore we have

$$I = \sum_{k,m,n,r,s=0}^{\infty} \frac{q^{\binom{k}{2} + \binom{m-k}{2} + \binom{n-k}{2} - \binom{m+n-2k}{2} + \binom{-k+(m+n+s-r)/2}{2}}}{(q;q)_k (q;q)_{m-k} (q;q)_{n-k} (q;q)_{-k+(m+n+s-r)/2}}$$

$$\times \frac{q^{\binom{k+(s+r-m-n)/2}{2} + \binom{-k+(m+n+r-s)/2}{2} + k - (r+m+n+s)/2}}{(q;q)_{k+(s+r-m-n)/2} (q;q)_{-k+(m+n+r-s)/2}}$$

$$\times (q;q)_{m+n-2k} q^{-(m+n+r+s)/2} t_1^r t_2^s t_3^m t_4^n.$$

Replace m and n by $m + k$ and $n + k$, respectively, to get

$$I = \sum_{k,m,n,r,s=0}^{\infty} \frac{q^{\binom{k}{2} + \binom{(m+n+s-r)/2}{2} + \binom{(r+s-m-n)/2}{2} + \binom{(m+n+r-s)/2}{2}}}{(q;q)_m (q;q)_n (q;q)_{(m+n+s-r)/2} (q;q)_{(r+s-m-n)/2}}$$

$$\times \frac{(q;q)_{m+n} q^{-mn-k-(m+n+r+s)/2}}{(q;q)_k (q;q)_{(m+n+r-s)/2}} t_1^r t_2^s t_3^{m+k} t_4^{n+k}.$$

Introduce the new summation indices

$$\alpha := (m+n+s-r)/2, \quad \beta := (m+n+r-s)/2,$$
$$\gamma := (r+s-m-n)/2 \tag{21.10.3}$$

instead of n, r, s. Our new summation indices are now k, m, α, β, and γ. Clearly

$$r = \beta+\gamma, \quad s = \alpha+\gamma, n = \alpha+\beta-m. \tag{21.10.4}$$

Thus

$$I = \sum_{k,m,\alpha,\beta,\gamma=0}^{\infty} \frac{q^{\binom{k}{2}+\binom{\alpha}{2}+\binom{\gamma}{2}+\binom{\beta}{2}+m(m-\alpha-\beta)-k-\alpha-\beta-\gamma}}{(q:q)_k(q:q)_m(q:q)_{\alpha+\beta-m}(q:q)_\alpha(q:q)_\gamma(q:q)_\beta}$$
$$\times (q:q)_{\alpha+\beta} t_1^{\beta+\gamma} t_2^{\alpha+\gamma} t_3^{m+k} t_4^{\alpha+\beta+k-m}.$$

By Euler's formula (12.2.25) the γ sum is $(-t_1t_2/q)_\infty$ while the k sum is $(-t_3t_4/q)_\infty$. This reduces I to a triple sum. Now replace $\alpha + \beta$ by p to see that I satisfies

$$\frac{I(t_1, t_2, t_3, t_4)}{(-t_1t_2/q, -t_3t_4/q; q)_\infty}$$
$$= \sum_{p=0}^{\infty} \frac{(t_1t_4/q)^p}{(q;q)_p} q^{\binom{p}{2}} \sum_{m=0}^{p} \frac{(q;q)_p (t_3/t_4)^m}{(q;q)_m(q;q)_{p-m}} q^{m(m-p)}$$
$$\times \sum_{\alpha=0}^{p} \frac{(q;q)_p (t_2/t_1)^\alpha}{(q;q)_\alpha(q;q)_{p-\alpha}} q^{\alpha(\alpha-p)}.$$

We now use the Poisson kernel for $\{h_n(x \mid q)\}$ to evaluate the above sum. This can be achieved by setting

$$t_1 = \sqrt{q}\,Te^\xi, \quad t_2 = -\sqrt{q}\,Te^{-\xi}, \quad t_3 = -R\sqrt{q}\,e^{-\eta}, \quad t_4 = R\sqrt{q}\,e^\eta.$$

This leads to

$$\frac{I(t_1, t_2, t_3, t_4)}{(-t_1t_2/q, -t_3t_4/q; q)_\infty}$$
$$= \sum_{p=0}^{\infty} \frac{(RT)^p q^{\binom{p}{2}}}{(q;q)_p} h_p(\sinh \xi \mid q) h_p(\sinh \eta \mid q)$$
$$= \frac{(-RTe^{\xi+\eta}, -RTe^{-\xi-\eta}, RTe^{\xi-\eta}, RTe^{\eta-\xi}; q)_\infty}{(R^2T^2/q; q)_\infty}$$
$$= \frac{(-t_1t_4/q, -t_2t_3/q, -t_1t_3/q, -t_2t_4/q; q)_\infty}{(t_1t_2t_3t_4/q^3; q)_\infty}.$$

and the proof of Theorem 21.10.1 is complete. □

Motivated by the construction of the Askey–Wilson polynomials, Ismail and Masson (Ismail & Masson, 1994) introduced rational functions

$$\varphi_n(\sinh \xi; t_1, t_2, t_3, t_4)$$
$$= {}_4\phi_3\left(\begin{matrix} q^{-n}, -t_1t_2q^{n-2}, -t_1t_3/q, -t_1t_4/q \\ -t_1e^\xi, t_1e^{-\xi}, t_1t_2t_3t_4q^{-3} \end{matrix} \,\middle|\, q, q \right). \tag{21.10.5}$$

Let

$$w(x) = w(x; t_1, t_2, t_3, t_4)$$

$$:= \prod_{j=1}^{4} \left(-t_j \left(x + \sqrt{x^2 + 1} \right), t_j \left(\sqrt{x^2 + 1} - x \right); q \right)_\infty . \qquad (21.10.6)$$

The biorthogonal rational functions which are analogues of the Askey–Wilson polynomials are

$$\varphi_n \left(\sinh \xi; t_1, t_2, t_3, t_4 \right) = {}_4\phi_3 \left(\begin{array}{c} q^{-n}, -t_1 t_2 q^{n-2}, -t_1 t_3/q, -t_1 t_4/q \\ -t_1 e^\xi, t_1 e^{-\xi}, t_1 t_2 t_3 t_4 q^{-3} \end{array} \bigg| q, q \right). \qquad (21.10.7)$$

Theorem 21.10.2 *The Ismail–Masson rational functions satisfy the orthogonality relation*

$$\int_{\mathbb{R}} w \left(x; t_1, t_2, t_3, t_4 \right) \varphi_m \left(x; t_1, t_2, t_3, t_4 \right)$$

$$\times \varphi_n \left(x; t_2, t_1, t_3, t_4 \right) d\mu(x) = g_n \delta_{m,n}, \qquad (21.10.8)$$

where μ is the probability measure with respect to which the h_n's are orthogonal, $w(x)$ is as in (21.10.6), and g_n is

$$g_n = \frac{1 + t_1 t_2 q^{n-2}}{1 + t_1 t_2 q^{2n-2}} \frac{\left(t_1 t_2 t_3 t_4 q^{-3} \right)^n \left(q, -q^2/t_3 t_4; q \right)_n}{\left(t_1 t_2 t_3 t_4 q^{-3}; q \right)_n}$$

$$\times \frac{\left(-t_1 t_3/q, -t_1 t_4/q, -t_2 t_3/q, -t_2 t_4/q, -t_3 t_4/q; q \right)_\infty}{\left(t_1 t_2 t_3 t_4 q^{-3} \right)_\infty / \left(-t_1 t_2 q^{n-1} \right)_\infty}. \qquad (21.10.9)$$

Proof Consider the integrals

$$J_{m,n} := \int_{\mathbb{R}} \frac{\varphi_n \left(x; t_1, t_2, t_3, t_4 \right)}{\left(-t_2(x + \sqrt{x^2 + 1}) , t_2 \left(\sqrt{x^2 + 1} - x \right); q \right)_m}$$

$$\times \prod_{j=1}^{4} \left(-t_j \left(x + \sqrt{x^2 + 1} \right), t_j \left(\sqrt{x^2 + 1} - x \right); q \right)_\infty d\psi(x).$$

We have

$$J_{m,n} = \sum_{k=0}^{n} \frac{\left(q^{-n}, -t_1 t_2 q^{n-2}, -t_1 t_3/q, -t_1 t_4/q; q \right)_k}{\left(q, t_1 t_2 t_3 t_4 q^{-3}; q \right)_k} q^k I \left(t_1 q^k, t_2 q^m, t_3, t_4 \right)$$

$$= \frac{\left(-t_3 t_4/q, -t_2 t_3 q^{m-1}, -t_2 t_4 q^{m-1}, -t_1 t_2 a^{m-1}, -t_1 t_3/q, -t_1 t_4/q; q \right)_\infty}{\left(t_1 t_2 t_3 t_4 q^{m-3}; q \right)_\infty}$$

$$\times {}_3\phi_2 \left(\begin{array}{c} q^{-n}, -t_1 t_2 q^{n-2}, t_1 t_2 t_3 t_4 q^{m-3} \\ t_1 t_2 t_3 t_4 q^{-3}, -t_1 t_2 q^{m-1} \end{array} \bigg| q, q \right).$$

Thus $J_{m,n}$ can be summed by the q-analog of the Pfaff–Saalschütz theorem and we get

$$J_{m,n}$$
$$= \frac{\left(-t_1 t_3/q, -t_1 t_4/q, -t_3 t_4/q, -t_1 t_2 q^{m-1}, -t_2 t_3 q^{m-1}, -t_2 t_4 q^{m-1}; q\right)_\infty}{(t_1 t_2 t_3 t_4 q^{m-3}; q)_\infty}$$
$$\times \frac{\left(-t_3 t_4 q^{-n-1}, q^{-m}; q\right)_n}{\left(t_1 t_2 t_3 t_4 q^{-3}, -q^{2-m-n}/t_1 t_2; q\right)_n}.$$

Clearly $J_{m,n} = 0$ if $m < n$ and

$$J_{n,n}$$
$$= \frac{\left(-t_1 t_3/q, -t_1 t_4/q, -t_3 t_4/q, -t_1 t_2 q^{n-1}, -t_2 t_3 q^{n-1}, -t_2 t_4 q^{n-1}; q\right)_\infty}{(t_1 t_2 t_3 t_4 q^{-3}; q)_\infty}$$
$$\times \frac{\left(-q^2/t_3 t_4, q\right)_n}{\left(-t_1 t_2 q^{n-1}; q\right)_n} q^{n(n-7)/2} \left(-t_1 t_2 t_3 t_4\right)^n.$$

Therefore

$$\int_{\mathbb{R}} \varphi_m \left(x; t_1, t_2, t_3, t_4\right) \varphi_n \left(x; t_2, t_1, t_3, t_4\right) w(x)\, d\mu(x)$$
$$= \frac{\left(-t_1 t_2 q^{n-2}, q^{-n}, -t_2 t_3/q, -t_2 t_4/q; q\right)_n}{\left(q, t_1 t_2 t_3 t_4 q^{-3}; q\right)_n} q^n J_{n,n} \delta_{m,n}.$$

Using the above evaluation of $J_{n,n}$ we establish the biorthogonality relation (21.10.8).
$\square$

The Sears transformation, (12.4.1), simply expresses the symmetry relation

$$\psi_n \left(\sinh \xi; t_1, t_2, t_3, t_4\right) = \psi_n \left(\sinh \xi; t_2/q, t_1 q, t_3, t_4\right), \qquad (21.10.10)$$

where

$$\psi_n \left(\sinh \xi; t_1, t_2, t_3, t_4\right) := t_1^{-n} \left(t_1 e^{-\xi}, -t_1 e^{\xi}\right)_n \varphi_n \left(\sinh \xi; t_1, t_2, t_3, t_4\right). \qquad (21.10.11)$$

One can rewrite ψ_n as an Askey–Wilson polynomial in a different variable. For example, the generating function of Theorem 15.2.2 gives rise to

$$\sum_{n=0}^{\infty} \frac{\left(t_1 e^{-\xi}, -t_1 e^{\xi}; q\right)_n}{\left(q, -q^2/t_3 t_4; q\right)_n} \varphi_n \left(\sinh \xi; t_1, t_2, t_3, t_4\right) t^n$$
$$= {}_2\varphi_1 \left(\begin{matrix} -t_1 t_4/q, -t_2 t_4/q^2 \\ t_1 t_2 t_3 t_4 q^{-3} \end{matrix} \Big| q, -t_1 t_3 t/q \right) \qquad (21.10.12)$$
$$\times {}_2\phi_1 \left(\begin{matrix} -q e^{-\xi}/t_4, q e^{\xi}/t_4 \\ -q^2/t_3 t_4 \end{matrix} \Big| q, -t_1 t_4 t/q \right).$$

The main term in the asymptotic expansion in Theorem 15.4.1 gives

$$
\varphi_n \left(\sinh \xi; t_1, t_2, t_3, t_4 \right)
$$

$$
= \frac{\left(-t_1 t_3/q, -t_2 t_3/q^2, -q e^{-\xi}/t_4, q e^{\xi}/t_4; q \right)_\infty}{\left(t_3/t_4, t_1 t_2 t_3 t_4 q^{-3}, t_1 e^{-\xi}, -t_1 e^{\xi}; q \right)_\infty} \tag{21.10.13}
$$

$$
\times \left(-t_1 t_4/q \right)^n \left[1 + o(1) \right],
$$

valid for $|t_3| < |t_4|$.

Exercises

21.1   Prove Theorem 21.9.3.

21.2   Prove Theorem 21.8.7.

21.3   Let $\{f_n(x)\}$ be a sequence of polynomials defined by

$$
f_n(\cosh \xi) = {}_{r+2}\phi_{r+1} \left(\begin{matrix} q^{-n}, a_1 e^{\xi}, a_1 e^{-\xi}, a_2, \ldots, a_r \\ a_1 b_1, a_1 b_2, \ldots, a_1 b_{r+1} \end{matrix} \,\middle|\, q, q^n b_1 b_2 \cdots b_{r+1} \right),
$$

$x_n(t) = \left(t q^{-2n} + q^{2n}/t \right)/2$. Prove that (Ismail, 2005a)

$$
\lim_{n \to \infty} q^{n^2 + n} \left(t a_1 b_1 b_2 \cdots b_r \right)^{-n} f_n \left(x_n(t) \right)
$$

$$
= \frac{A_q \left(q/ \left(a_1 b_1 b_2 \cdots b_{r+1} t \right) \right)}{(q; q)_\infty \prod_{j=1}^{r+1} \left(a_1 b_j; q \right)_\infty}
$$

This limit is uniform on compact subsets of $\mathbb{C} \smallsetminus \{0\}$.

22

The Riemann-Hilbert Problem for Orthogonal Polynomials

In this chapter a Riemann–Hilbert problem consists of finding an analytic function on $\mathbb{C} \setminus \Sigma$, where Σ is a collection of oriented curves, for which the boundary values on Σ (from both sides of the curves) are given. In order to have a unique solution one usually needs a normalization condition at a certain point in the complex plane and in this chapter this will be the point at infinity on the Riemann sphere $\overline{\mathbb{C}} = \mathbb{C} \cup \{\infty\}$. Initially the collection Σ of oriented curves will be the real line $\mathbb{R}$, the semi-axis $\mathbb{R}_+ = [0, \infty)$, or a closed interval, which without loss of generality can be taken to be $[-1, 1]$. The orientation is from left to right for these cases. The orientation of the curves is an indication of how the boundary values are taken: we define

$$f_+(z) = \lim_{z' \to z,\, +\text{ side}} f(z'), \quad f_-(z) = \lim_{z' \to z,\, -\text{ side}} f(z'), \quad z \in \Sigma,$$

where the $+$ side is on the left of the oriented curve and the $-$ side is on the right. This is well defined, except at endpoints of curves or at points of intersection of curves in Σ, where we have to impose extra conditions for the Riemann–Hilbert problem.

22.1 The Cauchy Transform

A typical scalar and additive Riemann–Hilbert problem is to find a function $f : \mathbb{C} \to \mathbb{C}$ such that

> 1. f is analytic in $\mathbb{C} \setminus \mathbb{R}$.
> 2. $f_+(x) = f_-(x) + w(x)$ when $x \in \mathbb{R}$.
> 3. $f(z) = \mathcal{O}(1/z)$ as $z \to \infty$.
> $\qquad\qquad\qquad\qquad\qquad$ (22.1.1)

Here $w : \mathbb{R} \to \mathbb{R}$ is a given function which describes the jump that f makes as it crosses the real axis.

Theorem 22.1.1 *Suppose that* $w \in L_1(\mathbb{R})$ *and* w *is Hölder continuous on* $\mathbb{R}$, *that is,*

$$|w(x) - w(y)| \le C|x - y|^\alpha, \qquad \text{for all } x, y \in \mathbb{R},$$

where $C > 0$ is a constant and $0 < \alpha \leq 1$. *Then the unique solution of the Riemann–Hilbert problem (22.1.1) is given by*

$$f(z) = \frac{1}{2\pi i} \int_{\mathbb{R}} \frac{w(t)}{t - z} \, dt,$$

which is the Cauchy transform or Stieltjes transform of the function w.

Proof Clearly this f is analytic in $\mathbb{C} \setminus \mathbb{R}$ and as $z \to \infty$ we have

$$\lim_{z \to \infty} z f(z) = \frac{-1}{2\pi i} \int_{\mathbb{R}} w(t) \, dt,$$

which is finite since $w \in L_1(\mathbb{R})$. Hence the first and third conditions of the Riemann–Hilbert problem are satisfied. We now show that

$$f_+(x) = \lim_{y \to 0+} f(x + iy) = \frac{1}{2} w(x) + \frac{i}{2\pi} \int_{\mathbb{R}} \frac{w(t)}{x - t} \, dt, \qquad (22.1.2)$$

where the second integral is a Cauchy principal value integral

$$\int_{\mathbb{R}} \frac{w(t)}{x - t} \, dt = \lim_{\delta \to 0} \int_{|t - x| > \delta} \frac{w(t)}{x - t} \, dt$$

which is called the *Hilbert transform* of w. Indeed, we have

$$f(x + iy) = \frac{1}{2\pi i} \int_{\mathbb{R}} w(t) \frac{(t - x) + iy}{(t - x)^2 + y^2} \, dt,$$

and therefore we examine the limits

$$\lim_{y \to 0+} \frac{y}{2\pi} \int_{\mathbb{R}} \frac{w(t)}{(t - x)^2 + y^2} \, dt \qquad (22.1.3)$$

and

$$\lim_{y \to 0+} \frac{1}{2\pi} \int_{\mathbb{R}} w(t) \frac{(t - x)}{(t - x)^2 + y^2} \, dt. \qquad (22.1.4)$$

For (22.1.3) we use the change of variables $t = x + sy$ to find

$$\lim_{y \to 0+} \frac{1}{2\pi} \int_{\mathbb{R}} \frac{w(x + sy)}{s^2 + 1} \, ds = \frac{w(x)}{2\pi} \int_{\mathbb{R}} \frac{ds}{1 + s^2} = \frac{w(x)}{2},$$

where the interchange of integral and limit can be justified by combining the continuity of w and Lebesgue's dominated convergence theorem. For (22.1.4) we write

$$\int_{\mathbb{R}} w(t) \frac{(t - x)}{(t - x)^2 + y^2} \, dt = \int_{|t - x| > \delta} w(t) \frac{(t - x)}{(t - x)^2 + y^2} \, dt$$

$$+ \int_{|t - x| \leq \delta} w(t) \frac{(t - x)}{(t - x)^2 + y^2} \, dt,$$

where $y \le \delta$. Clearly

$$\lim_{y \to 0+} \int_{|t-x|>\delta} w(t) \frac{(t-x)}{(t-x)^2 + y^2} \, dt = -\int_{|t-x|>\delta} \frac{w(t)}{x-t} \, dt,$$

which tends to

$$-\int_{\mathbb{R}} \frac{w(t)}{x-t} \, dt,$$

as $\delta \to 0$. So we need to prove that

$$\lim_{\delta \to 0} \lim_{y \to 0+} \int_{|t-x| \le \delta} w(t) \frac{(t-x)}{(t-x)^2 + y^2} \, dt = 0.$$

Observe that the symmetry implies

$$\int_{|t-x| \le \delta} \frac{(t-x)}{(t-x)^2 + y^2} \, dt = 0,$$

hence

$$\int_{|t-x| \le \delta} w(t) \frac{(t-x)}{(t-x)^2 + y^2} \, dt = \int_{|t-x| \le \delta} [w(t) - w(x)] \frac{(t-x)}{(t-x)^2 + y^2} \, dt.$$

If we estimate the latter integral, then the Hölder continuity gives

$$\left| \int_{|t-x| \le \delta} [w(t) - w(x)] \frac{(t-x)}{(t-x)^2 + y^2} \, dt \right| \le C \int_{|t-x| \le \delta} \frac{|t-x|^{\alpha+1}}{(t-x)^2 + y^2} \, dt$$

$$\le C \int_{|t-x| \le \delta} |t-x|^{\alpha-1} \, dt$$

$$= \frac{2C}{\alpha} \delta^{\alpha},$$

and this clearly tends to 0 as $\delta \to 0$ for every y. This proves (22.1.2). With the same method one also shows that

$$f_-(x) = \lim_{y \to 0+} f(x - iy) = -\frac{1}{2} w(x) + \frac{i}{2\pi} \int_{\mathbb{R}} \frac{w(t)}{x-t} \, dt, \qquad (22.1.5)$$

so that we conclude that

$$f_+(x) = f_-(x) + w(x), \qquad x \in \mathbb{R},$$

which is the jump condition of the Riemann–Hilbert problem.

To show uniqueness, assume that g is another solution of this Riemann–Hilbert problem. Then $f - g$ is analytic in $\mathbb{C} \setminus \mathbb{R}$ and on $\mathbb{R}$ we see that $(f - g)_+(x) = (f - g)_-(x)$ so that $f - g$ is continuous on $\mathbb{C}$. But then one can use Morera's theorem to conclude that $f - g$ is analytic on the whole complex plane. As $z \to \infty$ we have $f(z) - g(z) = \mathcal{O}(1/z)$ hence $f - g$ is a bounded entire function. Liouville's theorem then implies that $f - g$ is a constant function, but as it tends to 0 as $z \to \infty$, we must conclude that $f = g$. $\square$

Equations (22.1.2) and (22.1.5) are known as the Plemelj–Sokhotsky identities and should be compared with formula (1.2.10) in Chapter 1 (Perron–Stieltjes inversion formula).

22.2 The Fokas–Its–Kitaev Boundary Value Problem

The basic idea of the Riemann–Hilbert approach to orthogonal polynomials is to characterize the orthogonal polynomials corresponding to a weight function w on the real line via a boundary value problem for matrix valued analytic functions. This was first formulated in a ground-breaking paper of Fokas, Its and Kitaev (Fokas et al., 1992). The Riemann–Hilbert problem for orthogonal polynomials on the real line with a weight function w is to find a matrix valued function $Y : \mathbb{C} \to \mathbb{C}^{2 \times 2}$ which satisfies the following three conditions:

1. Y is analytic in $\mathbb{C} \setminus \mathbb{R}$.
2. (jump condition) On the real line we have

$$Y_+(x) = Y_-(x) \begin{pmatrix} 1 & w(x) \\ 0 & 1 \end{pmatrix}, \qquad x \in \mathbb{R}.$$

3. (normalization near infinity) Y has the following behavior near infinity

$$Y(z) = (I + \mathcal{O}(1/z)) \begin{pmatrix} z^n & 0 \\ 0 & z^{-n} \end{pmatrix}, \qquad z \to \infty. \tag{22.2.1}$$

A matrix function Y is analytic in z if each of its components is an analytic function of z. The boundary values $Y_+(x)$ and $Y_-(x)$ are defined as

$$Y_\pm(x) = \lim_{\epsilon \to 0+} Y(x \pm i\epsilon),$$

and the existence of these boundary values is part of the assumptions. The behavior near infinity is in the sense that

$$Y(z) \begin{pmatrix} z^{-n} & 0 \\ 0 & z^n \end{pmatrix} = \begin{pmatrix} 1 & 0 \\ 0 & 1 \end{pmatrix} + \begin{pmatrix} a(z) & b(z) \\ c(z) & d(z) \end{pmatrix}$$

where

$$|za(z)| \leq A, \quad |zb(z)| \leq B, \quad |zc(z)| \leq C, \quad |zd(z)| \leq D, \quad |z| > z_0, \ \Im z \neq 0.$$

Theorem 22.2.1 *Suppose that $x^j w \in L_1(\mathbb{R})$ for every $j \in \mathbb{N}$ and that w is Hölder continuous on $\mathbb{R}$. Then for $n \geq 1$ the solution of the Riemann–Hilbert problem*

(22.2.1) *for Y is given by*

$$Y(z) = \begin{pmatrix} P_n(z) & \dfrac{1}{2\pi i} \displaystyle\int_{\mathbb{R}} \dfrac{P_n(t)w(t)}{t-z}\, dt \\[2ex] -2\pi i \gamma_{n-1}^2 P_{n-1}(z) & -\gamma_{n-1}^2 \displaystyle\int_{\mathbb{R}} \dfrac{P_{n-1}(t)w(t)}{t-z}\, dt \end{pmatrix}, \qquad (22.2.2)$$

where P_n is the monic orthogonal polynomial of degree n for the weight function w and γ_{n-1} is the leading coefficient of the orthonormal polynomial p_{n-1}.

Proof Let us write the matrix Y as

$$Y = \begin{pmatrix} Y_{1,1}(z) & Y_{1,2}(z) \\ Y_{2,1}(z) & Y_{2,2}(z) \end{pmatrix}.$$

The conditions on Y imply that $Y_{1,1}$ is analytic in $\mathbb{C}\setminus\mathbb{R}$. The jump condition for $Y_{1,1}$ is $(Y_{1,1})_+ (x) = (Y_{1,1})_- (x)$ for $x \in \mathbb{R}$, hence $Y_{1,1}$ is continuous in $\mathbb{C}$ and Morera's theorem therefore implies that $Y_{1,1}$ is analytic in $\mathbb{C}$ so that $Y_{1,1}$ is an entire function. The normalization near infinity gives $Y_{1,1}(z) = z^n + \mathcal{O}\left(z^{n-1}\right)$, hence Liouville's theorem implies that $Y_{1,1}$ is a monic polynomial of degree n, and we denote it by $Y_{1,1}(z) = \pi_n(z)$.

Now consider $Y_{1,2}$. This function is again analytic in $\mathbb{C} \setminus \mathbb{R}$ and the jump condition is $(Y_{1,2})_+ (x) = w(x)\,(Y_{1,1})_- (x) + (Y_{1,2})_- (x)$ for $x \in \mathbb{R}$. Since $Y_{1,1}$ is a polynomial, this jump condition becomes

$$(Y_{1,2})_+ (x) = (Y_{1,2})_- (x) + w(x)\pi_n(x), \qquad x \in \mathbb{R}.$$

The behavior near infinity is $Y_{1,2}(z) = \mathcal{O}\left(z^{-n-1}\right)$. We can therefore use Theorem 22.1.1 to conclude that

$$Y_{1,2}(z) = \frac{1}{2\pi i} \int_{\mathbb{R}} \frac{\pi_n(t)w(t)}{t-z}\, dt$$

since $\pi_n(t)w(t)$ is Hölder continuous and in $L_1(\mathbb{R})$. The polynomial π_n is still not specified, but we haven't used all the conditions of the behavior near infinity: $Y_{1,2}(z) = \mathcal{O}\left(z^{-n-1}\right)$ as $z \to \infty$. If we expand $1/(t-z)$ as

$$\frac{1}{t-z} = -\sum_{k=0}^{n} \frac{t^k}{z^{k+1}} + \frac{1}{t-z}\frac{t^{n+1}}{z^{n+1}}, \qquad (22.2.3)$$

then

$$Y_{1,2}(z) = -\sum_{k=0}^{n} \frac{1}{z^{k+1}} \frac{1}{2\pi i}\int_{\mathbb{R}} t^k \pi_n(t)w(t)\, dt + \frac{1}{z^{n+1}}\frac{1}{2\pi i}\int_{\mathbb{R}} \frac{t^{n+1}\pi_n(t)w(t)}{t-z}\, dt,$$

so that π_n needs to satisfy the conditions

$$\int_{\mathbb{R}} t^k \pi_n(t)w(t)\, dt = 0, \qquad k = 0, 1, \ldots, n-1.$$

Hence π_n is the monic orthogonal polynomial P_n of degree n.

The reasoning for the second row of Y is similar with just a few changes. The function $Y_{2,1}$ is analytic in $\mathbb{C}\backslash\mathbb{R}$ and has the jump condition $(Y_{2,1})_+ (x) = (Y_{2,1})_- (x)$ for $x \in \mathbb{R}$. Hence $Y_{2,1}$ is an entire function. The behavior near infinity is $Y_{2,1}(z) = \mathcal{O}\left(z^{n-1}\right)$, which makes it a polynomial (not necessarily monic) of degree at most $n-1$. Let us denote it by $Y_{2,1}(z) = \pi_{n-1}(z)$. Next we look at $Y_{2,2}$ which is analytic in $\mathbb{C}\backslash\mathbb{R}$, satisfies the jump condition

$$(Y_{2,2})_+ (x) = (Y_{2,2})_- (x) + w(x)\pi_{n-1}(x), \qquad x \in \mathbb{R},$$

and behaves at infinity as $Y_{2,2}(z) = z^{-n} + \mathcal{O}\left(z^{-n-1}\right)$. Theorem 22.1.1 gives us that

$$Y_{2,2}(z) = \frac{1}{2\pi i}\int_{\mathbb{R}} \frac{\pi_{n-1}(t)w(t)}{t-z}\,dt.$$

Using the expansion (22.2.3) gives

$$Y_{2,2}(z) = -\sum_{k=0}^{n} \frac{1}{z^{k+1}}\frac{1}{2\pi i}\int_{\mathbb{R}} t^k \pi_{n-1}(t)w(t)\,dt$$
$$+ \frac{1}{z^{n+1}}\frac{1}{2\pi i}\int_{\mathbb{R}} \frac{t^{n+1}\pi_{n-1}(t)w(t)}{t-z}\,dt,$$

so that π_{n-1} needs to satisfy

$$\int_{\mathbb{R}} t^k \pi_{n-1}(t)w(t)\,dt = 0, \qquad k = 0,1,\ldots,n-2,$$

and

$$\int_{\mathbb{R}} t^{n-1}\pi_{n-1}(t)w(t)\,dt = -2\pi i. \tag{22.2.4}$$

This means that π_{n-1} is (up to a factor) equal to the monic orthogonal polynomial of degree $n-1$, and we write $\pi_{n-1}(x) = c_n P_{n-1}(x)$. Insert this in (22.2.4), then

$$-2\pi i = c_n \int_{\mathbb{R}} t^{n-1}P_{n-1}(t)w(t)\,dt$$
$$= c_n \int_{\mathbb{R}} P_{n-1}^2(t)w(t)\,dt = c_n/\gamma_{n-1}^2,$$

hence $c_n = -2\pi i \gamma_{n-1}^2$. $\square$

Remark 22.2.1 *The solution of the Riemann–Hilbert problem* (22.2.1) *for $n = 0$ is given by*

$$Y(z) = \begin{pmatrix} 1 & \dfrac{1}{2\pi i}\displaystyle\int_{\mathbb{R}} \dfrac{w(t)}{t-z}\,dt \\ 0 & 1 \end{pmatrix}. \tag{22.2.5}$$

This Riemann–Hilbert approach to orthogonal polynomials may not seem the most natural way to characterize orthogonal polynomials, but the matrix Y contains quite a lot of relevant information. First of all the first column contains the monic orthogonal polynomials of degrees n and $n-1$ and it also contains the leading coefficient γ_{n-1} of the orthonormal polynomial of degree $n-1$. Secondly the matrix Y also contains the functions of the second kind in the second column. The polynomials and the functions of the second kind are connected by the following identity

Theorem 22.2.2 *For every $z \in \mathbb{C}$ we have* $\det Y = 1$, *which gives*

$$p_n(z) \int_{\mathbb{R}} \frac{p_{n-1}(t)w(t)}{t-z}\, dt - p_{n-1}(z) \int_{\mathbb{R}} \frac{p_n(t)w(t)}{t-z}\, dt = -\frac{\gamma_n}{\gamma_{n-1}}, \qquad (22.2.6)$$

where $p_n = \gamma_n P_n$ are the orthonormal polynomials.

Proof The Riemann–Hilbert problem for the function $f(z) = \det Y(z)$ is

1. f is analytic in $\mathbb{C} \setminus \mathbb{R}$.
2. $f_+(x) = f_-(x)$ for $x \in \mathbb{R}$.
3. $f(z) = 1 + \mathcal{O}(1/z)$ as $z \to \infty$.

This means that f is an entire function which is bounded, hence Liouville's theorem implies that f is a constant function. Now $f(z) \to 1$ as $z \to \infty$, hence $f(z) = 1$ for every $z \in \mathbb{C}$. $\qquad\square$

This identity is also known as the Liouville–Ostrogradski formula. An important consequence of this result is that Y^{-1} exists everywhere in $\mathbb{C}$.

Corollary 22.2.3 *The solution (22.2.2) of the Riemann–Hilbert problem (22.2.1) is unique.*

Proof Suppose that X is another solution of the Riemann–Hilbert problem (22.2.1). Consider the matrix function $Z = XY^{-1}$, then

1. Z is analytic in $\mathbb{C} \setminus \mathbb{R}$.
2. $Z_+(x) = Z_-(x)$ for every $x \in \mathbb{R}$, since both X and Y have the same jump over $\mathbb{R}$.
3. $Z(z) = I + \mathcal{O}(1/z)$ as $z \to \infty$, since both X and Y have the same behavior near infinity.

Hence we see that Z is an entire matrix function for which the entries are bounded entire functions. Liouville's theorem then implies that each entry of Z is a constant function, and since $Z(z) \to I$ as $z \to \infty$ we must conclude that $Z(z) = I$ for every $z \in \mathbb{C}$. But this means that $X = Y$. $\qquad\square$

22.2.1 The three-term recurrence relation

The Riemann–Hilbert setting of orthogonal polynomials also enables us to find the three-term recurrence relation. We now use a subscript n and denote by Y_n the solution of the Riemann–Hilbert problem (22.2.1). Consider the matrix function $R = Y_n Y_{n-1}^{-1}$, then R is analytic in $\mathbb{C} \setminus \mathbb{R}$. The jump condition for R is $R_+(x) = R_-(x)$ for all $x \in \mathbb{R}$ since both Y_n and Y_{n-1} have the same jump matrix. Hence we conclude that R is analytic in $\mathbb{C}$. The behavior near infinity is

$$R(z) = [I + \mathcal{O}_n(1/z)] \begin{pmatrix} z & 0 \\ 0 & 1/z \end{pmatrix} [I + \mathcal{O}_{n-1}(1/z)]^{-1},$$

as $z \to \infty$. If we write

$$\mathcal{O}_n(1/z) = \frac{1}{z} \begin{pmatrix} a_n & b_n \\ c_n & d_n \end{pmatrix} + \mathcal{O}_n\left(1/z^2\right),$$

then this gives

$$R(z) = \begin{pmatrix} z - a_{n-1} + a_n & -b_{n-1} \\ c_n & 0 \end{pmatrix} + \mathcal{O}(1/z).$$

But since R is entire, we therefore must conclude that

$$R(z) = \begin{pmatrix} z - a_{n-1} + a_n & -b_{n-1} \\ c_n & 0 \end{pmatrix}, \qquad z \in \mathbb{C}.$$

Recall that $R = Y_n Y_{n-1}^{-1}$, so therefore we have

$$Y_n(z) = \begin{pmatrix} z - a_{n-1} + a_n & -b_{n-1} \\ c_n & 0 \end{pmatrix} Y_{n-1}(z). \qquad (22.2.7)$$

If we use (22.2.2), then the entry in the first row and column gives

$$P_n(z) = (z - a_{n-1} + a_n) P_{n-1}(z) + 2\pi i b_{n-1} \gamma_{n-2}^2 P_{n-2}(z). \qquad (22.2.8)$$

If we put $a_{n-1} - a_n = \alpha_{n-1}$ and $-2\pi i b_{n-1} \gamma_{n-2}^2 = \beta_{n-1}$, then this gives the three-term recurrence relation

$$P_n(z) = (z - \alpha_{n-1}) P_{n-1}(z) - \beta_{n-1} P_{n-2}(z),$$

which is (2.2.1).

In a similar way we can check the entry on the first row and second column and we see that

$$\tilde{Q}_n(z) = (z - \alpha_{n-1}) \tilde{Q}_{n-1} - \beta_{n-1} \tilde{Q}_{n-2}(z),$$

where

$$\tilde{Q}_n(z) = \int_{\mathbb{R}} \frac{P_n(t) w(t)}{z - t} \, dt$$

is a multiple of the function of the second kind, see Chapter 3. Hence we see that the function of the second kind satisfies the same three-term recurrence relation. The Wronskian (or Casorati determinant) of these two solutions is given by the determinant of Y_n, see (22.2.6).

We can also check the entry on the second row and first column in (22.2.7) to find

$$-2\pi i \gamma_{n-1}^2 P_{n-1}(z) = c_n P_{n-1}(z),$$

so that $c_n = -2\pi i \gamma_{n-1}^2$. We know that $\det Y_n = 1$, hence $\det R = b_{n-1}c_n = 1$ and therefore

$$b_{n-1} = \frac{-1}{2\pi i \gamma_{n-1}^2},$$

and thus

$$\beta_{n-1} = -2\pi i b_{n-1}\gamma_{n-2}^2 = \frac{\gamma_{n-2}^2}{\gamma_{n-1}^2}.$$

22.3 Hermite Polynomials

Consider the Riemann–Hilbert problem for Hermite polynomials where $Y : \mathbb{C} \to \mathbb{C}^{2\times 2}$ is a matrix valued function with the following properties:

1. Y is analytic in $\mathbb{C} \setminus \mathbb{R}$.
2. The boundary values Y_+ and Y_- exist on $\mathbb{R}$ and

$$Y_+(x) = Y_-(x) \begin{pmatrix} 1 & e^{-x^2} \\ 0 & 1 \end{pmatrix}, \qquad x \in \mathbb{R}. \qquad (22.3.1)$$

3. Near infinity we have

$$Y(z) = (I + \mathcal{O}(1/z)) \begin{pmatrix} z^n & 0 \\ 0 & z^{-n} \end{pmatrix}. \qquad (22.3.2)$$

Then

$$Y(z) = \begin{pmatrix} h_n(z) & \dfrac{1}{2\pi i}\displaystyle\int_{\mathbb{R}} \dfrac{h_n(t)}{t-z}e^{-t^2}\,dt \\ -2\pi i \gamma_{n-1}^2 h_{n-1}(z) & -\gamma_{n-1}^2 \displaystyle\int_{\mathbb{R}} \dfrac{h_{n-1}(t)}{t-z}e^{-t^2}\,dt \end{pmatrix},$$

where $h_n = 2^{-n} H_n$ are the monic Hermite polynomials.

22.3.1 A Differential Equation

Recall that the exponential of a matrix A is

$$e^A := \sum_{n=0}^{\infty} \frac{1}{n!} A^n.$$

This is well-defined, and the series converges in the operator norm when $\|A\| < \infty$.

We will show how to obtain the second order differential equation for Hermite polynomials from this Riemann–Hilbert problem. Consider the matrix

$$Z(z) = \begin{pmatrix} e^{z^2/2} & 0 \\ 0 & e^{-z^2/2} \end{pmatrix} Y(z) \begin{pmatrix} e^{-z^2/2} & 0 \\ 0 & e^{z^2/2} \end{pmatrix} = e^{\sigma_3 z^2/2} Y(z) e^{-\sigma_3 z^2/2},$$

where σ_3 is one of the Pauli matrices

$$\sigma_3 = \begin{pmatrix} 1 & 0 \\ 0 & -1 \end{pmatrix}. \qquad (22.3.3)$$

Then it is easy to check that

1. Z is analytic in $\mathbb{C} \setminus \mathbb{R}$.
2. The boundary values Z_+ and Z_- exist on $\mathbb{R}$ and

$$Z_+(x) = Z_-(x) \begin{pmatrix} 1 & 1 \\ 0 & 1 \end{pmatrix}, \qquad x \in \mathbb{R}.$$

3. Near infinity we have

$$Z(z) = e^{\sigma_3 z^2 / 2} \left(I + \mathcal{O}\left(\frac{1}{z}\right) \right) e^{-\sigma_3 z^2 / 2} z^{\sigma_3 n}, \qquad z \to \infty \qquad (22.3.4)$$

where $z^{n\sigma_3} := \exp\left(n \operatorname{Log} z\sigma_3 \right)$.

Now consider the auxiliary matrix function

$$\widehat{Z}(z) = \begin{cases} Z(z) & \Im z > 0, \\ Z(z) \begin{pmatrix} 1 & 1 \\ 0 & 1 \end{pmatrix}, & \Im z < 0, \end{cases}$$

then obviously $\widehat{Z}$ is analytic in $\mathbb{C} \setminus \mathbb{R}$. The boundary values on $\mathbb{R}$ are given by $\widehat{Z}_+(x) = Z_+(x)$ and $\widehat{Z}_-(x) = Z_-(x) \begin{pmatrix} 1 & 1 \\ 0 & 1 \end{pmatrix} = Z_+(x)$ for every $x \in \mathbb{R}$, hence $\widehat{Z}$ is analytic everywhere on $\mathbb{C}$ and therefore entire. But then the derivative $\widehat{Z}'$ is also entire, and in particular $\left(\widehat{Z}'\right)_+ (x) = \left(\widehat{Z}'\right)_- (x)$ for every $x \in \mathbb{R}$. This implies that

$$(Z')_+ (x) = (Z')_- (x) \begin{pmatrix} 1 & 1 \\ 0 & 1 \end{pmatrix}, \qquad x \in \mathbb{R}.$$

Writing $\mathcal{O}(1/z) = A_n/z + \mathcal{O}\left(1/z^2\right)$ in (22.3.4) then gives that near infinity we have

$$Z'(z) = e^{\sigma_3 z^2 / 2} \left(\sigma_3 A_n - A_n \sigma_3 + \mathcal{O}(1/z) \right) e^{-\sigma_3 z^2 / 2} z^{\sigma_3 n}, \qquad z-> \infty.$$

Consider the matrix function $Z'(z)Z(z)^{-1}$, which is well defined since $\det Z = 1$. Clearly $Z'Z^{-1}$ is analytic in $\mathbb{C} \setminus \mathbb{R}$ and on the real line $\left(Z'Z^{-1}\right)_+ (x) = \left(Z'Z^{-1}\right)_- (x)$ for every $x \in \mathbb{R}$ because Z' and Z have the same jump condition on $\mathbb{R}$. Hence $Z'Z^{-1}$ is an entire matrix valued function. Near infinity it has the behavior

$$Z'(z)Z(z)^{-1} = e^{\sigma_3 z^2 / 2} \left(\sigma_3 A_n - A_n \sigma_3 + \mathcal{O}(1/z) \right) e^{-\sigma_3 z^2 / 2}, \qquad z \to \infty,$$

hence by Liouville's theorem we have that

$$e^{-\sigma_3 z^2 / 2} Z'(z)Z(z)^{-1} e^{\sigma_3 z^2 / 2} = \sigma_3 A_n - A_n \sigma_3 = 2 \begin{pmatrix} 0 & b_n \\ -c_n & 0 \end{pmatrix},$$

where

$$A_n = \begin{pmatrix} a_n & b_n \\ c_n & d_n \end{pmatrix}.$$

But then

$$Z'(z) = \begin{pmatrix} 0 & 2b_n e^{z^2} \\ -2c_n e^{-z^2} & 0 \end{pmatrix} Z(z). \qquad (22.3.5)$$

The entry on the first row and first column in (22.3.5) gives

$$h'_n(z) = -4\pi i b_n \gamma_{n-1}^2 h_{n-1}(z).$$

If we compare the coefficient of z^n on both sides of this identity, then we see that $-4\pi i \gamma_{n-1}^2 b_n = n$, so that we get

$$h'_n(z) = n h_{n-1}(z), \qquad (22.3.6)$$

which is the lowering operator for Hermite polynomials, see (4.6.20). The entry on the second row and first column in (22.3.5) is

$$\pi i \gamma_{n-1}^2 \left(e^{-z^2} h_{n-1}(z) \right)' = c_n e^{-z^2} h_n(z).$$

Comparing the coefficient of z^n gives $-2\pi i \gamma_{n-1}^2 = c_n$, so that

$$\left(e^{-z^2} h_{n-1}(z) \right)' = -2 e^{-z^2} h_n(z), \qquad (22.3.7)$$

which is the raising operator for Hermite polynomials as in (4.6.21). Combining (22.3.6) and (22.3.7) gives the differential equation

$$h''_n(z) - 2z h'_n(z) = -2n h_n(z),$$

which corresponds to (4.6.23).

Observe that we can apply the same reasoning to the second column of (22.3.5) to find that the Hermite function of the second kind

$$Q_n(z) = e^{z^2} \int_{\mathbb{R}} \frac{h_n(t) e^{-t^2}}{z - t} \, dt$$

satisfies the same differential equation, namely

$$y''_n(z) - 2z y'_n(z) = -2n y_n(z). \qquad (22.3.8)$$

We just found that $-4\pi i \gamma_{n-1}^2 b_n = n$. The symmetry easily shows that the entry a_n in A_n is equal to zero, hence the recurrence relation (22.2.8) for monic Hermite polynomials becomes

$$h_{n+1}(z) = z h_n(z) - \frac{n}{2} h_{n-1}(z).$$

In terms of the usual Hermite polynomials $H_n(z)$, $H_n(z) = 2^n h_n(z)$, the above three-term recurrence relation becomes (4.6.27).

The procedure followed here proves that both $h_n(x)$ and $Q_n(x)$ satisfy (22.3.8) and that $\{h_n(x), Q_n(x)\}$ form a basis of solutions for (22.3.8).

22.4 Laguerre Polynomials

For orthogonal polynomials on the half line $\mathbb{R}_+ = [0, \infty)$ the Riemann–Hilbert problem requires a jump condition on the open half line $(0, \infty)$ and an additional condition which describes the behavior near the endpoint 0. If we do not impose this extra condition near the endpoint 0, then we lose the unicity of the solution, since we can add A/z^k to a given solution, where A is any 2×2 matrix and k is an integer ≥ 1. The extra condition will prevent this rational term in the solution.

Let us consider the Laguerre weight $w(x) = x^\alpha e^{-x}$ on $[0, \infty)$, where $\alpha > -1$. The appropriate Riemann–Hilbert problem is

1. Y is analytic in $\mathbb{C} \setminus [0, \infty)$.
2. (jump condition) On the positive real line we have
$$Y_+(x) = Y_-(x) \begin{pmatrix} 1 & x^\alpha e^{-x} \\ 0 & 1 \end{pmatrix}, \qquad x \in (0, \infty).$$

3. (normalization near infinity) Y has the following behavior near infinity
$$Y(z) = (I + \mathcal{O}(1/z)) \begin{pmatrix} z^n & 0 \\ 0 & z^{-n} \end{pmatrix}, \qquad z \to \infty.$$

4. (condition near 0) Y has the following behavior near 0
$$Y(z) = \begin{cases} \begin{pmatrix} \mathcal{O}(1) & \mathcal{O}(1) \\ \mathcal{O}(1) & \mathcal{O}(1) \end{pmatrix}, & \text{if } \alpha > 0, \\ \begin{pmatrix} \mathcal{O}(1) & \mathcal{O}(\log|z|) \\ \mathcal{O}(1) & \mathcal{O}(\log|z|) \end{pmatrix}, & \text{if } \alpha = 0, \qquad z \to 0. \\ \begin{pmatrix} \mathcal{O}(1) & \mathcal{O}(|z|^\alpha) \\ \mathcal{O}(1) & \mathcal{O}(|z|^\alpha) \end{pmatrix}, & \text{if } \alpha < 0, \end{cases}$$

$$(22.4.1)$$

Theorem 22.4.1 *The unique solution of the Riemann–Hilbert problem* (22.4.1) *for Y is given by*

$$Y(z) = \begin{pmatrix} \ell_n^\alpha(z) & \dfrac{1}{2\pi i} \displaystyle\int_0^\infty \dfrac{\ell_n^\alpha(t) t^\alpha e^{-t}}{t - z} \, dt \\ -2\pi i \gamma_{n-1}^2 \ell_{n-1}^\alpha(z) & -\gamma_{n-1}^2 \displaystyle\int_0^\infty \dfrac{\ell_{n-1}^\alpha(t) t^\alpha e^{-t}}{t - z} \, dt \end{pmatrix}, \qquad (22.4.2)$$

where $\ell_n^\alpha = (-1)^n n! L_n^{(\alpha)}$ is the monic Laguerre polynomial of degree n for the weight function $x^\alpha e^{-x}$ on $[0, \infty)$ and γ_{n-1} is the leading coefficient of the orthonormal Laguerre polynomial of degree $n - 1$.

Proof As in Theorem 22.2.1 it is clear that (22.4.2) satisfies the conditions (i)–(iii) of the Riemann–Hilbert problem (22.4.1). So we only need to verify that (22.4.2) also satisfies condition (iv) near the origin and that this solution is unique. Obviously $\ell_n^\alpha(z)$ and $\ell_{n-1}^\alpha(z)$ are bounded as $z \to 0$, hence the first column in (22.4.2) is $\mathcal{O}(1)$, as required. If $\alpha > 0$ then

$$\lim_{z \to 0} \int_0^\infty \frac{\ell_n^\alpha(t)t^\alpha e^{-t}}{t-z}\, dt = \int_0^\infty \ell_n^\alpha(t)t^{\alpha-1}e^{-t}\, dt,$$

which is finite since $\alpha - 1 > -1$, so that the second column of (22.4.2) is $\mathcal{O}(1)$ when $\alpha > 0$. If $-1 < \alpha < 0$ then we write

$$\int_0^\infty \frac{\ell_n^\alpha(t)t^\alpha e^{-t}}{t-z}\, dt = \int_0^\delta \frac{\ell_n^\alpha(t)t^\alpha e^{-t}}{t-z}\, dt + \int_\delta^\infty \frac{\ell_n^\alpha(t)t^\alpha e^{-t}}{t-z}\, dt,$$

where $\delta > 0$. As before, the second integral is such that

$$\lim_{z \to 0} \int_\delta^\infty \frac{\ell_n^\alpha(t)t^\alpha e^{-t}}{t-z}\, dt = \int_\delta^\infty \ell_n^\alpha(t)t^{\alpha-1}e^{-t}\, dt,$$

which is finite. Let $z = re^{i\theta}$, with $\theta \neq 0$, then in the first integral we make the change of variables $t = rs$ to find

$$\lim_{z \to 0} |z|^{-\alpha} \int_0^\delta \frac{\ell_n^\alpha(t)t^\alpha e^{-t}}{t-z}\, dt = \lim_{r \to 0} \int_0^{\delta/r} \frac{\ell_n^\alpha(rs)s^\alpha e^{-rs}}{s - e^{i\theta}}\, ds = \ell_n^\alpha(0) \int_0^\infty \frac{s^\alpha\, ds}{s - e^{i\theta}},$$

which is finite since $\alpha > -1$, showing that the second column of (22.4.2) is $\mathcal{O}\left(|z|^\alpha\right)$ whenever $\alpha < 0$. For $\alpha = 0$ we observe that

$$\left| \ell_n^0(t)e^{-t} - \ell_n^0(z)e^{-z} \right| \leq C_n |t - z|,$$

so that

$$\int_0^\delta \frac{\ell_n^0(t)e^{-t}}{t-z}\, dt = \int_0^\delta \frac{\ell_n^0(t)e^{-t} - \ell_n^0(z)e^{-z}}{t-z}\, dt + \ell_n^0(z)e^{-z} \int_0^\delta \frac{1}{t-z}\, dt.$$

Clearly

$$\left| \int_0^\delta \frac{\ell_n^0(t)e^{-t} - \ell_n^0(z)e^{-z}}{t-z}\, dt \right| \leq C_n \delta$$

and

$$\int_0^\delta \frac{1}{t-z}\, dt = \log(\delta - z) - \log(-z),$$

where log is defined with a cut along $(-\infty, 0]$. This shows that the second column of (22.4.2) is $\mathcal{O}(\log |z|)$ as $z \to 0$ and $z \notin [0, \infty)$.

To show that the solution is unique we first consider the function $f(z) = \det Y(z)$, with Y given by (22.4.2). Clearly f is analytic in $\mathbb{C} \setminus [0, \infty)$ and $f_+(x) = f_-(x)$ for

$x \in (0, \infty)$, hence f is analytic in $\mathbb{C} \setminus \{0\}$ and f has an isolated singularity at 0. By condition (iv) in (22.4.1) we see that

$$f(z) = \begin{cases} \mathcal{O}(1), & \text{if } \alpha > 0, \\ \mathcal{O}(\log|z|), & \text{if } \alpha = 0, \\ \mathcal{O}(|z|^\alpha), & \text{if } \alpha < 0, \end{cases}$$

hence, since $\alpha > -1$, the singularity at 0 is removable and f is an entire function. As $z \to \infty$ we have that $f(z) \to 1$, hence by Liouville's theorem $f(z) = \det Y(z) = 1$ for every $z \in \mathbb{C}$. Now let X be another solution of the Riemann–Hilbert problem (22.4.1). The matrix valued function XY^{-1} is analytic in $\mathbb{C} \setminus [0, \infty)$ and has the jump condition $(XY^{-1})_+ (x) = (XY^{-1})_- (x)$ for $x \in (0, \infty)$ because both X and Y have the same jump condition on $(0, \infty)$. Hence XY^{-1} is analytic on $\mathbb{C} \setminus \{0\}$ and each entry of XY^{-1} has an isolated singularity at the origin. Observe that

$$Y^{-1} = \begin{pmatrix} Y_{2,2} & -Y_{1,2} \\ -Y_{2,1} & Y_{1,1} \end{pmatrix}$$

so that

$$XY^{-1} = \begin{pmatrix} X_{1,1}Y_{2,2} - X_{1,2}Y_{2,1} & X_{1,2}Y_{1,1} - X_{1,1}Y_{1,2} \\ X_{2,1}Y_{2,2} - X_{2,2}Y_{2,1} & X_{2,2}Y_{1,1} - X_{2,1}Y_{1,2} \end{pmatrix},$$

and condition (iv) in (22.4.1) then gives

$$XY^{-1}(z) = \begin{cases} \begin{pmatrix} \mathcal{O}(1) & \mathcal{O}(1) \\ \mathcal{O}(1) & \mathcal{O}(1) \end{pmatrix}, & \text{if } \alpha > 0, \\ \begin{pmatrix} \mathcal{O}(\log|z|) & \mathcal{O}(\log|z|) \\ \mathcal{O}(\log|z|) & \mathcal{O}(\log|z|) \end{pmatrix}, & \text{if } \alpha = 0, \\ \begin{pmatrix} \mathcal{O}(|z|^\alpha) & \mathcal{O}(|z|^\alpha) \\ \mathcal{O}(|z|^\alpha) & \mathcal{O}(|z|^\alpha) \end{pmatrix}, & \text{if } \alpha < 0, \end{cases}$$

and since $\alpha > -1$ this means that each singularity is removable and hence XY^{-1} is an entire function. As $z \to \infty$ we have $XY^{-1}(z) \to I$, hence Liouville's theorem implies that $XY^{-1}(z) = I$ for every $z \in \mathbb{C}$, so that $X = Y$. $\qquad\square$

22.4.1 Three-term recurrence relation

The three-term recurrence relation can be obtained in a similar way as in Section 22.2.1. Consider the matrix function $R = Y_n Y_{n-1}^{-1}$, where Y_n is the solution (22.4.2) of the Riemann–Hilbert problem (22.4.1). Then R is analytic in $\mathbb{C} \setminus [0, \infty)$ and $R_+(x) = R_-(x)$ for $x \in (0, \infty)$, since both Y_n and Y_{n-1} have the same jump condition on $(0, \infty)$. Hence R is analytic in $\mathbb{C} \setminus \{0\}$ and has an isolated singularity at 0. Observe that R is equal to

$$\begin{pmatrix} (Y_n)_{1,1}(Y_{n-1})_{2,2} - (Y_n)_{1,2}(Y_{n-1})_{2,1} & (Y_n)_{1,2}(Y_{n-1})_{1,1} - (Y_n)_{1,1}(Y_{n-1})_{1,2} \\ (Y_n)_{2,1}(Y_{n-1})_{2,2} - (Y_n)_{2,2}(Y_{n-1})_{2,1} & (Y_n)_{2,2}(Y_{n-1})_{1,1} - (Y_n)_{2,1}(Y_{n-1})_{1,2} \end{pmatrix},$$

so that near the origin we have

$$
R(z) = \begin{cases}
\begin{pmatrix} \mathcal{O}(1) & \mathcal{O}(1) \\ \mathcal{O}(1) & \mathcal{O}(1) \end{pmatrix}, & \text{if } \alpha > 0, \\[12pt]
\begin{pmatrix} \mathcal{O}(\log|z|) & \mathcal{O}(\log|z|) \\ \mathcal{O}(\log|z|) & \mathcal{O}(\log|z|) \end{pmatrix}, & \text{if } \alpha = 0, \qquad z \to 0, \\[12pt]
\begin{pmatrix} \mathcal{O}(|z|^\alpha) & \mathcal{O}(|z|^\alpha) \\ \mathcal{O}(|z|^\alpha) & \mathcal{O}(|z|^\alpha) \end{pmatrix}, & \text{if } \alpha < 0,
\end{cases}
$$

hence the singularity at 0 is removable and R is an entire function. Near infinity we have

$$
R(z) = [I + \mathcal{O}_n(1/z)] \begin{pmatrix} z & 0 \\ 0 & 1/z \end{pmatrix} [I + \mathcal{O}_{n-1}(1/z)]^{-1}, \qquad z \to \infty,
$$

hence if we write

$$
\mathcal{O}_n(1/z) = \frac{1}{z} \begin{pmatrix} a_n & b_n \\ c_n & d_n \end{pmatrix} + \mathcal{O}_n\left(1/z^2\right),
$$

then Liouville's theorem implies that

$$
R(z) = \begin{pmatrix} z - a_{n-1} - a_n & -b_{n-1} \\ c_n & 0 \end{pmatrix},
$$

which gives

$$
Y_n(z) = \begin{pmatrix} z - a_{n-1} - a_n & -b_{n-1} \\ c_n & 0 \end{pmatrix} Y_{n-1}(z).
$$

Putting $a_{n-1} - a_n = \alpha_{n-1}$ and $-2\pi i b_{n-1}\gamma_{n-1}^2 = \beta_{n-1}$ then gives the three-term recurrence relation in the first row and first column.

22.4.2 A differential equation

To obtain the second order differential equation we need the complex function z^α which is defined as

$$
z^\alpha = r^\alpha e^{i\alpha\theta}, \qquad z = re^{i\theta}, \ \theta \in (-\pi, \pi).
$$

This makes z^α an analytic function on $\mathbb{C} \setminus (-\infty, 0]$ with a cut along $(-\infty, 0]$. Observe that

$$
[(-z)^\alpha]_+ = x^\alpha e^{-i\alpha\pi}, \qquad [(-z)^\alpha]_- = x^\alpha e^{i\alpha\pi}, \qquad x \in (0, \infty). \qquad (22.4.3)
$$

Consider the matrix

$$
Z(z) = Y(z) \begin{pmatrix} (-z)^{\alpha/2} e^{-z/2} & 0 \\ 0 & (-z)^{-\alpha/2} e^{z/2} \end{pmatrix}
$$
$$
= Y(z)(-z)^{\sigma_3 \alpha/2} e^{-\sigma_3 z/2},
$$

where σ_3 is the Pauli matrix (22.3.3), then Z is analytic in $\mathbb{C} \setminus [0, \infty)$. The boundary values Z_+ and Z_- exist on $(0, \infty)$ and if we take into account (22.4.3), then

$$Z_+(x) = Y_+(x) \begin{pmatrix} x^{\alpha/2} e^{-x/2} e^{-i\pi\alpha/2} & 0 \\ 0 & x^{-\alpha/2} e^{x/2} e^{i\pi\alpha/2} \end{pmatrix}$$

$$= Y_-(x) \begin{pmatrix} 1 & x^{\alpha} e^{-x} \\ 0 & 1 \end{pmatrix} \begin{pmatrix} x^{\alpha/2} e^{-x/2} e^{-i\pi\alpha/2} & 0 \\ 0 & x^{-\alpha/2} e^{x/2} e^{i\pi\alpha/2} \end{pmatrix}$$

$$= Z_-(x) \begin{pmatrix} x^{-\alpha/2} e^{x/2} e^{-i\pi\alpha/2} & 0 \\ 0 & x^{\alpha/2} e^{-x/2} e^{i\pi\alpha/2} \end{pmatrix} \begin{pmatrix} 1 & x^{\alpha} e^{-x} \\ 0 & 1 \end{pmatrix}$$

$$\times \begin{pmatrix} x^{\alpha/2} e^{-x/2} e^{-i\pi\alpha/2} & 0 \\ 0 & x^{-\alpha/2} e^{x/2} e^{i\pi\alpha/2} \end{pmatrix}$$

so that

$$Z_+(x) = Z_-(x) \begin{pmatrix} e^{-i\pi\alpha} & 1 \\ 0 & e^{i\pi\alpha} \end{pmatrix}, \qquad x \in (0, \infty).$$

Near infinity we have

$$Z(z) = \left(I + \frac{A_n}{z} + \mathcal{O}\left(1/z^2\right) \right) z^{\sigma_3 n} (-z)^{\alpha\sigma_3/2} e^{-z\sigma_3/2}, \qquad z \to \infty,$$

and near the origin we have

$$Z(z) = \begin{cases} \begin{pmatrix} \mathcal{O}(|z|^{\alpha/2}) & \mathcal{O}(|z|^{-\alpha/2}) \\ \mathcal{O}(|z|^{\alpha/2}) & \mathcal{O}(|z|^{-\alpha/2}) \end{pmatrix} & \text{if } \alpha > 0, \\[2ex] \begin{pmatrix} \mathcal{O}(1) & \mathcal{O}(\log|z|) \\ \mathcal{O}(1) & \mathcal{O}(\log|z|) \end{pmatrix} & \text{if } \alpha = 0, \qquad z \to 0. \\[2ex] \begin{pmatrix} \mathcal{O}(|z|^{\alpha/2}) & \mathcal{O}(|z|^{\alpha/2}) \\ \mathcal{O}(|z|^{\alpha/2}) & \mathcal{O}(|z|^{\alpha/2}) \end{pmatrix} & \text{if } \alpha < 0, \end{cases}$$

The advantage of using Z rather than Y is that the jump matrix for Z on $(0, \infty)$ is constant, which makes it more convenient when we take derivatives. Clearly Z' is analytic in $\mathbb{C} \setminus [0, \infty)$, and following the same reasoning as in Section 22.3.1 we see that

$$(Z')_+(x) = (Z')_-(x) \begin{pmatrix} e^{-i\pi\alpha} & 1 \\ 0 & e^{i\pi\alpha} \end{pmatrix}, \qquad x \in (0, \infty).$$

The behavior near infinity is given by

$$Z'(z) = \left(-\frac{1}{2}\sigma_3 - \frac{1}{2z} A_n \sigma_3 + \frac{1}{2z}(2n + \alpha)\sigma_3 + \mathcal{O}\left(1/z^2\right) \right)$$
$$\times z^{\sigma_3 n} (-z)^{\alpha\sigma_3/2} e^{-z\sigma_3/2},$$

and near the origin we have

$$
Z'(z) = \begin{cases} \begin{pmatrix} \mathcal{O}(|z|^{\alpha/2-1}) & \mathcal{O}(|z|^{-\alpha/2-1}) \\ \mathcal{O}(|z|^{\alpha/2-1}) & \mathcal{O}(|z|^{-\alpha/2-1}) \end{pmatrix} & \text{if } \alpha > 0, \\[2ex] \begin{pmatrix} \mathcal{O}(1) & \mathcal{O}(1/|z|) \\ \mathcal{O}(1) & \mathcal{O}(1/|z|) \end{pmatrix} & \text{if } \alpha = 0, \\[2ex] \begin{pmatrix} \mathcal{O}(|z|^{\alpha/2-1}) & \mathcal{O}(|z|^{\alpha/2-1}) \\ \mathcal{O}(|z|^{\alpha/2-1}) & \mathcal{O}(|z|^{\alpha/2-1}) \end{pmatrix} & \text{if } \alpha < 0, \end{cases} \qquad z \to 0.
$$

Let's now look at the matrix $Z'Z^{-1}$. This matrix is analytic in $\mathbb{C} \setminus [0,\infty)$ and $(Z'Z^{-1})_+ (x) = (Z'Z^{-1})_- (x)$ for $x \in (0,\infty)$ since both Z' and Z have the same jump matrix on $(0,\infty)$. Hence $Z'Z^{-1}$ is analytic in $\mathbb{C} \setminus \{0\}$ and has an isolated singularity at the origin. The behavior near the origin is

$$
Z'(z)Z^{-1}(z) = \begin{cases} \begin{pmatrix} \mathcal{O}(1/|z|) & \mathcal{O}(1/|z|) \\ \mathcal{O}(1/|z|) & \mathcal{O}(1/|z|) \end{pmatrix} & \text{if } \alpha \geq 0, \\[2ex] \begin{pmatrix} \mathcal{O}(|z|^{\alpha-1}) & \mathcal{O}(|z|^{\alpha-1}) \\ \mathcal{O}(|z|^{\alpha-1}) & \mathcal{O}(|z|^{\alpha-1}) \end{pmatrix} & \text{if } \alpha < 0, \end{cases} \qquad z \to 0,
$$

hence, since $\alpha > -1$ the singularity at the origin is at most a simple pole. Then $zZ'(z)Z^{-1}(z)$ is an entire function and the behavior near infinity is given by

$$
zZ'(z)Z^{-1}(z) = z \left(-\frac{1}{2}\sigma_3 - \frac{1}{2z}A_n\sigma_3 + \frac{1}{2z}(2n+\alpha)\sigma_3 + \mathcal{O}\left(1/z^2\right) \right)
$$
$$
\times \left(I - \frac{1}{z}A_n + \mathcal{O}\left(1/z^2\right) \right),
$$

hence Liouville's theorem gives

$$
zZ'(z)Z^{-1}(z) = -\frac{1}{2}\sigma_3 z + \frac{1}{2}(\sigma_3 A_n - A_n\sigma_3) + \frac{2n+\alpha}{2}\sigma_3, \qquad z \in \mathbb{C}.
$$

This means that

$$
zZ'(z) = \begin{pmatrix} -\frac{z-2n-\alpha}{2} & b_n \\ -c_n & \frac{z-2n+\alpha}{2} \end{pmatrix} Z(z), \qquad (22.4.4)
$$

where

$$
A_n = \begin{pmatrix} a_n & b_n \\ c_n & d_n \end{pmatrix}.
$$

Observe that

$$
Z(z) = \begin{pmatrix} (-z)^{\alpha/2}e^{-z/2}\ell_n^\alpha(z) & \dfrac{(-z)^{-\alpha/2}e^{z/2}}{2\pi i}\displaystyle\int_0^\infty \dfrac{\ell_n^\alpha(t)t^\alpha e^{-t}}{t-z}\,dt \\[3ex] e_n(-z)^{\alpha/2}e^{-z/2}\ell_{n-1}^\alpha(z) & e_n\,\dfrac{(-z)^{-\alpha/2}e^{z/2}}{2\pi i}\displaystyle\int_0^\infty \dfrac{\ell_{n-1}^\alpha(t)t^\alpha e^{-t}}{t-z}\,dt \end{pmatrix},
$$

where $e_n = -2\pi i \gamma_{n-1}^2$, hence if we look at the entry on the first row and first column of (22.4.4), then we get

$$z\left((-z)^{\alpha/2}e^{-z/2}\ell_n^\alpha(z)\right)'$$
$$= -\frac{z - 2n - \alpha}{2}(-z)^{\alpha/2}e^{-z/2}\ell_n^\alpha(z) + b_n e_n(-z)^{\alpha/2}e^{-z/2}\ell_{n-1}^\alpha(z),$$

which after simplification becomes

$$z\left[\ell_n^\alpha(z)\right]' = n\ell_n^\alpha(z) + b_n e_n \ell_{n-1}^\alpha(z). \tag{22.4.5}$$

In a similar way, the entry on the second row and first column of (22.4.4) gives

$$z e_n \left((-z)^{\alpha/2}e^{-z/2}\ell_{n-1}^\alpha(z)\right)'$$
$$= c_n(-z)^{\alpha/2}e^{-z/2}\ell_n^\alpha(z) + \frac{z - 2n - \alpha}{2}e_n(-z)^{\alpha/2}e^{-z/2}\ell_{n-1}^\alpha(z).$$

After simplification the factor $(-z)^{\alpha/2}e^{-z/2}$ can be removed, and if we check the coefficients of z^n in the resulting formula, then it follows that $c_n = -e_n$ and

$$z\left[\ell_{n-1}^\alpha(z)\right]' = -\ell_n^\alpha(z) + (z - n - \alpha)\ell_{n-1}^\alpha(z). \tag{22.4.6}$$

Elimination of ℓ_{n-1}^α from (22.4.5) and (22.4.6) gives the second order differential equation

$$z^2\left[\ell_n^\alpha(z)\right]'' + z(\alpha + 1 - z)\left[\ell_n^\alpha(z)\right]' = -\left[nz + b_n e_n - n(n + \alpha)\right]\ell_n^\alpha(z).$$

The left hand side contains z as a factor, hence we conclude that $b_n e_n = n(n + \alpha)$, and the differential equation becomes

$$z\left[\ell_n^\alpha\right]''(z) + (\alpha + 1 - z)\left[\ell_n^\alpha\right]'(z) = -n\ell_n^\alpha(z),$$

which corresponds to (4.6.16). If we recall that $\ell_n^\alpha = (-1)^n n! \, L_n^{(\alpha)}$, then (22.4.5) becomes

$$z\left[L_n^{(\alpha)}\right]'(z) = nL_n^{(\alpha)}(z) - (n + \alpha)L_{n-1}^{(\alpha)}(z),$$

which is (4.6.14). Formula (22.4.6) is

$$z\left[L_{n-1}^{(\alpha)}\right]'(z) = nL_n^{(\alpha)}(z) + (z - n - \alpha)L_{n-1}^{(\alpha)}(z).$$

Observe that we found that $-2\pi i \gamma_{n-1}^2 b_n = n(n + \alpha)$, so if we use this in the recurrence relation (22.2.8) then we see that

$$\ell_{n+1}^\alpha(z) = (z - \alpha_n)\,\ell_n^\alpha(z) - n(n + \alpha)\ell_{n-1}^\alpha(z).$$

If we evaluate (22.4.6) at $z = 0$ then we see that $\ell_n^\alpha(0) = (-1)^n(\alpha + 1)_n$. Use this in the recurrence relation to find that $\alpha_n = \alpha + 2n + 1$, so that

$$\ell_{n+1}^\alpha(z) = (z - \alpha - 2n - 1)\ell_n^\alpha(z) - n(n + \alpha)\ell_{n-1}^\alpha(z).$$

If we use the relation $\ell_n^\alpha = 2^n \, n! \, L_n^{(\alpha)}$ then this gives the recurrence relation (4.6.26).

22.5 Jacobi Polynomials

The next case deals with orthogonal polynomials on a bounded interval of the real line. Without loss of generality we can take the interval $[-1, 1]$. The Riemann–Hilbert problem now requires a jump condition on the open interval $(-1, 1)$ and extra conditions near both endpoints -1 and 1. Let us consider the Jacobi weight $w(x) = (1 - x)^\alpha(1 + x)^\beta$ on $[-1, 1]$, where $\alpha, \beta > -1$. The Riemann–Hilbert problem is then given by

1. Y is analytic in $\mathbb{C} \setminus [-1, 1]$.
2. (jump condition) On the open interval $(-1, 1)$ we have

$$Y_+(x) = Y_-(x) \begin{pmatrix} 1 & (1 - x)^\alpha(1 + x)^\beta \\ 0 & 1 \end{pmatrix}, \qquad x \in (-1, 1).$$

3. (normalization near infinity) Y has the following behavior near infinity

$$Y(z) = (I + \mathcal{O}(1/z)) \begin{pmatrix} z^n & 0 \\ 0 & z^{-n} \end{pmatrix}, \qquad z \to \infty.$$

4. (condition near ± 1) Y has the following behavior near 1

$$Y(z) = \begin{cases} \begin{pmatrix} \mathcal{O}(1) & \mathcal{O}(1) \\ \mathcal{O}(1) & \mathcal{O}(1) \end{pmatrix}, & \text{if } \alpha > 0, \\[2mm] \begin{pmatrix} \mathcal{O}(1) & \mathcal{O}(\log|z - 1|) \\ \mathcal{O}(1) & \mathcal{O}(\log|z - 1|) \end{pmatrix}, & \text{if } \alpha = 0, \qquad z \to 1. \\[2mm] \begin{pmatrix} \mathcal{O}(1) & \mathcal{O}(|z - 1|^\alpha) \\ \mathcal{O}(1) & \mathcal{O}(|z - 1|^\alpha) \end{pmatrix}, & \text{if } \alpha < 0, \end{cases}$$

Near -1 the behavior is

$$Y(z) = \begin{cases} \begin{pmatrix} \mathcal{O}(1) & \mathcal{O}(1) \\ \mathcal{O}(1) & \mathcal{O}(1) \end{pmatrix}, & \text{if } \beta > 0, \\[2mm] \begin{pmatrix} \mathcal{O}(1) & \mathcal{O}(\log|z + 1|) \\ \mathcal{O}(1) & \mathcal{O}(\log|z + 1|) \end{pmatrix}, & \text{if } \beta = 0, \qquad z \to -1. \\[2mm] \begin{pmatrix} \mathcal{O}(1) & \mathcal{O}(|z + 1|^\beta) \\ \mathcal{O}(1) & \mathcal{O}(|z + 1|^\beta) \end{pmatrix}, & \text{if } \beta < 0, \end{cases}$$

The unique solution of this Riemann–Hilbert problem is then given by

$$Y(z) = \begin{pmatrix} \tilde{P}_n^{(\alpha,\beta)}(z) & \dfrac{1}{2\pi i} \displaystyle\int_{-1}^1 \dfrac{\tilde{P}_n^{(\alpha,\beta)}(t)(1 - t)^\alpha(1 + t)^\beta}{t - z} \, dt \\[5mm] -2\pi i \gamma_{n-1}^2 \tilde{P}_{n-1}^{(\alpha,\beta)}(z) & -\gamma_{n-1}^2 \displaystyle\int_{-1}^1 \dfrac{\tilde{P}_{n-1}^{(\alpha,\beta)}(t)(1 - t)^\alpha(1 + t)^\beta}{t - z} \, dt \end{pmatrix},$$

$$(22.5.1)$$

where $\tilde{P}_n^{(\alpha,\beta)} = 2^n\, n!/(\alpha + \beta + n + 1)_n P_n^{(\alpha,\beta)}$ is the monic Jacobi polynomial and γ_{n-1} is the leading coefficient of the orthonormal Jacobi polynomial of degree $n - 1$. The proof is similar to the proof of Theorem 22.4.1 for Laguerre polynomials.

22.5.1 Differential equation

The three-term recurrence relation can be obtained in exactly the same way as before. The derivation of the differential equation is a bit different and hence we sketch how to obtain it from this Riemann–Hilbert problem. We need the complex functions $(z-1)^\alpha$ and $(z+1)^\beta$ which we define by

$$(z-1)^\alpha = |z-1|^\alpha e^{i\pi\alpha}, \qquad z = 1 + re^{i\theta}, \ \theta \in (-\pi,\pi),$$

so that $(z-1)^\alpha$ has a cut along $(-\infty, 1]$, and

$$(z+1)^\beta = |z+1|^\beta e^{i\pi\beta}, \qquad z = -1 + re^{i\theta}, \ \theta \in (-\pi,\pi),$$

so that $(z+1)^\beta$ has a cut along $(-\infty, -1]$. The function $(z-1)^\alpha(z+1)^\beta$ is now an analytic function on $\mathbb{C} \setminus (-\infty, 1]$. Observe that

$$[(z-1)^\alpha]_\pm = (1-x)^\alpha e^{\pm i\pi\alpha}, \qquad x \in (-\infty, 1), \tag{22.5.2}$$

and

$$[(z+1)^\beta]_\pm = (-1-x)^\beta e^{\pm i\pi\beta}, \qquad x \in (-\infty, -1). \tag{22.5.3}$$

Consider the matrix

$$Z(z) = Y(z) \begin{pmatrix} (z-1)^{\alpha/2}(z+1)^{\beta/2} & 0 \\ 0 & (z-1)^{-\alpha/2}(z+1)^{-\beta/2} \end{pmatrix}$$

$$= Y(z)(z-1)^{\sigma_3\alpha/2}(z+1)^{\sigma_3\beta/2},$$

where σ_3 is the Pauli matrix (22.3.3), then Z is analytic in $\mathbb{C} \setminus (-\infty, 1]$. This Z has a jump over the open interval $(-1, 1)$ but in addition we also created a jump over the interval $(-\infty, -1)$ by introducing the functions $(z-1)^{\pm\alpha/2}(z+1)^{\pm\beta/2}$. One easily verifies, using the jump condition of Y and the jumps (22.5.2)–(22.5.3), that

$$Z_+(x) = \begin{cases} Z_-(x) \begin{pmatrix} e^{i\pi\alpha} & 1 \\ 0 & e^{-i\pi\alpha} \end{pmatrix}, & x \in (-1,1), \\[4mm] Z_-(x) \begin{pmatrix} e^{i\pi(\alpha+\beta)} & 0 \\ 0 & e^{-i\pi(\alpha+\beta)} \end{pmatrix}, & x \in (-\infty, -1). \end{cases}$$

Observe that these jumps are constant on $(-1, 1)$ and $(-\infty, -1)$. Near infinity we have

$$Z(z) = \left(I + \frac{A_n}{z} + \mathcal{O}\left(1/z^2\right)\right) z^{\sigma_3 n}(z-1)^{\sigma_3\alpha/2}(z+1)^{\sigma_3\beta/2}, \qquad z \to \infty,$$

and near the points ± 1 we have

$$Z(z) = \begin{cases} \begin{pmatrix} \mathcal{O}(|z-1|^{\alpha/2}) & \mathcal{O}(|z-1|^{-\alpha/2}) \\ \mathcal{O}(|z-1|^{\alpha/2}) & \mathcal{O}(|z-1|^{-\alpha/2}) \end{pmatrix} & \text{if } \alpha > 0, \\[4mm] \begin{pmatrix} \mathcal{O}(1) & \mathcal{O}(\log|z-1|) \\ \mathcal{O}(1) & \mathcal{O}(\log|z-1|) \end{pmatrix} & \text{if } \alpha = 0, \qquad z \to 1, \\[4mm] \begin{pmatrix} \mathcal{O}(|z-1|^{\alpha/2}) & \mathcal{O}(|z-1|^{\alpha/2}) \\ \mathcal{O}(|z-1|^{\alpha/2}) & \mathcal{O}(|z-1|^{\alpha/2}) \end{pmatrix} & \text{if } \alpha < 0, \end{cases}$$

and

$$
Z(z) = \begin{cases}
\begin{pmatrix} \mathcal{O}(|z+1|^{\beta/2}) & \mathcal{O}(|z+1|^{-\beta/2}) \\ \mathcal{O}(|z+1|^{\beta/2}) & \mathcal{O}(|z+1|^{-\beta/2}) \end{pmatrix} & \text{if } \beta > 0, \\[2mm]
\begin{pmatrix} \mathcal{O}(1) & \mathcal{O}(\log|z+1|) \\ \mathcal{O}(1) & \mathcal{O}(\log|z+1|) \end{pmatrix} & \text{if } \beta = 0, \qquad z \to -1. \\[2mm]
\begin{pmatrix} \mathcal{O}(|z+1|^{\beta/2}) & \mathcal{O}(|z+1|^{\beta/2}) \\ \mathcal{O}(|z+1|^{\beta/2}) & \mathcal{O}(|z+1|^{\beta/2}) \end{pmatrix} & \text{if } \beta < 0,
\end{cases}
$$

We can again argue that Z' is analytic in $\mathbb{C} \setminus (-\infty, 1]$ with jumps

$$
(Z')_+(x) = \begin{cases}
(Z')_-(x) \begin{pmatrix} e^{i\pi\alpha} & 1 \\ 0 & e^{-i\pi\alpha} \end{pmatrix}, & x \in (-1,1), \\[3mm]
(Z')_-(x) \begin{pmatrix} e^{i\pi(\alpha+\beta)} & 0 \\ 0 & e^{-i\pi(\alpha+\beta)} \end{pmatrix}, & x \in (-\infty,-1).
\end{cases}
$$

The behavior near infinity is

$$
Z'(z) = \left[\left(I + \frac{A_n}{z} \right) \sigma_3 \left(\frac{n}{z} + \frac{\alpha}{2(z-1)} + \frac{\beta}{2(z+1)} \right) + \mathcal{O}\left(1/z^3\right) \right]
$$
$$
\times z^{\sigma_3 n}(z-1)^{\sigma_3\alpha/2}(z+1)^{\sigma_3\beta/2}, \qquad z \to \infty,
$$

and using

$$
\frac{1}{z-1} = \frac{1}{z} + \frac{1}{z^2} + \mathcal{O}\left(1/z^3\right), \qquad \frac{1}{z+1} = \frac{1}{z} - \frac{1}{z^2} + \mathcal{O}\left(1/z^3\right),
$$

this leads to

$$
Z'(z) = \left[\frac{2n+\alpha+\beta}{2z}\sigma_3 + \frac{1}{z^2}\left(-A_n + A_n\sigma_3\frac{2n+\alpha+\beta}{2} + \frac{\alpha-\beta}{2}\sigma_3 \right) \right.
$$
$$
\left. + \mathcal{O}\left(\frac{1}{z^3}\right) \right] z^{\sigma_3 n}(z-1)^{\sigma_3\alpha/2}(z+1)^{\sigma_3\beta/2}, \qquad z \to \infty.
$$

The behavior near ± 1 is

$$
Z'(z) = \begin{cases}
\begin{pmatrix} \mathcal{O}(|z-1|^{\alpha/2-1}) & \mathcal{O}(|z-1|^{-\alpha/2-1}) \\ \mathcal{O}(|z-1|^{\alpha/2-1}) & \mathcal{O}(|z-1|^{-\alpha/2-1}) \end{pmatrix} & \text{if } \alpha > 0, \\[2mm]
\begin{pmatrix} \mathcal{O}(1) & \mathcal{O}(1/|z-1|) \\ \mathcal{O}(1) & \mathcal{O}(1/|z-1|) \end{pmatrix} & \text{if } \alpha = 0, \qquad z \to 1, \\[2mm]
\begin{pmatrix} \mathcal{O}(|z-1|^{\alpha/2-1}) & \mathcal{O}(|z-1|^{\alpha/2-1}) \\ \mathcal{O}(|z-1|^{\alpha/2-1}) & \mathcal{O}(|z-1|^{\alpha/2-1}) \end{pmatrix} & \text{if } \alpha < 0,
\end{cases}
$$

and

$$
Z'(z) = \begin{cases}
\begin{pmatrix} \mathcal{O}(|z+1|^{\beta/2-1}) & \mathcal{O}(|z+1|^{-\beta/2-1}) \\ \mathcal{O}(|z+1|^{\beta/2-1}) & \mathcal{O}(|z+1|^{-\beta/2-1}) \end{pmatrix} & \text{if } \beta > 0, \\[2ex]
\begin{pmatrix} \mathcal{O}(1) & \mathcal{O}(1/|z+1|) \\ \mathcal{O}(1) & \mathcal{O}(1/|z+1|) \end{pmatrix} & \text{if } \beta = 0, \quad z \to -1. \\[2ex]
\begin{pmatrix} \mathcal{O}(|z+1|^{\beta/2-1}) & \mathcal{O}(|z+1|^{\beta/2-1}) \\ \mathcal{O}(|z+1|^{\beta/2-1}) & \mathcal{O}(|z+1|^{\beta/2-1}) \end{pmatrix} & \text{if } \beta < 0,
\end{cases}
$$

Now we look at the matrix $Z'Z^{-1}$. This matrix is analytic on $\mathbb{C} \setminus (-\infty, 1]$ and $(Z'Z^{-1})_+(x) = (Z'Z^{-1})_-(x)$ for $x \in (-1,1)$ and $x \in (-\infty,-1)$ since both Z' and Z have the same jumps on these intervals. Hence $Z'Z^{-1}$ is analytic in $\mathbb{C} \setminus \{-1,1\}$ and has isolated singularities at ± 1. The behavior near -1 and 1 implies that these singularities are simple poles, hence $(z^2-1)Z'(z)Z^{-1}(z)$ is an entire function and the behavior near infinity is given by

$$
(z^2-1)Z'(z)Z^{-1}(z) = \frac{2n+\alpha+\beta}{2}\sigma_3 z - A_n
$$
$$
+ \frac{2n+\alpha+\beta}{2}(A_n\sigma_3 - \sigma_3 A_n)
$$
$$
+ \frac{\alpha-\beta}{2}\sigma_3 + \mathcal{O}(1/z),
$$

hence if we set

$$
A_n = \begin{pmatrix} a_n & b_n \\ c_n & d_n \end{pmatrix},
$$

then Liouville's theorem implies that

$$
(z^2-1)Z'(z)
$$
$$
= \begin{pmatrix} \dfrac{2n+\alpha+\beta}{2}z - a_n + \dfrac{\alpha-\beta}{2} & -b_n(2n+\alpha+\beta+1) \\ c_n(2n+\alpha+\beta-1) & -\dfrac{2n+\alpha+\beta}{2}z - d_n - \dfrac{\alpha-\beta}{2} \end{pmatrix} Z(z).
$$
$$(22.5.4)$$

If we work out the entry on the first row and first column of (22.5.4) then we find

$$
(1-z^2)\left[\tilde{P}_n^{(\alpha,\beta)}(z)\right]'
$$
$$
= (-nz + a_n)\tilde{P}_n^{(\alpha,\beta)}(z) + b_n e_n(2n+\alpha+\beta+1)\tilde{P}_{n-1}^{(\alpha,\beta)}(z), \quad (22.5.5)
$$

where $e_n = -2\pi i\gamma_{n-1}^2$. Similarly, if we work out the entry on the second row and first column, then we can first check the coefficient of z^n to find that $c_n = e_n$, and with that knowledge we find

$$
(1-z^2)\left[\tilde{P}_{n-1}^{(\alpha,\beta)}(z)\right]'
$$
$$
= -(2n+\alpha+\beta-1)\tilde{P}_n^{(\alpha,\beta)}(z) + [(n+\alpha+\beta)z + d_n + \alpha - \beta]\tilde{P}_{n-1}^{(\alpha,\beta)}(z).
$$
$$(22.5.6)$$

If we eliminate $\tilde{P}_{n-1}^{(\alpha,\beta)}$ from (22.5.5) and (22.5.6) then we find

$$(1 - z^2)^2 \left[\tilde{P}_n^{(\alpha,\beta)}(z)\right]''$$

$$- (1 - z^2)\left[(\alpha + \beta + 2)z + \alpha - \beta + a_n + d_n\right]\left[\tilde{P}_n^{(\alpha,\beta)}(z)\right]'$$

$$= \left[-n\left(1 - z^2\right) - b_n e_n(2n + \alpha + \beta - 1)(2n + \alpha + \beta + 1)\right. \qquad (22.5.7)$$

$$-a_n\left(d_n + \alpha - \beta\right)$$

$$\left. + z\left[n\left(d_n + \alpha - \beta\right) - a_n(n + \alpha + \beta)\right] + n(n + \alpha + \beta)z^2\right]$$

$$\times \tilde{P}_n^{(\alpha,\beta)}(z).$$

The left hand side of this equation has $1 - z^2$ as a factor, so the right hand side should also have $1 - z^2$ as a factor, and since ± 1 are not zeros of $\tilde{P}_n^{(\alpha,\beta)}(z)$, the coefficient of z in the factor on the right hand side must be zero, which gives

$$n\left(d_n + \alpha - \beta\right) = a_n(n + \alpha + \beta) = 0.$$

Observe that $\det Y = 1$, hence since $Y = \left[I + A_n/z + \mathcal{O}\left(1/z^2\right)\right]z^{\sigma_3 n}$ we must have $\det\left[I + A_n/z + \mathcal{O}\left(1/z^2\right)\right] = 1$. This gives

$$\begin{vmatrix} 1 + a_n/z + \mathcal{O}\left(1/z^2\right) & b_n/z + \mathcal{O}\left(1/z^2\right) \\ c_n/z + \mathcal{O}\left(1/z^2\right) & 1 + d_n/z + \mathcal{O}\left(1/z^2\right) \end{vmatrix}$$

$$= 1 + \frac{a_n + d_n}{z} + \mathcal{O}\left(1/z^2\right),$$

so that $d_n = -a_n$. Solving for a_n then gives

$$a_n = \frac{n(\alpha - \beta)}{2n + \alpha + \beta}, \qquad d_n = \frac{-n(\alpha - \beta)}{2n + \alpha + \beta}. \qquad (22.5.8)$$

Put $z = \pm 1$ in the factor on the right hand side of (22.5.7), then we see that

$$b_n e_n = \frac{4n(n + \alpha + \beta)(n + \alpha)(n + \beta)}{(2n + \alpha + \beta - 1)(2n + \alpha + \beta)^2(2n + \alpha + \beta + 1)}. \qquad (22.5.9)$$

The factor $1 - z^2$ can now be canceled on both sides of (22.5.7) and we get

$$(1 - z^2)\left[\tilde{P}_n^{(\alpha,\beta)}(z)\right]'' - \left[(\alpha + \beta + 2)z + \alpha - \beta\right]\left[\tilde{P}_n^{(\alpha,\beta)}(z)\right]'$$

$$= -n(n + \alpha + \beta + 1)\tilde{P}_n^{(\alpha,\beta)}(z), \qquad (22.5.10)$$

which corresponds to the differential equation (4.2.6).

If we use the relation $\tilde{P}_n^{(\alpha,\beta)}(z) = 2^n n!/(\alpha + \beta + n + 1)_n P_n^{(\alpha,\beta)}(z)$, where $P_n^{(\alpha,\beta)}(z)$ is the usual Jacobi polynomial (see Chapter 4), and if we use (22.5.8)–(22.5.9), then (22.5.5) becomes (3.3.16) and (22.5.6) changes to

$$(2n + \alpha + \beta)\left[P_{n-1}^{(\alpha,\beta)}\right]'(z)$$

$$= -2n(n + \alpha + \beta)P_n^{(\alpha,\beta)}(z) + (n + \alpha + \beta)[(2n + \alpha + \beta)z + \alpha - \beta]P_{n-1}^{(\alpha,\beta)}(z).$$

Finally, we can use (22.5.8) and (22.5.9) in the recurrence relation (22.2.8) to find that

$$\tilde{P}_{n+1}^{(\alpha,\beta)}(z) = \left(z + \frac{\alpha^2 - \beta^2}{(2n+\alpha+\beta)(2n+\alpha+\beta+2)}\right)\tilde{P}_n^{(\alpha,\beta)}(z)$$
$$- \frac{4n(n+\alpha)(n+\beta)n+\alpha+\beta)}{(2n+\alpha+\beta-1)(2n+\alpha+\beta)^2(2n+\alpha+\beta+1)}\tilde{P}_{n-1}^{(\alpha,\beta)}(z).$$

If we use the relation $\tilde{P}_n^{(\alpha,\beta)}(z) = 2^n n!/(\alpha+\beta+n+1)_n P_n^{(\alpha,\beta)}(z)$ then this recurrence relation corresponds to (4.2.9).

22.6 Asymptotic Behavior

One of the main advantages of the Riemann–Hilbert approach for orthogonal polynomials is that this is a very useful setting to obtain uniform asymptotics valid in the whole complex plane. The idea is to transform the initial Riemann–Hilbert problem (22.2.1) in a few steps to another equivalent Riemann–Hilbert problem for a matrix valued function R which is analytic in $\mathbb{C} \setminus \Sigma$, where Σ is a collection of oriented contours in the complex plane. This new Riemann–Hilbert is normalized at infinity, so that $R(z) = I + \mathcal{O}(1/z)$ as $z \to \infty$, and the jumps on each of the contours Γ in Σ are uniformly close to the identity matrix:

$$R_+(z) = R_-(z)[I + \mathcal{O}(1/n)], \qquad z \in \Gamma.$$

One can then conclude that the solution R of this model Riemann–Hilbert problem will be close to the identity matrix

$$R(z) = I + \mathcal{O}(1/n), \qquad n \to \infty,$$

uniformly for $z \in \mathbb{C}$. By reversing the steps we can then go back from R to the original matrix Y in (22.2.2) and read of the required asymptotic behavior as $n \to \infty$.

The transformation from Y to R goes as follows:

1. Transform Y to T such that T satisfies a Riemann–Hilbert problem with a simple jump on $\mathbb{R}$ and such that T is normalized at infinity: $T(z) = I + \mathcal{O}(1/z)$ as $z \to \infty$. This step requires detailed knowledge of the asymptotic zero distribution of the orthogonal polynomials and uses relevant properties of the logarithmic potential of this zero distribution. The jump matrix on $\mathbb{R}$ will contain oscillatory terms on the interval where the zeros are dense.

2. Transform T to S such that S is still normalized at infinity but we deform the contour $\mathbb{R}$ to a collection Σ_S of contours such that the jumps of S on each contour in Σ_S are no longer oscillatory. This deformation is similar to a contour deformation in the steepest descent method for obtaining asymptotics of an oscillatory integral and hence this is known as a steepest descent method for Riemann–Hilbert problems. It was first developed by Deift and Zhou (Deift & Zhou, 1993).

3. Some of the jumps for S are close to the identity matrix. Ignoring these jumps, one arrives at a normalized Riemann–Hilbert problem for P with

jumps on $\Sigma_P \subset \Sigma_S$. This P is expected to be close to S as $n \to \infty$ and it will be called the *parametrix* for the outer region.

4. At the endpoints and at the intersection points of the contours in Σ_S the jumps for S will usually not be close to the identity matrix. Around these points z_k we need to make a local analysis of the Riemann–Hilbert problem. Around each endpoint or intersection point z_k we need to construct a *local parametrix* P_k, which is the solution of a Riemann–Hilbert problem with jumps on the contours in the neighborhood of the point z_k under investigation and such that this P_k matches the parametrix P on a contour Γ_k encircling z_k up to terms of order $\mathcal{O}(1/n)$.

5. Transform S to R by setting $R = SP^{-1}$ away from the points z_k, and $R = SP_k^{-1}$ in the neighborhood of z_k. This R will then be normalized at ∞ and it will have jumps on a collection of contours Σ_R which contains parts of the contours in $\Sigma_S \setminus \Sigma_P$ and the contours Γ_k encircling the endpoints/intersection points z_k. All these jumps are uniformly close to the identity matrix.

This looks like a reasonably simple recipe, but working out the details for a particular weight w (or a particular family of orthogonal polynomials) usually requires some work.

- The case where $w(x) = \exp\left(-nx^{2m}\right)$, with m an integer, has been worked out in detail by Deift (Deift, 1999). The case $m = 1$ gives uniform asymptotics for the Hermite polynomials which improves the Plancherel–Rotach asymptotics.
- The case where $w(x) = e^{-Q(x)}$ on $\mathbb{R}$, where Q is a polynomial of even degree with positive leading coefficient, has been worked out by Deift et al. (Deift et al., 1999b). The case $Q(x) = N\left(x^4 + tx^2\right)$ for parameter values $t < 0$ and $N > 0$ was investigated by Bleher and Its (Bleher & Its, 1999).
- Freud weights $w(x) = \exp\left(-|x|^\beta\right)$, with $\beta > 0$ are worked out in detail by Kriecherbauer and McLaughlin (Kriecherbauer & McLaughlin, 1999).
- The case where $w(x) = e^{-nV(x)}$ on $\mathbb{R}$, where V is real valued and analytic on $\mathbb{R}$ and

$$\lim_{|x| \to \infty} V(x)/\log(1 + x^2) = \infty,$$

has been worked out by Deift et al. (Deift et al., 1999a). An overview of the Riemann–Hilbert approach for this case and the three previous cases can be found in (Deift et al., 2001).

- The case where $w(x) = (1-x)^\alpha (1+x)^\beta h(x)$ on $[-1, 1]$, where h is a strictly positive real analytic function on $[-1, 1]$, has been worked out in detail in (Kuijlaars et al., 2004). The case where $h = 1$ gives strong asymptotics for Jacobi polynomials. See also Kuijlaars' lecture (Kuijlaars, 2003) for a readable exposition of this case.

- Generalized Jacobi weights of the form

$$w(x) = (1 - x)^\alpha (1 + x)^\beta h(x) \prod_{j=1}^{p} |x - x_j|^{2\lambda_j}, \qquad x \in [-1, 1],$$

where $\alpha, \beta, 2\lambda_j > -1$ for $j = 1, \ldots, p$, with $-1 < x_1 < \cdots < x_p < 1$ and h

is real analytic and strictly positive on $[-1, 1]$, were investigated by Vanlessen in (Vanlessen, 2003).

- Laguerre polynomials with α large and negative were investigated by Kuijlaars and McLaughlin (Kuijlaars & McLaughlin, 2001).
- The case where $w(x) = 1/\left(1 + x^2\right)^{N+n}$ on $\mathbb{R}$ was worked out by Gawronski and Van Assche (Gawronski & Van Assche, 2003). The corresponding orthogonal polynomials are known as relativistic Hermite polynomials.

22.7 Discrete Orthogonal Polynomials

The Riemann–Hilbert problem of Fokas–Its–Kitaev works whenever the orthogonal polynomials have a weight function which is sufficiently smooth. Recently, Baik and his co-authors (Baik et al., 2003) and (Baik et al., 2002) have formulated an *interpolation problem* which gives a characterization of discrete orthogonal polynomials which is similar to the Riemann–Hilbert problem. This interpolation problem is no longer a boundary value problem, but a problem in which one looks for a meromorphic matrix function Y for which the residues at a set of given poles satisfy a relation similar to the jump condition of the Riemann–Hilbert problem.

The Baik–Kriecherbauer–McLaughlin–Miller interpolation problem is to find a 2×2 matrix function Y such that

1. Y is analytic in $\mathbb{C} \setminus X_N$, where $X_N = \{x_{1,N}, x_{2,N}, \ldots, x_{N,N}\}$ is a set of real nodes.
2. (residue condition) At each node $x_{k,N}$ the first column of Y is analytic and the second column has a simple pole. The residue satisfies

$$\operatorname*{Res}_{z = x_{k,N}} Y(z) = \lim_{z \to x_{k,N}} Y(z) \begin{pmatrix} 0 & w_{k,N} \\ 0 & 0 \end{pmatrix}, \qquad (22.7.1)$$

where $w_{k,N} > 0$ are given weights.
3. (normalization) Near infinity one has

$$Y(z) = (I + \mathcal{O}(1/z)) \begin{pmatrix} z^n & 0 \\ 0 & z^{-n} \end{pmatrix}, \qquad z \to \infty.$$

Theorem 22.7.1 *The interpolation problem* (22.7.1) *has a unique solution when* $0 \leq n \leq N - 1$, *which for* $n = 0$ *is given by*

$$Y(z) = \begin{pmatrix} 1 & \sum_{k=1}^{N} \dfrac{w_{k,N}}{z - x_{k,N}} \\ 0 & 1 \end{pmatrix}, \qquad (22.7.2)$$

and for $1 \leq n \leq N - 1$ is given by

$$
Y(z) = \begin{pmatrix} P_{n,N}(z) & \displaystyle\sum_{k=1}^{N} \frac{w_{k,N} P_{n,N}(x_{k,N})}{z - x_{k,N}} \\ \gamma_{n-1,N}^2 P_{k,N}(z) & \gamma_{n-1,N}^2 \displaystyle\sum_{k=1}^{N} \frac{w_{k,N} P_{n-1,N}(x_{k,N})}{z - x_{k,N}} \end{pmatrix},
\tag{22.7.3}
$$

where $P_{n,N}$ are the monic discrete orthogonal polynomials on X_N for which

$$
\sum_{k=1}^{N} P_{m,N}(x_{k,N}) P_{n,N}(x_{k,N}) w_{k,N} = \frac{\delta_{m,n}}{\gamma_{n,N}^2}.
$$

Proof See (Baik et al., 2002) for a complete proof. □

   This interpolation problem can be transformed into a usual Riemann–Hilbert boundary value problem by removing the poles in favor of jumps on contours. This requires detailed knowledge of the asymptotic zero distribution of the discrete orthogonal polynomials. The resulting Riemann–Hilbert problem can then be analyzed asymptotically using the steepest descent method of Deift and Zhou. The Hahn polynomials were investigated in detail in (Baik et al., 2002) using this approach. The asymptotic results for Krawtchouk polynomials, obtained by Ismail and Simeonov (Ismail & Simeonov, 1998) using an integral representation and the classical method of steepest descent, can also be obtained using this Riemann–Hilbert approach, and then the strong asymptotics can be extended to hold everywhere (uniformly) in the complex plane.

22.8 Exponential Weights

This section is based on (Wang & Wong, 2005b) and states uniform Plancherel–Rotach asymptotics for orthogonal polynomials with exponential weights. Let

$$
w(x) = e^{-v(x)}, \quad v(x) = \sum_{r=0}^{2m} v_k x^k, \quad v_{2m} > 0, \quad m \geq 1.
\tag{22.8.1}
$$

   The Mhaskar–Rakhmanov–Saff (MRS) numbers r_n and s_n are determined by the equations

$$
\frac{1}{2\pi} \int_{r_n}^{s_n} \frac{v'(t)\,(t - r_n)}{\sqrt{(s_n - t)(t - r_n)}}\, dt = n, \quad \frac{1}{2\pi} \int_{r_n}^{s_n} \frac{v'(t)\,(s_n - t)}{\sqrt{(s_n - t)(t - r_n)}}\, dt = -n.
\tag{22.8.2}
$$

The existence of the MRS numbers for sufficiently large n has been established in (Deift et al., 1999a), where a convergent series representation in powers of $n^{-1/2m}$ has been given. Set

$$
c_n = (s_n - r_n)/2, \quad d_n = (r_n + s_n)/2.
\tag{22.8.3}
$$

The zeros of the corresponding orthogonal polynomials live in $[r_n, s_n]$. The numbers c_n and d_n give the radius and the center of $[r_n, s_n]$ and have the power series representations

$$c_n = n^{1/2m} \sum_{\ell=0}^{\infty} c^{(\ell)} n^{-\ell/2m}, \quad c^{(0)} = (m v_{2m} A_m)^{-1/2m}, \quad c^{(1)} = 0,$$

$$d_n = \sum_{\ell=0}^{\infty} d^{(\ell)} n^{-\ell/2m}, \qquad d^{(0)} = -v_{2m-1}/(2m v_{2m}), \tag{22.8.4}$$

where

$$A_m = \frac{(1/2)_m}{m!} = \prod_{j=1}^{m} \frac{2j-1}{2j}, \quad m \ge 1. \tag{22.8.5}$$

Set

$$\lambda_n(z) = c_n z + d_n, \quad V_n(z) = \frac{1}{n} v(\lambda_n(z)). \tag{22.8.6}$$

Clearly, V_n is a polynomial of degree $2m$ with leading term asymptotically equal to $1/(m A_m) > 0$ and all other coefficients tend to zero as $n \to \infty$. Let Γ_z be a simple, closed positively oriented contour containing $[-1, 1]$ and $\{z\}$ in its interior. Define the function $h_n(z)$ via

$$h_n(z) = \frac{1}{2\pi i} \int_{\Gamma_z} \frac{V_n'(\zeta)}{\sqrt{\zeta^2 - 1}} \frac{d\zeta}{\zeta - z}. \tag{22.8.7}$$

In order to state the Wang–Wong theorem, we need a few more notations. Set

$$\ell_n := \frac{1}{\pi} \int_{-1}^{1} \sqrt{1 - t^2} \, h_n(t) \ln |t| \, dt - V_n(0),$$

$$\psi_n(z) := \frac{1}{2\pi} (1 - z)^{1/2} (1 + z)^{1/2} h_n(z), \quad z \notin (-\infty, -1] \cup [1, \infty),$$

$$\xi_n(z) := -2\pi i \int_{1}^{z} \psi_n(\zeta) \, d\zeta, \quad z \in \mathbb{C} \setminus (-\infty, -1) \cup (1, \infty). \tag{22.8.8}$$

Define φ_n on $\mathbb{C} \setminus \mathbb{R}$ by

$$\varphi_n(z) = \begin{cases} -\frac{1}{2} \xi_n(z) & \text{for } \operatorname{Im} z > 0, \\ \frac{1}{2} \xi_n(z) & \text{for } \operatorname{Im} z < 0. \end{cases} \tag{22.8.9}$$

One can easily verify that

$$\begin{aligned} (\varphi_n)_+(x) &= (\varphi_n)_-(x) & x > 1, \\ (\varphi_n)_+(x) &= (\varphi_n)_-(x) - 2\pi i, & x < -1, \end{aligned} \tag{22.8.10}$$

hence we can analytically continue $e^{n\varphi_n(z)}$ to $\mathbb{C} \setminus [-1, 1]$. The function

$$\zeta_n(z) := \left[\frac{3}{2} \varphi_n(z) \right]^{2/3} \tag{22.8.11}$$

has the property $(\zeta_n)_+(x) = (\zeta_n)_-(x)$ for $x \in (1, 1)$, hence $\zeta_n(z)$ has an analytic continuation to $\mathbb{C} \setminus (-\infty, -1]$.

We now state the main result of this section.

Theorem 22.8.1 *Let* $\{P_n(x)\}$ *be monic polynomials orthogonal with respect to* w *of* (22.8.1) *and assume* $\int_{\mathbb{R}} w(x)\,dx = 1$. *Then*

$$P_n\left(c_n z + d_n\right) = \sqrt{\pi}\, c_n^n \exp\left(\frac{1}{2} n\ell_n + \frac{1}{2} nV_n(z)\right)$$

$$\left\{n^{1/6} Ai\left(n^{2/3}\zeta_n\right) \mathbf{A}(z,n) - n^{-1/6} A'i\left(n^{2/3}\zeta_n\right) \mathbf{B}(z,n)\right\},$$
(22.8.12)

where $\mathbf{A}(z,n)$ *and* $\mathbf{B}(z,n)$ *are analytic functions of* z *in* $\mathbb{C} \smallsetminus (-\infty, 1]$. *Moreover, when* z *is bounded away from* $(-\infty, -1]$, $\mathbf{A}(z,n)$ *and* $\mathbf{B}(z,n)$ *have the uniform asymptotic expansions*

$$\mathbf{A}(z,n) \sim \frac{\zeta_n^{1/4}(z)}{a(z)}\left[1 + \sum_{k=2m}^{\infty} \frac{A_k(z)}{n^{k/2m}}\right],$$

$$\mathbf{B}(z,n) \sim \frac{a(z)}{\zeta_n^{1/4}(z)}\left[1 + \sum_{k=2m}^{\infty} \frac{B_k(z)}{n^{k/2m}}\right],$$
(22.8.13)

and the coefficients $\{A_k(z)\}$ *and* $\{B_k(z)\}$ *are analytic functions in* $\mathbb{C} \smallsetminus [-\infty, -1]$. *The function* $a(z)$ *is*

$$a(z) = (z-1)^{1/4}/(z+1)^{1/4}.$$
(22.8.14)

The proof of Theorem 22.8.1 uses Riemann–Hilbert problem techniques. The case $v = x^4 + c$ was proved in (Rui & Wong, 1999).

23
Multiple Orthogonal Polynomials

Let μ be a given positive measure with moments $m_n (= \int_{\mathbb{R}} x^n \, d\mu(x))$. The nth degree monic orthogonal polynomial P_n is defined by requiring that

$$\int_{\mathbb{R}} P_n(x)x^k \, d\mu(x) = 0, \qquad k = 0, 1, \ldots, n-1, \qquad (23.0.1)$$

and the nth degree orthonormal polynomial $p_n = \gamma_n P_n$ is defined by taking γ_n from

$$\int_{\mathbb{R}} P_n(x)x^n \, d\mu(x) = \frac{1}{\gamma_n^2}, \qquad \gamma_n > 0. \qquad (23.0.2)$$

The system (23.0.1) is a linear system of n equations for the n unknown coefficients $a_{k,n}$ $(k = 1, \ldots, n)$ of the monic polynomial $P_n(x) = \sum_{k=0}^{n} a_{k,n} x^{n-k}$, with $a_{0,n} = 1$. This system of equations always has a unique solution since the matrix of the system is the Gram matrix

$$\begin{pmatrix} m_0 & m_1 & m_2 & \cdots & m_{n-1} \\ m_1 & m_2 & m_3 & \cdots & m_n \\ m_2 & m_3 & m_4 & \cdots & m_{n+1} \\ \vdots & \vdots & \vdots & \cdots & \vdots \\ m_{n-1} & m_n & m_{n+1} & \cdots & m_{2n-2} \end{pmatrix},$$

which is a positive definite matrix whenever the support of μ contains at least n points, see Chapter 2.

Multiple orthogonal polynomials are polynomials of one variable which are defined by orthogonality relations with respect to r different measures $\mu_1, \mu_2, \ldots, \mu_r$, where $r \geq 1$ is a positive integer. These polynomials should not be confused with multivariate or multivariable orthogonal polynomials of several variables. Other terminology is also in use:

- Hermite–Padé polynomials (Nuttall, 1984) is often used because of the link with Hermite–Padé approximation or simultaneous Padé approximation (de Bruin, 1985), (Bultheel et al., 2005), (Sorokin, 1984), and (Sorokin, 1990).
- Polyorthogonal polynomials is used in (Nikishin & Sorokin, 1991).

- The so-called d-orthogonal polynomials (Ben Cheikh & Douak, 2000a), (Ben Cheikh & Zaghouani, 2003), (Douak, 1999), (Douak & Maroni, 1995), and (Maroni, 1989) correspond to multiple orthogonal polynomials near the diagonal and d refers to the number of orthogonality measures (which we denote by r).
- Polynomials of simultaneous orthogonality is used in (Kaliaguine & Ronveaux, 1996).
- Multiple orthogonal polynomials are also studied as vector orthogonal polynomials (Kaliaguine, 1995), (Sorokin & Van Iseghem, 1997), and (Van Iseghem, 1987) and are related to vector continued fractions.

23.1 Type I and II multiple orthogonal polynomials

In this chapter we will often be using multi-indices in our notation. A multi-index $\vec{n} \in \mathbb{N}^r$ is of the form $\vec{n} = (n_1, \ldots, n_r)$, with each $n_j \geq 0$, and its size is given by $|\vec{n}| = n_1 + n_2 + \cdots + n_r$.

We distinguish between two types of multiple orthogonal polynomials. *Type I multiple orthogonal polynomials* are collected in a vector $(A_{\vec{n},1}, \ldots, A_{\vec{n},r})$ of r polynomials, where $A_{\vec{n},j}$ has degree at most $n_j - 1$, such that

$$\sum_{j=1}^{r} \int_{\mathbb{R}} x^k A_{\vec{n},j} \, d\mu_j(x) = 0, \qquad k = 0, 1, 2, \ldots, |\vec{n}| - 2, \qquad (23.1.1)$$

and

$$\sum_{j=1}^{r} \int_{\mathbb{R}} x^{|\vec{n}|-1} A_{\vec{n},j} \, d\mu_j(x) = 1. \qquad (23.1.2)$$

This gives a linear system of $|\vec{n}|$ equations for the $|\vec{n}|$ unknown coefficients of the polynomials $A_{\vec{n},j}$ ($j = 1, 2, \ldots, r$). We say that the index $\vec{n}$ is normal for type I if the relations (23.1.1)–(23.1.2) determine the polynomials $(A_{\vec{n},1}, \ldots, A_{\vec{n},r})$ uniquely. The matrix of the linear system is given by

$$M_{\vec{n}} = \left(M_{n_1}^{(1)} \quad M_{n_2}^{(2)} \quad \cdots \quad M_{n_r}^{(r)} \right),$$

where each $M_{n_k}^{(k)}$ is a $|\vec{n}| \times n_k$ matrix containing the moments of μ_k:

$$M_{n_k}^{(k)} = \begin{pmatrix} m_0^{(k)} & m_1^{(k)} & m_2^{(k)} & \cdots & m_{n_k-1}^{(k)} \\ m_1^{(k)} & m_2^{(k)} & m_3^{(k)} & \cdots & m_{n_k}^{(k)} \\ m_2^{(k)} & m_3^{(k)} & m_4^{(k)} & \cdots & m_{n_k+1}^{(k)} \\ \vdots & \vdots & \vdots & \cdots & \vdots \\ m_{|\vec{n}|-1}^{(k)} & m_{|\vec{n}|}^{(k)} & m_{|\vec{n}|+1}^{(k)} & \cdots & m_{|\vec{n}|+n_k-2}^{(k)} \end{pmatrix}.$$

Hence $\vec{n}$ is a normal index for type I if $\det M_{\vec{n}} \neq 0$. This condition gives some restriction on the measures $\mu_1, \ldots, \mu_r$. If all multi-indices are normal, then $(\mu_1, \ldots, \mu_r)$ is a *perfect system*.

A monic polynomial $P_{\vec{n}}$ is a *type II multiple orthogonal polynomial* if $P_{\vec{n}}$ is of degree $|\vec{n}|$ and

$$\int_{\mathbb{R}} P_{\vec{n}}(x)x^k \, d\mu_1(x) = 0, \qquad k = 0, 1, \ldots, n_1 - 1,$$

$$\int_{\mathbb{R}} P_{\vec{n}}(x)x^k \, d\mu_2(x) = 0, \qquad k = 0, 1, \ldots, n_2 - 1,$$

$$\vdots \qquad\qquad (23.1.3)$$

$$\int_{\mathbb{R}} P_{\vec{n}}(x)x^k \, d\mu_r(x) = 0, \qquad k = 0, 1, \ldots, n_r - 1.$$

The conditions (23.1.3) give a linear system of $|\vec{n}|$ equations for the $|\vec{n}|$ unknown coefficients of the monic polynomial $P_{\vec{n}}$. If this system has a unique solution, then we say that $\vec{n}$ is a normal index for type II. The matrix of this linear system is given by

$$\begin{pmatrix} \left[M_{n_1}^{(1)}\right]' \\ \left[M_{n_2}^{(2)}\right]' \\ \vdots \\ \left[M_{n_r}^{(r)}\right]' \end{pmatrix} = M_{\vec{n}}',$$

which is the transpose of the matrix for type I, and hence a multi-index is normal for type II if $\det M_{\vec{n}} \neq 0$. Clearly a multi-index is normal for type II if and only if it is normal for type I, hence we just talk of normal indices.

If $\det M_{\vec{n}} = 0$, so that $\vec{n}$ is not a normal index, then the system of equations (23.1.1), together with

$$\sum_{j=1}^{r} \int_{\mathbb{R}} x^{|\vec{n}|-1} A_{\vec{n},j}(x) \, d\mu_j(x) = 0, \qquad (23.1.4)$$

has non-trivial solutions $(A_{\vec{n},1}, \ldots, A_{\vec{n},r})$, which are all called type I multiple orthogonal polynomials for the index $\vec{n}$. Similarly, if $\det M_{\vec{n}} = 0$ then the system of equations (23.1.3) has solutions $P_{\vec{n}}$ where the degree is strictly less than $|\vec{n}|$, and these polynomials are all called type II multiple orthogonal polynomials. For a normal index the degree of the type II multiple orthogonal polynomial $P_{\vec{n}}$ is exactly equal to $|\vec{n}|$ (and we choose $P_{\vec{n}}$ to be monic), and for the type I multiple orthogonal polynomials the normalization (23.1.2) holds.

Corollary 23.1.1 *If $\vec{n}$ is a normal index, then the polynomial $A_{\vec{n}+\vec{e}_j,j}$ has degree exactly n_j for every $j \in \{1, 2, \ldots, r\}$.*

Proof The vector $\left(A_{\vec{n}+\vec{e}_j,1}, \ldots, A_{\vec{n}+\vec{e}_j,r}\right)$ satisfies the orthogonality relations (23.1.1) and (23.1.4). If $A_{\vec{n}+\vec{e}_j,j}$ has degree $< n_j$, then these are $|\vec{n}|$ homogeneous equations for the $|\vec{n}|$ coefficients of the polynomials $\left(A_{\vec{n}+\vec{e}_j,1}, \ldots, A_{\vec{n}+\vec{e}_j,r}\right)$, and the matrix of this linear system is $M_{\vec{n}}$. Since $\vec{n}$ is a normal index, we conclude that $M_{\vec{n}}$ is not singular, but then $\left(A_{\vec{n}+\vec{e}_j,1}, \ldots, A_{\vec{n}+\vec{e}_j,r} \right)$ is the trivial vector $(0, \ldots, 0)$. This is a contradiction since there is always a non-trivial vector of type I multiple orthogonal polynomials when $\vec{n} \neq \vec{0}$. $\qquad\square$

Corollary 23.1.2 *If $\vec{n}$ is a normal index, then for every $j \in \{1, 2, \ldots, r\}$ one has*

$$\int_{\mathbb{R}} x^{n_j-1} P_{\vec{n}-\vec{e}_j}(x)\, d\mu_j(x) \neq 0.$$

Proof If the integral vanishes, then $P_{\vec{n}-\vec{e}_j}$ satisfies the orthogonality conditions (23.1.3) for a type II multiple orthogonal polynomials with multi-index $\vec{n}$, so there is a polynomial of degree $\leq |\vec{n}| - 1$ satisfying the orthogonality conditions for index $\vec{n}$. This is in contradiction with the normality of $\vec{n}$. $\qquad\square$

23.1.1 Angelesco systems

An Angelesco system $(\mu_1, \ldots, \mu_r)$ consists of r measures such that the convex hull of the support of each measure μ_i is a closed interval $[a_i, b_i]$ and all open intervals $(a_1, b_1), \ldots, (a_r, b_r)$ are disjoint. Observe that the closed intervals are allowed to touch each other. Such a system was first introduced by Angelesco in 1919 (Angelesco, 1919) in the framework of algebraic continued fractions. Such a system is of interest because all the multi-indices are normal for the multiple orthogonal polynomials. Furthermore we can easily locate the sets where the zeros of the type II multiple orthogonal polynomials are.

Theorem 23.1.3 *If $P_{\vec{n}}$ is a type II multiple orthogonal polynomial of index $\vec{n} = (n_1, n_2, \ldots, n_r)$ for an Angelesco system $(\mu_1, \ldots, \mu_r)$, and if the support of each μ_i contains infinitely many points, then $P_{\vec{n}}$ has n_i zeros in the open interval (a_i, b_i) for each $i \in \{1, 2, \ldots, r\}$.*

Proof Let m_i be the number of sign changes of $P_{\vec{n}}$ in the open interval (a_i, b_i) and suppose that $m_i < n_i$ for some i with $1 \leq i \leq r$. Let q_{i,m_i} be the monic polynomial of degree m_i for which the zeros are the points in (a_i, b_i) where $P_{\vec{n}}$ changes sign, then

$$\int_{a_i}^{b_i} P_{\vec{n}}(x)\, q_{i,m_i}(x)\, d\mu_i(x) \neq 0$$

since the integrand does not change sign on $[a_i, b_i]$ and the support of μ_i contains infinitely many points. But the orthogonality (23.1.3) implies that this integral is 0. This contradiction means that $m_i \geq n_i$ for every $i \in \{1, 2, \ldots, r\}$. The intervals $(a_1, b_1), \ldots, (a_r, b_r)$ are all disjoint, so in total the number of sign changes of $P_{\vec{n}}$ on the real line is $\geq |\vec{n}|$. But since $P_{\vec{n}}$ is of degree $\leq |\vec{n}|$, we therefore have $m_i = n_i$ for each i. Each sign change therefore corresponds to a zero of multiplicity one. Hence $P_{\vec{n}}$ has degree $|\vec{n}|$, which implies that $\vec{n}$ is a normal index. $\qquad\square$

The polynomial $P_{\vec{n}}$ can therefore be factored as

$$P_{\vec{n}}(x) = q_{\vec{n},1}(x) q_{\vec{n},2}(x) \cdots q_{\vec{n},r}(x),$$

where each $q_{\vec{n},j}$ is a polynomial of degree n_j with its zeros on (a_j, b_j). The orthogonality (23.1.3) then gives

$$\int_{a_j}^{b_j} x^k q_{\vec{n},j}(x) \left(\prod_{i \neq j} q_{\vec{n},i}(x) \right) d\mu_j(x) = 0, \qquad k = 0, 1, \ldots, n_j - 1. \quad (23.1.5)$$

The product $\prod_{i \neq j} q_{\vec{n},i}(x)$ does not change sign on (a_j, b_j), hence (23.1.5) shows that $q_{\vec{n},j}$ is an ordinary orthogonal polynomial of degree n_j on the interval $[a_j, b_j]$ with respect to the measure $\prod_{i \neq j} |q_{\vec{n},i}(x)| \, d\mu_j(x)$. This measure depends on the multi-index $\vec{n}$. Hence many properties of the multiple orthogonal polynomials for an Angelesco system can be obtained from the theory of ordinary orthogonal polynomials.

23.1.2 AT systems

A Chebyshev system $\{\varphi_1, \ldots, \varphi_n\}$ on $[a, b]$ is a system of n linearly independent functions such that every linear combination $\sum_{k=1}^{n} a_k \varphi_k$ has at most $n - 1$ zeros on $[a, b]$. This is equivalent with the condition that

$$\det \begin{pmatrix} \varphi_1(x_1) & \varphi_1(x_2) & \cdots & \varphi_1(x_n) \\ \varphi_2(x_1) & \varphi_2(x_2) & \cdots & \varphi_2(x_n) \\ \vdots & \vdots & \cdots & \vdots \\ \varphi_n(x_1) & \varphi_n(x_2) & \cdots & \varphi_n(x_n) \end{pmatrix} \neq 0$$

for every choice of n different points $x_1, \ldots, x_n \in [a, b]$. Indeed, when $x_1, \ldots, x_n$ are such that the determinant is zero, then there is a linear combination of the rows that gives a zero row, but this means that for this linear combination $\sum_{k=1}^{n} a_k \varphi_k$ has zeros at $x_1, \ldots, x_n$, giving n zeros, which is not allowed.

A system $(\mu_1, \ldots, \mu_r)$ of r measures is an algebraic Chebyshev system (AT system) for the multi-index $\vec{n}$ if each μ_j is absolutely continuous with respect to a measure μ on $[a, b]$ with $d\mu_j(x) = w_j(x) \, d\mu(x)$, where μ has an infinite support and the w_j are such that

$$\left\{ w_1, x w_1, \ldots, x^{n_1 - 1} w_1, w_2, x w_2, \ldots, x^{n_2 - 1} w_2, \ldots, w_r, x w_r, \ldots, x^{n_r - 1} w_r \right\}$$

is a Chebyshev system on $[a, b]$.

Theorem 23.1.4 *Suppose $\vec{n}$ is a multi-index such that $(\mu_1, \ldots, \mu_r)$ is an AT system on $[a, b]$ for every index $\vec{m}$ for which $m_j \leq n_j$ $(1 \leq j \leq r)$. Then $P_{\vec{n}}$ has $|\vec{n}|$ zeros on (a, b) and hence $\vec{n}$ is a normal index.*

Proof Let $x_1, \ldots, x_m$ be the sign changes of $P_{\vec{n}}$ on (a, b) and suppose that $m < |\vec{n}|$.

We can then find a multi-index $\vec{m}$ such that $|\vec{m}| = m$ and $m_j \leq n_j$ for every $1 \leq j \leq r$ and $m_k < n_k$ for one index k with $1 \leq k \leq r$. Consider the interpolation problem where we want to find a function

$$L(x) = \sum_{j=1}^{r} q_j(x)\, w_j(x),$$

where q_j is a polynomial of degree $m_j - 1$ if $j \neq k$ and q_k is a polynomial of degree m_k, that satisfies the interpolation conditions

$$L(x_j) = 0, \qquad j = 1, 2, \ldots, m,$$
$$L(x_0) = 1, \qquad \text{for some other point } x_0 \in [a, b].$$

This interpolation problem has a unique solution since this involves the Chebyshev system with multi-index $\vec{m} + \vec{e}_k$. This function L has m zeros (by construction) and it is not identically zero, hence the Chebyshev property implies that L has exactly m zeros and each zero is a sign change. This means that $P_{\vec{n}}L$ has no sign changes on (a, b), and hence

$$\sum_{j=1}^{r} \int_a^b P_{\vec{n}}(x)\, q_j(x)\, d\mu_j(x) \neq 0.$$

But the orthogonality (23.1.3) implies that each term in the sum is zero. This contradiction implies that $m \geq |\vec{n}|$, but since $P_{\vec{n}}$ has degree $\leq |\vec{n}|$, we must conclude that $m = |\vec{n}|$ and hence $P_{\vec{n}}$ is a polynomial of degree $|\vec{n}|$ with all its zeros on (a, b). $\qquad\square$

We introduce a partial order relation on multi-indices by saying that $\vec{m} \leq \vec{n}$ whenever $m_j \leq n_j$ for every j with $1 \leq j \leq r$. The previous theorem then states that $\vec{n}$ is a normal index whenever $(\mu_1, \ldots, \mu_r)$ is an AT system on $[a, b]$ for every $\vec{m} \leq \vec{n}$.
There is a similar result for type I multiple orthogonal polynomials.

Theorem 23.1.5 *Suppose $\vec{n}$ is a multi-index such that $(\mu_1, \ldots, \mu_r)$ is an AT system on $[a, b]$ for every index $\vec{m}$ for which $\vec{m} \leq \vec{n}$. Then $\sum_{j=1}^{r} A_{\vec{n},j} w_j$ has $|\vec{n}| - 1$ sign changes on (a, b).*

Proof Let $x_1, \ldots, x_m$ be the sign changes of $\sum_{j=1}^{r} A_{\vec{n},j} w_j$ on (a, b) and suppose that $m < |\vec{n}| - 1$. Let π_m be the polynomial $\pi_m(x) = (x - x_1) \cdots (x - x_m)$, then $\pi_m \sum_{j=1}^{r} A_{\vec{n},j} w_j$ does not change sign on (a, b), hence

$$\sum_{j=1}^{r} \int_a^b A_{\vec{n},j}(x) \pi_m(x)\, d\mu_j(x) \neq 0.$$

But the orthogonality (23.1.1) implies that this sum is equal to zero. This contradiction shows that $m \geq |\vec{n}| - 1$. The sum $\sum_{j=1}^{r} A_{\vec{n},j} w_j$ is a linear combination of the Chebyshev system for the multi-index $\vec{n}$ and hence it has at most $|\vec{n}| - 1$ zeros on $[a, b]$. We therefore conclude that $m = |\vec{n}| - 1$. $\qquad\square$

Every $A_{\vec{n},j}$ has exactly degree $n_j - 1$ because otherwise $\sum_{j=1}^{r} A_{\vec{n},j} w_j$ is a sum involving a Chebyshev system with index $\vec{m} \leq \vec{n}$ and $\vec{m} \neq \vec{n}$, so that $|\vec{m}| < |\vec{n}|$, and in such a Chebyshev system the sum can have at most $|\vec{m}| - 1 < |\vec{n}| - 1$ zeros on $[a, b]$, which contradicts the result in Theorem 23.1.5.

23.1.3 Biorthogonality

In an AT system every measure μ_k is absolutely continuous with respect to a given measure μ on $[a, b]$ and $d\mu_k(x) = w_k(x) \, d\mu(x)$. In an Angelesco system we can define $\mu = \mu_1 + \mu_2 + \cdots + \mu_r$ and if all the intervals $[a_j, b_j]$ are disjoint, then each measure μ_k is absolutely continuous with respect to μ and $d\mu_k(x) = w_k(x) \, d\mu(x)$, with $w_k = \chi_{[a_k,b_k]}$ the characteristic function for the interval $[a_k, b_k]$, i.e.,

$$\chi_{[a_k,b_k]}(x) = \begin{cases} 1, & \text{if } x \in [a_k, b_k], \\ 0, & \text{if } x \notin [a_k, b_k]. \end{cases}$$

In case the intervals $[a_j, b_j]$ and $[a_{j+1}, b_{j+1}]$ are touching, with $b_j = a_{j+1}$, then one needs to be a bit more careful with possible Dirac measures at the common point $b_j = a_{j+1}$. If $\mu_j = \hat{\mu}_j + c_1 \delta_{b_j}$ and $\mu_{j+1} = \hat{\mu}_{j+1} + c_2 \delta_{a_{j+1}}$, where $\hat{\mu}_j$ and $\hat{\mu}_{j+1}$ have no mass at $b_j = a_{j+1}$, then the absolute continuity with respect to $\mu = \mu_1 + \mu_2 + \cdots + \mu_r$ still holds, but with

$$w_j = \chi_{(a_j,b_j)} + \frac{c_1}{c_1 + c_2} \chi_{\{b_j\}}$$

$$w_{j+1} = \chi_{(a_{j+1},b_{j+1})} + \frac{c_2}{c_1 + c_2} \chi_{\{a_{j+1}\}}.$$

Hence for an AT system and an Angelesco system we have $d\mu_k(x) = w_k(x) \, d\mu(x)$ for $1 \leq k \leq r$. For the type I multiple orthogonal polynomials we then define the functions

$$Q_{\vec{n}}(x) = \sum_{j=1}^{r} A_{\vec{n},j}(x) w_j(x). \tag{23.1.6}$$

The orthogonality (23.1.1) then becomes

$$\int_a^b Q_{\vec{n}}(x) x^k \, d\mu(x) = 0, \qquad k = 0, 1, \ldots, |\vec{n}| - 2, \tag{23.1.7}$$

and the normalization (23.1.2) becomes

$$\int_a^b Q_{\vec{n}}(x) x^{|\vec{n}|-1} \, d\mu(x) = 1. \tag{23.1.8}$$

The type II multiple orthogonal polynomials $P_{\vec{n}}$ and these type I functions $Q_{\vec{m}}$ turn out to satisfy a certain biorthogonality.

Theorem 23.1.6 *The following biorthogonality holds for type I and type II multiple orthogonal polynomials:*

$$\int_a^b P_{\vec{n}}(x)Q_{\vec{m}}(x)\,d\mu(x) = \begin{cases} 0, & \text{if } \vec{m} \leq \vec{n}, \\ 0, & \text{if } |\vec{n}| \leq |\vec{m}| - 2, \\ 1, & \text{if } |\vec{m}| = |\vec{n}| + 1, \end{cases} \qquad (23.1.9)$$

where $Q_{\vec{m}}$ is given by (23.1.6).

Proof If we use the definition (23.1.6), then

$$\int_{\mathbb{R}} P_{\vec{n}}(x)Q_{\vec{m}}(x)\,d\mu(x) = \sum_{j=1}^r \int_{\mathbb{R}} P_{\vec{n}}(x)A_{\vec{m},j}(x)\,d\mu_j(x).$$

Every $A_{\vec{m},j}$ has degree $\leq m_j - 1$, hence if $\vec{m} \leq \vec{n}$ then $m_j - 1 \leq n_j - 1$ and

$$\int_{\mathbb{R}} P_{\vec{n}}(x)A_{\vec{m},j}(x)\,d\mu_j(x) = 0$$

follows from the type II orthogonality (23.1.3). This proves the result when $\vec{m} \leq \vec{n}$.

The type II multiple orthogonal polynomial $P_{\vec{n}}$ has degree $\leq |\vec{n}|$, hence if $|\vec{n}| \leq |\vec{m}| - 2$ then the orthogonality (23.1.7) shows that the integral $\int_{\mathbb{R}} P_{\vec{n}}(x)Q_{\vec{m}}(x)\,d\mu(x)$ vanishes for $|\vec{n}| \leq |\vec{m}| - 2$.

Finally, if $|\vec{m}| = |\vec{n}| + 1$ then $P_{\vec{n}}$ is a monic polynomial of degree $|\vec{m}| - 1$ so that

$$\sum_{j=1}^r \int_{\mathbb{R}} P_{\vec{n}}(x)A_{\vec{m},j}(x)\,d\mu_j(x) = \int_{\mathbb{R}} x^{|\vec{m}|-1}Q_{\vec{m}}(x)\,d\mu(x) = 1,$$

where the last equality follows from (23.1.8). $\qquad\square$

Observe that Theorem 23.1.6 does not give the value of the integral of $P_{\vec{n}}Q_{\vec{m}}$ for all possible multi-indices $\vec{n}$ and $\vec{m}$, but the indices described by the theorem are useful in many situations.

23.1.4 Recurrence relations

Recall that monic orthogonal polynomials on the real line satisfy a three-term recurrence relation of the form

$$P_{n+1}(x) = (x - \alpha_n)\,P_n(x) - \beta_n\,P_{n-1}(x),$$

where $\alpha_n \in \mathbb{R}$ and $\beta_n > 0$. Multiple orthogonal polynomials also satisfy a finite order recurrence relation but, since we are working with multi-indices, there are several ways to decrease or increase the degree of the multiple orthogonal polynomials. Let $\{\vec{m}_k,\ k = 0, 1, \ldots, |\vec{n}|\}$ be a path from $\vec{0} = (0, 0, \ldots, 0)$ to $\vec{n} = (n_1, n_2, \ldots, n_r)$ with $\vec{m}_0 = \vec{0}$, $\vec{m}_{|\vec{n}|} = \vec{n}$, where in each step the multi-index $\vec{m}_k$ is increased by one at exactly one component, so that for some j with $1 \leq j \leq r$

$$\vec{m}_{k+1} = \vec{m}_k + \vec{e}_j.$$

For such a path we have $|\vec{m}_k| = k$ and $\vec{m}_k \leq \vec{m}_{k+1}$.

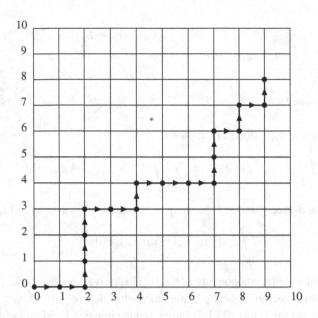

Fig. 23.1. A path from $(0,0)$ to $(9,8)$ for $r = 2$

Theorem 23.1.7 *Let* $(\pi(1), \pi(2), \ldots, \pi(r))$ *be a permutation of* $(1, 2, \ldots, r)$ *and let*

$$\vec{s}_j = \sum_{i=1}^{j} \vec{e}_{\pi(i)}, \qquad 1 \leq j \leq r.$$

Choose $k \in \{1, 2, \ldots, r\}$ *and suppose that all multi-indices* $\vec{m} \leq \vec{n} + \vec{e}_k$ *are normal. Then*

$$x P_{\vec{n}}(x) = P_{\vec{n}+\vec{e}_k}(x) + a_{\vec{n},0}(k) P_{\vec{n}}(x) + \sum_{j=1}^{r} a_j(\vec{n}) P_{\vec{n}-\vec{s}_j}(x), \qquad (23.1.10)$$

where $a_{\vec{n},0}(k)$ *and the* $a_j(\vec{n})$ *are real numbers.*

Observe that the right-hand side in (23.1.10) contains $r + 2$ terms. For $r = 1$ this reduces to the usual three-term recurrence relation.

Proof Let $\{\vec{m}_k, \ k = 0, 1, \ldots, |\vec{n}|\}$ be a path from $\vec{0}$ to $\vec{n}$ so that the last $r + 1$ multi-indices are $\vec{n} - \vec{s}_j = \vec{m}_{|\vec{n}|-j}$ for $j = 1, \ldots, r$ and $\vec{n} = \vec{m}_{|\vec{n}|}$. The polynomials $P_{\vec{m}_j}$ $(0 \leq j \leq |\vec{n}|)$ are monic and of degree j, hence they are a basis for the linear space of polynomials of degree $\leq |\vec{n}|$. Clearly $x P_{\vec{n}}(x) - P_{\vec{n}+\vec{e}_k}(x)$ is a polynomial of degree $\leq |\vec{n}|$, hence we can write

$$x P_{\vec{n}}(x) = P_{\vec{n}+\vec{e}_k}(x) + \sum_{j=0}^{|\vec{n}|} c_j(\vec{n}) P_{\vec{m}_j}(x). \qquad (23.1.11)$$

Multiply both sides of the equation by $Q_{\vec{m}_\ell}$ and observe that $\vec{m}_\ell \leq \vec{m}_j$ if and only if $\ell \leq j$, then Theorem 23.1.6 gives

$$\int_{\mathbb{R}} P_{\vec{m}_j}(x) Q_{\vec{m}_\ell}(x) \, d\mu(x) = 0, \qquad \ell \leq j.$$

Furthermore we observe that $|\vec{m}_j| < |\vec{m}_\ell|$ if and only if $j < \ell$, hence Theorem 23.1.6 also gives

$$\int_{\mathbb{R}} P_{\vec{m}_j}(x) Q_{\vec{m}_\ell}(x) \, d\mu(x) = 0, \qquad j \leq \ell - 2.$$

For $j = \ell - 1$ Theorem 23.1.6 gives

$$\int_{\mathbb{R}} P_{\vec{m}_{\ell-1}}(x) Q_{\vec{m}_\ell}(x) \, d\mu(x) = 1.$$

All this shows that

$$\int_{\mathbb{R}} x P_{\vec{n}}(x) Q_{\vec{m}_\ell}(x) \, d\mu(x) = c_{\ell-1}(\vec{n}), \qquad \ell = 1, 2, \ldots, |\vec{n}|. \qquad (23.1.12)$$

The left-hand side is of the form

$$\sum_{j=1}^{r} \int_{\mathbb{R}} P_{\vec{n}}(x) x A_{\vec{m}_\ell,j}(x) \, d\mu_j(x).$$

Observe that

$$\vec{m}_{|\vec{n}|-r} = (n_1 - 1, n_2 - 1, \ldots, n_r - 1)$$

and

$$\vec{m}_\ell \leq (n_1 - 1, n_2 - 1, \ldots, n_r - 1)$$

whenever $\ell \leq |\vec{n}| - r$. Hence, when $\ell \leq |\vec{n}| - r$ we see that $x A_{\vec{m}_\ell,j}$ is a polynomial of degree $\leq n_j - 1$, and hence by the orthogonality (23.1.3) we have

$$\sum_{j=1}^{r} \int_{\mathbb{R}} P_{\vec{n}}(x) x A_{\vec{m}_\ell,j}(x) \, d\mu_j(x) = 0, \qquad \ell \leq |\vec{n}| - r.$$

Using this in (23.1.12) implies that

$$c_{\ell-1}(\vec{n}) = 0, \qquad \ell \leq |\vec{n}| - r,$$

which gives

$$x P_{\vec{n}}(x) = P_{\vec{n}+\vec{e}_k}(x) + \sum_{j=|\vec{n}|-r}^{|\vec{n}|} c_j(\vec{n}) P_{\vec{m}_j}(x).$$

If we define $a_j(\vec{n}) = c_{|\vec{n}|-j}(\vec{n})$ for $1 \leq j \leq r$ and $a_{\vec{n},0}(k) = c_{|\vec{n}|}(\vec{n})$, then this gives the required recurrence relation (23.1.10). $\qquad \square$

Using (23.1.12) we see that the coefficients in the recurrence relation (23.1.10) are explicitly given by

$$a_j(\vec{n}) = \int_{\mathbb{R}} x P_{\vec{n}}(x) Q_{\vec{n}-\vec{s}_{j-1}}(x) \, d\mu(x), \qquad 1 \le j \le r, \qquad (23.1.13)$$

where $\vec{s}_0 = \vec{0}$. For $j = r$ we multiply both sides of (23.1.10) by $Q_{\vec{n}+\vec{e}_k}$. Theorem 23.1.6 then gives

$$\int_{\mathbb{R}} P_{\vec{n}+\vec{e}_k}(x) \, Q_{\vec{n}+\vec{e}_k}(x) \, d\mu(x) = 0$$

and

$$\int_{\mathbb{R}} P_{\vec{n}}(x) \, Q_{\vec{n}+\vec{e}_k}(x) \, d\mu(x) = 1,$$

so that

$$a_{\vec{n},0}(k) = \int_{\mathbb{R}} x P_{\vec{n}}(x) \, Q_{\vec{n}+\vec{e}_k}(x) \, d\mu(x). \qquad (23.1.14)$$

Observe that the coefficients $a_j(\vec{n})$ for $j < r$ do not depend on k.

Corollary 23.1.8 *If $k \ne \ell$ then*

$$P_{\vec{n}+\vec{e}_k} - P_{\vec{n}+\vec{e}_\ell} = d_{\vec{n}}(k, \ell) \, P_{\vec{n}}(x), \qquad (23.1.15)$$

where $d_{\vec{n}}(k, \ell) = a_{\vec{n},0}(\ell) - a_{\vec{n},0}(k)$.

Proof Subtract the recurrence relation (23.1.10) with k and with ℓ, then most terms cancel since the recurrence coefficients $a_j(\vec{n})$ with $j < r$ do not depend on k or ℓ. The only terms left give the desired formula. $\qquad \square$

The recurrence relation (23.1.10) is of order $r + 1$, hence we should have $r + 1$ linearly independent solutions. The type II multiple orthogonal polynomials are one solution. Other solutions are given by

$$S_{\vec{n},\ell}(x) = \int_{\mathbb{R}} \frac{P_{\vec{n}}(t)}{x - t} \, d\mu_\ell(t), \qquad 1 \le \ell \le r.$$

Indeed, we have

$$x S_{\vec{n},\ell}(x) = \int_{\mathbb{R}} P_{\vec{n}}(t) \, d\mu_\ell(t) + \int_{\mathbb{R}} \frac{t P_{\vec{n}}(t)}{x - t} \, d\mu_\ell(t).$$

Applying the recurrence relation (23.1.10) to the integrand in the last integral gives

$$x S_{\vec{n},\ell}(x) = S_{\vec{n}+\vec{e}_k,\ell}(x) + a_{\vec{n},0}(k) S_{\vec{n},\ell}(x) + \sum_{j=1}^{r} a_j(\vec{n}) S_{\vec{n}-\vec{s}_j,\ell}(x)$$

whenever $n_\ell > 0$.

The type I multiple orthogonal polynomials also satisfy a finite order recurrence relation.

Theorem 23.1.9 *Let π be a permutation on $(1, 2, \ldots, r)$ and let*

$$\vec{s}_j = \sum_{i=1}^{j} \vec{e}_{\pi(i)}, \qquad 1 \le j \le r.$$

Suppose that all multi-indices $\vec{m} \le \vec{n}$ are normal. Then

$$xQ_{\vec{n}}(x) = Q_{\vec{n}-\vec{e}_k}(x) + b_{\vec{n},0}(k) Q_{\vec{n}}(x) + \sum_{j=1}^{r} b_j(\vec{n}) Q_{\vec{n}+\vec{s}_j}(x), \qquad (23.1.16)$$

where $b_{\vec{n},0}(k)$ and the $b_j(\vec{n})$ are real numbers.

Proof Let $\{\vec{m}_j, \ j = 0, 1, 2, \ldots, |\vec{n}| + r\}$ be a path from $\vec{m}_0 = \vec{0}$ to $\vec{m}_{|\vec{n}|+r} = (n_1 + 1, n_2 + 1, \ldots, n_r + 1)$, such that $\vec{m}_{|\vec{n}|} = \vec{n}$, $\vec{m}_{|\vec{n}|+j} = \vec{n} + \vec{s}_j$ for $1 \le j \le r$ and $\vec{m}_{|\vec{n}|-1} = \vec{n} - \vec{e}_k$. Then we can write

$$xQ_{\vec{n}}(x) = \sum_{j=1}^{|\vec{n}|+r} \hat{c}_j(\vec{n}) Q_{\vec{m}_j}(x).$$

We don't need the index $j = 0$ since $Q_{\vec{0}} = 0$. Multiply both sides of this equation by $P_{\vec{m}_\ell}$ and integrate, to find that

$$\hat{c}_{\ell+1}(\vec{n}) = \int_{\mathbb{R}} xQ_{\vec{n}}(x) P_{\vec{m}_\ell}(x) \, d\mu(x).$$

This integral is 0 whenever $\ell + 1 \le |\vec{n}| - 2$, hence the expansion reduces to

$$xQ_{\vec{n}}(x) = \sum_{j=|\vec{n}|-1}^{|\vec{n}|+r} \hat{c}_j(\vec{n}) Q_{\vec{m}_j}(x).$$

For $j = |\vec{n}| - 1$ we have

$$\hat{c}_{|\vec{n}|-1}(\vec{n}) = \int_{\mathbb{R}} xQ_{\vec{n}}(x) P_{\vec{m}_{|\vec{n}|-2}}(x) \, d\mu(x)$$

$$= \int_{\mathbb{R}} x^{|\vec{n}|-1} Q_{\vec{n}}(x) \, d\mu(x)$$

$$= 1.$$

If we define $b_j(\vec{n}) = \hat{c}_{|\vec{n}|+j}(\vec{n})$ for $1 \le j \le r$ and $b_{\vec{n},0}(k) = \hat{c}_{|\vec{n}|}(\vec{n})$, then the required recurrence relation (23.1.16) follows. $\square$

Observe that the recurrence coefficients for type I are given by

$$b_j(\vec{n}) = \int_{\mathbb{R}} xQ_{\vec{n}}(x) P_{\vec{n}+\vec{s}_{j-1}}(x) \, d\mu(x), \qquad 1 \le j \le r, \qquad (23.1.17)$$

where $\vec{s}_0 = \vec{0}$, and that

$$b_{\vec{n},0}(k) = \int_{\mathbb{R}} xQ_{\vec{n}}(x) P_{\vec{n}-\vec{e}_k}(x) \, d\mu(x) \qquad (23.1.18)$$

is the only coefficient which depends on k.

Corollary 23.1.10 *If $k \neq \ell$ then*

$$Q_{\vec{n} - \vec{e}_k} - Q_{\vec{n} - \vec{e}_\ell} = \hat{d}_{\vec{n}}(k, \ell) \, Q_{\vec{n}}(x), \qquad (23.1.19)$$

where $\hat{d}_{\vec{n}}(k, \ell) = b_{\vec{n},0}(\ell) - b_{\vec{n},0}(k)$.

Theorem 23.1.9 implies that each component $A_{\vec{n},\ell}$ of the vector of type I multiple orthogonal polynomials satisfies the same recurrence relation

$$x A_{\vec{n},\ell}(x) = A_{\vec{n}-\vec{e}_k,\ell}(x) + \sum_{j=0}^{r} b_j \, (\vec{n}) \, A_{\vec{n}_j,\ell}(x) \qquad (23.1.20)$$

but with different initial conditions: $A_{\vec{n},\ell} = 0$ whenever $n_\ell \leq 0$. This gives r linearly independent solutions of the recurrence relation (23.1.16), which is of order $r + 1$. Yet another solution is given by

$$R_{\vec{n}}(x) = \int_{\mathbb{R}} \frac{Q_{\vec{n}}(t)}{x - t} \, d\mu(t),$$

because

$$x R_{\vec{n}}(x) = \int_{\mathbb{R}} Q_{\vec{n}}(t) \, d\mu(t) + \int_{\mathbb{R}} \frac{t Q_{\vec{n}}(t)}{x - t} \, d\mu(t)$$

and if we apply the recurrence relation (23.1.16) to the integrand of the last integral, then

$$x R_{\vec{n}}(x) = R_{\vec{n}-\vec{e}_k}(x) + \sum_{j=0}^{r} b_j \, (\vec{n}) \, R_{\vec{n}_j}(x),$$

whenever $|\vec{n}| \geq 2$.

The recurrence relation (23.1.10) gives a relation between type II multiple orthogonal with multi-indices ranging from $(n_1 - 1, n_2 - 1, \ldots, n_r - 1)$ to $(n_1, n_2, \ldots, n_r)$ and $\vec{n} + \vec{e}_k$. Another interesting recurrence relation connects type II multiple orthogonal polynomials $P_{\vec{n}}$ with type II multiple orthogonal polynomials with one multi-index $\vec{n} + \vec{e}_k$ and all contiguous multi-indices $\vec{n} - \vec{e}_j$ $(1 \leq j \leq r)$.

Theorem 23.1.11 *Suppose $\vec{n}$ and $\vec{n} + \vec{e}_k$ are normal indices. Then*

$$x P_{\vec{n}}(x) = P_{\vec{n}+\vec{e}_k}(x) + a_{\vec{n},0}(k) P_{\vec{n}}(x) + \sum_{j=1}^{r} a_{\vec{n},j} P_{\vec{n}-\vec{e}_j}(x), \qquad (23.1.21)$$

where

$$a_{\vec{n},0}(k) = \int_{\mathbb{R}} x P_{\vec{n}}(x) \, Q_{\vec{n}+\vec{e}_k}(x) \, d\mu(x), \qquad (23.1.22)$$

and

$$a_{\vec{n},\ell} = \frac{\int_{\mathbb{R}} x^{n_\ell} P_{\vec{n}}(x) \, d\mu_\ell(x)}{\int_{\mathbb{R}} x^{n_\ell - 1} P_{\vec{n}-\vec{e}_\ell}(x) \, d\mu_\ell(x)}. \qquad (23.1.23)$$

Proof Since $\vec{n}$ and $\vec{n} + \vec{e}_k$ are normal indices, both the polynomials $P_{\vec{n}}(x)$ and $P_{\vec{n}+\vec{e}_k}(x)$ are monic, and hence $xP_{\vec{n}}(x) - P_{\vec{n}+\vec{e}_k}(x)$ is a polynomial of degree $\leq |\vec{n}|$. By choosing $a_{\vec{n},0}(k)$ appropriately we can also cancel the term containing $x^{|\vec{n}|}$ so that $xP_{\vec{n}}(x) - P_{\vec{n}+\vec{e}_k}(x) - a_{\vec{n},0}(k)\,P_{\vec{n}}(x)$ is a polynomial of degree $\leq |\vec{n}| - 1$. It is easy to verify that this polynomial is orthogonal to polynomials of degree $\leq n_j - 2$ with respect to μ_j for $j = 1, 2, \ldots, r$. The linear space $\mathcal{A}$ which consists of polynomials of degree $\leq |\vec{n}| - 1$ which are orthogonal to polynomials of degree $\leq n_j - 2$ with respect to μ_j for $1 \leq j \leq r$ corresponds to the linear space $A \subset \mathbb{R}^{|\vec{n}|}$ of coefficients c of polynomials of degree $\leq |\vec{n}| - 1$, satisfying the homogeneous system of linear equations $\widetilde{M_{\vec{n}}}\mathbf{c} = \mathbf{0}$, where $\widetilde{M_{\vec{n}}}$ is obtained from the moment matrix $M'_{\vec{n}}$ by deleting r rows. The normalility of $\vec{n}$ implies that the rank of $\widetilde{M_{\vec{n}}}$ is $|\vec{n}| - r$ and hence the linear space A has dimension r. Each polynomial $P_{\vec{n}-\vec{e}_j}$ belongs to the linear space $\mathcal{A}$ and the r polynomials $P_{\vec{n}-\vec{e}_j}$ are linearly independent since if we set

$$\sum_{j=1}^{r} a_j\, P_{\vec{n}-\vec{e}_j} = 0,$$

then multiplying by $x^{n_\ell-1}$ and integrating with respect to μ_ℓ gives

$$a_\ell \int_{\mathbb{R}} x^{n_\ell-1}\, P_{\vec{n}-\vec{e}_\ell}(x)\, d\mu_\ell(x) = 0,$$

and by Corollary 23.1.2 this shows that $a_\ell = 0$ for $\ell = 1, 2, \ldots, r$. Hence we can write $xP_{\vec{n}}(x) - P_{\vec{n}+\vec{e}_k}(x) - a_{\vec{n},0}(k)P_{\vec{n}}(x)$ as a linear combination of this basis in $\mathcal{A}$, as in (23.1.21). If we multiply both sides of the equation (23.1.21) by $x^{n_\ell-1}$ and integrate with respect to μ_ℓ, then (23.1.23) follows. If we multiply both sides of (23.1.21) by $Q_{\vec{n}+e_k}$ and then use the biorthogonality (23.1.9), then (23.1.22) follows. $\qquad\square$

A similar recurrence relation for continuous multi-indices also holds for type I multiple orthogonal polynomials.

Theorem 23.1.12 *Suppose that $\vec{n}$ and $\vec{n} - \vec{e}_k$ are normal indices. Then*

$$xQ_{\vec{n}}(x) = Q_{\vec{n}-\vec{e}_k}(x) + b_{\vec{n},0}(k)\, Q_{\vec{n}}(x) + \sum_{j=1}^{r} b_{\vec{n},j}\, Q_{\vec{n}+\vec{e}_j}(x), \qquad (23.1.24)$$

where

$$b_{\vec{n},0}(k) = \int_{\mathbb{R}} xQ_{\vec{n}}(x)\, P_{\vec{n}-\vec{e}_k}(x)\, d\mu(x), \qquad (23.1.25)$$

and

$$b_{\vec{n},\ell} = \frac{\kappa_{\vec{n},\ell}}{\kappa_{\vec{n}+\vec{e}_\ell,\ell}}, \qquad (23.1.26)$$

and $\kappa_{\vec{n},\ell}$ is the coefficient of $x^{n_\ell-1}$ in $A_{\vec{n},\ell}$:

$$A_{\vec{n},\ell}(x) = \kappa_{\vec{n},\ell}\, x^{n_\ell-1} + \cdots.$$

Observe that the coefficients $\kappa_{\vec{n}+\vec{e}_j,j}$ $(1 \leq j \leq r)$ are all different from zero by Corollary 23.1.1.

23.2 Hermite–Padé approximation

Suppose we are given r functions with Laurent expansions

$$f_j(z) = \sum_{k=0}^{\infty} \frac{c_{k,j}}{z^{k+1}}, \qquad j = 1, 2, \ldots, r.$$

In type I Hermite–Padé approximation one wants to approximate a linear combination (with polynomial coefficients) of the r functions by a polynomial. We want to find a vector of polynomials $(A_{\vec{n},1}, \ldots, A_{\vec{n},r})$ and a polynomial $B_{\vec{n}}$, with $A_{\vec{n},j}$ of degree $\leq n_j - 1$, such that

$$\sum_{j=1}^{r} A_{\vec{n},j}(z) f_j(z) - B_{\vec{n}}(z) = \mathcal{O}\left(\frac{1}{z^{|\vec{n}|}}\right), \qquad z \to \infty.$$

Type II Hermite–Padé approximation consists of simultaneous approximation of the functions f_j by rational functions with a common denominator. We want to find a polynomial $P_{\vec{n}}$ of degree $\leq |\vec{n}|$ and polynomials $Q_{\vec{n},j}$ $(j = 1, 2, \ldots, r)$ such that

$$P_{\vec{n}}(z) f_1(z) - Q_{\vec{n},1}(z) = \mathcal{O}\left(\frac{1}{z^{n_1+1}}\right), \qquad z \to \infty,$$

$$\vdots$$

$$P_{\vec{n}}(z) f_r(z) - Q_{\vec{n},r}(z) = \mathcal{O}\left(\frac{1}{z^{n_r+1}}\right), \qquad z \to \infty.$$

If the functions f_j are of the form

$$f_j(z) = \int_{\mathbb{R}} \frac{d\mu_j(x)}{z - x},$$

then the coefficients $c_{k,j}$ in the Laurent expansion of f_j are moments of the measure μ_k

$$c_{k,j} = \int_{\mathbb{R}} x^k \, d\mu_j(x)$$

and the linear equations for the unknown coefficients of $(A_{\vec{n},1}, \ldots, A_{\vec{n},r})$ in type I Hermite–Padé approximation are the same as (23.1.1) so that these polynomials are the type I multiple orthogonal polynomials for the measures $(\mu_1, \ldots, \mu_r)$. In a similar way we see that the linear equations for the unknown coefficients of the polynomial $P_{\vec{n}}$ in type II Hermite–Padé approximation are the same as (23.1.3) so that the common denominator is the type II multiple orthogonal polynomial. The remaining ingredients in Hermite–Padé approximation can be described using the type I and type II multiple orthogonal polynomials. The polynomials $B_{\vec{n}}$ for type I

Hermite–Padé approximation is given by

$$B_{\vec{n}}(z) = \sum_{j=1}^{r} \int_{\mathbb{R}} \frac{A_{\vec{n},j}(z) - A_{\vec{n},j}(x)}{z - x} \, d\mu_j(x),$$

and the remainder is then given by

$$\sum_{j=1}^{r} A_{\vec{n},j}(z) \, f_j(z) - B_{\vec{n}}(z) = \sum_{j=1}^{r} \int_{\mathbb{R}} \frac{A_{\vec{n},j}(x)}{z - x} \, d\mu_j(x).$$

The polynomials $Q_{\vec{n},j}$ for type II Hermite–Padé approximation are given by

$$Q_{\vec{n},j}(z) = \int_{\mathbb{R}} \frac{P_{\vec{n}}(z) - P_{\vec{n}}(x)}{z - x} \, d\mu_j(x),$$

and the remainders are then given by

$$P_{\vec{n}}(z) \, f_j(z) - Q_{\vec{n},j}(z) = \int_{\mathbb{R}} \frac{P_{\vec{n}}(x)}{z - x} \, d\mu_j(x),$$

for each j with $1 \le j \le r$.

23.3 Multiple Jacobi Polynomials

There are various ways to define multiple Jacobi polynomials (Aptekarev et al., 2003), (Nikishin & Sorokin, 1991), and (Van Assche & Coussement, 2001). Two important ways are on one hand as an Angelesco system and on the other hand as an AT system.

23.3.1 Jacobi–Angelesco polynomials

Kalyagin (Kalyagin, 1979) and (Kaliaguine & Ronveaux, 1996) considered polynomials defined by a Rodrigues formula of the form

$$(x - a)^{\alpha} x^{\beta} (1 - x)^{\gamma} \binom{\alpha + \beta + \gamma + 3n}{n} P_{n,n}^{(\alpha,\beta,\gamma)}(x)$$
$$= \frac{(-1)^n}{n!} \frac{d^n}{dx^n} \left[(x - a)^{n+\alpha} x^{\beta+n} (1 - x)^{\gamma+n} \right], \quad (23.3.1)$$

where $a < 0$ and $\alpha, \beta, \gamma > -1$. This defines a monic polynomial $P_{n,n}^{(\alpha,\beta,\gamma)}$ of degree $2n$. Indeed, if we apply Leibniz' formula twice, then some calculus gives

$$\binom{\alpha + \beta + \gamma + 3n}{n} P_{n,n}^{(\alpha,\beta,\gamma)}(x)$$
$$= \sum_{k=0}^{n} \sum_{j=0}^{n-k} \binom{n+\alpha}{k} \binom{n+\beta}{j} \binom{n+\gamma}{n-k-j} (x - a)^{n-k} x^{n-j} (x - 1)^{k+j}.$$

By using integration by parts n times, one easily finds

$$\int_a^0 (x-a)^\alpha |x|^\beta (1-x)^\gamma P_{n,n}^{(\alpha,\beta,\gamma)}(x) x^k \, dx$$

$$= (-1)^n C_n(\alpha,\beta,\gamma) \int_a^0 \left(\frac{d^n}{dx^n} x^k \right) (x-a)^{n+\alpha} |x|^{\beta+n} (1-x)^{\gamma+n} \, dx$$

which is 0 whenever $k \le n-1$. In a similar way we see that

$$\int_0^1 (x-a)^\alpha x^\beta (1-x)^\gamma P_{n,n}^{(\alpha,\beta,\gamma)}(x) x^k \, dx = 0, \qquad 0 \le k \le n-1.$$

Hence $P_{n,n}^{(\alpha,\beta,\gamma)}$ is the multiple orthogonal polynomial with multi-index (n,n) for the Angelesco system (μ_1, μ_2), where $d\mu_1(x) = (x-a)^\alpha |x|^\beta (1-x)^\gamma \, dx$ on $[a,0]$ and $d\mu_2(x) = (x-a)^\alpha x^\beta (1-x)^\gamma \, dx$ on $[0,1]$. Observe that

$$\binom{\alpha+\beta+\gamma+3n}{n} P_{n,n}^{(\alpha,\beta,\gamma)}(x)$$

$$= x^n (x-a)^n \binom{n+\gamma}{n} F_1 \left(-n, -n-\alpha, -n-\beta, \gamma+1; \frac{x-1}{x-a}, \frac{x-1}{x} \right),$$

where F_1 is the Appell function defined in (1.3.36). The Rodrigues formula (23.3.1) only gives these Jacobi–Angelesco polynomials for diagonal multi-indices. For other multi-indices we can use

$$(x-a)^\alpha x^\beta (1-x)^\gamma \binom{\alpha+\beta+\gamma+3n}{n} P_{n+k,n}^{(\alpha,\beta,\gamma)}(x)$$

$$= \frac{(-1)^n}{n!} \frac{d^n}{dx^n} \left((x-a)^{n+\alpha} x^{n+\beta} (1-x)^{n+\gamma} P_{k,0}^{(\alpha+n,\beta+n,\gamma+n)}(x) \right),$$

where the extra polynomial $P_{k,0}$ is an ordinary orthogonal polynomial on $[a,0]$. There is a similar formula for $P_{n,n+k}^{(\alpha,\beta,\gamma)}$ which uses $P_{0,k}$, which is an ordinary orthogonal polynomial on $[0,1]$.

23.3.1.1 Rational approximations to π

One place where these polynomials appear is when one wants to approximate π by rational numbers (Beukers, 2000). Consider the case $\alpha = \beta = \gamma = 0$ and $a = -1$, and take the functions

$$f_1(z) = \int_{-1}^0 \frac{dx}{z-x}, \qquad f_2(z) = \int_0^1 \frac{dx}{z-x}.$$

Type II Hermite–Padé approximation to these functions gives

$$P_{n,n}(z)f_1(z) - Q_{n,n;1}(z) = \int_{-1}^{0} \frac{P_{n,n}(x)}{z-x}\, dx \qquad (23.3.2)$$

$$P_{n,n}(z)f_2(z) - Q_{n,n;2}(z) = \int_{0}^{1} \frac{P_{n,n}(x)}{z-x}\, dx \qquad (23.3.3)$$

Observe that for $z = i$ we have $f_1(i) = (2\log 2 - i\pi)/4$ and $f_2(i) = (-2\log 2 - i\pi)/4$, so if we evaluate the Hermite–Padé approximations at $z = i$, then

$$P_{n,n}(i)\frac{2\log 2 - i\pi}{4} - Q_{n,n;1}(i) = \int_{-1}^{0} \frac{P_{n,n}(x)}{i-x}\, dx$$

$$P_{n,n}(i)\frac{-2\log 2 - i\pi}{4} - Q_{n,n;2}(i) = \int_{0}^{1} \frac{P_{n,n}(x)}{i-x}\, dx.$$

Add these two expression together, then

$$P_{n,n}(i)\frac{i\pi}{2} + [Q_{n,n;1}(i) + Q_{n,n;2}(i)] = -\int_{-1}^{1} \frac{P_{n,n}(x)}{i-x}\, dx.$$

The Rodrigues formula now is

$$P_{n,n}(x) = \frac{1}{n!}\frac{d^n}{dx^n}\left(x^n\left(1-x^2\right)^n\right),$$

and if we expand $\left(1 - x^2\right)^n$ then this gives

$$P_{n,n}(x) = \sum_{k=0}^{n} \binom{n}{k}\binom{2k+n}{2k}(-1)^k\, x^{2k}.$$

Notice that this polynomial is not monic. When we evaluate this at $x = i$ then we get

$$P_{n,n}(i) = \sum_{k=0}^{n} \binom{n}{k}\binom{2k+n}{2k}$$

which obviously is a positive integer. For $Q_{n,n;1}$ we have

$$Q_{n,n;i}(i) = \int_{-1}^{0} \frac{P_{n,n}(x) - P_{n,n}(i)}{x-i}\, dx$$

$$= \sum_{k=0}^{n} \binom{n}{k}\binom{2k+n}{2k}(-1)^k \int_{-1}^{0} \frac{x^{2k} - i^{2k}}{x-i}\, dx$$

$$= \sum_{k=0}^{n}\sum_{j=0}^{2k-1} \binom{n}{k}\binom{2k+n}{2k}(-1)^{k+j}\frac{i^{2k-j-1}}{j+1}$$

and in a similar way

$$Q_{n,n;2}(i) = \sum_{k=0}^{n} \sum_{j=0}^{2k-1} \binom{n}{k} \binom{2k+n}{2k} (-1)^k \frac{i^{2k-j-1}}{j+1}$$

so that

$$Q_{n,n;1}(i) + Q_{n,n;2}(i) = -2i \sum_{k=0}^{n} \sum_{j=0}^{k-1} \binom{n}{k} \binom{2k+n}{2k} \frac{(-1)^j}{2j+1}.$$

This is i times a rational number, but if we multiply this by the least common multiple of the odd integers $3, 5, \ldots, 2n - 1$ then this gives i times an integer. All this gives the rational approximation

$$\frac{\pi}{2} \approx \frac{2i\left[Q_{n,n;1}(i) + Q_{n,n;2}(i)\right]}{P_{n,n}(i)}.$$

Unfortunately this rational approximation is not good enough to prove that π is irrational. A better approximation can be obtained by taking $2n$, adding (23.3.2)–(23.3.3), and then taking the nth derivative, which gives

$$P_{2n,2n}^{(n)}(z)\left[f_1(z) + f_2(z)\right] + \sum_{k=0}^{n-1} \binom{n}{k} P_{2n,2n}^{(k)}(z)\left[f_1 + f_2\right]^{(n-k)}(z)$$

$$- \left[Q_{2n,2n;1}^{(n)}(z) + Q_{2n,2n;2}^{(n)}(z)\right] = (-1)^n n! \int_{-1}^{1} \frac{P_{2n,2n}(x)}{(z-x)^{n+1}}\, dx,$$

and then to evaluate this at $z = i$. The right-hand side then becomes, using the Rodrigues formula,

$$(-1)^n \frac{(3n)!}{(2n)!} \int_{-1}^{1} \frac{x^{2n}(1-x^2)^{2n}}{(i-x)^{3n+1}}\, dx.$$

An even better rational approximation was given by Hata (Hata, 1993): if one replaces f_1 by

$$f_3(z) = \int_{-i}^{0} \frac{dx}{z-x}$$

then one takes $a = -i$ for the Jacobi–Angelesco polynomials, which gives complex multiple orthogonality. The resulting rational approximation to π then gives the upperbound 8.016 for the measure of irrationality, which is the best bound so far. Notice that neither Beukers (Beukers, 2000) nor Hata (Hata, 1993) mention multiple orthogonal polynomials, but their approximations implicitly use these Jacobi–Angelesco polynomials.

23.3.2 Jacobi–Piñeiro polynomials

Another way to obtain multiple Jacobi polynomials is to use several Jacobi weights on the same interval. It is convenient to take $[0, 1]$ as the interval, rather than $[-1, 1]$ as is usually done for Jacobi polynomials. These multiple orthogonal polynomials were first investigated by Piñeiro (Piñeiro, 1987) for a special choice of parameters. The idea is to keep one of the parameters of the Jacobi weight $x^\alpha (1-x)^\beta$ fixed and to change the other parameter appropriately for the r weights. Let $\beta > -1$ and choose $\alpha_1, \ldots, \alpha_r > -1$ so that $\alpha_i - \alpha_j \notin \mathbb{Z}$ whenever $i \neq j$. The measures $(\mu_1, \ldots, \mu_r)$ with $d\mu_i(x) = x^{\alpha_i}(1-x)^\beta \, dx$ on $[0, 1]$ then form an AT system. The polynomials $P_{\vec{n}}$ given by the Rodrigues formula

$$(-1)^{|\vec{n}|} \prod_{j=1}^{r} (|\vec{n}| + \alpha_j + \beta + 1)_{n_j} (1-x)^\beta P_{\vec{n}}^{(\vec{\alpha};\beta)}(x)$$

$$= \prod_{j=1}^{r} \left(x^{-\alpha_j} \frac{d^{n_j}}{dx^{n_j}} x^{n_j + \alpha_j} \right) (1-x)^{\beta + |\vec{n}|} \quad (23.3.4)$$

are monic polynomials of degree $|\vec{n}|$. The differential operators

$$x^{-\alpha_j} \frac{d^{n_j}}{dx^{n_j}} x^{n_j + \alpha_j}, \qquad j = 1, 2, \ldots, r$$

are commuting, hence the order in which the product in (23.3.4) is taken is irrelevant. Integration by parts shows that

$$\int_0^1 x^\gamma \prod_{j=1}^{r} \left(x^{-\alpha_j} \frac{d^{n_j}}{dx^{n_j}} x^{n_j + \alpha_j} \right) (1-x)^{|\vec{n}| + \beta} \, dx$$

$$= (-1)^{|\vec{n}|} \prod_{j=1}^{r} (\alpha_j - \gamma)_{n_j} \frac{\Gamma(\gamma + 1)\Gamma(|\vec{n}| + \beta + 1)}{\Gamma(|\vec{n}| + \beta + \gamma + 2)},$$

and this is 0 whenever $\gamma = \alpha_j + k$ with $0 \leq k \leq n_j - 1$ for every $j \in \{1, 2, \ldots, r\}$. Hence we have

$$\int_0^1 P_{\vec{n}}^{(\vec{\alpha},\beta)}(x) x^{\alpha_j + k}(1-x)^\beta \, dx = 0, \qquad k = 0, 1, \ldots, n_j - 1,$$

for $1 \leq j \leq r$, which shows that these are the type II multiple orthogonal polynomials for the r Jacobi weights $(\mu_1, \ldots, \mu_r)$ on $[0, 1]$. If we use Leibniz' rule several times, then

$$(1-x)^{-\beta} \prod_{j=1}^{r} \left(x^{-\alpha_j} \frac{d^{n_j}}{dx^{n_j}} x^{n_j + \alpha_j} \right) (1-x)^{|\vec{n}| + \beta}$$

$$= n_1! \cdots n_r! \sum_{k_1=0}^{n_1} \cdots \sum_{k_r=0}^{n_r} (-1)^{|\vec{k}|} \prod_{j=1}^{r} \binom{n_j + \alpha_j + \sum_{i=1}^{j-1} k_i}{n_j - k_j}$$

$$\times \binom{|\vec{n}| + \beta}{|\vec{k}|} \frac{|\vec{k}|! x^{|\vec{k}|}(1-x)^{|\vec{n}| - |\vec{k}|}}{k_1! \cdots k_r!},$$

which is a polynomial of degree $|\vec{n}|$ with leading coefficient

$$n_1!\cdots n_r!(-1)^{|\vec{n}|}\sum_{k_1=0}^{n_1}\cdots\sum_{k_r=0}^{n_r}\prod_{j=1}^{r}\binom{n_j+\alpha_j+\sum_{i=1}^{j-1}k_i}{n_j-k_j}\binom{|\vec{n}|+\beta}{|\vec{k}|}\frac{|\vec{k}|!}{k_1!\cdots k_r!},$$

which is equal to $(-1)^{|\vec{n}|}\prod_{j=1}^{r}(|\vec{n}|+\alpha_j+\beta+1)_{n_j}$. Another representation can be obtained by expanding

$$(1-x)^{|\vec{n}|+\beta}=\sum_{k=0}^{\infty}\binom{|\vec{n}|+\beta}{k}(-1)^k\,x^k,$$

then the Rodrigues formula implies that

$$(-1)^{|\vec{n}|}\prod_{j=1}^{r}(|\vec{n}|+\alpha_j+\beta+1)_{n_j}(1-x)^{\beta}P_{\vec{n}}^{(\vec{\alpha};\beta)}(x)$$

$$=\prod_{j=1}^{r}(\alpha_j+1)_{n_j}\,{}_{r+1}F_r\left(\begin{array}{c}-|\vec{n}|-\beta,\alpha_1+n_1+1,\ldots,\alpha_r+n_r+1\\\alpha_1+1,\ldots,\alpha_r+1\end{array}\middle|x\right).$$

$$(23.3.5)$$

This series is terminating whenever β is an integer.

One can obtain another family of multiple Jacobi polynomials by keeping the parameter α fixed and by changing the parameter β. If $(\mu_1,\ldots,\mu_r)$ are the measures given by $d\mu_k=x^\alpha(1-x)^{\beta_k}\,dx$ on $[0,1]$, where $\beta_i-\beta_j\notin\mathbb{Z}$ whenever $i\neq j$, then these multiple Jacobi polynomials are basiscally the Jacobi–Piñeiro polynomials $(-1)^{|\vec{n}|}P_{\vec{n}}(1-x)$ with parameters $\alpha_j=\beta_j$ $(j=1,2,\ldots,r)$ and $\beta=\alpha$.

23.3.2.1 *Rational approximations of $\zeta(k)$ and polylogarithms*

The polylogarithms are defined by

$$\mathrm{Li}_k(z)=\sum_{n=1}^{\infty}\frac{z^n}{n^k},\qquad |z|<1.$$

One easily finds that

$$\frac{(-1)^k}{k!}\int_0^1\frac{\log^k(x)}{z-x}\,dx=\mathrm{Li}_{k+1}(1/z).$$

Observe that

$$\mathrm{Li}_1(z)=-\log(1-z).$$

Simultaneous rational approximation to $\mathrm{Li}_1(1/z),\ldots,\mathrm{Li}_r(1/z)$ can be done using Hermite–Padé approximation and this uses multiple orthogonal polynomials for the system of weights $1,\log x,\ldots,(-1)^{r-1}/(r-1)!\,\log^{r-1}(x)$ on $[0,1]$. This is a limiting case of Jacobi–Piñeiro polynomials $P_{\vec{n}}^{(\vec{\alpha},\beta)}$ where $\beta=0$ and $\alpha_1=\alpha_2=\cdots=\alpha_r=0$. Indeed, if $n_1\geq n_2\geq\cdots\geq n_r$ then the polynomials defined by the

Rodrigues formula (23.3.4) are still of degree $|\vec{n}|$, but the orthogonality conditions are

$$\int_0^1 P_{\vec{n}}^{(\vec{0},0)}(x)\, x^k \log^{j-1}(x)\, dx = 0, \qquad k = 0, 1, \ldots, n_j - 1,$$

for $1 \le j \le r$. Observe that

$$\mathrm{Li}_k(1) = \sum_{n=1}^{\infty} \frac{1}{n^k} = \zeta(k),$$

whenever $k > 1$. The Jacobi–Piñeiro polynomials $P_{\vec{n}}^{(\vec{0},0)}(x)$ have rational coefficients, so Hermite–Padé approximation to these polylogarithms, evaluated at $z = 1$, gives rational approximations to $\zeta(k)$. In fact one gets simultaneous rational approximations to $\zeta(1), \ldots, \zeta(r)$. Unfortunately $\zeta(1)$ is the harmonic series and diverges, which complicates matters. However, if one combines type I and type II Hermite–Padé approximation, with some extra modification so that the divergence of $\zeta(1)$ is annihilated, then one can actually get rational approximations to $\zeta(2)$ and $\zeta(3)$ which are good enough to prove that both numbers are irrational. Apéry's proof (Apéry, 1979) of the irrationality of $\zeta(3)$ is equivalent to the following Hermite–Padé approximation problem (Van Assche, 1999): find polynomials (A_n, B_n), where A_n and B_n are of degree n, and polynomials C_n and D_n, such that

$$A_n(1) = 0$$
$$A_n(z)\mathrm{Li}_1(1/z) + B_n(z)\mathrm{Li}_2(1/z) - C_n(z) = \mathcal{O}\left(1/z^{n+1}\right), \qquad z \to \infty,$$
$$A_n(z)\mathrm{Li}_2(1/z) + 2B_n(z)\mathrm{Li}_3(1/z) - D_n(z) = \mathcal{O}\left(1/z^{n+1}\right), \qquad z \to \infty,$$

which is then evaluated at $z = 1$. Observe that the second and third line are each a type I Hermite–Padé approximation problem, but they both use the same vector (A_n, B_n) and hence lines two and three together form a type II Hermite–Padé problem with common denominator (A_n, B_n).

23.4 Multiple Laguerre Polynomials

For Laguerre polynomials there are also several ways to obtain multiple orthogonal polynomials (Aptekarev et al., 2003), (Nikishin & Sorokin, 1991), and (Van Assche & Coussement, 2001). For AT systems one can take the Laguerre weights $x^\alpha e^{-x}$ on $[0, \infty)$ and change the parameters α or one can keep the parameter α fixed and change the rate of exponential decrease at ∞.

23.4.1 Multiple Laguerre polynomials of the first kind

Consider the measures $(\mu_1, \ldots, \mu_r)$ given by $d\mu_j(x) = x^{\alpha_j} e^{-x}\, dx$ on $[0, \infty)$, where $\alpha_i - \alpha_j \notin \mathbb{Z}$ whenever $i \ne j$. The Rodrigues formula

$$(-1)^{|\vec{n}|}\, e^{-x}\, L_{\vec{n}}^{\vec{\alpha}}(x) = \prod_{j=1}^{r} \left(x^{-\alpha_j} \frac{d^{n_j}}{dx^{n_j}} x^{n_j + \alpha_j} \right) e^{-x} \qquad (23.4.1)$$

gives a polynomial $L_{\vec{n}}^{\vec{\alpha}}$ of degree $|\vec{n}|$ for which (use integration by parts)

$$\int_0^\infty L_{\vec{n}}^{\vec{\alpha}}(x)\, x^{\alpha_j}\, e^{-x}\, x^k\, dx = 0, \qquad k = 0, 1, \dots, n_j - 1,$$

for $j = 1, 2, \dots, r$, so that this is the type II multiple orthogonal polynomial for the AT system $(\mu_1, \dots, \mu_r)$ of Laguerre measures. Observe that

$$\frac{d}{dx}\left(x^{\alpha_k} e^{-x} L_{\vec{n}}^{\vec{\alpha}}(x)\right) = -x^{\alpha_k - 1} e^{-x} L_{\vec{n}+\vec{e}_k}^{\vec{\alpha}-\vec{e}_k}(x), \qquad k = 1, \dots, r, \qquad (23.4.2)$$

so that the differential operator

$$D_k = x^{-\alpha_k + 1} e^x D x^{\alpha_k} e^{-x}$$

acting on $L_{\vec{n}}^{\vec{\alpha}}(x)$ raises the kth component n_k of the multi-index by one and lowers the kth component α_k of the parameter $\vec{\alpha}$ by one. These differential operators are all commuting. Observe that the product of differential operators in (23.4.1) is the same as in (23.3.4) for the Jacobi–Piñeiro polynomials, but applied to a different function. An explicit expression as an hypergeometric function is given by

$$(-1)^{|\vec{n}|} e^{-x} L_{\vec{n}}^{\vec{\alpha}}(x) = \prod_{j=1}^r (\alpha_j + 1)_{n_j}$$

$$\times {}_rF_r\left(\begin{array}{c} \alpha_1 + n_1 + 1, \dots, \alpha_r + n_r + 1 \\ \alpha_1 + 1, \dots, \alpha_r + 1 \end{array}\middle| -x\right).$$

23.4.2 Multiple Laguerre polynomials of the second kind

If we take the measures $(\mu_1, \dots, \mu_r)$ with $d\mu_j(x) = x^\alpha e^{-c_j x}\, dx$ on $[0, \infty)$, where $\alpha > -1$, $0 < c_j$ and $c_i \neq c_j$ whenever $i \neq j$, then we get another AT system, and the corresponding type II multiple orthogonal polynomials are given by

$$(-1)^{|\vec{n}|} \prod_{j=1}^r c_j^{n_j} x^\alpha L_{\vec{n}}^{(\alpha,\vec{c})}(x) = \prod_{j=1}^r \left(e^{c_j x} \frac{d^{n_j}}{dx^{n_j}} e^{-c_j x}\right) x^{|\vec{n}|+\alpha}. \qquad (23.4.3)$$

The differential operators

$$D_k = x^{-\alpha + 1} e^{c_k x} D x^\alpha e^{-c_k x}$$

are again commuting operators and

$$\frac{d}{dx}\left(x^\alpha e^{-c_k x} L_{\vec{n}}^{(\alpha,\vec{c})}(x)\right) = -c_k x^{\alpha - 1} e^{-c_k x} L_{\vec{n}+\vec{e}_k}^{(\alpha-1,\vec{c})}(x). \qquad (23.4.4)$$

An explicit expression is given by

$$L_{\vec{n}}^{(\alpha,\vec{c})}(x) = \sum_{k_1=0}^{n_1} \cdots \sum_{k_r=0}^{n_r} \binom{n_1}{k_1} \cdots \binom{n_r}{k_r} \binom{|\vec{n}|+\alpha}{|\vec{k}|}$$

$$\times (-1)^{|\vec{k}|} \frac{|\vec{k}|!}{c_1^{k_1} \cdots c_r^{k_r}} x^{|\vec{n}|-|\vec{k}|}. \qquad (23.4.5)$$

23.4.2.1 Random matrices: the Wishart ensemble

Let M be a random matrix of the form $M = XX^T$, where X is a $n \times (n+p)$ matrix for which the columns are independent and normally distributed random vectors with covariance matrix Σ. Such matrices appear as sample covariance matrices when the $n + p$ columns of X are a sample of a multivariate Gaussian distribution in $\mathbb{R}^n$. The distribution

$$\frac{1}{Z_n} e^{-\text{Tr}(\Sigma^{-1}M)} (\det M)^p \, dM$$

for the $n \times n$ positive definite matrices M of this form gives the so-called Wishart ensemble of random matrices. This ensemble can be described using multiple Laguerre polynomials of the second kind (Bleher & Kuijlaars, 2004). The eigenvalues of M follow a determinantal point process on $[0, \infty)$ with kernel

$$K_n(x, y) = \sum_{k=0}^{n-1} p_k(x) \, q_k(y), \tag{23.4.6}$$

where $p_k(x) = P_{\vec{n}_k}(x)$ and $q_k(y) = Q_{\vec{n}_{k+1}}(y)$, and $\vec{n}_0, \vec{n}_1, \ldots, \vec{n}_n$ is a path from $\vec{0}$ to $\vec{n}$, the $P_{\vec{n}}$ are type II multiple Laguerre polynomials and the $Q_{\vec{n}}$ are type I multiple Laguerre polynomials (of the second kind). The parameters $\beta_1, \ldots, \beta_r$ for the multiple Laguerre polynomials of the second kind are the eigenvalues of the matrix Σ^{-1} and β_j has multiplicity n_j for $1 \leq j \leq r$. There is a Christoffel–Darboux type formula for kernels of the form (23.4.6) with multiple orthogonal polynomials, namely

$$(x - y) K_n(x, y) = P_{\vec{n}}(x) Q_{\vec{n}}(y) - \sum_{j=1}^{r} \frac{h_{\vec{n}}(j)}{h_{\vec{n}-\vec{e}_j}(j)} P_{\vec{n}-\vec{e}_j} Q_{\vec{n}+\vec{e}_j}(y) \tag{23.4.7}$$

where

$$h_{\vec{n}}(j) = \int_{\mathbb{R}} P_{\vec{n}}(x) \, x^{n_j} \, d\mu_j(x),$$

(Daems & Kuijlaars, 2004).

23.5 Multiple Hermite polynomials

If we consider the weights $w_j(x) = e^{-x^2 + c_j x}$ on $(-\infty, \infty)$, where $c_1, \ldots, c_r$ are r different real numbers, then

$$e^{-x^2} H_{\vec{n}}^{\vec{c}}(x) = (-1)^{|\vec{n}|} 2^{-|\vec{n}|} \left(\prod_{j=1}^{r} e^{-c_j x} \frac{d^{n_j}}{dx^{n_j}} e^{c_j x} \right) e^{-x^2} \tag{23.5.1}$$

defines a polynomial $H_{\vec{n}}^{\vec{c}}$ of degree $|\vec{n}|$. An explicit expression is

$$H_{\vec{n}}^{\vec{c}}(x) = (-1)^{|\vec{n}|} 2^{-|\vec{n}|}$$

$$\times \sum_{k_1=0}^{n_1} \cdots \sum_{k_r=0}^{n_r} \binom{n_1}{k_1} \cdots \binom{n_r}{k_r} (c_1)^{n_1-k_1} \cdots (c_r)^{n_r-k_r} (-1)^{|\vec{k}|} H_{|\vec{k}|}(x),$$

where H_n is the usual Hermite polynomial. Recall that the usual $H_n(x) = 2^n x^n + \cdots$ is an even polynomial, so that $H_{\vec{n}}^{\vec{c}}$ is a monic polynomial of degree $|\vec{n}|$, and

$$H_{\vec{n}}^{\vec{c}}(x) = x^{|\vec{n}|} - \frac{1}{2}\sum_{j=1}^{r} n_j c_j x^{|\vec{n}|-1} + \cdots .$$

The Rodrigues formula (23.5.1) and integration by parts give

$$g(t) = \int_{\mathbb{R}} e^{tx} H_{\vec{n}}^{\vec{c}}(x) e^{-x^2}\, dx = 2^{-|\vec{n}|}\sqrt{\pi}\,(t - c_1)^{n_1}\cdots(t - c_r)^{n_r}\, e^{t^2/4},$$

and hence

$$\int_{\mathbb{R}} x^k H_{\vec{n}}^{\vec{c}}(x) e^{-x^2+c_j x}\, dx = g^{(k)}(c_j) = 0, \qquad k = 0,1,\ldots,n_j - 1,$$

for $1 \le j \le r$, and

$$\int_{\mathbb{R}} x^{n_j} H_{\vec{n}}^{\vec{c}}(x)\, e^{-x^2+c_j x}\, dx = g^{(n_j)}(c_j) = 2^{-|\vec{n}|}\sqrt{\pi}\, n_j! \prod_{i \ne j}(c_j - c_i)^{n_i}\, e^{c_j^2/4},$$

$$\tag{23.5.2}$$

which indeed shows that these are multiple Hermite polynomials. If we use (23.5.2) then the recurrence coefficients in Theorem 23.1.11 are given by $a_{\vec{n},j} = n_j/2$ for $1 \le j \le r$, and by comparing the coefficient of $x^{|\vec{n}|}$ we also see that $a_{\vec{n},0}(k) = c_k/2$, so that the recurrence relation is

$$x H_{\vec{n}}^{\vec{c}}(x) = H_{\vec{n}+\vec{e}_k}^{\vec{c}}(x) + \frac{c_k}{2}\,H_{\vec{n}}^{\vec{c}}(x) + \frac{1}{2}\sum_{j=1}^{r} n_j H_{\vec{n}-\vec{e}_j}^{\vec{c}}(x). \tag{23.5.3}$$

23.5.1 Random matrices with external source

Recently a random matrix ensemble with an external source was considered by (Brézin & Hikami, 1998) and (Zinn-Justin, 1997). The joint probability density function of the matrix elements of the random Hermitian matrix M is of the form

$$\frac{1}{Z_N} e^{-\mathrm{Tr}(M^2 - AM)}\, dM$$

where A is a fixed $N \times N$ Hermitian matrix (the external source). Bleher and Kuijlaars observed that the average characteristic polynomial $p_N(z) = \mathrm{E}[\det(zI - M)]$ can be characterized by the property

$$\int_{\mathbb{R}} p_N(x) x^k e^{-(x^2 - c_j x)}\, dx = 0, \qquad k = 0,1,\ldots,N_j - 1,$$

where N_j is the multiplicity of the eigenvalue c_j of A, see (Bleher & Kuijlaars, 2004). This means that p_N is a multiple Hermite polynomial of type II with multi-index $\vec{N} = (N_1,\ldots,N_r)$ when A has r distinct eigenvalues $c_1,\ldots,c_r$ with multiplicities $N_1,\ldots,N_r$ respectively. The eigenvalue correlations and the eigenvalue density can be written in terms of the kernel

$$K_N(x,y) = \sum_{k=0}^{N-1} p_k(x)\, q_k(y),$$

where the q_k are the type I multiple Hermite polynomials and the p_k are the type II multiple Hermite polynomials for multi-indices on a path $\vec{0} = \vec{n}_0, \vec{n}_1, \ldots, \vec{n}_N = \vec{N}$. The asymptotic analysis of the eigenvalues and their correlations and universality questions can therefore be handled using asymptotic analysis of multiple Hermite polynomials.

23.6 Discrete Multiple Orthogonal Polynomials

Arvesú, Coussement and Van Assche have found several multiple orthogonal polynomials (type I) extending the classical discrete orthogonal polynomials of Charlier, Meixner, Krawtchouk and Hahn. Their work is (Arvesú et al., 2003).

23.6.1 Multiple Charlier polynomials

Consider the Poisson measures

$$\mu_i = \sum_{k=0}^{\infty} \frac{a_i^k}{k!}\, \delta_k,$$

where $a_1, \ldots, a_r > 0$ and $a_i \neq a_j$ whenever $i \neq j$. The discrete measures $(\mu_1, \ldots, \mu_r)$ then form an AT system on $[0, \infty)$ and the corresponding multiple orthogonal polynomials are given by the Rodrigues formula

$$C_{\vec{n}}^{\vec{a}}(x) = \left(\prod_{j=1}^{r} (-a_j)^{n_j} \right) \Gamma(x+1) \left(\prod_{j=1}^{r} a_j^{-x} \nabla^{n_j} a_j^x \right) \frac{1}{\Gamma(x+1)},$$

where the product of the difference operators $a_j^{-x} \nabla^{n_j} a_j^x$ can be taken in any order because these operators commute. We have the explicit formula

$$C_{\vec{n}}^{\vec{a}}(x) = \sum_{k_1=0}^{n_1} \cdots \sum_{k_r=0}^{n_r} (-n_1)_{k_1} \cdots (-n_r)_{k_r} (-x)_{|\vec{k}|} \frac{(-a_1)^{n_1-k_1} \cdots (-a_r)^{n_r-k_r}}{k_1! \cdots k_r!}.$$

They satisfy the recurrence relation

$$x C_{\vec{n}}^{\vec{a}}(x) = C_{\vec{n}+\vec{e}^k}^{\vec{a}}(x) + (a_k + |\vec{n}|)\, C_{\vec{n}}^{\vec{a}}(x) + \sum_{j=1}^{r} n_j a_j C_{\vec{n}-\vec{e}_j}^{\vec{a}}(x).$$

23.6.2 Multiple Meixner polynomials

The orthogonality measure for Meixner polynomials is the negative binomial distribution

$$\mu = \sum_{k=0}^{\infty} \frac{(\beta)_k c^k}{k!}\, \delta_k,$$

where $\beta > 0$ and $0 < c < 1$. We can obtain two kinds of multiple Meixner polynomials by fixing one of the parameters β or c and by changing the remaining parameter.

23.6.2.1 Multiple Meixner polynomials of the first kind

Fix $\beta > 0$ and consider the measures

$$\mu_i = \sum_{k=0}^{\infty} \frac{(\beta)_k c_i^k}{k!} \delta_k,$$

with $c_i \neq c_j$ whenever $i \neq j$. The system $(\mu_1, \ldots, \mu_r)$ is an AT system and the corresponding multiple orthogonal polynomials are given by the Rodrigues formula

$$M_{\vec{n}}^{\beta;\vec{c}}(x) = (\beta)_{|\vec{n}|} \left(\prod_{j=1}^{r} \frac{c_j}{c_j - 1} \right)^n_j$$

$$\times \frac{\Gamma(\beta)\Gamma(x+1)}{\Gamma(\beta + x)} \left(\prod_{j=1}^{r} c_j^{-x} \nabla^{n_j} c_j^x \right) \frac{\Gamma(|\vec{n}| + \beta + x)}{\Gamma(|\vec{n}| + \beta)\Gamma(x+1)}.$$

For $r = 2$ these polynomials are given by

$$M_{n,m}^{\beta;c_1,c_2}(x) = \left(\frac{c_1}{c_1 - 1} \right)^n \left(\frac{c_2}{c_2 - 1} \right)^m (\beta)_{n+m}$$

$$\times F_1 \left(-x; -n, -m; \beta; 1 - \frac{1}{c_1}, 1 - \frac{1}{c_2} \right),$$

where F_1 is the Appell function defined in (1.3.36).

23.6.2.2 Multiple Meixner polynomials of the second kind

If we fix $0 < c < 1$ and consider the measures

$$\mu_i = \sum_{k=0}^{\infty} \frac{(\beta_i)_k c^k}{k!} \delta_k,$$

with $\beta_i - \beta_j \notin \mathbb{Z}$ whenever $i \neq j$, then the system $(\mu_1, \ldots, \mu_r)$ is again an AT system and the corresponding multiple orthogonal polynomials are given by

$$M_{\vec{n}}^{\vec{\beta};c}(x) = \left(\frac{c}{c - 1} \right)^{|\vec{n}|} \prod_{j=1}^{r} (\beta_j)_{n_j}$$

$$\times \frac{\Gamma(x+1)}{c^x} \left(\prod_{j=1}^{r} \frac{\Gamma(\beta_j)}{\Gamma(\beta_j + x)} \nabla^{n_j} \frac{\Gamma(\beta_j + n_j + x)}{\Gamma(\beta_j + n_j)} \right) \frac{c^x}{\Gamma(x+1)}.$$

For $r = 2$ these polynomials are given by

$$M_{n,m}^{\beta_1,\beta_2;c}(x) = \left(\frac{c}{c-1}\right)^{n+m} (\beta_1)_n (\beta_2)_m$$

$$\times F_{1:0;1}^{1:1;2} \left(\begin{array}{c} (-x) : (-n); (-m, \beta_1 + n); \\ \\ (\beta_1) : - ; (\beta_2); \end{array} \quad \frac{c-1}{c}, \frac{c-1}{c}\right),$$

where

$$F_{\ell:m;n}^{p:q;k} \left(\begin{array}{c} \vec{a} : \vec{b}; \vec{c}; \\ \\ \vec{\alpha} : \vec{\beta}; \vec{\gamma} \end{array} \quad x, y\right) = \sum_{r=0}^{\infty} \sum_{s=0}^{\infty} \frac{\prod_{j=1}^{p} (a_j)_{r+s} \prod_{j=1}^{q} (b_j)_r \prod_{j=1}^{k} (c_j)_s}{\prod_{j=1}^{\ell} (\alpha_j)_{r+s} \prod_{j=1}^{m} (\beta_j)_r \prod_{j=1}^{n} (\gamma_j)_s} \frac{x^r y^s}{r! s!}$$

is a Kampé de Fériet series, (Appell & Kampé de Fériet, 1926).

23.6.3 *Multiple Krawtchouk polynomials*

Consider the binomial measures

$$\mu_i = \sum_{k=0}^{N} \binom{N}{k} p_i^k (1 - p_i)^{N-k} \delta_k,$$

where $0 < p_i < 1$ and $p_i \neq p_j$ whenever $i \neq j$. The type II multiple orthogonal polynomials for $\vec{n} \leq N$ are the multiple Meixner polynomials of the first kind with $\beta = -N$, and $c_i = p_i / (p_i - 1)$. This gives for $r = 2$ the explicit formula

$$K_{n,m}^{p_1,p_2;N}(x) = p_1^n p_2^m (-N)_{n+m} F_1 \left(-x; -n, -m; -N; \frac{1}{p_1}, \frac{1}{p_2}\right).$$

23.6.4 *Multiple Hahn polynomials*

If we consider the Hahn measure of (6.2.1) for $\alpha, \beta > -1$, then we fix one of the parameters α or β and change the remaining parameter. We will keep $\beta > -1$ fixed and consider the measures

$$\mu_i = \sum_{k=0}^{N} \frac{(\alpha_i + 1)_k (\beta + 1)_{N-k}}{k! (N - k)!} \delta_k,$$

with $\alpha_i - \alpha_j \notin \{0, 1, \ldots, N - 1\}$ whenever $i \neq j$. The case when α is fixed and the β_i are different can be obtained from this by changing the variable x to $N - x$.

The type II multiple orthogonal Hahn polynomials for $|\vec{n}| \leq N$ are given by the

Rodrigues formula

$$P_{\vec{n}}^{\vec{\alpha};\beta,N}(x) = \frac{(-1)^{|\vec{n}|}(\beta+1)_{|\vec{n}|}}{\prod\limits_{j=1}^{r}(|\vec{n}|+\alpha_j+\beta+1)_{n_j}} \frac{\Gamma(x+1)\Gamma(N-x+1)}{\Gamma(\beta+N-x+1)}$$

$$\times \left(\prod_{j=1}^{r}\frac{1}{\Gamma(\alpha_j+x+1)}\nabla^{n_j}\Gamma(\alpha_j+n_j+x+1)\right)\frac{\Gamma(\beta+N-x+1)}{\Gamma(x+1)\Gamma(N-x+1)}.$$

For $r = 2$ these polynomials are again given as a Kampé de Fériet series

$$P_{n,m}^{\alpha_1,\alpha_2;\beta,N}(x) = \frac{(\alpha_1+1)_n (\alpha_2+1)_m (-N)_{n+m}}{(n+m+\alpha_1+\beta+1)_n (n+m+\alpha_2+\beta+1)_m}$$

$$\times F_{2:0;2}^{2:1;3}\left(\begin{array}{c}(-x,\beta+n+\alpha_1+1):(-n);(-m,\beta+\alpha_2+n+m+1,\alpha_1+n+1);\\[4pt](-N,\alpha_1+1):-;(\alpha_2+1,\beta+n+\alpha_1+1);\end{array}1,1\right).$$

23.6.5 Multiple little q-Jacobi polynomials

As a last example we consider some basic multiple orthogonal polynomials which are q-analogs of the multiple Jacobi–Piñeiro polynomials. Little q-Jacobi polynomials are orthogonal polynomials with respect to the measure μ for which $d\mu(x) = w(x;a,b\,|\,q)\,d_q x$, where

$$w(x;a,b\,|\,q) = \frac{(qx;q)_\infty}{(q^{\beta+1}x;q)_\infty}x^\alpha,$$

with $\alpha,\beta > -1$. Again there are two kinds of multiple little q-Jacobi polynomials, by taking one of the two parameters α or β fixed and changing the remaining parameter (Postelmans & Van Assche, 2005).

23.6.5.1 Multiple little q-Jacobi polynomials of the first kind

Consider the measures

$$d\mu_i(x) = w(x;\alpha_i,\beta\,|\,q)\,d_q x,$$

where $\beta > -1$ is fixed and the $\alpha_i > -1$ are such that $\alpha_i - \alpha_j \notin \mathbb{Z}$ whenever $i \neq j$. The system $(\mu_1,\ldots,\mu_r)$ is then an AT system and the type II multiple orthogonal polynomials are given by a Rodrigues formula

$$P_{\vec{n}}(x;\vec{\alpha},\beta\,|\,q) = C(\vec{n},\vec{\alpha},\beta)\frac{(q^{\beta+1}x;q)_\infty}{(qx;q)_\infty}$$

$$\times\left(\prod_{j=1}^{r}x^{-\alpha_j}D_p^{n_j}x^{\alpha_j+n_j}\right)\frac{(qx;q)_\infty}{(q^{\beta+|\vec{n}|+1};q)_\infty},$$

where

$$C(\vec{n},\vec{\alpha},\beta) = (-1)^{|\vec{n}|}(1-q)^{|\vec{n}|}\frac{q^{\sum\limits_{j=1}^{r}(\alpha_j-1)n_j+\sum\limits_{1\le j\le k\le r}n_j n_k}}{\prod\limits_{j=1}^{r}(q^{\alpha_j/\beta+|\vec{n}|+1};q)_{n_j}}.$$

An explicit expression in terms of a basic hypergeometric series is

$$P_{\vec{n}}(x;\vec{\alpha},\beta\,|\,q) = C(\vec{n},\vec{\alpha},\beta)(1-q)^{-|\vec{n}|} q^{-\sum\limits_{j=1}^{r}\alpha_j n_j - \sum\limits_{j=1}^{r}\binom{n_j}{2}} \prod_{j=1}^{r}(q^{\alpha_j+1};q)_{n_j}$$

$$\times \frac{(q^{\beta+1}x;q)_\infty}{(qx;q)_\infty}\,{}_{r+1}\phi_r\left(\begin{array}{c} q^{-\beta-|\vec{n}|}, q^{\alpha_1+n_1+1}, \ldots, q^{\alpha_r+n_r+1} \\ q^{\alpha_1+1}, \ldots, q^{\alpha_r+1} \end{array}\bigg|\, q;q^{\beta+1}x\right).$$

23.6.5.2 *Multiple little q-Jacobi polynomials of the second kind*

Keep $\alpha > -1$ fixed and consider the measures

$$d\mu_i(x) = w\left(x;\alpha,\beta_i\,|\,q\right)d_q x,$$

where the $\beta_i > -1$ are such that $\beta_i - \beta_j \notin \mathbb{Z}$ whenever $i \neq j$. The system $(\mu_1,\ldots,\mu_r)$ is again an AT system and the type II multiple orthogonal polynomials are given by the Rodrigues formula

$$P_{\vec{n}}\left(x;\alpha,\vec{\beta}\,|\,q\right)$$

$$= \frac{C\left(\vec{n},\alpha,\vec{\beta}\right)}{(qx;q)_\infty x^\alpha}\left(\prod_{j=1}^{r}(q^{\beta_j+1}x;q)_\infty\, D_p^{n_j}\,\frac{1}{(q^{\beta_j+n_j+1}x;q)_\infty}\right)(qx;q)_\infty x^{\alpha+|\vec{n}|},$$

where

$$C\left(\vec{n},\alpha,\vec{\beta}\right) = (-1)^{|\vec{n}|}(1-q)^{|\vec{n}|}\,\frac{q^{(\alpha+|\vec{n}|-1)|\vec{n}|}}{\prod\limits_{j=1}^{r}\left(q^{\alpha+\beta_j+|\vec{n}|+1};q\right)_{n_j}}.$$

23.7 Modified Bessel Function Weights

So far all the examples are extensions of the classical orthogonal polynomials in Askey's scheme of hypergeometric orthogonal polynomials (and its q-analogue). These classical orthogonal polynomials all have a weight function w that satisfies *Pearson's equation*

$$[w(x)\sigma(x)]' = \tau w(x), \tag{23.7.1}$$

where σ is a polynomial of degree at most two and τ a polynomial of degree one, or a discrete analogue of this equation involving a difference or q-difference operator. There are however multiple orthogonal polynomials which are not mere extensions of the classical orthogonal polynomials but which are quite natural in the multiple setting. On one hand, we can allow a higher degree for the polynomials σ and τ in Pearson's equation (23.7.1), and this will typically give rise to an Angelesco system. The Jacobi–Angelesco polynomials are an example of this kind, where $w(x) = (x - a)^\alpha x^\beta(1-x)^\gamma$, for which $\sigma(x) = (x-a)x(1-x)$ is a polynomial of degree 3.

Another example are the *Jacobi–Laguerre polynomials,* for which

$$\int_a^0 P_{n,m}^{(\alpha,\beta)}(x)(x-a)^\alpha |x|^\beta e^{-x} x^k \, dx = 0, \qquad k = 0, \ldots, n-1,$$

$$\int_0^\infty P_{n,m}^{(\alpha,\beta)}(x)(x-a)^\alpha x^\beta e^{-x} x^k \, dx = 0, \qquad k = 0, \ldots, m-1,$$

where $a < 0$ and $\alpha, \beta > -1$. Here $w(x) = (x-a)^\alpha x^\beta e^{-x}$, which satisfies Pearson's equation (23.7.1) with $\sigma(x) = (x-a)x$ and τ a polynomial of degree two. Other examples have been worked out in (Aptekarev et al., 1997).

Another way to obtain multiple orthogonal polynomials with many useful properties is to consider Pearson's equation for vector valued functions or a system of equations of Pearson type. This corresponds to considering weights which satisfy a higher order differential equation with polynomial coefficients. Bessel functions are examples of functions satisfying a second order differential equation with polynomial coefficients. If we want positive weights, then only the modified Bessel functions are allowed.

23.7.1 Modified Bessel functions

Multiple orthogonal polynomials for the modified Bessel functions I_ν and $I_{\nu+1}$ were obtained in (Douak, 1999) and (Coussement & Van Assche, 2003). The modified Bessel function I_ν satisfies the differential equation (1.3.22) and has the series expansion in (1.3.17). The function is positive on the positive real axis for $\nu > -1$ and has the asymptotic behavior $I_\nu(x) = e^x / \sqrt{2\pi x} \, [1 + \mathcal{O}(1/x)]$ as $x \to +\infty$. We use the weights

$$w_1(x) = x^{\nu/2} I_\nu \left(2\sqrt{x} \right) e^{-cx}, \qquad w_2(x) = x^{(\nu+1)/2} I_{\nu+1} \left(2\sqrt{x} \right) e^{-cx},$$

on $[0, \infty)$, where $\nu > -1$ and $c > 0$. The system (w_1, w_2) then turns out to be an AT-system on $[0, \infty)$ (in fact it is a Nikishin system), hence every multi-index is normal. The multiple orthogonal polynomials on the diagonal (multi-indices (n, m) with $n = m$) and the stepline (multi-indices (n, m), where $n = m + 1$) can then be obtained explicitly and they have nice properties. Let $Q_{n,m}^{(\nu,c)}(x) = A_{n+1,m+1}(x)w_1(x) + B_{n+1,m+1}(x)w_2(x)$, where $(A_{n,m}, B_{n,m})$ are the type I multiple orthogonal polynomials, and define

$$q_{2n}^\nu(x) = Q_{n,n}^{(\nu,c)}(x), \qquad q_{2n+1}^\nu(x) = Q_{n+1,n}^{(\nu,c)}(x).$$

In a similar way we let $P_{n,m}^{(\nu,c)}$ be the type II multiple orthogonal polynomials and define

$$p_{2n}^\nu(x) = P_{n,n}^{(\nu,c)}(x), \qquad p_{2n+1}^\nu(x) = P_{n+1,n}^{(\nu,c)}(x).$$

Then we have the following raising and lowering properties:

$$\left[q_n^{\nu+1}(x) \right]' = q_{n+1}^\nu(x),$$

and
$$[p_n^\nu(x)]' = np_{n-1}^{\nu+1}(x).$$

The type I multiple orthogonal polynomials $\{q_n^\nu(x)\}$ have the explicit formula

$$q_n^\nu(x) = \sum_{k=0}^{n+1} \binom{n+1}{k} (-c)^k x^{(\nu+k)/2} I_{\nu+k}\left(2\sqrt{x}\right) e^{-cx},$$

and the type II multiple orthogonal polynomials have the representation

$$p_n^\nu(x) = \frac{(-1)^n}{c^{2n}} \sum_{k=0}^{n} \binom{n}{k} c^k k! \, L_k^\nu(cx),$$

where L_k^ν are the Laguerre polynomials. The type II multiple orthogonal polynomials satisfy the third order differential equation

$$xy'''(x) + (-2cx + \nu + 2)y''(x) + \left(c^2 x + c(n - \nu - 2) - 1\right) y'(x) = c^2 n y(x),$$

and the recurrence relation

$$x p_n^\nu(x) = p_{n+1}^\nu(x) + b_n p_n^\nu(x) + c_n p_{n-1}^\nu(x) + d_n p_{n-2}^\nu(x),$$

with recurrence coefficients

$$b_n = \frac{1}{c^2}\left[1 + c(2n + \nu + 1)\right], \quad c_n = \frac{n}{c^3}\left[2 + c(n + \nu)\right], \quad d_n = \frac{n(n-1)}{c^4}.$$

Multiple orthogonal polynomials for the modified Bessel functions K_ν and $K_{\nu+1}$ have been introduced independently in (Ben Cheikh & Douak, 2000b) and (Van Assche & Yakubovich, 2000), and were further investigated in (Coussement & Van Assche, 2001). The modified Bessel function K_ν satisfies the differential equation (1.3.22) for which they are the solution that remains bounded as $x \to \infty$ on the real line. An integral representation is

$$K_\nu(x) = \frac{1}{2}\left(\frac{x}{2}\right)^\nu \int_0^\infty \exp\left(-t - \frac{x^2}{4t}\right) t^{-\nu-1} \, dt.$$

We shall use the scaled functions

$$\rho_\nu(x) = 2x^{\nu/2} K_\nu\left(2\sqrt{x}\right),$$

and consider the weights

$$w_1(x) = x^\alpha \rho_\nu(x), \qquad w_2(x) = x^\alpha \rho_{\nu+1}(x),$$

on $[0, \infty)$, where $\alpha > -1$ and $\nu \geq 0$. The system (w_1, w_2) is an AT system (in fact, this system is a Nikishin system). If we put

$$Q_{n,m}^{(\alpha,\nu)}(x) = A_{n+1,m+1}(x)\rho_\nu(x) + B_{n+1,m+1}(x)\rho_{nu+1}(x),$$

where $(A_{n,m}, B_{n,m})$ are the type I multiple orthogonal polynomials, and define

$$q_{2n}^\alpha(x) = Q_{n,n}^{(\alpha,\nu)}(x), \qquad q_{2n}^\alpha(x) = Q_{n,n}^{(\alpha,\nu)}(x),$$

then the following Rodrigues formula holds:

$$x^\alpha q_{n-1}^\alpha(x) = \frac{d^n}{dx^n} \left[x^{n+\alpha} \rho_\nu(x) \right].$$

For the type II multiple orthogonal polynomials $P_{n,m}^{(\alpha,\nu)}$ we define

$$p_{2n}^\alpha(x) = P_{n,n}^{(\alpha,\nu)}(x), \qquad p_{2n+1}^\alpha(x) = P_{n+1,n}^{(\alpha,\nu)}(x),$$

and then have the differential property

$$[p_n^\alpha(x)]' = np_{n-1}^{\alpha+1}(x).$$

These type II multiple orthogonal polynomials have a simple hypergeometric representation:

$$p_n^\alpha(x) = (-1)^n \, (\alpha+1)_n \, (\alpha_\nu + 1)_n \, {}_1F_2 \left(\begin{matrix} -n \\ \alpha+1, \alpha+\nu+1 \end{matrix} \middle| x \right).$$

These polynomials satisfy the third order differential equation

$$x^2 y'''(x) + x(2\alpha + \nu + 3)y''(x) + [(\alpha+1)(\alpha+\nu+1) - x]y'(x) + ny(x) = 0,$$

and the recurrence relation

$$xp_n^\alpha(x) = p_{n+1}^\alpha(x) + b_n p_n^\alpha(x) + c_n p_{n-1}^\alpha(x) + d_n p_{n-2}^\alpha(x),$$

with

$$b_n = (n+\alpha+1)(3n+\alpha+2\nu) - (\alpha+1)(\nu-1)$$
$$c_n = n(n+\alpha)(n+\alpha+\nu)(3n+2\alpha+\nu)$$
$$d_n = n(n-1)(n+\alpha-1)(n+\alpha)(n+\alpha+\nu-1)(n+\alpha+\nu).$$

23.8 The Riemann–Hilbert problem for multiple orthogonal polynomials

In Chapter 22 it was shown that the usual orthogonal polynomials on the real line can be characterized by a Riemann–Hilbert problem for 2×2 matrices. In (Van Assche et al., 2001) it was shown that multiple orthogonal polynomials (of type I and type II) can also be described in terms of a Riemann–Hilbert problem, but now for matrices of order $r + 1$. Consider the following Riemann–Hilbert problem: determine an

$(r+1) \times (r+1)$ matrix function Y such that

1. Y is analytic in $\mathbb{C} \setminus \mathbb{R}$.
2. On the real line we have

$$Y_+(x) = Y_-(x) \begin{pmatrix} 1 & w_1(x) & w_2(x) & \cdots & w_r(x) \\ 0 & 1 & 0 & \cdots & 0 \\ 0 & 0 & 1 & & 0 \\ \vdots & \vdots & & \ddots & 0 \\ 0 & 0 & \cdots & 0 & 1 \end{pmatrix}, \qquad x \in \mathbb{R}.$$

3. Y has the following behavior near infinity

$$Y(z) = (I + \mathcal{O}(1/z)) \begin{pmatrix} z^{|\bar{n}|} & & & 0 \\ & z^{-n_1} & & \\ & & z^{-n_2} & \\ & & & \ddots \\ 0 & & & & z^{-n_r} \end{pmatrix}, \qquad z \to \infty.$$

$$(23.8.1)$$

Theorem 23.8.1 *Suppose that $x^j w_k \in L_1(\mathbb{R})$ for every j and $1 \leq k \leq r$ and that each w_k is Hölder continuous on $\mathbb{R}$. Let $P_{\bar{n}}$ be the type II multiple orthogonal polynomial for the measures $(\mu_1, \ldots, \mu_r)$ for which $d\mu_k(x) = w_k(x)\,dx$ on $\mathbb{R}$ and suppose that $\bar{n}$ is a normal index. Then the solution of the Riemann–Hilbert problem (23.8.2) is unique and given by*

$$\begin{pmatrix} P_{\bar{n}}(z) & \dfrac{1}{2\pi i} \displaystyle\int_{\mathbb{R}} \dfrac{P_{\bar{n}}(t)w_1(t)}{t-z}\,dt & \cdots & \dfrac{1}{2\pi i} \displaystyle\int_{\mathbb{R}} \dfrac{P_{\bar{n}}(t)w_r(t)}{t-z}\,dt \\[3mm] -2\pi i \gamma_1 P_{\bar{n}-\bar{e}_1}(z) & -\gamma_1 \displaystyle\int_{\mathbb{R}} \dfrac{P_{\bar{n}-\bar{e}_1}(t)w_1(t)}{t-z}\,dt & \cdots & -\gamma_1 \displaystyle\int_{\mathbb{R}} \dfrac{P_{\bar{n}-\bar{e}_1}(t)w_r(t)}{t-z}\,dt \\[3mm] \vdots & \vdots & \cdots & \vdots \\[3mm] -2\pi i \gamma_r P_{\bar{n}-\bar{e}_r}(z) & -\gamma_r \displaystyle\int_{\mathbb{R}} \dfrac{P_{\bar{n}-\bar{e}_r}(t)w_1(t)}{t-z}\,dt & \cdots & -\gamma_r \displaystyle\int_{\mathbb{R}} \dfrac{P_{\bar{n}-\bar{e}_r}(t)w_r(t)}{t-z}\,dt \end{pmatrix},$$

$$(23.8.2)$$

where

$$\frac{1}{\gamma_k} = \frac{1}{\gamma_k(\bar{n})} = \int_{\mathbb{R}} x^{n_k-1} P_{\bar{n}-\bar{e}_k}(t) w_k(t)\,dt.$$

Proof The function $Y_{1,1}$ on the first row and first column of Y is an analytic function on $\mathbb{C} \setminus \mathbb{R}$, which for $x \in \mathbb{R}$ satisfies $(Y_{1,1})_+ (x) = (Y_{1,1})_- (x)$, hence $Y_{1,1}$ is an entire function. The asymptotic condition shows that $Y_{1,1}(z) = z^{|\bar{n}|}[1 + \mathcal{O}(1/z)]$ as $z \to \infty$, hence by Liouville's theorem we conclude that $Y_{1,1}(z) = \pi_{|\bar{n}|}(z)$ is a monic polynomial of degree $|\bar{n}|$.

For the remaining functions $Y_{1,j+1}$ $(j = 1, 2, \ldots, r)$ on the first row the jump condition becomes

$$(Y_{1,j+1})_+ (x) = (Y_{1,j+1})_- (x) + w_j(x)\, \pi_{|\vec{n}|}(x), \qquad x \in \mathbb{R},$$

hence the Plemelj–Sokhotsky formulas give

$$Y_{1,j+1}(z) = \frac{1}{2\pi i} \int_{\mathbb{R}} \frac{\pi_{|\vec{n}|}(t)w_j(t)}{t - z}\, dt.$$

The condition near infinity is $Y_{1,j+1}(z) = \mathcal{O}\left(1/z^{n_j+1}\right)$ as $z \to \infty$. If we expand $1/(t - z)$ as

$$\frac{1}{t - z} = -\sum_{k=0}^{n_j-1} \frac{t^k}{z^{k+1}} + \frac{1}{t - z}\frac{t^{n_j}}{z^{n_j}},$$

then

$$Y_{1,j+1}(z) = -\sum_{k=0}^{n_j-1} \frac{1}{z^{k+1}} \int_{\mathbb{R}} t^k \pi_{|\vec{n}|}(t)w_j(t)\, dt + \mathcal{O}\left(\frac{1}{z^{n_j+1}}\right),$$

hence $\pi_{|\vec{n}|}$ has to satisfy

$$\int_{\mathbb{R}} t^k \pi_{|\vec{n}|}(t)w_j(t)\, dt = 0, \qquad k = 0, 1, \ldots, n_j - 1$$

and this for $j = 1, 2, \ldots, r$. But these are precisely the orthogonality conditions (23.1.3) for the type II multiple orthogonal polynomial $P_{\vec{n}}$ for the system $(\mu_1, \ldots, \mu_r)$, so that $\pi_{|\vec{n}|}(z) = P_{\vec{n}}(z)$.

The remaining rows can be handled in a similar way. The coefficients $\gamma_j(\vec{n})$ appear because the asymptotic condition for $Y_{j+1,j+1}$ is

$$\lim_{z \to \infty} z^{n_j} Y_{j+1,j+1}(z) = 1.$$

Observe that the coefficients $\gamma_j(\vec{n})$ are all finite since $\vec{n}$ is a normal index (see Corollary 23.1.2). $\qquad\square$

There is a similar Riemann–Hilbert problem for type I multiple orthogonal polynomials: determine an $(r+1) \times (r+1)$ matrix function X such that

1. X is analytic in $\mathbb{C} \setminus \mathbb{R}$.
2. On the real line we have

$$
X_+(x) = X_-(x) \begin{pmatrix} 1 & 0 & 0 & \cdots & 0 \\ -w_1(x) & 1 & 0 & \cdots & 0 \\ -w_2(x) & 0 & 1 & & 0 \\ \vdots & \vdots & & \ddots & 0 \\ -w_r(x) & 0 & \cdots & 0 & 1 \end{pmatrix}, \qquad x \in \mathbb{R}.
$$

3. X has the following behavior near infinity

$$
X(z) = (I + \mathcal{O}(1/z)) \begin{pmatrix} z^{-|\vec{n}|} & & & & 0 \\ & z^{n_1} & & & \\ & & z^{n_2} & & \\ & & & \ddots & \\ 0 & & & & z^{n_r} \end{pmatrix}, \qquad z \to \infty.
$$

$$(23.8.3)$$

Theorem 23.8.2 *Suppose that $x^j w_k \in L_1(\mathbb{R})$ for every j and $1 \le k \le r$ and that each w_k is Hölder continuous on $\mathbb{R}$. Let $(A_{\vec{n},1}, \ldots, A_{\vec{n},r})$ be the type I multiple orthogonal polynomials for the measures $(\mu_1, \ldots, \mu_r)$ for which $d\mu_k(x) = w_k(x)\,dx$ on $\mathbb{R}$ and suppose that $\vec{n}$ is a normal index. Then the solution of the Riemann–Hilbert problem (23.8.3) is unique and given by*

$$
\begin{pmatrix} \displaystyle\int_{\mathbb{R}} \frac{Q_{\vec{n}}(t)}{z-t}\,dt & 2\pi i A_{\vec{n},1}(z) & \cdots & 2\pi i A_{\vec{n},r}(z) \\[2ex] \dfrac{c_1}{2\pi i} \displaystyle\int_{\mathbb{R}} \frac{Q_{\vec{n}+\vec{e}_1}(t)}{z-t}\,dt & c_1 A_{\vec{n}+\vec{e}_1,1}(z) & \cdots & c_1 A_{\vec{n}+\vec{e}_1,r}(z) \\[2ex] \vdots & \vdots & \cdots & \vdots \\[2ex] \dfrac{c_r}{2\pi i} \displaystyle\int_{\mathbb{R}} \frac{Q_{\vec{n}+\vec{e}_r}(t)}{z-t}\,dt & c_r A_{\vec{n}+\vec{e}_r,1}(z) & \cdots & c_r A_{\vec{n}+\vec{e}_r,r}(z) \end{pmatrix},
$$

$$(23.8.4)$$

where

$$
Q_{\vec{n}}(x) = \sum_{j=1}^{n} A_{\vec{n},j}(x)\,w_j(x),
$$

and $1/c_j = 1/c_j(\vec{n})$ is the leading coefficient of $A_{\vec{n}+\vec{e}_j,j}$.

Proof For $1 \le j \le r$ the functions $X_{1,j+1}$ satisfy the jump condition $(X_{1,j+1})_+(x) = (X_{1,j+1})_-(x)$ for $x \in \mathbb{R}$, so that each $X_{1,j+1}$ is an entire function. Near infinity we have $X_{1,j+1}(z) = \mathcal{O}(z^{n_j-1})$, hence Liouville's theorem implies that each $X_{1,j+1}$

is a polynomial π_j of degree at most $n_j - 1$. The jump condition for $X_{1,1}$ is

$$(X_{1,1})_+ (x) = (X_{1,1})_- (x) - \sum_{j=1}^{r} w_j(x)\pi_j(x), \qquad x \in \mathbb{R},$$

hence we conclude that

$$X_{1,1}(z) = \frac{1}{2\pi i} \int_{\mathbb{R}} \sum_{j=1}^{r} w_j(t)\pi_j(t) \frac{dt}{z - t}.$$

If we expand $1/(z - t)$ as

$$\frac{1}{z - t} = \sum_{k=0}^{|\vec{n}|-1} \frac{t^k}{z^{k+1}} + \frac{1}{z - t} \frac{t^{|\vec{n}|}}{z^{|\vec{n}|}},$$

then

$$X_{1,1}(z) = \sum_{k=0}^{|\vec{n}|-1} \frac{1}{z^{k+1}} \int_{\mathbb{R}} t^k \sum_{j=1}^{r} w_j(t)\pi_j(t)\, dt + \mathcal{O}\left(1/z^{|\vec{n}|+1}\right),$$

hence the asymptotic condition $z^{|\vec{n}|} X_{1,1}(z) = 1 + \mathcal{O}(1/z)$ as $z \to \infty$ implies that

$$\int_{\mathbb{R}} t^k \sum_{j=1}^{r} w_j(t)\pi_j(t)\, dt = 0, \qquad k = 0, 1, \ldots, |\vec{n}| - 2,$$

and

$$\frac{1}{2\pi i} \int_{\mathbb{R}} t^{|\vec{n}|-1} \sum_{j=1}^{r} w_j(t)\pi_j(t)\, dt = 1.$$

But these are precisely the orthogonality conditions (23.1.1) and (23.1.2) for the type I multiple orthogonal polynomials for $(\mu_1, \ldots, \mu_r)$, up to a factor $2\pi i$, namely $\pi_j(x) = 2\pi i A_{\vec{n},j}(x)$. This gives the first row of X.

For the other rows of X one uses a similar reasoning, but now one has $X_{1+j,1}(z) = \mathcal{O}\left(z^{-|\vec{n}|-1}\right)$ and $X_{1+j,1+j}$ is a monic polynomial of degree n_j. These two properties explain that row $j + 1$ consists of type I multiple orthogonal polynomials with multi-index $\vec{n} + \vec{e}_j$ and that $X_{1+j,1+j} = c_j A_{\vec{n}+\vec{e}_j,j}$, where $1/c_j$ is the leading coefficient of $A_{\vec{n}+\vec{e}_j,j}$. Observe that all the $c_j(\vec{n})$ are finite since $\vec{n}$ is a normal index (see Corollary 23.1.1). □

There is a very simple and useful connection between the matrix functions X for type I and Y for type II. This relation can, with some effort, be found in Mahler's exposition (Mahler, 1968).

Theorem 23.8.3 (Mahler's relation) *The matrix X solving the Riemann–Hilbert problem (23.8.3) and the matrix Y solving the Riemann–Hilbert problem (23.8.2) are connected by*

$$X(z) = Y^{-T}(z),$$

where A^{-T} is the transpose of the inverse of a matrix A.

Proof We will show that Y^{-T} satisfies the Riemann–Hilbert problem (23.8.3), then unicity shows that the result holds. First of all it is easy to show that $\det Y(z) = 1$ for all $z \in \mathbb{C}$, hence Y^{-T} indeed exists and is analytic in $\mathbb{C} \setminus \mathbb{R}$. The behavior at infinity is given by

$$Y^{-T} = [I + \mathcal{O}(1/z)]^{-1} \begin{pmatrix} z^{|\vec{n}|} & & & 0 \\ & z^{-n_1} & & \\ & & z^{-n_2} & \\ & & & \ddots \\ 0 & & & z^{-n_r} \end{pmatrix}^{-1}$$

$$= [I + \mathcal{O}(1/z)] \begin{pmatrix} z^{-|\vec{n}|} & & & 0 \\ & z^{n_1} & & \\ & & z^{n_2} & \\ & & & \ddots \\ 0 & & & z^{n_r} \end{pmatrix},$$

which corresponds to the behavior in (23.8.3). Finally, the jump condition is

$$\left(Y^{-T}\right)_+(x) = \left(Y^{-T}\right)_-(x) \begin{pmatrix} 1 & w_1(x) & w_2(x) & \cdots & w_r(x) \\ 0 & 1 & 0 & \cdots & 0 \\ 0 & 0 & 1 & & 0 \\ \vdots & \vdots & & \ddots & 0 \\ 0 & 0 & \cdots & 0 & 1 \end{pmatrix}^{-T}$$

$$= \left(Y^{-T}\right)_-(x) \begin{pmatrix} 1 & 0 & 0 & \cdots & 0 \\ -w_1(x) & 1 & 0 & \cdots & 0 \\ -w_2(x) & 0 & 1 & & 0 \\ \vdots & \vdots & & \ddots & 0 \\ -w_r(x) & 0 & \cdots & 0 & 1 \end{pmatrix},$$

which corresponds to the jump condition in (23.8.3). □

Of course this also implies that $Y(z) = X^{-T}(z)$. For the entry in row 1 and column 1 this gives

$$P_{\vec{n}}(z) = \det \begin{pmatrix} c_1 A_{\vec{n}+\vec{e}_1,1}(z) & \cdots & c_1 A_{\vec{n}+\vec{e}_1,r}(z) \\ \vdots & \cdots & \vdots \\ c_r A_{\vec{n}+\vec{e}_r,1}(z) & \cdots & c_r A_{\vec{n}+\vec{e}_r,r}(z) \end{pmatrix},$$

which gives the type II multiple orthogonal polynomial $P_{\vec{n}}$ in terms of the type I multiple orthogonal polynomials $\left(A_{\vec{n}+\vec{e}_j,1}, \ldots, A_{\vec{n}+\vec{e}_j,r}\right)$ for $j = 1, 2, \ldots, r$.

23.8.1 Recurrence relation

Consider the matrix function $R_{\vec{n},k} = Y_{\vec{n}+\vec{e}_k} Y_{\vec{n}}^{-1}$, where $Y_{\vec{n}}$ is the matrix (23.8.2) containing the type II multiple orthogonal polynomials. Then $R_{\vec{n},k}$ is analytic in $\mathbb{C} \setminus \mathbb{R}$ and the jump condition is $(R_{\vec{n},k})_+(x) = (R_{\vec{n},k})_-(x)$ for $x \in \mathbb{R}$ since both

$Y_{\vec{n}+\vec{e}_k}$ and $Y_{\vec{n}}$ have the same jump matrix on $\mathbb{R}$. Hence $R_{\vec{n},k}$ is an entire matrix function and the behavior near infinity is

$$
R_{\vec{n},k}(z) = [I + \mathcal{O}_{\vec{n}+\vec{e}_k}(1/z)]
\begin{pmatrix}
z & & & & & & & \\
& 1 & & & & & & \\
& & \ddots & & & & & \\
& & & 1 & & & & \\
& & & & 1/z & & & \\
& & & & & 1 & & \\
& & & & & & \ddots & \\
& & & & & & & 1
\end{pmatrix}
[I + \mathcal{O}_{\vec{n}}(1/z)]^{-1},
$$

where the $1/z$ in the matrix is on row $k+1$. Liouville's theorem then implies that

$$
R_{\vec{n},k}(z) =
\begin{pmatrix}
z - a_0 & -a_1 & \cdots & & -a_k & & \cdots & -a_r \\
b_1 & 1 & & & & & & \\
\vdots & & \ddots & & & & & \\
& & & 1 & & & & \\
b_k & & & & 0 & & & \\
& & & & & 1 & & \\
\vdots & & & & & & \ddots & \\
b_r & & & & & & & 1
\end{pmatrix},
$$

where the $a_1, \ldots, a_r$ are constants depending on $\vec{n}$ and $b_1, \ldots, b_r$ are constants depending on $\vec{n} + \vec{e}_k$. This means that

$$
Y_{\vec{n}+\vec{e}_k}(z) =
\begin{pmatrix}
z - a_0 & -a_1 & \cdots & & -a_k & & \cdots & -a_r \\
b_1 & 1 & & & & & & \\
\vdots & & \ddots & & & & & \\
& & & 1 & & & & \\
b_k & & & & 0 & & & \\
& & & & & 1 & & \\
\vdots & & & & & & \ddots & \\
b_r & & & & & & & 1
\end{pmatrix}
Y_{\vec{n}}(z),
$$

and the entry on the first row and first column then gives the recurrence relation in Theorem 23.1.11. The entry in row $j + 1$ (for $j \neq k$) and the first column gives Corollary 23.1.8 but for another multi-index, and the entry in row $k + 1$ and the first column gives $-2\pi i \gamma_k (\vec{n} + \vec{e}_k) = b_k$.

23.8.2 Differential equation for multiple Hermite polynomials

Let us now take the multiple Hermite polynomials, where $w_j(x) = e^{-x^2 + c_j x}$ for $x \in \mathbb{R}$ for $1 \leq j \leq r$. Then each weight function w_j is actually an entire function on $\mathbb{C}$. Consider the matrix function

$$
Z(z) = E^{-1}(z) Y(z) E(z),
$$

where

$$E(z) = \begin{pmatrix} \exp\left(-\frac{r}{r+1}z^2\right) & & & \\ & \exp\left(\frac{1}{r+1}z^2 - c_1 z\right) & & \\ & & \ddots & \\ & & & \exp\left(\frac{1}{r+1}z^2 - c_r z\right) \end{pmatrix},$$

then Z is analytic on $\mathbb{C} \setminus \mathbb{R}$, it has the jump condition

$$Z_+(x) = Z_-(x) \begin{pmatrix} 1 & 1 & 1 & \cdots & 1 \\ 1 & 0 & \cdots & & 0 \\ & 1 & & & \vdots \\ & & \ddots & & 0 \\ & & & & 1 \end{pmatrix}, \qquad x \in \mathbb{R},$$

with a constant jump matrix, and the behavior near infinity is

$$E(z)Z(z)E^{-1}(z) = [I + \mathcal{O}(1/z)] \begin{pmatrix} z^{|\vec{n}|} & & & & 0 \\ & z^{-n_1} & & & \\ & & z^{-n_2} & & \\ & & & \ddots & \\ 0 & & & & z^{-n_r} \end{pmatrix}.$$

The derivative Z' is also analytic on $\mathbb{C} \setminus \mathbb{R}$, it has the same jump condition

$$Z'_+(x) = Z'_-(x) \begin{pmatrix} 1 & 1 & 1 & \cdots & 1 \\ 1 & 0 & \cdots & & 0 \\ & 1 & & & \vdots \\ & & \ddots & & 0 \\ & & & & 1 \end{pmatrix}, \qquad x \in \mathbb{R},$$

but the asymptotic condition is different. Observe that $E'(z) = L(z)E(z)$, where

$$L(z) = \begin{pmatrix} -\frac{2r}{r+1}z & & & \\ & \frac{2}{r+1}z - c_1 & & \\ & & \ddots & \\ & & & \frac{2}{r+1}z - c_r \end{pmatrix},$$

and $(E^{-1})'(z) = -L(z)E^{-1}(z)$, therefore the behavior near infinity is

$$E(z)Z'(z)E^{-1}(z) = \Big(-L(z)[1 + \mathcal{O}(1/z)] + [I + \mathcal{O}(1/z)]L(z) + \mathcal{O}(1/z) \Big)$$

$$\times \begin{pmatrix} z^{|\vec{n}|} & & & & 0 \\ & z^{-n_1} & & & \\ & & z^{-n_2} & & \\ & & & \ddots & \\ 0 & & & & z^{-n_r} \end{pmatrix}.$$

The matrix function $Z'(z)Z^{-1}(z)$ then turns out to be analytic in $\mathbb{C} \setminus \mathbb{R}$ with no jump on $\mathbb{R}$, so that it is an entire function, and hence $E(z)Z'(z)Z^{-1}(z)E^{-1}(z)$ is an entire matrix function. Observe that $L(z)$ is a matrix polynomial of degree one, hence the behavior near infinity and Liouville's theorem then give

$$E(z)Z'(z)Z^{-1}(z)E^{-1}(z) = 2 \begin{pmatrix} 0 & a_1 & a_2 & \cdots & a_r \\ -b_1 & 0 & 0 & \cdots & 0 \\ -b_2 & 0 & 0 & \cdots & 0 \\ \vdots & & & & \vdots \\ -b_r & 0 & 0 & \cdots & 0 \end{pmatrix},$$

where $a_1, \ldots, a_r$ and $b_1, \ldots, b_r$ are constants depending on $\vec{n}$. This gives the differential equation (in matrix form)

$$Z'(z) = 2 \begin{pmatrix} 0 & a_1 e^{z^2-c_1 z} & a_2 e^{z^2-c_2 z} & \cdots & a_r e^{z^2-c_r z} \\ -b_1 e^{-z^2+c_1 z} & 0 & 0 & \cdots & 0 \\ -b_2 e^{-z^2+c_2 z} & 0 & 0 & \cdots & 0 \\ \vdots & & & & \vdots \\ -b_r e^{-z^2+c_r z} & 0 & 0 & \cdots & 0 \end{pmatrix} Z(z).$$

The entry on the first row and first column gives

$$\left(H_{\vec{n}}^{\vec{c}}\right)'(z) = 2\sum_{j=1}^{r} a_{\vec{n},j} H_{\vec{n}-\vec{e}_j}^{\vec{c}}(z),$$

where the $a_{\vec{n},j}$ $(1 \le j \le r)$ are the coefficients appearing in the recurrence relation (23.1.21), which for multiple Hermite polynomials is equal to (23.5.3), so that $2a_{\vec{n},j} = n_j$. We therefore have

$$\left(H_{\vec{n}}^{\vec{c}}\right)'(z) = \sum_{j=1}^{r} n_j H_{\vec{n}-\vec{e}_j}^{\vec{c}}(z), \tag{23.8.5}$$

which can be considered as a lowering operation. The entry on row $j+1$ and the first column gives

$$\left(e^{-z^2+c_j z} H_{\vec{n}-\vec{e}_j}^{\vec{c}}(z)\right)' = -2e^{-z^2+c_j z} H_{\vec{n}}^{\vec{c}}(z), \qquad 1 \le j \le r, \tag{23.8.6}$$

which can be considered as r raising operators. If we consider the differential operators

$$D_0 = \frac{d}{dz}, \qquad D_j = e^{z^2-c_j z} \frac{d}{dz} e^{-z^2+c_j z}, \qquad 1 \le j \le r,$$

then the $D_1, \ldots, D_r$ are commuting operators and (23.8.5)–(23.8.6) give

$$\left(\prod_{j=1}^{r} D_j D_0\right) H_{\vec{n}}^{\vec{c}}(z) = \left(-2\sum_{j=1}^{r} n_j \prod_{i \ne j} D_i\right) H_{\vec{n}}^{\vec{c}}(z), \tag{23.8.7}$$

which is a differential equation of order $r+1$ for the multiple Hermite polynomials.

24

Research Problems

In this chapter we formulate several open problems related to the subject matter of this book. Some of these problems have already been alluded to in the earlier chapters, but we felt that collecting them in one place would make them more accessible.

24.1 Multiple Orthogonal Polynomials

In spite of the major advances made over the last thirty years in the area of multiple orthogonal polynomials, the subject remains an area with many open problems. We formulate several problems below that we believe are interesting and whose solution will advance our understanding of the subject.

Problem 24.1.1 *Consider the case when the measures $\mu_1, \ldots, \mu_r$ are absolutely continuous and $\mu_j'(x) = \exp(-v_j(x))$, and all the measures μ_j, $1 \le j \le r$ are supported on $[a, b]$. The problem is to derive differential recurrence relations and differential equations for the multiple orthogonal polynomials which reduce to the results in Chapter 3. Certain smoothness conditions need to be imposed on $v_j(x)$.*

Problem 24.1.2 *Evaluate the discriminants of general multiple orthogonal polynomials in terms of their recursion coefficients when their measures of orthogonality are as in Problem 24.1.1.*

The solution of Problem 24.1.2 would extend the author's result of Chapter 3 from orthogonal polynomials to multiple orthogonal polynomials, while the solution of Problem 24.1.1 would extend the work on differential equations to multiple orthogonal polynomials.

Problem 24.1.3 *Assume that μ_j, $1 \le j \le r$ are discrete and supported on $\{s, s + 1, s + 2, \ldots, t\}$, s, t are nonnegative integers, and t may be $+\infty$. Let $w_j(\ell) = \mu_j(\{\ell\})$ and $w_j(s - 1) = w_j(t + 1) = 0$, $1 \le j \le r$. Extend the Ismail–Nikolova–Simeonov results of §§6.3 and 6.4 to the multiple orthogonal polynomials.*

Problem 24.1.4 *Viennot developed a combinatorial theory of orthogonal polynomials when the coefficients $\{\alpha_n\}$ and $\{\beta_n\}$ in the monic form are polynomials in n or q^n. He gives interpretations for the moments, the coefficients of powers of x in*

$P_n(x)$, *and for the linearization coefficients in terms of statistics on combinatorial configurations. This work is in (Viennot, 1983). Further development in the special case of q-Hermite polynomials is in (Ismail et al., 1987). Extending this body of work to multiple orthogonal polynomials will be most interesting.*

Problem 24.1.5 *There is no study of zeros of general or special systems of multiple orthogonal polynomials available. An extension of Theorem 7.1.1 to multiple orthogonal polynomials would be a worthwhile research project. We may need to assume that μ_j is absolutely continuous with respect to a fixed measure α for all j. This assumes that all measures are supported on the same interval. The Hellman–Feynman techniques may be useful in the study of monotonicity of the zeros of multiple orthogonal polynomials like the Angelesco and AT systems, or any of the other explicitly defined systems of Chapter 23.*

24.2 A Class of Orthogonal Functions

As we pointed out in §14.5, the system of functions $\{\mathcal{F}_k(x)\}$ defined by (14.5.1) is a complete orthonormal system in $L^2(\mu, \mathbb{R})$, when $\{p_n(x)\}$ are complete orthonormal in $L^2(\mu, \mathbb{R})$ and $\{r_k(x)\}$ is complete orthonormal in L^2 weighted by a discrete measure with masses $\{\rho(x_k)\}$ at $x_1, x_2, \ldots$, and $\{u_n\}$ is a sequence of points on the unit circle.

Problem 24.2.1 *Explore interesting examples of the system $\{\mathcal{F}_n\}$ in (14.5.1) by choosing $\{p_n\}$ from the q-orthogonal polynomials in Chapters 13 and 15 and $\{r_n\}$ from Chapter 18. The special system $\{\mathcal{F}_n\}$ must have some interesting additional properties like addition theorems, for example.*

Problem 24.2.2 *The functions $\{\mathcal{F}_k(x)\}$ resemble bilinear forms in reproducing kernel Hilbert spaces. The problem is to recast the properties of $\{\mathcal{F}_k(x)\}$ in the language of reproducing kernel Hilbert spaces. The major difference here is that $\{\mathcal{F}_k(x)\}$ project functions in $L^2(\mu, \mathbb{R})$ on a weighted ℓ^2 space and functions in a weighted ℓ^2 space on $L^2(\mu, \mathbb{R})$.*

24.3 Positivity

Conjecture 24.3.1 *If*

$$\frac{\prod_{j=1}^{4}(1-t_j)^{-1}}{\sum_{1\leq i<j\leq 4}(1-t_i)(1-t_j)} = \sum_{k,\ell,m,n=0}^{\infty} E(k,\ell,m,n)\, t_1^k t_2^\ell t_3^m t_4^n,$$

then $E(k,\ell,m,n) \geq 0$.

The early coefficients in the power series expansion of

$$
\left[\sum_{1 \leq i < j \leq 4} (1 - t_i)(1 - t_j) \right]^{-1}
$$

are positive but the later coefficients do change sign. The factor $\prod_{j=1}^{4} (1 - t_j)^{-1}$ is an averaging factor that makes the early positive terms count more than the later terms.

Conjecture 24.3.2 ((Ismail & Tamhankar, 1979)) *Let*

$$
\begin{pmatrix} y_1 \\ y_2 \\ y_3 \\ y_4 \end{pmatrix} = \begin{pmatrix} \dfrac{1}{2} & 1 & \sqrt{12}\,i & -12 \\[2mm] \dfrac{1}{12} & \dfrac{1}{2} & 1 & -\sqrt{12}\,i \\[2mm] -\dfrac{i\sqrt{12}}{144} & \dfrac{1}{12} & \dfrac{1}{2} & 1 \\[2mm] -\dfrac{1}{144} & \dfrac{i\sqrt{12}}{144} & \dfrac{1}{12} & \dfrac{1}{2} \end{pmatrix} \begin{pmatrix} t_1 \\ t_2 \\ t_3 \\ t_4 \end{pmatrix},
$$

and let $G(k_1, k_2, k_3, k_4)$ be the coefficient of $t_1^{k_1} t_2^{k_2} t_3^{k_3} t_4^{k_4}$ in $y_1^{k_1} y_2^{k_2} y_3^{k_3} y_4^{k_4}$. Then

$$
\sum_{k_1=0}^{j_1} \sum_{k_2=0}^{j_2} \sum_{k_3=0}^{j_3} \sum_{k_4=0}^{j_4} G(k_1, k_2, k_3, k_4) \geq 0
$$

for all nonnegative integers j_1, j_2, j_3, j_4.

The equivalence of Conjectures 24.3.1 and 24.3.2 follows from the MacMahon Master Theorem.

Conjecture 24.3.3 *If*

$$
\frac{(1 - t_1 - t_2 - t_3 - t_4)^{-1}}{\sum_{1 \leq i < j \leq 4} (1 - t_i)(1 - t_j)} = \sum_{k, \ell, m, n = 0}^{\infty} F(k, \ell, m, n)\, t_1^k t_2^\ell t_3^m t_4^n,
$$

then $F(k, \ell, m, n) \geq 0$.

24.4 Asymptotics and Moment Problems

The problems formulated here deal with monic polynomials whose three-term recurrence relations have the form

$$
x P_n(x) = P_{n+1}(x) + \alpha_n P_n(x) + q^{-n} \beta_n P_{n-1}(x), \tag{24.4.1}
$$

and

$$
P_0(x) = 1, \quad P_1(x) = x - \alpha_n. \tag{24.4.2}
$$

We further assume

$$
q^{-n/2} \alpha_n \to 0, \quad \beta_n \to 1 \quad \text{as } n \to \infty. \tag{24.4.3}
$$

Set

$$x_n(t) = q^{-n/2}t - q^{n/2}/t. \tag{24.4.4}$$

Conjecture 24.4.1 *The limiting relation*

$$\lim_{n\to\infty} \frac{q^{n^2/2}}{t^n} P_n\left(x_n(t)\right) = A_q\left(1/t^2\right), \tag{24.4.5}$$

holds where A_q *is the function defined in* (21.7.3).

Conjecture 24.4.1 will imply that the zeros

$$x_{n,1} > x_{n,2} > \cdots > x_{n,n}$$

of $P_n(z)$ have the asymptotic property

$$\lim_{n\to\infty} q^{n/2} x_{n,k} = 1/\sqrt{i_k(q)}, \tag{24.4.6}$$

where

$$0 < i_1(q) < i_2(q) < \cdots$$

are the zeros of $A_q(z)$.

In a work in progress by Ismail, Li and Rahman, they establish Conjecture 24.4.1 when $\alpha_n = 0$ for all n. They also give the next two terms in the asymptotic expansion of $q^{n^2/2}t^{-n}P_n\left(x_n(t)\right)$ when $\beta_n = 1 + c_1 q^n + c_2 q^{2n} + o\left(q^{2n}\right)$, as $n \to \infty$.

A typical β_n in (24.4.1) is $1 + \mathcal{O}\left(q^{cn}\right)$, as $n \to \infty$, for $c > 0$. Moreover ζ_n, of (2.1.5), has the property that $\lim_{n\to\infty} q^{n(n+1)/2}\zeta_n$ exists. In all the examples we know of, if $\{P_n(x)\}$ is orthogonal with respect to a weight function w then $\lim_{n\to\infty} w\left(x_n(t)\right) t^{2n}q^{-n^2/2}$ exists. This leads to the following two conjectures.

Conjecture 24.4.2 *Assume that as* $n \to \infty$,

$$\beta_n = 1 + \mathcal{O}\left(q^{cn}\right), \quad c > 0,$$

$$\alpha_n = \mathcal{O}\left(q^{n(d+1/2)}\right), \quad d > 0,$$

and that $\{P_n(x)\}$ *is orthogonal with respect to a weight function* $w(x)$. *With* $\zeta_n = \int_{\mathbb{R}} P_n^2(x)w(x)\,dx$, *there exists* $\delta > 0$ *such that*

$$\lim_{n\to\infty} \sqrt{w\left(x_n(t)\right)} \frac{P_n\left(x_n(t)\right)}{\sqrt{\zeta_n}} q^{-n\delta} = f(t)A_q\left(1/t^2\right), \tag{24.4.7}$$

where the function $f(t)$ *is defined on* $\mathbb{C} \smallsetminus \{0\}$ *and has no zeros. Moreover* f *may depend on* w. *Furthermore* $A_q(t)$ *and* $1/f(t)$ *have no common zeros.*

Of course (24.4.6) will be an immediate consequence of (24.4.7). In the case of the polynomials $\{h_n(x\,|\,q)\}$, $\delta = 1/4$.

Note that w in Conjecture 24.4.2 is not unique.

Conjecture 24.4.3 *Under the assumptions in Conjecture 24.4.2, there exists δ such that*

$$\lim_{n \to \infty} w\left(x_n(t)\right) t^n q^{n\delta - n^2/4}$$

exists. Moreover δ is the same for all weight functions.

Recall the definitions of q-order, q-type and q-Phragmén–Lindelöf indicator in (21.1.18)–(21.1.20).

Conjecture 24.4.4 *Let A, B, C, D be the Nevanlinna functions of an indeterminate moment problem. If the order of A is zero, but A has finite q-order for some q, then A, B, C, D have the same q-order, q-type and q-Phragmén–Lindelöf indicator.*

24.5 Functional Equations and Lie Algebras

Let

$$w_1(x) = x^\alpha e^{-\phi(x)}, \quad x > 0, \ \alpha > -1, \tag{24.5.1}$$

$$w_2(x) = \exp(-\psi(x)), \quad x \in \mathbb{R}. \tag{24.5.2}$$

In §3.2, we proved that there exists linear differential operators $L_{1,n}$ and $L_{2,n}$ such that

$$L_{1,n} p_n(x) = A_n(x) p_{n-1}(x), \quad L_{2,n} p_{n-1}(x) = A_{n-1}(x) \frac{a_n}{a_{n-1}} p_n(x),$$

if $\{p_n(x)\}$ are orthonormal with respect to $e^{-v(x)}$.

Problem 24.5.1 *Assume that $\{p_n(x)\}$ is orthonormal on $\mathbb{R}$ with respect to $w_2(x)$. Then the Lie algebra generated by $L_{1,n}$ and $L_{2,n}$ is finite dimensional if and only if ψ is a polynomial of degree $2m$, in which case the Lie algebra is $2m + 1$ dimensional.*

Chen and Ismail proved the "if" part in (Chen & Ismail, 1997).

Problem 24.5.2 *Let $\{p_n(x)\}$ be orthonormal on $[0, \infty)$ with respect to $w_1(x)$. Then the Lie algebra generated by $xL_{1,n}$ and $xL_{2,n}$ is finite dimensional if and only if ϕ is a polynomial.*

The "if" part is Theorem 3.7.1 and does not seem to be in the literature.

Recall the Rahman–Verma addition theorem, Theorem 15.2.3. Usually group theory is the natural setup for addition theorems but, so far, the general Rahman–Verma addition theorem has not found its natural group theoretic setup. Koelink proved the special case $a = q^{1/2}$ of this result using quantum group theoretic techniques in (Koelink, 1994). Askey observed that the Askey–Wilson operator can be used to extend Koelink's result for $a = q^{1/2}$ to general a. Askey's observation is in a remark following Theorem 4.1 in (Koelink, 1994).

Problem 24.5.3 *Find a purely quantum group theoretic proof of the full Rahman–Verma addition theorem, Theorem 15.2.3.*

Koelink's survey article (Koelink, 1997) gives an overview of addition theorems for q-polynomials.

Koelink proved an addition theorem for a two-parameter subfamily of the Askey–Wilson polynomials in (Koelink, 1997, Theorem 4.1). His formula involves several $_8W_7$ series and contains the Rahman–Verma addition theorem as a nontrivial special case; see §5.2 of (Koelink, 1997).

Problem 24.5.4 *Find a nonterminating analogue of Theorem 4.1 in (Koelink, 1997) where all the special Askey–Wilson polynomials are replaced by $_8W_7$ functions.*

Floris gave an addition formula of the q-disk polynomials, a q-analogue of an addition theorem in (Koornwinder, 1978). Floris' result is an addition theorem in noncommuting variables and has been converted to a formula only involving commuting variables in (Floris & Koelink, 1997). Special cases appeared earlier in (Koornwinder, 1991); see (Rahman, 1989) for a q-series proof.

No addition theorem seems to be known for any of the associated polynomials of the classical polynomials.

Problem 24.5.5 *Find addition theorems for the two families of associated Jacobi polynomials, the Askey–Wimp and the Ismail–Masson polynomials.*

Recall Theorem 14.6.4 where we proved that the only solution to

$$f(x \oplus y) = f(x)f(y), \qquad (24.5.3)$$

is $\mathcal{E}_q(x; \alpha)$ if $f(x)$ has an expansion $\sum\limits_{n=0}^{\infty} f_n g_n(x)$, which converges uniformly on compact subsets of a domain Ω.

Problem 24.5.6 *Extend the definition of $\oplus$ to measurable functions and prove that the only measurable solution to (24.5.3) is $\mathcal{E}_q(x; \alpha)$.*

24.6 Rogers–Ramanujan Identities

The works (Lepowsky & Milne, 1978) and (Lepowsky & Wilson, 1982) contain a Lie theoretic approach to Rogers–Ramanujan and other partition identities. So far, this algebraic approach has not produced identities like (13.5.7) or (13.5.13) for m positive or negative; see Theorem 13.6.1.

Problem 24.6.1 *Find an algebraic approach to prove (13.5.13) for all integers m.*

As we pointed out in the argument preceeding Theorem 13.6.1, it is sufficient to establish (13.5.13) for $m = 0, 1, \ldots$, then use (13.6.6) and difference equation techniques to extend it for $m < 0$. Since m is now nonnegative, one needs to extend the techniques of (Lepowsky & Milne, 1978) and (Lepowsky & Wilson, 1982) to graded algebras where m will denote the grade.

Problem 24.6.2 *We believe that the quintic transformations in* (13.6.7) *are very deep and deserve to be understood better. Extending the above-mentioned algebraic approach to prove identities like* (13.6.7) *will be most interesting.*

Problem 24.6.3 *The partition identities implied by the first equality in* (13.6.7) *have not been investigated. A study of these identities is a worthwhile research project and may lead to new and unusual results.*

24.7 Characterization Theorems

Theorem 20.5.3 characterizes the Sheffer A-type zero polynomials relative to $\dfrac{d}{dx}$ and the Al-Salam–Chihara polynomials. Our first problem here deals with a related question.

Problem 24.7.1 *Characterize the triples* $\{r_n(x), s_n(x), \phi_n(x)\}$,

$$\phi_n(x) = \sum_{k=0}^{n} r_k(x) s_{n-k}(x),$$

when $\{r_n(x)\}$, $\{s_n(x)\}$ *and* $\{\phi_n(x)\}$ *are orthogonal polynomials.*

The ultraspherical and q-ultraspherical polynomials are examples of the ϕ_n's in the above problem.

Problem 24.7.2 *Characterize all orthogonal polynomial sequences* $\{\phi_n(x)\}$ *such that* $\{\phi_n(q^n x)\}$ *is also an orthogonal polynomial sequence.*

Theorem 20.5.5 solves Problem 24.7.2 under the added assumption $\phi_n(-x) = (-1)^n \phi_n(x)$. The general case remains open.

Problem 24.7.3 *Let* $\{x_n\}$, $\{a_n\}$, $\{b_n\}$ *be arbitrary sequences such that* $b_n \neq 0$, *for* $n > 0$ *and* $a_0 = b_0 = 1$. *The question is to characterize all monic orthogonal polynomials* $\{P_n(x)\}$ *which take the form*

$$b_n P_n(x) = \sum_{k=0}^{n} a_{n-k} b_k \prod_{j=1}^{k} (x - x_k), \tag{24.7.1}$$

where the empty product is 1.

Geronimus posed this question in (Geronimus, 1947) and, since then, this problem has become known as the "Geronimus Problem." He gave necessary and sufficient conditions on the sequences $\{a_n\}$, $\{b_n\}$ and $\{x_n\}$, but the identification of $\{P_n(x)\}$ remains ellusive. For example, the P_n's are known to satisfy (2.2.1) if and only if

$$a_{k+1}(B_{n-k} - B_{n+1}) = a_1 a_k(B_n - B_{n+1}) + \frac{\beta_n}{B_{n-1}} a_{k-1} + a_k(x_{n+1} - x_{n-k+1})$$

for $k = 0, 1, \ldots, n$, where $B_0 := 0$, $B_k = b_{k-1}/b_k$, $k > 0$. The problem remains open in its full generality, but some special cases are known. The case $x_{2k+1} = x_1$,

$x_{2k} = x_2$ for all k has been completely solved in (Al-Salam & Verma, 1982). The case $x_k = q^{1-k}$ is in (Al-Salam & Verma, 1988).

A polynomial sequence $\{\phi_n(x)\}$ is of Brenke type if there is a sequence $\{c_n\}$, $c_n \neq 0, n \geq 0$, and

$$\sum_{n=0}^{\infty} c_n \phi_n(x) t^n = A(t) B(xt), \tag{24.7.2}$$

where

$$A(t) = \sum_{n=0}^{\infty} a_n t^n, \quad B(t) = \sum_{n=0}^{\infty} b_n t^n, \tag{24.7.3}$$

$a_0 b_n \neq 0, n \geq 0$. It follows from (24.7.2) that

$$c_n \phi_n(x) = \sum_{k=0}^{n} a_{n-k} b_k x^k. \tag{24.7.4}$$

Chihara characterized all orthogonal polynomials which are of Brenke type in (Chihara, 1968) and (Chihara, 1971). In view of (24.7.1) and (24.7.4), this solves the Geronimus problem when $x_k = 0$ for $k > 0$.

A very general class of polynomials is the so-called Boas and Buck class. It consists of polynomials $\{\phi_n(x)\}$ having a generating function

$$\sum_{n=0}^{\infty} \phi_n(x) t^n = A(t) B(x H(t)).$$

where A and B are as in (24.7.3) and $H(t) = \sum_{n=1}^{\infty} h_n t^n$, $h_1 \neq 0$. Boas and Buck introduced this class because they can expand general functions into the polynomial basis $\{\phi_n(x)\}$, see (Boas & Buck, 1964). It does not seem to be possible to describe all orthogonal polynomials of Boas and Buck type. Moreover, some of the recently-discovered orthogonal polynomials (e.g., the Askey–Wilson polynomials) do not seem to belong to this class. On the other hand the q-ultraspherical, Al-Salam–Chihara and q-Hermite polynomials belong to the Boas and Buck class of polynomials.

Problem 24.7.4 *Determine subclasses of the Boas and Buck class of polynomials where all orthogonal polynomials within them can be characterized. The interesting cases are probably the ones leading to new orthogonal polynomials.*

One interesting subclass is motivated by Theorem 21.9.8.

Problem 24.7.5 *Determine all orthogonal polynomials $\{\phi_n(x)\}$ which have a generating function of the type*

$$\sum_{n=0}^{\infty} \phi_n(x) t^n = (1 - At)^{\alpha} (1 - Bt)^{\beta} B(x H(t)), \tag{24.7.5}$$

where B satisfies the conditions in (24.7.3). We already know that $H(t) = g(t)$
as defined in (21.9.31) leads to interesting orthogonal polynomials; see (Ismail &
Valent, 1998) and (Ismail et al., 2001).

Problem 24.7.6 *Characterize all orthogonal polynomials $\{\phi_n(x)\}$ having a gener-*
ating function

$$\sum_{n=0}^{\infty} \phi_n(x)t^n = A(t)\mathcal{E}_q(x; H(t)), \qquad (24.7.6)$$

where $H(t) = \sum_{n=1}^{\infty} h_n t^n$, $h_1 \neq 0$.

The special case $H(t) = t$ of Problem 24.7.6 has been solved in (Al-Salam, 1995)
and only the continuous q-Hermite polynomials have this property.

The next problem raises a q-analogue of characterizing orthogonal polynomial
solutions to (20.5.5).

Conjecture 24.7.7 *Let $\{p_n(x)\}$ be orthogonal polynomials satisfying*

$$\pi(x)D_q p_n(x) = \sum_{k=-r}^{s} c_{n,k}\, p_{n+k}(x),$$

for some positive integers r and s, and a polynomial $\pi(x)$ which does not depend
on n. Then $\{p_n(x)\}$ satisfies an orthogonality relation of the type (18.6.1), where w
satisfies (18.6.4) and u is a rational function.

Conjecture 24.7.8 *Let $\{p_n(x)\}$ be orthogonal polynomials and $\pi(x)$ be a polyno-*
mial of degree at most 2 which does not depend on n. If $\{p_n(x)\}$ satisfies

$$\pi(x)\mathcal{D}_q p_n(x) = \sum_{k=-1}^{1} c_{n,k}\, p_{n+k}(x), \qquad (24.7.7)$$

then $\{p_n(x)\}$ are continuous q-Jacobi polynomials, Al-Salam–Chihara polynomials,
or special or limiting cases of them. The same conclusion holds if $\pi(x)$ has degree
$s - 1$ and the condition (24.7.7) is replaced by

$$\pi(x)\mathcal{D}_q p_n(x) = \sum_{k=-r}^{s} c_{n,k} p_{n+k}(x), \qquad (24.7.8)$$

for positive integers r, s, and a polynomial $\pi(x)$ which does not depend on n.

In §15.5 we established (24.7.7) for continuous q-Jacobi polynomials and $\pi(x)$ has
degree 2. Successive application of the three-term recurrence relation will establish
(24.7.8) with $r = s$.

The Askey–Wilson polynomials do not have the property (24.7.7). The reason is
that, in general, $w\left(x; q^{1/2}\mathbf{t}\right)/w(x; \mathbf{t})$ is not a polynomial. On the other hand

$$\frac{w(x; q\mathbf{t})}{w(x; \mathbf{t})} = \prod_{j=1}^{4} \left(1 - 2xt_j + t_j^2\right) = \Phi(x),$$

say. Therefore there exists constants $c_{n,j}$, $-2 \leq j \leq 2$, such that

$$\Phi(x)\mathcal{D}_q^2 p_n(x; \mathbf{t}) = \sum_{j=-2}^{2} c_{n,j} p_{n+j}(x; \mathbf{t}). \qquad (24.7.9)$$

Conjecture 24.7.9 *Let* $\{p_n(x)\}$ *be orthogonal polynomials and* $\pi(x)$ *be a polynomial of degree at most 4. Then* $\{p_n(x)\}$ *satisfies*

$$\pi(x)\mathcal{D}_q^2 p_n(x) = \sum_{k=-r}^{s} c_{n,k} p_{n+k}(x) \qquad (24.7.10)$$

if and only if $\{p_n(x)\}$ *are the Askey–Wilson polynomials or special cases of them.*

The following two conjectures generalize the problems of Sonine and Hahn mentioned in §20.4.

Conjecture 24.7.10 *Let* $\{\phi_n(x)\}$ *and* $\{\mathcal{D}_q\phi_{n+1}(x)\}$ *be orthogonal polynomial sequences. Then* $\{\phi_n(x)\}$ *are Askey–Wilson polynomials, or special or limiting cases of them.*

Conjecture 24.7.11 *If* $\{\phi_n(x)\}$ *and* $\{\mathcal{D}_q^k\phi_{n+k}(x)\}$ *are orthogonal polynomial sequences for some* k, $k = 1, 2, \ldots$, *then* $\{\phi_n(x)\}$ *must be the Askey–Wilson polynomials or arise as special or limiting cases of them.*

If $\mathcal{D}_q$ is replaced by D_q in Conjectures 24.7.10–24.7.11, then it is known that $\{\phi_n(x)\}$ are special or limiting cases of big q-Jacobi polynomials.

The next two problems are motivated by the work of Krall and Sheffer, mentioned above Theorem 20.5.9.

Problem 24.7.12 *Let* $\{\phi_n(x)\}$ *be a sequence of orthogonal polynomials. Characterize all orthogonal polynomials* $Q_n(x)$,

$$Q_n(x) = \sum_{j=0}^{m} a_j(x) D_q^j \phi_n(x),$$

for constant m, $a_j(x)$ *a polynomial in* x *of degree at most* j. *Solve the same problem when* D_q *is replaced by* $\mathcal{D}_q$ *or* ∇.

Problem 24.7.13 *Let* $\{\phi_n(x)\}$ *be a sequence of orthogonal polynomials. Describe all polynomials* $Q_n(x)$ *of the form*

$$Q_n(x) = \sum_{j=0}^{m} a_j(x) D_q^{j+1} \phi_{n+1}(x),$$

which are orthogonal where $a_j(x)$ *and* m *are as in Problem 24.7.12. Again, solve the same problem when* D_q *is* $\mathcal{D}_q$ *or* ∇.

It is expected that the classes of polynomials $\{Q_n(x)\}$ which solve Problems 24.7.12–24.7.13 will contain nonclassical orthogonal polynomials.

24.8 Special Systems of Orthogonal Polynomials

Consider the following generalization of Chebyshev polynomials,

$$\Phi_0(x) = 1, \quad \Phi_1(x) = 2x - c\cos\beta, \tag{24.8.1}$$

$$2x\Phi_n(x) = \Phi_{n+1}(x) + \Phi_{n-1}(x) + c\cos(2\pi n\alpha + \beta)\,\Phi_n(x), \quad n > 0, \tag{24.8.2}$$

when $\alpha \in (0, 1)$ and is irrational.

This is a half-line version of the spectral problem of a doubly-infinite Jacobi matrix. This is a discrete Schrödinger operator with an almost periodic potential; see (Moser, 1981), (Avron & Simon, 8182), (Avron & Simon, 1982) and (Avron & Simon, 1983).

Problem 24.8.1 *Determine the large n behavior of $\Phi_n(x)$ in different parts of the complex x-plane. The measure of orthogonality of $\{\Phi_n(x)\}$ is expected to be singular continuous and is supported on a Cantor set.*

If n in (24.8.2) runs over all integers, then (24.8.2) becomes a spectral problem for a doubly infinite Jacobi matrix. Avron and Simon proved that if α is a Liouville number and $|c| > 2$, then the spectrum of (24.8.2) is purely singular continuous for almost all β; see (Avron & Simon, 1982). This model and several others are treated in Chapter 10 of (Cycon et al., 1987).

In a work in preparation, Ismail and Stanton have studied the cases of rational α.

We know that the Pollaczek polynomials $\left\{ P_n^\lambda(x; a, b) \right\}$ are polynomials in x. This fact, however, is far from obvious if $P_n^\lambda(x; a, b)$ is defined by (5.4.10).

Problem 24.8.2 *Prove that the right-hand side of (5.4.10) is a polynomial in $\cos\theta$ of degree n without the use of the three-term recurrence relation.*

Recently, Chu solved Problem 24.8.2 when $b = 0$.

As we noted in §5.5, Euler's formula (1.2.4) and the Chu–Vandermonde sum are the sums needed to prove directly the orthogonality of the polynomials $\{G_n(x; 0, b)\}$.

Problem 24.8.3 *Prove the orthogonality relation (5.5.18) directly using special functions and complex variable techniques.*

As we pointed out in Remark 5.5.1, it is unlikely that the integral and sum in (5.5.18) can be evaluated separately. So, what is needed is a version of the Lagrange inversion (1.2.4) or Theorem 1.2.3 where the sum is now an infinite sum plus an integral. One possibility is to carry out Szegő's proof of Theorem 5.4.2 until we reach the evaluation of the integral in (5.4.11). In the case where the measure of orthogonality has discrete part the integrals over the indented semicircles centered at ± 1 do not go to zero as the radii of the semicircles tends to zero. What is needed then is a careful analysis of the limits as the radii of the semicircles tend to zero possibly through deformation of the contour integral.

Problem 24.8.4 *The direct proof of orthogonality of $\{G_n(x; 0, b)\}$ used (1.2.4). The more general (1.2.5) has not been used to prove orthogonality relations for a specific* *system of orthogonal polynomials. The problem here is to find a specific system of orthogonal polynomials whose orthogonality can be proved using (1.2.5) to evaluate the integrals $\int_{\mathbb{R}} x^n p_n(x)\, d\mu(x)$.*

Askey and Ismail gave a q-extension of the polynomials $\{G_n(x; a, b)\}$ of §5.5 in Chapter 7 of (Askey & Ismail, 1984). They considered the polynomials

$$F_0(x; a, c) = 1, \quad F_1(x; a, c) = (c - a)\, x/(1 - q), \tag{24.8.3}$$

$$
x\,[c + 1 - q^n(a + 1)]\, F_n(x; a, c)
$$
$$
= \left(1 - q^{n+1}\right) F_{n+1}(x; a, c) + \left(c - aq^{n-1}\right) F_{n-1}(x; a, c). \tag{24.8.4}
$$

They proved that, in general, the polynomials $\{F_n\}$ are orthogonal with respect to a measure with a finite discrete part and an absolutely continuous part supported on $[-2\sqrt{c}/(1 + c), 2\sqrt{c}/(1 + c)]$. When $c = 0$ the discrete part becomes infinite and the continuous component disappears. In this case, the orthogonality measure has masses $\sigma_n(q)$ at $\pm x_n$, where

$$x_n = q^n \sqrt{\frac{b(1 - q)}{1 - q^n + b(1 - q)q^n}}, \tag{24.8.5}$$

$$\sigma_n(q) = \frac{b^n\,(1 - q^n)\, q^{n(n-1)}}{2(q; q)_n\, (aq^n/x_n^2; q)_\infty\, x_n^{2(n-1)}}\, [2 - q^n + b(1 - q)q^n],$$

and

$$c = a + b(1 - q), \tag{24.8.6}$$

so that $a = b(q - 1)$ in the present case. Set

$$
\alpha, \beta = \left[x(a + 1) \pm \sqrt{x^2(a + 1)^2 - 4a}\right]/(2a),
$$
$$
\mu, \nu = \left[x(c + 1) \pm \sqrt{x^2(c + 1)^2 - 4c}\right]/(2c). \tag{24.8.7}
$$

Askey and Ismail proved

$$\sum_{n=0}^{\infty} F_n(x; a, c)\, t^n = \frac{(t/\alpha, t/\beta; q)_\infty}{(t/\mu, t/\nu; q)_\infty}, \tag{24.8.8}$$

and used it to derive the representation

$$F_n(x; a, c) = \frac{\beta^n c^n}{(q; q)_n}\, (a/c; q)_n\, {}_3\phi_2 \left(\begin{matrix} q^{-n}, a\alpha\nu, a\alpha\mu \\ a/c, 0 \end{matrix} \,\middle|\, q, q \right). \tag{24.8.9}$$

When $c = 0$ we find

$$\sum_{n=0}^{\infty} F_n(x; a, 0)\, t^n = (t/\alpha, t/\beta; q)_\infty/(tx; q)_\infty, \tag{24.8.10}$$

from which it follows that

$$F_n(x; a, 0) = \frac{(a\alpha/x; q)_n}{(q; q)_n} x^n \, {}_1\phi_1\left(\begin{array}{c} q^{-n} \\ q^{1-n}\beta x \end{array} \bigg| \, q, -q\beta^2 a\right),$$

$$F_n(x; a, 0) = \frac{(-\alpha)^{-n} q^{n(n-1)/2}}{(q; q)_n} \, {}_1\phi_1\left(\begin{array}{c} q^{-n}, a\alpha/x \\ 0 \end{array} \bigg| \, q, q\alpha x\right),$$

(24.8.11)

and two similar formulas with α and β interchanged. Note that x_n solves $a\alpha q^n = x$ while $-x_n$ solves $a\beta q^n = x$. The orthogonality relation is

$$\sum_{k=0}^{\infty} \sigma_k(q) \left\{ F_m\left(x_k; a, 0\right) F_n\left(x_k; a, 0\right) + F_m\left(-x_k; a, 0\right) F_n\left(-x_k; a, 0\right) \right\}$$

$$= \frac{b^{n+1}(1 - q)^{n+1} q^{n(n-1)/2}}{(q; q)_n \left[1 - q^n + bq^n(1 - q)\right]} \delta_{m,n}. \quad (24.8.12)$$

Problem 24.8.5 *Prove (24.8.12) directly using special functions or function theoretic techniques.*

In a private communication, Dennis Stanton proved (24.8.12) when $m = n = 0$ using a version of q-Lagrange inversion from (Gessel & Stanton, 1983) and (Gessel & Stanton, 1986). The case of general m and n remains open.

As in §4.9, $\left\{\mu_{n,k}^{(\alpha,\beta)}\right\}$ are relative extrema of $\left|P_n^{(\alpha,\beta)}(x)\right|$. They occur at $\{z_{n,k}\}$, $-1 < z_{n,n-1} < \cdots < z_{n,1} < 1$.

Conjecture 24.8.6 ((Askey, 1990)) *We have $\mu_{n+1,k}^{(\alpha,\beta)} < \mu_{n,k}^{(\alpha,\beta)}$, $k = 1, 2, \ldots, n - 1$, if $\alpha > \beta > -1/2$.*

Wong and Zhang confirmed another conjecture of Askey's, namely that $\mu_{n+1,k}^{(0,-1)} > \mu_{n,k}^{(0,-1)}$. This was done in (Wong & Zhang, 1994a) and (Wong & Zhang, 1994b). A complete analysis of comparing $\mu_{n+1,k}^{(\alpha,\beta)}$ and $\mu_{n,k}^{(\alpha,\beta)}$ for $\alpha < \beta$ is an interesting open problem.

A polynomial f with integer coefficients is called irreducible if it is irreducible over the field of rational numbers $\mathbf{Q}$, that is if $f = gh$, g and h are polynomials with integer coefficients, then g or h must be a constant. Grosswald (Grosswald, 1978) devoted two chapters to the algebraic properties of the Bessel polynomials. The main problem is stated in the following conjectures.

Conjecture 24.8.7 *The Bessel polynomials $\{y_n(x)\}$ are irreducible.*

Conjecture 24.8.8 *The Galois group of a Bessel polynomial $y_n(x)$ is the full symmetric group on n symbols.*

Of course, Conjecture 24.8.7 implies Conjecture 24.8.8. There is ample evidence to support the validity of Conjecture 24.8.7. For example, it holds when the degree is of the form p^m, p is a prime. Also, Conjecture 24.8.7 has been verified for $n \leq 400$. With today's computing power one can probably verify it for a much larger range. For details and proofs, see Chapters 11 and 12 of (Grosswald, 1978).

24.9 Zeros of Orthogonal Polynomials

In this section, we discuss open problems involving monotonicity of zeros of orthogonal polynomials.

Problem 24.9.1 *Extend Theorem* 7.1.1 *to the case when*

$$d\alpha(x; \tau) = w(x; \tau)dx + d\beta(x; \tau)$$

where $\beta(x; \tau)$ is a jump function or a step function.

The case of purely discrete measures is of particular interest so we pose the problem of finding sufficient conditions on $d\beta(x; \tau)$ to guarantee the monotonicity of the zeros of the corresponding orthogonal polynomials when the mass points depend on the parameter τ. An example where such results will be applicable is the Al–Salam–Carlitz polynomials $U_n^{(a)}(x; q)$, where the point masses are located at $x = aq^n$, $x = q^n$, $n = 0, 1, \ldots$, Chihara (Chihara, 1978, pp. 195–198). The Al–Salam–Carlitz polynomials seem to possess many of the desirable combinatorial properties of a q-analogue of the Charlier polynomials and, as such, may be of some significance in Combinatorics. Additional examples of orthogonal polynomials with mass points depending on parameters are in (Askey & Ismail, 1984).

Problem 24.9.2 *Extend Theorem* 7.4.2 *to all zeros of $Q_N(x; \tau)$ and extend Theorem* 7.4.2 *to all positive zeros of $R_N(x; \tau)$.*

In Problem 24.9.2, we seek conditions on the coefficients $\lambda_n(\tau)$ and $\mu_n(\tau)$ which suffice to prove the monotonicity of all (positive) zeros of $Q_N(x; \tau)$ $(R_n(x; \tau))$. At the end of Section 3, we already indicated that the zeros of orthonormal polynomials strictly increase (or decrease) if the derivative of the corresponding Jacobi matrix is positive (negative) definite. We also indicated that we may replace "definite" by "semi-definite." However, we believe that definiteness or semi-definiteness is a very strong assumption and it is desirable to relax these assumptions.

One can combine Markov's theorem and quadratic transformation of hypergeometric functions to prove that the positive zeros $\{\zeta(\lambda)\}$ of an ultraspherical polynomial decrease as λ increases, $\lambda > 0$. The details are in Chapter 4 of Szegő (Szegő, 1975).

Recall that $N(n, N)$ is the number of integer zeros of $K_n(x; 1/2, N)$. The following conjectures are due to Krasikov and Litsyn, (Krasikov & Litsyn, 1996), (Habsieger, 2001a).

Conjecture 24.9.3 *For $2n - N < 0$, we have*

$$N(n, N) \leq \begin{cases} 3 & \textit{if } n \textit{ is odd} \\ 4 & \textit{if } n \textit{ is even}. \end{cases}$$

Conjecture 24.9.4 *Let $n = \binom{m}{2}$. Then the only integer zeros of $K_n\left(x; 1/2, m^2\right)$ are $2, m^2 - 2$ and $m^2/4$ for $m \equiv 2 \pmod{4}$.*

Bibliography

Abdi, W. H. (1960). On q-Laplace transforms. *Proceedings of the National Academy of Sciences of India, Section A*, 29, 389–408.

Abdi, W. H. (1964). Certain inversion and representation formulae for q-Laplace transforms. *Math. Zeitschr.*, 83, 238–249.

Abdi, W. H. (1966). A basic analogue of the Bessel polynomials. *Math. Nachr.*, 30, 209–219.

Ablowitz, M. J. & Ladik, J. F. (1976). Nonlinear differential-difference equations and Fourier analysis. *J. Mathematical Phys.*, 17(6), 1011–1018.

Abramowitz, M. & Stegun, I. A., Eds. (1965). *Handbook of mathematical functions, with formulas, graphs, and mathematical tables*, volume 55 of *National Bureau of Standards Applied Mathematics Series*. Superintendent of Documents, US Government Printing Office, Washington, DC. Third printing, with corrections.

Agarwal, R. P. (1969). Certain fractional q-integrals and q-derivatives. *Proc. Camb. Phil. Soc.*, 66, 365–370.

Ahmed, S., Bruschi, M., Calegro, F., Olshantsky, M. A., & Perelomov, A. M. (1979). Properties of the zeros of the classical orthogonal polynomials and of the Bessel functions. *Nuovo Cimento*, 49 B, 173–199.

Ahmed, S., Laforgia, A., & Muldoon, M. (1982). On the spacing of some zeros of some classical orthogonal polynomials. *J. London Math. Soc.*, 25(2), 246–252.

Ahmed, S. & Muldoon, M. (1983). Reciprocal power sums of differences of zeros of special functions. *SIAM J. Math. Anal.*, 14, 372–382.

Ahmed, S., Muldoon, M., & Spigler, R. (1986). Inequalities and numerical bounds for zeros of ultraspherical polynomials. *SIAM J. Math. Anal.*, 17, 1000–1007.

Akhiezer, N. I. (1965). *The Classical Moment Problem and Some Related Questions in Analysis*. Edinburgh: Oliver and Boyed.

Al-Salam, W. A. (1965). Characterization of certain classes of orthogonal polynomials related to elliptic functions. *Annali di Matematica Pura et Applicata*, 68, 75–94.

Al-Salam, W. A. (1966a). Fractional q-integration and q-differentiation. *Notices of the Amer. Math. Soc.*, 13(243).

Al-Salam, W. A. (1966b). q-analogues of Cauchy's formulas. *Proc. Amer. Math. Soc.*, 17, 616–621.

Al-Salam, W. A. (1966c). Some fractional q-integrals and q-derivatives. *Proc. Edinburgh Math. Soc., Ser. II*, 15, 135–140.

Al-Salam, W. A. (1990). Characterization theorems for orthogonal polynomials. In P. Nevai (Ed.), *Orthogonal Polynomials: Theory and Practice* (pp. 1–24). Dordrecht: Kluwer.

Al-Salam, W. A. (1995). Characterization of the Rogers q-Hermite polynomials. *Int. J. Math. Math. Sci.*, 18, 641–647.

Al-Salam, W. A. & Carlitz, L. (1965). Some orthogonal q-polynomials. *Math. Nachr.*, 30, 47–61.

Al-Salam, W. A. & Chihara, T. S. (1972). Another characterization of the classical orthogonal polynomials. *SIAM J. Math. Anal.*, 3, 65–70.

Al-Salam, W. A. & Chihara, T. S. (1976). Convolutions of orthogonal polynomials. *SIAM J. Math. Anal.*, 7, 16–28.

Al-Salam, W. A. & Chihara, T. S. (1987). q-Pollaczek polynomials and a conjecture of Andrews and Askey. *SIAM J. Math. Anal.*, 18, 228–242.

Al-Salam, W. A. & Ismail, M. E. H. (1977). Reproducing kernels for q-Jacobi polynomials. *Proc. Amer. Math. Soc.*, 67(1), 105–110.

Al-Salam, W. A. & Ismail, M. E. H. (1983). Orthogonal polynomials associated with the Rogers–Ramanujan continued fraction. *Pacific J. Math.*, 104(2), 269–283.

Al-Salam, W. A. & Ismail, M. E. H. (1994). A q-beta integral on the unit circle and some biorthogonal rational functions. *Proc. Amer. Math. Soc.*, 121, 553–561.

Al-Salam, W. A. & Verma, A. (1975a). A fractional Leibniz q-formula. *Pacific J. Math.*, 60(2).

Al-Salam, W. A. & Verma, A. (1975b). Remarks on fractional q-integrals. *Bul. Soc. Royal Sci. Liege*, 44(9-10).

Al-Salam, W. A. & Verma, A. (1982). On an orthogonal polynomial set. *Nederl. Akad. Wetensch. Indag. Math.*, 44(3), 335–340.

Al-Salam, W. A. & Verma, A. (1983). q-analogs of some biorthogonal functions. *Canad. Math. Bull.*, 26, 225–227.

Al-Salam, W. A. & Verma, A. (1988). On the Geronimus polynomial sets. In *Orthogonal polynomials and their applications (Segovia, 1986)*, volume 1329 of *Lecture Notes in Math.* (pp. 193–202). Berlin: Springer.

Alhaidari, A. D. (2004a). Exact L^2 series solution of the Dirac–Coulomb problem for all energies. *Ann. Phys.*, 312(1), 144–160.

Alhaidari, A. D. (2004b). L^2 series solution of the relativistic Dirac–Morse problem for all energies. *Phys. Lett. A*, 326(1-2), 58–69.

Alhaidari, A. D. (2004c). L^2 series solutions of the Dirac equation for power-law potentials at rest mass energy. *J. Phys. A: Math. Gen.*, 37(46), 11229–11241.

Alhaidari, A. D. (2005). An extended class of L^2 series solutions of the wave equation. *Ann. Phys.*, 317, 152–174.

Allaway, W. (1972). *The identification of a class of orthogonal polynomial sets*. PhD thesis, University of Alberta, Edmonton, Alberta.

Allaway, W. (1980). Some properties of the q-Hermite polynomials. *Canadian J. Math.*, 32, 686–694.

Alon, O. E. & Cederbaum, L. S. (2003). Hellmann–Feynman theorem at degeneracies. *Phys. Rev. B*, 68(033105), 4.

Andrews, G. E. (1970). A polynomial identity which implies the Rogers–Ramanujan identities. *Scripta Math.*, 28, 297–305.

Andrews, G. E. (1971). On the foundations of combinatorial theory V. Eulerian differential operators. *Studies in Applied Math.*, 50, 345–375.

Andrews, G. E. (1976a). On identities implying the Rogers–Ramanujan identities. *Houston J. Math.*, 2, 289–298.

Andrews, G. E. (1976b). *The Theory of Partitions*. Reading, MA: Addison-Wesley.

Andrews, G. E. (1981). Ramunujan's "lost" notebook. III. The Rogers–Ramanujan continued fraction. *Adv. in Math.*, 41(2), 186–208.

Andrews, G. E. (1986). *q-series: Their development and application in analysis, number theory, combinatorics, physics, and computer algebra*. Number 66 in CBMS Regional Conference Series. Providence, RI: American Mathematical Society.

Andrews, G. E. (1990). A page from Ramanujan's lost notebook. *Indian J. Math.*, 32, 207–216.

Andrews, G. E. & Askey, R. A. (1978). A simple proof of Ramanujan's summation $_1\psi_1$. *Aequationes Math.*, 18, 333–337.

Andrews, G. E. & Askey, R. A. (1985). Classical orthogonal polynomials. In C. Breziniski et al. (Ed.), *Polynômes Orthogonaux et Applications*, volume 1171 of *Lecture Notes in Mathematics* (pp. 36–63). Berlin Heidelberg: Springer-Verlag.

Andrews, G. E., Askey, R. A., & Roy, R. (1999). *Special Functions*. Cambridge: Cambridge University Press.

Andrews, G. E., Berndt, B. C., Sohn, J., Yee, A. J., & Zaharescu, A. (2003). On Ramanujan's continued fraction for $(q^2; q^3)_\infty/(q; q^3)_\infty$. *Trans. Amer. Math. Soc.*, 365, 2397–2411.

Andrews, G. E., Berndt, B. C., Sohn, J., Yee, A. J., & Zaharescu, A. (2005). Continued fractions with three limit points. *Advances in Math.*, 192, 231–258.

Angelesco, A. (1919). Sur deux extensions des fractions continues algébriques. *C.R. Acad. Sci. Paris*, 168, 262–263.

Anick, D., Mitra, D., & Sondhi, M. M. (1982). Stochastic theory of a data-handling system with multiple sources. *Bell System Tech. J.*, 61(8), 1871–1894.

Annaby, M. H. & Mansour, Z. S. (2005a). Basic fractional calculus. (To appear).

Annaby, M. H. & Mansour, Z. S. (2005b). Basic Sturm–Liouville problems. *J. Phys. A*. (To appear).

Anshelevich, M. (2004). Appell polynomials and their relatives. *Int. Math. Res. Not.*, (65), 3469–3531.

Anshelevich, M. (2005). Linearization coefficients for orthogonal polynomials using stochastic processes. *Ann. Prob.*, 33(1).

Apéry, R. (1979). Irrationalité de $\zeta(2)$ et $\zeta(3)$. *Astérisque*, 61, 11–13.

Appell, P. & Kampé de Fériet, J. (1926). *Fonctions Hypergéométriques et Hypersphérique; Polynomes d'Hermite*. Paris: Gauthier-Villars.

Aptekarev, A. I. (1998). Multiple orthogonal polynomials. *J. Comput. Appl. Math.*, 99, 423–447.

Aptekarev, A. I., Branquinho, A., & Van Assche, W. (2003). Multiple orthogonal polynomials for classical weights. *Trans. Amer. Math. Soc.*, 355, 3887–3914.

Aptekarev, A. I., Marcellán, F., & Rocha, I. A. (1997). Semiclassical multiple orthogonal polynomials and the properties of Jacobi–Bessel polynomials. *J. Approx. Theory*, 90, 117–146.

Arvesú, J., Coussement, J., & Van Assche, W. (2003). Some discrete multiple orthogonal polynomials. *J. Comput. Appl. Math.*, 153, 19–45.

Askey, R. A. (1970a). Linearization of the product of orthogonal polynomials. In *Problems in analysis (papers dedicated to Salomon Bochner, 1969)* (pp. 131–138). Princeton, NJ: Princeton Univ. Press.

Askey, R. A. (1970b). Orthogonal polynomials and positivity. In D. Ludwig & F. W. J. Olver (Eds.), *Studies in Applied Mathematics 6: Special Functions and Wave Propagation* (pp. 64–85). Philadelphia, PA: Society for Industrial and Applied Mathematics.

Askey, R. A. (1971). Orthogonal expansions with positive coefficients. II. *SIAM J. Math. Anal.*, 2, 340–346.

Askey, R. A. (1975a). *A note on the history of series*. Technical Report 1532, Mathematics Research Center, University of Wisconsin.

Askey, R. A. (1975b). *Orthogonal Polynomials and Special Functions*. Philadelphia, PA: Society for Industrial and Applied Mathematics.

Askey, R. A. (1978). Jacobi's generating function for Jacobi polynomials. *Proc. Amer. Math. Soc.*, 71, 243–246.

Askey, R. A. (1983). An elementary evaluation of a beta type integral. *Indian J. Pure Appl. Math.*, 14(7), 892–895.

Askey, R. A. (1985). Review of "a treatise on generating functions" by Srivastava and Manocha. Math. Rev., 85m:33016.

Askey, R. A. (1989a). Beta integrals and the associated orthogonal polynomials. In K. Alladi (Ed.), *Number theory, Madras 1987*, volume 1395 of *Lecture Notes in Math.* (pp. 84–121). Berlin: Springer.

Askey, R. A. (1989b). Continuous q-Hermite polynomials when $q > 1$. In D. Stanton (Ed.), *q-Series and Partitions*, IMA Volumes in Mathematics and Its Applications (pp. 151–158). New York: Springer-Verlag.

Askey, R. A. (1989c). Divided difference operators and classical orthogonal polynomials. *Rocky Mountain J. Math.*, 19, 33–37.

Askey, R. A. (1990). Graphs as an aid to understanding special functions. In R. Wong (Ed.), *Asymptotic and Computational Analysis* (pp. 3–33). New York: Marcel Dekker.

Askey, R. A. (2005). Evaluation of sylvester-type determinants using orthogonal polynomials. (To appear).

Askey, R. A. & Gasper, G. (1972). Certain rational functions whose power series have positive coefficients. *Amer. Math. Monthly*, 79, 327–341.

Askey, R. A. & Gasper, G. (1976). Positive Jacobi polynomial sums. II. *Amer. J. Math.*, 98, 109–137.

Askey, R. A. & Gasper, G. (1977). Convolution structures for Laguerre polynomials. *J. Analyse Math.*, 31, 48–68.

Askey, R. A. & Ismail, M. E. H. (1976). Permutation problems and special functions. *Canadian J. Math.*, 28, 853–874.

Askey, R. A. & Ismail, M. E. H. (1980). The Rogers q-ultraspherical polynomials. In E. Cheney (Ed.), *Approximation Theory III* (pp. 175–182). New York: Academic Press.

Askey, R. A. & Ismail, M. E. H. (1983). A generalization of ultraspherical polynomials. In P. Erdős (Ed.), *Studies in Pure Mathematics* (pp. 55–78). Basel: Birkhauser.

Askey, R. A. & Ismail, M. E. H. (1984). Recurrence relations, continued fractions and orthogonal polynomials. *Memoirs Amer. Math. Soc.*, 49(300), iv + 108 pp.

Askey, R. A., Ismail, M. E. H., & Koornwinder, T. (1978). Weighted permutation problems and Laguerre polynomials. *J. Comb. Theory Ser. A*, 25(3), 277–287.

Askey, R. A., Rahman, M., & Suslov, S. K. (1996). On a general q-Fourier transformation with nonsymmetric kernels. *J. Comp. Appl. Math.*, 68(1-2), 25–55.

Askey, R. A. & Wilson, J. A. (1979). A set of orthogonal polynomials that generalize the Racah coefficients or $6-j$ symbols. *SIAM J. Math. Anal.*, 10(5), 1008–1016.

Askey, R. A. & Wilson, J. A. (1982). A set of hypergeometric orthogonal polynomials. *SIAM J. Math. Anal.*, 13(4), 651–655.

Askey, R. A. & Wilson, J. A. (1985). Some basic hypergeometric orthogonal polynomials that generalize Jacobi polynomials. *Memoirs Amer. Math. Soc.*, 54(319), iv + 55 pp.

Askey, R. A. & Wimp, J. (1984). Associated Laguerre and Hermite polynomials. *Proc. Roy. Soc. Edinburgh*, 96A, 15–37.

Atakishiyev, N. M. (1994). A simple difference realization of the Heisenberg q-algebra. *J. Math. Phys.*, 35(7), 3253–3260.

Atakishiyev, N. M. (1996). On a one-parameter family of q-exponential functions. *J. Phys. A*, 29(10), L223–L227.

Atakishiyev, N. M. & Suslov, S. K. (1992a). Difference hypergeometric functions. In A. A. Gonchar & E. B. Saff (Eds.), *Progress in approximation theory (Tampa, FL, 1990)*, volume 19 of *Springer Ser. Comput. Math.* (pp. 1–35). New York: Springer.

Atakishiyev, N. M. & Suslov, S. K. (1992b). On the Askey–Wilson polynomials. *Constructive Approximation*, 8, 363–369.

Atkinson, F. V. (1964). *Discrete and Continuous Boundary Problems*. New York: Academic Press.

Atkinson, F. V. & Everitt, W. N. (1981). Orthogonal polynomials which satisfy second order differential equations. In *E. B. Christoffel (Aachen/Monschau, 1979)* (pp. 173–181). Basel: Birkhäuser.

Aunola, M. (2005). Explicit representations of Pollaczek polynomials corresponding to an exactly solvable discretization of the hydrogen radial Schrödinger equation. *J. Phys. A*, 38, 1279–1285.

Avron, J. & Simon, B. (1981/82). Almost periodic Schrödinger operators. I. Limit periodic potentials. *Comm. Math. Phys.*, 82(1), 101–120.

Avron, J. & Simon, B. (1982). Singular continuous spectrum for a class of almost periodic Jacobi matrices. *Bull. Amer. Math. Soc. (N.S.)*, 6(1), 81–85.

Avron, J. & Simon, B. (1983). Almost periodic Schrödinger operators. II. The integrated density of states. *Duke Math. J.*, 50(1), 369–391.

Azor, R., Gillis, J., & Victor, J. D. (1982). Combinatorial applications of Hermite polynomials. *SIAM J. Math. Anal.*, 13(5), 879–890.

Babujian, H. M. (1983). Exact solution of the isotropic Heisenberg chain with arbitrary spins: Thermodynamics of the model. *Nucl. Phys. B*, 215, 317–336.

Baik, J., Deift, P., & Johansson, K. (1999). On the distribution of the length of the longest increasing subsequence of random permutations. *J. Amer. Math. Soc.*, 12, 1119–1178.

Baik, J., Kriecherbauer, T., McLaughlin, K. T.-R., & Miller, P. D. (2002). Uniform asymptotics for polynomials orthogonal with respect to a general class of discrete weights and universality results for associated ensembles. math.CA/0310278 on http://arXiv.org.

Baik, J., Kriecherbauer, T., McLaughlin, K. T.-R., & Miller, P. D. (2003). Uniform asymptotics for polynomials orthogonal with respect to a general class of discrete weights and universality results for associated ensembles: announcement of results. *Internat. Math. Res. Notices*, 2003(15), 821–858.

Bailey, W. N. (1935). *Generalized Hypergeometric Series*. Cambridge: Cambridge University Press.

Balakrishnan, A. V. (1960). Fractional powers of closed operators and the semigroups generated by them. *Pacific J. Math.*, 10, 419–437.

Balawender, R. & Holas, A. (2004). Comment on "breakdown of the Hellmann–Feynman theorem: degeneracy is the key". *Phys. Rev. B*, 69(037103), 5.

Bank, E. & Ismail, M. E. H. (1985). The attractive Coulomb potential polynomials. *Constr. Approx.*, 1, 103–119.

Bannai, E. & Ito, T. (1984). *Algebraic Combinatorics I: Association Schemes*. Menlo Park: Benjamin/Cummings.

Baratella, P. & Gatteschi, L. (1988). The bounds for the error terms of an asymptotic approximation of Jacobi polynomials. In M. Alfaro, et al. (Ed.), *Orthogonal Polynomials and Their Applications*, volume 1329 of *Lecture Notes in Math.* (pp. 203–221). New York: Springer-Verlag.

Bateman, H. (1905). A generalization of the Legendre polynomials. *Proc. London Math. Soc.*, 3(2), 111–123.

Bateman, H. (1932). *Partial Differential Equations*. Cambridge: Cambridge University Press.

Bauldry, W. (1985). *Orthogonal polynomials associated with exponential weights*. PhD thesis, Ohio State University, Columbus.

Bauldry, W. (1990). Estimates of asymmetric Freud polynomials on the real line. *J. Approximation Theory*, 63, 225–237.

Baxter, R. J. (1982). *Exactly Solved Models in Statistical Mechanics*. London: Academic Press.

Beckermann, B., Coussement, J., & Van Assche, W. (2005). Multiple Wilson and Jacobi–Piñeiro polynomials. *J. Approx. Theory*. (To appear).

Ben Cheikh, Y. & Douak, K. (2000a). On the classical d-orthogonal polynomials defined by certain generating functions, I. *Bull. Belg. Math. Soc. Simon Stevin*, 7, 107–124.

Ben Cheikh, Y. & Douak, K. (2000b). On two-orthogonal polynomials related to the Bateman $j_n^{u,v}$-function. *Methods Appl. Anal.*, 7, 641–662.

Ben Cheikh, Y. & Douak, K. (2001). On the classical d-orthogonal polynomials defined by certain generating functions, II. *Bull. Belg. Math. Soc. Simon Stevin*, 8, 591–605.

Ben Cheikh, Y. & Zaghouani, A. (2003). Some discrete d-orthogonal polynomial sets. *J. Comput. Appl. Math.*, 156, 253–263.

Berezans'kiĭ, J. M. (1968). *Expansions in eigenfunctions of selfadjoint operators.* Translated from the Russian by R. Bolstein, J. M. Danskin, J. Rovnyak and L. Shulman. Translations of Mathematical Monographs, Vol. 17. Providence, RI: American Mathematical Society.

Berg, C., Chen, Y., & Ismail, M. E. H. (2002). Small eigenvalues of large Hankel matrices: the indeterminate case. *Math. Scand.*, 91(1), 67–81.

Berg, C. & Christensen, J. P. R. (1981). Density questions in the classical theory of moments. (French summary). *Ann. Inst. Fourier (Grenoble)*, 31(3), 99–114.

Berg, C. & Ismail, M. E. H. (1996). q-Hermite polynomials and classical orthogonal polynomials. *Canad. J. Math.*, 48, 43–63.

Berg, C. & Pedersen, H. L. (1994). On the order and type of the entire functions associated with an indeterminate Hamburger moment problem. *Ark. Mat.*, 32, 1–11.

Berg, C. & Valent, G. (1994). The Nevanlinna parameterization for some indeterminate Stieltjes moment problems associated with birth and death processes. *Methods and Applications of Analysis*, 1, 169–209.

Berndt, B. C. & Sohn, J. (2002). Asymptotic formulas for two continued fractions in Ramanujan's lost notebook. *J. London, Math. Soc.*, 65, 271–284.

Bethe, H. (1931). Zur theorie der metalle i. eigenverte und eigenfunctionen der linearen atomkette. *Zeitchrift für Physik*, 71, 206–226.

Beukers, F. (2000). A rational approach to π. *Nieuw Arch. Wisk. (5)*, 1(4), 372–379.

Biedenharn, L. & Louck, J. (1981). *The Racah–Wigner Algebra in Quantum Theory.* Reading: Addison-Wesley.

Bleher, P. & Its, A. (1999). Semiclassical asymptotics of orthogonal polynomials, Riemann–Hilbert problem, and universality in the matrix model. *Ann. of Math. (2)*, 150, 185–266.

Bleher, P. M. & Kuijlaars, A. B. J. (2004). Integral representations for multiple Hermite and multiple Laguerre polynomials. math.CA/0406616 at arXiv.org.

Boas, R. P. & Buck, R. C. (1964). *Polynomial expansions of analytic functions.* Second printing, corrected. Ergebnisse der Mathematik und ihrer Grenzgebiete, N.F., Bd. 19. New York: Academic Press Inc. Publishers.

Boas, Jr., R. P. (1939). The Stieltjes moment problem for functions of bounded variation. *Bull. Amer. Math. Soc.*, 45, 399–404.

Boas, Jr., R. P. (1954). *Entire functions.* New York: Academic Press Inc.

Bochner, S. (1929). Über Sturm–Liouvillesche polynomsysteme. *Math. Zeit.*, 29, 730–736.

Bochner, S. (1954). Positive zonal functions on spheres. *Proc. Nat. Acad. Sci. U.S.A.*, 40, 1141–1147.

Bonan, S., Lubinsky, D. S., & Nevai, P. (1987). Orthogonal polynomials and their derivatives. II. *SIAM J. Math. Anal.*, 18(4), 1163–1176.

Bonan, S. S. & Clark, D. S. (1990). Estimates of the Hermite and the Freud polynomials. *J. Approximation Theory*, 63, 210–224.

Bonan, S. S. & Nevai, P. (1984). Orthogonal polynomials and their derivatives. I. *J. Approx. Theory*, 40, 134–147.

Borzov, V. V., Damashinski, E. V., & Kulish, P. P. (2000). Construction of the spectral measure for deformed oscillator position operator in the case of undetermined moment problems. *Reviews in Math. Phys.*, 12, 691–710.

Bourget, J. (1866). Memoire sur le mouvement vibratoire des membranes circulaires (June 5, 1865). *Ann. Sci. de l'École Norm.*, Sup. III(5), 5–95.

Braaksma, B. L. J. & Meulenbeld, B. (1971). Jacobi polynomials as spherical harmonics. *Indag. Math.*, 33, 191–196.

Brenke, W. C. (1930). On polynomial solutions of a class of linear differential equations of the second order. *Bull. Amer. Math. Soc.*, 36, 77–84.

Bressoud, D. (1981). On partitions, orthogonal polynomials and the expansion of certain infinite products. *Proc. London Math. Soc.*, 42, 478–500.

Brézin, E. & Hikami, S. (1998). Level spacing of random matrices in an external source. *Phys. Rev. E*, 58, 7176–7185.

Broad, J. T. (1978). Gauss quadrature generated by diagonalization of H in finite L^2 bases. *Phys. Rev. A (3)*, 18(3), 1012–1027.

Brown, B. M., Evans, W. D., & Ismail, M. E. H. (1996). The Askey–Wilson polynomials and q-Sturm–Liouville problems. *Math. Proc. Cambridge Phil. Soc.*, 119, 1–16.

Brown, B. M. & Ismail, M. E. H. (1995). A right inverse of the Askey–Wilson operator. *Proc. Amer. Math. Soc.*, 123, 2071–2079.

Bryc, W. (2001). Stationary random fields with linear regressions. *Ann. Probab.*, 29, 504–519.

Bryc, W., Matysiak, W., & Szabłowski, P. J. (2005). Probabilistic aspects of Al-Salam–Chihara polynomials. *Proc. Amer. Math. Soc.*, 133, 1127–1134.

Bueno, M. I. & Marcellán, F. (2004). Darboux transformation and perturbation of linear functionals. *Linear Algebra and its Applications*, 384, 215–242.

Bultheel, A., Cuyt, A., Van Assche, W., Van Barel, M., & Verdonk, B. (2005). Generalizations of orthogonal polynomials. *J. Comput. Appl. Math.* (To appear).

Burchnal, J. L. & Chaundy, T. W. (1931). Commutative ordinary differential operators. II. The identity $p^n = q^m$. *Proc. Roy. Soc. London (A)*, 34, 471–485.

Burchnall, J. L. (1951). The Bessel polynomials. *Can. J. Math.*, 3, 62–68.

Bustoz, J. & Ismail, M. E. H. (1982). The associated classical orthogonal polynomials and their q-analogues. *Canad. J. Math.*, 34, 718–736.

Bustoz, J. & Suslov, S. K. (1998). Basic analog of Fourier series on a q-quadratic grid. *Methods Appl. Anal.*, 5(1), 1–38.

Butzer, P. L., Kilbas, A. A., & Trujillo, J. J. (2002a). Compositions of Hadamard-type fractional integration operators and the semigroup property. *J. Math. Anal. App.*, 269(2), 387–400.

Butzer, P. L., Kilbas, A. A., & Trujillo, J. J. (2002b). Fractional calculus in the Mellin setting and Hadamard-type fractional integrals. *J. Math. Anal. App.*, 269(1), 1–27.

Butzer, P. L., Kilbas, A. A., & Trujillo, J. J. (2002c). Mellin transform analysis and integration by parts for hadamard-type fractional integrals. *J. Math. Anal. App.*, 270(1), 1–15.

Carlitz, L. (1955). Polynomials related to theta functions. *Annal. Mat. Pura Appl. (4)*, 4, 359–373.

Carlitz, L. (1958). On some polynomials of Tricomi. *Boll. Un. Mat. Ital. (3)*, 13, 58–64.

Carlitz, L. (1959). Some formulas related to the Rogers–Ramanujan identities. *Annali di Math. (IV)*, 47, 243–251.

Carlitz, L. (1960). Some orthogonal polynomials related to elliptic functions. *Duke Math. J.*, 27, 443–459.

Carlitz, L. (1961). Some orthogonal polynomials related to elliptic functions. II. Arithmetic properties. *Duke Math. J.*, 28, 107–124.

Carlitz, L. (1972). Generating functions for certain Q-orthogonal polynomials. *Collect. Math.*, 23, 91–104.

Carnovale, G. & Koornwinder, T. H. (2000). A q-analogue of convolution on the line. *Meth. Appl. Anal.*, 7(4), 705–726.

Cartier, P. & Foata, D. (1969). *Problèmes combinatoires de commutation et réarrangements*, volume 85 of *Lecture Notes in Mathematics*. Berlin: Springer-Verlag.

Charris, J. A. & Ismail, M. E. H. (1987). On sieved orthogonal polynomials. V. Sieved Pollaczek polynomials. *SIAM J. Math. Anal.*, 18(4), 1177–1218.

Chen, Y. & Ismail, M. E. H. (1997). Ladder operators and differential equations for orthogonal polynomials. *J. Phys. A*, 30, 7817–7829.

Chen, Y. & Ismail, M. E. H. (1998a). Hermitean matrix ensembles and orthogonal polynomials. *Studies in Appl. Math.*, 100, 33–52.

Chen, Y. & Ismail, M. E. H. (1998b). Some indeterminate moment problems and Freud-like weights. *Constr. Approx.*, 14(3), 439–458.

Chen, Y. & Ismail, M. E. H. (2005). Jacobi polynomials from compatibility conditions. *Proc. Amer. Math. Soc.*, 133(2), 465–472 (electronic).

Chen, Y., Ismail, M. E. H., & Van Assche, W. (1998). Tau-function constructions of the recurrence coefficients of orthogonal polynomials. *Advances in Appl. Math.*, 20, 141–168.

Chihara, L. (1987). On the zeros of the Askey–Wilson polynomials, with applications to coding theory. *SIAM J. Math. Anal.*, 18(1), 191–207.

Chihara, T. S. (1962). Chain sequences and orthogonal polynomials. *Trans. Amer. Math. Soc.*, 104, 1–16.

Chihara, T. S. (1968). Orthogonal polynomials with Brenke type generating functions. *Duke Math. J.*, 35, 505–517.

Chihara, T. S. (1970). A characterization and a class of distribution functions for the Stieltjes–Wigert polynomials. *Canad. Math. Bull.*, 13, 529–532.

Chihara, T. S. (1971). Orthogonality relations for a class of Brenke polynomials. *Duke Math. J.*, 38, 599–603.

Chihara, T. S. (1978). *An Introduction to Orthogonal Polynomials*. New York: Gordon and Breach.

Chihara, T. S. (1982). Indeterminate symmetric moment problems. *J. Math. Anal. Appl.*, 85(2), 331–346.

Chihara, T. S. & Ismail, M. E. H. (1993). Extremal measures for a system of orthogonal polynomials. *Constructive Approximation*, 9, 111–119.

Christiansen, J. S. (2003a). The moment problem associated with the q-Laguerre polynomials. *Constr. Approx.*, 19(1), 1–22.

Christiansen, J. S. (2003b). The moment problem associated with the Stieltjes–Wigert polynomials. *J. Math. Anal. Appl.*, 277, 218–245.

Christiansen, J. S. (2005). Indeterminate moment problems related to birth and death processes with quartic rates. *J. Comp. Appl. Math.*, 178, 91–98.

Chudnovsky, D. V. & Chudnovsky, G. V. (1989). Computational problems in arithmetic of linear differential equations. Some Diophantine applications. In *Number theory (New York, 1985/1988)*, volume 1383 of *Lecture Notes in Math.* (pp. 12–49). Berlin: Springer.

Conrad, E. (2002). *Some continued fraction expansions of Laplace transforms of elliptic functions*. PhD thesis, Ohio State University, Columbus, OH.

Conrad, E. & Flajolet, P. (2005). The fermat cubic, elliptic functions, continued fractions, and a combinatorial tale. (To appear).

Cooper, S. (1996). The Askey–Wilson operator and the $_6\psi_5$ summation formula. Preprint.

Corteel, S. & Lovejoy, J. (2002). Frobenius partitions and the combinatorics of Ramanujan's $_1\psi_1$ summation. *J. Combin. Theory Ser. A*, 97(1), 177–183.

Coussement, E. & Van Assche, W. (2001). Some properties of multiple orthogonal polynomials associated with Macdonald functions. *J. Comput. Appl. Math.*, 133, 253–261.

Coussement, E. & Van Assche, W. (2003). Multiple orthogonal polynomials associated with the modified Bessel functions of the first kind. *Constr. Approx.*, 19, 237–263.

Cycon, H. L., Froese, R. G., Kirsch, W., & Simon, B. (1987). *Schrödinger operators with application to quantum mechanics and global geometry*. Berlin: Springer-Verlag.

Daems, E. & Kuijlaars, A. B. J. (2004). A Christoffel–Darboux formula for multiple orthogonal polynomials. math.CA/0402031 at arXiv.org.

Datta, S. & Griffin, J. (2005). A characterization of some q-orthogonal polynomials. *Ramanujan J.* (To appear).

de Boor, C. & Saff, E. B. (1986). Finite sequences of orthogonal polynomials connected by a Jacobi matrix. *Linear Algebra Appl.*, 75, 43–55.

de Branges, L. (1985). A proof of the Bieberbach conjecture. *Acta Math.*, 154(1-2), 137–152.

de Bruin, M. G. (1985). Simultaneous padé approximation and orthogonality. In C. Brezinski et al. (Ed.), *Polynômes Orthogonaux et Applications*, volume 1171 of *Lecture Notes in Mathematics* (pp. 74–83). Berlin: Springer.

de Bruin, M. G., Saff, E. B., & Varga, R. S. (1981a). On the zeros of generalized Bessel polynomials. I. *Nederl. Akad. Wetensch. Indag. Math.*, 43(1), 1–13.

de Bruin, M. G., Saff, E. B., & Varga, R. S. (1981b). On the zeros of generalized Bessel polynomials. II. *Nederl. Akad. Wetensch. Indag. Math.*, 43(1), 14–25.

Deift, P. (1999). *Orthogonal Polynomials and Random Matrices: a Riemann–Hilbert Approach*, volume 3 of *Courant Lecture Notes in Mathematics*. New York: New York University Courant Institute of Mathematical Sciences.

Deift, P., Kriecherbauer, T., McLaughlin, K. T.-R., Venakides, S., & Zhou, X. (1999a). Strong asymptotics of orthogonal polynomials with respect to exponential weights. *Comm. Pure Appl. Math.*, 52(12), 1491–1552.

Deift, P., Kriecherbauer, T., McLaughlin, K. T.-R., Venakides, S., & Zhou, X. (1999b). Uniform asymptotics for polynomials orthogonal with respect to varying exponential weights and applications to universality questions in random matrix theory. *Comm. Pure Appl. Math.*, 52(11), 1335–1425.

Deift, P., Kriecherbauer, T., McLaughlin, K. T.-R., Venakides, S., & Zhou, X. (2001). A Riemann–Hilbert approach to asymptotic questions for orthogonal polynomials. *J. Comput. Appl. Math.*, 133(1-2), 47–63.

Deift, P. & Zhou, X. (1993). A steepest descent method for oscillatory Riemann–Hilbert problems. asymptotics for the MKdV equation. *Ann. of Math. (2)*, 137(2), 295–368.

Derkachov, S. É., Korchemsky, G. P., & Manshov, A. N. (2003). Baxter $\mathbb{Q}$-operator and separation of variables for the open $sl(2, \mathbb{R})$ spin chain. *JHEP*, 10, 053. hep-th/0309144.

DeVore, R. A. (1972). *The Approximation of Continuous Functions by Positive Linear Operators*, volume 293 of *Lecture Notes in Mathematics*. Berlin: Springer-Verlag.

Diaconis, P. & Graham, R. L. (1985). The Radon transform on Z_2^k. *Pacific J. Math.*, 118(2), 323–345.

Diaz, J. B. & Osler, T. J. (1974). Differences of fractional order. *Math. Comp.*, 28, 185–202.

Dickinson, D. J. (1954). On Lommel and Bessel polynomials. *Proc. Amer. Math. Soc.*, 5, 946–956.

Dickinson, D. J., Pollack, H. O., & Wannier, G. H. (1956). On a class of polynomials orthogonal over a denumerable set. *Pacific J. Math*, 6, 239–247.

Dickson, L. E. (1939). *New Course on the Theory of Equations*. New York: Wiley.

Diestler, D. J. (1982). The discretization of continuous infinite sets of coupled ordinary linear differential equations: Applications to the collision-induced dissociation of a diatomic molecule by an atom. In J. Hinze (Ed.), *Numerical Integration of Differential Equations and Large Linear Systems (Bielefeld, 1980)*, volume 968 of *Lecture Notes in Mathematics* (pp. 40–52). Berlin-New York: Springer-Verlag.

Dijksma, A. & Koornwinder, T. H. (1971). Spherical harmonics and the product of two Jacobi polynomials. *Indag. Math.*, 33, 191–196.

Dilcher, K. & Stolarsky, K. (2005). Resultants and discriminants of Chebyshev and related polynomials. *Trans. Amer. Math. Soc.*, 357, 965–981.

Dixon, A. C. (1890). On the doubly periodic functions arising out of the curve $x^3 + y^3 - 3\alpha xy = 1$. *Quarterly J. Pure and Appl. Math.*, 24, 167–233.

Dominici, D. (2005). Asymptotic analysis of the Krawtchouk polynomials by the WKB method. (To appear).

Douak, K. (1999). On 2-orthogonal polynomials of Laguerre type. *Int. J. Math. Math. Sci.*, 22, 29–48.

Douak, K. & Maroni, P. (1992). Les polynômes orthogonaux 'classiques' de dimension deux. *Analysis*, 12, 71–107.

Douak, K. & Maroni, P. (1995). Une caractérisation des polynômes d-orthogonaux 'classiques'. *J. Approx. Theory*, 82, 177–204.

Drew, J. H., Johnson, C. R., Olesky, D. D., & van den Driessche, P. (2000). Spectrally arbitrary patterns. *Linear Algebra Appl.*, 308, 121–137.

Dulucq, S. & Favreau, L. (1991). A combinatorial model for Bessel polynomials. In C. Brezinski, L. Gori, & A. Ronveau (Eds.), *Orthogonal Polynomials and Their Applications* (pp. 243–249). J. C. Baltzer AG, Scientific Publishing Co.

Durán, A. J. (1989). The Stieltjes moments problem for rapidly decreasing functions. *Proc. Amer. Math. Soc.*, 107(3), 731–741.

Durán, A. J. (1993). Functions with given moments and weight functions for orthogonal polynomials. *Rocky Mountain J. Math.*, 23(1), 87–104.

Elbert, Á. & Siafarikas, P. D. (1999). Monotonicity properties of the zeros of ultraspherical polynomials. *J. Approx. Theory*, 97, 31–39.

Elsner, L. & Hershkowitz, D. (2003). On the spectra of close-to-Schwarz matrices. *Linear Algebra Appl.*, 363, 81–88.

Elsner, L., Olesky, D. D., & van den Driessche, P. (2003). Low rank perturbations and the spectrum of a tridiagonal pattern. *Linear Algebra Appl.*, 364, 219–230.

Erdélyi, A., Magnus, W., Oberhettinger, F., & Tricomi, F. G. (1953a). *Higher Transcendental Functions*, volume 2. New York: McGraw-Hill.

Erdélyi, A., Magnus, W., Oberhettinger, F., & Tricomi, G. F. (1953b). *Higher Transcendental Functions*, volume 1. New York: McGraw-Hill.

Even, S. & Gillis, J. (1976). Derangements and Laguerre polynomials. *Math. Proc. Camb. Phil. Soc.*, 79, 135–143.

Everitt, W. N. & Littlejohn, L. L. (1991). Orthogonal polynomials and spectral theory: a survey. In *Orthogonal polynomials and their applications (Erice, 1990)*, volume 9 of *IMACS Ann. Comput. Appl. Math.* (pp. 21–55). Basel: Baltzer.

Exton, H. (1983). *q-hypergeometric functions and applications*. Ellis Horwood Series: Mathematics and its Applications. Chichester: Ellis Horwood Ltd. With a foreword by L. J. Slater.

Faddeev, L. D. (1998). How algebraic Bethe Ansatz works for integrable models. In A. Connes, K. Gawedzki, & J. Zinn-Justin (Eds.), *Quantum symmetries/Symmetries quantiques* (pp. 149–219). Amsterdam: North-Holland. Proceedings of the Les Houches summer school, Session LXIV, Les Houches, France, August 1–September 8, 1995.

Favard, J. (1935). Sur les polynômes de Tchebicheff. *Comptes Rendus de l'Académie des Sciences, Paris*, 131, 2052–2053.

Faybusovich, L. & Gekhtman, M. (1999). On Schur flows. *J. Phys. A*, 32(25), 4671–4680.

Feldheim, E. (1941a). Contributions à la théorie des polynômes de Jacobi. *Mat. Fiz. Lapok*, 48, 453–504.

Feldheim, E. (1941b). Sur les polynômes généralisés de Legendre. *Bull. Acad. Sci. URSS. Sér. Math. [Izvestia Akad. Nauk SSSR]*, 5, 241–248.

Fernandez, F. (2004). Comment on "breakdown of the Hellmann–Feynman theorem: degeneracy is the key". *Phys. Rev. B*, 69(037101), 2.

Feynman, R. (1939). Forces in molecules. *Phys. Rev.*, 56, 340–343.

Fields, J. & Ismail, M. E. H. (1975). Polynomial expansions. *Math. Comp.*, 29, 894–902.

Fields, J. & Wimp, J. (1961). Expansions of hypergeometric functions in hypergeometric functions. *Math. Comp.*, 15, 390–395.

Fields, J. L. (1967). A uniform treatment of Darboux's method. *Arch. Rational Mech. Anal.*, 27, 289–305.

Filaseta, M. & Lam, T.-Y. (2002). On the irreducibility of the generalized Laguerre polynomials. *Acta Arith.*, 105(2), 177–182.

Flensted-Jensen, M. & Koornwinder, T. (1975). The convolution structure for Jacobi function expansions. *Ark. Mat.*, 11, 469–475.

Floreanini, R. & Vinet, L. (1987). Maximum entropy and the moment problem. *Bull. Amer. Math. Soc. (N.S.)*, 16(1), 47–77.

Floreanini, R. & Vinet, L. (1994). Generalized q-Bessel functions. *Canad. J. Phys.*, 72(7-8), 345–354.

Floreanini, R. & Vinet, L. (1995a). A model for the continuous q-ultraspherical polynomials. *J. Math. Phys.*, 36, 3800–3813.

Floreanini, R. & Vinet, L. (1995b). More on the q-oscillator algebra and q-orthogonal polynomials. *Journal of Physics A*, 28, L287–L293.

Floris, P. G. A. (1997). Addition formula for q-disk polynomials. *Compositio Math.*, 108(2), 123–149.

Floris, P. G. A. & Koelink, H. T. (1997). A commuting q-analogue of the addition formula for disk polynomials. *Constr. Approx.*, 13(4), 511–535.

Foata, D. (1981). Some Hermite polynomial identities and their combinatorics. *Adv. in Appl. Math.*, 2, 250–259.

Foata, D. & Strehl, V. (1981). Une extension multilinéaire de la formule d'Erdélyi pour les produits de fonctions hypergéométriques confluentes. *C. R. Acad. Sci. Paris Sér. I Math.*, 293(10), 517–520.

Fokas, A. S., Its, A. R., & Kitaev, A. V. (1992). The isomonodromy approach to matrix models in 2D quantum gravity. *Comm. Math. Phys.*, 147, 395–430.

Forrester, P. J. & Rogers, J. B. (1986). Electrostatics and the zeros of the classical orthogonal polynomials. *SIAM J. Math. Anal.*, 17, 461–468.

Frenzen, C. L. & Wong, R. (1985). A uniform asymptotic expansion of the Jacobi polynomials with error bounds. *Canad. J. Math.*, 37(5), 979–1007.

Frenzen, C. L. & Wong, R. (1988). Uniform asymptotic expansions of Laguerre polynomials. *SIAM J. Math. Anal.*, 19(5), 1232–1248.

Freud, G. (1971). *Orthogonal Polynomials*. New York: Pergamon Press.

Freud, G. (1976). On the coefficients in the recursion formulae of orthogonal polynomials. *Proc. Roy. Irish Acad. Sect. A (1)*, 76, 1–6.

Freud, G. (1977). On the zeros of orthogonal polynomials with respect to measures with noncompact support. *Anal. Numér. Théor. Approx.*, 6, 125–131.

Gabardo, J.-P. (1992). A maximum entropy approach to the classical moment problem. *J. Funct. Anal.*, 106(1), 80–94.

Garabedian, P. R. (1964). *Partial Differential Equations*. New York: Wiley.

Garrett, K., Ismail, M. E. H., & Stanton, D. (1999). Variants of the Rogers–Ramanujan identities. *Advances in Applied Math.*, 23, 274–299.

Gasper, G. (1971). Positivity and the convolution structure for Jacobi series. *Ann. Math.*, 93, 112–118.

Gasper, G. (1972). Banach algebras for Jacobi series and positivity of a kernel. *Ann. Math.*, 95, 261–280.

Gasper, G. & Rahman, M. (1983). Positivity of the Poisson kernel for the continuous q-ultraspherical polynomials. *SIAM J. Math. Anal.*, 14(2), 409–420.

Gasper, G. & Rahman, M. (1990). *Basic Hypergeometric Series*. Cambridge: Cambridge University Press.

Gasper, G. & Rahman, M. (2004). *Basic Hypergeometric Series*. Cambridge: Cambridge University Press, second edition.

Gatteschi, L. (1987). New inequalities for the zeros of Jacobi polynomials. *SIAM J. Math. Anal.*, 18, 1549–1562.

Gautschi, W. (1967). Computational aspects of three-term recurrence relations. *SIAM Rev.*, 9, 24–82.

Gawronski, W. & Van Assche, W. (2003). Strong asymptotics for relativistic Hermite polynomials. *Rocky Mountain J. Math.*, 33(2), 489–524.

Gelfand, I. M., Kapranov, M. M., & Zelevinsky, A. V. (1994). *Discriminants, Resultants, and Multidimensional Determinants*. Boston: Birkhäuser.

Geronimus, Y. L. (1946). On the trigonometric moment problem. *Ann. of Math. (2)*, 47, 742–761.

Geronimus, Y. L. (1947). The orthogonality of some systems of polynomials. *Duke Math. J.*, 14, 503–510.

Geronimus, Y. L. (1961). *Orthogonal polynomials: Estimates, asymptotic formulas, and series of polynomials orthogonal on the unit circle and on an interval*. Authorized translation from the Russian. New York: Consultants Bureau.

Geronimus, Y. L. (1962). *Polynomials orthogonal on a circle and their applications*, volume 3 of *Amer. Math. Soc. Transl.* Providence, RI: American Mathematical Society.

Geronimus, Y. L. (1977). *Orthogonal polynomials*, volume 108 of *Amer. Math. Soc. Transl.* Providence, RI: American Mathematical Society.

Gessel, I. M. (1990). Symmetric functions and P-recursiveness. *J. Comp. Theor.*, *Ser. A*, 53, 257–285.

Gessel, I. M. & Stanton, D. (1983). Applications of q-Lagrange inversion to basic hypergeometric series. *Trans. Amer. Math. Soc.*, 277(1), 173–201.

Gessel, I. M. & Stanton, D. (1986). Another family of q-Lagrange inversion formulas. *Rocky Mountain J. Math.*, 16(2), 373–384.

Gilewicz, J., Leopold, E., & Valent, G. (2005). New Nevanlinna matrices for orthogonal polynomials related to cubic birth and death processes. *J. Comp. Appl. Math.*, 178, 235–245.

Gillis, J., Reznick, B., & Zeilberger, D. (1983). On elementary methods in positivity theory. *SIAM J. Math. Anal.*, 14, 396–398.

Godoy, E. & Marcellán, F. (1991). An analog of the Christoffel formula for polynomial modification of a measure on the unit circle. *Bull. Un. Mat. Ital.*, 4(7), 1–12.

Goldberg, J. (1965). Polynomials orthogonal over a denumerable set. *Pacific J. Math.*, 15, 1171–1186.

Goldman, J. & Rota, G. C. (1970). On the foundations of combinatorial theory IV. Finite vector spaces and Eulerian generating functions. *Studies in Applied Math.*, 49, 239–258.

Golinskii, L. (2005). Schur flows and orthogonal polynomials on the unit circle. (To appear).

Gosper, R. W., Ismail, M. E. H., & Zhang, R. (1993). On some strange summation formulas. *Illinois J. Math.*, 37(2), 240–277.

Gould, H. W. (1962). A new convolution formula and some new orthogonal relations for inversion of series. *Duke Math. J.*, 29, 393–404.

Gould, H. W. & Hsu, L. C. (1973). Some new inverse series relations. *Duke Math. J.*, 40, 885–891.

Gray, L. J. & Wilson, D. G. (1976). Construction of a Jacobi matrix from spectral data. *Linear Algebra Appl.*, 14, 131–134.

Grenander, U. & Szegő, G. (1958). *Toeplitz Forms and Their Applications*. Berkeley, CA: University of California Press. Reprinted, Bronx, NY: Chelsea, 1984.

Grosswald, E. (1978). *The Bessel Polynomials*, volume 698 of *Lecture Notes in Mathematics*. Berlin: Springer.

Grünbaum, F. A. (1998). Variation on a theme of Stieltjes and Heine. *J. Comp. Appl. Math.*, 99, 189–194.

Grünbaum, F. A. & Haine, L. (1996). The q-version of a theorem of Bochner. *J. Comp. Appl. Math.*, 68, 103–114.

Grünbaum, F. A. & Haine, L. (1997). A theorem of Bochner, revisited. In *Algebraic aspects of integrable systems*, volume 26 of *Progr. Nonlinear Differential Equations Appl.* (pp. 143–172). Boston, MA: Birkhäuser Boston.

Gupta, D. P. & Masson, D. R. (1991). Exceptional q-Askey–Wilson polynomials and continued fractions. *Proc. Amer. Math. Soc.*, 112(3), 717–727.

Gustafson, R. A. (1990). A generalization of Selberg's beta integral. *Bull. Amer. Math. Soc. (N.S.)*, 22, 97–108.

Habsieger, L. (2001a). Integer zeros of q-Krawtchouk polynomials in classical combinatorics. *Adv. in Appl. Math.*, 27(2-3), 427–437. Special issue in honor of Dominique Foata's 65th birthday (Philadelphia, PA, 2000).

Habsieger, L. (2001b). Integral zeroes of Krawtchouk polynomials. In *Codes and association schemes (Piscataway, NJ, 1999)*, volume 56 of *DIMACS Ser. Discrete Math. Theoret. Comput. Sci.* (pp. 151–165). Providence, RI: Amer. Math. Soc.

Habsieger, L. & Stanton, D. (1993). More zeros of Krawtchouk polynomials. *Graphs Combin.*, 9(2), 163–172.

Hadamard, J. (1932). *Le Problème de Cauchy et les Équations aux Dérivées Partielles linéares Hyperboliques*. Paris: Hermann.

Hahn, W. (1935). Über die Jacobischen Polynome und zwei verwandte Polynomklassen. *Math. Z.*, 39, 634–638.

Hahn, W. (1937). Über höhere Ableitungen von Orthogonalpolynomen. *Math. Z.*, 43, 101.

Hahn, W. (1949a). Beiträge zur Theorie der Heineschen Reihen. Die 24 Integrale der Hypergeometrischen q-Differenzengleichung. Das q-Analogon der Laplace-Transformation. *Math. Nachr.*, 2, 340–379.

Hahn, W. (1949b). Über Orthogonalpolynome, die q-Differenzengleichungen genügen. *Math. Nachr.*, 2, 4–34.

Hata, M. (1993). Rational approximations to π and other numbers. *Acta Arith.*, 63(4), 335–349.

Heisenberg, W. (1928). Zur theorie des feromagnetismus. *Zeitchrift für Physik*, 49, 619–636.

Heller, E. J. (1975). Theory of J-matrix Green's functions with applications to atomic polarizability and phase-shift error bounds. *Phys. Rev.*, 22, 1222–1231.

Heller, E. J., Reinhardt, W. P., & Yamani, H. A. (1973). On an equivalent quadrature calculation of matrix elements of $(z - P^2/2m)$ using an L^2 expansion technique. *J. Comp. Phys.*, 13, 536–549.

Hellmann, E. (1937). *Einfuhrung in die Quantamchemie*. Vienna: Deuticke.

Hendriksen, E. & van Rossum, H. (1988). Electrostatic interpretation of zeros. In *Orthogonal Polynomials and Their Applications (Segovia, 1986)*, volume 1329 of *Lecture Notes in Mathematics* (pp. 241–250). Berlin: Springer.

Hilbert, D. (1885). Über die discriminante der in endlichen abbrechenden hypergeometrischen reihe. *J. für die reine und angewandte Matematik*, 103, 337–345.

Hille, E. (1959). *Analytic Theory of Functions*, volume 1. New York: Blaisdell.

Hirschman, I. I. & Widder, D. V. (1955). *The convolution transform*. Princeton, NJ: Princeton University Press.

Hisakado, M. (1996). Unitary matrix models and Painléve III. *Mod. Phys. Lett. A*, 11, 3001–3010.

Hong, Y. (1986). On the nonexistence of nontrivial perfect e-codes and tight $2e$-designs in Hamming schemes $H(n, q)$ with $e \geq 3$ and $q \geq 3$. *Graphs Combin.*, 2(2), 145–164.

Horn, R. A. & Johnson, C. R. (1992). *Matrix Analysis*. Cambridge: Cambridge University Press.

Hwang, S.-G. (2004). Cauchy's interlace theorem for eigenvalues of Hermitian matrices. *Amer. Math. Monthly*, 111(2), 157–159.

Ibragimov, I. A. (1968). A theorem of Gabor Szegő. *Mat. Zametki*, 3, 693–702.

Ihrig, E. C. & Ismail, M. E. H. (1981). A q-umbral calculus. *J. Math. Anal. Appl.*, 84, 178–207.

Ismail, M. E. H. (1977a). Connection relations and bilinear formulas for the classical orthogonal polynomials. *J. Math. Anal.*, 57, 487–496.

Ismail, M. E. H. (1977b). A simple proof of Ramanujan's $_1\psi_1$ sum. *Proc. Amer. Math. Soc.*, 63, 185–186.

Ismail, M. E. H. (1981). The basic Bessel functions and polynomials. *SIAM J. Math. Anal.*, 12, 454–468.

Ismail, M. E. H. (1982). The zeros of basic Bessel functions, the functions $J_{v+ax}(x)$, and associated orthogonal polynomials. *J. Math. Anal. Appl.*, 86(1), 1–19.

Ismail, M. E. H. (1985). A queueing model and a set of orthogonal polynomials. *J. Math. Anal. Appl.*, 108, 575–594.

Ismail, M. E. H. (1986). Asymptotics of the Askey–Wilson polynomials and q-Jacobi polynomials. *SIAM J. Math Anal.*, 17, 1475–1482.

Ismail, M. E. H. (1987). The variation of zeros of certain orthogonal polynomials. *Advances in Appl. Math.*, 8, 111–118.

Ismail, M. E. H. (1989). Monotonicity of zeros of orthogonal polynomials. In D. Stanton (Ed.), *q-Series and Partitions*, volume 18 of *IMA Volumes in Mathematics and Its Applications* (pp. 177–190). New York: Springer-Verlag.

Ismail, M. E. H. (1993). Ladder operators for q^{-1}-Hermite polynomials. *Math. Rep. Royal Soc. Canada*, 15, 261–266.

Ismail, M. E. H. (1995). The Askey–Wilson operator and summation theorems. In M. Ismail, M. Z. Nashed, A. Zayed, & A. Ghaleb (Eds.), *Mathematical Analysis, Wavelets and Signal Processing*, volume 190 of *Contemporary Mathematics* (pp. 171–178). Providence, RI: American Mathematical Society.

Ismail, M. E. H. (1998). Discriminants and functions of the second kind of orthogonal polynomials. *Results in Mathematics*, 34, 132–149.

Ismail, M. E. H. (2000a). An electrostatic model for zeros of general orthogonal polynomials. *Pacific J. Math.*, 193, 355–369.

Ismail, M. E. H. (2000b). More on electronic models for zeros of orthogonal polynomials. *Numer. Funct. Anal. and Optimiz.*, 21(1-2), 191–204.

Ismail, M. E. H. (2001a). An operator calculus for the Askey–Wilson operator. *Ann. of Combinatorics*, 5, 333–348.

Ismail, M. E. H. (2001b). Orthogonality and completeness of Fourier type systems. *Z. Anal. Anwendungen*, 20, 761–775.

Ismail, M. E. H. (2003a). Difference equations and quantized discriminants for q-orthogonal polynomials. *Advances in Applied Math.*, 30(3), 562–589.

Ismail, M. E. H. (2003b). A generalization of a theorem of Bochner. *J. Comp. and Appl. Math.*, 159, 319–324.

Ismail, M. E. H. (2005a). Asymptotics of q-orthogonal polynomials and a q-Airy function. *Internat. Math. Res. Notices*, 2005(18), 1063–1088.

Ismail, M. E. H. (2005b). Determinants with orthogonal polynomial entries. *J. Comp. Appl. Anal.*, 178, 255–266.

Ismail, M. E. H. & Jing, N. (2001). q-discriminants and vertex operators. *Advances in Applied Math.*, 27, 482–492.

Ismail, M. E. H. & Kelker, D. (1976). The Bessel polynomial and the student t-distribution. *SIAM J. Math. Anal.*, 7(1), 82–91.

Ismail, M. E. H. & Letessier, J. (1988). Monotonicity of zeros of ultraspherical polynomials. In M. Alfaro, J. S. Dehesa, F. J. Marcellán, J. L. R. de Francia, & J. Vinuesa (Eds.), *Orthogonal Polynomials and their Applications (Proceedings, Segovia 1986)*, number 1329 in Lecture Notes in Mathematics (pp. 329–330). Berlin, Heidelberg: Springer-Verlag.

Ismail, M. E. H., Letessier, J., Masson, D. R., & Valent, G. (1990). Birth and death processes and orthogonal polynomials. In P. Nevai (Ed.), *Orthogonal polynomials (Columbus, OH, 1989)* (pp. 229–255). Dordrecht: Kluwer.

Ismail, M. E. H., Letessier, J., & Valent, G. (1988). Linear birth and death models and associated Laguerre and Meixner polynomials. *J. Approx. Theory*, 55, 337–348.

Ismail, M. E. H. & Li, X. (1992). Bounds for extreme zeros of orthogonal polynomials. *Proc. Amer. Math. Soc.*, 115, 131–140.

Ismail, M. E. H., Lin, S. S., & Roan, S. S. (2005). Bethe ansatz equations of the XXZ model and q-Sturm–Liouville problems. *J. Math. Phys.* (To appear).

Ismail, M. E. H. & Masson, D. R. (1991). Two families of orthogonal polynomials related to Jacobi polynomials. In *Proceedings of the U.S.-Western Europe Regional Conference on Padé Approximants and Related Topics (Boulder, CO, 1988)*, volume 21 (pp. 359–375).

Ismail, M. E. H. & Masson, D. R. (1994). q-Hermite polynomials, biorthogonal rational functions, and q-beta integrals. *Trans. Amer. Math. Soc.*, 346, 63–116.

Ismail, M. E. H. & Muldoon, M. (1988). On the variation with respect to a parameter of zeros of Bessel functions and q-Bessel functions. *J. Math. Anal. Appl.*, 135, 187–207.

Ismail, M. E. H. & Muldoon, M. (1991). A discrete approach to monotonicity of zeros of orthogonal polynomials. *Trans. Amer. Math. Soc.*, 323, 65–78.

Ismail, M. E. H. & Mulla, F. S. (1987). On the generalized Chebyshev polynomials. *SIAM J. Math. Anal.*, 18(1), 243–258.

Ismail, M. E. H., Nikolova, I., & Simeonov, P. (2004). Difference equations and discriminants for discrete orthogonal polynomials. (To appear).

Ismail, M. E. H. & Rahman, M. (1991). Associated Askey–Wilson polynomials. *Trans. Amer. Math. Soc.*, 328, 201–237.

Ismail, M. E. H. & Rahman, M. (1998). The q-Laguerre polynomials and related moment problems. *J. Math. Anal. Appl.*, 218(1), 155–174.

Ismail, M. E. H., Rahman, M., & Stanton, D. (1999). Quadratic q-exponentials and connection coefficient problems. *Proc. Amer. Math. Soc.*, 127, 2931–2941.

Ismail, M. E. H., Rahman, M., & Zhang, R. (1996). Diagonalization of certain integral operators II. *J. Comp. Appl. Math.*, 68, 163–196.

Ismail, M. E. H. & Ruedemann, R. (1992). Relation between polynomials orthogonal on the unit circle with respect to different weights. *J. Approximation Theory*, 71, 39–60.

Ismail, M. E. H. & Simeonov, P. (1998). Strong asymptotics for Krawtchouk polynomials. *J. Comput. Appl. Math.*, 100, 121–144.

Ismail, M. E. H. & Stanton, D. (1988). On the Askey–Wilson and Rogers polynomials. *Canad. J. Math.*, 40, 1025–1045.

Ismail, M. E. H. & Stanton, D. (1997). Classical orthogonal polynomials as moments. *Canad. J. Math.*, 49, 520–542.

Ismail, M. E. H. & Stanton, D. (1998). More on orthogonal polynomials as moments. In B. Sagan & R. Stanley (Eds.), *Mathematical Essays in Honor of Gian-Carlo Rota* (pp. 377–396). Boston, MA: Birkhauser.

Ismail, M. E. H. & Stanton, D. (2000). Addition theorems for q-exponential functions. In *q-Series from a Contemporary Perspective*, volume 254 of *Contemporary Mathematics* (pp. 235–245). Providence, RI: American Mathematical Society.

Ismail, M. E. H. & Stanton, D. (2002). q-integral and moment representations for q-orthogonal polynomials. *Canad. J. Math.*, 54, 709–735.

Ismail, M. E. H. & Stanton, D. (2003a). Applications of q-Taylor theorems. In *Proceedings of the Sixth International Symposium on Orthogonal Polynomials, Special Functions and their Applications (Rome, 2001)*, volume 153 (pp. 259–272).

Ismail, M. E. H. & Stanton, D. (2003b). q-Taylor theorems, polynomial expansions and interpolation of entire functions. *J. Approx. Theory*, 123, 125–146.

Ismail, M. E. H. & Stanton, D. (2005). Ramnujan's continued fractions via orthogonal polynomials. *Adv. in Math.* (To appear).

Ismail, M. E. H., Stanton, D., & Viennot, G. (1987). The combinatorics of the q-Hermite polynomials and the Askey–Wilson integral. *European J. Combinatorics*, 8, 379–392.

Ismail, M. E. H. & Tamhankar, M. V. (1979). A combinatorial approach to some positivity problems. *SIAM J. Math. Anal.*, 10, 478–485.

Ismail, M. E. H. & Valent, G. (1998). On a family of orthogonal polynomials related to elliptic functions. *Illinois J. Math.*, 42(2), 294–312.

Ismail, M. E. H., Valent, G., & Yoon, G. J. (2001). Some orthogonal polynomials related to elliptic functions. *J. Approx. Theory*, 112(2), 251–278.

Ismail, M. E. H. & Wilson, J. (1982). Asymptotic and generating relations for the q-Jacobi and the $_4\phi_3$ polynomials. *J. Approx. Theory*, 36, 43–54.

Ismail, M. E. H. & Wimp, J. (1998). On differential equations for orthogonal polynomials. *Methods and Applications of Analysis*, 5, 439–452.

Ismail, M. E. H. & Witte, N. (2001). Discriminants and functional equations for polynomials orthogonal on the unit circle. *J. Approx. Theory*, 110, 200–228.

Ismail, M. E. H. & Zhang, R. (1988). On the Hellmann–Feynman theorem and the variation of zeros of special functions. *Adv. Appl. Math.*, 9, 439–446.

Ismail, M. E. H. & Zhang, R. (1989). The Hellmann–Feynman theorem and zeros of special functions. In N. K. Thakare (Ed.), *Ramanujan International Symposium on Analysis (Pune, 1987)* (pp. 151–183). Delhi: McMillan of India.

Ismail, M. E. H. & Zhang, R. (1994). Diagonalization of certain integral operators. *Advances in Math.*, 109, 1–33.

Ismail, M. E. H. & Zhang, R. (2005). New proofs of some q-series results. In M. E. H. Ismail & E. H. Koelink (Eds.), *Theory and Applications of Special Functions: A Volume Dedicated to Mizan Rahman*, volume 13 of *Developments in Mathematics* (pp. 285–299). New York: Springer.

Jackson, F. H. (1903). On generalized functions of Legendre and Bessel. *Trans. Royal Soc. Edinburgh*, 41, 1–28.

Jackson, F. H. (1903–1904). The application of basic numbers to Bessel's and Legendre's functions. *Proc. London Math. Soc. (2)*, 2, 192–220.

Jackson, F. H. (1904–1905). The application of basic numbers to Bessel's and Legendre's functions, II. *Proc. London Math. Soc. (2)*, 3, 1–20.

Jacobson, N. (1974). *Basic Algebra I*. San Francisco: Freeman and Company.

Jin, X.-S. & Wong, R. (1998). Uniform asymptotic expansions for Meixner polynomials. *Constr. Approx.*, 14(1), 113–150.

Jin, X.-S. & Wong, R. (1999). Asymptotic formulas for the zeros of the Meixner polynomials. *J. Approx. Theory*, 96(2), 281–300.

Jones, W. B. & Thron, W. (1980). *Continued Fractions: Analytic Theory and Applications*. Reading, MA: Addison-Wesley.

Joni, S. J. & Rota, G.-C. (1982). Coalgebras and bialgebras in combinatorics. In *Umbral calculus and Hopf algebras (Norman, Okla., 1978)*, volume 6 of *Contemp. Math.* (pp. 1–47). Providence, RI: American Mathematical Society.

Jordan, C. (1965). *Calculus of Finite Differences*. New York: Chelsea.

Kadell, K. W. J. (1987). A probabilistic proof of Ramanujan's $_1\psi_1$ sum. *SIAM J. Math. Anal.*, 18(6), 1539–1548.

Kadell, K. W. J. (2005). The little q-Jacobi functions of complex order. In M. E. H. Ismail & E. H. Koelink (Eds.), *Theory and Applications of Special Functions: A Volume Dedicated to Mizan Rahman*, volume 13 of *Developments in Mathematics* (pp. 301–338). New York: Springer.

Kaliaguine, V. (1995). The operator moment problem, vector continued fractions and an explicit form of the Favard theorem for vector orthogonal polynomials. *J. Comput. Appl. Math.*, 65, 181–193.

Kaliaguine, V. & Ronveaux, A. (1996). On a system of "classical" polynomials of simultaneous orthogonality. *J. Comput. Appl. Math.*, 67, 207–217.

Kalnins, E. G. & Miller, W. (1989). Symmetry techniques for q-series: Askey–Wilson polynomials. *Rocky Mountain J. Math.*, 19, 223–230.

Kaluza, T. (1933). Elementarer Bewies einer Vermutung von K. Friedrichs und H. Lewy. *Math. Zeit.*, 37.

Kalyagin, V. A. (1979). On a class of polynomials defined by two orthogonality relations. *Mat. Sb. (N.S.)*, 110(152), 609–627. Translated in Math. USSR Sb. **38** (1981), 563–580.

Kamran, N. & Olver, P. J. (1990). Lie algebras of differential operators and Lie-algebraic potentials. *J. Math. Anal. Appl.*, 145, 342–356.

Kaplansky, I. (1944). Symbolic solution of certain problems in permutations. *Bull. Amer. Math. Soc.*, 50, 906–914.

Karlin, S. & McGregor, J. (1957a). The classification of birth and death processes. *Trans. Amer. Math. Soc.*, 86, 366–400.

Karlin, S. & McGregor, J. (1957b). The differential equations of birth and death processes and the Stieltjes moment problem. *Trans. Amer. Math. Soc.*, 85, 489–546.

Karlin, S. & McGregor, J. (1958). Many server queuing processes with Poisson input and exponential service time. *Pacific J. Math.*, 8, 87–118.

Karlin, S. & McGregor, J. (1959). Random walks. *Illinois J. Math.*, 3, 66–81.

Karlin, S. & McGregor, J. (1961). The Hahn polynomials, formulas and an application. *Scripta Mathematica*, 26, 33–46.

Kartono, A., Winata, T., & Sukirno (2005). Applications of nonorthogonal Laguerre function basis in helium atom. *Appl. Math. Comp.*, 163(2), 879–893.

Khruschev, S. (2005). Continued fractions and orthogonal polynomials on the unit circle. *J. Comp. Appl. Math.*, 178, 267–303.

Kibble, W. F. (1945). An extension of theorem of Mehler on Hermite polynomials. *Proc. Cambridge Philos. Soc.*, 41, 12–15.

Kiesel, H. & Wimp, J. (1996). A note on Koornwinder's polynomials with weight function $(1 - x)^\alpha (1 + x)^\beta + m\delta(x + 1) + n\delta(x - 1)$. *Numerical Algorithms*, 11, 229–241.

Knopp, K. (1945). *Theory of Functions*. New York, NY: Dover.

Knuth, D. E. & Wilf, H. S. (1989). A short proof of Darboux's lemma. *Appl. Math. Lett.*, 2(2), 139–140.

Koekoek, R. & Swarttouw, R. (1998). *The Askey-scheme of hypergeometric orthogonal polynomials and its q-analogues*. Reports of the Faculty of Technical Mathematics and Informatics 98-17, Delft University of Technology, Delft.

Koelink, E. (1997). Addition formulas for q-special functions. In M. E. H. Ismail, D. R. Masson, & M. Rahman (Eds.), *Special functions, q-series and related topics (Toronto, ON, 1995)*, volume 14 of *Fields Inst. Commun.* (pp. 109–129). Providence, RI: Amer. Math. Soc.

Koelink, H. T. (1994). The addition formula for continuous q-Legendre polynomials and associated spherical elements on the $SU(2)$ quantum group related to Askey–Wilson polynomials. *SIAM J. Math. Anal.*, 25(1), 197–217.

Koelink, H. T. (1999). Some basic Lommel polynomials. *J. Approx. Theory*, 96(2), 345–365.

Koelink, H. T. (2004). Spectral theory and special functions. In R. Álvarez-Nodarse, F. Marcellán, & W. Van Assche (Eds.), *Laredo Lectures on Orthogonal Polynomials and Special Functions* (pp. 45–84). Hauppauge, NY: Nova Science Publishers.

Koelink, H. T. & Swarttouw, R. (1994). On the zeros of the Hahn–Exton q-Bessel function and associated q-Lommel polynomials. *J. Math. Anal. Appl.*, 186(3), 690–710.

Koelink, H. T. & Van Assche, W. (1995). Orthogonal polynomials and Laurent polynomials related to the Hahn–Exton q-Bessel function. *Constr. Approx.*, 11(4), 477–512.

Koelink, H. T. & Van der Jeugt, J. (1998). Convolutions for orthogonal polynomials from Lie and quantum algebra representations. *SIAM J. Math. Anal.*, 29(3), 794–822.

Koelink, H. T. & Van der Jeugt, J. (1999). Bilinear generating functions for orthogonal polynomials. *Constr. Approx.*, 15(4), 481–497.

Konovalov, D. A. & McCarthy, I. E. (1994). Convergent J-matrix calculation of the Poet-Temkin model of electron-hydrogen scattering. *J. Phys. B: At. Mol. Opt. Phys.*, 27(14), L407–L412.

Konovalov, D. A. & McCarthy, I. E. (1995). Convergent J-matrix calculations of electron-helium resonances. *J. Phys. B: At. Mol. Opt. Phys.*, 28(5), L139–L145.

Koornwinder, T. H. (1972). The addition formula for Jacobi polynomials, I. Summary of results. *Indag. Math.*, 34, 188–191.

Koornwinder, T. H. (1973). The addition formula for Jacobi polynomials and spherical harmonics. *SIAM J. Appl. Math.*, 25, 236–246.

Koornwinder, T. H. (1974). Jacobi polynomials, II. An analytic proof of the product formula. *SIAM J. Math. Anal.*, 5, 125–237.

Koornwinder, T. H. (1975). Jacobi polynomials, III. An analytic proof of the addition formula. *SIAM J. Math. Anal.*, 6, 533–543.

Koornwinder, T. H. (1977). Yet another proof of the addition formula for Jacobi polynomials. *J. Math. Anal. Appl.*, 61(1), 136–141.

Koornwinder, T. H. (1978). Positivity proofs for linearization and connection coefficients for orthogonal polynomials satisfying an addition formula. *J. London Math. Soc.*, 18(2), 101–114.

Koornwinder, T. H. (1981). Clebsch–Gordan coefficients for $SU(2)$ and Hahn polynomials. *Nieuw Arch. Wisk. (3)*, 29(2), 140–155.

Koornwinder, T. H. (1982). Krawtchouk polynomials, a unification of two different group theoretic interpretations. *SIAM J. Math. Anal.*, 13(6), 1011–1023.

Koornwinder, T. H. (1984). Orthogonal polynomials with weight function $(1 - x)^\alpha (1 + x)^\beta + m\delta(x + 1) + n\delta(x - 1)$. *Canadian Math Bull.*, 27, 205–214.

Koornwinder, T. H. (1985). Special orthogonal polynomial systems mapped onto each other by the Fourier–Jacobi transform. In A. Dold & B. Eckmann (Eds.), *Polynômes Orthogonaux et Applications (Proceedings, Bar-le-Duc 1984)*, volume 1171 of *Lecture Notes in Mathematics* (pp. 174–183). Berlin: Springer.

Koornwinder, T. H. (1990). Jacobi functions as limit cases of q-ultraspherical polynomials. *J. Math. Anal. Appl.*, 148, 44–54. Appendix B.

Koornwinder, T. H. (1991). The addition formula for little q-Legendre polynomials and the $SU(2)$ quantum group. *SIAM J. Math. Anal.*, 22(1), 295–301.

Koornwinder, T. H. (1993). Askey-Wilson polynomials as zonal spherical functions on the $SU(2)$ quantum group. *SIAM J. Math. Anal.*, 24(3), 795–813.

Koornwinder, T. H. (2005a). *Lowering and raising operators for some special orthogonal polynomials*. Contemporary Mathematics. American Math. Soc.: Providence, RI. (To appear).

Koornwinder, T. H. (2005b). A second addition formula for continuous q-ultraspherical polynomials. In M. E. H. Ismail & E. Koelink (Eds.), *Theory and Applications of Special Functions*, volume 13 of *Developments in Mathematics* (pp. 339–360). New York: Springer.

Koornwinder, T. H. & Swarttouw, R. F. (1992). On q-analogues of the Fourier and Hankel transforms. *Trans. Amer. Math. Soc.*, 333(1), 445–461.

Korepin, V. E., Bogoliubov, N. M., & Izergin, A. G. (1993). *Quantum Inverse Scattering Method and Correlation Functions*. Cambridge Monographs on Mathematical Physics. Cambridge: Cambridge University Press.

Krall, H. L. (1936a). On derivatives of orthogonal polynomials. *Bull. Amer. Math. Soc.*, 42, 424–428.

Krall, H. L. (1936b). On higher derivatives of orthogonal polynomials. *Bull. Amer. Math. Soc.*, 42, 867–870.

Krall, H. L. (1938). Certain differential equations for Tchebychev polynomials. *Duke Math. J.*, 4, 705–719.

Krall, H. L. & Frink, O. (1949). A new class of orthogonal polynomials. *Trans. Amer. Math. Soc.*, 65, 100–115.

Krall, H. L. & Sheffer, I. M. (1965). On pairs of related orthogonal polynomial sets. *Math. Z.*, 86, 425–450.

Krasikov, I. (2001). Nonnegative quadratic forms and bounds on orthogonal polynomials. *J. Approx. Theory*, 111, 31–49.

Krasikov, I. (2003). Bounds for zeros of the Laguerre polynomials. *J. Approx. Theory*, 121, 287–291.

Krasikov, I. & Litsyn, S. (1996). On integral zeros of Krawtchouk polynomials. *J. Combin. Theory Ser. A*, 74(1), 71–99.

Kriecherbauer, T. & McLaughlin, K. T.-R. (1999). Strong asymptotics of polynomials orthogonal with respect to Freud weights. *Internat. Math. Res. Notices*, 1999(6), 299–333.

Kuijlaars, A. B. J. (2003). Riemann–Hilbert analysis for orthogonal polynomials. In E. Koelink & W. Van Assche (Eds.), *Orthogonal Polynomials and Special Functions*, volume 1817 of *Lecture Notes in Mathematics* (pp. 167–210). Berlin: Springer.

Kuijlaars, A. B. J. & McLaughlin, K. T.-R. (2001). Riemann–Hilbert analysis for Laguerre polynomials with large negative parameter. *Comput. Methods Funct. Theory*, 1, 205–233.

Kuijlaars, A. B. J., McLaughlin, K. T.-R., Van Assche, W., & Vanlessen, M. (2004). The Riemann–Hilbert approach to strong asymptotics for orthogonal polynomials on $[-1, 1]$. *Adv. Math.*, 188(2), 337–398.

Kuijlaars, A. B. J. & Van Assche, W. (1999). The asymptotic zero distribution of orthogonal polynomials with varying recurrence coefficients. *J. Approx. Theory*, 99(1), 167–197.

Kulish, P. P. & Sklyanin, E. K. (1982). *Quantum spectral transform method. Recent developments*, volume 151 of *Lecture Notes in Physics*. Berlin-New York: Springer.

Kulish, P. P. & Sklyanin, E. K. (1991). The general $u_q[sl(2)]$ invariant XXZ integrable quantum spin chain. *J. Phys. A: Math. Gen.*, 24, 435–439.

Kwon, K. H. (2002). *Orthogonal polynomials I.* Lecture notes, KAIST, Seoul.

Kwon, K. H., Kim, S. S., & Han, S. S. (1992). Orthogonalizing weights of Tchebychev sets of polynomials. *Bull. London Math. Soc.*, 24(4), 361–367.

Kwon, K. H., Lee, D. W., & Park, S. B. (1997). New characterizations of discrete classical orthogonal polynomials. *J. Approx. Theory*, 89(2), 156–171.

Kwon, K. H., Lee, J. K., & Yoo, B. H. (1993). Characterizations of classical orthogonal polynomials. *Results Math.*, 24(1-2), 119–128.

Kwon, K. H. & Littlejohn, L. L. (1997). Classification of classical orthogonal polynomials. *J. Korean Math. Soc.*, 34(4), 973–1008.

Kwon, K. H., Littlejohn, L. L., & Yoo, B. H. (1994). Characterizations of orthogonal polynomials satisfying differential equations. *SIAM J. Math. Anal.*, 25(3), 976–990.

Labelle, J. & Yeh, Y. N. (1989). The combinatorics of Laguerre, Charlier, and Hermite polynomials. *Stud. Appl. Math.*, 80(1), 25–36.

Laforgia, A. (1985). Monotonicity properties for the zeros of orthogonal polynomials and Bessel functions. In C. Bresinski, A. P. Magnus, P. Maroni, & A. Ronveaux (Eds.), *Polynômes Orthogonaux et Applications*, number 1171 in Lecture Notes in Mathematics (pp. 267–277). Berlin: Springer-Verlag.

Laforgia, A. & Muldoon, M. E. (1986). Some consequences of the Sturm comparison theorem. *Amer. Math. Monthly*, 93, 89–94.

Lancaster, O. E. (1941). Orthogonal polynomials defined by difference equations. *Amer. J. Math.*, 63, 185–207.

Landau, H. J. (1987). Maximum entropy and the moment problem. *Bull. Amer. Math. Soc. (N.S.)*, 16(1), 47–77.

Lanzewizky, I. L. (1941). Über die orthogonalität der Fejér–Szegöschen polynome. *C. R. Dokl. Acad. Sci. URSS (N. S.)*, 31, 199–200.

Leonard, D. (1982). Orthogonal polynomials, duality, and association schemes. *SIAM J. Math. Anal.*, 13(4), 656–663.

Lepowsky, J. & Milne, S. C. (1978). Lie algebraic approaches to classical partition identities. *Adv. in Math.*, 29(1), 15–59.

Lepowsky, J. & Wilson, R. L. (1982). A Lie theoretic interpretation and proof of the Rogers–Ramanujan identities. *Adv. in Math.*, 45(1), 21–72.

Lesky, P. (1989). Die Vervollständigung der diskreten klassischen Orthogonalpolynome. *Österreich. Akad. Wiss. Math.-Natur. Kl. Sitzungsber. II*, 198(8-10), 295–315.

Lesky, P. (2001). Orthogonalpolynome in x und q^{-x} als Lösungen von reellen q-Operatorgleichungen zweiter Ordnung. *Monatsh. Math.*, 132(2), 123–140.

Lesky, P. A. (2005). *Eine Charakterisierung der klassischen kontinuierlichen-, diskreten- und q-Orthogonalpolynome.* Åchen: Shaker Verlag.

Letessier, J. & Valent, G. (1984). The generating function method for quadratic asymptotically symmetric birth and death processes. *SIAM J. Appl. Math.*, 44, 773–783.

Levit, R. J. (1967a). A variant of Tchebichef's minimax problem. *Proc. Amer. Math. Soc. 18 (1967), 925–932; errata, ibid.*, 18, 1143.

Levit, R. J. (1967b). The zeros of the Hahn polynomials. *SIAM Rev.*, 9, 191–203.

Lew, J. S. & Quarles, Jr., D. A. (1983). Nonnegative solutions of a nonlinear recurrence. *J. Approx. Theory*, 38(4), 357–379.

Lewis, J. T. & Muldoon, M. E. (1977). Monotonicity and convexity properties of zeros of Bessel functions. *SIAM J. Math. Anal.*, 8, 171–178.

Li, X. & Wong, R. (2000). A uniform asymptotic expansion for Krawtchouk polynomials. *J. Approx. Theory*, 106(1), 155–184.

Li, X. & Wong, R. (2001). On the asymptotics of the Meixner–Pollaczek polynomials and their zeros. *Constr. Approx.*, 17(1), 59–90.

Lomont, J. S. & Brillhart, J. (2001). *Elliptic polynomials*. Boca Raton, FL: Chapman & Hall / CRC.

Lorch, L. (1977). Elementary comparison techniques for certain classes of Sturm–Liouville equations. In G. Berg, M. Essén, & A. Pleijel (Eds.), *Differential Equations (Proc. Conf. Uppsala, 1977)* (pp. 125–133). Stockholm: Almqvist and Wiksell.

Lorentzen, L. & Waadeland, H. (1992). *Continued Fractions With Applications*. Amsterdam: North-Holland Publishing Co.

Louck, J. D. (1981). Extension of the Kibble–Slepian formula for Hermite polynomials using Boson operator methods. *Advances in Appl. Math.*, 2, 239–249.

Lubinsky, D. S. (1987). A survey of general orthogonal polynomials for weights on finite and infinite intervals. *Acta Applicandae Mathematicae*, 10, 237–296.

Lubinsky, D. S. (1989). *Strong Asymptotics for Extremal Errors and Polynomials Associated with Erdős-Type Weights*, volume 202 of *Pitman Research Notes in Mathematics*. Harlow: Longman.

Lubinsky, D. S. (1993). An update on orthogonal polynomials and weighted approximation on the real line. *Acta Applicandae Mathematicae*, 33, 121–164.

Luke, Y. L. (1969a). *The special functions and their approximations, Vol. I*. Mathematics in Science and Engineering, Vol. 53. New York: Academic Press.

Luke, Y. L. (1969b). *The special functions and their approximations. Vol. II*. Mathematics in Science and Engineering, Vol. 53. New York: Academic Press.

MacMahon, P. (1915–1916). *Combinatory Analysis*, volume 1 and 2. Cambridge: Cambridge University Press. Reprinted, New York: Chelsea 1984.

Magnus, A. (2003). MAPA3072A special topics in approximation theory 1999–2000: Semi-classical orthogonal polynomials on the unit circle. http://www.math.ucl.ac.be/~magnus/.

Magnus, A. P. (1988). Associated Askey–Wilson polynomials as Laguerre–Hahn orthogonal polynomials. In M. Alfaro et al. (Ed.), *Orthogonal Polynomials and Their Applications*, volume 1329 of *Lecture Notes in Mathematics* (pp. 261–278). Berlin: Springer-Verlag.

Mahler, K. (1968). Perfect systems. *Compositio Math.*, 19, 95–166.

Makai, E. (1952). On monotonicity property of certain Sturm–Liouville functions. *Acta Math. Acad. Sci. Hungar.*, 3, 15–25.

Mandjes, M. & Ridder, A. (1995). Finding the conjugate of Markov fluid processes. *Probab. Engrg. Inform. Sci.*, 9(2), 297–315.

Maroni, P. (1985). Une caractérisation des polynômes orthogonaux semi-classiques. *C. R. Acad. Sci. Paris Sér. I Math.*, 301(6), 269–272.

Maroni, P. (1987). Prolégomènes à l'étude des polynômes orthogonaux semi-classiques. *Ann. Mat. Pura Appl. (4)*, 149, 165–184.

Maroni, P. (1989). L' orthogonalité et les récurrences de polynômes d'ordre supérieur à deux. *Ann. Fac. Sci. Toulouse*, 10, 105–139.

Marshal, A. W. & Olkin, I. (1979). *Inequalities: Theory of Majorization and Its Applications*. New York: Academic Press.

Mazel, D. S., Geronimo, J. S., & Hayes, M. H. (1990). On the geometric sequences of reflection coefficients. *IEEE Trans. Acoust. Speech Signal Process.*, 38, 1810–1812.

McBride, E. B. (1971). *Obtaining Generating Functions*. New York: Springer-Verlag.

McCarthy, P. J. (1961). Characterizations of classical polynomials. *Portugal. Math.*, 20, 47–52.

McDonald, J. N. & Weiss, N. A. (1999). *A Course in Real Analysis*. New York: Wiley.

Mehta, M. L. (1979). Properties of the zeros of a polynomial satisfying a second order linear partial differential equation. *Lett. Nuovo Cimento*, 26, 361–362.

Mehta, M. L. (1991). *Random Matrices*. Boston: Academic Press, second edition.

Meixner, J. (1934). Orthogonale polynomsysteme mit einer besonderen Gestalt der erzeugenden funktion. *J. London Math. Soc.*, 9, 6–13.

Mhaskar, H. (1990). Bounds for certain Freud polynomials. *J. Approx. Theory*, 63, 238–254.

Mhaskar, H. (1996). *Introduction to the Theory of Weighted Polynomial Approximation*. Singapore: World Scientific.

Miller, W. (1968). *Lie Theory and Hypergeometric Functions*. New York: Academic Press.

Miller, W. (1974). *Symmetry Groups and Their Applications*. New York: Academic Press.

Milne, S. C. (2002). Infinite families of sums of squares formulas, Jacobi elliptic functions, continued fractions, and Schur functions. *Ramanujan J.*, 6, 7–149.

Milne-Thomson, L. M. (1933). *The Calculus of Finite Differences*. New York: Macmillan.

Młotkowski, W. & Szwarc, R. (2001). Nonnegative linearization for polynomials orthogonal with respect to discrete measures. *Constr. Approx.*, 17(3), 413–429.

Moak, D. (1981). The q-analogue of the Laguerre polynomials. *J. Math. Anal. Appl.*, 81(1), 20–47.

Moser, J. (1981). An example of a Schröedinger equation with almost periodic potential and nowhere dense spectrum. *Comment. Math. Helv.*, 56(2), 198–224.

Mullin, R. & Rota, G.-C. (1970). On the foundations of combinatorial theory: III. Theory of binomial enumeration. In B. Harris (Ed.), *Graph Theory and Its*

Applications (Proc. Advanced Sem., Math. Research Center, Univ. of Wisconsin, Madison, Wis., 1969) (pp. 167–213 (loose errata)). New York: Academic Press.

Nassrallah, B. & Rahman, M. (1985). Projection formulas, a reproducing kernel and a generating function for q-Wilson polynomials. *SIAM J. Math. Anal.*, 16, 186–197.

Nehari, Z. (1952). *Conformal Mapping.* New York: McGraw-Hill.

Nehari, Z. (1961). *Introduction to Complex Analysis.* Boston, MA: Allyn and Bacon.

Nelson, C. A. & Gartley, M. G. (1994). On the zeros of the q-analogue exponential function. *J. Phys. A*, 27(11), 3857–3881.

Nelson, C. A. & Gartley, M. G. (1996). On the two q-analogue logarithmic functions: $\ln_q(w)$, $\ln\{e_q(z)\}$. *J. Phys. A*, 29(24), 8099–8115.

Nevai, P. (1979). Orthogonal polynomials. *Memoirs Amer. Math. Soc.*, 18(213), v + 185 pp.

Nevai, P. (1986). Géza Freud, orthogonal polynomials and Christoffel functions. A case study. *J. Approx. Theory*, 48, 3–167.

Nikishin, E. M. & Sorokin, V. N. (1991). *Rational Approximations and Orthogonality*, volume 92 of *Translations of Mathematical Monographs*. Providence, RI: Amer. Math. Soc.

Novikoff, A. (1954). *On a special system of polynomials.* PhD thesis, Stanford University, Stanford, CA.

Nuttall, J. (1984). Asymptotics of diagonal Hermite–Padé polynomials. *J. Approx. Theory*, 42, 299–386.

Olver, F. W. J. (1974). *Asymptotics and Special Functions.* New York, NY: Academic Press.

Osler, T. J. (1970). Leibniz rule for fractional derivatives generalized and an application to infinite series. *SIAM J. Appl. Math.*, 18, 658–674.

Osler, T. J. (1972). A further extension of the Leibniz rule to fractional derivatives and its relation to Parseval's formula. *SIAM J. Math. Anal.*, 3, 1–16.

Osler, T. J. (1973). A correction to Leibniz rule for fractional derivatives. *SIAM J. Math. Anal.*, 4, 456–459.

Parlett, B. N. (1998). *The Symmetric Eigenvalue Problem*, volume 20 of *Classics in Applied Mathematics*. Philadelphia, PA: Society for Industrial and Applied Mathematics (SIAM). Corrected reprint of the 1980 original.

Pastro, P. I. (1985). Orthogonal polynomials and some q-beta integrals of Ramanujan. *J. Math. Anal. Appl.*, 112, 517–540.

Periwal, V. & Shevitz, D. (1990). Unitary-matrix models as exactly solvable string theories. *Phys. Rev. Lett.*, 64, 1326–1329.

Piñeiro, L. R. (1987). On simultaneous approximations for a collection of Markov functions. *Vestnik Mosk. Univ., Ser. I*, (2), 67–70. Translated in *Moscow Univ. Math. Bull.* 42 (1987), no. 2, 52–55.

Pollaczek, F. (1949a). Sur une généralisation des polynômes de Legendre. *C. R. Acad. Sci. Paris*, 228, 1363–1365.

Pollaczek, F. (1949b). Systèmes de polynômes biorthogonaux à coefficients réels. *C. R. Acad. Sci. Paris*, 228, 1553–1556.

Pollaczek, F. (1956). *Sur une généralisation des polynômes de Jacobi*, volume 131 of *Memorial des Sciences Mathematique*. Paris: Gauthier-Villars.

Pólya, G. & Szegő, G. (1972). *Problems and Theorems in Analysis*, volume 1. New York: Springer-Verlag.

Pólya, G. & Szegő, G. (1976). *Problems and Theorems in Analysis*, volume 2. New York: Springer-Verlag.

Postelmans, K. & Van Assche, W. (2005). Multiple little q-Jacobi polynomials. *J. Comput. Appl. Math.* (To appear).

Potter, H. S. A. (1950). On the latent roots of quasi-commutative matrices. *Amer. Math. Monthly*, 57, 321–322.

Pruitt, W. (1962). Bilateral birth and death processes. *Trans. Amer. Math. Soc.*, 107, 508–525.

Pupyshev, V. I. (2000). The nontriviality of the Hellmann–Feynman theorem. *Russian J. Phys. Chem.*, 74, S267–S278.

Qiu, W.-Y. & Wong, R. (2000). Uniform asymptotic formula for orthogonal polynomials with exponential weight. *SIAM J. Math. Anal.*, 31(5), 992–1029.

Qiu, W.-Y. & Wong, R. (2004). Asymptotic expansion of the Krawtchouk polynomials and their zeros. *Comp. Meth. Func. Theory*, 4(1), 189–226.

Rahman, M. (1981). The linearization of the product of continuous q-Jacobi polynomials. *Canad. J. Math.*, 33(4), 961–987.

Rahman, M. (1984). A simple evaluation of Askey and Wilson's q-beta integral. *Proc. Amer. Math. Soc.*, 92(3), 413–417.

Rahman, M. (1989). A simple proof of Koornwinder's addition formula for the little q-Legendre polynomials. *Proc. Amer. Math. Soc.*, 107(2), 373–381.

Rahman, M. (1991). Biorthogonality of a system of rational functions with respect to a positive measure on $[-1, 1]$. *SIAM J. Math. Anal.*, 22, 1421–1431.

Rahman, M. (2001). The associated classical orthogonal polynomials. In J. Bustoz, M. E. H. Ismail, & S. K. Suslov (Eds.), *Special Functions 2000: Current Perspective and Future Directions*, Nato Science Series (pp. 255–279). Dordrecht: Kluwer Academic Publishers.

Rahman, M. & Tariq, Q. (1997). Poisson kernels for associated q-ultrasperical polynomials. *Methods and Applications of Analysis*, 4, 77–90.

Rahman, M. & Verma, A. (1986a). Product and addition formulas for the continuous q-ultraspherical polynomials. *SIAM J. Math. Anal.*, 17(6), 1461–1474.

Rahman, M. & Verma, A. (1986b). A q-integral representation of Rogers' q-ultraspherical polynomials and some applications. *Constructive Approximation*, 2, 1–10.

Rainville, E. D. (1960). *Special Functions*. New York: Macmillan.

Ramis, J.-P. (1992). About the growth of entire functions solutions of linear algebraic q-difference equations. *Ann. Fac. Sci. Toulouse Math. (6)*, 1(1), 53–94.

Reuter, G. E. H. (1957). Denumerable Markov processes and associated semigroups on l. *Acta Math.*, 97, 1–46.

Rivlin, T. J. & Wilson, M. W. (1969). An optimal property of Chebyshev expansions. *J. Approximation Theory*, 2, 312–317.

Rogers, L. J. (1893). On the expansion of certain infinite products. *Proc. London Math. Soc.*, 24, 337–352.

Rogers, L. J. (1894). Second memoir on the expansion of certain infinite products. *Proc. London Math. Soc.*, 25, 318–343.

Rogers, L. J. (1895). Third memoir on the expansion of certain infinite products. *Proc. London Math. Soc.*, 26, 15–32.

Rogers, L. J. (1907). On the representation of certain asymprortic series as convergent continued fractions. *Proc. Lond. Math. Soc. (2)*, 4, 72–89.

Roman, S. & Rota, G.-C. (1978). The umbral calculus. *Advances in Math.*, 27, 95–188.

Rota, G.-C., Kahaner, D., & Odlyzko, A. (1973). On the foundations of combinatorial theory. VIII. Finite operator calculus. *J. Math. Anal. Appl.*, 42, 684–760.

Routh, E. (1884). On some properties of certain solutions of a differential equation of the second order. *Proc. London Math. Soc.*, 16, 245–261.

Rudin, W. (1976). *Principles of Mathematical Analysis*. New York: McGraw-Hill, third edition.

Rui, B. & Wong, R. (1994). Uniform asymptotic expansion of Charlier polynomials. *Methods Appl. Anal.*, 1(3), 294–313.

Rui, B. & Wong, R. (1996). Asymptotic behavior of the Pollaczek polynomials and their zeros. *Stud. Appl. Math.*, 96(3), 307–338.

Rui, B. & Wong, R. (1999). A uniform asymptotic formula for orthogonal polynomials associated with $\exp(-x^4)$. *J. Approx. Theory*, 98(1), 146–166.

Saff, E. B. & Totik, V. (1997). *Logarithmic Potentials With External Fields*. New York: Springer-Verlag.

Saff, E. B. & Varga, R. S. (1977). On the zeros and poles of Padé approximants to e^z. II. In E. B. Saff & R. S. Varga (Eds.), *Padé and Rational Approximations: Theory and Applications* (pp. 195–213). New York: Academic Press, Inc.

Sarmanov, I. O. (1968). A generalized symmetric gamma-correlation. *Dokl. Akad. Nauk SSSR*, 179, 1279–1281.

Sarmanov, O. V. & Bratoeva, Z. N. (1967). Probabilistic properties of bilinear expansions of Hermite polynomials. *Theor. Probability Appl.*, 12, 470–481.

Scheinhardt, W. R. W. (1998). *Markov-modulated and feedback fluid queues*. PhD thesis, Faculty of Mathematical Sciences, University of Twente, Enschede.

Schlosser, M. (2005). Abel–Rothe type generalizations of Jacobi's triple product identity. In M. E. H. Ismail & E. H. Koelink (Eds.), *Theory and Applications of Special Functions: A Volume Dedicated to Mizan Rahman*, volume 13 of *Developments in Mathematics* (pp. 383–400). New York: Springer.

Schur, I. (1929). Einige Sätze über Primzahlen mit Anwendungen auf Irreduzibilitätsfragen, I. *Sitzungsber. Preuss. Akad. Wissensch. Phys.-Math. Kl.*, 23, 125–136.

Schur, I. (1931). Affektlose Gleichungen in der Theorie der Laguerreschen und Hermiteschen Polynome. *J. Reine Angew. Math.*, 165, 52–58.

Schützenberger, P. M. (1953). Une interprétation de certains solutions de l'equation fonctionnelle: $f(x + y) = f(x)f(y)$. *C. R. Acad. Sci. Paris*, 236, 352–353.

Schwartz, H. M. (1940). A class of continued fractions. *Duke J. Math.*, 6, 48–65.

Selberg, A. (1944). Bemerkninger om et multiplet integral. *Norsk Mat. Tidsskr.*, 26, 71–78.

Sericola, B. (1998). Transient analysis of stochastic fluid models. *Perform. Eval.*, 32, 245–263.

Sericola, B. (2001). A finite buffer fluid queue driven by a Markovian queue. *Queueing Syst.*, 38(2), 213–220.

Sharapudinov, I. I. (1988). Asymptotic properties of Krawtchouk polynomials. *Mat. Zametki*, 44(5), 682–693, 703.

Sheen, R.-C. (1987). Plancherel–Rotach-type asymptotics for orthogonal polynomials associated with $\exp(-x^6/6)$. *J. Approx. Theory*, 50(3), 232–293.

Sheffer, I. M. (1939). Some properties of polynomial sets of type zero. *Duke Math. J.*, 5, 590–622.

Shilov, G. E. (1977). *Linear Algebra*. New York: Dover.

Shohat, J. & Tamarkin, J. D. (1950). *The Problem of Moments*. Providence, RI: American Mathematical Society, revised edition.

Shohat, J. A. (1936). The relation of the classical orthogonal polynomials to the polynomials of appell. *Amer. J. Math.*, 58, 453–464.

Shohat, J. A. (1938). Sur les polynômes orthogonèaux généraliséès. *C. R. Acad. Sci.*, 207, 556–558.

Shohat, J. A. (1939). A differential equation for orthogonal polynomials. *Duke Math. J.*, 5, 401–417.

Siegel, C. L. (1929). Über einige anwendungen diophantischer approximationery. Abh. der Preuss. Akad. der Wissenschaften. Phys-math. Kl. Nr. 1.

Simon, B. (1998). The classical moment as a selfadjoint finite difference operator. *Adv. Math.*, 137, 82–203.

Simon, B. (2004). *Orthogonal Polynomials on the Unit Circle*. Providence, RI: American Mathematical Society.

Sklyanin, E. K. (1988). Boundary conditions for integrable quantum systems. *J. Phys. A: Math. Gen.*, 21(10), 2375–2389.

Slater, L. J. (1964). *Generalized Hypergeometric Series*. Cambridge: Cambridge University Press.

Slepian, D. (1972). On the symmetrized Kronecker power of a matrix and extensions of Mehler's formula for Hermite polynomials. *SIAM J. Math. Anal.*, 3, 606–616.

Sneddon, I. N. (1966). *Mixed boundary value problems in potential theory*. Amsterdam: North-Holland.

Sonine, N. J. (1880). Recherches sur les fonctions cylindriques et le développement des fonctions continues en series. *Math. Ann.*, 16, 1–80.

Sonine, N. J. (1887). On the approximate computation of definite integral and on the rational integral functions occurring in this connection. *Warsaw Univ. Izv.*, 18, 1–76. Summary in *Jbuch. Fortschritte Math.* 19, 282.

Sorokin, V. N. (1984). Simultaneous Padé approximants for finite and infinite intervals. *Izv. Vyssh. Uchebn. Zaved., Mat.*, (8), 45–52. Translated in *J. Soviet Math.* 28 (1984), no. 8, 56–64.

Sorokin, V. N. (1986). A generalization of Laguerre polynomials and convergence of simultaneous Padé approximants. *Uspekhi Mat. Nauk*, 41, 207–208. Translated in *Russian Math. Surveys* 41 (1986), 245–246.

Sorokin, V. N. (1990). Simultaneous Padé approximation for functions of Stieltjes type. *Sibirsk. Mat. Zh.*, 31(5), 128–137. Translated in *Siberian Math. J.* 31 (1990), no. 5, 809–817.

Sorokin, V. N. & Van Iseghem, J. (1997). Algebraic aspects of matrix orthogonality for vector polynomials. *J. Approx. Theory*, 90, 97–116.

Srivastava, H. M. & Manocha, H. L. (1984). *A Treatise on Generating Functions*. Chichester: Ellis Horwood Ltd.

Srivastava, H. M. & Singhal, J. P. (1973). New generating functions for Jacobi and related polynomials. *J. Math. Anal. Appl.*, 41, 748–752.

Stanley, R. P. (1978). Generating functions. In G. C. Rota (Ed.), *Studies in Combinatorics*. Washington, DC: Mathematical Association of America.

Stanley, R. P. (1985). Reconstruction from vertex-switching. *J. Combin. Theory Ser. B*, 38(2), 132–138.

Stanton, D. (2001). Orthogonal polynomials and combinatorics. In J. Bustoz, M. E. H. Ismail, & S. K. Suslov (Eds.), *Special functions 2000: current perspective and future directions (Tempe, AZ)*, volume 30 of *NATO Sci. Ser. II Math. Phys. Chem.* (pp. 389–409). Dordrecht: Kluwer Acad. Publ.

Stieltjes, T. J. (1885a). Sur les polynômes de Jacobi. *Comptes Rendus de l'Academie des Sciences, Paris*, 100, 620–622. Reprinted in Œuvres Complètes, volume 1, pp. 442–444.

Stieltjes, T. J. (1885b). Sur quelques théorèmes d'algèbre. *Comptes Rendus de l'Académie des Sciences, Paris*, 100, 439–440. Reprinted in Œuvres Complètes, volume 1, pp. 440–441.

Stieltjes, T. J. (1894). Recherches sur les fractions continues. *Annal. Faculté Sci. Toulouse*, 8, 1–122.

Stone, M. H. (1932). *Linear Transformations in Hilbert Space and Their Application to Analysis*. Providence, RI: American Mathematical Society.

Suslov, S. K. (1997). Addition theorems for some q-exponential and trigonometric functions. *Methods and Applications of Analysis*, 4, 11–32.

Szász, O. (1950). On the relative extrema of ultraspherical polynomials. *Bol. Un. Mat. Ital. (3)*, 5, 125–127.

Szász, O. (1951). On the relative extrema of Hermite orthogonal functions. *J. Indian Math. Soc.*, 25, 129–134.

Szegő, G. (1926). Beitrag zur theorie der thetafunktionen. *Sitz. Preuss. Akad. Wiss. Phys. Math. Kl.*, XIX, 242–252. Reprinted in Collected Papers, (R. Askey, ed.), Volume I, Boston: Birkhauser, 1982.

Szegő, G. (1933). Über gewisse Potenzenreihen mit lauter positiven Koeffizienten. *Math. Zeit.*, 37, 674–688.

Szegő, G. (1950a). Conformal mapping of the interior of an ellipse onto a circle. *Amer. Math. Monthly*, 57, 474–478.

Szegő, G. (1950b). On certain special sets of orthogonal polynomials. *Proc. Amer. Soc.*, 1, 731–737.

Szegő, G. (1950c). On the relative extrema of Legendre polynomials. *Bol. Un. Mat. Ital. (3)*, 5, 120–121.

Szegő, G. (1968). An outline of the history of orthogonal polynomials. In D. T. Haimo (Ed.), *Orthogonal Expansions and Their Continuous Analogues* (pp. 3–11). Carbondale: Southern Illinois Univ. Press.

Szegő, G. (1975). *Orthogonal Polynomials*. Providence, RI: American Mathematical Society, fourth edition.

Szegő, G. & Turán, P. (1961). On monotone convergence of certain Riemann sums. *Pub. Math. Debrencen*, 8, 326–335.

Szwarc, R. (1992a). Connection coefficients of orthogonal polynomials. *Canad. Math. Bull.*, 35(4), 548–556.

Szwarc, R. (1992b). Convolution structures associated with orthogonal polynomials. *J. Math. Anal. Appl.*, 170(1), 158–170.

Szwarc, R. (1992c). Orthogonal polynomials and a discrete boundary value problem. I. *SIAM J. Math. Anal.*, 23(4), 959–964.

Szwarc, R. (1992d). Orthogonal polynomials and a discrete boundary value problem. II. *SIAM J. Math. Anal.*, 23(4), 965–969.

Szwarc, R. (1995). Connection coefficients of orthogonal polynomials with applications to classical orthogonal polynomials. In *Applications of hypergroups and related measure algebras (Seattle, WA, 1993)*, volume 183 of *Contemp. Math.* (pp. 341–346). Providence, RI: Amer. Math. Soc.

Szwarc, R. (1996). Nonnegative linearization and quadratic transformation of Askey–Wilson polynomials. *Canad. Math. Bull.*, 39(2), 241–249.

Szwarc, R. (2003). A necessary and sufficient condition for nonnegative product linearization of orthogonal polynomials. *Constr. Approx.*, 19(4), 565–573.

Takhtadzhan, L. A. (1982). The picture of low-lying excitations in the isotropic Heisenberg chain of arbitrary spins. *Phys. Lett. A*, 87, 479–482.

Takhtadzhan, L. A. & Faddeev, L. D. (1979). The quantum method of the inverse problem and the heisenberg xyz model. *Russ. Math. Surveys*, 34, 11–68.

Terwilliger, P. (2001). Two linear transformations each tridiagonal with respect to an eigenvasis of the other. *Linear Algebra and its Applications*, 330, 149–203.

Terwilliger, P. (2002). Leonard pairs from 24 points of view. *Rocky Mountain J. Math.*, 32(2), 827–888.

Terwilliger, P. (2004). Leonard pairs and the q-Racah polynomials. *Linear Algebra Appl.*, 387, 235–276.

Titchmarsh, E. C. (1964). *The Theory of Functions*. Oxford: Oxford University Press, corrected second edition.

Toda, M. (1989). *Theory of Nonlinear Lattices*, volume 20 of *Springer Series in Solid-State Sciences*. Berlin: Springer-Verlag, second edition.

Todd, J. (1950). On the relative extrema of the Laguerre polynomials. *Bol. Un. Mat. Ital. (3)*, 5, 122–125.

Tracy, C. A. & Widom, H. (1999). Random unitary Matrices, permutations and Painlève. *Commun. Math. Phys.*, 207, 665–685.

Tricomi, F. G. (1957). *Integral Equations*. New York: Interscience Publishers. Reprinted, New York: Dover Publications, Inc., 1985.

Underhill, C. (1972). *On the zeros of generalized Bessel polynomials.* Internal note, University of Salford.

Uvarov, V. B. (1959). On the connection between polynomials, orthogonal with different weights. *Dokl. Acad. Nauk SSSR*, 126, 33–36.

Uvarov, V. B. (1969). The connection between systems of polynomials that are orthogonal with respect to different distribution functions. *USSR Computat. Math. and Math. Phys.*, 9, 25–36.

Valent, G. (1994). Asymptotic analysis of some associated orthogonal polynomials connected with elliptic functions. *SIAM J. Math. Anal.*, 25, 749–775.

Valent, G. (1995). Associated Stieltjes–Carlitz polynomials and a generalization of Heun's differential equation. *J. Comp. Appl. Math.*, 57, 293–307.

Valent, G. & Van Assche, W. (1995). The impact of Stieltjes' work on continued fractions and orthogonal polynomials: additional material. In *Proceedings of the International Conference on Orthogonality, Moment Problems and Continued Fractions (Delft, 1994)*, volume 65 (pp. 419–447).

Van Assche, W. (1994). Presentation at orthogonal polynomials and special functions meeting. Delft.

Van Assche, W. (1999). Multiple orthogonal polynomials, irrationality and transcendence. In B. C. Berndt et al. (Ed.), *Continued Fractions: from analytic number theory to constructive approximation*, volume 236 of *Contemporary Mathematics* (pp. 325–342). Providence, RI: Amer. Math. Soc.

Van Assche, W. (2004). Difference equations for multiple Charlier and Meixner polynomials. In S. Elaydi et al. (Ed.), *New Progress in Difference Equations* (pp. 547–555). London: Taylor & Francis.

Van Assche, W. & Coussement, E. (2001). Some classical multiple orthogonal polynomials. *J. Comput. Appl. Math.*, 127, 317–347.

Van Assche, W., Geronimo, J. S., & Kuijlaars, A. B. J. (2001). Riemann–Hilbert problems for multiple orthogonal polynomials. In J. Bustoz et al. (Ed.), *Special Functions 2000: current perspectives and future directions (Tempe, AZ)*, volume 30 of *NATO Sci. Ser. II Math. Phys. Chem.* (pp. 23–59). Dordrecht: Kluwer Acad. Publ.

Van Assche, W. & Yakubovich, S. B. (2000). Multiple orthogonal polynomials associated with Macdonald functions. *Integral Transforms Spec. Funct.*, 9, 229–244.

Van der Jeugt, J. (1997). Coupling coefficients for Lie algebra representations and addition formulas for special functions. *J. Math. Phys.*, 38(5), 2728–2740.

Van der Jeugt, J. & Jagannathan, R. (1998). Realizations of $su(1, 1)$ and $U_q(su(1, 1))$ and generating functions for orthogonal polynomials. *J. Math. Phys.*, 39(9), 5062–5078.

Van Diejen, J. F. (2005). On the equilibrium configuration of the BC-type Ruijsenaars–Schneider system. *J. Nonlinear Math. Phys.*, 12, 689–696.

Van Doorn, E. & Scheinhardt, W. R. W. (1966). Analysis of birth-death fluid queues. In *Proceedings of the Applied Mathematics Workshop*, volume 5 (pp. 13–29). Taejon: Korea Advanced Institute of Science and Technology.

Van Iseghem, J. (1987). Vector orthogonal relations, vector QD-algorithm. *J. Comput. Appl. Math.*, 19, 141–150.

Vanlessen, M. (2003). Strong asymptotics of the recurrence coefficients of orthogonal polynomials associated to the generalized Jacobi weight. *J. Approx. Theory*, 125, 198–237.

Varga, R. S. (2000). *Matrix iterative analysis*, volume 27 of *Springer Series in Computational Mathematics*. Berlin: Springer-Verlag, expanded edition.

Vatsaya, S. R. (2004). Comment on "breakdown of the Hellmann–Feynman theorem: degeneracy is the key". *Phys. Rev. B*, 69(037102), 2.

Verma, A. (1972). Some transformations of series with arbitrary terms. *Ist. Lombardo Accad. Sci. Lett. Rend. A*, 106, 342–353.

Viennot, G. (1983). Une théorie combinatoire de polynômes orthogonaux generaux. Université de Québec à Montréal. Lecture Notes.

Vinet, L. & Zhedanov, A. (2004). A characterization of classical and semiclassical orthogonal polynomials from their dual polynomials. *J. Comput. Appl. Math.*, 172(1), 41–48.

Wall, H. S. (1948). *Analytic Theory of Continued Fractions*. Princeton, NJ: Van Nostrand.

Wall, H. S. & Wetzel, M. (1944). Quadratic forms and convergence regions for continued fractions. *Duke Math. J.*, 11, 89–102.

Wallisser, R. (2000). On Lambert's proof of the irrationality of π. In F. Halter-Kock & R. Tichy (Eds.), *Algebraic Number Theory and Diophantine Analysis* (pp. 521–530). New York: de Gruyter.

Wang, Z. & Wong, R. (2003). Asymptotic expansions for second-order linear difference equations with a turning point. *Numer. Math.*, 94(1), 147–194.

Wang, Z. & Wong, R. (2005a). Linear difference equations with transition points. *Math. Comp.*, 74(250), 629–653.

Wang, Z. & Wong, R. (2005b). Uniform asymptotics for orthogonal polynomials with exponential weights—the Riemann–Hilbert approach. (To appear).

Wang, Z. & Wong, R. (2005c). Uniform asymptotics of the stieltjes–wigers polynomials via the Riemann–Hilbert approach. (To appear).

Watson, G. N. (1944). *A Treatise on the Theory of Bessel Functions*. Cambridge: Cambridge University Press, second edition.

Weber, M. & Erdélyi, A. (1952). On the finite difference analogue of Rodrigues' formula. *Amer. Math. Monthly*, 59(3), 163–168.

Weinberger, H. F. (1956). A maximum property of Cauchy's problem. *Ann. of Math. (2)*, 64, 505–513.

Wendroff, B. (1961). On orthogonal polynomials. *Proc. Amer. Math. Soc.*, 12, 554–555.

Westphal, U. (1974). An approach to fractional powers of operators via fractional differences. *Proc. London Math. Soc. (3)*, 29, 557–576.

Whittaker, E. T. & Watson, G. N. (1927). *A Course of Modern Analysis*. Cambridge: Cambridge University Press, fourth edition.

Widder, D. V. (1941). *The Laplace Transform*. Princeton, NJ: Princeton University Press.

Wilf, H. S. (1990). *Generating Functionology*. Boston: Academic Press.

Wilson, J. A. (1977). Three-term contiguous relations and some new orthogonal polynomials. In *Padé and Rational Approximation (Proc. Internat. Sympos., Univ. South Florida, Tampa, Fla., 1976)* (pp. 227–232). New York: Academic Press.

Wilson, J. A. (1980). Some hypergeometric orthogonal polynomials. *SIAM J. Math. Anal.*, 11(4), 690–701.

Wilson, J. A. (1982). Hypergeometric series recurrence relations and properties of some orthogonal polynomials. (Preprint).

Wilson, J. A. (1991). Orthogonal functions from Gram determinants. *SIAM J. Math. Anal.*, 22(4), 1147–1155.

Wilson, M. W. (1970). Nonnegative expansions of polynomials. *Proc. Amer. Math. Soc.*, 24, 100–102.

Wimp, J. (1965). On the zeros of confluent hypergeometric functions. *Proc. Amer. Math. Soc.*, 16, 281–283.

Wimp, J. (1985). Some explicit Padé approximants for the function ϕ'/ϕ and a related quadrature formula. *SIAM J. Math. Anal.*, 10, 887–895.

Wimp, J. (1987). Explicit formulas for the associated Jacobi polynomials and some applications. *Can. J. Math.*, 39, 983–1000.

Wimp, J. & Kiesel, H. (1995). Non-linear recurrence relations and some derived orthogonal polynomials. *Annals of Numerical Mathematics*, 2, 169–180.

Wintner, A. (1929). *Spektraltheorie der unendlichen Matrizen*. Leipzig: S. Hirzel.

Witte, N. S. & Forrester, P. J. (2000). Gap probabilities in the finite and scaled Cauchy random matrix ensembles. *Nonlinearity*, 13, 1965–1986.

Wong, R. & Zhang, J.-M. (1994a). Asymptotic monotonicity of the relative extrema of Jacobi polynomials. *Canad. J. Math.*, 46(6), 1318–1337.

Wong, R. & Zhang, J.-M. (1994b). On the relative extrema of the Jacobi polynomials $P_n^{(0,-1)}(x)$. *SIAM J. Math. Anal.*, 25(2), 776–811.

Wong, R. & Zhang, J.-M. (1996). A uniform asymptotic expansion for the Jacobi polynomials with explicit remainder. *Appl. Anal.*, 61(1-2), 17–29.

Wong, R. & Zhang, J.-M. (1997a). Asymptotic expansions of the generalized Bessel polynomials. *J. Comput. Appl. Math.*, 85(1), 87–112.

Wong, R. & Zhang, J.-M. (1997b). The asymptotics of a second solution to the Jacobi differential equation. *Integral Transform. Spec. Funct.*, 5(3-4), 287–308.

Wong, R. & Zhao, Y.-Q. (2004). Uniform asymptotic expansions of the Jacobi polynomials in a complex domain. *Proc. Roy. Soc. Lon. Ser. A.* (To appear).

Wong, R. & Zhao, Y.-Q. (2005). On a uniform treatment of Darboux's method. *Constr. Approx.*, 21, 225–255.

Yamani, H. A. & Fishman, L. (1975). J-matrix method: Extension to arbitrary angular momentum and to Coulomb scattering. *J. Math. Physics*, 16, 410–420.

Yamani, H. A. & Reinhardt, W. P. (1975). L^2 discretization of the continuum: radial kinetic energy and Coulomb Hamiltonian. *Phys. Rev. A*, 11, 1144–1155.

Yee, A. J. (2004). Combinatorial proofs of Ramanujan's $_1\psi_1$ summation and the q-Gauss summation. *J. Combin. Thy., Ser. A*, 105, 63–77.

Yoo, B. H. (1993). *Characterizations of orthogonal polynomials satisfying differential equations*. PhD thesis, Korea Advanced Institute of Science and Technology, Taejon, Korea.

Zeng, J. (1992). Weighted derangements and the linearization coefficients of orthogonal sheffer polynomials. *Proc. London Math. Soc. (3)*, 65(1), 1–22.

Zhang, G. P. & George, T. F. (2002). Breakdown of the Hellmann–Feynman theorem: degeneracy is the key. *Phys. Rev. B*, 66(033110), 4.

Zhang, G. P. & George, T. F. (2004). Extended Hellmann–Feynman theorem for degenerate eigenstates. *Phys. Rev. B*, 69(167102), 2.

Zinn-Justin, P. (1997). Random Hermitian matrices in an external field. *Nuclear Phys. B*, 497, 725–732.

Zygmund, A. (1968). *Trigonometric Series: Vols. I, II*. London: Cambridge University Press, second edition.

Index

Author index